XI International Conference on Computational Heat, Mass and Momentum Transfer (ICCHMT 2018)

MATEC Web of Conferences Volume 240 (2018)

Cracow, Poland
21 - 24 May 2018

Editors:

Slawomir Gradziel
Stanislaw Lopata
Tomasz Sobota
Wieslaw Zima

ISBN: 978-1-5108-7565-4

Printed from e-media with permission by:

Curran Associates, Inc.
57 Morehouse Lane
Red Hook, NY 12571

Some format issues inherent in the e-media version may also appear in this print version.

For additional information, please contact EDP Sciences – Web of Conferences
at the address below.

EDP Sciences – Web of Conferences
17, Avenue du Hoggar
Parc d'Activité de Courtabœuf
BP 112
F-91944 Les Ulis Cedex A
France

Phone: +33 (0) 1 69 18 75 75
Fax: +33 (0) 1 69 28 84 91

contact-edps@webofconferences.org

Additional copies of this publication are available from:

Curran Associates, Inc.
57 Morehouse Lane
Red Hook, NY 12571 USA
Phone: 845-758-0400
Fax: 845-758-2633
Email: curran@proceedings.com
Web: www.proceedings.com

TABLE OF CONTENTS

Preface

Computational techniques and methods are evolutionary field from macroscopic-level to nano-level, using continuum and discrete mechanics. It is necessary for scientist and engineers to come together every two years in a relaxed environment to discuss new ideas and developments in the area of computational methods and applications, especially in the topic of fluid dynamics, heat and mass transfer in nonreactive and reactive processes. Also, it is good opportunity for new scientist and engineer to explore the art of computational methods.

The main topics of the ***XI International Conference on Computational Heat, Mass and Momentum Transfer*** **ICCHMT 2018** were the following: Advanced Numerical Methods, Aeronautical and Space Applications, Bio-Fluidics and Biomedical Engineering, Bio-Inspired Flow and Heat Transfer, Complex Chemical Reaction Modeling, Compressible Flows, Convection and Buoyancy-Driven Flows, Double Diffusive Convection, Fluid Flow and Heat Transfer in Biomedical Devices and Biotechnology, Fluid Machinery, Granular Flows, Heat and Mass Transfer in Manufacturing and Materials Processing, Heat and Mass Transfer in Nuclear Applications, Heat and Mass Transfer in Particle-Laden Flows, Heat Exchangers / Heat Pipe, Internal Flow and Heat Transfer, Micro / Nano Heat and Mass Transfer, Mixing Devices and Phenomena, Multi-Phase Flows, Optimization in Thermal Engineering, Reactive Flows and Combustion, Steam and Gas Turbines, Technology for Renewable Energy Sources, Thermal Flow Visualization, Thermal Fluid Machinery, Transport Phenomena in Porous Media, Waste Management and Waste Disposal, and Any Other Related Topics.

In this Proceedings, selected papers covering the above topics, mainly related to heat, mass and momentum transfer, heat exchangers, fluid dynamics, nanofluids, renewable energy, and modeling of the energy and industrial processes are presented. Proceedings contains a total of 116 papers.

Tradition is that the engineers and scientists come together to share experiences every two years on the international conference. In 2018 **the Institute of Thermal Power Engineering of the Cracow University of Technology** fell the honour of organizing the conference in Cracow, Poland. We hope that this meeting was productive not only for the exchange of experiences, ideas and changes in methods and applications computational in the topics of fluid dynamics and heat and mass transfer, but also was a good opportunity for young engineers and scientists to learn about new trends in computational methods.

We hope, that the readers of this Proceedings will find the presented papers useful and stimulating for their studies in the area on heat, mass and momentum transfer realized in various processes.

Kraków, Poland
November 2018

Editors:
Sławomir Grądziel
Stanisław Łopata
Tomasz Sobota
Wiesław Zima

Opening address

The 11th edition of the International Conference on Computational Heat, Mass and Momentum Transfer (ICCHMT 2018) took place on 21-24 May 2018 in Kraków (Poland). The conference was organized by the Institute of Thermal Power Engineering of the Cracow University of Technology in cooperation with the University of Calgary (Canada), the Sapienza University of Rome (Italy) and the Paris Saclay University LMT Cachan (France).

The topics of the conference included issues related to numerical methods in solving Navier-Stokes equations, numerical modelling of single- and multiphase flows, modelling of conjugated heat transfer problems in solids and liquids, and nanofluid heat transfer. The aim of the conference was to present the results of scientific research, exchange experiences, as well as to build a platform for cooperation between the scientific community and industry.

Representatives of the world's leading scientific institutions involved in computer modelling of heat and mass transfer processes were invited to participate in the conference. These issues find a practical application in the design and modelling of heat exchangers, both micro-scale and industrial energy systems, air conditioning and refrigeration systems, as well as many others.

The venue of the conference was the Galaxy hotel at Gęsia Street in Cracow, Poland. During the plenary sessions, world-renowned specialists in the field of numerical modelling of heat and mass transfer processes delivered their lectures. A total of 24 thematic sessions and 3 poster sessions were held.

The chairpersons of the ICCHMT2018 were: Prof. Jan Taler (Cracow University of Technology), Prof. Wen-Quan Tao (Xi'an Jiaotong University, China), Prof. Abdulmajeed Mohamad (University of Calgary, Canada) and Prof. Ali Cemal Benim (Duesseldorf University of Applied Sciences, Germany).

ICCHMT 2018 was held under the honorary patronage of:
- Ministry of Energy (Poland),
- President of the City of Kraków Jacek Majchrowski,
- Committee on Thermodynamics and Combustion of the Polish Academy of Sciences.

Committee of the XI International Conference on Computational Heat, Mass and Momentum Transfer

Honorary Committee:

Prof. Jan Kazior	Cracow University of Technology	Poland
Prof. Tadeusz Tatara	Cracow University of Technology	Poland
Prof. Wen-Quan Tao	Xi'an Jiaotong University	China
Prof. Ulrich Gross	TU Bergakademie Freiberg	Germany
Prof. Shen Yu	Chinesse Academy of Sciences	China
Prof. Pradeep Bansal	Satya International Ltd./Bansal Engineering Associates	USA
Prof. Afshin Ghajar	Oklahoma State University	USA
Prof. Henrik Lund	Aalborg University	Denmark
Prof. Tadeusz Bohdal	Koszalin University of Technology	Poland
Prof. Simeon Oka	University of Belgrade	Serbi

Steering Committee:

Prof. Abdulmajeed Mohamad	University of Calgary	Canada
Prof. Jan Taler	Cracow University of Technology	Poland
Prof. Rachid Bennacer	ENS Cachan Paris-Saclay	France
Prof. Sang-Ho Suh	Soongsil University	South Korea

International Committee:

Bayazitoglu Y.	Rice University	USA
Benim A. C.	Duesseldorf Univ. Applied Sciences	Germany
Bennacer R.	ENS	France
Biringen S.	University of Colorado	USA
Bontoux P.	CNRS, les universités d'Aix Marseille	France
Cheng P.	Shanghai Jiaotong Univ.	China
Chikh S.	University USTHB	Algeria
Chorowski M.	Wroclaw Univ. of Sci. & Tech.	Poland
Chung J. T.	Korea Univ.	Korea
Corcione M.	Sapienza Univ. Roma	Italy
Costa V.	University of Aveiro	Portugal
Cruchaga M.	Univ. de Santiago de Chile	Chile
De Lemos M.	Instituto Tecnologico de Aeronautica	Brazil
Duic N.	Univ. of Zagreb	Croatia
Epple B.	Technical Univ. of Darmstadt	Germany
Ferreira J.	ANSYS	Germany
Fue-Sang Lien	Univ. of Waterloo	Canada
Ghajar A.	Oklahoma State University	USA
Gnutek Z.	Wrocław University of Tech.	Poland
Gross U.	TU Bergakademie Freiberg	Germany
Günes D.	Bilgi University	Turkey
Hamed M. S.	McMaster Univ.	Canada
He Y-L	Jiaotong University	China
Jasak H.	Univ. of Zagreb	Croatia
Jin Y.	Zhejiang Sci-Tech Univ.	China
Joos F.	Helmut Schmidt University	Germany
Kakac S.	TOBB Univ. of Econ and Tech	Turkey
Kanna R.P.	Velammal College of Engi. and Tech.	India
Kim H. D.	Andong Nat'l Univ.	Korea
Kim S. J.	KAISTDaejeon,	Korea
Kim Y. J.	Sungkyunkwan Univ.	Korea
Klemeš J. J.	Brno University of Technology	Czech Republic
Kinoue Y.	Saga Univ.	Japan
Lewis R. W.	Swansea University	Great Britan
Li J.	Xi'an Jiaotong Univ.	China
Lund H.	Aalborg University	Denmark
Merlen A.	Univ. Sciences et Technologies de Lille	France
Nagano Y.	Nagoya Inst. of Tech.	Japan
Nagatama G.	Kyushu Inst. of Tech.	Japan
Ocłoń P.	Cracow University of Tech.	Poland
Ohara T	Tohoku University	Japan
Oka S.	VINČA Institute of Nuclear Sciences	Serbia
Olsen J.E.	SINTEF	Norway

Orfi J.	KSU	Saudi Arabia
Pasqualotto E.	ALSTOM	Switzerland
Pop I.	University of Cluj-Napoca	Romania
Saeid N.	Institut Teknologi Brunei	Brunei
Stanek W.	Silesian University of Technology	Poland
Suh S.H.	Soongsil University	Korea
Sunden B.	Lund University	Sweden
Sheremet M. A.	Tomsk State Univ.	Russia
Suryan A.	College of Eng. Trivandrum	India
Syed K.J.	ALSTOM	Switzerland
Tao WQ	Xian Jiaotong University	China
Taler J.	Cracow University of Tech.	Poland
Taler D.	Cracow University of Tech.	Poland
Rao R.V.	SV NIT	India
Vallati A.	Sapienza Univ.	Italy
Viskanta R.	Purdue University	USA
Wang B.C.	Univ. of Manitoba	Canada
Wang G.	Beijing Inst. Of Tech.	China
Watanabe S.	Kyushu Univ.	Japan
Wiedermann A.	MAN Diesel&Turbo	Germany
Yamamoto K.	Nagoya University	Japan
Yan J.	KTH Royal Institute of Technology	Sweden
Younis L. B.	SNC.LAVALIN	Canada

Local Organizing Committee:

Bohdan Węglowski DSc PhD	Cracow Univ. of Tech. Poland
Dawid Taler DSc PhD	Cracow Univ. of Tech. Poland
Piotr Duda DSc PhD	Cracow Univ. of Tech. Poland
Wiesław Zima DSc PhD	Cracow Univ. of Tech. Poland
Stanisław Łopata DSc PhD	Cracow Univ. of Tech. Poland
Sławomir Grądziel DSc PhD	Cracow Univ. of Tech. Poland
Piotr Wais PhD	Cracow Univ. of Tech. Poland
Tomasz Sobota PhD	Cracow Univ. of Tech. Poland
Magdalena Jaremkiewicz PhD	Cracow Univ. of Tech. Poland
Anna Korzeń PhD	Cracow Univ. of Tech. Poland
Artur Cebula PhD	Cracow Univ. of Tech. Poland
Paweł Ocłoń DSc PhD	Cracow Univ. of Tech. Poland
Damian Muniak PhD	Cracow Univ. of Tech. Poland
Piotr Dzierwa PhD	Cracow Univ. of Tech. Poland
Marcin Trojan PhD	Cracow Univ. of Tech. Poland
Karol Majewski PhD	Cracow Univ. of Tech. Poland
Marzena Nowak-Ocłoń PhD	Cracow Univ. of Tech. Poland
Piotr Cisek PhD	Cracow Univ. of Tech. Poland
Marcin Pilarczyk PhD	Cracow Univ. of Tech. Poland
Karol Kaczmarski MSc	Cracow Univ. of Tech. Poland
Marek Majdak MSc	Cracow Univ. of Tech. Poland
Monika Rerak MSc	Cracow Univ. of Tech. Poland
Vincenzo Spena	DIAEE Sapienza University of Rome, Italy
Elisa Ricci	DIAEE Sapienza University of Rome, Italy

edp sciences

MATEC Web of Conferences

Statement of Peer review

In submitting conference proceedings to MATEC Web of Conferences, I certify to the Publisher that I adhere to the **Policy on Publishing Integrity** of the journal in order to safeguard good scientific practice in publishing.

1. All articles have been subjected to peer review administered by the proceedings editors.
2. Reviews have been conducted by expert referees, who have been requested to provide unbiased and constructive comments aimed, whenever possible, at improving the work.
3. Proceedings editors have taken all reasonable steps to ensure the quality of the materials they publish and their decision to accept or reject a paper for publication has been based only on the merits of the work and the relevance to the journal.

Title, date and place of the conference

XI International Conference on Computational Heat, Mass and Momentum Transfer **ICCHMT 2018, May 21-24 2018, Cracow, Poland**

Proceedings editor(s):

Slawomir Grądziel

Stanislaw Łopata

Tomasz Sobota

Wieslaw Zima

Date and editor's signature

14 November 2018

MATEC Web of Conferences **240**, 01001 (2018) https://doi.org/10.1051/matecconf/201824001001
ICCHMT 2018

Numerical investigation of boiling and forced convection heat transfer in inclined porous enclosure using modified enthalpy formulation

Omar Rafae Alomar[1,*], *Rafie Rushdy* Mohammed[2], and *Karam Hashim* Mohammed[1]

[1] Northern Technical University, Engineering Technical College of Mosul, Cultural Group Street, Mosul, Iraq
[2] Northern Technical University, Mosul Technical Institute, Cultural Group Street, Mosul, Iraq

Abstract—Two-Phase flow in an inclined rectangular porous media, under unsteady-state condition, has been numerically investigated in this article, based on the modified h-formulation of Two-Phase Mixture Model (TPMM). The governing equations have been discretised using Finite Volume Method (FVM) and solved iteratively in a SIMPLE-like manner. Effects of various parameters on the flow and temperature fields have been investigated, which clearly demonstrated that the inclination angle strongly affect the boiling initiation. Recirculating flow has been observed when the inclination angle $\theta > 0^\circ$. The results clearly indicated that the operating conditions and the porous medium properties have significant effects on the initiation and termination of phase change process. A closer inspection of the results reveals that the presence of a critical inclination angle θ depending on the value of K^* and Re_l, which correspond to a maximum values of vapour volume. The modified h-formulation requires significantly less computational time as compared with the existing H-formulation of TPMM.

Keywords: porous media; modified enthalpy formulation; TPMM; phase change process; sloping evaporator; Local Thermal Equilibrium model

Nomenclature

$\mathbf{b}$	Body force vector per unit mass, $\mathrm{m/s^2}$
$\tilde{\mathbf{b}}$	Normalized body force per unit mass $= \mathbf{b}/g$
D	Capillary diffusion coefficient, $\mathrm{m^2/s}$
f	Hindrance function
Fr	Froude number $= u_{in}/u_g = u_{in}/\sqrt{gW} = Re_{in}/Re_g$
h_{fg}	Latent heat of vaporization $= h_{v,sat} - h_{l,sat}$, J/kg
k_{rl}, k_{rv}	Relative permeability for liquid & vapour, respectively
n	Exponent of expression for relative permeabilities
p_c	Capillary pressure, Pa
Pe	Peclet number $= u_{in}W/\alpha = RePr$
Q^*	Normalized heat flux $= \dot{q}''W/\mu h_{fg}$
Re	Reynolds number $= u_{in}W/\nu$
Re_g	Gravitational Reynolds number $= u_g W/\nu_l$
u_g	Gravitational velocity vector $\sqrt{gW}$, m/s
W	Height of the duct, m
Greek Symbols	
β	Isobaric expansion coefficient, $\mathrm{K^{-1}}$
θ	Inclination angle of the evaporator
$\Delta\rho$	Difference in densities $= (\rho_l - \rho_v)$, $\mathrm{kg/m^3}$
γ	Advection correction coefficient
Γ	Diffusion coefficient for enthalpy, kg/ms
ε	Porosity
λ	Relative mobility
σ	Surface tension, N/m
$\tilde{\sigma}$	Normalized surface tension coefficient $= \rho_l \sigma W/\mu_l^2$
Subscripts	
in	Inlet
l	Liquid
s	Solid
sat	Saturation
v	Vapour
Superscripts	
*	Dimensionless

1 Introduction

Two-phase flow and analysis of boiling in porous media is an important topic, which spans a broad spectrum of engineering disciplines [1–3]. Based on Two-Phase Mixture Model (TPMM) proposed by Wang and co-authors [4 – 6], which consists of mass, momentum and energy conservation equations in the modified form, three distinct regions can be observed, if a liquid is injected into a porous evaporator. They are: sub-cooled liquid region, two-phase region and superheated gaseous region. Owing

* Correspondence author: Omar Rafae Alomar, email: sedrarasha@yahoo.com; omar.alomar@ntu.edu.iq

to the generality and the simplicity of the numerical formulation, TPMM of Wang [6] has been most often used for the simulation of two-phase flow within porous media [6–19].

However, a major problem that is encountered when applied TPMM [6] during the simulation of two-phase flow within porous media is the occurrence of drastic, non-physical change in the predicted temperature distribution over an extremely short distance that results primarily due to the presence of discontinuity in the effective diffusion coefficient across the interfaces. In order to avoid the aforementioned problem, a new smoothing algorithm has been developed by Alomar et al [20]. The proposed algorithm has been also employed by Alomar et al. [21–24] for the successful prediction of complete phase change process of liquid water inside a divergent [21], an annular [22] and constant cross-sectional area [23] porous evaporator as well as porous channel [24] using both Local Thermal Equilibrium (LTE) and Non-Equilibrium (LTNE) conditions.

After critically analysed the drawbacks of the original enthalpy (h) formulation [3], a modified formulation has been developed by Ray and Alomar [25] that can easily accommodate substantial density variations in the single phase regions as compared to the popular modified H-formulation [6]. Most importantly, the modified h-formulation [25] does not require any artificial definition of the modified volumetric (H) enthalpy as the dependent variable for the energy conservation equation [5, 6] while successfully eliminating the identified problems. It has been demonstrated that the modified h-formulation requires significantly less computational time as compared to the H-formulation [6] and hence it is recommended for the simulation of two-phase flow inside porous media. Later, the modified h-formulation [25] has been employed by Alomar et al. [26] in order to test the performance of non-staggered and staggered grid layouts for the simulation of complete evaporation process inside porous evaporator. The results indicated that the solutions obtained by using non-staggered grid layout are identical with that obtained using staggered grid layout. After that Alomar et al. [27] performed a numerical study on the complete evaporation process inside anisotropic porous media using the same formulation. The results indicated that the anisotropy of permeability and solid thermal conductivity have significant effect on the complete evaporation process.

Another interesting feature of boiling and force convection effects is the influence of porous evaporator inclination. As a continuation of the previous research efforts toward a complete understanding of boiling and force convection in porous media, the two-phase flow through a porous enclosure with a finite wall heat sources has been numerically investigated in the present article to include the effect of inclination angle of the evaporator. The mathematical formulation is based on the modified h-formulation [25] along with LTE model.

2 Mathematical Formulation

2.1 Problem Description

Schematic representation of the problem geometry is shown in Fig. 1 along with its dimensions. The problem under investigation is incomplete phase change process of liquid water in an inclined porous channel, under unsteady-state condition, which has been considered as a demonstrative example. The model has been applied for a two-parallel plate channel ($L \times W$), filled with a homogeneous as well as isotropic porous medium of known properties. The channel has unheated inlet and exit lengths of l_i and l_e located on the bottom wall, respectively. The length of heated section l_h, where an asymmetric heat source $\dot{q}_w''^{*}$ has been applied on the bottom wall of the channel, whereas the inlet and exit lengths are considered well insulated. On the other hand, the upper plate of the channel is considered at constant temperature. An external pressure gradient drives the sub-cooled water at the inlet, with a low temperature T_{in} $(T_{in} < T_{sat})$ and uniform velocity u_{in}, to flow through the channel. The flowing water is heated as it passes through the heated surface of the channel. When the heat flux is sufficiently increased, boiling occurs at the heated section and thus a two-phase zone is formed. For numerical simulations, the modified h-formulation [25] along with LTE model has been used. Further, all the properties have been assumed to be constant for both liquid and vapour phases.

2.2 Governing Conservation Equations

In the present study, the LTE model has been employed for the present simulations[2]. The existing model considers unsteady, two-dimensional incomplete evaporation process within an isotropic porous medium due to localized asymmetric heating in an inclined porous channel. In order to construct a modified h-formulation [25] applicable under very low flow velocity, Darcy's law is introduced. All applicable equations have been made dimensionless using the definitions in Table 1 in order to reduce the number of parameters those affect final results[3]. By adopting the modified h-formulation [25], the conservation equations of mass, momentum and energy by including the transient term and the effects of inclination angle are given as:

$$\varepsilon \frac{\partial \rho^*}{\partial t^*} + \nabla^* \cdot \left(\rho^* \mathbf{u}^*\right) = 0 \tag{1}$$

$$\rho^* \mathbf{u}^* = -\frac{K^*}{\nu^*}\left[\nabla^* p^* - \rho_k^* \mathbf{b}^*\right] \tag{2}$$

[2] This model assumes that the solid and the fluid phases are in thermal equilibrium with each other and they coexist at the same temperature.

[3] The dimensionless parameters have been carefully chosen such that they retain all governing equations along with expressions for the mixture variables in the same form of modified h-formulation [25].

$$(1-\varepsilon)\rho_s^* C_{ps}^* \frac{\partial T^*}{\partial t^*} + \varepsilon \frac{\partial(\rho^* h^*)}{\partial t^*} + \nabla^* \cdot \left[\gamma_h \rho^* \mathbf{u}^* h^*\right] = \nabla^* \cdot \left[\Gamma_h^* \nabla^* h^*\right] + \nabla^* \cdot \left(f \frac{K^* \Delta\rho^* h_{fg}^*}{\nu_v^*} \mathbf{b}^* \right) \quad (3)$$

where symbols with superscript '*' denote the dimensionless values and h is the enthalpy. Further definitions of dimensionless mixture properties in the two-phase region, appearing in Eq. (1), (2), and (3) are provided in Tables 1 and 2.

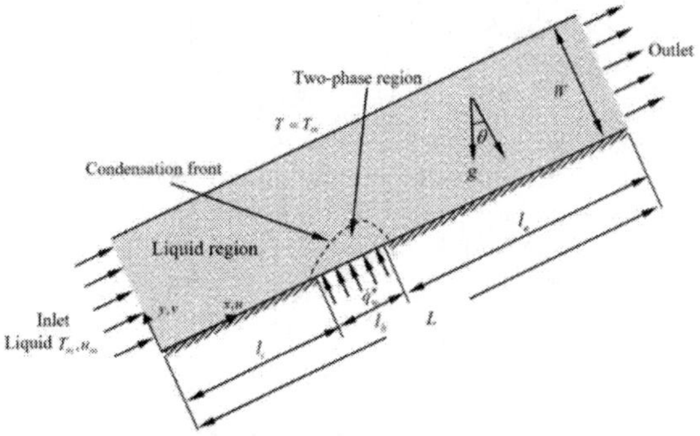

Fig. 1. Schematic representation of the phase change problem.

Table 1. Definitions of Dimensionless Variables [20, 25].

Dimensionless variables	Expressions
Time	$t^* = u_{in} t/W$
Length	$x^* = x/W$; $y^* = y/W$
Density	$\rho^* = \rho/\rho_l$; $\rho_l^* = 1$
Velocity	$u_j^* = u_j/u_{in}$
Pressure	$p^* = p/\rho_l u_{in}^2$
Temperature	$T^* = T C_{pl}/h_{fg}$
Enthalpy	$h^* = h/h_{fg}$; $h_{fg}^* = 1$
Specific Heat	$C_p^* = C_p/C_{pl}$
Dynamic Viscosity	$\mu^* = \mu/\mu_l Re_l$; $\mu_l^* = 1/Re_l$
Thermal Conductivity	$k^* = k/\rho_l C_{pl} W u_{in} = k/k_l Pe_l$; $k_l^* = 1/Pe_l$
Body Force per unit Mass	$\mathbf{b}^* = \tilde{\mathbf{b}}\, gW/u_{in}^2 = \tilde{\mathbf{b}}\,/Fr^2$
Isobaric Expansion	$\beta^* = \beta\, h_{fg}/C_{pl}$
Wall Heat Flux	$\dot{q}_w''^* = \dot{q}_w''/\rho_l u_{in} h_{fg} = Q_w^*/Re_l$
Darcy Number	$K^* = K/W^2$

It may be noted here that according to TPMM, the momentum balance given in Eq. (2) considers only the body force term and Darcy's law [4–6] for the additional pressure drop due to the presence of a porous matrix. Since Darcy's law is used to model the flow in porous media, Eq. (2) is valid only for low mass flow rate applications[4], where the fluid Reynolds number based on the permeability of the porous medium, $Re_K = \rho u \sqrt{K}/\mu$, is less than one [17]. In Eqs. (2) and (3), the permeability of the porous medium K (or Darcy number $K^* = K/W^2$) can be evaluated either from a separate detailed study or by using empirical correlation where K is expressed as a function of the porosity ε and average pore diameter of porous matrix [28, 29]. In Eq. (2), ρ_k^* is the dimensionless kinetic density, which according to the Boussinesq approximation and it could be expressed using the linear expansions of single phase densities as a function of local temperature around their saturation temperature [6]:

$$\rho_k^* = \rho_l^*\left[1-\beta_l^*\left(T^*-T_{sat}^*\right)\right]\lambda_l + \rho_v^*\left[1-\beta_v^*\left(T^*-T_{sat}^*\right)\right]\lambda_v \quad (4)$$

Additionally, $\mathbf{b}^* = \tilde{\mathbf{b}}_j/Fr^2$ ($\tilde{\mathbf{b}} = \mathbf{b}/g$) is the normalised body force vector per unit mass, where $\tilde{b}_x = -\sin\theta$ and $\tilde{b}_y = -\cos\theta$ have been specified since the phase change process in an inclined porous media has been considered.

Table 2. Definitions of Dimensionless Properties for Two-Phase Mixture [6, 20, 25].

Variables	Expressions
Density	$\rho^* = \rho_l^* s + \rho_v^*(1-s)$
Mass Velocity	$\rho^* \mathbf{u}^* = \rho_l \mathbf{u}_l^* + \rho_v \mathbf{u}_v^*$
Enthalpy	$\rho^* h^* = h_l^* \rho_l^* s + (1-s)\rho_v^* h_v^*$
Dynamic Viscosity	$\mu^* = \frac{1}{Re_l}\left[\frac{\rho_l^* s + \rho_v^*(1-s)}{k_{rl}/\nu_l^* + k_{rv}/\nu_v^*}\right]$
Kinematic Viscosity	$\nu^* = \left[k_{rl}/\nu_l^* + k_{rv}/\nu_v^*\right]^{-1}$
Advection Correction Coefficient	$\gamma_h = \frac{\left[\rho_l^* s + \rho_v^*(1-s)\right]\left(h_{v,sat}^* - \lambda_l h_{fg}^*\right)}{\rho_l^* h_{l,sat}^* s + \rho_v^* h_{v,sat}^*(1-s)}$
Effective Diffusion Coefficient	$\Gamma_h^* = \frac{\rho^* h_{fg}^*}{\rho_v^*\left(h_{v,sat}^* - h^*\right) + \rho_l^*\left(h^* - h_{l,sat}^*\right)} D^* + k_{eff}^* \frac{dT^*}{dh^*}$
Capillary Diffusion Coefficient	$D^* = \frac{\left(\varepsilon K^*\right)^{1/2}\sigma^*}{\mu_l^*} \frac{k_{rv}k_{rl}}{\left(\nu_v^*/\nu_l^*\right)k_{rl} + k_{rv}}\left[-J'\right]$
Surface tension	$\sigma^* = \sigma/\rho_l u_{in}^2 W = \tilde{\sigma}/Re_l^2$
Relative Mobilities	$\lambda_l(s) = \frac{k_{rl}/\nu_l^*}{k_{rl}/\nu_l^* + k_{rv}/\nu_v^*}$; $\lambda_v(s) = \frac{k_{rv}/\nu_v^*}{k_{rl}/\nu_l^* + k_{rv}/\nu_v^*}$
Hindrance Function	$f = \frac{k_{rl}k_{rv}/\nu_v^*}{k_{rl}/\nu_l^* + k_{rv}/\nu_v^*}$
Relative Permeabilities	$k_{rl} = s^n$, $k_{rv} = (1-s)^n$; $n=1$
Capillary Pressure Function	$J = 1.417(1-s) - 2.120(1-s)^2 + 1.263(1-s)^3$

The expression of Γ_h^* in Eq. (3) is presented in Table 2, which contains the effective thermal conductivity of the combined medium k_{eff}^* in its dimensionless form. A simple estimation for k_{eff}^* can be obtained from a separate numerical and/or experimental study [30 – 33]. In the present investigation, it has been evaluated according to the parallel arrangement model as [6, 20]:

$$k_{eff}^* = (1-\varepsilon)k_s^* + \varepsilon\left[k_l^* s + (1-s)k_v^*\right] \quad (5)$$

where k_l^*, k_v^* and k_s^* are the thermal conductive of liquid, vapour and solid, respectively. The velocities of individual phases can be evaluated using the following relations [6]:

$$\rho_l \mathbf{u}_l^* = \lambda_l \rho^* \mathbf{u}^* + \mathbf{j}^* \quad (6a)$$

$$\rho_v \mathbf{u}_v^* = \lambda_v \rho^* \mathbf{u}^* + \mathbf{j}^* \quad (6b)$$

where $\mathbf{j}^*$ is the total diffusive flux, which can be written as:

[4] Since the convective and the diffusive effects have been neglected.

$$\mathbf{j}^* = -\rho_l^* D^* \nabla s + f \frac{K^* \Delta\rho^*}{\nu_v^*} \mathbf{b}^* \quad (7)$$

The combined temperature T^* in the single phase region and the liquid saturation s in the two-phase zone can be calculated from the solution of h^* as shown in Table 3. It is evident from Table 1 that the dimensional temperatures can be retrieved from their dimensionless values using only the properties of working fluid [20].

Table 3. Supplementary relations between enthalpy, temperature, and liquid saturation [25].

h^*	T^*	s	dT^*/dh^*
$h^* \le h_{l,sat}^*$	$T^* = T_{sat}^* + \frac{h^* - h_{l,sat}^*}{C_{pl}^*}$	1	$1/C_{pl}^* = 1$
$h_{l,sat}^* < h^* \le h_{v,sat}^*$	T_{sat}^*	$\frac{\rho_v^*(h_{v,sat}^* - h^*)}{\rho_v^*(h_{v,sat}^* - h^*) + \rho_l^*(h^* - h_{l,sat}^*)}$	0
$h_{v,sat}^* < h^*$	$T^* = T_{sat}^* + \frac{h^* - h_{v,sat}^*}{C_{pv}^*}$	0	$1/C_{pv}^*$

2.3 Boundary Conditions

For the present investigation, the following conditions have been employed at different boundaries of the computational domain in order to solve Eqs. (1) – (3):

i) At the inlet $t^* = 0$, the porous medium has been assumed to be at constant temperature T_{in}^* and the liquid velocity is uniform:

$$h^* = h_{l,sat}^* + C_{pl}^*\left(T_{in}^* - T_{sat}^*\right) \quad (8a)$$

$$u^* = u_{in}^* \quad (8b)$$

ii) At the inlet $x^* = 0$, the porous medium has been assumed to be saturated with sub-cooled liquid water that enters the evaporator with uniform velocity u_{in}^* and specified constant temperature of the fluid $T_{in}^* < T_{sat}^*$ [5]. Therefore, the prescribed boundary conditions at the inlet may be summarized as:

$$h^* = h_{l,sat}^* + C_{pl}^*\left(T_{in}^* - T_{sat}^*\right) \quad (9a)$$

$$\frac{\partial p^*}{\partial x^*} = -\frac{\mu_l^*}{K^*} u_{in}^* + \rho_k^* b_x^* \quad (9b)$$

$$u^* = u_{in}^* \quad (9c)$$

$$v^* = 0 \quad (9d)$$

iii) At the evaporator outlet, i.e., at $x^* = L^*$ and for $0 \le y^* \le 1$, the axial diffusion in Eq. (3) has been neglected and hence the second derivatives of h^* in the flow direction has been set zero[6]. Therefore, h^* at the outlet has been obtained by linearly extrapolating the values at the neighbouring interior nodes that ensure $d^2h^*/dx^{*2} = 0$ [7]. On the other hand, the flow has been assumed to be hydro-dynamically fully-developed at the exit[8]. Therefore, the exit boundary condition for the flow may be written as:

$$\frac{\partial p^*}{\partial x^*} = -\frac{\mu_l^*}{K^*} u_{in}^* + \rho_k^* b_x^* \quad (10a)$$

$$\frac{\partial u^*}{\partial x^*} = \frac{\partial v^*}{\partial x^*} = 0 \quad (10b)$$

iv) At $y^* = 0$ (bottom wall) for $0 \le x^* \le L^*$ impermeable condition has been imposed[9]. As far as the energy conservation equation is concerned, the specified heat flux boundary condition has been employed at the walls, where a vanishing local heat flux has been enforced at the unheated segments, i.e., for $0 \le x^* \le l_i^*$ and $l_i^* + l_h^* \le x^* \le L^*$. The boundary conditions may be formulated as:

$$-\Gamma_h^* \frac{\partial h^*}{\partial y^*} = f \frac{K^* \Delta\rho^* h_{fg}^*}{\nu_v^*} b_y^* \quad (11a)$$

$$\frac{\partial p^*}{\partial y^*} = -\rho_k^* b_y^* \quad (11b)$$

$$v^* = 0 \quad (11c)$$

v) At the position of the wall that is heated with constant heat flux, $\dot{q}_{w,x}^{\prime\prime *}$, i.e., for $l_i^* \le x^* \le l_i^* + l_h^*$ and $y^* = 0$. Therefore, the wall boundary condition may be written as follows:

$$-\Gamma_h^* \frac{\partial h^*}{\partial y^*} - f \frac{K^* \Delta\rho^* h_{fg}^*}{\nu_v^*} b_y^* = \dot{q}_{w,x}^{\prime\prime *} \quad (12)$$

vi) At $y^* = 1$ (top wall) for $0 \le x^* \le L^*$ impermeable condition has been imposed and the wall has been kept at constant temperature T_{in}^*.

3 Numerical Solution

The Finite Volume Method (FVM) on staggered grid layout has been used in order to discretise the conservation equations (1), (2), and (3). According to FVM, the entire computational domain is first divided into a number of non-overlapping control volumes. In the present study, only uniform control volumes have been used for all considered cases. The required pressure gradient and the kinetic density at the cell faces have been obtained by employing the second order Central Differencing Scheme (CDS) and

[5] Since sub-cooled liquid enters the channel, $\rho_{in}^* = 1$ has been set.

[6] For shorter unheated exit length, or even in its absence, as long as the flow is assumed to be thermally fully-developed for constant wall heat flux, the present condition would always hold, indicating linear rise in temperature, although the latter condition may never be met.

[7] This condition imply that the gradient of h^* in the axial direction is assumed to remain constant at the exit, which, would allow finite heat transfer at the exit. As will be shortly apparent from the predicted results, owing to the presence of unheated exit length, heat transfer at the exit of the evaporator turns out to be negligible and hence $dh^*/dx^* = 0$ is observed for all tested cases.

[8] This is quite reasonable for the present investigation in the presence of an unheated exit length filled with a porous medium.

[9] By imposing a vanishing normal velocity component all along the axial extent of the domain.

by linearly interpolating the adjacent nodal values, respectively. For the energy conservation, the first order Upwind Differencing Scheme (UDS) and second order CDS have been used for expressing the convective terms and the diffusive terms in Eq. (3), respectively. Discretised equations for all the nodes, along with boundary conditions, exhibit a tri-diagonal matrix structure and hence they have been solved by the Thomas algorithm [34]. In addition, diffusivities for the enthalpy Γ_h^* at cell faces have been obtained using the harmonic mean approximation, which ensures the balance of diffusive energy flux. On the other hand, the advection correction coefficient γ_h and the hindrance coefficient f at the cell faces have been obtained using linear interpolation from the adjacent nodal values.

In the present study, the smoothing algorithm proposed by Alomar et al. [20, 35] along with the recommended smoothing parameters has been employed for Γ_h^* to avoid the non-physical jump in the predicted temperature distributions[10] and to obtain the converged solutions for all considered cases. A grid independence study has been carried out and the results show that a mesh of $N_{CV} = 180 \times 300$ control volumes produce a grid-independent solutions. In this regards, all subsequent have been performed using this mesh size. Since the mixture properties, listed in Table 2, are strong functions of the liquid saturation, both s and dT^*/dh^* has been iteratively updated from the previous solution of h^*, according to the relations provided in Table 3. The time step size for all cases studied is 0.5s and the flow time is $t = 2\times10^4$s. Owing to the strong non-linearity in the governing conservation equations and the related constitutive relations arising out of the interdependence of dependent variable and since during discretisation, the discretised equations have been further under-relaxed and a relaxation factor of 0.1 has been generally used for most of the simulations in order to achieve convergence with the convergence criterion set to 10^{-5}.

4 Results and Discussion

For all studied cases, the length and height of the evaporator/channel have been kept fixed at $L = 1000$mm ($L^* = 10$) and $W = 100$mm ($W^* = 1$), respectively. The unheated starting and exit lengths have been set as $l_i = 0.4L$ ($L_i^* = 4$) and $l_e = 0.5L$ ($L_e^* = 5$)[11]. Therefore, the length of the heated section on the bottom plate has been taken as $l_h = 0.1L$ ($L_h^* = 1$). Sub-cooled liquid water has been used as the working fluid and its properties are given in Table 4, where the saturation temperature T_{sat} of water has been considered to be $100\,^\circ$C [20, 36, 37]. Water enters the domain with uniform velocity and temperature $T_{in} = 20^\circ C$. The inlet velocity u_{in} has been allowed to vary from 4×10^{-6} m/s to 45×10^{-6} m/s for the entire study. The inlet velocity has been carefully chosen in order to satisfy the criterion of the fluid Reynolds number $Re_K < 1$ for the velocity field of both liquid and vapour phases. With these inputs and the properties of water in Table 4, the Re_l has been assumed to vary from $Re_l = 1.5$ to $Re_l = 14.5$.

For the porous medium, the porosity has been assumed to be around 0.35. For the present demonstrative examples, the permeability of the porous medium has been assumed to follow the Carman-Kozeny type relation [28, 29]. Based on this, the permeability has been varied from $1\times10^{-11}\,\mathrm{m}^2$ to $7\times10^{-11}\,\mathrm{m}^2$. In dimensionless form, Darcy number K^* has been varied between 1×10^{-9} and 1×10^{-8}. On the other hand, the properties of solid phase, i.e., thermal conductivity k_s, density ρ_s and specific heat Cp_s have been considered 0.95 W/mK, 2645 kg/m^3 and 879 J/kg K, respectively, assuming a porous matrix made of glass bead system [36, 37]. At the heated section, it has been assumed that heat is added from an external source, supplying a constant heat flux $q_w'' = 3000 W/m^2$ ($Q_w^* = 4.75\times10^{-1}$). In addition, the acceleration due to gravity vector has been assumed to act in both negative x- and y-directions and hence $\tilde{b}_y = -\sin\theta$ and $\tilde{b}_y = -\cos\theta$ have been considered for all the cases. The inclination angle of the channel has been considered to vary from 0° to 90° for all cases studied.

The gravitational Reynolds number Re_g, required for the calculation of the Froude number Fr, and the normalized surface tension $\tilde{\sigma}$ have been taken as 3.4×10^5 and 7.25×10^7, respectively, based on properties of water in Table 4, gravity $g = 9.81\,\mathrm{m/s^2}$ and the distance between the parallel plate of the channel $W = 100$ mm. Finally, for the calculation of relative permeabilities, n=1 has been adopted without any exception.

Table 4 Properties of Water [10, 36 and 37].

Property	Liquid	Vapour
Density, ρ (kg/m^3)	957.85	0.5978
Specific Heat, C_p (J/kgK)	4190.2	2029.0
Dynamic Viscosity, μ (kg/ms)	2.79×10^{-4}	1.202×10^{-5}
Thermal Conductivity, k (W/mK)	0.68	0.0248
Saturation Enthalpy, h (J/kg)	419.02×10^3	2676.05×10^3
Latent Heat of Evaporation, h_{fg} (J/kg)	957.85	0.5978
Surface Tension Coefficient, σ (N/m)	2257.03×10^3	
Density, ρ (kg/m^3)	0.0589	

[10] The non-physical jump occurs primarily due to the discontinuity in the effective diffusion coefficient across the saturated liquid condition.

[11] The exit length has been carefully chosen in order to avoid the influence of exit boundary condition on the internal solution.

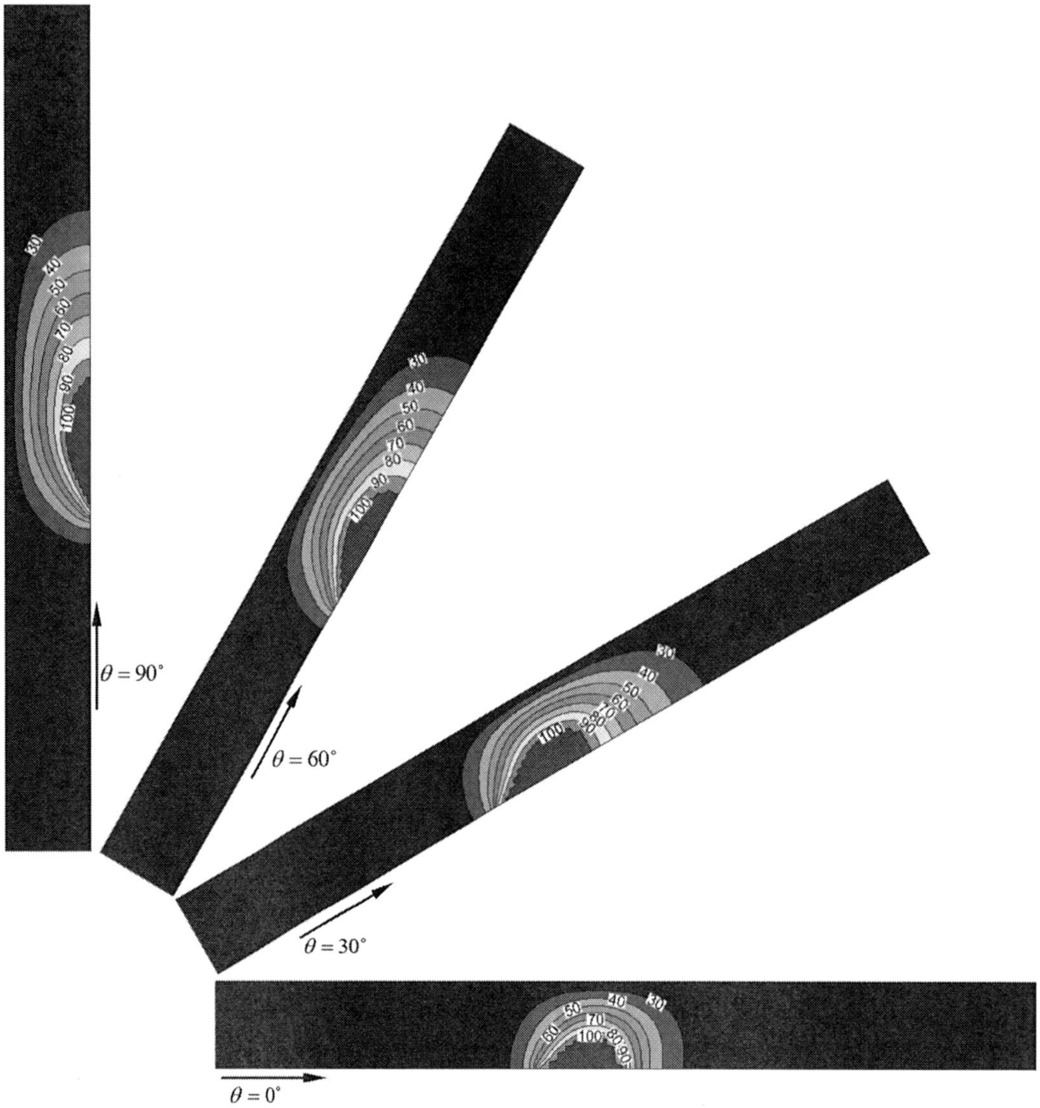

Fig. 2. Combined temperature distributions for various inclination angles with $K^* = 5 \times 10^{-9}$, $Re_l = 1.5$ and $t^* = 0.845$.

The results for different inclination angle θ of the channel, including the four particular cases presented in Figs. 2 – 5, show that the two-phase zone develops inside the channel and covers the heated surface, where the temperature remains constant at $T_{sat} = 100°C$. Wang [6] denoted the interface that separates the sub-cooled liquid region from the two-phase zone as a condensation front. The condensation front is appreciably inclined toward the outlet due to incoming flow of sub-cooled liquid. In order to identify the interface separating the sub-cooled liquid and the two-phase regions, isotherm is presented for $T_{sat} = 100 - 10^{-5}$ °C. This isotherm is, however, marked as that corresponding to $T_{sat} = 100°C$ in all relevant figures, where the negative sign refer to the saturation isotherm from the sub-cooled liquid zone. It is evident from the Figs. 2 and 3 that the two-phase zone is extended in both axial and transverse directions as compared with the horizontal case due to more energy added to the domain associated with increasing the buoyancy force. The buoyancy force in the transverse direction increases and reaches its maximum value at $\theta = 0^\circ$ and then the condensation front is moved considerably towards the cooled wall. Most importantly, the axial and transverse diffusions remain present even in the two-phase region as there exists a finite gradient of h^*, although the temperature remains unchanged at saturation condition. Since h^* increases upward and in the downstream direction due to heating and buoyancy effect, energy is transferred in the backward direction owing to the axial diffusion [20].

https://doi.org/10.1051/matecconf/201824001001

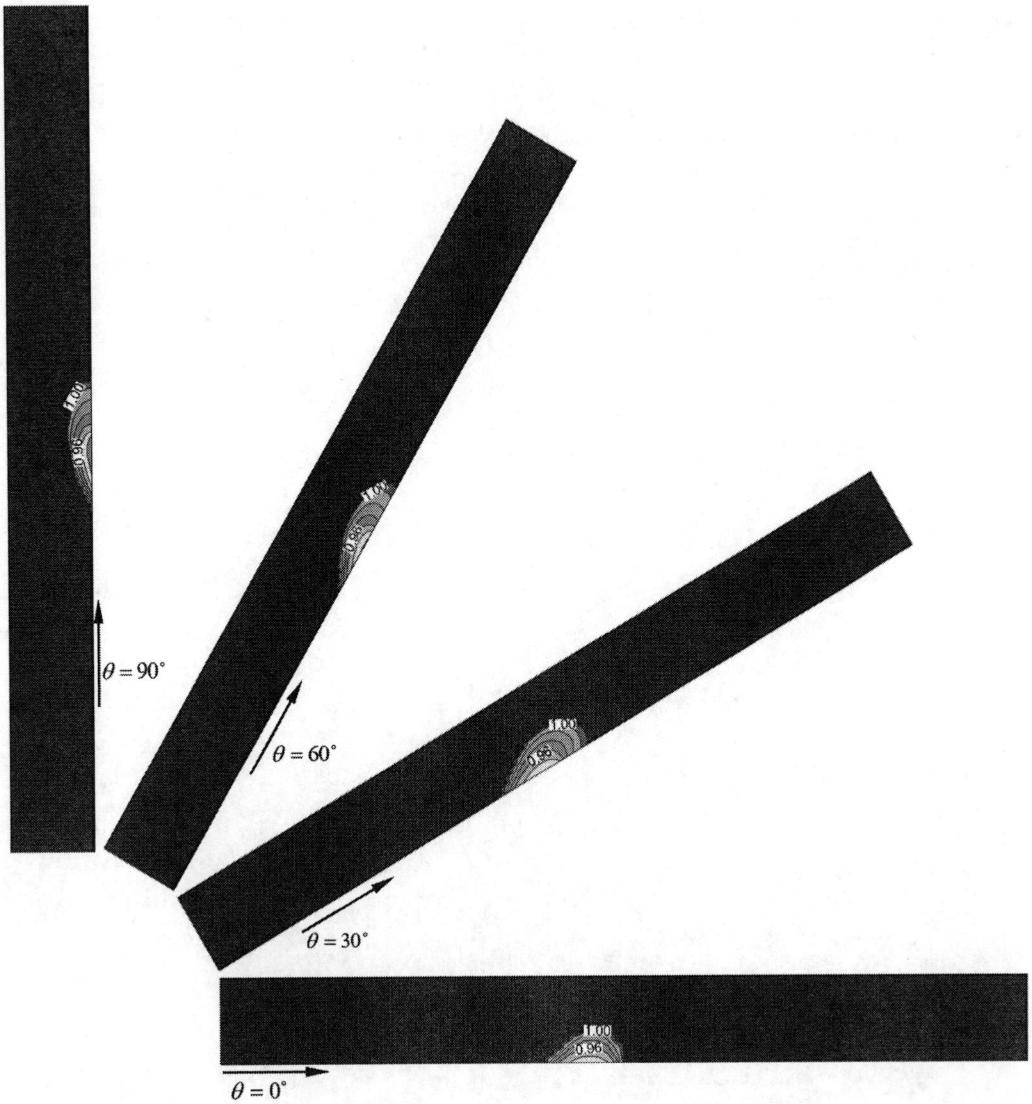

Fig. 3. Liquid saturation distributions for various inclination angles with $K^* = 5\times10^{-9}$, $Re_l = 1.5$ and $t^* = 0.845$.

The liquid velocity indicates parallel flow of the sub-cooled liquid at the inlet but deflected movement towards the upper surface of the channel when the liquid approaches the condensation front, so as to bypass the vapour region surrounding the bottom heated surface, as shown in Fig. 4 for $\theta = 0^\circ$. When the channel is inclined from $\theta = 0^\circ$, recirculating flow has been observed as shown in Fig. 4. For inclined and vertical positions, ascending forced flow assists buoyancy-induced flow near the heated wall and the heated zone is more extended and is stretched toward the outlet. When $\theta > 0^\circ$, the buoyancy force in the longitudinal direction becomes higher and evacuated heat from the heat source is increased. The vapour velocity vectors presented in Fig. 5 show a primarily upward movement upon the vapour generated on the heated surface due to local buoyancy. The $Re_{K,\max}$ for liquid and vapour phases is evaluated according to the maximum velocity in the respective phase in order to ensure that the assumption of Darcy flow is still valid.

In order to further demonstrate the effect of inclination angle of the channel, results have been generated considering a variation in Re_l and K^*. These results are summarized in Figs. 6 – 9. The effect of K^* on the temperature, liquid saturation, velocity vectors for both liquid and vapour phases is presented in Fig. 6 when $\theta = 90^\circ$ and $Re_l = 1.5$. It is evident from Fig. 6 that K^* has significant effects on temperature distribution, particularly in the two-phase region. An increase in K^* leads to a decrease in the size of two-phase region due to increase the effective diffusion coefficient in this region. As

K^* is decreased, the two-phase region initiates earlier and the size of this region is extended in the axial and transverse directions, as may be observed from Figs. 6(a) and 6(b). An early initiation of phase change processes for $K^* = 10^{-9}$ as compared to $K^* = 10^{-8}$ may also be explained by the reduced capillary diffusion affecting the effective diffusion coefficient in the two-phase region. As K^* is increased, the volume of vapour is decreased by increasing θ and the recirculating flow is increased, as shown in Figs. 6 – 9.

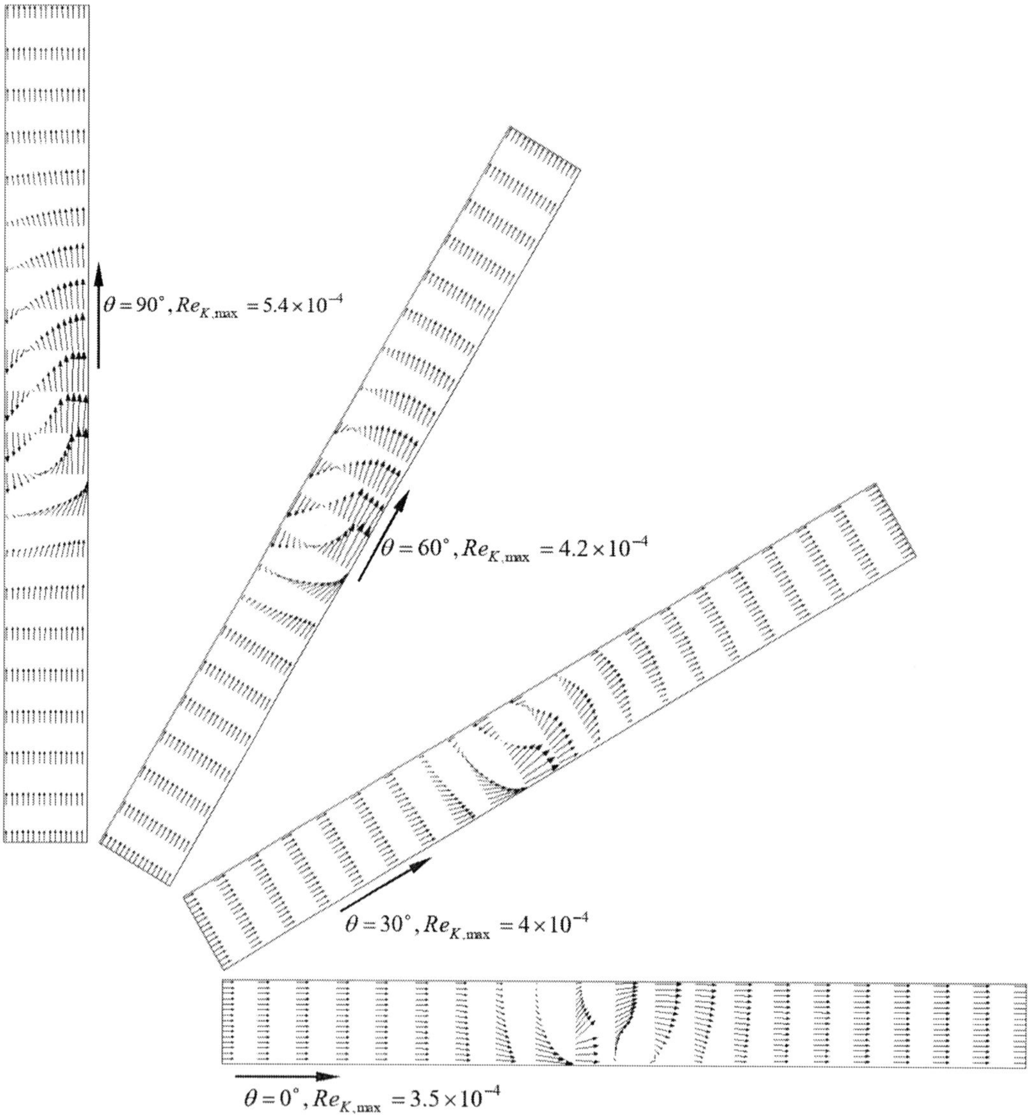

Fig. 4. Liquid velocity vectors for various inclination angles with $K^* = 5 \times 10^{-9}$, $Re_l = 1.5$ and $t^* = 0.845$.

The effects of Re_l on the temperature distribution with different values of inclination angles are presented in Figs. 7 – 9. For the same inclination angle, a low Re_l initiates the liquid evaporation earlier and increases the size of two-phase region towards the upper wall of the channel. This trend is reasonable and can be explained by the fact that when Re_l decreases, the residence time of fluid inside the channel is increased and less energy is required to evaporate the liquid phase. For large values of Re_l, the buoyancy effects are not important. For low Re_l, the volume of vapour is reduced by increasing θ, as may be observed in Figs. 7 – 9. Most importantly, it is evident from the Figs. 7 – 9 that a marginal change in Re_l leads to a substantial change in the predicted temperature distribution. Thus, the inlet flow condition requires to be satisfactorily designed in order to achieve the desired objective.

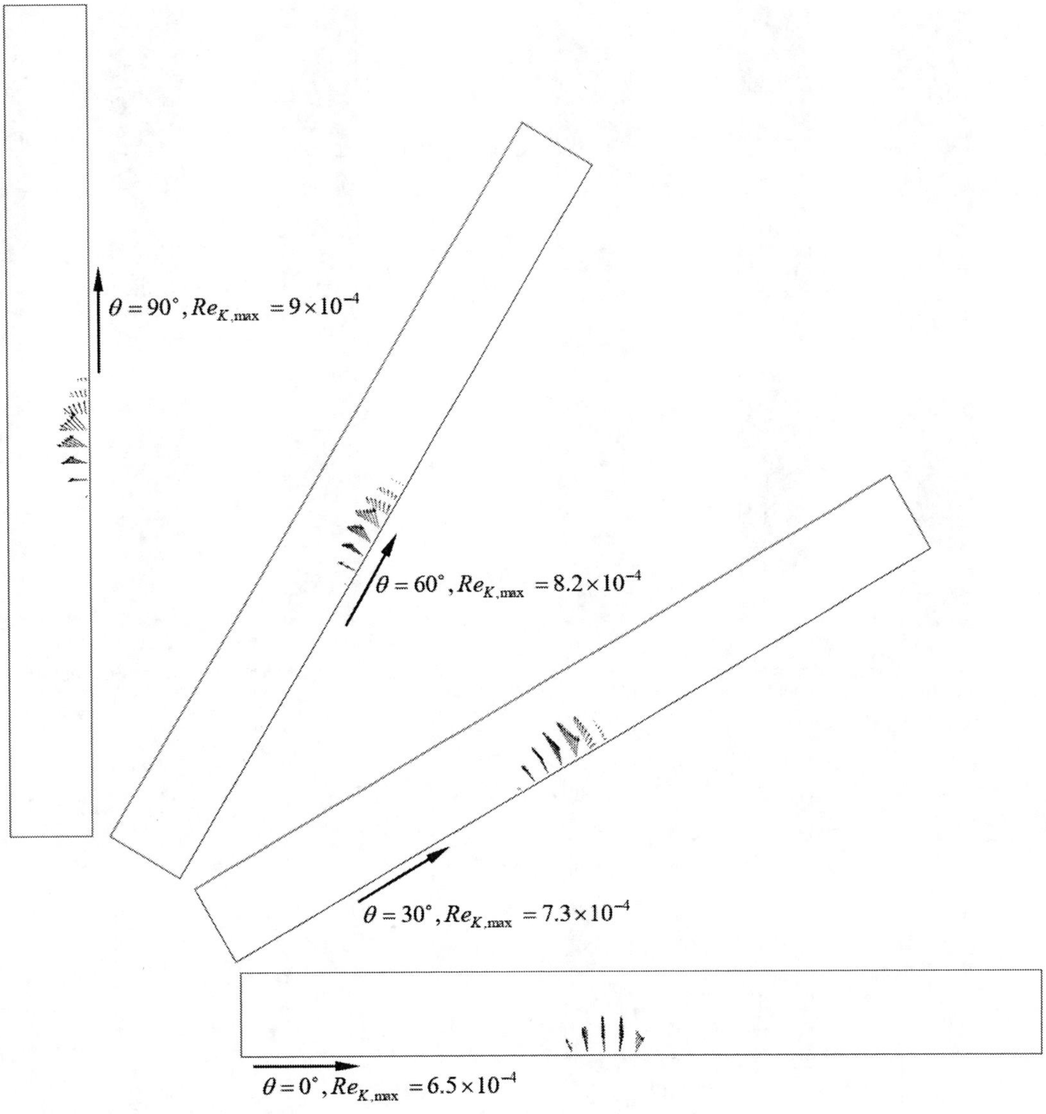

Fig. 5. Vapour velocity vectors for various inclination angles with $K^* = 5\times10^{-9}$, $Re_l = 1.5$ and $t^* = 0.845$.

 https://doi.org/10.1051/matecconf/201824001001

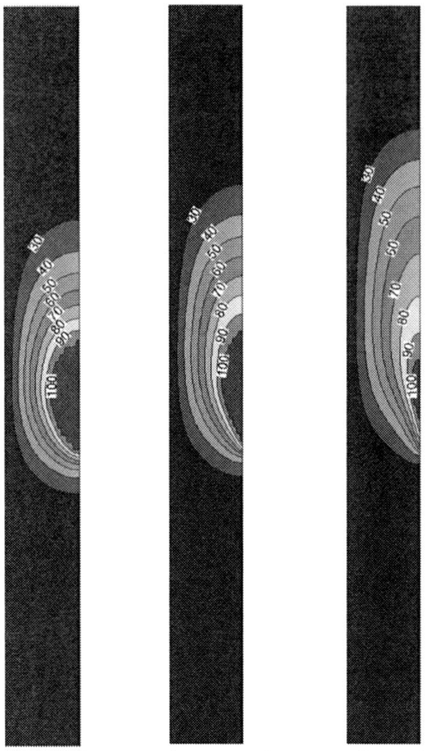

$K^* = 10^{-9}$ $K^* = 5\times10^{-9}$ $K^* = 10^{-8}$

(a) Combined temperature distributions

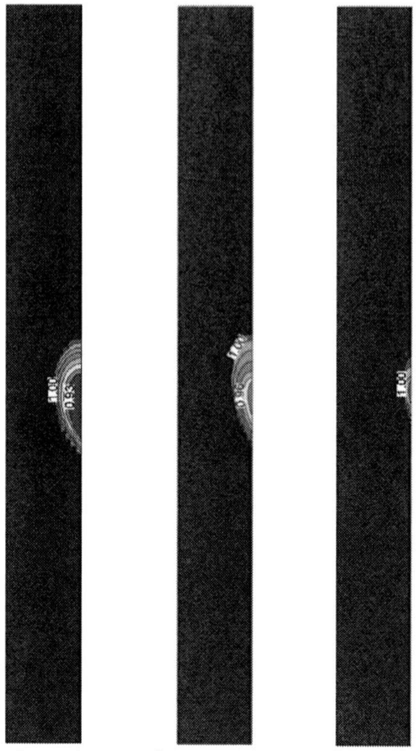

$K^* = 10^{-9}$ $K^* = 5\times10^{-9}$ $K^* = 10^{-8}$

(b) Liquid saturation distributions

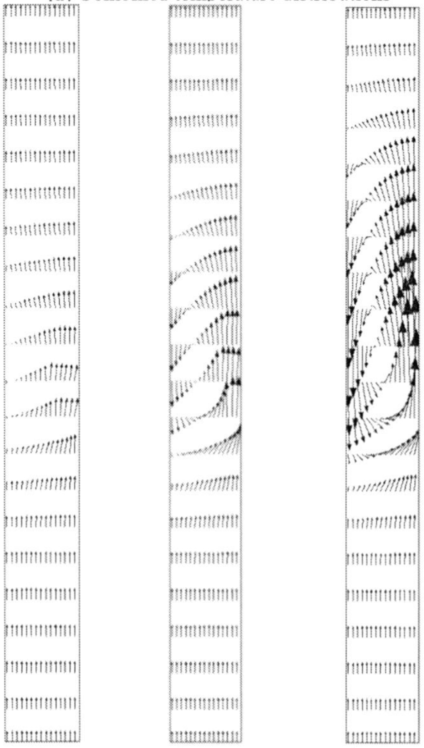

$K^* = 10^{-9}$ $K^* = 5\times10^{-9}$ $K^* = 10^{-8}$

(c) Liquid velocity vectors

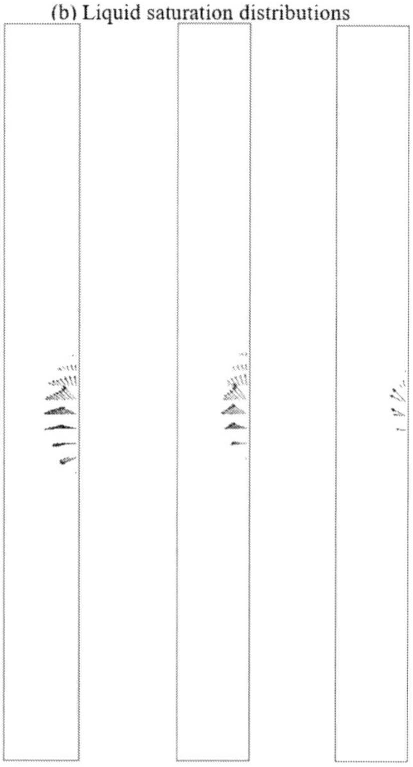

$K^* = 10^{-9}$ $K^* = 5\times10^{-9}$ $K^* = 10^{-8}$

(d) Vapour velocity vectors

Fig. 6. Effects of K^* on (a) combined temperature distribution, (b) liquid saturation distribution, (c) liquid velocity vectors and (d) vapour velocity vectors for $\theta = 90^\circ$, $Re_l = 1.5$ and $t^* = 0.845$.

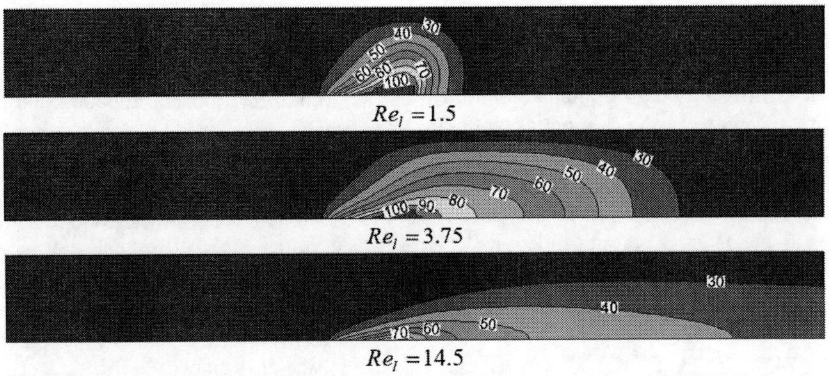

Fig. 7. Combined temperature distributions for various Re_l with $\theta = 30°$, $K^* = 10^{-8}$ and $t^* = 0.845$.

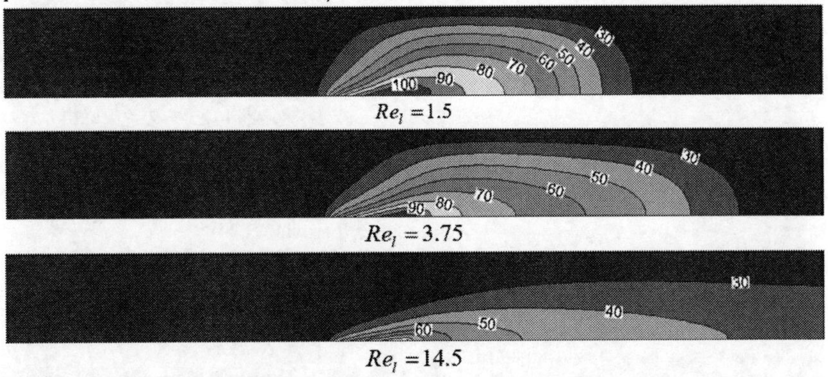

Fig. 8. Combined temperature distributions for various Re_l with $\theta = 60°$, $K^* = 10^{-8}$ and $t^* = 0.845$.

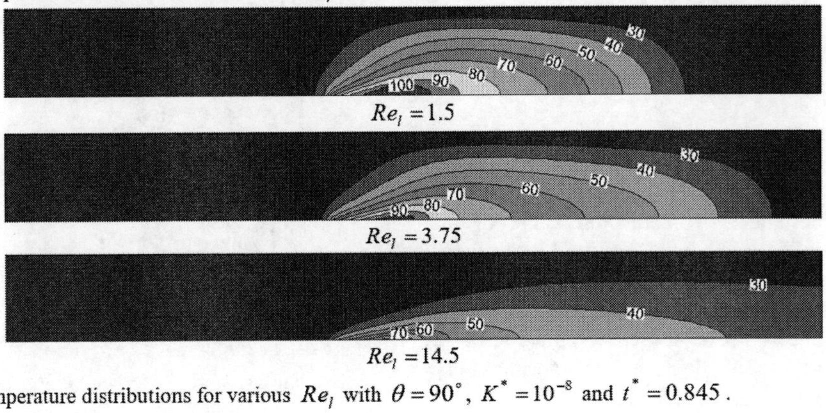

Fig. 9. Combined temperature distributions for various Re_l with $\theta = 90°$, $K^* = 10^{-8}$ and $t^* = 0.845$.

In order to demonstrate the usefulness of the modified h-formulation [25] for multi-dimensional problems, under unsteady-state condition, the solutions of two-phase flow obtained using the current formulation has been compared with those obtained using the H-formulation [9] under the same conditions. The comparison is presented in terms of the evaporated volume fraction ev as a function of the inclination angle, as shown in Fig. 10. The ev has been used to describe the importance of boiling and it has been defined in dimensional form as [9] [12]:

$$ev = \frac{W^*}{L^*} \int_0^1 \int_0^{L^*/W^*} (1 - s)\, dx^* dy^* \qquad (13)$$

It may be mentioned here that all reported computations have been carried out on a desktop personal computer (Intel Core i7 2.80 GHz CPU with 4 cores, each with 4 GB DDR3 1333 MHz RAM made by Hynix). The typically required computation time on this machine for the present study has been recorded to vary from 1 hour to 2 hours. It is evident from Fig. 10 that the results obtained from modified h- formulation provide an excellent agreement with that obtained by using H-formulation [9] for all tested cases. It is clear from the figure that the volume of vapour

[12] Najjari and Nasrallah [9] pointed out that the evaporated volume fraction is a direct measure of the boiling strength and the extent of the resulting two-phase and vapour regions.

decreases for a higher K^* and Re_l as well as for an increasing inclination angle θ. It has been found that the *h*-formulation requires significantly less computation time. The average absolute difference in the evaporated volume fraction has been found to be 0.05%. Particularly for these cases, the modified *h*-formulation requires approximately 45% less computational time as compared to the existing *H*-formulation [6] and hence it is recommended for the simulation of the two-phase flow inside porous media.

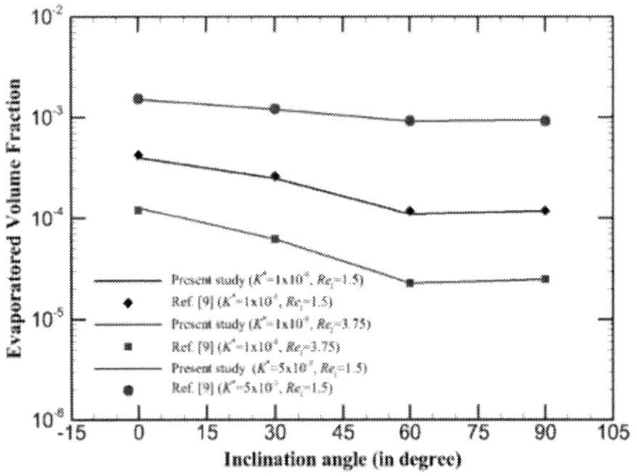

Fig. 10. Comparison of evaporated volume fraction determined in the present study and results of Najjari and Nasrallah [9] for various θ at different K^* and Re_l for $Q_w^* = 4.75 \times 10^{-1}$ and $t^* = 0.845$.

5 Conclusions

Numerical simulations of two-phase flow of water inside an inclined porous evaporator, based on the modified *h*-formulation [25], have been investigated in this study in order to assess the effect of inclination angle of the evaporator. The flow has been assumed to be unsteady as well as two-dimensional. All the governing equations have been discretised using the Finite Volume Method (FVM) and the resultant coupled set of discretised equations have been solved iteratively using SIMPLE-like algorithm [34]. Numerical simulations of incomplete phase change process have been obtained for similar geometry with identical heat flux value of Najjari and Nasrallah [9] for the purpose of comparison. The effects of inclination angle, inlet Reynolds number, Darcy number and total evaporated volume fraction have been investigated. The results obtained from modified *h*- and *H*-formulation [6] have been compared. The major conclusions from the present study may be summarised as follows:

1. The results clearly demonstrated that the boiling depends strongly on inclination angle of the channel.
2. Recirculating flow has been observed except when the channel in a horizontal position.
3. The results also indicated that the presence of a critical inclination angle θ depending on the value of K^* and Re_l which correspond to a maximum values of vapour volume. The results show that Re_l and K^* have a strong effect on the initiation and termination of phase change process.
4. Nearly identical solutions of the evaporated volume fraction predicted by both modified *h*-formulation and the existing H -formulation that used by Najjari and Nasrallah [9] for all tested cases demonstrate the consistency of the modified h -formulation. It has been also observed that the modified h -formulation requires significantly less computation time as compared to that for the H -formulation [6].

References

1. C.Y. Wang and P. Cheng, *Advances in Heat Transfer*, Academic Press, NewYork **30**, 93–196 (1997)
2. J. Bear, *Dynamics of Fluids in Porous Media* (Elsevier, New York, 1972)
3. A.E. Scheidegger, *The Physics of Flow Through Porous Media* (3rd Ed.. University of Toronto Press. Toronto, 1974)
4. C.Y. Wang and C. Beckermann, Int. J. Heat Mass Transfer **36**, 11, 2747–2758 (1993)
5. C.Y. Wang, C. Beckermann and C. Fan, Heat Transfer, Part A **26**, 375–398 (1994)
6. C.Y. Wang, Numer. Heat Transfer, Part B **32,** 85–105 (1997).
7. T.S. Zhao and Q. Liao, J. Thermophys. Heat Transfer **13**, 3, 302–307, (1999)
8. T.S. Zhao, P. Cheng and C.Y. Wang, Chem. Eng. Sci. **55**, 2653–2661 (2000)
9. M. Najjari and S.B. Nasrallah, J. Porous Media **6**, 1, 71–81 (2003)
10. M. Najjari and S.B. Nasrallah, J. Porous Media **8**, 1, 1–12 (2005)
11. M. Najjari and S.B. Nasrallah, J. Porous Media **9**, 7, 671–681 (2006).
12. M. Najjari and S.B. Nasrallah, Int. J. Therm. Sci. **47**, 825–833 (2008)
13. M. Najjari and S.B. Nasrallah, Drying Technol. **27**, 336–343 (2009)
14. H.Y. Li, K.C. Leong, L.W. Jin and J.C. Chai, Int. J. Therm. Sci. **49**, 1115–1127 (2010)
15. H.Y. Li, K.C. Leong, L.W. Jin and J.C. Chai, Int. J. Heat Mass Transfer **53**, 5209–5222 (2010)
16. H.Y. Li, K.C. Leong, L.W. Jin and J.C. Chai, Int. J. Therm. Sci. **49**, 2363–2375 (2010)
17. G.P. Peterson and C.S. Chang, Numer. Heat Transfer, Part A **31**, 2, 113–130 (1997)
18. J.X. Shi and J.H. Wang, Transp. Porous Media. **87**, 703–716 (2011)
19. F. Lindner, C. Mundt and M. Pfitzner, Transp. Porous Media., **106**, 201–220 (2014)
20. O. R. Alomar, M.A.A. Mendes, D. Trimis, S. Ray, Int. J. Therm. Sci. **86**, 408-420 (2014)

21. O. R. Alomar, M.A.A. Mendes, D. Trimis, S. Ray, Int. J. of Material, Mechanics and Manufacturing. **2**, 3, 223–229 (2014)
22. O. R. Alomar, D. Trimis, S. Ray, *Numerical simulation of complete liquid-vapour phase change process inside an annular porous evaporator* (10th international conference on Freiberg-St. Petersburg Colloquium of Young Scientists, 2015)
23. O. R. Alomar, M.A.A. Mendes, D. Trimis, S. Ray, Int. J. Therm. Sci. **94**, 228–241 (2015)
24. O. R. Alomar, M.A.A. Mendes, D. Trimis, S. Ray, Int. J. Therm. Sci. **112**, 222–241 (2017).
25. S. Ray, O. R. Alomar, Int. J. Therm. Sci. **105**, 123–136 (2016)
26. O. R. Alomar, M. A.A. Mendes, S. Ray, D. Trimis, Int. J. Therm. Sci. **129**, 56–72 (2018)
27. O. R. Alomar, R. R. Mohammed. M.A.A. Mendes, S. Ray , D. Trimis, Int. J. Therm. Sci. **135**, 1–16 (2019)
28. D.A. Nield, A. Bejan, *Convection in Porous Media* (Third ed., Springer, New York, 2006)
29. M. Kaviany, *Principles of Heat Transfer in Porous Media* (Second Ed. Springer-Verlag, New York, 1995)
30. M.A.A. Mendes, S. Ray, D. Trimis, Int. J. Heat Mass Transfer. **66**, 412-422 (2013)
31. M.A.A. Mendes, S. Ray and D. Trimis, Int. J. Therm. Sci. **79**, 260-265 (2014)
32. M.A.A. Mendes, S. Ray and D. Trimis, Int. J. Heat Mass Transfer. **75**, 224-230 (2014)
33. M.A.A. Mendes, S. Ray and D. Trimis, *Proc. 22nd National & 11th ISHMT-ASME Heat and Mass Transfer Conf.* (Indian Institute of Technology, Kharagpur, India, December 28–31, 2013)
34. S.V. Patankar, *Numerical Heat Transfer and Fluid Flow* (Hemisphere, New York, 1980)
35. O. R. Alomar, *Modeling and Simulation of Complete Liquid-Vapor Phase Change Process inside Porous Media* (Cuvillier Verlag, 2016)
36. F.P. Incropera, D.P. Dewitt, T.L. Bergman, A.S. Lavine, *Fundamentals of Heat and Mass Transfer* (John Wiley & Sons, 2007)
37. C. Borgnakke, R.E. Sonntag, *Fundamentals of Thermodynamics* (John Wiley & Sons, 2009)

Influence of forced convection on the evaporation and internal dynamics inside of an array of salt solution droplets

Omar Al-Rawi[1,2,*], and *Mark* Wilson[1]

[1] School of Mechanical Engineering, University of Leeds, Leeds, LS2 9JT, UK
[2] State Company for Oil Projects, Ministry of Oil, Baghdad, Iraq

Abstract. The effects of a gentle forced air convection on the internal dynamics of an array of multiple pinned sessile salt solution droplets are investigated via fully-coupled transient ALE finite element analysis. Results highlight the competition between the shear-induced circulation within the droplets and the gravity-driven flow in the droplets arising from increasing liquid density in regions of high water evaporation. At low air speeds, gravity effects dominate, resulting in a non-uniform concentration distribution. However, at higher speeds the shear-induced circulation within the droplets becomes sufficient to mix the liquid within the droplets via a 3D flow pattern, resulting in greater concentration uniformity. In addition, the shielding effect of leading droplets on downstream droplets is explored for various air speeds, with results showing differences in average concentration levels.

Nomenclature

$\bar{c}, \bar{C}$	mol/m^3	solute, vapour concentration
j_r	mol/m^{2}s	total molar flux
ρ_l, ρ_v	kg/m^3	liquid, vapour water density
$\mathbf{n}$		unit surface normal vector
$\mathbf{T}$	Pa	stress tensor
D_s, D_v	m^2/s	solute, vapour diffusivity

1 Introduction

In recent decades, the dynamics inside multiple droplets evaporating close to each other has attracting interest in many industrial and scientific applications such as surface coatings [1], inkjet printing [2], and pharmaceuticals [3]. Experimental and numerical studies show that the dynamics inside a droplet having a pinned contact line depend on the evaporation conditions and the environment surrounding the droplet [4]. An outward flow towards the edge of the droplet producing a ring-like deposit is observed, which is caused by either the replenishment of liquid lost in this area or by edge enhanced evaporation [4, 5]. In contrast, a flow towards the centre of the droplet inhibiting the ring-like deposit is generated due to either centre enhanced evaporation or surface tension gradients causing Marangoni flow [6].

In droplets of evaporating aqueous solutions, another flow mechanism is observed. Kang *et al.* [7], Savino and Monti [8], Lee *et al.* [9], and Pradhan and Panigrahi [10, 11] found that due to the concentration gradients inside the droplet, buoyancy-driven convection is induced where the liquid moves along the droplet surface towards the contact line and towards the droplet apex in the central region of the droplet. Further observations reveal that this internal flow field is significantly affected by the solute concentration inside evaporating droplets [12].

The interaction between adjacent evaporating droplets via their shared vapour space affects the evaporation behaviour, and hence the flow patterns inside these droplets [13]. Davoust and Theisen characterised experimentally the evaporation rate of a population of droplets in a microsystem [14, 15]. Recently, Pradhan and Panigrahi [10] observed how the fluid convection inside an evaporating droplet is strongly affected by the influence of an adjacent droplet and the distance between them.

The importance of vapour/air convection, i.e. natural and forced, in the evaporation of sessile droplets has also been explored. Experimental work by Kelly-Zion *et al.* [16] examined the crucial role of natural convection in increasing the evaporation rates of various liquids (3-methylpentane, hexane, cyclohexane, and heptane) on an aluminium substrate. Saada *et al.* [17] observed numerically the effect of buoyant vapour convection on evaporation rates for different droplet radii. The experimental findings for salt solution droplet evaporation show the importance of gas convection around these droplets, which is significantly influenced by the salt concentration, and neglecting this mechanism leads to underestimation of evaporation rates [18, 19].

Numerically, Raghuram *et al.* [20] found that the flow-field inside the droplet induced by Marangoni effects is greatly affected by the forced convective environment. Doursat *et al.* [21] studied the evolution of the mass and heat transfer during the evaporation of multiple pure water droplets under forced convective environment. Chakraborty et al. [22] analysed the feasibility of using an array of evaporating droplets to dissipate the heat released from microprocessors. They noticed that providing a fan to generate forced air flow in order to remove the diffusive vapour between the

* Corresponding author: o.alrawi@hotmail.com

droplets would improve the evaporation rates of these droplets.

Though the above studies illustrate the importance of considering convection in the evaporation of droplets in pure water or binary solvent droplets, it appears that there are no prior studies of the effects of forced convection on the internal dynamics during evaporation of neighbouring salt solution droplets. This is therefore the focus of the present work.

2 Mathematical model

The system considered is a semi-infinite regular array of sessile salt solution droplets, over which a laminar airflow passes, as shown in Figure 1. The repeating pattern of the geometry is exploited to reduce the computational domain to a unit flow consisting of two droplets aligned parallel with the airflow, with periodic conditions along the side boundaries.

2.1 Assumptions

Droplets of aqueous NaCl solution with initial volume 2.1 μl and 1M concentration are used to investigate the effect of the forced convective environment on the solute distribution inside the droplets. For this type of salt solution, the experimental observations of Kang *et al.* [7] show that, in the absence of an overlying airflow, the flow inside the droplet is driven by convection due to density gradients as a results of increasing salinity, and Marangoni effects are negligible. The system is also assumed to be in thermal equilibrium and thermal effects are neglected. Hence surface tension gradients are ignored here. The effect of gravity is captured via the Boussinesq approximation.

The contacts lines of the droplets are assumed to be pinned, with an initial contact angle of 90°, but the free surfaces otherwise evolve in a fully transient manner in response to local mass loss due to water evaporation. Both liquid and air are modelled as incompressible.

2.2 Governing equations

Based on the above assumptions, the species transport, continuity and momentum equations are:

$$\frac{\partial \bar{c}}{\partial t} + \mathbf{u}.\nabla\bar{c} = \nabla.(D\nabla\bar{c}) \tag{1}$$

$$\nabla.\mathbf{u} = 0 \tag{2}$$

$$\rho\left(\frac{\partial \mathbf{u}}{\partial t} + (\mathbf{u}.\nabla)\mathbf{u}\right) = \nabla.[-p\mathbf{I} + \mu(\nabla\mathbf{u} + (\nabla\mathbf{u})^{T})] - \Delta\rho g\mathbf{k} \tag{3}$$

where $\bar{c}$ is the molar concentration field and D is the diffusion coefficient in the relevant fluid.

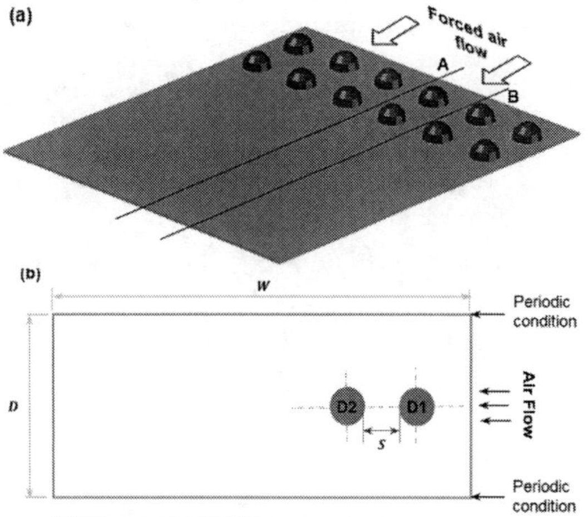

Key	Description
W	Width of a unit cell
D	Depth of a unit cell
D1	Lead droplet
D2	Downstream droplet
S	Downstream droplet distance from the lead droplet in air flow direction

Fig. 1. Sessile droplet array considered (a), and plan view of the representative computational domain (b).

2.3 Initial and boundary conditions

The droplets are initially hemispherical, with a uniform solute concentration of 1M. During evaporation, the solute concentration $\bar{c}$ increases at the droplet free surfaces, and diffuses to the inner regions of the droplets. At the free surfaces, the condition on the solute concentration is therefore

$$-D_s\nabla\bar{c}\cdot\mathbf{n} = -\frac{\bar{c}\,\mathbf{j}_{\Gamma}}{\rho_l}\bar{M}_{H_2O} \tag{4}$$

where

$$\mathbf{j}_{\Gamma} = -D_v\nabla\bar{C} + \mathbf{u}\bar{C} \tag{5}$$

is local total molar flux of water vapour at the free surface and $\bar{C}$ is the molar water vapour concentration. At the droplet surfaces, the concentration of water vapour is assumed to be saturated, with an adjustment for the solute presence via Raoult's law [10], hence $\bar{C} = 0.93$ mol/m^3 at the free surfaces. At the outer boundaries, the vapour concentration is also fixed in terms of saturation conditions, but assuming a relative humidity of 60% and ambient temperature of 20°C. The velocity and stress conditions at the free surfaces are given by:

$$\mathbf{u}_v - \mathbf{u}_l = \bar{M}_{H_2O}\left({}^1/_{\rho_v} - {}^1/_{\rho_l}\right)(\mathbf{j}_\Gamma \cdot \mathbf{n})\mathbf{n} \quad (6)$$

$$\mathbf{n} \cdot \mathbf{T}_v - \mathbf{n} \cdot \mathbf{T}_l = \sigma(\nabla \cdot \mathbf{n})\mathbf{n} \quad (7)$$

where $\mathbf{T}$ is the stress tensor, and the subscripts *v* and *l* on **u** and **T** refer to the values of these entities on the vapour and liquid sides of the interface respectively. Condition (6) accounts for the increase in specific volume of water as it evaporates, resulting in a Stefan flow, while condition (7) incorporates both the capillary pressure and shear stress of the airflow acting on the liquid.

At fluid-solid boundaries, the no-slip and no-flux conditions are applied, while the airflow inlet is a fully-developed laminar flow with entrance length $0.06ReD_H$. At the outlet, $-\mathbf{n}.\, D_v \nabla C = 0$ and atmospheric pressure constraints are applied. As noted above, periodic conditions are applied along the sides of the domain, which is chosen to have a streamwise length of 60 mm, a spanwise width of 20 mm and a height of 10mm.

2.4 Numerical method

Equations (1)-(3) were solved subject to the above conditions via a 3D transient Arbitrary Lagrangian-Eulerian finite element approach implemented in COMSOL Multiphysics®, v. 5.3. An unstructured quadratic tetrahedral mesh was used, where nodes on the free surfaces were displaced according to the local evaporative mass loss, with a velocity given by

$$\dot{\mathbf{x}}_\Gamma = (\mathbf{u}_l \cdot \mathbf{n})\mathbf{n} - \frac{(\mathbf{j}_\Gamma \cdot \mathbf{n})\bar{M}_{H_2O}}{\rho_l}\mathbf{n} \quad (8)$$

Nodes in the interior of the liquid and vapour regions were moved via the Winslow mesh deformation method, subject to equation (8) at the free surface and zero mesh displacement at the outer boundaries.

3 Results and discussion

The forced convection air speeds considered here are U = 0.08, 0.2 and 2.0 m/s. The value 0.08 m/s was chosen as it was the lowest non-zero speed for which mesh-independent results could be obtained with the available computational resources. Below this value, the velocities on the free surfaces are very small, and this is an area of parameter space that is currently being further explored. The speed was limited to 2 m/s to ensure that the air flow remained laminar, and that no displacement of the droplets should be expected [23]. Note that, 3 mm is considered here as a separation distance between the two droplets.

3.1 Validation against experiments

Before using the model to explore the effect of forced convection on the salt solution droplets, its predictions for natural convection (i.e. no imposed airflow) were first tested against the experimental results of Pradhan and Panigrahi [10], with all parameters specified as in Ref. [10]. Figure 2 shows predictions of the velocity

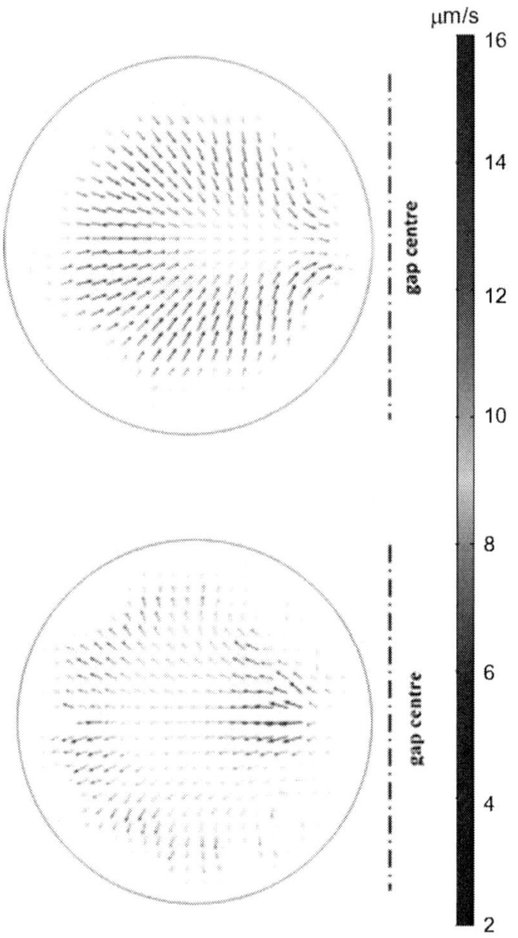

Fig. 2. Numerical results for the velocity field on horizontal places inside the left-hand droplet of two adjacent droplets. The planes shown are at heights (a) 50 µm and (b) 350 µm above the substrate.

vectors in horizontal planes at heights of 50 and 350 µm above the substrate. These show the same general flow patterns as the corresponding experimental velocity fields.

3.2 Distribution of solute inside droplets

In a natural convection environment, the salt concentration increases along the droplet free surfaces due to the water evaporation from those surfaces. This induces a gravity-driven flow due to the increase in density with increasing salinity, and a non-uniform distribution of solute concentration within the droplets.

The presence of a gentle forced air convection changes the flow and salt distribution inside droplets, as a result of the shear-induced flow within the droplet caused by the passing air. Figure 3 shows the resulting velocity field in the lead (i.e. upstream) droplet on a central vertical streamwise plane, along with the velocity field on the interior side of the free surface (recall that the velocity field has different values on the interior and exterior sides of the free surface, as expressed in equation 6). Note the difference in order of magnitude of the scale bars ($\times10^{-4}$, $\times10^{-3}$ and $\times10^{-2}$ m/s). In each case the maximum speed of the flow induced by the air is approximately 1% of the air speed.

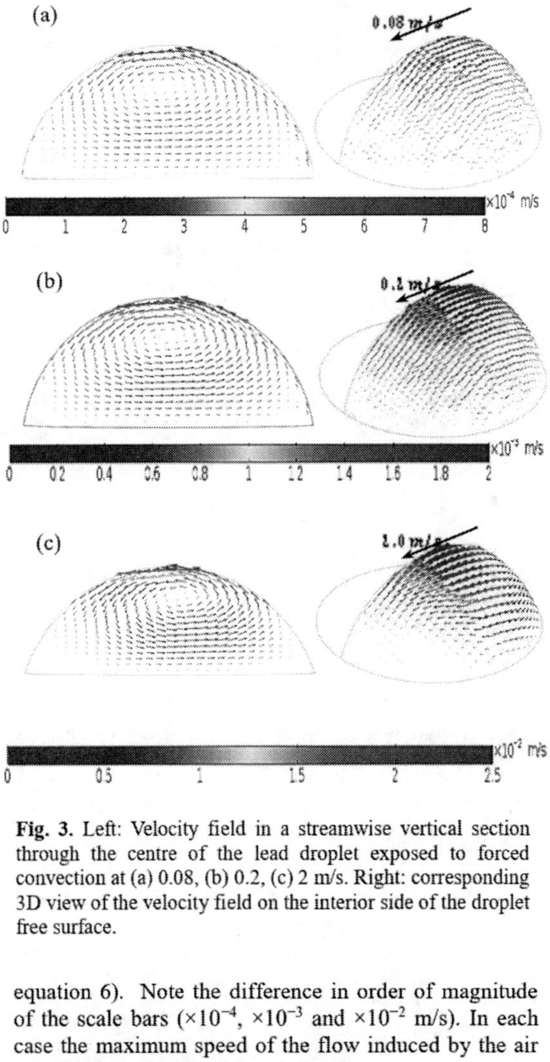

Fig. 3. Left: Velocity field in a streamwise vertical section through the centre of the lead droplet exposed to forced convection at (a) 0.08, (b) 0.2, (c) 2 m/s. Right: corresponding 3D view of the velocity field on the interior side of the droplet free surface.

Figure 4 shows the distribution of the solute concentration on the vertical centre plane passing through both droplets, at time 300 s. At the lowest speed, the concentration distribution in the upstream droplet (the rightmost image in Figure 4a) looks very similar to that in droplets under natural convection, since the shear- induced flow is very small. A clear non-uniformity is observable. Generally, it can be noticed that the salt concentration is higher near the upstream sections of the of droplet surfaces, since the highest evaporative flux occurs at these regions. The salt concentration gradually decreases towards the apex region of the droplets surface, and then increases gradually towards the downstream contact-line region, where there is also enhanced evaporation as a result of the sharp vapour concentration gradients there (and all along the contact lines).

Considering the downstream droplet (left-hand droplet in Figure 4a), it is very noticeable that the concentration level in this droplet is significantly less than in the upstream droplet. This is because the water vapour evaporated from the upstream droplet is transported downstream to the other droplet, raising the vapour concentration levels there and hence reducing the local concentration gradients and the associated evaporative flux given by equation (5). When the forced convection speed is increased to 0.2 m/s, the greater magnitude of the shear-driven circulation pushes the regions of non-uniform concentration closer to the contact lines. The average concentration level is also raised a little by the convection-enhanced evaporation.

At the maximum speed considered, see Figure 4(c), the shear-induced flow within the droplets dominates the gravity-driven flow, and is sufficient to ensure effective mixing of the solute. This results in much more uniform concentration fields in the two droplets (which is still observed when the contour scales are more focused than shown in Figure 4). As expected, the faster air flow increases the evaporation rates of the droplets. However, there is proportionally bigger increase in the evaporation rate of the upstream droplet, and consequently a proportionally bigger difference in the average concentration levels of the two droplets compared to the 0.2 m/s case. While the upstream droplet is exposed to 'fresh' air, the downstream droplet sits in the wake of the upstream droplet and consequently experiences both a shielding effect and raised surrounding vapour levels from the downstream water vapour transport.

3.2 Distribution of water vapour inside the domain

For a more visual depiction of the 'reach' of the upstream droplet, Figure 5 shows 3D views of the vapour concentration field plotted on the bottom surfaces of the domain and the droplet surfaces. These clearly show how increasing the air speed affects the vapour space between the two droplets, and how the downstream droplet is influenced by the shielding effect of the lead droplet and by the downstream transport from the lead droplet. Consequently, different evaporative behavior is expected between the two droplets. It is worth recalling that periodic conditions are applied in this problem, to represent the transverse array of droplets shown in Figure 1. At zero or low air speed, transverse diffusive transport of water vapour will affect the evaporation rates of the droplets in the domain, particularly at small domain widths. As the droplet wakes show, increasing the air speed will make this transverse diffusive transport within the array negligible.

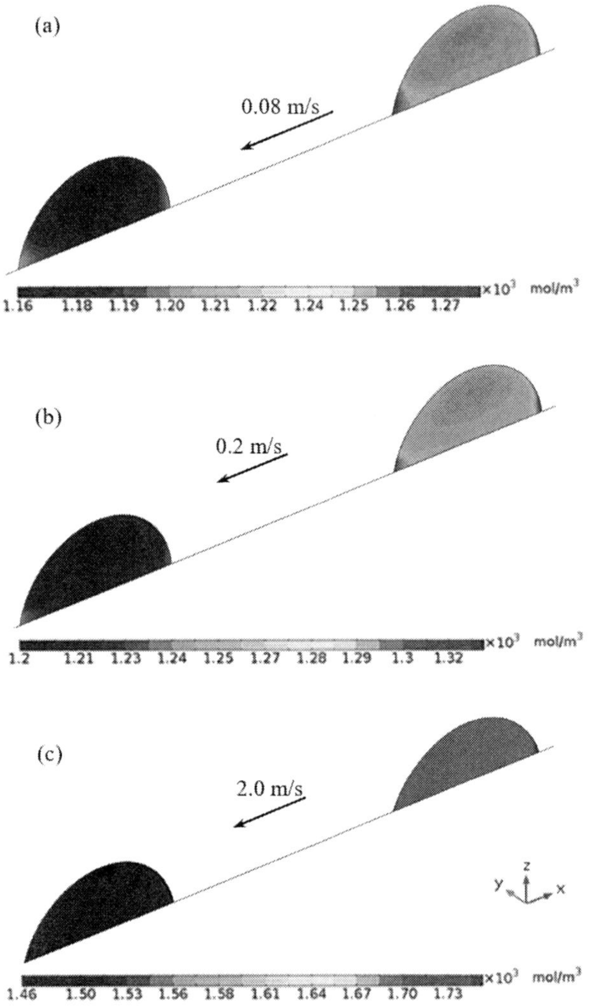

Fig. 4. Contour plot of the solute concentration field (after 300 s) at a droplet separation distance of 3mm and various air speeds.

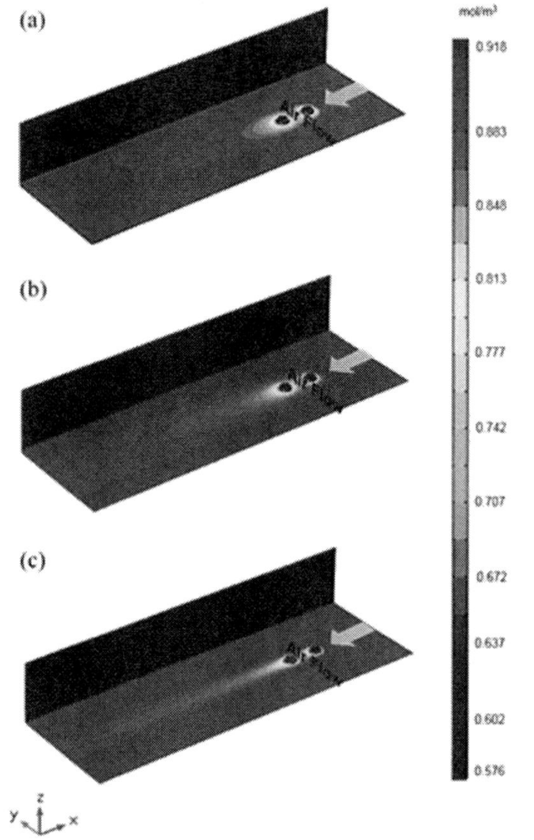

Fig. 5. Vapour concentration on the bottom surfaces of the domain at air speed (a) 0.08, (b) 0.2, (c) 2 m/s and at 3 mm separations.

4 Conclusions

The key observations from this fully-coupled 3D transient finite element study are as follows:

- The forced convective airflow over sessile salt solution droplets produced a circulation flow within the droplets with a maximum speed approximately 1% of the air speed.
- As air speed increased, the shear-induced circulation becomes sufficient to mix the solution effectively, producing much more uniform concentration distributions in the droplets. Strong variations in concentration still exist near contact lines, but these regions become more localised as air speed increases.
- Downstream vapour transport from the upstream droplet significantly affects the evaporative behaviour of the downstream droplet by raising the surrounding vapour concentration levels and hence reducing concentration gradients.

Omar Al-Rawi thanks the Higher Committee for Education Development in Iraq for funding this work.

References

1. Kimura, M., et al., Long-Range Ordering of Diblock Copolymers Induced by Droplet Pinning, Langmuir, **19**(23), pp. 9910-9913 (2003)
2. Park, J. and J. Moon, Control of Colloidal Particle Deposit Patterns within Picoliter Droplets Ejected by Ink-Jet Printing, Langmuir, **22**(8), pp. 3506-3513(2006)
3. Vehring, R., Pharmaceutical Particle Engineering via Spray Drying, Pharmaceut. Res., **25**(5), pp. 999-1022 (2008)
4. Fischer, B.J., Particle Convection in an Evaporating Colloidal Droplet, Langmuir, **18**(1), pp. 60-67 (2002)
5. Deegan, R.D., et al., Capillary flow as the cause of ring stains from dried liquid drops, Nature, **389**(6653), pp. 827-829 (1997)

6. Hu, H. and R.G. Larson, Marangoni Effect Reverses Coffee-Ring Depositions, J. Phys. Chem. B, **110**(14), pp. 7090-7094 (2006)
7. Kang, K.H., et al., Evaporation-induced saline Rayleigh convection inside a colloidal droplet, Phys. Fluids, **25**(4), pp. 042001 (2013)
8. Savino, R. and R. Monti, Buoyancy and surface-tension-driven convection in hanging-drop protein crystallizer, J. Cryst. Growth, **165**(3), pp. 308-318 (1996)
9. Lee, S.J., J. Hong, and Y.-S. Choi, Evaporation-Induced Flows inside a Confined Droplet of Diluted Saline Solution, Langmuir, 30(26), pp. 7710-7715 (2014)
10. Pradhan, T.K. and P.K. Panigrahi, Influence of an adjacent droplet on fluid convection inside an evaporating droplet of binary mixture, Colloid. Surface. A, 500, pp. 154-165 (2016)
11. Pradhan, T.K. and P.K. Panigrahi, Evaporation induced natural convection inside a droplet of aqueous solution placed on a superhydrophobic surface, Colloid. Surface. A, **530**, pp. 1-12 (2017)
12. Xu, G., et al., Effect of Salt Concentration on the Motion of Particles near the Substrate in Drying Sessile Colloidal Droplets, Langmuir, **33**(3), pp. 685-695 (2017)
13. Pradhan, T.K. and P.K. Panigrahi, Deposition pattern of interacting droplets, Colloid. Surface. A, **482**, pp. 562-567 (2015)
14. Davoust, L. and J. Theisen, Evaporation Rate of Drop Arrays within a Digital Microsystem, Procedia Engineering, **47**, pp. 1-4 (2012)
15. Davoust, L. and J. Theisen, Evaporation rate of drop arrays within a digital microfluidic system, Sensor. Actuat. B-Chem., **189**, pp. 157-164 (2013)
16. Kelly-Zion, P.L., et al., Evaporation of sessile drops under combined diffusion and natural convection, Colloid. Surface. A, **381**(1–3), pp. 31-36 (2011)
17. Ait Saada, M., S. Chikh, and L. Tadrist, Numerical investigation of heat and mass transfer of an evaporating sessile drop on a horizontal surface, Phys. Fluids, **22**(11): Art. No. 112115 (2010)
18. Misyura, S.Y., Free convection and vapor diffusion of droplet aqueous solutions, Chem. Eng. Res. Des., **126**(Supplement C), pp. 153-160 (2017)
19. Misyura, S.Y., Evaporation of a sessile water drop and a drop of aqueous salt solution, Sci. Rep., **7**(1), pp. 14759 (2017)
20. Raghuram, S., et al., Numerical study of Marangoni convection during transient evaporation of two-component droplet under forced convective environment, Int. J. Heat Mass Tran., **55**(25–26), pp. 7949-7957 (2012)
21. Doursat, C., et al., Droplet evaporation on a solid surface exposed to forced convection: Experiments, simulation and dimensional analysis, Int. J. Heat Mass Tran., **113**, pp. 1234-1245 (2017)
22. Chakraborty, S., M.A. Rosen, and B.D. MacDonald, Analysis and feasibility of an evaporative cooling system with diffusion-based sessile droplet evaporation for cooling microprocessors, Appl. Therm. Eng., **125**, pp. 104-110 (2017)
23. Fan, J., M.C.T. Wilson, and N. Kapur, Displacement of liquid droplets on a surface by a shearing air flow, J. Colloid Interf. Sci., **356**(1), pp. 286-292 (2011)

https://doi.org/10.1051/matecconf/201824001003

Numerical investigation for convective heat transfer and friction factor under pulsating flow conditions

Erman Aslan[1,*], *Mert* Ozsaban[2], *Guven* Ozcelik[3], and *Hasan Riza* Guven[1]

[1]Mechanical Engineering Department, Istanbul University-Cerrahpasa, Avcilar Campus, 34320 Avcilar, Istanbul, Turkey
[2]Mechanical Engineering Department, Recep Tayyip Erdogan University, Zihni Derin Campus, 53100 Merkez, Rize, Turkey
[3]Mechanical Engineering Department, Istanbul Arel University, Tepekent Campus, 34537 Buyukcekmece, Istanbul, Turkey

Abstract. For pulsating flow, the behaviours of the convective heat transfer and friction factor for a periodic corrugated channel are investigated numerically. The finite volume method is used in the numerical study. Three different Reynolds Averaged Numerical Simulation based turbulence models, namely the k-ω model, the Shear Stress Transport (SST) model and the transition SST model are used and compared. The results are also compared with the previous experiments for non-pulsating flow. Analyses are conducted for air flow through a corrugated channel which has sharp corrugation peaks with an inclination angle of 30° and a 5mm minimum channel height. Reynolds number is changed in the range 6294 to 7380, while keeping the Prandtl number constant at 0.70. A sinusoidal pulsatile flow condition which is F=400 and u_A^*=0.5 is used. Variations of the Nusselt number and the friction factor with the Reynolds number are studied. Non-pulsating flow results and pulsating flow results are compared with each other.

Nomenclature

D_h: hydraulic diameter, m
f: friction factor, frequency 1/s
F: nondimensionalized frequency
H: channel height, m
h: cycle average heat transfer coefficient, W/(m^{2}K)
h: axially local heat transfer coefficient, W/(m^{2}K)
k: thermal conductivity, W/(mK)
Nu: Nusselt number
p: Pressure, Pa
Re: Reynolds number
S: pitch (axial length of cycle), m
T: temperature, K
t: time, s
u: velocity, m/s
x: axial coordinate, m

Greek symbols

θ: contact angle, °
ρ: density, kg/m^3
μ: dynamic viscosity, kg/(m.s)
ν: kinematic viscosity, m^2/s

Sub- and Superscripts:

A: amplitude
min: minimum
max: maximum
0: uniform inlet
$*$: nondimensionalized

1 Introduction

Lately, because of the increasing demand by industries including power generation [1] and heat recovery [2] for more efficient and compact exchangers [3, 4], exploration of heat transfer enhancement methods has gained great momentum. To this purpose, two techniques can generally be identified, namely, the active and passive techniques. Active techniques include surface vibration, fluid vibration, electrostatic fields, boundary layer injection, boundary layer suction, electro-hydrodynamics, etc. for increasing heat transfer. Thus, active techniques require more costs and attention.

In passive techniques, besides using fluid additives, the applied approaches generally involve a modification of the channel in a way to increase the heat transfer rate. A special category of heat transfer enhancement is the use of bluff bodies that increase heat transfer coefficient by turbulence generation and vortex shedding in their wake [5, 6]. A further category is the insertion of structures that interrupt the wall boundary layer and create recirculation and turbulence near the wall [7]. Passive methods may also include categorical changes in the flow arrangement, such as the use of jet impingement techniques [8, 9] instead of wall-congruent flow, which is, however, out of the present scope. Enhancement of the radiative heat transfer [10] is also not included in the present scope. The present focus is on the enhancement of heat transfer by forced convection. In the present investigation, a combination of passive and active techniques is considered. The use of corrugated channels

* Corresponding author: erman.aslan@istanbul.edu.tr

and pulsatile inlet flow conditions belong to the passive and active categories, respectively.

Experimental investigations of forced convection in corrugated channels were performed by Sparrow and Hossfeld [11], Snyder et al. [12], Bilen et al. [13] and Nilpueng and Wongwises [14], who investigated the velocity and temperature fields experimentally. Computational investigation of force forced convection in corrugated channel in turbulent regimes were also performed to get convective heat transfer coefficient and pressure drop by Ciafalo et al. [15], Mirzaei et al. [16].

Pulsating flows are encountered in many engineering applications and also in nature and biological systems. One of the engineering applications is pulsating flow in pipe and channels. Ray et al. [17] studied analytically and numerically to find development length of sinusoidally pulsating laminar pipe flows in moderate and high Reynolds number regimes. Hydrodynamics of piston-driven laminar pulsating flow for developing flow was studied numerically at Re=1000 by Aygun and Aydin [18]. Five different values of the frequencies were considered. They found that the friction coefficient increases with frequency. In second paper of Aygun and Aydın, hydrodynamics of piston-driven laminar pulsating flow for fully developed flow was investigated numerically, analytically and experimentally [19]. Again five different frequency values were determined like in first paper [18]. The near/off-wall flow reversals are at F=105, 226 and 402, where F is the non-dimensional frequency. Numerical studies of flow and heat transfer in circular tube under pulsating flow condition were carried out in the laminar regime by Chattopadhyay et al [20]. The frequency interval was taken from 0 to 20Hz, and amplitude was determined to be greater than one. It was observed that pulsation has no positive effect in augmenting heat transfer within the investigated range of pulsation frequency and amplitude. A computational investigation of the pulsating flow and heat transfer in a straight circular tube, for the turbulent flow regime was presented by Benim et al. [21], where no significant enhancement of the time-averaged heat transfer rate by pulsations was observed for the considered conditions.

A perturbation approach was used to obtain analytical analytical expression for the velocity and temperature distribution and transient Nusselt number for the problem of forced convection, in a parallel plates or a circular tube by Nield and Kuznetsov [22]. Pulsating flow conditions were formed via applied pressure gradient that fluctuates with small amplitude harmonically in time. It was observed that, Nusselt number alters with frequency. In another investigation, performance of closed-loop pulsating heat pipes with uniform and altering tube diameter was studied experimentally by Tseng et al. [23]. Metha and Khandekar [24] investigated heat transfer of single-phase pulsating laminar flow in a square mini-channel, experimentally. According to the results, heat transfer enhancement due to periodic pulsating flow depends on frequency. With increasing frequency, heat transfer rate increases. The effect on flow and heat transfer characteristics for pulsating laminar flow in a heated square cylinder was investigated numerically by Yu et al. [25], where pulsation amplitude and frequency were changing in the range 0.2 - 0.8[-] and 0 - 20Hz, respectively. It was found [24] that when the pulsating frequency was about the natural vortex shedding frequency, heat transfer was also substantially enhanced. Convection heat transfer in a corrugated channel was investigated by Jafari et al. [26] for pulsating flow conditions using the Lattice Boltzmann Method. The dimensionless frequency of pulsating velocity and oscillating amplitude were studied at a wide range (Strouhal number, from 0.05 to 1, and amplitude, from 0 to 0.25, where amplitude is equal to "0" represents the steady flow). It was found that the best effect of flow pulsation on convection rate was achieved for a special value of the oscillation frequency, which was depending on the Reynolds number. Pulsating flows also occur in biological systems, such as the blood flow in human aorta under physiological, or surgery conditions, and have been investigated by different researchers [27].

In the present study, convective heat transfer and friction factor characteristics in a corrugated channel with pulsating flow conditions are investigated numerically using the finite volume method. Only one type corrugation geometry is used which has sharp corrugation peak. One sinusoidal flow condition is considered. Three different RANS based turbulence models, namely the k-ω, the Shear Stress Transport (SST) and the transition SST, are used.

2 Problem definition

A symmetrically corrugated channel with sharp corrugation peaks is considered. The schematic of channel with the geometry parameters is shown in Figure 1. Contact angle (θ) is 30°, pitch or axial length of cycle (S) is 17.32mm, channel half height (b) is 7.5mm and height of corrugation (a) is 5mm.

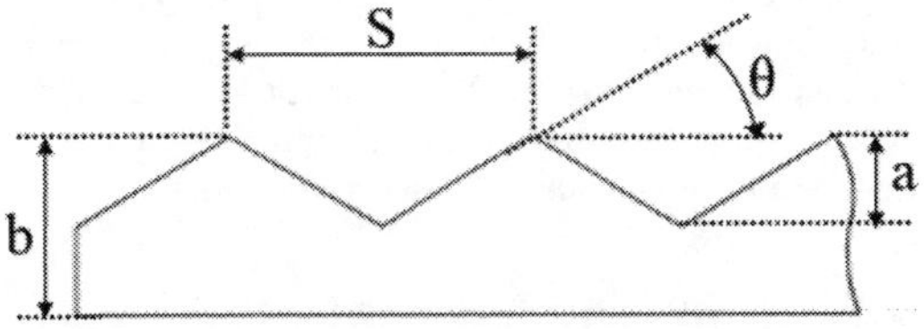

Fig. 1. Schematic figure of corrugation channel with the representative parameters.

According to these parameters maximum channel height is defined as;

$$H_{max} = 2b \quad (1)$$

while the minimum channel height is given by;

$$H_{min} = 2(b-a) \quad (2)$$

The hydraulic diameter is defined as;

$$D_h = H_{min} + H_{max} \quad (3)$$

In the numerical study, Reynolds number (Re) is varied from 6294 to 7380, and Prandtl number is 0.7.

In the numerical calculatios, the flow and heat transfer were assumed to be two-dimensional. Material properties of air are assumed to be constant. Figure 2 shows a schematic of typical solution domain with boundary types. At the inlet, a sinuosidial velocity profile in time is applied for getting pulsating flow conditions. Sinusodial velocity profile is defined as;

$$u(t) = u_0\left[1 + u_A^* \sin(2\pi f t)\right] \quad (4)$$

where u_0 is the uniform inlet velocity, f is frequency, t is time and u_A^* is nondimensionalized amplitude. The nondimensionalized amplitude is defined as;

$$u_A^* = \frac{u_A}{u_0} \quad (5)$$

where u_A is amplitude. Nondimensionalized frequency F is defined as;

$$F = \frac{D_h^2 f}{\upsilon} \quad (6)$$

where υ is the kinematic viscosity. The Reynolds number is defined as

$$\mathrm{Re} = \frac{\rho u_0 D_h}{\mu} \quad (7)$$

where ρ is density and μ is dynamic viscosity. One sinusoidal pulsating flow condition which is F=400, u_A^*=0.5 solved numerically.

An outflow boundary condition is imposed at the outlet for the momentum equations. No-slip boundary condition is applied at walls. For the energy equation, constant temperature (300K) is imposed at the inlet. A constant heat flux (616 W/m^2) is applied at the walls. Zero temperature gradient is used at the outlet.

3 Modelling and Assumptions

The time-dependent incompressible Navier-Stokes equations coupled with the continuity equation, the energy equation and the turbulence model equations are solved using the finite volume based commercial CFD code Ansys-Fluent [28]. For turbulence modelling, three models, namely the *k-ω* model [29], the SST model [30] and the transition SST model [31] are employed. For modelling, the near-wall turbulence model, no wall functions approach [32] is used. The applied turbulence equations are capable of representing the near-wall turbulence, provided that the grid resolution is sufficiently fine.

For discretizing convective terms a second-order upwinding procedure is used [28]. A second-order implicit scheme is used for integration in time [28]. For pressure velocity coupling PISO algorithm is used [28]. Default under-relaxation factors (pressure: 0.3, momentum: 0.7, turbulence quantities: 0.8 and energy: 1.0) are used [28]. As the converge criteria, a residual value 10^{-6} is required for all equations except energy equations. For energy equation, residual value 10^{-8} is used. At the inlet boundary, the turbulence intensity is kept at 4% and the length scale was assumed to be one third of the hydraulic diameter.

A structured meshing strategy based on quadrilateral finite volumes is used. A grid independence study is performed for finding an adequate grid resolution for all turbulence models. The Nusselt number and the friction factor are taken as the quantities to be monitored. The grid independence study was performed for steady-state flow conditions. Doing so, the maximum temporal flow rate of the pulsating flow is assumed to be the flow rate in the steady-state. This means that the velocity $u_{\max} = u_0 + u_A$, is prescribed at the inlet, where u_0 is taken from the maximum Reynolds number. In order to cope with the requirements of turbulence models for modelling the near-wall turbulence, without using wall-functions, an all grids, it is ensured that the distance of first near-wall cell fulfill the condition of $y^+ < 1$ and at least three cells are contained in the region for $y^+ < 5$. For all turbulence models, and four cases, the grid independence mesh number is 192000. In the unsteady calculation, time step size has been chosen in such a way that the cell Courant number takes a value about unity.

Two parameters of interest for the present study are (1) Nusselt number, (2) friction factor. In numerical calculations, cycle averaged Nusselt number is calculated as

$$Nu = \frac{hD_h}{k} \quad (8)$$

where h is the cycle averaged heat transfer coefficient and k is the thermal conductivity. In numerical calculations, firstly, local heat transfer coefficients (h_x) at the previous three cycles are calculated with using the wall and fluid time-averaged temperatures. Finally, the cycle averaged heat transfer coefficient is obtained using the local heat transfer coefficients.

$$h = \frac{1}{3S}\int_0^{3S} h_x dx \quad (9)$$

In numerical simulations, two sections are specified for calculating the pressure gradient (dP/dx). The first section is placed at the beginning of the previous three cycles from the last cycle. Second section is located on the end of the previous cycle from the last cycle. Friction factor is calculated using pressure gradient as.

$$f = -\frac{dP}{dx} D_h \Big/ \frac{1}{2}\rho u_0^2 \quad (10)$$

4 Results

Figure 2 shows the axial velocity distribution predicted by (a) *k-ω*, (b) SST and (c) transition SST turbulence model. These predictions are obtained for the highest Reynolds number (Re=7380), for F=400 and u_A^*=0.5

The flow develops through the channel and the flow characteristics are observed to reach a certain periodicity for the last cycles of the corrugated channels for three turbulence models. Therefore, one can say that the flow is rather fully developed, in this sense, for the last cycles of the channel. For the lower Reynolds number, the flow is observed to reach the fully-developed state earlier.

The displayed plots show the velocity distribution in the cycle before the last, where the velocity field can be assumed to be fully developed. In all predictions, a jet-like forward flow in the central region can be observed, which is surrounded by a recirculation zone.

The predictions of the three turbulence models are qualitatively almost the same. Still, some differences can be observed. One can see that the positive axial velocities in the central region are predicted to be larger by the k-ω model, in comparison to the predictions of the other turbulence models. The transition SST model predicts the lowest values of the axial velocity in the central region, whereas the prediction of the SST model is in between (Fig. 2).

The predicted time-averaged temperature distribution by (a) *k-ω*, (b) SST and (c) transition SST turbulence model is shown in Figure 3 for the highest Reynolds number (Re=7380) for F=400 and u_A^*=0.5. The displayed cycle in Figure 3 is the same as the one in Figure 2, i.e. the second form the last.

The flow is gradually and continuously heated up through the channel by the applied constant heat flux at the wall. It is clearly seen in Figure 2, that the onset and the growth of the circulation zones promote the mixing of cold fluid from the core with the hot fluid near to the boundary layer. This induces higher temperature gradients near the corrugated walls. Thus, the net heat transfer rate from the corrugated wall to the fluids get enhanced. The *k-ω* turbulence model produce higher temperature is than the other turbulence models, where the temperature increase is smallest for the transitional SST model.

Figure 4 shows the relationship between the Nusselt and Reynolds numbers for the F=400 and u_A^*=0.5. The curves predicted by the three turbulence models are compared with each other and with the experimental results for the non-pulsating conditions [33]. One can see that Nu increases for increasing Re, In SST and transition SST turbulence models, the experimental steady-state Nu is lower than for pulsating case. In a previous study [33], the SST model, was found to perform better than other turbulence models. A superior performance of the SST model in predicting convective heat transfer was also observed by other authors [34, 35]. Therefore, we can say that pulsating flow enhances the convective heat transfer.

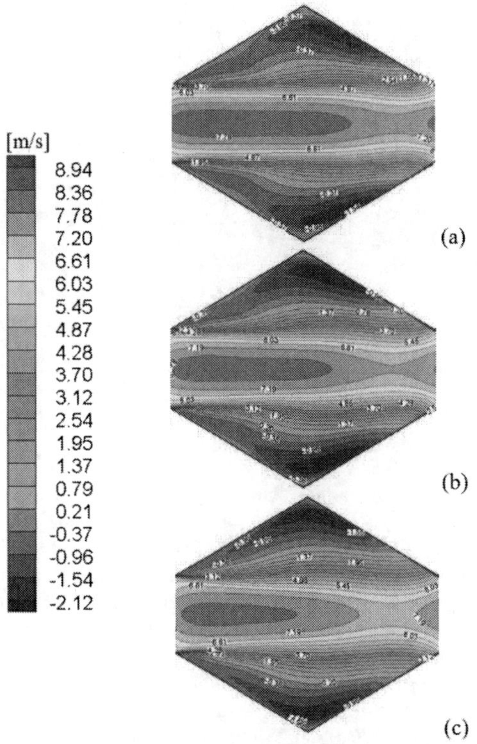

Fig. 2. Time-averaged axial velocity [m/s] contours for (a) *k-ω*, (b) SST, (c) transition SST.

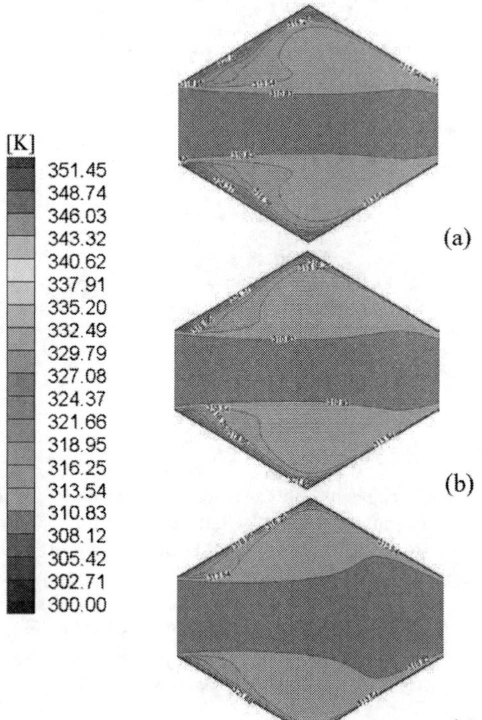

Fig. 3. Time-averaged temperature [K] contours for (a) *k-ω*, (b) SST, (c) transition SST.

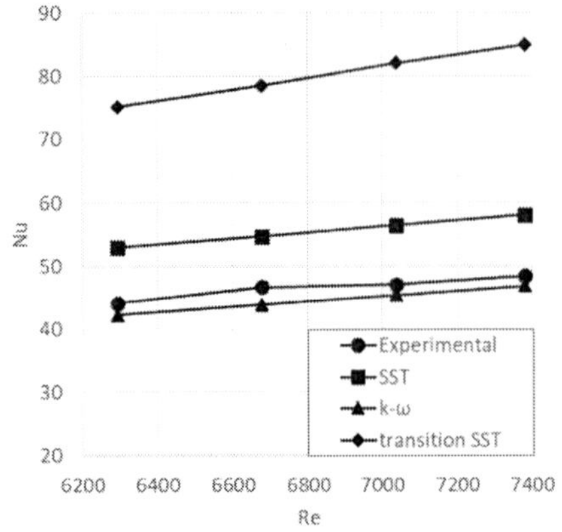

Fig. 4. Nusselt number as function of Re, for F=400, u_A^*=0.5.

Figure 5 shows the relationship between the friction factor and Reynolds number for the F=400 and u_A^*=0.5. One can see that f has an increasing trend with increasing Re, except for the transition SST model. Again, SST model was predicting the best results than the other turbulence models, according to previous study [33]. Friction factor prediction of the SST model is lower than experimental results which is get from non-pulsating flow. Thus, we can say that, with using pulsating flow conditions friction factor decreases.

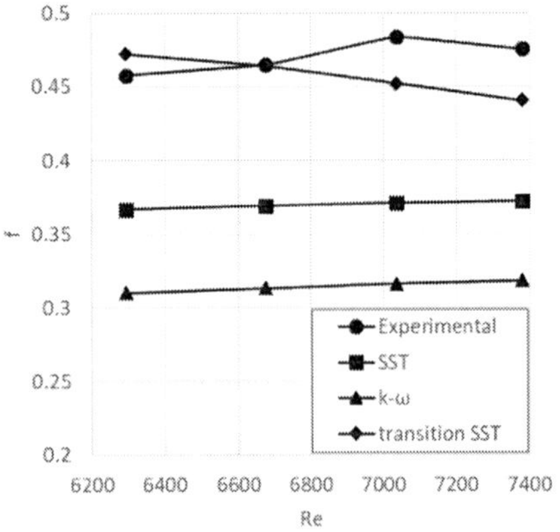

Fig. 5. Friction factor as function of Re, for F=400, u_A^*=0.5.

5 Conclusions

The present paper presents a computational study for a corrugated channel under pulsating flow conditions. One pulsating flow condition, which is F=400 and u_A^*=0.5, is applied. The pulsating numerical results are compared with each other, and the non-pulsating experimental results [33]. Three turbulence modes are used, namely, the k-ω model, the SST model and the transition SST model. The major conclusions to be drawn are:

- Pulsating flow condition have a positive effect to enhance heat transfer.
- Nusselt number increases with increasing Reynolds number.
- The SST and transition SST turbulence models produce the highest Nusselt numbers, which are also higher than the non-pulsating experimental results. However, the k-ω turbulence model produces the lowest Nusselt numbers that are also lower than the non-pulsating experimental results
- Pulsation decreases the friction factor.
- Friction factor increases with increasing Reynolds number for the k-ω and SST turbulence models. However, for the transitional SST turbulence model friction factor decreases with increasing Reynolds number.
- In the future work, the new pulsating cases will be investigated for understanding the amplitude and frequency effects of Nusselt number and friction factor, and for validating the numerical results.

References

1. P. Oclon, S. Lopata, M. Nowak, A.C. Benim, Progress in Computational Fluid Dynamics **15**, 5, pp. 290-306 (2015)
2. D.G. Ebling, A. Krumm, B. Pfeiffelmann, J. Gottschald, J. Bruchmann, A.C. Benim, M. Adam, R. Labs, R. R. Herberz, A. Stunz, Journal of Electric Materials **45**, 7, pp. 3433-3439 (2016)
3. A. Jokar, S.J. Eckels, M.H. Hosni, Heat Transfer Engineering **31**, 1, pp. 3-16 (2010)
4. A.M. Maqableh, A.F. Khadrawi, M.A. Al-Nimr,, S.A. Ammourah, A.C. Benim, Progress in Computational Fluid Dynamics **11**, 5, pp. 318-328 (2011)
5. A.C. Benim, H. Chattopadhyay, A. Nahavandi, International Journal of Thermal Sciences **50**, 10, pp. 1973-1983 (2011)
6. I. Taymaz, E. Aslan, A.C. Benim, Thermal Science **19**, 2, pp. 537-547 (2015)
7. S. Bhattacharyya, H. Chattopadhyay, A. Guin, A.C. Benim, Heat Transfer Engineering, published online, doi: 10.1080/01457632.2018.1474593 (2018)
8. H. Chattopadhyay, A.C. Benim, Journal of Heat Transfer – Transactions of the ASME **133**, 10, Article No: 104502, doi:10.1115/1.4004075 (2011)
9. S.D. Farahani, F. Kowsary, Heat Transfer Engineering **39**, 10, pp. 901-913 (2018)
10. A.C. Benim, Computer Methods in Applied Mechanics and Engineering **67**, 1, pp. 1-14 (1988)
11. E.M. Sparrow, L.M. Hossfeld, Int. J. Heat Mass Trans. **27**, pp. 1715-1723 (1984)
12. B. Snyder, K.T. Li, R.A. Wirtz, Int. J. Heat Mass Trans., **36**, pp. 2965-2976 (1993)
13. K. Bilen, M. Cetin, H. Gul, T. Balta, Appl. Thermal Eng. **29**, pp. 753-761 (2009)
14. K. Nilpueng, S. Wongwises, Exp. Thermal Fluid Sci. **15**, pp. 290-306 (2006)

15. M. Ciofalo, J. Stasiek, M.W. Collins, *In: Proc. 2nd Int. Symp. Eng. Turbulence Modelling and Measurements* (Florence, Italy, pp. 293-292, 1993)
16. M. Mirzaei, A. Sohankar, L. Davidson, F. Innings, Int. J. Heat Mass Trans. **76**, pp. 432-446 (2014)
17. S. Ray, B. Unsal, F. Durst, Int. J. Heat Fluid Flow **37**, pp. 167-176 (2012)
18. C. Aygun, O. Aydin, Nucl. Eng. Design **274**, pp. 164-171 (2014)
19. C. Aygun, O. Aydin, Nucl. Eng. Design **274**, pp. 172-180 (2014)
20. H. Chattopadhyay, F. Durst, S. Ray, Int. Com. Heat Mass Transfer **33**, pp. 475-481 (2006)
21. A.C. Benim, M. Cagan, D. Gunes, Int. J. of Thermal Sciences **43**, 8, pp. 725-732 (2004)
22. D.A. Nield, A.V. Kuznetsov, Int. J. Thermal Sciences **46**, pp. 551-560 (2007)
23. C.Y. Tseng, K.S. Yang, K.H. Chien, M.S. Jeng, C.C. Wang, Exp. Therm. Fluid Sciences **54**, pp. 85-92 (2014)
24. B. Metha, S. Khandekar, Int. J. Thermal Sci. **91**, pp. 157-166 (2015)
25. J.Y. Yu, W. Lin, X.T. Zhen, Heat Mass Trans. **50**, pp. 849-864 (2014)
26. M. Jafari, M. Farhadi, K. Sedighi, Int. Commun. Heat Mass. Trans. **45**, pp. 146-154 (2013)
27. A. Assmann, A.C. Benim, F. Gül, P. Lux, P. Akhyari, U. Boeken, F. Joos, P. Feindt, A. Lichtenberg, Journal of Biomechanics **45**, 1, pp. 156-163 (2012)
28. *Ansys Fluent 14.5 User's Guide* (Ansys Inc., Canonsburg, PA , 2012)
29. D.C. Wilcox, *Turbulence Modelling for CFD* (DCW Industries, Inc., La Canada, California, 1988)
30. F.R. Menter, AIAA Journal **32**, pp. 1598-1605 (1994)
31. F.R. Menter, R. Langtry, S. Volker, Flow Turbulence Combustion **77**, pp. 277-303 (2006)
32. P.A. Durbin, B.A. P. Reif, *Statistical Theory and Modelling for Turbulent Flows*, 2nd ed. (Wiley, Chichester, 2011)
33. E. Aslan, I. Taymaz, Y. Islamoglu, M. Engin, I. Colpan, G. Karabas, G. Ozcelik, Progress in Computational Fluid Dynamics **18**, 1, pp. 33-45 (2018)
34. A.C. Benim, D. Brillert, M. Cagan, ASME Paper No. GT2004-54151, in: Proc. ASME Turbo Expo 2004, Vienna, Austria, June 14-17, 2004, Vol. 4, pp. 453-460, doi:10.1115/GT2004-54151 (2004)
35. A.C. Benim, K. Ozkan, M. Cagan, D. Gunes, International Journal of Numerical Methods for Heat & Fluid Flow, **17**, 3, pp. 284-301 (2007)

Study of a hydraulic system converting energy from sea waves near the coast

Silvio Barbarelli*, *Mario* Amelio, *Gaetano* Florio, and *Nino Michele* Scornaienchi

Department of Mechanical, Energy and Management Engineering (DIMEG)
University of Calabria – Ponte P. Bucci, cube 44/c, I87036 Rende, Italy

Abstract. This paper proposes a system able to recover energy from sea waves. The system is made up essentially of a floating structure, connected to a piston pump. The pump, constrained to move vertically under the action of the waves, pushes water into a reservoir maintained at constant pressure, from which it flows in a Pelton turbine. A methodology was developed for the preliminary sizing of the proposed system. This methodology takes the height of the waves and the pressure of the reservoir as reference parameters and then provides the main geometric data of the machines. A case study developed in the present work, demonstrates the possibility of designing miniaturized components of the system, able anyway to provide acceptable energy output with contained installation costs.

Nomenclature

A_{buoy}	buoyant area	[m^2]
b	viscous damping coefficient	[kg/s]
A_P	piston area	[m^2]
C_o	pump stroke	[m]
c_o	spouting velocity	[m/s]
CWR	capture width ratio	[%]
D_{buoy}	characteristic size of the buoy	[m]
D_n	nozzle diameter	[m]
D_P	piston diameter	[m]
D_T	turbine diameter	[m]
F_A	Archimedes's force	[N]
F_w	weight force	[N]
F_p	pressure force	[N]
g	gravity	[m/s^2]
H_m	reservoir pressure	[m]
H_w	wave height	[m]
J	average wave power	[kW/m]
k	specific turbine speed	[-]
m_{add}	added mass	[kg]
m_b	buoyant mass	[kg]
N_{cy}	wave frequency	[min^{-1}]
n_T	turbine rotational speed	[rpm]
P_P	pump power	[kW]
P_{System}	system power	[kW]
P_T	turbine power	[kW]
Q_P	pump flow rate	[m^3/s]
Q_T	turbine flow rate	[m^3/s]
T_e	wave period	[s]
u	peripheral velocity	[m/s]
V_P	pump volume	[m^3]
V_z	sunk buoy volume	[m^3]
z	vertical position of the buoy	[m]
z_A	pressure sinking of the buoy	[m]
z_w	weight sinking of the buoy	[m]

Greek letter

β	deflection turbine blade angle	[°]
φ	velocity reduction coefficient	[-]
η_{el}	electrical efficiency	[-]
η_P	pump efficiency	[-]
η_T	turbine efficiency	[-]
ρ	water density	[kg/m^3]
ψ	velocity reduction coefficient	[-]

Keywords: Wave Energy, Point Absorber, Piston Pump, Pelton Turbine, Sizing Procedure, Specific Power, Pressure Control.

1 Introduction

Renewable energy sources are considered the way to unlock energy transition positively affecting the National Power Grid performance [1] by supplying natural driven energy such as solar, wind and marine [2]. Among marine resources, ocean waves are one of the most promising sources of renewable energy with an estimated potential of 2 TW around the world [3].

Waves are generated by bodies moving on or near the surface of the ocean, winds, seismic disturbances, and lunar and solar gravitational fields. The most common types of waves considered for engineering analysis are those created by the wind or swells [4]. Point absorbers

* Corresponding author: silvio.barbarelli@unical.it

are floating wave energy converters (WEC's) whose dimensions are much smaller than the wave length. These devices can oscillate, under the waves action, with one or more degrees of freedom. A point absorber is able to absorb wave energy from a wave front greater than the physical dimensions of the device itself [5]. After assessing the waves energy available at a particular site, an appropriate WEC can be designed which can be located offshore or inshore [6], according to the seabed characteristics. Furthermore, it can be coupled to storage systems to solve the mismatch between energy production and actual use [7].

Anyway, wave energy conversion technology is in the pre-commercial stage. The cost of a wave energy converter increases with its size. Therefore, the geometry optimization of the system has a significant role in the design process to make for an economically feasible system. A solution should be provided by the design of miniaturized components of the energy transformation mechanism, by leaving the dimensions of the point absorber relatively large.

In the present paper, the authors, inspired by the work of Paula B. Garcia-Rosa [8] propose a system whose point absorber (floating structure) is connected directly to a piston pump, which in turn delivers water to a pressurized reservoir.

The reservoir feeds a small Pelton turbine, able to provide power, determined by the available water flow and by the reservoir pressure, this last depending on the wave height.

The sizing methodology of the system is described, by pointing out the off-design operations and the management of the pressurized reservoir.

A study case demonstrated that the components can be designed with small dimensions, consequently containing the installation costs [9].

2 Waves motion

The waves motion constitutes a huge quantity of energy, currently exploited only in minimal part.

According to linear wave theory, by indicating with H_w and T_e height and period, the average power, transported by a wave per meter wave front is described by the following equation:

$$J = \frac{\rho g^2}{32\pi} \frac{T_e H_w^2}{1000} \quad (1)$$

while the capture width ratio (*CWR*) of the point absorber is defined as:

$$CWR = 100 \frac{P_{av}}{J D_{buoy}} \quad (2)$$

P_{av} and D_{buoy} being the average power and diameter of the point absorber.

As an example, in the next figure the potential of energy related to the Mediterranean sea, near to Pantelleria island, is reported.

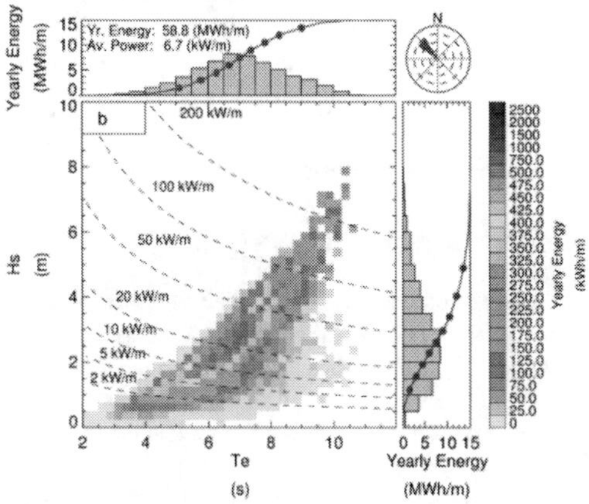

Fig. 1. Yearly average energy as a function of the wave height H_w and of the period T_e near Pantelleria island (Italy).

From the reading of the graph, a global resource of 60 MWh/m results with an average power of 6.7 Kw/m. The wave status of the sea is very variable determined by the factors that transfer energy to the wave like winds, seismic disturbances, and waves generated by the lunar and solar gravitational fields. Moreover, the seabed bathymetry or generic obstacles, influences shape and extension of the waves.

About that, various models are proposed in the literature which more or less fit the physical phenomenon [10-11]. Here, for simplicity, the "fully developed sea model" is taken into consideration. The characteristics of this wave model are reported in Tab. 1.

Table 1: Wave characteristics.

Wind velocity (km/h)	Fetch (km)	Last (h)	Wave height (m)	Wave length (m)	Period (s)
20	24	2.75	0.33	10.6	3.2
40	176	11.50	1.80	39.70	6.2
60	660	27.50	5.10	89.2	9.1
80	1682	50.00	10.30	158.6	12.4

From Tab. 1 it is possible to determine a relationship between the cycles number per minutes and the height of the wave.

$$N_{cy} = 11.92 H_w^{-0.38} \quad (3)$$

For completeness, the next table reports the Douglas scale for determining the sea state

Table 2: State of the sea (wind sea).

Degree	Height (m)	Description
0	no wave	Calm (Glassy)
1	0–0.10	Calm (rippled)
2	0.10–0.50	Smooth
3	0.50–1.25	Slight
4	1.25–2.50	Moderate
5	2.50–4.00	Rough

6	4.00–6.00	Very rough
7	6.00–9.00	High
8	9.00–14.00	Very high
9	14.00+	Phenomenal

3 Description of the system

The technical scheme of Fig. 2 illustrates the proposed system. The buoy (point absorber) can be put near the shore where the seabed allows the working. This buoy is connected to a piston pump at which transfers the wave motion. The piston pump delivers water in a pressurized tank partially filled by air, whose pressure can be maintained constant or changed to the desired value thanks an auxiliary compressor, able to restore the air losses. The hydraulic circuit closes on a Pelton turbine connected to an electrical generator. In more detail, the buoy is more extended in the horizontal plane and it is free to oscillate in the vertical direction (see Fig. 2), by following the wave crest.

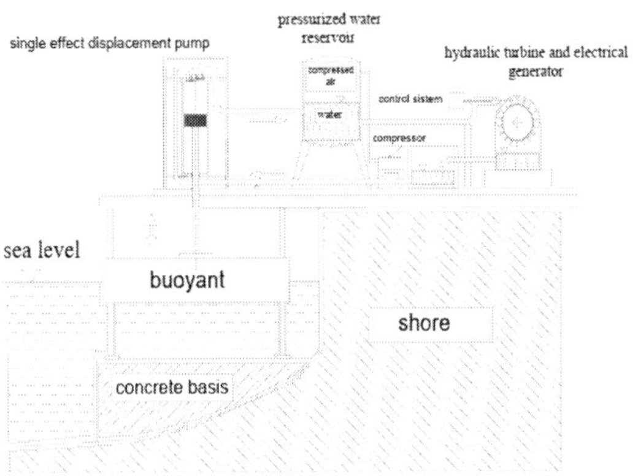

Fig. 2. System layout.

4 Mathematical model of the buoy oscillations

Making reference to Fig 3, the dynamic equation of the motion is given by the following equation [12]:

$$(m_b + m_{add})\ddot{z} + b\dot{z} = F_p + F_w - F_A \qquad (4)$$

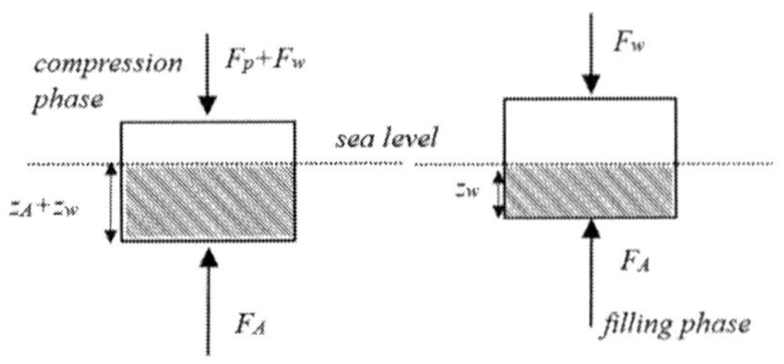

Fig. 3. Scheme of the forces on the buoy.

where:

- $F_A = \rho g V_z$ is the Archimedes's thrust
- $F_p = \dfrac{\pi D_p^2}{4}\rho g H_m$ is the pressure force acted by the piston pump
- $F_w = g m_b$ is the weight force

The solution of the differential equation will provide the vertical motion of the buoy, and consequently the excursion of the piston pump. It is worth pointing out that the piston pump will charge the water reservoir during the rising of the buoy (compression phase, see Figs. 3 and 4). Instead, during the filling phase (see Figs. 3 and 4), the piston pump is not connected to the reservoir and so the pressure force stops acting.

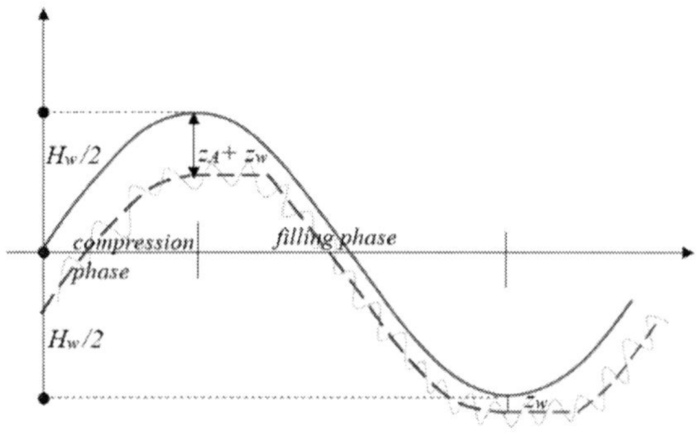

Fig. 4. Path of the buoy.

Fig. 4 illustrates the path of the buoy (red dashed line) under the action of a wave with height H_w, and the inertia secondary oscillations, derived from eq. 2, which overlap on the red dashed line.

During the compression phase, the buoy sinks under the sea level of $z_A + z_w$ because of the reservoir pressure and its weight. During the filling phase, the buoy instead sinks by z_w. So the total excursion of the piston pump is:

$$C_o = H_w/2 - z_A - z_w + H_w/2 + z_w = H_w - z_A \qquad (5)$$

5 Sizing of the system

Figure 5 summarizes the sizing procedure, starting from the wave height (H_w) , the buoy area (A_{buoy}), the reservoir pressure (H_m), the specific speed (k) and the rated rotational speed (n_T) of the Pelton Turbine, which in Fig. 5 are highlighted in the orange boxes. All the other devices/parameters consequently proportioned, are depicted in the blue boxes in the same figure.

In more detail, the height of the wave determines the number of cycles per minute N_{cy} (see eq. 3) and also the stroke C_o of the pump, while the piston area derives from the equilibrium equation of the buoy. So that, the displacement of the pump V_P is found and consequently its flow rate Q_P. The latter has to be equal to the turbine flow rate Q_T. Now, by considering a fixed pressure of the reservoir $\rho g H_m$, and a specific speed k, besides a rated rotational turbine speed n_T, the Pelton turbine can be sized, i. e. its diameter D_T and its nozzle diameter D_n.

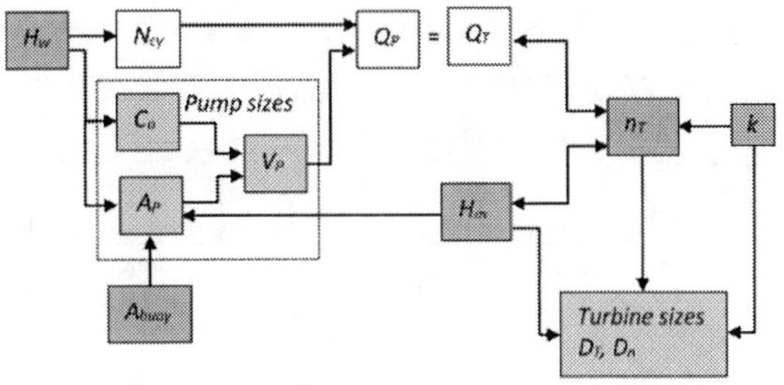

Fig. 5. Sizing procedure scheme.

In the following, the steps of the procedure above illustrated are explained in more detail.

5.1 Piston pump and buoy sizing

The sizing of these two devices is strictly correlated because of the equilibrium equation of the buoy, connected directly to the piston pump.

The first parameter considered is the flow rate elaborated by the pump, which is:

$$Q_P = V_P \frac{N_{cy}}{60} = A_P C_o \frac{N_{cy}}{60} = \frac{\pi}{4} D_P^2 C_o \frac{N_{cy}}{60} \quad (6)$$

The piston area is given by imposing the equilibrium in the steady condition. From eq. 4, by neglecting the weight of the buoy, it results that:

$$\frac{\rho g H_m}{\eta_P} A_p = \rho g z_A A_{buoy} \quad (7)$$

By simplifying:

$$A_p = \eta_P z_A \frac{A_{buoy}}{H_m} \quad (8)$$

5.2 Pelton choice

The choice of the Pelton starts from the pressure level of the reservoir equal to $\rho g H_m$.

Taking into account the specific speed:

$$k = \frac{1}{53} n_T \frac{\sqrt{Q}}{(gH_m)^{0.75}} \quad (9)$$

whose optimal range is [0 ÷ 0.18], the Pelton nominal flow rate can be determined as:

$$Q_T = 2809 k^2 \frac{H_m^{\frac{3}{2}}}{n_T^2} \quad (10)$$

Under the maximum efficiency conditions it results that:

$$u \cong 0.5 c_0 = 0.5 \varphi \sqrt{2gH_m} \quad (11)$$

where

$c_0 = \varphi \sqrt{2gH_m}$ is the spouting velocity of the jet, u is the peripheral velocity and φ a velocity reduction coefficient.

The peripheral velocity depends on the Pelton diameter D_T as follows:

$$u = \frac{\pi D_T n_T}{60} \quad (12)$$

So that the turbine diameter is expressed as:

$$D_T = \frac{30}{\pi} \frac{\varphi}{n_T} \sqrt{2gH_m} = 42.2 \frac{\varphi}{n_T} \sqrt{H_m} \quad (13)$$

In an analog way, the nozzle diameter D_n, can be determined by considering that the turbine flow rate can be expressed as:

$$Q_T = \frac{\pi D_n^2}{4} c_0 \quad (14)$$

Then, by involving the specific speed of eq. 9 it results that:

$$D_n = \frac{28.41}{n_T} \frac{k}{\sqrt{\varphi}} \sqrt{H_m} \quad (15)$$

5.3 Specific power of the system

The waves energy is converted into electrical power, through the pump and the turbine working.

The pump power, by considering the pressure of the reservoir H_m and the provided flow rate Q_P, is expressed by the following equation:

$$P_P = \frac{\rho g Q_P H_m}{\eta_P} = \rho g \frac{\eta_P}{\eta_P} z_A \frac{A_{buoy}}{H_m} C_o \frac{N_{cy}}{60} H_m = \rho g z_A A_{buoy} (H_w - z_A) \frac{11.92 H_w^{-0.38}}{60} \quad (16)$$

The power supplied by the pump is equal to that employed by the Pelton turbine for producing electrical energy, reduced by the turbine losses, so that one obtains:

$$P_T = \eta_T \rho g z_A A_{buoy} (H_w - z_A) \frac{11.92 H_w^{-0.38}}{60} \quad (17)$$

Instead the specific turbine power is:

$$\frac{P_T}{A_{buoy}} \cong \eta_T 2000 z_A (H_w - z_A) H_w^{-0.38} \quad (18)$$

By deriving the specific power with respect to z_A a maximum value can be found for:

$$z_A = \frac{H_w}{2} \tag{19}$$

The specific power of the system expressed in kW (see Fig. 6) considering the generator electrical loss, in the best condition, is given by:

$$\frac{P_{system}}{A_{buoy}} \cong \frac{1}{2}\eta_{el}\eta_T H_w^{1.62} \tag{20}$$

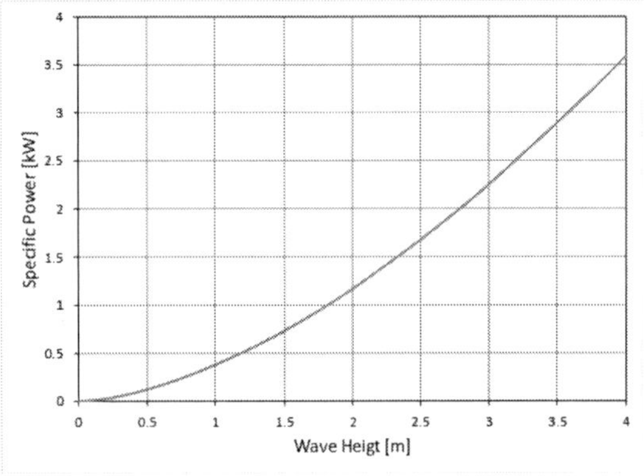

Fig. 6. Specific power of the system.

6 Case study

In this section, a system sizing is proposed with the aim to obtain small dimensions of all the components.

A key design parameter is the wave height. This parameter determines the size of the piston pump connected to the buoy, because the optimal stroke of the pump is equal to the wave design half height (eq. 19).

Fig. 7 reports an example of the probability distribution chart of the wave heights for the near shore region of Rio de Janeiro [12].

Making reference to Fig. 7, by choosing a wave height of 1 m, the system will be able to work for 37% of the time at partial load and for 63 % of the remaining time at full load (stroke of the pump equal to maximum value of 0.5 m).

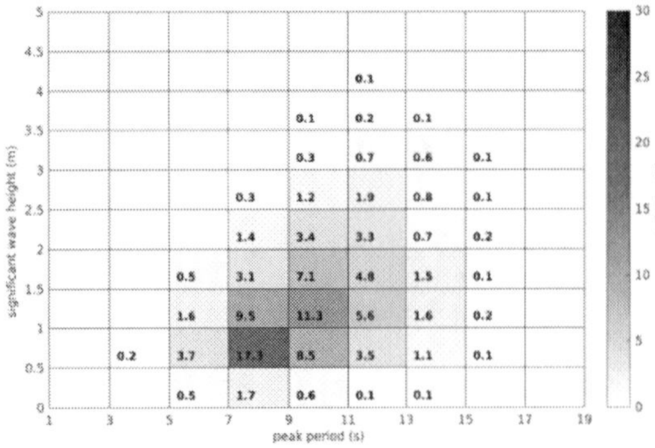

Fig. 7. Joint Probability distribution (%) for the near shore region of Rio de Janeiro.

By considering that of Fig. 7 as a typical distribution, in a first rough configuration the design wave height is set equal to 1.

All the system input parameters are reported in the next table:

Table 3: Input parameters of the system.

Parameter	Symbol	Value
Wave nominal height	H_w	1 *m*
Buoy area	A_{buoy}	20 *m*
Specific speed	*k*	0.09
Pressure reservoir	H_m	160 *m*
Turbine rotational speed	n_T	3000 *rpm*

These inputs allow all the components of the system to be sized. In fact, by applying eqs. 5, 8, 13, 15, 19, the following values can be obtained:

Table 4: Output parameters of the system.

Parameter	Symbol	Value
Piston pump stroke	C_o	0.5 *m*
Piston pump diameter	D_P	0.28 *m*
Piston pump volume	V_P	30 *l*
Turbine diameter	D_T	0.168 *m*
Turbine nozzle diameter	D_n	0.011 *m*

The flow rate elaborated by the system, is equal to 5.1 l/s (see eqs. 6, 10 and 14), while the nominal power amounts to 7kW.

In these conditions, the *CWR* is equal to 33%, *J* being equal to 4.7 kW/m (see eq. 1) and D_{buoy} (characteristic size of the buoy) equal to 4.5 m.

6.1 Off design operations

The Pelton turbine has the possibility of changing its flow rate without efficiency losses, by moving the needle of the nozzle so that the flowing area varies from zero (no wave) to the maximum value (design wave). All the intermediate waves (under 1 m) imply proportional strokes of the piston pump, flow rates and consequent powers up to the design value of 7 kW.

When the pump starts working, under the waves action, the nozzle feeding the turbine is closed. The water level of the reservoir rises until the pressure reaches the design value of 16 bar (160 m water column).

For guaranteeing this pressure value, an air mass of 20 kg at environment temperature, is compressed by the remaining part of water (2 m^3), occupying a volume of 1 m^3.

At this point, the nozzle of the turbine opens and the flow rate incoming is discharged on the turbine itself.

The successive increasing of the flow rate delivered by the pump, determined by waves height increases, would try to augment the air pressure. At purpose, the needle of the nozzle moves opening the nozzle area more, allowing the turbine flow rate to rise, so restoring the pressure level.

When the nozzle area is completely open, successive increases of the flow rate, for waves higher than 1 m (design value), cannot be managed by the nozzle opening, but by greater pressure levels. For this reason the air mass is further compressed, occupying a minor volume, and the jet velocity, investing the Pelton turbine, augments as in eq. 11.

In this case is worth pointing out that the efficiency will not be the maximum. In fact, the efficiency expression is equal to:

$$\eta_T = 2\varphi^2 \frac{u}{c_o}(1-\frac{u}{c_o})(1+\psi\cos\beta) \qquad (21)$$

which finds a maximum value for u/c_o = 0.5 if the velocity co remains to the design value, as in eq. 11.

Increasing flow rates imply higher values of co and a decay of efficiency plus or minus consistent. This inconvenience can be overcome, by changing the rotational regime [13] so that the peripheral velocity u changes opportunely for maintaining unchanged the ratio u/co.

Finally, when the wave motion stops and the sea is still, the reservoir does not receive more water. In this condition the reservoir starts emptying. The water mass, flowing through the nozzle, lowers its level and the reservoir pressure decreases. Under a certain level, the turbine nozzle is closed and the reservoir remains filled partly by water and partly by air, by waiting until the sea becomes wavy again.

7 Conclusions

A hydraulic system made up of buoy, piston pump, pressurized reservoir and Pelton turbine has been sized, by following a procedure which aims to optimize all the geometrical parameters, starting from the design height wave, the reservoir pressure and the specific speed of the Pelton turbine.

The reservoir is conceived to operate at the same pressure for waves with height under 1 m (design value). For higher wave heights, the pressure of the reservoir together with the rotational regime of the turbine have to increase for maintaining the best efficiency conditions.

The system is conceived to be not excessively expensive, the design choices being addressed toward miniaturized components.

Anyway, the specific power of such a system is estimated equal to 0.38 kW/m2, so that, by consider a point absorber with area equal to 20 m2, the nominal power is equal about to 7 kW with a capture width ratio of 33%.

References

1. M. Noussan, R. Roberto, B. Nastasi, Energies **11**, 650 (2018)
2. S. Barbarelli, M. Amelio, T. Castiglione, G. Florio, N.M. Scornaienchi, G. Lo Zupone, A. Cutrupi, Energy Convers Manag **87**, 1124–33 (2014)
3. B. Drew, A.R. Plummer, and M. N. Sahinkaya. *Proceedings of the IMechEA: Journal of Power and Energy*, **223**, 8, 887–902 (2009)
4. M.E. McCormick, *Ocean Wave Energy Conversion*, (Dover Publications, New York, NY, USA, 2007)
5. G. Thomas, *Ocean Wave Energy: Current Status and Future Perspectives*, J. Cruz, Ed. (Springer, Berlin, Germany, 2008)
6. S. Barbarelli, G. Florio, G. Lo Zupone, N.M. Scornaienchi, Renew Energy **129** Part A, 183-200, (2018)
7. D. Astiaso Garcia, F. Barbanera, F. Cumo, U. Di Matteo, B. Nastasi, Energies **9**, 963 (2016)
8. P.B. Garcia-Rosa, J.P.V.S. Cunha, F. Lizarralde, *American Control Conference Hyatt Regency Riverfront* (St. Louis, MO, USA, 2009)
9. G. Lo Zupone, M. Amelio, S. Barbarelli, G. Florio, N.M. Scornaienchi, A. Cutrupi, Appl Energy **185**, 1292–1302 (2017)
10. A. Guissard, C. Baufays, P. Sobieski, Remote Sensing of Environment **48**, 1, 25-38 (1994)
11. P. Zwolan, K. Czaplewski, Scientific Journals, Maritime University of Szczecin **32**, 104, 186–190 (2012)
12. M. Shadman, S.F. Estefen, C.A. Rodriguez, I.C.M. Nogueira, Renew Energy **115**, 533-546 (2018)
13. S. Barbarelli, G. Florio, M. Amelio, N.M. Scornaienchi, Appl Energy **224**, 717-730 (2018)

https://doi.org/10.1051/matecconf/201824001005

Heat transfer and flow field in a circular twisted channel

Suvanjan Bhattacharyya[1,*], *Himadri* Chattopadhyay[2], *Arnab* Banerjee[3], and *Ali Cemal* Benim[4]

[1]Deparement of Mechanical and Aeronautical Engineering, University of Pretoria, South Africa
[2]Mechanical Engineering Department, Jadavpur University, Kolkata, West Bengal, India
[2]Mechanical Engineering Department, MCKV Institute of Engineering, Liluah, Howrah, West Bengal, India
[4]Center of Flow Simulation (CFS), Dept. Mechanical and Process Engineering, Düsseldorf University of Applied Sciences, Germany

Abstract. In the present paper, along with experimental study, computational fluid dynamics analysis is performed, using the transition SST model which can predict the change of flow regime from laminar through transition to turbulent. The differential governing equations are discretized by the finite volume method. The investigations are conducted for Reynolds numbers ranging from 100 to 50,000 covering laminar, transitional and turbulent regimes, and for three length and three pitch ratios. The predictions are observed to show a good agreement with the measurements and published correlations of other authors. The analysis indicates that the large length ratio and small pitch ratio yields a higher heat transfer rate with relatively low performance penalty. The transition from laminar to turbulent regime is observed between Reynolds numbers of 2,500 to 3,500 for all cases. For almost all investigated cases the performance factors are greater than unity.

1 Introduction

Heat exchanger is a device facilitating heat transfer between two or more fluids [1]. It is used in a vast number of industrial applications, such as thermal power plants, chemical processing plants, air conditioning equipment, refrigerators, radiator for space vehicles as well as automobiles, extending up to micro-scale applications [2,3]. Obviously, for improving the efficiency of energy utilization, enhancement of heat transfer in such applications play an important role. To this goal, two techniques can generally be identified, namely, the active and passive techniques. Active techniques include surface vibration, fluid vibration etc. for increasing heat transfer [4]. In passive techniques, the applied methods usually involve a modification of the channel geometry to enhance heat transfer. A special category herein is the use of bluff bodies that increase heat transfer by vortex shedding [5,6]. Passive methods may also include basically different arrangements, such as impinging jet [7,8] instead of wall-parallel flow.

A further passive technique that attracted considerable attention is the insertion of structures such as twisted tapes that induce a swirling motion. Indeed, industrial heat transfer equipment are generally operated in turbulent/swirl flow conditions where their performance in terms of energy transfer rate is high, compared with laminar flow by virtue of the high degree of turbulence in turbulence/swirl flow. Heat transfer characteristics under transitional flow conditions in most of thermal problems are of considerable interest. Predicting transition of laminar regime to turbulent in heat transfer augmentation techniques will be highly useful to design any heat transfer equipment. There is constant thrive in the studies on transition from laminar to turbulent flow.

Many researchers performed surveys on passive enhancement method of this kind, which were very inspiring, such as channel with a cylinder roughness [9] and compound heat transfer enhancement of a convergent-divergent tube with a twisted tape [10].Swirling flow may increase the heat transfer through the boundary layer by interruption or thinning and also increase turbulence intensity [11]. Tube with wire coil inserts [12] can generate vortex and also can develop secondary flow. Tube insert technology – i.e., insertion of tape inside a channel - was used in some previous work. Bhattacharyya et al. [13,14] and Saha et al. [15] studied experimentally center-clearing twisted tapes with artificial rib roughness and achieved considerable improvement. Meng et al. [16] performed experiments on elliptical tube type swirl generator. Some earlier research [17,18] showed that vortex geometry may effect heat transfer and be utilized to design heat exchangers. Chen et al. [19] studied experimentally heat transfer augmentation techniques of a dimpled tube. Vicente et al. [20] reported on thermohydraulic performance of a helically dimpled tubes for laminar and transition flow. Mengna et al. [21] studied on compound heat transfer enhancement of a converging–diverging tube with swirl generator inserts. Sivashanmugam et al. [22] reported on experiments on thermohydraulic characteristics of laminar stream in circular channel fitted with screw type tape. Additional information on similar vortex originator might be acquired from [23–27].

In the present paper, a numerical and experimental study of forced convection in circular twisted channel is

* Corresponding author: suvanjanmechanical@gmail.com

presented. Air is used as working medium. At the prevailing relatively small/moderate temperature differentials, constant material properties are assumed A sketch of the device is presented in Figure 1, where the circular twisted enhancer can be recognized.

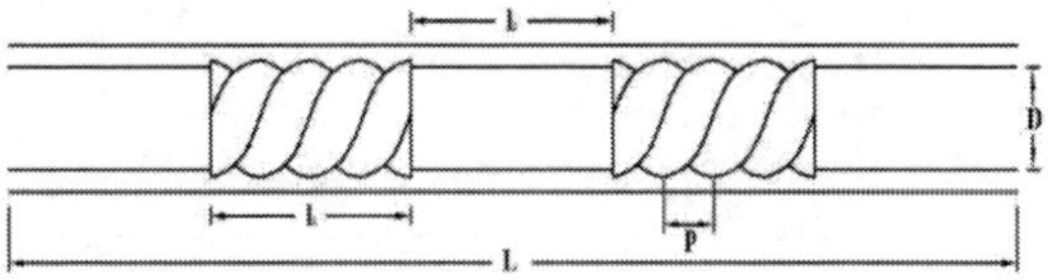

Fig. 1. Sketch of geometry.

2 Experimental rig

Figure 2 demonstrates the schematic figure and original look of the experimental setup. Labeled portions show the significant parts of the test rig in which experiment is prosecuted. Air is considered as the working medium while carrying out the experiment. The experiment is conducted in a non-air conditioned laboratory and usually the temperature in the room is varied between 27°C - 33°C.

The atmospheric air was sucked by a 7.0 kW blower. This air is then crossed throughout a PVC (Polyvinyl chloride) made pipe and finally comes to a rotameter. The rotameter is utilized to assess the mass flow rate of the employed fluid (Pr = 0.707). With higher blower speed the mass flow rate fluctuation decreases. A bypass valve is installed in between the blower and the rotameter. The foremost intention of this bypass valve is to pass the excess air through this valve and maintain a low mass flow rate. The rotameter can assess flow rate varying in between 120 to 520 LPH. An error of 0.0442% is present in the rotameter. U-tube manometer is utilized to assess the pressure drop along the test section. Three circular twisted channels of different length ratios (H = l/L = 0.2, 0.25 and 0.30), and three different pitch ratios (S = p/d = 0.22, 0.36, and 0.45) were used for evaluation.

3 Test section

For validation purpose the experiment is executed in a smooth tube and then the twisted parts are installed which are used to enrich the heat transfer rate. A circular corrugated tube having 22 mm inside diameter and length of 2.0 m is used for the experiment purpose. The channel as depicted in Figure 1 is made up of brass. In order to assess the pressure drop throughout the experiment two pressure taps are placed in the experimental setup. The two pressure taps are placed at inlet and outlet of the test section. Net distance between the two pressure taps is 2.10 m. The pressure taps are placed in such a way such that they can only assess the pressure values of the fully developed flow. The pressure drop is measured using a differential U-tube manometer ranging from 0 -150 mm Hg. The differential manometer is calibrated to ensure no error.

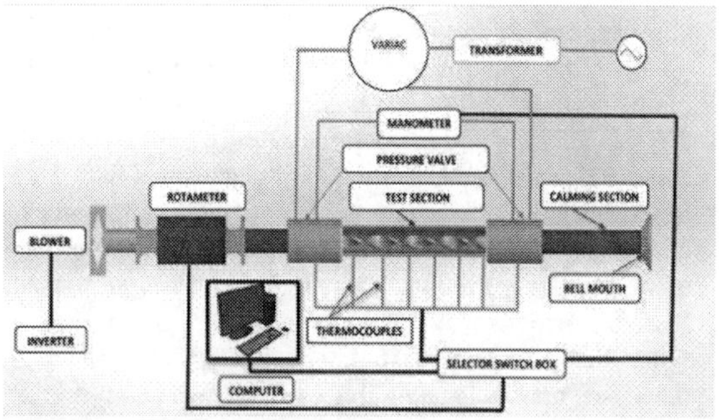

(a)

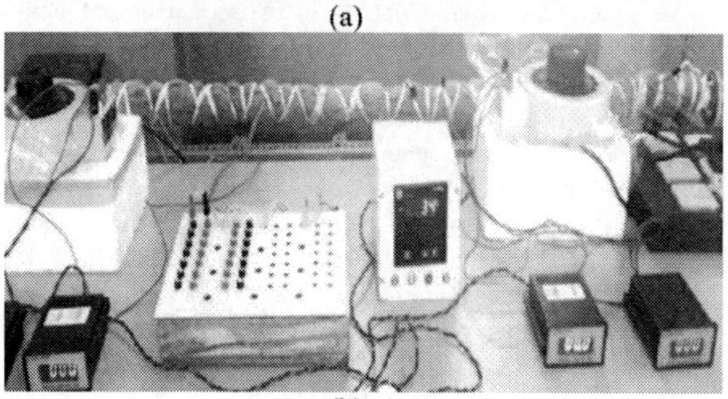

(b)

Fig. 2. Schematic diagram (a) and photographic view (b) of experimental setup.

A total of 28 thermocouples were fitted on the tube outside wall by brazing. The thermocouples are installed at 7 locations. The positions of thermocouples were 10 cm, 60 cm, 1.00 m, 1.35 m, 1.6 m, 1.75 m and 1.90 m over the wall of corrugated tube. Each position contains 4 thermocouples.

The test section was electrically heated up by nichrome heater wire providing uniform wall heat flux boundary condition. Porcelain bead types of insulation were used on nichrome heater wire, so no direct interaction of the nichrome heater wire with the tube wall. A 1.5 KW variac is used in which the variations of voltage and current are 10.0 – 30.0 V and 1.0 – 3.0 A respectively. There is an error of 0.072% in the current and voltage compared to the nominal value. In order to shrink the heat loss the brass corrugated tube is insulated using glass-wool having thermal conductivity 0.0037 W/mK that are tightly insulated by fabricated ropes. The total insulation thickness is 140 mm, and approximately 99% of the heat loss is prevented using such insulations.

A data accumulating system is used for recording all the measurements. The mass flow rate, inlet, outlet and bulk temperatures, differential pressure readings are taken and recorded in the data accumulating system. The output data from the data accumulating system is provided as an input into the personal computer which helps in data reduction.

4 Experimental procedure

The investigational setup is turned on and the voltage is setup at 20V to obtain the required heat flux. The setup is allowed to avail steady state by providing a time allowance of approximately 1 hour and 45 minutes. For most of the experiments the upmost mass flow rate is maintained at 12 kg/s. When the standard deviation of the temperature and mass flow rate are less than 0.03°C and 0.011 kg/s respectively then it is assumed that steady state is attained. As soon as steady state is attained, the data accumulated system receives a set of 110 readings of all the temperatures, pressures and mass flow rates. First the experiment is conducted in the plain channel. After that the circular twisted channel was placed in place of plain channel. The experiment is conducted using three circular twisted channels of different length ratios (H = l/L = 0.2, 0.25 and 0.30), and three different pitch ratios (S = p/d = 0.22, 0.36, and 0.45). The experiment is conducted by the variation of mass flow rate for both plain channel and twisted channel. A total of 40 experiments are executed for plain channel and 75 experiments for twisted channel. Uncertainties of some major experimental variables are given in Table 1. The uncertainty in Reynolds number, friction factor and Nusselt number were estimated as ± 4.17 %, ± 6.66 % and ± 7.42 %, respectively following the procedure of Bhattacharyya and Saha [11,24].

5 Modelling

In the current numerical study, three-dimensional convective heat transfer (HT) augmentation of air through circular twisted channel as shown in Figure 1 is simulated. For modelling purposes, an unstructured grid with non-uniform grid density distribution was created, in order to observe the flow and heat transfer more accurately. The regions in the channel close to the channel wall are meshed into smaller control volumes in order to achieve a more precise prediction. Different grid sizes were tested as part of grid independence study. As result, a grid with 777,235 cells are selected as the numerical grid. The surface grid is shown in Figure 3.

In the present investigation, circular twisted channels of four different length ratios (H = l/L = 0.2, 0.25 and 0.30), and three different pitch ratios (S = p/d = 0.22, 0.36, and 0.45) are investigated. Fluid enters the channel at an inlet temperature (T_i) of 300 K. At the wall, a uniform wall temperature (T_w) boundary condition is employed with T_w=500 K, along with no slip. The range of Reynolds number considered is from 100 to 50,000. At the inlet, a uniform velocity profile is used with a turbulent intensity of 5%, even in the laminar regime following Abraham et al [28].

The governing equations are discretized on a non-uniform unstructured grid using finite volume method within the framework of the general purpose CFD code Ansys Fluent [29]. Buoyancy effects as well as radiation [30] are neglected. The SST model of turbulence [29] was successfully used in a wide spectrum of applications [28,31]. The transitional version of the SST model [29] is used as the turbulence model. Kadiyala et al. [32] used this model for studying the effect of jets on moving surface, Oclon et al. for predicting fouling in a compact heat exchanger [33] and Bhattacharyya et al. [34] for enhancement purposes. The velocities and pressures

Table 1. Experimental uncertainties of major parameters.

Name of variables	Errors
Flow velocity	0.18%
Voltage on the heater	0.15%
Electrical resistance	0.33%
Heat transfer coefficient	1.19%
Current on the heater	0.23%
Ambient temperature	0.10%
Electrical power on the heater	0.17%
Average temperature	0.60%

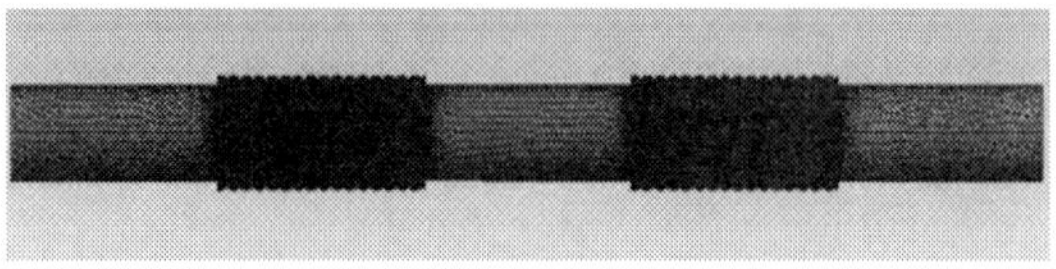

Fig. 3. Surface grid (detail).

were predicted using the SIMPLE scheme. For convective terms, a second order upwind scheme [29] is used. The discretized equations are then solved iteratively using in double precision. The convergence criteria for continuity, momentum and energy are set at 10^{-4}, 10^{-5}, and 10^{-7} respectively. The convergence criterion for the other four turbulence quantities was also fixed at 10^{-4} [29]. The Reynolds number of air flow in the channel is calculated from, where U, D_h, ν denote the bulk velocity, hydraulic diameter and kinematic viscosity, respectively

$$Re = UD_h / \nu \quad (1)$$

The heat transfer coefficient (h) is, then, obtained from where q_w, T_w, T_b denote the wall heat flux, wall temperature and bulk fluid temperature, respectively.

$$h = q_w /(T_w - T_b) \quad (2)$$

The convective heat transfer coefficient is then used to obtain Nusselt number (Nu) as, where k denotes the thermal conductivity

$$Nu = hD_h / k \quad (3)$$

The friction factor is determined from the measured values of pressure drop (Δp), across the test section length using the following relation (L is tube length).

$$f = 2D_h \Delta p / \rho L U^2 \quad (4)$$

To evaluate the effect of heat transfer enhancement under given pumping power, the performance evaluation criteria η (Bhattacharyya et al. [35]) is determined as

$$\eta = Nu\, f_0^{0.33} / Nu_0\, f^{0.33} \quad (5)$$

where the subscript 0 denotes the plain tube.

The field synergy principle has a special application. To reveal the variation of the included angle between velocity and temperature Guo et al. [36] employed

$$\beta_m = \cos^{-1}\left(\int |\vec{u}||\vec{\nabla}T| \cos\beta dV / \int |\vec{u}||\vec{\nabla}T| dV\right) \quad (6)$$

5 Results

In order to inspect the heat transfer augmentation of the horizontal heated twisted channel, the Nusselt numbers of the channel with twisted geometry were evaluated as shown in Figure 4 for selected cases. It is observed that the Nusselt number raises with a raise in the Reynolds number (Re). This is indeed expected. Based on the Nusselt numbers, it is evident that the channel with twisted geometry have higher convective heat transfer (HT) compared that of the plain channel [1]. Interruption of the fluid flow (starting from the inlet of the experimental rig) owing to the occurrence of the twist leads to secondary flows and enhances fluid mixing [37]. It is also clear from Figure 4 that the Nusselt numbers are higher for small pitch ratio (S = p/d). This is likely because a small pitch ratio promotes flow recirculation and separation, which underside of the wake swirls to the upper side, interacting with the thermal boundary layer. As the fluid passes, the twisted portion enhances mixing between the core and surface. The Nu rises by 51% compared to the plain channel [1].

The friction penalty was used to assess the characteristics of fluid flow in this study. Figure 5 shows the variations of the friction factor (f) with the Reynolds number. It can be viewed that the 'f' reduces with a raise in Re. The 'f' are higher for the twisted channel evaluated with those for the plain channel, which is likely due to the increase in the contact surface area and reduced free flow, which increases the flow velocity. It is apparent that the friction factors are higher for the twisted channel small pitch ratio (S = p/d) evaluated against the other pitch ratios and plain channel [1]. The fluid is transported toward the channel walls due to the generation of secondary flow, which deforms the profiles of axial velocity along the length of the channel.

Turbulent Intensity at the exit of the channel is shown in Figure 6. Turbulent dies out for low Reynolds number (Re = 100) and increases with increasing Reynolds number. At the entry of the tube the TI is given 5% [38]. From the Figure 6, one can see that the highest level of turbulent intensity is about 14%-15% at Reynolds number of 40,000 - 50,000.

Figure 7 shows the variations of the thermal performance with the Re of the twisted channel. It can be distinguished that the thermal performance is more than unity for most of the time in varied Re. At higher Re (Re = 35000 – 50000) for the case of H = 0.2 and S = 0.45, the thermal performance decreases to a value less than 1.0. $\eta \geq 1.0$ indicates that the heat transfer for the channel with twisted geometry is higher than that for the plain one for the same pumping power. Figure 7 also shows that the thermal performance of the channel with twisted geometry increases with an increase of length ratio (H = l/L) and decrease in pitch ratio (S = p/d). The highest thermal performance is obtained for channel with H = 0.3 and S = 0.22. The lowest thermal performance is obtained for the channel with H = 0.2 and S = 0.45.

The synergy principle analysis [36] helps to realize the actual enhancement caused due to the presence of the twisted channel. Major part of the flow domain is redistricted to synergy angle close to 90^0. This is the reason the average synergy angle for each case is lies in the range of 87^0 to 90^0 as seen in Figure 8. The general nature of the synergy angle is to increase with increase in Reynolds's number. It is also evident that as the flow reaches turbulence there is a clear increase in the synergy angle. This could be used to explain why we see such a dramatic drop in the efficiency at the turbulent region. Again, it is visible that the increase in length ratio with small pitch ratio decreases the synergy sufficiently to provide good heat transfer enhancement which is also supported by the evidence in the efficiency.

5 Conclusions

In this work, the heat transfer and fluid flow characteristics of turbulent flow through a heat exchanger mounted with circular twisted channel were investigated experimentally and numerically. It is observed the twisted geometry enhances heat transfer considerably. Heat transfer also increases with an increase of length ratio and decrease in pitch ratio. The

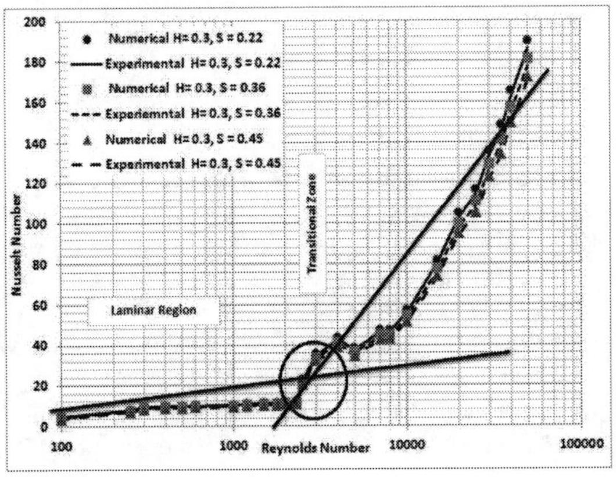

Fig. 4. Nusselt number as function of Reynolds number, (a) at constant value of length ratio (H = 0.3) with different pitch ratio (S = 0.22, 0.36, 0.45).

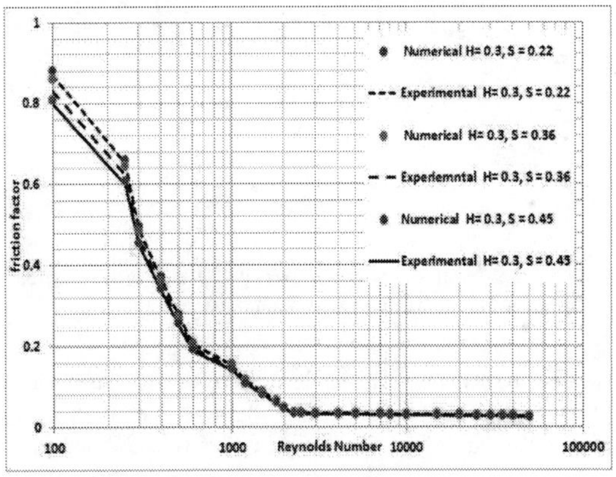

Fig. 5. Friction factor as function of Reynolds number, (a) at constant value of length ratio (H = 0.3) with different pitch ratio (S = 0.22, 0.36, 0.45).

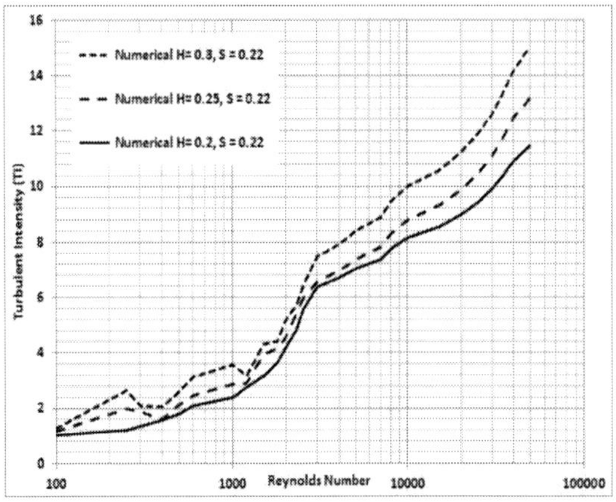

Fig. 6. Variation of turbulent intensity (TI) with Re.

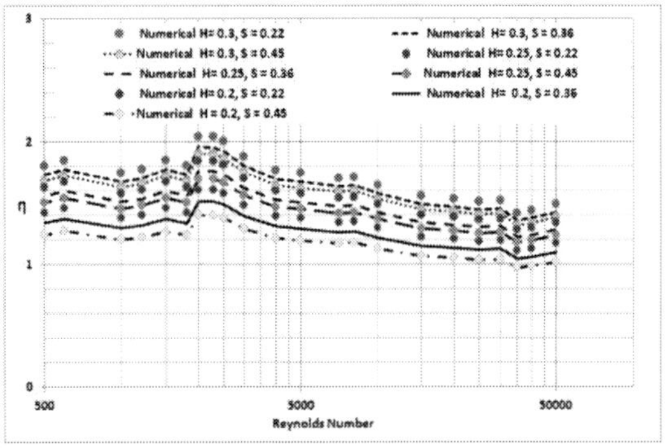

Fig. 7. Thermal enhancement factor.

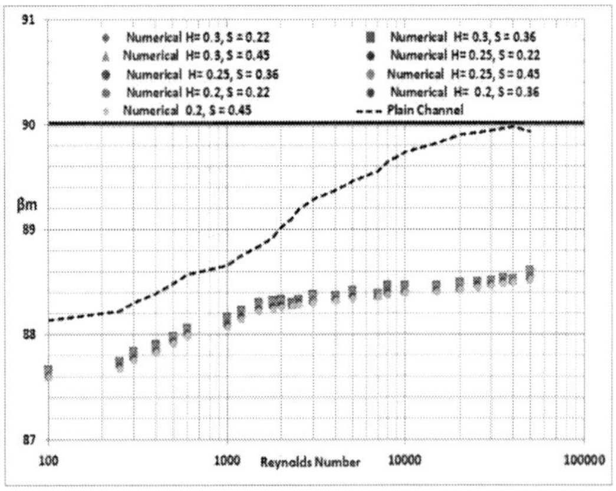

Fig. 8. Variation of average synergy angle with Reynolds number for different length and pitch ratio.

twisted geometry with the large length ratio and small pitch ratio leads to the highest heat transfer augmentation, where the Nu is higher by 51% compared with those for the plain channel. The friction factor decreases as the Re increases, but increases increasing length ratio and decreasing pitch ratio. In spite of the high friction penalty, H=0.3 and S=0.22 yields the maximum thermal performance, with a value of 2.08.

References

1. K.R. Shah, D.P. Sekulic, *Fundamentals of Heat Exchanger Design* (Wiley, Chichester, UK, 2003)
2. A.M. Maqableh, A.F. Khadrawi, M.A. Al-Nimr, S. A. Ammourah, A.C. Benim, Progress in Computational Fluid Dynamics, **11**, 5, pp.318-328 (2011)
3. D.G. Ebling, A. Krumm, B. Pfeiffelmann, J. Gottschald, J. Bruchmann, A.C. Benim, M. Adam, R. Labs, R.R. Herbertz, A. Stunz, Journal of Electric Materials, **45**, 7, pp.3433-3439 (2016)
4. A.C. Benim, M. Cagan, D. Gunes, International Journal of Thermal Sciences, **43**, 8, pp.725-732 (2004)
5. A.C. Benim, H. Chattopadhyay, A. Nahavandi, International Journal of Thermal Sciences, **50**, 10 pp.1973-1983 (2011)
6. I. Taymaz, E. Aslan, A.C. Benim, **19**, 2, pp. 537-547 (2015)
7. H. Chattopadhyay, A.C. Benim, Journal of Heat Transfer – Transactions of the ASME, **133**, 10, Article Number: 104502, 5 pages (2011)
8. A.C. Benim, K. Ozkan, M. Cagan, D. Gunes, International Journal of Numerical Methods for Heat & Fluid Flow, **17**, 3, pp.284-301 (2007)
9. N. Mahir, Z. Altaç, Heat Transfer Eng., **38**, 16, pp.1367 – 1381 (2016)
10. M. Hong, X. Deng, K. Huang, Z. Li, Chinese J. Chemical Eng., **15**, 6, pp.814-820 (2007)
11. A.C. Benim, M.P. Escudier, A. Nahavandi, A.K. Nickson, K.J. Syed, F. Joos, International Journal of Numerical Methods for Heat & Fluid Flow, **20**, 3, pp.348-370 (2010)
12. P.K. Rout, S.K. Saha, ASME J Heat Transfer, **135**, 2, Article Number: 021901 (2013)
13. S. Bhattacharyya, S. Saha, S.K. Saha, Exp. Thermal Fluid Science, **44**, pp.727-735 (2013)
14. S. Bhattacharyya, S.K. Saha, Experimental Thermal and Fluid Science, **42**, pp. 154-162 (2012)
15. S.K. Saha, S. Bhattacharyya, P.K. Pal, Exp. Therm.Fluid Sci., **41**, pp.121-129 (2012)
16. J.A. Meng, X.G. Liang, Z.J. Chen, Z.X. Li, Experimental Thermal and Fluid Science, **29**, pp.457-465 (2005)
17. S.K. Saha, ASME Journal Heat Transfer, **132**, 8, Article Number: 081701, 12 pages (2010)
18. A. Mwesigye, T. Bello-Ochende, J. P. Meyer, International Journal of Thermal Sciences, **99**, pp.238-257 (2016)
19. J. Chen, H. Müller-Steinhagen, G.G. Duffy, Applied Thermal Engineering, **21**, pp.535–547 (2011)
20. P.G. Vicente, A. Garcia, A. Viedma, International Journal of Heat and Mass Transfer, **45**, pp.5091–5105 (2002)
21. H. Mengna, D. Xianhe, H. Kuo, L. Zhiwu, Chinese Journal of Chemical Engineering, **15**, pp.814–820 (2007)

22. P. Sivashanmugam, S. Suresh, Applied Thermal Engineering, **26,** 16, pp.1990-1997 (2006)
23. G. Biswas, H. Chattopadhyay, A. Sinha, Heat Transfer Engineering, **33,** 4, pp.406-424 (2012)
24. P. Kumar, F. Topin, M. Miscevic, P. Lavieille, L. Tadrist, Numerical Heat and Mass Transfer in Porous Media, Advanced Structured Materials, **27,** pp.181–208 (2012)
25. B. Li, B. Feng, Y.L. He, W.Q. Tao, Appl. Thermal Eng., **26,** pp.2336-2344 (2006)
26. S.K. Saha, S. Bhattacharyya, G.L. Dayanidhi, Heat Trans. Research, **43,** 3, pp. 207-227 (2012)
27. S. Bhattacharyya, H. Chattopadhyay, S. Bandyopadhyay, International J. Heat and Technology, **34,** 3, pp. 401-406 (2016)
28. J.P. Abraham, E.M. Sparrow, J.C.K. Tong, Int. Journal of Heat and Mass Transfer, **52,** 3-4, pp. 557–563 (2009)
29. www.ansys.com
30. A.C. Benim, Computer Methods in Applied Mechanics and Engineering", **67,** 1, pp.1-14 (1988)
31. A. Assmann, A.C. Benim, F. Gül, P. Lux, P. Akhyari, U. Boeken, F. Joos, P. Feindt, A. Lichtenberg, Journal of Biomechanics, **45,** 1, pp.156-163 (2012)
32. P.K. Kadiyala, H. Chattopadhyay, Heat Transfer Engineering, doi: 10.1080/01457632.2017.1288045, pp. 1-9 (2017)
33. P. Ocłoń, S. Łopata, M. Nowak, A.C. Benim, Progress in Computational Fluid Dynamics, **15,** 5, pp.290-306 (2015)
34. S. Bhattacharyya, H. Chattopadhyay, A.C. Benim, Progress in Computational Fluid Dynamics, **17,** 3, pp. 193-197 (2017)
35. S. Bhattacharyya, H. Chattopadhyay, A. Swami, Md. K. Uddin, Int. J. Heat and Technology, **34,** 4, pp.727-733 (2016)
36. Z.Y. Guo, W.Q. Tao, R.K. Shah, International Journal of Heat Mass Transfer, **48,** pp.1797-1807 (2005)
37. S. Bhattacharyya, H. Chattopadhyay, A. Guin, A.C. Benim, Heat Tran. Eng., doi: 10.1080/01457632.2018.1474593 (2018)
38. S. Bhattacharyya, H. Chattopadhyay, A.C. Benim, Progress in Computational Fluid Dynamics, **17,** 6, pp. 390-396 (2017)

Natural convection melting of nano-enhanced phase change material in a cavity with finned copper profile

Nadezhda Bondareva, and *Mikhail* Sheremet*

Laboratory on Convective Heat and Mass Transfer, Tomsk State University, 36 Lenin Avenue, 634050 Tomsk, Russia

Abstract. Present study is devoted to numerical simulation of heat and mass transfer inside a cooper profile filled with paraffin enhanced with Al_2O_3 nanoparticles. This profile is heated by the heat-generating element of constant volumetric heat flux. Two-dimensional approximation of melting process is described by the Navier-Stokes equations in non-dimensional variables such as stream function, vorticity and temperature. The enthalpy formulation has been used for description of the heat transfer. The influence of volume fraction of nanoparticles and intensity of heat generation on melting process and natural convection in liquid phase has been studied.

1 Introduction

Cooling and maintenance of temperature has a significant role in electronic industry and power engineering. During the operating of the semiconductor appliances and accumulators a large amount of energy generates. To dissipate this energy the metal profiles with high thermal conductivity are used. The presence of fins on the profile increases heat exchange with the environment.

Paraffin is widely used in application in the field of heat storage and energy absorption. Phase transition is accompanied by either absorption or release of energy. As a rule, paraffin waxes has a low melting temperature and it is suitable for operating of electronic systems. Many experimental and numerical researches are devoted to melting of paraffin. Phase change materials such as paraffin wax in spite of high latent heat have a relatively low thermal conductivity. Therefore, it can be used in conjunction with metal profiles, metal foams, graphite matrixes and others. One of the ways to enhance heat transfer inside the system containing paraffin is the adding of nanoparticles to paraffin [1, 2]. Including of nanoparticles with high heat conductivity allows to increase the intensity of heat exchange [3–6]. It was shown experimentally that thermal conductivity increases with a growth of loadings of nanoparticles and temperature [3, 5]. Bayat et al. analyzed numerically heat transfer inside the heat sink filled with PCM enhanced by nanoparticles of aluminum oxide and cooper oxide [7]. The authors showed that the volume fraction of nanoparticles is reflected in time of reaching critical temperature. The liquid volume fraction rises faster with adding nanoparticles.

At the same time the effect of particles on the viscosity of the suspension and influence of nanoparticles size play an important role in the fluid motion and phase transition [8]. The combination of these factors determines the heat transfer modes inside the material. Numerical investigations of melting process of ice with Cu nanoparticles inside in the thermal storage unit were carried out by Jourabian et al. [9, 10]. The domains under consideration represent a two-dimensional cross-section model of heated tube with constant temperature inside round [9] or square [10] shells. It was shown that a growth of particle volumetric concentration reduces the influence of heater position, due to a raise of viscosity and as a consequence due to the weakening of convection heat transfer. It should be noted that for the case of low Rayleigh numbers ($Ra = 10^5$ and $Ra = 2\cdot 10^5$) the addition of nanoparticles leads to an intensification of melting in the lower part of the region and has little effect on the melting rate at the top [9, 10], whereas in the case of high Rayleigh number $Ra = 1.25\cdot 10^7$ the melting rate rises both above the heater and under the heater [11].

Another factor affecting the heat transfer is the change in the latent energy of suspension melting. An inclusion of solid particles of greater density leads to a reduction of the latent heat.

Present study is devoted to a numerical investigation of heat transfer inside the passive cooling system based on nano-enhanced phase-change material.

2 Physical and mathematical model

Figure 1 shows the geometry of considered two-dimensional domain. It represents a rectangular cavity with a copper radiator filled with pure n-octadecane, the bottom wall of which is in contact with a rectangular source of constant volumetric heat generation. The heater has thermophysical properties of silicon. The initial temperature of the whole system is the same as the ambient temperature. Heat exchange with the

*Corresponding author: Michael-sher@yandex.ru

https://doi.org/10.1051/matecconf/201824001006

environment is carried out through the upper wall of the region, other borders of the system are considered to be thermally insulated. Fluid motion is assumed to be laminar. The thermophysical properties of paraffin, such as thermal conductivity, viscosity and density both in solid paraffin and in melt are considered to be constant and independent of temperature. The Boussinesq approximation for describing the buoyancy force effect inside the molten region, is applied.

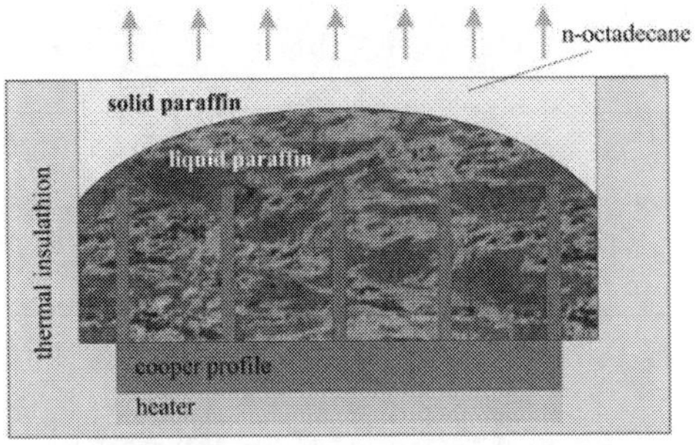

Fig. 1. Physical model.

Governing equations in primitive variables velocity and pressure can be written as:

$$\frac{\partial u}{\partial x}+\frac{\partial v}{\partial y}=0 \tag{1}$$

$$\frac{\partial u}{\partial t}+u\frac{\partial u}{\partial x}+v\frac{\partial u}{\partial y}=-\frac{1}{\rho_{nm}}\frac{\partial p}{\partial x}+\frac{\mu_{nm}}{\rho_{nm}}\left(\frac{\partial^2 u}{\partial x^2}+\frac{\partial^2 u}{\partial y^2}\right) \tag{2}$$

$$\frac{\partial v}{\partial t}+u\frac{\partial v}{\partial x}+v\frac{\partial v}{\partial y}=-\frac{1}{\rho_{nm}}\frac{\partial p}{\partial y}+\frac{\mu_{nm}}{\rho_{nm}}\left(\frac{\partial^2 v}{\partial x^2}+\frac{\partial^2 v}{\partial y^2}\right)+ \frac{(\rho\beta)_{nm}}{\rho_{nm}}g\left(T-T_m\right) \tag{3}$$

The heat conduction equation within the solid phase and energy equation inside the liquid phase of paraffin can be written as follows:

$$\frac{\partial h}{\partial t}=\left(k_s\right)_{nm}\left(\frac{\partial^2 T}{\partial x^2}+\frac{\partial^2 T}{\partial y^2}\right) \tag{4}$$

$$\frac{\partial h}{\partial t}+u\frac{\partial h}{\partial x}+v\frac{\partial h}{\partial y}=\left(k_l\right)_{nm}\left(\frac{\partial^2 T}{\partial x^2}+\frac{\partial^2 T}{\partial y^2}\right) \tag{5}$$

For the heat source and the copper profile, the heat transfer can be described by the following heat conduction equations:

$$\rho_1 c_1\frac{\partial T}{\partial t}=k_1\left(\frac{\partial^2 T}{\partial x^2}+\frac{\partial^2 T}{\partial y^2}\right) \tag{6}$$

$$\rho_2 c_2\frac{\partial T}{\partial t}=k_2\left(\frac{\partial^2 T}{\partial x^2}+\frac{\partial^2 T}{\partial y^2}\right)+Q \tag{7}$$

In present work, the hydrodynamics equations were solved using the stream function and vorticity variables defined as: $u=\partial\psi/\partial y$, $v=-\partial\psi/\partial x$ and $\omega=\partial v/\partial x-\partial u/\partial y$. Dimensionless variables were correlated to dimensional ones through the relations:

$$X=x/H,\ Y=y/H,\ U=u/V_0,\ V=v/V_0,$$
$$V_0=\sqrt{g\beta\left(T_h-T_F\right)H},\ \tau=H/V_0,\ \Omega=\omega H/V_0, \tag{8}$$
$$\Theta=\left(T-T_F\right)/\left(T_h-T_F\right),\ \Psi=\psi/\left(V_0 H\right)$$

Taking into account these relations (8) the governing equations (1)–(5) can be written as follows:

$$\frac{\partial^2\Psi}{\partial X^2}+\frac{\partial^2\Psi}{\partial Y^2}=-\Omega \tag{9}$$

$$\frac{\partial\Omega}{\partial\tau}+U\frac{\partial\Omega}{\partial X}+V\frac{\partial\Omega}{\partial Y}=\frac{\mu_{nm}/\mu_f}{\rho_{nm}/\rho_f}\sqrt{\frac{Pr}{Ra}}\nabla^2\Omega+ +\frac{(\rho\beta)_{nm}/(\rho\beta)_f}{\rho_{nm}/\rho_f}\frac{\partial\Theta}{\partial X} \tag{10}$$

$$\zeta(\varphi)\left[\frac{\partial\Theta}{\partial\tau}+U\frac{\partial\Theta}{\partial X}+V\frac{\partial\Theta}{\partial Y}\right]+\frac{\rho_{nm}L_{nm}}{\rho_f L_f}\cdot Ste\cdot\left[\frac{\partial\varphi}{\partial\tau}+ +U\frac{\partial\varphi}{\partial X}+V\frac{\partial\varphi}{\partial Y}\right]=\frac{\xi(\varphi)}{\sqrt{Ra\cdot Pr}}\nabla^2\Theta \tag{11}$$

In energy equation the smoothing function φ, reflecting the melt fraction, and the auxiliary functions $\zeta(\varphi)$ and $\xi(\varphi)$ were introduced:

$$\varphi=\begin{cases}0, & T<T_m-\eta\\ \dfrac{T-\left(T_m-\eta\right)}{2\eta}, & T_m-\eta\le T\le T_m+\eta\\ 1, & T>T_m+\eta\end{cases} \tag{12}$$

$$\zeta(\varphi)=\frac{\left(\rho_s c_s\right)_{nm}}{\rho_l c_l}+\varphi\left(\frac{\left(\rho_l c_l\right)_{nm}}{\rho_l c_l}-\frac{\left(\rho_s c_s\right)_{nm}}{\rho_l c_l}\right),$$
$$\xi(\varphi)=\frac{\left(k_s\right)_{nm}}{k_l}+\varphi\left(\frac{\left(k_l\right)_{nm}}{k_l}-\frac{\left(k_s\right)_{nm}}{k_l}\right) \tag{13}$$

Dimensionless energy equation for the heater can be written as:

$$\frac{\partial\Theta}{\partial\tau}=\frac{\alpha_2/\alpha_0}{\sqrt{Ra\cdot Pr}}\left(\nabla^2\Theta+Os\right) \tag{14}$$

Here the Ostrogradsky number (*Os*) characterizes the intensity of volumetric heat generation of the heater.

The thermophysical properties of the nanomaterial are determined using the following correlations [7]:

$$
\begin{aligned}
&(\rho\beta)_{nm} = (1-\Phi)(\rho\beta)_f + \Phi(\rho\beta)_{np} \\
&(\rho c)_{nm} = (1-\Phi)(\rho c)_f + \Phi(\rho c)_{np} \\
&\rho_{nm} = (1-\Phi)\rho_f + \Phi\rho_{np} \\
&c_{nm} = (\rho c)_{nm} / \rho_{nm}
\end{aligned} \tag{15}
$$

The correlation for the effective thermal conductivity can be taken for the liquid paraffin as [5]:

$$
\begin{aligned}
k_{nm} = &\frac{k_{np} + 2k_f - 2(k_f - k_{np})\Phi}{k_{np} + 2k_f + (k_f - k_{np})\Phi} k_f + \\
&+5\cdot 10^4 \beta_\lambda \Phi \rho_f c_f \sqrt{\frac{\kappa T}{\rho_{np} d_{np}}} f(T,\Phi)
\end{aligned} \tag{16}
$$

and in the solid material:

$$
k_{nm} = \frac{k_{np} + 2k_m - 2(k_f - k_{np})\Phi}{k_{np} + 2k_m + (k_f - k_{np})\Phi} k_m \tag{17}
$$

Here $\beta_\lambda = 8.4407(100\Phi)^{-1.07304}$, and $\kappa = 1.381\cdot 10^{-23}$ J/K is the Boltzmann constant, the function $f(T,\Phi)$ determines as:

$$
\begin{aligned}
f(T,\Phi) = &\left(2.817\cdot 10^{-2}\Phi + 3.917\cdot 10^{-3}\right)\frac{T}{T_0} + \\
&+\left(-3.0669\cdot 10^{-2}\Phi - 3.91123\cdot 10^{-3}\right)
\end{aligned} \tag{18}
$$

Here $T_0 = 273$ K and $T = 320$ K, latent heat is $L_{nm} = \dfrac{(1-\Phi)\rho_m L_m}{\rho_{nm}}$.

The nanofluid viscosity was defined as

$$
\mu_{nm}/\mu_f = 0.983\exp(12.959\Phi) \tag{19}
$$

The dimensionless parameters were defined as $Pr = \nu\rho_l c_l / k_l$ is the Prandtl number; $Ste = L_f/(c_l \Delta T)$ is the Stefan number; $Ra = g\beta\Delta T H^3/(\nu\alpha_l)$ is the Rayleigh number; $Os = qH^2/(k_2\Delta T)$ is the Ostrogradsy number; $Bi = \bar{h}H/k$ is the Biot number.

2.1. Initial and boundary conditions

At the initial time, the temperature of system was set to equal the ambient temperature $\Theta = \Theta_0$.

Boundary conditions for the governing equations are:

- condition at the interface between the copper profile and the phase change material

$$
\Theta_m = \Theta_1,\ k_m \frac{\partial \Theta_m}{\partial Y} = k_1 \frac{\partial \Theta_1}{\partial Y} \tag{20}
$$

- condition at the interface between the copper profile and the heat source:

$$
\Theta_1 = \Theta_2,\ k_1 \frac{\partial \Theta_1}{\partial Y} = k_2 \frac{\partial \Theta_2}{\partial Y} \tag{21}
$$

- the convective boundary condition is set at the upper border

$$
\frac{\partial \Theta}{\partial Y} = -Bi(\Theta - \Theta_0) \tag{22}
$$

- Adiabatic boundary conditions for temperature were set at other borders:

$$
\frac{\partial \Theta}{\partial \bar{n}} = 0 \tag{23}
$$

On the solid boundaries of the area occupied by the melted paraffin the no-slip boundary conditions have been used for velocity in the form: $\Psi = 0$, $\Omega = -\nabla^2\Psi$.

2.2. Numerical procedure

To solve the dimensionless equations (9)–(11) with corresponding initial and boundary conditions (20)–(23) the finite difference method was used [12–14]. The difference elliptic equation for stream function was solved by using the successive over relaxation method. The energy and vorticity equations were solved by using the Samarskii's local one-dimensional scheme. This algorithm was verified and the results were in good agreement with experimental and numerical data of other authors.

3 Results and Discussion

Figure 2 shows isotherms for different nanoparticles volume fractions and dimensionless time. It can be seen that at initial stage of heating (upper row) the presence of nanoparticles promotes the rapid melting of the paraffin, taking into account the phase front position. The movement of the interphase occurs mostly due to heat conduction. At the same time, a growth of time (bottom row) leads to more intensive melting in the case of low nanoparticles concentration due to convective flow intensification, while a rise of nanoparticles volume fraction results in a growth of the melt viscosity and as a result convective flow attenuates. Therefore, for high Φ one can find less intensive melting.

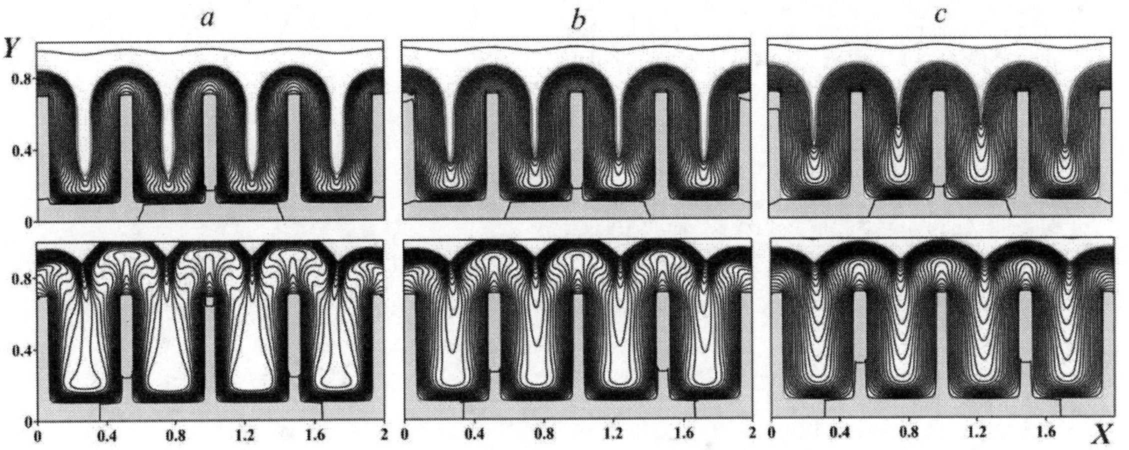

Fig. 2. Isotherms for $Os = 0.007$ at $\tau = 636.3$ (upper row) and $\tau = 848.4$ (bottom row): *a*) $\Phi = 0$, *b*) $\Phi = 0.04$, *c*) $\Phi = 0.08$.

Figure 3 presents the mean Nusselt number time curves. Initial increase in the average Nusselt number relates with heating of the cavity due to heat conduction. As soon as the temperature of profile surface reaches the fusion temperature the melting process begins with absorbing a large amount of energy. The Nusselt number is growing rapidly, until the entire fins surface is covered by a layer of melt. Further, the phase front is removed from the copper profile, where temperature gradually increases, the mean Nusselt number is changing insignificantly. With the appearance and development of natural convection, the interphase boundary moves upward, vertical velocity increases and the upper part of the melt region warms up faster, while at the bottom part the descending flow of the liquid cools the spaces between the fins.

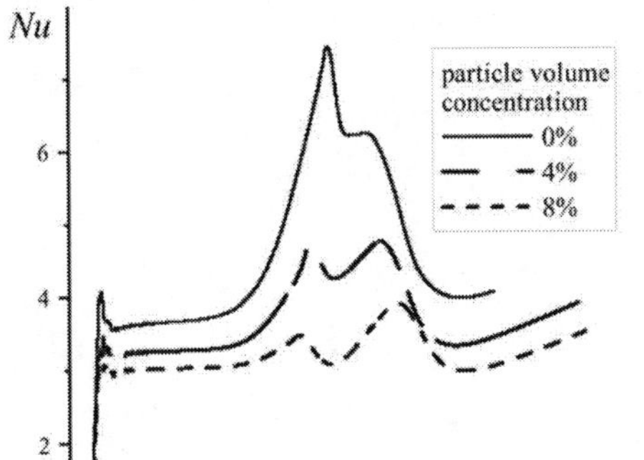

Fig. 3. Nusselt number time curves for $Os = 7{\cdot}10^{-3}$.

When convective mode is formed in the melt, the cold flows fall down, as a result the cooling of the plate surface and increasing the heat transfer are occurs. With adding of nanoparticles, this effect becomes less pronounce. Enlargement of viscosity in nano-enhanced material leads to weakening of convective flows. The development of the circulation flow with the expansion of the melt region promotes melting, with addition of particles and an increase in viscosity, the melting rate decreases.

Figure 4 shows the evolution of the liquid volume fraction. An inclusion of nanoparticles inside the PCM results in more intensive melting of the medium at the initial and middle level of the process, while in the case of whole melting an addition of nanoparticles rise the time for complete melting.

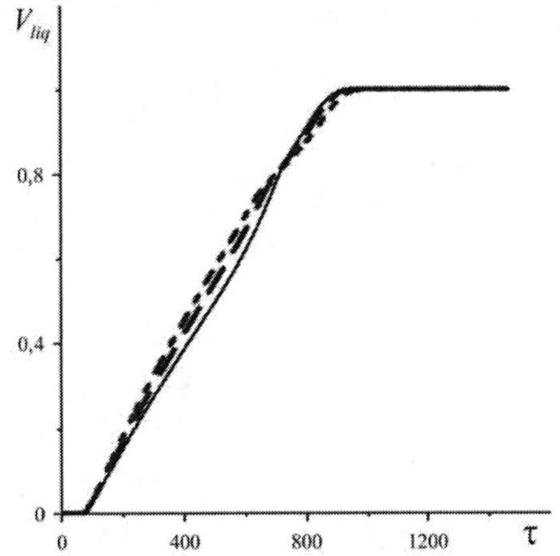

Fig. 4. Liquid volume fraction time curves for $Os = 7{\cdot}10^{-3}$.

4 Conclusions

Numerical investigation of conjugate natural convection during the melting process of nano-enhanced phase change material inside the cooling system with heat sink was carried out. Distributions of isotherms were received for different nanoparticles volume fraction and dimensionless time. It was shown that an increase in the nanoparticles concentration leads to the melting process accelerates at the initial and middle level of the considered process due to a rise of the thermal conductivity, while with the growth of the melt volume and the development of natural convective heat transfer,

*Corresponding author: Michael-sher@yandex.ru

the phase front moves upward faster in the case of pure paraffin. It should be noted that the maximum value of average Nusselt number is observed for the case of $\Phi = 0$.

Acknowledgements

This work was conducted as a government task of the Ministry of Education and Science of the Russian Federation (Project Number 13.9724.2017/8.9).

References

1. C.J Ho, J.Y. Gao, Int. J. Heat Mass Transfer **62**, (2013).
2. S.S. Sebti, M. Mastiani, H. Mirzaei, A. Dadvand, S. Kashani, S.A. Hosseini, Applied Physics & Engineering **14** (2013).
3. L.-W. Fan, Z.-Q. Zhu, Y. Zeng, Q. Lu, Z.-T. Yu, Int. J. Heat Mass Transfer **79** (2014).
4. N.S. Dhaidan, J.M. Khodadadi, T.A. Al-Hattab, S.M. Al-Mashat, Int. J. Heat Mass Transfer **66** (2013).
5. R.S. Vajjha, D.K. Das, Int. J. Heat Mass Transfer **52** (2009).
6. A. Ebrahimi, A. Dadvand, Alexandria Engineering Journal **54** (2015).
7. M. Bayat, M.R. Faridzadeh, D. Toghraie, Thermal Science and Engineering Progress **5** (2018).
8. Y.M.F. El Hasadi, J.M. Khodadadi, J. Heat Transfer **135** (2013).
9. M. Jourabian, M. Farhadi, A.A.R. Darzi, Applied Mathematical Modelling **37** (2013).
10. M. Jourabian, M. Farhadi, K. Sedighi, Computers and Mathematics with Applications **67** (2014).
11. A.J. Chamkha, A. Doostanidezfuli, E. Izadpanahi, M. Ghalambaz, Advanced Powder Technology **28** (2017).
12. N.S. Bondareva, M.A. Sheremet, Thermophysics and Aeromechanics **23** (2016).
13. N.S. Bondareva, M.A. Sheremet, Int. J. Mechanical Sciences **134** (2017).
14. N.S. Bondareva, M.A. Sheremet, Int. J. Thermal Sciences **115** (2017).

Determination of distribution of heat-conducting material concentration in protective layer of TPS panel

Łukasz Brodzik[1,*], *Andrzej* Frąckowiak[1]

[1]Poznań University of Technology, Chair of Thermal Engineering, Piotrowo 3 street, 60-965 Poznań, Poland

Abstract. Paper presents the problem of heating the damaged insulation of an orbiter. Changes of the insulation's thermal properties, made by adding conductive material of high value of specific heat in a form of a dope to the protective layer, were examined. The analysis was made using insulating material LI900 and beryllium dope. In the paper was investigated the effect of the dopant's share on the temperature distribution on the aluminum surface. This influence was depended on the size of the damage, defined by parameter D, as well as a constant percentage volume fraction.

Nomenclature

T – temperature,
ρ – density,
p - pressure,
c – specific heat,
k – thermal conductivity,
γ – concentration of the addition

1 Introduction

The object of the research described in this paper is high temperature reusable surface insulation (HRSI) of an orbiter. In extreme cases, the damage of the insulation material results in melting of the orbiter aluminum structure, leading directly to its destruction. Under normal conditions, when the thermal protection system is fully operative, a significant increase in the temperature of the orbiter structure takes place only after the vehicle has landed. When a malfunction occurs, the temperature increase takes place earlier, already at the flight stage. Contrary to many concepts discussed in literature, the paper does not focus on increasing of the thermal durability. The LI900 surface insulation is a well-explored material [1, 2, 3], therefore that kind of thermal barrier in analysis was used.

2 Formulation of the heat transfer problem

The numerical simulations were made in the FreeFem++ environment with the assumed flight duration of the shuttle of t_2 = 2000 s and the heat flux [4]. Each axisymmetric model used in the research had 3 layers, i.e. aluminum structure of the vehicle, NOMEX SIP and the LI900 insulation. This simplification is due to the negligible effect of thin layers, Fig. 1.

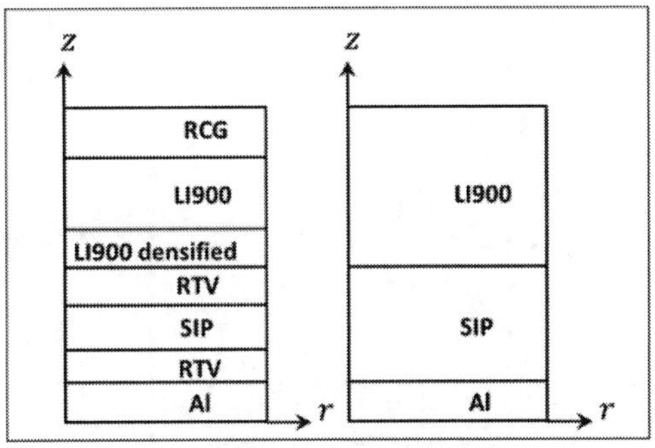

Fig. 1. Geometry of model.

The purpose of the analysis was to determine such a distribution of the concentration as to reduce the temperature increase on the surface of the aluminum structure. For this reason, the dopant was used at constant concentration throughout the entire insulation volume. The filings were oriented parallel to the outer surface of the insulation. In this way, in the insulation part of the heat flux was accumulated and conducted in the transverse direction of the main stream. Significant differences in maximum temperature should occur before 2000s when the orbiter is in flight.

In the investigations, a material loss of the diameter D of 1, 2, 3 and 4 cm was assumed, Fig. 2. The additional conductive material, as a dopant, in the insulation was beryllium. The specifications of the HRSI tile material and additional material were taken from literature [5].

The first law of thermodynamics and Fourier's law was used in carrying out the analysis of the heat flow in non-stationary conditions for axisymmetric heat conduction model.

* Corresponding author: lukasz.brodzik@put.poznan.pl

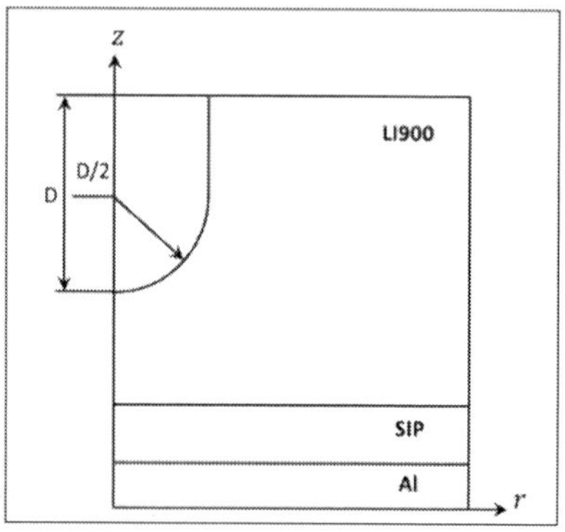

Fig. 2. Geometry of model.

Therefore, for the cylindrical coordinate system in space Ω_1 the following formula was used (Fig. 3):

$$\rho c_p\left(\frac{\partial T}{\partial t}\right)=\frac{1}{r}\frac{\partial}{\partial r}\left(k_r r\frac{\partial T}{\partial r}\right)+\frac{\partial}{\partial z}\left(k_z\frac{\partial T}{\partial z}\right) \qquad (1)$$

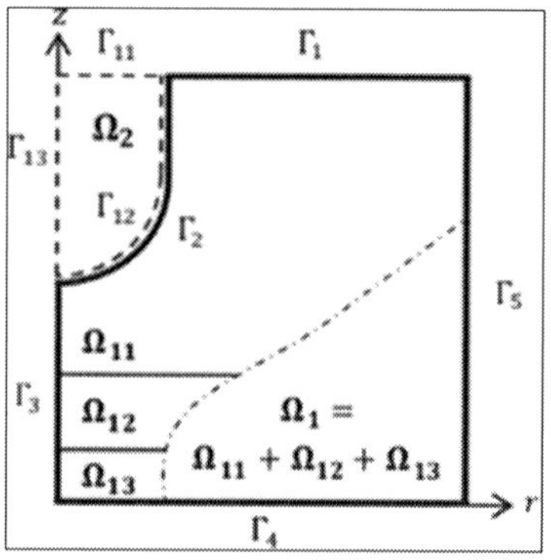

Fig. 3. Geometry of model.

In place of damage (Γ_{12} edge) problem of radiation was solved using P1 approximation. This model assumes plane-parallel symmetry, that matter is in local thermodynamic equilibrium, scattering is coherent and isotropic. This model is one of approximations to solve Radiative Transfer Equation (RTE).

3 Results

For the smallest considered diameter D, the addition of beryllium causes, that the maximum temperature on the surface of the underlying structure is less than the maximum working temperature of this material, Fig. 4.The increase of temperature in function of damage magnitude for 1% dopant of beryllium is described by the following polynomial

$$T_{max} = 0.195D^2 - 1.79D + 145.5 \qquad (2)$$

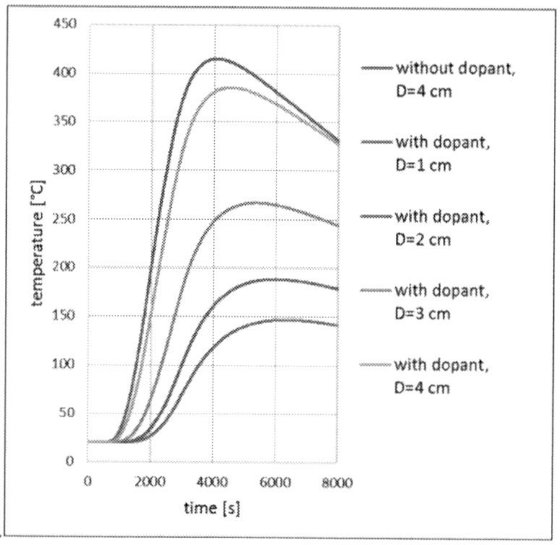

Fig. 4. Comparison of the distribution of temperature on the surface of the orbiter structure for a damaged tile, γ =1%.

Results are presented in Tab. 1.

Table 1. Results for models with 1% dopant of beryllium.

D [cm]	T_{max} [°C]	t [s]
1	147	6305
2	188	5975
3	267	5350
4	386	4525

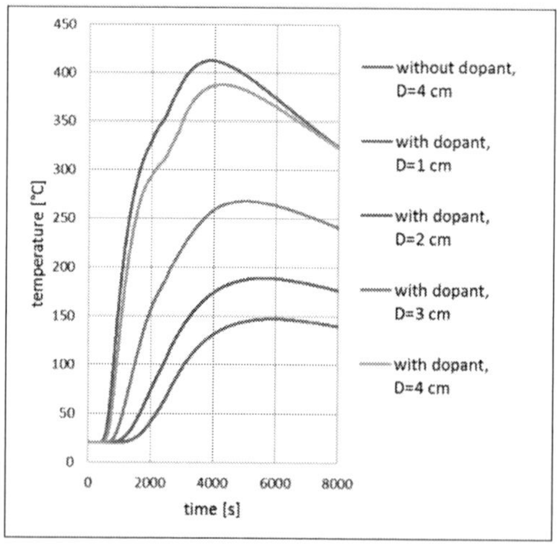

Fig. 5. Comparison of the distribution of temperature on the surface of SIP for a damaged tile, γ =1%.

Very similar distribution of temperature is observed on the surface of SIP, on merging with insulation, Fig. 5.

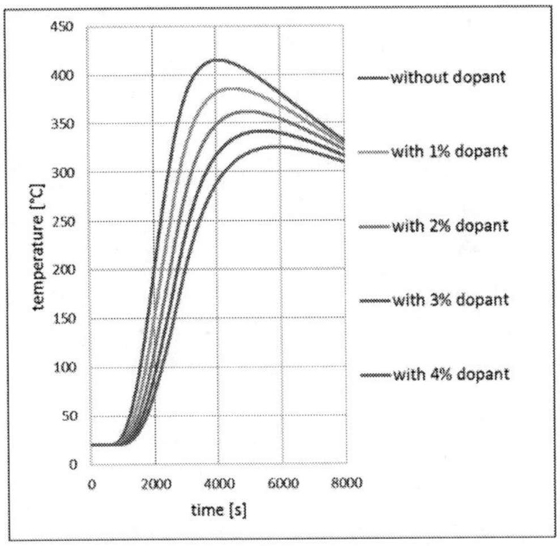

Fig. 6. Comparison of the distribution of temperature on the surface of the orbiter structure for a damaged tile, D=4 cm.

The Fig. 6 shows, what is the effect of adding an increasingly large dopant of beryllium. The increase of temperature in function of dopant magnitude for 4 cm of damage diameter is described by the following polynomial

$$T_{max} = 1.75\gamma^2 - 29.05\gamma + 413.25 \quad (3)$$

Results are presented in Tab. 2.

Table 2. Results for models with different dopant of beryllium.

γ [%]	T_{max} [°C]	t [s]
1	386	4525
2	362	4970
3	342	5425
4	325	5880

It is possible to analyze how big delay in the occurrence of maximum temperature, as well as its reduction is provided by beryllium filings. There were compared for the same type of tile damage, D = 4cm, 4 variants of models. The first variant was a model, as before, with an insulation thickness of 77 mm. The next two models contain insulation with dopant of beryllium equal to 1% of volume and insulation with an increased thickness equal to 86.868 mm. Both insulations had the same weight of 0.2909 kg. In the basic configuration, the insulation had a weight of 0.2579 kg. The last option is a model in which dopant of so-called ideal material has been added. It had the properties of hydrogen characterized by a very high value of specific heat. The results are presented in the Fig.7.

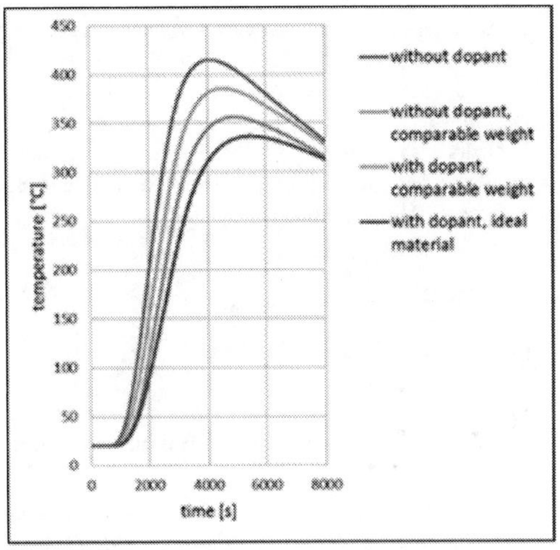

Fig. 7. Comparison of the distribution of temperature on the surface of the orbiter structure for a damaged tile, using the same weight, D=4 cm.

Maximum temperature for model with thicker insulation is 357°C obtained after 4905s. For the corresponding model with a dopant of beryllium, these parameters are respectively 386°C and 4525s. The results related to the temperature distribution in the insulation including ideal material show, that using this method there can be delay the occurrence of its maximum additionally by almost 10 minutes.

4 Conclusions

The additional conductive layer should be made from a material characterized by a high value of specific heat and high melting point. These properties may turn out useful not only for future orbital flights but perhaps also for interplanetary flights. Such a material should also have a relatively high density. During the research, nickel was a good candidate in this respect ensuring good insulator properties. For the adopted distribution of the heat flux for the two considered additions of nickel during the flight, the structure did not exceed its admissible operating temperature range, hence was not damaged. Moreover, the insulation with an additional material should ensure limited heat conduction towards the orbiter structure. It should also be characterized by a greater conduction in the transverse direction (compared to the main direction of the heat flux).

It is important that the new material should be contained in an insulator in a low admixture value. This will provide a small increase in weight of the tile. Considered material, despite its small weight and relatively good thermal properties, proved to be useless in the proposed method. As a result, still thicker insulator layer exhibits more favorable heat transfer. The analysis proved, that the effect of the method used is similar to increasing the thickness of the insulator. However, there is the possibility of using such a material, which in practice proves to be more effective in reducing heat flow. Even ensuring a small reduction in

https://doi.org/10.1051/matecconf/201824001007

maximum temperature in relation to the thicker insulation without an additional conductive layer but maintaining the same weight will bring big benefits. Using a modified material, the thickness of the insulator in the tile can be reduced. There may be tens of thousands of such tiles. Finally, it would reduce weight, and also cost of space missions.

References

1. Yuen W. W., Takara E., Cunnington G., *Combined conductive/radiative heat transfer in high porosity fibrous insulation materials: theory and experiment*, (The 6th ASME-JSME Thermal Engineering Joint Conference, 2003)
2 Balasubramaniam R., Callister S., *Materials Science And Engineering: Indian Adaptation* (Wiley India Pvt, 2009)
3 Dobrzański A. Leszek, *Engineering materials and material design. Fundamentals of materials science and metallurgy*, (WNT, 2006)
4. Ng Heok Wei, *Thermomechanical analysis of a damaged thermal protection system*, (doctoral thesis, University of Michigan, 2007)
5. Wiliams S. D., Curry M. Donald, *Thermal Protection Materials-Thermophysical Property Data*, (NASA Houston, 1992)

Numerical modeling of heat transfer in the fixed-matrix regenerator working in the Electric Thermal Storage heating system

Piotr Cisek[1,*], *Dawid* Taler[2], and *Grzegorz* Cisek[3]

[1]Institute of Thermal Power Engineering, Cracow University of Technology, 31-864 Cracow, Poland
[2]Faculty of Environmental Engineering, Cracow University of Technology, 31-155 Cracow, Poland,
[3]Department of Telecommunications, AGH University of Science and Technology, 30-962 Cracow, Poland

Abstract. The study presents the concept of Electric Thermal Storage (ETS) central heating system. Thermal Energy Storage (TES) is carried out in the fixed-matrix regenerator. The energy conservation equations, determined for the discharge period of the regenerator operation, are implemented in MATLAB numerical procedures based on the Finite Volume Method (FVM). In the model pressure drops within the system are calculated, both for the airflow through the inner tubes, and between the tubes. The flow distribution calculations show that the assumption of even air flow distribution would not be justified. Subsequently, the values of heat transfer coefficients are determined for the four distinct heat transfer surfaces, for the variable axial coordinate z and during the time of the system operation. The use of two different criterion equations is considered, for determining the mean Nusselt number Nu_m for fluid flow through the concentric annular duct, as well as for the local Nusselt number Nu_z calculated for the fluid flow through a circular or non-circular ducts. The most appropriate approach is selected by comparing the calculation results with experimental data. Taking into account the relative error, *RMSE*, and *MAPE* values calculated, it may be concluded that the Taler correlation – for non-circular ducts – gives results closer to the experimental data obtained.

1 Introduction

The presented study is related to sensible heat storage carried out in regenerative heat exchangers, also called as regenerators. Regenerative heat exchangers are classified as fixed-matrix and rotary regenerators. The fixed-matrix or fixed-bed regenerator is a periodic-flow heat transfer device with high thermal capacity matrix through which the hot fluid stream and cold fluid stream pass alternately [1, 2]. Specific parameters determine regenerators performance, i.e., geometry, thermal and mechanical properties of the solid matrix, working fluid properties, as well as a charge and discharge time-length periods [3].

Surface geometries used for the compact fixed-matrix regenerators are, i.e., quartz pebbles, steel, copper, or lead shots, copper wool, packed fibers, powders, randomly packed woven screens, and crossed rods, among others. Fixed-matrix regenerators application involves both high temperature, 800°C – 1 100°C, heat recovery systems [4–6], and low-temperature applications, like space heating in HVAC installations [7–9].

As shown by Ismail et al. [10], two broad groups of models of heat transfer between the working fluid and the regenerator bed are commonly used. The first group includes models in which the regenerator bed and the fluid is considered as one phase. Therefore they are called single-phase models. The second group comprises of models where the bed and the fluid domains are treated as a separate phase, hence the name - two-phase models. In the second group, three basic models may be distinguished, the Schumann's model, the continuous solid phase model, and the concentric dispersion model, that assumes a thermal gradient in the solid particles [10].

In the paper numerical investigation of heat transfer between fixed matrix and working fluid during cooling period is presented. Thus, heat balance equations for the regenerator's matrix and working fluid domains are derived for the concentric dispersion model. The heat balance equations are formulated for the novel design of fixed-bed regenerator that works as a Thermal Energy Storage (TES) unit in Electric Thermal Storage (ETS) central heating system. The governing system of differential equations with formulated boundary and initial conditions is solved using a Finite Volume Method. In the study, an appropriate correlation for determining heat transfer coefficients is also chosen. The numerical model is verified by comparing the computation results with experimental data.

* Corresponding author: cisekpiotr@mech.pk.edu.pl

2 Test stand

The heating system with the regenerator that works as TES is shown in Fig. 1. Two periods may be distinguished during the TES unit operation. In the first stage, when the off-peak period of power system load occurs, the regenerators matrix is heated up with low-priced energy. Twenty four electrical heaters, 500W of power output each, are installed in the TES unit bed what gives 12 kW of power output in total. The charging period is extended during the time to achieve uniform matrix temperature throughout the entire bed. The bed temperature may reach even 700°C and is measured in three various cross-sections of the regenerator bed. The test facility is equipped with the plate fin-and-tube heat exchanger, as shown in Fig. 1, being investigated and tested in a broad range of previous papers [11–18].

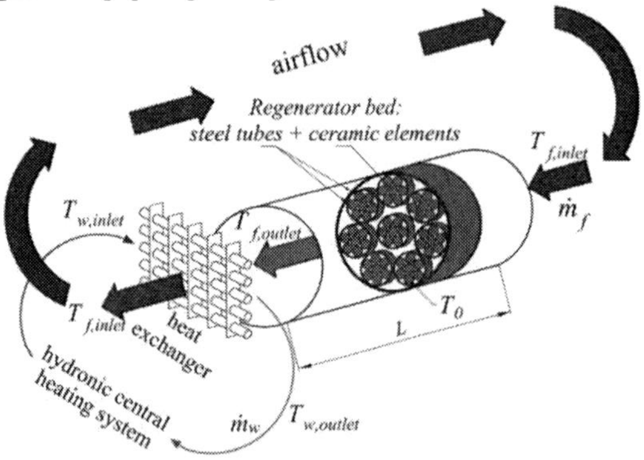

Fig. 1. The central heating system with fixed-matrix regenerator working as a TES unit.

Subsequently, during the discharge period, when electricity price is high (daytime, and on-peak periods), the stored heat is gradually released and used for space heating. Thus, according to Fig. 1, the regenerator matrix is cooled by the air that is circulating in a closed circuit. The centrifugal fan, located at the regenerator inlet section, controls the airflow. When leaving the TES unit, hot air is cooled in the plate fin–and–tube heat exchanger and flows back through the air duct to the regenerator inlet. Hot water, heated up in the heat exchanger, feeds the central heating system which consists of two radiators located in the adjacent room. Thus, in the discharge period, the heat exchanger performs the function of a low–temperature water boiler.

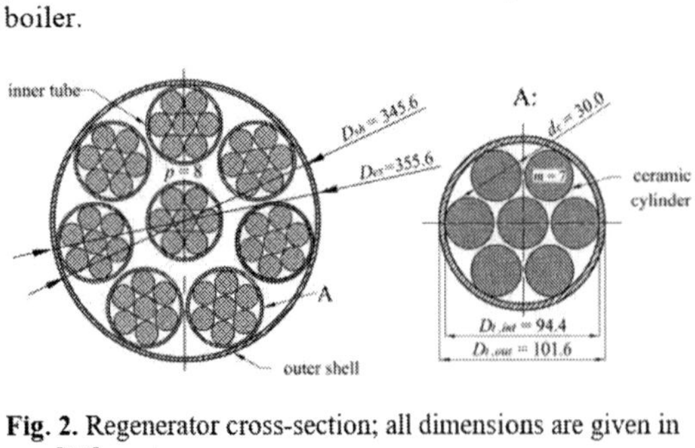

Fig. 2. Regenerator cross-section; all dimensions are given in mm [19].

The regenerator geometry is shown in Fig. 2. The bed consists of eight steel tubes with the length of 2.0 m being filled with ceramic cylinders arranged one by one in seven rows in each tube. There are 64 cylinders in one row. The total mass of the ceramic elements is 200,7 kg. The ceramic cylinders of diameter $d_c = 30$ mm and 30 mm height, are made of refractory concrete, which thermal properties are as follows: specific heat, $c_c = 790$ J/(kg·K); thermal conductivity, $k_c = 6.4$ W/(m·K); density, $\rho_c = 2700$ kg/m^3. Other bed parameters may be found in [19].

The experimental setup scheme along with measurement points marked is shown in Fig. 1. The air mass flow rate $\dot{m}_f$ is determined using thermal mass flow meter installed in the air duct, while the water mass flow rate $\dot{m}_w$ is measured using turbine flow meter installed at the plate fin-and-tube heat exchanger outlet. The air temperatures are measured at the regenerator inlet $T_{f,inlet}$, and outlet section $T_{f,outlet}$. The water temperature measurements are carried out both at the heat exchanger inlet and outlet, determining $T_{w,inlet}$, and $T_{w,outlet}$ respectively. Temperature measurements are performed using pre-calibrated T-type thermocouples (Cu-CuNi).

3 Model formulation

In the previous studies, an analytical solution [8, 20, 21] and lumped-parameter model [19] of the regenerator bed cooling is presented. The models assumed that the bed is a porous medium, where radial heat conduction is neglected. The two-dimensional model, shown in this study, takes into account the bed geometry. The energy conservation equations are formulated for the solid as ceramic cylinders, steel tubes of the regenerator core and for the outer shell, as well as for the fluid - air flowing in the air gaps across the bed and between the tubes.

When taking into account the simplifying assumptions, that are listed below, the energy conservation equations for the fluid and solid domains may be derived. The following assumptions are made when formulating the model equations:

1. Heat conduction in the air in the axial and radial direction is neglected.
2. Contact between the neighboring solids is neglected so that there is no heat conduction between adjacent ceramics rows as well as between ceramics and steel tubes.
3. Air temperature is a function of time t and axial coordinate z and is uniform at a given regenerator cross-section.
4. Physical properties of the ceramics and steel are considered as temperature independent.
5. There is no heat transfer between the system and the surroundings.

According to the abovementioned assumptions, the heat balance equations for solid and fluid domains are written in the following form:

❖ for a single row of ceramic cylinders:

$$\rho_c c_c \frac{\partial T_c}{\partial t} = \frac{1}{r}\frac{\partial}{\partial r}\left(r k_c \frac{\partial T_c}{\partial r}\right) + \frac{\partial}{\partial z}\left(k_c \frac{\partial T_c}{\partial z}\right) \quad (1)$$

where c subscript denotes the ceramics parameter and $f{,}int$ stands for the air flowing in the inner tubes.

❖ for the airflow inside the inner tubes, contacting with the ceramic cylinders

$$\rho_f c_f \left(\frac{\partial T_{f,int}}{\partial t} + w_{f,int} \frac{\partial T_{f,int}}{\partial z}\right) = \frac{1}{r}\frac{\partial}{\partial r}\left(r k_f \frac{\partial T_{f,int}}{\partial r}\right) + \frac{\partial}{\partial z}\left(k_f \frac{\partial T_{f,int}}{\partial z}\right) \quad (2)$$

where the air velocity inside the inner tubes is denoted as $w_{f,int}$.

❖ for the steel tubes of the regenerator core

$$\rho_{st} c_{st} \frac{\partial T_t}{\partial t} = \frac{1}{r}\frac{\partial}{\partial r}\left(r k_{st} \frac{\partial T_t}{\partial r}\right) + \frac{\partial}{\partial z}\left(k_{st} \frac{\partial T_t}{\partial z}\right) \quad (3)$$

where subscript t indicates the tubes parameters, and st stands for steel.

❖ for the air flowing between the tubes

$$\rho_f c_f \left(\frac{\partial T_{f,out}}{\partial t} + w_{f,out} \frac{\partial T_{f,out}}{\partial z}\right) = \frac{1}{r}\frac{\partial}{\partial r}\left(r k_f \frac{\partial T_{f,out}}{\partial r}\right) + \frac{\partial}{\partial z}\left(k_f \frac{\partial T_{f,out}}{\partial z}\right) \quad (4)$$

Air velocity in the space between the tubes is given as $w_{f,out}$.

❖ for the regenerator outer shell

$$\rho_{st} c_{st} \frac{\partial T_{sh}}{\partial t} = \frac{1}{r}\frac{\partial}{\partial r}\left(r k_{sh} \frac{\partial T_{sh}}{\partial r}\right) + \frac{\partial}{\partial z}\left(k_{sh} \frac{\partial T_{sh}}{\partial z}\right) \quad (5)$$

The system of partial differential equations, along with boundary conditions, is solved using Finite Volume Method [22, 23]. The primary goal of the calculations is to determine the heat transfer fluid temperature at the regenerator outlet, for the axial coordinate $z = L$, where L is the regenerator length. Input parameters of the model are the total air mass flow rate $\dot{m}_f$, air temperature at the inlet section $T_{f,inlet}$, and initial temperature T_0 of the whole bed.

3 Modeling and analysis

When determining the airflow distribution within the bed, pressure drop calculations in the system are performed. Basic principles of fluid mechanics state that the fluid flows at a higher rate through the channel, wherein encounters lower pressure drop. In this case, the flow resistance is determined, according to [3], for the subsequent parts of the regenerator, inlet and outlet sections, interior of the steel tubes, and space between the tubes. Based on the estimated pressure drops, it is possible to determine the air flow rates in the inner tubes $\dot{m}_{f,int}$ and in the space between the tubes $\dot{m}_{f,out}$ with the constraint that its sum must be equal to the total mass flow rate $\dot{m}_f$

$$\dot{m}_f = \dot{m}_{f,out} + p\,\dot{m}_{f,int} \quad (6)$$

where p is the number of steel tubes in the regenerator core, for the considered regenerator $p = 8$.

Acquiring information on air distribution is necessary when calculating heat transfer coefficients, and therefore modeling convective heat transfer in the wall-fluid system for the considered bed geometry. An exemplary experimental dataset, collected during the test stand operation, is shown in Fig. 3 and 4. As it is shown in Fig. 4, air flow rate is constant during the system operation. Therefore, for $\dot{m}_f$ = 0.07891 kg/s values of $\dot{m}_{f,int}$ and $\dot{m}_{f,out}$ are calculated, and equals to $\dot{m}_{f,int}$ = 0.00175 kg/s and $\dot{m}_{f,out}$ = 0.06491 kg/s respectively. Thus, less than 17.5% of the total air flow $\dot{m}_f$ goes through the inner tubes. In this case, the crucial role is played by linear pressure losses. The side surfaces roughness of the ceramic cylinders, approx. 0.3 mm, results in significantly higher pressure losses due to the liquid layers friction on the surface of the cylinder. For stainless steel, the roughness value is approx. 0.045 mm.

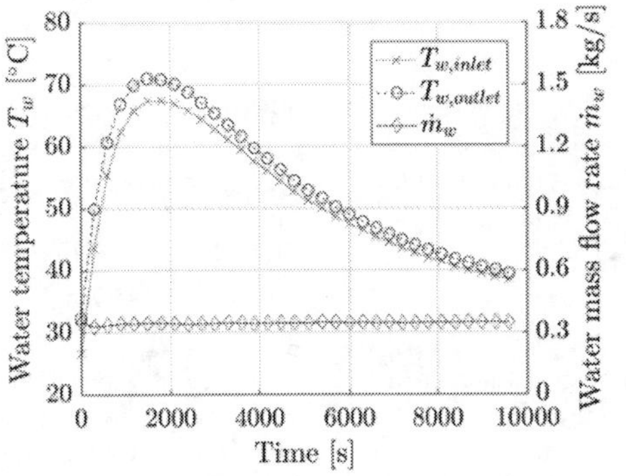

Fig. 3. The water temperature measured at the heat exchanger inlet $T_{w,inlet}$ and outlet $T_{w,outlet}$, along with the water mass flow rate $\dot{m}_w$ circulating in the central heating system.

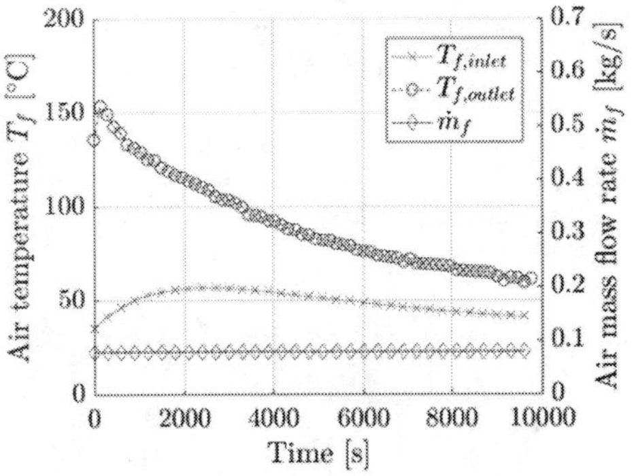

Fig. 4. The air temperature measured at both regenerator inlet $T_{f,inlet}$, and outlet $T_{f,outlet}$, along with the air mass flow rate $\dot{m}_f$.

It can be seen in Fig. 4 that at the initial time outlet air temperature $T_{f,outlet}$ is more than 15°C lower than for the next points marked. In this case, a dynamic temperature measurement error occurs which is related to the thermal inertia of the thermocouples [24–26].

After determining the flow distribution within the regenerator bed, calculation of heat transfer coefficients is performed. In the study, two approaches for determining the heat transfer coefficients for the air flow through the regenerator bed are compared. The first methodology is presented by Gnielinski in [3]. Herein a chapter devoted to heat transfer in concentric annular ducts may be found. In the second approach, the use of modified Petukhov and Kirillov correlation for the fluid flow through a circular or non-circular duct is considered. The modification is proposed by Taler in [27, 28].

3.1 Heat transfer in concentric annular ducts

For the case of airflow through the tubes filled with ceramics, it is assumed that each of the cylinder rows is contacting with a ring-shaped air layer of equivalent diameter equal to 35.7 mm. While for the steel tubes in the space between the tubes it is 122.2 mm. The air mass flow rate through the single annular duct is an equal part of the flow rates $\dot{m}_{f,int}$ and $\dot{m}_{f,out}$. So that, for the inside of inner tubes, the mass flow rate per a single row of ceramic cylinders is equal to ⅐ of $\dot{m}_{f,int}$, and for the single tube it is ⅛ of $\dot{m}_{f,out}$.

In the case of thermally and hydrodynamically developing flow in a concentric annular duct, the mean Nusselt number in the laminar region may be calculated from Martin correlation:

$$\mathrm{Nu}_{\mathrm{m,lam}} = \left(\mathrm{Nu}_{\mathrm{m,1}}^{\ 3} + \mathrm{Nu}_{\mathrm{m,2}}^{\ 3} + \mathrm{Nu}_{\mathrm{m,3}}^{\ 3}\right)^{1/3} K \tag{7}$$

where $\mathrm{Nu}_{m,1}$, $\mathrm{Nu}_{m,2}$ and $\mathrm{Nu}_{m,3}$ are the mean Nusselt numbers, calculated for Re < 2300, for boundary conditions specified in [3]. Moreover, K is a correction factor taking into account the effect of changing the fluid physical properties with the temperature

$$K = \left(\frac{T_f + 273}{T_c + 273}\right)^n \tag{8}$$

where the exponent n is $n = 0$ when the air is cooling, and $n = 0.45$ when it is heated up.

For fully turbulent flow, for $\mathrm{Re} \geq 10^4$, the Gnielinski equation for determining the mean Nusselt number during the air flow through the annular duct takes the following form:

$$\mathrm{Nu}_{\mathrm{m,turb}} = \frac{\left(\frac{\lambda}{8}\right)\mathrm{Re}\,\mathrm{Pr}}{k_1 + 12.7\sqrt{\frac{\lambda}{8}}\left(\mathrm{Pr}^{2/3} - 1\right)}\left[1 + \left(\frac{d_h}{L}\right)^{2/3}\right] F_{ann} \tag{9}$$

with

$$k_1 = 1.07 + \frac{900}{\mathrm{Re}} - \frac{0.63}{1 + 10\,\mathrm{Pr}} \tag{10}$$

where λ is a friction factor an annular duct, Pr is a Prandtl number, and F_{ann} is a correction factor [3]. Hydraulic diameter d_h is a calculated as

$$d_h = d_o - d_i \tag{11}$$

where d_o is the outer diameter, and d_i is the inner diameter of the annuli. In the above equations, the ranges of validity are $10^4 < \mathrm{Re} < 10^6$; $0.6 < \mathrm{Pr} < 1000$; $0 \leq d_h/L \leq 1$.

For the transition region between laminar and fully developed turbulent flow, for $2300 \leq \mathrm{Re} \leq 10^4$, the mean Nusselt number may be represented by the equation

$$\mathrm{Nu}_{\mathrm{m,trans}} = [(1-\gamma)\,\mathrm{Nu}_{\mathrm{m,lam}}\big|_{\mathrm{Re}=2300} + \gamma\,\mathrm{Nu}_{\mathrm{m,turb}}\big|_{\mathrm{Re}=10^4}]K \tag{12}$$

with

$$\gamma = \frac{\mathrm{Re} - 2300}{10^4 - 2300} \quad \text{and} \quad 0 \leq \gamma \leq 1. \tag{13}$$

The drawback of formulae (12) is the adoption of the Reynolds number limit value of Re = 2300, and Re = 10^4, which marks the boundaries of the transitional flow region. So that, the first derivative dNu/dRe in the transition point is no longer continuous. Consequently, when changing the flow region, from the laminar to the transitional, at Re = 2300, and from the transitional to the turbulent region, at Re = 10^4, a rapid change in the course of the function Nu = f(Re) occurs. For this reason, a new correlation is proposed by Taler [27, 28] that is valid for the transitional and the turbulent flow region.

3.2 Heat transfer in non-circular ducts

As the second case, the correlation for determining the local value of the Nusselt number Nu_z for the fluid flow in the plain tubes is used. The relationship is proposed by Taler [27, 28] for a wide range of Reynolds number values varying between $2300 < \mathrm{Re} < 10^6$. The Taler correlation is a modification of Petukhov and Kirillov equation initially formulated for a turbulent flow range. The coefficients in the Taler equation are determined experimentally by comparing a large amount of experimental data available in the literature. For undeveloped velocity and temperature profiles, assuming a constant wall temperature, the criterion equation takes the following form

$$\mathrm{Nu}_z = \mathrm{Nu}_z\big|_{\mathrm{Re}=2300} + \frac{\frac{\lambda}{8}(\mathrm{Re}-2300)\mathrm{Pr}^{1.008}}{1.08 + 12.39\sqrt{\frac{\lambda}{8}}\left(\mathrm{Pr}^{2/3} - 1\right)}\left[1 + \frac{1}{3}\left(\frac{d_h}{z}\right)^{2/3}\right] K \tag{14}$$

The Nusselt number $\mathrm{Nu}_{z|\mathrm{Re}=2300}$ in Eq. (14) is calculated from Eq. (15) for Re = 2300

$$\mathrm{Nu}_z\big|_{\mathrm{Re}=2300} = [\mathrm{Nu}_{z,1}^{\ 3} + 0.7^3 + \left(\mathrm{Nu}_{z,2} - 0.7\right)^3 + \mathrm{Nu}_{z,3}^{\ 3}]^{1/3} K \tag{15}$$

where $\mathrm{Nu}_{z,1}$, $\mathrm{Nu}_{z,2}$ and $\mathrm{Nu}_{z,3}$ are the local Nusselt numbers, calculated for Re = 2300, for boundary conditions specified in [27]. In the equations mentioned above, λ is a friction factor a plain tube, calculated from Taler equation [29], z is the axial coordinate, changing

 https://doi.org/10.1051/matecconf/201824001008

from 0 to L, and d_h is the hydraulic diameter calculated as 4 times the flow area divided by the wetted perimeter of the conduit. For $z = 0$ is the point from which heating or cooling of the fluid begins. So that, it is a coordinate of the regenerator inlet section.

3.3 Determination of heat transfer coefficients

The heat transfer coefficients are determined for the four different contact surfaces:

- h_c for the lateral surface of ceramic cylinders;
- $h_{t,int}$ for the inner surface of the tubes filled with the ceramic cylinders;
- $h_{t,out}$ for the outer surface of the tubes;
- h_{sh} for the internal surface of the outer shell.

Values of the relevant heat transfer coefficients are determined in the model for each time-step and for each of the N control volumes of the solid body being in contact with the fluid.

Figures 5 and 6 show the time courses of the heat transfer coefficient values determined for various axial coordinates z, where $z = 0$ is the inlet, and $z = 2$ m is the outlet cross-section of the regenerator bed.

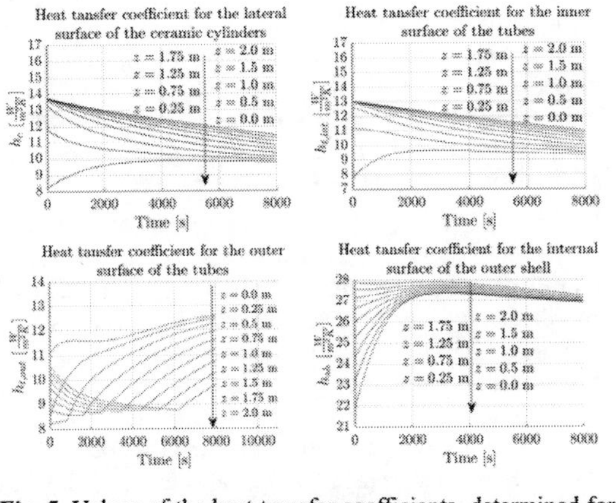

Fig. 5. Values of the heat transfer coefficients, determined for annular duct flow, for the variable axial coordinate z and during the time of the system operation.

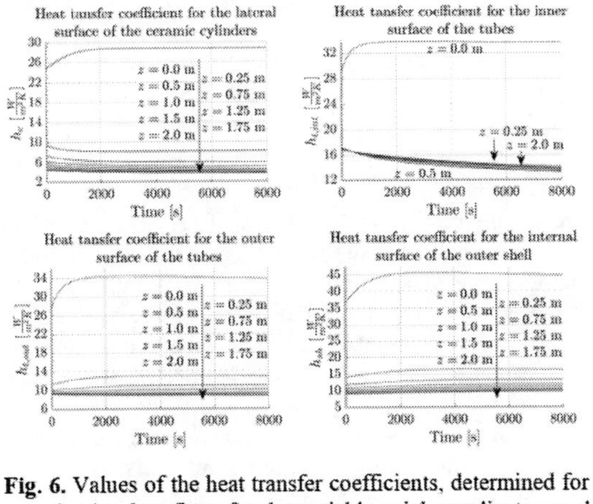

Fig. 6. Values of the heat transfer coefficients, determined for non-circular duct flow, for the variable axial coordinate z and during the time of the system operation.

Figure 5 presents the calculation results when the annular flow is assumed. The h_c and $h_{t,int}$ values are the lowest in the inlet section, for $z = 0$, and rises with increasing z, reaching the highest values at $z = 2$ m, so at the outlet section. The results shown in Fig. 6 are entirely different. According to Taler correlation, in the inlet cross-section the highest values of heat transfer coefficients are obtained, which decrease in the range $z = 0 \div 0.25$ almost three times, and then set at a similar level. In the inlet section, severe disturbances and changes in the flow direction are expected. Thus, it may result in an intensification of heat transfer until the flow stabilizes. It can be supposed that the h_c and $h_{t,int}$ values obtained in this case are closer to reality.

The time course of $h_{t,out}$ shown in Fig. 5 reveals the fundamental drawback of averaging the heat transfer coefficient values in the transitional region, according to the Eq. (12). A rapid change in the course of the $h_{t,out}$ values, may be noticed when changing from transitional to the laminar flow region. As it can be seen in Fig. 6, Taler's approach is free of this drawback.

3.4 Experimental validation

Experimental verification of the proposed heat transfer correlations is carried out with comparing the temperature $T_{f,outlet}$, obtained from measurements, with mass-average temperature $\bar{T}_f\big|_{z=L}$, determined from Eq. (16) for the outlet cross-section of the regenerator, for $z = L$

$$\bar{T}_f\Big|_{z=L} = \frac{p\,\dot{m}_{f,int}\,T_{f,int}\big|_{z=L} + \dot{m}_{f,out}\,T_{f,out}\big|_{z=L}}{\dot{m}_f} \tag{16}$$

where $T_{f,int}$, and $T_{f,out}$ are the air temperatures determined, respectively, within the steel tubes, and in the space between the tubes.

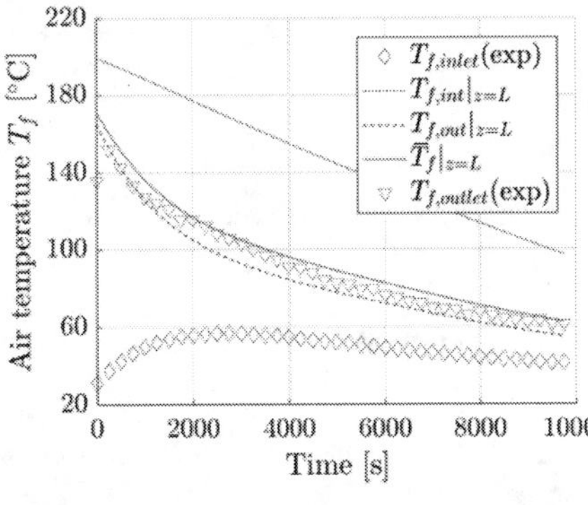

Fig. 7. Calculation results when determining heat transfer coefficients for an annular duct flow – according to Gnielinski.

In Figs 7 and 8, computation results of the air outlet temperature along with its experimental values are presented. To better compare the numerical and experimental results, the relative error is calculated, for each of the time-step number n, and for the both considered cases

$$\varepsilon_f^n = \frac{\bar{T}_f^n\big|_{z=L} - T_{f,outlet}^n}{T_{f,max} - T_{f,min}} \cdot 100\% \qquad (17)$$

where $T_{f,max}$ and $T_{f,min}$ are the maximal and minimal air temperatures determined within the system.

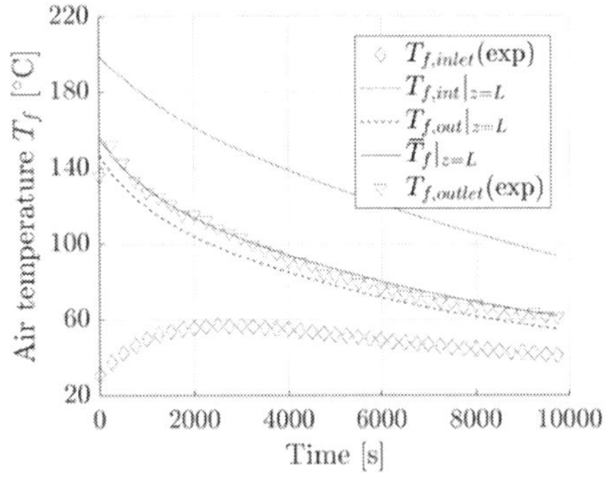

Fig. 8. Calculation results when determining heat transfer coefficients for a non-circular duct flow – according to Taler.

For the uncertainty analysis, the Root Mean Square Error (*RMSE*) and the Mean Absolute Percentage Error (*MAPE*) are also calculated

$$RMSE = \sqrt{\frac{1}{N_t}\sum_{n=1}^{N_t}\left(\bar{T}_f^n\big|_{z=L} - T_{f,outlet}^n\right)^2} \qquad (18)$$

$$MAPE = \frac{1}{N_t}\sum_{n=1}^{N_t}\frac{\left|\bar{T}_f^n\big|_{z=L} - T_{f,outlet}^n\right|}{T_{f,outlet}^n} \cdot 100\% \qquad (19)$$

where N_t is the total number of time steps.

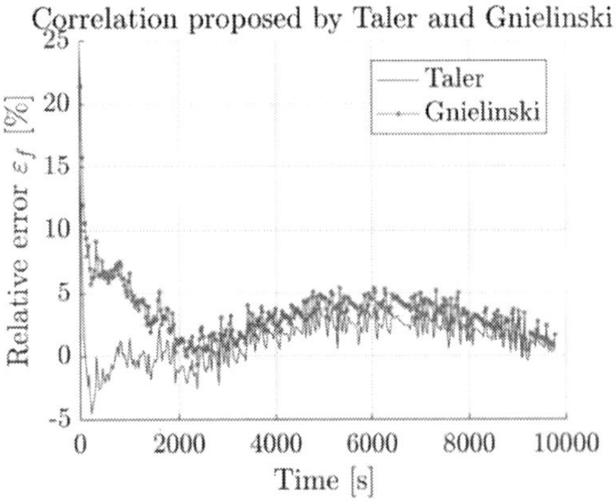

Fig. 9. The values of ε_f determined from Eq. (17) for the results obtained from the model and compared to the experimental ones.

For the results shown in Fig. 7, when using Eqs (7) – (13), *RMSE* = 5,77°C, and *MAPE* = 5,43%. When considering Taler correlation, Eqs (14) – (15), *RMSE* = 2,71°C, and *MAPE* = 2,66%, see Fig. 8. As it is shown, the *RMSE* and *MAPE* values are more than 2 times lower for the Taler correlation. Based on the *RMSE* and *MAPE* values, as well as maximum values of the relative error, presented in Fig. 9, it may be concluded, that a good agreement between the experiment and calculation results is achieved when using Taler approach. A significant relative error for an initial time, exceeding 16% in both cases, is caused by a dynamic temperature measurement error. In the remaining time period, ε_f is not exceeding ±5%.

4 Conclusions

The study presents the concept of Electric Thermal Storage (ETS) central heating system. Thermal Energy Storage (TES) in the system is carried out in the fixed-matrix regenerator. In the article construction of the regenerator is presented. The test stand built in the laboratory of the Institute of Thermal Power Engineering at the Cracow University of Technology is used to test the proposed ETS central heating system.

The energy conservation equations, formulated for the two-dimensional model, takes into account the regenerator bed geometry. After the model validation, the developed heat balance equations are implemented in MATLAB numerical procedures based on the Finite Volume Method. In the model pressure drops within the system are calculated, both for the airflow through the inner tubes, and between the tubes. The flow distribution calculations show that in the tubes filled with ceramics flow resistance is much higher than for the flow between the tubes. Therefore, approx. 17.5% of the total flow rate $\dot{m}_f$ goes through the eight tubes filled with the ceramic cylinders. Thus, the assumption of even air flow distribution would not be justified.

Subsequently, the values of heat transfer coefficients h_c, $h_{t,int}$, $h_{t,out}$, and h_{sh} are determined for the four distinct heat transfer surfaces, for the variable axial coordinate z and during the time of the system operation. The use of two different criterion equations is considered: first one, proposed by Gnielinski, for determining the mean Nusselt number Nu_m for fluid flow through the duct of the annular cross-section; the second one is recommended by Taler for the fluid flow through a circular or non-circular ducts. The Taler correlation is further modified for determining local Nusselt numbers Nu_z. The most appropriate approach is selected by comparing the calculation results with experimental data. Taking into account the relative error, *RMSE*, and *MAPE* values calculated, it may be concluded that better results are obtained in this case for the Taler correlation.

The numerical model may be used in the design and performance assessment of the presented accumulative heating systems. The development of mathematical procedures that allows determining the temperature of the air leaving the TES unit during the time, with the selected parameters of the system, i.e., mass flow rate, number of tubes in the core, shape, and material from which bed particles are made, etc. The numerical procedures may support the design and selection of devices at the planning stage of installation capacity. Moreover, the model may also be useful during the implementation of a model-based control system that is planned to be developed in the future.

References

1. D.L. Zunft, Chapter in L.F. Cabeza (Ed.), *Advances in Thermal Energy Storage Systems. Methods and Applications* (Woodhead Publishing Series in Energy **66**, 2015)
2. K. Thulukkanam, *Heat Exchanger Design Handbook* (2nd Ed., CRC Press, 2013)
3. Verein Deutscher Ingenieure VDI-Gesellschaft Verfahrenstechnik und Chemieingenieurwesen (GVC) (Ed.), *VDI Heat Atlas* (2nd Ed., Springer, 2010).
4. Y. Jian, Q. Falcoz, P. Neveu, F. Bai, Y. Wang, Z. Wang, Appl. Energy **139**, 30–42 (2015).
5. M.A. Al-Nimir, M.K. Abu-Qudasi, M.D. Mashaqi, Energy Convers. Manag. **37**, 23–30 (1996)
6. A. Mawire, M. McPherson, R.R. van den Heetkamp, S.J.P. Mlatho, Appl. Energy **86**, 1246–1252 (2009)
7. P.M. Cuce, S. Riffat, Renew. Sust. Energ. Rev. **47**, 665–682 (2015)
8. D. Taler, P. Cisek, Archives of Thermodynamics **4**, 55–67 (2013)
9. J. Sacharczuk, D. Taler, Archives of Thermodynamics **3**, 281–295 (2014)
10. K.A.R. Ismail, R. Stuginsky Jr, Appl. Ther. Eng. **19**, 757–788 (1999)
11. A. Cebula, D. Taler, Archives of Thermodynamics **30**, 45–66 (2009)
12. D. Taler, P. Ocłoń, Int. J. Therm. Sci. **84**, 309–322 (2014)
13. D. Taler, P. Ocłoń, Chem. Eng. Process. **83**, 1–11 (2014)
14. D. Taler, Energy Convers. Manag. **96**, 452–462 (2015)
15. A. Korzeń, D. Taler, Int. J. Therm. Sci. **92**, 188–198 (2015)
16. J. Taler, P. Ocłoń, D. Taler, M. Nowak-Ocłoń, Chapter in: M. Salim Newaz Kazi (Ed.), *Heat Transfer Studies and Applications* (InTech 2015)
17. D. Taler, Energy Convers. Manag. **159**, 232–243 (2018)
18. D. Taler, A. Korzeń, Int. J. Numer. Method. H. **28**, 81–91 (2018)
19. P.Cisek, D. Taler, Energ. Buildings **130**, 747–760 (2016)
20. D. Taler, K. Kaczmarski, J. Therm. Sci. **25**, 549–557 (2016)
21. D. Taler, K. Kaczmarski, Technical Transactions. Mechanics **1-M**, 231–242 (2016)
22. J. Taler, P. Duda, *Solving Direct and Inverse Heat Conduction Problems* (Springer, 2006)
23. A. Cebula, D. Taler, Chapter in: R.B. Hetnarski (Ed.), *Encyclopedia of thermal stresses* **4**: F-G, 1645–1658 (Springer, 2014)
24. M. Majdak, M. Jaremkiewicz, Measurement Automation Monitoring **62**, 284–287 (2016)
25. M. Jaremkiewicz, D. Taler, T. Sobota, Int. J. Therm. Sci. **87**, 241–250 (2015)
26. M. Jaremkiewicz, Heat Mass Transf. **53**, 887–897 (2017)
27. D. Taler, Int. J. Therm. Sci. **108**, 108–122 (2016)
28. D. Taler, *Numerical modelling and experimental testing of heat exchangers* (Springer, 2019)
29. D. Taler, Int. J. Therm. Sci. **105**, 109–122 (2016)

Experimental and numerical studies of heat and mass transfer in low-temperature heat accumulator with phase transformations of accumulating material

Valery Gorobets[1], *Ievgen* Antypov[1], *Viktor* Trokhaniak[1], and *Yurii* Bohdan[2,*]

[1]Heat and Power Engineering Department, Education and Research Institute of Energetics, Automation and Energy Efficiency, National University of Life and Environmental Sciences of Ukraine, Heroyiv Oborony st., 15, Kyiv, 03041, Ukraine
[2]Ships Power Plant Operation Department, Marine Engineering Faculty, Kherson State Maritime Academy, Ushakova avenue, 20, Kherson, 73000, Ukraine

Abstract. Accumulation of thermal energy is produced with the aim of storing at certain times, when there is an overabundance of this energy and its further use in other periods of time when there is a deficit thermal energy. Thermal energy storage may be carried out under heating of any material (water, solid materials etc.) or by using the phase or chemical transformation of the material (melting and crystallization processes, direct and reverse chemical reaction). Thermal accumulators with phase or chemical transformations are allowed to concentrate a large amount of energy in a relatively small volume of accumulating material. In this paper an experimental study and numerical modeling of heat and mass transfer in the heat accumulator during phase transformations of the accumulating material are presented. The experimental plant consists of a chamber filled with paraffin. In experimental studies, the changes of the temperature distribution in heat accumulating material and tube bundles have investigated. Numerical simulation of melting and solidification of heat accumulating material during the heated and cooled of tube bundles were performed. As the results of research, the basic laws of melting and crystallization processes in heat accumulator during phase transformations of heat accumulating material were determined.

1 Introduction

Analysis of different methods of heat accumulation showed that the most promising type of heat accumulators are heat accumulators of periodic action with phase or chemical transformations of accumulation material [1-3]. Accumulators of this type give an opportunity to provide a high density of accumulated energy per unit of mass and a stable temperature of the coolant at the exit from the heat accumulator. The works [1-14] provide detailed information on the research of heat and mass transfer processes that occur during the phase transformations of the accumulation material. From the analysis of these studies it can be concluded that it is necessary to intensify the processes of heat and mass transfer with "charge" and "discharge" of accumulators and to reduce losses in storage of thermal energy in such accumulators [4-9]. Given the complexity of processes in heat-storage material for such devices, it is necessary to conduct a theoretical and experimental investigations of heat and mass transfer processes in order to optimally locate heat sources in the accumulative material. Such issues are insufficiently studied and require further research [10-14]. In particular, the processes of heat and mass transfer during phase transformations of accumulation material around several horizontally placed cylindrical heat sources are insufficiently investigated.

2 Analysis and modelling

On the basis of the analysis of the structures of existing accumulation devices using a package of COMSOL Multiphysics 3.5a a new axonometric model of the heat-storage phase of the phase transition was proposed and developed [15]. We assume that the coolant enters the accumulator and moves along the system of heating pipes, giving or absorbing at the same time the thermal energy that is accumulated or taken away from the accumulation material. As a heat-accumulating material T3 paraffin with a phase transition temperature $T_f = 54 \div 56$ °C was selected.

In the numerical study of these processes, we consider that: 1) the accumulator case is insulated and thermal losses in the environment are absent; 2) the initial temperature throughout the volume of the accumulation material is the same; 3) the temperature of the surface of cylindrical heat sources is based on the condition of the thermal balance between the heat received by the heat accumulation material and transferred to the material from heat sources. At the

* Corresponding author: bohdanyurii09@gmail.com

boundary of the phase separation, heat is taken into account, which is absorbed (released) during melting (crystallization) of the accumulation material.

The mathematical model of heat and mass transfer processes in phase transformations of the accumulation material includes the Navier-Stokes equation and the convective heat transfer equation.

$$\left.\begin{aligned}\frac{\partial\rho w_x}{\partial\tau}+w_x\frac{\partial\rho w_x}{\partial x}+w_y\frac{\partial\rho w_x}{\partial y}=-\frac{\partial p}{\partial x}+\rho g_x+\frac{\partial}{\partial x}\left(\mu\frac{\partial w_x}{\partial x}\right)+\frac{\partial}{\partial y}\left(\mu\frac{\partial w_x}{\partial y}\right),\\ \frac{\partial\rho w_y}{\partial\tau}+w_x\frac{\partial\rho w_y}{\partial x}+w_y\frac{\partial\rho w_y}{\partial y}=-\frac{\partial p}{\partial y}+\rho g_y+\frac{\partial}{\partial x}\left(\mu\frac{\partial w_y}{\partial x}\right)+\frac{\partial}{\partial y}\left(\mu\frac{\partial w_y}{\partial y}\right)\end{aligned}\right\} \quad (1)$$

$$\frac{\partial\rho}{\partial\tau}+\frac{\partial\rho w_x}{\partial x}+\frac{\partial\rho w_y}{\partial y}=0 \quad (2)$$

$$\frac{\partial T}{\partial\tau}+w_x\frac{\partial T}{\partial x}+w_y\frac{\partial T}{\partial y}=\frac{1}{\rho c_p}\frac{\partial}{\partial x}\left(\lambda\frac{\partial T}{\partial x}\right)+\frac{1}{\rho c_p}\frac{\partial}{\partial y}\left(\lambda\frac{\partial T}{\partial y}\right)\pm Q_s(x,y) \quad (3)$$

In this case, the initial and boundary conditions have the form:

- at the initial moment of time:

$$T(x,y,\tau=0)=T_0, \quad (4)$$

- on the surface of the case:

$$\left.\frac{dT}{dn}\right|_{s,c}=0, \quad (5)$$

- on the surface of cylindrical heat sources:

$$T_s(x,y)=T_{s,t}, \quad (6)$$

In carrying out numerical calculations of heat transfer in the region of the solid phase of the accumulating material, the equation (3) was used, in which it was assumed that $w_x = 0$, $w_y = 0$ and the convective elements were absent.

To test the validity of numerical simulation results, a test module (Figure 1) is developed that is made of steel which it is filled the solid accumulative material. This module consist of a horizontally oriented container (housing) in the form of a parallelepiped with height H = 240 mm, length L = 500 mm and width W = 300 mm. To ensure a uniform accumulation of heat in an array of accumulation material, a tube bundle with a chess or corridor arrangement consisting of 8 steel tubes (d = 21.3x2.8 mm) located horizontally parallel to the lower wall of the module casing with a relative radius around the heat sources, which is not exceeds $R \leq 40$ mm. We assume that the operating time of the apparatus is $\tau \approx 28800$ s (8 hours).

The mass of the accumulative material (paraffin) in the working chamber is M_{hm} = 26.1 kg. The thermophysical properties of the accumulative material were determined by the results of laboratory tests, and the properties of the coolant (water) – based on known table values.

The following parameters were chosen for the investigated factors of influence during the accumulation process: temperature and velocity of heat carrier. As parameters of optimization were selected: coefficient of useful use of mass or accumulated material and the degree of heating of material for the lower levels in volume of accumulator.

Fig. 1. Appearance of the experimental module: 1 - collapsible case; 2 - face cap; 3 - distribution collector; 4 - tube bundle; 5 - thermal insulation panel.

3 Results and discussions

As a result of the numerical simulation (Figure 2a, b) and the experimental studies (Figure 2c, d) of heat and mass transfer processes during phase transformations of the accumulation material around several cylindrical heat sources were found the existence of three melting stages of the heat-accumulating material. In the first stage of melting, the main mechanism of heat transfer is conductive heat transfer. In the second stage of melting, the convective heat transfer is dominant, which is accompanied by intense melting of the accumulation material. The third stage of melting is characterized by overheating of the molten mass of the material with slow melting of the solid mass of the material located in the bottom of the container.

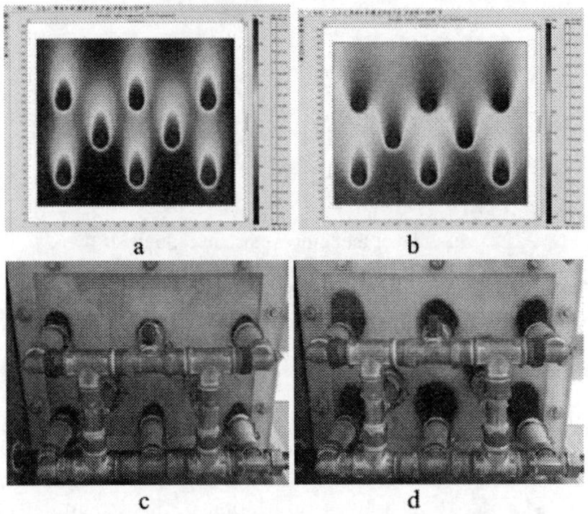

Fig. 2. The temperature fields (a, b - numerical simulation) and the change of the melting profiles (c, d - experimental research) in the system «solid - melt" in the cross section of accumulation material for heat accumulator at the different moments of time: a, c - 30 min; b, d - 180 min.

On the showed Figure 2a,b present temperature fields that has the same scale of the color map with the temperature range from 295 to 350 K with step 5 K.

On the basis of analysis of the performed researches it was found that for the second phase of melting, the maximum value of the mass increase of the molten substance per unit time is characteristicfor the second phase of melting. The third phase is characterized by a slower growth of the mass of the melt in the lower part of the working volume, where there is the appearance of zones of low melting point of the accumulation material, hereinafter referred to as "stagnant zones", the temperature of which is 10-12 % lower than in the region of intense circulation fluid flows (see Fig. 3). In this case, a large amount of heat is spent on overheating of the molten material volume, which is located on the top of the heat accumulator. The duration of such "overheating" is about 25 % of the total time when heat accumulator is in "charge" mode.

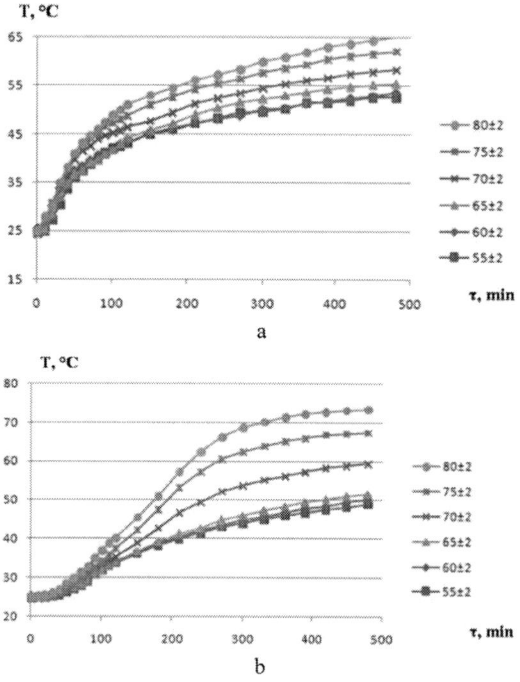

Fig. 3. Dependence of the temperature in volume of the accumulative material on time at different values of the temperature of the coolant at the inlet: a - in the "stagnant zones"; b - in the melt of the material.

The individual curves in Figure 3 showed the dependence of temperature of volume of accumulating material on time for different temperatures of inlet coolant in range from 55 to 80 °C with step 5 °C.

In order to maximize the heating of the accumulative material in the lower areas of the container and to determine the optimal distances between adjacent cylindrical heat sources and the bottom of accumulator, a series of model calculations was conducted. In the process of numerical computing the dimensions and geometry of the location of heat sources in the volume of accumulation material is changed. As a result of numerical calculations, an optimal distance (radius) between heat sources was found, which made it possible to reduce the weight and dimensions of heat accumulator and increase its heat accumulation capacity. For different geometric parameters of the cylindrical heat transfer surface, the value of the optimal distance (radius) R between the heat sourcesthe bottom of accumulatoris obtained.

The analysis of the results of numerical simulation gives the following conclusion: placing the lower row of heating elements from the bottom and lateral walls of the accumulator case should be carried out at a distance not exceeding the value of the optimal distance (radius) R, which are shown in Table 1.

Table 1. The value of the optimal distance (radius) between the heat sources and the bottom and lateral walls in the volume of the accumulative material, depending on the diameter of the cylindrical sources

Parameter	Nominal diameter (inches)					
	½	¾	1	1 ¼	1 ½	2
External diameter of heat exchangers (mm)	21,25	26,75	33,50	42,25	48,00	60,00
Optimal radius R (mm)	30,00	35,00	45,00	57,00	70,00	85,00

Experimental studies have confirmed the results of numerical modeling on which the necessity of placing the bottom row of heating elements on the bottom and lateral walls of the battery casing, the axial distance which is equal to the optimal radius R. It is established that the excess of this value by 20 % leads to a decrease in the mass of accumulation material by 7 %. Choosing the optimal distance between the cylindrical sources and the bottom wall of case leads to minimization of stagnant zones during the accumulation of thermal energy and reducing the time of "charging" the heat accumulator by 15-20 %.

A comparison of the results of numerical simulation (curves 1.3 in Figure 4) and experimental data (curves 2.4) for melting processes (curves 1.2 - "charge" of the accumulator) and crystallization (curves 3.4 - "discharge" of the accumulator) is carried out. The analysis of the obtained results is showed that the difference between the numerical and experimental data for the temperatures during melting (curves 1 and 2) of the heat-accumulating material does not exceed 3 %, and during its crystallization – no more than 5 % (curves 3 and 4 in Figure 4).

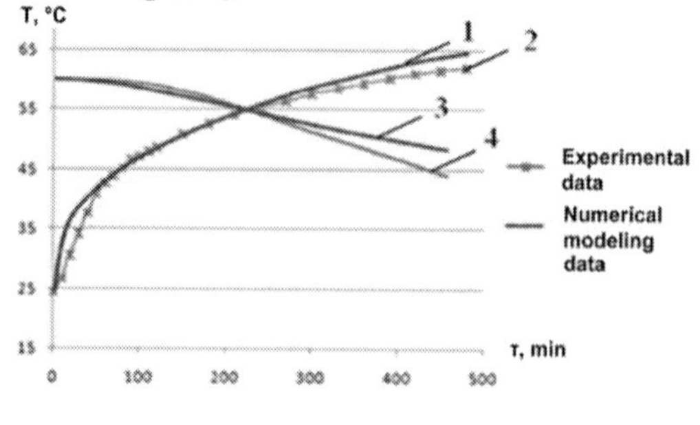

Fig. 4. Comparison of the results of numerical simulation (1, 3) with experimental data (2, 4) of the thermal characteristics of the heat-accumulating material in the process of "charging" and "discharging" the heat accumulator

4 Conclusions

1. Analysis of the obtained temperature fields and melting profiles of the accumulation material in the system "solid-melt" for different moments of time revealed the presence of three melting phases.
2. In the first phase, the heat transfer from the heat source to the phase separation by the main transfer mechanism is conductive heat transfer with uniform melting of the heat accumulating material along the perimeter of the cylindrical heat source.
3. In the second phase, the convective heat transfer is dominant. This period is characterized by the presence of intense convective flows, which leads to an increase in the efficiency of the melting process and to increase the melting speed of the material placed above the heat sources.
4. The third phase of melting is characterized by overheating of liquid mass of accumulation material, slowing melting processes in the lower part of the battery and reducing the growth of the mass of the melt. The duration of the "overheating" stage is about 25 % of the total time of work for heat accumulator in the "charge" mode. In the lower part of the working volume there are "stagnant zones", the temperature of which is 12 % lower than in the area of intense circulation of fluid.
5. As a result of the numerical calculations of the processes of heat and mass transfer in the heat accumulator of phase transition on the basis of paraffin, an optimal distance (radius) R between cylindrical heat sources is established in which the heat accumulation capacity of the accumulator will be maximal. Experimental investigations have confirmed that the bottom row of heating elements in the accumulating material should be located at a distance not exceeding the optimal radius R, which minimizes the number of stagnant zones and reduces the time of "charging" the accumulator by 15-20 %.

Acknowledgments

Publication is based on the research provided by the grant support of the State Fund for Fundamental Research of Ukraine, the Ministry of Education and Science of Ukraine (project N F64/11-2016).

References

1. R. Velraja, R. Seeniraja, B. Hafnerb, C. Faberb, K. Schwarzerb, Sol. Energy, **65**(3), pp. 171–180 (1999).
2. Q. Zhang, Y. Huo, Z. Rao, Science Bulletin, **61**(5), pp. 391–400 (2016).
3. Iu. Nasieka, V. Strelchuk, V. Naseka, Yu. Stubrov, S. Dudnik, V. Gritsina, O. Opalev, K. Koshevoy, V. Strel'nitskij, V. Tkach, M. Boyko, I. Antypov, J. Cryst. Growth, **491**, pp. 103–110 (2018).
4. J. Pereira da Cunha, P. Eames, Appl. Energ., **177**, pp. 227–238 (2016).
5. L. Liu, D. Su, Y. Tang, G. Fang, Renew. Sust. Energ. Rev., **62**, pp. 305–316 (2016).
6. L. Fan, J. Khodadadi, Renew. Sust. Energ. Rev., **15**(1), pp. 24–46 (2011).
7. F. Agyenima, N. Hewitta, P. Eamesb, M. Smyth, Renew. Sust. Energ. Rev., **14**(2), pp. 615–628 (2010).
8. A. Regin, S. Solanki, J. Saini, Renew. Energ., **31**, pp. 2025–2041 (2006).
9. S. Jegadheeswaran, D. Pohekar, Renew. Sust. Energ. Rev., **13**(9), pp. 2225–2244 (2009).
10. S. Kuboth, A. König-Haagen, D. Brüggemann, Energies, **10**, 274 (2017).
11. Y. Huo, Z. Rao, Int. J. Heat Mass Tran., **86**, pp. 197–206 (2015).
12. S. Kalaiselvam, M. Veerappan, A. Arul, S. Iniyan, Int. J. Therm. Sci., **47**(7), pp. 858–874 (2008).
13. A. Sharma, V. Tyagi, R. Chen, D. Buddhi, Renew. Sust. Energ. Rev., **13**, pp. 318–345 (2009).
14. V.G. Gorobets, V.V. Treputnev, Teplofizika Vysokikh Temperatur, **33** (4), pp. 588–593 (1995).
15. I. Antypov, Scientific herald of the National University of Life and Environmental Sciences of Ukraine, **224**, pp. 208–213 (2015).

Analysis of heat transfer in a coated bed of an adsorption chiller

Karolina Grabowska[1,*], *Marcin* Sosnowski[1], *Jaroslaw* Krzywanski[1], *Karol* Sztekler[2], *Wojciech* Kalawa[2], *Anna* Zylka[1], and *Wojciech* Nowak[2]

[1] Faculty of Mathematics and Natural Science, Jan Dlugosz University, al. Armii Krajowej 13/15; PL42-200 Częstochowa, Poland
[2] AGH University of Science and Technology, al. A. Mickiewicza 30; PL30-059 Cracow, Poland

Abstract. Adsorption chillers can be a promising part of sustainable development concept of the global economy due to the utilization of low grade thermal energy sources for cooling production. Therefore, research aiming at improving their performance i.e. Coefficient of Performance (COP) by optimizing the heat and mass transfer condition in the adsorption beds are crucial. Innovative modification of the sorbent layer structure are proposed in the paper in order to improve the heat transfer characteristics in the heat exchanger boundary layer. The analysis of desorption conditions in the parametric model of a coated adsorption bed construction is presented in the paper. The computational fluid dynamics with conjugate heat transfer analysis are used to determine the crucial input parameters for further analytical calculations. The heat transfer condition in novel coated design and a conventional fixed bed are compared in the paper. The developed computational model consisted of three subdomains representing heating water, heat exchanger material (copper) and sorbent (silica gel).

Nomenclature

(in) – inlet
(out) – outlet
λ - Thermal Conductivity; W/(m*K)
T_w – average temperature of hot water; K
T_s – average temperature of silica gel; K
T_c – average temperature of coated layer; K

1 Introduction

Industrial waste heat, solar power and heat from cogeneration are ecological sources suitable to power adsorption chillers [1-3]. It is the best strategy to reduce consumption of conventional energy, because the compressor air conditioning devices are mostly very high electrical energy consuming systems. Therefore research on improving adsorption cooling technology efficiency is a crucial task to sustainable economic development.

Currently the low coefficient of performance (COP) of adsorption chillers limits the dissemination of this technology. High thermal resistance in the vicinity of a heat exchanger surface, which is the result of the high porosity of a fixed sorbent layer is responsible for the low heating power (HP) of the bed [4]. Therefore the crucial area of chiller optimization is to improve the heat transfer coefficient between the adsorbent bed and the heat transfer surface.

In a paper [5] to improve the heat transfer conditions, polydispersive sorbent structures were investigated. The porous media was mixed with metal filings which are characterized by higher thermal conductivity.

A new approach to bed modification is using a coated bed construction [6] where the sorbent layer is glued to the heat exchanger surface. The glue with good thermal parameters, fills the gaseous spaces between particles and reduces the thermal contact resistance of the heat exchanger boundary layer. The optimal parameters of components used to build the novel coated bed construction was defined in [7]. Another important concept of improving the adsorption chiller efficiency is further developing of the heat exchanger surface. Finned tubes of various shapes are investigated in [8-10]. Complementing the heat transfer experimental studies in the sorbent bed is modeling research based on computational fluid dynamics (CFD) and artificial intelligence (AI) methods [11-13].

A validated CFD model was used to optimize the adsorption cooling cycle based on activated carbon – ethanol pair in [14]. In paper [15] silica gel-water working pair was analyzed numerically. The innovative discretization method, which can be used in numerical modeling of flow within granular materials like porous media was presented in [16, 17].

The goal of this paper is to develop a numerical model of heat transfer in the adsorption bed of modified coated construction.

* Corresponding author: k.grabowska@ujd.edu.pl

2 Analysis and modelling

1.1 Experimental study

In order to develop the adsorption bed model, thermal parameters of the porous media layer were defined experimentally. Measurements of heat conductivity λ was carried out with the use of a LFA 457 MicroFlash® apparatus by Netzsch company which performs precise thermal diffusivity measurements based on the Laser Flash Methods (LFM).

The thermal conductivity of fixed and coated sorbent samples were examined. Silica gel was used as a sorbent and an epoxy resin was used as a glue material. The results of experimental tests are presented in Table 1. As it can be seen from table, fixed layer of porous media is characterized by significantly lower thermal conductivity than the coated layer.

Table 1. Thermal conductivity of sorbent samples

Samples	λ (W/m*K)	Specific Heat (J/kg*K)	Bulk Density /Density (g/cm3)
fixed sorbent layer	0.18	924	0.8
coated sorbent layer	0.48	700	0.8

1.2 Numerical geometry

The numerical geometry of the adsorption bed was built with a basic component repeated 60-times, one after another along the axis, which is presented in Figure 1.

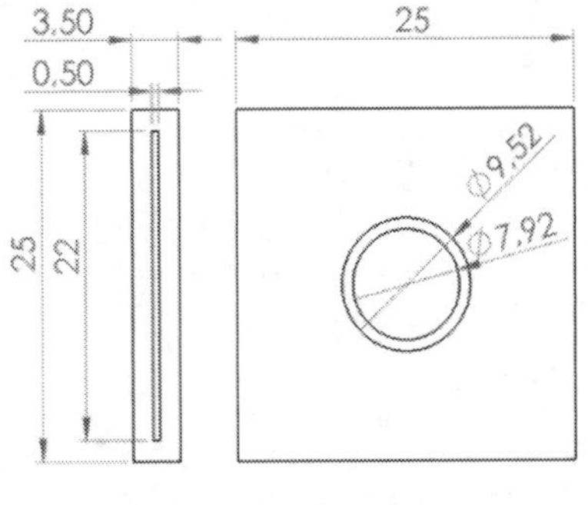

Fig. 1. The basic component of the adsorption bed model.

A 3D model of the adsorption bed was created with the use of the above geometry and it was used to perform the numerical simulations. The adsorption bed model consisted of three computational subdomains representing respectively heating water, heat exchanger material (copper) and silica gel or coated silica gel layer dependent on the analyzed case, such as presented in Figure 2. The temperature distribution was analyzed in a computational domain composed of heat exchanger (finned cooper pipe) with a length of 210 mm and a width of 125 mm (5 components with a width of 25mm each).

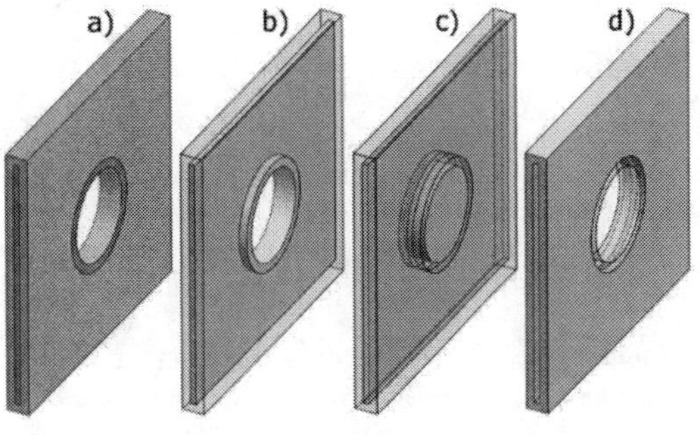

Fig. 2. Computational domains: a) entire domain, b) metal, c) water, d) silica gel or coated silica gel.

This geometry is characterized by the repeatability of the layout of finned tubes, therefore in order to minimize calculation time, the model was reduced by using periodicity condition. The simulation were carried out on a model reduced to the form as presented in Figure 3.

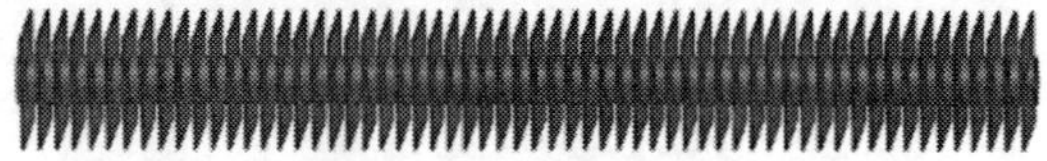

Fig. 3. Computational module of the heat exchanger.

1.3 Numerical analysis

The ANSYS Fluent commercial code was used to perform numerical simulations. The CFD solver was configured as pressure-based and the analysis were carried out in steady state. The solver employs an algorithm which belongs to a general class of methods called the projection method, where the constraint of mass conservation of the velocity field is achieved by solving a pressure equation. The pressure equation is derived from the continuity and the momentum equations in such a way that the velocity field, corrected by the pressure, satisfies the continuity. The solution process involves iterations in order to solve the nonlinear and coupled governing equations.

The mass-flow-inlet boundary condition with mass flow rate equal 0.02 kg/s, was specified at the inlet of the heating water. Moreover the k-ε turbulence model was selected with the standard wall function and the turbulence intensity was 5%. The outflow boundary condition was defined on the surface of water outflow.

The desorption process of the adsorption bed working cycle was simulated. Improvement of heat transfer conditions during desorption process will allow to decrease the hot water temperature which is necessary to well prepared the bed to the subsequent adsorption cycle. Therefore the carried out numerical simulations were focused on desorption process only, because the optimization of heat transfer in the adsorbent bed will allow to utilize the low grade thermal energy sources effectively. The temperature of the inlet hot water was 343 K. Two cases were analysed: first was for coated bed construction and the second was for conventional bed with packed silica gel.

3 Results and discussion

Heat transfer between adjacent subdomains was modelled as conjugate heat transfer. The remaining walls were defined as adiabatic.

All figures are presented with the same temperatures range in order to compare the differences in the obtained temperature fields. The obtained temperature distributions in the computational domain of the heat exchanger are depicted for two cases in Figure 4.

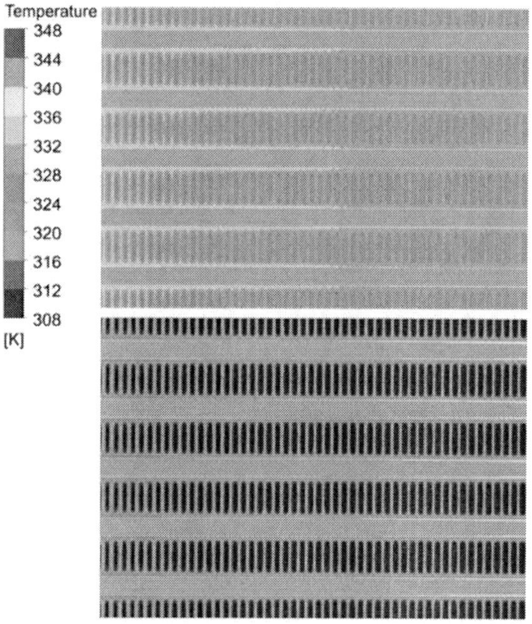

Fig. 4. The temperature distribution in the in longitudinal section of whole computational domain; top - case 1 with glue, bottom - case 2 without glue.

The intensification of the heat transfer between the heat exchanger surface and the coated bed is clearly visible. In consequence desorption in a coated bed will be performed more efficiency. Heat transfer in the conventional fixed bed is strongly limited by very low thermal conductivity of silica gel and high porosity of sorbent layer.

In Table 2, numerically determined average temperatures in particular subdomains of adsorption bed are presented.

Table 2. Comparison of Numerically Determined Temperatures (w-water, s-silica gel, c-coated layer).

Average variable	Coated bed	Fixed bed
T_w(in)	343	343
T_w(out)	341.13	339.34
T_s(in)	322.40	302.43
T_s(out)	319.50	296.75
T_w	342.20	341.10
T_s	321.12	303.76
T_c	328.10	-

Higher average temperatures in the silica gel volume are noticed for the coated bed construction, which confirms the positive effect of the coated layer on the desorption process.

The comparison of two cases reveals, that in the coated sorbent, higher temperatures during the desorption process at the inlet and outlet of the computational domain was noticed, respectively. It has been confirmed that the use of a glued layer results in a more intense heat transfer from water to the volume of silica gel. High thermal resistance, which is observed in fixed bed leads to the increase of desorption time, thus increasing of the energy input for the cooling production and lowering the COP of adsorption chiller.

4 Conclusions

The numerical analysis of heat transfer in two adsorption beds construction is presented in this paper. Numerical simulations under the same flow conditions were carried out, which allowed to indication of the crucial impact of the bed component's thermal parameters.

The influence of a coated sorbent layer on heat transfer intensification is investigated. The average temperature of silica gel volume in the coated bed case was 321.12 K and in the fixed bed case it was only 303.76 K. The obtained temperature distributions confirm that the glued layer of sorbent contributed to the intensification of heat transfer from the hot water to the adsorption bed volume.

Numerical methods can be very useful in the optimization of adsorption bed parameters, because they allow determination of variables that are very difficult to define experimentally, such as temperature changes in the silica gel volume. However, empirical validation on a physical adsorption bed model is necessary in order to precisely define the boundary conditions of the numerical model.

Acknowledgments

Scientific work was performed within the project: "The development of innovative technology of adsorption chiller NETI®, using special, glued construction of the adsorption beds" (number: POIR.01.01.01-00-1659/15) it was partially supported by National Science Centre of Poland (Narodowe Centrum Nauki) grant number 2017/01/X/ST8/00019 as well as subsidies granted by the Faculty of Mathematics and Natural Sciences of Jan Dlugosz University in Czestochowa. The support is gratefully acknowledged.

References

1. B. B. Saha, S. Koyama, T. Kashiwagi, A. Akisawa, K. C. Ng, H. T. Chua, 2003, Int. J. Refrig. **26**, 7, pp. 749-757 (2003)
2. K. Sztekler, W. Kalawa, W. Nowak, S. Stefański, J. Krzywanski, K. Grabowska, *Proc. Experimental Fluid Mechanics 2017* (N. J. Dančová Petra, ed., Technical University of Liberec, pp. 650-653, 2017)

3. J., Krzywanski, K. Grabowska, F. Herman, P. Pyrka, M. Sosnowski, T. Prauzner, W. Nowak, Energ Convers Manage, **153**, pp. 313-322 (2017)
4. D. C. Wang, Z. Z. Xia, J. Y. Wu, R. Z. Wang, H. Zhai, W.D. Dou, Int J Refrig, **28**, 7, pp. 1073-1083 (2005)
5. A.A. Askalany, S. K. Henninger, M. Ghazy, B. B. Saha, Appl Therm Eng **110**, pp. 695-702 (2017)
6. Y. Jiang, T.S. Ge, R. Z. Wang, L. M. Hu, Int J Refrig **51**, pp. 169-179 (2015)
7. K. Grabowska, J. Krzywanski, W. Nowak, M. Wesolowska, Energy **151**, pp. 317-323 (2018)
8. Y. I. Aristov, I. S. Glaznev, I. S. Girnik, Energy **46**, 1, pp. 484-492 (2012)
9. Q. W. Pan, R.Z. Wang, L.W. Wang, D. Liu, Int J Refrig **67**, pp. 336-344 (2016)
10. A. Rezk, R. K. Al-Dadah, S. Mahmoud, A. Elsayed, Appl Therm Eng **53**, 2, pp. 278-284 (2013)
11. J. Krzywanski, W. Nowak, Int J Heat Mass Transfer **94**, pp. 327-334 (2016)
12. A. Gnatowski, M. Sosnowski, A. Pajdak, J. Sobczyk, Adv Sci Technol Res J. **12**, 2, (2018)
13. M. Sosnowski, J. Krzywanski, K. Grabowska, R. Gnatowska, *International Conference on Experimental Fluid Mechanics 2017* (P. Dančová, and J. Novosád, eds., pp. 589-594, Mikulov, 2017)
14. S. Jribi, T. Miyazaki, B.B. Saha, S. Koyama, S. Maeda, T. Maruyama, Int J Refrig **74**, pp. 345-353 (2017)
15. K. Grabowska, M. Sosnowski, J. Krzywanski, K. Sztekler, W. Kalawa, A. Zylka, W. Nowak, J. Therm. Sci. **27**, 5, pp. 421-426 (2018)
16. M. Sosnowski, *International Conference on Experimental Fluid Mechanics 2017* (P. Dančová, and N. J., eds., pp. 582-588, Mikulov, 2017)
17. M. Sosnowski, *Engineering Mechanics 2018,* (Svratka, pp. 801–804, 2018)

The application of Picard method in identification of the heat transfer coefficient in flow boiling in a minichannel

Mirosław Grabowski[1], *Sylwia* Hożejowska[2,*], *Anna* Pawińska[2], and *Mieczysław E.* Poniewski[1]

[1]Warsaw University of Technology, Płock Campus, ul. Łukasiewicza 17, 09-400 Płock, Poland
[2]Kielce University of Technology, Faculty of Management and Computer Modelling, Al.1000-lecia PP 7, 25-314 Kielce, Poland

Abstract. This paper summarizes the procedure of experimental determination of flow boiling heat transfer coefficient in the rectangular minichannel 2 mm wide, 1.8 mm deep and 193 mm long. In this experiment, the basic thermal and flow parameters were measured and the temperature distribution on the foil insulating the heater was recorded using an infrared camera. A high speed video camera observed the vapor patterns forming during the boiling process. The void fraction was determined with the image processing technique. Stationary heat transfer process in the insulating foil, the heater and the flowing liquid was described using two-dimensional Laplace's equation (for the foil), the Poisson equation (for the heater) and energy equation (for the liquid) – all complemented with an appropriate system of boundary conditions. The system of the differential equations with adopted boundary conditions gives the three consecutive inverse problems in elements of the test section containing minichanel. Trefftz method was used to determine two dimensional temperature distribution of the foil and the heater. The liquid temperature was calculated using the Picard-Trefftz method. The values of the experimentally obtained heat transfer coefficients with the values of these coefficients calculated on the basis of correlations known from the literature were also compared.

1 Introduction

Recent years have been abundant with experimental and theoretical works on thermal processes occurring in flow boiling in mini- and micro channels. A wide and comprehensive review of the accomplishments in the field is presented in [1], where much has been said about mathematical models and their application to mathematical analysis. The most general model, that is, two-fluid four-field model that at least theoretically allows continuous modelling of flow changes, has been discussed in [2]. This model comprises a system of several partial differential equations, even when related to developed nucleate boiling. Additionally, a suitable set of boundary conditions has to be defined. Solving such a complex system of equations is not always possible, especially in the case of inverse problems. Even commercial CFD software may be inefficient in such a case. Therefore, there is a need to develop numerical and other methods for solving inverse problems especially coupled inverse problems in areas in contact, with different physical parameters.

The methods presented here: 1) the Trefftz method and 2) the Picard -Trefftz method, belong to the class of semianalytical methods for solving partial differential equations. The Trefftz method was first described by Trefftz [3] and was addressed to linear equations. This method was generalized by many authors [4-8] who expanded its applicability. The paper presents application of the Trefftz method and the Picard-Trefftz method to an approximate solution of the inverse heat conduction problems for stationary flow of a refrigerant in rectangular and horizontal minichannel. The temperature distribution of the elements of the minichannel insulating foil, heating foil and boiling liquid was described with the Laplace equation, Poisson equation and stationary energy equation, respectively. The Trefftz method was used to obtain two dimensional temperature distributions of the insulating foil and of the heater. Two dimensional fluid temperatures was calculated with the use of the Picard - Trefftz method. The heat transfer coefficient at the heater-liquid interface was determined based on the boundary condition of the third kind. Local values of the experimental heat transfer coefficient were compared with the values obtained with correlations of developed experimentally by other researchers.

2 Experiment

2.1 Experimental stand

Figure 1 shows the flow boiling apparatus used. Its main part is the measuring section, Fig.1 #1. Containing the milled in the body, a 193 mm long minichannel with a rectangular cross-section (2 mm wide, 1.8 mm deep). A 2 mm thick transparent lid with a sodium glass visor closes the minichannel on one side, and a heater in the

*Corresponding author: ztpsf@tu.kielce.pl

form of a flat strip made of Fecral 135 retaining alloy with dimensions of 2x0.1 mm, sealed with insulating foil (3M SCOTCH S 33+ insulating tape) is mounted on the other side. Czaki TP-201 type K thermocouples, 0.5 mm in diameter, and Kobold pressure sensors with a range of 0 - 2.5 bar (accuracy ± 1%) were also mounted in the body. The both sensor types are shown schematically in Fig.1, #P and #T. The heater is powered from the programmable DC power supply by TDK Lambda, Fig.1 #2.

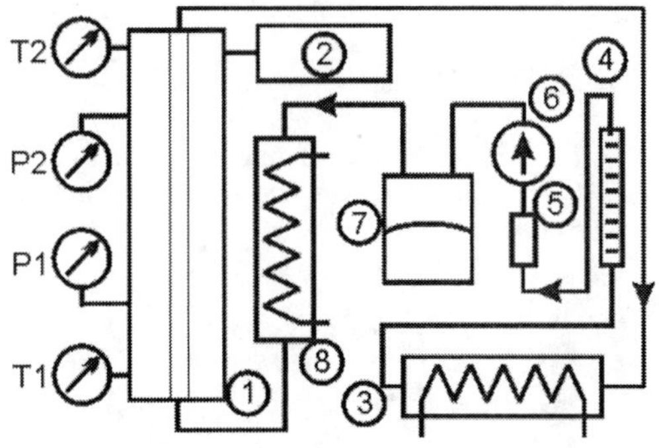

Fig.1. Flow loop: 1 – test section with the minichannel (described in detail in Figure 2), 2 – DC power supply for the heater, 3 – cooler, 4 – rotameter, 5 – filter, 6 – pump, 7 – pressure control, 8 – preheater.

Vapor structures forming due to a phase change during the liquid heating are observed and recorded using a Vision Research Phantom 710 high speed camera with a resolution of 1280x800 pixels at 7530 frames per second.

The two-dimensional temperature distribution on the external surface of the heater is measured and recorded using the FLIR SC 7600 Series infrared camera with a recording resolution of 640x512 pixels and recording frequency of up to 100Hz.

The experimental set-up was designed to:

- displace/force the test liquid at a given temperature (Fig.1, heater #8) and a given pressure through the flow loop (Fig.1, pump #6, pressure set #7), with a maximum volumetric flow rate up to $1.5 \cdot 10^{-7}$ m^3s^{-1} (Fig.1, rotameter #4);
- preheat the liquid flowing into the minichannel using the preheater (Fig.1 #8), and then overheat it above the saturation temperature in the minichannel located in the measurement section (Fig.1 #1) using the heater (not visible in the diagram);
- measure and record the following parameters: liquid pressure and temperature at the inlet and outlet of the minichannel, Fig.1a, manometers P and thermocouples T;
    - volumetric flow rate of the tested liquid, Fig.1, rotameter #4;
    - thermal power supplied to the heating foil, Fig.1, current and voltage supplied by the DC power supply #2;
    - concurrently and thermographically visualize the heating foil temperature distribution and provide images of two-phase flow patterns.

2.2 Experimental results

The experimental setup described above and in [9] enabled to collect the folowing parameters:the flow of the liquid at a set rate, temperature and pressure at the inlet to the minichannel, to heat the liquid in the minichannel to the set temperature, to measure and record the temperature and pressure at the inlet and outlet of the minichannel, liquid volumetric flow rate, and the electric current supplied to the heating foil, and to record the flow structures and two-dimensional temperature distributions on the outer surface of the insulation foil.

The temperature distributions recorded with the thermal imaging camera on the insulating foil were particularly important for further calculations, Fig. 2.

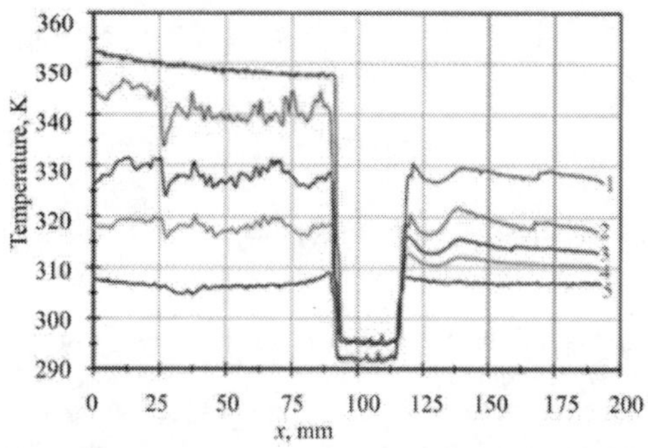

Fig. 2. Exemplary temperature profiles on the minichannel side in contact with ambient surroundings as a function of the distance from the channel inlet. Volumetric flow rate for all cases was $1.41 \cdot 10^{-7}$ m^3/s. Case1: heat flux 154086 W/m^2, Case 2: heat flux 102634 W/m^2, Case 3: heat flux 55699W/m^2, Case 4: heat flux 24194 W/m^2, Case 5: heat flux 0 W/m^2.

They formed the basis for determining heat transfer coefficients in the minichannel using the Picard-Trefftz method. Observations of the two-phase flow structures helped determine the void fraction, The method of determining the void faction was described in detail in [10].

The minichannel was split into three domains related to the heat supply method used. In the domain Ω_I, the heat supplied to the liquid was generated by the electric heater. Due to design reasons, it was not possible to record the temperature field in the foil insulating of the domain Ω_{II}. The domain Ω_{III} was quasi-adiabatic and all thermal and flow parameters were recorded there, just as in Ω_I, Fig 3.

3 Mathematical model and methods

The presented mathematical model is a modification of the approaches from [9,11]. Two dimensions were taken into account in the mathematical model: one in the direction of the flow (x) and one in the perpendicular direction (y) referred to the foil thickness (δ_F), to the heater (δ_H) and to the depth of the minichannel (δ_M). Assumption that the phenomena occurring at the side surfaces of the measurement module do not affect the thermodynamic parameters of its central part allows limiting the considerations to this central part only (along its length L).

Due to the measurement module design this mathematical model concerns heat transfer in only first Ω_I region where in the section with a length of $L_I = 93$ mm there is the heater heating the liquid flowing into the minichannel.

It was assumed that the insulating foil temperature T_F and the heater temperature T_H satisfy the Laplace and Poisson equations in the form, [9,11]

$$\Delta T_F = 0 \text{ for } 0 < x < L_1,\ 0 < y < \delta_F \tag{1}$$

$$\Delta T_H = -\frac{q_V}{\lambda_H} \text{ for } 0 < x < L_1,\ \delta_F < y < \delta_F + \delta_H \tag{2}$$

where $\Delta = \frac{\partial^2}{\partial x^2} + \frac{\partial^2}{\partial y^2}$, L_I – length of the domain Ω_I, q_V – volumetric heat flux supplied to the heater, λ_H – the thermal conductivity of the heater. For equations (1) and (2) the assumption was that:

- at the foil-heater interface the temperature and heat fluxes were equal,
- at the foil - external environment contact the temperature of the foil was equal to the temperature measurements (conducted by infrared thermography) and

$$\lambda_F \frac{\partial T_F}{\partial y} = -q_{loss} \text{ for } y = 0,\ 0 \le x \le L_1 \tag{3}$$

where q_{loss} – the heat loss to the surroundings was determined as in [12],

- the remaining boundaries are isolated, Fig. 3.

It was assumed that the fluid flow in the minichannel was steady state and laminar (Reynolds number $Re < 2000$) with a constant volumetric flow rate, [11]. Additionally, the fluid velocity in the minichannel has only one non-zero component $u(y)$ with parabolic profile, parallel to the flow direction and determined by analogy with [11].

The fluid temperature T_f in the domain $\Omega_{I,f}$ satisfied the energy equation

$$\lambda_f \Delta T_f - u(y) c_p \rho_f \frac{\partial T_f}{\partial x} = 0 \tag{4}$$

where λ_f – the thermal conductivity of the fluid, c_p –specific heat, ρ_f – fluid density, $u(y)$ – liquid velocity.

Adequate boundary conditions were adopted for equation (4), Fig. 3:

$$T_f(x, \delta_F + \delta_H) = T_H(x, \delta_F + \delta_H) \text{ for } 0 \le x \le L_1 \tag{5}$$

$$T_f(0, y) = T_{in} \text{ for } \delta_F + \delta_H \le y \le \delta_F + \delta_H + \delta_M \tag{6}$$

$$T_f(L_1, y) = T_{sat}(L_1) \text{ for } \delta_F + \delta_H \le y \le \delta_F + \delta_H + \delta_M \tag{7}$$

where T_{in}– the liquid temperature at the minichannel inlet, T_{sat} –the saturation temperature depends on linearly changing pressure.

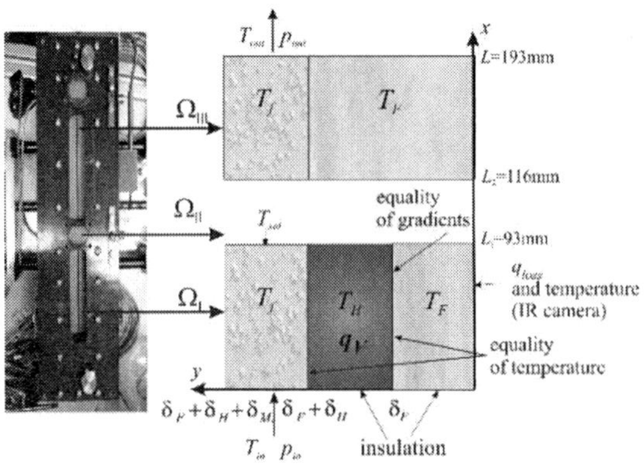

Fig. 3. Test section and its schematic illustrating the underlying model assumptions (pictorial view, not to scale).

For the considered bubbly and bubbly-slug flow the heat flux generated in the heater is transferred to the liquid phase in the proportion relative to the void fraction $\varphi(x)$. Therefore, the heat transfer coefficient at the foil-liquid interface was determined using the Robin condition in the form

$$\alpha_{2D}(x) = (1 - \varphi(x)) \frac{-\lambda_H \frac{\partial T_H}{\partial y}(x, \delta_F + \delta_H)}{T_H(x, \delta_F + \delta_H) - T_{ave}(x)} \tag{8}$$

where the reference temperature T_{ave} was the average fluid temperature along the minichannel depth

$$T_{ave}(x) = \frac{1}{0.5\delta_M} \int_{\delta_F + \delta_H}^{\delta_F + \delta_H + 0.5\delta_M} T_f(x, y)\, dy \tag{9}$$

The main idea of the Picard–Trefftz method is a decomposition of the differential operator from equation (4) into the Laplace operator and the remaining component of equation (4). Next, the Picard's iterations were used where in each iteration a linear combination of the Trefftz functions for the Laplace operator was used, [13].

Equation (4) with the conditions (5) – (7) can be written using operator notation

$$\Delta T_f + NT_f = 0 \text{ for } (x, y) \in \Omega_{1,f} \tag{10}$$

$$BT_f = g(x, y) \text{ for } (x, y) \in \partial\Omega_{1,f} \tag{11}$$

where differential operator N is defined as

$$N = -u(y)\frac{c_p \rho_f}{\lambda_f}\frac{\partial}{\partial x} \tag{12}$$

and operator B describes boundary conditions (5)- (7).

In the Picard iterations the following procedure was used to approximate the solutions to equation (4):
- in the first step, for $k = 1$:

$$\Delta T_f^{(1)} = 0 \text{ for } (x, y) \in \Omega_{1,f} \tag{13}$$

$$BT_f^{(1)} = g(x, y) \text{ for } (x, y) \in \partial\Omega_{1,f} \tag{14}$$

- in subsequent steps, for $k > 1$:

$$\Delta T_f^{(k)} = -NT_f^{(k-1)} \text{ for } (x, y) \in \Omega_{1,f} \tag{15}$$

$$BT_f^{(k)} = g(x, y) \text{ for } (x, y) \in \partial\Omega_{1,f} \tag{16}$$

In each step the Trefftz method was used to determine the approximate solutions to equations (13) and (15) as in [13]. The unknown solutions equations (13) and (15) was approximating in the form of a linear combination of the Trefftz functions $w_j(x,y)$ for Laplace equation, i.e.

$$T_f^{(k)} = \sum_{j=1}^{M^{(k)}} a_j^{(k)} w_j(x, y) + T_{sol}^{(k)}(x, y) \tag{17}$$

where $T_{sol}^{(k)}(x, y)$ is a particular solution to the Poisson's equation (15) and is given by formula

$$T_{sol}^{(k)} = -\Delta^{-1}\left[NT_f^{(k-1)}\right] \tag{18}$$

and the particular solution equation (13) is equal to zero, $T_{sol}^{(1)}(x, y) = 0$. In (18) operator Δ^{-1} is calculated according to the formula

$$\Delta^{-1}(x^m y^l) = m!\,l! \sum_{k=0}^{[m/2]} (-1)^k \frac{x^{m-2k}\, y^{l+2+2k}}{(m-2k)!\,(l+2+2k)!} \tag{19}$$

where the symbol [.] denotes the floor function.

The coefficients of the linear combinations in (17) were determined based on the adopted boundary conditions, following the method described in [14].

4 Results and discussion

Numerical calculations were performed using the Trefftz method and the Picard –Trefftz method for the two-phase flow of ethanol in a 1.8 mm deep and 2.0 mm wide horizontal minichannel. The following experimental data were used in calculations: inlet pressure 10.9 kPa (average), liquid temperature at the inlet 306 K (average), volume flow rate $8.2 \cdot 10^{-8}$ m^3 s^{-1} and five heat flux densities q_w in the range of 99.2 ÷ 154.1kWm^{-2}. The thermal imaging camera helped to measure the temperature distribution of the insulating foil. In further calculations, the local void fraction determined, in the manner described in [10], at the distances of 0 m, 0.02m, 0.04 m, 0.06 m, and 0.08 m from the minichannel inlet was approximated with a quadratic function, Fig. 4.

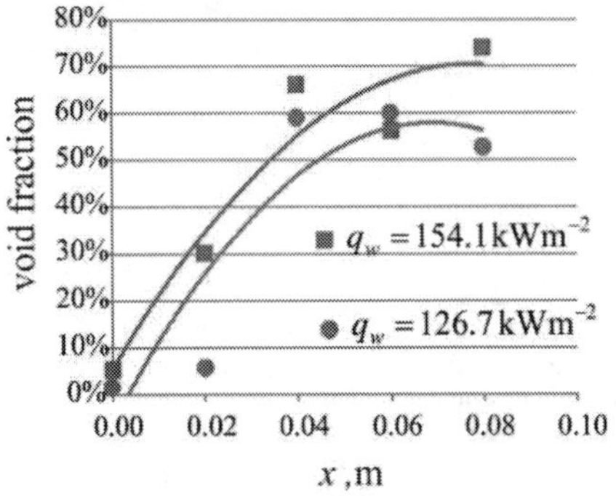

Fig. 4. The void fraction approximated by quadratic function.

As follows from the presented mathematical approach two-dimensional temperature distributions in the insulating foil and the heater were determined first using the Trefftz method with six harmonic polynomials (Trefftz functions for Laplace equation). In the next stage, the liquid temperature distribution in the minichannel was determined by the Picard-Trefftz method with four iterative steps and two harmonic polynomials at each step.

The values of local heat transfer coefficients obtained from (8) are shown in Fig. 5. It was observed that when the heat flux supplied to heating foil grows, the fraction of the vapour phase increases, leading to the decrease of the heat transfer coefficient.

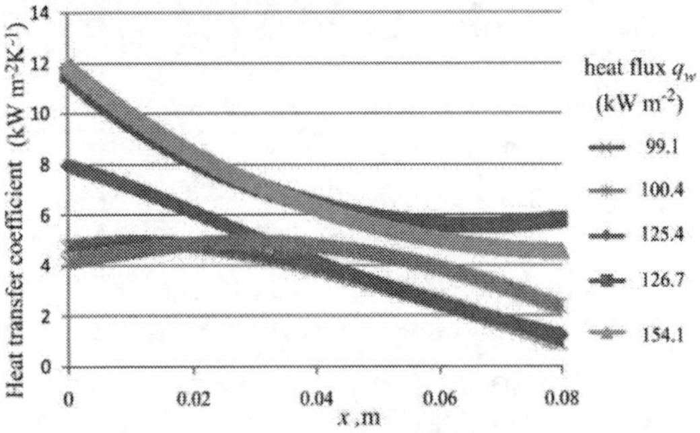

Fig. 5. Heat transfer coefficients for different heat fluxes as a function of the minichannel length obtained using the Picard-Trefftz method.

This paper compared the values of the experimental heat transfer coefficient α_{2D} calculated from formula (8) with the values α_{TP} calculated based correlations known from the literature:

-Kandlikar and Balasubramanian correlation for low Reynolds numbers [15]:

$$\alpha_{TP} = \left(0.6683 \cdot Co^{-0.2} + Bo^{0.7} F_{Fl}\right)\left(1-X\right)^{0.8} \alpha_{SP} \quad (20)$$

where fluid-surface parameter F_{Fl} and the heat transfer coefficient α_{SP} for a single phase flow were determined as in [15]:

-Bohdal and Charun [16]:

$$\alpha_{TP} = 19.2 Re^{0.8} Bo^{0.6} \left(\frac{\rho_g}{\rho_l}\right)^{0.5} \frac{\lambda_f}{D} \quad (21)$$

- Gungor and Winterton [17]:

$$\alpha_{TP} = E \cdot 0.023 Re_l^{0.8} Pr_l^{0.4} \frac{\lambda_f}{D} \quad (22)$$

$$E = 1 + 3000 Bo^{0.86} + 1.12 \left(\frac{X}{1-X}\right)^{0.75} \left(\frac{\rho_l}{\rho_g}\right)^{0.41}$$

- Tran and Wambsganss [18]:

$$\alpha_{TP} = 840000 \left(Bo^2 We_l\right)^{0.3} \left(\frac{\rho_g}{\rho_l}\right)^{0.4} \quad (23)$$

The following symbols were adopted in correlations (20) – (23) the following symbols means: Bo–boiling number, Co–convection number, X–vapour quality, D–minichannel internal perimeter, Pr– Prandtl number, Re–Reynolds number, We – Weber number, subscripts l and g refers to the liquid and gas phase respectively.

Figure 6 compares the values of heat transfer coefficient calculated based on correlations (20) – (23) and the values of experimental coefficient α_{2D} calculated from (8) for q_w= 125.2 kWm^{-2}.

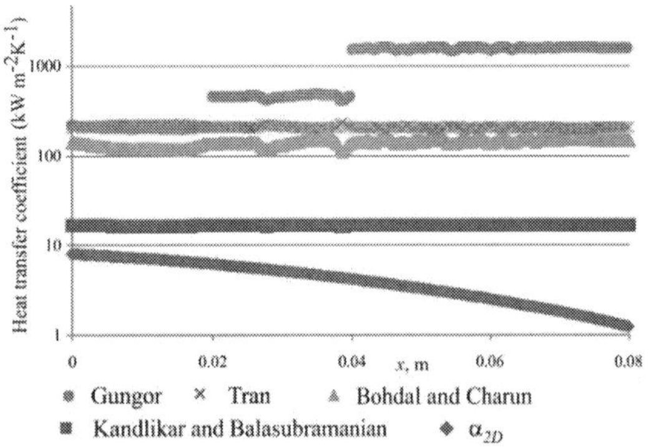

Fig.6. Local heat transfer coefficients calculated based on correlations (20)-(23) and formula (11) for q_w = 125.2 kWm^{-2}.

5 Conclusions

The proposed 2D stationary mathematical approach was constructed for the heat transfer in three adjacent elements of the test section. Trefftz method was used to determine two dimensional temperature distribution of the insulating foil and the heater. The liquid temperature was calculated using the Picard-Trefftz method. The heat transfer coefficients obtained from the two-dimensional approach were compared with correlations known from the literature.

The heat transfer coefficients calculated from the Kandlikar and Balasubramanian correlations [15] were most similar to those calculated experimentally from (8), Fig.5. With other correlations the differences between the experimental and correlation-based results were more pronounced. These considerable differences may result from the fact that the correlations in question have been developed for the flows with higher Reynolds numbers than those in our experiment. From the descriptions of experiments performed by the cited researchers, it appears that heat losses to the environment are often omitted, with a comment added that they can have significant value. This inaccuracy in determining the heat flux transmitted to the fluid may also explain obtaining by the cited authors coefficient values consistently higher compared to those presented in these study, where heat losses are included in the calculation procedure.

This opens a new field of research devoted to the modification of the existing correlations or the development of original correlation equation.

References

1. Z. Guo, D.F. Fletcher, B.S. Haynes, J. Comput. Multiphase Flows **6** (2014)
2. R.T. Lahey Jr., *Proc. XIII US National Congress on Applied Mechanics*, June 21-26, University of Florida, USA (1998)
3. E. Trefftz, *Proc. II Int.Congress of Applied Mechanics*, Zurich, 131–137 (1926)
4. O.C. Zienkiewicz, D.W. Kelly, P. Bettes, Int. J. Numer. Meth. Eng. **11** (1977)
5. J. Jirousek 1978, Comput. Meth. App. Mech. Eng. **14** (1978)
6. I. Herrera, *Boundary methods. An algebraic theory*, (Pitman, Boston–London–Melbourne, 1984)
7. Z.-C. Li, T.-T. Lu, H.-Y. Hu, A. H.-D. Cheng., *Trefftz and Collocation Methods*, (WIT Press, Southampton 2008)
8. K. Grysa, A. Maciąg, Int. J. Heat Mass Transfer **100**, (2016)
9. S. Hożejowska, M. Grabowski, EPJ Web of Conf. **114**, 02041 (2016)
10. K. Płaczkowski, M. Poniewski, M. Grabowski, S. Alabrudziński, App. Mech. Materials **797** (2015)
11. S. Hożejowska, R. Kaniowski, M.E. Poniewski, Exp. Thermal Fluid Sci.**78** (2016)
12. M. Piasecka, K. Strąk, B. Maciejewska, Heat Transfer Eng. **38** (2016)
13. K. Grysa, A. Maciąg, A. Pawińska, Int. J. Heat Mass Transfer **55** (2012)

14. S. Hożejowska, M. Piasecka, M. Poniewski, Int. J. Therm. Sci. **48** (2009)
15. S.G. Kandlikar, P. Balasubramanian, Heat Transfer Eng. **25**, 3 (2004)
16. T. Bohdal, H. Charun, K.Dutkowski, 2008, Annals of Environment Protection **10** (2008)
17. K.E. Gungor, R.H.S. Winterton, Chem. Eng. Res. Des. **65** (1987)
18. T. Tran, M.W. Wambsganss, D.M. France, Int. J. Multiphase Flow **22** (1996)

The application of Fourier transform to the identification of temperature distribution in HFE-7100 flow boiling in an annular minigap

Sylwia Hożejowska [1,*], and *Magdalena* Piasecka[2]

[1]Kielce University of Technology, Faculty of Management and Computer Modelling, Al. 1000-lecia PP 7, 25-314 Kielce, Poland
[2]Kielce University of Technology, Faculty of Mechatronics and Mechanical Engineering, Al. 1000-lecia PP 7, 25-314 Kielce, Poland

Abstract. In the paper the results of investigations into HFE-7100 flow boiling heat transfer in a cylindrical minigap a 1 mm wide created between the external glass pipe and the copper pipe were discussed. A cartridge heater located axially heated fluid flowing along the minigap. The cooling fluid temperature and pressure at the inlet and the outlet to/from the minigap and temperature of the heater in 18 points were measured. Two-dimensional mathematical model for heat transfer coefficient determination was proposed. It was assumed that in the test section the heat transfer process was in steady state and the fluid flow was laminar. The temperature of the metal pipe near the heater was assumed to satisfy Laplace's equation. The problem formulated in this way was solved by two methods: the Fourier transform and the Trefftz method. The working fluid temperature was calculated depending on the flow type: for single phase flow with boiling incipience and for two-phase flow, respectively. The heat transfer coefficient at the fluid – copper pipe interface was calculated due to the Robin condition. Local heat transfer coefficient values obtained from the Fourier transform and from the Trefftz method were similar. Results were presented and discussed.

1 Introduction

Heat exchangers with minigaps have attracted wide attention from the heat transfer issues due to their prominent ability to remove heat flux from confined spaces. Highly efficient heat transfer system is of great importance to the operations and performance of high power components of heat exchangers. Miniaturization of heat transfer exchangers help to decrease size of these devices.

Recently, a lot of works have been published to address the issues of heat transfer in flow boiling in minigaps of different geometry. Previous authors' studies mostly have been performed to evaluate the flow boiling heat transfer in rectangular minichannels, as [1-5]. This paper shows the results of research on flow boiling heat transfer during refrigerant flow along an annular minigap [6-8]. Cylindrical geometry of a minigap in other researchers' works can be find in [9-11].

Currently, the research often abandons the experimental studies in favour of numerical simulations, utilising commercial software, where the experimental results serve the purpose of validation of numerical calculations only, such as, for example, in [12, 13]. Application of commercial software for solving an inverse heat conduction problem (IHCP), as contemplated in this paper, is limited, which results from the following facts:

- too small set of available experimental data which have to be implemented into the codes,
- necessity to solve instantaneously energy equations in the annular minigap and the heated surface with an incomplete set of closure equations (unknown boundary condition on the unheated side of the channel).

In this paper, two methods are used for solving the IHCP: the Fourier transform and the Trefftz method [8].

2 Experimental background

The schematic diagram of the experimental stand is presented in Fig. 1.

The flow loop in which a working fluid (HFE-7100) circulates and the data acquisition system consists of: the test section (1), a tube-type heat exchanger (5), a compensating tank (6), a deaerator (7), a filter (8), a gear pump (9) and a mass flow meter (10). The data acquisition system comprises: two data acquisition stations (2, 3), a pc computer (4) with a specialist software and a high speed camera.

The schematic diagram of the test section is shown in Fig. 2. A minigap (3) 1 mm wide was created between a glass pipe (1) and a copper pipe (2) positioned along the same axis. A cartridge heater (4) was located inside

* Corresponding author: ztpsf@tu.kielce.pl

the pipe axially. To fix thermoelement sensors (6) with a spacing of approximately 10 mm in the flow line longitudinal slots was made in the inside surface of the pipe. Thermoelement wires run in the gap between the pipe and the heater in the thermal paste. The sensors of thermoelements to measure the working fluid temperature (6) and pressure meters (7) for fluid pressure control at the inlet and the outlet to/from the minigap, were placed in the module headers (5).

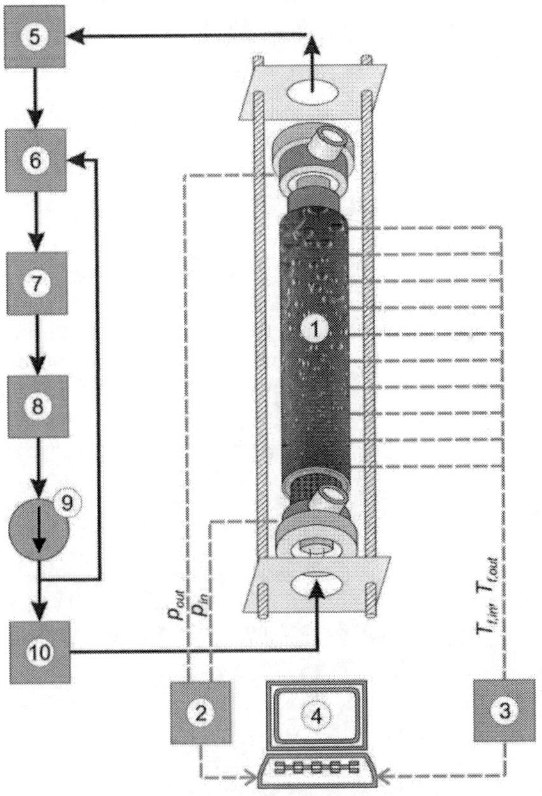

Fig.1. The schematic diagram of the experimental stand: 1 - the test section, 2,3 - data acquisition stations, 4 - a pc computer, 5 - a tube-type heat exchanger, 6 - a compensating tank, 7 - a deaerator, 8 - a filter, 9 - a gear pump, 10 - a mass flow meter.

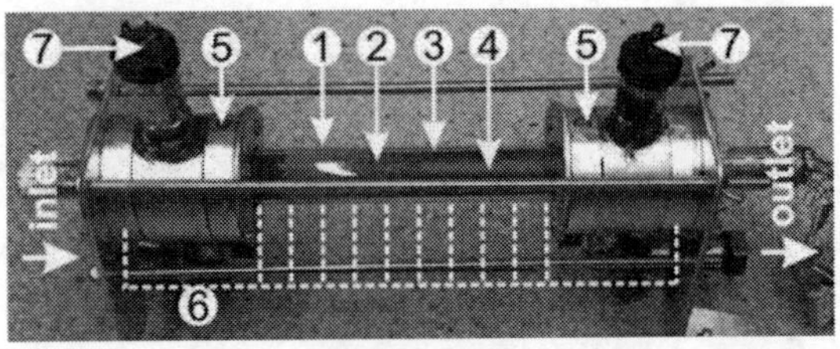

Fig. 2. The schematic diagram of the test section: 1 - a glass pipe, 2 - a copper pipe, 3 - a minigap, 4 - a cartridge heater, 5 - a module header,6 - thermoelements, 7 - a pressure meter.

During the experiment, there was a HFE-7100 (3M) flow in the flow loop along the minigap. The heat flux supplied to the heater in the test section was increased gradually, by adjustment of heating power by autotransformer with controlled current intensity, until developed nucleate boiling occurs. The signals from the thermocouples and the pressure meters, the power supplied to the heater and mass flow rate were recorded by data acquisition stations.

3 Mathematical model and numerical solution

In a two – dimensional mathematical model it was assumed the steady state in the test section and the laminar flow of the fluid (Reynolds number < 2100).

Scheme of the test section is presented in Fig. 3. The heat transfer process in the test module is axisymmetric i.e. the temperature distribution of the elements in the module is only dependent on two variables: z (along the flow direction) and r (referring to the thickness of the cartridge heater and the copper pipe). It was assumed that the thin layer of the thermal conductive filler (~10^{-4} m) and the size of the thermocouples placed in it were omitted. Additionally, it was assumed that the surface has direct contact with the surface of the heater. All the thermal properties were stationary and independent on temperature.

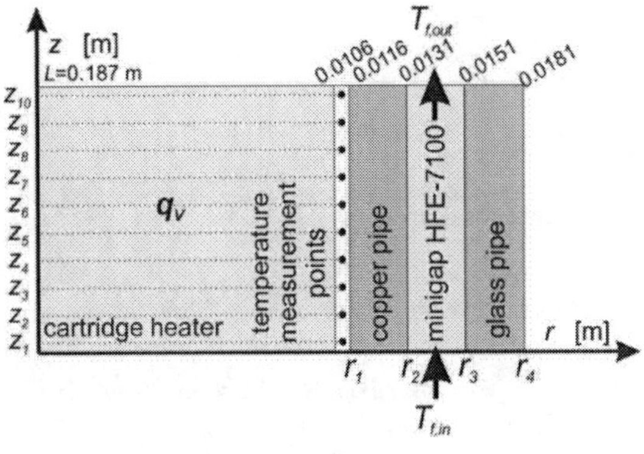

Fig. 3. Scheme of the test section (pictorial view, not to scale).

The temperature of the copper pipe T_{cp} was assumed to satisfy Laplace's equation

$$\frac{1}{r}\left(r\frac{\partial T_{cp}}{\partial r}\right)+\frac{\partial^2 T_{cp}}{\partial z^2}=0 \tag{1}$$

for which we imposed proper boundary conditions: the layers of the cartridge heater and the copper pipe are in perfect thermal contact while the walls perpendicular to them are isolated, i.e.

$$\lambda_{cp}\frac{\partial T_{cp}}{\partial r}=\frac{q_V r_1}{2} \quad \text{for } r=r_1 \tag{2}$$

$$T_{cp}(r_1,z)=T_{approx}(z) \tag{3}$$

$$\frac{\partial T_{cp}}{\partial z}=0 \text{ for a) } z=0 \text{ and b) } z=L \tag{4}$$

where λ_{cp} – the thermal conductivity of the cooper pipe, L – length of the minigap, q_V – volumetric heat flux supplied to the cartridge heater, T_{approx} – a polynomial approximating the temperature measurements of the cooper pipe.

The IHCP described by equations (1) – (4) was solved by two methods, based on the Fourier transform in the first approach and on the Trefftz method in the second one.

The working fluid temperature T_f was calculated depending on the flow type:

- for single phase flow and in the subcooled boiling region (small share of the vapour phase in the mixture), the fluid temperature changes linearly in the minichannel from the inlet temperature to the outlet temperature;
- for two-phase flow in the saturated boiling region (when it dominates the participation of the vapour phase in the mixture) - the fluid temperature is equal to local saturation temperature; local saturation temperature is determined on the basis of the linear distribution of pressure along the length of the minigap from the inlet to the outlet.

If the temperature distributions of the copper pipe and the refrigerant are known, we can calculate the heat transfer coefficient α at the fluid – pipe interface due to the Robin condition

$$\alpha(z)=\frac{0.5q_V r_1^2 r_2^{-1}}{T_{cp}(r_2,z)-T_f(z)} \tag{5}$$

3.1 Fourier transform

In the solution of IHCP described by equations (1) - (4) the Fourier transform was used, where the integral-transform $\hat{T}_{cp,F}$ and the inversion formula for the function $T_{cp,F}$ in the finite interval $0 \le z \le L$ were defined by the formulas respectively

$$\hat{T}_{cp,F}(r,\beta_i)=\int_0^L T_{cp,F}(r,z)K(\beta_i,z)dz \tag{6}$$

$$T_{cp,F}(r,z)=\sum_{i=0}^{\infty}\hat{T}_{cp,F}(r,\beta_i)K(\beta_i,z) \tag{7}$$

The transform kernel K is given by analogy to [14]

$$K(\beta_i,z)=\begin{cases}\dfrac{1}{L} & \text{for } i=0\\ \sqrt{\dfrac{2}{L}}\cos\beta_i z & \text{for } i=1,2,\ldots\end{cases} \tag{8}$$

and the eigenvalues $\beta_i = {\pi i}/{L}$ for i= 1,2,….

The solution of equation (1) has the form

$$T_{cp,F}(r,z)=\sum_{i=0}^{\infty}\left(a_i I_0(\beta_i r)+b_i K_0(\beta_i r)\right)K(\beta_i,z) \tag{9}$$

where $a_0=\frac{1}{L}\int_0^L T_{approx}(z)dz$, b_0=0, I_0 – modified Bessel function of the first kind, K_0 – modified Bessel functions of the second kind. The coefficients a_i and b_i for i=1,2,…, were calculated from the system of equations

$$\begin{cases}a_i I_0(\beta_i r_1)+b_i K_0(\beta_i r_1)=\int_0^L K(\beta_i,z)T_{approx}(z)dz\\ a_i\beta_i I_0(\beta_i r_1)+b_i\beta_i K_0(\beta_i r_1)=\dfrac{q_V r_1}{2\lambda_{cp}}\int_0^L K(\beta_i,z)dz\end{cases} \tag{10}$$

The function T_{cp} satisfies Laplace's equation (1) and the assumed boundary conditions (2)-(4) exactly.

3.2 Trefftz method

The idea of the Trefftz method consists in an approximation of an unknown solution of equation (1) using a linear combination of functions (called T-functions), which satisfy this equation exactly. First of all, T-function for the equation (1) should be defined. Solving Laplace's equation (1) by the Fourier method (separation of variables method), we obtain two solutions depends on Bessel functions of the first J_0 and second Y_0 kind

$$\text{a) } J_0(\theta r)e^{\theta z}, \qquad \text{b) } Y_0(\theta r)e^{\theta z} \tag{11}$$

Expanding the functions (11) into a power series in regard to powers of the θ parameter, yields T-functions of the first $u_n(x,y)$ and second $w_n(x,y)$ kind for equation, [15]

$$u_n(r,z)=\sum_{k=0}^{[n/2]}\frac{(-1)^k}{(k!)^2(n-2k)!}\left(\frac{r}{2}\right)^{2k}z^{n-2k} \tag{12}$$

$$w_n(r,z)=u_n(r,z)\cdot\ln r-\sum_{k=0}^{[n/2]}\frac{(-1)^n(0.5r)^{2n}H_n}{(k!)^2(n-2k)!}z^{n-2k} \tag{13}$$

where H_n – harmonic numbers and the symbol [.] denotes the floor function.

Knowledge of the T-function given by equations (12) and (13) allows for approximating the unknown solution of Laplace's equation (1) in the form of a linear combination, [15]

$$T_{cp,T}(r,z)=\sum_{n=0}^{N}c_n u_n(r,z)+\sum_{m=0}^{M}d_m w_m(r,z) \tag{14}$$

Unknown coefficients c_n, d_m in (14) were determined by minimizing the error functional that described the mean square error with which the function $T_{cp,T}$ satisfied the assumed boundary conditions (2)-(4), analogously as in [1]. The function $T_{cp,T}$ satisfies Laplace's equation exactly and the assumed boundary conditions (2)-(4) approximately. Approximate fulfillment of boundary conditions by the calculated approximation is a characteristic feature of the Trefftz method.

4 Results and discussion

The calculations were performed with the Fourier transform and the Trefftz method for the following experimental parameters: pressure at the inlet 220 kPa (average), inlet liquid subcooling 56 K (average), constant mass flow rate of 0.017 kg/s) and five heat fluxes q_w in the range of 8.8÷17.4 kW/m^2.

Figure 4 shows the heat transfer coefficient as a function of the minigap inlet at the subcooled boiling region, calculated using the Fourier transform.

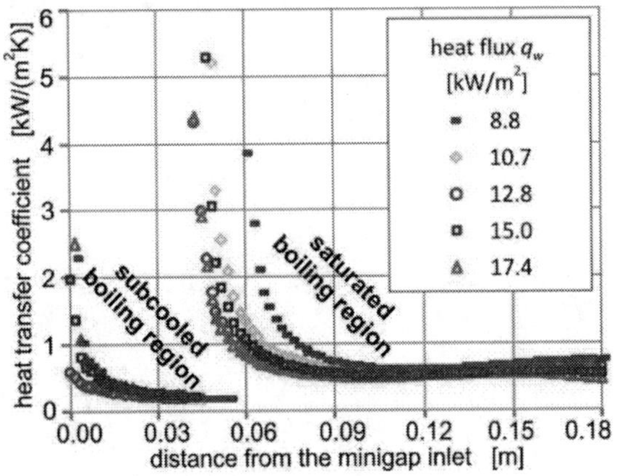

Fig. 4. The heat transfer coefficient α_F vs. the distance from the minigap inlet z.

For expected improvement of accuracy, measurement temperature of the cartridge heater – copper pipe interface was approximated with a third degree polynomial. In the Trefftz method the same number of T-functions of the first and second kind were taken. The gradient of the copper pipe temperature T_{cp} was calculated based on the known foil temperature distribution.

Similar values of the heat transfer coefficient have been obtained based on the Trefftz method, which is confirmed by the values of the maximum relative differences. The maximum relative differences *MRD* between heat transfer coefficient values obtained using the Trefftz method (α_T) and Fourier transform (α_F) were calculated from the formula

$$MRD = \max\left\{\frac{\|\alpha_T-\alpha_F\|_2}{\|\alpha_T\|_2}, \frac{\|\alpha_T-\alpha_F\|_2}{\|\alpha_F\|_2}\right\} \tag{15}$$

where L^2 – norm is defined in in measurements points z_{data} as

$$\|\alpha\|_2 = \sqrt{\sum_{z_{data}} \alpha^2(z_{data})} \tag{16}$$

For the considered heat fluxes, the *MRD* ranges from 0.21 to 1.46 and only for heat flux 12.8 kW/m^2 *MRD* is equal to 12.10.

By analogy to [4,16,17], the mean relative error of the heat transfer coefficient (*MRE*) was determined only for the smallest and the largest value of heat flux. In the *MRE* calculations it was adopted that:

- uncertainty of temperature measurement: ΔT_{data} =0.9 K, [4],
- accuracy of the copper pipe temperature approximation: $\Delta T_{cp} = \left|\frac{\partial T_{cp}}{\partial z}\Delta z\right| + \Delta T_{data}$ (average: 0.93 K);
- uncertainty of thermocouple location: $\Delta z = 10^{-4}$ m,
- uncertainty of radius: $\Delta r = 10^{-4}$ m,
- accuracy of the volumetric heat flux supplied to the cartridge heater Δq_V (average: 222 kW/m^3), [17];
- accuracy of the fluid temperature determination: ΔT_f=0.9 K in the subcooled boiling region and $\Delta T_f = \Delta T_{sat}$ in the saturated boiling region, [4].

Both used numerical methods gave the *MRE* less than 9.4%, except q_w=12.8 kW/m^2, where *MRE* was equal to 16.7% for the Fourier transform.

It was observed that local heat transfer coefficient values decreased with increasing distance from the minigap inlet at all heat fluxes supplied to the heater at the subcooled and the saturated boiling regions. In the subcooled boiling region, heat transfer coefficients were in several hundred kW/(m^{2}K), which is much lower than those obtained for the saturated boiling region. The highest values of the heat transfer coefficient are in the order of several thousand kW/(m^{2}K) and occur at the beginning of the saturated boiling region. The highest value reached 5.5 kW/(m^{2}K).

It should be underlined that the heat transfer coefficient obtained with the use of the Fourier transform and by the Trefftz method were similar. Values of the maximum relative differences between results were in the range: 0.21-1.22, the smallest values *MRD* (0.21) were obtained for the smallest heat flux, for higher heat fluxes *MRD* values were higher, and the highest one (1.22) was found for the highest heat flux.

MRE was very similar for both used methods. The *MRE* for Trefftz method did not exceed 9.4% in the entire experiment.

5 Conclusions

In this work, the application of two numerical methods for the identification of temperature distribution in flow boiling of HFE-7100 in an annular minigap was proposed. The two-dimensional temperature distribution of the copper pipe was determined from the Fourier transform and from the Trefftz method. Using the Fourier transform, a solution satisfying the governing differential equation and the adopted boundary conditions exactly. In the Trefftz method, the obtained approximation of the copper pipe temperature has been satisfying exactly only the governing differential equation – the boundary conditions are satisfied by the function approximately. Using the Trefftz method, as opposed to the Fourier transform, the IHCP may be solved also in regions with more complex shapes, with a lower number of boundary conditions, whether in a continuous form, or in a discrete form.

As the results of the mathematical modelling of the temperature distributions in the test section was known, the heat transfer coefficient at the fluid – wall interface was calculated due to the Robin condition. The results were presented as the heat transfer coefficient versus the distance from the minigap inlet. It was observed that local heat transfer coefficient values decreased with increasing distance from the minigap inlet at all heat fluxes supplied to the heater at the subcooled and the saturated boiling regions. In the subcooled boiling region, heat transfer coefficients were in several hundred kW/(m^{2}K). The highest value in the order of several thousand kW/(m^{2}K) occurred at the beginning of the saturated boiling region. The values of the heat transfer coefficient obtained with the use of the Fourier transform and by the Trefftz method were similar.

References

1. S. Hożejowska, M. Piasecka, Heat Mass Transf. **50**, 1053-1063 (2014)
2. B. Maciejewska, M. Piasecka, Heat Mass Transf. **53**(4), 1211–1224 (2017)
3. K. Strąk, M. Piasecka, B. Maciejewska, Int. J. Heat Mass Transf. **117**, 375–387(2018)
4. B. Maciejewska, M. Piasecka, Int. J. Heat Mass Transf. **107**, 925–933 (2017)
5. T. Musiał, M. Piasecka, S. Hożejowska, EPJ Web of Conf. **143**, 02077 (2017)
6. M. Piasecka, S. Hożejowska, T. Musiał, 2017, E3S Web of Conf. **13**, 02002 (2017)
7. M. Piasecka, S. Hożejowska, A. Piasecki, *Proc. 9th World Conf. ExHFT-9*, June 12-15, Brasil, **PT30** (2017)
8. S. Hożejowska, T. Musiał, M. Piasecka, EPJ Web of Conf. **114**, 02042 (2016)
9. G. Boye, J. Schmidt, F. Beyrau, Adv. Mech. Eng. **7** (6),1-14 (2015)
10. F.C. Hsieh, K.W. Li, Y.M. Lie, C.A. Chen, T.F. Lin, Int. J. Heat Mass Transf. **51**, 3763-3775 (2008)
11. C.P. Yin, Y.Y. Yan, T.F. Lin, B.C Yang, Int. J. Heat Mass Transf. **43**, 1885-1896 (2000)
12. Z. Guo, B.S. Haynes, D.F. Fletcher, J. Comput. Mult. Flows **8**, 61–82 (2016)
13. J. Yao, M.K. Patel, Y. Yao, P.J. Mason, Adv. Appl. Math. Mech. **1**, 231–241 (2009)
14. M. N. Ozisik, *Boundary Value Problems of Heat Conduction* (International Textbook Company, Scranton, Pennsylvania, 1968)
15. S. Futakiewicz, *The heat functions method for solving direct and inverse heat conduction problems*, (Ph.D. Thesis in Polish, Poznan University of Technology, 1999)
16. M. Piasecka, Metrol. Meas. Syst. XX (2), 205–216 (2013)
17. S. Hożejowska, M. Piasecka, M.E. Poniewski, Int. J. Therm. Sci. **48,** 1049–1059 (2009)

An inverse geometric problem in steady state heat conduction - the solution and stability analysis

Leszek Hożejowski*

Faculty of Management and Computer Modelling, Kielce University of Technology, al.Tysiąclecia P. P. 7, 25-314 Kielce, Poland

Abstract. The paper addresses a numerical method for boundary identification in a problem governed by Laplace's equation. The proposed numerical procedure for discrete reconstruction of the unknown boundary from the given temperature data is based on the Trefftz method. In contrast to the procedures described in the reference papers, the present approach requires significantly less and easier computation. The paper undertakes analysis of the resistance of the solution to small perturbations of the prescribed temperature condition at the unknown part of the boundary. We define and then estimate a sensitivity factor which allows quantitative assessment of the relationship between temperature measurement errors and boundary identification errors, even if the exact solution is not known. The included numerical examples demonstrate the effectiveness of the proposed method for boundary reconstruction and present the analysis of numerical stability using a sensitivity factor.

1 Introduction

There are different kinds of inverse geometry problems like shape and design optimization or identification of defects in materials. In this paper, however, we address a problem which consists in identification of the unknown part of a domain boundary. More precisely, given a problem described by a differential equation, a numerical reconstruction of the unknown boundary has to be achieved from the available boundary condition. There have been different propositions of the efficient solution algorithms to this problem. One method, however, seems very suitable for approaching the problem, namely the Trefftz method and the method of fundamental solutions (MFS) which are mathematically equivalent though differ in formulation [1]. In recent years some researchers proposed their solution procedures based on the Trefftz method (or collocation Trefftz method) and concerning the problems governed by the Laplace equation [2-6], biharmonic equation [7] or the modified Helmholtz equation [8]. Also, there appeared some successful applications of MSF to boundary identification [9-11].

The present study employs the Trefftz method for discrete reconstruction of the unknown boundary. Unlike in [2-4, 7, 8], we recover the boundary points one after the other. In terms of computation this means solving one equation at a time (in contrast to solving a large system of equations). A new approach to stability analysis proposed in the paper allows to predict boundary identification errors, calculated on the basis of the known measurement errors.

2 Two dimensional boundary identification problem

In this section a certain inverse geometry problem will be set in terms of a mathematical model. Besides, a solution algorithm and a new approach to stability analysis will be proposed.

2.1 Problem formulation

Consider a problem in a bounded two-dimensional region Ω. Assume that its boundary $\partial\Omega$ is composed of two disjoint curves γ_0 and γ_1:

$$\partial\Omega = \gamma_0 \cup \gamma_1 \qquad (1)$$

such that γ_0 is known and γ_1 unknown. The governing equation has a form

$$\Delta T = 0 \quad \text{in}\,\Omega \qquad (2)$$

where Δ denotes the Laplace operator so the problem can be interpreted as describing steady state heat conduction. We solve (2) subject to

$$T = T_0 \quad \text{on} \quad \gamma_0 \qquad (3)$$

$$\frac{\partial T}{\partial \mathbf{n}} = q_0 \quad \text{on} \quad \gamma_0 \qquad (4)$$

$$T = T_1 \quad \text{on} \quad \gamma_1 \qquad (5)$$

* Corresponding author: hozej@tu.kielce.pl

where $\partial/\partial\mathbf{n}$ denotes differentiation along the unit outward normal $\mathbf{n}$. Conditions (3) and (4) mean that both the temperature and its gradient along γ_0 are known while equation (5) indicates that the unknown part of the boundary, γ_1, has to be inferred from the given temperature data. Assuming that T_1 is known from measurement as a set of discrete data (which better corresponds to the real situation), condition (5) should be rewritten as

$$T = T_1^m \quad \text{at} \quad P_m \in \gamma_1 \quad (m = 1, 2, \ldots M) \tag{6}$$

2.2 Solution method

The problem described by (2) – (4) and (6) will be solved using the Trefftz method which consists in approximating the solution by a linear combination of the basis functions (named T-complete functions) which satisfy the differential equation (2). Then we have

$$T \approx \sum_1^N c_n V_n \tag{7}$$

where V_n denote the T-complete functions and c_n – the scalar coefficients which we specify to minimize the functional

$$\Phi = \left\| \sum_1^N c_n V_n \big|_{\gamma_0} - T_0 \right\|^2 + \left\| \sum_1^N c_n \frac{\partial V_n}{\partial \mathbf{n}} \bigg|_{\gamma_0} - q_0 \right\|^2 \tag{8}$$

In the above, $\|\cdot\|$ denotes L^2 norm. Note that finding c_n from the minimum of (8) requires solving a system of linear algebraic equations.

Discrete reconstruction of γ_1 should be interpreted as: given the x-coordinate of the point P_m, find its y-coordinate. However, depending on a particular problem, we would sometimes prefer a formulation in polar coordinates.

Since one has to calculate y_m from (6), where y_m denotes the y-coordinate of P_m, we propose to solve M nonlinear equations (one after the other), each with one unknown:

$$T(x_m, y_m) = T_1^m, \quad m = 1, 2, \ldots, M \tag{9}$$

This is in contrast to other solution algorithms based on the Trefftz method [2-4, 7,8] which include both the coefficients c_n of T-complete functions and the sought coordinates y_m of P_m's in a system of simultaneous nonlinear equations whose solution might be problematic, especially if M is very large.

Numerical methods for solving nonlinear equations usually require an initial guess. In our problem, the y-coordinate of the point joining γ_0 and γ_1 could make the initial guess for the first equation (corresponding to $m = 1$) while in subsequent steps the previously obtained y_m seems a reasonable initial point for the $(m+1)$-th equation.

2.3 Stability analysis

Since our task is to recover γ_1 from the temperature data T_1, the problem of solution sensitivity comes down to the question: If T_1 is perturbed with a small error into $T_1 + \Delta T_1$, how will the position of the recovered boundary points P_m change in comparison with that obtained from the original data?

In the reference papers addressing a boundary identification problem, both computation and further analysis is based on synthetic data coming from numerical simulations. Then, stability analysis consists in disturbing the inputs with simulated random errors and observing if the boundary identification errors have more or less the same magnitude as input errors. Such an approach seems insufficient. In fact, the crucial problem lies in comparison of the two ratios: $\Delta T_1/T_1$ and $\Delta y/y$ which represent relative errors of temperature measurement and boundary identification, respectively. We can write

$$\frac{\frac{\Delta T}{T}}{\frac{\Delta y}{y}} = \frac{y}{T}\frac{\Delta T}{\Delta y} \approx \frac{y}{T}\frac{\partial T}{\partial y} \tag{10}$$

where the latter comes from replacing the differential quotient by the partial derivative. For simplicity, we will write "=" instead of "≈". Let $\varepsilon_{T,y}$ denote the right-hand side of (10):

$$\varepsilon_{T,y} = \frac{y}{T}\frac{\partial T}{\partial y} \tag{11}$$

$\varepsilon_{T,y}$ is known as a partial elasticity of the function T with respect to y and is defined locally (at a point). Equation (10) referred to γ_1 gives

$$\left|\frac{\Delta y}{y}\right| = \left|\varepsilon_{T,y}\right|^{-1} \cdot \left|\frac{\Delta T_1}{T_1}\right| \quad \text{on } \gamma_1 \tag{12}$$

Formula (12) is crucial to further analysis. It says that a 1% perturbation of the temperature will result in $\left|\varepsilon_{T,y}\right|^{-1}$ times larger percent error of the boundary identification. In view of this interpretation, $\left|\varepsilon_{T,y}\right|^{-1}$ could be called a sensitivity factor. Evaluating $\varepsilon_{T,y}$ is not problematic because the Trefftz method allows for obtaining the derivative $\partial T/\partial y$ by analytical differentiation of (7). On the other hand, $\Delta T_1/T_1$ can be treated as given because it denotes either a simulated error or a measurement error, usually known by the experimenter. As a consequence, we are able to effectively estimate $|\Delta y/y|$. Besides, one should note that the evaluation of errors using a sensitive factor does not need the exact analytical solution of Laplace's equation which serves only for simulating temperature measurements.

3 Numerical examples

For two-dimensional Laplace's equation the T-complete functions in (x,y)-coordinates are

$$\left\{1, Re(x+iy)^n, Im(x+iy)^n \ : \ n=1,2,\dots\right\} \qquad (13)$$

where *Re* and *Im* denote real and imaginary part of a complex number, respectively, and i stands for the imaginary unit. We recognize the functions (13) as harmonic polynomials.

In further considerations, the choice of an analytical solution to Laplace's equation will allow for numerical simulation of the required temperature T_0 and T_1 and the temperature gradient q_0.

3.1 Example 1

Let the problem domain be

$$\Omega \ : \ 0 \le x \le 1, \quad f(x) \le y \le g(x) \qquad (14)$$

Its upper boundary, $g(x)$, will be treated as unknown and has tobe determined. In this example we choose $f(x)=(x-1)(x+1)^{-1}$, $g(x)=2(2+x\sin 10x)^{-1}$ and $T(x,y)=e^x \cos y$. Besides, we take $N = 17$ T-complete functions in expansion (7) and perform calculations with the simulated measurements at $M = 40$ points.

Using the unperturbed temperature data, the algorithm recovers the upper boundary with very high accuracy (errors less than 0.0016%). Figure 1 presents the results of boundary identification from the noisy data, with randomly generated errors ranging from -5% to 5% and normally distributed. For the purpose of stability analysis, we evaluated the sensitivity factor which turned out to vary (depending on x) from 0.2 to 1.6.

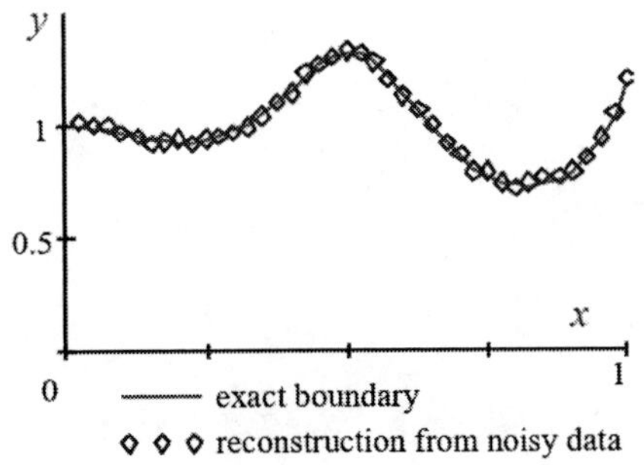

Fig. 1. Reconstructed part of the unknown boundary (γ_1) in Example 1.

Visual inspection of the graph in Fig. 1 shows barely noticeable influence of the introduced noise on the boundary reconstruction. This should not be a surprise since the sensitivity factor in this case is relatively small.

Figure 2 presents the true boundary identification errors and compares them to those estimated from equation (12).

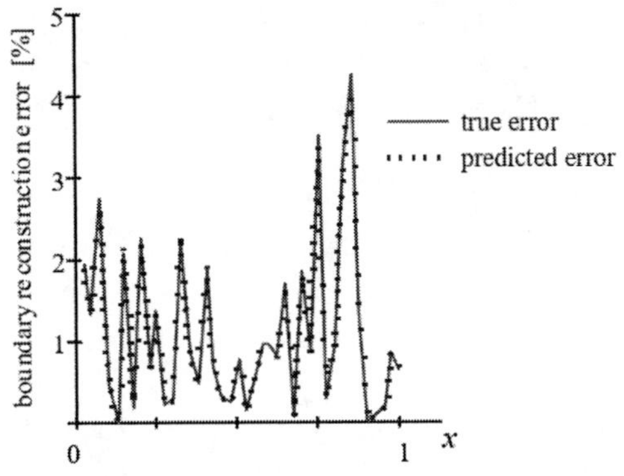

Fig. 2. Boundary reconstruction errors in Example 1 (for noisy data).

The graph shows very good agreement between the true and predicted errors. The result also proves (though not explicitly) that we succeeded in evaluating the sensitivity factor with high accuracy.

3.2 Example 2

This example differs from the previous one only in a choice of an upper boundary to be determined. Here we take $g(x)=1(2+x\sin 10x)^{-1}$, obtained by dividing the previous $g(x)$ by 2. All the other data, including the simulated errors, are remained the same.

The computation using errorless data yields the reconstruction having errors at most 0.0003%. In comparison to the previous example, the use of noisy data significantly perturbed the recovered boundary points locations. Such result was expected, since the magnitude of the sensitivity factor was considerably greater than in Example 1 (for details, see Tab.1).

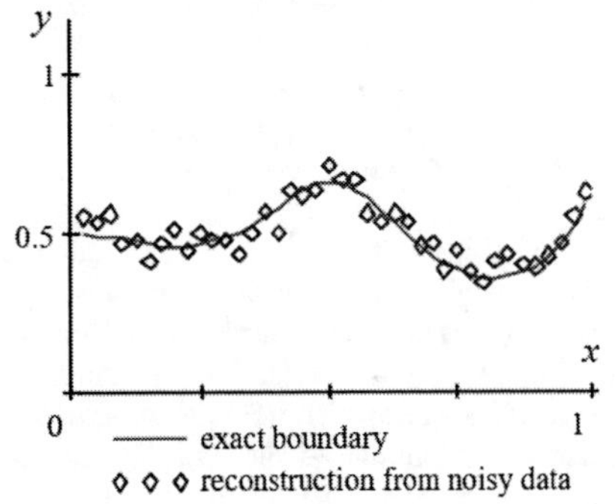

Fig. 3. Reconstructed part of the unknown boundary (γ_1) in Example 2.

Similarly to the previous case, we present the percentage boundary reconstruction errors, both true and predicted (see: Fig. 4).

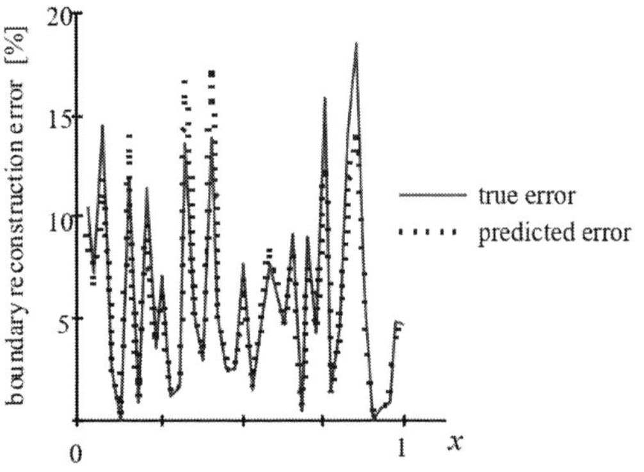

Fig. 4. Boundary reconstruction errors in Example 2 (for noisy data).

The results shown in Fig. 4 confirm what we could predict by analyzing the estimated sensitivity factor (2.1 ÷ 7.3), namely that the boundary reconstruction errors in this example must be a few times larger than in the previous example.

A tabular summary of the computation results concerning the stability analysis of both numerical examples in contained in Tab.1.

Table 1. Boundary reconstruction analysis - accuracy and stability.

	Sensitivity factor	Maximum boundary reconstruction error	
		exact input data	noisy input data
Example 1	0.2 ÷ 1.6	0.0016%	4.2%
Example 2	2.1 ÷ 7.3	0.0003%	18.7%

Generally, when a sensitivity factor grows, its estimation from noisy data becomes less accurate. This explains why the predicted errors shown in Fig. 4 are considerably underestimated or overestimated at some points. Nevertheless, in this and other similar cases, formula (12) still provides some useful information which can be indicative of the true accuracy of boundary reconstruction. Finally, we note that insufficient accuracy of the solution presented in Example 2 is not implied by imperfection of the used computational algorithm. A high sensitivity factor indicates problems where accurate boundary reconstructions from noisy data are very problematic or, in extreme cases, unavailable.

The stability analysis in problems described in polar coordinates goes similar to the analysis shown above. The only difference is that given such a problem, we seek for the radial coordinates of some unknown boundary points so the proper sensitivity coefficient contains the derivative of T with respect to r.

4 Conclusions

The paper presents an application of the Trefftz method for discrete reconstruction of an unknown part of the domain boundary in a heat conduction problem. In terms of computation, the algorithm outputs every recovered point by solving a single nonlinear equation with one unknown. Therefore, with a growing number of the unknown boundary points, the computation does not become more complicated or problematic, unlike in the related algorithms which employ large systems of nonlinear equations. Moreover, the presented method gives accurate boundary reconstructions, as shown in the included examples. The use of what was defined as a sensitivity factor, allows for credible estimation of the boundary identification errors without knowing the true boundary.

References

1. J.T. Chen, C.S. Wu, Y.T. Lee, K.H. Chen, Comput. Math. Appl. 53, 6 (2007)
2. C.M. Fan, H.F. Chan, Numer. Heat Transfer, Part B **59**, 1 (2011)
3. C.M. Fan, H.F. Chan, C.L. Kuo, W. Yeigh, Eng. Anal. Boundary Elem. **36**, 1 (2012)
4. A. Karageorghis, D. Lesnic, L. Marin, Inverse Prob. Sci. Eng. **22**, 3 (2014)
5. L. Hożejowski, J. Theor. Appl. Mech.-Pol. **54**, 3 (2016)
6. L. Hożejowski, Journal of Applied Mathematics and Computational Mechanics **15**, 2 (2016)
7. H.F. Chan, C.M. Fan, J. Mech **29**, 02 (2013)
8. C.M. Fan, Y.C. Liu, H.F. Chan, S.S. Hsiao, Inverse Prob. Sci. Eng. **22**, 2 (2014)
9. L. Marin, A. Karageorghis, CMC **10**, 3 (2009)
10. L. Marin, A. Karageorghis, D. Lesnic, Eng. Anal. Boundary Elem. **35** (2011)
11. Y. Sun, F. Ma, Inverse Prob. Sci. Eng. **23**, 6 (2015)

https://doi.org/10.1051/matecconf/201824001014

Numerical analysis of lifted hydrogen flame

Sohail Iqbal[1], *Björn* Pfeiffelmann[1], *Ali Cemal* Benim[1,*], and *Franz* Joos[2]

[1]Center of Flow Simulation (CFS), Dep. Mechanical and Process Engineering, Düsseldorf University of Applied Sciences, Germany
[2]Laboratory of Turbomachinery, Helmut Schmidt University, Hamburg, Germany

Abstract. A numerical analysis of a turbulent lifted H_2/N_2 flame is presented. As combustion mechanisms, a large spectrum is considered including single-step and two-step, as well as detailed mechanisms. As combustion models, various models are considered that treat turbulence-chemistry interaction in different ways, including the Eddy Dissipation Concept and the Laminar Flamelet Method. For turbulence modelling Reynolds Averaged Numerical Simulation and Large Eddy Simulation approaches are used. Results are compared with measurements.

1 Introduction

Power generation by thermal machinery, including the gas turbine [1] and steam turbine [2] plants, largely depends on the conversion of the chemically bound energy by the combustion process. Parallel to the efforts for exploiting new energy sources, as well as recovery techniques [3], combustion continues to play an important role in power generation, also through the renewable energies, considering the significance of biomass in the latter.

Combustion of hydrogen and hydrogen-blend fuels plays an especially important role in clean and efficient energy supply, climate protection and resource efficiency. On the one hand, hydrogen represents an attractive alternative to storing excess energy in power generation from wind power or photovoltaics. On the other hand, instead of combustion [4,5] the gasification of biomass and coal is a good possibility for an efficient and clean power generation. The gasification product, the synthesis gas contains, in addition to carbon monoxide and small amounts of methane, significant amounts of hydrogen. In addition, there is a rapidly growing interest in hydrogen production using nuclear power plants, using the nuclear power for electrolysis, thermochemical cycles or hybrid approaches. From an environmental point of view, the subsequent combustion of hydrogen is most welcome because it produces no carbon dioxide when burned.

Use of hydrogen or hydrogen blend gases as fuel in combustion systems represents a great challenge. Hydrogen is extremely reactive and, compared to other gases, has very different material properties, so that even relatively small proportions of hydrogen can greatly alter the combustion properties of the gas mixture. Therefore, the combustion of fuel gas mixtures with hydrogen fractions requires new combustion chamber concepts. To this purpose development of accurate prediction procedures are required. This is the main motivation for the present contribution.

In power generation by gas turbines, the state-of-the-art technology for the land-based gas turbine combustors is the so-called "lean premixed combustion", due to its high potential for achieving very low nitrogen oxide emissions. A principal problem in premixed combustion systems is the so-called "flashback" [6]. This means an undesirable flame propagation upstream, towards the inner parts of the burner, which cause serious damage [6]. A classical cause of flashback is an imbalance between the local flow velocity and flame speed, in the favor of the latter, which may be caused by a change/disturbance in the operation of the burner. This aspect makes the use of hydrogen fuel in premixed combustion to become an especially great challenge, since the high reactivity of hydrogen considerably increases the flame speed and flashback propensity [6].

The counterpart of flashback is the so-called "blow-off", i.e. loss of the flame, which is also an undesirable scenario, which can happen when the flow velocity is higher than the flame speed. The precursor of blow-off is the flame lift-off which starts as the flame root leaves the rim. If the jet velocity is further increased, the flame moves downstream to a position where it stabilizes, thus, a lifted flame results. Lift-off height is the distance between the lifted flame base and the jet exit. When the jet velocity reaches the blow-off velocity, the reaction cannot be sustained and the flame is extinguished. Both stability limits (flashback & blow-off) are important for operation of combustion systems and have safety implication for handling fuels, in our case, hydrogen.

For a lifted flame, a turbulent mixing region exists which is very similar to that for the unignited jet, where an accurate prediction of the subsequent ignition is crucial for the prediction of lift-off height and the blow-off behavior of the flame. The prediction of the lifted flame is, thus, highly demanding. The scope of the

* Corresponding author: alicemal@prof-benim.com

present investigation is the prediction of the lifted flame by various modelling strategies.

Besides experimental studies [7,8], computational investigations [8,9] were performed for lifted hydrogen flames. In the previous modelling work, either the 2nd generation Eddy Dissipation Concept (EDC) in combination with Reynolds Averaged Numerical Simulation (RANS) [8] or a subgrid joint probability density method in combination with Large Eddy Simulation [9] was used as the modelling strategy. In the present investigation, different combustion modelling approaches will be assessed within RANS and LES formulations for turbulence.

2 Modelling

For the computational investigations, the general-purpose finite volume method based open source CFD code OpenFOAM [10] is used, where a pressure-correction scheme (PIMPLE which is a combination of PISO and SIMPLE schemes) was applied for treating the velocity-pressure coupling. A reacting mixture of ideal gases is considered along with a Newtonian behavior and negligible buoyancy effects. Radiative heat transfer [11] is neglected as gas radiation is obviously weaker compared to radiation encountered in flames containing liquid droplets or solid particles such as pulverized coal flames [4,5].

Flow turbulence is described either within a RANS [12] or an LES [13] framework. For RANS the SST model [14] is used as the turbulence model, which has been successfully used, so far, in a wide range of applications [15-19]. For LES, the Smagorinsky model [13,20-22] is employed as the subgrid-scale model, assuming a constant value of 0.1 for the subgrid-scale model constant C_S [13].

For scalars, a gradient-diffusion approximation is used, along with the assumption of constant turbulent Schmidt numbers of value 0.7.

A general, reversible chemical reaction involving N species can be expressed as [23]

$$\sum_{i=1}^{N} \nu'_i M_i \xleftrightarrow{k_f, k_b} \sum_{i=1}^{N} \nu''_i M_i \qquad (1)$$

where M_i, ν_i denote the species, the stoichiometric coefficients, whereas k_f and k_b stay for the rate coefficients of the forward and backward reaction, respectively.

According to the law of mass action of chemical kinetics, the molar conversion rate of a species i ($R_{C,i}$) is expressed as [23]

$$R_{C,i} = \left(\nu''_i - \nu'_i\right)\left(k_f \prod_{j=1}^{N} C_j^{n'_j} - k_b \prod_{j=1}^{N} C_j^{n''_j}\right) \qquad (2a)$$

where C_i denote the species concentrations. The exponents n_i are equal to ν_i for elementary reactions. For global reactions n_i are empirically determined and may be different from ν_i.

After Arrhenius [23], the rate coefficient k_f is expressed by

$$k_f = A\,T^m \exp\left(-E_a / R_u T\right) \qquad (2b)$$

In (2b), T is the thermodynamic temperature, E_a is the activation energy, R_u is the universal gas constant, A is the so-called pre-exponential factor, m is the so-called temperature exponent. For irreversible reactions, k_b is zero, for reversible reactions, given k_f, k_b can be obtained with the help of the equilibrium constant K_C ($K_C = k_f/k_b$) [23].

The source terms of species transport equations depend on expressions as given by Eq. (2). Obviously, in a turbulent flow, where averaged/filtered equations are solved, such highly non-linear source terms lead to severe closure problems. Thus, models are needed to consider turbulence-chemistry interaction.

2.1 Turbulence-chemistry interaction (turbulence model)

Following models are considered:

2.1.1 Chemical kinetics controlled flame (C)

It is assumed that turbulent mixing is infinitely fast compared to the chemical reaction rate, so that the resultant averaged/filtered conversion rate is determined purely by chemical kinetics. Expressions such as Eq. (2) are used as they are, plugging in the averaged/filtered quantities, neglecting the fluctuations.

2.1.2 Turbulent mixing controlled flame (T)

It is assumed that the chemical kinetics is infinitely fast compared to the turbulent mixing rate, so that the resultant averaged/filtered conversion rate is determined purely by small-scale turbulent mixing. A model based on this assumption is the so-called Eddy Dissipation Concept [24]. Here the averaged/filtered molar conversion rate ($R_{T,i}$) is calculated from

$$R_{T,i} = \nu_i\, \alpha\, \rho \frac{\varepsilon}{k} \min\left(\min_R \left(\frac{Y_R}{\nu'_R M_R} \right), \beta \frac{\sum_P Y_P}{\sum_j^N \nu''_j M_j} \right) \qquad (3)$$

where, ρ, k, ε denote the time-averaged mixture density, turbulence kinetic energy and its dissipation rate, respectively. The terms Y_i and M_i stay for the averaged/filtered mass fraction and molar mass of the species i. The subscripts R and P denote the each if the reacting species, and the sum of the product species, respectively. The coefficients α and β are the empirical model constants. The standard model constants [24] are used. Since an improved version of this model was proposed later (see below), the abbreviation "EDC" will be reserved to denote the improved version of the Eddy Dissipation Concept, which will be presented below. The present version will be denoted by the abbreviation "T".

2.1.3 Turbulence & chemistry controlled flame (TC)

A rather simple method for considering of both effects is the following combination of the previously mentioned two methods, where the resulting averaged/filtered conversion rate is obtained from

$$R_i = min\left(R_{C,i}, R_{T,i}\right) \tag{4}$$

2.1.3.1 Eddy Dissipation Concept (EDC)

The initial Eddy Dissipation Concept [24] assumes an infinitely fast chemistry. An improved version of the Eddy Dissipation Concept was proposed, later, again by Magnussen [25] where, chemistry and turbulence effects are considered in a combined manner (in a more sophisticated form compared to the "TC" approach expressed by Eq. (4)) Here, the averaged/filtered source/sink term of the species transport equation for a species i (S_i), is modelled as

$$S_i = \rho \frac{\gamma^* \psi}{1-\gamma^* \Psi} \frac{1}{\tau^*}\left(Y_i - Y_i{}^*\right) \tag{5}$$

where γ^* and τ^* denote the mass fraction occupied by the fine structures and the corresponding time scale, ψ representing the reacting fraction of the fine structures (constant estimated by stoichiometry). The terms Y_i and Y^*_i represent the averaged/filtered and fine structure values of the species mass fractions. Assuming adiabatic, isobaric perfectly stirred reactors for the fine structures, the fine structure mass fractions are obtained from

$$\frac{dY_i{}^*}{dt} = R_i(p,T,Y_i{}^*) + \frac{1}{\tau^*}\left(Y_i^0 - Y_i{}^*\right) \tag{6}$$

where Y^0_i denotes the surrounding value of the mass fraction and R_i is the reaction rate obtained from chemical kinetics (Eq. (2)). The two key parameters γ^* and τ^* are modelled based on turbulence quantities.

2.1.3.2 Laminar Flamelet Model (LFM)

The Laminar Flamelet Method is an elegant way of combining turbulence and chemistry effects [26]. In the present study, a flamelet model based on the mixture fraction (Z) and the reaction progress variable (C) is adopted, which was originally proposed by Pierce and Moin [27]. Within this concept, all thermo-chemical variables (φ) are expressed functions of Z and C as

$$\phi = \phi(Z, C) \tag{7}$$

Such functional relationships are obtained by performing 1D laminar flame calculations, once for all, and made available in tabulated form (flamelet libraries) for the subsequent CFD analysis.

Assuming Z and C are independent, with the help of presumed probability density functions (PDFs), the average values of the thermo-chemical variables are obtained. In the present work, a single-delta PDF is assumed for C, where a beta PDF is used for Z.

The model requires the solution of modelled transport equations for three quantities, i.e. for Z, its variance and the reaction progress variable [28].

The elegance of the approach lies therein that the number of equations to be solved does not increase with the complexity of the underlying reaction mechanism, since species transport equations are not solved during the CFD calculation, but extracted out of the libraries, which were prepared by detailed chemistry calculations in a preceding step.

Thus, LFM is very convenient to use with a detailed reaction mechanism, and, it is normally used, as such. In the present work, the flamelet libraries are constructed using the FlameMaster code [29].

2.2 Reaction mechanisms

The combustion reactions consist of elementary reactions between a large number of species, which are expressed by the so-called "detailed mechanisms" [23]. However, except for LFM, the computational overhead increases with the number of species considered in the mechanism. Thus, simplified mechanisms comprising a small number of mechanisms are required for practical purposes. These are, however, less universal compared to the detailed mechanisms and necessitate a careful evaluation and validation for the specific purpose. The following global and detailed mechanisms are considered in the present study:

2.2.1 Single-step global mechanisms

Here, the mechanism of Varma et al. [30] (VCB) and Marinov et al. [31] (MWP) are considered. Note that the reaction of MWP is a reversible one, whereas the reaction of VCB is an irreversible reaction.

2.2.2 Two-step global mechanisms

Here, the mechanism of Hsu and Jemcov [32] (HJ) is considered. Both reactions are formulated as reversible reactions.

2.2.3 Detailed reaction mechanisms

A detailed mechanism that is considered is the one suggested by Conaire et al. [33] (CCSPW).

A further detailed mechanism considered is the full GRI Mech 3.0 [34]. Referring to the definitions above (Eq. (2)), the single-step and two-step global reaction mechanisms [30-32] are summarized in Table 1.

Table 1. Global mechanisms (units: cm^3, mol, s, kcal, K).

	A	m	E_a	n'_{H2}	$n'_{O2/OH}$
	VCB [30]:				
$H_2 + ½ O_2 \rightarrow H_2O$	$3.53 \cdot 10^{12}$	0	7292	1.1	1.1
	MWP [31]:				
$H_2 + ½ O_2 \leftrightarrow H_2O$	$1.80 \cdot 10^{13}$	0	34979	1	0.5
	HJ [32]:				
$H_2 + O_2 \leftrightarrow 2\ OH$	$2.50 \cdot 10^{26}$	-3	23900	1	1
$H_2 + 2OH \leftrightarrow 2\ H_2O$	$8.10 \cdot 10^{35}$	-5	18800	1	2

The detailed reaction mechanism of Conaire et al. [33] is presented in Table 2 (note that the concentration exponents, n_i, are not explicitly given in Table 2, as the reactions are elementary reactions)

The applied models with different combinations are summarized in Table 3 (note that for the combustion model TC, the backward reaction is disabled, if the reaction mechanism is reversible).

Table 2. Detailed mechanism [33] (units: cm^3, mol, s, kcal, K).

	Reaction	A	m	E_a
	H_2/O_2 Chain Reactions			
1	$H + O_2 = O + OH$	$1.91 \cdot 10^{14}$	0.00	16.44
2	$O + H_2 = H + OH$	$5.08 \cdot 10^{4}$	2.67	6.292
3	$OH + H_2 = H + H_2O$	$2.16 \cdot 10^{8}$	1.51	3.43
4	$O + H_2O = OH + OH$	2.95×10^{6}	2.02	13.40
	H_2/O_2 Dissociation/Combination Reactions			
5[a]	$H_2 + M = H + H + M$	$4.57 \cdot 10^{19}$	-1.40	105.1
6[a]	$O + O + M = O_2 + M$	$6.17 \cdot 10^{15}$	-0.50	0.00
7[a]	$O + H + M = OH + M$	$4.72 \cdot 10^{18}$	-1.00	0.00
8[b]	$H + OH + M = H_2O + M$	$4.50 \cdot 10^{22}$	-2.00	0.00
	Formation and Consumption of HO_2			
9[c]	$H + O_2 + M = HO_2 + M$	$3.48 \cdot 10^{16}$	-0.41	-1.12
	$H + O_2 = HO_2$	$1.48 \cdot 10^{12}$	0.60	0.00
10	$HO_2 + H = H_2 + O_2$	$1.66 \cdot 10^{13}$	0.00	0.82
11	$HO_2 + H = OH + OH$	$7.08 \cdot 10^{13}$	0.00	0.30
12	$HO_2 + O = OH + O_2$	$3.25 \cdot 10^{13}$	0.00	0.00
13	$HO_2 + OH = H_2O + O_2$	$2.89 \cdot 10^{13}$	0.00	-0.50
	Formation and Consumption of H_2O_2			
14	$HO_2 + HO_2 = H_2O_2 + O_2$	$4.20 \cdot 10^{14}$	0.00	11.98
	$HO_2 + HO_2 = H_2O_2 + O_2$	$1.30 \cdot 10^{11}$	0.00	-1.629
15[d]	$H_2O_2 + M = OH + OH + M$	$1.27 \cdot 10^{17}$	0.00	45.5
	$H_2O_2 = OH + OH$	$2.95 \cdot 10^{14}$	0.00	48.4
16	$H_2O_2 + H = H_2O + OH$	$2.41 \cdot 10^{13}$	0.00	3.97
17	$H_2O_2 + H = H_2 + HO_2$	$6.03 \cdot 10^{13}$	0.00	7.97
18	$H_2O_2 + O = OH + HO_2$	$9.55 \cdot 10^{6}$	2.00	3.97
19	$H_2O_2 + OH = H_2O + HO_2$	$1.00 \cdot 10^{12}$	0.00	0.00
	$H_2O_2 + OH = H_2O + HO_2$	$5.80 \cdot 10^{14}$	0.00	9.56
a	Efficiency factors: H_2O =12.0; H_2 = 2.5			
b	Efficiency factors: H_2O =12.0; H_2 = 0.73			
c	Troe param.: a=0.5, $T^{***}=10^{-30}$, $T^{**}=10^{30}$, $T^{*}=10^{100}$ Efficiency factors: H_2O =14.0; H_2 = 1.3			
d	Troe param.: a=0.5, $T^{***}=10^{-30}$, $T^{**}=10^{30}$ Efficiency factors: H_2O =12.0; H_2 = 2.5			

Table 3. Summary of applied models.

Case	Reaction mechanism	Combustion model	Turbulence model
T	VCB [30]	T	RANS
V-C	VCB [30]	C	RANS
V-TC	VCB [30]	TC	RANS
V-EDC	VCB [30]	EDC	RANS
M-C	MWP [31]	C	RANS
M-TC	MWP [31]	TC	RANS
M-EDC	MWP [31]	EDC	RANS
H-C	HJ [32]	C	RANS
H-TC	HJ [32]	TC	RANS
H-EDC	HJ [32]	EDC	RANS
C-C	CCSPW[33]	C	RANS
C-EDC	CCSPW[33]	EDC	RANS
LES-LFM	GRI [34]	LFM	LES

3 The test case

The considered test case is the atmospheric, vitiated coflow burner, which was investigated by Cabra et al. [8], which is depicted in Figure 1

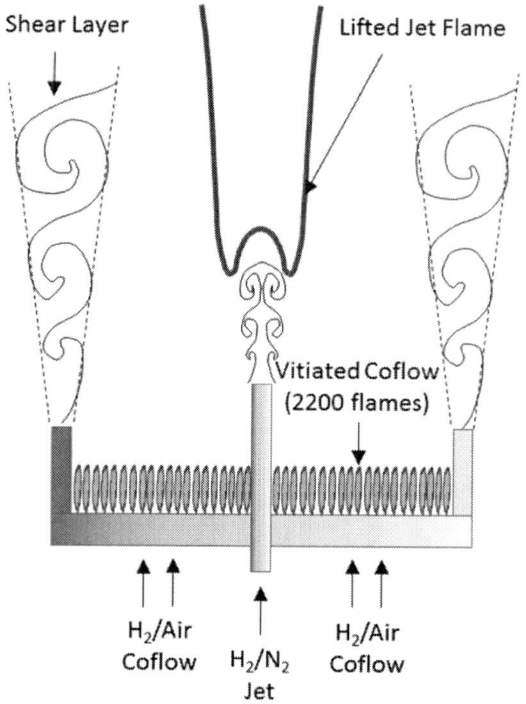

Fig. 1. The investigated flame [8].

This consists of a central H_2/N_2 turbulent jet with a coaxial flow of hot combustion products from a lean premixed H_2/Air flame. The central jet exit diameter is d=4.57 mm. The coflow is stabilized on a perforated disk with 87% blockage and an outer diameter of D=210 mm. The central jet extends 70 mm above the surface of the perforated disk. The boundary conditions are summarized in Table 4 (X_{N2}=1-X_{H2}-X_{O2}-H_{H2O}).

Table 4. Inlet velocity (V), temperature (T), mole fract. (X_i).

	Central Jet	Coflow
V (m/s)	107	3.5
T (K)	305	1045
X_{H2}	0.2537	0
X_{O2}	0	0.1474
X_{H2O}	0	0.0989

4 Results

The grids are generated using a combination of rectangular/hexahedral and triangular/tetrahedral cells with a fine discretization near the central jet, and decreasing resolution towards outer parts.

Grid independence is ensured by detailed grid studies that are not discussed here in detail. For all models with RANS turbulence modelling (Table 3), a 2D-axisymmetric grid is used, utilizing the axisymmetry of the time-averaged fields. For LES turbulence model (GRI-LFM, Table 3), a 3D grid is used. The 2D ($2.2 \cdot 10^4$ cells) and 3D ($1.1 \cdot 10^6$ cells) are displayed in Figure 2.

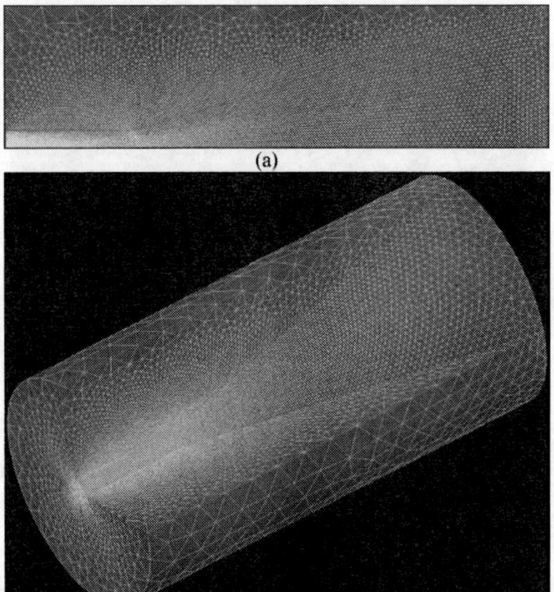

(b)

Fig. 2. Grids, (a) 2D grid, (b) 3D grid.

Predicted fields of temperature (in K), Z, C by LES-LFM (Table 3), in a longitudinal plane through the middle of the jet flame at an instant of time are presented in Figure 3. Development of vortex structures in the edges of the jet, as well as lifting effects can be recognized.

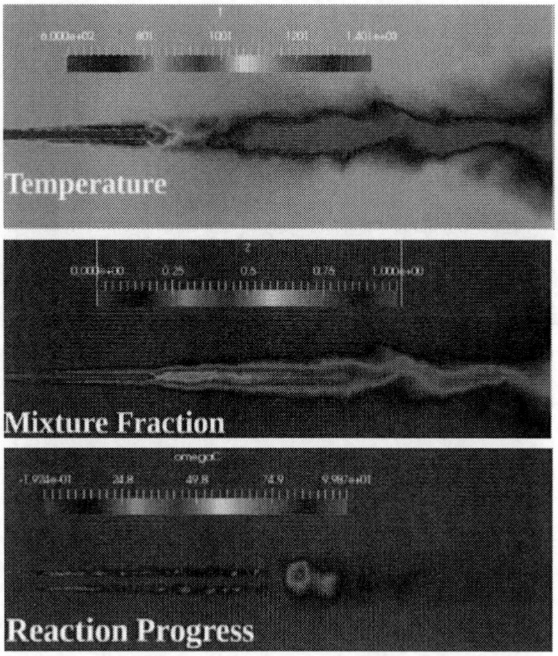

Fig. 3. Predicted fields of temperature (in K), mixture fraction and reaction progress variable in a longitudinal plane through middle of jet flame at an instant of time by LES-LFM (Table 3).

Predicted temperature field by C-EDC (Table 3) is displayed in Figure 4. The lifting of the flame can be seen by inspecting the temperature field.

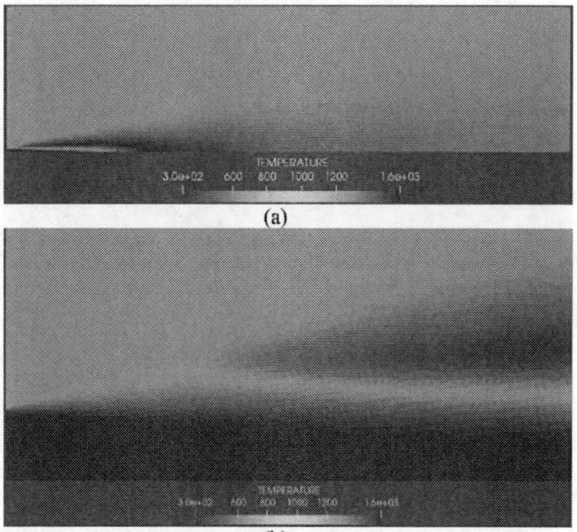

(b)

Fig. 4. Predicted temperature field (in K) by C-EDC (Tab. 3) (a) complete domain, (b) detail near central jet.

Predicted radial temperature profiles at x/d=1 (x: axial distance measured from nozzle outlet) are compared with measurements in Figure 5. M-C, M-TC, M-EDC, C-C, C-EDC and LES-LFM perform similarly and quite well.

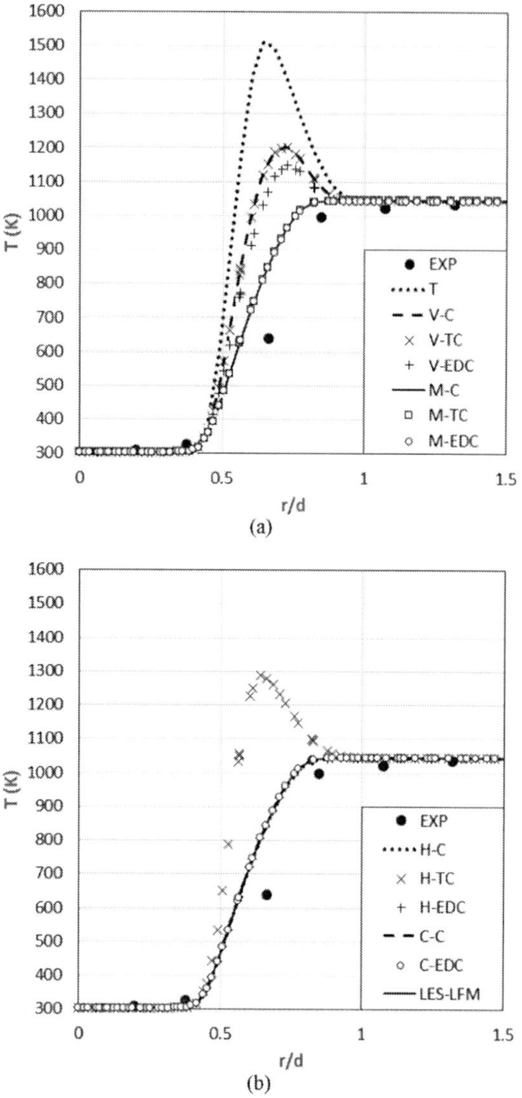

Fig. 5. Radial temperature profiles at z/d=1 (a) 1-Step reaction mechanism, (b) 2-Step & detailed reaction mechanism.

4 Conclusions

A numerical analysis of a turbulent lifted H_2/N_2 flame is presented. As combustion mechanisms, a large spectrum is considered including single-step and two-step, as well as detailed mechanisms. As combustion models, various models are considered that treat turbulence-chemistry interaction in different ways, including the EDC and the LFM. For turbulence modelling RANS and LES approaches are used. Results are compared with measurements. The mechanism of Marinov et al. [31] is observed to perform quite well, for its being a single-step mechanism. The Hsu and Jemcov mechanism [32] shows, an inferior performance compared to the mechanisms of Varma et al. [30] and Marinov et al. [31], although it is a two-step mechanism [32], whereas the former [31,32] are single-step ones.

References

1. A. C. Benim, D. Brillert, M. Cagan, *Investigation into the computational analysis of direct-transfer pre-swirl systems for gas turbine cooling*, ASME Paper No. GT2004-54151, in: Proc. ASME Turbo Expo 2004, **4**, pp. 453-460 (2004)
2. A. C. Benim, M. Geiger, S. Doehler, M. Schoenenberger, H. Roemer, *Modelling the flow in the exhaust hood of steam turbines under consideration of turbine-exhaust hood interaction*, (in: Proceed. 1st European. Conf. Turbomachinery - Fluid Dynamic and Thermodynamic Aspects: Computational. Methods, Erlangen, Germany, Book Series: VDI Ber., Vol.1185, pp.343-357, VDI Verlag, Duesseldorf, 1995)
3. D. G. Ebling, A. Krumm, B. Pfeiffelmann, J. Gottschald, J. Bruchmann, A. C. Benim, M. Adam, R. Labs, R. R. Herbertz, A. Stunz, Journal of Electronic Materials, **45**, 7, pp.3433-3439 (2016)
4. A. C. Benim, B. Epple, B. Krohmer, Progress in Comput. Fluid Dynamics, **5**, 6, pp.345-361 (2005)
5. J. P. Kim, U. Schnell, G. Scheffknecht, A. C. Benim, Progress in Computational Fluid Dynamics, **7**, 6, pp.337-346 (2007)
6. A. C. Benim, K. J. Syed, *Flashback Mechanisms in Lean Premixed Gas Turbine Combustion*, (Academic Press, Cambridge, MA, USA, 2015)
7. C. N. Markides, E. Mastorakos, E., Proceedings of the Combustion Institute, **30**, pp.883-891 (2005)
8. R. Cabra, T. Myhrvold, J. Y. Chen, R. W. Dibble, A. N. Karpetis, R. S. Barlow, Proc. of the Combustion Institute, **29**, pp.1881-1888 (2002)
9. W. P. Jones, S. Navarro-Martinez, Combustion and Flame, **150**, pp.170-187 (2007)
10. www.openfoam.com
11. A. C. Benim, Computer Methods in Applied Mechanics and Engineering, **67**, 1, pp.1-14 (1988)
12. P. A. Durbin, B. A. P. Reif, *Statistical Theory and Modeling for Turbulent Flows*, 2nd ed. (Wiley, New Jersey, USA, 2010)
13. P. Sagaut, *Large Eddy Simulation for Incompressible Flows*, 3rd ed. (Springer, Berlin, 2005)
14. F. Menter T. Esch. S. Kubacki, "Transition modelling based on local variables", in: *Engineering Turbulence Modeling and Experiments 5*, W.Rodi, N. Fueye (Eds.) (Elsevier, Amsterdam, 2002) pp.555-564
15. A. C. Benim, M. Cagan, D. Gunes, International Journal of Thermal Sciences, **43**, 8, pp.725-732 (2004)
16. A. C. Benim, K. Ozkan, M. Cagan, D. Gunes, Int. J. Num. Methods for Heat & Fluid Flow, **17**, 3, pp.284-301 (2007)
17. H. Chattopadhyay, A.C. Benim, Journal of Heat Transfer – Transactions of the ASME, **133**, 10, Article Nr.: 104502, 5 pages (2011)

18. A. Assmann, A. C. Benim. F. Gül, P. Lux, P. Akhyari, U. Boeken, F. Joos, P. Feindt, A. Journal of Biomechanics, **45,** 1, pp.156-163 (2012)
19. P. Oclon, S. Lopata, M. Nowak, A. C. Benim, Progress in Comput. Fluid Dynamics, **15,** 5, pp.290-306 (2015)
20. A. C. Benim, A. Nahavandi, K. J. Syed, Progress in Computational Fluid Dynamics – An International Journal, **5,** 8, pp.444-454 (2005)
21. A. C. Benim. M. P. Escudier. A. Nahavandi, A. K. Nickson, K. J. Syed, F. Joos, International Journal of Numerical Methods for Heat & Fluid Flow, **20,** 3, pp.348-370 (2010)
22. A. C. Benim, H. Chattopadhyay, A. Nahavandi, International Journal of Thermal Sciences, **50,** 10, pp. 1973-1983 (2011)
23. Turns, S. R., *An Introduction to Combustion*, 3rd ed. (McGrawHill, NY, USA, 2012)
24. B. F. Magnussen, B. H. Hjertager, *On mathematical models of turbulent combustion with special emphasis on soot formation and combustio*", Proc.16th Symp. (Int.) Combustion (The Combust. Inst., Pittsburgh, PA, 1976) pp.719-729
25. B. F. Magnussen, *On the structure of turbulence and generalized eddy dissipation concept for chemical reactions on turbulent flow* (Proc. 19th AIAA Aerospace Meeting, January 12-15, 1981, St. Louis, USA, pp.1-6 (1981) doi: 10.2514/6.1981-42)
26. N. Peters, *Turbulent Combustion*, (Cambridge University Press, Cambridge, UK, 2002)
27. C. D. Pierce, P. Moin, J. Fluid Mech., **504**, pp.73–97 (2004)
28. T. Poinsot, D. Veynante, *Theoretical and Numerical Combustion*, 2nd ed. (R. T. Edwards Inc., Philadelphia, PA, USA, 2005)
29. H. Pitsch, *A C++ computer program for 0-D and 1-D laminar flamelet calculations* (Technical Report, RWTH Aachen, Germany, 1998)
30. A. K. Varma, A. U. Chatwani, F. V. Bracco, Combustion and Flame, **64**, pp.233-236 (1986)
31. N. M. Marinov, C. K. Westbrook, W. J. Pitz, *Detailed and global chemical kinetics model for hydroge*" (Report UCRL-JC-120677, Lawrence Livermore National Lab., California, USA, 1995)
32. K. Hsu, A. Jemcov, AIAA Paper 2000-2478 (2000)
33. M. O. Conaire, H. J. Curran, J. M. Simmie, W. J. Pitz, C. K. Westbrook, International Journal of Chemical Kinetics, **36**, pp.603-622 (2004)
34. http://combustion.berkeley.edu/gri-mech/

Experimental and numerical investigation on the impact of vacuum level on effective thermal conductivity and microstructure of carbon fibre reinforced epoxy composites manufactured by vacuum bag method

Michał Kubiś[1,*], *Mirosław* Seredyński[1], *Łukasz* Cieślikiewicz[1], *Tomasz* Wiśniewski[1], and *Anna* Boczkowska[2]

[1] Institute of Heat Engineering, Faculty of Power and Aeronautical Engineering, Warsaw University of Technology, 21/25 Nowowiejska Str., 00-665 Warsaw, Poland
[2]Faculty of Materials Science and Engineering, Warsaw University of Technology,
02-507 Warsaw, Poland

Abstract. The physical properties of epoxy based carbon reinforced composites are highly anisotropic due to their directional structure and dependent on the manufacturing process parameters. Thermal conductivity was found to be dependent on the void volume fractions, which appear as a result of the insufficient vacuum level. In the proposed paper the multi-scale computational model of heat transfer across the carbon fiber-epoxy resin composite is proposed. The meso-scale effective thermal conductivities are determined with analytical formulae for isotropic and anisotropic media, the latter takes into account thermal resistance at the interface of fibres and epoxy resin. Proposed model is utilized to determine the effective thermal conductivity in the direction perpendicular to plies of composite. The influence of void fractions and the thickness of the composite on the effective thermal conductivity is investigated. The numerical outcomes underestimate the real variation in conductivity, which can be caused by change in carbon volume fraction of samples manufactured at different vacuum levels which was not considered in numerical computation.

1 Introduction

Epoxy based carbon reinforced composites, especially those with long fibers are well known for their high strength to density ratio as well as strong anisotropy of their properties both mechanical and physical [1, 2]. It is a result of layered structure and fibers alignment e.g. unidirectional, braided or fabric. On the other hand, microstructure quality is the second factor that is listed to have great impact on effective thermal conductivity. Void volume fraction and fiber volume fraction are two parameters that in minority affects overall properties of composites [3]. They could vary amongst different manufacturing methods and be altered by manufacturing process parameters e.g. vacuum level in the case of methods based on a vacuum bag.

Effective thermal conductivity of composites became significant factor as their structures became designing to work in environment of higher temperature gradients e.g. engine casing, battery casing, composite tooling [4] as well as wing shields in aeroplane which surface heats up to more than 100°C during the flight. According to complex alignment of fibers in carbon fabrics braids infiltrated with epoxy resin their effective thermal conductivity are not obvious to predict.

Classical theoretical models: effective medium approximation methods [5, 6, 7, 9, 10] or based on micromechanics [11, 12, 13, 14] are well known and very simple, and quick to apply. Nevertheless, their scope of application is limited to simplified cases due to composites parameters e.g. particle arrangement, their shape, restricted volume fractions or considering of interfacial contact resistance.

Effective thermal conductivity could be computed also based on simulation e.g. FEM, FDM, Element Free Galerkin, Lattice Boltzman Method, off-lattice Monte Carlo as well as Molecular Dynamics [15]. The great advantage of simulation approach is their ability to accurate characterization of details of composite microstructure at different scales: micro-scale, meso-scale and macro-scale. However, simulation of multiscaled microstructure requires to determine a large representative elementary volume to evaluate effective thermal conductivity accurately. Hence great computing power demand is the bottleneck of these methods especially if dimensions of microstructure elements vary in orders of magnitude. Wherefore multiscale modeling based of

* Corresponding author: mkubis@itc.pw.edu.pl

integrating of above-mentioned methods could provide efficient solution for that inconvenience.

As the input to the model several micro-scale properties are needed. Among them are thermal conductivity of a single carbon fiber in parallel and radial directions, and thermal conductance at the interface of fibre and matrix material. In both cases literature review indicates that there is a wide range of values reported by many authors. Thermal conductivity in parallel direction can be as high as 4.4-75 W/(m·K) for PAN-based carbon fibres [16, 17, 18] and 0.5-25 W/(m·K) in radial direction [19, 20].

Values of thermal conductance showed in literature, or its reciprocial, the interface resistance, R_k, cover wide range. Hua et al. [20] coupled the LFA experimental analysis and simplified numerical simulations to predict effective thermal conductivity along fibers and interface resistance. The procedure of tuning of both parameters was based on finding the best correlation between experimental and numerically predicted signal curve. The considered thermal resistances were in the range 10^{-6} and 10^{-4} m^2K/W. The best fitted value determined was 10^{-5} m^2K/W.

Bui et al. [21] calculated the effective thermal conductivity of the carbon nanotube-polystyrene composite. They estimated the interfacial resistance by tuning the results given with Molecular Dynamics simulations and experimental findings. The calculated resistance between carbon nanotube and polymer was $2.21 \cdot 10^{-8}$ m^2K/W.

The objective of this paper was to investigate the impact of manufacturing parameter on composite structure. Therefore three types of composites were manufactured with vacuum assisted resin infusion method at three different levels of vacuum. Then their physical properties were measured: thermal diffusivity a (LFA 447 Nanoflash, Netzsch) and specific heat c_p (DSC 404 Pegasus, Netzsch). Density ρ was calculated based on geometry and mass measurement. Thermal conductivity in through the thickness direction was determined according to formula: $k=a \cdot c_p \cdot \rho$. Microstructures of composites were investigated with scanning electron microscopy. Quantitive 2D image analysis technique was utilized to determine void volume fractions (Micrometer program [22, 23]). Based on the same images, geometry of the composite structure was developed and model of medium approximation integrated with numerical simulation were applied for effective thermal conductivity calculation. Both results were compared and discussed.

2 Structure of the composite

Six layers of plain fabric (200g/m^2) made of carbon fibers were infiltrated with Axson epoxy system consist of a resin Epolam 2031 and a hardener Epolam 2031. SEM image shows exemplary cross section of composite where yarns/tows are interlaced and positioned perpendicular to each other and denoted as numbers 1 and 2 (Fig.1). Single carbon fibers within yarns/tows are the brithest elements of image infiltrated with resin that is pictured herein as a darker shades of gray and the black areas are voids.

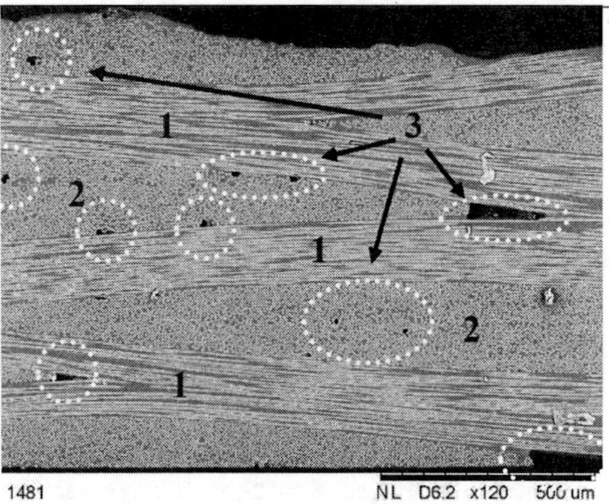

Fig. 1. Exemplary SEM image of composite microstructure. 1-fiber tows transversal to normal of crossection, 2-fibre tows parallel to normal of crosssection, 3-voids.

Three different types of samples were made at different vacuum levels applied during manufacturing process. Value of vacuum levels presented in table 1 referred to atmospheric pressure. Detailed observations of bunches of carbon fibers, done with a scanning electron microscope in the cross-sections normal and parallel to fibers, revealed very good penetration of bunches with epoxy resin. Although small amount of voids appeared as a result of entrapped air during infiltration process. Each type of manufactured samples was investigated.

Experimental results of effective thermal conductivity in through the thickness direction were shown in table 1, as well as void volume fraction and manufacturing parameters.

Table 1. Experimental **results** of effective thermal conductivity and void volume fraction at various manufacturing conditions.

Vacuum level below atmospheric pressure [kPa]	Volume fraction of voids [%]	Effective thermal conductivity [W/(m·K)]
-90	1,6	0.54
-60	2,5	0.50
-30	3,5	0.45

3 Analysis and modelling

3.1 Geometry

Based on detailed measurements of the composite geometry, its threedimensional model is developed. The composite has the multi-scale structure. At the meso-scale the discernible components are yarn, warp and the matrix, epoxy resin. At the micro-scale, both the yarn and warp are bunches of parallel carbon fibres infiltrated with the epoxy resin. For the sake of analysis simplification both yarn and warp are treated as continuous with the same structure and properties..

The simplified geometry of warp and yarn (Fig. 2) reflects the real one observed with the scanning microscope (Fig. 1). The computational domain is the cuboid which dimensions are L_x, L_y and L_z, measured in directions along X, Y and Z axes of the Cartesian coordinate system, respectively. At the plane perpendicular to fibers the shape of cross-section of the bunch is an ellipse with diameters d and D, the former is measured in the Z direction and the latter in the X or Y direction, dependent on the configuration (Fig. 2b). Measured diameters *d* and *D* were equal to 0.2 mm and 2.373 mm. Volume representing the single bunch is formed by swapping the ellipse along a path having the shape of the cosine function, which period is equal to doubled length of the domain, $2L_x$ or $2L_y$, dependent on its orientation. The amplitude is equal to 0.22 mm. For considered case the half distance of the yarn/warp structure period, namely L_x or L_y, is equal to 2.5 mm. The total height of the composite, L_z, is equal to 0.6mm. Three layers of weaves are taken into account (Fig. 2). The distance between the horizontal axes of two corresponding layers of bunches is equal to L_z = 0.6 mm. The intermediate layer is positioned in the midway between them and it is additionally shifted in the X and Y directions by the half of the cavity width, namely $L_x/2$ or $L_y/2$.

The volumetric fraction of all bunches in the whole domain is 94%.

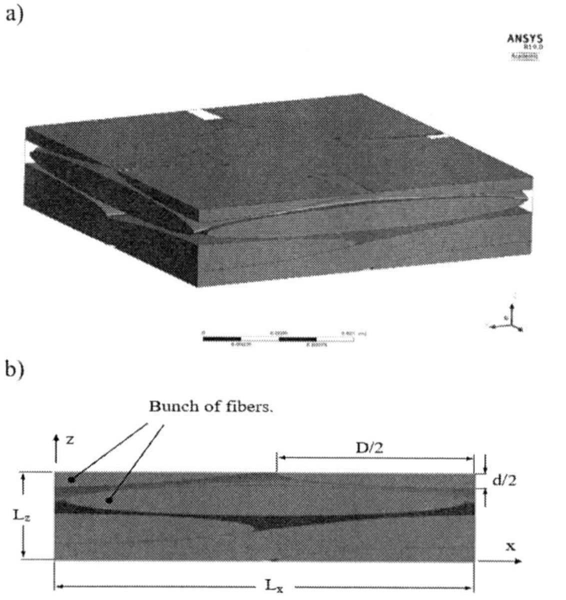

Fig. 2. Isometric view of the considered part of the composite (a) and the side view with dimensions of the macro-structures (b). Bunches of fibers are denoted with pale blue color and the filler, not marked here, fills the rest of the box.

3.2 Mathematical model

In the domain presented in Fig. 1, the stationary heat conduction equation is solved

$$\nabla(k\nabla T) = 0 \tag{1}$$

where the thermal conductivity, *k*, is evaluated in each mesoscopic zone (yarn, warp or epoxy resin) of the domain. Due to the non-uniform structure of all zones, their effective mesoscopic thermal conductivities are calculated with the averaging formula. The thermal resistance at the interface of various materials, i.e. fibers and epoxy, is taken into account. The matrix material, namely epoxy resin, has also nonuniform structure, due to presence of entrapped bubbles of air. The averaged thermal conductivity of the matrix material is determined on the assumption, the entrapped gas is uniformly distributed in the epoxy resin, present outside and inside bunches of fibers.

The effective thermal conductivity of the matrix material, km, which is considered as the uniform mixture of the entrapped gas (voids) and epoxy resin, is constant and isotropic. It is derived with the Maxwell model [24],

$$k_m = k_e\left[\frac{k_v+2k_e+2\phi_v(k_v-k_e)}{k_v+2k_e-\phi_v(k_v-k_e)}\right] \tag{2}$$

as the function of thermal conductivities of epoxy, k_e, and voids, k_v, and volume fraction of voids, ϕ_v.

The effective thermal conductivity of bunch of carbon fibers filled with epoxy is highly anisotropic and can be expressed as the diagonal tensor according to its main directions, parallel or perpendicular to fibers. For the former orientation the average thermal conductivity is given with the mixture law

$$k_{beff}^p = \phi_f k_f^p + (1-\phi_f)k_m \tag{3}$$

where ϕ_f is the volume fraction of fibers in the yarn or warp. Based on the analysis of photos of the microstructure it is estimated to be equal to 0.55, so the volume fraction of carbon fibers in the whole composite is 51.7%. The thermal conductivity of the matrix material, k_m, takes into account presence of voids in between carbon fibers (eq. 2).

The effective thermal conductivity of the composite, where the thermal resistance, related to imperfect contact of particles and matrix material or the Kapitza resistance [25], should not be neglected, the Hasselmann-Johnson relation can be used. This was originally derived for spherical inclusions and next was generalized for cases involving sets of parallel fibers immersed in the matrix material [7],

$$k_{beff}^n = k_m\left[\frac{(\alpha_1-\alpha_2-1)\phi_f+(1+\alpha_1+\alpha_2)}{(1-\alpha_1+\alpha_2)\phi_f+(1+\alpha_1+\alpha_2)}\right] \tag{2}$$

where

$$\alpha_1 = \frac{k_f^n}{k_m} \; and \; \alpha_2 = \frac{k_f^n}{ah_C}.$$

where a is the radius of fiber and hc is the effective thermal conductance, reciprocial of the thermal resistance, R_k. Based on the analysis of microstructure of the composite crossection perpendicular to warp, the average radius of single carbon fiber was estimated to be 3.5 µm.

Both yarn and warp are interlaced and crimped, so fibers are not perfectly straight, parallel to axes of the Cartesian coordinate system. For general case the components of the thermal conductivity tensor must be projected orthogonally on the directions of the fixed Cartesian coordinate system, related to orientation of the composite. To fix the idea the bunch of fibers shown in Fig. 2b, parallel to X axis is considered. The components of thermal conductivity in the global coordinate system are functions of squared sine and cosine functions of angle of fibers inclination, $\theta(x)$, with respect to the X axis [26]. For considered geometry and arrangement of bunches, the maximum value for θ is estimated to be approximately 6.28°, so variation in the thermal conductivity with changes of the inclination angle of bunch of fibers can be neglected in this preliminary study, and the diagonal components of the tensor of conductivity are simplified to the form

$$k_{b,x} \approx k^{p}_{b,eff}$$

$$k_{b,y} \approx k^{n}_{b,eff}$$

$$k_{b,z} \approx k^{n}_{b,eff} \tag{3}$$

The thermal conductivities of all substances are listed in the table 2.

Table 2. Material properties of the components of the composite.

	Units	Value
Thermal conductivity of epoxy resin, k_e	W/(m·K)	0.19 [26]
Thermal conductivity of voids, k_v	W/(m·K)	0.01
Thermal conductivity of the fiber, parallel to axis, k_f	W/(m·K)	30 [16]
Thermal conductivity of the fiber, perpendicular to axis, k_f	W/(m·K)	25 [27]

3.3 Mesh

The domain presented in Fig. 2a was spatially discretized with control volume grid refined close to curved surfaces. The maximum size of the control volume (c.v.) face is 0.02 mm, total number of control volumes is 2.31 mln. The general quality of mesh was good, but due to geometry constraints, some highly skewed control volumes were created close to edges common for bunches and matrix material. Views of the meshed cross-section of the composite, meshed surface of bunches are shown in Fig. 3.

a)

b)

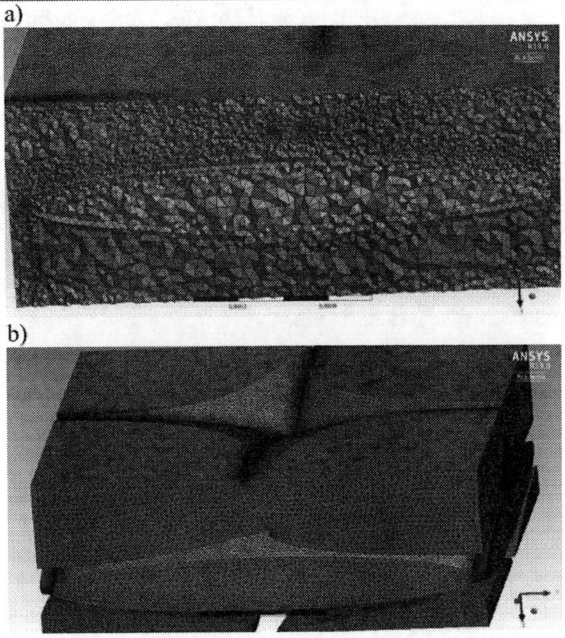

Fig. 3. View of the meshed cross-section of the domain (a) and meshed surface of bunches (b).

The isothermal conditions were imposed at two opposite external surfaces *Z=const*, namely temperature 300K on the Zmin and 301K on Z_{max}. At other external surfaces the adiabatic condition was imposed. The stationary heat conduction problem was solved with the ANSYS Fluent software. Gradients of temperature were derived with the Least Squares Cell Based scheme and warped-face gradient correction was utilized to avoid nonphysical oscillations. The underrelaxation coefficient was set to 0.8.

4 Results and discission

Five samples of each type of composites were prepared for measurement of their physical properties. Hence experimental values of effective thermal conductivity showed in table 1 were determined as an average results of five samples. The values vary between 0.45 - 0.54 [W/(m·K]) along with decreasing vacuum level below atmospheric (-30 kPa to -90 kPa). Void volume fractions were obtained according to quantitative image analysis technique on the basis of analysis of 30 SEM images of 5 different samples for each composite type.

At the first step of setting simulation conditions, the thermal resistance at the surface of carbon fibers was approximated. This was estimated for the volume fraction of entrapped gas in the epoxy resin, equal to 1.6%. The calculated with the Maxwell formula (eq. 2) effective thermal conductivity of epoxy resin with entrapped gas is 0.186 W/(m·K).

The thermal conductivity of the bunch, predicted in the normal direction to fibers with the H-J formula (eg. 4), as a function of the radial thermal conductivity, k^{n}_{f} and interfacial

https://doi.org/10.1051/matecconf/201824001015

thermal resistance, R_k, is presented in Fig. 4. It is evident that the interfacial resistance has considerable impact on the effective thermal conductivity. The red line represents the value of the experimentally determined through-thickness thermal conductivity of the composite. The total effective conductivity of the composite is an average value of conductivities of the epoxy resin, km, and bunch of fibers, in normal direction to fibers, $k_{b,eff}^n$. Since k_m is lower than $k_{b,eff}^n$, the latter must be higher than the measured thermal conductivity, so allowable values of the $k_{b,eff}^n$ there are on the left side of the red line in Fig. 4.

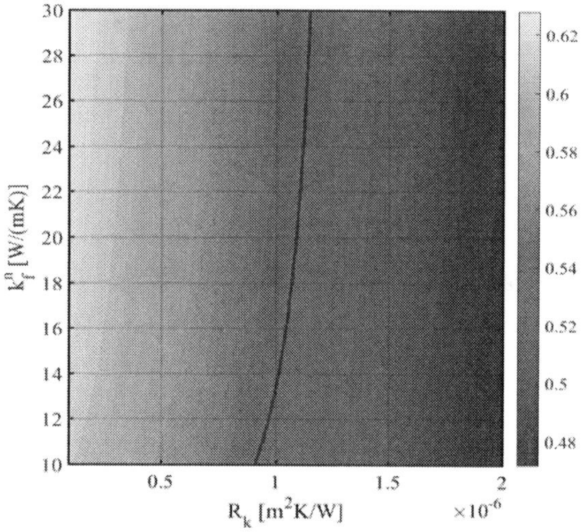

Fig. 4. Effective thermal conductivity of the bunch of fibers immersed in epoxy resin, perpendicular to fibers, predicted with the H-J model (eq. 4). The isoline of conductivity equal to 0.54 W/(m·K) is shown with red line.

A series of simulations was performed for the set of thermal interfacial resistances, in the range 10-8 to 10-6 m^{2}K/W and constant radial thermal conductivity of fibers, k_f^n, equal to 25 W/(m·K). As a result, the heat flux density, q, was determined on two opposite surfaces, Zmin and Zmax, where temperature boundary conditions were imposed. The effective thermal conductivities were determined with the formula:

$$k_{z,eff} = q\frac{L_z}{\Delta T} \qquad (4)$$

where ΔT is the imposed temperature difference, equal to 1K. Calculated through-thickness effective conductivities, as functions of the interfacial resistance, as well as thermal conductivities of bunches, $k_{b,eff}^n$, are plotted in Fig. 5. The best fitted resistance, for assumed parameters, is close to 6·10^{-6} m^{2}K/W (Fig. 5).

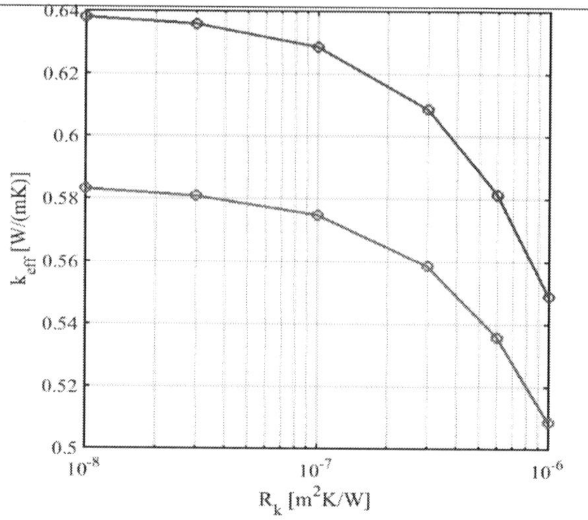

Fig. 5. Effective thermal conductivity of bunch of fibers infiltrated with epoxy (blue line) and predicted numerically conductivity of the composite in the Z direction (red line) as functions of thermal resistance.

Temperature distributions in the two crossections of the domain, perpendicular to the Y axis, crossing the half and the one-fourth of the domain width and on two side surfaces are presented in Fig. 6. The best fitted thermal resistance was adopted here. The temperature distribution in the cross-sections is only slightly disturbed with the presence of epoxy resin, due to high fractions of bunches and immediate contact of bunches.

The impact of the void fraction of entrapped gas in the epoxy resin, present outside and inside bunches, on the overall effective thermal conductivity of the composite in direction normal to tows is next examined (Fig. 7).

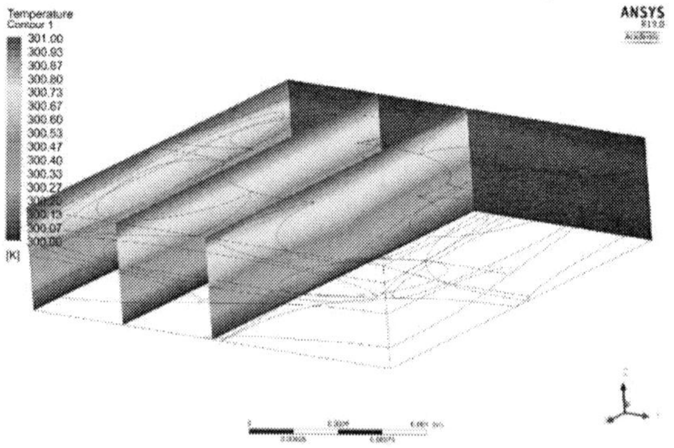

Fig. 6. Temperature map on two cross-sections of the domain and on two side walls.

Both numerical and experimental outcomes show the conductivity decreases with increasing volume fraction of voids. Numerical results underestimate the changes of conductivity. This can be induced by a change of carbon volume fraction as a result of higher compression forces introduced to the composite surface resulted from higher vacuum applied during manufacturing process. This was resulted in change of carbon fibre to matrix ratio which wasn't taken into consideration at numerical computations. However REV geometry was based on the microstructure of composite manufactured at the lower vacuum level (90 kPa below atmospheric preassure), which possessed the lower void volume fraction. Hence discrepancy of results in that case is minor.

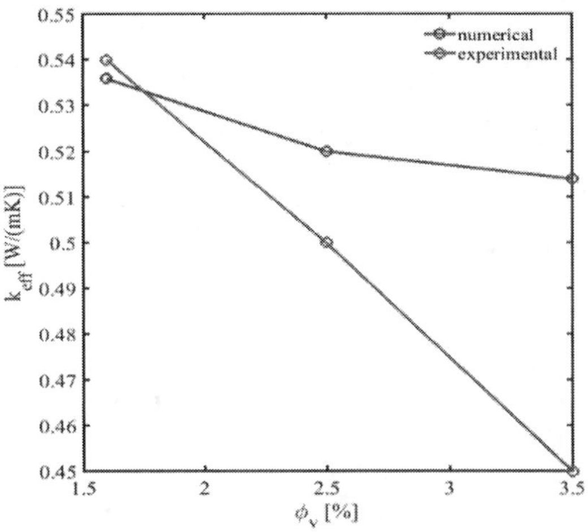

Fig. 7. Comparison of effective thermal conductivity in direction normal to yarns determined experimentally and predicted with numerical simulation.

5 Conclusions

In the proposed paper the computational model of heat transfer across the complex carbon fibre-epoxy composite, is proposed. The three dimensional geometry reflects the meso-structure of the composite components. The non-uniform micro-structure of warps, yarns and matrix material including anisotropic thermal conductivity of fibers, voids present in epoxy resin, and interfacial resistances on the surfaces of fibres is taken into account. The effective meso-scale thermal conductivities are determined with analitical formulae and next applied in the numerical meso-scale model. The interfacial resistance is estimated on the basis of simulations results, which assures the best fitting of the calculated through-thickness effective thermal conductivity with the determined experimentally one. For that determined resistance, the effect of void fractions in the epoxy resin on overall through-thickness thermal conductivity is investigated numerically. Numerical simulations underestimate this relation. However this could be caused by slight difference between modelled microstructure and the real one. It is exept composite manufactured at the lower vacuum level which, herein, was a benchmark for REV geometry development. Hence both experimentaly and computational results in this case are close. This slight alteration should be take into consideration in further works.

References

1. A. Boczkowska and G. Krzesiński, *Kompozyty i techniki ich wytwarzania*, OWPW (2016).
2. S.H. Goodman, H. Dodiuk-Kenig, *Handbook of thermosets plastics*, Wiliam Andrew (2014).
3. SAE International, *Composites material handbook Vol. 2, CMH-17* (2016).
4. P. Czarnocki, A. Boczkowska, W. Frączek, P. Chabera, et al., , SAE Int. J. Mater. Manf., vol. **11**, 2, (2018).
5. J.C. Maxwell-Garnett, Philos. Trans. Roy. Soc. London Ser. A, vol. **203**, 385 (1904).
6. H. Fricke, J.Phys. Chem., vol. **57**, 9, 934 (1953).
7. D.P.H. Hasselman, L.F. Johnson, J. Compos. Mater., vol. **21** 6, 508 (1987).
8. D. Burggeman, Ann. Phys., vol. **24**, 7, 636 (1935).
9. A.G. Every, Y. Tzou, D.P.H. Hasselman, R. Raj, Acta Metall. Mater., vol. **40,** 1, 123–129 (1992).
10. R. Landauer, J. Appl. Phys., vol. **23** 7, 779 (1952).
11. Z. Hashin, S. Shtrikman, J. Appl. Phys., vol. **33**, 10, 3125 (1962).
12. T. Mori, K. Tanaka, Acta Metall., vol. **21** ,5, 571 (1973).
13. Y. Benveniste, J. Appl. Phys., vol. **61**, 8, 2840 (1987).
14. Y. Agari, M. Tanaka, S. Nagai, T. Uno, J. Appl. Polym. Sci., vol. **34** , 4, 1429 (1987).
15. S. Zhai, P. Zhang, Y. Xian, J. Zeng, B. Shi, Int. J. Heat Mass Transf., vol. **117**, 358 (2018).
16. Z. Han, A. Fina, Prog. Polym. Sci., vol. **36,** 914 (2011).
17. J. X. Wei, "*Report: Carbon fiber thermal conductivity measurement and analysis*", Beijing Institute Mater. Sci. Technol. Div. (1989)
18. I. Hatta, T. Yamane, S. Katayama, M. Todoki, J. Wide Bandgap Mater., vol. **7**, 294 (2000).
19. S. W. Beckwith, *Composites reinforcement fibers : III - The carbon and graphite families (Part B)*, in SAMPE 49th International Symposium and Exhibition, vol. **47**, 44 (2011).
20. N. Hu, S. W. Chiang, J. Yi, X. Li, J. Li, H. Du, C. Xu, Y. He, B. Li, F. Kang, Compos. Sci. Technol., vol. **110**, 69 (2015).
21. K. Bui, B.P. Grady, D.V. Papavassiliou, Chem. Phys. Lett., vol. **508**, 248 (2011).
22. T. Wejrzanowski, M. Lewandowska, K.J. Kurzydłowski, Image Anal.Stereol., vol. **29**, 1 (2010).
23. T. Wejrzanowski, W.L. Spychalski, K. Rożniatowski, K.J. Kurzydłowski, Int. J. Appl. Math. Comput. Sci., vol. **18**, 33 (2008).
24. J.C. Maxwell, *A Treatise on Electricity and Magnetism. 3rd ed., chapter 9*, 1, New York, Dover;(1954).
25. K. Pietrak, T.S. Wiśniewski,, Journal of Power Technologies, vol. **95**,14 (2015).
26. A. Dasgupta, R.K. Agarwal, S.M. Bhandarkar, Compos. Sci. Technol., vol. **56**, 209 (1996).
27. Y.S. Song, J.R. Youn, Carbon, vol. **44**, 710 (2006)

Turbulence models impact on the flow and thermal analyses of jet impingement

Tomasz Kura[1,*], *Elżbieta* Fornalik-Wajs[1], *Jan* Wajs[2] and *Sasa* Kenjeres[3]

[1]AGH University of Science and Technology, Faculty of Energy and Fuels,
Department of Fundamental Research in Energy Engineering, al. Mickiewicza 30, 30-059 Krakow, Poland
[2]Gdansk University of Technology, Faculty of Mechanical Engineering,
Department of Energy and Industrial Apparatus, ul. Narutowicza 11/12, 80-233 Gdansk, Poland
[3]Delft University of Technology, Faculty of Applied Sciences,
Department of Chemical Engineering, van der Maasweg 9, 2629 HZ, Delft, The Netherlands

Abstract. Accurate numerical reconstruction of heat and mass transfer processes in particular applications, such a jet impingement, is difficult to obtain even with the use of modern computational methods. In the proposed paper, the flow and thermal phenomena occurring during single minijet impingement on the flat, concave and convex, heated surfaces were considered. Problem of impingement on non-flat surface, still not common and purely described in the literature, can be of big importance in engineering applications, such as the heat exchangers. Numerical analyses, based on the mass, momentum and energy conservation laws, were conducted with the OpenFOAM software. Focus was placed on the proper model construction, in which turbulence and boundary layer modelling was crucial, due to their significance in the heat transfer processes. Analysis of results obtained by RANS models focused mostly on the comparison of turbulent and hydrodynamics parameters.

1 Introduction

Modern world and its technical development is connected strictly with the trials to improve the ways of energy resources utilization. One of the trends is to apply methods known from one branch of engineering in other. An example can be a jet impingement. It was being used in the cooling systems of electronics or in some metallurgical processes. Recently, however, it was proposed in the novel construction of cylindrical heat exchanger [1], which resulted in very promising values of transferred heat rates and lower than in other similar devices hydraulic resistance [2-3].

In abovementioned device, about 1000 orifices were generating minijets with the core diameter of ~1 mm. Due to complexity of the system, experimental investigation of it can be very difficult to perform, especially, when the detailed flow phenomena in the small scale is the main goal of research. Until now only general data was available, then, concerning macroscale parameters, such as already mentioned pressure losses or overall heat transfer efficiency. Understanding of the phenomena occurring in the device requires more detailed analysis, concerning also microscale.

For such cases numerical methods are very helpful. However, even now they demonstrate lack of accuracy in some scientific problems. Jet impingement unfortunately is one of the examples. According to Zuckerman [4] proper predicting of flow and thermal parameters in the case of jet impingement and its numerical modelling depends significantly on the type of turbulence model chosen for the simulation. In his paper it can be found, that the best results for RANS simulations can be obtained with v^2-f model, which is the four-equation, enhanced k-ε type, model. While hydrodynamics of the phenomena is predicted with decent level of accuracy by RANS methods, heat transfer causes problems. Hadžiabdić [5] in his doctoral thesis confirmed this statement, moreover he concluded, that even LES methods exhibited lack of accuracy when used to predict heat transfer in jet impingement.

Before analyzing complex systems, it is essential to identify the advantages and disadvantages of possible numerical methods to be applied. The following paper regards the analysis of single minijet that impinge various surfaces – flat, convex and concave. Their geometry was chosen in the basis of previous research [6], as well as data from [1-3]. While the available information for the flat case is generally broad, non-flat impingement is still not a popular and widely described scientific topic. Correlations do not exist, which can be used in the problems regarded in [1-3]. The goal of paper was to identify the turbulence model impact on the results and describe the differences between impingement on various surfaces – since such knowledge would be essential to correctly analyze arrays of minijets, existing for example in heat exchangers [1-3].

* Corresponding author: kura@agh.edu.pl

https://doi.org/10.1051/matecconf/201824001016

2 Mathematical model and geometry

Steady-state, single-phase, two-dimensional axisymmetric analyses were performed. High resolution, second order discretization schemes were applied to provide sufficient results. The conservation laws of mass (Equation 1), momentum (Equation 2) and energy (Equation 3) were applied, using Reynolds averaging approach. All variables marked with overline, such as $\overline{( )}$, would represent time-averaged value. They were coupled with various turbulence models, both high- (k-ε and v^2-f) and low-Reynolds (SST k-ω) types:

$$\frac{\partial \overline{u_i}}{\partial x_i} = 0, \quad (1)$$

$$\rho \frac{\partial}{\partial x_j}\left(\overline{u_j}\,\overline{u_i}\right) = -\frac{\partial \overline{p}}{\partial x_j} + \frac{\partial}{\partial x_j}\left(2\mu S_{ij} - \rho\overline{u_i' u_j'}\right), \quad (2)$$

$$\frac{\partial}{\partial x_j}\left(\overline{u_j}\left(\rho\overline{E} + \overline{p}\right)\right) = \frac{\partial}{\partial x_j}\left(\lambda_{ef}\frac{\partial \overline{T}}{\partial x_j}\right), \quad (3)$$

where u is the velocity, m/s; ρ is the density, kg/m^3; p is the pressure, Pa; μ is the dynamic viscosity, Pa·s; S_{ij} is the strain rate tensor, 1/s, $\overline{u_i' u_j'}$ is the Reynolds stress term, m^2/s^2; E is the total energy, J; T is the temperature, K; λ_{ef} is the effective thermal conductivity, W/(m·K).

Figures 1, 2 and 3 represent the characteristic dimensions of analyzed flat, concave and convex geometries, respectively, as well as location and type of boundary conditions. For better graphical representation, their scale varies between figures.

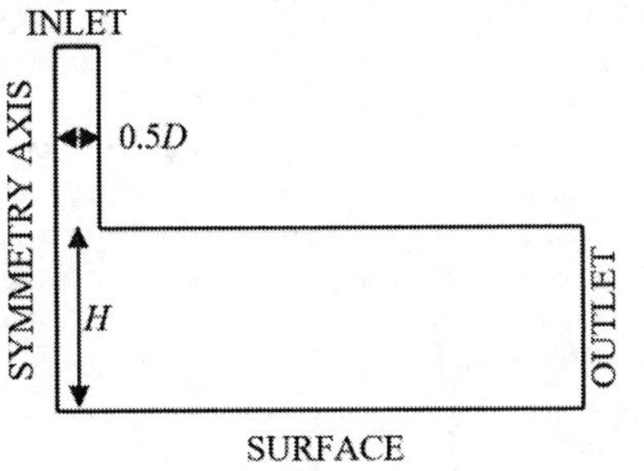

Fig. 1. Flat type geometry boundary conditions.

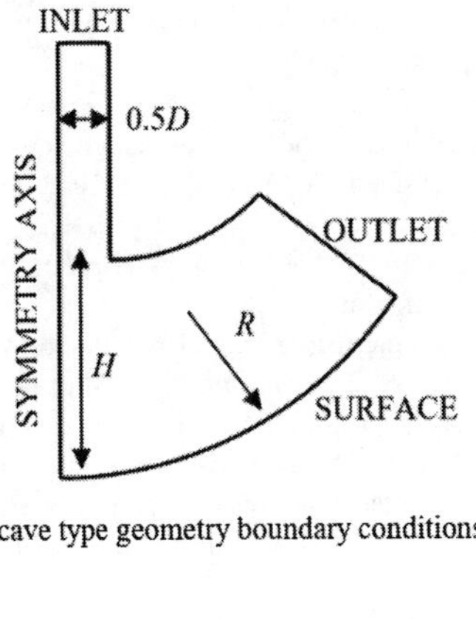

Fig. 2. Concave type geometry boundary conditions.

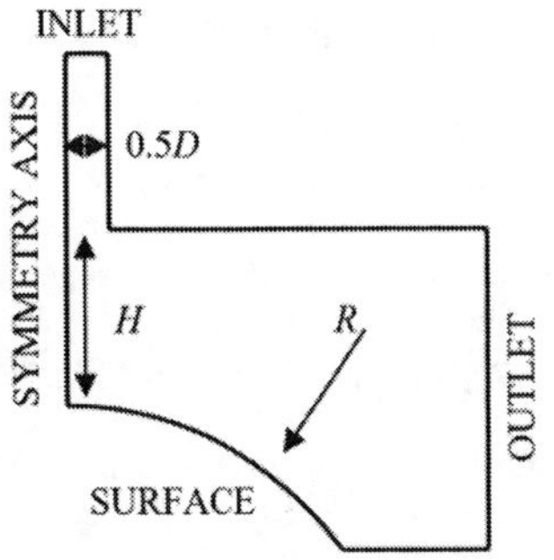

Fig. 3. Convex type geometry boundary conditions.

As it can be seen, all cases were axisymmetric. Moreover, values of H, which was the height between orifice exit and stagnation point, and R, which was the radius of curvature, were chosen to be multiplications of orifice diameter D. Multiplication factors used in the paper are listed in Table 1.

3 Flat surface, validation case

Numerical analyses of jet impingement could be compared for example with the ERCOFTAC benchmark cases, describing the experimental work by Cooper et al. [7] and Yan [8]. They represented the situation, in which air is impinging the flat surface, heated with constant heat flux. Orifice to surface distance H was equal to two diameters of the orifice, $2D$. Heat flux at the surface was equal to 1000 W/m^2. Reynolds number, defined as:

$$\mathrm{Re} = \frac{\rho u D}{\mu}, \quad (4)$$

at the exit of the orifice was equal to 23000, and, in addition, the flow there was fully developed, which was achieved with the mapped inlet boundary condition. Parameter D in Equation 4 is the jet impingement characteristic length, denoting the orifice diameter.

Table 1 presents the boundary conditions for validation case. They were also used for non-flat cases, from Section 4.

Table 1. Boundary and geometry conditions, validation and non-flat cases [6-10].

Inlet Reynolds number	Surface heat flux, W/m^2	Inlet temperature, K	*H/D*	*R/D*
23000	**1000**	**293**	**2**	**4**

Numerous publications were based on those results, among which the papers by Behnia et al. [9-10] were chosen as the reference case. Both papers confirmed, that v^2-f model presents the best performance when analyzing jet impingement. Moreover, the main drawback of standard k-ε model, which is overproduction of turbulence kinetic energy in the stagnation zone, was also confirmed. Results obtained with v^2-f model were compared with standard k-ε and SST k-ω models. First two were used in low-Reynolds mode, while SST k-ω

was used in the combination with wall functions, to check their performance.

Initially, for the validation case, it was important to fulfil the mesh requirements regarding jet impingement simulations. Mesh construction process was time consuming, required not only knowledge of phenomena but also the trial-error method. Regular mesh independence checks were performed as well. The final space division was chosen in the basis of the comparison of obtained results with [7-10]. Table 2 presents number of mesh elements that were chosen to analyze the process.

Table 2. Number of mesh elements for analyze cases.

Mesh 1	Mesh 2	Mesh 3
100 000	**170 000**	**250 000**

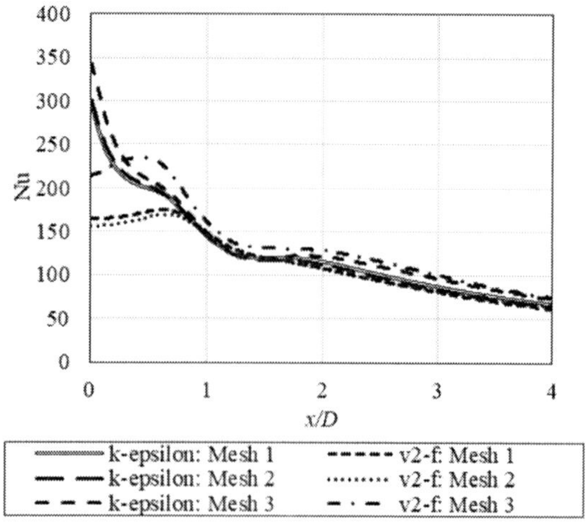

Fig. 4. Impact of mesh elements number on results of local Nusselt number.

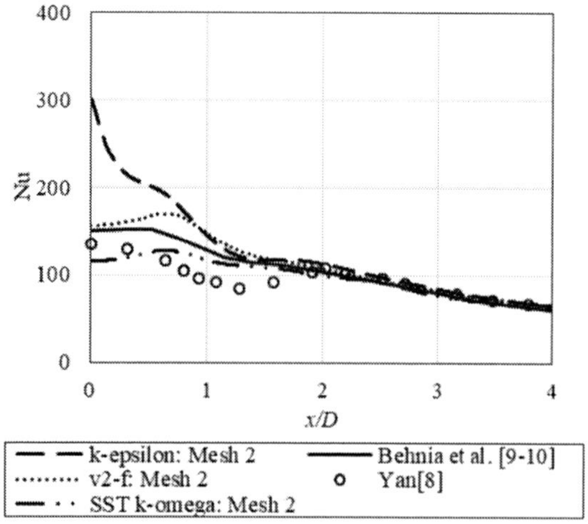

Fig. 5 Comparison of local Nusselt number values, obtained with various turbulence models, with benchmark data [8-10].

Figure 4 represents the impact of elements number on the results – which are represented by local Nusselt number distribution at the distance x/D along the heated surface. The Nusselt number, used as the performance indicator, was defined as follows:

$$\text{Nu} = \frac{\alpha D}{\lambda}, \tag{5}$$

where α is the convective heat transfer coefficient, W/(K·m^2) and λ is the thermal conductivity, W/(K·m). As can be seen, depending on the mesh size and settings, very different results were obtained.

Figure 5 shows selected results from Figure 4, compared with benchmark data, experimental results by Yan [8] and numerical results by Behnia et al. [9-10]. Also results obtained with SST k-ω are presented there, to check the performance of wall-functions. Only the values calculated with Mesh 2 were included, as they exhibited the best agreement with [8-10]. Moreover, results presented in Section 4 were also obtained with utilization of Mesh 2 construction process.

As it can be noticed, far from the stagnation region all presented results are almost the same. However, large discrepancies occur in the stagnation zone. The k-ε model overpredicted the heat transfer significantly – but this effect was expected. On the other hand results obtained with OpenFOAM v^2-f model also did not reflect the ones obtained by Behnia et al. [9-10]. Especially, in the region, where the distance from the stagnation point is slightly higher than the orifice radius. The answer for that discrepancy can be found in the paper of Billard and Lawrence [11]. From theoretical point of view they analyzed evolution of the v^2-f models, because, depending on particular implementations, various results could be obtained. They described, that some models were adjusted to use very robust Dirichlet boundary condition at the wall for the elliptic relaxation function, f. It led to the possibility of their usage in the segregated solvers, commonly applied in many commercial codes or OpenFOAM. However, such method led to omitting one term in the relaxation equation and, as a result, another overprediction – of the velocity scale, v^2. It explains the issue with Nusselt number values, presented in Figure 5. In the papers by Behnia et al. [9-10], their own solver was used, so they were able to use the original v^2-f model, not having such drawbacks. Still, though, the difference between their results and the experiment can be noticed that may possibly be never fully avoided, as described in [4].

Another reason of presented differences may lay in the characteristics of the fully developed flow. Behnia et al. in [9] included Figure 14, in which the impact of different orifice-exit velocity and turbulence profiles on the Nusselt number distribution at the impinged surface was very significant.

Another conclusion can be related with results obtained with SST k-ω model. In that case the heat transfer was underpredicted in the stagnation region and proper far from it. However, because of the wall functions consideration, which are just a simplification of actual heat and mass transfer processes, this model would not be analyzed and described in the next sections.

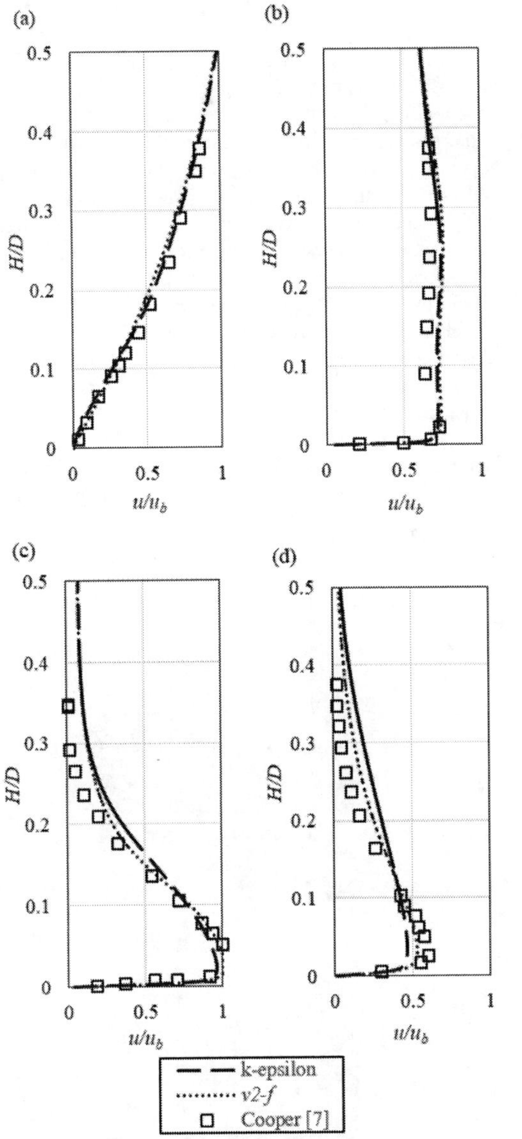

Fig. 6. Velocity magnitude of the jet at various radial locations, normalized by bulk velocity u_b = 34.5 m/s. (a) x/D=0, (b) x/D = 0.5, (c) x/D=1, (d) x/D = 2.5.

Apart from the thermal parameters, also a flow prediction by particular numerical models is very important for results validation. Figure 6 presents obtained numerically and compared with experimental [5] profiles of the velocity in various locations far from the stagnation point, x/D. At the bulk Reynolds number in the orifice equal to 23000, bulk air velocity u_b was equal to 34.5 m/s.

Analysis of Figure 6 leads to conclusion, that in contrary to thermal parameters, flow behavior was predicted well by the v^2-f model implemented in OpenFOAM and used in presented studies. Differences between it and standard k-ε are also clearly visible. Profiles of velocity obtained in [9-10] were very similar, including the noticeable difference between the experimental and numerical results starting at ratio H/D higher than 0.2, for x/D = 1 and 2.5.

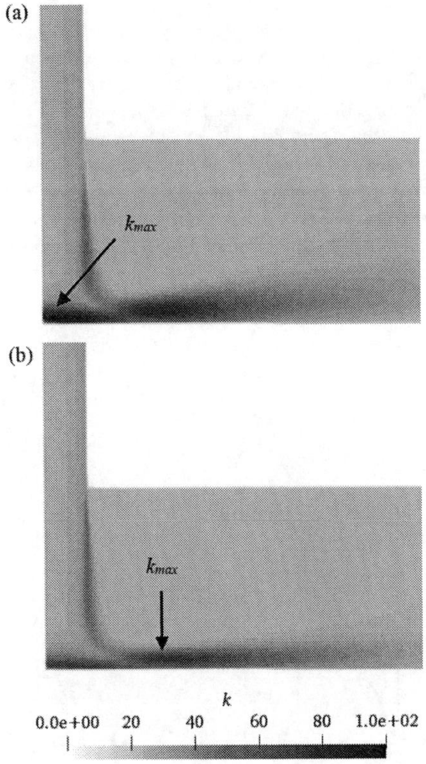

Fig. 7. Flat surface impingement. Turbulence kinetic energy, k, distribution. (a) k-ε model, (b) v^2-f model. Arrows indicate location of maximum values.

In Figure 7, the turbulence kinetic energy distribution for validation case is presented. Its budget can be written as follows:

$$\rho\frac{\partial u_i k}{\partial x_i}-\frac{\partial}{\partial x_i}\left[\left(\mu+\frac{\mu_t}{\sigma_k}\right)\frac{\partial k}{\partial x_i}\right]=\rho\left(\underbrace{\mu_t S_{ij}}_{production}-\underbrace{\varepsilon}_{dissipation}\right), \quad (6)$$

where μ_t is the eddy viscosity; Pa·s, ε is turbulence dissipation rate, m^2/s^3. Two terms, production and dissipation, are emphasized, as they were used in Section 4 for data presentation. The k-ε model was characterized by overproduction of turbulence kinetic energy in the stagnation region. Its maximum was located there. On the other hand, usage of v^2-f model caused the maximum to move outside this region. It reflects the real-life situation that can be observed in [5, 9-10]. As mentioned at the beginning of Section 3, the most common implementation of v^2-f model is not able to limit the extensive production of turbulence in stagnation zone properly. That is the reason of relatively high values of turbulence kinetic energy, visible in Figure 7(b), which do not occur in [5, 9-10]. Nevertheless, the v^2-f model was chosen for the next analyses, presented in Section 4.

4 Non-flat surfaces

Before analyzing arrays of jets, that impinge non-flat surfaces, it is important to define the influence of the surface shape on the flow behavior. In [6] Authors tried to define the critical curvature radius to orifice diameter ratio *R*/*D*, for which the curvature effect plays a role. It was concluded, that depending on the type of surface and mentioned ratio, the difference in heat transfer occurs between the particular non-flat and flat surface impingement. Moreover, with increasing ratio, the values tended to vary less – since the curvature of stagnation zone was almost negligible. However, [6] did not contained hydrodynamic data, which also is very important.

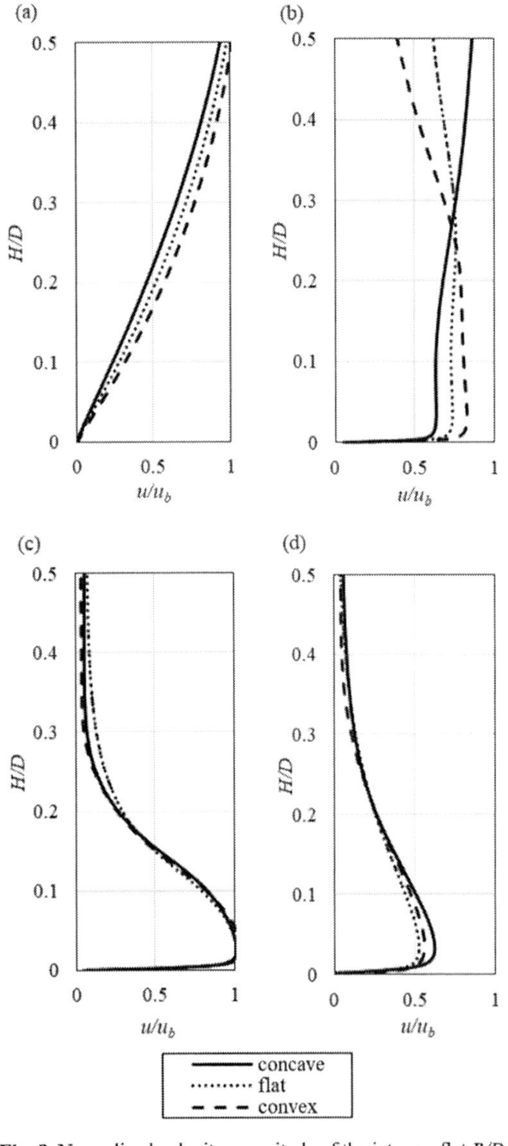

Fig. 8. Normalized velocity magnitude of the jet, non-flat *R*/*D* = 4 and flat surface. *H*/*D* = 2. (a) *x*/*D*=0, (b) *x*/*D* = 0.5, (c) *x*/*D*=1, (d) *x*/*D* = 2.5.

In [6] it was noticed, that for ratio of surface curvature and orifice diameter equal to 4, the curvature effect is noticeable. Therefore, this ratio was selected for representation in the following studies. In Figure 8, comparison of normalized (in the same manner, as in Figure 5) velocity profiles, depending on the type of surface: flat, concave or convex, is presented. For the radial distance *x*/*D* higher than 1 only slight differences occur. However, when analyzing results in Figure 8(a) and 8(b), it can be seen, that depending on the type of surface, the flow behaves in different way. Therefore, the stagnation zone is the area, where the most noticeable discrepancies takes place. It is especially visible in Figure 8(b), for height values $H/D \geq 0.25$. To be able to provide data described above, it was important to propose the method of comparison between flat and non-flat surfaces. For that purpose, the distance *x* was measured as the straight chord connecting the stagnation point and particular point on the curve. Moreover it was measured, that difference between its length and the length of the arc connecting both points was negligible.

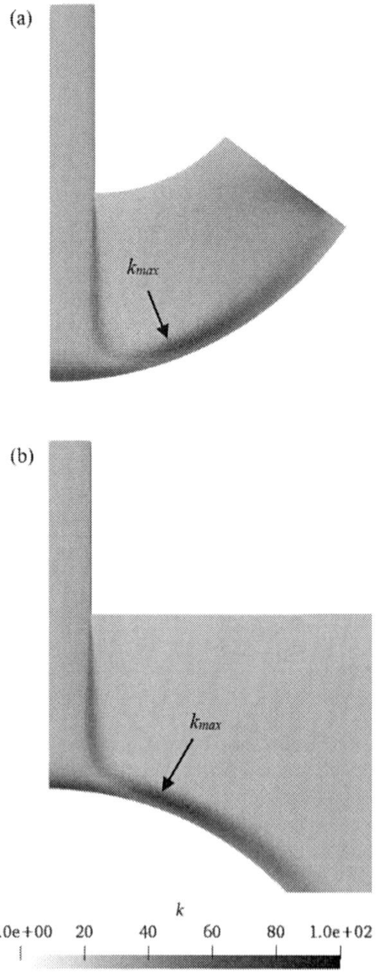

Fig. 9. Distribution of turbulence kinetic energy, *k*. (a) concave surface, (b) convex surface. *R*/*D* = 4. *H*/*D* = 2. Arrows indicate maximum values, their locations are listed in Table 3.

In Figure 9, distribution of the turbulence kinetic energy during impingement on concave, Figure 9(a) and convex, Figure 9(b), surfaces is shown. Its maxima are located outside the stagnation zone, as in the flat case.

Table 3. Distance of turbulence kinetic energy maxima from stagnation point, x/D.

concave surface	flat surface	convex surface
1.5039	**1.4676**	**1.3775**

In Table 3, locations of turbulence kinetic energy maxima are listed, in relation with orifice diameter D. They were established for flat and non-flat cases in the same manner, as data from Figure 8. Differences between each case exist – it is an important aspect to mention, because it may influence the results when the jets array impingement occurs, such as in [1-3]. Multiple jets might strongly interfere with each other.

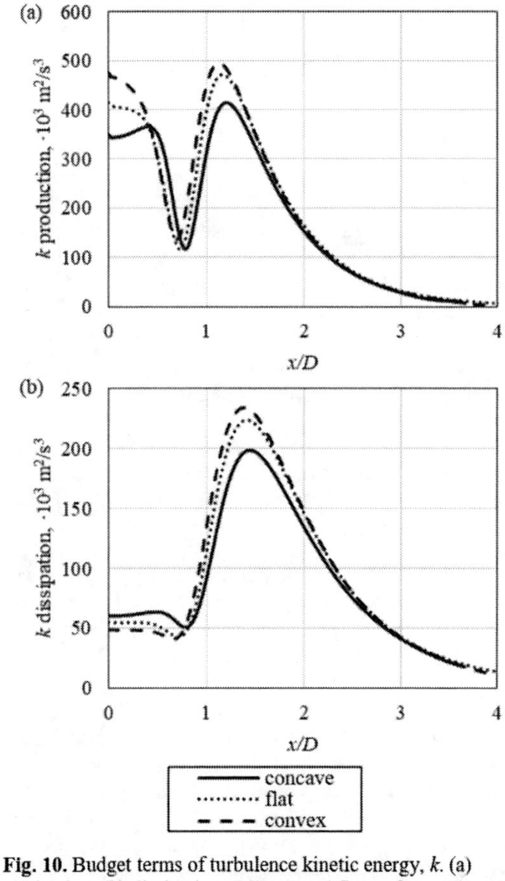

Fig. 10. Budget terms of turbulence kinetic energy, k. (a) production, (b) dissipation. $R/D = 4$ and flat surface. $H/D = 2$.

In Figure 10, two terms of turbulence kinetic energy budget are presented, along the curvature or segment distant from stagnation point, at height normal to impinged surface, where maxima from Table 3 occurred. In the stagnation zone, the highest production exists in the convex case, followed by flat and concave ones. For the dissipation, situation is opposite. However, in the regions close to the highest values of turbulence kinetic energy, located at the distances x/D mentioned in Table 3, for both production and dissipation highest values were obtained for convex surface. In general, behavior presented in plots for each situation is quite similar.

5 Summary

In this paper, thermal and hydrodynamic analyses of jet impingement on flat and non-flat surfaces were presented. Both boundary conditions and scope of interest were based on [1-3, 5-8]. Mesh and software configurations were determined by ERCOFTAC data [7-8]. Velocity profiles and turbulence kinetic energy budgets comparison revealed important differences. In Authors opinion those variations would matter, when the whole jets array impinging flat and non-flat surfaces would be analyzed. Different turbulence model were considered, however, as it was proved in [4], v^2-f has given the best results. While RANS models, presented in the following paper, can reveal important information, they should be verified and extended also by more extensive methods, such as LES approach.

The present work was supported by the Polish Ministry of Science and Higher Education and by the PLGrid Infrastructure.

References

1. J. Wajs, D. Mikielewicz, E. Fornalik-Wajs, *Cylindrical jet heat exchanger dedicated to heat recovery, especially from low temperature waste sources* (Patent PL224494, 2013 (in Polish))
2. J. Wajs, D. Mikielewicz, E. Fornalik-Wajs, M. Bajor, Arch. of Thermodynamic **36(4)**, 48-63 (2015)
3. J. Wajs, D. Mikielewicz, E. Fornalik-Wajs, M. Bajor, Heat Tr. Eng. doi:10.1080/01457632.2018.1442369
4. N. Zuckerman, N. Lior, J. Heat Transfer **127**(5), 544-552 (2005)
5. M. Hadžiabdić, *LES, RANS and combined simulation of impinging flows* (PhD Thesis, Delft University of Technology, 2006)
6. T. Kura, E. Fornalik-Wajs, J. Wajs, Arch. of Thermodynamics **39**(1), 147-166 (2018) doi: 10.1515/aoter-2018-0008
7. D. Cooper, D.C. Jackson, B.E. Launder, G.X. Liao, Int. J. Heat Mass Transfer **36**, 2675-2684 (1993)
8. X. Yan, *A preheated-wall transient method using liquid crystals for the measurement of heat transfer on external surfaces and in ducts* (PhD Thesis, University of California, 1993)
9. M. Behnia, S. Parneix, Y. Shabany, P.A. Durbin, Int. J. of Heat and Fluid Flow **20**(1), 1-9 (1999)
10. M. Behnia, S. Parneix, P.A. Durbin, Int. J. of Heat and Mass Transfer **41**(12), 1845-1855 (1998)
11. F. Billard, D. Laurence, Int. J. of Heat and Fluid Flow **33**(1), 45-58 (2012)

Modelling microwave heating of an oil palm mesocarp

Ming Law [1,*], and *Jessie* Chang[1]

[1]Department of Mechanical Engineering, Faculty of Engineering and Science, Curtin University Malaysia, CDT 250, 98009, Miri, Sarawak, Malaysia

Abstract. An oil palm mill often uses steam in sterilization and fruit detachment processes. Consequently, a large amount of wastewater is produced. To reduce the wastewater, microwave (MW) heating of oil palm fruits has been studied and positive results were reported in open literature. Nevertheless, MW heating of oil palm fruits requires proper control to avoid overheating, which deteriorates the oil quality. MW heating efficiency depends not only on the electromagnetic strength, but also on the complex permittivity of oil palm fruits. This study reports the MW heating model of an oil palm mesocarp. The three-dimensional model solved Maxwell's Electromagnetic waves equation, diffusion equation of moisture content, and heat conduction equation. The model was validated with mesocarp's experimental data of moisture content and temperatures. The electromagnetics, moisture, and temperature distributions of the mesocarp were studied. The simulation results showed that the volume- and surface-averaged temperature were similar, thus the surface temperature might be used as a good approximation to the volumetric temperature. Besides, the model also showed that radiation and convection were the main heat loss mechanisms. This validated model can be used for designing a microwave heating reactor for oil palm fruits.

1 Introduction

A typical oil palm fruit consists of three major parts: mesocarp (flesh of the fruit), shell and kernel. Palm oil and kernel oil can be obtained from the mesocarp and the kernel respectively. The usages of palm and kernel oils are very extensive. Palm oil is widely used in food products such as cooking oil, margarine and non-dairy creamers as well as feedstock for biofuel. In addition palm oil has a longer shelf life compared to other vegetable oils. On the other hand, palm kernel oil is a raw material for the production of non-food products such as soaps, detergents, toiletries, cosmetics and candles.

Palm and kernel oils are extracted from oil palm fruits through physical and mechanical processes in a palm oil milling process. Sterilization is the first stage of the milling process where fresh fruit bunches (FFB) are exposed to heat rendering operation with high-pressure steam operated at 15-45 psi for about 90mins in horizontal cylinder autoclaves [1]. Sterilization aims to soften the fruits which allow easy detachment from the FFB and to improve the quality of fruits by deactivating the enzymatic activities. The sterilized fruits are then detached from the FFB through the threshing process in rotating thresher. Palm oil mill effluent (POME) is generated when hot steam is used in the sterilization process. According to Kamyab et al. [2], POME is one of the major environmental pollutants to contaminate the aquatic environment. In addition, POME requires an expensive and challenging treatment process before disposal [3]. Thus there is an urge to look for alternatives. Microwave (MW) heating technology could be a potential energy to replace the conventional sterilization process [4, 5]. This is because not only the MW heating is a fast process, it also reduces energy consumption and yields a better quality of palm oils compared to the conventional sterilization process. Furthermore, POME can be eliminated as hot steam is not required in the MW heating process. In addition, the better efficiency on fruit detachment using MW heating was also observed [6].

To date, there are a limited modelling works on MW heating of oil palm fruits, compared to other materials, such as potatoes, apples and woods. Puangsuwan et al. [7] simulated the MW heating of a oil plam fruit (OPF). The model did not account for the shrinkage effects on the OPF. Law et al. [8] modelled the MW heating of oil palm kernels and the effects of OPKs arrangement in a MW reactor were studied. Recently, the effects of MW heating of 180W on mechanical strength of OPKs, mesocarp and empty fruit bunches were extensively investigated by Law et al. [9]. The authors verified that MW heating could aid the detachedment of oil palm fruits from bunches. However, prolonged MW heating will cause the bunches to be harden and making the detachment more difficult. The present work is an extension of the previous work [9] and the attention is paid on the thermal behavior of oil palm mesocarp subjected to 100W and 300W.

* Corresponding author: m.c.law@curtin.edu.my

2 Mathematical modelling

Figures and tables, as originals of good quality and well contrasted, are to be in their final form, ready for reproduction, pasted in the appropriate place in the text. Try to ensure that the size of the text in your figures is approximately the same size as the main text (10 point). Try to ensure that lines are no thinner than 0.25 point.

2.1. Computational domain

The computational domain of the model is shown as shown in Figure 1.

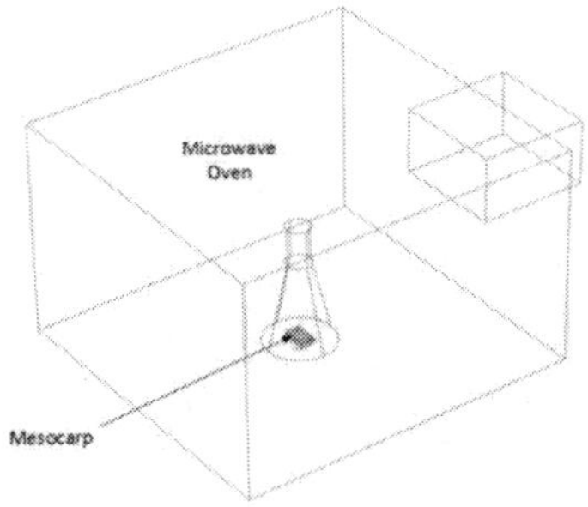

Fig. 1. The computational domain of the model.

2.2 Assumptions

The following assumptions were made in the current study:
1) The ambient air inside the MW oven remained in isothermal condition during the heating of mesocarp.
2) Only liquid water was considered and the effect of crude palm oil was not accounted for.
3) Moisture evaporation occurs on the mesocarp's surface only.
4) Mesocarp was assumed to be an isotropic material and the shrinkage due to the moisture lost was not considered.

2.3 Governing equations

The governing equations are represented as follows:
MW can be modelled by Maxwell's equation in frequency domain as

$$\nabla\cdot(\nabla\times\mathbf{E})-\frac{\omega^2}{c_o}\mu\varepsilon\mathbf{E}=0 \quad (1)$$

where complex permittivity, $\varepsilon=\varepsilon_0\left(\varepsilon_r'-j\varepsilon_r''\right)$.

The conservation of moisture molar concentration (mol/m^3)

$$\frac{\partial c}{\partial t}=\nabla\cdot(D\nabla c) \quad (2)$$

The moisture content percentage based on dry basis (% kg/kg (d.b.)) can be related with c , via

$$\text{mc}=\frac{c\text{M}_{\text{H}_2\text{O}}}{\rho}\times 100\% \quad (3)$$

where $\text{M}_{\text{H}_2\text{O}}$ is the water molecular weight.

The conservation of heat energy

$$\rho c_p\frac{\partial T}{\partial t}=\nabla\cdot(k\nabla T)+\dot{Q}_{MW} \quad (4)$$

The MW heat source is defined as

$$\dot{Q}_{MW}=\pi f\varepsilon_o\varepsilon_r''|\mathbf{E}|^2 \quad (5)$$

2.3.1 Boundary conditions

Electromagnetic boundary condition:
Two MW powers, 100 W and 300 W, were applied at the inlet port of the MW oven. Intermittent heating mode was used and controlled by the oven processor, the 'on' and 'off' periods for both power level are as follows:

$$P(t)=\begin{cases}100\text{ W} & t\le 10\text{s}\\ 0\text{ W} & 10<t\le 30\text{s}\end{cases} \quad (6)$$

$$P(t)=\begin{cases}300\text{ W} & t\le 13\text{s}\\ 0\text{ W} & 13<t\le 30\text{s}\end{cases} \quad (7)$$

The period of each cycle is 30s.

Impedance boundary condition was used for the metallic wall of the oven, i.e.

$$\frac{1}{\mu}\mathbf{n}\times(\nabla\times\mathbf{E})-\frac{jk_o}{\eta}\mathbf{n}\times(\mathbf{n}\times\mathbf{E})=0 \quad (8)$$

where $k_o=\omega\sqrt{\varepsilon_o\mu_o}$, μ_o is the permeability of free space (1.257×10^{-6}H/m) and $\eta=\sqrt{\mu/\varepsilon}$.

Boundary condition of the moisture content of the oil pam mesocarp:

The moisture content was lost through evaporation and mass transfer mechanism such that,

$$-\mathbf{n}\cdot(D\nabla c)=\dot{c}_{evap}(c-c_e)+h_m(c_e-c) \quad (9)$$

The evaporation rate was described by a 1st order Arrhenius rate [8],

$$\dot{c}_{evap}=-AT\exp(-E_a/RT) \quad (10)$$

Boundary condition of the energy conservation:

Heat losses of the mesocarp occurred through convection, radiation and moisture evaporation, such that

$$-\mathbf{n}\cdot k\nabla T=h_t(T_\infty-T)+\dot{Q}_{rad}+\dot{Q}_{evap} \quad (11)$$

h_t was assumed to be 5 W/(m2.K) for natural convection in a MW oven [10]. Heat lost through radiation was described as

$$\dot{Q}_{rad}=\gamma_T\sigma_T\left(T_\infty^4-T^4\right) \quad (12)$$

Heat lost via evaporation was denoted as

$$\dot{Q}_{evap} = \dot{c}_{evap} H_{evap} \tag{13}$$

The enthaphy of evaporation was as follows [11]

$$H_{evap} = -0.0469T^2 - 13.493T + 52202 \tag{14}$$

Initial conditions and input parameters:

The initial moisture content of the mesocarp was experimentally determined as 25.9% kg/kg d.b. and the room temperature was 33°C. Zero electric field strength was assumed inside the oven before the heating occurred. The input parameters of the modeling works are listed in Table 1.

Table 1. Parameters used in the model.

Property	Value
ρ (kg/m³)	995 kg/m³
E_a (J/mol)	43000
A (1/s)	0.005
c_p (J/(kg.K))	2816
k (W/(m.K))	0.458
h_t (W/(m².K))	5
ε_r'	$3.47 + 0.0704\text{mc}$
ε_r''	$0.486 + 0.0402\text{mc}$
h_m (m/s)	2.06×10^{-6} @ 100W 2.06×10^{-5} @ 300W

3 Results and discussion

Figure 2 and Figure 3 show the experimental data and predicted moisture contents of the mesocarp, subject to 100 and 300 W. Close agreement is found for both cases. Comparisons between the predicted temperature and experimental data are shown in Figure 4 and Figure 5. The tempature data was obtained at the centre of the top surface of the mesocarp using a fibre-optic temperature sensor. Good agreement is obtained for the 100 W case until around t = 300s. Higher discrepancy was observed at the later stage of the MW heating as seen in Figure 4. On the other hand, good agreement was obtained for the case of MW power of 300W as shown in Figure 5. To main the crude palm oil quality, it is important to keep the temperature of the mesocarp to below 100 °C. Thus, when MW power of 300 W is used, it is advisable that the heating time should not be more than 100 s. Nevertheless, as MW often results in non-uniform heating, it is essential to gain a more thorough understanding of the thermal response of the mesocarp before an optimum MW reactor can be designed.

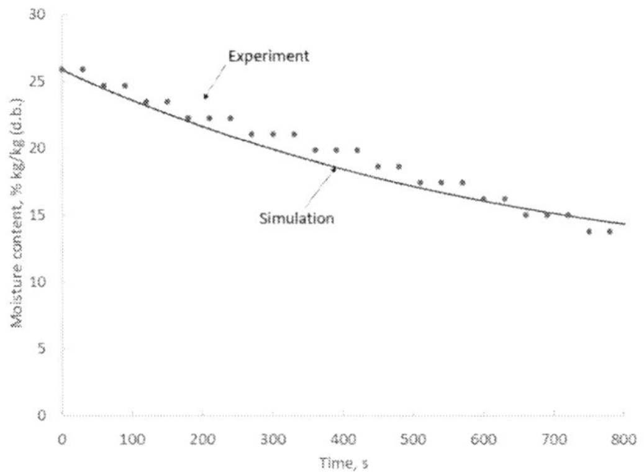

Fig. 2. Moisture content of the mesocarp sample subjected to 100 W MW power.

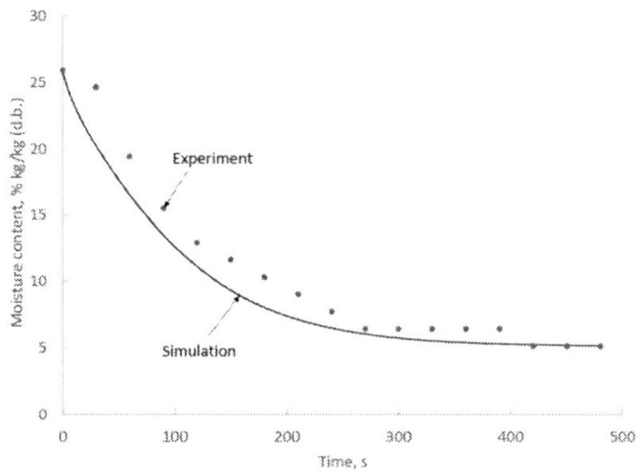

Fig. 3. Moisture content of the mesocarp sample subjected to 300 W of MW power.

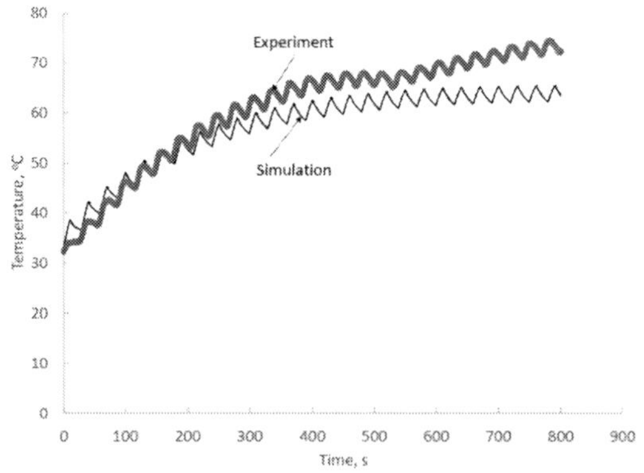

Fig. 4. Comparison between the experimental and simulation-predicted temperatures of the sample subjected to 100 W of MW power.

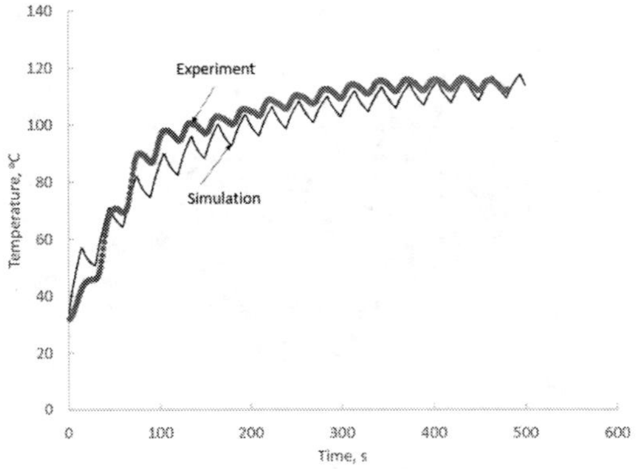

Fig. 5. Comparison between the experimental and simulation-predicted temperatures of the sample subjected to 300 W of MW power.

As MW heating often results in non-uniform temperature profile [8], Figure 6 shows the volume-, surface-averaged temperatures and also the maximum and minimum temperatures of the mesocarp sample subject to 300W. As there is only a small difference between volume- and surface-averaged temperatures, so surface temperature could be used as a good approximation for mesocarp sample. Similar readings are found for both volume- and surface-averaged temperatures because the mesocarp has a relatively high thermal conductivity. Higher difference between volume- and surface-averaged temperatures is expected if a material possesses a lower thermal conductivity. This result has an important implication in the temperature measurement because surface temperature can be obtained cheaply using an infra-red camera and the internal temperature profiles are usually obtained by using a more expensive magnetic resonance imaging (MRI) device. Hence, the result shows that surface temperature can be used as a good approximation to the volumetric-averaged temperature.

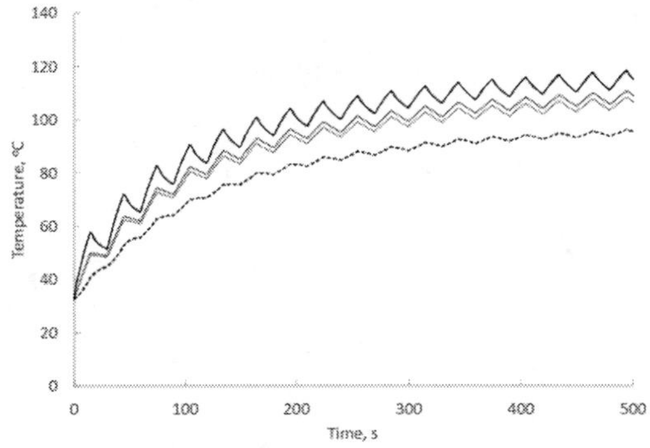

Fig. 6. Comparison between the volume- and surface-averaged temperatures of the mesocarp sample subject to a MW power of 300 W.

In addition, the simulation shows that under 300 W MW irradiation, the mesocarp sample experiences a temperature difference of around 20 °C. Neverthess, the temperature of the mesocarp is affected by the MW electric field strength, E. In the current simulation, the mesocarp is placed at the centre of the MW oven but it is not at the location where the strongest electric field strength is found as can be seen in Figure 7. Hence, the temperature difference could be higher if the E field is increased. The distribution of the E field inside a MW oven is as a result of constructive and destructive interference [8]. Constructive interference results in temperature hot-spot and destructive interference causes lower temperature in a sample.

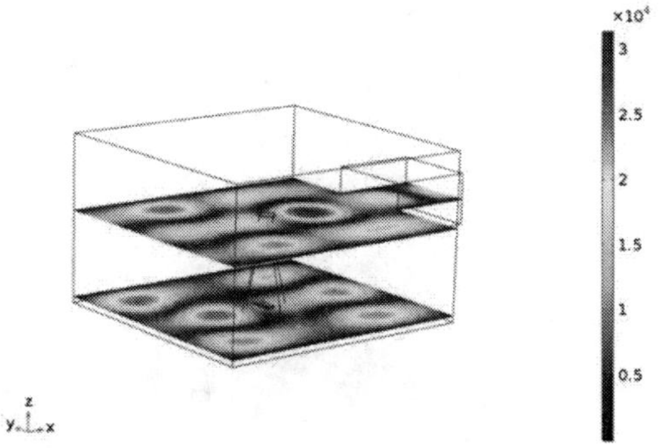

Fig. 7. The electric field strength [V/m] distribution in the MW oven on xy plane.

Figure 8 shows the MW power density distribution in the mesocarp at t = 60s and 360s. Higher power density is found at the top part of the sample due to stronger electric field strength. As the electric field is absorbed as it passes through the sample, electric field is generally weaker at the bottom part of the sample. The MW power density absorbed is decreasing as moisture is lost during the drying process.

Figure 9 shows the internal temperature distribution of the mesocarp. The temperature and the MW power density distributions are not the same because the mesocarp's temperature is also affected by other cooling effects such as convection, radiation and evaporation. The result shows that the maximum temperature is found in the middle, internal region of the mesocarp. Comparing the experimental temperature measurement in Figure 5 and the predicted temperature profile in Figure 9, both results are in good agreement.

Figure 10 shows the surface temperature profile of the mesocarp at t = 60s and 360s. Hot spot is found at the centre of the sample. The temperature difference at t = 60s is around 9 °C while it is around 14 °C when t = 360s. We expect that the temperature non-uniformity will be more severe when a large quantity of oil palm fruits are heated inside MW oven. Hence, there is a need to carry MW heating for an actual size of an oil palm fruitlet.

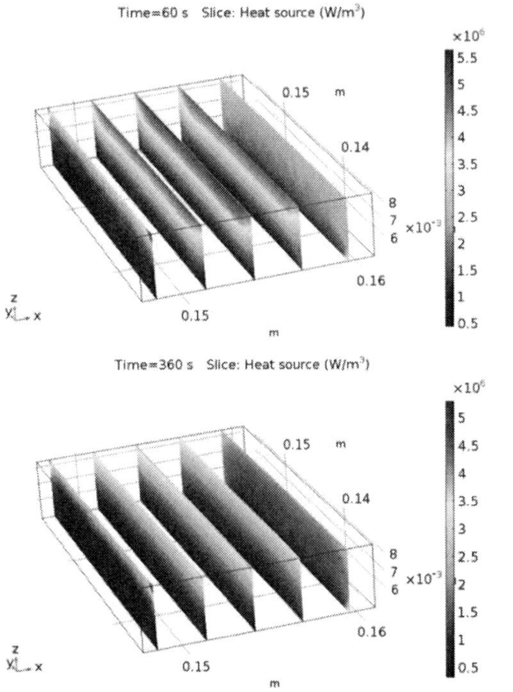

Fig. 8. The distribution of MW power density in the mesocarp sample at t = 60s (top) and 360s (bottom)

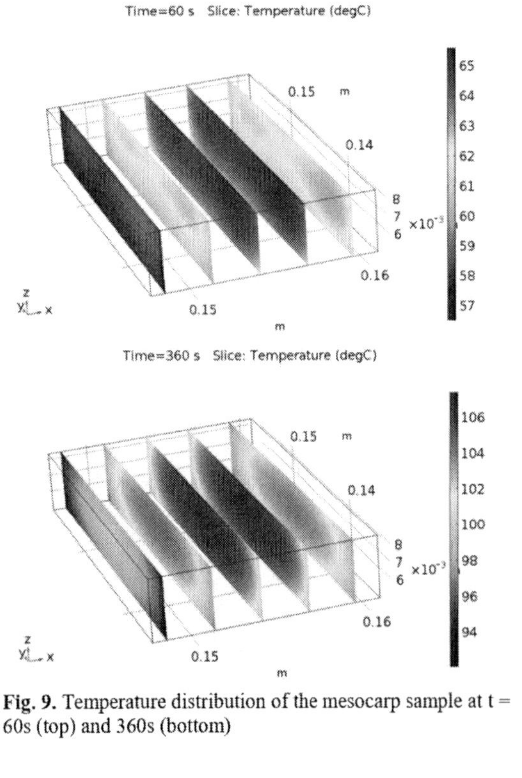

Fig. 9. Temperature distribution of the mesocarp sample at t = 60s (top) and 360s (bottom)

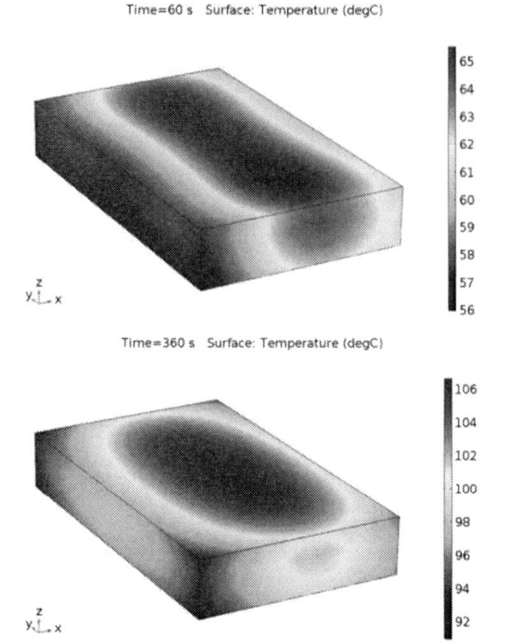

Fig. 10. Surface temperature distribution at t = 60s (top) and 360s (bottom)

The heating of the mesocarp is affected by four main mechanisms, namely, MW heating, convection, evaporation and radiation. Figure 11 shows the heat power contributed by each of these mechanisms. From the result, MW heating is clearly the dominant heat source. As it depends on dielectric loss factor, which in turn depends on moisture content, hence the heat source decreases as the moisture content evaporates. The simulation result shows that convection is a main heat sink, followed by radiation and evaporation. Each of these heat sink is not more than 0.5W. The contribution of evaporation decreases as the mesocarp is dried.

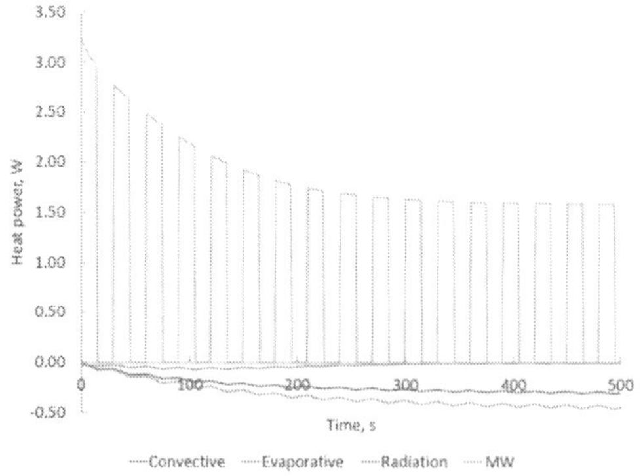

Fig. 11. Heat lost / gained by the mesocarp sample.

4 Conclusions

The current study discussed the model development of an oil palm mesocarp sample subject to MW heating. The model was validated against experimental data for the moisture content and temperature history.

The model result showed that radiation and convection are dominant cooling mechanisms to reduce the surface temperature, whereby evaporation is significant when sample has high moisture content. Temperature non-uniformity increased during MW heating. Finally, surface temperature measurement can be used as an approximation for the volumetric-averaged temperature for mesocarp.

This research is supported by Malaysian Ministry of Higher Education, through the fundamental research grant schemes (FRGS/2/2014TK06/CURTIN/02/1; FRGS/2/2014/TK06/CURTIN/03/1).

References

1. K. Sivasothy, M. Rohaya, *International Planters Conference, Kuala Lumpur* (2000)
2. H. Kamyab, M. Din, A. Keyvanfar, M. Majid, A. Talaiekhozani, A. Shafaghat, C. Lee, L. Shiun, H. Ismail, Energy Procedia, **75**, 2400-2408, (2015)
3. Y. Madaki, L. Seng, 2013, IJSET, **2**, 1138 – 1155, (2013)
4. C. Vincent, R. Shamsudin, A. Baharuddin, J Food Eng, **143**,123 - 131, (2014)
5. M. Chow, A. Ma, J Microw Power Electromagn Energy, **40**, pp. 165 – 173, (2007)
6. N. Sukaribin, K. Khalid, Ind Crops Prod., **30**, 179-183, (2009)
7. K. Puangsuwan, M. Chongcheawchamnan, C. Tongurai, J Microw Power Electromagn Energy, **49**, 100-111, (2015)
8. M. Law, E. Liew, S. Chang, Y. Chan, C. Leo, Appl Therm Eng., **98**, 702-726, (2016)
9. M. Law, J. Chang, Y. Chan, D. Pui, K. You, Drying Technol, (2018), DOI: 10.1080/07373937.2018.1439057
10. P. Verboven, A. Datta, N. Anh, N. Scheerlinck, B. Nicolaï, J Food Eng, **59**, 181-190, (2003)
11. C. Hii, C. Law, C.L., Law, M.C., 2013, Appl Therm Eng., **54**, 264-271, (2013)

On modeling of the initial stage of nonstationary nucleate boiling for the high heat fluxes

Anatoliy Levin*, and *Polina* Khan

Melentiev Energy System Institute SB RAS, 130, Lermontov St., Irkutsk, 664033, Russia

Abstract. During recent years, there have been made significant achievements in the numerical description of the bubble boiling, particularly, in the calculation of the bubble growth dynamics, the nucleation density, and the bubble boiling threshold [1, 2]. However, the numerical prediction is mostly based on the empirical correlations, the accuracy of which does not exceed 20%. Boiling is a complicated process, where each parameter affects not only the general outcome but other parameters, too. The non-stationary heat release is most difficult for modeling, because many of the existing researches are based on analytical expressions for the fluid temperature distribution. The basic stages of the explosive boiling are: (i) heating of the wall to the nucleation temperature; (ii) nucleation of the isolated bubbles; (iii) merging of bubbles and coverage of the entire surface by the steam phase; (iv) heating of the surface to the Leidenfrost temperature and transition to the boiling crisis. This paper presents results of the experimental study for the initial stages of the explosive boiling (i) and (ii), as well as an attempt to simulate them in order to clarify whether the existing approaches can be extended to the case of the nonstationary heat release.

Nomenclature

D	bubble diameter, m
E	electric field intensity
f	bubble nucleation frequency
h_{lg}	latent heat, J/kg
J	electric current density
m	fraction of the bubble covered by the subcooled liquid
n	nucleations per area, 1/m^2
q''	heat flux, W/m^2
q'''	heat source, W/m^3
V	voltage
δ	thickness of the layer, m
ϵ	permittivity
σ	conductivity

Subscripts

c	conduction
e	evaporation
i	initial
s	superheated
sat	saturation
ml	microlayer
-	heat absorption
+	heat release

1 Introduction

The present research considers the initial stage of the bubble boiling without bubble agglomerations. In the case of high heat fluxes the subcooled liquid is located at the small distance from the heater surface. Figure 1 shows the comparison of the experimental bubble diameter with the theoretical value of the overheated layer thickness, $\sqrt{\alpha t}$, where α is the thermal diffusivity and t is the time passed since the wall achieved the saturation temperature. Analysis of this figure shows that the bubble diameter can be estimated on the basis of the overheated layer thickness, taking into account the liquid bulk temperature, wall overheat, and the heat flux. However, it was revealed in [3] that no single approximation of this kind can cover all possible regimes of non-stationary heat release. We propose the following approach: the ratio between the overheated layer thickness and the bubble maximum diameter is assumed to be constant [2], and the non-stationary overheated layer thickness is obtained numerically by taking into account not only the conductive heat transfer but also the evaporative heat transfer. The coupling between the temperature field and the averaged bubble diameter in homogeneous flow is modeled with the help of COMSOL Multiphysics (ver. 5.3) software.

* Corresponding author: lirt@mail.ru

The calculation results are compared with the experimental data obtained in a vertical channel with a cylindrical heater (120 mm long, 12 mm in diameter and 1 mm thick). The stepwise heat flux was used with the maximum up to 1.0 MW/m^2.

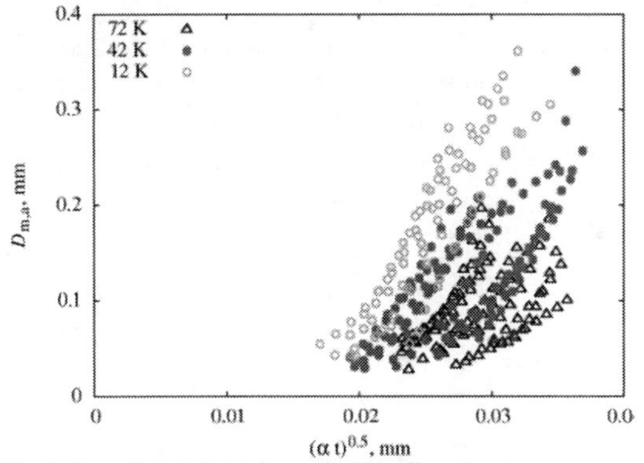

Fig. 1. Experimental maximum bubble diameters versus theretical overheated layer thickness.

2 The heat balance based model for nucleate boiling

The expression for the maximum diameter, given in [5,6], is based on determining the heat balance for a bubble, part of which is submerged in a superheated liquid layer and another part is surrounded by the liquid subcooled to the saturation temperature:

$$D_m = 1.21AB^{-0.5}, \quad (5)$$

where $A = \frac{(1-m)k_f(T_w-T_S)}{\rho_g h_{fg}\sqrt{\pi a_g}}$, $B = \frac{m\varphi C(T_s - T_0)}{1-\rho_g/\rho_f}$, k is the thermal conductivity. The factor m is the fraction of the bubble surface area covered by the subcooled liquid at the temperature T_0, as proposed in [5]. Following [4], φ (characterizes the effect of liquid velocity) and C (a function of system pressure) for the present condition are equal to 1 and 65, correspondingly. The cooling area fraction m of 0.3 was indicated as the best one for approximating the experiment data in [4] and 0.5 for the data in [4]. By definition, m is related to the location of T_s isotherm as follows:

$$m = 1 - x_s/D_m \quad (7)$$

where x_s is the thickness of the superheated liquid layer.

For the boiling stages of short duration, we can assume that the ratio between the thickness of the superheated layer and the maximum bubble diameter is a value of small variability. It means that during nonstationary boiling simulation we can calculate the temperature field and easily obtain the bubble diameter via Eq. (7). In [3] we employed the homogenous representation of the liquid wall layer without taking into consideration the influence made by the nucleate boiling on the temperature distribution. It was shown that in this case the value of m stayed within the range 0.5–0.9. In the present work we enhance the existing axisymmetric model of the nonstationary heat release into the boundary liquid layer with the dependencies for the main heat fluxes arising from bubble boiling.

3 Numerical simulation

To calculate the heat balance, it is required to know the description of the temperature field. According to the measured wall temperatures and bubble diameters, the temperature gradient near the heater surface exceeds 0.5K/μm. Such detail is possible only if we use numerical modeling. The water temperature profile, the heater surface temperature, and the radial heat flux were calculated with the help of the COMSOL software.

We used a 2D axisymmetric model (Fig.2), which comprised a 55 mm long steel heater with inner diameter of 10 mm and outer diameter of 12 mm, the surrounding water layer was 3 mm thick. To obtain a finer resolution for the superheated layer, we applied a mesh of about 180000 elements with 2 boundary layers near heater-water interface: the first layer is 0.05 mm thick, and has the step of 0.001 mm, the second layer is 0.27 mm thick, and has the step of 0.005 mm (Fig. 3). The remaining part of the model is meshed with the step of 0.1 mm. In order to obtain an overheated layer with a sufficient resolution, we used a mesh with a total number of elements of about 180,000 and two boundary layers near the interface between the heater and water regions: the first 0.05 mm thick layer has a mesh step of 0.001 mm and the second 0.27 mm thick layer has a mesh step of 0.005 mm. The rest of the model is covered with a 0.1 mm mesh.

The water flows into the channel with uniform velocity of u_{in}=0.2 m/s and flows out at atmospheric pressure. No-slip boundary conditions were set at the outer wall and the heater surface. To obtain the temperature field relevant to experimental bubbles data, the resistive heat release inside the heater was calculated by solving the problem of electrical conductivity with the boundary conditions based on the voltage history of each individual experiment. The bulk temperature T_b was specified at the water inlet and at the outer wall. The zero normal heat flux condition was used at the water outlet.

The laminar fluid flow problem

$$\rho\frac{\partial \mathbf{u}}{\partial t} - \rho(\mathbf{u}\cdot\nabla)\mathbf{u} = \nabla\cdot[-p\mathbf{I} + \mu(\nabla\mathbf{u} + (\nabla\mathbf{u})^{\mathrm{T}})] \quad (8)$$

$$\rho\nabla\cdot\mathbf{u} = 0 \quad (9)$$

coupled with the heat transfer in fluid problem

$$\rho C_p\frac{\partial T}{\partial t} + \rho C_p\mathbf{u}\cdot\nabla T - \mathrm{k}\nabla\cdot\nabla T = 0 \quad (10)$$

was solved for the subcooled water flow domains. Here T denotes transposition. The heat transfer in solid in presence of resistive heat

$$\rho C_p\frac{\partial T}{\partial t} - \mathrm{k}\nabla\cdot\nabla T = \frac{1}{2}\mathbf{J}\cdot\mathbf{E} \quad (11)$$

was coupled with the direct electric current problem

$$\nabla \cdot \mathbf{J} = 0 \quad (12)$$

$$\mathbf{J} = \left(\sigma + \epsilon_0 \epsilon_r \frac{\partial}{\partial t}\right) \mathbf{E} \quad (13)$$

$$\mathbf{E} = -\nabla V \quad (14)$$

The changes in the temperature profile that took place beyond the first 50 mm of the heater length and the first 3 mm of the heater thickness, as well as the gravitation effect, were checked numerically and found negligible. The heater surface temperature measured by thermocouples showed good agreement (<5%) with numerical modeling until the moment of the fully developed nucleate boiling.

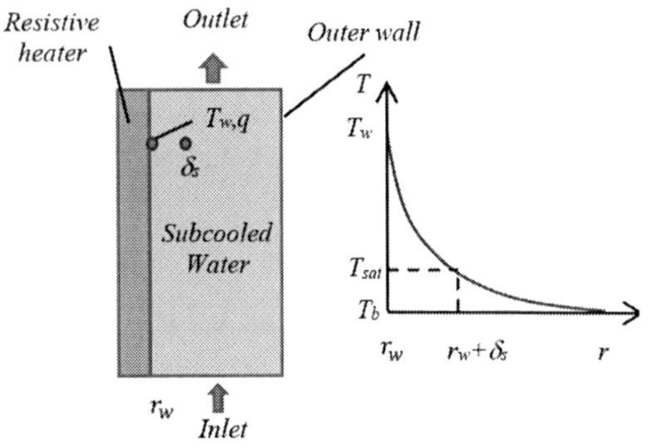

Fig. 2. Model for the numerical simulation.

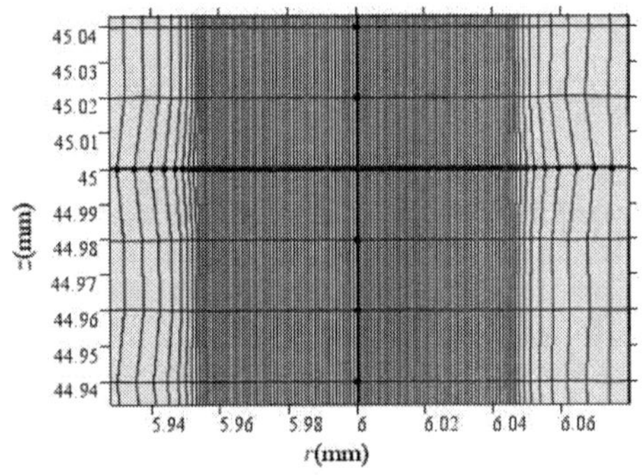

Fig. 3. Mesh near the heater-water interface.

In addition to the heat fluxes described by the heat transfer equations (10-11), this paper also introduces evaporative heat transfers associated with the life cycle of bubbles (Fig. 4). In this case, the bubbles, as well as the vapor cavities of a specific localization, are not considered, and the sequence of events of the bubble life cycle is not modeled. Heat fluxes are averaged over the bubble lifetime , $\frac{1}{f} \approx 0.2\,ms$, and over the area of its influence of size $D_{\pm}$ in the plane of the heater surface. Since the axisymmetric model cannot be used for representation of an individual bubble, we model a solid ring of bubbles at the height of 45 mm from the beginning of the channel. Modeling the ring of bubbles, rather than a continuous coating of the heater surface, makes it possible to estimate the effect of convection and thermal conductivity along the channel on the temperature field, and to determine the dimensions of the long-term area of influence over the entire period of the nucleate boiling, which lasted about 2.5 ms. The limitations of the axisymmetric model do not allow for the distribution of heat in the angular direction. Due to this fact, in this work we obtained a high estimate of the influence of evaporative heat transfer on the change in the thickness of the superheated layer and the coefficient m.

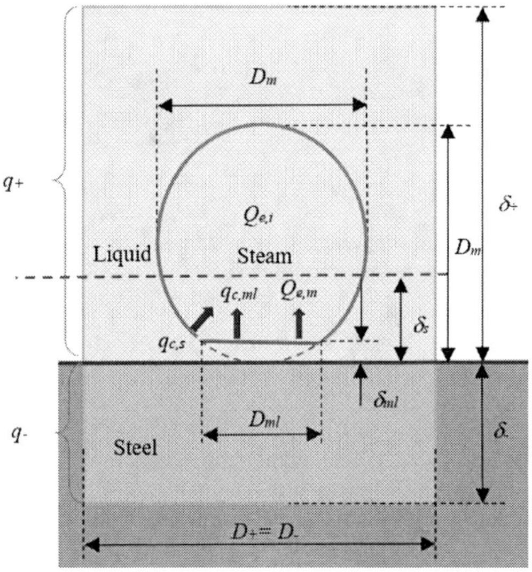

Fig. 4. The bubble and the heat fluxes associated with it: $Q_{e,ml}$ – microlayer evaporation heat, $q_{c,ml}$ – microlayer conduction heat flux, $q_{c,s}$ – superheated layer conduction heat flux, $Q_{e,i}$ – initial evaporation heat.

In addition to estimating the local heat fluxes in the area of bubble influence, it would be interesting to measure the contribution of all bubbles to the total heat transfer from the heater surface. To do this, we need to know the nucleation centers density. When processing video data, for each frame we calculated the corresponding time from the start of the heating and wrote out the coordinates and sizes of all the bubbles in the frame. Based on the scale of the frames (0.01 mm/pixel along the x axis and 0.02 mm/pixel in the y axis) and the angle of inclination between the heater surface and the filming direction, we obtained the area of the frame (8.5 mm^2) and the nucleation centers density, n, in each frame. The obtained values are approximated by the equation

$$n = 0.243 * \exp((t - 6.14)/0.77) \quad (15)$$

where the time from the start of heating is given in milliseconds. The nucleation centers density at the end of the nucleate boiling period reaches about 5 mm^2. Therefore, the distance between the centers of nucleation is about 0.44 mm. With a maximum observed bubble diameter of this period up to 0.2 mm, the distance between the centers of the bubbles is approximately 2.2 diameters. When studying the sequence of frames, it can be seen that the bubbles appear in the same places – nucleation centers, the number of which becomes bigger as the temperature of the heater increases. Throughout their life, bubbles are spherical and touch the surface of the heater. They do not slide along the surface, but collapse, or decrease, remaining at the nucleation point. For each nucleation center, a sequence of frames corresponding to the local maxima of the bubble diameters was chosen. The volume-averaged maximum diameters, for all the nucleation centers in the field of view, were approximated by the formula

$$D_m = 0.45 - 0.17 * t + 0.017 * t^2 \quad (16)$$

where the time from the start of heating is given in milliseconds, whereas the maximum bubble diameter is measured in millimeters. For each measurement of the local maximum of the bubble diameter, its lifetime was determined as the period of time elapsed since the previous measurement of the local maximum of the diameter in the same nucleation center. We also measured the inverse of the lifetime value – the nucleation frequency. The average nucleation frequency was $f = 4.67$ kHz. In the model, the dependency of the nucleation frequency on the start of the heating was represented as

$$f = 16.6 - 1.6\,\mathrm{t} \quad (17)$$

Here the frequency is expressed in kHz, and the time from the start of heating – in milliseconds. The approximations presented here for the nucleation centers, the nucleation frequency, and the bubble diameter, as well as the results of calculations of the temperature field with allowance for evaporative heat transfer, were used in the same experiment with an initial temperature of $T_b = 90$ ° C and an average growth rate of the heater surface temperature of 5.6 K/ms.

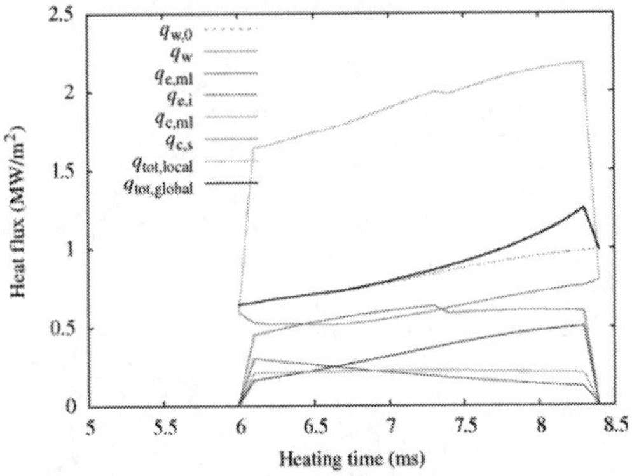

Fig. 5 Heat fluxes associated with a bubble.

The size of the influence area of an individual bubble can be estimated based on its lifetime. Due to forced convection with a maximum flow rate of 0.2 m/s, the temperature distortions associated with the formation and collapse of bubbles move 40 mkm. However, it should be taken into account that at the distance of one bubble diameter from the wall, the water flow velocity does not exceed 0.05 m / s. In this case, the estimation of the influence distance due to forced convection is reduced to 10 μm. The distance of influence due to thermal conductivity, calculated by the formula $\sqrt{\alpha t}$, amounts to about 5 μm for water at the saturation temperature of 102 ° C and to δ_{-}=30 μm for steel. Therefore, the maximum diameter of the influence area of a bubble during its lifetime is stipulated by the thermal conductivity of steel, and is equal to D_{-}= D_{+}=0.23 mm, which corresponds to the influence area $A_{-} = A_{+} = \frac{\pi D_{+}^2}{4} = 4.15 \cdot 10^{-8} m^2$. Judging by the size of the temperature perturbations, which are visible on the frames following the bubble collapse, the thickness of the influence area is about one and a half maximum diameters, $\delta_{+}(\tau) = \frac{3}{2} D_m(\tau)$.

The heat flux per unit surface of the heater is considered to be composed of four components: the bubble and the microlayer evaporation heat, both averaged over the lifetime and the influence area; the thermal conductivity through the microlayer; the thermal conductivity through the superheated layer.

$$q''_{lg} = q''_{e,i} + q''_{e,ml} + q''_{c,ml} + q''_{c,s}. \quad (18)$$

This is the local heat flux in the influence area of a bubble. When estimating the contribution of bubbles to the overall heat transfer, it is necessary to use data on the density of nucleation centers:

$$q''_{lg,n} = q''_{lg} A_{+} n. \quad (19)$$

The heat transfer through the dry spot and the rewetting effect are neglected, since it is known that in this series of experiments bubbles were spherical and the diameter of a dry spot was much smaller than that of a bubble. The heat flux of the initial evaporation, $q_{e,i}$, is determined as the evaporation heat released during the bubble growth, averaged over the influence area and the bubble lifetime

$$Q_{e,i} = h_{lg} \rho_g \frac{\pi D_m^3}{6} \quad (20)$$

$$q''_{e,i} = Q_{e,i} f / \mathrm{A}_{+} \quad (21)$$

The absorption of heat by a bubble due to evaporation of the microlayer is determined by the volume of the microlayer, $\frac{2}{3} \delta_{ml} \frac{\pi D_{ml}^2}{4}$, where the coefficient 2/3 is responsible for the wedge-shaped narrowing.

$$Q_{e,ml} = h_{lg} \rho_l \delta_{ml} \frac{\pi D_{ml}^2}{6} \quad (22)$$

$$q''_{e,ml} = Q_{e,ml} f / \mathrm{A}_{+} \quad (23)$$

The heat flux due to the thermal conductivity of the microlayer is assumed to be constant within the radius of the microlayer $\frac{D_{ml}}{2}$ from the center of a bubble, and is averaged over the whole bubble influence area.

In accordance with Fig. 4, the radius of the microlayer is defined as follows:

$$\frac{D_{ml}}{2} = (D_m \delta_{ml} - \delta_{ml}^2)^{0.5} \quad (24)$$

$$q''_{c,ml} = \frac{k_l (T_w - T_{sat})}{\delta_{ml}} \frac{\pi D_{ml}^2}{4} / \mathrm{A}_{+} \quad (25)$$

As shown in some papers, including [6–8], the microlayer thickness reaches its maximum value at the stage of bubble growth. Further, the area and thickness of the microlayer decrease due to its evaporation. Therefore, the calculations for a constant-sized microlayer that were carried out in this work are the estimates aimed at evaluation of the relative contribution of each component of the heat flux. The microlayer thickness, similar to the formula obtained in [8], is determined by the expression

$$\delta_{ml} = 0.25\sqrt{\frac{\nu_l}{3f}} \qquad (26)$$

Here, the bubble growth time is estimated as 1/3 of the nucleation period, and the coefficient 0.25 is chosen so that the thickness of the microlayer does not exceed 1.5 µm.

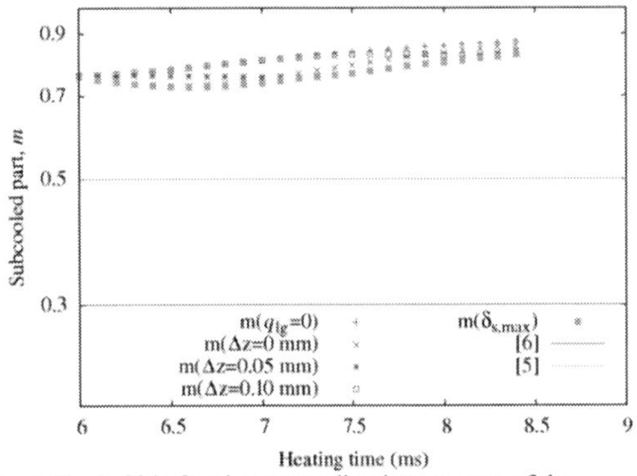

Fig. 6 The bubble fraction protruding into an area of the subcooled liquid.

The heat flux through the superheated layer is considered proportional to the wall overheating $(T_w - T_{sat})$, but the layer thickness varies from δ_{ml} to δ_s. This heat flux is integrated over the heater area under the corresponding part of a bubble and is averaged over the entire influence area.

$$q''_{c,s} = \frac{1}{A_+}\int_{r_w+\delta_{ml}}^{r_w+\delta_s} k_l(T_w - T_{sat})\,\pi\left(\frac{D_m}{r-r_w} - 2\right)dr \qquad (27)$$

The obtained value q''_{lg} is the heat flux, additionally transferred from the heater into the liquid. Heat transfer is realized in the form of a volumetric heat source, q'''_{lg}. With respect to z, this source is determined in the range $45 - \frac{D_+}{2} < z < 45 + \frac{D_+}{2}$. In the thickness of the heater, the heat is absorbed to the depth δ_- . In a liquid, heat is released linearly at the distance up to $\delta\_+ = 1.5\ D_m$ and reaches its maximum at the center of a bubble. The heat release integral q'''_{lg} is equal to zero over all r .

$$q'''_{lg} = \begin{cases} -\dfrac{q''_{lg}}{h_i} & r_w - \delta_- < r < r_w \\ \dfrac{8}{3}\dfrac{q''_{lg}(\mathrm{r} - \mathrm{r_w})}{D_m^2} & r_w < r < r_w + \dfrac{D_m}{2} \\ \dfrac{4}{3}q''_{lg}\dfrac{r_w + 3\dfrac{D_m}{2} - \mathrm{r}}{{D_m}^2} & r_w + \dfrac{D_m}{2} < r < r_w + \delta_+ \end{cases}$$

It follows from Fig. 5 that in the neighborhood of nucleation the effect of heat flux on the heating of the wall layer is inhomogeneous both with respect to time and to the contribution of an individual bubble. It is impossible to isolate the biggest heat flux among the ones we have considered, whereas the heat flux in the bubble neighborhood $q''_{tot,local} = q''_w + q''_{lg}$ is twice bigger than the calculated value for a homogeneous heat flux, $q''_{w,0}$.

4 Conclusions

Taking into account the contribution of interphase transitions to heat exchange, the change in the thickness of the superheated layer is proportional to the change in the maximum bubble diameters. The fraction of the bubble protruding into the region of the subcooled liquid, as a result of the introduction of heat fluxes associated with the bubble into the model, decreased by no more than 0.1, remaining substantially higher than the values indicated by other authors (Fig. 6).

Acknowledgements

This research was performed at the Melentiev Energy Systems Institute SB RAS using the High-Temperature Circuit Multiple Access Research Center and was supported by the Basic Research Program of the SB RAS No. AAAA-A17-117030310443-5.

References

1. Mohanty R.L., and Das M.K., Renew. Sustain. Energy Rev., 78, pp. 466–494 (2017).
2. Hoang N.H., Song C.H., Chu I.C., and Euh D. J., Int. J. Heat Mass Trans., 112, pp. 454-464 (2017).
3. Levin A., and Khan P., MATEC Web of Conference, **115**, STS-33, 8014 (2017).
4. Unal, H. C., International Journal of Heat and Mass Transfer, **19(6)**, 643–649 (1976).
5. Hoang, N. H., Chu, I. C., Euh, D. J., & Song, C. H., International Journal of Heat and Mass Transfer, 94, 174–179 (2016).
6. Guion, A., Langewisch, D., & Buongiorno, J. International Conference on Multiphase Flow, 1–13. (2013).
7. Hansch, S., Narayanan, C., Kcboux, S., Giuhlui, G., & Walker, S. International Topical Meeting on Nuclear Reactor Thermal Hydraulics 2015, NURETH 2015, 6(August 2015), 4848–4861. (2015).
8. Cooper, M. G., Lloyd, A. J. P., Division, C. A., & Company, T. B. International Journal of Heat and Mass Transfer, 12, 895–913. (1969).

Influence of vacuum level on insulation thermal performance for LNG cryogenic road tankers

Lisowski Edward[1], *Lisowski* Filip[2*]

[1] Cracow University of Technology, Institute of Applied Informatics
[2] Cracow University of Technology, Institute of Machine Design

Abstract. In this paper, thermal properties of materials for evacuated insulation systems of double-walled cryogenic tanks were discussed. Than the comparison of insulation variants for LNG cryogenic road tankers was presented. The use of several layers of insulation made of materials such as aerogel and fiberglass or the use of multilayer isolation (MLI) has been compared to the use of perlite powder. The average heat flux through the tank walls and insulation system has been compared under different vacuum levels and in the absence of vacuum. The comparative analysis was performed by applying transient thermal analysis using finite element method.

1 Introduction

Typical cryogenic road tankers consist of inner vessel and outer jacket. The evacuated annular space in between is fully or partially filled by insulation system. The thermal performance of insulation system depends on applied materials and vacuum level. The vacuum level should be constantly controlled during tanks exploitation and improved in case of pressure increase. This paper discussed the influence of vacuum level on thermal performance of different insulation systems for LNG cryogenic road tankers.

2 Insulation materials for LNG cryogenic road tankers

Insulation materials for cryogenic applications can be divided into three groups: foams, bulk-fill and layered [1]. Typical foams including polyurathene or polystyrene have thermal conductivity in the range of 35-55 mW/(mK) [2]. However foams are rigid and flammable [3], which is not favorable for application in mobile tanks.

The bulk–fill powders insulations have been mainly used in stationary or large transportation containers. The best known natural origin microporous material is an expanded perlite with bulk density from 50 to 300 kg/m^3 [4] and thermal conductivity coefficient 1-2 mW/(mK) under vacuum pressure around 0,1 Pa. [5]. Other well known powders are: glass bubbles, silica aerogels or composites. The glass bubbles have a slightly lower heat conduction coefficient than them perlite or aerogel powders. In turn, aerogel powders with microporous cellular structure have high compressive strength, but at high production costs [4]. However, the powder materials have the tendency to settle and compact due to vibration and thermal cycling loads [6].

The best thermal performance, as low as 0,03 mW/(mK), can be obtained in the laboratory conditions for multilayer insulation [7]. MLI, also known as superinsulation, requires vacuum level below 10^{-3} Pa to be fully effective [8]. It consists of a large number of highly reflecting shields separated by spacers having low conductivity. When the pressure distribution is very homogeneous, the local thermal conductivity between the radiation shields is strongly suppressed [9]. MLI is sensitive to mechanical compression and edge effects. The effectiveness of MLI depends on vacuum level. In case of sudden loss of vacuum, thermal performance can degrade rapidly causing a dangerous pressure increase inside the inner tank [10]. Furthermore, the steps of evacuation, heating, and vacuum retention are expensive and time consuming [6].

The other layered insulation systems that incorporate aerogel or fiberglass composite blankets can remain effective even at lower vacuum level [11]. In that case the conduction is greater and those systems will not be as efficient as MLI at higher vacuum pressure.

The aerogels are flexible, highly porous materials, which have low density and high porosity, that makes them very good insulators. However, compressive loads may affect the thermal resistance due to their structure [12].

An another material, developed as an alternative to commonly used perlite is the micro–fiberglass material Cryo-Lite®. Its low density minimizes tanker weight, eliminates problem with settling or compaction and offers fast vacuum pump down rates with minimal outgassing [8]. Due to its low opacity Cryo-Lite® blankets should be used with radiation shields.

Since the effectiveness of MLI is about 50% worse at atmospheric pressure, system involving combination of MLI and aerogels [13] or MLI and fiberglass blankets

* Corresponding author: filip.lisowski@mech.pk.edu.pl

were also tested [8]. Experimental results showed that combination of different insulation materials can improve insulation system effectiveness and it was found that it is significant whether a particular material is on the warm or cold side.

3 Modelling of the heat transfer through tanker insulation

In order to compare thermal performacne of different insulating materials for application in LNG cryogenic road tanker, heat transfer through vary evacuated insulation systems and temperature distribution were examined by applying transient thermal analysis using Ansys software. Two dimensional models within cross–section of the tank walls and insulation system, analogous as shown in the Fig. 1 were accepted. The study included four different insulation materials in several variants as given in table 1. Each variant was tested in three levels of the residual gas pressure (CVP): 10^{-1} Pa, 10^{-3} Pa and 100 kPa.

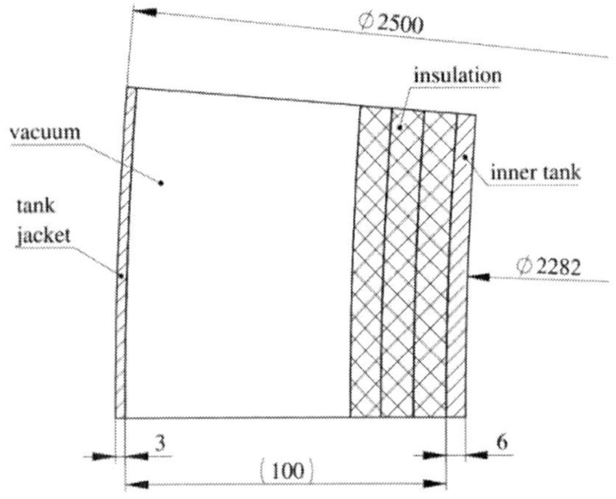

Fig. 1. Cross–section of the exemplary insulation system

Table 1. Variants of the tested insulation systems

Insulation material	Number of layers	Insulation thickness [mm]
Cryogel® Z	2	20
Cryogel® Z	6	60
Cryogel® Z	10	100
Cryo-Lite®	1	25
Cryo-Lite®	2	50
Cryo-Lite®	5	100
MLI	40	16
Perlite	–	100

3.1 Material properties

A stainless steel 1.4301 was accepted as a material of inner vessel and outer jacket. The coefficient of thermal conductivity in the function of temperature for steel 1.4301 is shown in the Fig. 3, whereas for insulating materials such as Cryo-Lite® and Cryogel® Z in the Fig. 2 (vendor data). MLI insulation consisted of 40 layers of double–aluminized Mylar and polyester net spacer. An expanded perlite accepted for analysis has bulk density of 200 kg/m3. The coefficients of thermal conductivity in the function of pressure for MLI and perlite are given in table 2.

Table 2. Coefficients of thermal conductivity for MLI and perlite in the function of pressure [8].

Material	k [mK/m·K]	CVP [Pa]
MLI	0,028	10^{-3}
MLI	0,072	10^{-1}
MLI	35	10^{5}
Perlite	0, 95	10^{-3}
Perlite	1	10^{-1}
Perlite	44	10^{5}

3.2 Assumptions for numerical analysis

– the model is in a state of thermal equilibrium,
– the internal surface of inner tank exchanges heat with liquid by convection with accepted convection coefficient 337 W/(m2K) at 77 K [14],
– the external surface of tank jacket exchanges heat with environment by convection with accepted convection coefficient 8,89 W/(m2K) at 293 K [14],
– heat is transferred through insulation layers only by conduction with accepted coefficient of thermal conductivity,
– for residual gas pressure of 10-1 Pa and 100 kPa, heat is transferred by conduction and by radiation,
– for residual gas pressure of 10-3 Pa, heat is transferred through residual gas only by radiation,
– for layered insulation systems, external layer of insulation is covered with an aluminum radiation shield with emissivity ε=0,03 [9],
– surface to surface model of radiation was accepted.

Coefficient of thermal conductivity of residual gas under pressure of 10^{-1} Pa was assumed equal to 0,01 W/(mK) [4]. For the case of no vacuum, the assumed gas pressure of 100 kPa corresponds to atmospheric pressure. Coefficient of thermal conductivity in the function of temperature for dry air was accepted as presented in the Fig. 4.

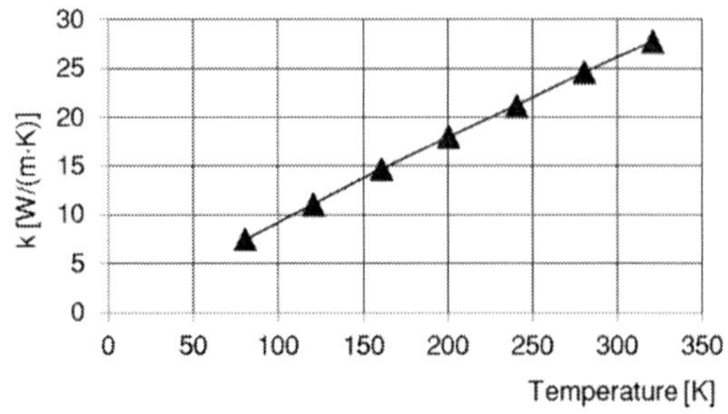

Fig. 4. Thermal conductivity of dry air at 100 kPa [15]

https://doi.org/10.1051/matecconf/201824001019

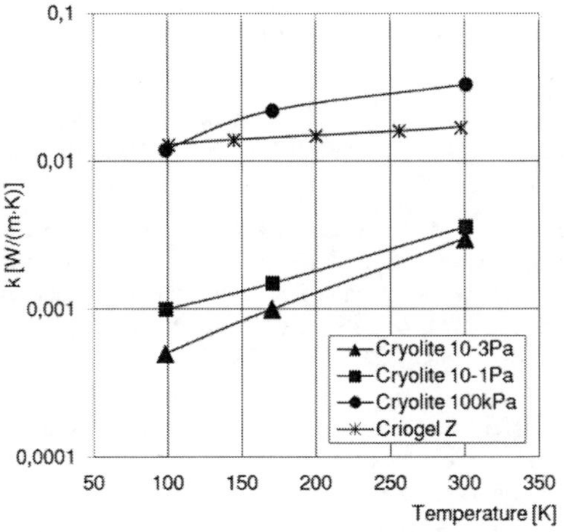

Fig. 2. Thermal conductivity of Cryo-Lite® and Cryogel® Z

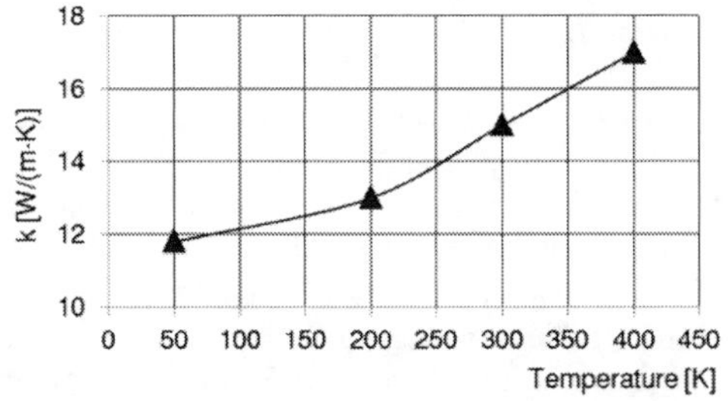

Fig. 3. Thermal conductivity of stainless steel 1.4301 [16]

3.3 Simulation results

The comparative results of the transient thermal analysis are presented in diagrams as well as in the maps of heat flux and temperature distribution. The exemplary results of heat flux and temperature distribution in the cross-section of insulation system involving Cryo-Lite® under vacuum pressure of 10^{-1} Pa are shown in the Fig. 5. and Fig. 6. Average heat flux in the cross-section of the inner tank wall for insulation systems containing Cryo-Lite® and Cryogel® insulation blankets are summarized in the Fig. 7 and Fig. 8. Fig. 9 presents results for insulation systems including MLI, perlite and only residual gas.

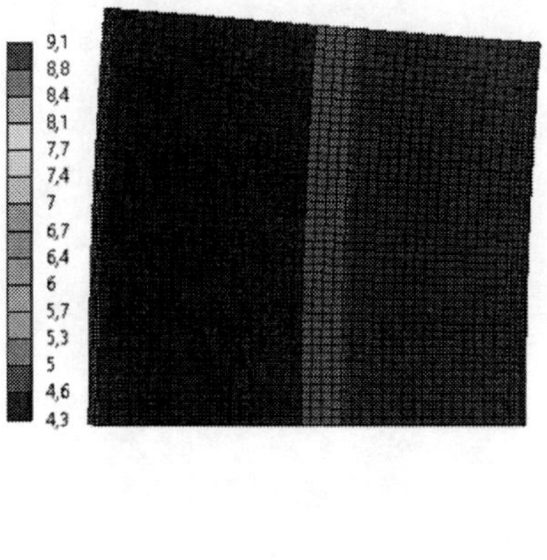

Fig. 5. Heat flux distribution in the cross section of insulation system with Cryo-Lite® under vacuum pressure of 10^{-1} Pa

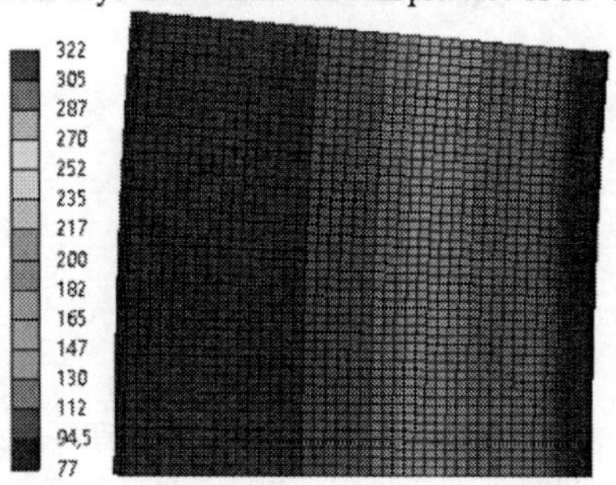

Fig. 6. Temperature distribution in the cross section of insulation with Cryo-Lite® under vacuum pressure of 10^{-1} Pa

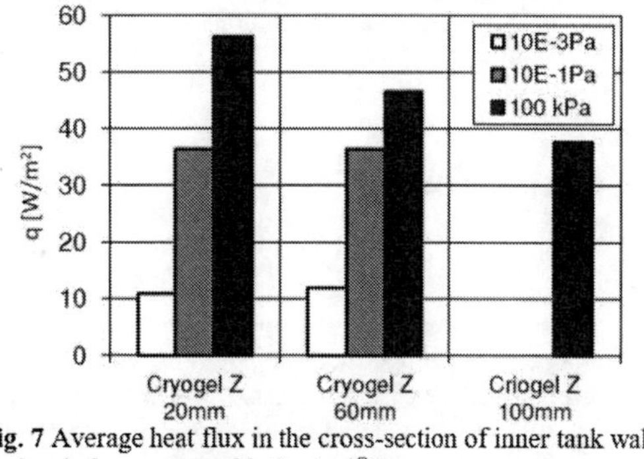

Fig. 7 Average heat flux in the cross-section of inner tank wall for insulation system with Cryogel®Z

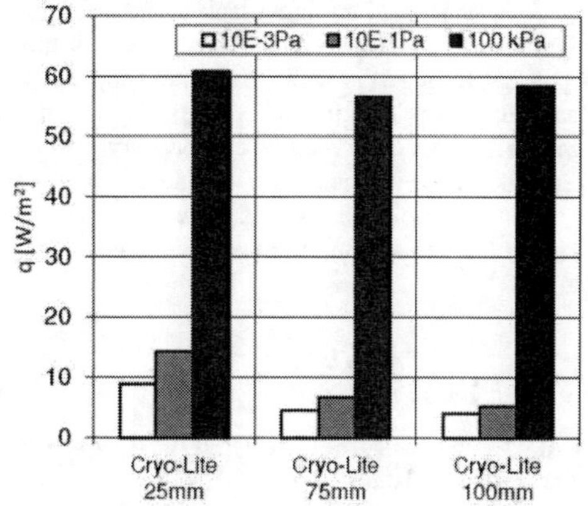

Fig. 8. Average heat flux in the cross-section of inner tank all for insulation system with Cryo-Lite®

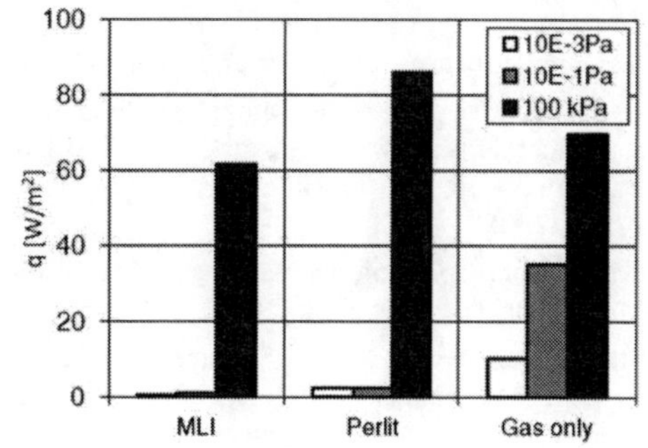

Fig. 9. Average heat flux in the cross-section of inner tank wall for insulation systems with MLI, perlite and gas only

4 Results and discussion

The results of thermal analysis of 2D models show that multilayer insulation (MLI) system in comparison with other tested cryogenic insulation systems is the best performer in the high vacuum (HV) range under residual gas pressure of 10^{-3} Pa and in the soft vacuum (SV) range under pressure of 10^{-1} Pa. Bulk-fill expanded perlite is the second, followed by the fiberglass and aerogel blankets. Cryogel® aerogel insulation blanket is the best performer in the range of no vacuum (NV) under pressure of 100 kPa. The insulation system with Cryo-Lite® fiberglass blankets had slightly worse thermal performance than insulation systems with MLI and perlite in the soft and high vacuum, but it was better in no vacuum range.

5 Conclusions

The liquefied natural gas is condensated at - 162 °C. In order to maintain a liquid state of gas, it is important to keep its proper temperature level throughout the storage period. Therefore the thermal insulation of LNG cryogenic tanks plays an important role. Its effectiveness depends on applied materials and pressure of residual gas. The presented analysis has shown that already the usage of vacuum significantly reduce heat leakage to the inner tank. In addition the use of modern insulating materials allows to control the heat flux passing through the walls of the tank and insulation system. However, it should be noticed, that the thermal analysis of the whole structure of LNG cryogenic road tankers should be extended to include the analysis of heat flux passing both through the tank walls and through the internal supports between inner tank and outer jacket.

REFERENCES

1. J.E Fesmire, S.D Augustynowicz, Cryogenic Thermal Insulation Systems, 16th Thermal and Fluids Analysis Workshop, Orlando, (2005)
2. M. Chorowski, Cryogenics Fundamentals and Applications, IPPU Masta, Gdańsk (in Polish) (2005)
3. A. Bahadori, Thermal Insulation Handbook for Oil, Gas and Petrochemical Industries, Elsevier (2014)
4. U. Bünger, G. Owren, Development potentials for small mobile tanks with vacuum powder insulations, International Journal of Hydrogen Energy, pp. 259-279, **4** (1998)
5. T.M. Flynn, Cryogenic engineering. Second Edition, Marcell Dekker, New York (2005)
6. S.D Augustynowicz, J.E. Fesmire, J.P.Wikstrom, Cryogenic Insulation Systems, 20th International Congress of Refrigeration, Sydney (1999)
7. J.E. Fesmire, Standardization in cryogenic insulation systems testing and performance data, Physics Procedia, pp. 1089-1097, **67** (2015)
8. A. Kogan, , J.E. Fesmire W. Johanson, J. Minnick, Cryogenic Vacuum Insulation for Vessels and Piping, NASA Technical Report, Nr. KSC-2010-126 (2010)
9. H. Reiss, A coupled numerical analysis of shield temperatures, heat losses and residual gas pressures in an evacuated super-insulation using thermal and fluid networks Part I: Stationary conditions., Cryogenics, pp. 259-279, **44** (2004).
10. G.F. Xie, X.D. Li, R.S. Wang, Study on the heat transfer of high-vacuum-multilayer-insulation tank after sudden, catastrophic loss of insulating vacuum, Cryogenics, pp. 682– 687, **50** (2010).
11. B. E. Coffman, J.E. Fesmire, S. White, G. Gould, S. Augustynowicz, Aerogel blanket insulation materials for cryogenic applications, Advances in Cryogenics Engineering AIP Conf. Proc., pp. 913-920, **1218** (2010)
12. A. Hoseini, A. Melekian, M. Bahrami, Deformation and thermal resistance study of aerogel blanket insulation material under unaxial compression, Energy and Buildings, pp. 228-237, **130** (2016)
13. W. L. Johnson, J. A. Demko, J. E. Feasmire, Analysis and testing of multilayer and aerogel insulation configurations, Advances in Cryogenics Engineering *AIP Conf. Proc.*, pp. 780-787, **1218** (2010).
14. W. Czyżycki, Modeling of heat flow through multilayer internal supports of cryogenic vessels, Technical Transactions, Mechanics, pp.27-34, **2-M** (2015)
15. N.B. Vargaftik., L.P. Filippov, A.A. Tarzimanov, Totskii E.E., Handbook of thermal conductivity of liquids and gases, CRC Press. (2005)
16. W. Czyżycki, Heat flow modelling on thermal insulation of cryogenics tanks using SolidWorks Simulation package, Technical Transactions, Mechanics, pp.29-36, **4-M** (2011)

Development of hybrid model based on Lattice Boltzmann Method and Cellular Automata devoted for phase transformation – simulation of heat flow with consideration of enthalpy

Łukasz Łach [1,*], *Robert* Straka[1], and *Dmytro* Svyetlichnyy[1]

[1]AGH University of Science and Technology, Faculty of Metals Engineering and Industrial Computer Science, al. Mickiewicza 30, Kraków, Poland

Abstract. In heat treatment of materials, the phase transformation is an important phenomenon, which determines the final microstructure. The microstructure of different materials described by such parameters as morphology, grain size, phase fraction and their spatial distribution, largely effects on the mechanical and functional properties of final products. The subject of the work is a development of a hybrid model based on CA and Lattice Boltzmann method (LBM) for modeling of the diffusion phase transformation. The model has a modular structure and simulates three basic phenomena: diffusion, heat flow and phase transformation. The objective of the paper is a presentation of module of the hybrid model for simulation of heat flow with considering of enthalpy of transformation. This is one of the stages in the development of the model and obtained results will be used in a combined solution of heat transfer and diffusion during the modeling of diffusion phase transformations. Lately, the model will be extended to three dimensions and will use hybrid computational systems (CPU and GPU). CA and LBM are used in the model as follows. LBM is used for modeling of heat flow, while CA is used for modeling of microstructure evolution during the phase transformation. The main factors considered in the model are the enthalpy of transformation and heat transfer. The paper presents the results of the modeling of the new phase growth determined by different values of overcooling affecting on different values in the enthalpy of transformation. The heat flow is simulated and the results for some modeling variants are shown. Examples of simulation results obtained from the modeling are presented in the form of images, which present the growth of new phase and temperature distributions.

Introduction

1 fundamental principle of materials engineering is that the microstructure affects on the physical properties and behaviour of a material. Microstructure is formed through a variety of different processes and in the most cases it is created during the phase transformation under effects of the temperature and/or pressure. Microstructures can be created through deformation or processing of the material (e.g. rolling, welding) and also artificially by combining different materials to form a composite material. Because the microstructure can be described as a composition of phases presented in a material (their morphology, distribution and internal structure), the phenomenon of the phase transformation plays an important role in the formation of the microstructure. Many of important material properties can be engineered by control of its solid state phase transformations, so experimental and modeling studies related to this type of the transformations are intensively realized. Thus, a modeling of the phase transformations is a primordial task for development of new materials with new properties, for improving the performance of existing materials, or designing new processes. Recently, many analytical and numerical methods are used for modeling of phase transformation, and various aspects of transformation are considered, for example nucleation of grains of new phase [1,2], grain growth [3,4], diffusion [5–7], *etc.* New numerical models of phase transformations are created based on the cellular automata (CA), finite difference method (FDM), phase field (PF) and others methods. Two-dimensional cellular automata model was used by An *et al.* [8] to simulate the ferrite-austenite transformation in binary low-carbon steels, while the Varma *et al.* [9] used two-dimensional diffusion based model to describe transformation of austenite into ferrite and pearlite under continuous cooling conditions. Three dimensional CA model was developed for example by Svyetlichnyy *et al.* [10] and austenite into ferrite and perlite transformation was modeled. FDM method was used for example by Wang *et al.* [11] for the verification of two models of phase transitions described by various mechanisms of the new phase growth and by Singh [12] for modeling the allotropic transformation of titanium as a result of high temperature oxidation. PF method was used for study the

* Corresponding author: lach@agh.edu.pl

cubic to monoclinic martensitic phase transformation in nickel–titanium (NiTi) shape memory alloys [13] and for modeling of martensitic transformation [14].

Taking into account the modeling of the diffusion phase transformations and existing models, it can be seen that a lot of limitations exists for example in relation to the choice of the shape of growing grains and the introduction of variable grain growth rate, and only some models allow for accurately modeling of diffusion phenomena and there are a few comprehensive studies, which combine the final microstructure with the actual conditions of its formation. Therefore, the objective of the work is a development of a new hybrid model based on Lattice Boltzmann Method and Cellular Automata for modeling of the diffusion phase transformations. The model will provide completely new opportunities and will allow to comprehensive modeling of the diffusion phase transformations in many processes, which consist of a heating, annealing and cooling.

The paper presents the structure of the hybrid model and its basic modules – the main attention is paid on the temperature and microstructure evolution modules, and the simulation of heat flow with consideration of the enthalpy of transformation. It is one of the stages in development of the model. Examples of the simulation results of the heat flow modeling using the LBM and CA are also presented.

2 Hybrid LBM and CA model – heat flow module

The part of the structure of the hybrid model of the phase transformation that relates to heat flow is shown in Fig. 1. The whole model is based on two modeling methods. The first method is the LBM method and the second one is cellular automata. In the main part, the model is based on the LBM method, which is a relatively new method and it is just beginning to be widely used in modeling. Taking into account that this method has got excellent numerical stability and constitutive versatility, it can play an essential role as a simulation tool for modeling of different phenomena and processes. Recently, LBM has been used for different issues, for example: modeling and simulation of liquid jet breakup [15], modeling dendritic solidification under forced and natural convection [16], modeling of thermal conduction in composites with thermal contact resistance [17], modeling of transport phenomena in fuel cells and flow batteries [18], modeling of the additive layer manufacturing [19]. Lattice Boltzmann method consider flows to be composed of a collection of pseudo-particles that are represented by a velocity distribution function, so it can be applied for modeling of different variants of flows, as well as phase transformations in flows. Taking into account the advantages of the LBM method and its applicability in the flow modeling, the heat flow model with consideration of enthalpy is based on this method. One of the most effective methods for modeling of microstructure evolution during various processes is cellular automata. Models based on cellular automata allow to take into account several simultaneously occurring phenomena, bringing modeling closer to real processes. CA allow for modeling of dynamic processes occurring in the materials, for example solidification [20], static [21] and dynamic recrystallization [22], phase transformation [10]. CA can be also used for modeling of phenomena in complex processes, such as additive layer manufacturing of orthopaedic implants [23], flat rolling [24] and shape rolling [25]. Taking into account the advantages of cellular automata, this method is used in the microstructure evolution model for modeling of nucleation and grain boundary movement. Since the basic structure of the LBM method is the same as in the synchronous automata (LBM originates from the variant of CA application known as Lattice Gas Automata), it is also an ideal platform for realizing combinations of these two methods in one modeling solution. The model can be also considered as a combined solution for heat flow, diffusion, and grain growth. The main modules considered in the model and also their interactions are shown in Fig.2. Taking into account this scheme, the main module considered in the paper is the temperature module and its connection with the microstructure module, marked by dashed rectangle. As can be seen on the scheme, the temperature influences directly on the nucleation of new grains and grain boundary movements, while the enthalpy of transformation is associated with the temperature. The next section presents the results of modeling for the analyzed issues. The diffusion module and its connection with the microstructure module have been developed at the previous stage of work (unpublished manuscript).

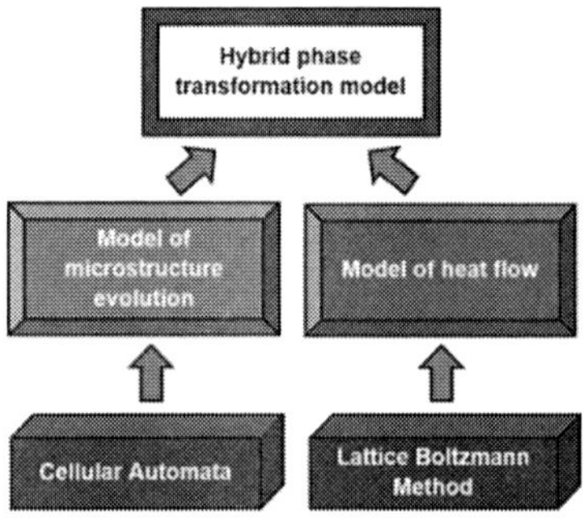

Fig. 1. Structure of hybrid model in part related to heat flow.

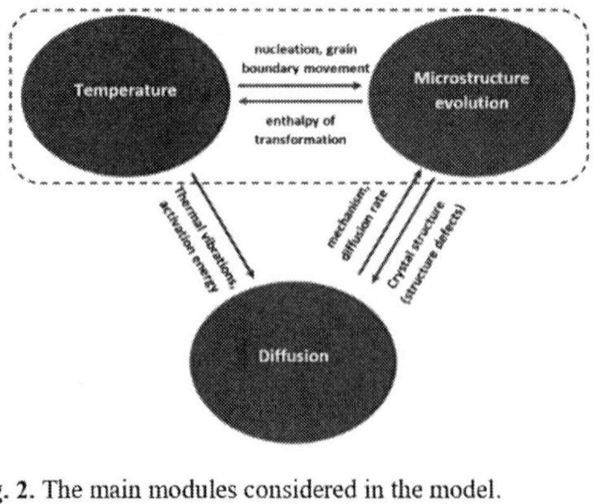

Fig. 2. The main modules considered in the model.

3 Analysis and modeling

On this stage of model development, several variants of heat flow with consideration of enthalpy were modeled using LBM algorithm. The results obtained from the heat flow model were directly linked with the microstructure evolution model, based on cellular automata. Because of there are no applications based on CA and LBM for solution the problem mentioned above, the one dimensional model was developed at first to obtain the preliminary results and verify the correctness of the proposed approach. The 1D Fourier equation describing the heat flow was used in the calculations:

$$\frac{1}{\alpha}\frac{\partial T}{\partial t} = \frac{\partial^2 T}{\partial x^2} \tag{1}$$

where: T – temperature, α - thermal diffusivity.

The lattice Boltzmann – BGK (Bhatnagar–Gross–Krook) model with single-relaxation-time was used for modeling. Then, evolution of distribution function can be described by the following expression:

$$f_k(x+\Delta x, t+\Delta t) - f_k(x,t) = -\frac{\Delta t}{\tau}\left[f_k(x,t) - f_k^{eq}(x,t)\right] \tag{2}$$

where: $f_k(x,t)$ and $f_k^{eq}(x,t)$ are the particle and equilibrium distribution functions at (x,t) and τ is the single-relaxation-time parameter that controls the rate of solution approach to local equilibrium f_k^{eq}.

The equilibrium distribution function f_k^{eq}:

$$f_k^{eq} = w_k \phi(x,t) \tag{3}$$

where w_k stands for the weighting factor in the direction k and ϕ is the dependent variable (temperature T in the presented task).

According to the equation (2) two steps can be considered:
collision:

$$f_k(x, t+\Delta t) = f_k(x,t)[1-\omega] + \omega f_k^{eq}(x,t) \tag{4}$$

where: $\omega = \Delta t/\tau$.
streaming:

$$f_k(x+\Delta x, t+\Delta t) = f_k(x, t+\Delta t) \tag{5}$$

Fig. 3 shows D1Q2 scheme, which was applied for modeling. The lattice velocities e_i and corresponding lattice factors for D1Q2 scheme are given as: $e_1 = 1$, $e_2 = -1$, $w_1 = 1/2$, $w_2 = 1/2$.

Fig. 3. D1Q2 LBM lattice.

The initial states for nodes in the calculation grid and their update during the transformation are presented in Fig. 4. Taking into account the type of state, the node can be in one of three states: Austenite (A), Interface (I) and Ferrite (F). Dark red nodes represents the austenite, yellow - interface and gray - ferrite.

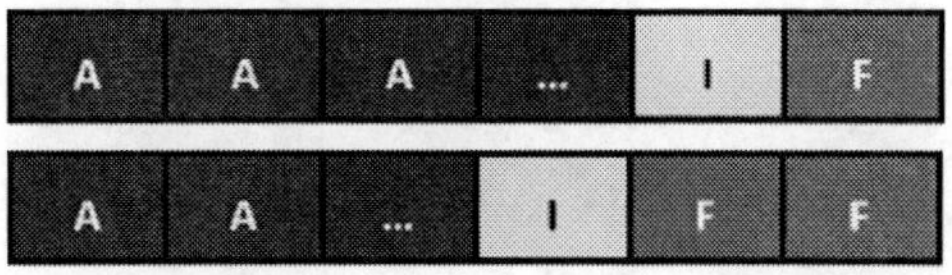

Fig. 4. The changes of states during the transformation.

The calculation algorithm used in the heat flow model includes the following repeated steps:
1. Calculation of a boundary velocity v: $v = k_v(T_p - T_i)$, where: k_v – velocity coefficient, T_p – temperature of phase transformation, T_i - interface temperature,
2. Determination of the quantity, mass or volume of the transformed material: $\Delta\varphi = v\Delta t$.
3. Determination of a fraction of new phase in interface node: $\varphi_t = \varphi_{t-1} + \Delta\varphi_i$.
4. Fraction checking:
4a) if $\varphi_t < 1 \Rightarrow$ go to point 5,
4b) if $\varphi_t \geq 1$ then:
– $\varphi_t' = 1$,
– change the state of node from I to F and the neighboring node from A to I,
– set the value of the fraction for the new I:
$\varphi = \varphi_t - \varphi_t'$, $\Delta\varphi_{i+1} = \varphi_t - 1$,
– the $\Delta\varphi$ for the old I node:
$\Delta\varphi_N = 1 - \varphi_{t-1} = \Delta\varphi_i - \Delta\varphi_{i+1}$
5. Calculation of heat source during the transformation: $Q = k_q\Delta\varphi$, where: k_q – specific enthalpy,
6. Calculation of new temperature: $T = \Sigma f_i + Q$,
7. Determination of the equilibrium distribution functions f^{eq} (3),
8. Collision - determination f^{out} (Eq.4),
9. Streaming - transfer of the distribution function to the neighboring nodes, (Eq.5)
10. Consideration of boundary conditions.

4 Results and discussion

The parameters used for modeling in the first case were as follows: number of nodes - *nodes* = 100, number of time steps - *steps* = 300, $\Delta x = 1$, $\Delta t = 1$, $k_v = 0.006$, $k_q = 25$. For the calculations the following relationship between the relaxation parameter (ω) and thermal diffusivity (α) was used: $\omega = 1/\left(\frac{\alpha}{\Delta t \cdot \left(\frac{\Delta x}{\Delta t}\right)^2} + 0.5\right)$.
Bounce back boundary conditions were applied for the edges. The initial temperature for all nodes was set to 750°C, while the temperature of phase transformation was equal to 800°C. The initial states for the nodes were the same as in Fig. 4. Fig. 5 shows three different variants of interface movement and growth of new phase depending on the value of thermal diffusivity α (time is on y axis). The thermal diffusivity has an influence on the temperature distribution and for higher value of α, the heat transfer is faster and affects on the rate of phase transformation as can be seen in the presented images.

Fig. 5. Growth of new phase depending on the thermal diffusivity α: a) 0.25, b) 0.5, c) 1.

The next presented results were obtained with the same base parameters as in the first case, and the effect of change of k_q coefficient on temperature and rate of phase transformation was studied. The calculations were realized with the same $\alpha = 0.5$. Fig. 6 shows the growth of new phase for three different values of k_q. For lower values of k_q, the rate of phase tansformation is higher, which is related with the fact that the temperature in the interface grows slower and this has an impact on higher values of boundary velocity. The temperature distributions after 200 time steps are presented in Fig. 7. As can be seen, a higher value of k_q gives a higher temperature.

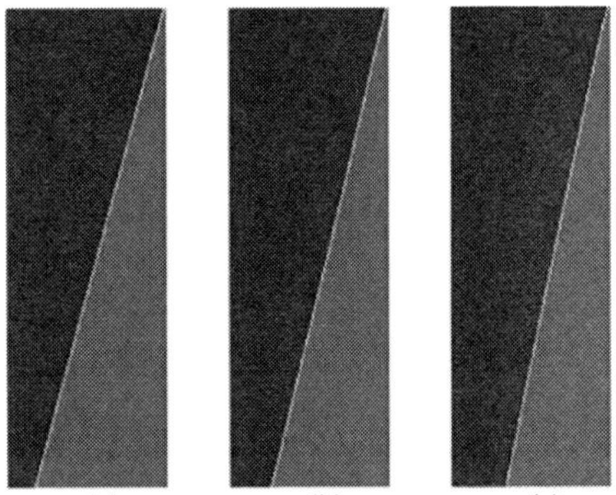

Fig. 6. Growth of new phase depending on k_q: a) 5, b) 10, c) 15.

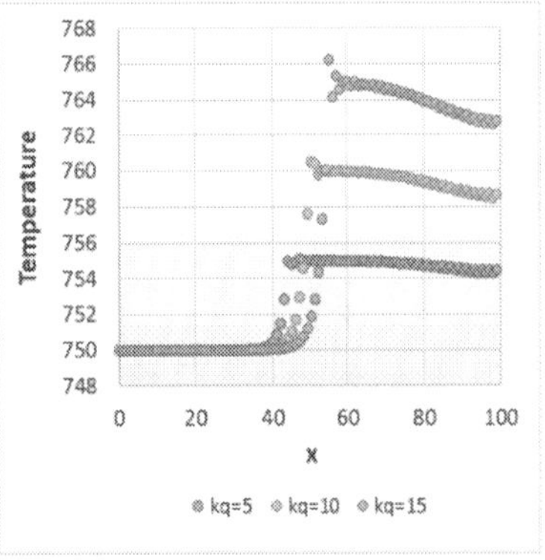

Fig. 7. Temperature distributions after 200 time steps for different value of k_q: a) 5, b) 10, c) 15.

The influence of coefficient k_v was modeled in the last variant. The base simulation parameters were the same as in the first variant, $\alpha = 0.5$ and $k_q = 25$. The simulation results for different value of k_v are presented in Fig. 8. Because k_v is used for the calculation of velocity, which determines the quantity of the transformed matter, this coefficient influences on the rate of the transformation.

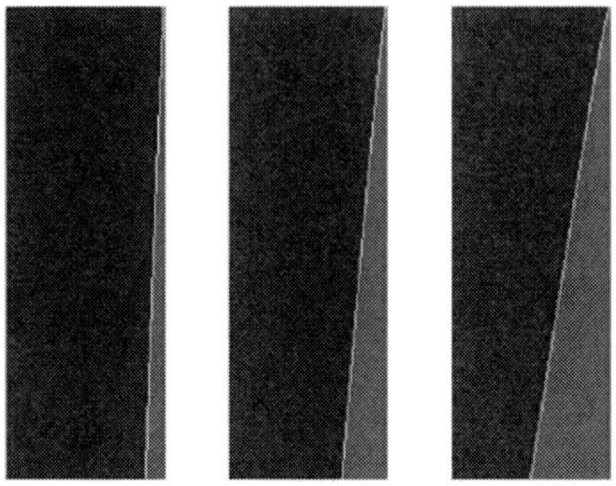

Fig. 8. Growth of new phase depending on k_V: a) 0.001, b) 0.003, c) 0.006.

Taking into account the presented results it can be seen that the new phase grows due to the movement of phase interface, while the α, k_v and k_q coefficients effect on the velocity of interface movement. Determination of the proper and optimal value of k_q and k_v will be important during the modeling in relation to the real conditions. The appropriate values of these coefficients will be determined based on the results of experimental studies.

5 Conclusions

A new hybrid model of diffusion phase transformation in the part related to heat flow was presented in the paper. A connection was established between the two main modules considered in the model: temperature and microstructure evolution module. The use of Lattice Boltzmann Method and Cellular Automata in the model creates completely new possibilities for modeling of phase transformations and in comparison with commonly used approaches significantly simplifies interface and interaction between different parts of the model. The modeling scheme and calculation algorithm were presented and examples simulation results of heat flow with consideration of enthalpy of transformation were shown. The hybrid model in the part related to diffusion has been developed at the previous stage of work (manuscript is submitted to a journal), currently allowing for 1D simulations and now it is possible to combine it with the developed and presented model of heat flow. 2D model of heat flow is also being developed at the current stage of model development and modeling results will be presented in the extended version of the paper. At the final stage of research, the model will be 3D and will use hybrid computational systems (CPU and GPU).

Acknowledgments: Support of the Polish National Science Centre is greatly appreciated (Grant No. 2016/21/D/ST8/01690).

References

[1] T.W. Heo, L.-Q. Chen, JOM. **66**, 1520 (2014)
[2] D. Zhang, Y. Yin, J. Zhou, Z. Tu, Arch. Foundry Eng. **17**, 155 (2017)
[3] L. Feng, B. Jia, C. Zhu, G. An, R. Xiao, X. Feng, Chinese Phys. B. **26**, 80504 (2017)
[4] S.E. Offerman, N.H. van Dijk, J. Sietsma, S. Grigull, E.M. Lauridsen, L. Margulies, H.F. Poulsen, M.T. Rekveldt, S. van der Zwaag, Science. **298**, 1003 (2002)
[5] M. Chen, H. Alimadadi, S. Molin, L. Zhang, N. Ta, P. V. Hendriksen, R. Kiebach, Y. Du, J. Electrochem. Soc. **164**, F1005 (2017)
[6] E. Gamsjäger, F.D. Fischer, J. Svoboda, J. Eng. Mater. Technol. **125**, 22 (2003)
[7] K. Nakajima, M. Apel, I. Steinbach, Acta Mater. **54**, 3665 (2006)
[8] D. An, S. Pan, L. Huang, T. Dai, B. Krakauer, M. Zhu, ISIJ Int. **54**, 422 (2014)
[9] M.R. Varma, R. Sasikumar, S.G.K. Pillai, P.K. Nair, Bull. Mater. Sci. **24**, 305 (2001)
[10] D.S. Svyetlichnyy, A.I. Mikhalyov, ISIJ Int. **54**, 1386 (2014)
[11] G.W. Wang, Z. Liu, J. Zhang, Adv. Mater. Res. **753–755**, 932 (2013)
[12] N. Singh, doi:10.13140/RG.2.1.3625.6483.
[13] Y. Zhong, T. Zhu, ACTA Mater. **75**, 337 (2014)
[14] A. Malik, G. Amberg, A. Borgenstam, J. Ågren, Model. Simul. Mater. Sci. Eng. **21**, 085003 (2013)
[15] S. Saito, Y. Abe, K. Koyama, Phys. Rev. E. **96**, 13317 (2017)
[16] M. Eshraghi, M. Hashemi, B. Jelinek, S. Felicelli, Metals (Basel) **7**, 474 (2017)
[17] C. Xie, J. Wang, D. Wang, N. Pan, M. Wang, Commun. Comput. Phys. **17**, 1037 (2015)
[18] A. Xu, W. Shyy, T. Zhao, Acta Mech. Sin. **33**, 555 (2017)
[19] D. Svyetlichnyy, M. Krzyzanowski, R. Straka, L. Lach, W.M. Rainforth, Int. J. Numer. Methods Heat Fluid Flow. **28**, 31 (2018)
[20] X. Zhan, Y. Wei, Z. Dong, J. Mater. Process. Technol. **208**, 1 (2008)
[21] D.S. Svyetlichnyy, J. Nowak, Ł. Łach, *ACRI 2012* (Lecture Notes in Computer Science, Sirakoulis G.C., Bandini S., 494 2012)
[22] W. Shen, L. Zhang, C. Zhang, Y. Xu, X. Shi, J. Mater. Eng. Perform. **25**, 2065 (2016)
[23] M. Krzyzanowski, D. Svyetlichnyy, G. Stevenson, W.M. Rainforth, Int. J. Adv. Manuf. Technol. **87**, 519 (2016)
[24] Ł. Łach, J. Nowak, D. Svyetlichnyy, J. Mater. Process. Technol. **255**, 488 (2018)
[25] Ł. Łach, D.S. Svyetlichnyy, Mater. Res. Innov. **18**, S6-295 (2014)

Development and validation of an inverse method for identification of thermal characteristics of a short laser pulse

Piotr Łapka[1,*], *Karol* Pietrak[1], *Małgorzata* Kujawińska[2], *Marcin* Malesa[2], and *Łukasz* Oniszk[2]

[1]Institute of Heat Engineering, Warsaw University of Technology, 21/25 Nowowiejska St., 00-665 Warsaw, Poland
[2]Institute of Micromechanics and Photonics, Warsaw University of Technology, 8 Sw. A. Boboli St., 02-525 Warsaw, Poland

Abstract. The paper presents development and validation of an inverse method for the identification of thermal characteristics of a short single laser pulse which stroke in a metal sample. The inverse method was applied to find unknown power of the laser pulse, the dimensionless shape parameter of the super-Gaussian function which describes the beam spatial profile as well as beginning and end times of the exposition of the metal sample to the laser pulse. The proposed inverse algorithm was based on the Levenberg-Marquardt technique as well as on temporal and spatial distributions of temperature on the rear surface of the sample, i.e., the opposite to the irradiated one, measured using the experimental stand. The performed investigations showed that the problem was ill-posed but good accuracy was obtained. The low sensitivity of registered temperature to changes in both power and duration of the pulse affected the retrieving accuracy most significantly. Moreover, the dependence of solution of the inverse problem on the initial guess was observed. The accuracy was also affected by low temporal resolution (500 Hz, with the exposure time from 0.2 to 1 ms) of the IR camera. This resolution affected the temporal sampling of measured temperatures. Despite these problems, the method was able to retrieve unknown pulse parameters with 20-25% accuracy.

1 Introduction

The high-power lasers (1-100 kW) are applied in many engineering fields, e.g., in material processing for cutting, drilling, heat treating, etching, machining and welding [1], in electro-optical systems for optical pumps for solid state lasers, fibre amplifiers, LIDAR, target illuminators and designators [2] as well as in optical data storage and in the free-space optical communication [2]. They are also utilized in the medicine [3] (e.g., in dermatology, ophthalmology, stomatology and surgery) and cosmetics [3]. The defence industry is another area of wide applications of the high-power lasers [4]. They may also serve as heat sources in materials characterization [5]. The interactions of high-power beams with matter are very complex and depend on the laser power and duration of the exposition [6]. The process of absorption of the laser radiation by matter involves many phenomena which are difficult to describe and model [7]. Heating, melting, evaporation, formation of clouds of the evaporated material, shock waves generation, ionization and formation of plasma are examples of these phenomena [7, 8].

In various high-power laser beam applications the beam intensity distribution and spot size are formed in an optic system. These two parameters as well as the laser power are very important because they determine how the laser beam will interact with the target. However, due to imperfections, manufacturing tolerances and other effects like input laser distortions, optical distortion, heating, overall instabilities and non-linear effects the beam profile might deviate from the assumed one [9]. These processes affect the beam profile changing it from the desired Gaussian one to super-Gaussian or even more complex form. Therefore, monitoring, diagnostic and optimization of the laser beam power, temporal and spatial distribution of beam intensity as well as spot size are very important for repeatable operation of many devices. Among the facilities for beam profile analysis are the scanning aperture profilers which use slits, knife-edges, or pinholes with single large area detectors, or camera-based profilers using CCD's or photodiode arrays. Most of the available measuring devices cannot be applied for direct analysis of the high-power lasers in industrial environment and their response time and mounting is inadequate for fast or on-line measurements [9]. For example, the high sensitivity of the camera profilers requires the laser power to be reduced by many orders of magnitude using beam sampling or optical attenuation [10]. Recently a new profiling technique which based on the Rayleigh scatter from the beam have been proposed [10]. This method overcomes the power obstacle and allows dynamic measurement and diagnostic of the beam parameters as well as determination of M^2 parameters of laser beams with power from 1kW to 100 kW. However, most of instrumentation for the beam monitoring is very expensive, fragile and requires great care in its usage. Recently, Kujawińska et al. [11] proposed a new approach to measure laser beam characteristics which is

* Corresponding author: piotr.lapka@itc.pw.edu.pl

based on the maps of temperature and displacements acquired at the rear surface of the heated metal sample with the aid of the high-speed infrared thermography and the 3D Digital Image Correlation (DIC) method [12], respectively. The thermographic method registers the temporal and spatial variations of temperature while the 3D DIC method allows to track sample deformation resulting from thermal stresses induced by significant sample heating by the laser pulse. Moreover, the proposed approach utilizes inverse thermal analysis to find unknown laser beam parameters [13] in which the surface heat flux resulting from the laser interaction with the sample is retrieved. The developed method is very simple and therefore may be easily applied in the industrial or field conditions with a relatively low cost compared to the other methods.

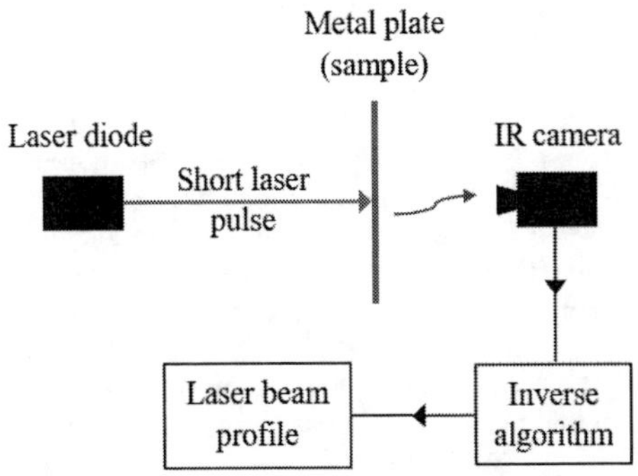

Fig. 1. Schematic of the identification method of thermal parameters of a short laser pulse.

In this paper, development and validation of an inverse thermal method for the identification of thermal characteristics of a short single super-Gaussian laser pulse which incident on a metal sample is presented. The method was applied to find four unknown laser pulse parameters, i.e., its power, the dimensionless shape parameter as well as beginning and end times of the exposition. The inverse algorithm based on the Levenberg-Marquardt (LM) technique [14] which used the temporal and spatial distributions of temperature on the rear surface of the sample, i.e., the opposite to the irradiated one. The temperature profiles were recorded by a high-speed IR camera. The developed method is a first step in the implementation of the new and more general approach to measure laser beam characteristics proposed in [11]. The second step will be related to the incorporation into the recovering procedure the maps of displacements registered by the 3D DIC method.

2 Method for the identification of the short laser pulse parameters

In the developed method, temporal and spatial distributions of temperature on the rear side (opposite to irradiated one) of the metal plate heated by the laser beam are utilized [11] in the identification of the laser pulse parameters. The schematic of the method is presented in Figure 1. A short pulse is emitted by the laser diode and hits the front surface of a metal plate. The thermal response of the sample at the rear surface serves as the input for the inverse algorithm which allows to retrieve the beam parameters.

In order to increase the thermal response on the rear side of the plate the thermal resistivity of the sample along its thickness should be as low as possible [11]. The temperature in the region of the sample irradiated by the laser beam might increase substantially, therefore high thermal stresses might also appear. Considering both thermal and structural aspects of the problem, 1 mm thick aluminum sheet metal (AW2017A T4) was selected among others available on the market [11]. Moreover, the thickness of the sample in the irradiated region was reduced to 0.5 mm by circular milling as shown in Figure 2 to lower its thermal resistivity.

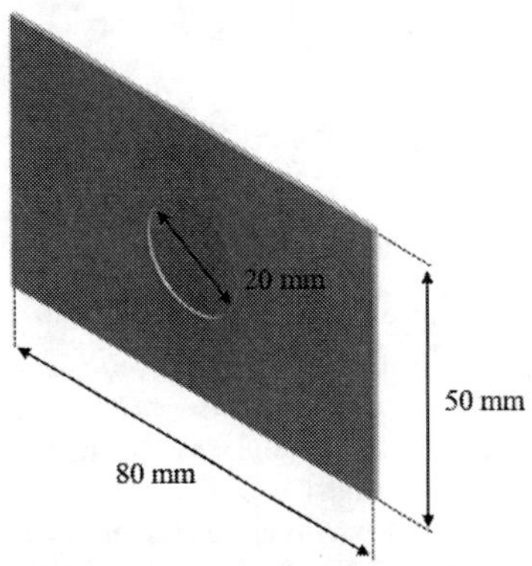

Fig. 2. Aluminium plate with main dimensions – front view.

3 Model of the forward problem

The non-destructive character of the proposed identification method was assumed. The duration of the laser beam was very short and expected maximal temperatures in the sample were below the melting point of aluminium (T_m = 883.0 K) – see [11]. Therefore, only heat conduction in the plate was considered, as described by the following equation:

$$\rho c(T)\frac{\partial T}{\partial t}=\nabla\cdot\left[\lambda(T)\nabla T\right] \quad (1)$$

with boundary conditions given by following formulas (see Figure 2):
- for the circular hollow:

$$-\lambda\nabla T\cdot\mathbf{n}_w=q_l+h\left(T_\infty-T_w\right)+(1-r)\sigma\left(T_\infty^4-T_w^4\right) \quad (2)$$

- for side walls:

$$-\lambda\nabla T\cdot\mathbf{n}_w=0 \quad (3)$$

- for other walls:

$$-\lambda\nabla T\cdot\mathbf{n}_w=h\left(T_\infty-T_w\right)+(1-r)\sigma\left(T_\infty^4-T_w^4\right) \quad (4)$$

and with the following initial condition:

$$T(t=0)=T_{init} \quad (5)$$

where: c is the specific heat, h – heat transfer coefficient, **n** – normal vector, r – surface reflectivity, t – time, T – temperature, w – wall, λ – thermal conductivity, ρ – density, σ – Stefan-Boltzmann constant and ∞ – surroundings. The profile of the heat flux, generated at the specimen surface, was assumed to be rectangular and super-Gaussian with respect to time and space, respectively, and was given by the following relationship:

$$q_l(t)=\begin{cases}(1-r)Pg_0\exp\left[-2\left(\dfrac{R}{R_0}\right)^p\right] & \text{for } t\in\langle t_s,t_e\rangle \\ 0.0 & \text{for } t\notin\langle t_s,t_e\rangle\end{cases} \tag{6}$$

with the following normalization constant:

$$g_0=p2^{\frac{2}{p}}\left[2\pi R_0^2\Gamma\left(\frac{2}{p}\right)\right]^{-1} \tag{7}$$

where: p is the dimensionless shape coefficient of the super-Gauss function, P – laser power, R – distance from the beam centre to a given location, R_0 – distance over which the Gaussian profile ($p = 2$) decreases to e^{-2} of its axial value, t_s and t_e – initial and end time of the laser pulse, respectively, and Γ – Euler gamma function.

The developed model was implemented applying the ANSYS Workbench environment. The geometry and mesh were generated in the ANSYS DesignModeler and ANSYS Meshing, respectively. The grid consisted of almost 198 thousand of elements. The mesh skewness and orthogonal quality were below 0.7 and above 0.3, respectively. These parameters were within acceptable range. The forward problem was solved applying the control volume method [15] based solver ANSYS Fluent and its customization interface, i.e., user-defined functions.

4 Experimental measurements

The experimental stand is ready for more complex measurements which include monitoring of both temperature and 3D displacements at the investigated samples (see Figure 3). In tests the sample was illuminated by single pulse or train of 5 pulses of the Nd:YAG laser of the following parameters: wavelength of 1.06 μm, frequency of 10 Hz and impulse duration in the range of 0.2-1 ms. The energy of a single pulse varied from 2 up to 5 J and was focused on the sample by a lens into a circular area of diameter equal to 12 mm. The effects of the interaction between the laser beam and the solid sample have been observed in the form of both the temperature and displacement maps provided by the IR camera and 3D DIC [12] systems, respectively, captured at the rear surface of the sample. The setup (Figure 3A) consisted of one infrared camera (FLIR SC 7500, spectral range of 1.5-5.1 μm, temperature range of 5-300°C, accuracy of ±1°C) used for monitoring of the temperature and two fast CMOS cameras (Optronis CR3000×2) equipped with 100 mm focal length lenses used for the 3D DIC measurements. The sampling rate of the IR camera was 500 Hz. To ensure sufficient amount of illumination for the 3D DIC measurements and to avoid additional heating of the sample, two 100 W LED reflectors had been used for illumination of the field of view (see Figure 3B). The IR camera was mounted at a heavy free-standing tripod and directed at a slight angle towards the illumination axis to avoid reflections from the sample (Figure 3B). The samples have been mounted in a sample holder made of constructional aluminum profiles. The holder fixed shorter edges of the sample while the longer ones were left unconstrained. It has been bolted to the optical table to ensure its stability.

5 Inverse algorithm

In the presented version of the inverse algorithm single laser pulses were investigated. Their parameters were retrieved based on the temporal and spatial variations of temperature on the rear surface of the irradiated sample (thermal problem only, see Figure 4). These distributions were acquired applying the high-speed IR camera. The unknown pulse parameters were: the power P, dimensionless shape coefficient p as well as initial and end times of the pulse t_s and t_e, respectively. These unknown parameters were represented by the following vector: $\mathbf{q} = [q_1, q_2, q_3, q_4]$, where q_1 refers to the laser power P, q_2 to the dimensionless shape coefficient p, q_3 to the initial time of the pulse t_s and q_4 to the end time t_e.

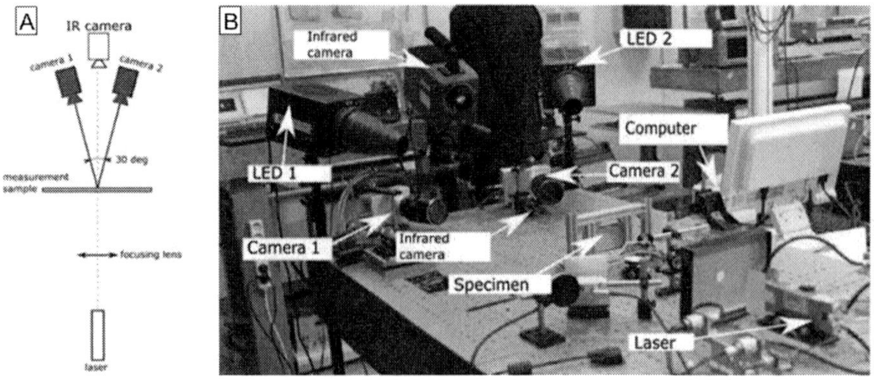

Fig. 3. Experimental stand: A) the schematic and B) the photo of the setup for the multimodal IR and 3D DIC monitoring of the solid sample under short single or multiple laser pulse.

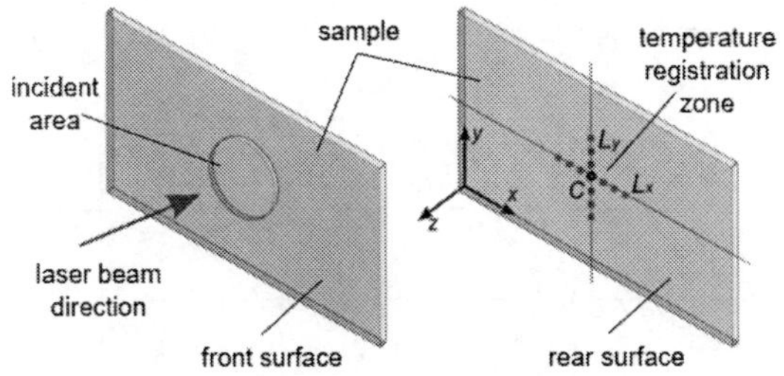

Fig. 4. Schematic of performed experiments.

The inverse problem referred to minimization of the following last squares norm [14]:

$$\mathbf{S}=\mathbf{r}^T\mathbf{r} \tag{8}$$

where the residual vector **r** was defined as a difference between temperatures obtained from the forward problem model ($\mathbf{T}^m$) and from the experimental measurements ($\mathbf{T}^e$), i.e.,

$$\mathbf{r}=\mathbf{T}^m(\mathbf{q})-\mathbf{T}^e \tag{9}$$

The norm **S** given by eq. (8) depended on the unknown vector **q**. Therefore, the task of the inverse algorithm was to find vector **q** which minimized this norm.

The LM technique [14] was applied to solve the defined inverse problem. This method uses the adaptive damping factor for regularization. Preliminary, the calibration of the method was necessary to find the initial value of this factor and additionally to select the stopping criteria. These parameters are case-dependent and were found by extensive testing of the developed inversed algorithm. The assumed values of the damping factor and stopping criteria are discussed later in this paper. Moreover, before the inverse method was applied to find unknown parameters of the laser pulse a thorough sensitivity analysis of the proposed method was performed and described by Pietrak et al. [13]. The investigations carried out in [13] answered the question about both the spatial locations on the sample surface and time moments in which the measured quantities (temperatures) were most sensitive to the changes in the unknown laser beam parameters.

The conclusion drawn from the investigation carried out in [13] was that the temperature acquisition should last long enough to include the cool down period at the rear surface of the sample. This was related to the shifting of the maximum temperature at the rear surface of the sample with respect to the maximum temperature at the front surface. Therefore, in the developed inverse algorithm the temperature history in the point C at the middle of the rear surface of the sample (Figure 4) was applied. These data were denoted by vector $\mathbf{T}_t$. Moreover, the analysis carried out in [13] showed that the inverse procedure should also use the spatial temperature distribution data along the lines L_x and L_y shown in Figure 4 (denoted by vectors $\mathbf{T}_x$ and $\mathbf{T}_y$, respectively) from at least one time instant (frame). These data are required to allow proper identification of the p-parameter, which describes the flatness of the super-Gaussian profile. The utilized frame should be collected just before the laser pulse ended when the temperature at the rear surface of the sample approaches maximum value. If necessary, additional frames might be also collected with a delay of a few milliseconds from the first frame.

After thorough testing of the developed inverse algorithm applying experimental data, a single frame collected close to the moment of maximum temperature at the rear surface of the sample was used in the procedure as suggested by Pietrak et al. [13]. Moreover, in order to improve convergence of the power and temporal parameters (i.e., P, t_s and t_e) the information about time variation of temperature at the point C ($\mathbf{T}_t$) was repeated six times in the vectors $\mathbf{T}^e$ and $\mathbf{T}^m$ which resulted in their following final form:

$$\mathbf{T}^e=\left[\mathbf{T}_x^e,\mathbf{T}_y^e,\mathbf{T}_t^e,\mathbf{T}_t^e,\mathbf{T}_t^e,\mathbf{T}_t^e,\mathbf{T}_t^e,\mathbf{T}_t^e\right]^T \tag{10}$$

$$\mathbf{T}^m=\left[\mathbf{T}_x^m,\mathbf{T}_y^m,\mathbf{T}_t^m,\mathbf{T}_t^m,\mathbf{T}_t^m,\mathbf{T}_t^m,\mathbf{T}_t^m,\mathbf{T}_t^m\right]^T \tag{11}$$

The temperature history data from point C (denoted as $\mathbf{T}_t$) were repeated several times in the residual vector in order to remove disparity between temporal and spatial resolution of the experimental data applied in the inverse algorithm. The former resolution was significantly lower than the latter one, i.e., 1 ms vis 20×20 pixel, where 1 pixel corresponded to 0.41 mm, while the laser pulse duration varied from 0.2 to 1 ms. As mentioned earlier, the spatial temperature distribution data along lines L_x and L_y at the rear surface of the sample (vectors $\mathbf{T}_x$ and $\mathbf{T}_y$) were taken from the thermogram map for a single time frame. The instant of the collection was close to the time in which maximal temperature at the rear surfaces in point C was attained, i.e., approximately 5 ms after the beginning of the laser impulse (the exact delays are not known because the sampling in the camera was not synchronized with the laser triggering).

In the adopted LM technique [14], the unknown vector **q** has been found iteratively according to the following formula:

$$\mathbf{q}^{j+1}=\mathbf{q}^j+\left(\pi\mathbf{J}^T\cdot\mathbf{J}+\mu^j\mathbf{\Omega}\right)^{-1}\cdot\mathbf{J}^T\left(\mathbf{T}^e-\mathbf{T}^m\right) \tag{12}$$

where superscript j denotes j-th iteration and the symbol **J** is the sensitivity (Jacobian) matrix, μ – adaptive regularization coefficient and $\mathbf{\Omega}$ – diagonal amplification matrix. The iterative loop started from the supplied initial guess $\mathbf{q}^0$. The elements of the Jacobian matrix were approximated by the following finite differences formula:

$$J_{ik}=\frac{\partial T_i^m}{\partial q_k}\cong\frac{T_i^m(q_k-\Delta q_k)-T_i^m(q_k)}{\Delta q_k} \tag{13}$$

where the subscript i is the index of the element of temperature vectors $\mathbf{T}^e$ and $\mathbf{T}^m$, $k\in 1\div 4$ – index of the unknown parameter, Δq_k – perturbation of k-th unknown parameter. In practice, single computation of the Jacobian matrix required five runs of the numerical model of direct problem, i.e., one for the current values of unknown parameters and four times with perturbed value of each unknown parameter separately. The perturbation vector $\Delta\mathbf{q}$ was taken as 1/10 of the initial guess vector $\mathbf{q}^0$. The matrix $\boldsymbol{\Omega}$ was assumed after Marquardt [16] in the following form:

$$\boldsymbol{\Omega}=\mathbf{J}^T\cdot\mathbf{J} \tag{14}$$

The starting value of μ was estimated equal to $\mu^0 = 50$. Then in each iteration of the LM algorithm this coefficient was modified in the following way. If the norm $\mathbf{S}^{j+1}$ for the current iteration was smaller or equal to the norm $\mathbf{S}^j$ for the previous iteration the regularization coefficient was decreased by the factor of 0.3. Otherwise μ was increased by 0.3^{-1}.

The following three stopping criteria were applied in the developed inverse algorithm:

$$\mathbf{S}\left(\mathbf{q}^{j+1}\right)<\varepsilon_1 \tag{15}$$

$$\left\|\mathbf{J}^T\left(\mathbf{q}^j\right)\left[\mathbf{T}^e-\mathbf{T}^m\left(\mathbf{q}^j\right)\right]\right\|<\varepsilon_2 \tag{16}$$

$$\left\|\mathbf{q}^{j+1}-\mathbf{q}^j\right\|<\varepsilon_3 \tag{17}$$

where the symbol ||.|| denotes Euclidean norm. The values of constants ε_1, ε_2 and ε_3 were set equal to $\varepsilon_1 = 5$, $\varepsilon_2 = 5$ and $\varepsilon_3 = 0.002$, determined for power in kW, temporal parameters in ms and least squares norm **S** defined by eq. (8). After the one of these stopping criteria was attained the unknown parameters were assumed found.

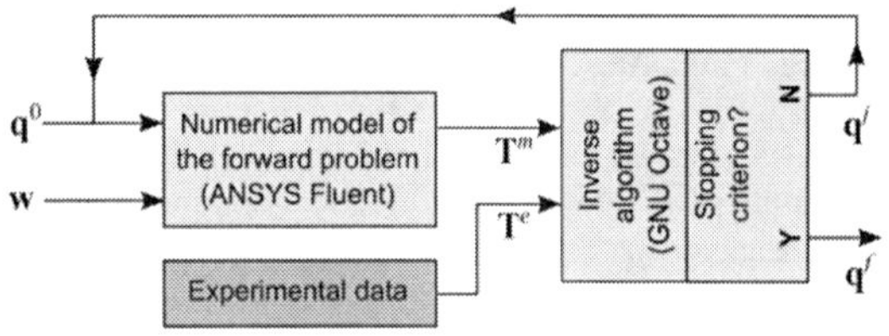

Fig. 5. Implementation of the inverse procedure ($\mathbf{q}^f$ denotes vector of estimated parameters, **w** – vector of known system parameters).

The inverse algorithm was developed applying the GNU Octave and ANSYS Fluent software. The former was the master program and the latter was the slave, invoked with a journal file modified beforehand with the use of an in-house GNU Octave routine. The GNU Octave allowed for external solution of the forward problem (in the ANSYS Fluent), modification of input parameters to the thermal model (modification of input files which were then read by UDF macros in the ANSYS Fluent), utilization of the output parameters from the thermal model (reading of output files from ANSYS Fluent which were written by UDF macros), development of the inverse problem solution method by applying build-in libraries, iterative execution of the overall inverse algorithm and analysis of the obtained results. The general schematic of the solution methodology is presented in Figure 5.

Table 1. Thermophysical parameters of the aluminum alloy AW2017A T4 [10].

Temperature (°C)	**Thermal conductivity** (W/K/m)	**Specific heat** (J/kg/K)	**Density** (kg/m³)
30	122.6	902.0	
50	125.0	911.0	
100	133.7	943.0	
150	151.9	1046.0	2700.0
200	154.9	1039.0	
250	130.0	832.0	
300	159.6	995.0	

Table 2. Initial and boundary conditions as well as the laser beam parameters for the considered cases.

	Case 1	**Case 2**	**Case 3**	**Case 4**
Initial temperature T_{init} (K)	303.89	302.80	303.55	303.73
Surroundings temperature T_∞ (K)	303.89	302.80	303.55	303.73
Heat transfer coefficient h (W/m²/K)	5.0			
Surface reflectivity r (-)	0.95			
Number of pulses	1	1	5	5
Laser power P (W)	21000			
Shape coefficient p (-)	2			
Beam radius R_0 (m)	0.006			
Pulse duration $t_e - t_s$ (ms)	0.2	1.0	0.2	0.2

6 Retrieving of the laser beam parameters

The aluminium sample used in the experiments had the following geometrical dimensions: 80.0×50.0×1.0 mm (width×length×thickness of the plate, respectively) and 2.0×0.5 mm (diameter×thickness of the milling, respectively). Its thermophysical properties (aluminium alloy AW2017A T4) are given in Table 1 [11]. Four cases, e.g., two single and two multiple laser pulses were used in the investigations presented in this paper. During the experiments some of the laser beam parameters, i.e., the laser power, the beam spatial profile as well as beginning and end time of the exposition were known. Therefore, the measured data were used for the validation of both the forward problem and proposed inverse algorithm. The initial and boundary conditions as well as laser pulse parameters for considered cases are given in Table 2. The value of heat transfer coefficient at sample surfaces was assumed for natural convection in the case of horizontally oriented plates [17], while the value of surface reflectivity was taken for polished aluminium surface [17]. The initial and surroundings temperatures were measured during the experiments.

Before application of the forward problem model in the inverse procedure its correctness and accuracy was proven. In the first step in the verification procedure convergence studies for time step and grid sizes were carried out. The forward problem was solved applying three different time step sizes, i.e., the base one ($\Delta t = 10^{-5}$ s), two times smaller and two times larger one as well as two additional meshes, i.e., the coarse and finer one with 10% less and 10% more elements, respectively. The spatial and temporal temperature distributions obtained for three time steps and three meshes were almost identical with the relative differences below 1%. Then in the second step, the validation using the experimental measurements conducted on the stand described above was carried out. The numerical and experimental results were found to agree very well. Detailed analysis showed that absolute differences between predicted and measured data were below 0.4 K and 0.1 K for the peak and cooling period, respectively. The comparison proved that the forward model gave physically-accurate results which matched the measurements for single and multiple pulses.

The results of the retrieving of four unknown laser beam parameters using experimental data are presented in Table 3 for cases defined in Table 2. For cases which involved multiple pulses (case 3 and 4), the data for the first pulse only were examined with the presented inverse algorithm. In Table 3 the obtained results are compared with the real laser pulse parameters which were known during experiments. The temporal variation of temperature in point *C* and spatial distribution of temperature along lines L_x and L_y for initial guess and final estimation as well as measured in the experiments for cases 2 and 4 are shown in Figure 6 and 7.

The tests carried out showed that the considered problem is ill-posed. However, relatively good retrieving accuracy of the unknown parameters of the laser pulse have been obtained – see Table 3. The main reason of ill-conditioning is very low sensitivity of the measured temperature to the variation of the end time of the pulse. Inaccurate retrieval of the end time of the exposition to the laser pulse results in decrease in the accuracy of the laser power fitting. Decrease or increase of the laser pulse duration was related to decrease or increase of the amount of energy absorbed by the plate. The same effect was obtained by decreasing or increasing of the laser power. Therefore, the end time of the exposition and the laser power were mutually dependent parameters and there was no unique solution of the problem.

To correctly retrieve the end time of the pulse (pulse duration was on the order of 10^{-1} ms) as high as possibly temporal resolution of temperature registration was required (1 ms resolution was available in the measurements). The available temporal resolution the IR camera (much higher than duration of pulse) limited the accuracy of the exposition end time fitting.

Table 3. Results of the retrieving of four unknown parameters with comparison against real parameters of the pulse.

Case	Quantity	P (kW)	p (-)	t_s (ms)	t_e (ms)	t_e–t_s (ms)
1	q_i^0	10.0	1.2	3.7	3.8	0.1
	q_i^e	21.0	2.0	-	-	0.2
	q_i^f	15.38	2.33	3.636	3.863	0.227
	Δq_i	5.62	0.33	-	-	0.027
	$\Delta q_i/q_i^e$ (%)	**26.76**	**16.5**	-	-	**13.45**
2	q_i^0	13.0	5	4	4.5	0.5
	q_i^e	21.0	2.0	-	-	1
	q_i^f	18.54	2.39	3.745	4.749	1.004
	Δq_i	2.46	0.39	-	-	0.004
	$\Delta q_i/q_i^e$ (%)	**11.71**	**19.5**	-	-	**0.40**
3	q_i^0	10.0	1.2	3.7	3.8	0.1
	q_i^e	21.0	2.0	-	-	0.2
	q_i^f	17.04	2.30	3.519	3.718	0.201
	Δq_i	3.96	0.3	-	-	0.001
	$\Delta q_i/q_i^e$ (%)	**18.86**	**15.0**	-	-	**0.5**
4	q_i^0	10.0	1.2	3.7	3.8	0.1
	q_i^e	21.0	2.0	-	-	0.2
	q_i^f	15.75	2.38	3.607	3.822	0.215
	Δq_i	5.25	0.38	-	-	0.015
	$\Delta q_i/q_i^e$ (%)	**25.0**	**19.0**	-	-	**7.5**

The accuracy of the retrieved parameters was found dependent on the initial guess. Good matches were obtained when the inverse algorithm started from the low laser power and short duration of the laser pulse. Moreover, the initial starting time of the exposition to the laser pulse should be closed to its real value. It should be noted that in practical applications the pulse duration is typically known and the main objective is to verify the spatial distribution of pulse energy. Moreover, for the specimen with a higher value of the thermal resistivity along its thickness (e.g., for thicker plate or sample made of material with the low value of thermal diffusivity) the conditioning of the inverse problem under consideration will be decreased. Such sample will have higher ability to damp thermal response on the rear surface of the sample.

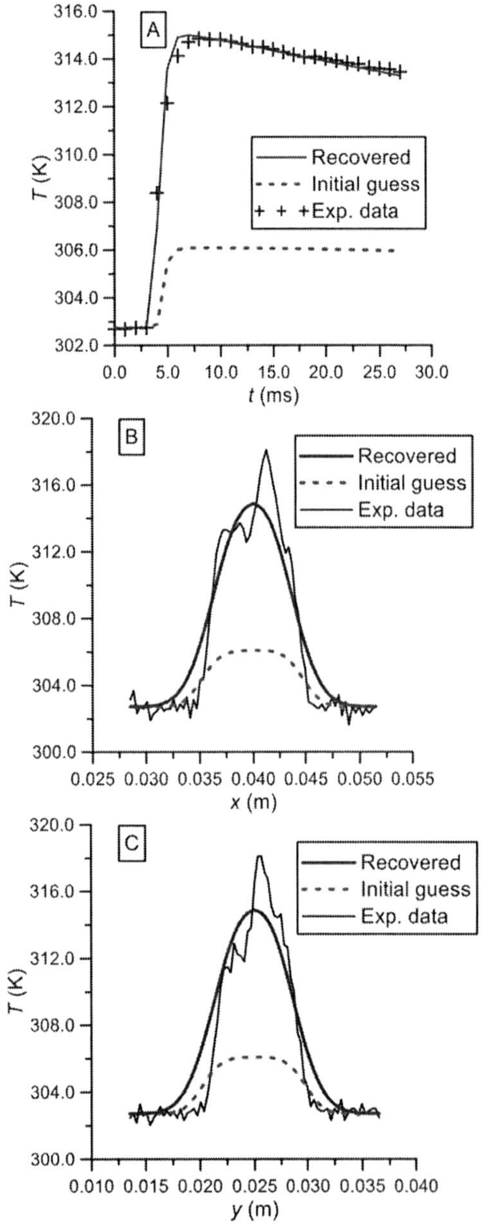

Fig. 6. A) Temporal temperature variation in point C as well as B) and C) distribution of temperature along lines L_x and L_y, respectively for approximately 5 ms after heating started for case 2 and for initial guess, final estimation and experimental measured.

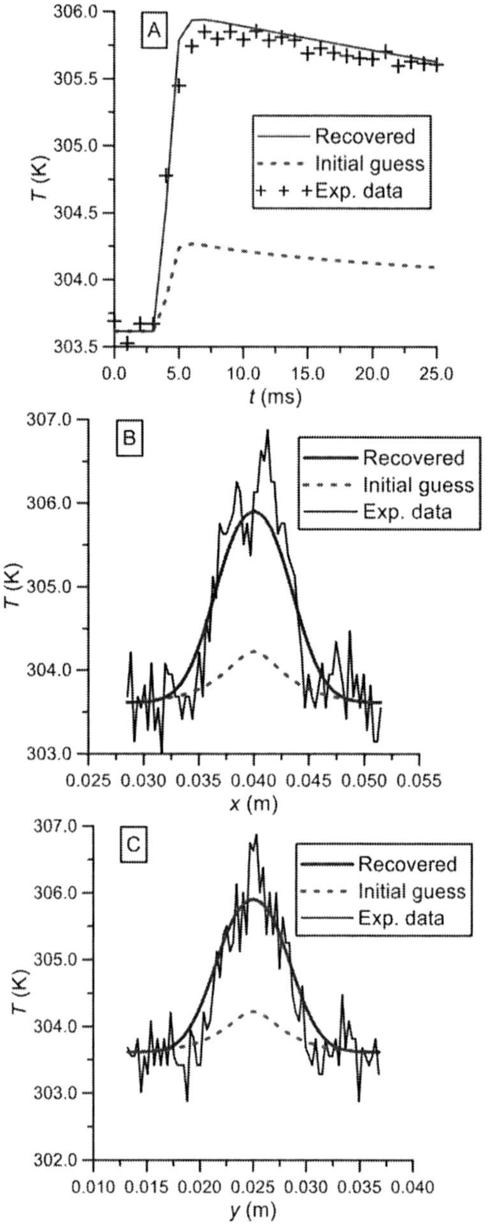

Fig. 7. A) Temporal temperature variation in point C as well as B) and C) distribution of temperature along lines L_x and L_y, respectively for approximately 5 ms after heating started for case 4 and for initial guess, final estimation and experimental measured.

Despite the aforementioned problems and limitations, the accuracy of the proposed retrieving method was on the level of 20-25%. The limitations of the quality of the experimental data were related to damping of the temperature response on the rear surface of the sample (effect of material type and thickness of the sample) and relatively low resolution of the IR camera (the frame rate of 500 Hz at 20×20 pixels, while the exposure times were in the range of 0.2-1 ms). The method recognized shorter (case 1, 3 and 4) and longer impulse (case 2).

Moreover, the spatial distribution of the laser pulse matched well to the real data – see Figure 6 and 7. The under- or over-prediction of the pulse duration resulted in the over- or under-prediction of the laser power, respectively. However, the amount of absorbed energy was predicted very well.

The proposed inverse procedure has also disadvantages related to its iterative character and required numerical computation of the sensitivity matrix at each iteration of the solution procedure. The Jacobian is non-linear and depends on the unknown parameters and should be updated in each iteration. Moreover, the determination of the Jacobian matrix applying finite difference method required $N+1$ model evaluations, where N was the number of unknowns. Therefore, computations are quite expensive especially in the case of large grids. Fortunately, this problem may be solved by applying parallel computation of the forward problem.

7 Conclusions

The paper presents a method of identification of thermal characteristics of the short single super-Gaussian laser pulse. The proposed method is based on the thermal imaging technique and utilizes the LM algorithm for retrieval of four unknown parameters of the laser pulse, i.e., the laser power, dimensionless shape parameter of the super-Gaussian function which describes the beam spatial profile as well as beginning and end times of the exposition of the sample to the laser pulse. The method involves thin metal plate illuminated by the investigated laser source as well as temporal and spatial distributions of temperature on the rear surface of the sample, i.e., the opposite to the irradiated one. The solution of the inverse problem was carried out applying the commercial software GNU Octave and ANSYS Fluent. Even though the data from the 3D DIC method were not yet considered in the retrieval procedure, it was shown that accurate determination of four considered parameters is possible, with 20-25% of accuracy.

The tests carried out showed that the considered inverse problem is ill-posed. The main reason for this behavior was a very low sensitivity of the measured temperature to the variation of the end time of the exposition to the laser pulse. Moreover, this parameter was found to influence the measurement data in the same way as the laser power. Therefore, these two parameters were mutually dependent and contributed to the solution non-uniqueness. The conditions under which the influence of ill-conditioning may be minimized were identified and described in this paper. The first one was related to required high temporal resolution of temperature registration. The second one referred to the proper form of the input data (vectors $\mathbf{T}^e$ and $\mathbf{T}^m$). The third one concerned the proper selection of the initial guess. The last one was related to the thermal properties and thermal resistivity of the aluminum sample. As shown in this paper, consideration of these conditions allows to obtain satisfying results. The method correctly recognized differences between the shorter and longer impulse duration. Moreover, the predicted spatial distribution of the laser pulse was found to correspond well to the experimental data. The under- or over-prediction of the pulse duration resulted in the over- or under-prediction of the laser power, respectively. However, the amount of absorbed energy was predicted correctly.

This work was supported by the National Centre for Research and Development (Poland) under grant no. DOB-16-6/1/PS/2014 and the statutory funds of Faculty of Power and Aeronautical Engineering and Faculty of Mechatronics of Warsaw University of Technology.

References

1. A.K. Dubey, V. Yadava, Int. J. Mach. Tools Manu. **48**, 609 (2008)
2. A. Kozłowska, P. Łapka, M. Seredyński, M. Teodorczyk, E. Dąbrowska-Tumańska, Appl. Therm. Eng. **91**, 279 (2015)
3. Q. Peng, A. Juzeniene, J. Chen, L. O Svaasand, T. Warloe, K.-E. Giercksky, J. Moan, Rep. Prog. Phys. **71**, 56701 (2008)
4. G.P. Perram, M.A. Marciniak, M. Goda, Proc. SPIE Int. Soc. Opt. Eng. **5414**, 1 (2004)
5. K. Pietrak, T.S. Wiśniewski, J. Power Technol. **94**, 270 (2014).
6. M. von Allmen, A. Blatter, *Laser-beam interactions with materials: Physical principles and applications* (Springer-Verlag, Berlin, 1995)
7. J. Marczak, K. Jach, A. Sarzynski, Proc. SPIE Int. Soc. Opt. Eng. **5146**, 215 (2003)
8. M. Kujawinska, K. Kustron, K. Siedlecki, M. Malesa, Proc. SPIE Int. Soc. Opt. Eng. **10436**, 104360H (2017)
9. O. Aharon, Proc. SPIE Int. Soc. Opt. Eng. **8963**, 89630M (2014)
10. J.L. Guttman, Laser Tech. J. **12**, 20 (2015)
11. M. Kujawińska, P. Łapka, M. Malesa, K. Malowany, M. Prasek, J. Marczak, Proc. SPIE Int. Soc. Opt. Eng. **10159**, 1015916 (2016)
12. M.A. Sutton, J.J. Orteu, H. Schreier, *Image correlation for shape, motion and deformation measurements* (Springer, New York, 2009)
13. K. Pietrak, P. Łapka, M. Kujawińska, Int. J. Therm. Sci. **134**, 585 (2018)
14. M.N. Ozisik, H.R.B. Orlande, *Inverse heat transfer* (Taylor&Francis, New York, 2000)
15. P. Łapka, P. Furmański, Heat Mass Transfer **44**, 937 (2008)
16. D.W. Marquardt, J. Soc. Ind. Appl. Math. **11**, 431 (1963)
17. Y.A. Cengel, A.J. Ghajar, *Heat and mass transfer, Fundamentals and applications* (McGraw-Hill, New York, 2015)

Preliminary mathematical and numerical transient models of convective heating and drying of a brick

Piotr Łapka[1,*], *Michał* Wasik[1], *Piotr* Furmański[1], *Mirosław* Seredyński[1], *Łukasz* Cieślikiewicz[1], *Karol* Pietrak[1], *Michał* Kubiś[1], *Tomasz S.* Wiśniewski[1], and *Maciej* Jaworski[1]

[1]Institute of Heat Engineering, Warsaw University of Technology, 21/25 Nowowiejska St., 00-665 Warsaw, Poland

Abstract. The paper presents the initial approach to mathematical and numerical modelling and optimization of heat and mass transfer in elements of the masonry wall. The considered single moist brick is placed in the channel through which the dry air is flowing. Only one wall of the brick is in contact with the flowing air and through this boundary heat and moisture are exchanged. The non-equilibrium mathematical model is formulated for general case with three phases of moisture present in the porous building material, i.e., water vapour in the moist air filling the pores, liquid water (bound water) adsorbed at the surface of the solid component of the material and free liquid water either in a discontinuous form (funicular) or continuous one (pendular). However, at this stage of development of the numerical model the moisture in the form of motionless liquid water and water vapour which reside in the ceramic material are considered. Moreover, the moisture and heat diffusion in the porous hydroscopic material of the brick as well as moisture and heat convection in the flowing air are assumed. The numerical model is developed with the aid of the commercial software ANSYS Fluent and its advanced customisation functionalities like the User-Defined Function, User-Defined Scalar and User-Defined Memory. Subsequently, investigation of the influence of the mass transfer coefficient between the free liquid water and moist air in the porous building material are carried out.

1 Introduction

In the masonry walls and buildings foundations moisture problems are very often encountered. There are many sources of the excessive moisture in the building masonry structures. The natural disasters like floods, storms and hurricanes, high level of ground water and lack or worn out of the moist insulation are typical examples. The consequences of the excessive presence of the moisture in the walls are very serious. The damp deteriorates the physical and chemical conditions of the walls [1, 2] and increases heat gains or losses from or to the surroundings and respective cooling or heating costs [3, 4]. What is the most important the presence of the moisture in the masonry walls may lead to growth of mildew and microorganisms which may cause health problems of the occupants [5]. Transfer phenomena in the building material are very slow and moisture might be accumulated for long time, especially in historical buildings [6, 7]. Therefore, renovation of the damp walls is very important.

There are many methods of drying of the walls and then protecting them against the water re-penetration. One of the techniques of protecting the walls against the moisture relies on the drying of the damp walls and then on the creation of the hydrophobic (waterproof) membrane by using special fluid mixture which is injected through several boreholes and then penetrates the masonry walls. The moisture insulation created in this way is very effective and durable. The effectiveness of the fluid penetration and hence the creation of the waterproof membrane is directly connected to the temperature and dryness of the wall, i.e., the higher temperature (but not higher than 62°C due to cross-link above this temperature of the silicones used for membrane forming) and the lower water content in the wall the better penetration and more durable moisture membrane. On the other hand, the main drawback of this method is substantial consumption of the energy required for heating and drying the walls. Therefore, to increase both energetic and membrane formation efficiencies of this method heating and drying processes of the masonry walls should be optimized. The development of advanced and reliable mathematical and numerical models of heat and mass transfer in the masonry materials are the first steps which should be undertaken in the optimization process.

The problem of combined heat, air and moisture transfer is common not only in building materials [8] but also in many engineering areas, e.g., in textile industry [9, 10], protective clothing [11-14], wood drying [15], food processing [8], granular material drying [8] or in composite membrane [8]. Although, wide range of application area has provided many similar models,

* Corresponding author: piotr.lapka@itc.pw.edu.pl

general-purposed accurate models have not been developed so far.

The combined heat and moisture transfer models may be divided into two general groups, i.e., the equilibrium [16] and non-equilibrium models [9, 10]. In the field of moisture transfer modelling in buildings or building materials two main groups of models are distinguished, i.e., building energy simulation models (BES) and heat, air and moisture transport models (HAM). Some commercial software, e.g., TRNSYS, ESP-r and EnergyPlus are qualified as BES models. They contain simplified models of moisture transfer and should not be used for accurate simulations. For example, engineering software like WUFI or Delphin belongs to the second group. Both computer programs solve combined heat and mass transport problems but implemented models are simplified. In order to achieve better performance HAM models are combined with Computational Fluid Dynamics (CFD) models and new group of models, i.e., CFD-HAM models is proposed. A several equilibrium CFD-HAM models based on two conservation equations in the dried materials, i.e., the energy and moisture balance equations were developed [8, 15-17]. These models predict total moisture content present in three phases (i.e., water vapour, bounded water and free liquid water) and with additional relations allows to find fractions of each phase.

Below the general non-equilibrium mathematical model of combined heat, moisture and air transfer in the porous building materials is formulated. Then the model in a simplified form is implemented in the ANSYS Fluent and combined with the air flow model around the dried brick. In the last step several test simulations are carried out.

2 Mathematical model of transport phenomena in the porous building material

The considered building material is assumed to be hygroscopic porous media. The moisture which reside in the brick consists of three phases, i.e., water vapour in the moist air filling the pores, liquid water (bound water) adsorbed at the surface of the solid component of the material and free liquid water either in a discontinuous form (funicular) or continuous one (pendular). The equilibrium curve relating the moisture content in the building material to the relative humidity in the moist air does not distinguish between the water adsorbed at the surface of the solid component and the free liquid. Therefore, it is convenient to treat these both forms as the liquid phase. Therefore, the sum of the volume fraction of all constituents in the considered system satisfies the following condition:

$$\varepsilon_s + \varepsilon_l + \varepsilon_g = 1 \tag{1}$$

where: subscripts g, l and s denote moist air, liquid water and solid component, respectively and ε is the volume fraction. Moreover, the sum of volume fraction of moist air and liquid water equals to the volume fraction of the pores, i.e.,

$$\varepsilon_l + \varepsilon_g = \varepsilon_p \tag{2}$$

where: subscript p denotes pores in the brick. The volume fraction of the dry air and water vapour are calculated from the following relationships:

$$\varepsilon_a = \varepsilon_g \rho_a \tag{3}$$

$$\varepsilon_v = \varepsilon_g \rho_v \tag{4}$$

where: subscripts a and v denote dry air and water vapour, respectively and ρ is the density.

2.1 Balance of the moisture

The amount of the water vapour and liquid water in the porous material varies due to sorption/desorption and condensation/evaporation.

If the liquid water is below certain critical volume fraction $\varepsilon_{l,max}$ the liquid is in the funicular state (separated droplets) and is motionless. Its concentration chances only due to evaporation and condensation phenomena. The balance equation for liquid water is then in the following form:

$$\frac{\partial}{\partial t}(\rho_l \varepsilon_l) = -\dot{m}_{lv} \tag{5}$$

where: $\dot{m}_{lv}$ is the volumetric intensity of liquid evaporation/condensation expressed in kg/s/m^3.

If the liquid water is above $\varepsilon_{l,max}$ separated droplets join together and the liquid flow due to capillary forces appears (pendular state). Its concentration changes due to diffusive and convective motion as well as evaporation and condensation phenomena. The balance equation for liquid water is then in the following form:

$$\frac{\partial}{\partial t}(\rho_l \varepsilon_l) + \nabla \cdot \mathbf{j}_l = -\dot{m}_{lv} \tag{6}$$

where: $\mathbf{j}_l$ is the mass flux of the liquid water and is expressed as follows:

$$\mathbf{j}_l = \zeta \rho_l \frac{K_{lg}}{\mu_l} \nabla p_c - \zeta \rho_l \frac{K_{lg}}{\mu_l} \nabla p_g + \zeta \rho_l \frac{K_{lg}}{\mu_l} \rho_l \varepsilon_l \mathbf{g} \tag{7}$$

In the eq. (7) $\mathbf{g}$ is the specific gravity, K_{lg} denotes the permeability of the porous medium for liquid water, when the water vapour is present in the pores, p_c stands for the capillary pressure, which is a difference between gas and liquid pressures in the pores, p_g is the pressure of the gas in the pores, μ_l is the liquid water dynamic viscosities and ζ accounts for mutual interaction of liquid and gaseous phases during their flow in the material. In eq. (7) the diffusive term was related to the capillary pressure, while the convective term was dependent on the gas pressure in the pores and gravitational acceleration applying the Darcy's law.

The water vapour amount in the pores changes due to diffusion in the moist air and convective motion with the moist air as well as evaporation and condensation. Its balance equation is following:

$$\frac{\partial}{\partial t}\left(\rho_v \varepsilon_g\right) + \nabla \cdot \mathbf{j}_v = \dot{m}_{lv} \tag{8}$$

where the diffusive-convective mass flux of the water vapour $\mathbf{j}_v$ is expressed as follows:

$$\begin{aligned} \mathbf{j}_v = & -\frac{\rho_v \zeta K_{gl}}{\mu_g} \nabla p_g + \rho_v \frac{\zeta K_{gl}}{\mu_g} \rho_g \mathbf{g} + \\ & -\rho_g D_{v-a,ef} \nabla \left(\frac{\rho_v}{\rho_g} \right) \end{aligned} \tag{9}$$

In the eq. (9) $D_{v\text{-}a,ef}$ is the effective water vapour diffusivity in building material and K_{lg} denotes the permeability of the porous medium for water vapour, when the liquid water is present in the pores.

The liquid water evaporation appears if the water vapour partial pressure in the pores of the building material in lower than the local water vapour saturation pressure while the vapour condensation from the humid air in the pores appears if the local water vapour partial pressure is higher than the water vapour saturation pressure. Moreover, the intensity of the evaporation or condensation depends on the difference between the densities of the water vapour in the air and at the surface of evaporation/condensation and is given by following formula:

$$\dot{m}_{lv} = \begin{cases} h_m a_s \dfrac{\varepsilon_l}{\varepsilon_p} \left(\rho_{v,sat} - \rho_v \right) & \text{– for evaporation} \\ h_m a_s \left(\rho_{v,sat} - \rho_v \right) & \text{– for condensation} \end{cases} \tag{10}$$

where: a_s denotes the pores area for unite volume of the porous medium, h_m is mass transfer coefficient between liquid water and moist air in the porous material and $\rho_{v,sat}$ stands for vapour density for the saturation conditions. The ratio of $\varepsilon_l/\varepsilon_p$ denotes the part of the pores area (volume) occupied by liquid water. According to eq. (10) for evaporation $\dot{m}_{lv}$ has positive value while for condensation negative.

2.2 Balance of the dry air

The amount of dry air in the pores changes due to diffusion in the moist air and convective motion with the moist air. Its balance equation is following:

$$\frac{\partial}{\partial t}\left(\rho_a \varepsilon_g\right) + \nabla \cdot \mathbf{j}_a = 0 \tag{11}$$

where the diffusive-convective mass flux of the dry air $\mathbf{j}_a$ is expressed as follows:

$$\begin{aligned} \mathbf{j}_a = & -\frac{\rho_a \zeta K_{gl}}{\mu_g} \nabla p_g + \rho_a \frac{\zeta K_{gl}}{\mu_g} \rho_g \mathbf{g} + \\ & +\rho_g D_{v-a,ef} \nabla \left(\frac{\rho_a}{\rho_g} \right) \end{aligned} \tag{12}$$

2.2 Balance of energy

The thermal equilibrium between different components and phases in the building material is assumed, i.e., constituents are in the same temperature. Then the energy balance equation in the brick is following:

$$\frac{\partial}{\partial t} H + \nabla \cdot (\mathbf{j}_l h_l + \mathbf{j}_v h_v + \mathbf{j}_a h_a) = -\nabla \cdot \mathbf{q} \tag{13}$$

where: h is the specific enthalpy, while heat flux and enthalpy per unit volume are defined in the following way:

$$\mathbf{q} = -k_{ef} \nabla T \tag{14}$$

$$H = \rho_s \varepsilon_s h_s + \rho_l \varepsilon_l h_l + \rho_v \varepsilon_g h_v + \rho_a \varepsilon_g h_a \tag{15}$$

In eq. (14) k_{ef} is the effective thermal conductivity of the medium, which is dependent on the moisture content. The specific enthalpies of medium components are based on the triple point temperature as the reference one and can be expressed as:

$$h_a = c_{p,a}\left(T - T_{ref}\right) \tag{16a}$$

$$h_l = c_w\left(T - T_{ref}\right) \tag{16b}$$

$$h_s = c_s\left(T - T_{ref}\right) \tag{16c}$$

$$h_v = \Delta h_{vap} + c_{p,v}\left(T - T_{ref}\right) \tag{16b}$$

where: c and c_p are the specific heat and specific heat at constant pressure, respectively and Δh_{vap} is the latent heat of evaporation.

Noting the relations for the heat flux, enthalpy and specific enthalpies, eq. (14), (15) and (16), respectively, and balances of the liquid water, water vapour and dry air, eq. (5) or (6), (8) and (11), respectively, and neglecting the differences between the specific heats of different moisture phases (in comparison to the latent heat), the energy equation, eq. (13), reduces to the following equations for pendular and funicular form of the liquid water, respectively:

$$\begin{aligned} & (\rho c)_{ef} \frac{\partial}{\partial t} T + \left(\mathbf{j}_v c_v + \mathbf{j}_a c_a\right) \cdot \nabla T = \\ & \nabla \cdot \left(k_{ef} \nabla T\right) - \dot{m}_{lv} \Delta h_{par} \end{aligned} \tag{17a}$$

$$\begin{aligned} & (\rho c)_{ef} \frac{\partial}{\partial t} T + \left(\mathbf{j}_l c_l + \mathbf{j}_v c_v + \mathbf{j}_a c_a\right) \cdot \nabla T = \\ & \nabla \cdot \left(k_{ef} \nabla T\right) - \dot{m}_{lv} \Delta h_{par} \end{aligned} \tag{17b}$$

where the effective volumetric specific heat of the moist material is defined as:

$$\left(\rho c\right)_{ef} = \varepsilon_s \rho_s c_s + \varepsilon_l \rho_l c_l + \varepsilon_g \rho_a c_{p,a} + \varepsilon_g \rho_v c_{p,v} \quad (18)$$

3 Numerical model

Development of the numerical model consists of three steps, i.e., generation of the computational geometry, generation of the mesh and implementation of equations in the software. The numerical model was developed in the ANSYS CFD environment. The geometry and mesh were created in the ANSYS DesignModeler and ANSYS Meshing, while the mathematical model was implemented and the simulations were carried out in the ANSYS Fluent.

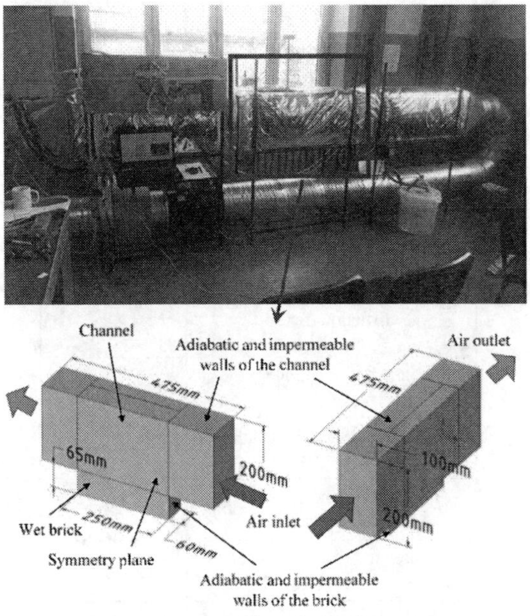

Fig. 1. Experimental stand (upper) and computational geometry (lower) with basic dimensions and boundary conditions.

The computational domain was prepared in relation to the experimental stand which is developed in parallel – see Figure 1. It consists of a part of a channel at the top through which dry air is flowing and wet material at the bottom as shown in Figure 1. Due to symmetry a half of the channel and sample are considered. The assumed basic dimensions and boundary conditions are presented in Figure 1. Subsequently, the structural mesh with over 3.6 million of elements was generated applying the sweep method. The grid for the air region was refined close to the channel walls and contact surface between the air and moist brick as well as in the region which refers to walls of the brick. The maximum aspect ratio of the mesh was 98.9.

Before implementation the mathematical model of transport phenomena in the porous building material in the ANSYS Fluent following assumptions were introduced:

- Influence of the convection due to gravitational acceleration as well as moist air and capillary pressure variations in the porous brick was neglected in the all mass balance equations, i.e., in eq. (5) or (6), (8) and (11).
- Influence of diffusive and convective motion of the liquid water as well as convection of water vapour and dry air were neglected in the energy equations, eqs (17).

These assumptions will be released in the future work.

With the above introduced assumptions the governing equations in the brick were cast in the following form for both pendular and funicular form of the liquid water:

$$\frac{\partial}{\partial t}\left(\rho_l \varepsilon_l\right) = -\dot{m}_{lv} \quad (19)$$

$$\frac{\partial}{\partial t}\left(\rho_v \varepsilon_g\right) = \rho_g D_{v-a,ef} \nabla\left(\frac{\rho_v}{\rho_g}\right) + \dot{m}_{lv} \quad (20)$$

$$\frac{\partial}{\partial t}\left(\rho_a \varepsilon_g\right) = \rho_g D_{v-a,ef} \nabla\left(\frac{\rho_a}{\rho_g}\right) \quad (21)$$

$$\left(\rho c\right)_{ef} \frac{\partial}{\partial t} T = \nabla \cdot \left(k_{ef} \nabla T\right) - \dot{m}_{lv} \Delta h_{par} \quad (22)$$

In the flowing air standard turbulent convective-diffusive balance equations for water vapour and dry air as well as energy were assumed. The air flow was turbulent and the velocity field was found using SST k-ω turbulence model. This model exactly resolves transport phenomena in the boundary layer, which is important for accurate modelling of transport of the moisture and heat from the brick to the flowing air. Moreover, the mesh in the air fulfilled constrain of y+ below 1 at all surfaces.

The governing equations, eq. (19)-(22) were implemented in the ANSYS Fluent applying advanced customisation functionalities like the User-Defined Function, User-Defined Scalar and User-Defined Memory. For the liquid fraction the scalar equation was implemented only in the solid region which represented the brick. For water vapour and dry air the single scalar equation for each constituent was implanted in both solid and liquid region which represented the brick and flowing air, respectively. Similarly, for the temperature the single scalar equation was implemented in both solid and liquid region. The transient and convective terms in these scalar equations were implemented using DEFINE_UDS_UNSTEADY and DEFINE_UDS_FLUX macros, respectively. The mass diffusivity and thermal conductivity as well as source terms were implemented applying DEFINE_DIFFUSIVITY and DEFINE_SOURCE macros, respectively.

Applying the single scalar equation for solid and liquid region for water vapour and dry air densities as well as for temperature have a lot of benefits. It significantly simplified treatment of conditions at the top wall of the brick which was in contact with the flowing air. In this way mass and heat convection at the top surface of the brick was resolved by the solution of the scalar equations in the fluid region in which turbulent air flow occurred.

4 Results

The simulations were carried out in two steps. In the first step the flow filed in the channel in the steady-state was found by simulating the air flow in the channel with periodic boundary conditions. Then in the second step only scalar balance equations were solved in the transient simulation with steady flow field obtained in the previous step. The boundary at the inlet to the computational domain were following: the dry air and water vapour densities were equal to 1.225 and 0.00968 kg/m^3, respectively, while the mixture velocity and temperature were 10 m/s and 23.8°C, respectively. The initial conditions for the transient simulation were following: the flow field from the periodic simulation of the air flow in the channel with the parameters the same as at the inlet to the channel, while the brick was at 23.8°C and pores were fully filled with the liquid water which volume fraction was 13%. The dry air and water vapour densities in the brick were equal to 0.

The thermo-physical parameters used in the simulations are given in Table 1. The closing relationships for the system of governing equations, eq. (19)-(22), were following [18]:

- Pores area for unite volume:

$$a_s = \frac{6}{d_{av}(1-\varepsilon_s)} \tag{23}$$

- Effective thermal conductivity of the brick and moist air:

$$k_{ef} = k_s + 0.0047W \tag{24a}$$

$$k_g = \frac{\rho_a k_a + \rho_v k_v}{\rho_a + \rho_v} \tag{24b}$$

- Volumetric moisture content:

$$W = \varepsilon_l \rho_l + \varepsilon_v \rho_g \tag{25}$$

- Vapour mass diffusivity in the brick and dry air:

$$D_{v-a,ef} = \frac{2.61\cdot 10^{-5} M_v \left[1-\left(\frac{W}{W_{cap}}\right)\right]}{C_{dry} BT \left[0.503\left(1-\frac{W}{W_{cap}}\right)^2 + 0.497\right]} \tag{26a}$$

$$D_{v-a} = 2.23\cdot 10^{-5}\left(\frac{T}{273.13}\right)^{1.75} \tag{26b}$$

- Water vapour density for saturation conditions:

$$\rho_{v,sat} = \frac{p^*_{v,sat} M_v}{BT} \tag{27}$$

Table 1. Thermophysical parameters of dry brick, liquid water, water vapour and dry air

Property	Value
Brick porosity ε_p	0.14
Capillary moisture content W_{cap} (kg/m^3)	130.0
Average pore diameter d_{av} (m)	$1.6\cdot 10^{-5}$
Brick density ρ_s (kg/m^3)	1800.0
Brick thermal conductivity k_s (W/m/K)	0.77
Brick specific heat c_s (J/kg/K)	880.0
Liquid water density ρ_l (kg/m^3)	1000.0
Liquid water conductivity k_l (W/m/K)	0.62
Liquid water specific heat c_l (J/kg/K)	4190.0
Liquid water surface tension σ (N/m)	$7.2\cdot 10^{-4}$
Water vapour resistance diffusion factor C_{dry}	24.79
Water vapour conductivity k_v (W/m/K)	0.018
Water vapour specific heat c_v (J/kg/K)	1864.0
Water vapour molecular mass M_v (kg/kmol)	18
Latent heat of evaporation Δh_{vap} (J/kg)	$2.6\cdot 10^6$
Dry air thermal conductivity k_g (W/m/K)	0.026
Dry air specific heat c_a (J/kg/K)	1005.0
Brick permeability for moist air K_g (m^2/s)	$1.15\cdot 10^{-12}$
Minimal saturation s_{min}	0.25
Universal gas constant B (J/mol/K)	8.314

- Modified saturation pressure:

$$p^*_{v,sat} = p_{v,sat}\exp\left(-\frac{p_c}{\rho_l BT}\right) \tag{28}$$

- Saturation pressure:

$$p_{v,sat} = 614.3\exp\left(17.06\frac{T-273.15}{T-40.25}\right) \tag{29}$$

- Capillary pressure:

$$p_c = \frac{\sigma}{\sqrt{K_g\left(1-\varepsilon_s\right)}}\begin{bmatrix}1.417\left(1-S\right)+ \\ -2.120\left(1-S\right)^2+ \\ 1.263\left(1-S\right)^3\end{bmatrix} \tag{30}$$

- Modified saturation:

$$S = \frac{s - s_{\min}}{1 - s_{\min}} \tag{31}$$

- Saturation:

$$s = \frac{\varepsilon_l}{1-\varepsilon_s} \tag{32}$$

where: B denotes the universal gas constant, C_{dry} is the water vapour resistance diffusion factor, d_{av} stand for the average pore diameter, K_g denotes the permeability of the porous building material for moist air, M_v is the molecular mass of vapour, s_{min} is the minimal saturation for water in the pendular form, W_{cap} denotes the capillary moisture content and σ is the water surface tension.

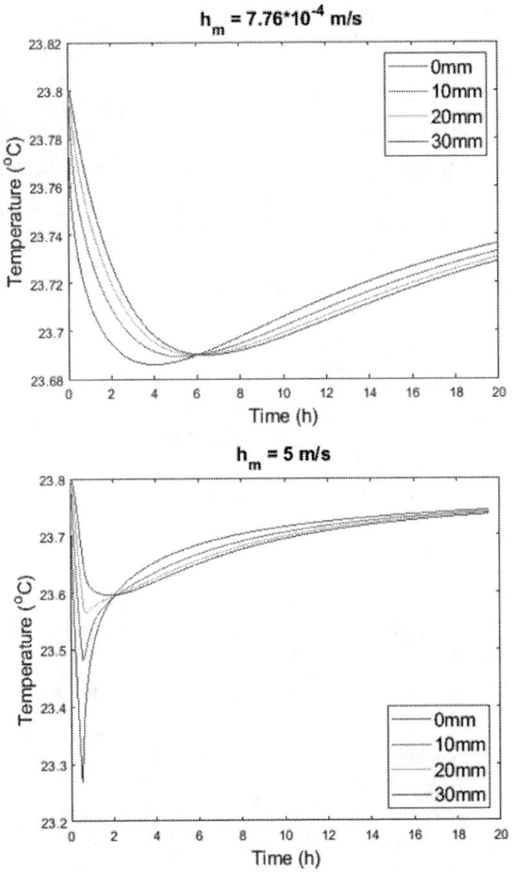

Fig. 2. Temporal variations of brick temperature at the brick top surface and various depths for two values of the mass transfer coefficient between liquid water and moist air in the porous material

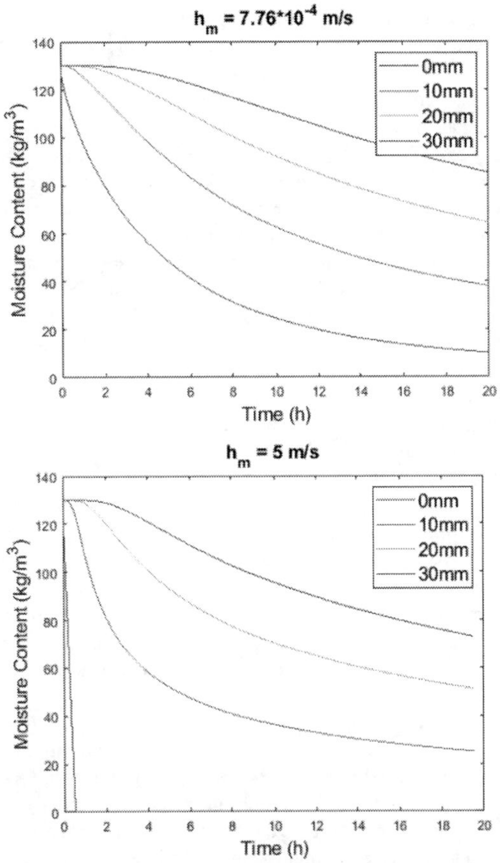

Fig. 3. Temporal variations of moisture content at the brick top surface and various depths for two values of the mass transfer coefficient between liquid water and moist air in the porous material

In this paper the influence of the mass transfer coefficient between liquid water and moist air in the porous material (evaporation speed) was analysed. Two values of this coefficient were assumed, i.e., $7.76 \cdot 10^{-4}$ and 5 m/s. The obtained temperature temporal variations at the brick top surface and at different depths inside the brick body are presented in Figure 2. For higher value of mass transfer coefficient the drying process was faster than for the lower one – see Figure 3. The temperature drops for the higher coefficient at different depths in the brick reached maximum after 30 to 60 minutes from the beginning of drying and temperature at the sample surface decreases below 23.3°C (lower part of Figure 2). The temperature drops for the lower coefficient reached maximum after 4 to 6 hours and temperature at the sample top surface decreases to 23.69°C (upper part of Figure 2). These values of temperature drops were lower than reported in the literature. This was probably caused by neglection of the capillary liquid moisture transport, which upholds evaporation. The second factor which might affect temperature drops was high mass and heat transfer rate between the air and brick. High speed of flowing air (10 m/s) resulted in the intensive convective mass and heat transport at the brick surface and lower temperature drops inside the brick.

The shapes of the temporal temperature curves presented in Figure 2 partially agree with data presented in the literature. The real drying curves are divided in three periods, i.e., initial temperature drop, temperature plateau and temperature increase in the final period. In Figure 2 during the first period of drying process, temperature falling rates were high as a consequence of the intensive evaporation of the liquid water. Unfortunately, the second period of drying process with constant temperature was not observed in presented simulations. Only for the deepest monitors in the brick (see curves for 30 mm in Figure 2) the flattened temperature curves might be found. The temperature after reaching the lowest values started to rise. The temperature plateau in the second period is partially caused by the capillary pressure gradient, which results in transport of liquid water to the drier regions and upholds isothermal evaporation. However, capillary movement of the liquid water was not accounted for in the present model, hence the constant temperature periods were not observed or were very short. In the last period of the drying process the predicted temperature growths were slowed by evaporation in the internal parts of the brick. These growths were slower than presented in literature because of more intensive evaporation inside the brick.

Temporary variations of the local moisture content at the brick top surface and various depths inside the brick were presented in Figure 3. For higher value of the mass transfer coefficient between liquid water and moist air in the porous building material the drying process was faster and water amount decreased more rapidly. At first water evaporated from the surface of the brick and then from the deeper regions inside the body of the brick. Although the surface was dried out or almost dried the internal parts of the brick were still wet.

7 Conclusions

In the paper complex non-equilibrium mathematical model of a drying process in the porous building materials was formulated. Three phases of moisture were considered to reside in the brick, i.e., water vapour in the moist air filling the pores, liquid water (bound water) adsorbed at the surface of the solid component of the material and free liquid water either in a discontinuous form (funicular) or continuous one (pendular). The model accounted for moisture diffusion and convection due to gravitational acceleration as well as gradients of moist air and capillary pressures inside the material. The model in a simplified form, i.e., with motion of the liquid water and convection of moist air as well as capillary forces in the porous building material neglected, was implemented in the commercial software ANSYS CFD with the aid of its advanced customisation functionalities like the User-Defined Function, User-Defined Scalar and User-Defined Memory. Subsequently, the model was coupled with heat and fluid flow model in air. Then the model was applied to initial investigations of drying of a single moist brick placed in the channel with flowing dry air.

The model allowed to predict variations of temperature and moisture content in the brick. However, the obtained results did not exhibit the same character of the drying temperature curves as these presented in the literature. Neglecting the capillary movement of the liquid water in the brick and assuming high heat and mass transfer rates between the flowing air and the brick top surface might be sources of this discrepancy.

The future work will be related to the implementation of the full mathematical model in the ANSYS Fluent with the free liquid water motion accounted for, determination of missing thermophysical parameters of the brick as well as its verification and validation. The complete numerical model will be used for parametric investigations and optimisation of the drying phenomena in the building materials.

This work was supported by the National Centre for Research and Development (Poland) under grant no. POIR.04.01.02-00-0099/16 *Development of innovative technology of drying and moisture sealing of masonry walls, DryWall* and the statutory funds of Faculty of Power and Aeronautical Engineering of Warsaw University of Technology.

References

1. P.B. Lourenço, E. Luso, M.G. Almeida, Build. Environ. **41**, 223 (2006)
2. N. Karagiannis, M. Karoglou, A. Bakolas, M. Krokida, A. Moropoulou, Constr. Build. Mater. **137**, 441 (2017)
3. N. Mendes, F.C. Winkelmann, R. Lamberts, P.C. Philippi, Energy Build. **35**, 631 (2003)
4. O.F. Osanyintola, C.J. Simonson, Energy Build. **38**, 1270 (2006)
5. R.C. Shoemaker, D.E. House, Neurotoxicol. Teratol. **28**, 573 (2006)
6. L. Falchi, D. Slanzi, E. Balliana, G. Driussi, E. Zendri, Build. Environ. **131**, 117 (2018)
7. A.M. Gueli, V. Garro, M. Liuzzo, G. Margani, S. Pasquale, G. Politi, G. Stella, Measurement **118**, 289 (2018)
8. T. Defraeye, Appl. Energy **131**, 323 (2014)
9. J. Fan, X. Wen, Int. J. Heat Mass Transf. **45**, 4045 (2002)
10. X. Cheng, J. Fan, Int. J. Therm. Sci. **43**, 665 (2004)
11. P. Łapka, P. Furmański, T.S. Wiśniewski, Int. J. Numer. Methods Heat Fluid Flow **27**, 1078 (2017)
12. P. Łapka, P. Furmański, Przem. Chem. **96**, 343 (2017)
13. P. Łapka, P. Furmański, Heat Mass Transf. **54**, 2461 (2018)
14. P. Furmański, P. Łapka, Int. J. Heat Mass Transf. **114**, 1331 (2017)
15. G.H. dos Santos, N. Mendes, Int. J. Heat Mass Transf. **52**, 2390 (2009)

https://doi.org/10.1051/matecconf/201824001022

16. M. Van Belleghem, M. Steeman, H. Janssen, A. Janssens, M. De Paepe, Build. Environ. **81**, 340 (2014)
17. M. Steeman, M. Van Belleghem, M. De Paepe, A. Janssens, Build. Environ. **45**, 2202 (2010)
18. M. Kaviany, *Principles of heat transfer in porous media* (Springer, New York, 1995)

Thermal and water transfer in cementitious porous medium: thermal building and durability

Xiaoyan Ma[1,*], *Farid* Benboudjema[1], and *Rachid* Bennacer [1,2]

[1]LMT / ENS Cachan / CNRS / Paris-Saclay University, 61 avenue du Président Wilson, 94235 Cachan, France
[2]EPMI, F-95092 Cergy Pontoise, France

Abstract. Cementitious materials are typical porous medium, which are widely used in civil engineering and thermal buildings. The thermal and hydrate properties of the material are fundamental to predict accurately the energy needs and the reliable duration of the structures in a long term. Due to the complicated coupling, the thermal capacity, conductivity and source term in energy equation are affected by the humidity distribution and the microstructure evolution of the porous materials. The coupling relation of thermal and hydrate properties are summarized in this work, the coupled drying model is discussed, and the experiment is performed to verify the simulation results for RH (Relative humidity) and ML (Mass loss). Two approaches are adopted to estimate the humidity/water content profile in the porous medium and to analyse the temperature effect. In addition, the equivalent hydrate capacity and conductivity are identified as function of humidity. The significance of this work lies in predicting the energy demand, the long-term thermal behaviour, taking into account the change of the thermal cementitious materials properties during their service time. Such inner structure is consequence of the system durability where the temperature, water content and humidity will act on the cracks settling within the material.

Nomenclature

c	Heat capacity
D_{eff}	Effective diffusivity $[m^2/s]$
k_{la}	Relative permeability
q	Source term
L_v	Latent heat [J/kg]
M	Location at (x,y,z)
M_v	Molecular mass $[kg/mol]$
p	Pressure$[Pa]$
RH	Relative humidity
S	Saturation degree
Φ	Porosity [%]
ρ	Density $[kg/m^3]$
μ	Dynamic viscosity $[Pa \cdot s]$
λ	Heat conductivity $[W/(m \cdot K)]$
μ	Dynamic viscosity $[Pa \cdot s]$

Sub/superscript

vs	Saturation vapour
app	Apparent
l	Liquid water
v	Water vapour

1 Introduction

The performance of thermal building is dependent on the properties of wall materials and their evolution overtime. The properties control is a key point in predicting accurately the energy needs over the time. Such cementitious materials are typical kind of complex porous medium, which are widely used in civil engineering. The building walls are subjected to various ambient conditions. During the lifetime of building concrete structure, some undesirable pathologies will appear after a long induction period, and eventually resulting in irreversible structure damages [1]. The most important factors that induce such structure changes (pathologies) are surrounding temperature and humidity.

Therefore, to understand the thermal properties which are related to the humidity and water contents and structure change, we have to be accurate and master the mechanisms of moisture transport inside the structure. However, the difficulties lie in the complexity of cementitious porous media, the strong coupling between temperature effect and humidity/water sorption, and the inaccurate energy source term induced by the phase change (evaporation/condensation process).

First of all, thermal conductivity as the main heat transfer parameter is necessary and often subject of measurement for various types of cement-based

*Corresponding author: ma@lmt.ens-cachan.fr

composites. However, very rare reference quantify the dynamic dependence of thermal conductivity on moisture and water content. It is complex coupling depending on the strong interlink between temperature and humidity conditions and the diffusion phenomena. The coupling phenomena also includes local aspect as total pore volume, pore distribution and connectivity. The porous structure (cementitious material) thermal conductivity varies in a very significant way due to the difference in values between air and cement stone (depending on the amount and the type of aggregates). In common service conditions, cementitious composites always contain certain amount of water. The thermal conductivity of water is $0.6\ W/(mK)$, which is more than twenty times higher of the air. Therefore, heterogeneous water contents in the pores compete with the effect of air, so the thermal conductivity of such composite material will exhibit important change over space and between the identified values by different authors.

Pavlik and Rovnanikova measured the thermal conductivity of a cement-based composite material [2], focusing on the dependence of thermal conductivity to the moisture content from fully dry state to totally saturated state by adopting an impulse technique. Results show that the thermal conductivity increase with the increase of ambient humidity under the same temperature condition. Dell'Isola *et al.* demonstrated that the effective thermal conductivity of some common building materials increases with temperature and moisture content [3].

Kim *et al.* proposed Eq. (1) of thermal conductivity for concrete [4], deducing from the regression analysis results. In which, the aggregate volume fraction and moisture condition are revealed as the main factors affecting the concrete conductivity.

$$\lambda_R = 0.8 \times [1.62 - 1.54 \times (W/C)] + 0.2 \times RH \quad (1)$$

In which, λ_R is a modification index associated with average relative humidity (RH) and water-to-cement ratio (W/C). Bristow provided equivalent reliable measurements of soil thermal properties as a function of water content [5].

As consequence of change in thermal conductivity, it will modify the density and heat capacity as well. Hillel studied the thermal diffusivity varying with water content [6]. It indicated that the thermal diffusivity is small at low water contents, and it increases as water content increases, reaches a maximum, and then slowly decreases as water content continues to rise to saturation state. In addition, the thermal conductivity of humid air is also function of temperature, pressure, and mole fraction of water. Beirão *et al.* performed research on humid air, and it indicated that thermal conductivity increase with temperature under the same pressure for same humidity, and specific capacity of moist air increase with temperature under all humidity cases [7]. Bednarska and Koniorczyk also verified the negative influence of moisture on thermal properties of material. The values of thermal conductivity coefficient increase with the increase of liquid content in pore [8].

Another coupling of heat and mass during drying or wetting process is the local energy source term, representing the evaporation or condensation. The phase change induces the energy loss or gained accompanied with the local moisture and water contents transfer process. Such energy source term in porous medium is related to the water changing quantity. Such energy source at location M, i.e. q(M), is the latent heat of the evaporation/condensation multiplied by the local water content change. The transient heat transfer equation with source term is given by Eq. (2).

$$(\rho c)_M \frac{\partial T}{\partial t} = \nabla(\lambda(M) \cdot \nabla T) + q(M) \qquad (2)$$

Heat capacity and thermal conductivity are function of the solid material, fluid, porosity, local water content, and HR. Such inhomogeneity could be expressed as:

$(\rho c)_M$ or $\lambda(M) \sim \text{function}(porosity, water\ content, gas, HR)$

The water evaporation/condensation quantify the source term and it takes major weight because the latent heat value is $2260(kJ \cdot kg^{-1})$, which is much higher than the sensible heat 4182 $(J \cdot kg^{-1} \cdot K^{-1})$. The question (2) could be rewritten without the q(M) term by adopting the technique of apparent heat capacity.

It is the strategy of variable heat capacity with temperature in order to incorporate the released or absorbed energy during the phase change. The partial differential equation is expressed by Eq. (3).

$$\left(\rho C_p(M)\right)^{app} \frac{\partial T}{\partial t} = \nabla(\lambda(M) \cdot \nabla T) \qquad (3)$$

The advantage of this method is to avoid the complexity in transform the latent heat term and simplify the calculation in source term. Therefore, to estimate this energy, we need to quantify the water that transformed in phase during the drying or wetting process. On the other hand, the drying process is quite slow in comparison to the time needed to reach temperature equilibrium. So in most studies, the drying or wetting of cementitious materials are considered as isotherm process.

In this work, aiming to obtain the local and global information of the hydrate state inside the porous material, we firstly performed the experimental trials. The inner relative humidity and the global mass loss are traced, which will be taken as the reference for the simulation results. Secondly, the analytical solution for hydrate state prediction is put forward. It is a simplified method to estimate the humidity profile because the solution is based on linear partial differential equation with constant diffusion coefficient instead of non-linear model. Afterwards the non-linear drying model for porous medium is presented and computational results for relative humidity and mass loss are obtained by CAST3M programming. Results are compared with the experimental one. At last, the equivalent diffusion coefficient and equivalent hydrate capacity is calculated as function of humidity. This information is fundamental for optimizing the hydrate and thermal properties and in the long-term predicting the durability of the cementitious materials.

2 Experiment and analytical solutions

For the experiment part, the porous specimen is concrete B11, which has been widely studied and properties are mostly know [9]. The geometry is prismatic with size of $7 \times 7 \times 28\ cm^3$, which is subjected to drying inside the climate chamber on two lateral faces, and the other four lateral faces are well enclosed by aluminium. The climate chamber (see Fig. 1a) is automatic regulation at temperature of 38 ± 0.5 °C, and relative humidity of 32 ± 1%. The evolution of mass is recorded by balance, and humidity sensors are fixed at the preserved holes in the samples, which are 1.5, 3.0 and 6.0 cm to the drying surface, respectively (see arrows), as illustrated in Fig. 1b. The relative humidity of the fixed points are recorded continuously. Further details for the devices and protocol can be referred to previous work [10].

a)

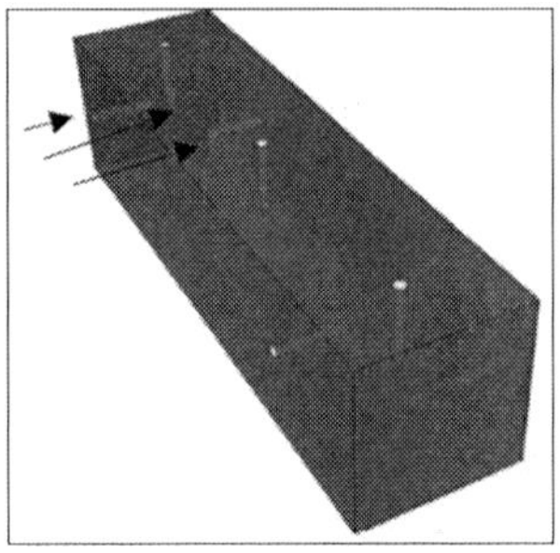

b)

Fig. 1. The climate chamber used for drying a); and prismatic concrete block with sensor locations b).

As a consequence of the pressure difference inside the domain, the complexity of the non-linear coupling of vapour diffusion and liquid water displacement feature the local phenomena. In order to have a global idea of the humidity distribution inside the porous medium, an analytical method with constant diffusion coefficient will be applied.

We will illustrate the physical tendency, to avoid the complexity by using a simplified linear model. A constant property without phase change is considered in order to use the existing analytical solution of the humidity distribution. The constant diffusion coefficient is used instead of variation function of humidity and temperature and no liquid displacement is considered. The resulting analytical solution of the internal relative humidity (RH) as function of dimensionless time and relative length are plotted in Fig. 2a. It shows the initial dimensionless humidity was unity as initial condition and the used boundary condition is zero concentration ($x = 1.0, \mathrm{RH} = 0$).

The relative humidity decrease fast at the vicinity of the external boundary. The central positions (such as $x \leq 0.2$) exhibit a slowly decrease over the considered time. The corresponding three experimental dimensional positions of the humidity sensors are chosen to have a clearer visualization of the local humidity time evolution, as showed in Fig. 2b.

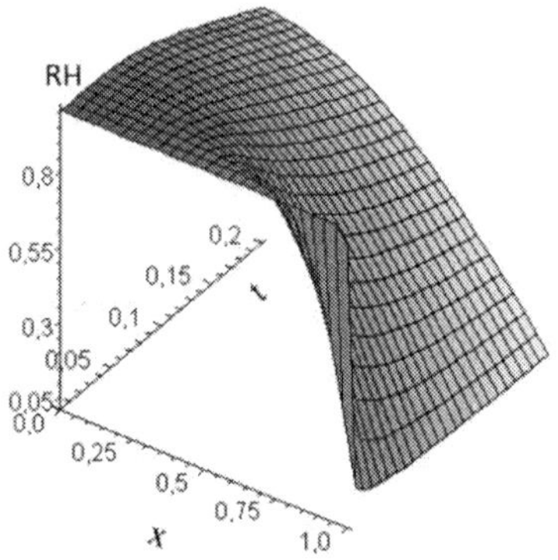

a)

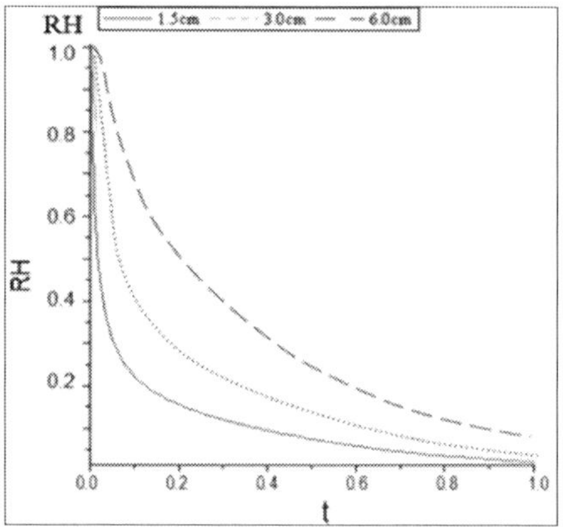

b)

Fig.2. Analytical solution for dimensionless humidity versus dimensionless time and space in 3D a); and RH time evolution at different positions to drying surface b).

3 Drying model

The drying model is fundamental for simulating the drying phenomenon in porous medium, and quite a few of literatures can be found in previous researches [11, 12]. For the drying model, it is assumed that the water transport in the concrete induced from the combination of the diffusion of water vapour and the permeation of liquid water. If the concrete porous network is saturated, the latter mechanism plays predominant role, whereas for a low degree of saturation, the movement of water in the form of vapour diffusion cannot be neglected. The simplified approach involves water transfer coefficient that is a function of saturation degree or the specific water [13,14]. The liquid water flow can be expressed by Darcy's law, which relates the fluid mass flow through the medium to the pressure gradient of this fluid. The water vapour flow can be expressed by Fick's law, representing the flow of water in form of vapour due to a gradient in vapour pressure in the considered porous area. The two driving forces expressed as pressure gradient, could be related to the relative humidity by Kelvin equation. The relation between the capillary forces and the relative humidity in concrete is expressed in Eq (4),

$$p_c = \frac{-\rho_l\, R\, T\, ln(HR)}{M_w} \quad (4)$$

Deducing from the mass balance of water and the equilibrium driving forces, the final drying model is given in Eq. (5).

$$\rho_l\left[\Phi\frac{\partial S_l}{\partial HR}\left(1-\frac{\rho_v}{\rho_l}\right)\right]\frac{\partial HR}{\partial t} = \mathrm{div}\left[\left(k_{la}\frac{1}{HR}\frac{\rho_l^2 RT}{\mu_l M_v} + D_{eff}\frac{M_v p_{vs}}{RT}\right)grad(HR)\right] \quad (5)$$

Among which, the desorption isotherm on the left-hand of the equation (5) links the saturation degree (S_l) to the relative humidity and contribute to the water storing capacity. The most widely used expression is given by Eq. (6) [15], where a_{vg} and b_{vg} are two parameters that can be obtained by fitting experimental results of the desorption isotherm for the tested material.

$$S_l = \left(1 + \left(-a_{vg} ln(HR)\right)^{b_{vg}}\right)^{-c_{vg}} \quad (6)$$

On the right hand side of the Eq. (5), it suggests that the diffusion coefficient term (or hydrate conductivity) is associated with temperature, both for liquid permeation and vapour diffusion term. k_{la} is the apparent permeability of liquid, which can be expressed by intrinsic water permeability k_0 multiplied by relative permeability k_{rl} ($0 < k_{rl} < 1$). This latter term can be determined by the Mualem empirical relation shown in Eq. (8), where S_l is the saturation degree, p_{krl} is a fitting parameter [11].

$$k_{la} = k_{rl} \cdot k_0 \quad (7)$$

$$k_{rl} = S_l^{p_{krl}}\left(1 - \left(1 - S_l^{\frac{1}{a_{vg}}}\right)^{a_{vg}}\right)^2 \quad (8)$$

For the diffusion coefficient of vapour, the effective diffusivity D_{eff} can be expressed by Eq. (9) and (10) [16].

$$D_{eff} = d_{rl} D_0 \quad (9)$$

$$d_{rl} = \Phi^{a_{mq}}\,(1 - S_l)^{b_{mq}} \quad (10)$$

The apparent diffusion is related to the relative diffusion function d_{rl} and to the diffusion coefficient of water in air D_0. In a similar way, the diffusion of water vapour is highly dependent on the saturation degree. In addition, the tortuosity of the concrete porosity also affects these parameter a_{mq} and b_{mq}.

What is worth noticing is that in the drying model, some parameters are not only function of humidity, but also function of temperature. In Eq. (5), despite of the apparent temperature, there is some other parameters affected by temperatures. The following Eqs. (11-13) illustrate the dependency to temperature, for the diffusion coefficient of water in air D_0 , the saturation vapor pressure P_{vs}, and the dynamic viscosity of liquid water μ_l. This also indicates the complex coupling between moisture and heat transfer, as well as the identification of parameters affecting the thermal property and hydrate property.

$$D_0 = 2.17 \times 10^{-5} \,.\, \frac{P_{atm}}{P_g} \,.\, \left(\frac{T}{T_0}\right)^{2.88} \quad (11)$$

$$P_{vs} = P_{atm} . e^{\frac{M_v . L_v}{R}\left(\frac{1}{T_0} - \frac{1}{T}\right)} \quad (12)$$

$$\mu_l = \mu_0 \,.\, e^{-1.94 - 4.8\frac{T_0}{T} + 6.74\left(\frac{T_0}{T}\right)^2} \quad (13)$$

4 Results and discussion

Computational results of the drying model and the experimental results are compared in Fig. 3. Both the relative humidity evolution at the point of 1.5 cm (to drying surface) and relative mass loss (ML) are presented. The relative humidity provide a local information of the moisture distribution, while the mass loss is the integral of water content over the domain, meaning the global water loss during the drying process. The local humidity exhibit the decrease over time as consequence of the applied boundary condition of 32%. This evolution has the same tendency as what was presented in the previous linear-simplified approach in Fig. 2b. During the test duration of 160 days, the simulation results Eqs. (5-12) match well with the experiment data using all the identified properties of the different contributing phenomenon.

In comparison to the first linear simplified case, this numerical model, which is more close to the reality, uses a variable volumetric capacity in the transient term and variable hydrate conductivity (humidity and water diffusion). We illustrate in the figures below (Fig. 4) the profile of the hydrate capacity and hydrate conductivity versus relative humidity in the present drying model.

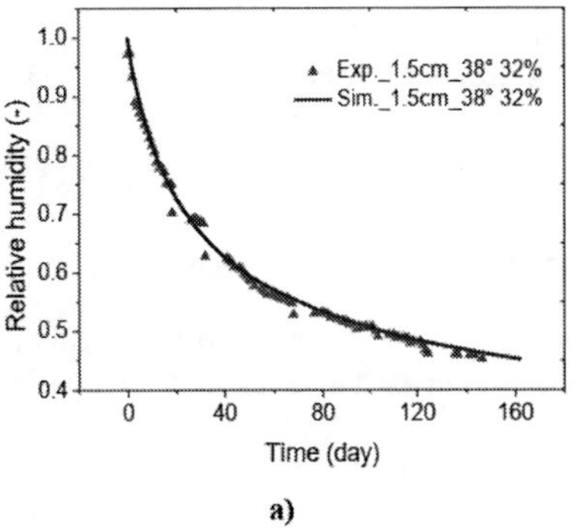

a)

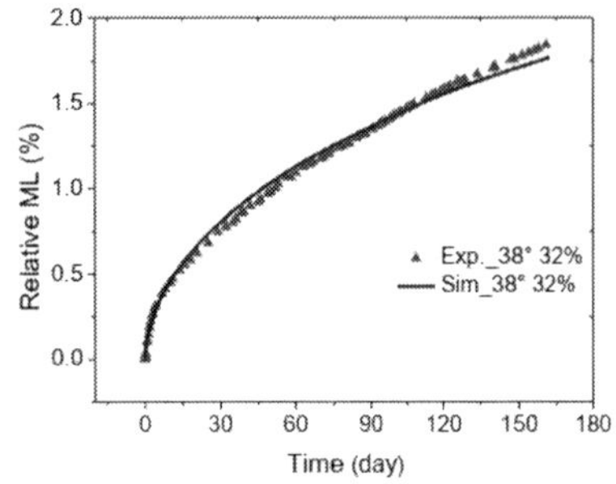

b)

Fig. 3. Experimental and simulation results of mass loss a); and relative humidity evolution b).

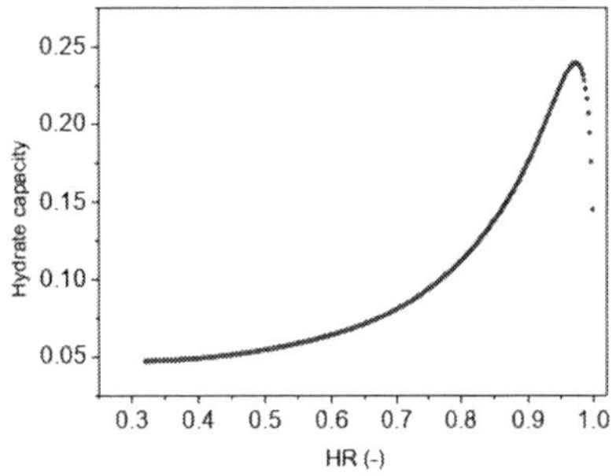

a)

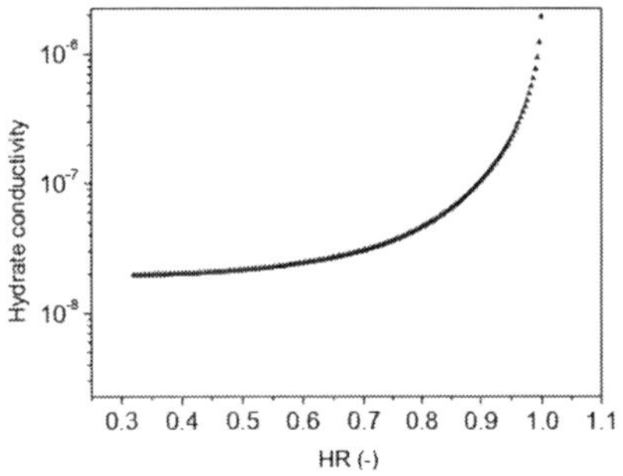

b)

Fig.4. Profile of equivalent hydrate capacity a); hydrate conductivity b) versus RH.

It demonstrates that both of the parameters are highly dependent to the water content level. This is obtained under the assumption of isotherm process, but it is worth noting that the humidity has effect on thermal properties as well, inducing the heat and mass transfer process.

5 Conclusion

This work emphasized on the hydrate transfer properties, and it is fundamental and necessary for accurately estimating the real local density, heat capacity, thermal conductivity and the heat source term in the heat transfer equation. The analysis can be applied to optimizing the hydrate and thermal properties and in the long-term predicting the durability of the cementitious materials. This information also allows to resolve the drying model and to predict the temperature history and the exchange between the wall and the living area (air). In further level, it will be able to predict the durability of the inner structure and cracks within the material, with taking into account the change in properties and especially the permeability (water redistribution).

In short, the practical meaning of predicting the real distribution of hydrate state over space and time, including the evaporation or condensation, is to estimate accurately the real local density, heat capacity and thermal conductivity, as well as the source term in the heat transfer equation.

Acknowledgments

This ThermaSMART project has received funding from the European Union's Horizon 2020 -MSCA-RISE-2017 program under grant agreement No. 778104.
The first author acknowledges China Scholarship Council for the financial support, No. 201608120052.

References

1. H. Wang and J.E. Gillott, Cem. and Concr. Res., **21** 647-654 (1991)
2. Z. Pavlik, and P. Rovnanikova, Int. J. Thermophys, **27** (4), 1228–1240 (2006)
3. M. Dell'Isola , F. R. D'Ambrosio Alfano, G. Giovinco, and E. Ianniello, Int. J. Thermophys, **33** (8–9), 1674–85 (2012)
4. K. Kim, J. Sang-eun, K. Jin-keun, and Y.Sungchul,. Cem. Concr. Res. 33, 363–71 (2003)
5. K. L. Bristow, Agricultural and Forest Meteorology **89** (2), 75–84 (1998)
6. D. Hillel, *Fundamentals of Soil Physics*. Academic Press, New York (1980)
7. S.G.S. Beirão, A.P.C. Ribeiro, M.J.V. Lourenço, F.J.V. Santos, and C.A. Nieto de Castro, Int. J. Thermophys. **33**, 8–9 (2012)
8. D. Bednarska , and K. Marcin, AIP Conf. Proc. (June), 1577–78 (2017).
9. A. Hilaire, PhD Thesis, ENS Cachan (2014)
10. X. Ma, A. Darquennes, F. Benboudjema, G. Nahas, R. Bennacer, *Int. Conf. Mater. Energy*, (2017)
11. Y. Mualem, Water Resour. Res., **12**(3), 513–522 (1976)
12. Z. Zhang, M. Thiery, Baroghel-Bouny, Cem. and Concr. Res., **89**, 257-268 (2016)
13. Z. Bažant, L. Najjar, Cem. and Concr. Res., **1**(5), 461–473 (1971)
14. B. Perrin, V. B. Bouny, L. Chemloul, Mater. Struct., **31**(4), 235–241 (1998)
15. M. A. van Genuchten, Soil Science Society of America Journal, **44**(5), 892–898 (1980)
16. R. Millington, J. Quirk, Trans. Faraday Soc., **57**,1200–1207 (1961)

Heat transfer coefficient determination using the FEM with time-dependent Trefftz-type basis functions in subcooled flow boiling in a minichannel

Beata Maciejewska[1,*], and *Magdalena* Piasecka[2]

[1] Faculty of Management and Computer Modelling,
[2] Faculty of Mechatronics and Mechanical Engineering,
Kielce University of Technology, Al. Tysiaclecia Panstwa Polskiego 7, 25-314 Kielce, Poland

Abstract. Results concerning flow boiling heat transfer in a vertical minichannel of 1.7 mm depth were shown. The channel was asymmetrically heated by a thin foil. Its surface temperature was recorded continuously in points by thermocouples. Measurements were carried out in 0.01 s intervals. The objective of the numerical calculations was to determine the heat transfer coefficient on the heated foil–fluid contact surface in the minichannel from the Robin boundary condition. Both the foil and fluid temperatures were the result of solving the nonstationary two-dimensional problem in the foil and flowing fluid. The problem was solved by using the FEM combined with Trefftz-type basis functions. The values of the time-dependent local heat transfer coefficient were presented and discussed.

1 Introduction

Miniaturization of devices is being progressively applied in cooling technologies designed to prevent exceeding operating temperatures. A compact heat exchanger responds to the issues of transfer in confined spaces in which conventional size channels cannot be used. Extensive efforts to recognize boiling phenomena in minichannels include theoretical analyses based on experimental measurements. The results can be applied to the construction of compact heat exchangers and using them in heating, cooling and thermoregulation applications.

In order to estimate the intensity of heat transfer accompanying flow boiling in a mini heat exchanger device, the main experimental data are needed for the heat transfer coefficient identification: the heated wall temperature, the temperature gradient and the temperature of the fluid flowing along a minichannel. These quantities were obtained by solving the inverse problem [1] in the heated wall and in the flowing fluid. The method proposed by Trefftz [2] can be used to solve the inverse problems. This approach is based on an approximation of the unknown solution by a linear combination of functions that exactly satisfy the differential equation.

Details of the method based on the Trefftz functions can be found in [3]–[12]. In this work, to solve the non-stationary two-dimensional problem in the flowing fluid the time-dependent Trefftz functions for the Fourier–Kirchhoff equation were determined. These functions were used to construct the nonstationary FEM basis functions.

2 Experiment

The test section with two parallel minichannels (Fig. 1) is the main element of flow loop realized in the experimental stand, which view is shown in Fig. 2.

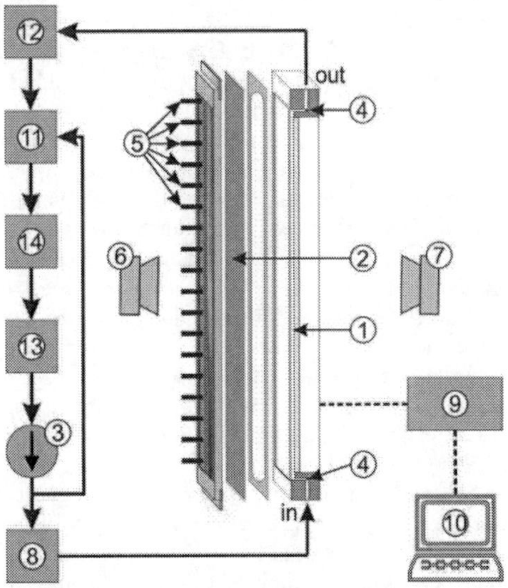

Fig. 1. The main elements of the experimental stand: 1 - a minichannel, 2 - a heated foil, 3 - a gear pump; 4,5 - thermocouples, 6 - an infrared camera, 7 - a high speed camera, 8 - a Coriolis mass flow meter, 9 - data acquisition stations, 10 - a pc computer, 11 - a compensating tank/ a pressure regulator, 12 - a heat exchanger, 13 - a filter, 14 - a deaerator.

* Corresponding author: beatam@tu.kielce.pl

Fig. 2. The view of the experimental stand.

The components of the test section are two 26.5 mm wide, 180 mm long and 1.7 mm deep minichannels (1). A 0.1 mm thick foil (2), made of Haynes-230 alloy is heated while FC-72 Fluorinert flowing along the minichannels thanks to the gear pump (3). In the selected minichannel temperature of the outer side of the foil is measured with 18 points by K-type thermocouples (5). Two other thermocouples control fluid temperature at the inlet and outlet of the minichannels (4). Foil temperature of the second minichannel is monitored by an infrared camera (6). The opposite surface of the heated foil is observed through the glass using high-speed camera (7), simultaneously. A mass flow meter (8) is used to control the fluid flow in the flow loop. Pressure at the minichannel inlet and outlet, current supplied to the foil and voltage drop across a foil were also monitored continuously.

The supply and control system contains: an inverter welder, a shunt, an ammeter and a voltmeter. The data acquisition system consists of two data acquisition stations (9) and a computer (10) with a appropriate software.

During an experiment series, there was a gradual increase in the electric power supplied to the foil. All measurements were carried out in 0.01 s intervals.

3 Analysis and modelling

The objective of the numerical calculations, refer to an experiment, was to determine the heat transfer coefficient during FC-72 flow boiling in a minichannel from the Robin boundary condition:

$$\alpha(x,t)=\frac{-\lambda_F\dfrac{\partial T_F(x,\delta_F,t)}{\partial y}}{T_F(x,\delta_F,t)-T_f(x,\delta_F+\delta_T,t)} \tag{1}$$

where λ_F – the thermal conductivity coefficient of the foil, δ_F – the foil thickness, T_F – the foil temperature, T_f – the fluid temperature, δ_T – the thickness of the thermal boundary layer, x – the spatial variable referred to the flow direction, y – the spatial variable referred to the direction perpendicular to the flow direction (referring to the thickness of the heated foil).

Both the foil and fluid temperatures were the result of solving the inverse nonstationary two-dimensional problem in two neighboring domains: the heated foil and flowing fluid. It was assumed that nonstationary temperature distribution in the heated foil $(x,y)\in\Omega_F$ for $t>0$ was described by the heat equation

$$\nabla^2 T_F-\frac{\rho_F\, c_{p\_F}}{\lambda_F}\frac{\partial T_F}{\partial t}=-\frac{I\cdot\Delta U}{A\cdot\delta_F\cdot\lambda_F} \tag{2}$$

whereas in the thermal boundary layer of fluid ($(x,y)\in\Omega_T$) for $t>0$ by the Fourier–Kirchhoff equation

$$\nabla^2 T_f-w_x(y)\frac{\partial T_f}{\partial x}-\frac{\rho_f\, c_{p\_f}}{\lambda_f}\frac{\partial T_f}{\partial t}=0 \tag{3}$$

The boundary conditions have the form:

$$T_f(x,y,0)=T_{0\_f}(x,y) \tag{4}$$

$$T_F(x,y,0)=T_{0\_F}(x,y) \tag{5}$$

$$T_F(x_p,0,t_k)=T_p(t_k)\ \text{for}\ p=1,2,\ldots,P,\ k=1,2,\ldots K \tag{6}$$

$$\frac{\partial T_F}{\partial x}(0,y,t)=0 \tag{7}$$

$$\frac{\partial T_F}{\partial x}(L,y,t)=0 \tag{8}$$

$$\lambda_F\frac{\partial T_F}{\partial y}(x,0,t)=q_{loss}(t) \tag{9}$$

$$T_f(x,\delta_F,t)=T_F(x,\delta_F,t) \tag{10}$$

$$T_f(0,y,t)=T_f^{in}(t) \tag{11}$$

$$T_f(L,y,t)=T_f^{out}(t) \tag{12}$$

where $\Omega_F = \{(x,y) \in R^2 : 0 < x < L, \quad 0 < y < \delta_F\}$, $\Omega_T = \{(x,y) \in R^2 : 0 < x < L, \quad \delta_F < y < \delta_F + \delta_T\}$, L – the minichannel length, I – the electrical current, ΔU – the voltage drop, A – the surface area of the heated foil, P – the number of measurements, K – the number of time intervals, T_p – the foil temperature measured by thermocouples at the boundary $y = 0$, q_{loss} – the heat loss to the surroundings [13], w_x – component of vector fluid velocity parallel to the minichannel heated surface [14], ρ_F, ρ_f – the density of the foil and fluid, respectively, c_{p_F}, c_{p_f} – specific heat of the foil and fluid, respectively, λ_F, δ_F, δ_T – defined as for Eq. (1). The boundary conditions are shown in Fig. 3.

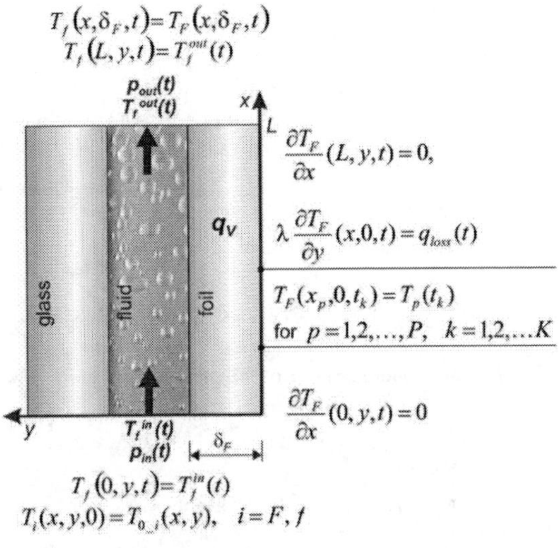

Fig. 3. Boundary conditions.

The solution method of Eqs. (2-3) presented in this article is based on an approach described in [15] concerning to the one-dimensional heat equation.

The problem described in Eqs. (3), (4), (10), (11) and (12) in the thermal boundary layer of fluid Ω_T was solved in the time subintervals by means of the FEM with time-dependent basis functions. This base functions are a linear combinations of Trefftz functions, strictly satisfying the governing equation, and have the following properties in nodes:

$$f_{jk}(x_i, y_i, t_i) = \delta_{ki}, \; i = 1,2,\ldots,lw \tag{13}$$

where δ_{ki} – the Kronecker delta, j – the element number, k – the basis function number in the j-th element, lw – number of nodes in the j-th element.

The Trefftz functions are coefficients at partial derivatives in expansion of the differential equation solution in the Taylor series, in which the derivative $\frac{\partial^2 T_f}{\partial y^2}$ is replaced by the formula resulting from the Fourier–Kirchhoff equation:

$$\frac{\partial^2 T_f}{\partial y^2} = w_x(y)\frac{\partial T_f}{\partial x} + \frac{\rho_f c_{p\_f}}{\lambda_f}\frac{\partial T_f}{\partial t} - \frac{\partial^2 T_f}{\partial x^2} \tag{14}$$

The fluid temperature in each time – space element Ω_T^j is approximated by a linear combination of time-dependent Trefftz-type basis functions $f_{jk}(x,y,t)$:

$$\tilde{T}_f^j(x,y,t) = \sum_{k=1}^{lw} T_f^n f_{jk}(x,y,t) \tag{15}$$

where n is the node number in the entire domain Ω_T, T_f^n denotes the approximate values of the fluid temperature in nodes j, k, lw are defined as for Eq. (13).

The unknown temperature values at nodes T_f^n were calculated by minimizing the functional, similar to this described in [16]. It represents the mean square error of the approximate solution on the boundary, in the initial time and along common edges of neighboring elements.

The set of Eqs. (2) and (5-9) was solved like in [16].

4 Results and discussion

The study shows the results obtained for the subcooled boiling region and the most important concern was to the heat transfer coefficient identification. The coefficient was found by solving the inverse heat conduction problem by using the FEM with time-dependent Trefftz-type basis functions. The calculations were performed using two minutes time-dependent temperature measurements but only selected results - from 1 s to 100 s - were presented. This data were selected to present the results obtained for the subcooled boiling region.

The relationship between the heat transfer coefficient and the distance from the minichannel inlet, were determined for three times: t_{10} = 10 s, t_{50} = 50 s and t_{80} = 80 s and shown in Fig. 4a. Figure 4b presents the heat transfer coefficient vs. time calculated on the basis of temperature measurements from six selected thermocouples (T2, T5, T8, T11, T14 and T17) with the increasing heat flux being supplied to the foil. Results obtained for the points where listed above thermocouples were selected as representative for the subcooled boiling region data.

When analysing the dependence transfer coefficient vs. the distance from the minichannel inlet shown in Fig. 4a, it can be noticed that the highest values of the heat transfer coefficient up to 1.6 kW/(m^{2}K) are achieved near the outlet of the minichannel and the lowest are obtained at its inlet. It is worth mentioning that presented results concerns first part of experiment when the lowest heat flux was supplied to the heated foil.

The results in Fig. 4b are presented as the heat transfer coefficient vs. time. The local heat transfer coefficients determined on the basis of the foil temperature measured at six selected points measured between 1 s and 100 s were shown in this graph. It was observed that local heat transfer coefficient values increased with increasing distance from the minichannel and achieved the highest values up to 1.6 kW/(m²K), at the minichannel outlet and the lowest were near the minichannel inlet approx. 0.2 kW/(m²K).

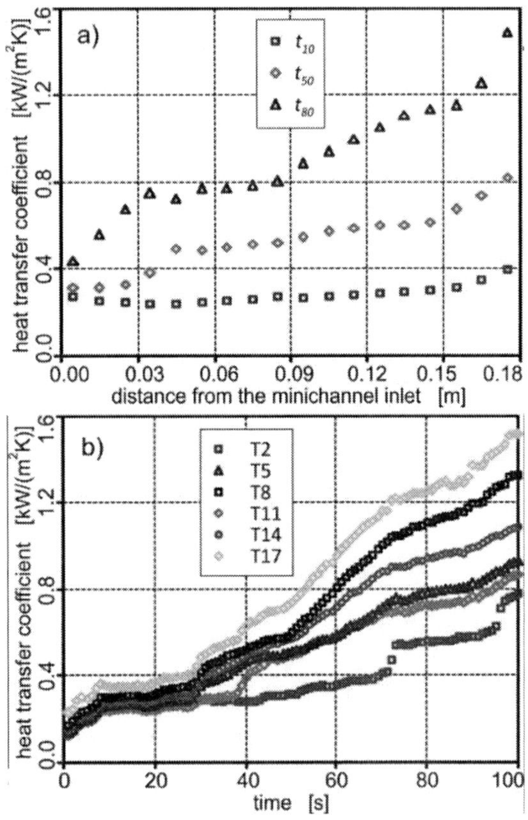

Fig. 4. Heat transfer coefficient vs.:
a) the distance from the minichannel inlet for selected time intervals : t_{10} = 10 s, t_{50} = 50 s and t_{80} = 80 s,
b) time calculated on the basis of temperature measurements from selected thermocouples: T2, T5, T8, T11, T14 and T17 ; experimental parameters: heat flux density from 1.8 kW/m² to 46 kW/m², average inlet pressure 122 kPa, average mass flow rate 0.012 kg/s.

5 Conclusions

The method of solving the time-dependent inverse heat transfer problem in flowing fluid using the FEM with time-dependent Trefftz-type basis functions were discussed in this paper. The objective of the numerical calculations, refer to an experiment, was to determine the heat transfer coefficient during FC-72 flow in an asymmetrically heated minichannel. Only results concerned subcooled boiling region was discussed The heated foil surface temperature was recorded continuously in 18 points by thermocouples. Other experimental parameters necessary for calculations were also monitored in 0.01 s intervals. The heat transfer coefficient, on the heated foil–fluid in the minichannel, was obtained from the Robin boundary condition. It was assumed that the nonstationary temperature distribution in the flowing fluid was described by the Fourier–Kirchhoff equation. The unknown temperature values at nodes were computed by minimizing the functional which describes the mean square error of the approximate solution on the boundary, in the initial time and along common edges of adjacent elements. The auxiliary nonstationary temperature distribution in the heated foil, described by the heat equation, was obtained by Trefftz method. The values of the time-dependent local heat transfer coefficient were discussed. It was confirmed that under the subcooled boiling, local heat transfer coefficients achieved rather values up to 1.6 kW/(m²K), at the minichannel outlet. The lowest values of the coefficient were obtained at the minichannel inlet.

References

1. M. N. Özişik, H. R. B. Orlande, *Inverse heat transfer: fundamentals and applications. Fundamentals and Applications* (Taylor & Francis, New York, 2000)
2. E. Trefftz, International Kongress für Technische Mechanik (1926)
3. I. Herrera, Numer. Methods Partial Differ. Equ. **16**, 6 (2000)
4. S. Y. Reutskiy, Eng. Anal. Bound. Elem. **28**, 1 (2004)
5. B. Movahedian, B. Boroomand, S. Soghrati, Eng. Anal. Bound. Elem., **37** (2013)
6. S. Blasiak, A. Pawinska, Int. J. Heat Mass Transf., **90** (2015)
7. K. Grysa, A. Maciag, Int. J. Heat Mass Transf., **100** (2016)
8. K. Grysa, B. Maciejewska, J. Theor. Appl. Mech. **51**, 2 (2013)
9. B. Maciejewska, J. Theor. Appl. Mech. **55**, 1 (2017)
10. B. Maciejewska, M. Piasecka, Heat Mass Transf., **53** (2017)
11. B. Maciejewska, K. Strak, M. Piasecka, Int. J. Numer. Methods Heat Fluid Flow **28**, 1 (2018)
12. B. Maciejewska, K. Strąk, M. Piasecka, Procedia Eng., **157** (2016)
13. M. Piasecka, K. Strąk, B. Maciejewska, Heat Transf. Eng. **38**, 3 (2017)
14. S. Hożejowska, J. Theor. Appl. Mech. **53**, 4 (2015)
15. M. Cialkowski, J. Therm. Sci. **11**, 2 (2002)
16. B. Maciejewska, M. Piasecka, Int. J. Heat Mass Transf., **107** (2017)

Effect of baffle shape in heat transfer for jet impingement on a solid block

Muthukannan Marimuthu[1], *Uthayakumar*[2], *Rajesh* Kanna P[3], *Paweł* Ocłoń, *[4], *Jan* Taler[5], and *Dawid* Taler[6]

[1] Department of Mechanical Engineering, Kalasalingam University, Anand Nagar,Krishnan Koil,Tamilnadu,India 626126
[2] Department of Mechanical Engineering, Kalasalingam University, Anand Nagar,Krishnan Koil,Tamilnadu,India 626126
[3]Department of Mechanical Engineering, College of Engineering and Computing, Alghurair University,, Dubai. UAE.
[4,5,6] Institute of Thermal Power Engineering, Cracow University of Technology, Kraków, Poland.

Abstract. The numerical solution solution is obtained for fluid flow and heat transfer in a confined impinging slot on a solid block with the presence of baffles. In order to consider the effect of baffle shape the rectangular and semi circular baffles are considered and for the effect for Reynolds number the Reynolds number is varied from 100 to 300 with the step of 50. The present study reveals the vital impact of Baffle shape and Reynolds number (Re) on the fluid flow and heat transfer characteristics over a wide range. It is finally added that the presence of baffle improves the Nusselt number. The Nusselt number increases with the increase of Reynolds number. The present study proved that, the primary peak of Nusselt number occurs nearer to the reattachment length. The secondary peak of Nusselt number occurs nearer to the baffle. It is observed that for semi circle baffle the velocity attains maximum one compared to rectangular baffle.

1 Introduction

The jet impingement cooling technique is used in industries like turbine blade cooling, rocket launcher cooling, electrical, metal annealing, and textile drying and electronic equipments in order to confiscate a huge amount of heat. The common types of impingement are confined and unconfined jets. The design and fabrication of confined jets are very complicated compared to unconfined jets. There are many numerical and experimental investigations are conducted to find out the effect of confinement wall on jet impingement. The experimental and numerical predictions are mostly related to turbulent jet impingement to have fast cooling. But in the case of cooling of electronic equipments the area need to be cooled is small and also the velocity. In order to consider the present situation, the laminar regime is considered for the entire computational domain.
Habib et al[1] experimentally investigated that there was increase in average nusselt number with the increase of Reynolds number and block height. Hussein M.Maghrabie et al [2] numerically investigated the heat transfer and pressure drop of in line array of heated obstacles cooled by jet impingement in cross flow. They concluded that there was a significant change in jet impingent cross flow and normal flow of jet impingement. WithadaJedsadaratanachai etal [3] predicted numerically V baffles can induce a drastic increase in Nusselt number and higher friction.

Yang etal [4] investigated experimental as well as numerically in array of jet impingement on a concave surfaces.The finally concluded that,increase in jet diameter hole will not improve the heat transfer significantly. Eiamsa-ard[5] et al numerically examined the fluid flo and heat transfer chararteristics in the presence of triangular wavy baffles.They concluded that for 30° and 45° baffles endow with more nusselt number compared to 60° baffles. Alsanossi M.Aboghrara [6] examined analysis of solar air heater with jet impingement on corrugated absorber plate. They exposed that corrugated plate is the strong function of heat transfer enhancement. In previous series of studies performed by author [7&8]concentrated more on fluid flow structure behavior of different fluids like air, water and nano fluid jet impingement on a solid block .In the present paper the author revealed a numerical study report on the effect of imposing the different shape of baffles on jet impingement on a solid block. The primary aim of this study is to elucidate the enhancement in heat transfer by incorporating baffles by (i.e. increasing area).

2 Problem Descriptions

A The schematic diagram for single slot jet impingement on a solid block in the presence of baffles are shown in figure 1. The air jet come out from a nozzle of width W, with a uniform velocity (U), impinges normal to a

* Corresponding author: poclon@mech.pk.edu.pl

computational domain which is called impingement plate at a distance of L_Y from the nozzle. The horizontal and vertical velocity are considered as U and V are respectively. Since the computational domain is symmetry about vertical or y-axis, it is prudent to model and simulate only right hand of the computational domain thereby saving precious computing power and time.

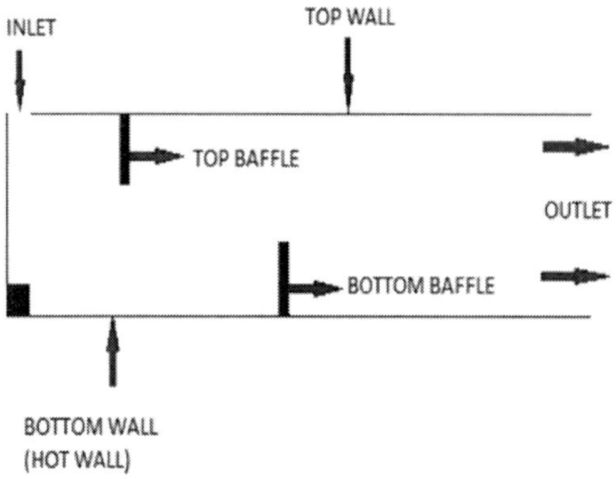

Fig. 1(a). Schematic diagram of computational domain with rectangular deflector

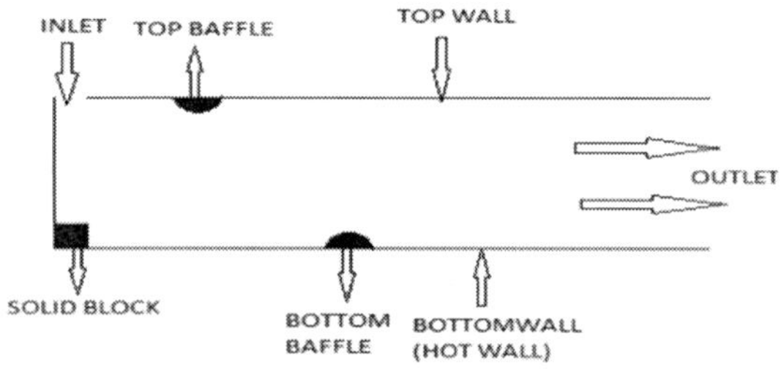

Fig. 1(b). Schematic diagram of computational domain with baffle shaped deflector

The inlet velocity of air (U) is calculated by using the formula: Re = ($\rho UD_h/\mu$) where ρ denotes the density of the fluid; D_h the hydraulic diameter, and μ-dynamic viscosity of the fluid. The deflectors are provided at the top and bottom walls at equal distance having same area. Dimensionless velocity (U&V) are determined by $U = u / U_{in}$, $V = v / U_{in}$ where u and v are the dimensional velocities and U_{in} is considered as the reference velocity and equal to unity.The aspect ratio (AR) is defined as length of the domain in Y direction(L_Y) to hydraulic diameter(D_h)

2.1 Boundary Conditions

The numerical study had a uniform wall temperature at the bottom wall and top wall is considered as an adiabatic one. Uniform velocity is considered as inlet and pressure outflow is considered at outlet. Symmetry condition is adopted by considering right hand of the computational domain inorder to reduce the computational time.

2.2 Assumptions

The following assumptions were made while solving this problem:

- Flow was taken as laminar i.e. Reynolds number varies from 50 to 300 with the step of Re=50 only since the presence of the block induces turbulence at low Reynolds number also.
- Incompressible fluid.
- The case solved here is that of non-conjugate heat transfer i.e. the effect of the conductivity of the slab material is not taken into account.
- The geometry at the outlet is assumed to be so long that the flow eventually gets fully developed.

2.3 Commercial Package Used

The continuity equation, momentum equation and energy equation are solved by commercial package ANSYSCFD15.0.

$$\frac{\partial U}{\partial X}+\frac{\partial V}{\partial Y}=0 \qquad \text{Continuity Equation}$$

X Momentum Equation

$$U\left(\frac{\partial U}{\partial X}\right)+V\frac{\partial U}{\partial Y}=-\frac{\partial P}{\partial X}+\frac{1}{Re}\left(\frac{\partial^2 U}{\partial X^2}+\frac{\partial^2 U}{\partial Y^2}\right)$$

Y Momentum Equation

$$U\frac{\partial V}{\partial X}+V\frac{\partial V}{\partial Y}=-\frac{\partial P}{\partial Y}+\frac{1}{Re}\left(\frac{\partial^2 V}{\partial X^2}+\frac{\partial^2 V}{\partial Y^2}\right)$$

Energy equation:

$$U\frac{\partial\theta}{\partial X}+V\frac{\partial\theta}{\partial Y}=\frac{1}{RePr}\left(\frac{\partial^2\theta}{\partial X^2}+\frac{\partial^2\theta}{\partial Y^2}\right)$$

The non-dimensional variables considered for this problem are,

$X = Lx/D_h$, $Y = Ly/D_h$, $U = u/U_{in}$, $V = v/U_{in}$ $P = p/\rho U_{in}^2$

Re = ($\rho uD_h/\mu$) and $Nu = (-\frac{\partial\theta}{\partial Y})_{y=0}$

where $\theta_f = \frac{T_f-T_c}{T_h-T_c}$ is the non dimensional temperature in the fluid region and

$\theta_s = \frac{T_w-T_c}{T_h-T_c}$ is the non dimensional temperature in the solid region.

The term θ_s is always 1 for the non-conjugate case, i.e., the effect of the slab thickness is neglected. Here the temperature of the wall is assumed as the temperature of the hot fluid (T_w=T_h).

2.4 Validation of the problem

In order to prove our algorithm the validation work has been done against sivasamy et al paper[4]. The centers of the primary and secondary vortexes for aspect ratio 5 for Reynolds number from 100 to 400, the same for Reynolds number 200 and aspect ratio 2 to 5 has been used as the parameters for validation.

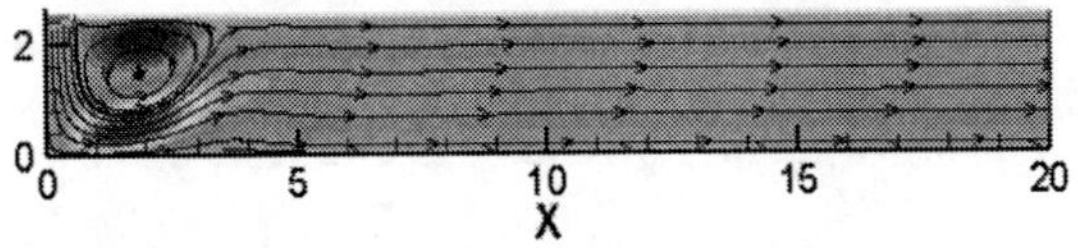

Fig. 2. X-velocity contours and streamlines for aspect ratio 5 and Re=100

Table 1 compares the center of primary vortex in x coordinate produced by present algorithm with sivasamy et al result. Table 1 compares the center of primary vortex in y coordinate produced by present algorithm with sivasamy et al result.

Table 1 Centres for primary vortex in X &Y-coordinate.

Re	Primary Vortex Coordinates			
	X		Y	
	Present	Sivasamy et al	Present	Sivasamy et al
100	1.90	1.90	1.43	1.41
200	3.12	3.07	1.52	1.49
300	4.08	4.05	1.60	1.56
400	4.86	4.84	1.64	1.63
500	5.43	5.42	1.71	1.67

The figure 3 shows that,the coefficient of friction in bottom wall is compared with sivasamy et al and it is having good agreement also.

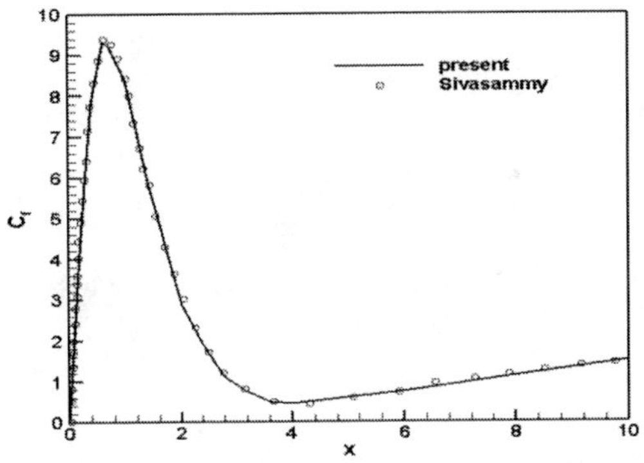

Fig 3. Comparison of Bottom wall friction (Cf) with sivasamy et al

The coefficient of friction in bottom wall is compared with sivasamy et al and it is having good agreement also.

For validation of flow field investigation, the extraction of vortex centre and coefficient of friction is extracted along the computational domain and it is compared with sivasamy et al and it is having good agreement also.

For validation of heat transfer result, the nusselt number calculated along the computational domain and compared with the sivasamy et al (2008) and it is also having good conformity also.

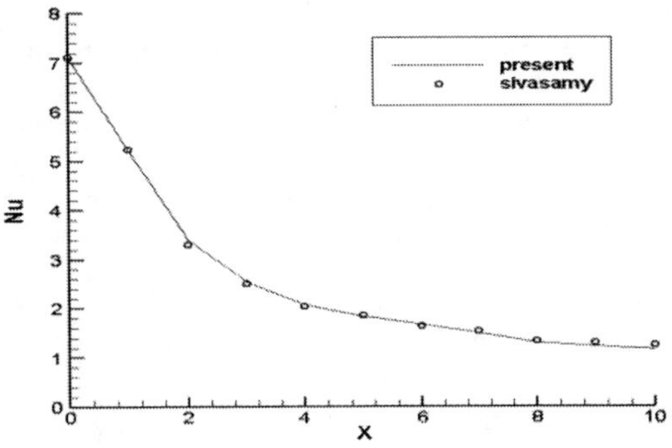

Fig 4. Validation of the bottom wall Nusselt number against Sivasamy et al [2008]

2.4. Grid Independence Study

Table 2 Grid Independence Study.

Grid level	Total No of nodes	Average Nu Number for the bottom wall
Grid 1	259246	8.31
Grid 2	274584	10.5
Grid 3	289584	11.367
Grid 4	304584	12.31
Grid 5	319922	12.3768
Grid 6	335260	12.4092

Table 2 reveals the prediction of nusselt number for same aspect ratio and Reynolds number. From the table it is concluded that there was not a substantial change in Nusselt Number after Grid Number 4.So Grid Number 4 is considered as the grid selction for the particular aspect ratio.

3 Numerical Approach and Procedures

In the present numerical investigation the two dimensional laminar incompressible jet is considered for the present study . The single phase model is adopted in the present simulation for the sake of simplicity. No slip condition and no penetration conditions are applied for the velocity components on the solid wall. The present numerical study concentrates on heat transfer investigation in the presence of solid block with different shape of deflector. The algorithm used for solving is the semi implicit method for pressure linked equation algorithm. The numerical investigations based on finite volume method are simulated by ANSYS FLUENT 15.0 CFD code. The Tecplot 9.0 (postprocessor) is used to interpret the result, for obtaining the contour plots of X velocity, Y velocity and streamline contour for flow field.

4 Results and Discussion

The present author has already done heat transfer investigation in jet impingement on a solid block. Numerical simulations of heat transfer characteristics of jet impingement in the presence of baffles are presented. In the present investigation rectangular and semi circle (Dimple) shaped baffles are adopted. The influence of rectangular and dimple shaped baffles are presented interms of velocity of fluid and Nusselt number.

The following figures shows the temperature contour for different Reynolds number in the presence of dimple shaped baffles

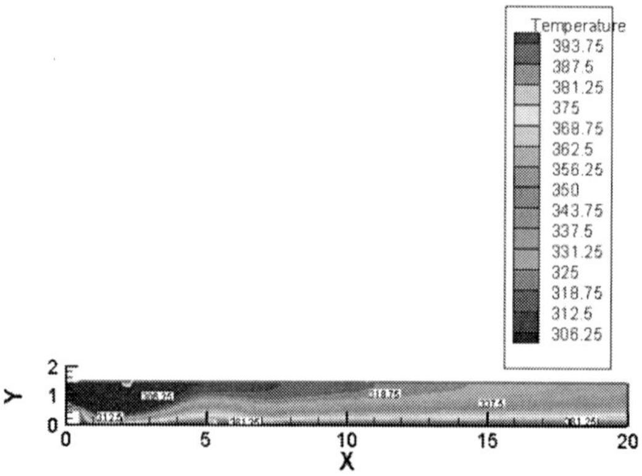

Fig. 5(a). Temperature contour in the presence of dimple baffle for Re=150 for AR=3

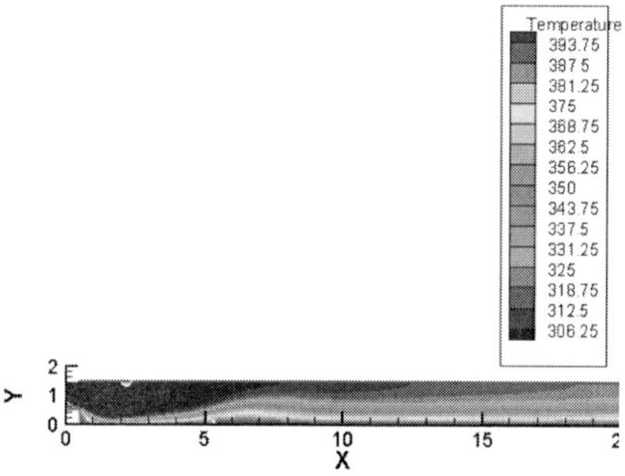

Fig. 5(b). Temperature contour in the presence of dimple baffle for Re=250 for AR=3

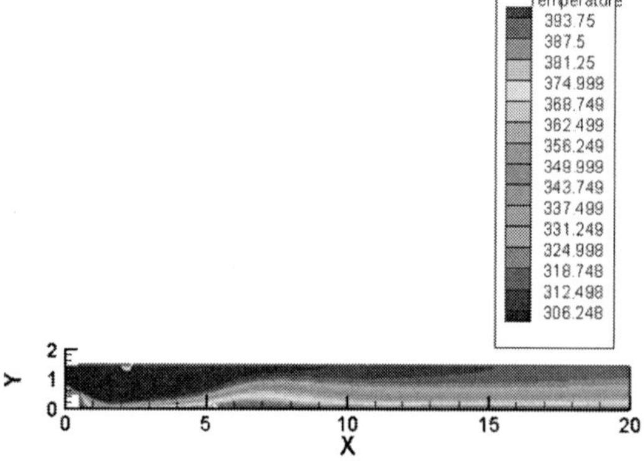

Fig. 5(c). Temperature contour in the presence of dimple baffle for Re=300 for AR=3

From figure 5 it is inferred that the average surface temperature increases with the increase of Reynolds number. When the Reynolds number increase the thermal

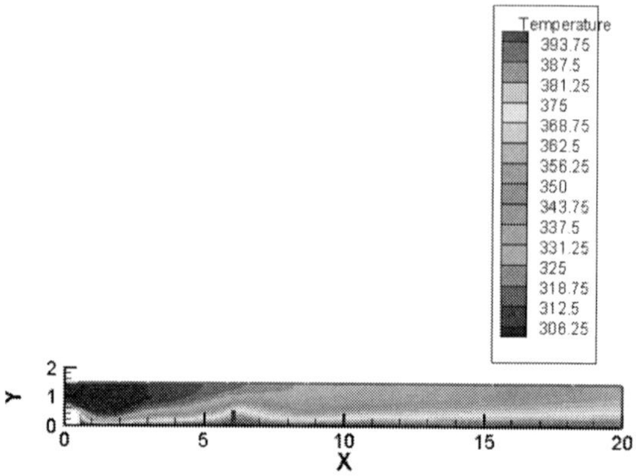

Fig 6(a). Temperature contour in the presence of rectangular baffle for Re=100 for AR=3

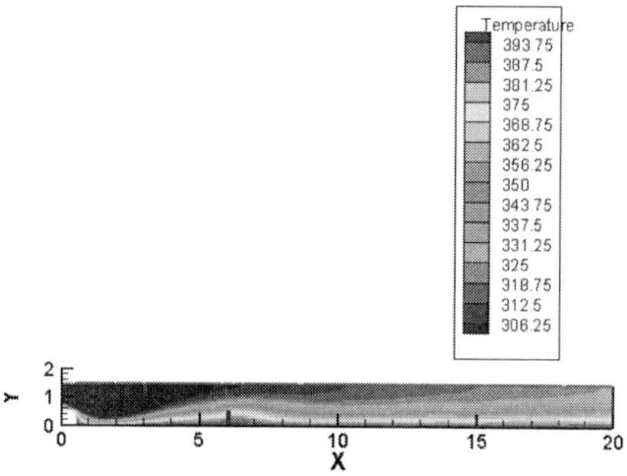

Fig. 6(b). Temperature contour in the presence of rectangular baffle for Re=200 for AR=3

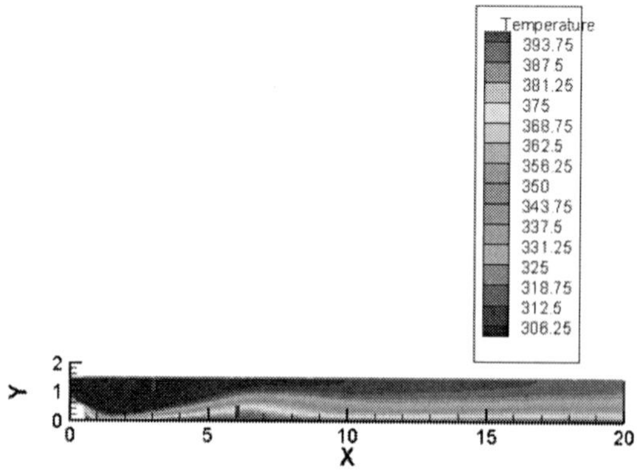

Fig.6(c). Temperature contour in the presence of rectangular baffle for Re=300 for AR=3

The figure 6 shows the temperature contour for different Reynolds number in the presence of rectangular baffles. From figure 5 & 6, it is concluded that the vortex induced by the baffle increases mixing and thus enhances the heat transfer. The figure 5 & 6 shows that there was a reduction in thickness of shear layer with the increase of Reynolds number.

The influence of baffle on the heat transfer can be determined by comparing the heat transfer characteristics downstream of the baffle with the presence of the baffle and without the presence of baffle. It is noticed that peak horizontal velocity in the presence of baffle is higher than peak horizontal velocity in the absence of baffle.

The followings table shows the peak Nusselt number with the presence of rectangular baffle & without baffle.

Table 3. Comparison of Peak Nusselt number between with and with out baffle in the computational domain.

Re Number	Peak Nu without baffle	Peak Nu with baffle
100	9.881	11.4216
200	14.856`	16.2422
300	20.4421	21.245

Figure 7(a)&7(b) Local Nusselt number in the downstream location in the presence of rectangular baffle.

For Reynolds number 100 nearly 15.59% improvement in Nusselt number and similary for 9.33% and 3.92% increment in for Re=200 & Re=300 respectively.

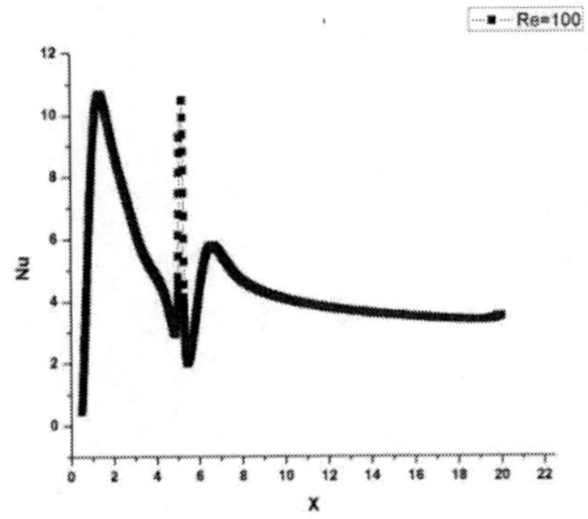

Fig. 7(a)

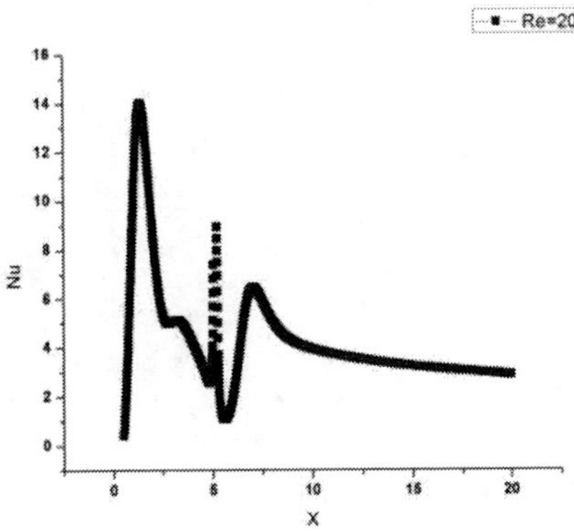

Fig. 7(b)

The figure 7(a) and 7(b)shows the local nusselt number in the downstream location.In the presence of rectangular baffle there are three peak nusselt number are identified in the downstream location. This will improve the average nussely number in the donstream location.Primary peak occurs at the reattachment of fluid in the computational domain.Seondary peak occurs due to the baffle and tertiary peak occurs due to high inertia force which cause flow sepeartion.

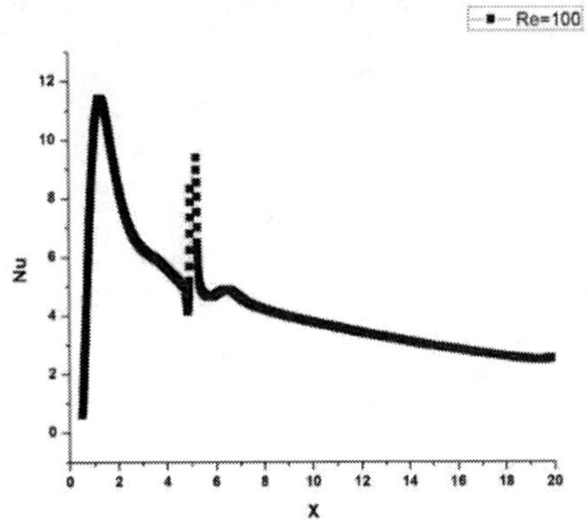

Fig 8(a)

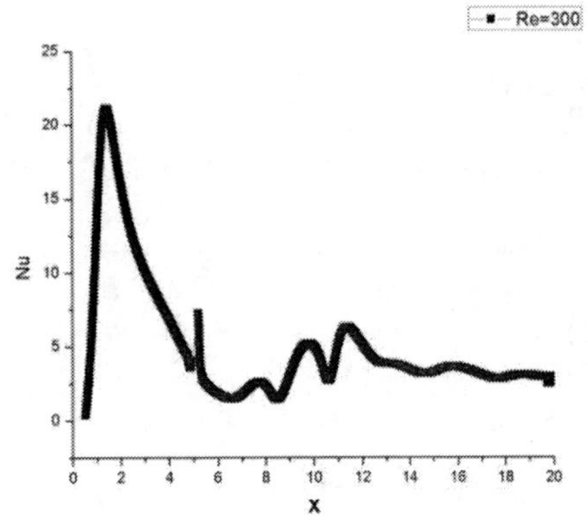

Fig 8(b)

Figure 8(a)&8(b) Local nusselt number in the downstream location in the presence of dimple baffle .

In case of rectangular shaped baffles the form drag/resistance is more due to large pressure drop where as loss due to friction will be more in dimple shaped baffle.The overall loss is combination of these two.

Table 4. Peak Nusselt number for various shape deflectors

Reynolds Number	Peak Nusselt number	
	Rectangular	Dimple
100	7.02	8.16
150	10.52	12.95
200	13.46	14.02
250	19.57	21.25
300	23.70	23.58

The table 4 shows that the presence of dimple shaped baffle produces high heat transfer compared to rectangular baffle.From the table 4 it is confirmed that the percentage of increase in nusselt number decreases with the increase of Reynolds number.

5 Conclusion

In the present numerical analysi of heat transfer, while the jet impinges on a block, is investigated using commercial CFD codes. The computed thermal fields in terms of the Nusselt number ranging from 2 to 5 with the Reynolds number ranging from 100 to 300 are analyzed in which the flow comes under the laminar regime.

Due to the presence of baffles, there was increase in nusselt number .The nusselt number profile along the downstream location is plotted for various Reynolds number.

The Nusselt number attains first peak value nearer to the reattachement length and second peak occurs nearer to the baffle and third peak occurs due to the higher Reynolds number

References

1. M.A. Habib, A.M. Mobarak, M.A. Sallak, E.A. Abdel Hadi, R.I. Affify, J Heat Trans-T ASME. **116**, 363–368 (1994)
2. Hussein M. Maghrabie,M. Attalla, H. E. Fawaz, M.Khalil, AEJ, **56,** 3, 285-296 (2017)
3. WithadaJedsadaratanachai,NuthvipaJayranaiwachira, PongjetPromvonge, Chin. J. Chem. Eng. **23,** 2, 342-349 (2015)
4. Liu Haiyong,Liu Cunliang, Wu Wenming, ,CJA, 1616-1628 (2015).
5. S.Eiamsaard, S.Pattanapipat. Promvonge, JMST, **27,** 7, 2199-2208 (2013)
6. Alsanossi M. Aboghrara , B.T.H.T. Baharudin, M.A. Alghoul, Nor Mariah Adam, A.A. Hairuddin, Husam A. Hasan, **10**, 111-120 (2017).
7. Muthukannan.M, Rajeshkanna.P, Ankit Bajpai, Jeyakumar.S, AJSE, **39**, 8077–8098 (2016),
8. Muthukannan.M, Rajeshkanna.P, Ankit Bajpai, Jeyakumar.S, JAFM, **9**, **4**, 1679-1694 (2016)
9. Sivasamy A, Selladurai V Rajeshkanna P, Int. J. Numer. Methods Fluids, **55**, 965–983 (2007)

MATEC Web of Conferences **240**, 01026 (2018) https://doi.org/10.1051/matecconf/201824001026
ICCHMT 2018

Numerical study of the effect of heat transfer on solid phase formation during decompression of CO_2 in pipelines

Sergey Martynov, *Wentian* Zheng, and *Haroun* Mahgerefteh*

Department of Chemical Engineering, University College London, WC1E 7JE, London, UK

Abstract. CO_2 solid phase formation accompanying rapid decompression of high-pressure CO_2 pipelines may lead to blockage of the flow and safety valves, presenting significant hazard for safe operation of the high-pressure CO_2 storage and transportation facilities. In this study, a homogeneous equilibrium flow model, accounting for conjugate heat transfer between the flow and the pipe wall, is applied to study the CO_2 solid formation in a 50 mm internal diameter and 37 m long pipe for various initial thermodynamic states of CO_2 fluid and wide range of discharge orifice diameters. The results show that the rate of CO_2 solid formation in the pipe is limited by heat transfer at the pipe wall. The predicted amounts of solid CO_2 are discussed in the context of venting of CO_2 pipelines.

1 Introduction

Despite the fact that carbon dioxide (CO_2) at low concentrations is commonly considered as a safe substance, accidental failure of facilities transporting large quantities of CO_2 at high pressures for carbon capture and storage or utilization [1,2] may cause significant harm to personnel and local populations as a result of explosion overpressure and the asphyxiate nature of the CO_2 in the ensuing dispersion cloud [3,4].

Blockage of safety valves by solid CO_2 which may form as a result of near-isentropic decompression of high-pressure CO_2 to the level below 5.18 bar (triple point of CO_2) is considered as one of major causes of failure of CO_2 storage and transport facilities [5,6]. Also, accumulation of solid CO_2 in pipelines and vessels may increase the risk of blockage and overfilling of units at later stages of operation [7–11]. In particular, our recent studies showed that solid CO_2 may form during decompression of pipelines initially filled with CO_2 at 60-80 bar pressures [12,13]. Given that solid CO_2 accumulation in safety valves and vented sections of pipelines presents a risk for the system integrity, designing venting equipment and procedures that minimize the amounts of solid phase formed becomes critically important. This requires development and application of mathematical models of pipeline decompression accounting for multiphase nature of the decompression flow and thermodynamic properties of CO_2 fluid in various states, including the triple point.

Recent studies showed that Homogeneous Equilibrium Mixture (HEM) model predicts well the pressure and temperature measured in the pipeline Full Bore Rupture test [13] and amounts of solid CO_2 formed in the pipeline in the orifice discharge tests [12]. Using this model it was demonstrated that duration of the pipeline decompression to the triple point where CO_2 solids can form in the pipe, scales with the pipeline length [14], while the amount of solid CO_2 formed depends on the history of decompression flow [12]. The latter depends on a number of factors, including the initial thermodynamic state of the fluid, the discharge hole diameter, and the rate of heat transfer at the pipe wall. Despite the fact that understanding the above effects is crucial for preventing/minimizing the CO_2 solids formation in process equipment, they have not been systematically studied. The present study is focused on the application of the HEM flow model to quantify the amount of CO_2 solid phase that may form under various scenarios of pipeline decompression..

* Corresponding author: h.mahgerefteh@ucl.ac.uk

2 Theory

In the present study, in order to predict the history of decompression of a CO_2 pipeline (Figure 1) accounting for spatial variations in the flow along the pipe, a set of quasi-one-dimensional HEM-based mass, momentum and energy conservation equations, is applied [15]:

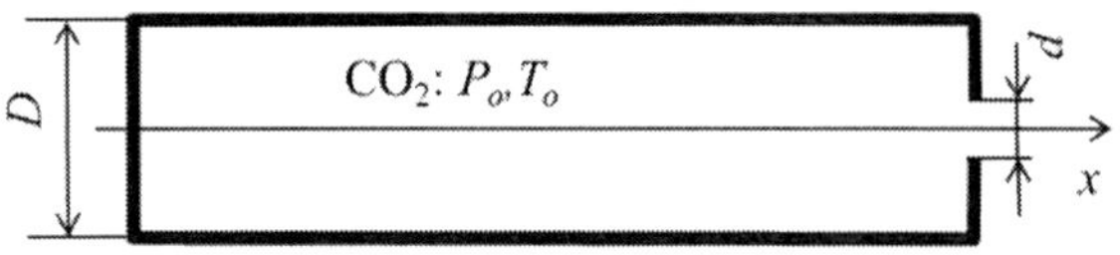

Fig. 1. Schematics of a pipeline filled with CO2 at pressure Po and temperature To prior to decompression through an orifice of diameter d.

$$\frac{\partial}{\partial t}\rho A+\frac{\partial}{\partial x}\rho uA=0 \tag{1}$$

$$\frac{\partial}{\partial t}\rho uA+\frac{\partial}{\partial x}(\rho u^2+p)A=p\frac{\partial A}{\partial x}-2f\rho u^2\frac{A}{D} \tag{2}$$

$$\frac{\partial}{\partial t}\left(u^2/2+e\right)\rho A+\frac{\partial}{\partial x}\left[\rho\left(u^2/2+e\right)+p\right]uA= =\left(-2f\rho u^3+4q\right)\frac{A}{D} \tag{3}$$

where p is the pressure, u is the velocity, D and $A=\pi D^2/4$ are respectively the pipe diameter and cross-sectional area of the flow; f is the Fanning friction factor, calculated using Chen's correlation [16], and q is the heat flux at the pipe wall.

In the above equations, the HEM density and specific internal energy are calculated knowing the fluid phase composition and properties of saturated vapor, liquid and solid phases as functions of pressure. The thermodynamic properties of CO2 in liquid and vapor states are calculated using the GERG 2004 equation of state (EoS) [17], while properties of solid CO2 are predicted using the extended Peng-Robinson EoS [18].

The heat flux, q, is defined by Newton's cooling law for single-phase and solid-vapor turbulent flows (with the Dittus-Boelter correlation [19] adopted to predict the heat transfer coefficient), and using Rohsenow's correlation for nucleate boiling [20] for vapor-liquid flows.

The set of equations (1) - (3) is closed by the boundary conditions specified at both ends of the pipeline (Figure 1), and using a lumped heat capacity model predicting evolution of the pipe wall temperature [14].

For the numerical solution of the governing equations of the model, Godunov's finite volume method combined with a fractional stepping time-integration scheme is applied [21].

At any time level, the total amount of CO_2 solid phase formed in the pipe is obtained by numerically integrating the resolved solid phase density profiles along the pipeline. The time integration procedure is terminated when the pressure at the discharge end of the pipe reaches 1 bar.

3 Results and discussion

This section is aimed at evaluation of the impact of key parameters of decompression process on the solid CO_2 formation in pipelines. For this purpose, the above described model is applied to predict the amount of solid CO_2 formed in a pipe for various initial states of CO_2 fluid and various discharge orifice diameters. For the sake of example, the study is performed for a medium-scale mild steel pipeline of 50 mm internal diameter, 5 mm wall thickness and 37 m length (http://www.co2quest.eu/).

3.1 The effect of initial conditions

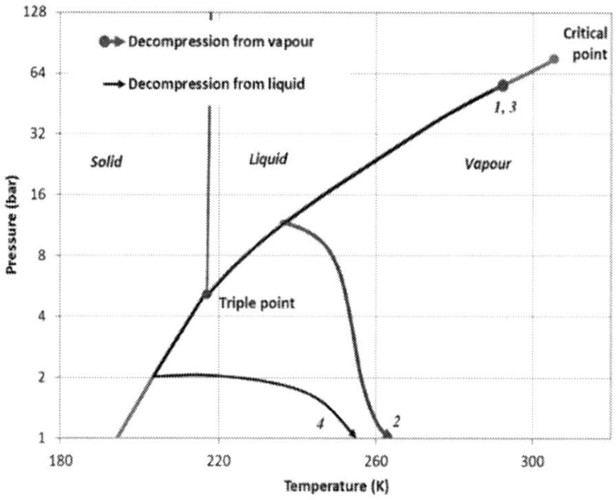

Fig. 2. Pressure-temperature phase diagrams of CO_2, showing the decompression trajectories predicted by the flow model for the fluid initially in the vapor state (trajectory *1*-*2*) and in the liquid state (trajectory 3-4).

In order to demonstrate the impact of initial conditions on the amount of solid CO_2 that could form in the pipe, the study was performed for two cases involving depressurization of the pipeline, initially at 54.4 bar pressure and 18 °C, via 6 mm orifice. In the first case the pipeline was initially filled with saturated liquid, while in the second case the pipe contained compressed saturated gas.

In Figure 2 evolution of the fluid pressure and temperature predicted by the model at location in the middle of the pipe, are plotted in the phase diagram of

CO_2. As can be seen in Figure 2, the liquid phase decompression path (trajectory *3-4*) crosses the triple point, and hence corresponds to a scenario where CO_2 solids can be expected to form in the pipe. At *ca.* 2 bar, the model predicts complete sublimation of the solid phase, and the trajectory deviates from the sublimation line into the vapor phase region. On the other hand, the decompression from the compressed gas state (trajectory *1-2*) follows the saturation line only till *ca.* 12 bar pressure, where the fluid turns to vapor, as can be explained by heating from the pipe wall. As such, the model indicates that decompression of a compressed CO_2 gas state doesn't lead to CO_2 solids forming in the pipe.

3.2 The effect of release orifice diameter

Figure 3 shows the effect of *d/D* ratio on the mass of solid CO_2 formed in the pipe upon its decompression to the triple point, obtained based on the HEM flow model predictions, in comparison with estimates using the thermodynamic method assuming isentropic decompression [12].

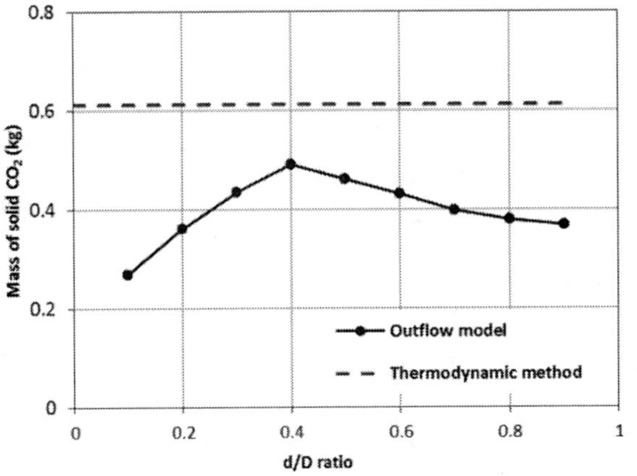

Fig. 3. The impact of the release orifice diameter on the amount of CO_2 solid phase formed in the pipeline, initially in liquid state at 18 ^{o}C and 54.4 bar.

As can be seen in Figure 3, predictions by the thermodynamic method are insensitive to the orifice diameter. Also, the thermodynamic method systematically (by *ca.* 15-50 %) overestimates the mass of solid phase in comparison with the predictions by the decompression flow model. This discrepancy can be directly attributed to non-isentropic nature of the fluid expansion process, involving conjugate heat transfer between the fluid and the pipe wall. Figure 3 also shows that the mass of solid CO_2 predicted by the outflow model scales nearly linearly with d/D in the range from 0.1 to 0.4. This indicates that using small ratios d/D can be more advantageous for use in design of decompression systems where the solid CO_2 pose hazard for the system operation and integrity.

4 Conclusions

The results obtained using the HEM pipeline outflow model showed that decompression of CO_2 from initial vapor state is characterized by significant heating of the flow by the pipe wall. This heating may result in complete evaporation of liquid phase before the fluid decompression to the triple point, leading to no CO_2 solids formed in the pipe. This contrasts to scenarios of decompression of pipes containing CO_2 in a form of compressed liquid, which evaporates only partially upon decompression to the triple point, potentially leading to large amount of solid CO_2 forming in the pipe, creating a hazard for the system operation and integrity.

The amounts of solid CO_2 calculated based on the simulations using the HEM flow model were compared with predictions using the thermodynamic model. While the latter gives conservative estimate of the amount of solid CO_2, the HEM decompression flow model resolves the effect of the orifice diameter and hence can become useful, *e.g.* for design of CO_2 venting systems.

Acknowledgments

The authors gratefully acknowledge the financial support provided by the European Union 7th Framework Program FP7-ENERGY-2012-1-2STAGE under grant agreement number 309102. The paper reflects only the authors' views and the European Union is not liable for any use that may be made of the information contained therein.

References

1 IPCC, *Carbon Dioxide Capture and Storage*, (Cambridge University Press, Cambridge, UK, 2005)

2 Styring P, Jansen D, de Coninck H, Reith H, Armstrong K, *Carbon Capture and Utilisation in the Green Economy* (The Centre for Low Carbon Futures and CO2 Chem Publishing, 2011)

3 Harper P, Wilday J, Bilio M, *Assessment of the Major Hazard Potential of Carbon Dioxide (CO_2)*, 2011.

4 M. van der Voort, A. van den Berg, D. Roekaerts, M., Xie, and P. de Bruijn, 2012, Shock Waves, **22**, pp. 129–140.

5 D. Huang, H. Quack, and G. Ding, 2007, Appl. Therm. Eng., **27**, pp. 1911–1922.

6 W. Clayton, and M. Griffin, 1994, Process Saf. Prog., **13**(4), pp. 202–209.

7 DNV, 2010, *Design and Operation of CO2*

Pipelines. Recommended Practice DNV-RP-J202.

8 Chart Inc., 2014, *CO_2 Storage Tank - Product Manual*, New Prague, MN 56071 USA.

9 M. Kim-E, 1981, "The Possible Consequences of Rapidly Depressurizing a Fluid," Massachusetts Institute of Technology.

10 Emerson Climate Technologies, 2015, *Commercial CO_2 Refrigeration Systems. Guide for Subcritical and Transcritical CO_2 Applications*.

11 D. Huang, G. Ding, H. Quack, 2008, Front. Energy Power Eng. China, **2**(4), pp. 494–498.

12 S. Martynov, W. Zheng, H. Mahgerefteh, S.Brown, J. Hebrard, D. Jamois, C. Proust, 2018, Ind. Eng. Chem. Res., **57**, pp. 7054–7063.

13 W. Zheng, H.Mahgerefteh, S. Martynov, S. Brown, 2017, Ind. Eng. Chem. Res., **56**, pp. 10491–10499.

14 S. Martynov, S., Brown, H., Mahgerefteh, V., Sundara, S., Chen, Y., Zhang, 2014, Process Saf. Environ. Prot., **92**(1), pp. 36–46.

15 S. Brown, S. Martynov, H. Mahgerefeth, 2015, Comput. Fluids, **120**, pp. 46–56.

16 N. Chen, 1979, Ind. Eng. Chem. Fundam., **18**(3), pp. 296–297.

17 O. Kunz, W. Wagner, 2012, J. Chem. Eng. Data, **57**(11), pp. 3032–3091.

18 Martynov, S., Brown, S, Mahgerefteh, H., 2013, Greenh. Gases Sci. Technol., **3**(2), pp. 136–147.

19 F. Dittus, and L.Boelter, 1930, University of California Press, Berkeley, California, p. 443.

20 Incropera F, De Witt D, *Fundamentals of Heat and Mass Transfer* (John Wiley & Sons, New York, 1985)

21 LeVeque R. J, *Finite Volume Methods for Hyperbolic Problems*, Cambridge Univ. Press, 2002)

Natural convection in a rotating cavity partially filled with a porous medium under the effect of a local heater

Stepan Mikhailenko, and *Mikhail* Sheremet*

Laboratory on Convective Heat and Mass Transfer, Tomsk State University, 36 Lenin Avenue, 634050 Tomsk, Russia

Abstract. The combined effect of uniform rotation and porous layer on natural convective heat transfer in a square cavity with a local heater has been studied numerically. The local thermal equilibrium model with the Brinkman-extended Darcy law has been used for description of the heat transfer and fluid flow within the porous layer. Governing equations with corresponding boundary conditions formulated using the non-dimensional stream function and vorticity variables have been solved by the finite difference method. The effects of Taylor number and porous layer thickness on fluid flow and heat transfer have been examined. It has been found a non-linear effect of the porous layer thickness on the heat transfer rate.

1 Introduction

Rotating cavities filled with porous material can be met in different engineering applications including heat exchangers, solar collectors, chemical reactors and others. Numerical and experimental analysis of convective heat transfer within rotating cavities has been performed by different authors [1–6]. Thus, Zhao et al. [1] have numerically analyzed natural convection in a fluid saturated porous horizontal cylinder under the effect of rotation about its axis. They have found that rotation significantly decreases the overall heat transfer. Jin et al. [2] have studied experimentally and numerically the heat transfer performance from discrete heat sources in a rotating cavity. It has been revealed that due to thermal inertia, rotation leads to a steady temperature field inside the solid zone and unsteady flow and temperature fields within the fluid zone, even in the low rotation speed range. Linear stability of a rotating fluid-saturated porous layer heated from below and cooled from above has been investigated by Malashetty et al. [3]. The obtained results have shown the competition between the processes of rotation and thermal diffusion that causes the convective instability to set in an oscillatory mode rather than stationary. The influence of local thermal non-equilibrium conditions on the linear and non-linear thermal instability in a nanofluid saturated rotating porous layer has been examined by Agarwal and Bhadauria [4]. They found that the system tends to be stabilized by the presence of rotational component in the porous medium and the onset of convection is delayed. The influence of rotation on the onset of convection in an anisotropic horizontal porous layer with internal heat source has been analyzed by Bhadauria et al. [5].

The objective of the present work is to analyze numerically natural convection of water inside a rotating square cavity partially filled with a porous material under the effect of local heater of constant temperature using local thermal equilibrium conditions. We believe that this work is noteworthy, and its usefulness for different engineering systems as discussed above motivated us.

2 Analysis and modelling

A square cavity with walls of length H with two opposite isothermally cold walls $\left(\bar{x}=0, \bar{x}=H\right)$ rotates counterclockwise with a constant angular velocity ω_0 (see Fig. 1). A heater of constant high temperature is located on the wall $\bar{y}=0$. A homogeneous porous layer covers the heated element. The following assumptions are made in the study:

– the Brinkman-extended Darcy model is used for the porous layer;

– the Boussinesq approximation is applied for description of the buoyancy force effect;

– the fluid phase and the solid matrix of the porous medium are described using the local thermal equilibrium conditions;

– the interface between the clear fluid and porous medium is permeable and fluid can penetrates through this border;

– properties of the fluid and porous medium are homogeneous and isotropic;

– boundaries of the cavity are assumed to be impermeable.

The equations governing the unsteady Brinkman natural convection flow and heat transfer inside the rotating porous cavity using the above mentioned assumptions can be written in the form including the the local thermal equilibruim inside the porous layer.

* Corresponding author: Michael-sher@yandex.ru

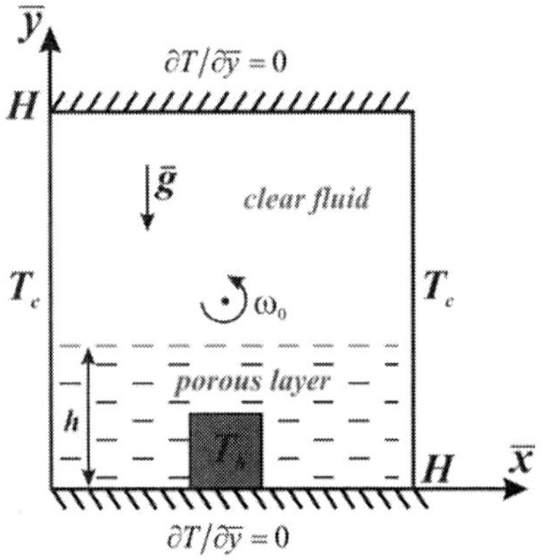

Fig. 1. Physical model and coordinate system.

Taking into account the dimensionless stream function (ψ) and vorticity (ω) variables as well as above-mentioned approaches, the governing equations can be formulated in the following form:

– for the clear fluid layer [7]:

$$\frac{\partial^2 \psi}{\partial x^2} + \frac{\partial^2 \psi}{\partial y^2} = -\omega \tag{1}$$

$$\frac{\partial \omega}{\partial \tau} + u\frac{\partial \omega}{\partial x} + v\frac{\partial \omega}{\partial y} = \frac{1}{\sqrt{Ta}}\left(\frac{\partial^2 \omega}{\partial x^2} + \frac{\partial^2 \omega}{\partial y^2}\right) + \frac{Ra}{Pr \cdot Ta}\left[\frac{\partial \theta}{\partial x}\cos(\tau) - \frac{\partial \theta}{\partial y}\sin(\tau)\right] \tag{2}$$

$$\frac{\partial \theta}{\partial \tau} + u\frac{\partial \theta}{\partial x} + v\frac{\partial \theta}{\partial y} = \frac{1}{Pr \cdot \sqrt{Ta}}\left(\frac{\partial^2 \theta}{\partial x^2} + \frac{\partial^2 \theta}{\partial y^2}\right) \tag{3}$$

– for the porous layer [8]:

$$\frac{\partial^2 \psi}{\partial x^2} + \frac{\partial^2 \psi}{\partial y^2} = -\omega \tag{4}$$

$$\varepsilon\frac{\partial \omega}{\partial \tau} + u\frac{\partial \omega}{\partial x} + v\frac{\partial \omega}{\partial y} = \frac{\varepsilon}{\sqrt{Ta}}\left(\frac{\partial^2 \omega}{\partial x^2} + \frac{\partial^2 \omega}{\partial y^2} - \varepsilon\frac{\omega}{Da}\right) + \frac{\varepsilon^2 \cdot Ra}{Pr \cdot Ta}\left[\frac{\partial \theta}{\partial x}\cos(\tau) - \frac{\partial \theta}{\partial y}\sin(\tau)\right] \tag{5}$$

$$\eta\frac{\partial \theta}{\partial \tau} + u\frac{\partial \theta}{\partial x} + v\frac{\partial \theta}{\partial y} = \frac{\alpha_{pm}/\alpha_f}{Pr \cdot \sqrt{Ta}}\left(\frac{\partial^2 \theta}{\partial x^2} + \frac{\partial^2 \theta}{\partial y^2}\right) \tag{6}$$

Following the geometry in Fig. 1, we assume that the initial and boundary conditions for these equations are

$\psi = 0,\ \omega = 0,\ \theta = 0$ at $\tau = 0$

$\psi = 0,\ \omega = -\dfrac{\partial^2 \psi}{\partial x^2},\ \theta = 0$ at $x = 0, 1$

$\psi = 0,\ \omega = -\dfrac{\partial^2 \psi}{\partial y^2},\ \dfrac{\partial \theta}{\partial y} = 0$ at $y = 0, 1$

$\psi = 0,\ \omega = -\dfrac{\partial^2 \psi}{\partial n^2},\ \theta = 1$ at the heater surface

$$\begin{cases} \psi|_{pm} = \psi|_f \\ \left.\dfrac{\partial \psi}{\partial y}\right|_{pm} = \left.\dfrac{\partial \psi}{\partial y}\right|_f \end{cases} \quad \begin{cases} \omega|_{pm} = \omega|_f \\ \left.\dfrac{\partial \omega}{\partial y}\right|_{pm} = \left.\dfrac{\partial \omega}{\partial y}\right|_f \end{cases}$$

$$\begin{cases} \theta|_{pm} = \theta|_f \\ \dfrac{k_{pm}}{k_f}\left.\dfrac{\partial \theta}{\partial y}\right|_{pm} = \left.\dfrac{\partial \theta}{\partial y}\right|_f \end{cases} \quad \text{at the fluid-porous interface}$$

Here Ra is the Rayleigh number, Pr is the Prandtl number, Ta is the Taylor number and Da is the Darcy number, which are defined as

$$Ra = \frac{\rho g \beta \Delta T H^3}{\mu \alpha},\quad Pr = \frac{\mu}{\rho \alpha},\quad Da = \frac{K}{H^2},\quad Ta = \frac{\rho^2 \omega_0^2 H^4}{\mu^2} \tag{7}$$

The physical quantity of interest is the average Nusselt number at the heater surface, which can be defined as

$$\overline{Nu} = -\frac{1}{l}\int_0^l \frac{\partial \theta}{\partial \mathbf{n}} d\zeta \tag{8}$$

Equations (1)–(6) with corresponding initial and boundary conditions were solved using an in-house computational code developed by means of the finite difference method of the second order accuracy [7–9]. The developed mathematical model and numerical technique have been verified comprehensively using the data of other authors [7–9].

The grid independency test has been performed for the present formulation (see Fig. 1) at $Ra = 10^5$, $Ta = 10^4$, $Da = 10^{-3}$, $Pr = 6.82$, $\varepsilon = 0.398$, $\gamma = h/H = 0.5$. Three uniform grids are tested. Figure 2 presents the effect of the grid parameters on the profiles of temperature (Fig. 2*a*) and vertical velocity (Fig. 2*b*) at cross-section $y = 0.5$ and $\phi = 0$ or 2π.

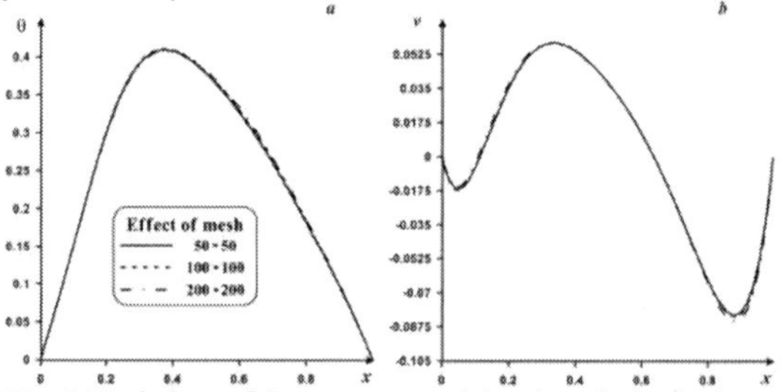

Fig. 2. Variations of the temperature (*a*) and vertical velocity (*b*) along the cross-section $y = 0.5$ versus the mesh parameters.

Using the results presented in Fig. 2, the uniform grid of 100×100 points has been chosen for the present study.

3 Results and Discussion

Analysis has been carried out for the following values of the governing parameters: Rayleigh number ($Ra = 10^5$), Taylor number ($Ta = 10^3 - 10^6$), Prandtl number ($Pr = 6.82$), Darcy number ($Da = 10^{-3}$), porosity ($\varepsilon = 0.398$), and porous layer height ($\gamma = 0.25$–0.75). Particular efforts have been focused on the effects of the

Taylor number and porous layer height on streamlines and isotherms as well as average Nusselt number and fluid flow rate inside the cavity.

Figure 3 shows streamlines during complete revolution with an angle step $\pi/2$ for $Ta = 10^4$, $\gamma = 0.5$. It should be noted that the presented streamlines have been obtained for the steady state, when the following revolutions repeat the previous ones. For description of these isolines of stream function it is possible to analyze the evolution of convective vortices formed near the heater. In the case of $\phi = 0$ one can find a development of small vortex from the left side of the heater and large vortex from the right side of the heater. A rotation of the cavity for $\phi = \pi/2$ leads to a growth of the sizes of the left convective cell, while the right one decreases in sizes. The following rotation illustrates the development and attenuation of mentioned convective cells. It is interesting to note that major convective cell core is located inside the clear fluid layer, where circulation occurs without the resistance from the solid matrix of porous medium.

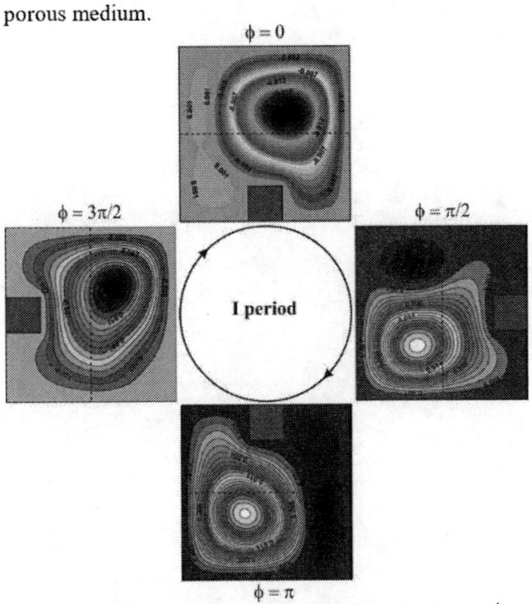

Fig. 3. Streamlines during complete revolution for $Ta = 10^4$, $\gamma = 0.5$.

Figure 4 demonstrates the isotherms inside the cavity during the complete revolution in accordance to the mentioned streamlines. It is worth noting that major heat transfer inside the porous layer is a heat conduction, while inside the clear fluid one can find more essential distortion of isolines of temperature due to intensive convective circulation. It is well-known that rotation can lead to the convective flow attenuation. Such effect of rotation can be found in Fig. 5.

The effects of the Taylor number and porous layer thickness on the average Nusselt number at the heated element and fluid flow rate during one complete revolution are presented in Fig. 5. As has been mentioned above, high angular velocity illustrates an essential attenuation of convective flow and heat transfer. At the same time, for $Ta = 10^6$ oscillations for the average Nusselt number and fluid flow rate vanish, while in the case of low angular velocity $Ta = 10^3$ one can find intensive circulation and high heat transfer rate.

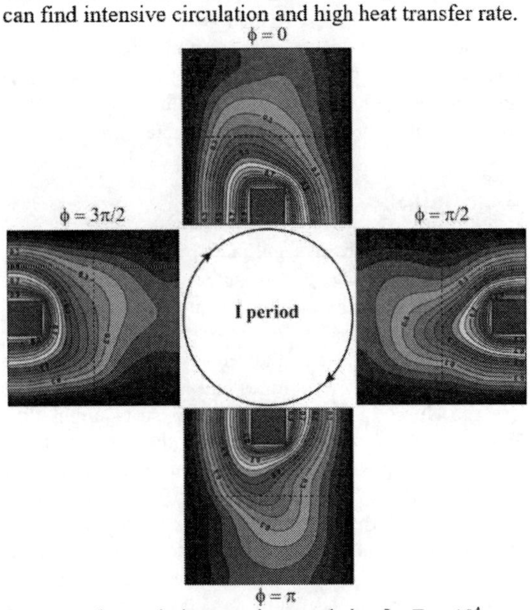

Fig. 4. Isotherms during complete revolution for $Ta = 10^4$, $\gamma = 0.5$.

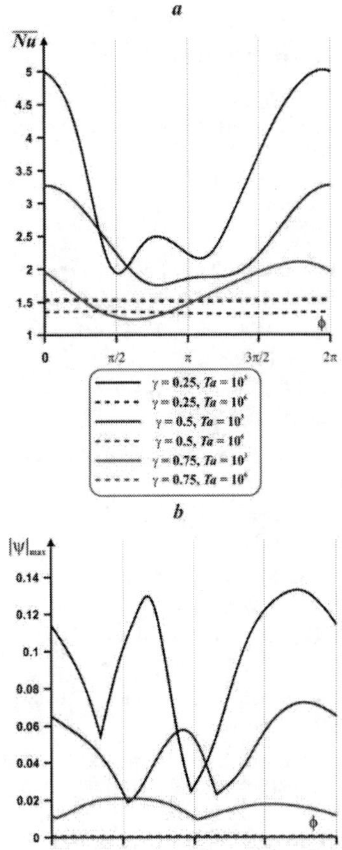

Fig. 5. Variations of average Nusselt number (*a*) and fluid flow rate (*b*) with rotation angle during one complete revolutions for different values of porous layer thickness and Taylor number.

A growth of the porous layer thickness results in the reduction of the heat transfer rate in the case of low Taylor number, while in the case of high Taylor number a rise of δ leads to weak heat transfer enhancement. At the same time, high porous layer thickness illustrates low amplitude of the heat transfer rate oscillations. The same changes can be found in distributions of the fluid flow rate with ϕ.

4 Conclusions

Natural convection in a partially porous square cavity with a local heater under the effect of uniform rotation has been studied numerically using the local thermal equilibrium approach. Particular efforts have been focused on the effects of the Taylor number and porous layer thickness on streamlines and isotherms as well as average Nusselt number at the heater surface and fluid flow rate. It has been found that in the case of steady state the complete revolution can be considered as an evolution of the convective cells near the heater. A growth of the Taylor number results in the convective flow and heat transfer attenuation. It should be noted that for low Taylor numbers a rise of γ leads to the heat transfer rate reduction, while for high values of Taylor number a rise of γ results in weak heat transfer enhancement.

Acknowledgements

This work was conducted as a government task of the Ministry of Education and Science of the Russian Federation (Project Number 13.9724.2017/8.9).

References

1. M. Zhao, L. Robillard, M. Prud'homme, Heat Mass Transfer **31** (1996).
2. L.F. Jin, K.W. Tou, C.P. Tso, Experimental Heat Transfer **18** (2005).
3. M.S. Malashetty, M. Swamy, S. Kulkarni, Physics of Fluids **19** (2007).
4. S. Agarwal, B.S. Bhadauria, Transport Porous Media **90** (2011).
5. B.S. Bhadauria, A. Kumar, J. Kumar, N.C. Sacheti, P. Chandran, Transport Porous Media **90** (2011).
6. A. Alhusseny, A. Turan, Heat Mass Transfer **51** (2015).
7. S.A. Mikhailenko, M.A. Sheremet, Numerical Heat Transfer A **72** (2017).
8. M.A. Sheremet, T.A. Trifonova, Transport Porous Media **101** (2014).
9. I.A. Aleshkova, M.A. Sheremet, Int. J. Heat Mass Transfer **53** (2010).

Comparison of predictive methods for flow boiling heat transfer in conventional channels and minichannels - the effect of reduced pressure

Dariusz Mikielewicz[1], *Blanka* Jakubowska[1,*]

[1]Gdansk University of Technology, Faculty of Mechanical Engineering, Department of Energy and Industrial Apparatus, Narutowicza 11/12, 80-233 Gdansk, Poland

Abstract. In the paper are presented the results of follow on studies from [1]–[3] using authors own model to predict heat transfer coefficient during flow boiling. The model has been tested against a large selection of experimental data collected from various researchers to investigate the sensitivity of the in-house developed model. The collected experimental data came from various studies from literature and were conducted for the full range of quality variation and a wide range of mass velocity and saturation temperatures. In the work are presented the results of calculations obtained using the in-house developed semi empirical model on selected experimental flow boiling data of the refrigerants: R134a, R1234yf, R600a, R290, NH3, CO2, R236fa, R245fa, R152a and HFE7000. In the present study the particular attention was focused on the influence of reduced pressure on the predictions of the theoretical model. The results of calculations were to test the sensitivity of the flow boiling model with respect to selection of the appropriate two-phase flow multiplier, which is one of the distinctive elements of the in-house model. The main purpose of this paper however is to show the effect of the reduced pressure on the predictions of heat transfer during flow boiling.

1 Introduction

Nowadays, there is an increasing interest in refrigerants featuring low Global Warming Potential (GWP). The reason of this concern can be attributed to the growing number of regulations and laws prohibiting the use of some of synthetic refrigerants [4]. According to this regulations the new fluids used in e.g. air-conditioning and refrigeration applications cannot be manufactured with fluorinated greenhouse gases having GWP greater than 150 [5]. Within the Montreal Protocol [5], most of the substances used in refrigeration system had been regulated due to its Ozone Depletion Potential (ODP). Consequently one of the working fluids most extensively used in medium evaporation temperatures, i.e. refrigerant R134a has a 100-year GWP=1430 [6], needs to be replaced by more environmentally friendly fluids. Previous studies have considered R152a [7] and the natural refrigerant CO_2 [8] as possible replacements for R134a. However, R152a is a flammable refrigerant and for the CO_2 system, the working pressure is significantly higher than for R134a system. Apart from that drawback CO_2, as compared to other contemporary fluids is a relatively safe one. It is non-toxic, non-flammable, non-explosive, cheap and can be coupled with most metals and plastics. Design of evaporators for use of the CO_2 requires the exact determination of the heat transfer coefficient during flow boiling. Available in the literature empirical correlations give different results as compared to the results obtained experimentally. There is hardly any robust and recommended correlation for the purpose of calculation of carbon dioxide two-phase heat transfer, despite some devoted contributions [9]. Moreover, the traditional refrigerants can be also replaced by other natural refrigerants such as R600a and R290. These working fluids do not exhibit a harmful impact on ozone layer and have a negligible GWP. Furthermore, the mixture of R290 and R600a can be one of the most appropriate refrigerants to replace R134a [10]. Recently, the new fluid R1234yf has been believed to be the promising candidate as an alternative of R134a. Its ODP=0, whereas GWP=4 and its thermophysical properties are similar to those of R134a. There are many available experimental investigations into flow boiling heat transfer of all mentioned above working fluids in the literature. Unfortunately, the results published for some of these fluids are still inconsistent and difficult in mathematical modelling [3]. In the literature there are many empirical correlations for modelling of flow boiling heat transfer. As mentioned earlier, in case of CO_2 most of well-known models does not exhibit a good consistency with experimental data. It was authors intention to show the performance of their own approach in predicting of flow boiling heat transfer coefficient on the example of the data collected from literature using the in-house developed model [11]–[14]. Based on the evidence of comparisons with experimental data a

* Corresponding author: blanka.jakubowska@pg.edu.pl

correction incorporating the effect of reduced pressure has been postulated to the authors own model to provide the best consistently of the predictions with the experimental data.

2 The model

The versatile semi-empirical model for calculations of flow boiling and flow condensation due to J. Mikielewicz [11] and the final version due to D. Mikielewicz et al. [12]–[14] has been tested for a significant number of experimental data and has returned satisfactory results for the case of the flow boiling process for numerous fluids. The fundamental hypothesis of the model is the fact that heat transfer during flow boiling with bubble generation can be modeled as a sum of two contributions constituting the total energy dissipation in the flow, namely the energy dissipation due to the shearing flow without the bubbles and dissipation resulting from the bubble generation. The final version of the model reads:

$$\frac{\alpha_{TPB}}{\alpha_{LO}} = \sqrt{R_{MS}^{n} + \frac{1}{1+P}\left(\frac{\alpha_{Pb}}{\alpha_{LO}}\right)^2} \tag{1}$$

Equation (1) also includes the empirical correction P defined by equation (2). Occurring in the equation (1) the two-phase multiplier is raised to the power n (n=0.76 for turbulent flow, n=2 for laminar flow).

$$P = 2.53\cdot 10^{-3}\cdot \mathrm{Re}^{1.17}\cdot Bo^{0.6}\cdot (R_{MS}-1)^{-0.65} \tag{2}$$

In calculations tested was the sensitivity of the developed model to the selection of the two-phase flow multiplier. For that purpose four models were introduced into equation (1), namely modified correlation due to Muller-Steinhagen and Heck [16], Friedel [17], Tran [18] and taking into account of the non-adiabatic effect [14]. The relationships, which describes the two-phase multiplier using during this analyse has been show in Table 1.

Table 1. Two-phase multiplier correlations.

Autor(s)	Equation
Muller-Steinhagen and Heck [16]	(3) $R_{MS} = \left[1+2\cdot\left(\frac{1}{f_1}-1\right)\cdot x\cdot Con^{m}\right]\cdot(1-x)^{\frac{1}{3}} + x^3\cdot\frac{1}{f_{1z}}$ Turbulent flow $f_1=(\rho_L/\rho_G)(\mu_L/\mu_G)^{0.25}$, $f_{1z}=(\mu_G/\mu_L)(\lambda_L/\lambda_G)^{1.5}(c_{pL}/c_{pG})$ Laminar flow $f_1=(\rho_L/\rho_G)(\mu_L/\mu_G)$ $f_{1z}=(\lambda_G/\lambda_L)$ m=0 for flow in conventional channels m = -1 for flow in minichannels.
Friedel [17]	$R_F = E + \frac{3.24\cdot F\cdot H}{Fr^{0.045}\cdot We^{0.035}}$ (4) $E = (1-x)^2 + x^2\left(\frac{\rho_l f_v}{\rho_v f_l}\right)$ $F = x^{0.78}(1-x)^{0.2224}$ $H = \left(\frac{\rho_l}{\rho_v}\right)^{0.91}\left(\frac{\mu_v}{\mu_l}\right)^{0.19}\left(1-\frac{\mu_v}{\mu_l}\right)^{0.7}$
Tran [18]	(5) $R_T = 1+\left[4{,}3\cdot\frac{(dp/dz)_{GO}}{(dp/dz)_{LO}}\right]\times$ $\left[N_{conf}\cdot x^{0,875}\cdot(1-x)^{0,875} + x^{1,75}\right]$ $N_{conf} = \sqrt{\frac{\sigma}{g(\rho_l-\rho_g)\cdot d_h^2}}$
non-adiabatic effects [14]	(6) $R_B = \begin{cases} R_{MS}\cdot\left(1-\frac{B}{2}\right) & for\ 0.1 < x \le 1 \\ R_{MS}\sqrt{1+\left(\frac{8\cdot\alpha_{Pb}\cdot d_w}{\lambda_l\cdot \mathrm{Re}\cdot \mathrm{Pr}\cdot\xi_0\cdot R_{MS}}\right)^2} & for\ 0 \le x \le 0.1 \end{cases}$ $B = \frac{2\cdot q\cdot\frac{\rho_l}{\rho_v}}{f_r\cdot G(s-1)\cdot r}$, $s = \sqrt[3]{\frac{\rho_l}{\rho_v}}$

It was expected that the accuracy of model predictions could be improved by some modifications to the empirical correction P, here by incorporation of the reduced pressure effect. The modified empirical correction P yields:

$$P = 2.53\cdot 10^{-3}\cdot \mathrm{Re}^{1.17}\cdot Bo^{0.6}\cdot (R_{MS}-1)^{-0.65}\cdot\left(\frac{p_{sat}}{p_{kr}}\right)^{a} \tag{7}$$

In the following part, the basic model and its subsequent modifications, which have been selected for discussion, will be analysed with respect to predictions of the heat transfer coefficient. These models are denoted respectively as: model I – using equation (3), model II – using equation (4), model III – using equation (5) and model IV – using equation (6). Moreover, exponent a, present in the modified two-phase flow multiplier in equation (7) was adjusted to the available data bank for flow boiling.

3 Results and discussion

The value of the confinement number Con and reduced pressure for all considered refrigerants are presented in Table 2. Using the Kew and Cornwell [19] criterion, the available data bank was divided into conventional size channels and minichannels.

Table 2. List of experimental data.

Author(s)	Fluid	*Con*	P_{sat}/P_{kr}
Docoloumbier et al. [20]	CO_2	**1.611 – 1.412**	0.359 – 0.472
Pamitran et al. [21]		0.241–0.483	0.485 – 0.61
Mastrullo et al. [22]		0.113–0.138	0.382 – 0.548
Yoon et al. [23]		0.083–0.099	0.472 – 0.61
Choi et al. [24]		0.415	0.61

Oh et al. [25]		0.071– 0.274	0.413 0.777
Dang et al. [26]		0.274	0.69
Kim et al. [27]		0.096 – 0.16	0.413 – 0.777
Wu et al. [28]		**0.526 – 0.78**	0.136 – 0.472
Cho et al. [29]		0.048– 0.149	0.472 – 0.777
Zhao et al. [30]		0.223	0.19
Choi et al. [24]	NH_3	**0.75 – 1.5**	0.0455
Anwar [31]		**1.167**	0.1125
Copetti et al. [32]	R600a	**0.537**	0.0884
Anwar [31]		**0.855**	0.1024
Choi et al. [24]	R290	**0.5 – 0.889**	0.111 – 0.154
Wang et al. [33]		0.08 – 0.272	0.0328 – 0.105
Anwar [31]		0.021	0.273
Anwar [31]	R152a	**0.639 – 0.656**	0.14 – 0.162
Lu et al. [34]	R134a	0.234	0.129
Satioh et al. [35]		0.444	0.151
Copetti et al. [32]		0.328	0.149
Kundu et al. [36]		0.132	0.094
Xu et al. [37], [38]		0.205 – 0.38	0.142 – 0.202
Mancin et al. [39]		0.238	0.189
Tibirçá and Ribatski [40]		0.325 – 0.371	0.149 – 0.257
Owhaib et al. [41]		**0.645**	0.213
Shiferaw et al. [42]		**0.732**	0.195
Martin-Callizo et al. [43]		**1.224**	0.218
Consolini and Thome [44]		**1.579**	0.195
Mahmoud et al. [45]		**0.731 – 1.65**	0.145 – 0.197
Ong and Thome [46]		**0.782**	0.195
Anwar et al. [47]		0.499 – **1.267**	0.174 – 0.2
Del Col et al. [4]	R1234yf	**0.736**	0.238
Satioh et al. [35]		0.414	0.151
Choi et al. [24]		0.285 – 0.567	0.111 – 0.133
Diani et al. [48]		0.218	0.232
Anwar et al. [47]		0.456 – 0.475	0.213 – 0.244
Ong and Thome [46]	R236fa	**0.826**	0.104
Tibirçá and Ribatski [40]	R245fa	0.441 0.425	0.050 0.071
Ong and Thome [46]		**0.986**	0.050
Anwar [31]		**1.592**	0.049
Own experimental research	HFE700	0,361 – 0,396	0.034 – 0.080

It can be concluded that the transition from conventional size channels to minichannels takes place on average at channel diameter smaller than 1.5 mm, i.e. Con > 0.5. It should be also noted that the analysis of the parameters from Table 2. indicates the fact that collected for scrutiny experimental research covers a full range of quality and a relatively wide range of mass velocity. Belyaev et al. [49] tried to confirm the hypothesis that in case of high value of reduced pressure, the two-phase flow structures in small diameter channels are similar to those occurring in conventional size diameter. Based on their [49] study they observed that when reduced pressure is greater than 0.4 then is no differences between heat transfer during flow boiling in minichannels and conventional channels.

Figures 1 to 4 show the results of calculations of heat transfer coefficient for all data bank obtained using the described earlier flow boiling model described by Eq. (1), (2) and (6), supplemented by different definitions of the two-phase flow multiplier (3), (4) and (5).

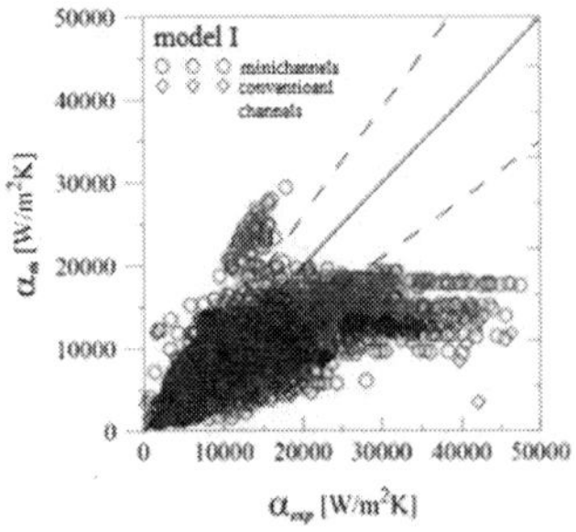

Fig. 1. Comparison of the test results αexp with predictions αth using Muller Steinhagen and Heck model, Eq. (3).

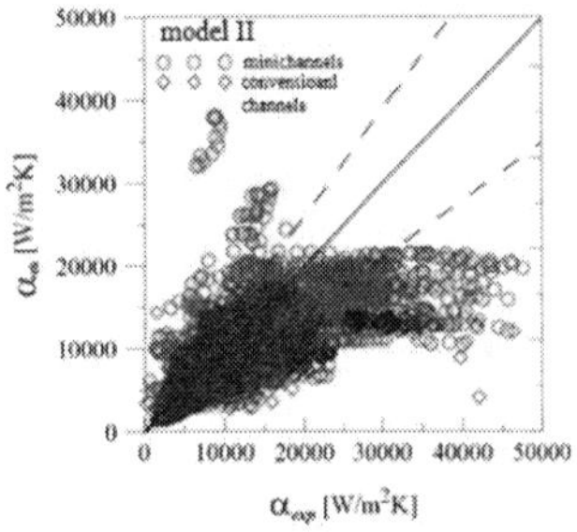

Fig. 2. Comparison of the test results α_{exp} with predictions α_{th} using Friedel model Eq. (4).

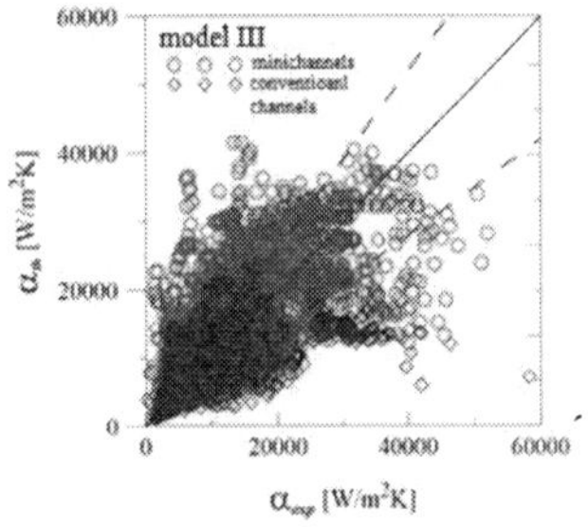

Fig. 3. Comparison of the test results α_{exp} with predictions α_{th} using Tran et al. model Eq. (5).

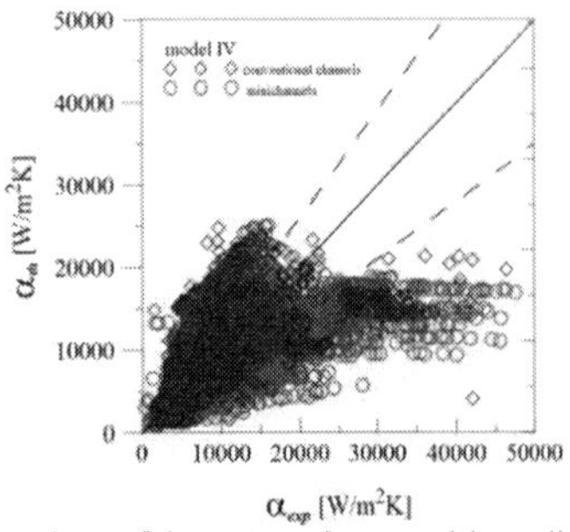

Fig. 4. Comparison of the test results α_{exp} with predictions α_{th} using Eq. (6).

The modification to the empirical correction described by Eq. (2) includes the effect of reduced pressure. The new version of the correction P is presented by Eq. (7), where the reduced pressure is raised to power a. The value of the exponent a was adjusted using the regression analysis. The results of calculations, which were obtained with taking into account of the reduced pressure are presented in Figures 5 to 8. The information about approximating error AE and correlation factors R^2 is given in Table 3.

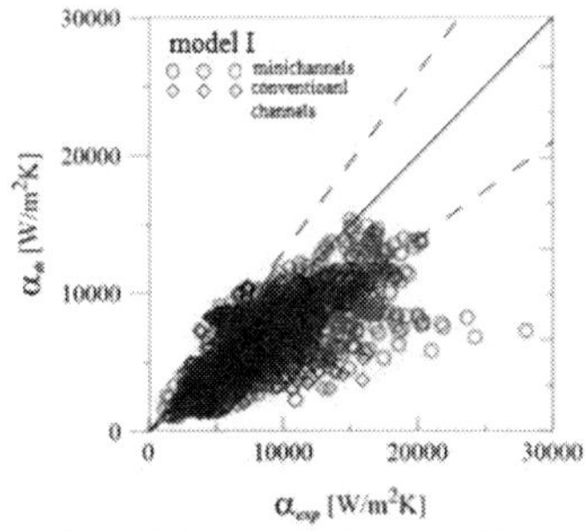

Fig. 5. Comparison of the test results α_{exp} with predictions α_{th} using Eqs. (3), (7) and a=1 for synthetic refrigerants: R134a, R1234yf, R236fa, R245fa, R152a, HFE7000.

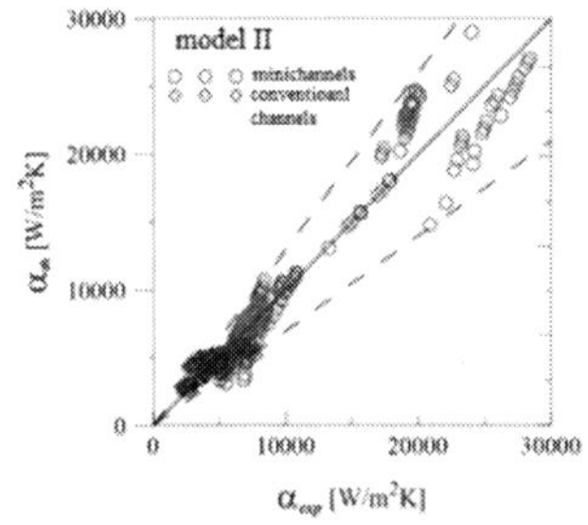

Fig. 6. Comparison of the test results α_{exp} with predictions α_{th} using Eqs. (4), (7) and a= -0.86 for natural refrigerants: R600a, NH_3, R290.

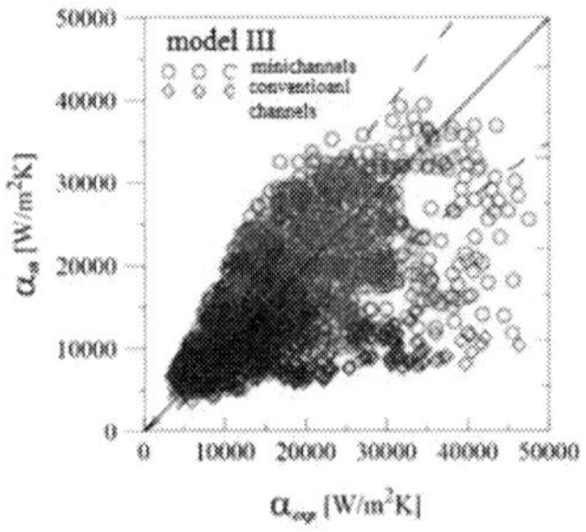

Fig. 7. Comparison of the test results α_{exp} with predictions α_{th} using Eqs. (5), (7) and a= 1 for CO_2.

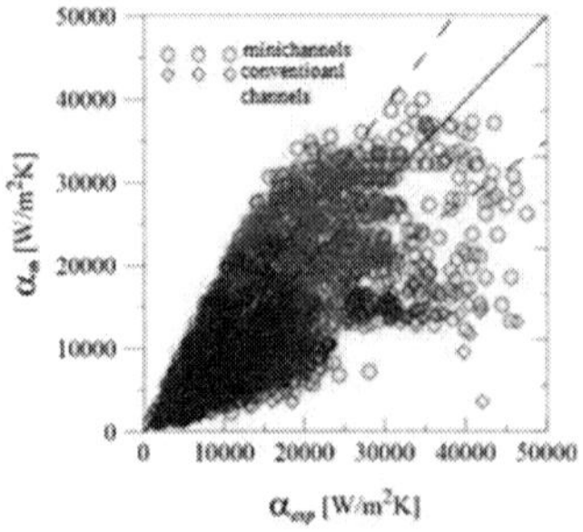

Fig. 8. Comparison of the test results α_{exp} with predictions α_{th} for final version of correlation.

Table 3. Values of exponent *a*, *AE* and R^2.

Model	*a*	data	*AE* [%]	R^2
Model I	0	all	33.11	0.426
		natural	37.41	0.548
		synthetic	30.20	0.239
		CO_2	38.46	0.243
Model II		all	32.64	0.344
		natural	33.45	0.344
		synthetic	24.63	0.184
		CO_2	34.18	0.297
Model III		all	50.17	0.115
		natural	30.34	0.142
		synthetic	52.10	0.050
		CO_2	31.25	0.349
Model IV		all	44.72	0.221
		natural	46.50	0.483
		synthetic	49.00	0.072
		CO_2	40.34	0.216
Model I	-0.86	natural	32.40	0.513
	1	synthetic	27.62	0.275
	1	CO_2	37.74	0.256
Model II	-0.86	natural	28.60	0.338
	1	synthetic	23.36	0.250
	1	CO_2	33.84	0.320
Model III	-0.86	natural	25.88	0.150
	1	synthetic	24.33	0.201
	1	CO_2	31.08	0.362
Model IV	-0.86	natural	43.34	0.415
	1	synthetic	47.46	0.090
	1	CO_2	41.00	0.196
All data presented in Figure 8			28.00	0.490

4. Conclusions

The paper presents the analysis of the results of flow boiling calculations using an authors' own model to predict heat transfer coefficient. The model was studied in several way, i.e., it was used as original one, in a modified version where the two-phase flow multiplier was changed and also in modified version where into the empirical correction *P* was included the reduced pressure effect (the value of exponent a was modeled). The results show that change the model which describing two-phase multiplier is significant especially in case of CO^2, where the best compliance with experimental data obtained using Tran correlation. The results also show that taking into account appropriate two-phase multiplier model and reduced pressure effects can significantly contribute to the convergence with experimental data compared to original model. In authors' opinion, the proposed method to calculate the heat transfer coefficient is a reliable tool in engineering calculations e.g. designing heat exchangers [50, 51].

Acknowledgements
The results presented in the paper have been partially funded from the project 2017/25/B/ST8/00755 by the National Science Centre, Poland.

References

1 D. Mikielewicz, B. Jakubowska, Arch. Thermodyn., **35**, 97–114 (2014)
2 D. Mikielewicz, B. Jakubowska, Arch. Thermodyn., **37**, 89–106 (2016)
3 D. Mikielewicz, B. Jakubowska, Polish Marit. Res., **24**, 141–148 (2017)
4 D. Del Col, S. Bortolini, D. Torresin, A. Cavallini, *Procedings of 23rd IIR International Congress of Refrigeration*, Prague, Czech Republic (2011)
5 United Nations Environment Program (UNEP) Montreal Protocol on Substances that Deplete the Ozone Layer, *Final Act. United Nations*, New York, USA, (1997)
6 M.P. Simmonds, W. J. Elliott, J. Mar. Biol. Assoc. UK, **89**, 203-210 (2009)
7 M. Ghodbane, *Int. Proceedings of the international congress and exposition,* Detroit, USA (1999)
8 G. Lorentzen, Revival of carbon dioxide as refrigerant, Int. J. Refrig., **17**, 292–301 (1994)
9 L. Cheng, G. Ribatski, J. R. Thome, Int. J. Heat Mass Transf. **51**, 125–135 (2008)
10 S. Wongwiesess, N. Chimres, Energ. Convers. Manag., **46**, 85–100 (2005)
11 J. Mikielewicz, Int. J. Heat Transf., **17**, 1129–1134 (1973)
12 D. Mikielewicz, J. Mikielewicz, J. Tesmar, Int. J. Heat Mass Transf., **50**, 3949–3956 (2007)
13 D. Mikielewicz, J. Mikielewicz, Heat Transf. Eng., **32**, 1173–1181 (2011)
14 D. Mikielewicz, R. Andrzejczyk, B. Jakubowska, J. Mikielewicz, Heat Transf. Eng., **37**, 1158–1171 (2016)
15 H. Müller-Steinhagen, K. Heck, Chem. Eng. Prcoess, **20**, 197–308 (1986)
16 D. Mikielewicz, Heat Transf. Eng., **31**, 276–287 (2010)
17 L. Friedel, *European Two-Phase Flow Group Meeting*, Paper E2, Ispra, Italy (1979)
18 T. N. Tran, M. C. Chyu, M. W. Wambsganss, Int. J. Refrig., **26**, 1739–1754 (2000)
19 P. A. Kew, K. Cornwell, Appl. Therm. Eng., **17**, 705–715 (1997)
20 M. Docoulombier, S. Colasson, B. J. P. Haberschil, Exp. Therm. Fluid Sci., **35**, 597 – 611 (2011)
21 A. S. Pamitran, K.-I. Choi, J.-T. Oh, Nasruddin, Int. Multiph. Flow, **37**, 794–801 (2011)
22 R. Mastrullo, A. W. Mauro, A. Rosato, G. P. Vanoli, Int. J. Refrig., **33**, 1068–1085 (2010)

23 S. H. Yoon, E. S. Cho, Y. W. Hwang, M. S. Kim, K. Min, Y. Kim, Int. J. Refrig, **27**, 111–119 (2004)
24 K.-I. Choi, A. S. Pamitran, C.-Y. Oh, J.-T. Oh, Int. J. Refrig., **30**, 1336–1346 (2007)
25 H.-K. Oh, C. H. Son, Appl. Therm. Eng., **31**, 163–172 (2011)
26 C. Dang, N. Haraguchi, E. Hihara, Int. J. Refrig, **33**, 655–663 (2010)
27 Y. J. Kim, J. M. Cho, M. . Kim, Int. J. Refrig., **31**, 771–779 (2008)
28 J. Wu, T. Koettig, C. Franke, D. Helmer, T. Eisel, F. Haug, J. Bremer, Int. J. Heat Mass Transf., **54**, 2154–2162 (2011)
29 J. M. Cho, M. S. Kim, Int. J. Refrig., **30**, 986–994 (2007)
30 X. Zhao, P. K. Bansal, Int. J. Refrig., **30**, 937–945 (2007)
31 Z. Anwar, *Flow boiling heat transfer, pressure drop and dryout characteristics of low GWP refrigerants in a vertical mini-channel*, Stockholm, Sweden (2014)
32 J. B. Copetti, M. H. Macaganan, F. Zinani, Int. J. Refrig., **33**, 325–334 (2013)
33 C. C. Wang, C. S. Chiang, D. C. Lu, Exp Therm. Fluid Sci, **15**, 395–405 (1997)
34 M.-C. Lu, J.-R. Tong, W.C-C, Int. J. Heat Mass Transf., **65**, 545–551 (2013)
35 S. Satioh, C. Dang, Y. Nakamura, E. Hihara, Int. J. Refrig., **33**, 1846–1853 (2011)
36 A. Kundu, R. Kumar, A. Gupta, Exp. Therm. Fluid Sci., **57**, 344–352 (2014)
37 Y. Xu, X. Fang, G. Li, D. Li, Int. J. Heat Mass Transf., **80**, 597–606 (2015)
38 Y. Xu, X. Fang, G. Li, D. Li, Int. J. Heat Mass Transf., **92**, 1143–1157 (2016)
39 S. Mancin, A. Diani, L. Rossetto, Energy Procedia, **45**, 6098–615 (2014)
40 C. B. Tibirçá, G. Ribatski, Int. J. Heat Mass Transf., **53**, 2459–2468 (2010)
41 W. Owhaib, *Experimental Heat Transfer , Pressure Drop and Flow Visualization of R-134a in Vertical Mini/Micro Tubes*, Stockholm, Sweden (2007)
42 D. Shiferaw, T. G. Karayiannis, D. B. R. Kenning, Int. J Therm. Sci., **48**, 331–341 (2009)
43 C. Martin-Callizo, R. Ali, B. Palm, *Heat Transfer Conference Proceedings,* Edinburgh, UK (2007)
44 L. Consolini, J. R. Thome, Microfluid Nanofluid, **6**, 731–746 (2009)
45 M. M. Mahmoud, D. B. R. Kenning, T. G. Karayiannis, *7th Int. Conference in Enhanced, Compact and Ultra-compact Heat Exchangers: From Microscale Phenomena to Industrial Applications*, Heredia, Costa Rice (2009)
46 C. I. Ong, J. R. Thome, Exp. Therm. Fluid Sci., **33**, 651–663 (2009)
47 Z. Anwar, B. Palm, R. Khodabanden, Exp. Therm. Fluid Sci., **66**, 137–149 (2015)
48 A. Diani, S. Mancin, L. Rossetto, Exp. Therm. Fluid Sci., **66**, 127–136 (2015)
49 A. V Belyaev, A. N. Varava, A. V Dedov, A. T. Komov, Int. J. Heat Mass Transf., **110** 360–373 (2017)
50 R. Andrzejczyk, T. Muszyński, App. Therm. Eng., **136**, 237-251 (2018)
51 J. Wajs, D. Mikielewicz, B. Jakubowska, Energy, **157**, 853-861 (2018)

https://doi.org/10.1051/matecconf/201824001029

Leidenfrost evaporation of water droplet

Tadeusz Orzechowski*

Kielce University of Technology, Al. Tysiąclecia PP 7, 25-314 Kielce, Poland

Abstract. The present paper examines the behaviour of a single large droplet levitating over a hot surface, unsteady mass of the droplet, and heat transfer. It was assumed that the evaporation from the upper surface of the droplet is negligibly small compared with the amount discharged from its lower surface, and the heat transfer coefficient is the power function of droplet orthogonal projection onto the heating surface. Based on the photographic documentation, the dependence of the droplet projection on time was approximated. A heat balance was written in the form of a non-linear first order differential equation. The solution to the equation was included. The analytical function of droplet mass change in time was used to determine the exponent of the power dependence of the heat transfer coefficient on the orthogonal droplet projection onto the heating surface. The comparison showed the method proposed in the study could be applied to analyse the behaviour of a water drop levitating above a surface at the temperature higher than the Leidenfrost point.

1 Introduction

In the mid-1980s, the thermal load on components amounted to 100 W/cm^2 [1]. The only possible solution was to use cooling that employed change of phase processes in the cooling agent [2,3,4]. Another important application area of phase-change cooling includes evaporation processes in refrigeration and air conditioning [5,6]. In thermal treatment of metals, controlled temperature changes are decisive for the formation of near-surface microstructures [7].

Surface spray cooling is another important issue [8]. Various investigations conducted into the subject focus on the physics of the phenomenon, which depends strongly on the liquid and its components [9] and [10], thermal and flow conditions [11,12], interaction with the surface [13], radiation heat transmission through the water film [14], etc. Also, when the basic liquid is modified with additives such as nanoparticles, significant changes in the nature of heat transfer are observed [15]. A brief review of their impact on the Leidenfrost temperature in heat transfer applications is delivered in [16].

Physical processes involved in droplet evaporation are closely related to the surface. It was shown that droplets located on low conductivity materials get colder and evaporate at about half the rate of similar droplets on surfaces with high thermal conductivity [17]. The evaporation characteristics of sessile water droplets on smooth hydrophobic and structured superhydrophobic heated surfaces were experimentally investigated in [18]. It was proved that surfaces with high wettability enhance heat transfer coefficient and critical heat flux in pool boiling [19]. Clear differences are observed between nucleate pool boiling and boiling in the droplet. Intense nucleate boiling in the droplet is accompanied by continuous changes in its shape and the area of contact with the substrate. Those phenomena are caused by vapour bubbles that flow inside and depart. In this case, local changes in the surface temperature are observed, which results in a lower value of the heat transfer coefficient [20]. Only at the wall superheat above 10 K, bubbles visibly change the droplet circumference. As a result, the surface is locally dried up, and the density of the removed heat is considerably reduced. Surface temperature grows and the inflowing liquid evaporates abruptly, which makes the whole process unstable. This range is termed a transition boiling regime [21]. The final stage of this regime is the formation of a stable vapour film, which supports the droplet above the heating surface [22,23], while intense internal currents inside the droplet are observed [24]. The first observation of that effect was reported by Johann Gottlob Leidenfrost in *A Tract About Some Qualities of Common Water* in 1756, and the phenomenon has been named after him.

In many technological processes, the element that needs to be cooled has the initial temperature that is higher than Leidenfrost temperature. The amount of heat carried away and the rate of cooling are the parameters that need to be precisely controlled [25]. That refers to technological processes [26,27], and to maintaining the thermal stability of the elements cooled [28]. To solve the task thus posed, it is necessary to have possibly complete information on the physical processes that occur in droplet evaporation, which despite numerous investigations have not been fully explained.

The aim of this study is to provide the analysis of heat transfer between a large water droplet at atmospheric pressure, and a hot surface at a temperature higher than Leidenfrost temperature.

* Corresponding author: todek@tu.kielce.pl

https://doi.org/10.1051/matecconf/201824001029

2 Experimental set-up

A large droplet of water was placed on the heating surface with temperature higher than the Leidenfrost point. Measurements taken for this study concerned the droplet mass changes in time. Electronic weighing scales, with the maximum capacity of 500 g, was used. The scales readability is 0.02 g, and sensitivity is 0.001 g. The interface is used for direct communication with the computer.

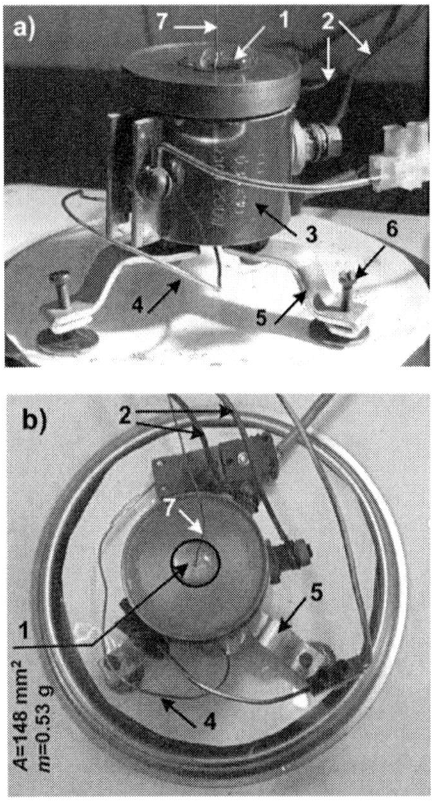

Fig. 1. Photograph of the experimental set-up: a) side view, b) top view 1- droplet, 2 - electrical power supply, 3 - heater, 4 - thermocouple, 5- support pin, 6 - screw, 7 - thermocouple.

Investigations into droplet evaporation at a temperature above the Leidenfrost point were conducted at the stand, the main component of which was a cylindrical copper heater (see Fig. 1). The cylinder upper surface was shaped like a bowl having a very large radius, which allowed centric, stable positioning of the liquid droplet under investigation, as shown in Fig.1b. Directly under the surface, in the symmetry axis, k-type sheathed thermocouple was placed. The heater was installed on the cylinder outer surface and supplied with electric power from the autotransformer by suspended conductors. The output power of the heater was set by voltage, adjusted according to the voltmeter readings. The detailed description of the measurement procedure, together with the analysis of uncertainties were presented in [29] and [30].

3 Results and discussion

Large mass droplets show instabilities of various types, which result from complex convective movements inside the droplet [31-34].

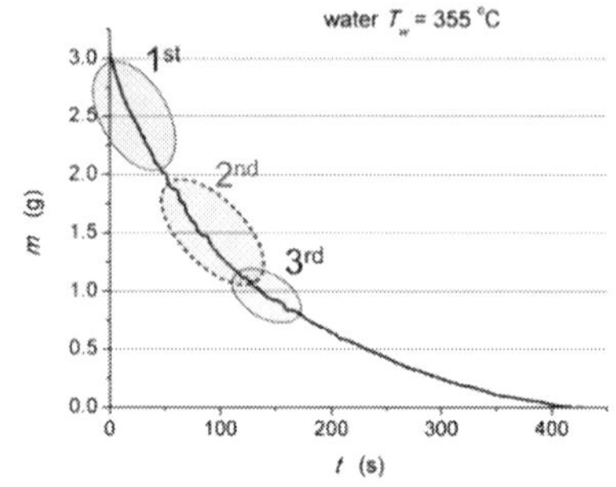

Fig. 2. Evaporation of the water droplet on the wall of the temperature T_w=355 °C.

In Fig.2, three zones with different droplet behaviour are marked. In the first zone under the surface of the droplet having the mass of as much as approx. 2g, vapour lingers forming large bubbles. They grow very slowly and relatively seldom depart freely through the droplet upper surface.

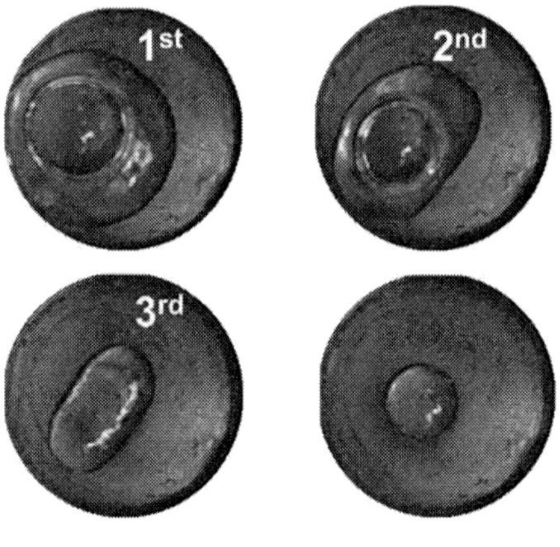

Fig. 3. Droplet shape evolution at various stages of its evaporation acc. Fig. 2.

Due to the measurement system inertia, relatively small changes in weight are recorded during this stage. As the droplet mass decreases, the area of its surface is also reduced, which leads to the deterioration of vapour accumulation. The frequency of bubble departure clearly rises, which is reflected in the readings of the scales in the second zone. For droplets having the mass of approx. 1g, vapour bubbles departing from the upper surface are not observed. Small changes in the droplet weight in third zone result from a chaotic droplet movement over the surface. Small drops weighing less than 0.7 g take on stable shapes, as shown in Figure 3.

Photographic imaging makes it possible to measure droplet projection onto the heating surface. Digital photo camera was suspended directly above the droplet to record the projection of its surface. Then, the footage was converted into individual frames. Using software tools, droplet surface area was estimated in the selected frames. For the series of concern, images are presented in Fig.4.

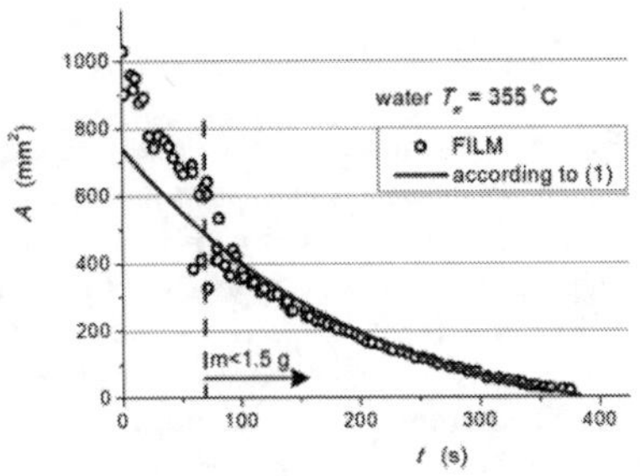

Fig. 4. Droplet projection onto the heating surface.

The literature provides different formulas for converting droplet orthogonal projection onto the heating surface. The correlations are usually time-dependent power formulas. One of them has the following form [35]:

$$\frac{A(t)}{A_0} = C_B\left(a^t - C_A\right) \tag{1}$$

where A_0 is the surface projection at the initial instant, whereas C_A, $C_A>0$ and $a<1$ are constants under the assumed and pre-set conditions of the experiment.

Using the correlation above (1), droplet surface area was calculated as a function of time, as shown in Fig.4. The fitting of the curve from Eq. (1) was conducted in such a way so that it could be matched, as accurately as possible, to the results obtained from photographic imaging. The comparison of the profiles of both curves shows fairly good congruence for small mass droplets. For large mass droplets, i.e. above 1.5g, a clear discrepancy between the curves is observed.

4 Heat transfer analysis

Assuming that evaporation from the upper surface of the droplet is small compared with the amount of heat carried away from the bottom surface, the overall heat transfer coefficient comprises the convective and radiative part, i.e. $\alpha=\alpha_c+\alpha_r$. Then, it possible to write energy balance equation, which implies a change in droplet mass m in time t:

$$\alpha(T_w - T_d)A = -\left(c_p(T_s - T_d) + h_{fg}\right)\frac{\mathrm{d}m}{\mathrm{d}t} \tag{2}$$

where T_w, T_d and T_s are surface, droplet and saturation temperatures respectively, A – orthogonal droplet projection onto the heating surface, c_p – specific heat, and h_{fg} phase change enthalpy at saturation temperature, here at atmospheric pressure.

To obtain solution to Eq. (2), it is necessary to provide a correlation for droplet projection onto the heating surface and heat transfer coefficient. Experimental data indicate exponential dependence of heat transfer coefficient on the surface. For that reason, the following form of the function was assumed [30]:

$$\alpha = C_\theta A^n \tag{3}$$

where n is a constant exponent, and coefficient $C_\theta=C(T_w)$ is different for individual measurement curves, and varies with the wall temperature T_w.

The equation (2) together with the adopted assumptions (1) and (3) can be integrated, and its solution has the following form:

$$m_0 - m = M_0\left(C_B C_A\right)^{n+1}\frac{1}{\ln a} * \left[\int\frac{(z-1)^{n+1}}{z}dz - \int\frac{(z-1)^{n+1}}{z}dz\bigg|_{t=0\to z=1/C_A}\right] \tag{4}$$

where for the sake of convenience, a new variable is introduced:

$$z = \frac{a^t}{C_A} \tag{5}$$

At t=0, droplet mass m_0 is known and M_0 is a flux of mass evaporated from the droplet surface at t=0 defined as follows:

$$M_0 = \frac{T_w - T_d}{h_{fg} + c_p(T_s - T_d)}C_\theta(A_0)^{n+1} \tag{6}$$

The integral in the equation above can be presented as an infinite series:

$$\int\frac{(z-1)^{n+1}}{z}dz = \sum_{i=1}^{j}\frac{(i-1)!}{\sum_{k=1}^{i}(n+1+k)}\frac{(z-1)^{n+1+i}}{z^i} + \frac{(j-1)!}{(n+2)(n+3)\ldots(n+1+j)}\int\frac{(z-1)^{n+j}}{z^{j+1}}dz \tag{7}$$

which is a fast-converging series. The first term of the sum on the right-hand side of the equation (7) is a partial sum, whereas the second term is the series remainder. A solution with accuracy below 1% is obtained already when only ten terms are taken into account, i.e. $j\leq10$.

Shape disturbances essentially alter droplet projection onto the heating surface, which can be seen in Figs. 3 and 4. The disturbances are caused by lingering and successive departing of large vapour bubbles. Therefore, further analysis was limited to droplets having a mass of $m \leq 1.5$g. Using correlation (1) for orthogonal projection vs. time, the surface area of the droplet was calculated, as shown in Fig.4.

The following values of coefficients were assumed: C_A=0.11, C_B=1.16, and a=0.9943. C_A and C_B were calculated from the dependence given in [35]. The base of power was selected following the criterion of the minimum mean quadratic error between the measured areas and areas computed acc. Eq. (1).

The accuracy of the approximation can be evaluated by comparing residuals between measurement points and calculated points at the same time instant. The greatest differences between computed and measured quantities are found for the initial period for small time values, where clear spherical shape aberration is visible, as can be seen in Fig. 3.

The solution (4) to the problem of concern determines the dependence of the droplet mass on time. To plot the curve, all the coefficients used before are needed. Three of them, namely C_A, C_B and a, result from the fitting of the correlation for droplet orthogonal projection on the heating surface. The remaining quantities that need to be specified are coefficient C_θ and power exponent n. They are related to the heat transfer coefficient, the magnitude of which is proposed to be expressed by power dependence on the instantaneous droplet projection area A. The values of both quantities mentioned above are decisive for magnitude of the mass flux that is evaporated. Therefore, they need to be selected in such a way so that the curve of droplet mass change in time, calculated acc. Eq. (4), would closely coincide with the curve determined experimentally, as shown in Fig.5.

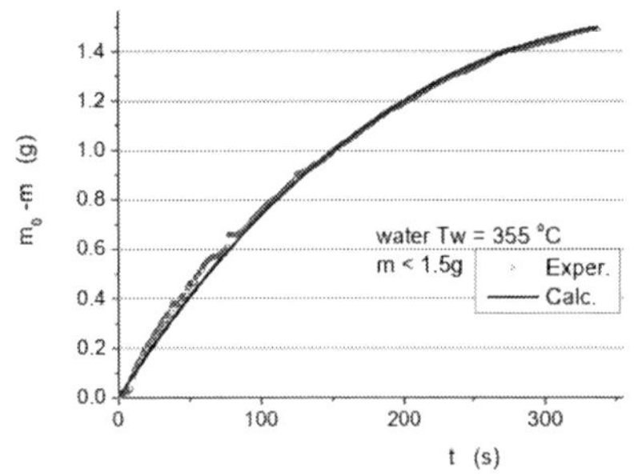

Fig. 5. Measured and calculated instantaneous droplet mass.

Both coefficients were determined acc. data from Fig. 5 following the criterion of the minimum mean quadratic error. Approximated values C_θ=19.1 and n=-0.3 make it possible to calculate local values of the heat transfer coefficient acc. formula (3), as shown in Fig. 6.

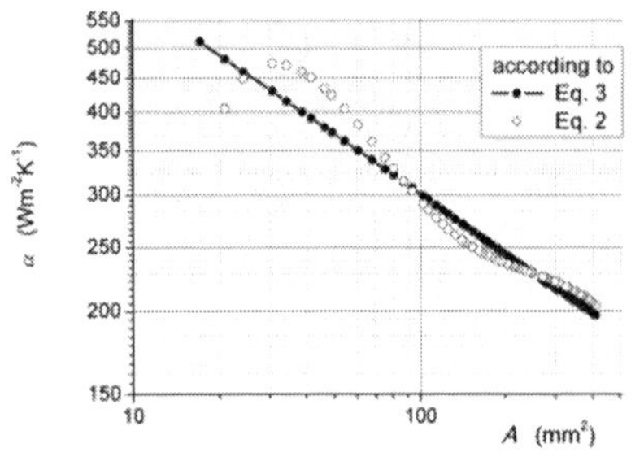

Fig. 6. Local values of the coefficient of heat transfer from the bottom water droplet surface for the orthogonal projections onto the heating surface at constant temperature T_w=355°C.

For the sake of comparison, the results of direct calculations from Eq. (2) are also given in that figure. The computational procedure was discussed in study [30].

5 Conclsions

A large droplet of water was placed on the heating surface with temperature higher than the Leidenfrost point. The analysis of results was carried out employing the droplet heat balance, under the assumption that heat transfer results only from droplet interaction with the heating surface. To make calculations, the dependence, available in the literature, for the droplet orthogonal projection as the function of time was applied (see Eq. 1). On the basis of the author's investigations and photographic imaging the following values were selected: coefficients C_A=0.11, C_B=1.16, and also a=0.9943. The values were calculated for the droplet with mass below 1.5 g and heating surface temperature T_w=355°C at the condition of the minimum mean quadratic error.

The comparison of the measured values of mass loss in time and the computed ones made it possible to determine coefficients C_θ and n. They specify instantaneous values of the heat transfer coefficient acc. Eq. (3). They were determined on the basis of the data in Fig. 5 for the criterion of the minimal mean quadratic error.

The determined values C_θ=19.1 and n=-0.3 make it possible to compute local values of the heat transfer coefficients acc. formula (3), which was shown in Fig. 6. For the sake of comparison, the results of direct computations acc. Eq. (2) were also given in the figure. In the procedure, polynomial approximation of the mass flux carried away from the droplet surface is used. That causes fluctuation visible in Fig. 6, which is difficult to eliminate, especially at the boundaries of the region. When analytical solution (Eq. 4) is applied, this inconvenience can be avoided. Also, the adopted power dependence of instantaneous values of the heat transfer coefficient on droplet orthogonal projection onto the heating surface is confirmed.

References

1. M. Aamir, Q. Liao, W. Hong, Z. Xun, S. Song, M. Sajid, Heat Mass Transfer, **53** (2017)
2. S. Hożejowska, J. Theor. Appl. Mech. **53**(4) (2015)
3. T. Orzechowski, Heat Mass Transfer, **52** (2016)
4. T. Orzechowski, EPJ Web of Conferences, **143** (2017)
5. T. Orzechowski, Z. Skrobacki, EPJ Web of Conferences, **114** (2016)
6. P.O. Rašković, G.D. Vučković, M.V. Vukić, Therm. Science, **12** (4) (2018)
7. K.H.M. Abdalrahman, S. Sabariman, E. Specht, Int. J. Heat Mass Tran. **78** (2014)
8. M. Gradeck, N. Seiler, P. Ruyer, D. Maillet, Exp. Therm. Fluid Sci. **47** (2013)

9. P. Keller, A. Bader, C. Hasse, Int. J. Heat Mass Tran. **67** (2013)
10. Y. Wang, Y. Jiang, W. Chen, B. Zhou, Appl. Therm. Eng. **123** (2017)
11. V.E. Nakoryakov, S.Y. Misyura, S.L. Elistratov, Int. J. Heat Mass Tran. **55** (2012)
12. K. Strąk, M. Piasecka, B. Maciejewska, Int. J. Heat Mass Tran. **117** (2017)
13. M. Gao, P. Kong, L-X. Zhang, J-N. Liu, Int. Commun. Heat Mass Tran. **88** (2017)
14. G.V. Kuznetsov, K.Y. Osipov, M.V. Piskunov, R.S. Volkov, Int. J. Heat Mass Tran. **117** (2018)
15. A.M. Ali, W. Arshad, H.M. Ali, H. M. Ali, M.A. Nasir, J. Thermal Science, **21**(5) (2017)
16. S. Wciślik, Chem. Pap. **71** (2017)
17. T. Log, Appl. Therm. Eng. **93** (2016)
18. S. Dash, S.V. Garimella, Phys. Rev. E, **89** (2014)
19. X. Chen, H. Qiu, Int. J. Heat Mass Tran. **88** (2015)
20. Y. Wang, H. Fu, Q. Huang, Y. Cui, Y. Sun, L. Jiang, Energy, **93** (2015)
21. Y. Mitsutake, S. Illias, K. Tsubaki, M.N. Hasan, M. Monde, Procedia Eng. **105** (2015)
22. T. Orzechowski, EPJ Web of Conferences, **143** (2017)
23. T. Orzechowski, S. Wciślik, Exp. Therm. Fluid Sci. **59** (2014)
24. A. Al-Sharafi, B.S. Yilbas, H. Ali, Appl. Therm. Eng. **128** (2018)
25. Y.Y. Puzina, J. Phys. Conf. Ser. **891** (2017)
26. N. Shahidzadeh-Bonn, S. Rafa, A. Azouni1, D. Bonn, J. Fluid Mech. **549** (2006)
27. X. Liu, J. Liu, R. Xue, L. Chen, Y. Hou, Y, Exp. Therm. Fluid Sci. **92** (2018)
28. D. Liu, F-Y. Zhao, H-X. Yang, G-G. Tang, Energy, **83** (2015)
29. T. Orzechowski, T, S. Wciślik, Energ. Convers. Manage. **76** (2013)
30. T. Orzechowski, T, S. Wciślik, Int. J. Heat Mass Tran. **73** (2014)
31. D. Brutin, B. Sobac, F. Rigollet, C. Le Niliot, Exp. Therm. Fluid Sci. **35** (2011)
32. M. He, H. Qiu, H, Int. J. Therm. Sci. **100** (2016)
33. S. Wciślik, J. Phys.: Conf Ser, **745** (2016)
34. R. Kaniowski, R. Pastuszko, EPJ Web of Conf. **143**, 02050 (2016)
35. G. Paul, P.K. Das, I. Manna, Exp. Therm. Fluid Sci. **60** (2015)

A thermal parametric study of non-evaporative spray cooling process

Abdessalam Otmani[1], *Hocine* Mzad[2,*], and *Kamel* Bey[2]

[1]Laboratory of Industrial Mechanics, Badji Mokhtar University of Annaba, DZ-23000, Algeria
[2]Department of Mechanical Engineering, Badji Mokhtar University of Annaba, P.O. Box 12, DZ-23000, Algeria

Abstract. Ordinary water spray cooling is connected with very high temperatures where heat transfer during evaporation plays a key role. However, during cooling without phase change, the behaviour of the spray cooling parameters is rarely considered. The purpose of this paper is to study the influence of spray hydrodynamic parameters on heat transfer without liquid phase change during the cooling of an aluminium 3003-H18 plate at a temperature of 92 °C. First of all, the flow rate was varied from 0.497 up to 1 l/min. Then, the inlet pressure varied from 0.7 to 2.1 bars. The influence of nozzle-to-target distance is also tested since the simulations were carried out in a wide height range, from 100 mm to 505 mm. The present simulation was achieved using the version 5.2 of COMSOL Multiphysics code.

1 Introduction

Optimizing the hydrodynamic parameters of the jet during nozzle design helps to obtain better dispersion and economy in water consumption. The nozzle regulates the flow rate, atomizes the liquid into droplets, and disperses the droplets in a specific pattern. The proper tip nozzle must be selected for the desired job; no single nozzle can meet all the spraying needs.

The thermal performance during spray injection of liquid onto surfaces is normally represented by the associated heat transfer coefficient or the Nusselt number, which is a measure of the cooling rate due to forced convection. Characterization of the heat transfer coefficient is important for determining the required cooling capacity and corresponding internal properties of the product. For this reason, a vast amount of work has been devoted to parameterize or model the heat transfer coefficient during spray cooling in different flow regimes [1, 2].

A correlation was established for the quantification of water jet dispersion on the totality of a plane surface in the function of the spray nozzle dimensions and hydrodynamic parameters. Subsequently, numerical simulations based on the above experimental results were conducted to investigate the effect of the water-spray characteristics on the heat transfer mechanisms involved during cooling of steel plates at high temperatures [3, 4].

A model to predict the heat and mass transfer in spray cooling was presented with consideration of droplet-film impaction, film formation, film motion, bubble boiling, droplet-bubble interaction, bulk air convection and radiation [5].

If multiple nozzles are used in a conveyorized application, the overlapping liquid distribution pattern of the nozzles needs to be considered because the process may depend strongly on the relative local volume flux of the spray. In order to understand the fluid behaviour in the overlap region, a hydrodynamic study was proposed using a FORTRAN programming based on experimental correlations of pulverized water jets [6].

The effects of droplet velocity non-uniformity, Sauter mean diameter distribution non-uniformity, droplet number non-uniformity, and heating power on the fluid film thickness, fluid film velocity, and surface temperature distribution were investigated, and then the surface temperature non-uniformity correlations in non-boiling regime and nucleate boiling regime were correlated [7].

The heat transfer characteristics of air water spray impingement cooling of stationary steel plate were experimentally investigated. The controlling parameters taken during the experiments were fluid pressure, water flow rate, nozzle tip to target distance and mass impingement density. The effects of the controlling parameters on the cooling rates were critically examined during spray impingement cooling [8].

* Corresponding author: h_mzad@yahoo.fr

 https://doi.org/10.1051/matecconf/201824001030

A computational fluid dynamics (CFD) assessed the potential and performance of evaporative cooling by water spray system with a hollow-cone nozzle configuration [9].

Nizetic et al. [10] proposed an alternative cooling technique for photovoltaic (PV) panels that includes a water spray application over panel surfaces. Both sides of the PV panel were cooled simultaneously, to investigate the total water spray cooling effect on the PV panel performance in circumstances of peak solar irradiation levels. Furthermore, it was also possible to decrease panel temperature from an average 54 °C (non-cooled PV panel) to 24 °C in the case of simultaneous front and backside PV panel cooling.

The purpose of this study is to highlight, by simulation under COMSOL Multiphysics, the influence of hydrodynamic parameters, namely the nozzle height, water flow rate and spraying pressure in order to evaluate the impact of nozzle characteristics and operating conditions on cooling performance of the spray system.

2 Modelling and Simulation

In the simplest case, a single nozzle is employed to distribute liquid in a defined pattern to cover a surface or fill a spatial volume in a prescribed manner. Hydraulic atomizers depend on the internal geometry and liquid pressure alone to produce a desired pattern. The exact definition of the spray characteristics and the specification of the operation parameters are the first two steps to be taken when a spray nozzle is being designed. Using the commercial software COMSOL Multiphysics 5.2 we achieved a specific application, coupled between heat transfer and fluid flow interface, for the simulation of a cooling process beneath liquid sprayer. The objective is to cool an aluminium 3003-H18 plate (Tab. 1) with 6 mm thickness and an area of 1200 mm^2, initially at the temperature of 92 °C.

A single nozzle fixed at the centre upon the plate, delivers the simulated water spray in an ambience at atmospheric pressure and a temperature of 20 °C. The nozzle geometry and meshing are shown on figure 1.The modelling section is executed in two steps, the fluid flow model and heat transfer model.

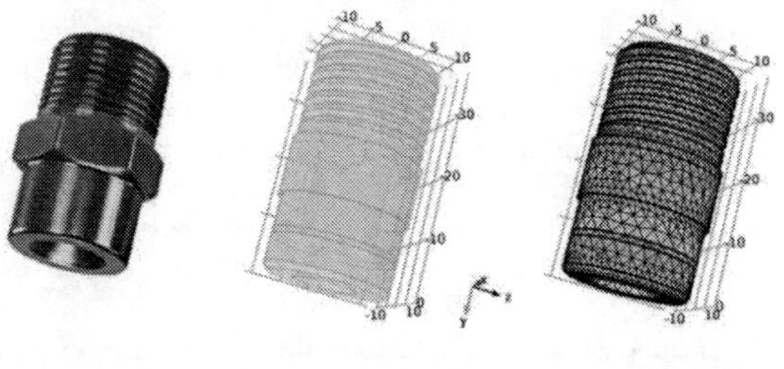

Fig. 1. BETE WL6 nozzle.

Table 1. Aluminium 3003-H18 Properties.

Parameter	Value
Specific heat capacity, C_p	890 J/(kg.K)
Density, ρ	2.8 g/cm^3
Thermal conductivity, k	180 W/(m.K)
Thermal expansion coeff., α	2.3E-5 1/K
Melting point, T_m	640 °C

The resolution of the Navier-Stokes equations was achieved for laminar fluid flow and stationary regime. The simulation was realized in two steps. The first one is the simulation of the laminar flow around the aluminium plate, inside and outside the nozzle. The obtained results were used in the second step, which is the simulation of the laminar bubbly flow (Fig. 2). Using a two-phase Euler-Euler model, a momentum balance equation and a continuity equation describe the dynamics of each phase.

Therefore, a combination of the momentum and the continuity equations gives:

$$\frac{\partial \rho_g \phi_g}{\partial t} + \nabla.\left(\phi_g \rho_g \left(u_l + u_{slip}\right)\right) = \nabla.\left(\frac{\tilde{\mu} \rho_g}{\rho_l} \nabla \phi_g\right) - m_{gl} \tag{1}$$

The gas phase transport equation is expressed as:

$$\frac{\partial \rho_g \phi_g}{\partial t} + \nabla.\left(\phi_g \rho_g u_g\right) = -m_{gl} \tag{2}$$

The gas velocity u_g is the sum of the following velocities:

$$u_g = u_l + u_{slip} + u_{drift} \tag{3}$$

The liquid volume fraction is calculated from:

$$\phi_l = 1 - \phi_g \tag{4}$$

The drift velocity expression is:

$$u_{drift} = -\frac{\tilde{\mu}}{\rho_l} \frac{\nabla \phi_g}{\phi_g} \tag{5}$$

Where $\tilde{\mu}$ is an effective viscosity causing the drift.

A diffusive term is introduced in the gas transport equation (Eq. 2), which is implemented in the fluid-flow model.

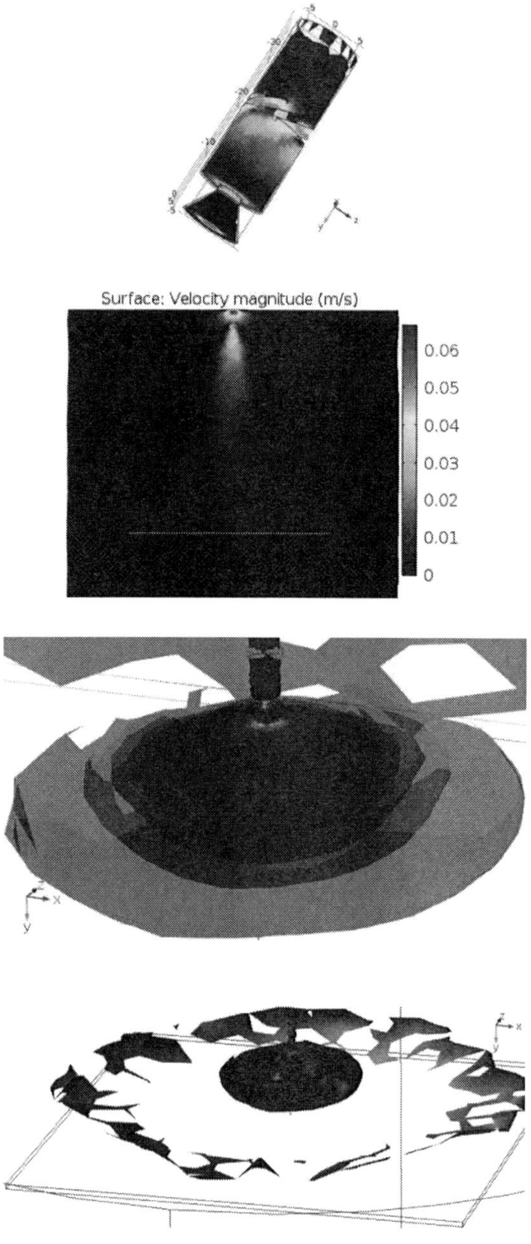

Fig. 2. Simulation of water spray dispersion.

The temperature distribution in a three-dimensional space within the considered solid is described by the equation of the heat conduction:

$$\rho C_p \frac{\partial T}{\partial t} + \rho C_p u \nabla T + \nabla q = 0 \tag{6}$$

By solving the heat equation, we evaluate the maximum plate temperature and the average convection heat flux variation.

There is no liquid evaporation, because the maximum plate temperature is below the saturation temperature of water.

The initial water supply temperature was 20 °C. The heat transfer in fluid flow is solved using the equation:

$$\rho C_p \left(\frac{\partial T}{\partial t} + u.\nabla T \right) + \nabla.(q + q_r) = Q_p + Q_{Vd} \tag{7}$$

Equation in which:

$$Q_p = \alpha_p T \left(\frac{\partial p}{\partial t} + u.\nabla p \right) \tag{8}$$

$$\alpha_p = -\frac{1}{\rho} \frac{\partial \rho}{\partial T} \tag{9}$$

and,

$$Q_{Vd} = \tau : \nabla u \tag{10}$$

The above term represents viscous heat dissipation in the fluid. For the gas density estimation, the model use the ideal gas law, i.e. the thermal expansion coefficient $\alpha_p = 1/T$.

The grid generation was executed resulting in a free tetrahedral meshing which consists of 41640 domain elements. A finer grid is applied for the nozzle and extra-coarse grid for the aluminium plate and the controlled volume of fluid. The chosen computational time stepping is $\Delta t = 1$ min.

3 Discussion of the results

The purpose of this simulation is to highlight the influence of three important hydrodynamic parameters, involved in water spraying cooling process, on heat transfer behaviour of an aluminium plate at a temperature of 92 °C.

Figure 3 illustrate the effect of water mass-flow rate on the temperature evolution, the tested flow values are 0.497, 0.7 and 1 litre/min. The results show a rapid decrease of the plate temperature during the first five minutes. The rest of time there is a very slow decrease of the temperature until t = 60 min, this is due to the insignificant temperature difference between the plate and the impinging water. To disclose the influence of flow variation on heat transfer, an enlargement (Fig. 3) makes it possible to distinguish the flow rate increase effect on the cooling stretch.

Convection heat dissipation is also influenced by water-spray flow, since we notice a decrease in convective heat flux which means the internal energy fall and consequently the diminution of wall temperature (Fig. 4).

From the graphical results obtained on figure 5, it appears that the optimum cooling pressure is 2.1 bars among those applied namely 0.7 bars, 1.9 bars and 2.1 bars. In the current context, it is the pressure that offers the best heat transfer resulting in a drop of the plate temperature, at t = 5 min the temperature value recorded is 53 °C. The decrease in temperature is explained by the heat flux released by forced convection with an average value of approximately 900 W/m^2 during the entire cooling period (Fig. 6). For example, at P = 0.7 bars corresponds a heat release of 17 W/m^2, which represents a derisory value compared to the 1100 W/m^2 evacuated at the optimal pressure P = 2.1 bars.

Figures 7 and 8 aims to investigate the influence of the nozzle-to-target distance on the cooling spray performance, to be done, we considered four different positions: 100, 191, 330 and 505 mm. The role of this parameter appears clearly on the graph since we note the significant influence of the nozzle position on the plate temperature evolution (Fig. 7). In this study, the value H = 505 mm is the optimum height and it is the one that ensures the fastest cooling. The effect of the height also appears on the convection heat flux (Fig. 8), indeed for H = 505 mm the significant surface covered by the spray enhances the convection heat transfer. Obviously, the minimum heat flux is recorded for H = 100 mm.

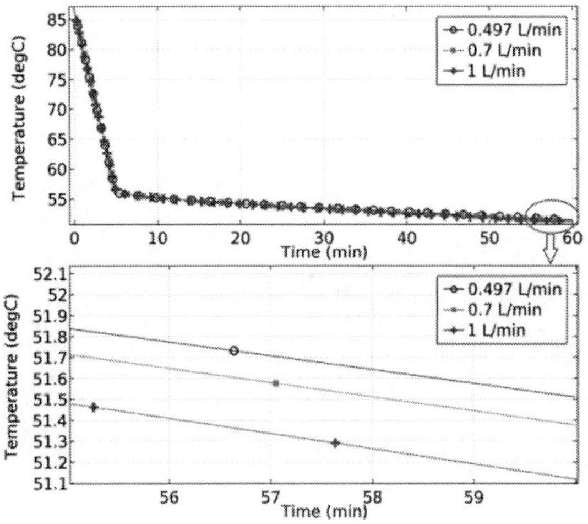

Fig. 3. Plate temperature decrease profile under flow rate variation.

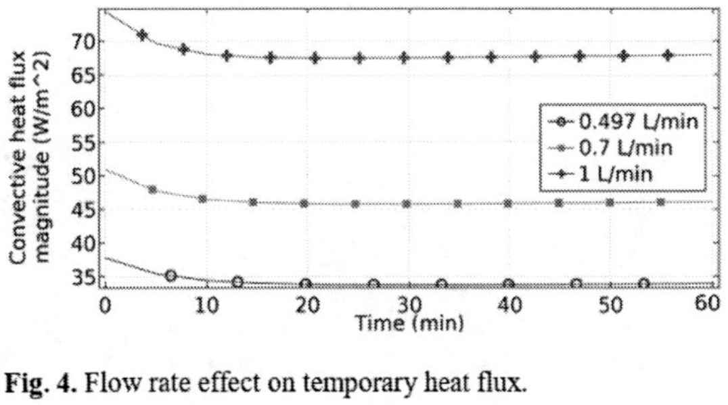

Fig. 4. Flow rate effect on temporary heat flux.

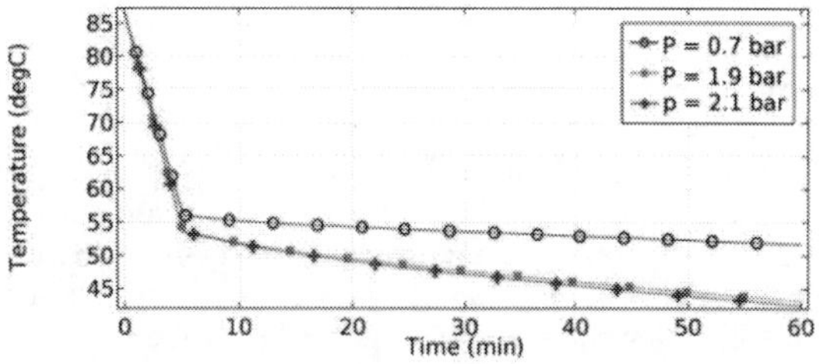

Fig. 5. Evolution of plate temperature under spray pressure influence.

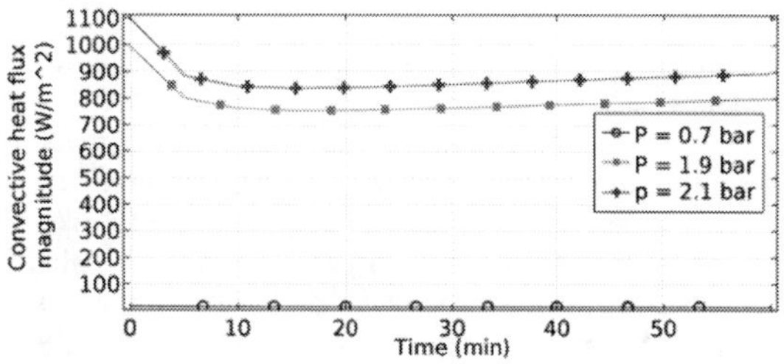

Fig. 6. Influence of spraying pressure on convection heat transfer.

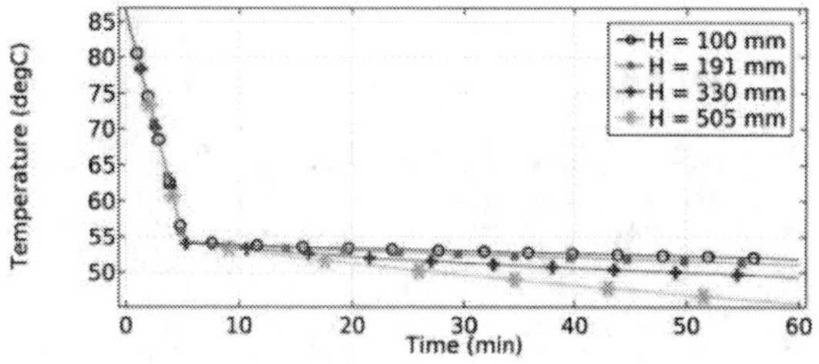

Fig. 7. Effect of nozzle height on plate temperature.

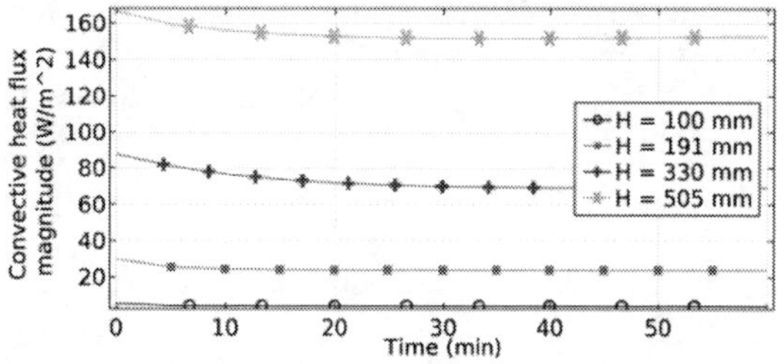

Fig. 8. Convection heat flux evolution with nozzle height.

4 Sum-up and conclusion

This paper deals with influence of spray cooling process parameters on the thermodynamic behaviour of fluid flow and an aluminium plate at the temperature of 92 °C. Unlike previous studies devoted solely to intensive cooling of highly heated surfaces, the present study makes it possible to observe the evolution of thermal parameters during laminar water jet impingement cooling of surface at a temperature below the saturation temperature of the cooling fluid.

The results of the simulation discussed in this paper signal the need to provide more insight for parametric analysis during spray cooling process involving the applications in such case of surface temperature, below 100 °C.

Increase in electrical efficiency depends primarily on cooling techniques. It is well known that a decrease in the PV panel temperature will lead to an increase in electrical efficiency. Besides, high temperatures reduce the time-life of the photovoltaic system. In order to increase the average efficiency in conventional PV panels it is necessary to have more efficient backside thermal dissipation. The backside surface is usually of aluminium made and the analysis presented in this study considering aluminium 3003-H18 cooling allows instantaneously the control and monitoring of photovoltaic panels temperature.

References

1. B.W. Webb, C.F. Ma, Advances in Heat Transfer, **26**, 105-217 (1995)
2. J. Kim, Int. J. of Heat and Fluid Flow, **28**, 753-767 (2007)
3. M. Tebbal, H. Mzad, Forsch. im Ingenieurwesen, **68**, 126-132 (2004)
4. H. Mzad, M. Tebbal, Heat Mass Transfer, **45**, 287-295 (2009)
5. R. Zhao, W.L. Cheng, Q.N. Liu, H.L. Fan, Heat and Mass Transfer, **46**, 821-829 (2010)
6. H. Mzad, M. Elguerri, Atomization and Sprays, **22**, 447-460 (2012)
7. W.L. Cheng, F.Y. Han, Q.N. Liu, R. Zhao, Int. J. Heat Mass Transfer, **55**, 5357-5366 (2012)
8. H. Mzad, R. Khelif, Procedia Engineering, **157**, 106-113 (2016)
9. H. Montazeri, B. Blocken, J.L.M. Hensen, Building and Environment, **83**, 129-141 (2015)
10. S. Nizetic, F. Grubisic-Cabo, I. Marinic-Kragic, A.M. Papadopoulos, Energy, **111**, 211-225 (2016)

Numerical study of the quenching of a laminar premixed hydrogen flame

Björn Pfeiffelmann, and *Ali Cemal* Benim*

Center of Flow Simulation (CFS), Department of Mechanical and Process Engineering, Düsseldorf University of Applied Sciences, Münsterstr. 156, D-40476 Düsseldorf, Germany

Abstract. A numerical analysis of the quenching of a laminar, premixed hydrogen-air flame is presented. A global and a detailed reaction mechanism are considered. First, one-dimensional flame propagation is analyzed and the models are validated based on the predicted flame speed. Subsequently, the quenching near a solid wall of a duct is analyzed, within a two-dimensional, steady-state formulation. Finally, propagation of a flame front through a quenching mesh, within an unsteady, two-dimensional analysis is considered. It is observed that the global mechanism does not predict a quenching of the flame by the mesh, whereas the detailed mechanism does.

1 Introduction

Power generation by thermal machinery, including the gas turbine [1] and steam turbine [2] plants, largely depends on the conversion of the chemically bound energy by the combustion process. Parallel to the efforts for exploiting new energy sources, as well as recovery techniques [3], combustion continues to play an important role in power generation, also through the renewable energies, considering the significance of biomass in the latter.

Combustion of hydrogen and hydrogen-blend fuels plays an especially important role in clean and efficient energy supply, climate protection and resource efficiency. On the one hand, hydrogen represents an attractive alternative to storing excess energy in power generation from wind power or photovoltaics [4]. On the other hand, instead of combustion [5,6] the gasification of biomass and coal is a good possibility for an efficient and clean power generation [7]. The gasification product, the synthesis gas contains, in addition to carbon monoxide and small amounts of methane, significant amounts of hydrogen. Additionally, there is a rapidly growing interest in hydrogen production using nuclear power plants, using the nuclear power for electrolysis, thermochemical cycles or hybrid approaches [8]. From an environmental point of view, the subsequent combustion of hydrogen is most welcome because it produces no carbon dioxide when burned.

Combustion of hydrogen has relevance also for safety engineering. Hydrogen has a high diffusivity and its leakage is more likely to be an issue compared to other gases. Hydrogen plays also a special role in conjunction with the safety of nuclear power plants. In nuclear reactors, metal components are cooled by light or heavy water. As consequence, the cooled metal parts corrode at some rate in the presence of intense radiation, which is accompanied by hydrogen gas production. Existence of hydrogen gas added to the liquid or steam coolant induces excess pressure in the steam line and has further negative effects such as deteriorating the heat transfer and the phase change processes, as well as embrittlement and swelling of the nuclear metals, thus, reduction in the power output and shortening of the reactor life. Thus, as a part of normal reactor operation, slightly radioactive hydrogen-steam mix is released from a cooler part of the line and captured for further treatment, which creates a safety issue in case of an accident.

Use of hydrogen or hydrogen blend gases as fuel in combustion systems represents a great challenge. Hydrogen is extremely reactive and, compared to other gases, has very different material properties, so that even relatively small proportions of hydrogen can greatly alter the combustion properties of the gas mixture.

Therefore, the combustion of fuel gas mixtures with hydrogen fractions requires new combustion chamber concepts. In power generation by combustion, the modern technology that is, in the meantime, state-of-the-art for land-based gas turbine combustors is the so-called "lean premixed combustion", due to its high potential for achieving very low nitrogen oxide emissions. A principal problem in premixed combustion systems is the so-called "flashback" [9]. This means an undesirable flame propagation upstream, towards the inner parts of the burner, which cause serious damage [9].

A classical cause of flashback is an imbalance between the local flow velocity and flame speed, in the favor of the latter, which may be caused by a change/disturbance in the operation of the burner. This

* Corresponding author: alicemal@prof-benim.com

aspect makes the use of hydrogen fuel in premixed combustion to become an especially great challenge, since the high reactivity of hydrogen considerably increases the flame speed and flashback propensity [9]. Thus, effective measures are needed to prevent an undesirable flame propagation in the combustion of hydrogen containing fuels. This is equally important as far as the above-mentioned safety issues are considered.

A measure to guard against undesirable flame propagation is the use of a so-called "quenching mesh" [9]. Here, although it is not the only cause, a primary cause that prevents the flame propagation can be considered to be the heat loss to the confining walls. A decrease of the temperature in the flame front retards the combustion reactions, and, thus, the flame propagation. A termination of the combustion reactions, as a result, is called "quenching" and the temperature that this occurs is called "quenching temperature". The range, within which the flame is quenched near the solid wall is termed as "quenching distance".

Thus, a solid mesh in the flow path, with mesh sizes smaller than the quenching distance can completely extinguish an upward propagating flame front and prevent flashback, which is the main idea behind the concept of "quenching mesh".

Although experimental values for quenching distances can be found in the literature [10], there is a need for a better understanding of the phenomenon with respect to the influence of different conditions such the geometry, mixture composition, fluid dynamic and thermodynamic conditions [11]. Until recently a number of further experimental [12] and theoretical [13] studies were performed on the subject. Computational investigations [14,15] are, however, still rare and bound to simplifications leaving room for further exploration.

In the present study, the flow, heat and mass transfer processes for a hydrogen flame in the presence of a quenching mesh is investigated computationally. In the present study, a laminar flame is considered, first, concentrating on the chemical kinetics, and, excluding the additional challenges related with the modelling of turbulence [16-20], and of turbulence-chemistry interactions [21-22], at the present stage.

2 Modelling

In the present study, a hydrogen-air mixture under atmospheric pressure is considered, which is mixed in a stoichiometric ratio (i.e. the equivalence ratio φ is unity, φ=1.0) [23].

The continuity equation, the Navier-Stokes equations, the energy equation and species transport equations are solved for subsonic and laminar flow of the chemically reacting mixture, assuming an ideal gas behavior [23,24]. Buoyancy is neglected. The radiative heat transfer [25] is also omitted, as gas radiation is obviously weaker compared to radiation encountered in flames containing liquid droplets or solid particles such as pulverized coal flames [5]. Considering the multi-component diffusion, the Soret and Dufour effects are also neglected, in the species transport and energy equations, respectively [23,24].

For the computational modelling, the finite volume method based, general-purpose CFD code ANSYS Fluent 18.0 [26] is used as the basic software platform. For the velocity-pressure coupling, the SIMPLEC [27] and PISO [28] schemes are used for steady-state and unsteady formulations, respectively. For unsteady calculations, a second-order accurate backward differencing scheme is used for the time discretization [29], whereas the time step-size is selected to assure the cell Courant number to be smaller than unity. For the discretization of the convective terms, a second-order accurate upwind scheme [30] is used for structured grids, whereas for the structured grids the QUICK scheme [31] is used, which is formally third-order accurate. In all cases grid independence studies are performed and grid independence is ensured.

The chemical kinetics software Cantera [32], is also employed in one-dimensional flame calculations for analyzing reaction mechanisms and verifying their implementation in Fluent [26].

2.1 Material properties

Laminar flame propagation, which is the key process in the present problem strongly depends on the material properties. Therefore, an accurate modelling in this respect is attempted. For all species, the specific heat capacities are represented by a pair of fourth order polynomials of temperature [33], one for the low (300K<T<1000K) and the other for the high (1000K<T<5000K) temperature range. The viscosities, thermal conductivities, and the multi-diffusion diffusion coefficients of all species are calculated according to the kinetic theory [33-35]. The mixture properties are calculated according to the mixing laws [23], based on the local composition.

2.2 Reaction mechanisms

Reaction rate coefficients are expressed by the Arrhenius equation [23]. For a general, reversible reaction between the two general reactants M_1 and M_2 (to form M_3 and M_4)

$$n_1 M_1 + n_2 M_2 \leftrightarrow n_3 M_3 + n_4 M_4 \quad (1)$$

where n_1, n_2, n_3, n_4 denote the stoichiometric coefficients, the rate constant of the forward reaction (k_f) is expressed as

$$k_f = A\, T^n \exp(-E_a/R_u T)\, [M_1]^{n1}\, [M_2]^{n2} \quad (2)$$

In (2), T is the thermodynamic temperature, E_a is the activation energy, R_u is the universal gas constant, A is the so-called pre-exponential factor, n is the so-called temperature exponent and the brackets indicate molar concentrations of the respective components. The rate constant of the backward reaction (k_b) of the reversible reaction is calculated based on the equilibrium constant of the reaction [23]. Obviously, detailed reaction mechanisms [36,37] that consist of elementary reactions are accurate and comprehensive, but, computationally not very convenient due to the usually large number of species involved that cause long computing times. Global and reduced mechanisms [36,37] are computationally more attractive as they comprise a much smaller number of species, allowing much shorter computational times. The penalty paid for this advantage is, nevertheless, a loss in the accuracy and/or universality as well as comprehension.

However, under certain conditions, for specific purposes, the global/reduced mechanisms can also deliver sufficiently accurate results. In the present study, two reaction mechanisms are selected, which, both, consist of reversible reactions. These are the single-step, global mechanism of Marinov et al. [38] and the detailed mechanism of Conaire et al. [39], which are summarized in Table 1 and Table 2, respectively.

Table 1. Global H_2/O_2 mechanism of Marinov et al. [38] (units: cm^3, mol, s, kcal, K)

	Reaction	A	n	E_a
1	$2\,H_2 + O_2 = 2\,H_2O$	$1.80\cdot10^{13}$	0.00	34.98

2.3 Wall quenching

Quenching of the flame by a wall is a complex phenomenon. Basically, two mechanisms have been stated, so far, to be responsible for this to occur [40]: I) Heat loss to the cold wall, II) Disturbance of the radical pool by the existence of the wall.

The opinion on the relative importance of these effects varies among the researchers [40]. The influence of the details of flow field in the wall boundary layer via flame stretch and curvature is, nevertheless, is addressed by neither of the two criteria. In the present study, mainly, the heat loss effect will be investigated. The modelling of the influence of the wall on the radical pool in the boundary layer is not a straightforward issue and will be investigated in the future studies.

3 Results

3.1 One-dimensional laminar flame

For the flashback phenomenon, the flame speed is a key parameter. As a basis validation, the propagation speed of an undisturbed (planar, unstretched) one-dimensional flame ($S_L{}^0$) is calculated with the global (Table 1, [38]) and detailed (Table 2, [39]) reaction mechanisms.

The predictions for a stoichiometric flame under standard conditions are compared with measurements in Table 3. One can see that the predicted flame speed by the detailed mechanism is well in the range of measured values, whereas the global mechanism overpredicts only slightly. The predicted profiles of temperature are presented in Figure 1. One can observe that the temperature profile is predicted much sharper by the global mechanism, compared to the detailed one.

Table 2. Detailed H_2/O_2 mechanism of Conaire et al. [39] (units: cm^3, mol, s, kcal, K)

	Reaction	A	n	E_a
	H_2/O_2 Chain Reactions			
1	$H + O_2 = O + OH$	$1.91\cdot10^{14}$	0.00	16.44
2	$O + H_2 = H + OH$	$5.08\cdot10^{4}$	2.67	6.292
3	$OH+H_2 = H + H_2O$	$2.16\cdot10^{8}$	1.51	3.43
4	$O + H_2O = OH + OH$	$2.95x10^{6}$	2.02	13.40
	H_2/O_2 Dissociation/Combination Reactions			
5[a]	$H_2 + M = H + H + M$	$4.57\cdot10^{19}$	-1.40	105.1
6[a]	$O + O + M = O_2 + M$	$6.17\cdot10^{15}$	-0.50	0.00
7[a]	$O + H + M = OH + M$	$4.72\cdot10^{18}$	-1.00	0.00
8[b]	$H+OH+M = H_2O+M$	$4.50\cdot10^{22}$	-2.00	0.00
	Formation and Consumption of HO_2			
9[c]	$H +O_2+M = HO_2 +M$	$3.48\cdot10^{16}$	-0.41	-1.12
	$H + O_2 = HO_2$	$1.48\cdot10^{12}$	0.60	0.00
10	$HO_2 + H = H_2 + O_2$	$1.66\cdot10^{13}$	0.00	0.82
11	$HO_2 + H = OH + OH$	$7.08\cdot10^{13}$	0.00	0.30
12	$HO_2 + O = OH + O_2$	$3.25\cdot10^{13}$	0.00	0.00
13	$HO_2 +OH = H_2O +O_2$	$2.89\cdot10^{13}$	0.00	-0.50
	Formation and Consumption of H_2O_2			
14	$HO_2+HO_2 = H_2O_2+O_2$	$4.20\cdot10^{14}$	0.00	11.98
	$HO_2+HO_2 = H_2O_2+O_2$	$1.30\cdot10^{11}$	0.00	-1.629
15[d]	$H_2O_2+M=OH+OH+M$	$1.27\cdot10^{17}$	0.00	45.5
	$H_2O_2 = OH + OH$	$2.95\cdot10^{14}$	0.00	48.4
16	$H_2O_2 +H = H_2O +OH$	$2.41\cdot10^{13}$	0.00	3.97
17	$H_2O_2 +H = H_2 +HO_2$	$6.03\cdot10^{13}$	0.00	7.97
18	$H_2O_2 +O = OH +HO_2$	$9.55\cdot10^{6}$	2.00	3.97
19	$H_2O_2+OH=H_2O+HO_2$	$1.00\cdot10^{12}$	0.00	0.00
	$H_2O_2+OH=H_2O+HO_2$	$5.80\cdot10^{14}$	0.00	9.56

a Efficiency factors: H_2O =12.0; H_2 = 2.5
b Efficiency factors: H_2O =12.0; H_2 = 0.73
c Troe param.: a=0.5, $T^{***}=10^{-30}$, $T^{**}=10^{30}$, $T^{*}=10^{100}$
Efficiency factors: H_2O =14.0; H_2 = 1.3
d Troe param.: a=0.5, $T^{***}=10^{-30}$, $T^{**}=10^{30}$
Efficiency factors: H_2O =12.0; H_2 = 2.5

Table 3. Predicted and measured $S_L{}^0$ (m/s) for stoichiometric flame under standard conditions.

	Global mech. (Table 1)	Detailed mech. (Table 2)	Exp. [38]
$S_L{}^0$ (m/s)	2.8	2.3	2.0–2.5

3.2 Flame stabilized in a channel

As next step, a steady-state, two-dimensional formulation is applied. A laminar, stoichiometric, premixed H_2/Air flame stabilized in the entrance region of a channel is considered. The channel has an inlet height of 6 mm, and a length of 8 mm. For enhancing flame stabilization, a diffuser-like, slightly diverging channel is considered, with an expansion angle of 4°. The boundaries of the solution domain are the symmetry plane, channel wall, inlet and outlet boundaries. At the

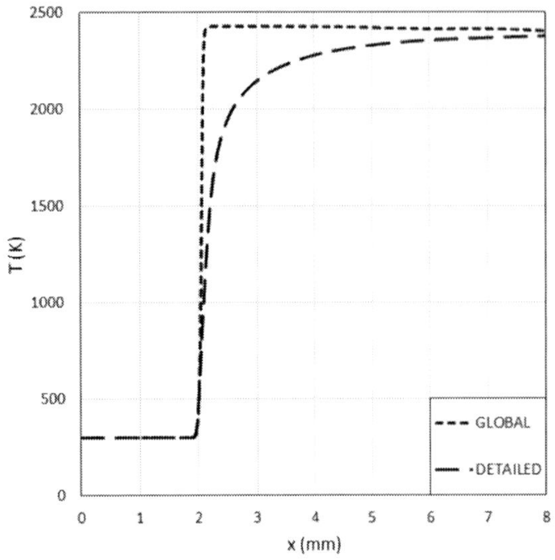

Figure 1. Temperature profiles for 1D flame.

inlet the stoichiometric mixture enters into the channel with a velocity slightly higher than the laminar flame speed, and a temperature of 300 K. The wall temperature is also set to 300 K. The predicted fields of the static temperature (in K) by the two reaction mechanisms are shown in Figure 2.

The temperature boundary layer near the cold diffuser wall can be seen in the figure. The low temperatures lead to a retardation of the combustion reactions and a local quenching. The present configuration is different from the experimental configurations at which the quenching distances are measured [10,40].

However, one can still see the qualitative behavior of different reaction mechanisms in predicting the quenching distance. One can see in the figure that the quenching effect by the cold wall is predicted to a stronger extent by the detailed mechanism (Fig. 2b) compared to the global one (Fig. 2a).

An indication for the implied quenching distance can be deduced by applying a rather simple criterion suggested by Zeldovich (cited in [15]), based on a single-step global mechanism, where the local flame temperature at which quenching shall occur (T_q) is estimated by

$$T_q = T_{ad} - \frac{T_{ad}^2}{E_a / R_u} \tag{3}$$

where T_{ad} denotes the adiabatic flame temperature. The implied quenching distances (δ_q) that result by applying this criterion to both solutions (Fig. 2) at the flame front are presented in Table 4 (using the activation energy, E_a, of the global mechanism reaction from Table 1, for the both cases). One can see that the quenching distance implied by the detailed mechanism is in the range of experimental values that have been usually observed.

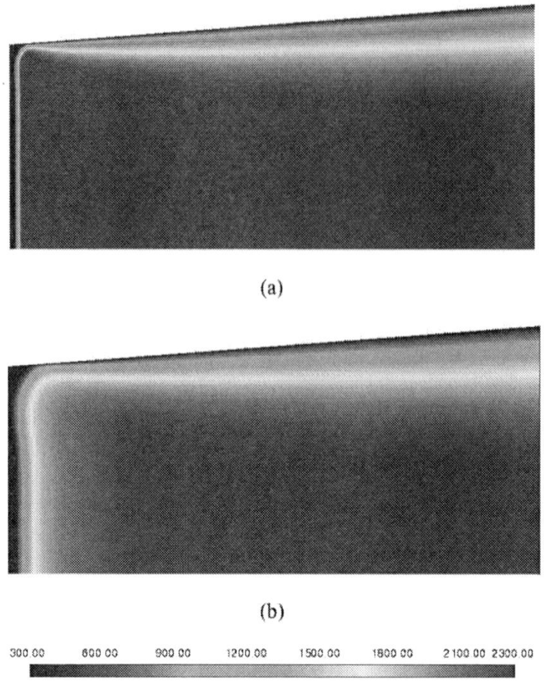

Figure 2. Predicted fields of static temperature (in K) for flame stabilized near the inlet of a diverging duct (a) global mechanism (Table 1), (b) detailed mechanism (Table 2)

Table 4. Implied quenching distances.

	Global mechanism (Table 1)	Detailed mechanism (Table 2)
δ_q (mm)	0.1	0.6

3.3 Flame passing a quenching mesh

As a further step, flame propagation through a quenching mesh is analyzed within an unsteady, two-dimensional formulation. A quenching mesh consisting of cylindrical wires is considered. The configuration is the same as the one analyzed by Kudriakov et al. [15], using a different global mechanism. The wire diameter (D) is assumed to be 0.3, which is also taken to be the spacing between the wires. Using the symmetries in the configuration, the solution domain is defined as shown in Figure 3.

The initially non-reacting, quiescent stoichiometric H_2/Air mixture at atmospheric pressure and 300 K in the domain is instantaneously ignited and the flame propagation in the domain is calculated. The initial ignition zone is also indicated in the figure.

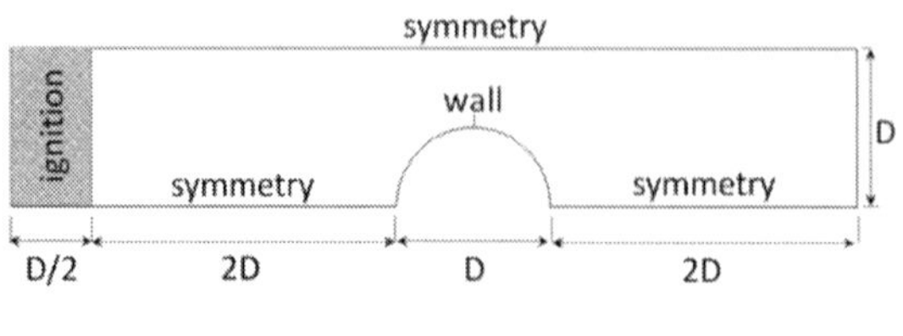

Figure 3. Solution domain

 https://doi.org/10.1051/matecconf/201824001031

The cylinder wall is kept at 300 K during the process. It is arguable that the cylinder does not remain at its initial temperature but get heated during the process. However, using an estimated convective heat transfer coefficient based on the laminar flame speed, material properties of steel and applying the block capacity method (which is applicable as the Biot number is well below 0.1) one can calculate the estimated temperature change of the wire to be less than 1K within the time required the flame front to pass over the wire. Thus, the assumption of a constant wall temperature can be seen to be realistic.

In generating the computational mesh, principally an equidistant mesh distribution is applied. The mesh size is set to be 0.005 mm everywhere and in some places (e.g. on the cylinder surface) slightly smaller. Looking at the flame structure shown in Fig. 1, one can deduce that this mesh size provides a sufficient resolution to resolve the flame front (this was additionally ensured by a grid independence study for the present case). The mesh consists mainly of triangular cells, which are complemented by a layer of quadrangles along the cylinder wall. The total number of cells is above 40,000, which is nearly four times finer than the one used by Kudriakov et al. [15].

The temperature fields calculated by the global (Table 1) and detailed reaction (Table 2) mechanism for the instant of 125 μs after ignition are presented (in K) in Figure 4. According to the global mechanism (Fig. 4a) the flame front propagates towards the quenching mesh (from left to right) and can pass through the mesh, indicating that the mesh (which has a gap width of 0.3 mm) is not able to protect the unburnt mixture behind the mesh from ignition. However, according to the detailed reaction mechanism (Fig. 4b) reactions slow down and the flame gets quenched, implying that the quenching mesh is able to protect the unburnt mixture.

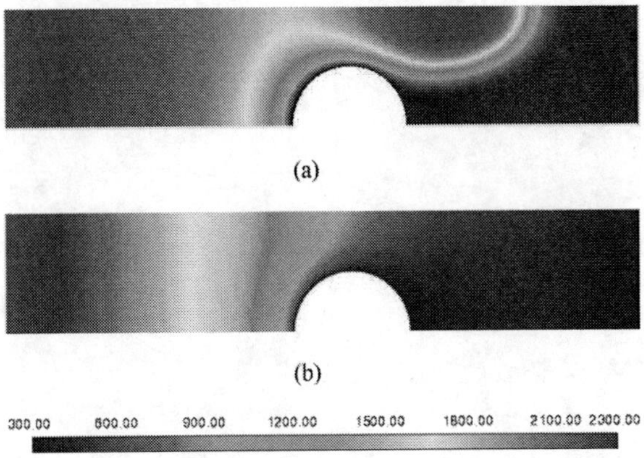

Figure 4. Predicted temperature fields 125 μs after ignition (a) global mechanism (Table 1) (b) detailed mechanism (Table 2)

4 Conclusions

A numerical analysis of the quenching of a premixed hydrogen-air flame is presented. The analysis is confined to laminar flames. As combustion mechanism, a global, and a detailed reaction mechanism are considered. First, one-dimensional flame propagation is analyzed and the computational models are validated by comparing the predicted flame speed with the measured values. Subsequently, the quenching near a solid wall of a duct is analyzed, within a two-dimensional, steady-state formulation. Finally, propagation of a flame front through a quenching mesh with a gap width of 0.3 mm is investigated, within an unsteady, two-dimensional analysis. It is observed that the global mechanism does not predict a quenching of the flame, by the mesh, whereas the detailed mechanism predicts a quenching. More detailed studies will be performed in the future work.

References

1. A. C. Benim, D. Brillert, M. Cagan, ASME Paper No. GT2004-54151, in: Proc. ASME Turbo Expo 2004, Vienna, Austria, **4**, pp. 453-460 (2004)
2. A. C. Benim, M. Geiger, S. Doehler, M. Schoenenberger, H. Roemer, *Modelling the flow in the exhaust hood of steam turbines under consideration of turbine-exhaust hood interaction*, (in: Proceed. 1st European. Conf. Turbomachinery - Fluid Dynamic and Thermodynamic Aspects: Computational. Methods, Erlangen, Germany, Book Series: VDI Berichte, Vol.1185, pp.343-357, VDI Verlag, Duesseldorf, 1995)
3. D. G. Ebling, A. Krumm, B. Pfeiffelmann, J. Gottschald, J. Bruchmann, A. C. Benim, M. Adam, R. Labs, R. R. Herbertz, A. Stunz, Journal of Electronic Materials, **45**, 7, pp.3433-3439 (2016)
4. M. Götz, J. Lefebvre, F. Mörs, A. McDaniel Koch, F. Graf, S. Bajohr, R. Reimert, Th. Kolb, Renewable Energy, **85**, pp.1371-1390 (2016)
5. A. C. Benim, B. Epple, B. Krohmer, Progress in Computational Fluid Dynamics, **5**, 6, pp.345-361 (2005)
6. J. P. Kim, U. Schnell, G. Scheffknecht, A. C. Benim, Progress in Computational Fluid Dynamics – An International Journal, **7**, 6, pp.337-346 (2007)
7. S. De, A. K. Agarwal, V. S. Moholkar, B. Thallada (Eds.), *Coal and Biomass Gasification*, (Springer, Berlin, Germany, 2018)
8. IAEA, "Hydrogen Production Using Nuclear" *Energy*, IAEA Nuclear Energy Series No. NP-T-4.2, Internat. Atomic Energy Agency, Vienna (2013)
9. A. C. Benim, K. J. Syed, *Flashback Mechanisms in Lean Premixed Gas Turbine Combustion*, (Academic Press, Cambridge, MA, USA, 2015)
10. B. Lewis, G. von Elbe, *Combustion, Flames and Explosions of Gases*, 3rd ed., (Academic Press, Orlando, USA 1987)
11. H. Philips, Proc. Roy. Soc., **7**, pp.129-135 (1963)
12. S. Y. Yang, S. H. Chung, H. J. Kim, Nuclear Engineering and Design, **224**, pp.199-206 (2003)

13. S. W. Hong, J. H. Song, Journal of Nuclear Sci. Techn., **50**, pp.1213-1219 (2013)
14. Z. B. Song, L. J. Wei, Z. Z. Wu, Z. Z., Comb. Sci. and Techn., **180**, pp.264-278 (2008)
15. S. Kudriakov, E. Studer, C. Bin, International J. Hydrogen Energy, **36**, pp.2555-2559 (2011)
16. J. L. Xia, B. L. Smith, A. C. Benim, J. Schmidli, G. Yadigaroglu, Computers & Fluids, **26**, 8, pp.811-823 (1997)
17. A. C. Benim, A. Nahavandi, K. J. Syed, K.J., Progress in Computational Fluid Dynamics – An International Journal, **5**, 8, pp.444-454 (2005)
18. A. C. Benim, M. P. Escudier, A. Nahavandi, A. K. Nickson, K. J. Syed, F. Joos, International Journal of Numerical Methods for Heat & Fluid Flow, **20**, 3, pp.348-370 (2010)
19. H. Chattopadhyay, A.C. Benim, Journal of Heat Transfer – Transactions of the ASME, **133**, 10, Article Nr.: 104502, 5 pages (2011)
20. P. Oclon, S. Lopata, M. Nowak, A. C. Benim, Progress in Comput. Fluid Dynamics, **15**, 5, pp.290-306 (2015)
21. A. C. Benim, S. Iqbal. W. Meier, F. Joos, A. Wiedermann, Applied Thermal Engineering, **110**, pp.202-212 (2017)
22. S. Iqbal, A. C. Benim, S. Fischer, F. Joos, D. Kluß, A. Wiedermann, Journal of Thermal Science, **25**, 5, pp.460-469 (2016)
23. S. R. Turns, *An Introduction to Combustion*, 3rd ed., (McGrawHill, New York, USA, 2012)
24. R. B. Bird, W. E. Stewart, E. N. Lightfoot, *Transport Phenomena*, 2nd ed., (Wiley, New York, USA, 2002)
25. A. C. Benim, Computer Methods in Applied Mechanics and Engineering, **67**, 1, pp.1-14 (1988)
26. www.ansys.com
27. J. P. Van Doormaal, G. D. Raithby, Num. Heat Trans., **7**, pp.147-163 (1984)
28. R. I. Issa, J. Computational Physics, **62**, pp.66-82 (1986)
29. H. K. Versteeg, W. Malalasekera, *An Introduction to Computational Fluid Dynamics – The Finite Volume Method*, 2nd ed., (Pearson, London, UK, 2007)
30. T. J. Barth, D. C. Jespersen, *The design and application of upwind schemes on unstuctured meshes*, AIAA-89-0366 (1989)
31. B. P. Leonard, Comput. Meths. Appl. Mechanics and Eng., **8**, pp.59-98 (1979)
32. www.cantera.org
33. R. J. Kee, F. M. Rupley, J. A. Miller, *The Chemkin thermodynamic data base*, Sandia Report, SAND87-8215B (1991)
34. R. A. Shevla, *Estimated viscosities and thermal conductivities of gases at high temperatures*, NASA Technical Report R-132 (1962)
35. R. C. Reid, J. M. Prausnitz, B. E. Poling, *The Properties of Gases and Liquids*, 4th ed., (McGraw-Hill, New York, USA, 1987)
36. P. Gerlinger, *Numerische Verbrennungssimulation*, (Springer, Berlin, Germany. 2005)
37. V. P. Zhukov, International Scholarly Research Network, Article ID 475607, 11 pages (2012)
38. N. M. Marinov, C. K. Westbrook, W. J. Pitz, *Detailed and global chemical kinetics model for hydroge*", Report No. UCRL-JC-120677, Lawrence Livermore National Laboratory, California, USA (1995)
39. M. O. Conaire, H. J. Curran, J. M. Simmie, W. J. Pitz, C. K. Westbrook, International Journal of Chemical Kinetics, **36**, pp.603-622 (2004)
40. H. C. Barnett, R. R. Hibbard (Eds.), *Basic considerations in the combustion of hydrocarbon fuels with air*, NACA Report 1300 (1959)

Numerical analysis of liquid jet impingement cooling of a thermoelectric generator

Björn Pfeiffelmann[1,*], *Ali Cemal* Benim[2], and *Franz* Joos[2]

[1]Center of Flow Simulation (CFS), Dept. Mechanical and Process Engineering, Düsseldorf University of Applied Sciences, Germany
[2]Laboratory of Turbomachinery, Helmut Schmidt University, Hamburg, Germany

Abstract. Designs of heat exchangers are quite often disconnected to the performance of thermoelectric generators (TEG). In this work, the TEG and the heat exchanger are numerical modelled simultaneously in a computational fluid dynamics (CFD) environment (OpenFOAM) to maximize the output power of the system while minimize the hydraulic power required. A preliminary work was done where the modelling of the heat exchanger, a single laminar slot jet, and the modelling of a 16 element TEG are validated. The considered heat exchanger is a laminar slot jet consists of a linear array of discrete heat sources which accord with the geometry of a thermoelectric generator. The considered 16 element TEG is modelled using the temperature dependent material properties which require a solution of a system of nonlinear differential equations, namely the conservation of energy and the conservation of electric current. The conjugate heat transfer OpenFOAM solver chtMultiRegionFoam is extended by an additional differential equation for the solid region to model the conservation of current. The conservation of energy is expanded by additional source terms based on Peltier/Thomson effect and Joule heat. To simplify the calculation, interface and 1D resistor load boundary conditions are developed and implemented. The heat exchanger and the TEG model, both, are validated by comparisons with measurements, where a good agreement is observed.

Nomenclature

$\vec{E}$ Electric field intensity vector, V/m²
$\vec{J}$ Electric current density vector, A/m²
R Resistance, Ω

Greek Symbols
α Seebeck coefficient, V/K
σ Electric conductivity, S/m
φ Electric scalar potential, V

Subscripts

C cold
H hot
n n-type material
p p-type material

1 Introduction

One of the key answers to increase the overall efficiency of industrial processes is the waste heat recovery. The energy emission by industrial processes is recognized as low-grade waste heat in many circumstances. This implies that the temperature is often lower than 600K. It is nearly impossible to recover this kind of waste heat with conventional heat to electricity conversion systems, like steam or gas turbines. A conversion process which can deal with low-grade waste heat is the Organic Rankine Cycle (ORC) [1-3], but it is rather complex and the involved turbine contain rotary parts [4] which, in general, needs a high-maintenance effort.

The thermoelectric generator (TEG) is a waste heat recovery system, which doesn't contain rotary parts and, thus, needs a low maintenance effort. The today's efficiency of around 5% for the commercial TEGs is a drawback of this technology. Nevertheless, there is an ongoing process in the material research to increase the efficiencies of TEGs and cost optimization of the TEG systems to implement them in various processes optimally with a minimum amortization time. The heat management system around the TEG is a further necessary development, e.g. heat source or cooling, which would decrease the TEG efficiency significantly, if not working in an optimal manner. Even though the TEG cooling system should provide an adequate cooling to achieve a large temperature difference across the TEG and to allow the TEG to work efficiently, the cooling costs should be low to achieve high overall system efficiency. There are diverse approaches to cool a TEG. One approach is the use of the natural convection, where the cooling does not need any power, and, thus, does not influence the system efficiency negatively [5-8]. The alternative approach is the use of forced convection. It is important to minimize the cooling costs for this approach, in order to preserve high system efficiency. For forced convection, there are mostly two working

* Corresponding author: bjoern.pfeiffelmann@hs-duesseldorf.de

fluids, which are normally considered for practical applications, i.e. air [9] and water [6, 10-13]. Though an amount of electric energy needs to be invested to cool by the forced convection, it is generally observed [9-13] that a higher overall TEG system efficiency can still be realized in comparison to the natural convection systems.

A constant temperature boundary condition is mostly assumed for the hot and cold side surfaces of the TEG in the practical development of TEG systems, which represents an ideal condition for an optimal behaviour of the TEG. A one-dimensional model for the prediction the performance and efficiency of a TEG can be derived with this assumption, under the consideration of temperature independent [14] and temperature dependent [15] material properties. However, the constant boundary temperature assumption of these simplified, one-dimensional models cannot manage the behaviour of most practical fluid systems that reveal multidimensional, irregular geometries, spatial variations in the flow parameters. Hence, multi-dimensional methods [16, 17] are required for being able to perform realistic simulations of practical systems. This is also the goal of the present study, where a prediction instrument for TEG performance with ability to three-dimensional applications is developed.

One of the early investigations on the computational modelling of thermo-electricity is done by Antonova and Looman [16], who modelled the thermo-electric phenomena by using the Finite Element Method (FEM), within the ANSYS software package. Although, the suggested procedure is capable to model in three dimensions, the presented application on the TEG was only two-dimensional. A three-dimensional investigation of a TEG system was presented by Chen et al. [17], where the TEG model was implemented in the commercial computational fluid dynamics software ANSYS Fluent based on the Finite Volume Method (FVM). Nevertheless, the validation of the model in comparison to measurements and other predictions were not very suitable [17]. Baskaya et. al. [18] studied the application of TEG in a rather complex practical case, i.e. to a condensing combi-boiler using the commercial FVM based software ANSYS Fluent. However, their study was indirectly related with the modelling of thermo-electric processes. In the study, six TEGs were considered in the boiler, and a simulation of the hot and cold side of each TEG was performed, without considering thermoelectric effects. Thus, TEGs were modelled simply as heat-conducting parts. However, this work [18] demonstrates the importance of multi-dimensional effects in TEG applications and is therefore a significant contribution. Rabari et al. [15] presented a one-dimensional, analytical model of a TEG, as well as a two-dimensional computational study of a TEG considering temperature dependent material properties based on an in-house code using the FEM. However, the enhancement of the model to include three-dimensional effects and, beyond this, its coupling to the fluid domain do not seem to be straightforwardly feasible, and were not put as a possible future scope [15], which limits the perspectives for the application of the model in complex practical situations. Hu et al. [19] presented a further, recent study on TEG modelling. They presented a study of different operation conditions for a TEG using the commercial software package COMSOL Multiphysics®. The used TEG model, which is based on the FEM, is a part of the general-purpose software COMSOL Multiphysics®. Thus, three-dimensional applications of the model and its coupling with different physical phenomena are possible.

There are two main goals targeted by the present work. The first is the development of a fully three-dimensional TEG model, by properly accounting for all relevant thermo-electric effects, which can readily be coupled with thermo-fluid dynamics of the heating and cooling mediums as well as with the structural mechanics of the solid parts. The second part is a validation study of a laminar slot jet as a preliminary work for the coupling of the thermo-fluid dynamics with the TEG model. The coupling is part of the future work.

A distinguishing feature of the present work is that the model is implemented in a general-purpose, open source C++ toolbox for field operation and manipulation (OpenFOAM) [20], which is principally based on an unstructured FVM that can inherently cope with complex three-dimensional geometries. Coupled with the fluid dynamics and structural mechanics modules of OpenFOAM, a complete multi-physics model of the TEG system can be obtained. Furthermore, the open source platform allows a great amount of freedom in implementing different modelling aspects.

Based on the implemented models for thermo-electricity in OpenFOAM, the before developed solver entitled tegFoam [21] is extended by the ability to handle temperature dependent material properties. The extended solver is again validated by comparisons with the experiments and numerical predictions of Hu et al. [19], who used a commercial package, for a three-dimensional problem. The three-dimensionality results from the spatial distribution of the combined TEG array, whereas each TEG element has a one-dimensional behaviour for its own, in accordance with Hu et al. [19]. Independent from this validation test case, the present model can inherently consider multi-dimensional field distributions within each TEG module.

Furthermore the steady state solver for buoyant, turbulent fluid flow and solid heat conduction with conjugate heat transfer between solid and fluid regions (chtMultiRegionSimpleFoam) is used to model the confined liquid laminar slot jet investigated experimental and numerical by Schafer et. al. [22,23].

2 Modelling

2.1 Validation of extended solver tegFoam

As already mentioned above, the present study deals with the modelling of the thermo-electric processes in the solid phase of the TEG module. The corresponding governing equations are presented in this section. For a steady state conservation of energy and charge of a solid thermoelectric material with temperature independent material properties, the governing equations can be expressed as [24]

$$\nabla \cdot (T\,\alpha(T)\,\vec{J}) - \nabla \cdot (k(T)\,\nabla T) = \vec{J} \cdot \vec{E} \qquad (1)$$

$$\nabla \cdot (\sigma(T)\,\alpha(T)\,\nabla T) + \nabla \cdot (\sigma(T)\,\nabla\varphi) = 0 \qquad (2)$$

with the electric field intensity vector ($\vec{E}$) and electric current density vector ($\vec{J}$) given as

$$\vec{E} = -\nabla\varphi \qquad (3)$$

$$\vec{J} = \sigma(T)\,(\vec{E} - \alpha\nabla T) \qquad (4)$$

Here, the variables T and φ are the resulting fields of temperature and the electric scalar potential, respectively. The coefficients α, k, σ represent the material properties, i.e. the Seebeck coefficient, thermal conductivity, and electric conductivity, respectively.

The differential governing equations are discretized by the FVM by a 2nd order accurate central differencing scheme. The discretized equations (Eq. 1, 2) are solved subsequently, within an iterative procedure based on successive substitution. In the present applications, equidistant, rectangular grids are used. For problems, where the material properties exhibit substantial local variations, the details of the interpolation procedure used for the calculation of diffusion coefficients on cell faces play an important role for the accuracy for a given grid. In the tegFoam solver, the diffusion coefficients are interpolated to the cell face centers by linear interpolation. Higher accuracy could be obtained, if the harmonic mean, which was suggested by Patankar [25], were used. This would enable, in comparison to linear interpolation, the same level of accuracy with a coarser mesh, or a better accuracy with the same mesh. This scheme is not implemented in the tegFoam solver yet. However, grid independence studies have always been performed to ensure a minimal discretization error, while using the linear interpolation scheme.

The applied boundary conditions shall also be discussed in this section. The formulation and implementation of the boundary conditions of the energy conservation equation (Eq. 1) in terms of temperature are quite straightforward to formulate and apply. Principally, boundary conditions of all three kinds can be used. In a coupled treatment of the TEG with the heat exchangers, the temperature distributions on the TEG boundary surfaces will directly result from the solution of the conjugate heat transfer problem, which may also indicate a spatially non-uniform distribution of temperature on these surfaces. Note that the p-type and n-type legs of the TEG are not thermally connected and are subject to hot-side and cold-side temperatures as boundary conditions of their own, which may also differ from one leg to the other.

The formulations of the boundary and interface conditions for the conservation of charge (Eq. 2) are not very straightforward and deserve special attention. They formulate the electric connection between the materials, which ensures that the electric circuit is closed. A special situation arises here, since the legs are connected electrically, which needs to be described by an adequate formulation. The applied formulations for the boundary and interface conditions are described by Pfeiffelmann et al. [21] directly in terms of the discretized equations.

2.2 Preliminary study – slot jet

In the second part of the present study a laminar slot jet will be considered, which is a preliminary work for coupling the tegFoam solver with thermo-fluid dynamics. The continuity equation, the Navier-Stokes equations and the energy equation are solved for laminar, incompressible flow of a Newtonian fluid with constant material properties. For the prevailing small dimensions, turbulent flow is unlikely and assumption of laminar flow is realistic [26-28], in contrast to the majority of applications in forced convection [29,36]. Buoyancy is neglected. The radiative heat transfer [37] is also omitted currently, as the temperatures are rather low.

For the computational modelling, the steady state solver chtMultiRegionSimpleFoam and the transient solver chtMultiRegionFoam, is used out of the basic software platform OpenFoam [20]. For the velocity-pressure coupling, the SIMPLEC [38] and PISO [39] schemes are used for steady-state and unsteady formulations, respectively. For unsteady calculations, a first-order accurate backward differencing scheme is used for the time discretization, whereas the time step-size is selected to assure the cell Courant number to be smaller than unity. For the discretization of the convective terms, a second-order accurate upwind scheme is used. In all cases grid independence studies are performed and grid independence is ensured.

3 Results

3.1 Validation of extended solver tegFoam

As a validation test case, a 16 element Bi2Te3 TEG module is considered, which was experimentally and computationally investigated by Hu et al. [19]. This module consists of 8 p-type and 8 n-type semiconductor legs as illustrated in Figure 1.

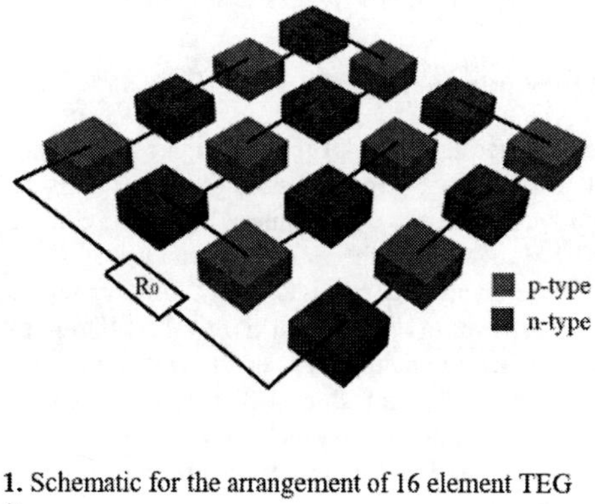

Fig. 1. Schematic for the arrangement of 16 element TEG circuit.

This module was measured and modelled for three cases, where temperatures of 403 K, 453 K and 503 K are applied to the hot surface, while keeping the cold surface temperature constantly at 303K. In [19], temperature dependent material properties (λ, α, σ) are considered by means of the following cubic polynomial fits. These polynomials are also adopted in the present study.

p-type Bi_2Te_3

$$\alpha_p = -7.0396 \cdot 10^{-4} + 6.3023 \cdot 10^{-6}\, T - 1.3273 \cdot 10^{-8} T^2 + 8.3335 \cdot 10^{-12} T^3 \left(\frac{V}{K}\right)$$

$$\sigma_p = 4.0733 \cdot 10^{5} - 2.1322 \cdot 10^{3}\, T + 4.1324\, T^2 - 2.7159 \cdot 10^{-3}\, T^3 \left(\frac{S}{m}\right)$$

$$k_p = 1.5775 - 2.1177 \cdot 10^{-3}\, T - 3.3157 \cdot 10^{-6} T^2 + 1.5888 \cdot 10^{-8}\, T^3 \left(\frac{W}{mK}\right)$$

n-type Bi_2Te_3

$$\alpha_n = 3.9458 \cdot 10^{-4} - 3.8211 \cdot 10^{-6}\, T + 7.3415 \cdot 10^{-9} T^2 - 3.9819 \cdot 10^{-12} T^3 \left(\frac{V}{K}\right)$$

$$\sigma_n = 3.1964 \cdot 10^{5} - 1.3972 \cdot 10^{3}\, T + 2.3388\, T^2 - 1.2792 \cdot 10^{-3}\, T^3 \left(\frac{S}{m}\right)$$

$$k_n = 2.4815 - 4.5128 \cdot 10^{-3}\, T - 4.5960 \cdot 10^{-6} T^2 + 2.1889 \cdot 10^{-8}\, T^3 \left(\frac{W}{mK}\right)$$

The presently predicted electrical power and efficiency as a function of current are compared to the measurements [19] and the COMSOL Multiphysics® predictions [19] in Figure 2 and Figure 3, respectively. It can be observed that the present predictions and the numerical results of Hu et al. [19] for the power show a quite good agreement (Fig. 2). For efficiency (Fig. 3), the present predictions and the numerical results of Hu et al. [19] show a very good agreement. However, both predictions show deviations from the measurements. As Hu et al. [19] argued, this might be caused by errors in predicting the heat flux by neglecting the parasitic radiative heat flow and imperfect soldering.

3.2 Preliminary study – Slot jet

The confined laminar slot jet investigated experimentally and numerically by Schafer et al. [22,23] is considered in the present study as a preliminary work for coupling the tegFoam solver with thermo-fluid dynamics.

In the symmetric slot jet, water flows through a 12.7 mm x 3.18 mm (L_h x B) slot nozzle and impinges on a bottom plate containing five square heat sources (L_h x L_h = 12.7 mm x 12.7 mm) in a 15.87 mm wide channel.

The first heat source was centered directly below the nozzle exit, and the remaining heat sources are placed downstream

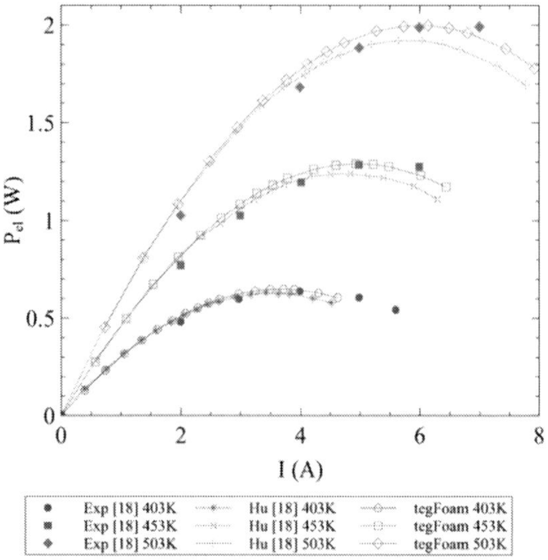

Fig. 2. Electric power (in W) against current (in A) for the modelled 16 element TEG, with T_C=303Km T_H=403K, 453K, 503K. Comparison of the present predictions (tegFoam) with the numerical (Hu) and experiments (Exp) of Hu et al. [19].

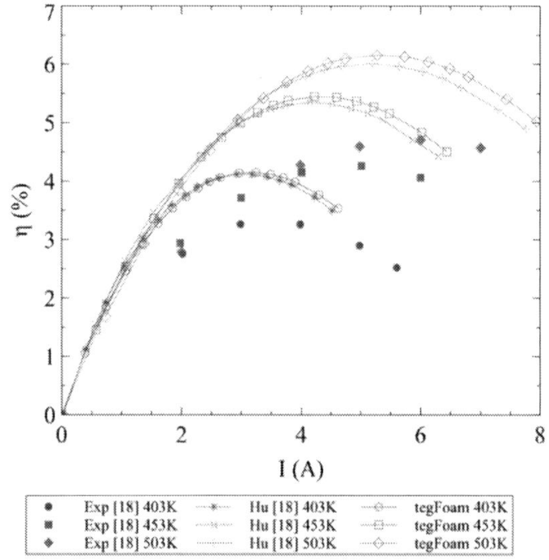

Fig. 3. Efficiency (in %) against current (in A) for the modelled 16 element TEG, with T_C=303Km T_H=403K, 453K, 503K. Comparison of the present predictions (tegFoam) with the numerical (Hu) and experiments (Exp) of Hu et al. [19].

with a spacing of 3.18 mm. The discrete heat sources are cooled by water with a Prandtl number of 7.4. Two different channel height by nozzle width H_C/B are considered, H_C/B=1.5 and H_C/B=1. Simulations are performed in the range of Reynolds numbers from 650 to 4000.

The Nusselt number $Nu_L=hL_h/\lambda$ is averaged over each discrete heat source, where $h=q/(T_h-T_o)$ is the heat transfer coefficient, q the local heat flux, T_e the temperature of the fluid at the nozzle exit and T_h the

https://doi.org/10.1051/matecconf/201824001032

local temperature on the heater surface. The other walls are assumed as adiabatic. For the outlet a constant pressure boundary condition and for the inlet a fully developed velocity profile of a duct flow is assumed.

The presently predicted average Nusselt number for the central heater with H/B of 1.5 and 1 is shown in Table 1.

The present prediction and the experimental results of Schafer et al. are in good agreement for H/B=1.5 and even better for H/B=1. The uncertainties in Re_L and Nu_L for the experiments of Schafer et al. [22] are ± 7 and ± 3 percent, respectively. The computational model slightly underpredicts the values of the experiments.

Tab. 1. Average Nusselt number Nu_L at the central heater with H/B = 1.5. Comparison of the present predictions (chtMRF) with the experimental (Exp) results of Schafer et al. [22]

Re_L	Nu_L		
	Exp [22]	Numerical	% Error
650	64.0	57.9	9.5
920	79.2	69.4	12.4
1000	86.1	74.0	14.1
1300	97.2	83.5	14.1
1800	107.0	98.9	7.6
1900	119.8	102.0	14.9
2570	130.3	117.7	9.7
2600	138.2	118.5	14.3
3300	153.4	133.8	12.8
3600	153.4	139.5	9.1
4000	167.5	146.4	12.6

Conclusions

The new solver for OpenFOAM, tegFoam, was extended to model thermoelectric effects with temperature dependent material properties. The solver was validated for 16 element TEG modules.

The results obtained by the developed solver is observed to show a good agreement for current and voltage and a fairly good agreement for heat flow with the experiments [19] and an overall good agreement with the predictions of the commercial finite element solver COMSOL Multiphysics® [19]. It is expected that the accuracy will be increased by the implementation of radiation [37], as well as the harmonic interpolation scheme [25] in the new solver, which will be performed in the future work.

Additionally, numerical simulations have been performed to determine flow and heat transfer of a confined laminar slot jet driven by liquids. The prediction of OpenFoam solver chtMultiregionFoam shows a good agreement with the experiments performed by Schafer et al. [22].

The further modelling step will be the combined modelling of the cold side laminar slot jet together with the TEG module.

References

1. L. Wang, A. P. Roskilly, R. Wang, Heat Transfer Engineering, **35**, 11-12, pp.1028-1034 (2014)
2. A. Kaya, M. Lazova, Ö. Bagci, S. Lecompte, B. Ameel, M. DePaepe, Heat Transfer Engineering, **38**, 11-12, pp.1018-1033 (2017)
3. Y. Dai, D. Hu, Y. Wu, Y. Gao, Y. Cao, Heat Transfer Engineering, **38**, 11-12, pp.990-999 (2017)
4. A. C. Benim, M. Geiger, S. Doehler, M. Schoenenberger, H. Roemer, *Modelling the flow in the exhaust hood of steam turbines under consideration of turbine-exhaust hood interaction*, in: Proceed. 1st European. Conf. Turbomachinery - Fluid Dynamic and Thermodynamic Aspects: Computational. Methods, Erlangen, Germany, March 1-3, 1995, Book Series: VDI Ber., Vol.1185, pp.343-357 (VDI Verlag, Duesseldorf, 1995)
5. C. Wang, C. Hung, W. Chen, Energy, **39**, 1, pp.236-245 (2012)
6. P. Mayer, R. Ram, Nanos. Micros. Thermophysical Eng., **10**, (2), pp.143-155 (2006)
7. G. Li, G. Zhang, W. He, J. Ji, S. Lv, X. Chen, H. Chen, Energy Conversion and Management, **112**, pp.191-198 (2016)
8. R. Y. Nuwayhid, A. Shihadeh, N. Ghaddar, Energy Conversion and Management, **46**, 9-10, pp.1631-1643 (2005)
9. M. F. Remeli, A. Date, B. Orr, L. C. Ding, B. Singh, N. D. Affandi, A. Akbarzadeh, Energy Conversion and Management, **111**, pp.147-157 (2016)
10. A. Rezania, L. A. Rosendahl, Journal of Electronic Materials, **41**, 6, pp.1298-1304 (2012)
11. Ding, L. C., Meyerheinrich, N., Tan, L., Rahaoui, K., Jain, R., Akbarzadeh, A., Energy Procedia, **110**, pp. 32-37 (2017)
12. R. Stobart, M. Wijewardane, Z. Yang, Applied Thermal Eng., **112**, pp.1433-1444 (2017)
13. D. G. Ebling, A. Krumm, B. Pfeiffelmann, J. Gottschald, J. Bruchmann, A. C. Benim, M. Adam, R. Labs, R. R. Herbertz, A. Stunz, Journal of Electronic Materials, **45**, 7, pp.3433-3439 (2016)
14. S. W. Angrist, *Direct Energy Conversion*, 3rd ed. (Cambridge University Press, Cambridge, 1996)
15. R. Rabari, S. Mahmud, A. Dutta, M. Biglarbegian, Heat Transfer Engineering, **36**, 17, pp.1458-1471 (2015)
16. E. E. Antonova, D. C. Looman, *Finite elements for thermoelectric device in ANSYS* (Proc. 24th International Thermoelectric Conference, Clemson, SC, USA, pp.200-203, 2005)
17. M. Chen, A. Rosendahl, T. A. Condra, International Journal of Heat and Mass Transfer, **54**, 1-3, pp.345-355 (2011)
18. S. Baskaya, S. Karaaslan, T. Calisir, M. Z. Yilmazoglu, T. O. Yilmaz, Heat Transfer Engineering, **36**, 14-15, pp.1292-1302 (2015)

19. X. Hu, H. Takazawa, K. Nagase, M. Ohta, A. Yamamoto, J. Electronic Mat., **44,** 10, pp.3637-3645 (2015)
20. The OpenFOAM Foundation. Retrieved July 27, from http://www.openfoam.org/ (2018)
21. B. Pfeiffelmann, A. C. Benim, F. Joos, Heat Transfer Engineering, **40,** 17 (2019)
22. D. Schafer, F. P. Incropera, S. Ramadhyani, Journal of Electronic Packaging, **113**, pp.359-366 (1991)
23. D. Schafer, S. Ramadhyani, F. P. Incropera, Numerical Heat Transfer, Part A: Applications, **22**, pp.121-141 (1992)
24. C. Goupil, C. (Ed.), *Continuum Theory and Modelling of Thermoelectric Elements* (Wiley-VCH, Weinheim, Germany, 2016)
25. S. V. Patankar, *A numerical method for conduction in composite materials, flow in irregular geometries and conjugate heat transfer* (Proc. 6th Int. Heat Transfer Conf., Vol. 3, pp. 297-302, Toronto, 1978)
26. A. M. Maqableh, A. F. Khadrawi, M. A. Al-Nimr, S. A. Ammourah, A. C. Benim, Progress in Computational Fluid Dynamics - An International Journal, **11**, 5, pp.318-328 (2011)
27. A. M. Maqableh, S. A. Ammourah, A. F. Khadrawi, M. A. Al-Nimr, A. C. Benim, Canadian Journal of Physics, **90, 1**, pp.83-89 (2012)
28. S. A. Ammourah, A. C. Benim, A. M. Maqableh, A. F. Khadrawi, M. A. Al-Nimr, JP Journal of Heat and Mass Transfer, **7**, 1, pp.17-34 (2013)
29. A. C. Benim, A. Nahavandi, *A computational analysis of turbulent swirling flows*, in: K. Hanjalic, Y. Nagano. M. J. Tummers (Eds.) *Turbulence, Heat and Mass Transfer*, vol. 4, pp.715-722 (Begell House, Danbury, Connecticut, 2003)
30. A. C. Benim, M. Cagan, A. Nahavandi, E. Pasqualotto, *RANS predictions of turbulent flow past a circular cylinder over the critical regime* (Proc. 5th IASME/WSEAS International Conference on Fluid Mechanics and Aerodynamics, Athens, Greece, August 25-27, 2007, pp.232-237, 2007)
31. A. C. Benim, D. Brillert, M. Cagan, ASME Paper No. GT2004-54151, in: Proc. ASME Turbo Expo 2004, **4**, pp. 453-460 (2004)
32. A. C. Benim, M. Cagan, D. Gunes, International Journal of Thermal Sciences, **43,** 8, pp.725-732 (2004)
33. A. C. Benim, K. Ozkan, M. Cagan, D. Gunes, Int. J. Num. Methods for Heat & Fluid Flow, **17,** 3, pp.284-301 (2007)
34. H. Chattopadhyay, A.C. Benim, Journal of Heat Transfer – Transactions of the ASME, **133,** 10, Article Nr.: 104502, 5 pages (2011)
35. A. C. Benim, H. Chattopadhyay, A. Nahavandi, International Journal of Thermal Sciences, **50,** 10, pp. 1973-1983 (2011)
36. P. Oclon, S. Lopata, M. Nowak, A. C. Benim, Progress in Comput. Fluid Dynamics – An International Journal, **15,** 5, pp.290-306 (2015)
37. A. C. Benim, Computer Methods in Applied Mechanics and Engineering, **67,** 1, pp.1-14 (1988)
38. J. P. Van Doormaal, G. D. Raithby, Numerical Heat Transfer, **7**, pp.147-163 (1984)
39. R. I. Issa, Journal of Computational Physics, **62**, pp.66-82 (1986)

The flow boiling heat transfer coefficient determination in a minichannel used the FEM combined with Trefftz functions

Magdalena Piasecka[1*], *Kinga* Strąk[2], and *Beata* Maciejewska[3]

[1,2] Faculty of Mechatronics and Mechanical Engineering
[3] Faculty of Management and Computer Modelling
Kielce University of Technology, Al. 1000-lecia P.P. 7, 25-314 Kielce, Poland

Abstract. The method of solving the inverse heat conduction problem, by means of the FEM with Trefftz-type basis functions, during flow boiling in a minichannel was shown. This basis functions were constructed with using the Hermite interpolation and Trefftz functions. The aim of the numerical calculations was to determine the local heat transfer coefficient on the basis of experimental data in a horizontally oriented minichannel. The refrigerant flowing along the minichannel (HFE-649 or HFE-7100) was heated by a thin enhanced plate by vibration-assisted laser texturing. The temperature on an outer smooth side of the plate was detected by means of infrared thermography. On the heated wall–fluid contact surface in the minichannel the heat transfer coefficient was obtained from the Robin boundary condition. It was assumed that the temperature distribution in the heated plate was described by the Poisson equation. The unknown values of temperature and temperature derivatives at nodes were computed by minimizing the functional which describes the mean square error of the approximate solution on the boundary and along common edges of neighbouring elements. The results were presented as the heated plate temperature and heat transfer coefficient versus the minichannel length.

1 Introduction

Heat transfer in small channels has been investigated intensively over the past years, particularly for application as cooling electronic components. Due to change of the state during the flow boiling in small channels, it is possible to meet opposing needs at the same time, i.e. receipt a heat flux as large as likely at small temperature difference between the heating surface and the saturated liquid. It is known from the literature that many factors influence flow boiling heat transfer in minichannels. One of them is using boiling liquids of different physical properties.

Ramakrishnan et al. [1] performed the pool boiling using HFE-649 and HFE-7100 fluids in microporous and microfinned heat sinks. Enhanced heat sinks performed better than bare die in experiments using HFE-649 and HFE-7100 under subcooled conditions. Konishi et al. [2] presented a study to FC-72 flow boiling CHF mechanisms encountered at different orientations relative to Earth's gravity when the fluid is supplied as a two-phase mixture. Mancin et al. [3] discussed the effects of various parameters on the single-phase liquid and flow boiling of R134a and R1234ze (E) inside a 5 pores per linear inch copper foam. Dutkowski [4] presented studies on boiling of the refrigerant R134a in minichannels. Mikielewicz et al. [5] performed flow boiling heat transfer and flow condensation for data collected from literature for such fluids as R404a, R600a, R290, R32, R134a, R1234yf and other. Bohdal et al. [6] conducted the investigation of flow structures during HFE-7100 refrigerant condensation in pipe minichannels.

The data from the experiments using several refrigerants were the basis for the analysis of steady-state flow boiling heat transfer in minichannels with smooth and enhanced heated surface performed by the authors of this paper in their previous works [7–10].

The numerical calculations involves determining the heat transfer coefficient by solving the boundary inverse heat conduction problem [11] using the FEM with the Trefftz functions and the Hermite interpolation [12]. These functions exactly satisfying the governing equation and are suitable for solving both direct and inverse problems. More information on the Trefftz function can be found in [13–18].

2 Experiment

The test section with a single rectangular minichannel (1, Fig. 1, Fig. 2) is the essential part of the main flow loop at the experimental stand shown in Fig. 1.

The refrigerant (HFE-649 or HFE-7100, 3M) flowing along the minichannel was heated by a thin plate enhanced by vibration-assisted laser texturing in contact with the fluid. Infrared thermography was used to determine changes in the temperature on the outer smooth side of the plate. The other side of the channel

* Corresponding author: tmpmj@tu.kielce.pl

with a glass panel allowed observing the vapour-liquid flow patterns on the enhanced heated plate side. K-type thermocouples and pressure meters were installed in the inlet and outlet of the minichannel. The research equipment comprised: a data and image acquisition system with digital cameras, a computer with special software, a lighting system and the power supply system.

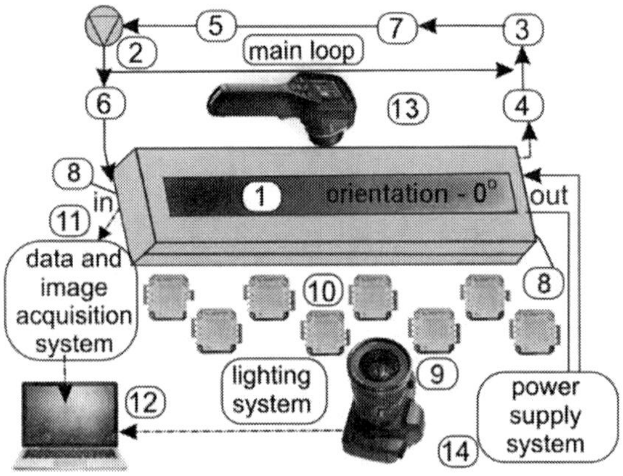

Fig. 1. The schematic diagram of the main systems of the experimental setup, 1- a test section with a minichannel, 2 -a gear pump, 3 - a compensating tank/a pressure regulator, 4 - a tube-type heat exchanger, 5 - a filter, 6 - a mass flowmeter, 7 - a deaerator, 8 - a pressure converter, 9 - a quick shot camera, 10 - a high power LEDs, 11 - a data acquisition station, 12 - a PC, 13 - an infrared camera, 14 - a power supply system.

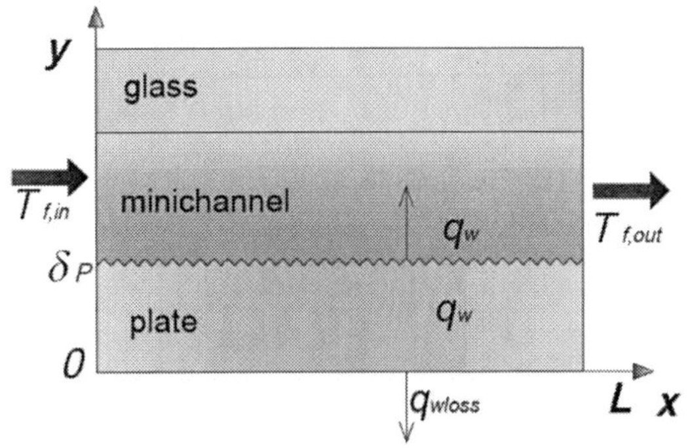

Fig. 2. Schematic diagram of the main elements of the test section with a minichannel.

3 Analysis and modelling

The aim of the numerical calculations, based on an experimental data, was to determine the heat transfer coefficient during FC-72 flow boiling in a minichannel from the Robin boundary condition:

$$h(x)=\frac{-\lambda_P \dfrac{\partial T_P(x,\delta_P)}{\partial y}}{T_P(x,\delta_P)-T_f(x)} \tag{1}$$

where λ_P – the thermal conductivity coefficient of the plate, δ_P – the plate thickness, T_P – the plate temperature, T_f –the fluid temperature calculated from the assumption of the linear dependence between the the fluid inlet temperature and the fluid outlet temperature in the minichannel, x – the spatial variable related to the flow direction, y – the spatial variable referred to the perpendicular direction to the flow direction and associated with the plate thickness (see Fig. 2).

The plate temperature was the solution of the inverse two-dimensional heat conduction problem in the plate. The plate temperature was described by Poisson's equation:

$$\nabla^2 T_P = -\frac{q_w}{\delta_P \cdot \lambda_P} \text{ for } (x,y)\in\Omega_P , \tag{2}$$

The boundary conditions have the form:

$$T_P(x_k,0) = T_k \text{ for } k = 1,2,\ldots,K \tag{3}$$

$$\frac{\partial T_P}{\partial x}(0,y) = 0, \tag{4}$$

$$\frac{\partial T_P}{\partial x}(L,y) = 0, \tag{5}$$

$$\lambda_P \frac{\partial T_P}{\partial y}(x,0) = q_{wloss} , \tag{6}$$

where heat flux is defined as $q_w = \frac{I\cdot\Delta U}{A}$, I – current intensity, ΔU – the voltage drop accross the plate, A – the surface area of the plate, $\Omega_P = \{(x,y)\in R^2 : 0<x<L,\ \ 0<y<\delta_P\}$ L – the length of a minichannel, K – the number of measurements, T_k – the plate temperature measured by infrared thermography at the boundary $y=0$, q_{wloss} – the heat loss to the surroundings, λ_P, δ_P – defined as for Eq. (1).

The solution method for Eq. (2) is a generalisation of the approach shown in [19] regarding to the one-dimensional heat conduction equation and using the Lagrange interpolation.

The inverse problem (Eqs. 2-6) in the heated plate Ω_P divided into elements Ω_P^j was solved by the FEM with Trefftz-type basis functions $f_{jk}(x,y)$, $g_{jk}(x,y)$, $h_{jk}(x,y)$ based on the Hermite interpolation [20], where j – the element number, k– the basis function number in the j-th element.

The basis functions $f_{jk}(x,y)$, $g_{jk}(x,y)$, $h_{jk}(x,y)$ are a linear combinations of Trefftz functions and they have the following properties in nodes (x_i, y_i)

$$f_{jk}(x_i,y_i)=\delta_{ki}, \quad \frac{\partial f_{jk}}{\partial x}(x_i,y_i)=0, \quad \frac{\partial f_{jk}}{\partial y}(x_i,y_i)=0$$

$$g_{jk}(x_i,y_i)=0,\ \frac{\partial g_{jk}}{\partial x}(x_i,y_i)=\delta_{ki},\ \frac{\partial g_{jk}}{\partial y}(x_i,y_i)=0,$$

$$h_{jk}(x_i,y_i)=0,\ \frac{\partial h_{jk}}{\partial x}(x_i,y_i)=0,\ \frac{\partial h_{jk}}{\partial y}(x_i,y_i)=\delta_{ki},$$

$$i=1,2,...,N \qquad (7)$$

where j – the element number, k– the basis function number in the j-th element, N – number of nodes in the j-th element.

The plate temperature in each element Ω_P^j is approximated by a linear combination of Trefftz-type basis functions and has the form:

$$T_P^j(x,y)=u(x,y)+\sum_{k=1}^{N}\left((a_n-u(x_n,y_n))f_{jk}(x,y)+\right.$$
$$\left.(b_n-u_x'(x_n,y_n))g_{jk}(x,y)+(c_n-u_y'(x_n,y_n))h_{jk}(x,y)\right), \qquad (8)$$

where n is the node number in the whole domain Ω_P, $u(x,y)$ – particular solution of equation Eq. (2), a_n– value of the unknown temperature function at the n-th node of the domain Ω_P, b_n –value of the partial derivative of the unknown temperature function with respect to x at the n-th node of the domain Ω_P, c_n – value of the partial derivative of the unknown temperature function with respect to y at the n-th node of the domain Ω_P, j, k, N– defined as for Eq. (7).

The unknown coefficients a_n, b_n, c_n of linear combination of Eq. (8) were calculated, as in [21], by minimizing the functional whichdescribes the mean square error of the approximate solution at the domain boundaries and along common edges of neighboring elements.

4 Results and discussion

The analysis includes the data from the experiments performed using two refrigerants: HFE-649 and HFE-7100. The experiments were recorded for two heat fluxes supplied to the heated plate under similar stable thermal and flow conditions (at a stationary condition).

Experimental parameters are presented in Table 1.

The results are presented as:

- the heated plate temperature measured by infrared thermography and the distance from the minichannel inlet - Fig. 3,
- the heat transfer coefficient and the distance from the minichannel inlet - Fig. 4,

for two selected values of heat flux: 48 and 65 kW/m^2.

The study focused on heat transfer coefficient identification only in the region of subcooled boiling. The coefficient values were determined by solving the inverse heat conduction problem using the FEM with Trefftz-type basis functions based on the Hermite interpolation. The results obtained on the basis of heated surface temperatures at the inlet and outlet of the minichannel (about 10% from each side) were ignored due to the occurrence of the largest measurement errors when using infrared thermography, as reported in [22].

Table 1. Experimental parameters.

Heat flux density q_w kW/m^2	**Average mass flux G kg /(m^2·s)**	**Average inlet pressure p_{in} kPa**	**Average inlet liquid sub-cooling ΔT_{sub} K**
48	361	108	45
65	428	145	52

Analysis of temperature distributions on the heated surface, shown in Fig. 3, indicated the highest plate temperatures were achieved for HFE-7100 fluid and for the highest heat flux (65 kW/m^2) from 0.08 m to 0.16 m distance from the minichannel inlet, Fig. 3b. For lower heat flux (48 kW/m^2), the highest temperature distribution was obtained for HFE-649 fluid (Fig. 3a) at all distance of the minichannel length. For higher heat fluxes was obtained the highest temperature distributions for both fluids.

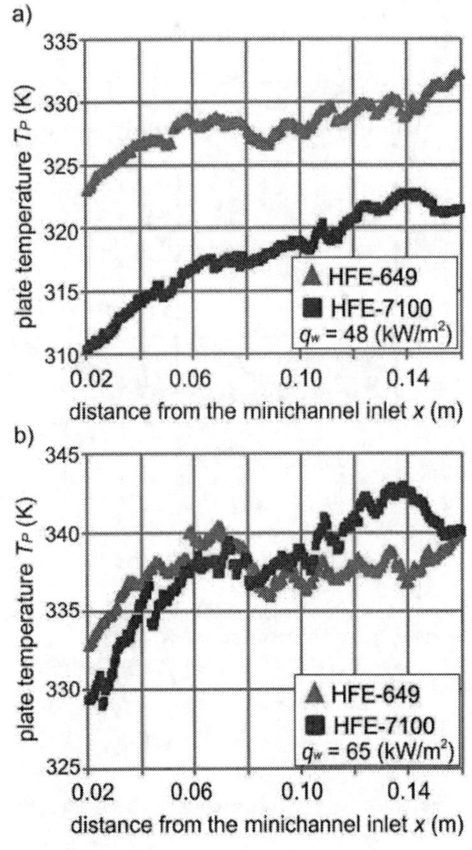

Fig.3. The plate temperature vs. the minichannel length, a) q_w = 48 kW/m^2, b) q_w = 65 kW/m^2; experimental parameters are shown in Table 1.

In the subcooled boiling region heat transfer coefficient values were in the range 1.7 kW/(m^2·K) to 2.6 kW/(m^2·K) for two selected values of the heat flux, Fig. 4. The highest values of the heat transfer coefficient were recorded for HFE-7100 fluid at lower heat flux (48 kW/m^2), Fig. 4a. It was also observed that for heat flux of 65 kW/m^2 higher values of the heat transfer coefficient were recorded for HFE-649 fluid at the distance from 0.05 m from the minichannel inlet to its outlet, Fig. 4b. The heat transfer values of coefficients calculated for both fluids slightly fluctuated with the distance from the minichannel inlet.

It can be noticed that the coefficient values were relatively low in comparison to those at the saturated boiling region, discussed in [7,23,24].

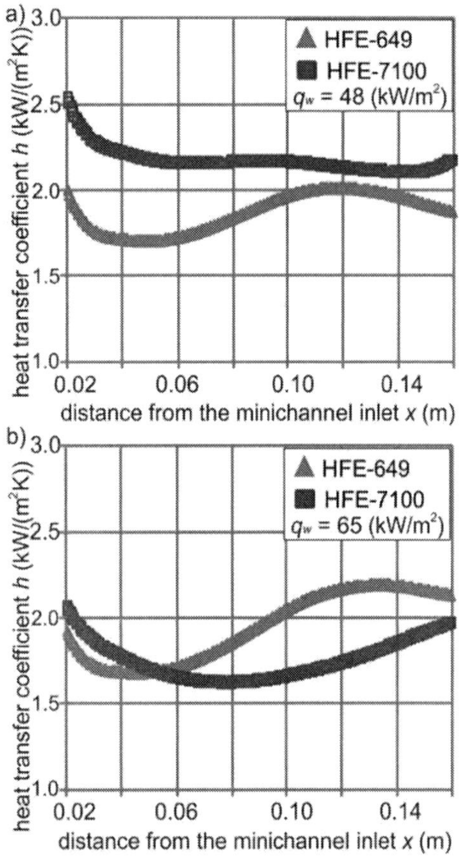

Fig. 4. The heat transfer coefficient vs. the minichannel length, at the subcooled boiling region, a) q_w = 48 kW/m^2, b) q_w = 65 kW/m^2; experimental parameters are shown in Table 1.

5 Conclusions

In this results of flow boiling heat transfer in a minichannel at subcooled boiling region were shown. During experiments two refrigerants (HFE-649 or HFE-7100, 3M) flowing in a single minichannel were heated by a thin enhanced plate. Infrared thermography was used to determine changes in the temperature on its outer side. The method of solving the inverse heat conduction problem by means of the FEM with Trefftz-type basis functions was proposed. This basis functions were constructed with using the Hermite interpolation and Trefftz functions. The heat transfer coefficient, on the heated wall–fluid contact surface in the minichannel, was obtained from the Robin boundary condition. The numerical calculations were performed in the subcooled boiling region. The results were presented as the heated plate temperature and heat transfer coefficient versus the channel length. The calculated heat transfer coefficient values were in the range 1.7 to 2.6 kW/(m^2·K). The highest values of the heat transfer coefficient were recorded for HFE-7100 fluid at lower heat flux. The heat transfer coefficients calculated for both fluids slightly fluctuated with the distance from the minichannel inlet.

References

1. B. Ramakrishnan, S. H. Bhavnani, J. Gess, R. W. Knight, D. Harris, R. W. Johnson, IEEE Semicond. Therm. Meas. Manag. Symp. 6892210 (2014)
2. C. Konishi, I. Mudawar, M. M. Hasan, Int. J. Heat Mass Transf., **65** (2013)
3. S. Mancin, A. Diani, L. Doretti, L. Rossetto, Int. J. Heat Mass Transf., **74** (2014)
4. K. Dutkowski, Int. J. Heat Mass Transf. **53**, 5–6 (2010)
5. D. Mikielewicz, R. Andrzejczyk, B. Jakubowska, J. Mikielewicz, Arch. Thermodyn. **35**, 3 (2014)
6. T. Bohdal, M. Sikora, K. Widomska, A. M. Radchenko, Arch. Thermodyn. **36**, 4 (2015)
7. B. Maciejewska, K. Strąk, M. Piasecka, Procedia Eng., **157** (2016)
8. K. Strąk, M. Piasecka, B. Maciejewska, Int. J. Heat Mass Transf., **117** (2018)
9. M. Piasecka, Metrol. Meas. Sys. **XX**, (2) 205–216 (2013)
10. M. Piasecka, K. Strąk, B. Maciejewska, Heat Transf. Eng. **38**, 3 (2017)
11. J. V. Beck, B. Blackwell, Jr., C. R. St. Clair, *Inverse Heat Conduction* (Wiley – Interscience Publ., New York, 1985)
12. D. Kincaid, W. Cheney, *Numerical Analysis: Mathematics of Scientific Computing* (3rd Ed., Brooks/Cole Publishing Company, Belmont, California, 2002)
13. I. Herrera, Numer. Methods Partial Differ. Equ. **16**, 6 (2000)
14. A. Maciąg, J. Theor. Appl. Mech. **49**, 1 (2011)
15. K. Grysa, B. Maciejewska, J. Theor. Appl. Mech. **51**, 2 (2013)
16. S. Hozejowska, R. Kaniowski, M. E. Poniewski, Int. J. Numer. Methods Heat Fluid Flow **24**, 4 (2014)
17. B. Maciejewska, K. Strak, M. Piasecka, Int. J. Numer. Methods Heat Fluid Flow **28**, 1 (2018)

18. B. Maciejewska, M. Piasecka, Int. J. Heat Mass Transf., **107** (2017)
19. M. Cialkowski, J. Therm. Sci. **11**, 2 (2002)
20. B. Maciejewska, J. Theor. Appl. Mech. **55**, 1 (2017)
21. B. Maciejewska, M. Piasecka, Heat Mass Transf. **53**, 4 (2017)
22. D. Michalski, K. Strak, M. Piasecka, EPJ Web Conf. **143**, 02075 (2017)
23. M. Piasecka, K. Strąk, B. Grabas, Arch. Metall. Mater. **62**, 4 (2017)
24. M. Piasecka, K. Strąk, B. Maciejewska, B. Grabas, J. Phys. Conf. Ser. **745**, 032123 (2016)

Modelling of propane emissions from a tank containing a liquefied phase

Zdzisław Salamonowicz[1,*] and *Radosław* Makowski[1]

[1]Faculty of Fire Safety Engineering, The Main School of Fire Service, Słowackiego 52/54 St., 01-629 Warsaw, Poland

Abstract. During gas-phase release from the tank containing liquefied gas, the phenomenon of boiling, heat transfer from moist air to the tank, heat exchange between shell, liquid, and vapor, and mass loss can be observed. The aim of this paper is modelling the thermal response of the tank containing liquefied gas during jet emission of the vapor phase. The model takes into account heat exchange between air, tank's shell, liquid phase, vapor phase and mass balance. The proposed model predicts: pressure inside the tank, tank's shell temperature in part with liquid and vapor, the temperature of the liquid phase and vapor phase, and mass loss from the tank. The results of the theoretical model were compared with experimental results. An experiment based on the typical home container for LPG, containing 10 kg of propane (27 dm^3 volume) was conducted. In general, the proposed model predicts well the changes in measured parameters - temperatures and pressure - during vapor phase emission from a tank containing liquefied gas.

1 Introduction

The process of storing and transporting liquefied gases in pressure tanks, from the safety point of view, requires the development of emergency scenarios and risk analysis. Storage tanks for liquefied medium under pressure can be found both in the refinery, energy and transport industries. A special way, as an alternative to natural gas, is the use of propane-butane in home tanks and portable cylinders.

The analysis of phenomena during the opening of a safety valve in a tank containing a liquefied medium is conducted by many researchers [1-7, 11] and is a complex issue requiring the consideration of many variables [8-12]. During the sudden opening of the safety valve or damage to the tank in the upper part, the vapor phase is emitted with the simultaneous process of the liquid boiling. Due to the low boiling point, this process can be very intensive, even with two-phase emission [23]. The modelling of the evaporation of the vapor phase from the tank is based on the mass and energy balance of the liquid-vapor-tank-environment system. The most important issues are related to heat exchange with the environment including condensation of steam with further freezing and mass exchange related to the boiling process and emission through the outlet.

2 Modelling and experiments

2.1. Model

The proposed model of jet outflow of propane from the tank presents changes in thermodynamic parameters during emergency release. The model allows to simulate the change of the following system parameters: temperature of the gas phase, temperature of the liquid phase, temperature of the upper part of the tank wall (in contact with the gas phase), temperature of the lower part of the tank wall (in contact with the liquid phase), mass of the gas phase and mass of the liquid phase. The model analyses the change of the above parameters during jet outflow and immediate ignition. The isentropic-adiabatic discharge takes place through an outlet of a given diameter. The algorithm, depending on the pressure in the tank, includes the throttled and unthrottled flow. Numerical calculations were made using Matlab's - mathematical computation environment.

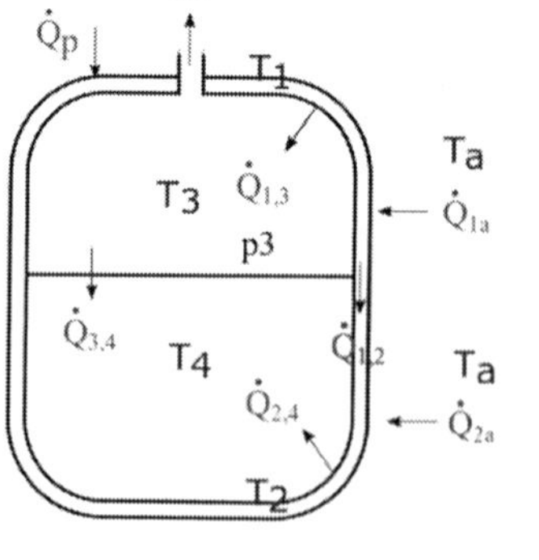

Fig. 1. Propane-tank-environment system.

* Corresponding author: zsalamonowicz@sgsp.edu.pl

Boundary conditions:

- in the initial stage, the liquid phase and the gas phase are in equilibrium,
- the tank-environment system is also in equilibrium, i.e. the initial temperatures: liquid phase, gas phase, tank (both upper and lower part) are equal to the ambient temperature,
- the ambient temperature is constant,
- the flow of heat flux through the walls of the tank is treated as flow through an infinitely thin shell (with the flow of heat the thickness of the tank is not taken into account),
- in calculations, the tank is treated as a homogeneous cylinder,
- the tank's shell has been divided into two parts by an imaginary line, reflecting the contact of the lower and upper parts of the tank with the liquid phase and the gas phase respectively,
- temperatures of the: liquid phase, gas phase, lower and upper part of the tank's shell are constant in the entire volume,
- coefficients used in the model are treated as constant values, their changes are not including changes of the thermodynamic parameters,
- values which are treated as variables in time are: heat exchange coefficients, mass streams: outflow and evaporation, liquid and gas phase density, height of both phases, heat exchange surfaces, heat flux from the environment, heat flux delivered from a fire, heat flux of liquid phase vaporization, flux of heat discharged into the environment from the gas phase, pressure in the tank.

The analysis of parameter changes over time results from the solution of the following system of equations:

$$\begin{cases} Q_1 = \dot{Q}_{1a}t - \dot{Q}_{1,3}t - \dot{Q}_{1,2}t + \dot{Q}_p t \\ Q_2 = \dot{Q}_{2a}t + \dot{Q}_{1,2}t - \dot{Q}_{2,4}t \\ Q_3 = \dot{Q}_{1,3}t - \dot{Q}_{3,4}t - \dot{Q}_{str}t \\ Q_4 = \dot{Q}_{2,4}t + \dot{Q}_{3,4}t - \dot{Q}_{lg}t \\ m_3 = m_{30} - \dot{M}_{str}t + \dot{M}_{lg}t \\ m_4 = m_{40} - \dot{M}_{lg}t \end{cases} \quad (1)$$

which considered in detail takes the form:

$$\begin{cases} \frac{dT_1}{dt} = \frac{h_{1a}A_{1a}(T_a-T_1)t - h_{1,3}A_{1,3}(T_1-T_3)t - h_{st}A_{st}(T_1-T_2)t + \dot{Q}_p t}{m_1 c_{st}} \\ \frac{dT_2}{dt} = \frac{h_{2a}A_{2a}(T_a-T_2)t + h_{st}A_{st}(T_1-T_2)t - h_{2,4}A_{2,4}(T_2-T_4)t}{m_2 c_{st}} \\ \frac{dT_3}{dt} = \frac{h_{1,3}A_{1,3}(T_1-T_3)t - h_{3,4}A_{3,4}(T_3-T_4)t - \dot{Q}_{str}t}{m_3 c_3} \\ \frac{dT_4}{dt} = \frac{h_{2,4}A_{2,4}(T_2-T_4)t + h_{3,4}A_{3,4}(T_3-T_4)t - \dot{Q}_{lg}t}{m_4 c_4} \\ \frac{dm_3}{dt} = -\dot{M}_{str}t + \dot{M}_{lg}t \\ \frac{dm_4}{dt} = -\dot{M}_{lg}t \end{cases} \quad (2)$$

The mass flow rate (mstr) was calculated including the throttled outflow criterion according to the formula [27]:

$$p_3 \geq p_a \left(\frac{\gamma+1}{2}\right)^{\frac{\gamma}{\gamma-1}} \quad (3)$$

The throttled outflow was calculated from the equation:

$$\dot{M}_{str} = A_h p_3 \left[\frac{\gamma M}{RT_3}\left(\frac{2}{\gamma+1}\right)^{\frac{\gamma+1}{\gamma-1}}\right]^{\frac{1}{2}} \quad (4)$$

whereas the unthrottled outflow was calculated by the equation:

$$\dot{M}_{str} = A_h \sqrt{2\rho_3 p_3 \left(\frac{\gamma}{\gamma-1}\right)\left[\left(\frac{p_a}{p_3}\right)^{\frac{2}{\gamma}} - \left(\frac{p_a}{p_3}\right)^{\frac{\gamma+1}{\gamma}}\right]} \quad (5)$$

The heat flux from the jet fire is described by the formula [24]:

$$\dot{Q}_p = \frac{A_{34}\eta\tau_a \dot{M}_{str} H_{comb}}{4\pi x^2} \quad (6)$$

The heat flow needed to evaporate the boiling liquid is calculated on the basis [21]:

$$\dot{Q}_{lg} = A_{34}\mu_4 h_{lg}\left[\frac{(\rho_4-\rho_3)g_z}{\sigma}\right]^{0,5}\left[\frac{c_{p4}(T_2-T_{sat})}{C_{sf}h_{lg}Pr_l^n}\right]^3 \quad (7)$$

The mass flow of evaporation:

$$\dot{M}_{lg} = \frac{\dot{Q}_{lg}}{h_{lg}} \quad (8)$$

The initial masses of gaseous and liquid phases were calculated from the system of equations:

$$\begin{cases} m_{30} + m_{40} = m_{netto} \\ \frac{m_{30}}{\rho_{30}} + \frac{m_{40}}{\rho_{40}} = \pi r^2 H_{zb} \end{cases} \quad (9)$$

The Peng-Robinson equation [28] was used to calculate the liquid-vapor balance:

$$p = \frac{RT}{V-b} - \frac{a\alpha(T)}{V(V+b)+b(V-b)} \quad (10)$$

Heat exchange coefficients between the environment and the upper and lower tank parts were calculated from equations [26]:

$$h_{1a} = 1{,}31(T_1 - T_a)^{\frac{1}{3}} \quad (11)$$

$$h_{2a} = 1{,}31(T_2 - T_a)^{\frac{1}{3}} \quad (12)$$

The heat exchange coefficient between the tank wall and the gas phase was calculated based on the equation [29]:

$$h_{13} = \frac{k_3 C Ra_3^{\frac{1}{3}}}{H_g} \quad (13)$$

where:

$$Ra_3 = \frac{g_z\beta_3(T_1-T_3)H_g^3}{\alpha_3 \nu_3} \quad (14)$$

and the coefficient of volumetric thermal expansion of the gas phase:

$$\beta_3 = \frac{2}{T_1+T_3} \quad (15)$$

Similarly, for the heat exchange coefficient between the tank wall and the liquid phase, the following equation was used:

$$h_{2,4} = \frac{k_4 C Ra_4^{\frac{1}{3}}}{H_c} \quad (16)$$

The coefficient of heat exchange along the steel wall [27]:

$$h_{st} = \frac{k_{st}}{\sqrt{k_{st}g}\left(\frac{1}{\sqrt{h_{1a}+h_{13}}} + \frac{1}{\sqrt{h_{2a}+h_{24}}}\right)} \quad (17)$$

The heat transfer coefficient between the liquid and gas phases was calculated from the relationship between the number of Nussetel, Jakob and Prandtl [27]:

$$h_{3,4} = \frac{k_4 \cdot Ja^2}{d \cdot C_{mb}{}^3 \cdot Pr^l} \quad (18)$$

2.2 Experiments

Standard cylinders with a water capacity of 27 dm^3 containing 10 kg of propane were used for the tests. A head with a valve and inputs for the measurement elements was designed and made, in order to insert a thermocouple and a pressure sensor inside the cylinder. During the experiments, the temperature of the liquid phase, vapor phase, temperature of the outer part of the tank wall corresponding to the liquid and vapor phase, the pressure in the tank, the mass of the system and the outside temperature were recorded. Figure 2 shows the research stand in terms of visible and infrared radiation.

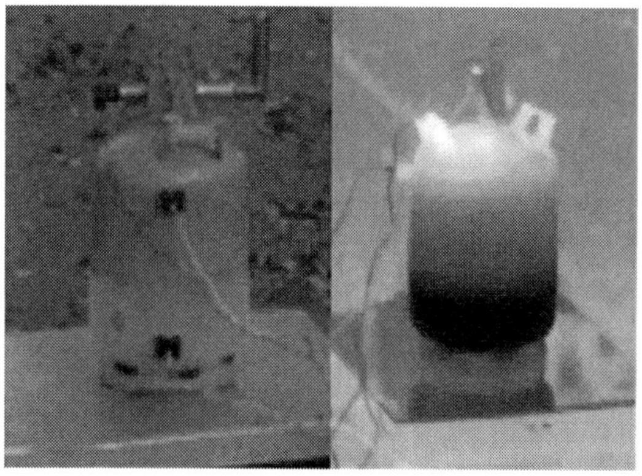

Fig. 2. Research stand – VIS and IR image.

3 Results and discussion

During the outflow of the gas phase from the system which was in equilibrium, the pressure was decreasing in the tank. At lower pressure, the saturation temperature is also lower, as a result of the pressure drop, the liquid is at a temperature above the saturation temperature and starts to boil. The boiling intensity is proportional to the difference between the temperature of the liquid phase and the saturation temperature at the specific pressure. The liquid phase absorbs heat from the system while evaporating. The heat absorbed by the system from the environment is lower than the heat needed to evaporate the liquid phase. Therefore, during the jet outflow, the decrease of temperature in both phases and both parts of the tank can be observed. Figures 3-6 present comparisons of changes in individual temperatures obtained from the proposed model and experimental data. The results are shown for four diameters of the outlet: 6, 5, 4 and 2 mm.

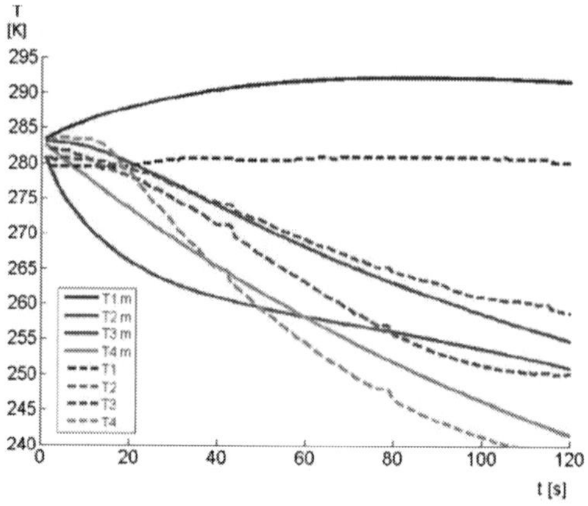

Fig. 3. Comparison of model temperature changes with experimental results for outflow diameter d = 6 mm.

According to the above results, only the temperature of the upper part of the tank (T_1) increases during the jet fire. This is the result of the thermal radiation of the flame (with an initial stream value <750 W) only on this part of the tank when the outflow is directed vertically upwards. The lowest values are obtained for the temperature of the liquid phase (T_4), and therefore the heat required to evaporate the liquid phase is high. The temperature of the lower part of the tank (T_2) decreases in a similar way to the temperature of the liquid phase. The decrease in T_2 is slightly smaller than T_4. The liquid phase absorbs large amounts of heat from the lower part of the shell of the tank. A slightly different course is characterized by gas phase temperature (T_3). Initially, its drop is more rapid than in the case of the liquid phase, then it becomes milder and finally, the temperature of the liquid phase is much lower. The rapid decrease in the initial phase is the result of the discharge of a large amount of heat from the gas phase to the environment. As the time passes, the pressure in the tank drops, so that the flow and speed of the outflow are also reduced.

Temperature measurements carried out during the jet fire tests for outflow through a diameter of 6 mm and an ambient temperature of approx. 280-285K confirm the tendency of temperature changes obtained in the model. In the case of the temperature of the upper part of the tank (T_1), an increase of approx. 10K was noted, which is not confirmed by experimental studies. The difference is due to the fact that the calculation assumes that the flame during combustion is stable, its shape does not change and the influence of wind is excluded [5, 13-17]. The temperature of the lower part of the tank (T_2) obtained as a result of the simulation is very similar to the temperature measured in reality. The model temperature of the liquid phase (T_4) maintains a similar trend to the temperature of the liquid phase during the experimental test. The temperature T_4 obtained from the model is characterized by a constant decrease in time, but in experimental measurement, the drop in temperature T_4 in the first minute is more distinct than in the subsequent time of the experiment. In the case of the gas phase temperature (T_3), convergence in the simulation and actual results were also achieved, however, in the initial phase a difference in the course of these temperatures is visible. The analysis of calculations in the model was adopted for the simulation time of 120 s. The model includes only transition boiling, film boiling was not included. During the film boiling, the evaporation heat flux

decreases significantly. Hence, after reaching the appropriate difference between the temperature of the liquid and the saturation temperature, the temperature drop of the liquid phase slows down and as a consequence the drop of temperatures T_1, T_2 slows down as well and even the increase of the gas phase temperature appears.

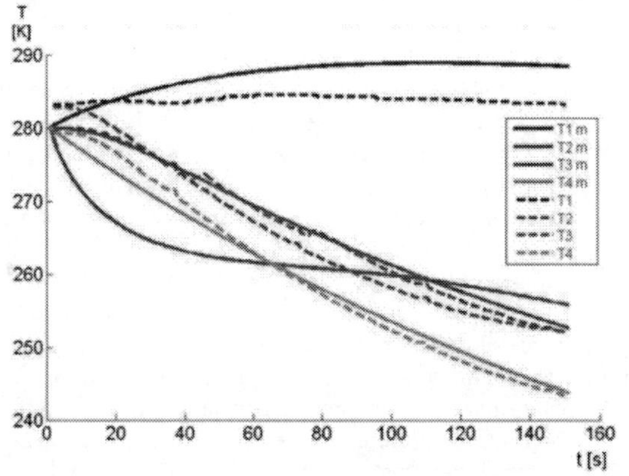

Fig. 4. Comparison of model temperature changes with experimental results for outflow diameter d = 5 mm.

Experimental temperatures for outflow through a diameter of 5 mm and an ambient temperature of approx. 280-285K also confirm the tendency of temperature changes obtained in the model. For this outflow, the results obtained from the model are even more similar to the empirical results. The temperatures of the liquid phase (T_4) and the upper part of the tank (T_2) are almost identical throughout the simulation. The temperature of the gas phase (T_3) obtained as a result of the simulation temporarily differs significantly from the values actually measured, however, it should be noted that the general trend is maintained, and also the temperature values in the final simulation phase are similar. The model temperature of the upper part of the tank (T_1) as well as in Figure 3 is also characterized by an upward tendency, while the temperature T_1 during the experiment remains rather unchanged.

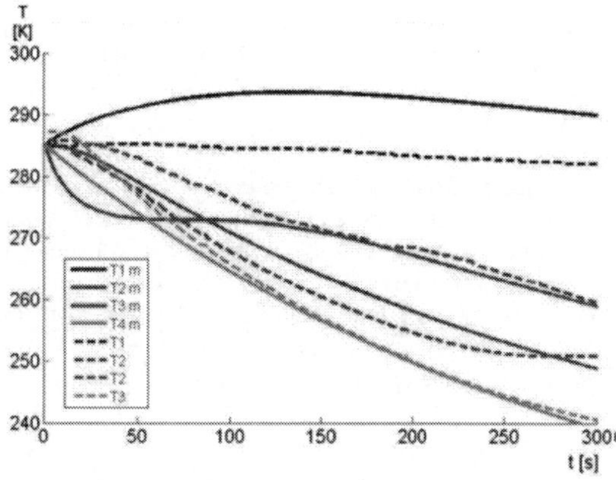

Fig. 5. Comparison of model temperature changes with experimental results for outflow diameter d = 4 mm.

When comparing the results obtained by the Matlab software with experimental results for a 4 mm diameter outflow, it can be seen that the liquid phase temperatures (T_4) in both cases are very similar throughout the simulation. The experimental temperature of the lower part of the tank (T_2), differs from calculations. Influence of external factors, such as air humidity or wind speed might explain those differences. The temperature of the upper part of the tank (T_1) does not differ from the previous cases. As a result of the calculations, a higher temperature was obtained. In the case of gas phase temperature (T_3), the differences between calculations and actual results are more visible.

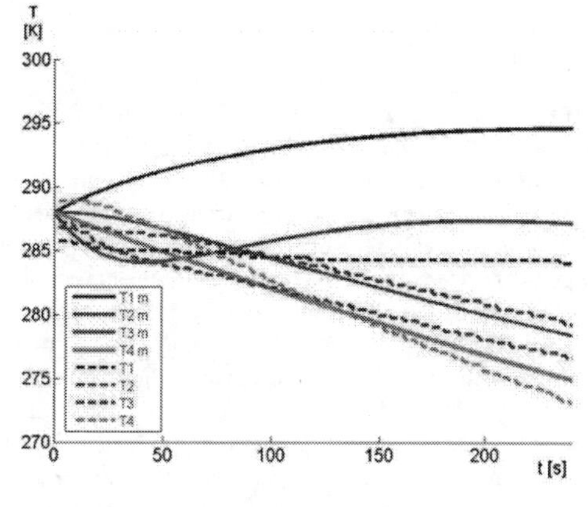

Fig. 6. Comparison of model temperature changes with experimental results for outflow diameter d = 2 mm.

In case of 2 mm diameter outflow, it can be observed again that the temperature of the liquid phase (T_4) and the temperature of the lower part of the tank (T_2) obtained during the simulation is very similar to the experimental results. However, a significant difference was noted for the gas phase temperature (T_3) and the temperature of the upper part of the tank (T_1). The calculated temperature T_1 in the final phase is about 10 K higher than the actual temperature, which is a big difference but comparable to the previous results. Although the temperature of the gas phase (T_3) obtained as a result of the simulation exceeds the temperature T_3 measured during the experiment also by about 10 K, however, it is a very big difference compared to the results for larger diameters. In the case of a 2 mm outflow, the jet was much smaller than in the previous tests, and therefore the evaporation rate was also smaller, as a result of which less heat was absorbed for the liquid evaporation. In the experiment, temperatures T_4 and T_2 decreased slower than in previous cases. As a result of a significant increase in temperature T_1 during the simulation, an excessive increase in temperature T_3 was obtained in comparison to the experimental results.

The results of the temperature change of the liquid phase (T_4) obtained in the model are very similar to the experimental data. In the experiment, during the first seconds, the temperature stays at a constant level, after which in the initial phase the drop is more rapid and in the final stage it is milder. The temperature of the lower part of the tank (T_2) looks similar. In the case of gas phase temperature (T_3), the results differ more. The downward trend is maintained and in the final phase the temperature difference is insignificant, however, in the initial phase, the model has a much faster temperature drop, while in the final part it is slower. In the case of the upper tank temperature (T_1), the difference between the model and the experiment is the most visible. In the model, the temperature increases by less than 10 K, while in the experiment the temperature is rather constant. The model assumes that the flame is stable with a fixed shape and wind speed is not taken into account. In fact, the heat flux coming from the fire was smaller than the one used in the model.

4 Conclusions

Taking into account the obtained results of computational simulations and comparing them with the results of experimental research, the first conclusion concerns the temperature course during the jet outflow without ignition. Under these conditions, all analyzed temperatures drop, i.e. the liquid phase, the gas phase, the upper part of the tank's shell and the lower part of the tank's shell. The temperature of the liquid phase decreases the most, while the decrease in the temperature of the upper part of the tank is the smallest. The drop in temperature of the liquid phase results from a large amount of heat necessary to evaporate the liquid phase. Consequently, the temperature of the lower part of the tank and the temperature of the gas phase is also significantly reduced. As the time passes, the evaporation stream decreases, therefore less heat is required to evaporate the liquid phase, as a result, the temperature of the liquid phase and the lower part of the tank are stabilized, and the temperature of the gas phase even begins to grow slightly. The heat required for evaporation has the greatest effect on the temperature distribution in the system. The first impression while observing the phenomenon of jet fire is that the situation should change dramatically in comparison to the jet outflow without ignition. In practice, it turns out that it is not. The temperature of the liquid phase, the temperature of the gas phase and the temperature of the lower part of the tank are still decreasing. Only the temperature of the upper part of the tank remains more or less constant, but it also does not grow. The thermal radiation from the flame only affects the temperature of the upper part of the tank.

The proposed model of the jet outflow explains the phenomena occurring during a jet fire by applying the principles of thermodynamics and the criterion equations. According to the results obtained from simulations carried out in the Matlab computing environment, the temperature of the liquid phase drops the most during the jet outflow. The slightly smaller decrease is recorded for the temperature of the lower part of the tank and the temperature of the gas phase. The temperature of the upper part of the tank decreases in the smallest degree. During a jet fire, the course of changes in the temperature of the liquid phase and the temperature of the lower part of the tank remains almost unchanged. The decrease in the gas phase temperature is slightly slower, while the temperature of the upper part of the tank slightly increases, which differs from the experimental results. The increase in the temperature of the upper part of the tank is the result of the assumptions regarding the stability and constancy of shape of the flame throughout the simulation and the exclusion of the influence of atmospheric conditions. After performing a series of simulations for jet fires at the outflow with different diameters, it can be concluded that as the diameter increases, temperature changes more rapidly, i.e. they decrease more. The most noticeable difference occurs when comparing the temperature of the liquid phase, while the smallest differences are visible for the temperature of the upper part of the tank. The influence of the ambient temperature on the rate of temperature changes is small, but it should be taken into account that assuming that the system is in equilibrium, the initial temperatures of particular phases and parts of the tank are higher. The proposed calculation model allows simulation of various fire conditions.

Summing up, the results obtained by simulations with experimental tests, it was found that the trends regarding temperature changes of the respective phases and part of the tank as well as the mass loss are correct. In addition, the courses of calculation of temperature changes: liquid phase and the lower part of the tank are very similar to real results. Some differences can be seen in the course of the gas phase temperature, while the temperature of the upper part of the tank in each case is greater during simulation. This difference results from the assumptions that the flame is stable and its shape is unchanged in time. The influence of atmospheric conditions, especially wind speed, is also not taken into account. These assumptions also affect the course of changes in the gas phase temperature.

References

1. M. Bi, J. Ren, B. Zhao, W. Che, Journal of Hazardous Materials **192**, p. 874-879 (2011)

2. X. Zhang, H. Tao, Z. Zhang, J. Liu, X. Liu, Applied Thermal Engineering **132**, p. 801-807 (2018)
3. M. Gomez-Mares, M. Zarate, J. Casal, Fire Safety Journal **43(8)**, p. 583-588 (2009)
4. N. Gopalaswami, Y. Liu, D.M. Laboureur, B. Zhang, M. Sam Mannan, Journal of Loss Prevention in the Process Industries **41**, p. 365-375 (2016)
5. C. Tao, Y. Shen, R. Zong, Applied Thermal Engineering **91**, p. 884-887 (2016)
6. D.M. Laboureur, N. Gopalaswami, Y. Liu, B. Zhang, M. Sam Mannan, Journal of Loss Prevention in the Process Industries **41**, p. 355-364 (2016)
7. B. Zhang, Y. Liu, D. Laboureur, D. Mannan, Ind. Chem. Res. **54 (37)**, p. 9251-9256 (2015)
8. P.S. Cumber, M. Spearpoint, Fire Safety Journal **41 (3)**, p. 215-228 (2006)
9. X.L. Zhang, L.H. Hu, Q. Wang, X.C. Zhang, Y. Jiang, Applied Thermal Engineering **110**, p. 11-114 2017)
10. I.W. Ekoto, A.J. Ruggles, L.W. Creitz, J.X. Li, Int. J. Hydrogen Energy **39**, p. 20570-20577 (2014)
11. K. Zhou, J. Liu, J. Jiang, Applied Thermal Engineering **106**, p. 634-639 (2016)
12. K. Zhou, J. Jiang, J. Heat Transfer **138** (2015)
13. M. Gomez-Mares, M. Munoz, J. Casal, *Axial temperature distribution in vertical jet fires*, J. of Hazardous Materials, **172**, p. 54-60 (2009)
14. X. Zhang, L. Hu, W. Zhu, X. Zhang, L. Yang, Applied Thermal Engineering **73**, p. 15-22 (2014)
15. A. Palacios, M. Munoz, R.M. Darbra, J. Casal, Fire Safety Journal **51**, p. 93-101 (2012)
16. A. Palacios, J. Casal, Fuel **90**, p. 824-833 (2011)
17. L. Hu, Q. Wang, F. Tang, M. Delichatsios, X. Zhang, Fuel **106**, p. 779-786 (2013)
18. M. Gomez-Mares, M. Munoz, J. Casal, Experimental Thermal and Fluid Science **34**, p. 323-329 (2010)
19. D.Y. Krian, D.P. Mishra, Fuel **86**, p. 1545-1551 (2007)
20. S.R. Gollahalli, T.A. Brzustowski, H.F. Sullivan, Transactions to the CSME 1975, **3(4)**, p. 205-214 (1975)
21. Y. A. Cengel, *Heat and Mass Transfer*, McGraw Hill (2006)
22. H. Clewell, Energy Systems Laboratory, ESL-TR-83-03, (1983)
23. P. Cumber, Journal of Hazardous Materials, A89 (2002)
24. *Guidelines for Evaluating the Characteristics of Vapor Cloud Explosions, Flash Fires and BLEVE's* (American Institute of Chemical Engineers, 1994)
25. J. P. Holman, *Heat transfer* (McGraw Hill, 1992)
26. D. Incropera, L. Bergman, *Fundamentals of Heat and Mass Transfer* (Willey, 2006)
27. A. F. Mills, *Heat Transfer* (Prentice Hall, 1998)
28. R. H. Perry and D. W. Green, *Perry's Chemical Engineers' Handbook* (The McGraw-Hill Companies, 1999)
29. P. K. Ramskill, *The development of ENGULF to model a multi-component liquid in a fire engulfed tank* (UKAEA Safety and Reliability Directorate, 1987)

Assessment of gas permeability through porous skeletal media with anisotropic internal structure

Grzegorz Wałowski[1], and *Gabriel* Filipczak[2,*]

[1]Institute of Technology and Life Sciences, Renewable Energy Department, Biskupińska Street 67, 60-463 Poznań, Poland
[2]Opole University of Technology, Faculty of Mechanical Engineering, Department of Process Engineering, Mikołajczyka Street 5, 45-271 Opole, Poland

Abstract. The results of experimental studies in the field of gas permeability of porous materials referring to gas flow have been shown. Coal chars with anisotropic porous-slotted structure have been tested. The studies have been carried out on a special test bench that allows for the measurement of gas permeability in terms of three flow orientations related to symmetrical specimens in the cubic form. The results of the measurements indicate that there is a considerable effect of the flow directivity on the permeability of carbonizers, which results from their integral anisotropic structure. A permeability coefficient for such materials has been defined and the experimental assessment of the value of this coefficient against gas flow and the total pressure drop on a porous medium has been made. The usefulness of computation methods characterising the hydrodynamics of gas flow through porous materials has been made, and the possibility of a numerical mapping of the flow geometry of solid materials with tortuous structure has been indicated.

1 Introduction

Gas flow through media with porous structures takes place in many process areas. It is most often associated with filtration or gas flow through filling layers as a porous medium with a loose composition. This type of flow is also included in the technological processes involving the thermal processing of coal and also during the migration of natural gas (e.g. methane) through natural formations, or the flow of reactive gas through various kinds of coal char, such as coke, activated carbon etc. [1-3]. In some technological processes, the operation of apparatus occurring in them depends on the type and design of gaseous phase dispensers, which are very often made from porous materials [4, 5].

In each case, the recognition of gas flow conditions through a porous media carries significant problems with the description of hydrodynamics and the assessment of the mechanism of gas flow through porous media, especially in the context of their varied internal structure. On the other hand, the knowledge of these mechanisms allows us to assess the process conditions associated with hydrodynamics of gas flow through such types of material, and, consequently, the detailed description of hydrodynamic phenomena accompanying the flow of gas through porous media.

Despite the widespread presence of gas flow in porous materials, in the literature on this area there is still a lack of an unequivocal interpretation and a clear indication of the nature of the hydrodynamic phenomena accompanying this flow. The main reason for this is seen in the very complex and diverse structure of porous materials, which due to the variable flow conditions entails many difficulties in describing these phenomena, but very often also due to the varying scale of the process - from the porous solid grain to the porous layer bed. Although a literature does not lack models that incorporate structural features of porous materials in its description, especially in terms of homogenization theory [6-9], the problem of the influence of the anisotropic structure on permeability of porous materials is still not sufficiently recognized. This situation becomes even more complicated with respect to the hydrodynamics assessment of gas flow through backbone (skeletal) porous materials. At the same time, it is difficult to find information concerning the hydrodynamics of gas flow through such kinds of porous materials. An exemplary comparison of some literature position [10-18] referring on hydrodynamics of gas flow through selected porous beds is characterized in Table 1.

The scope of this work involves the presentation of the investigation results in the field of gas permeability through porous material with skeletal structure. Experimental results have been reported and gas permeability assessment of various kinds of solid (backbone) porous materials has been made. Evaluation of process conditions related to hydrodynamics of gas flow through porous materials with an anisotropic internal structure is carried out. Additionally, the possibilities of numerical modelling techniques to describe the flow structure of porous materials with a skeletal structure are also demonstrated.

*Corresponding author: g.filipczak@po.opole.pl

Table 1. Selected studies referring to investigations upon hydrodynamics of gas flow through porous media.

Author	Kind of gas	Deposit characteristics				
		Kind of material	Deposit height H, m	Mean diameter of deposit D, m	Equivalent diameter of grain / pores d_ε, m	Porosity of sample / deposit ε
Blicharski et. al. [10]	air	rock	0.0254	0.0254	-	0.49
Brauer [11]	air	formed spheres of uniform size	0.50	-	0.0024	0.32
Dyga, Płaczek [12]	air	aluminium foam, 20PPI	1.00	0.020	0.00345	0.93
Ergun [13]	CO_2, N_2, CH_4, H_2	coke	0.95	0.724	-	0.42
Hehlmann [14]	air	fine coke	0.20	-	0.0049	0.61
Mertas et al. [15]	N_2	coke yielded grains	0.08	0.020	-	-
Skotniczy [16]	air	glass balls	0.04	0.110	0.0050	-
Warpechowski, Jopkiewicz [17]	air	coke	0.95	0.380	0.0330	0.42
Wong et al. [18]	air	biochar with clay	0.07	0.070	$3 \cdot 10^{-6}$	0.10

2 Quality of process conditions

Porous medium is characterised by a given porosity and its flow structure depends not only on this porosity, but also on the size (diameter) of tubules (pores) and their shape - at the given length of the pore. Another specific feature of porous materials is associated with their ability to store and transport of fluids as the results of internal and external forces. A study by Aksielrud and Altszuler [19] contains a statement that gas flow through porous media with the pores size of millimetres and less is dominated by the process phenomena resulting from the flow hydrodynamics of viscous fluid whereas in flows through structures with a very small pore size, e.g. tenths of a micrometre, these phenomena are restrained by the physiochemical and diffusive mechanism influences that take place at their interface.

This is confirmed by other studies [20, 21], where the disparity between these phenomena decreases in the conditions when a high intensity of gas movement is maintained. Nonetheless, in each case, the gas flow mechanism is closely related to the geometric structure of porous layer. Therefore, it depends on the configuration and the pore size, as well as on their shape and sinuosity. In the general assumption, two major cases of fluid flow in porous media could be distinguished, as it is shown in Fig. 1. The first one occurs for the case of the flow through a grained layer, while the other one concerns a porous material with a solid (skeletal) structure with empty pores.

The first case (Fig. 1a) involves the flow in the space between the grains and we can assume that the space resulting from the porosity of packed bed layer (ε) is completely available for the fluid flow. Whereas, in the second case (Fig. 1b), the flow occurs only in the area of the pores and channels of open and interconnected, with a surface of flow much smaller than that required by absolute porosity of the material.

An additional complexity of hydrodynamics stems from the fact that skeletal structures are compact (rigid) deposits that are not able to be relaxed during pressure increase in the system. As a consequence, the flow conditions through such structures will be different from each other and the flow deviation will be greater than greater the porous sinuosity and content of closed or blinded pores. Lambe and Whitman [22], among others, point out that the measure of this deviation can be coefficient of pressure drop which value corresponds to the bed porosity (ε), the hydraulic diameter of the pores (d_ε), and the really flow path (L_ε), which results from the sinuosity of the tubular pores.

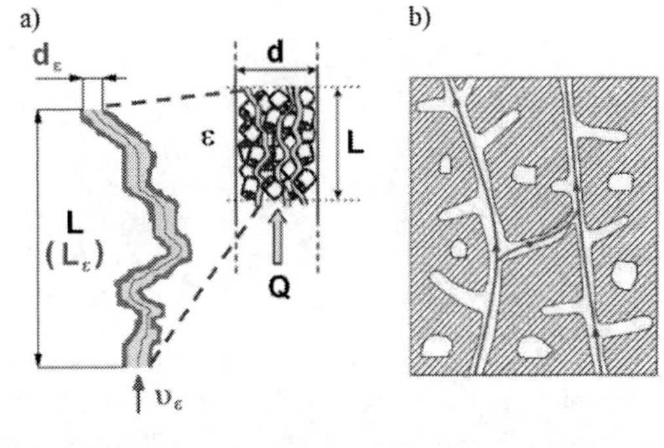

Fig. 1. Flow scheme through a porous bed: a) grain layer with a tortuous channels [22]; b) rigid skeletal structure with open flow channels and blind pores and closed for the flow [23].

The basis for a detailed analysis of fluid flow in porous media is still Darcy law. In its original form, this law describes the conditions for permeability of various kinds of grained bed by reference to the filtration mechanism during laminar flow of water through a sand layer, which constitutes a model grained medium (Fig. 1a). If the variability of liquid properties is take into account, the velocity through a porous bed will be proportional to the change in density (ρ) and inversely proportional to the change in viscosity (η). Then the Darcy equation describing the permeability (Q) of the porous bed takes the following formula

$$Q = K A_o \frac{\rho \hat{g}}{\eta} \frac{\Delta h}{L} \quad (1)$$

where Δh denotes pressure drop, L- height of porous medium, A_o- layer bed cross-section and g- gravitational acceleration.

This formula remains one of the characteristics of the contemporary description of this phenomenon, although it refers (from definition) only to laminar flow.

The K coefficient in equation (1) describes the so-called permeability of a porous medium, and its value, as shown by the Darcy model, is characteristic of a given porous medium. Since this coefficient (by definition) has a surface dimension, its value from a hydrodynamic point of view – as characteristic dimension – is very often considered as a certain geometric feature that characterizes the total permeability of the porous material. On the other hand, the value of such permeability depends not only on filtration characteristics of the porous medium (its structure, particle size, their density, porosity etc.), but also on the physical properties of fluid, especially its viscosity. As a rule, this factor does not depend on the shape and size of the bed itself.

Of course, the Darcy model also applies to the description of pressuring flows. Then for equation (1) we will get

$$Q = K A_o \frac{\Delta P}{\eta L} \quad (2)$$

Therefore

$$K = \eta \frac{Q}{A_o} \frac{L}{\Delta P} \quad (3)$$

The last equation shows that for a given volumetric flow rate (Q) the permeability of a porous bed could be determined by means of experiments if the fluid properties (η) and the geometrical parameters of the flow system (A_o) are known. The pressure drop on the bed is then experimental value.

From physics phenomena point of view, the permeability of the porous medium can be subordinated to various hydrodynamic criteria, depending on the structure of the medium, the type of fluid (single- and multi-phase) and the method of flow forced (gravity, pressuring).

Numerous publications in this respect [1, 23, 24-27], describing this process both from the research and analytical side, *de facto* relates to the filtration phenomenon and are generally associated with the laminar flow of liquids through a granular bed, according to Darcy's law [28]. Naturally, this does not exhaust the many other cases of fluid flow through porous media. For example, for the transitional and turbulent liquid flow Forchheimer study can be distinguished [29-30] or Ergun [13] and Brinkman's research [31]. A more advanced description of flow for a spatial system of capillaries in the form of a bundle of sinuous tubules may be found in the studies by Kozeny-Carman [1, 32].

A characteristic comparison between the ranges of the applicability of Darcy, Forchheimer and Brinkman models is shown in Figure 2, which taken from the Hansen study [33] From the sequence of function describing the gas permeability versus a pressure drop in a porous medium it follows that outside of the Darcy area (A) where for laminar flow linear dependence of this function exists, in other areas referring to the transitional and turbulent flow (B and C), this function is nonlinear. In addition, if fluid motion is influenced only by the diffusion phenomenon (close to the smallest permeability value), in the laminar region the description of the permeability is also nonlinear. It follows that the greater turbulence of flow, the inertia forces has a more significant effect on flow resistance. The scope of applicability of the fundamental models for the granular layers resulting from these premises relative to the Reynolds number is characterized in Table 2.

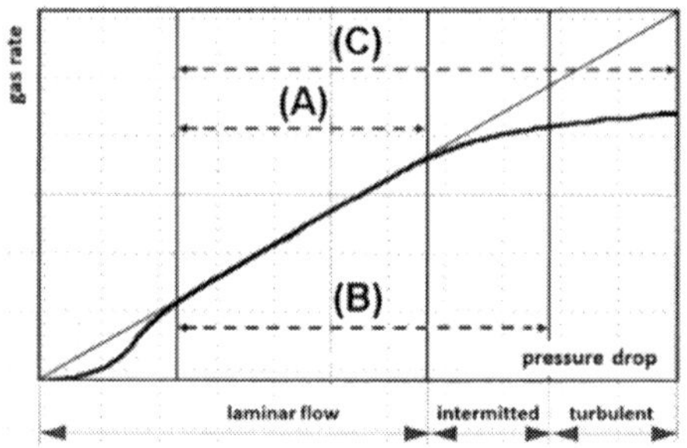

Fig. 2. General characteristics of flow conditions through porous beds: (A) - Darcy area, (B) - Forchheimer area, (C) - Brinkman area, acc. [33].

Table 2. The scope of significance of flow models through granular layers, acc. [29, 30].

Hydrodynamics models			
laminar flow		intermitted	turbulent
Re<1-3	3<Re<15	15<Re<120	Re>120
pre-Darcy area	Darcy law		
	Forcheimer formula		
	Brinkman formula		

In each case, the Reynolds number is determined by the equivalent diameter (d_ε) of flow channels and the mean fluid velocity (v_f) in these channels, according to equation

$$\mathrm{Re} = \frac{v_\varepsilon d_\varepsilon \rho_f}{\eta_f} \quad (4)$$

An exemplary comparison of a few characteristic models used in the calculation of the coefficient of permeability for different beds is shown in Table 3.

It is evident that these models are very varied and in many cases are far from Darcy assumptions, and very often – like to Slichter and Dullien models – do not take into account the fluid properties. Not all of them, such as ASTM and Szczelkaczew, take into account the porosity parameter as a characteristic feature of each porous material. On the other hand, there is no reference to porous skeletal (rigid) structures, which considerably limits the possibility of using these types of models to describe hydrodynamics of gas flow through porous skeletal forms. The self-made tests are described below.

3 Experimental studies and analysis of measurement results

The tests on permeability of porous materials were carried out for a many different types of material, for which the average porosity was within the limits (22-56) %. Most of them were coal chars (cokes) derived from miscellaneous thermal processing of bituminous coal, and materials such as partially melted gangue (including volcanic rock), natural and synthetic pumice, and porous polyamide with symmetrical spatial structure.

The tested materials were different in structure and their internal form was evaluated on the basis of a scanning image – Fig. 3. Based on the following, the porosity of each material was determined and its effective value resulting from the openings of the pores.

Experimental studies were carried out on a specially designed set-up that is currently being patented [37]. An essential element of this measuring system is a flow module (Fig. 4a), in which a specimen of porous material (1) is placed.

The specimens were cube shaped (Fig. 4b), and the module flow channel design allowed for the measurements of permeability for each of the three main flow directions (X, Y, Z) by rotating the cubic specimen in a selected plane of the measuring cell.

Experimental set-up was equipped with a rotameter to measure of gas flow and two pressure gauges for assessment of pressure drop (at known temperature of gas). For a measuring of permeability a gas volume flow rate was assumed, resulting from allowable pressure drop on the porous specimen. Of course, the scope of research included also determination of the apparent density of porous deposits as well as the overall porosity of each tested specimen.

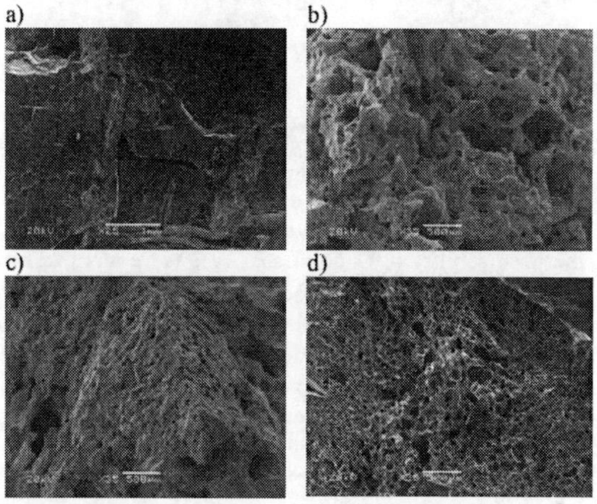

Fig. 3. Exemplary SEM images of the coal char specimen forms: a) *in situ* coal char, b) coke, c) *in situ* gangue, d) *ex situ* coal char (×35 zoom).

Table 3. Coefficient of permeability in the assessment of various research sources.

Author	Model	Remarks
ASTM (*American Society for Testing and Materials*) [34]	$K_{ASTM}=\frac{2\eta L Q_g P_{wyl}}{A_o(P_{wl}^2-P_{wyl}^2)}=\frac{\eta L Q_g P_{wyl}}{A_o \Delta P_{zm}}\frac{2}{(P_{wl}+P_{wyl})}$	oil and gas sands and shales
Brinkman, [33]	$K_B=\frac{v_o\eta}{\frac{\Delta P}{L}-(\rho\beta v_o^2)-\eta v_o}$	air cooling of packed fruit
Darcy [23, 28]	$K_{Da}=\frac{v_o\eta L}{\Delta P_{zm}}$	sand-ground (water)
Dullien [23]	$K_{Du}=\frac{d_e^2}{180}\left(\frac{\varepsilon^3}{(1-\varepsilon)^2}\right)$	granular materials
Ergun [13]	$K_E=\frac{1}{4}\left(\frac{1{,}75}{\sqrt{150}}\frac{\rho_f v_o^2}{\sqrt{\varepsilon^3}}\frac{L}{\Delta P}+\sqrt{\left(\frac{1{,}75^2}{150}\frac{\rho_f v_o^4}{\varepsilon^3}\frac{L^2}{\Delta P^2}\right)+\left(4v_o\eta\frac{L}{\Delta P}\right)}\right)^2$	pressure drop in gas flow
Forchheimer [30, 33]]	$K_F=\frac{v_o\eta}{\frac{\Delta P}{L}-(\rho\beta v_o^2)+C}$	granular and porous materials
Slichter [35]	$K_S=7{,}8\varepsilon^{3{,}26}d_e^2$	symmetrical spherical bed (gas)
Szczełkaczew [36]	$K_{Sz}=\frac{v_o\eta}{g\rho}$	grained medium - flow through a straight pipe (fluid)

https://doi.org/10.1051/matecconf/201824001035

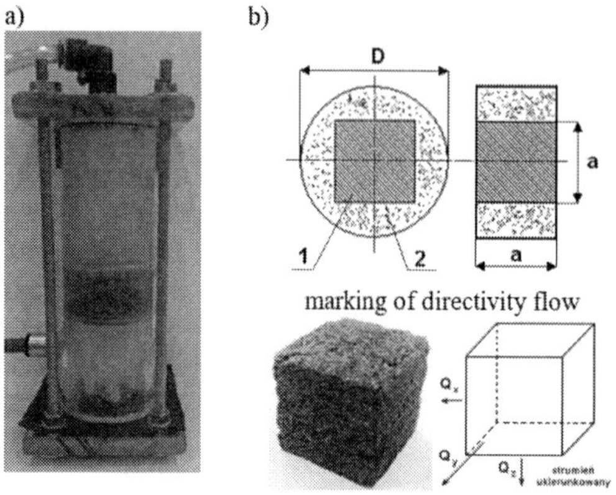

Fig. 4. Gas permeability meter: a) flow module channel, b) measurement cell: 1 - specimen of porous material, 2 - sealing frame with diameter (*D*) and the size of the sample (*a*).

An example of measurement results characterizing the permeability (Q) of the char *in situ* versus pressure drop (ΔP_{zm}) on the porous bed are shown in Fig. 5. From distribution of experimental points it can be seen that the permeability of this kind of material depends significantly on the direction (X, Y, Z) of gas flow. This indicates a distinct asymmetry of permeability flow rate to a given flow direction and consequently to the apparently anisotropic structure of the material to which the permeation tensor is an anisotropic tensor. Similar results have also been obtained for other such materials.

The point of reference for such asymmetry may be a porous polyamide which is a material with spherical sintered bed of particles with a diameter of approximately 0.1 mm. The experimental data presented in Fig. 5b show that the polyamide permeability characteristics are practically independent of the flow direction of the gas, which indicates the symmetrical structure of the polyamide. At the same time, despite a significantly lower porosity than carbonate *in situ*, this polyamide is characterized by similar gas flow characteristics. These results confirm the perception that higher permeability of coal char results more from their porous-slit structure than from porosity scale.

It is interesting to note that for the porous polyamide the permeability characteristics are also non-linear, which in the scope of conducted measurements proves the advantage of the turbulent gas flow.

This is confirmed by experiment results shown in Fig. 6, which characterize the gas permeability of coke with a multiplied volume of cubic specimen – a=30×30×30, 40×40×40 and 50×50×50 mm. Also, the results of these measurements show that the "reciprocal distance" for gas permeability between multiple-volume specimens is not transferred to a proportional displacement of the permeability characteristics. This is very important in terms of geometry scale transfer, and at the same time it confirms the thesis that in case of porous skeletal materials their porosity scale does not form the dominant factor in the description of gas permeability hydrodynamics.

This observation confirms the measurement results shown in Fig. 7, referring to the average permeability of various materials for airflow in two orthogonal directions (Y, Z). In each case (and this is confirmed by the results of studies for other systems – [38]), materials with a relatively large porosity such as pumice or cokes are characterised by a very low permeability, which demonstrates that their internal structure is largely closed for fluid flow.

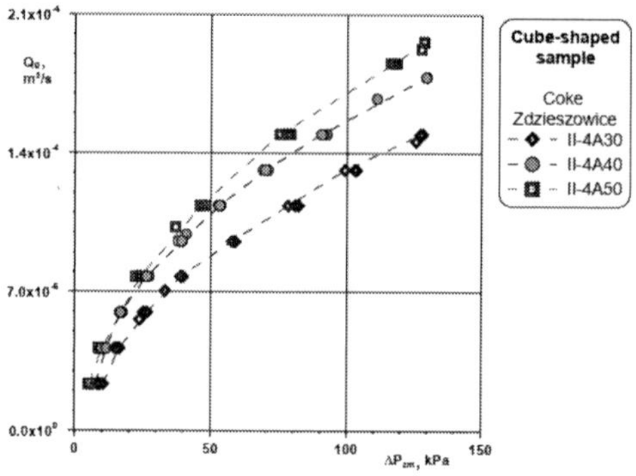

Fig. 6. Characteristics of permeability of coke with multiplied specimen size (*a*=30, 40 and 50 mm) – air flow.

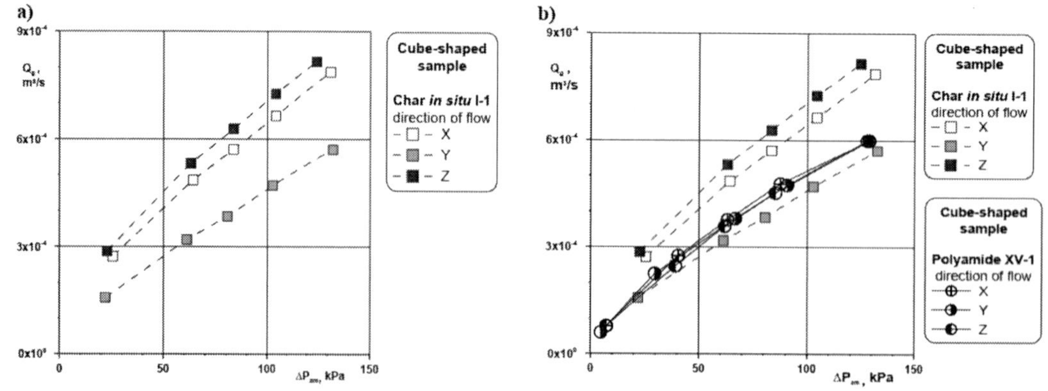

Fig. 5. Experimental points characterizing asymmetric flow of air through a deposit of carbonate *in situ* (a) and comparison of this asymmetry to sintered polyamide (b); cubic shape sample with dimension *a*=20 mm.

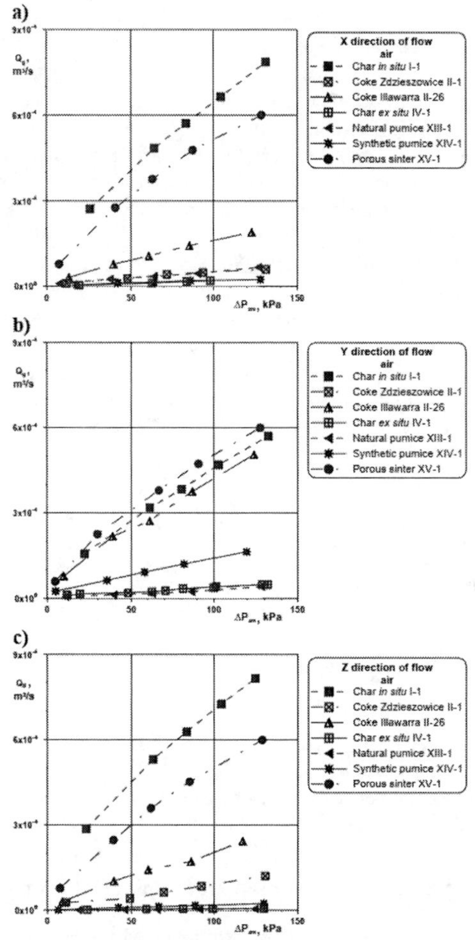

Fig. 7. Permeability of porous materials in regard of the gas flow direction: a) flow direction X, b) Y, c) Z.

A comparative statement of calculation value of the permeability coefficient, calculated with respect to the models shown in Table 3, has been illustrated in Fig. 8. As can be seen, in all analysed cases, very different results are obtained, which are characteristic both in terms of the calculation method used and the kind of material. There are three characteristic trends distinguish oneself between the value of the permeability coefficient and the pressure increase: decreasing, unchanging and increasing.

According to Darcy model, declining trend indicates a greater impact of gas choking than it would be by rise the pressure value.

In turn, the apparent trend for Szczelkaczew model results directly from the increase in the calculated gas flow velocity (with its properties), which is proportionally related to the value of this coefficient. Furthermore, by definition of this method, the values of the permeability coefficient also have values much higher than for the other models. In that above context, invariant trends in the values of the permeability coefficient are directly related to the form of the computational models (Tab. 3).

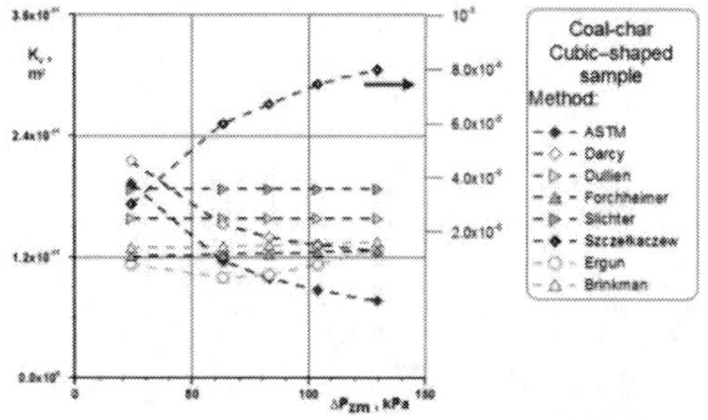

Fig. 8. Coefficient of permeability applied to the different calculation methods.

For comparison, on Fig. 9 are presented the results of calculations according to the method of ASTM (Tab. 3), put together for chosen chars from deposits of hard coals in the area of world. These results are pointing for very strong tying together the structure of these chars from of them gas permeability. It proves about distinct anisotropic of this group of porous materials.

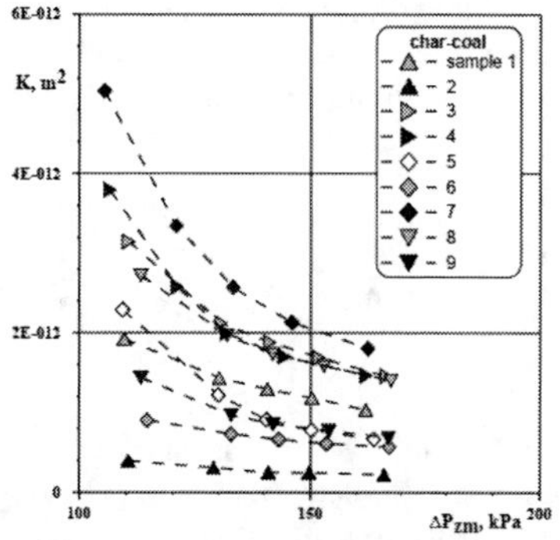

Fig. 9. Coefficient of permeability applied to ASTM method (cubic samples): ▲ (1) Czech Republic-a, ▲ (2) Czech Republic-b, ► (3) Australia-a, ► (4) Australia-b, ◊ (5) Poland-a, ♦ (6) Poland-b, ♦ (7) Poland-c, ▼ (8) USA, ▼ (9) Canada.

Due to the great discrepancies and simultaneously in view of numerous limitations in use of models known in the literature, both in the aspect of the structure of porous materials and in the approach to hydrodynamics evaluation of gas flow through such materials, an attempt was made to develop an own model for estimating the permeability coefficient.

From the assumption, definition of model is referred to an alternative approach that results from pressure drop connected with local flow resistance. With reference of the loss coefficient of flow (ξ_ε) and gas velocity (v_g) to the conditions associated with the mean porosity value (ε), this could be written as

$$\Delta P = \xi_\varepsilon \frac{\rho_g v_\varepsilon}{2} \tag{4}$$

Transformation of this equation leads to expression

$$Q_g = \sqrt{A_o^2 \frac{2\,\Delta P}{\xi_\varepsilon \rho_g}} = \sqrt{\frac{2A_o^2}{\xi_\varepsilon}}\sqrt{\frac{\Delta P}{\rho_g}} \tag{5}$$

in which the effective surface area of gas flow (A_o) results from the material porosity. In terms of the overall cross-section (A) of the flow module channel, it yields to $A_o = \varepsilon A$.

It can be easily noticed that the component $\sqrt{\frac{2A_o^2}{\xi_\varepsilon}}$ of equation (5) has a surface dimension, which in hydrodynamic terms allows it to be treated as the equivalent value of the permeability coefficient. Therefore (by definition)

$$K_V = \sqrt{\frac{2A_o^2}{\xi_\varepsilon}} \tag{6}$$

Then, the gas flow rate is described by the following relation

$$Q_g = K_V \sqrt{\frac{\Delta P}{\rho_g}} \tag{7}$$

If the hydrodynamic parameters are known (flow rate, pressure drop, material porosity and type of gas of course), the permeability coefficient value may be determined by means of experiments. Then relationship (7) can be written as

$$K_V = \frac{Q_g}{\sqrt{\frac{\Delta P_{zm}}{\rho_g}}} \tag{8}$$

The coefficient described in this way may be (from definition) used to determine the permeability with respect to each direction of gas flow (X, Y, Z).

In contrast to the some other shown above models (Tab. 3), formula (8) does not include the impact of gas viscosity on the permeability, but it provides an experimental value, and therefore accounts for all features and parameters resulting from the hydrodynamics of gas flow through porous materials.

The exemplary results of the permeability coefficient determined on Eq. (8) for *in situ* coal char are shown in Fig. 10. It is possible to observe the varying intensity of gas flow relative to a given flow direction, indicating a strong anisotropy of this type of material. Similar results were obtained for other carbonates [38].

The characteristic result of the study on the permeability coefficient is the monotonous nature of the change in its value, together with the increase in pressure. This indicates an appropriately proportional increase in permeability with increasing pressure. In relation to well-known models, this effect much better corresponds to the flow phenomena of porous structures. Of course, such a change is equivalent to an increase in the gas stream.

From equation (8), it is also apparent that the increase of the value of permeability coefficient is more caused by the gas stream then relatively pressure drop. This may be facilitated by the slit-porous of coal-char structure and the possible internal volume deformation of such a skeleton. Experimental data confirm that in context of comparison of the permeability coefficient to other calculation methods (Fig. 8 and 9) the nature of change such defined coefficient much better reflects the mechanism of this phenomenon. On the other hand, as is apparent from the experimental data [38], for many other materials is observed very low permeability, what has its origin in the different structure of these materials, with a relatively high share of blinded pores and channels closed for flow (Fig. 3).

Errors and statistics. For all cases the measurement errors are connected with measure of gas flow rate and pressure drop. With the confidence level of 0.99 for flow gauge calibration, the mean relative error for gas flow rate was ±5.3%. The average error for manometer gauge was ±8.7%. As a result, the mean estimation error of permeability coefficient is about ±12% (Fig. 10).

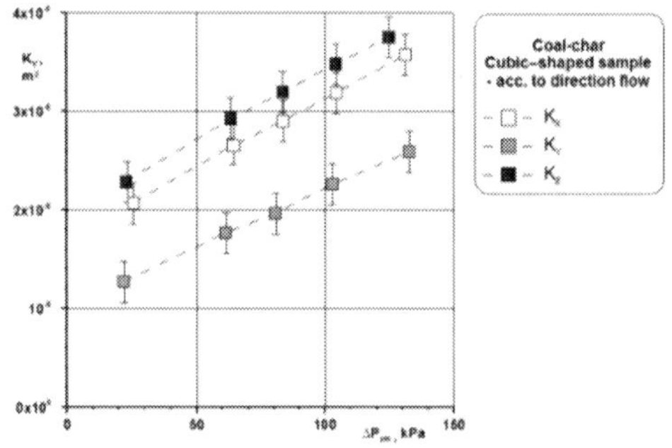

Fig. 10. Gas permeability coefficients for an *in situ* coal char, according to Eq. (8).

3.1 Numerical view of skeletal material porosity

Taking into account a very complicated process features of skeletal porous material (including their permeability and anisotropy), an attempt was made to create an integrated area of a tortuous multi-layer network by numerical means. For this purpose there was prepared a relevant simulation algorithm by means of the ANSYS programme, using the so-called Geometry, Mesh and Fluent computational blocks, the details of which are included in other own studies [38, 39].

The analysis included a configuration corresponding to a network of microchannels arranged randomly, when in extreme cases the network channels partially overlapped, creating a quasi-fractal network of curvilinear microchannel (Fig. 11a, b).

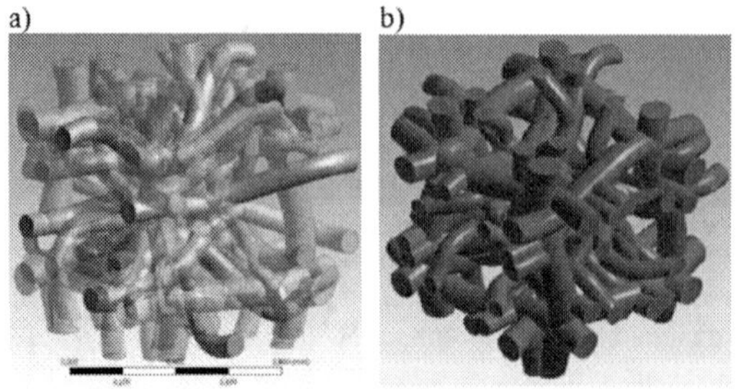

Fig. 11. Modelling of the geometry of microchannel network with curvilinear microchannels (a) and quasi-fractal network of tortuous microchannels (b); a) ε=3.8%, b) ε=39.5%.

On this basis, a fundamental cellular structure as quasi-fractal network of tortuous microcircuits with open for flow pores was modelled, as shown in Fig. 12. This network corresponds to the actual structure of the porous material (Fig. 3), analogous to the quasi-fractal network of tortuous microchannels [39]. Obtained by this way structure model of skeletal porous material allows for further calculations, in accordance with appropriate numerical methods for assessing hydrodynamic of gas flow through such materials.

Fig. 12. Numerical projection of structure of the skeletal porous material (a= 20 mm, d_e= 0.123 mm, ε= 54.3%).

3 Conclusions

The recognition of the gas permeability problem of porous media has shown that in literature there is very little information on hydrodynamics of gas flow through solid (skeletal) porous materials. In this respect, appropriate experimental studies of gas permeability of such group of materials have been made, and hydrodynamic phenomena resulting from pressure drop of gas flow have been evaluated. These studies were supplemented with numerical interpretation simulating the internal structure of the materials tested, taking into account the spatial distribution of the flow channels.

It has been found that the scale of permeability of the skeleton porous materials is determined by characteristic parameters such as the size of effective porosity for gas flow and also the anisotropic structure of these materials. Both these quantities have a considerable effect on the permeability coefficient, what was taken into account during the theoretical assessment of the issue.

Based on the local pressure drop a simply model for identification of the permeability coefficient of porous materials is proposed. The investigation results show that such definite coefficient properly reflects trends in the changes of hydrodynamic parameters at gas flow through skeletal porous media.

References

1. M. Błaszczyk, *Study of migration processes of petroleum substances in porous structures (in Polish)* (PhD Thesis, Łódź University of Technology, 2014)
2. K. Stańczyk at al. *Hydrogen-oriented underground coal gasification for Europe (HUGE). Final report.* (Research Found for Coal and Steel. EUR 25044 EN Project, 2012)
3. P.L. Younger P.L., Mine Water Environ. **30**, 127 (2011)
4. M. Sadatomi, J. Multiphase Flow **8**, 6, 641 (1982)
5. J.L. Williams, Catalysis Today **69**, 3 (2001)
6. J.L. Auriault, T. Strzelecki, J. Bauer, S. He S., Eur. J. Mech. A/Solids **9**, 4, 373 (1990)
7. J.L. Auriault J.L., P. Royer, Soil Mechanics and Porous Media **3117**, Serie II, 431 (1993)
8. O. Coussy, *Mechanics and Physics of Porous Solids* (John Wiley, 2007)
9. D. Łydżba D., Studia Geotechnica et Mechanica **XIII** (3-4), 51 (1991)
10. J. Blicharski, R. Smulski, AGH Drilling Oil Gas **29**, 1, 89 (2012)
11. H. Brauer, *Grundlagen der Einphasen - und Mehrphasen Strömungen* (Verlag Säuerländer, 1971)
12. R. Dyga, M. Płaczek, Inż. Ap. Chem. **52**, 4, 300 (2013)
13. S. Ergun, Chemical Eng. Progress, **48**, 2, 89 (1952)
14. J. Hehlmann at al., Rocznik Ochrony Środowiska, Koszalin **11**, 281 (1999)
15. B. Mertas at al., Karbo **2**, 163 (2013)
16. P. Skotniczy, Prace Instytutu Mech. Górotworu PAN, Kraków, **10**, 1-4, 103 (2008)
17. K. Warpechowski, A. Jopkiewicz, Archiwum Odlewnictwa PAN **2**, 5, 124 (2002)
18. J.T. Wong, Z. Chen, C.W. Ng, M.H. Wong, Environ. Science and Pollution Research Int. **23**, 8, 7126 (2016), DOI 10.1007/s11356-015-4871-2
19. G. Aksielrud, M.A. Altszuler, *Mass transfer in porous body (in Polish)* (WNT Warszawa, 1987)
20. G. Dulniew, V. Novikov, *Prociessy pierenosa w nieodnorodnych sriedach.* (Eniergoatomizdat, 1991)
21. A.P. Mozhaev, Journal of Engineering Physics and Thermophysics **75**, 2, 371 (2002)
22. T. Lambe, R.V.Whitman R., *Soil mechanics (in Polish)*, (Arkady, Warszawa 1978)
23. T. Strzelecki at al., *Modelling of flows through porous media (in Polish)* (Wyd. Edukacyjne Wrocław, 2008)
24. A.L. Dullien, *Porous media: fluid transport and pore structure* (Elsevier, 1991)
25. T. Piecuch, Rocznik Ochrona Środowiska **11**, 299 (2009)
26. Z. Orzechowski, J. Prywer, R. Zarzycki, *Fluid mechanics in engineering and environmental protection (in Polish)* (WNT Warszawa, 2009)
27. W. Sobieski (Ed.), *Granular porous media (in Polish)* (UW-M Olsztyn, 2016)
28. B. Bębenek, H. Bębenek, *Energy losses in fluid flows (in Polish)* (Wyd. Politechniki Krakowskiej, Kraków 1987)
29. A.M. Amao, *Mathematical model for Darcy Forchheimer flow with applications to well performance analysis* (Master's Thesis, Texas University, Misato, 2007)
30. M. Peszyńska, A. Trykozko, W. Sobieski, Mathematical Sciences and Applications **32**, 463 (2010)
31. F. Franchi, B. B. Straughan, Math. Methods in the Applied Sciences **19**, 1335 (1996)
32. Z. Kembłowski, *Theoretical basis of chemical and process engineering (in Polish)* (WNT Warszawa, 1985)
33. T.E. Hansen, *Flow in micro porous silicon carbide* (Technical Univ. of Denmark, Copenhagen, 2007)
34. *Standard Test Method for Permeability of Rocks by Flowing Air.* ASME D 4525-90, Reapproved (2001)
35. P. Popielski, *Model of mechanical scouring in terms of the finite element method (in Polish)* (PhD dissertation, Warszawa University of Technology, 2000)
36. S. Pisarczyk, *Soil mechanics (in Polish)* (Politechnika Warszawska, 2005)
37. G. Filipczak, E. Krause, G. Wałowski, Patent PL 225 980
38. G. Wałowski, *Hydrodynamics of gas flow through porous deposits (n Polish)* (PhD Thesis, OUTech Opole, 2015)
39. G. Wałowski, G. Filipczak, E. Krause, Modelowanie inżynierskie **20**, 51, 1236 (2014)

Review of existing and development of a novel mathematical model for supercritical CO_2 cycles working fluid

*Marcin Wołowicz**, *Jarosław Milewski*, and *Piotr Lis*

Institute of Heat Engineering, Warsaw University of Technology, Nowowiejska 21 Street 00-665 Warsaw, Poland

Abstract. The paper aims to compare the models of working fluids against experimental data for carbon dioxide close to its critical conditions. Fortunately, most of the work is already done and published where the authors compared the models based on the equation of the state (EoS). There are a few other models which were not investigated, thus we would like to add a few new results here and focus only on near-critical properties where the biggest deviation between experimental and calculated properties can be observed. The area of interest was pressure range of 7.39 – 20 MPa and temperature range of 304-340 K just above fluid critical point (7.39 MPa, 304.25 K). Model validation was performed for density and heat capacity as one of the most important parameters in preliminary cycle analysis.

1 Introduction

The paper aims to compare the models of working fluids against experimental data for carbon dioxide close to its critical conditions. Fortunately, most of the work is already done and published by Zhao et al [1], where the authors compared the models based on the equation of the state (EoS) as follows:

1. Peng Robinson (PR)
2. Peng–Robinson combined with the Boston-Mathias alpha function (PR-BM)
3. Soave–Redlich–Kwong (SRK)
4. Benedict–Webb–Rubin modified by Starling and Nushiumi (BWRS)
5. Lee–Kesler–Plöcker (LKP)
6. Span–Wagner (SW)

They concluded that in terms of pure-component behavior, among the six studied EoS, the Span-Wagner EoS is the one that leads to the most accurate predictions of the CO_2 in the near–critical and supercritical regions. In terms of cycle efficiency, it has been observed that the influence of the EoS choice is limited but not necessarily negligible. However, from a component–sizing point of view, they found that certain deviations can be considered as high, leading to economic issues due to over or underestimation of equipment costs (at level of 10%). Thus, the SW EoS is recommended for evaluating the performances of a SC-CO_2 Brayton cycle in order to avoid inaccurate predictions in terms of equipment sizing and optimization.

There are a few other models which were not investigated in [1], thus we would like to add a few new results here and focus only on near-critical properties where the biggest deviation between experimental and calculated properties can be observed. The area of interest was pressure range of 7.39 – 20 MPa and temperature range of 304-340 K just above fluid critical point (7.39 MPa, 304.25 K). Model validation was performed for density and heat capacity as one of the most important parameters in preliminary cycle analysis.

The following fluid models not included in Zo study were considered for modeling super critical CO_2 cycles focusing on near-critical point area:

1. Klimeck (GERG2008)
2. Cubic-Plus-Association (CPA)
3. Benedict-Webb-Rubin-Lee-Starling (BWR-LS)
4. Peng-Robinson-Wong-Sandler (PRWS)
5. Predictive Redlich-Kwong-Soave (PRKS)
6. Redlich-Kwong-Soave (RKS) with Boston-Mathias alpha function (RKS-BM)
7. Redlich-Kwong-Soave-Wong-Sandler (RKSWS)
8. Schwarzentruber-Renon (SR-POLAR)

The Span–Wagner EoS was used as a reference as it was proofed to be the most accurate in previous studies. Investigation was done using Aspen One® simulation software.

2 Supercritical cycles

Supercriticality is a hybrid state, where distinct liquid and gas phases do not exist. A supercritical fluid is dense, like a liquid, but expands to fill a volume the way a gas does. Small changes in temperature near the critical point—about 31°C cause big changes in density, similar to evaporating where the liquid changes to a vapor. The

* Corresponding author: marcin.wolowicz@itc.pw.edu.pl

density change, however, is more than two orders of magnitude smaller that the density change when, e.g., water becomes steam at atmospheric pressure. Similarly, it takes a lot of energy to increase the temperature a small amount when the fluid is near the critical point, much the way the heat of vaporization requires great amount of energy. Consequently, a large spike in heat capacity occurs near the critical point of CO_2.

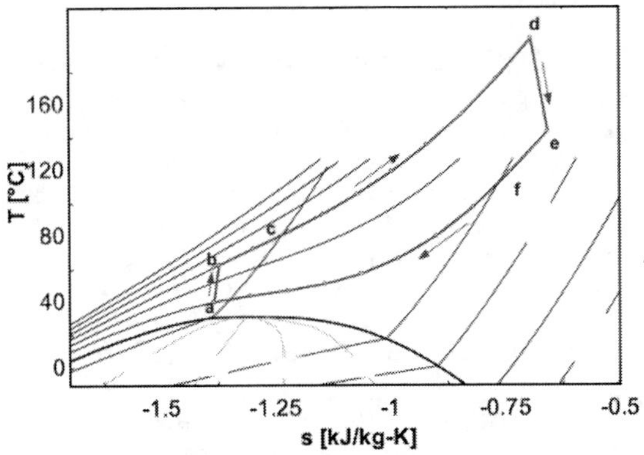

Fig. 1. Carbon dioxide Brayton cycle T-S chart [2].

Both supercritical and transcritical cycles are presented in Fig. 1 and Fig. 2 respectively. It can be seen from T-S chart, that the plot of the working fluid of a supercritical cycle is located in the supercritical area throughout all the points of the cycle. In case of a transcritical cycle a part of the plot is below the critical point, nonetheless most of the points are situated over the saturation line, therefore both cycles are usually related to supercritical cycles in the literature.

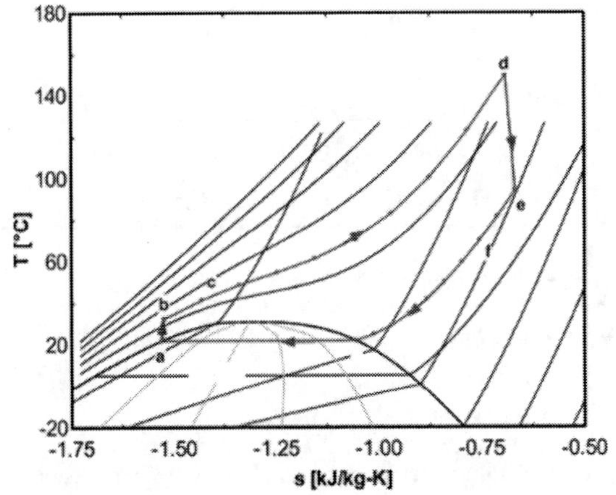

Fig. 2. Carbon dioxide transcritical cycle T-S chart [2].

Both cycles consists of four processes, namely: compression (pumping) (a-b), isobaric heat supply (b-d), expansion (d-e) and isobaric heat rejection (e-a). It is worth to note that the transcritical cycle has considerably lower low cycle temperature.

Since carbon dioxide used as a working fluid in both supercritical and transcritical power cycles operates mostly in the supercritical area, the ideal gas model cannot be applied to obtain the working parameters and the cycle performance. On the contrary, carbon cycle behavior deviates substantially from an ideal gas behavior particularly near the critical point. Therefore, the best suited real gas equation of state has to be selected to calculate CO_2 parameters.

Obviously, all the equations of state designed to describe a real gas behavior that have been developed in the past are burdened with error. However, there are some of them that are usually recommended in the literature. Oh et al. [4] proposed the Lee-Kesler-Plocker equation of state as the most accurate and used it for all computer calculations. A comparison made by Mazzoccoli in 2012 [4] proved that out of 17 examined equations of state, the Span-Wagner [5] equation has the lowest error in the density prediction in every region analyzed. Wang et al. [6] performed an analysis which aimed at developing an improved density correlation for supercritical CO_2. An experimental apparatus was designed to measure accurately the density of supercritical CO_2 in temperature range from 303 to 473 K, the pressure from 3 to 60 MPa.

Then, the relative error between the experimental and calculated density values for a number of frequently used equation of state was evaluated. The high-precision equations, such as the MBWR, Lee-Kesler, Span-Wagner, and Sun-Kiselev-Ely equations of state, proved to provide the most accurate values. However, the maximum relative error can reach approximately 20% in the near critical region, thus the authors proposed an error correction term to the original equations. The corrected Lee-Kesler, MBWR, Span-Wagner, and Sun-Kiselev-Ely equations can precisely predict the density of supercritical CO_2, and the average relative error can be restricted to within 1.5%. Besides, the new polynomial density calculation formula with 28 parameters was proposed based on the experimental data. This formula can restrict the average relative error of density to 2.0% across the whole studied range.

3 Theory

Selected EoS can be divided into three categories similar as in paper [1]:

1. *Helmoholtz-type*: GERG2004
2. *Virial-type*: BWR-LS
3. *Cubic-type*: PRWS, PRKS, CPA, RKS-BM, RKSWS, SR-POLAR

3.1 Klimeck (GERG2004)

The GERG2004 equation of state is based on a multi-fluid approximation explicit in the reduced Helmholtz free energy. This model was developed under the auspices of the Groupe Européen de Recherches

Gazières (GERG) in cooperation with the German technical and Scientific Association on Gas and Water and European natural gas companies (E.ON Ruhrgas, Germany; Enagas, Spain; Gasunie, The Netherlands; Gaz de France, France; Snam Rete Gas, Italy; and Statoil, Norway). The model of carbon dioxide is based on work done by Klimeck [14] and is valid for temperature range of 216 – 900 K and pressure up to 300 MPa. EoS contains 22 terms. As it is based on SW EoS the prediction accuracy is expected to be similar.

3.2 Benedict-Webb-Rubin-Lee-Starling (BWR-LS)

The Benedict-Webb-Rubin-Lee-Starling EoS [15] is a generalization by Lee and Starling of the Virial EoS for pure fluids by Benedict, Webb and Rubin. The equation is used for non-polar components, and can manage hydrogen-containing systems.

3.3 Peng-Robinson-Wong-Sandler (PRWS)

This model uses the Peng-Robinson equation-of-state for pure compounds. The mixing rules are the predictive Wong-Sandler rules. The binary parameter PRWSK is used for the kit. Several alpha functions can be used in the Peng-Robinson-Wong-Sandler equation-of-state model for a more accurate description of the pure component behavior. The pure component behavior and parameter requirements are described in Peng-Robinson, and in Peng-Robinson Alpha Functions.

3.4 Predictive Redlich-Kwong-Soave (PRKS)

This model uses the Redlich-Kwong-Soave equation-of-state for pure compounds. The mixing rules are the predictive Holderbaum rules, or PSRK method. Several alpha functions can be used in the PSRK equation-of-state model for a more accurate description of the pure component behavior. The pure component behavior and parameter requirements are described in Redlich-Kwong-Soaveand in Soave Alpha Functions.

3.5 Cubic-Plus-Association (CPA)

The CPA method represents the Cubic-Plus-Association EOS model developed by Kontogeorgis and co-workers (Kontogeorgis, Voutsas, Yakoumis, Tassios, IECR 1996). The model combines the SRK cubic EOS with an association term similar to that of SAFT, as present in the PC-SAFT model. Mixing rules apply to the cubic, whereas combining rules are used for the association term. The model's applicability covers the VLE and VLLE of mixtures containing hydrocarbons and polar/associating chemicals such as water, alcohols, glycols, esters, and organic acids.

3.6 Redlich-Kwong-Soave (RKS) cubic equation of state with Boston-Mathias alpha function (RKS-BM)

The RKS-BM EoS is the Redlich-Kwong-Soave equation-of-state with the Boston-Mathias alpha function. It is recommended for hydrocarbon processing applications, such as gas-processing, refinery, and petrochemical processes. Its results are comparable to those of the Peng-Robinson-Boston-Mathias equation-of-state.

3.7 Redlich-Kwong-Soave-Wong-Sandler (RKSWS)

This EoS uses the Redlich-Kwong-Soave equation-of-state for pure compounds. The predictive Wong-Sandler mixing rules are used. Several alpha functions can be used in the Redlich-Kwong-Soave-Wong-Sandler EoS model for a more accurate description of the pure component behavior. The pure component behavior and parameter requirements are described in Redlich-Kwong-Soave, and in Soave Alpha Functions.

3.8 Schwarzentruber-Renon (SR-POLAR)

The Schwartzentruber-Renon EoS is based on Redlich-Kwong EoS and can be used to model chemically non-ideal systems with the same accuracy as activity coefficient property methods. This EoS is recommended for highly non-ideal systems at high temperatures and pressures, such as in methanol synthesis and supercritical extraction applications.

4 Experimental data

The experimental data used for working fluid EoS validation is reported in Table 1.

Table 1. Database for model validation.

property	temperature range (K)	pressure range (MPa)	data points	reference
heat capacity	290-417	0-90	284	[8]-[10]
density	217-340	0-90	362	[11]

The area of interest was pressure range of 7.39 – 20.00 MPa and temperature range of 304-340 K just above fluid critical point (7.39 MPa, 304.25 K).

5 Results and discussion

Mean absolute percentage error (MAPE) was used to compare models quality by calculating differences between EoS and experimental data. Table 2 contains MAPE results for considered EoS. The Span-Wagner EoS results were added for reference as this EoS is considered to be the most accurate for supercritical CO_2 applications [1]. Values in bold shows the most accurate equation for each property.

Table 2. Database for model validation.

EoS	MAPE(c_p) (%)	MAPE(ρ) (%)
GERG'04	**2.67%**	0.29%
BWR-LS	11.09%	13.97%
CPA	26.66%	27.15%
PRWS	14.49%	8.60%
PRKS	14.26%	17.21%
RKS-BM	12.84%	15.36%
RKSWS	14.26%	17.21%
SR-POLAR	12.84%	15.36%
SW (reference)	2.68%	**0.22%**

Results confirm that Helmoholtz-type EoS is the most accurate choice for supercritical CO_2 applications. GERG'04 EoS shows almost the same accuracy as SW EoS. It was expected as GERG'04 EoS is based on SW EoS with minor modifications by Klimeck [13-14]. Virial-type as well cubic-type EoS accuracy is not acceptable and in most cases exceed 10%.

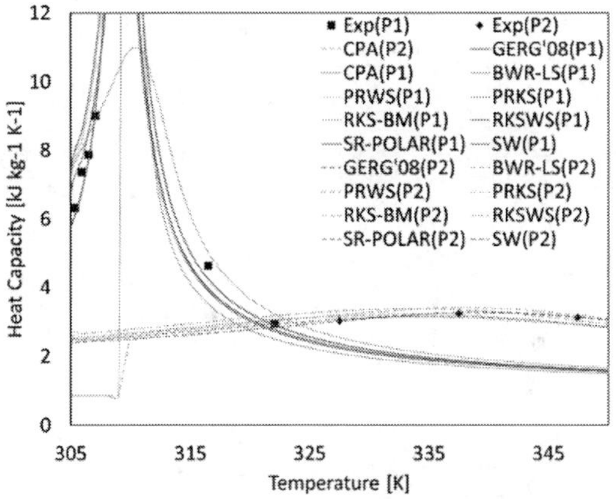

Fig. 3. Overview of different EoS results for isobaric heat capacity.

Graphical comparison between different EoS and experimental data is shown on Figure 3 and Figure 4. Two different isobars for each property were considered as follows: heat capacity (P1 = 8.3 MPa, P2 = 15.8 MPa), density (P1 = 7.4 MPa, P2 = 9 MPa). At lower pressures near the critical point peak in heat capacity can be observed.

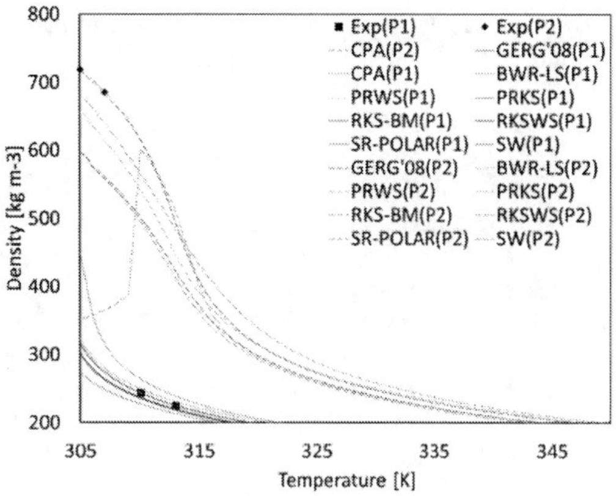

Fig. 4. Overview of different EoS results for density.

6 Conclusions

The paper investigates correctness of usage of different EoS in supercritical carbon dioxide applications. Eight different EoS available in popular commercial simulation software was tested. The area of interest selected for the study was pressure range of 7.39 – 20.00 MPa and temperature range of 304-340 K just above fluid critical point (7.39 MPa, 304.25 K) where the strongest deviation from ideal gas behavior can be observed. When considering pure component Helmoholtz-type EoS occurs to be the most accurate. The study confirms the statement by Zaho [1] that SW and GERG2008 should give similar results as both have the same theoretical foundations.

Based on the study it is recommended to use thermodynamic properties databases provided by The National Institute of Standards and Technology (NIST-REFPROP) or Groupe Européen de Recherches Gazières (GERG) as they include accurate Helmoholtz-type EoS for carbon dioxide.

Acknowledgements

This scientific work was financially supported as a research project number 2015/19/D/ST8/02780 by the National Science Centre, Poland.

References

1. Q. Zhao, M. Mecheri, T. Neveux, R. Privat, J.-N. Jaubert, Selection of a proper equation of state for the modeling of a supercritical CO_2 brayton cycle: consequences on the process design, Ind. Eng. Chem. Res., **56**, pp. 6841-6853 (2017)
2. Y. Chen, P. Lundqvist, P. Platell, Theoretical research of carbon dioxide power cycle application in automobile industry to reduce vehicle's fuel consumption, Applied Thermal Engineering, **25**(14–15), pp. 2041 – 2053 (2005)
3. C. Oh, T. Lillo, W. Windes, T. Totemeier, R. Moore, Development of a supercritical carbon

dioxide brayton cycle: Improving VHTR efficiency and testing material compatibility, INL/EXT-06-01271 (2004)

4. M. Mazzoccoli, B. Bosio, E. Arato, , Analysis and comparison of equations-of-state with p-rho-t experimental data for CO_2 and CO_2-mixture pipeline transport, Energy Procedia **23**, pp. 274 – 283, The 6th Trondheim Conference on CO_2 Capture, Transport and Storage (2012)
5. R. Span and W. Wagner, A new equation of state for carbon dioxide covering the fluid region from the triple point temperature to 1100K at pressures up to 800MPa, J. Phys. Chem. Ref. Data **25**(6), 1509-1596 (1996)
6. Z. Wang, B. Sun, and L. Yan, 2014, Improved density correlation for supercritical CO_2, Chem. Eng. Tech. **38** (1) 75-84 (2015)
7. Y. Kim, C. Kim, and D. Favrat, Transcritical or supercritical CO_2 cycles using both low- and high-temperature heat sources, Energy **43**(1), pp. 402 – 415, 2nd International Meeting on Cleaner Combustion (2012)
8. L. B. Koppel, J. M. Smith, Thermal Properties of Carbon Dioxide in the Critical Region, J. Chem. Eng. Data, **5**, pp. 437 (1960)
9. L. Dordian, J.-Y. Coxam, J. R. Quint, J.-P. E. Grolier, E. W. Lemmon, and S. G. Penoncello, Isobaric Heat Capacities of Carbon Dioxide and Argon Between 323 and 423 K and at Pressures up to 25 MPa, J. Supercrit. Fluids **8**, pp. 228 (1995)
10. G. Ernst, G. Maurer, E. Wiederuh, Flow calorimeter for the accurate determination of the isobaric heat capacity at high pressures results for carbon dioxide, J. Chem. Thermodynamics, **21, pp.** 53-65 (1989)
11. W. Duschek, R. Kleinrahm, W. Wagner, Measurement and correlation of the (pressure, density, temperature) relation of carbon dioxide, J. Chem. Thermodynamics, **32**, pp. 827-840 (1990)
12. O. Kunz, W. Wagner, The GERG-2008 Wide-Range Equation of State for Natural Gases and Other Mixtures: An Expansion of GERG-2004, J. Chem. Eng. Data, **57** (11), pp. 3032–3091 (2012)
13. O. Kunz, R. Klimeck, W. Wagner, M. Jaeschke, The GERG-2004 Wide-Range Equation of State for Natural Gases and Other Mixtures, Düsseldorf (2007)
14. R. Klimeck,, Entwicklung einer Fundamentalgleichung für Erdgase für das Gas- und Flüssigkeitsgebiet sowie das Phasengleichgewicht. Dissertation, Fakultät für Maschinenbau, Ruhr-Universität Bochum, (2000)
15. M.R. Brulé, C.T. Lin, L.L. Lee, and K.E. Starling, J. AIChE, **28**, pp. 616 (1982)

The possibility of simplified modelling of radiation heat transfer within a steel porous charge

Rafał Wyczółkowski[1,*], *Tomasz* Wyleciał[1], and *Marek* Gała[2]

[1]Czestochowa University of Technology, Faculty of Production Engineering and Materials Technology, Department of Industrial Furnaces and Environmental Protection, Al. Armii Krajowej 19, 42-200 Czestochowa, Poland
[2]Czestochowa University of Technology, Faculty of Electrical Engineering, Institute of Electric Power Engineering, Al. Armii Krajowej 17, 42-200 Czestochowa, Poland

Abstract. The article refers to the problem of calculating the effective thermal conductivity k_{ef} of a steel porous charge. In proposed approach for each heat transfer mechanism, which occurs in the considered medium, the corresponding thermal resistance is assigned. The model values of the k_{ef} coefficient were determined twice for the same input data (geometric dimensions, emissivity, temperature, thermal conductivity of steel and gas). The difference between successive calculations depended on the use of two different methods to determine the thermal radiation resistance R_{rd}. In the first approach the resistance R_{rd} was calculated using the exact method and in the second approach the simplified, approximate method was used. The discrepancy between the obtained results of k_{ef} in both approaches provides the evidence to use the approximate method to determine the resistance R_{rd}. A bundle of square steel sections were used to demonstrate the challenge. In the exact method, data regarding the temperature distribution within a single profile was used. These data were obtained based on experimental research using a guarded hot plate apparatus. The calculations were performed using a temperature range between 200°C and 800°C for two sections: 60×60mm and 80×80mm and three emissivities: 0.5, 0.7 and 0.9.

1 Introduction

Heat treatment operations of steel products have significant bearing on all key performance metrics of the plant, i.e.: productivity, energy consumption, product quality and emission of pollutants. This causes that heat treatment parameters should be selected very carefully. In the metallurgical industry for control and design of heat treatment parameters special numerical models have been used for more than three decades [1-3]. Such models predict spatial and temporal changes in the temperature of the charge. One of the challenges of this prediction is having required knowledge about the thermal properties of the heated charge. This issue becomes particularly complicated when the charge with a porous structure (e.g bundles of long elements or wire coils) is treated. Such kind of charge is a granular medium with the gas phase filling the gaps in between the steel elements. Therefore, during the heat treatment the following complex processes take place simultaneously: conduction in steel, conduction in gas, contact conduction, free convection and thermal radiation. As a result, the effective thermal conductivity k_{ef} becomes the key thermal property of the porous charge. This parameter is commonly used in the theory of porous media [4, 5].

The effective thermal conductivity of the porous material can be calculated by applying the model which is based on the analysis of thermal resistance for individual modes of heat transfer [6, 7]. If the high-temperature process is applied to the above scenario, one of the heat transfer modes to be considered in the k_{ef} model is thermal radiation. The thermal resistance of radiation R_{rd} in this model can be calculated in two ways: exact and approximate. In the exact approach, resistance R_{rd} is determined using the radiosity method. The main challenge comes down to solving the system of equations. Obtaining a solution by this method requires knowledge on the temperature values of all the surfaces that close the space of radiative heat transfer. To obtain information on changes in the temperature distribution within the heated porous charge, experimental investigations are necessary. Therefore, this is a significant drawback of this method. This problem is eliminated in the approximate method as this utilises one simple equation in which only the average temperature is applied.

This paper compares the two methods of calculating the effective thermal conductivity of a porous charge. The first, thermal radiation resistance is determined using the exact method, the second the approximated method. The analysis was carried out for the porous charge of a steel square section bundle which is illustrated in Fig. 1. Due to the high porosity of this charge, the fraction of thermal radiation in the total heat transfer is particularly high in this case.

* Corresponding author: rwyczolkowski@wip.pcz.pl

Fig. 1. Bundle of square sections as an example of a steel porous charge.

2 Analysis and modelling

When analyzing the thermal radiation in the area of the bundle, the process within each profile needs to be considered. This is radiation heat transfer in a three-surface enclosure as illustrated in Fig 2. It is assumed that the temperature of the bottom surface A_1 is T_1, the temperature of the top surface A_2 is T_2 and the temperature of the lateral surface A_3 is T_3 and $T_1 > T_3 > T_2$.

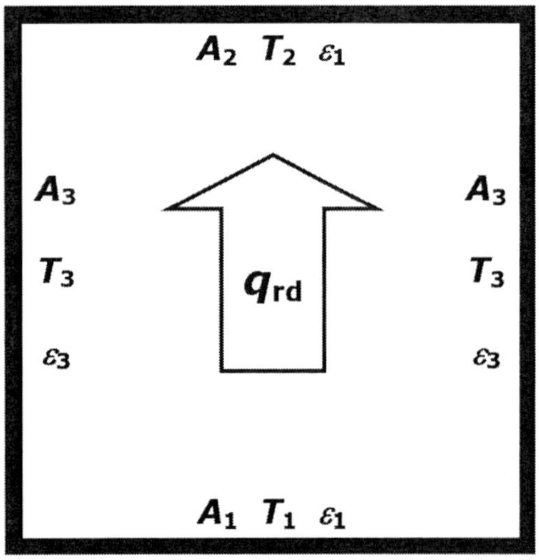

Fig. 2. The layout of the surfaces considered in the analysis of radiation heat transfer in the square section.

Since the enclosure is square the following relationship takes place: $A_1 = A_2 = 0.5\,A_3$. It is also assumed that the surfaces of the enclosure are opaque, diffuse, grey and its radiative properties are the same and expressed by the emissivity ε ($\varepsilon_1 = \varepsilon_2 = \varepsilon_3$). The net heat flux of radiation q_{rd} in this system is [8]:

$$q_{rd} = J_1 - J_2 \tag{1}$$

were J_1 and J_2 represents radiosity of the surfaces A_1 and A_2 respectively. The radiation resistance R_{rd} for this system is expressed by the following relationship:

$$R_{rd} = \frac{T_1 - T_2}{q_r} = \frac{T_1 - T_2}{J_1 - J_2} \tag{2}$$

Therefore, the challenge in calculating the resistance R_{rd} in the exact method is to determine the radiosities J_i of all surfaces in the considered enclosure, which are described by the following equations:

$$J_1 = E_1 - F_{2\text{-}1}\rho J_2 - F_{3\text{-}1}\rho J_3 \tag{3a}$$

$$J_2 = E_2 - F_{1\text{-}2}\rho J_1 - F_{3\text{-}2}\rho J_3 \tag{3b}$$

$$J_3 = E_3 - F_{1\text{-}3}\rho J_1 - F_{2\text{-}3}\rho J_2 - F_{3\text{-}3}\rho J_3 \tag{3c}$$

where $\rho = 1 - \varepsilon$ is surface reflectivity.

The emissive power per unit area of each surface E_i is described by the equation [8]:

$$E_i = \varepsilon \sigma T_i^4 \tag{4}$$

where σ is Stefan-Boltzmann constant.

The view factors $F_{i\text{-}j}$ used in equations (3) for the surface layout under consideration have the following values:

$$F_{1\text{-}2} = F_{2\text{-}1} = F_{3\text{-}3} = 0.4 \tag{5a}$$

$$F_{1\text{-}3} = F_{2\text{-}3} = 0.6 \tag{5b}$$

$$F_{3\text{-}1} = F_{3\text{-}2} = 0.3 \tag{5c}$$

In order to determine each radiosity, the system of equations (3) comes down to the matrix form:

$$\mathbf{E} = \mathbf{C} \times \mathbf{J} \tag{6a}$$

where

$$\mathbf{E} = \begin{bmatrix} E_1 \\ E_2 \\ E_3 \end{bmatrix} \tag{6b}$$

$$\mathbf{C} = \begin{bmatrix} 1 & -F_{2\text{-}1}\rho & -F_{3\text{-}1}\rho \\ -F_{1\text{-}2}\rho & 1 & -F_{3\text{-}2}\rho \\ -F_{1\text{-}3}\rho & -F_{2\text{-}3}\rho & 1 - F_{3\text{-}3}\rho \end{bmatrix} \tag{6c}$$

$$\mathbf{J} = \begin{bmatrix} J_1 \\ J_2 \\ J_3 \end{bmatrix} \tag{6d}$$

In the approximate approach, the resistance R_{rd} is calculated using a single equation. In this equation there is only one temperature T_m and its value is the average for the whole enclosure:

$$R_{rd} = \frac{X_{rd}}{4\sigma\, T_m^3} \quad (7)$$

where X_{rd} is a dimensionless coefficient with the value depending on the emissivity and the shape, as well as the relative position of the surfaces that represent the boundaries for the space [9]. For square enclosure $X_{rd} = 1/\varepsilon$.

The effective thermal conductivity of the analyzed charge is calculated using the definition of conduction resistance for the flat layer with dimension l [10]:

$$k_{ef} = \frac{l}{R_{to}} \quad (8)$$

where R_{to} is the total thermal resistance of the considered medium.

For the section bundle the l represents the sum of section dimension in the direction of heat flow and the width of the gap between the individual sections. While resistance R_{to} is the sum of the section thermal resistance R_{st} and thermal resistance of the gap R_{gp}:

$$R_{to} = R_{st} + R_{gp} \quad (9)$$

The methodology for calculating the section thermal resistance R_{st} is described in the publication [11]. While calculating this parameter, the R_{rd} resistance needs to be taken into account. Whereas the gap thermal resistance R_{gp} is calculated by the polynomial [12]:

$$R_{gp} = \left(B_1 t^2 + B_2 t + B_3\right) \cdot 10^{-4} \quad (10)$$

The values of coefficients B_i from this polynomial depend on the surface physical state of the adjacent sections.

3 Results and discussion

The calculations presented below were performed for two square sections: 60 mm and 80 mm with the wall thickness of 3 mm and three emissivities: 0.5, 0.7 and 0.9. The temperatures of the individual section surfaces required for the calculation of the resistance R_{rd} in the exact method were obtained from experimental research. These measurements were taken while determining the effective thermal conductivity of the section bundles using the guarded hot plate apparatus [13]. During these experiments, temperatures t_1 and t_2 were captured using thin jacked thermocouples within the selected sections. Measurements points were located on opposite surfaces of the section, perpendicular to the direction of the heat flow. The results of the measurements were used to calculate the temperature differences $\Delta t = t_1 - t_2$, which corresponds to the mean temperature $t_m = 0.5\cdot(t_1 + t_2)$. The values of the parameter Δt in the function of the temperature t_m obtained for the sections 60 and 80mm are shown in Fig. 3.

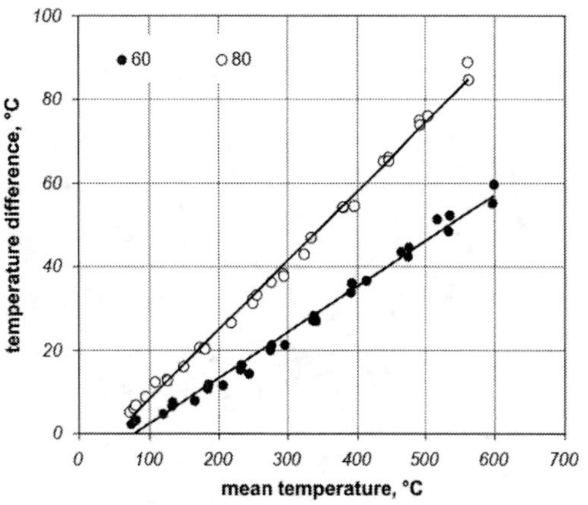

Fig. 3. Temperature differences within the 60 and 80mm sections obtained from the experimental research.

The data presented in Fig. 3 was then approximated by linear regression equations. The following relationships were obtained for each section:

$$\Delta t_{60} = 0.1094\, t_m - 8.45 \quad (11a)$$

$$\Delta t_{80} = 0.1681\, t_m - 8.94 \quad (11b)$$

The equations (11) were used to determine the temperatures for each surfaces of the section (T_1, T_2 and T_3) which are necessary for calculating the R_{rd} resistance in the exact method:

$$T_1 = (t_m + 273) + 0.5\Delta t \quad (12a)$$

$$T_2 = (t_m + 273) - 0.5\Delta t \quad (12b)$$

$$T_3 = (t_m + 273) \quad (12c)$$

Fig. 4 presents the results of radiation resistance calculations for the considered scenarios, which were obtained by both methods (exact and approximate). It was observed that the results for both sections using the same method (Figs. 4a and 4b) were very similar with only a very minor difference of 0.1% However, there are significant differences between the exact and the approximate methods for the same sections, with the exact method producing higher values. These differences in the results of both methods depend on the emissivity. When the emissivity decreases greater differences are observed.

In order to analyze the deviations in detail, the percentage excess of radiation resistance ΔR_{rd} were calculated:

$$\Delta R_{rd} = \frac{R_{rd-ex} - R_{rd-ap}}{R_{rd-ap}} \cdot 100\% \quad (13)$$

where $R_{rd\text{-}ex}$ is the resistance obtained using the exact method, whereas $R_{rd\text{-}ap}$ is the resistance obtained in the approximate method.

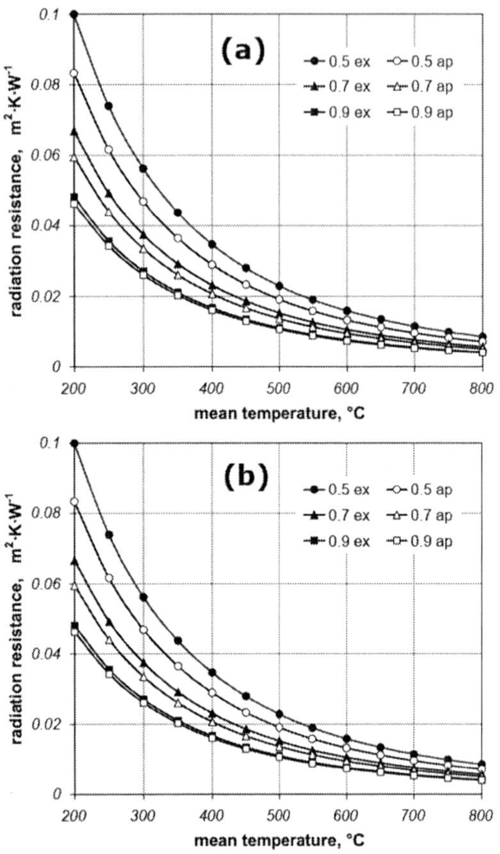

Fig. 4. Resistance R_{rd} calculated by exact and approximate method for: a) section 60 mm, b) section 80 mm.

The results of the ΔR_{rd} parameter obtained for the 80mm section are presented in Fig. 5. It can be noted that this parameter is almost constant for each emissivity in a temperature function. Very similar results were obtained for the 60 mm section. This implies that for a given emissivity the average value of this parameter can be applied.

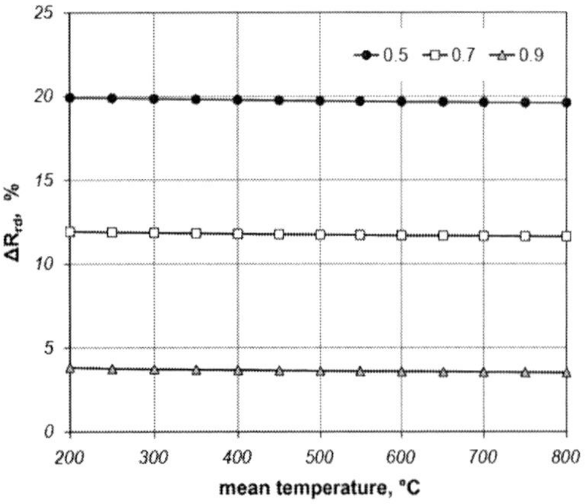

Fig. 5. Percentage excess of resistance R_{rd} for the 80 mm section.

Thus, the difference in R_{rd} resistance between the exact and the approximate method is reduced linearly with the increase in emissivity, reaching zero for the black body surface. However, for a square enclosure with an emissivity of 0.5, the $R_{rd\text{-}ex}$ value is about 20% greater than the $R_{rd\text{-}ap}$ value.

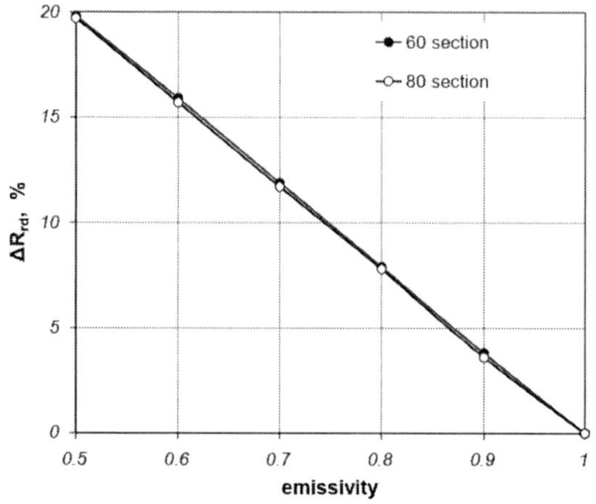

Fig. 6. The percentage excess of radiation resistance for the considered sections versus emissivity.

In line with the purpose of this article, it was decided to evaluate how the observed difference in $R_{rd\text{-}ex}$ and $R_{rd\text{-}ap}$ values affects the value of the effective thermal conductivity of the square section bundle. In these calculations it was assumed that the thermal conductivity of steel k_s and the thermal conductivity of gas phase k_g change with temperature in the following relationships:

$$k_s = 1.2\cdot10^{-8}t^3 - 3.2\cdot10^{-5}t^2 - 1.2\cdot10^{-2}t + 51.3 \quad (14)$$

$$k_g = -2.88\cdot10^{-8}t^2 + 8.05\cdot10^{-5}t + 0.024 \quad (15)$$

Equation (14) describes changes in thermal conductivity of low-alloy steel with carbon content of 0.2%, while equation (15) describes changes in thermal conductivity of air. These equations were determined through approximation of the literature data [14, 15].

The calculations of the coefficient k_{ef} were performed for the same scenarios as for the radiation resistance. To determine the gap thermal resistance R_{gp}, in equation (10), the following values of B_i coefficients were used:

- $B_1 = 2.31\cdot10^{-5}$;
- $B_2 = -0.0534$;
- $B_3 = 58.16$.

Calculations results for the effective thermal conductivity for the analyzed bundles are presented in Fig. 7. The obtained k_{ef} values for the 60 mm sections are within the 3.9÷8.8 W/(m^2·K) range, whereas for the 80 mm sections the k_{ef} values are within the 3.5÷11.3 W/(m^2·K) range. This coefficient increases as a function of temperature, and the dynamic of it increases with the increase in the emissivity. At the same time, higher values were obtained when the radiation resistance $R_{rd\text{-}ap}$ was taken into account in the calculations.

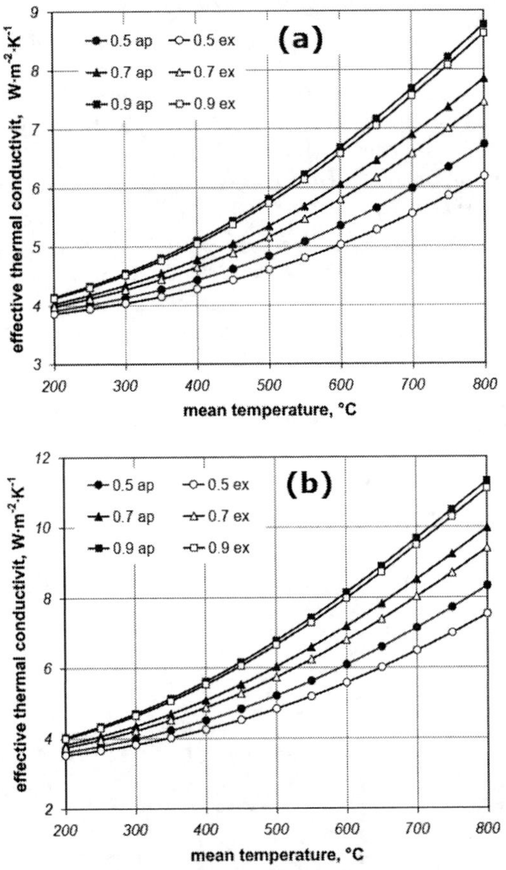

Fig. 7. The coefficient k_{ef} calculated with both the exact and the approximate methods of the radiation resistance, results for: a) 60 mm section bundle, b) 80 mm section bundle.

In order to demonstrate the influence of the radiation resistance value on the effective thermal conductivity, the percentage difference between the k_{ef} values obtained for both scenarios was calculated. This parameter is defined as follows:

$$\Delta k_{ef} = \frac{k_{ef-ap} - k_{ef-ex}}{k_{ef-ex}} \cdot 100\% \tag{16}$$

where $k_{ef\text{-}ap}$ denote the k_{ef} value calculated taking into account the resistance $R_{rd\text{-}ap}$, and $k_{ef\text{-}ex}$ denote the k_{ef} value calculated taking into account the resistance $R_{rd\text{-}ex}$. The results of the calculations for the Δk_{ef} for each bundle are presented in Fig. 8.

The value of the Δk_{ef} for the 60mm section bundle in 200°C temperature was not greater than 1.5%. However, when the temperature increased to 800°C the obtained values were between 1.7% and 8.7%. The highest values were observed for the emissivity of 0.5. For the emissivity of 0.7 the maximum value was 5.3%. A similar trend was observed for the 80mm sections. For 200° the maximum value obtained was 2.6%, whilst for 800°C and the successive emissivities obtained values were 10.6%, 6.1% and 1.9% respectively.

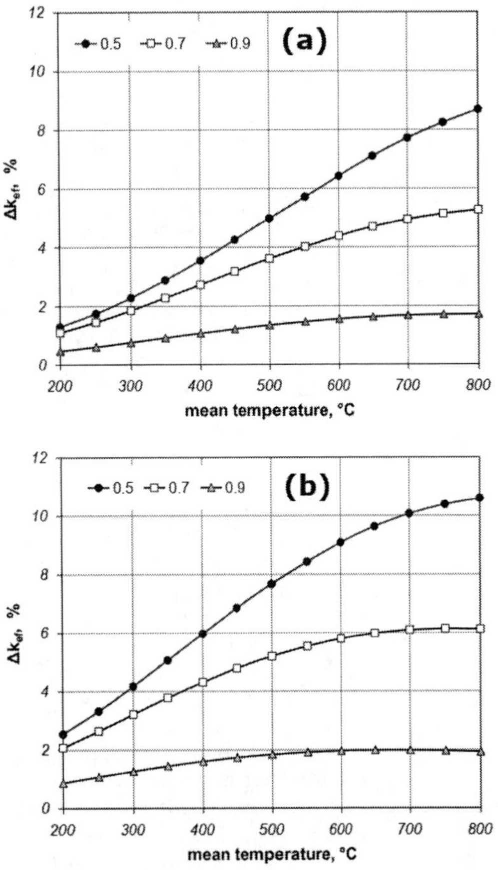

Fig. 8. Values of the Δk_{ef} parameter for: a) 60 mm section bundle, b) 80 mm section bundle.

The values of the Δk_{ef} parameter averaged for the whole temperature range obtained for all the analyzed scenarios are summarized in Table 1.

Table 1. The averaged values of Δk_{ef} for each bundles.

Emissivity	Sections 60 mm	Sections 80 mm
0.5	5.0%	7.2%
0.7	3.4%	4.7%
0.9	1.2%	1.6%

As it can be seen, the use of the approximate method for calculations the radiation resistance in the model of the effective thermal conductivity, in relation to the entire temperature range, exceeds the maximum value of the k_{ef} coefficient by approximately 7%. However, this applies to the emissivity of 0.5, while the surface of steel profiles subjected to heat treatment is usually characterized by the emissivity above 0.7. Therefore, the overestimation of the value of k_{ef} using the approximate method will be less than 5%. It can be stated that for industrial needs, the accuracy of calculations at this level is sufficient.

However, in a situation where more accurate calculations of the k_{ef} coefficient are required, the radiation resistance can be calculated using the approximate method. In order to apply this approach, the use of an appropriate correction term is required, which takes into account the linear impact of the emissivity:

$$R_{rd-cr} = (1.39 - 0.39\varepsilon)\frac{X_{rd}}{4\sigma T_m^3} \quad (17)$$

Due to the use of equation (17), the mathematical model of the effective thermal conductivity avoids solving the system of equations (6a), however, the results of these calculations will not be overstated as observed with equation (7).

4 Conclusions

The problem presented in this article is related to the optimization of heat treatment of the square steel profiles heated in the form of bundles. Due to the porous structure of this charge, its basic thermal property is effective thermal conductivity. This parameter quantifies the ability of the bundle to transmit heat as a result of the complex processes of conduction in steel and air, contact conduction, thermal radiation and free convection. The calculations of the k_{ef} coefficient were performed for two scenarios, which differ in the way of their modeling (exact and approximate) for radiation resistance. It has been shown that with the approximate method of determining resistance R_{rd}, the results of the k_{ef} calculations are overstated by an average of about 5%. In order to eliminate this discrepancy an updated version of the approximate equation was proposed, in which a corrective term was added. This term takes into account the linear impact of emissivity on the value of radiation resistance R_{rd}.

References

1. T.R. Rao, G.J. Barth, J.R. Miller, Iron and Steel Eng. **60** (1983)
2. Y. Jaluria, Int. J. Num. Met. Eng. **25** (1988)
3. S. Sahay, R. Mehta, K. Krishnan, Mat. Man. Proc. **22** (2007)
4. M. Kaviany, *Principles of heat transfer in porous media* (2nd ed. Springer–Verlag, New York, 1995)
5. W van Antwerpen, C.G. du Toit, P.G. Rousseau, Nuc. Eng. Des. **240** (2010)
6. D. Kunii, J.M. Smith, AIChE J., **6** (1960)
7. P. Zehner, E.U. Schlunder, Chem. Ingr. Tech. **42** (1970)
8. F.M. Modest, *Radiative heat transfer* (3rd ed. Academic Press, Amsterdam, 2013)
9. R. Wyczolkowski, D. Urbaniak, J. Ther. Heat Trans. **30** (2016)
10. Y.A. Cengel, *Heat and mass transfer, a practical approach* (3rd ed. McGraw-Hill, New York, 2007)
11. R. Wyczolkowski, Proc. Eng. **157** (2016)
12. R. Wyczolkowski, H. Radomiak, T. Wylecial, *Proceedings of 10th International Conference on Computational Heat, Mass and Momentum Transfer* (ICCHM2T 2017, May 28-June 1, Seoul, Korea, Paper Number 160149)
13. R. Wyczolkowski, A. Benduch, Civ. Env. Eng. Rep. **14** (2014)
14. T. Senkara, *Thermal calculations of furnaces in metallurgy* (Wydawnictwo Śląsk, Katowice, 1981 - in Polish)
15. K. Raźnjević, *Thermal tables with diagrams* (WNT, Warsaw, 1966 - in Polish)

Heat transfer study on a hybrid smooth and spirally corrugated tube

Chen Yang[1], *Min-rui* Chen[1], *Jin-yuan* Qian[1, 2, *], *Zan* Wu[2], *Zhi-jiang* Jin[1], and *Bengt* Sunden[2]

[1]Institute of Process Equipment, College of Energy Engineering, Zhejiang University, Hangzhou, 310027, PR China
[2]Department of Energy Sciences, Lund University, P.O. Box 118, SE-22100 Lund, Sweden

Abstract. Corrugated tubes are widely used in a range of applications for heat transfer enhancement. The spirally corrugated tube has a better heat transfer performance than the smooth tube. In this paper, the heat transfer performance of a hybrid smooth and six-start spirally corrugated tube is studied. With a validated numerical model, the effects of the corrugation part length on the vortex in the downstream smooth tube are studied for a range of high Reynolds numbers, where the existence of the corrugation part can turn out the secondary flow and enhance heat transfer. Meanwhile, it is found that in the smooth part, the fluid flow part with whirling can reach a maximum length, even if the length of the corrugation part continuously increases. Thus a series of critical corrugation lengths can be obtained. This work can reveal the enhanced heat transfer mechanism of the hybrid smooth and spirally corrugated tube and be of interest to researchers in heat transfer issues of corrugated tubes.

1 Introduction

Spirally corrugated tubes, owning the merits of better thermal performance, have been widely used in coaxial heat exchangers and have attracted the attention of researchers on heat transfer characteristics.

Mimura and Iaozaki [1] compared the corrugated tubes with smooth tubes on heat transfer and pressure drop performance under the effects of spiral structures. Others confirmed the influence spiral corrugation exerts on enhancing heat transfer performance through experimental works and numerical simulation [2-7]. Taking account of corrugation depth (DR), pitch ratio (PR) and Reynolds number, Promthaisong *et al.* [8] discovered that heat transfer enhancement was induced by vortex flows generated only around the wall corrugation.

Nowadays, a wide range of researchers have turned their research priority to multi-start spirally corrugated tubes given their exclusive structure and good heat transfer performance. Lazim *et al.* [9] presented a numerical analysis on thermal characteristics and pressure drop performance. Taking four-start spirally corrugated tubes as an example, they concluded that more corrugations can enhance flow mixing. Kareem *et al.* [10-11] performed numerical simulation on two-start and three-start spirally corrugated tubes and indicated that corrugation profiles determined thermal performance, which had a better effect at high *Re*. Balla [12] conducted a numerical study and studied the effects of corrugation characteristics on heat transfer performance. Jin *et al.* [13-14] also investigated six-start spirally corrugated tubes and obtained a criterion correlation for flow resistance as well as heat transfer characteristics. Sun *et al.* [15] and Vocale *et al.* [16] applied the field synergy theory to analyse the augmentation effect. Promthaisong *et al.* [17] made a conclusion that main swirl flow and five-secondary swirl flow enhanced heat transfer process of five-start spiral corrugated tube.

2 Analysis and modelling

2.1 Geometric model

The copper smooth tube experiences a rolling operation in a rotational stretching way, passing through six processing mechanisms successively. As a result, the structure of six-start spirally corrugated tube is shown as Fig. 1, including geometric parameters.

* Corresponding author: qianjy@zju.edu.cn

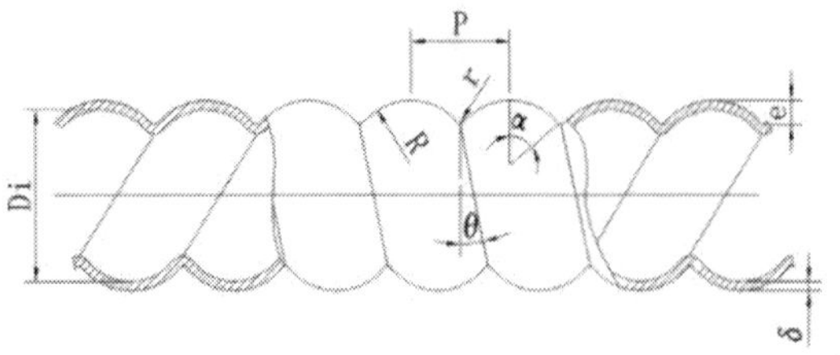

Fig. 1. Structural parameters of six-start spirally corrugated tube.

To exclude the influence of spiral corrugation parameters, including inter diameter D_i, pitch p and corrugation depth e, we set these variables as constant value, which is elaborated in Table 1.

Table 1. Constant values of variables.

Inter diameter D_i (mm)	Pitch p (mm)	Corrugation depth e (mm)	Reynolds number Re
27	60	4.5	10000

The hybrid tubes we studied are consisted of six-head spirally corrugated tube and smooth tube. They start from corrugation part, the length of which differs from each other, while the whole length is 1 m. The flow path model can be seen from Fig. 2.

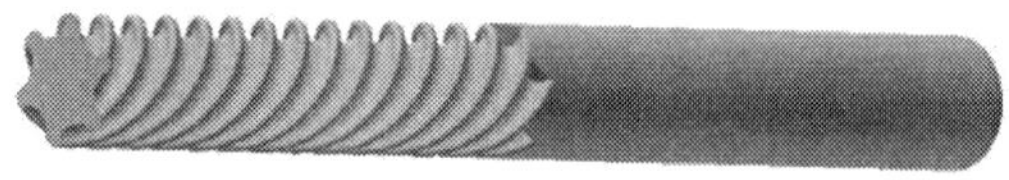

Fig. 2. Flow path model of the hybrid tube.

2.2 Mesh and boundary condition

To ensure the accuracy and reliability of solving results, it is of utter necessity to verify the grid independence. We adopt tetrahedral mesh in corrugation part while multi-zone method in smooth part. Furthermore, boundary layer mesh is applied. Eventually, the grid size is controlled in 1 mm^3 when the Nusselt number tends to reach its asymptotic value. When grid numbers increases from 3246214 to 3460198, difference of Nu is between the acceptable range of 5%. Simultaneously, the results of assessed y+ coincide with the recommended values ranging from 30 to 100.

The heat transfer medium in the hybrid smooth and six-start spirally corrugated tube is water. We set inlet boundary as velocity inlet, the temperature of which is 363 K while outlet boundary as pressure outlet, the value is 0 Pa. The Realizable k-ε turbulence model, combined with standard wall function method is adopted in the simulation, applying pressure implicit solver, SIMPLE algorithm and the second order upwind scheme. The values of residuals in the convergence conditions are less than 1×10^{-6}.

2.3 Reliable verification

Compare the calculated Nusselt number with the value of the empirical formula at Re ranging from 1×10^4 to 2.5×10^4 in smooth tube, the errors are in the allowable range and the largest one is +4.63 %.

3 Results and discussion

3.1 Heat transfer performance

In this paper, we research the velocity field distribution of a hybrid smooth and six-start spirally corrugated tube.

On the basis of the previous studies, the successive corrugation structure can induce a significant degree of swirl perpendicular to the mainstream direction. As a result, the secondary flow velocity V_{xy} is boosted and heat transfer is enhanced. To figure out the influence, the lengths of the corrugation part are 100 mm, 200 mm, 300 mm, corresponding to Tube 1, Tube 2 and Tube 3, respectively. The temperatures at various positions of these three tubes, together with the smooth tube are calculated to analyse the heat transfer performance, which is depicted in Fig. 3.

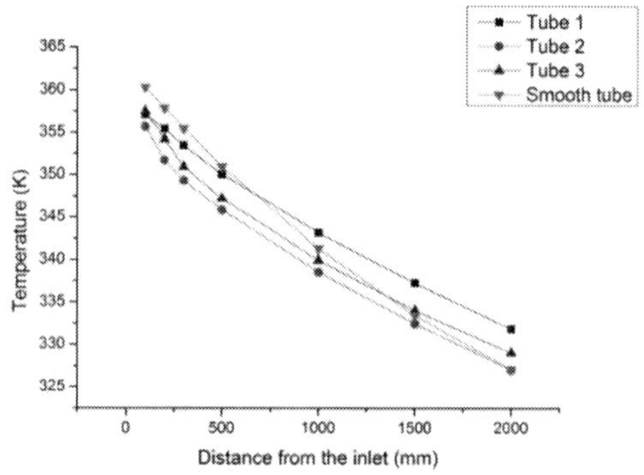

Fig. 3. Temperatures at various positions of four tubes.

3.2 Velocity analysis

For further investigation of the flow situation, cross sections of tube 1 are created in z=0.01, 0.05, 0.1 m (the end of the corrugation part). As is shown in Fig. 4, the velocity contour reflects the secondary flow velocity distribution that fluid of low velocity is around the center position while high at the corrugation. The secondary flow derives from the flowing fluid under the influence of the centrifugal force. The different centrifugal force values result from the variance in axial velocity. Fluid particles flowing in the tube core and close to the tube wall generate the velocity disparity. The former have a higher axial velocity and thus experience a larger centrifugal force. As a result, fluid in the tube core region is pushed towards the outside wall of the tube, where it drives the fluid towards the inside wall of the tube.

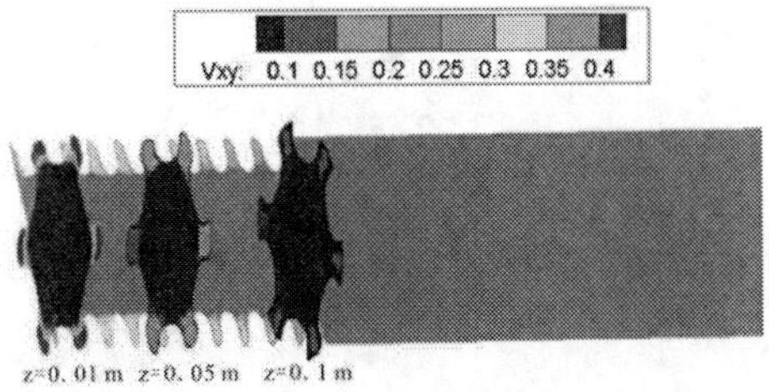

Fig. 4. Secondary flow velocity distribution in 100 mm corrugation part.

The scope of the centered fluid shrinks as the fluid flows away from the inlet, and the secondary flow at the corrugation is becoming less intensive at the same time. In other words, the secondary flow at the cross section tends to reduce velocity variance, which can be seen in Fig. 5.

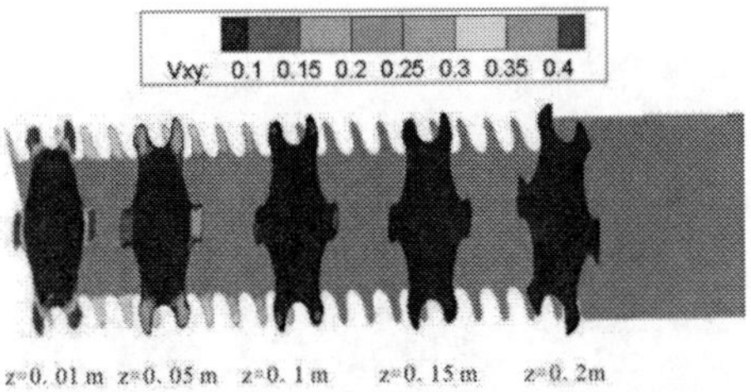

Fig. 5. Secondary flow velocity distribution in 200 mm corrugation part.

In tube 2, cross sections are generated in z=0.15 and 0.2 m. Compared with Fig. 6, it can be concluded that the secondary flow keeps developing at the corrugation part. The cross section of tube 3 in z=0.25 m is illustrated in Fig. 6, and the velocity contours almost remain constant, which means the secondary flow gets fully developed 0.2 m away from the inlet.

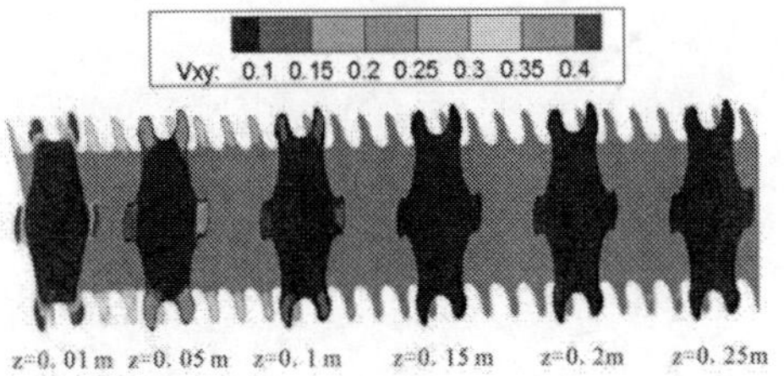

Fig. 6. Secondary flow velocity distribution in 300 mm corrugation part.

The spirally corrugated tube plays a significant role in generating secondary flow, which gets diminished in smooth tube. The distance from the interface between the corrugation part and smooth part to where secondary flow gets fully vanished is recorded. When the cross section is 0.01 m away from the interface, the secondary flow velocity vectors are intensive and the fluid particles around the wall are faster than the other part. As the fluid flows through the smooth tube, the secondary velocity vectors tends to distribute evenly and the velocity values of the whole cross section gets slower. As is shown in Fig. 7, the vector distribution characteristics of three tubes are comparable under the same working condition, in other words, secondary flow dies away following a specific regular pattern even if the length of the corrugation part continuously increases.

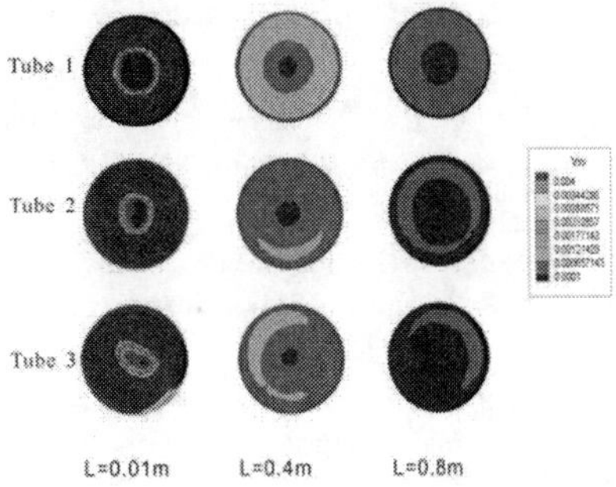

Fig. 7. Comparison of three tubes at same distance from the interface.

No matter how long the corrugation part is, the secondary flow vanishes at the position of 0.8 m away from the interface.

4 Conclusion

In this study, a novel structure hybrid smooth and six-start spirally corrugated tube is proposed, in order to discover the appropriate length of the corrugation part. Simulation is conducted with fixed geometric parameters under *Re*=10,000. The main conclusions are drawn as follows:

Firstly, the corrugation structure gives rise to the generation of secondary flow and enhances heat transfer as a result. Besides, once the corrugation part is substituted by smooth tube, the secondary flow will experience attenuation and finally vanishes no matter how long the corrugation part is or whether the secondary flow has been fully developed. The key to maintaining secondary flow is the corrugation structure.

This work is supported by the Key project of Natural Science Foundation of Zhejiang Province, China through Grant No. LZ17E050002, and the Fundamental Research Funds for the Central Universities through Grant No. 2018QNA4013.

References

1. K. Mimura, A. Isozaki, Desalination, **22,** 1, pp. 131-139, (1977)
2. S. Garimella, R. N. Christensen, SHAAE Trans, **99,** 1, pp. 1205-1216, (1993)
3. J.E. Walker, G.A. Whan, R.R. Rothfus, AIChE J., **3,** 4, pp. 484-489, (1957)
4. H.M. Li, K.S. Ye, Y.K. Tan, S.J. Deng, Heat Transf, **25,** 63, pp. 75-80, (1982)
5. S. Garimella, V. Chandrachood, R.N. Christensen, D. E. Richards, ASHRAE T., **94**, pp. 1119-1131, (1988)

6. T.S.Ravigururajan, A.E. Bergles, J. Heat Transf., **116,** 1, pp. 54-57, (1994)
7. Y.X. Li, J.H. Wu, H. Wang, L.P. Kou, X.H. Tian, Energy Proc., **17(Part A)**, pp. 791-800, (2012)
8. P. Promthaisong, W. Jedsadaratanachai, S. Eiamsa-Ard, Numer. Heat Tranf. A-Appl., **69,** 6, pp. 607-629, (2016)
9. T.M. Lazim, Z.S. Kareem, M.N.M. Jaafar, S. Abdullah, A.F. Abdulwahid, Int. Rev. Modell. Simul., **7,** 6, pp. 970, (2014)
10. Z.S. Kareem, M.N.M. Jaafar., T.M. Lazim, S. Abdullah, A.F. Abdulwahid, Alexandria Eng. J., **54,** 3, pp. 415-422, (2015)
11. Z.S. Kareem, S. Abdullah, T.M. Lazim, M.N.M. Jaafar, A.F. Abdulwahid, Chem. Eng. Sci., **134**, pp. 746-757, (2015)
12. H.H. Balla, Case Stud. Therm. Eng., **9(C)**, pp. 79-89, (2017)
13. Z.J. Jin, B.Z. Liu, F.Q. Chen, Z.X. Gao, X.F. Gao, J.Y. Qian, Int. J. Heat Mass Transf., **103**, pp. 1198-1207, (2016)
14. Z.J. Jin, F.Q. Chen , Z.X. Gao , X.F. Gao , J.Y. Qian, Int. J. Heat Mass Transf., **108**, pp. 1011-1025, (2017)
15. M. Sun, M. Zeng, Appl. Therm. Eng., **129**, pp. 1-11, (2018)
16. P. Vocale, F. Bozzoli, A. Mocerio, S. Rainieri, ICHMT 9th, Napoli, Italy. May 28-June 1, (2017)
17. P. Promthaisong, W. Jedsadaratanachai, V. Chuwattanakul, S. Eiamsa-Ard, Int. Conf. Chem. Mater. Process, pp.020005, (2017)

Thermal performance and redesign of beam stoppers under superficial heat transfer from intense ion beams in matter

André Pilan Zanoni[1,*], *Jose Antonio* Briz Monago[1], and *Marco* Calviani[1]

[1]CERN - European Organization for Nuclear Research, Geneva 23, Switzerland

Abstract. In the Linear accelerator 3 (Linac 3) and Low Energy Ion Ring (LEIR) at CERN dedicated beam stoppers intercept ion beams for machine and personnel protection whenever required. The interaction of accelerated ions and the stopper induces through-thickness heat deposition in the stopper material as deep as a few hundreds of micrometers. Following a multi-layered geometrical discretization this paper discusses the thermal and structural performance of the stoppers under different ion beams (Pb, Xe, Ar). Design considerations based on the projected range of different ions in target materials (aluminum, steel and copper alloys) are discussed. An improved design for ion stoppers based on thermo-structural performance and projected range is proposed.

Nomenclature

A: number of nucleons per ion
c_p: specific heat capacity
e: elementary charge
E_a: energy per nucleon in MeV/n
E_{ion}: energy per ion in MeV
I: current
l: smallest element size
n_l: number of layers
n_{ion}: number of ions
q: ion charge
$Q_{ion,max}$: maximum deposited heat density per ion
S_m: medium equivalent stress amplitude
S_a: average equivalent stress amplitude
S_s: endurance limit (Soderberg fatigue criterion)
S_{LF}: yield limit (Soderberg fatigue criterion)
T_0: initial temperature
T_{ad}: adiabatic temperature
α : thermal diffusivity
ρ: density
σ_x, σ_y: beam sizes at 1 σ in horizontal and vertical directions respectively
σ_e : equivalent von-Mises stress
σ_y (T): temperature-dependent yield strength
μ: pulse length

1 Introduction

The Linear accelerator 3 (Linac 3) and the Low Energy Ion Ring (LEIR) at CERN prepare ion beams to be injected in the Proton Synchrotron (PS) and for the subsequence accelerator chain. In order to protect personnel in case of accidental intrusions in the machine beam stoppers are installed in the accelerator line (Fig. 1).

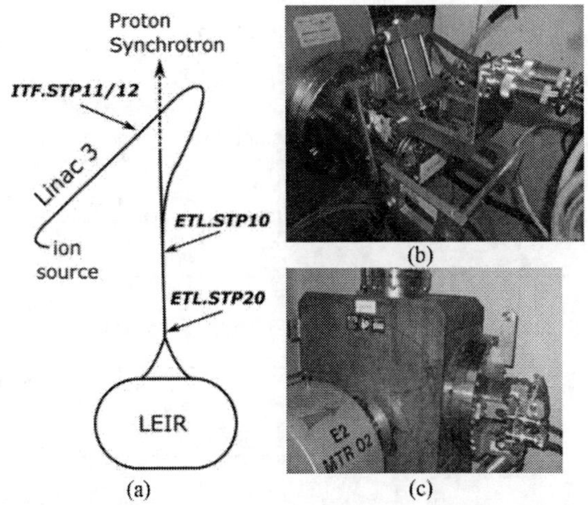

Fig. 1. (a) Schematic layout of Linac 3 and LEIR. (b) Beam stopper ITF.STP11/12. (c) Beam-stopper type ETL.STP10 and ETL.STP20.

Beam stoppers consist of a material block in vacuum that rests out of the beam line during beam operation. By means of a pneumatic actuator the material block closes – when required – the beam aperture preventing any circulating or unexpected beam to impact downstream areas. The current design of the beam stoppers in Linac 3 and LEIR dates from the early 1990s with limited design documentation available.

When an ion beam at energies of a few tens of MeV/n hits the material block heat is shallowly deposited in the block reaching only hundreds of micrometers through-thickness. This heat is the result of the heat transferred from the ion beam momentum, ionization and nuclear interaction processes between

* Corresponding author: andre.pilan.zanoni@cern.ch

beam and matter [1]. For safely designing beam stoppers the material block must withstand the thermal gradients induced by the beam-matter interaction.

The first goal of this paper is to validate the thermal and structural performance of those stoppers through numerical simulation. In this analysis we opt for multi-layered discretization of the beam-impact area instead of the time and computationally consuming models with solid elements [2]. Next we propose an improved beam stopper design by analyzing the thermo-structural performance of different targets under ion beam impacts and the projected range of different ions in matter.

2 Analysis and modeling

2.1 Beam operation

The Linac 3 injects ion beams in LEIR which splits the beam in bunches accelerated from 4.2 to 72 MeV/n [3]. A typical ion is lead, although xenon is also employed in the CERN physics program [4]. In the past the facility ran other ions such as oxygen, indium, argon and helium [5].

Table 1 shows critical beam scenarios for the analysis of the ion beam stoppers at injection energies of 4.2 MeV/n. The lead ion refers to $^{208}Pb^{54+}$, the xenon ion to $^{129}Xe^{39+}$ and the argon ion to $^{40}Ar^{11+}$. σ_x and σ_y correspond to beam sizes at 1 σ following a planar Gaussian distribution in horizontal and vertical directions respectively.

Table 1. Beam parameters for ion beam stoppers at injection [5].

Beam stopper	Ion	Current [μA]	Ion energy* [MeV]	σ_x [mm]	σ_y [mm]
ITF	Pb	50	874	4.3	6.9
	Xe	50	542	4.3	6.9
	Ar	62	168	4.3	6.9
ETL	Pb	50	874	5.9	10.3
	Xe	50	542	5.9	10.3
	Ar	62	168	5.9	10.3

*$E_{ion}=A.E_a$

The beam stoppers can receive beam at injection energies over four continuous weeks following two scenarios. In the first scenario the beam has a 200 μs long pulse repeated each 0.2 s. The second scenario considers 600 μs long pulses spaced by 0.4 s.

In rare accidental scenarios an extracted pulse of lead ions at 72 MeV/n from LEIR might hit the stopper ETL.STP20. The accidental pulse has two bunches, each 200 ns long spaced by 354 ns. The number of ions per pulse is 3×10^9 with $\sigma_x = 5.0$ mm and $\sigma_y = 4.5$ mm.

2.2 Beam-matter interaction

We simulate the interaction between ion beams and the stopper block using the Monte Carlo FLUKA code [1, 6], a particle transport and interaction code. The block is divided in energy bins in which the energy deposited is normalized to one ion.

At low energies the ions come to a complete stop in the block material due to strong ionization losses [1, 6].

2.3 Geometrical and thermal modeling

The material block in the stopper ITF.STP11/12 is a 10 mm thick stainless steel plate. In ETL.STP10 and 20 the absorbing block is made of aluminum alloy Antico 100 with Ø200 mm and 210 mm length. None of them is actively cooled, with only radiation to vacuum chambers and heat conduction through supports being the only heat sinks. Figure 2 and Fig. **3** present the geometrical model and meshing of both absorbers.

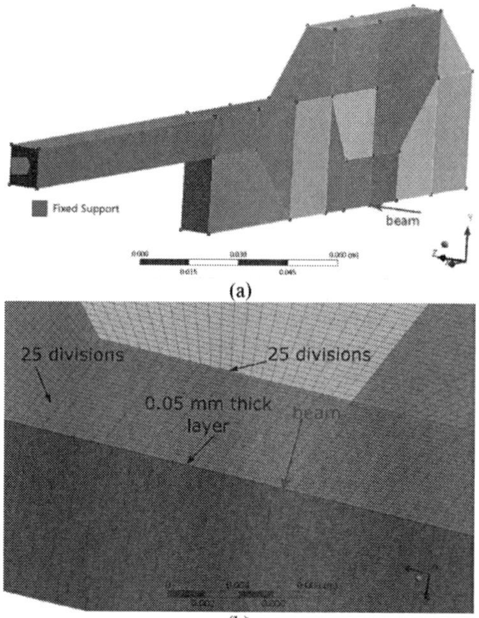

(a)

(b)

Figure 2. ITF absorber: (a) geometrical model divided in mesh zones. The XZ plane is used for symmetry. (b) Meshing near the beam impact.

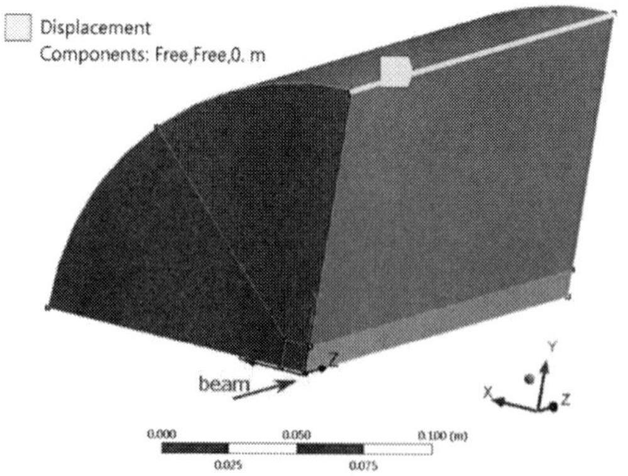

(a)

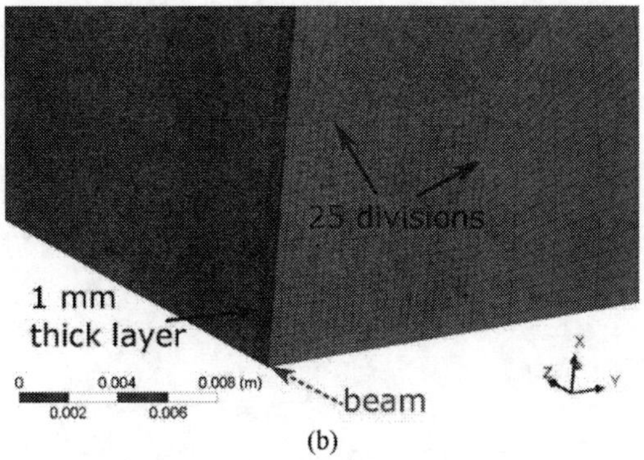

(b)

Fig. 3. ETL absorber: (a) geometrical model divided in mesh zones. The XZ and ZY planes are used for symmetry. The yellow edge is not allowed to displace in Z direction. (b) Meshing near the beam impact.

The projected range of 4.2 MeV/n lead ions is 20 μm in stainless steel [Fig. **4**(a)] and 46 μm in Antico 100. To discretize the very tight energy deposition through-thickness we opt for a multi-layered discretization of the area impacted by the beam following [Fig. **4**(b)] [2].

In ANSYS 17.1 we model this area with layered shell elements and couple them with regular solid elements. This method allows accurate layer-wise application of heat generation, greatly reducing the total amount of elements [2]. We use up to 30 layers for the multi-layered discretization.

Steady-state and transient-thermal models are solved in ANSYS 17.1 [7]. To guarantee correct initial time-step resolution we implement initial time steps smaller than $l^2/(4\alpha)$ [8].

For the beam injection scenarios we simulate the four-week beam operation through a steady-state analysis in which the heat generation is averaged by the pulse repetition rate (Fig. **5**). We use the steady-state temperatures as the initial condition to a transient-thermal analysis for modeling one steady pulse averaging the heat generation by the pulse length.

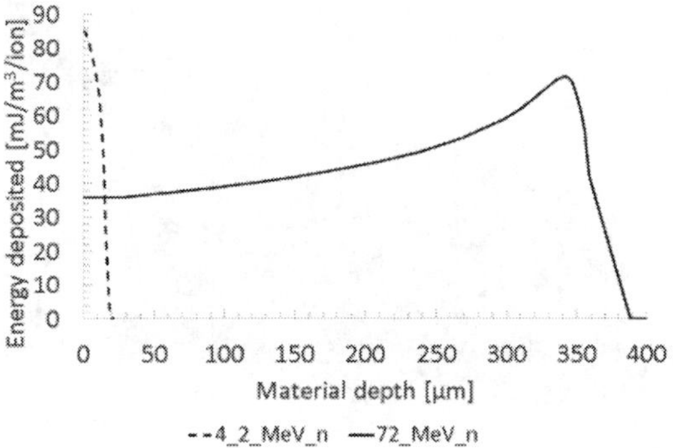

(a)

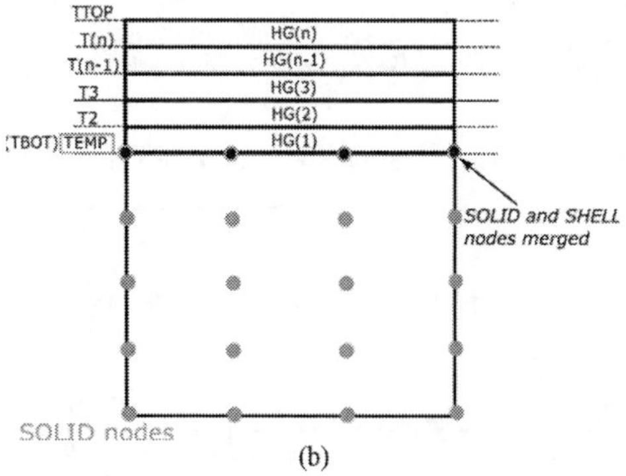

(b)

Fig. 4. (a) Projected range of lead ions at 4.2 MeV/n and 72 MeV/n in stainless steel [1, 6]. (b) Schematic representation of the multi-layered discretization of the geometrical model [2].

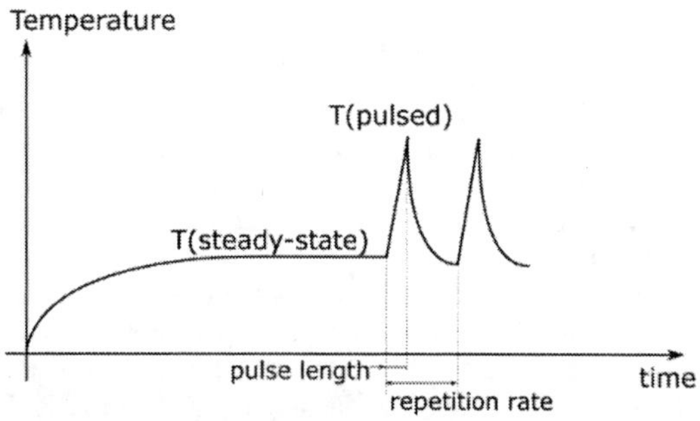

Fig. 5. Schematic representation of the steps for thermal modeling of beam injection scenarios.

For both stopper blocks a radiation emissivity of 0.1 from the outer core surfaces is implemented. For the ITF stopper block we implement a total heat conductance through the screws of 300 W/(m^{2}K), comparable to the one adopted in Ref. [9]. The initial temperature of both stoppers is assumed to be homogeneous at 22°C.

The equivalent stresses induced by the thermal gradients are calculated using the von-Mises yield criterion [10]. Considering the temperature-dependent yield strength as the structural limit, $\sigma_e/\sigma_y(T) < 1$ indicates the thermal deformation is reversible, otherwise it is permanent.

3 Results and discussion

3.1 Projected range of ions in matter

Table 2 presents the results of the range of argon, xenon and lead ions at 4.2 and 72 MeV/n in the aluminum alloy Antico 100 and stainless steel. Note that the range in matter is inversely proportional to the material density.

The range of the light argon ion in the ITF stainless steel block is 1 mm, shorter than the block thickness of 10 mm. In the ETL block out of Antico 100 the longest ion range (2.56 mm) is still much shorter than the total block length (210 mm). Both blocks completely stop the primary ion beams having overdimensioned block lengths.

In the thermal analysis we adjust the thickness of the layered-element shell according to each ion-matter scenario.

Table 2. Projected range of ions in matter via SRIM [11].

Ion	Ion energy [MeV/n]	Range in stainless steel [µm]	Range in Antico 100 [µm]
Lead	4.2	20	46
Xenon	4.2	17	39
Argon	4.2	17	39
Lead	72	382	980
Xenon	72	433	1100
Argon	72	1000	2560

3.2 Heat diffusivity and model verification

Because of long energy deposition rates of hundreds of microseconds and the tight projected ranges heat can easily diffuse in the material during beam impacts. In order to quantify submillimetric heat diffusion in the stopper blocks we compare the maximum adiabatic temperature with the maximum actual temperature in the block for one beam impact.

We calculate the adiabatic temperature at the position of maximum heat density deposition per ion ($Q_{\mathrm{ion,max}}$) in matter calculated via FLUKA using Eq. 1. In order to validate the heat import to the thermal model in ANSYS 17.1 we run a transient analysis of one beam impact in the material block with nearly zero diffusivity by setting the material thermal conductivity to 10^{-10} W/(mK).

$$T_{\mathrm{ad}} = \frac{Q_{\mathrm{ion,max}} n_{\mathrm{ion}}}{\rho c_p} + T_0 \quad (1)$$

The number of ions is calculated through Eq. 2.

$$n_{ion} = \frac{I\mu}{e|q|} \quad (2)$$

We perform another transient analysis of one beam pulse impact now with the actual material thermal conductivity. We compare the maximum temperature of both transient analyses.

Table 3 presents the results for lead ions at injection energy in the ITF and ETL blocks with a pulse length of 200 µs. Note that the hypothetical adiabatic temperatures serve only to present the effect of heat diffusion and validating thermal results. The adiabatic temperatures in both models have a maximum difference of 1.7%.

Table 3. Maximum adiabatic and actual temperatures of lead ions in ITF and ETL stoppers. Initial block temperature: 22°C.

Stopper block	Adiabatic [°C]		Actual [°C]
	FLUKA model	Thermal model	Thermal model
ITF (steel)	47.2	47.0	33.5
ETL (Al)	39.5	39.2	24.4

Turning on heat diffusion effects we estimate the actual peak temperature rise in the ITF steel block is 2.2 times smaller than the adiabatic results. Heat diffusion is stronger in the ETL aluminum-alloy block being 7.2 times smaller than the adiabatic result.

3.3 Thermo-structural performance

We discuss the performance of ion beam stoppers at injection and extraction energies.

3.3.1 Injection

Table 4 presents the maximum temperature in the stopper blocks for one beam pulse after steady state for different ions and beam-pulse setups. Although the block receives repeated beam during four continuous weeks thermal steady state is reached in few minutes. Fig. 6 shows the steady-state temperature distribution in the ITF and ETL blocks under repeated lead ion impacts with 0.2 s pulse repetition rate.

Table 4. Maximum temperatures for injection scenarios. Pulse length in brackets.

Stopper block	Ion	0.2 s (200 µs)		0.4 s (600 µs)	
		sty.-st.	Pulsed	sty.-st.	pulsed
ITF	Pb	31.9	43.1	37.3	57.1
	Xe	30.5	37.0	34.7	48.6
	Ar	32.8	38.5	38.0	49.7
ETL	Pb	22.3	24.7	22.4	30.3

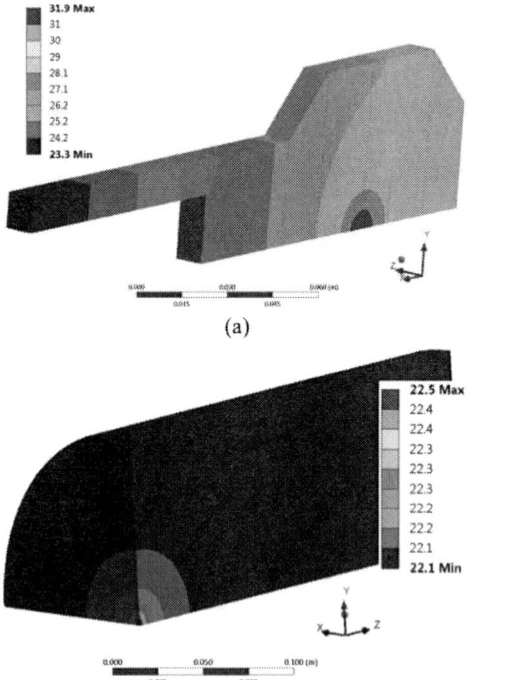

(b)

Fig. 6. Steady-state results for injection scenarios of lead ions with a repetition rate of 0.2 s in °C: (a) ITF block. (b) ETL block.

https://doi.org/10.1051/matecconf/201824001039

We skip the analysis on the ETL stopper block under xenon and argon ions as only very small temperature variations are expected.

The ion beam scenario with 0.4 s pulse repetition rate and 600 μs pulse length induces the highest stresses in both ITF and ETL blocks. The maximum equivalent stress in the ITF steel block is 86.7 MPa corresponding to max $\sigma_e/\sigma_y(T) = 0.55$. In ETL the maximum equivalent stress is 19.0 MPa with max $\sigma_e/\sigma_y(T) = 0.08$. We take temperature-dependent yield strengths from [12] for stainless steel and [13] for Antico 100.

3.3.2 Extraction

We combine the two bunches of an extracted ion beam in one single pulse with 400 ns pulse length. Because of this short pulse length heat deposition is faster than heat diffusion in the block. We calculate a difference of 1.0% between the peak adiabatic temperature and the actual peak temperature from the thermal model.

A single lead pulse induces a maximum temperature of 59.4°C and a maximum equivalent stress of 78.1 MPa in the ETL block, corresponding to max $\sigma_e/\sigma_y(T) = 0.33$. The maximum temperature and stresses are 45.5°C and 50.4 MPa for xenon and 28°C and 13.1 MPa for argon.

4 Redesign of the ion beam stoppers

Although the absorbing blocks of both ITF and ETL stoppers withstand all possible beam impact scenarios we investigate a compact stopper design for both ITF and ETL lines with thinner blocks.

4.1 Beam stopper design

With the objective of harmonizing the design of beam stoppers with those of similar devices at CERN we study the compatibility of a small size beam stopper design from ELENA (Extra Low Energy Antiproton) [14] adapted to Linac 3 and LEIR (Fig. 7).

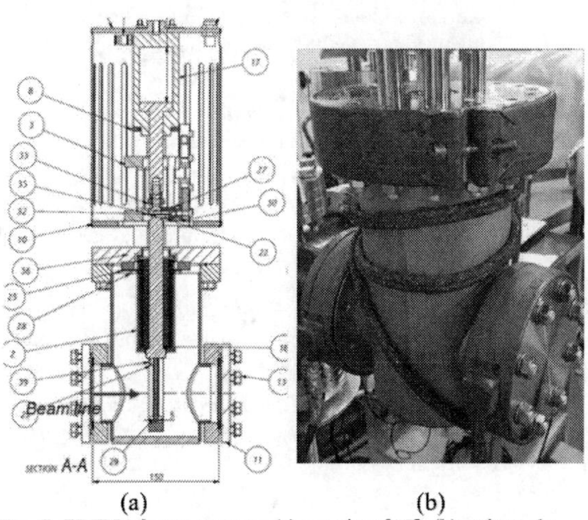

(a) (b)

Fig. 7. ELENA beam stopper: (a) cut view [15]. (b) unit ready for installation at company CINEL, Italy.

The absorbing block consists of a stainless steel plate with Ø83 mm and 5 mm thickness. However an extracted lead ion beam already induces a maximum temperature of 78.3°C and a maximum equivalent stress of 154 MPa and max $\sigma_e/\sigma_y(T) = 1.01$, exceeding the yield strength of stainless steel.

Maintaining the same ELENA-type beam stopper design the plate can be replaced by any geometry with Ø 83 mm and a maximum thickness of 20 mm.

4.2 Proposed materials for block absorbers

The absorber material must have good thermal conductivity to attenuate thermal gradients caused by continuous beam impacts at injection energies over four weeks. In order to allow the LEIR machine to run a wide range of ions at extraction energies the absorbing block must completely stop the lightest ion, which is helium.

From a thermo-structural perspective aluminum is a strong material candidate. However the project range of helium in aluminum is 20.6 mm, which is longer than the allowed thickness for the ELENA-type stopper.

Another alternative is a copper-based material. We choose CuCr1Zr, a copper-alloy with superior mechanical resistance and similar thermal properties as pure copper [16]. Knowing the projected range of helium ions ($^4He^{2+}$) at 72 MeV/n in this alloy is 7.5 mm we analyze the thermo-structural performance of a 10 mm thick CuCr1Zr block under both injection (4.2 MeV/n) and extraction (72 MeV/n) beam scenarios.

4.3 Performance of a CuCr1Zr block

Considering thermal cooling through radiation to the vacuum chamber ($\varepsilon = 0.1$) and poor heat conductance through supports of 25 W/(m^{2}K) [17] we analyze the beam injection scenario of lead ions continuously hitting the block over four weeks. We choose the worst-case beam scenario with 0.4 s pulse repetition rate and 600 μs pulse length. For some pulses after steady state the minimum and maximum temperatures in the block are 106.7 and 135.0°C respectively. This temperature distribution yields a maximum equivalent stress of 73.2 MPa and max $\sigma_e/\sigma_y(T) = 0.28$.

In order to assess the fatigue behavior of the new block we apply the Soderberg criterion (Eq. 3) [18]. The material is safe if $F > 1$.

$$F = \frac{1}{\left(\frac{S_m}{S_S} + \frac{S_a}{S_{LF}}\right)} > 1 \qquad (3)$$

The number of cycles over four weeks is 1.2×10^7 for a beam pulse with the shortest repetition rate. We consider $S_m = S_a = 36.6$ MPa, $S_S = 90$ MPa for this cycle range [19], and $S_{LF} = 0.7\sigma_y(T) = 185$ MPa at T = 140°C. The plate does not reach any fatigue limit as $F = 5.1$.

An extracted beam pulse of lead ions from LEIR induces a maximum temperature of 86°C and max $\sigma_e/\sigma_y(T) = 0.49$ (Fig. 8). The maximum energy deposition, temperature and equivalent stresses in the stopper lie in the beam axis. The temperature peaks at 320 μm.

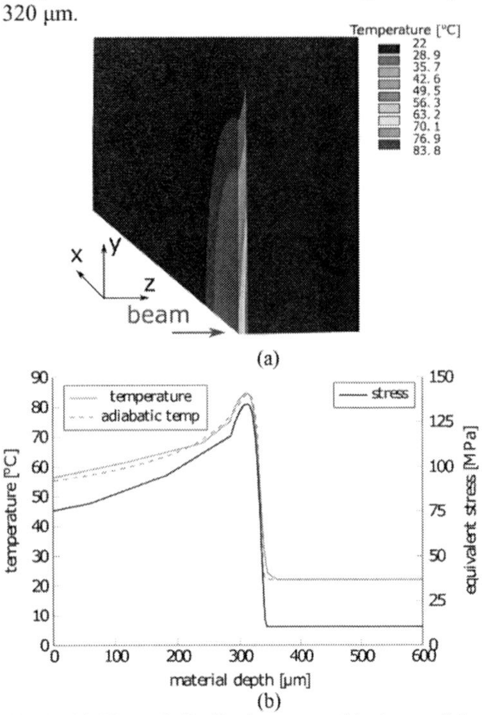

Fig. 8. (a) Thermal distribution stopper block out of CuCr1Zr after one extracted ion beam pulse from LEIR [2]. (b) Thermal and equivalent-stress distribution in the beam axis [2].

5 Conclusions

We validated the thermo-structural performance of beam stoppers under shallow energy deposition from intense ion beams in stopper targets at Linac 3 and LEIR. Taking into account short ranges of ion in matter from 17 μm to 2.56 mm we discretized the beam-impact zone with multi-layered elements. To verify this method we calculated a maximum temperature difference of less than 1.7% between the multi-layered and adiabatic-temperature calculations.

We analyzed the stoppers under long-term ion injection at 4.2 MeV/n and one-pulse extraction scenario at 72 MeV/n. For all ion beam configurations none of the targets reach structural limits. The most critical scenario was the impact of injected lead ions in the ITF stainless steel block over four continuous weeks inducing max $\sigma_e/\sigma_y(T) = 0.55$.

Although the ion stoppers withstand all beam configurations we proposed a compact beam stopper for both Linac 3 and LEIR, replacing the 210 mm long block at ETL. The new stopper block consists of a copper-alloy disc with Ø83 mm and 10 mm thickness. It withstands continuous impacts of 4.2 MeV/n lead ions with 0.4 s pulse repetition rate and 600 μs pulse length over four weeks as well as one extracted lead ion at 72 MeV/n with a pulse length of 400 ns. We designed the stopper block thicker than the range of the lightest ion in the copper alloy, allowing Linac 3 and LEIR to run any ion with mass number from 4 (helium) to 208 (lead).

The authors want to thank Richard Scrivens for beam operation data and Sven de Man for the design data on the ELENA-type beam stoppers.

References

[1] A. Ferrari, P. R Sala, A. Fasso, J. Ranft, *FLUKA: A multi-particle transport code* (Program version 2005, Stanford, CA 94309, 2005)

[2] A. Pilan Zanoni, Phys. Rev. Accel. Beams, **21**, 5, p.054801 (2018)

[3] S. Gilardoni, D. Manglunki et al, *Fifty years of the CERN Proton Synchrotron* (CERN, 2013)

[4] D. Manglunki et al, *CERN's Fixed Target Primary Ion Programme* (International Particle Accelerator Conference, Busan, Korea, 2016)

[5] R. Scrivens, "Beam Parameters from Linac3 to LEIR for Beam stoppers," 2017.

[6] T. Boehlen et al, Nuclear Data Sheets, **120**, pp.211-214 (2014)

[7] ANSYS Academic Research, Release 17.1 (manual).

[8] R. Stout, P. D. Billings and PEON Semiconductors, *Accuracy and time resolution in thermal transient finite element analysis*, (ANSYS-TM Users Conference, 2002)

[9] J. J. Scialdone, C. H. Clatterbuck, J. L. Wall, *Thermal conductance of two interface materials and their applications in space systems*, (NASA Technical Memorandum 104554, March 1992)

[10] R. v. Mises, Nachrichten von der Gesellschaft der Wissenschaften zu Goettingen, Mathematisch-Physikalische Klasse, **1913**, 4, pp.582-592 (1913)

[11] J. F. Ziegler, M. D. Ziegler, J. P. Biersack, Nuclear Instruments and Methods in Physics Research Section B: Beam Interactions with Materials and Atoms, **268**, 11-12, pp.1818-1823 (2010)

[12] J. F. Dempsey et al, Challenges in Mechanics of Time-Dependent Materials and Processes in Conventional and Multifunctional Materials, **2**, pp.7-15 (2013)

[13] Pechinery group, *Aluminium - semi-finished products* (Paris: Edition Pechiney Rhenalu, 1997)

[14] D. Horvath, *FLUKA calculations for AD beam stopper design* (CERN - EDMS 1529712, 2015)

[15] *Drawing H1AB0100 (company CINEL, Italy)*, 2015-2017.

[16] D. Kupferinstitut, *Kupferdatenblatt CuCr1Zr*, (Deutsches Kupferinstitut, Duesseldorf, Germany, 2005)

[17] J. E. Fontenot Jr, *The Thermal Conductance of Bolted Joints* (Louisiana State University (LSU Historical Dissertations and Theses) 1968)

[18] E. Iurzolla, I criteri di resistenza, Cortina, 1991.

[19] *CES Selector 2017*, CERN.

Numerical study of the effect of forced convective flow on dropwise condensation by thermal LBM simulation

Shaofei Zheng[1,*], *Ferdinand* Eimann[1], *Tobias* Fieback[1], and *Ulrich* Gross[1]

[1]Institute of Thermal Engineering, TU Bergakademie Freiberg, Gustav-Zeuner-Str. 7, 09599 Freiberg, Germany

Abstract. The enhancement mechanism of forced convective flow on dropwise condensation over a cold spot is numerically investigated by two-dimensional hybrid thermal lattice Boltzmann (LB) model based on the Shan-Chen pseudopotential LB model. After validating the present LB model, dropwise condensation over a cold spot as the nucleation region is simulated. The well-known power law for the growth of a single condensing droplet is demonstrated. Finally, the simulation of dropwise condensation considering the convection flow or not is carried out in the constant contact radius (CCR) mode. Using the CCR model, the effect of contact angle can be also investigated. The result of streamline field indicates that the forced convectional flow complicates the internal flow of droplet and main flow. The dragging force from main flow changes the size of two symmetric vortices inside the droplet. And the channel flow is also strongly influenced by the suction effect caused by condensation at the three phase contact line. By comparison, the heat transfer enhancement of the superimposed flow is not worth mentioning. The present study illustrates the mechanisms of dropwise condensation under forced convectional flow.

1 Introduction

Because of the complicated dynamic behavior of droplets, dropwise condensation becomes to be a typical case of multiphase flow involving vapor-liquid phase change, free surface flow, coupled heat and mass transfer, multi-scale feature, and interfacial behaviors. Therefore, even though various experimental and theoretical investigations have been carried out, dropwise condensation has been rarely studied numerically [1, 2].

Over the past two decades, compared to the conventional CFD based multiphase models, the LB model has been developed into a powerful simulation method in modeling multiphase flow undergoing phase change. Theories and applications of LB model for multiphase flow and phase-change heat transfer have been deeply reviewed and summarized by Li and his co-workers [3]. For multiphase flow, a popular model used in LB model is the pseudopotential LB method proposed by Shan and Chen [4, 5] because of its computational efficiency and straightforward algorithm. In this model, the phase separation is realized spontaneously by the introduced short-range intermolecular interaction based on a potential function. To alleviate the weaknesses of the original Shan-Chen model, several improvements [6-11] have been carried out. For phase-change heat transfer, the thermal pseudopotential LB model is developed by devising the temperature distribution function [12-16]. The temperature distribution function is used to solve the macroscopic energy equation. The hybrid thermal LB model, which solves the temperature equation by the finite-difference scheme, is devised for avoiding the error term in energy equation recovered. Using the hybrid thermal LB model, Li and his co-authors have made several successful works [17-18].

Using the hybrid thermal LB model, the present paper is focused on the coupling effect between the fully developed laminar flow and dropwise condensation. After validating the present LB model, dropwise condensation considering the convective flow or not are simulated using the same properties. The main objective is to understand the mechanism of dropwise condensation in the forced convective laminar flow.

2 Model description

The multiphase single-component pseudopotential LB model proposed by Shan and Chen [4, 5] is employed to simulate the multiphase flow for the density and velocity fields. The energy equation is discretized by the finite difference scheme for the temperature field.

2.1 The multiphase single-component pseudopotential LB model

Based on the Bhatnagar-Gross-Krook (BGK) collision operator [19], the discretized LB equation is given below for the flow fields:

* Corresponding author: shaofei.zheng@ttd.tu-freiberg.de

$$f_\alpha\left(\boldsymbol{x}+\boldsymbol{e}_\alpha\delta_t,t+\delta_t\right)-f_\alpha\left(\boldsymbol{x},t\right)= \\ -\frac{1}{\tau_f}\left(f_\alpha\left(\boldsymbol{x},t\right)-f_\alpha^{eq}\left(\boldsymbol{x},t\right)\right)+\Delta f_\alpha\left(\boldsymbol{x},t\right) \tag{1}$$

where $f_\alpha(\boldsymbol{x},t)$ denotes the particle distribution function, $\boldsymbol{x}$ is the particle position, t is the time, $\boldsymbol{e}_\alpha$ represents the discrete velocity along the α-th direction, τ_f is called as the non-dimensional relaxation time, the equilibrium density distribution function $f_\alpha^{eq}(\boldsymbol{x},t)$ can be written as below:

$$f_\alpha^{eq}\left(\boldsymbol{x},t\right)=w_\alpha\rho\left[1+\frac{\boldsymbol{e}_\alpha\cdot\boldsymbol{u}}{c_s^2}+\frac{\left(\boldsymbol{e}_\alpha\cdot\boldsymbol{u}\right)^2}{2c_s^4}-\frac{\boldsymbol{u}^2}{2c_s^2}\right] \tag{2}$$

where w_α and c_s are the weight coefficients and lattice sound speed, respectively. In this work, two-dimensional 9-velocity scheme (D2Q9) is chosen. The lattice quantities (density ρ and velocity $\boldsymbol{u}$) are given as:

$$\rho=\sum_\alpha f_\alpha \qquad \rho\boldsymbol{u}=\sum_\alpha \boldsymbol{e}_\alpha f_\alpha \tag{3}$$

In Eq. (1), $\Delta f_\alpha(\boldsymbol{x},t)$ is the body force term. In LB model, the hydrodynamic equations are correctly recovered by incorporating the body force. According to the former works [8, 20], the exact difference method (EDM) developed by Kupershtokh et al. [21] shows better stability and accuracy. Therefore, the EDM forcing scheme is used in this work:

$$\Delta f_\alpha\left(\boldsymbol{x},t\right)=f_\alpha^{eq}\left(\rho,\boldsymbol{u}+\Delta\boldsymbol{u}\right)-f_\alpha^{eq}\left(\rho,\boldsymbol{u}\right) \tag{4}$$

where $\Delta\boldsymbol{u}=\boldsymbol{F}\delta_t/\rho$ is the effect of body force in velocity during each time step. For EDM forcing scheme, the actual fluid velocity $\boldsymbol{v}$ is defined by:

$$\boldsymbol{v}=\boldsymbol{u}+\frac{\delta_t\boldsymbol{F}}{2\rho} \tag{5}$$

In pseudopotential LB model, the short-range intermolecular interaction force is significant and responsible for the phase transition. The mixed scheme for the interparticle interaction force is used in this work:

$$F_{\text{int}}\left(\boldsymbol{x}\right)=-\beta\psi\left(\boldsymbol{x}\right)\sum_\alpha G\left(\boldsymbol{x}+\boldsymbol{e}_\alpha\right)\psi\left(\boldsymbol{x}+\boldsymbol{e}_\alpha\right)\boldsymbol{e}_\alpha \\ -\frac{1-\beta}{2}\sum_\alpha G\left(\boldsymbol{x}+\boldsymbol{e}_\alpha\right)\psi^2\left(\boldsymbol{x}+\boldsymbol{e}_\alpha\right)\boldsymbol{e}_\alpha \tag{6}$$

where β is the weighting factor depending on the non-ideal equation of state. The Green function $G(\boldsymbol{x}+\boldsymbol{e}_\alpha)$ denotes the interaction strength, which can be found in [20].

The pseudopotential $\psi(\boldsymbol{x})$ is determined by the local density and equation of state and given by:

$$\psi\left(\boldsymbol{x}\right)=\sqrt{\frac{2\left(p_{EOS}-\rho c_s^2\right)}{c_0 g}} \tag{7}$$

The pseudopotential $\psi(\boldsymbol{x})$ is linked to the temperature fields through the non-ideal equation of state. The Peng-Robinson (P-R) equation of state is adopted in this work:

$$p_{EOS}=\frac{\rho RT}{1-b\rho}-\frac{a\varphi\left(T\right)\rho^2}{1+2b\rho-b^2\rho^2} \tag{8}$$

where $a = 0.45724R^2T_c^2/p_c$ and $b = 0.0778RT_c/p_c$, $\varphi(T)$ can be found in [6]. In this work, we set $a = 2/49$, $b = 2/21$ and $R=1$. In present work, the saturation temperature is set as $T_{sat} = 0.86T_c$, which has the coexistence liquid density $\rho_l = 6.5$ and the coexistence vapor density $\rho_v = 0.38$. After numerical test, the weighting factor in Eq. (6) is used as $\beta = 1.18$ which can give stable and accurate results in a wide temperature range. It should be mentioned that the lattice unit (l.u.) is used for all quantities here. And the gravity is free.

2.2 Energy equation

After neglecting the viscous heat dissipation, the governing equation for the temperature field can be derived from the local balance law of entropy [17]:

$$\partial_t T= \\ -\mathbf{v}\cdot\nabla T+\frac{1}{\rho c_v}\nabla\cdot\left(\lambda\nabla T\right)-\frac{T}{\rho c_v}\left(\frac{\partial p_{EOS}}{\partial T}\right)_\rho\nabla\cdot\mathbf{v} \tag{9}$$

with the thermal conductivity λ and the specific heat at constant volume c_V. The second-order Runge-Kutta method is employed to implement the time discretization:

$$T^{t+\delta_t}=T^t+\frac{\delta_t}{2}\left(h_1+h_2\right) \tag{10}$$

$$h_1=K(T^t),\ h_2=K(T^t+\frac{\delta_t}{2}h_1) \tag{11}$$

where $K(T)$ is the right-side term of Eq. (9). Based on the second-order isotropic difference schemes, the spatial gradient and Laplacian of and arbitrary quantity ϕ can be calculated as:

$$\partial_i\phi\left(\boldsymbol{x}\right)\approx\frac{1}{c_s^2\delta_t}\sum_\alpha w_\alpha\phi\left(\boldsymbol{x}+\boldsymbol{e}_\alpha\delta_t\right)\boldsymbol{e}_{\alpha i} \tag{12}$$

$$\nabla^2\phi\left(\boldsymbol{x}\right)\approx\frac{2}{c_s^2\delta_t^2}\sum_\alpha w_\alpha\left[\phi\left(\boldsymbol{x}+\boldsymbol{e}_\alpha\delta_t\right)-\phi\left(\boldsymbol{x}\right)\right] \tag{13}$$

3 Model validations

To validate the capability of present model for simulating vapor-liquid phase change, several tests are carried out.

3.1 Phase transition

The 2D phase transition process at a constant temperature is simulated. A square domain (150 × 150 l.u.) with all periodic boundary conditions is initialized with the local density $\rho = \rho_c + \Delta\rho$, where $\Delta\rho$ is a small random perturbation parameter. During simulation, the phase transition is achieved automatically and the droplet arises after reaching the equilibrium state. Then the corresponding vapor and liquid densities can be determined. Over a wide temperature range, the liquid and vapor densities from the simulation show a good agreement with the theoretical Maxwell construction, as shown in Fig. 1.

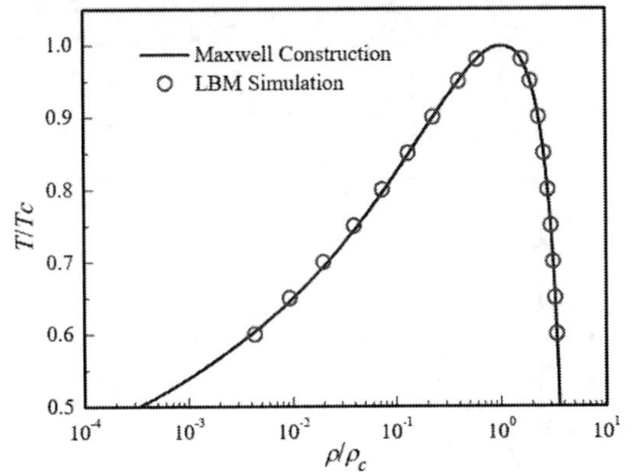

Fig. 1. Coexistence curve from the simulation and P-R equation of state.

3.2 Laplace test

For Laplace test, a droplet with radius R is initialized at the center of the computational domain having all periodic boundary conditions. Once the simulation reaches to the equilibrium state, the pressure difference Δp across the droplet interface can be quantified. The simulations are carried out for different droplet radius at three different temperatures, and the results are plotted in Fig. 2. According to the Laplace law, the pressure difference Δp is the linear function of the inverse of $1/R$, in which the coefficient represents the surface tension. Obviously, the simulation results follow Laplace law.

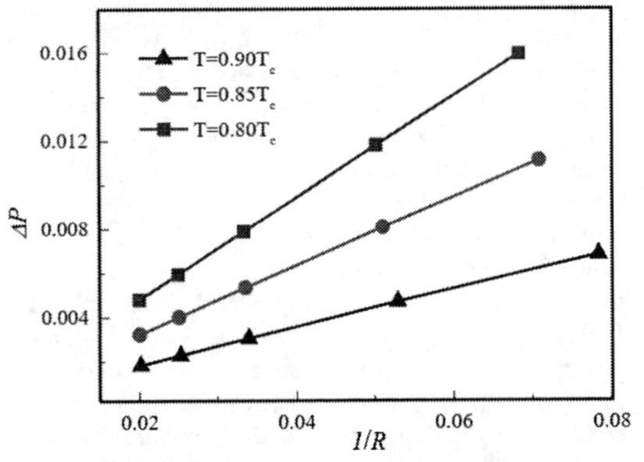

Fig. 2. Laplace test for the liquid droplet at different temperatures.

3.3 Wettabilit test

The equilibrium contact angle (wetting condition) on solid wall is implemented through the geometric formulation proposed by Ding and Spelt [22]. Using the second-order central difference scheme, the discretized form for numerical application is written as:

$$\rho_{i,0} = \rho_{i,2} + \tan\left(\pi/2 - \theta_w^a\right)\left|\rho_{i+1,1} - \rho_{i-1,1}\right| \quad (14)$$

with the prescribed static contact angle θ_w^a, $\rho_{i,0}$ denotes the density of the ghost layer $(i,0)$ beneath the boundary lattices, where the first and second indexes represent the coordinate tangential and normal to the boundary. For $\theta_w^a = 60^\circ$ and 120°, the numerical simulation obtain the equilibrium contact angle $\theta_w = 60.6^\circ$ and 121.5° respectively. Fig. 3 shows the equilibrium shape of liquid droplet over the solid wall.

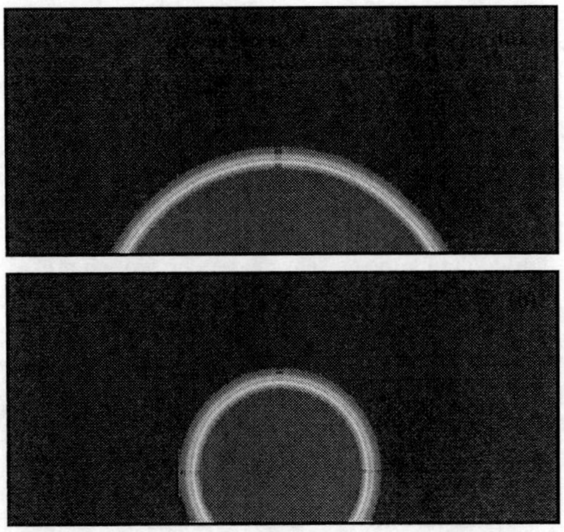

Fig. 3. Contact angles obtained by the geometric formulation approach. (a) $\theta_w^a = 60^\circ$, (b) $\theta_w^a = 120^\circ$.

3.4 2D droplet evaporation

Evaporation of a stationary droplet is considered to verify the accuracy of the present model in simulating liquid-vapor phase change. For droplet evaporation, the classic D2 law states that the square of the droplet radius reduces linearly over time. The simulation is examined on a 200 l.u. × 200 l.u. computational domain with a droplet (radius R = 30 l.u.) placed at the center. At the initial time step, the droplet temperature is set at the saturation temperature $0.86T_c$ and a uniform temperature $1.0T_c$ is given to the vapor surrounding the droplet. All boundaries are employed at the periodic boundary condition and constant temperature condition ($1.0T_c$). The kinematic viscosity is chosen as $\upsilon = 0.15$ ($\tau_f = 0.95$). The specific heat at a constant volume is c_v = 5.0. According to the D2 law, the thermal conductivity λ should be constant, and the evaporation rate depends linearly on λ. Two cases are carried out: Case A with λ = 2/3, Case B with λ = 3/3. Fig. 4 plots the square of the droplet radius normalized by the initial radius during the evaporation process for these two cases. As expected, the linear relationship between $(R/R_0)^2$ and t is clearly observed for two cases, and the increasing λ accelerates the evaporation.

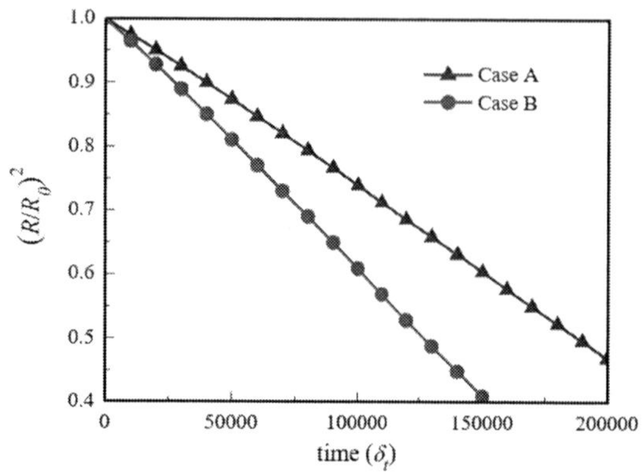

Fig. 4. The square of the non-dimensional droplet radius versus time (R_0 is the initial droplet radius).

4 Results and discussion

4.1 Dropwise condensation on plate

After validating the present model, the growth process of a single droplet is simulated in a rectangle domain ($L_x \times L_y = 800 \times 200$ l.u.). The thermal conductivity is taken as $\lambda = \rho c_V \chi$ with $c_V = 5.0$ and $\chi = 0.04$. The kinematic viscosity υ is set to 0.05. The whole domain is initialized as the vapor having the saturation temperature T_{sat}. A cold spot of 30 l.u. as a nucleation region is placed at the center of bottom wall and is fixed at $T_w = 0.7T_c$. The rest of the bottom wall is at $T_w = 0.83T_c$. The periodic boundary condition is applied in the x-direction. For the bottom surface, the non-slip boundary condition including the forcing term is used. And the convective boundary condition proposed by Lou et al. [23] is specified at the top boundary. The prescribed static contact angle is $\theta_w^a = 90°$ and the contact angle hysteresis is considered with a hysteresis window (0°, 90°) [24].

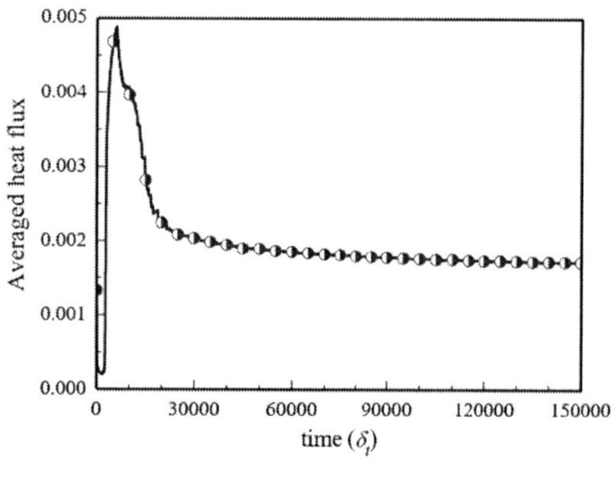

Fig. 5. Variation of the averaged heat flux on the cold spot during condensation.

For showing the heat transfer during condensation, the variation of the averaged heat flux over the cold spot with time is plotted in Fig. 5. At the start of condensation, the heat flux increases dramatically because of nucleation. Once the first droplet forms, the increasing conduction resistance of droplet decreases the heat flux. Finally, the heat flux goes to a stable state. It can be concluded that the heat transfer of dropwise condensation is mainly attributed by the small size droplet. Additionally, Fig. 6 presents the time evaluation of droplet radius during condensation. As known, a single droplet growth in condensation can be described by the power law ($R \sim t^a$). Based on a theoretical analysis, the power law exponent $a = 0.5$ for a 2D droplet growth has been demonstrated by Ashrafi and Moosavi [25]. As seen in Fig. 6, the present simulated droplet radius obey a power law with $a = 0.407$. The difference with the theoretical value falls into the normal difference range presented in reference [25].

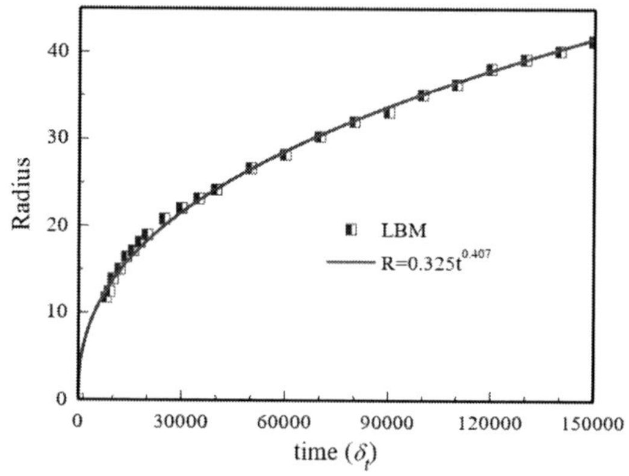

Fig. 6. Comparison of the simulated droplet radius during time with the power law.

4.2 The effect of forced convective flow

In this part, the effect of forced convection flow for dropwise condensation will be investigated. Two cases are considered for comparision: Case A without convection flow, Case B with convection flow. For these two cases, the simulations are carried out in the rectangle domain having the same grid size ($L_x \times L_y = 800 \times 200$ l.u.). The same properties are used: the kinematic viscosity $\upsilon = 0.05$, the thermal conductivity $\lambda = \rho c_V \chi$ with $c_V = 8.0$ and $\chi = 0.06$. The boundary conditions for Case A are listed below: the periodic boundary condition in x-direction, the non-slip boundary condition for the bottom solid wall and the top open boundary described by the convective boundary condition. For Case B, different boundary conditions are used to include the convective flow. The top and bottom are considered as a solid wall implementing the non-slip boundary condition. Following the parabolic velocity profile along y-direction, the velocity inlet boundary condition is implemented. As before, the convective outlet boundary condition is employed at the right boundary. These two cases have the same thermal boundary condition, which set the saturation temperature T_{sat} for all boundaries in spite of a cold spot 80 l.u. located in the center of bottom wall. As the nucleation region, the temperature of cold spot is set as T_w, and the sub-cooling $\Delta T = T_{sat} - T_w$ is chosen as $0.16T_c$. For Case B, the parabolic velocity profile is initialized on the whole computational domain, and the first 20 000 steps are carried out without condensation so that the velocity distribution can reach its equilibrium state.

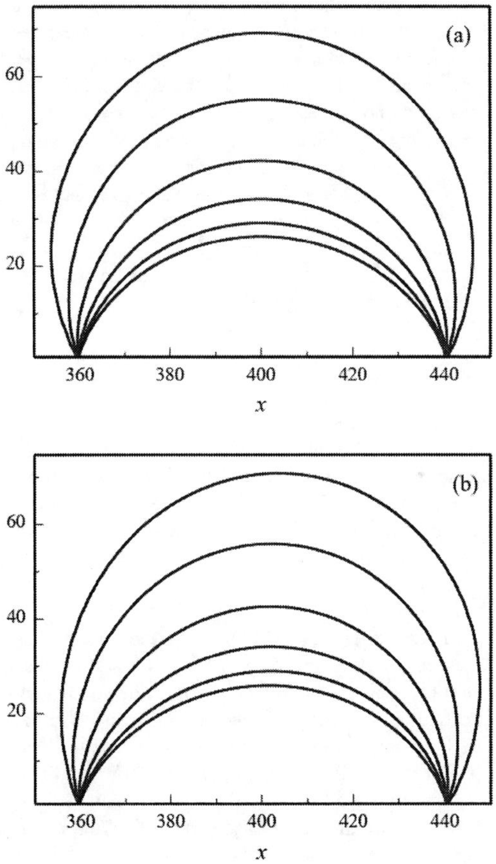

Fig. 7. Evolution of the droplet profile during condensation. (a) Case A, (b) Case B.

At the beginning of condensation, the wettability effect is not added so the liquid can cover the cold spot freely. When the whole cold spot is covered by the liquid film, the wettability effect is taken into consideration with a hysteresis window of (0°, 180°) for contact angle. Therefore, the droplet grows in the constant contact radius (CCR) mode, which implies that the contact angle increases gradually whereas the contact line is pinned on the solid wall. Using this way, the effect of contact angle can be also researched and the droplet basis can always keep at the constant temperature T_w. In this work, the averaged inlet velocity is $U = 0.025$. Thus, the Reynolds number is $Re = UL_y/\upsilon = 100$ which corresponds to the laminar flow. Fig. 7 gives the evolution of droplet profile during condensation for two cases. In Case A, the contact angle is increased owing to the constant contact radius. Due to the shearing effect of external flow, the deformation of droplet can be observed and the deformation is enhanced with the growth of droplet.

In order to deeply understand the mechanism of dropwise condensation, it is necessary to study the fluid dynamics. Fig. 8 shows the streamline field inside and outside of the condensing droplet for two cases. As shown in the left side, three different contact angles (CA) for Case A are presented, which is corresponded by Case B in right side at the same condensation time. For Case A, the x-coordinate of showing region span from 320 to 480. In order to display the flow field in the channel for Case B, the range from 310 to 560 is used. That is why the droplet in right side looks smaller than the left side although the droplet size from two cases keeps close at the same condensing time. From the left side, two symmetric vortices (namely, Marangoni flow) inside the droplet can be clearly found in different CA because of the temperature gradient. Also it can be seen that most of the vapor flow goes to the three phase contact line and condensate. However, two symmetric vortices are replaced by two different size vortices in Case B. The left vortex is narrowed by the flow dragging force having the opposite direction, while the same direction of the flow dragging force expand the right vortex. An interesting phenomenon is the suction effect caused by the condensation at the three phase contact line. The pressure difference induced by the suction effect forces the main flow going to the bottom wall, and the main flow is strongly compressed under a big size droplet. At the small contact angle CA = 69.4°, the strong suction effect brings out a strong compression for the main flow. Additionally, the lee-side vortex structure is also replaced by the suction flow in the back region of droplet, and the region of suction flow is enlarged with the increasing of droplet height.

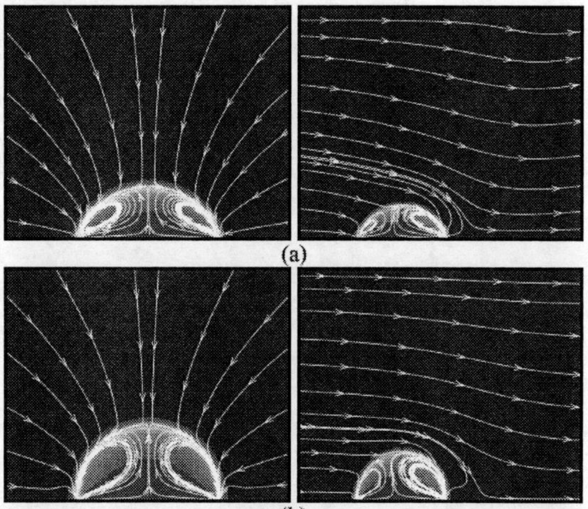

Fig. 8. Streamline field inside and outside of the droplet for Case A (left side) and Case B (right side). (a) CA = 69.4°, (b) CA = 90.8°, (c) CA = 118.3°.

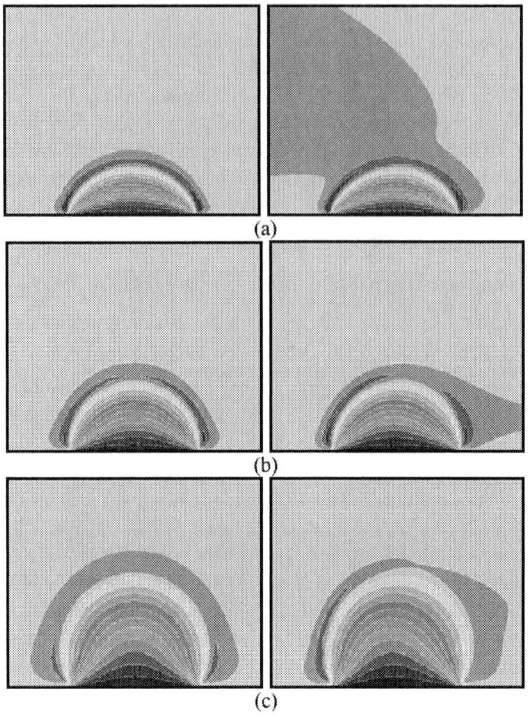

Fig. 9. Temperature distribution inside and outside of the droplet for Case A (left side) and Case B (right side). (a) CA = 69.4°, (b) CA = 90.8°, (c) CA = 118.3°.

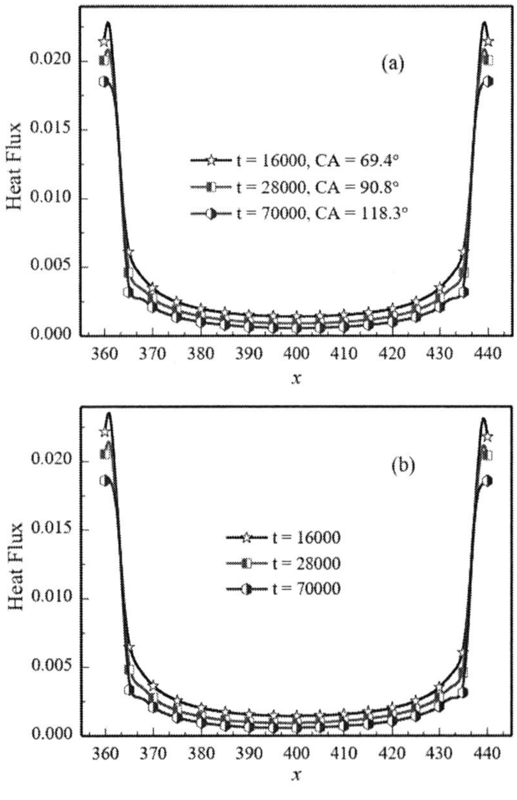

Fig. 10. Local heat flux over the cold spot. (a) Case A, (b) Case B.

About the effect of convection flow in heat transfer, Fig. 9 and Fig. 10 present the temperature distribution and the local heat flux over the cold spot. From Fig. 9, it looks that the convection flow has no influence for the heat transfer inside the droplet. Also it can be found that the maximum temperature gradient locates at the three phase contact line, which can also explain that three phase contact line is the main place for condensation. From the heat flux shown in Fig. 10, the heat flux is decreased with increasing the contact angle, which is attributed to the variation of conduction resistance. Comparing the local heat flux for two cases, the heat transfer is slightly enhanced by the forced convectional flow. Therefore, it can conclude that the effect of forced convectional flow for heat transfer can be neglected compared with the latent heat released by condensation.

5 Conclusions

In present work, the 2D hybrid thermal LB model is used to simulate dropwise condensation. The fluid flow and temperature field is solved by the pseudopotential LB model and finite difference scheme, respectively. In order to avoid the drawbacks of the original pseudopotential LB model, some recent improvements from literature are utilized in the present work. For validation, some tests are also carried out.

Based on the CCR model, the effect of forced convective flow for dropwise condensation in different contact angles is investigated. For two symmetric vortices inside the droplet in dropwise condensation without convection flow, superimposed flow narrows the left vortex while enlarges the right one. The maximum temperature gradient can explain that most of the vapor flow goes to the three phase contact line and condensates. Condensation at three phase contact line brings out a suction effect. The suction effect is responsible for the flowing down of the main flow. Compared with the latent heat of condensation, the contribution of forced convective flow for heat transfer is no worth of mention. Therefore, the conclusion is that the superimposed convective flow complicates the fluid dynamics for the main flow and internal flow of droplet whereas the enhancement effect for heat transfer is trivial.

The authors want to thank the financial support from China Scholarship Council (CSC) and German Research Foundation (DFG, Grant No. Gr1060/18).

References

1. J. W. Rose, Proc. Inst. Mech. Eng., part A: J. Power and Energy **216**, 115-128 (2002).
2. A. A. Mohamad, *Lattice Boltzmann Method: Fundamentals and Engineering Applications with Computer Codes* (Springer, London, 2011).
3. Q. Li, K. H. Luo, Q. J. Kang, Y. L. He, Q. Chen, Q. Liu, Prog. Energy Combust. Sci. **52**, 62-105 (2016).
4. X. Shan, H. Chen, Phys. Rev. E **47**, 1815-1820 (1993).
5. X. Shan, H. Chen, Phys. Rev. E **49**, 2941-2948

(1994).
6. P. Yuan, L. Schaefer, Phys. Fluids **18**, 042101 (2006).
7. A. L. Kupershtokh, D. I. Karpov, D. A. Medvedev, C. P. Stamatelatos, V. P. Charalambakos, E. C. Pyrgioti, D. P. Agoris, IET Sci. Meas. Technol. **1**, 303-311 (2007).
8. Q. Li, K. H. Luo, X. J. Li, Phys. Rev. E **86**, 016709 (2012).
9. X. Shan, Phys. Rev. E **73**, 047701 (2006).
10. M. Sbragaglia, R. Benzi, L. Biferale, S. Succi, K. Sugiyama, F. Toschi, Phys. Rev. E **75**, 026702 (2007).
11. Q. Li, K. H. Luo, Appl. Therm. Eng. **72**, 56-61 (2014).
12. R. Zhang, H. Chen, Phys. Rev. E **67**, 066711 (2003).
13. G. Hazi, A. Markus, Int. J. Heat Mass Transfer **52**, 1472-1480 (2009).
14. S. Gong, P. Cheng, Int. J. Heat Mass Transfer **55**, 4923-4927 (2012).
15. M. R. Kamali, J. J. J. Gillissen, H. E. A. van den Akker, Phys. Rev. E **88**, 033302 (2013).
16. Q. Li, P. Zhou, H. J. Yan, Phys. Rev. E **96**, 063303 (2017).
17. Q. Li, Q. J. Kang, M. M. Francois, Y. L. He, K. H. Luo, Int. J. Heat Mass Transfer **85**, 787-796 (2015).
18. Q. Li, P. Zhou, H. J. Yan, Langmuir **32**, 9389-9396 (2016).
19. Y. H. Qian, D. D'Humieres, P. Lallemand, Europhysics Letters **17**, 479-484 (1992).
20. S. Gong, P. Cheng, Computers & Fluids **53**, 93-104 (2012).
21. A. L. Kupershtokh, D. A. Medvedev, D. I. Karpov, Comput. Math. Appl. **58**, 965-974 (2009).
22. H. Ding, P. D. M. Spelt, Phys. Rev. E **75**, 046708 (2007).
23. Q. Lou, Z. Guo, B. Shi, Phys. Rev. E **87**, 063301 (2013).
24. L. Wang, H. B. Huang, X. Y. Lu, Phys. Rev. E **87**, 013301 (2013).
25. A. Ashrafi, A. Moosavi, J. Appl. Phys. **120**, 124901 (2016).

Numerical study of heat and mass flow in layered heat storage

Bartosz Ziegler[1,*], *Jędrzej* Mosiężny[1], and *Paweł* Czyżewski[1]

[1]Poznań University of Technology, Chair of Thermal Engineering, ul. Piotrowo 3, 60-965 Poznań, Poland

Abstract. An analysis of argon gas flow in exemplary configuration of layered bed thermal energy storage is presented. The analysis incorporates URANS model with conjugate heat transfer between gas and solid storage core. The aim of this analysis was to identify key factors limiting exergy efficiency of this storage type and determine some details about storage transient behaviour. Three full cycles of storage loading and unloading having 17 hours physical time in total are simulated, with calculation of exergetic efficiency for each of the cycles. Conclusions regarding this storage type feasibility for indirect power storage in pumped heat systems are made.

1 Introduction and motivation

This work refers to article provided by authors [1] and awaiting publication.

Many energy markets experience rapid increase in renewable energy sources share, motivated by need to reduce CO_2 emissions and related climate changes [2]. This solution suffers from general mismatch between renewable (mainly solar and wind) production and energy demand of the grid. Overproduction should be stored, otherwise it is wasted. Stored surplus can be then used when demand rises above production. Modern EES (electrical energy storage) methods are seen as solution to face this problem.

Considered EES methods must meet many economical and operational often conflicting requirements such as: meeting peak energy demand, flattening RE generation fluctuations, support of energy system management [3] and provide flexibility for introduction of new forms of modern economy like smart grids, electrical vehicle grids and distributed power generation.

Economic viability determined by daily electricity price ratio is fundamental. There are many important parameters of ES methods characterizing their utility such as overall efficiency, energy density, power density, specific energy, specific power, power rating and rated energy capacity. For commercial feasibility, system electric efficiency must be at least as high, as the average ratio of low energy price (when surplus of energy production occurs) to the high price (when there is a shortage of energy).

EES based on indirect storage of energy in the form of heat (Thermal Energy Storage – TES), offers economically favorable investment costs with relatively high energy density, especially comparing to batteries, considered as potential best solution for modern ES approaches [4]. Overall efficiency of TES systems is impacted most by compression/expansion inefficiencies and thermal store exergy losses [5]. Resulting efficiencies are usually too low, when referring to daily electricity price fluctuations in most energy markets. Therefore a concept called pumped heat energy storage (PHES) is believed to meet this obstacle reaching more than 70% overall efficiency. PHES concept is based on heat pump thermodynamic cycle principle during loading and heat engine cycle during discharging of storage tank. The diagram of one of the possible PHES configurations is shown in figure 1.

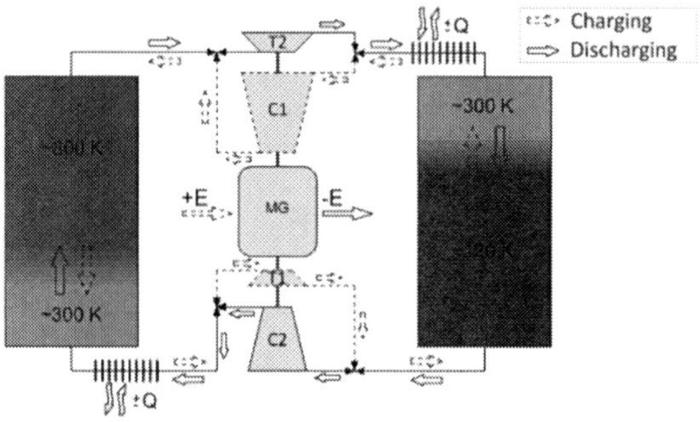

Fig. 1. Scheme of PHES.

G.F Frate et al. proved that PHES is promising ES technology and founded that the main approach to boost PHES performance is thermal integration. They achieved best round-trip efficiency results for working fluid R1233zd at 110°C, but for large industry scale applications it is crucial to prove economical reasonable gas [6].

Roskosch and Atakan investigated PHES with Rankine cycle (ORC cycle). They conclude that overall efficiency decreases deeply with increasing storage temperature and achieved 70% of overall efficiency in most favorable case [7].

Most of performed work on PHES take advantage of expensive working fluid like ORC, ammonia or refrigerants. It involves high costs of their purchase and further recycling or disposal. It impacts highly operating

* Corresponding author: bartosz.ziegler@put.poznan.pl

costs and final profitability. Therefore it is necessary to investigate less sophisticated working fluid to face this obstacle. Because of its thermodynamic characteristics, availability and safety reasons, argon seems to be very promising working fluid for utilization in PHES systems.

Small number of papers describe existing prototypes of PHES in the context of heat transfer and storage parts of the system [8]. This paper concerns the CFD calculation of fluid and heat flow in fixed bed heat storage reservoir with argon as working fluid.

2 Heat storage model

The numerical model used in the analysis represents a slice of a 32 layer storage, with layers of solid material with holes arranged in hexagonal pattern. Every second layer has the hole pattern shifted in a way that n^{th} layer hole axes are positioned exactly in the middle of next layer's holes. Such arrangement makes the jet created in each hole to impinge on the surface of next layer, intensifying the heat transfer between argon gas and heat storing solid. A scheme of the arrangement, with smallest possible part of the arrangement that can be extracted from it with pattern symmetries is shown on figure 2. Such cut of the 32 layer storage was investigated numerically with a transient conjugate heat transfer analysis.

2.1 Physical model

Parameters of the storage core solid, are generic values of 500 $J{*}kg^{-1}K^{-1}$ specific heat capacity, 8000 $kg{*}m^{-3}$ density, and 50 $Wm^{-1}K^{-1}$ thermal conductivity. This parameters are close to ingot iron parameters, which is one of the considered core material candidates because of its low cost and relatively high thermal conductivity compared to ceramic materials.

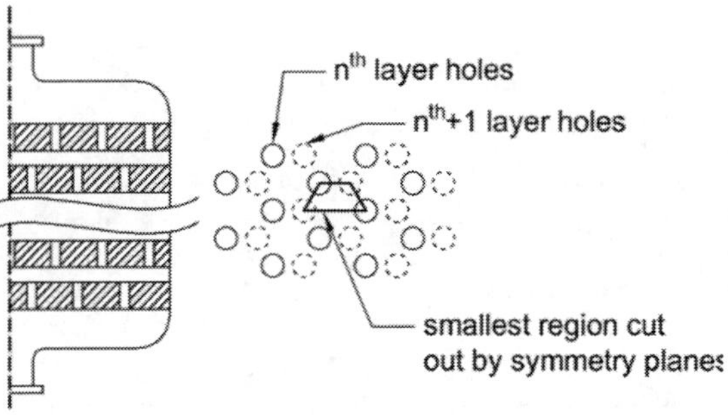

Fig. 2. Layered heat storage scheme.

The gas used in the simulation was argon, modelled as incompressible ideal gas model where its density is modelled as:

$$\rho = \frac{p_{ref}}{RT} \tag{1}$$

Where: ρ, R, T – are density, individual gas constant and temperature respectively, but p_{ref} is a constant reference pressure. Such model is suitable for current analysis, because in the presented storage device, gas flow is very slow and therefore pressure changes insignificantly in the fluid domain (absolute pressure changes by less than 0.3% throughout the analysis domain). Therefore it would be unjustified to use fully compressible flow model. The storage operating pressure was set to 2 bar, and this was the value of static pressure used for outlet boundary condition. Inlet boundary was set to constant flow velocity of 0,611 m/s for hot (800K) side inlet during charging process and 0,23 m/s inlet at the cold (300K) side inlet during storage discharge. Values of the velocities are matching a mass flow of $3.2x10^{-3}$ kg/s for both charging and discharging stages of storage operation.

A geometry of two layers was created (shown on figure 3) and meshed with conformal periodicity on its maximum and minimum z coordinate. This mesh was then copied 16 times and merged into single, two zone mesh. Because of the shift between solid layers, no practical topology was determined for all hexa structured mesh. Therefore, solid region and contacting with it region of gas, are meshed as a structured multiblock meshes with a layer of pyramid and tetra elements connecting structured regions (see figure 4).

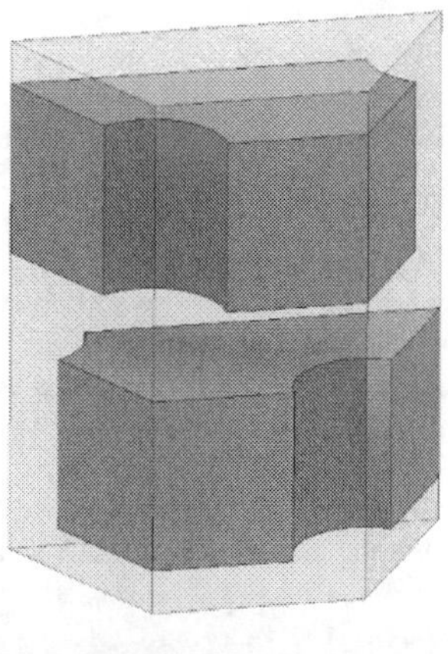

Fig. 3. Repeatable geometry element.

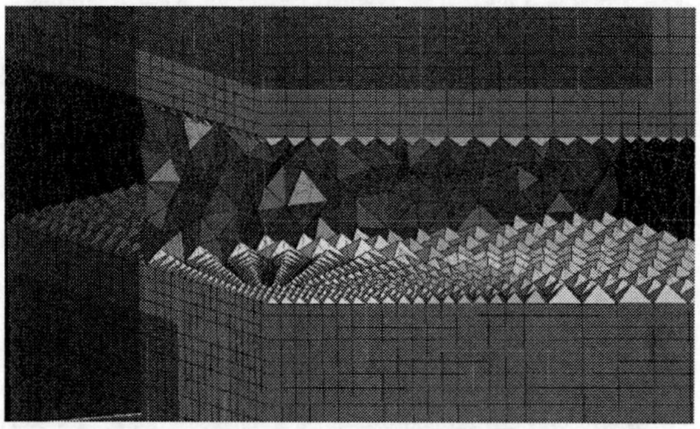

Fig. 4. Coarse mesh detail view with exposed unstructured region joining two multiblock structured fluid regions.

2.2 Numerical setup

The analysis was performed with Fluent 18.1 software, with the use of pressure based coupled solver (coupled momentum and pressure-based continuity equations, turbulence model and energy equations solved in segregated manner). Two equation k-ω-SST turbulence model was used. Second order bounded implicit scheme was chosen for time integration.

Meshes of two resolutions were prepared. A fine mesh, with boundary layers first element y+ H 1 (based on average velocities in the holes) having in total 14.2 million and second having y+ of around 15, having in total 2.1 million elements. Complete simulation was done on the coarse mesh.

Convective terms were discretized using first order upwind schemes, this together with coarse mesh approach was aimed at obtaining a highly diffusive setup, which would damp out small scale flow structures and allow for practical computation time.

The need for highly diffusive convective schemes and coarse mesh approach, was discovered after first attempts on computing the flow on the fine mesh. For this case, flow field oscillations are characterised by many orders of magnitude smaller time scales, compared to heat accumulation process in the solid. Therefore allowing for those oscillations to be resolved, requires impractically small time steps to be used when compared to the simulated process time of 17 hours. On the fine mesh, time step convergence was possible with given setup for time steps of the order of 10-20 milliseconds. After simulating first minute of the flow, it was estimated that the analysis would take around 32 years on 120 cores (almost 34 million CPU hours). Switching to coarse mesh resulted in numerical damping of the velocity field oscillations and allowed for two orders of magnitude higher timesteps. At the beginning of each storage operation stage, the time step was set to 1e-04 seconds and then gradually increased to 2.5 s which was the timestep size for major part of each charge/discharge stage. The simulation with that setup was finished in 13 days on 48 cores. To assess the inaccuracy of that approach, average heat fluxes absorbed by first five layers of the storage in first 50 seconds of the charging cycle (so in the part of cycle when highest temperature gradients are present so highest numerical errors are expected) where calculated on both meshes. Those average heat fluxes where 526.8 W and 479.2 W for fine and coarse meshes respectively. This 9% heat flux reduction, was found small enough to accept this approach because of its preliminary character aimed to assess how the storage core behaves and not to perform a detailed design.

3 Results and discussion

During analysis runtime, for each time step area averaged pressure and mass flow averaged temperature was recorded on inlet and outlet boundaries. Those data were later used to calculate inlet and outlet exergy fluxes as:

$$\dot{B} = \dot{m}\left[C_p(T - T_{ref}) - T_{ref}\left(C_p ln\frac{T}{T_{ref}} - Rln\frac{p}{p_{ref}}\right)\right] \tag{2}$$

Where: $\dot{B}$ – exergy flux, T_{ref} – reference temperature of 300K, p_{ref} – reference (operating) pressure of 2 bar, C_p – specific heat capacity at constant pressure, R – individual gas constant.

Exergy fluxes integrated in time, where then used to calculate amounts of exergy absorbed and recovered from the storage in each charge/discharge cycle to assess exergetic efficiency.

The time dependence of mass flow averaged temperatures at both storage ends is presented on figure 5. Constant temperature parts, represent prescribed values when given side acts as an inlet. Constant temperature of 800 K was assigned for hot side inlet for charging processes, and 300 K was assigned for cold side inlet for discharging processes. It can be seen, that except first cycle when hot gas is ingested into completely cold storage, outlet temperatures behave almost linearly, hence the heat flux exchanged with the storage drops in that almost linear manner throughout entire process. This observation suggests that the storage accumulates heat in its whole volume, instead of having a moving temperature front. This conclusion was also supported by recorded temperature profiles which are detailed in related article [1].

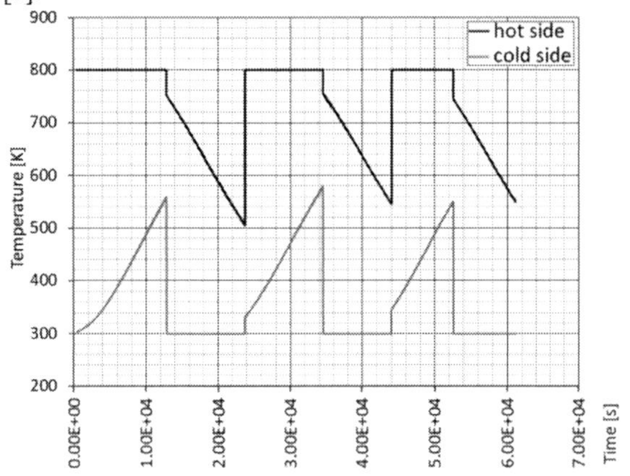

Fig. 5. Storage hot and cold side, mass flow averaged temperatures during whole process.

As can be also seen from figure 5, times of the subsequent processes are not equal. The storage heat capacity and mass flow, where designed for a 3.5 hours charge/discharge processes, that was the time of the first storage charging. During first discharge however, the outlet temperature was decreasing faster than expected and the discharge was stopped after 3 hours when outlet temperature dropped to almost 500 K. Second charge was also performed for 3 hours. After that it was decided (because first cycle exergy efficiency was calculated as 52.2% which is quite low) that the condition for process end should be the outlet temperature reaching 550K, which is halfway between charge/discharge inlet range instead of using prescribed stage time. This adaptive process allowed to increase the exergy efficiencies of subsequent cycles from 52.2% for first cycle, through 60.7% for second to 66.8% for the last operating cycle.

4 Conclusions

Main purpose of this study was to find qualitative results for gas-solid thermal interaction in conjugate heat transfer of the generic layered bed thermal storage. Moreover, limiting factors for the CFD pipeline of such devices.

The study in question combines two processes with large discrepancies in time scale: flow fluctuations occurring in order of milliseconds and heating of solids measured in hours. Numerical approach capable of time accurate capturing of flow fluctuations requires high density meshes with proper boundary layer resolution and leads to extremely slow computation and impractical walltime in order of magnitude of years. Coarsening the mesh resolution both in fluid and solid leads to filtering out the small time scale fluid flow fluctuations which leads to stabilisation of the numerical calculation and reducing the computation walltime to order of magnitude of hours. This is achieved at the cost of underestimation of conjugate heat flux.

It was also found, that the lower part of storage capacity is utilized during the thermal cycling, the higher exergy efficiency is obtained. The most useful conclusion however, is that the biggest temperature drop on the way of heat transfer between gas and solid is in the inner part of the gas boundary layer (closest to the solid wall, where the heat transfer coefficient is limited by molecular thermal conductivity of the gas). This leads to a conclusion, that in order to increase the thermal storage exergetic efficiency, layers (especially those nearest to its ends) should be of a much higher specific surface (gas-solid interface surface per solid mass). It is further suggested that the use of granular beds, or porous materials in applications requiring high exergy efficiency might be more feasible.

This research was supported in part by PLGrid Infrastructure.

References

1. J. Mosiężny, B. Ziegler, P. Czyżewski, Int. J. Numer. Methods Heat Fluid Flow (to be published)
2. R.A. Betts, C.D. Jones, J.R. Knight, R.F. Keeling, J.J. Kennedy, Nat. Clim. Chang. **6,** 806–810 (2016)
3. H. Saboori, R. Hemmati, S.M.S. Ghiasi, S. Dehghan, Renew. Sustain. Energy Rev. **79,** 1108–1121 (2017)
4. X. Luo, J. Wang, M. Dooner, J. Clarke, Appl. Energy. **137,** 511–536 (2015)
5. F. Ni, H.S. Caram, Appl. Therm. Eng. **84,** 34–44 (2015)
6. Frate, M. Antonelli, U. Desideri, Appl. Therm. Eng. **121,** 1051–1058 (2017)
7. D. Roskosch, B. Atakan, Energy Procedia. **129,** 1026–1033 (2017)
8. A. Smallbone, V. Jülch, R. Wardle, A.P. Roskilly, Energy Convers. Manag. **152,** 221–228 (2017)

Experimental investigation on straight and u-bend double tube heat exchanger with active and passive enhancement methods

Rafal Andrzejczyk [1,*], *Tomasz* Muszynski [1], and *Przemysław* Kozak [1]

[1] Gdansk University of Technology, Faculty of Mechanical Engineering, Narutowicza 11/12, 80-233 Gdansk, Poland

Abstract. Authors in this work want to demonstrate the possibility to increase the heat transfer efficiency by using simple wire coil inserts to create turbulent flow in the boundary layer as well as air blowing into the annulus of the pipe. In the study, Wilson plot approach was applied in order to estimate heat transfer coefficients for all heat exchanger (HX) configurations. The study focuses on experimental values of heat transfer coefficient (HTC) and pressure drops. The primary objectives of the work are to: I. Provide an experimental comprehensive database for HTC and pressure drops; II. Analysis effect of different flow conditions e.g.. water mass flow rate, the void fraction on heat transfer and hydraulic performance of tested elements. III. Compare influences of both passive and active methods at the efficiency of simple heat exchangers constructions; IV. Validation experimental results with selected experimental models from the open literature.

1 Introduction

In recent years, investigations aimed at increasing heat transfer efficiency are of primary importance [1,2]. Reducing the energy requirement of energy-intensive systems and also maximization of energy utilization are common subjects of research [3]. For this reason, numerical and experimental studies are conducted in order to create more efficient heat exchangers for thermal energy recovery and storage systems [4,5]. In spite of strong progress in various fields of mechanical and materials engineering, simple constructions, such as pipe in pipe, U-type or coil heat exchangers are still very popular. The growing, negative industry impact on environment strongly accelerates the search for new solutions in the field of heat transfer engineering [6]. They are many papers in open literature focuses on heat transfer enhancement techniques in plain heat exchangers constructions. Those techniques as generally divided into passives and active techniques. In passive techniques, there is no required external power. Sheikholeslami et al. [7] presented comprehensive literature review of passive techniques, such as various turbulators (coiled tubes, extended surfaces (fin, winglet),rough surfaces(corrugated tube, rib) and swirl flow devices such as twisted tape, conical ring, snail entry turbulator, vortex rings, coiled wire). They noted that wire coils turbulator provide the best performance increase while maintaining reasonable increase of pressure drop. They also concluded that passive techniques are more popular as heat transfer techniques because of the simple manufacturing process and a possibility to employed to existing heat exchanger construction. Moosavi et al. [8] experimentally tested the possibility to optimize heat transfer and pressure drop characteristics via air bubble injection inside a shell and coiled tube heat exchanger. In the paper, it was emphasized that the air flow rate and injection side play a key role in the possibility to enhance heat transfer in shell coil heat exchanger. Authors noted that the amount of augmentation and curves behavior was completely different when airflow was injected into the shell side and coil side of the heat exchanger.

It should be noted that in all types of heat exchangers heat and fluid flow can be complicated and difficult to predict by means of literature correlations [9], especially in two-phase flow conditions [10–12]. What is more important not many studies concerned at augmentation of heat transfer in case of U-bend double pipe heat exchanger (XH) However latest reports showed the advantage of wire coil turbulators and air bubble injections compare to others heat transfer enhancement techniques. Authors in this study were concentrated on the possibility to combine active and passive techniques to increase the heat performance of plain heat exchangers constructions.

2 Experimental setup

In this study four configurations of heat exchangers (HXs) have been used: straight double tube HX, straight double tube HX with helicoidal turbulization (SDTHXT), U-bend double tube HX, U-bend double tube HX with helicoidal turbulization (UBDTHXT). Also experimentally test were done in two-phase flow

* Corresponding author: rafal.andrzejczyk@pg.edu.pl

regime (air-water mixture) for all heat exchangers configurations. The heat exchangers were made of copper pipes with the following diameters: d = 10 mm , D = 18 mm and wall thickness 1 mm. The total length of the exchanger, in each configuration, was constant and was 430 mm. The helical turbulent element is made of brass wire with a diameter of 2.4 mm and a pitch of 11 mm. What is more, turbulator was located in the jacket of the heat exchanger. Construction of experimental setup (see Fig.1) provides energy balance in water-water configuration The water inlet and outlet temperature were measures by K-type thermocouples in the first class of accuracy. All thermocouples were connected to the CHY510 meter made in the ITS-90 standard. Pressure difference was measured only at the shell side by PELTRON pressure transducer. The volume flow of water was measured on the shell side of the heat exchanger with a ROL 16 rotameter in class 2.5 and on the inner tube side with a Meterc water meter (T30 / 90). The volume flow of air was measured with ROL 06 rotameter in class 2.5 .

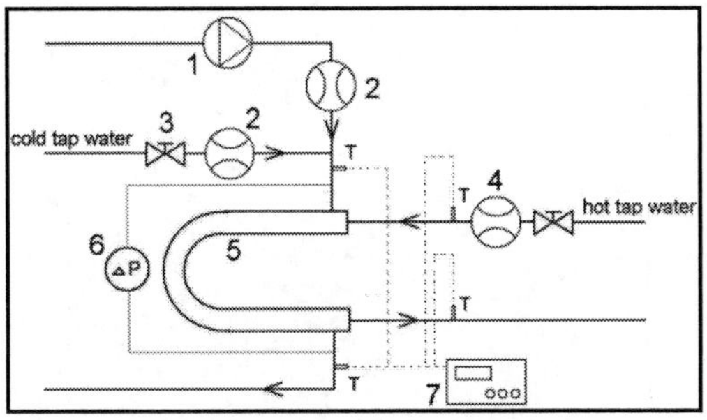

Fig. 1. Experimental setup: 1- compressor, 2 –rotameter, 3 – valve, 4- water flow meter, 5- experimental module, 6- pressure drop transducer, 7- data acquisition.

Table 1. Uncertainties of selected parameters.

Parameter	Measuring range	Systematic error
T [°C]	10-60	Max. error = +/-2,8K
m [kg/s]	0,000056-0,094	Maximum error = 2,5%
ΔP [kPa]	0-200	Maximum error = 0,5 kPa
P_{el} [W]	100-3000W	Maximum error = ± 5%

To supply resistance heater M10-522-10 250/4 autotransformer was used and Bemco 6907C wattmeter to measure the actual electrical power of the heater. At table 1 the uncertainties of selected parameters were presented. To estimate heat loss to ambient the infrared photos(Flir E6) of insulation were used. Each time during experiments whole of tested heat exchanger was covered with 10 mm of polyurethane insulation. Results of calculations shown that heat losses were not larger than 1%, so it was neglected in calculations procedure.

3 Experimental procedure

The heat flux in water-water configuration was calculated from energy balance. There were used followed equations:

$$\dot{Q}_h = \dot{m}_h \cdot c_p \cdot \left(T_{h,in} - T_{h,out}\right) \quad (1)$$

$$\dot{Q}_c = \dot{m}_c \cdot c_p \cdot \left(T_{c,out} - T_{c,in}\right) \quad (2)$$

A number of transfer units was calculated from the formula:

$$NTU = \frac{U \cdot A}{C_{min}} \quad (3)$$

where the minimum value of the heat capacity was obtained from the next dependence:

$$C_{min} = f(C_h, C_c) \quad (4)$$

The overall heat transfer coefficient U was calculated as below:

$$U = \frac{\dot{Q}_{av}}{\left(\Delta T_{log} \cdot A\right)} \quad (5)$$

where: $\dot{Q}_{av} = \frac{\dot{Q}_h + \dot{Q}_c}{2}$

Heat transfer effectiveness ε [-] was calculated from correlations 6-9:

$$\varepsilon = \frac{\dot{Q}_{av}}{\dot{Q}_{max}} \quad (6)$$

$$\dot{Q}_{max} = C_{min} \cdot \Delta T_{max} \quad (7)$$

$$\Delta T_{max} = T_{c,wlot} - T_{z,wlot} \quad (8)$$

$$\dot{Q} = \dot{Q}_c = \dot{Q}_h \quad (9)$$

where $\dot{Q}_{max}$ [W] is a maximum heat flux that can be received in the tested heat exchanger. At the shell side the Reynolds number at single phase regime was calculated as the mass flow rate through equivalent diameter:

$$Re = \frac{G_{sh} \cdot D_e}{\mu_L} \quad (10)$$

In case of two-phase flow regime, the liquid Reynolds number was used. This number was calculated as follows:

$$Re_L = \frac{4 \cdot \dot{m}_w}{\pi \cdot \sqrt{1 - \varphi} \cdot \mu_L \cdot D} \quad (11)$$

To the calculation of void fraction Chisholm correlation was used [13]:

$$\varphi = \left[1 + \left(\frac{V_G}{V_L} \right) \cdot \left(\frac{1-x}{x} \right) \cdot \frac{\rho_G}{\rho_L} \right]^{-1} \quad (12)$$

Equivalent diameter depends on the volume of the annulus divided by heat exchanger length and circumference.

$$De = \frac{4 \cdot V_{sh}}{\pi \cdot D \cdot L} \quad (13)$$

The volume available for the flow of fluid in the annulus, V_{sh} can be calculated knowing geometrical dimensions of the heat exchanger.

$$V_{sh} = \frac{\pi}{4} D^2 L - \left(\frac{\pi}{4} d^2 L + \frac{\pi}{4} d_c^2 L \right) \quad (14)$$

The measured pressure drop is the sum of friction pressure drop, expansion and contraction losses due to the headers at both ends of the test section. To better compare experimental results for different heat exchanger configurations with different heat transfer enhancement techniques (passive and active) also the performance factor was calculated. Authors were also used similar calculation methodology in their previous study [14]:

$$PF = \frac{\frac{Nu}{Nu_R}}{\left(\frac{f}{f_R} \right)^{0.33}} \quad (15)$$

Heat transfer coefficient of shell fluid was calculated based on Wilsons plot method [15] and then the Nusselt number at the shell side was also calculated.

$$Nu = \frac{\alpha_{sh} \cdot D_e}{\lambda} \quad (16)$$

It was assumed Nusselt number for straight double HX as a reference value. The friction factor was calculated as below [7]:

$$f = \frac{\Delta P}{\left(\frac{\rho_L \cdot w^2}{2} \right) \cdot \left(\frac{L}{D_e} \right)} \quad (17)$$

4 Results and discussion

As can be seen in figures: 2,3 pressure drops are significantly larger for heat exchangers with helicoidal turbulator. It should be also noted that in two-phase flow regime pressure drops were always larger than in single phase.

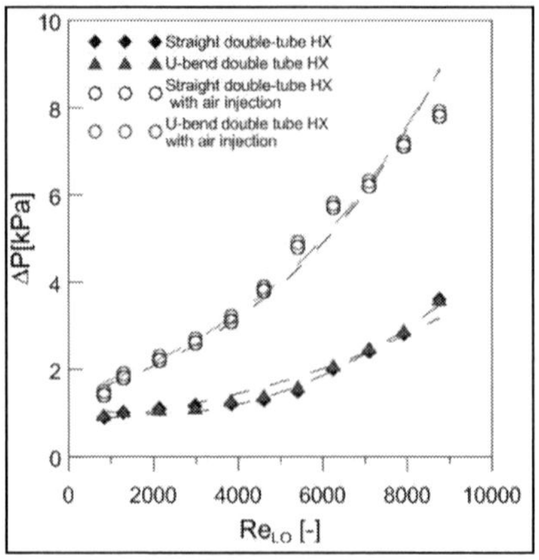

Fig. 2. Hydraulic characteristic at shell side for SDTHX and UBDTHX in single and two-phase flow regimes.

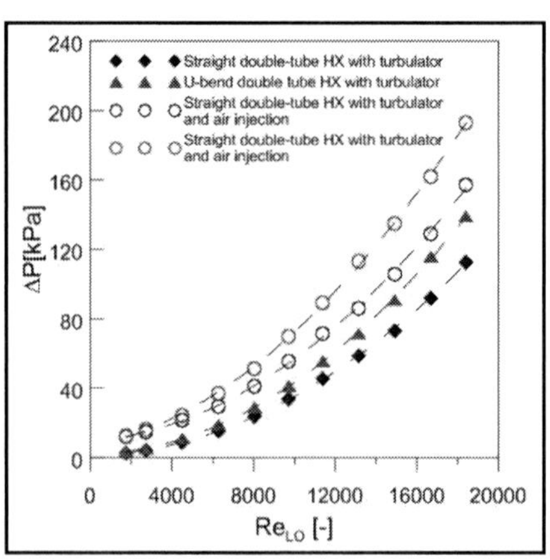

Fig. 3. Hydraulic characteristic at shell side for SDTHXT and UBDTHXT in single and two-phase flow regimes.

In case of single phase flow regime, (see Fig.4) the U-bend double heat exchanger (U-BDTHX) has larger values of the Nusselt number at a whole range of Reynolds numbers. However, the difference is no larger than 50%. Outcomes were compared with very well-known Dittus-Boelter correlation. It should be noted that experimental results for straight double tube HX (SDTHX) have a good fit with that correlation. The situation is quite opposite in case of those heat exchangers working at two-phase flow conditions (air-water mixture), see Fig.5. This time SDTHX has larger values of Nusselt numbers at a whole range of liquid Reynolds numbers (the same water mass flow rate as in single phase regime). The Nusselt numbers are almost two-times larger compared to the single phase in case of SDTHX and more than 50% larger in case of U-BDTHX. The experimental results for the two-phase flow regime were compared with Kim et al. correlations [16]. Both calculations methodology provided results with fairly good accuracy. However, the literature correlations overestimate Nuselt numbers in case of U-BDTHX and underestimate Nusselts number for

https://doi.org/10.1051/matecconf/201824002001

SDTHX. It should be also emphasized that HXs working at two-phase flow-regime are characterized larger thermal power as well as larger pumping power consumptions Despite the fact that, ε-NTU analysis (Fig.6) shows that all of the designs presented are of similar efficiency. However in two-phase flow regime constructions are characterized by higher NTU values, which ultimately results in a maximum efficiency of 58% for UBDTHX and 55% for SDTHX. In case of single phase flow regime, (see Fig.7) the U-bend double heat exchanger with turbulator (U-BDTHXT) has larger values of the Nusselt number at a whole range of Reynolds numbers.

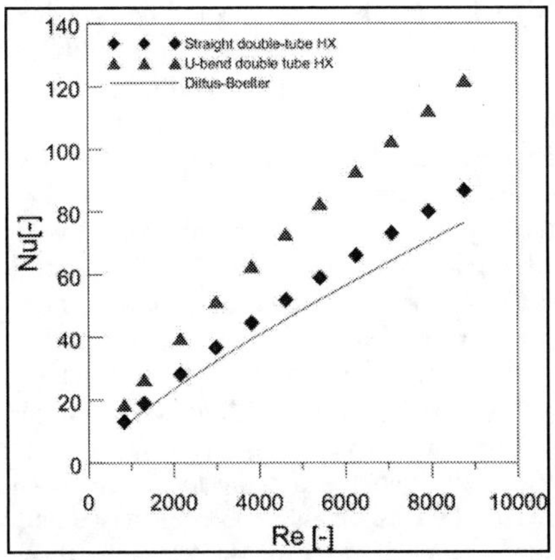

Fig. 4. Nusselt number at shell side for SDTHX and UBDTHX in single phase flow regime.

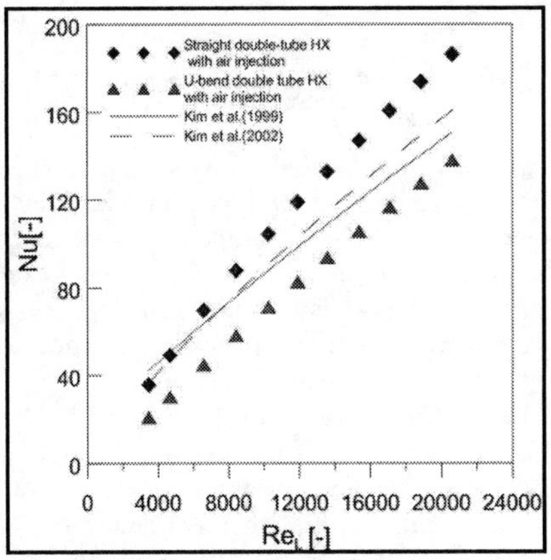

Fig. 5. Nusselt number at shell side for SDTHX and UBDTHX in two-phase flow regime.

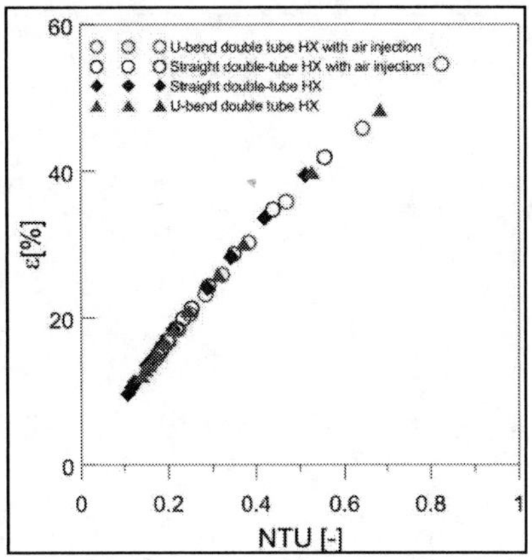

Fig. 6. The dependence of heat transfer efficiency as a function of the NTU for SDTHX and UBDTHX in single and two-phase flow regimes.

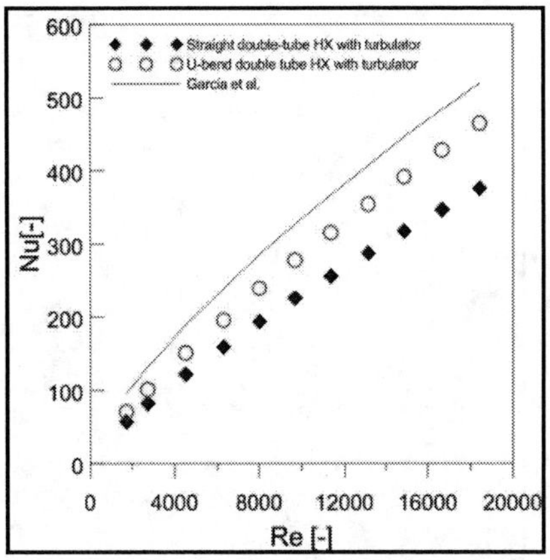

Fig. 7. Nusselt number at shell side for SDTHX and UBDTHX with helicoidal turbulator in single phase flow regime.

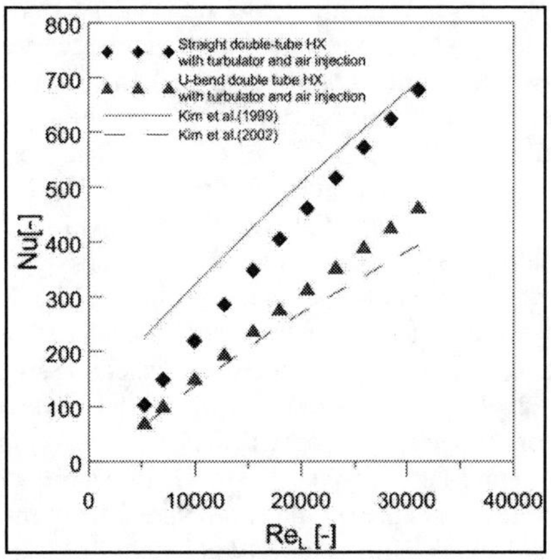

Fig. 8. Nusselt number at shell side for SDTHX and UBDTHX with helicoidal turbulators in two-phase phase flow regime.

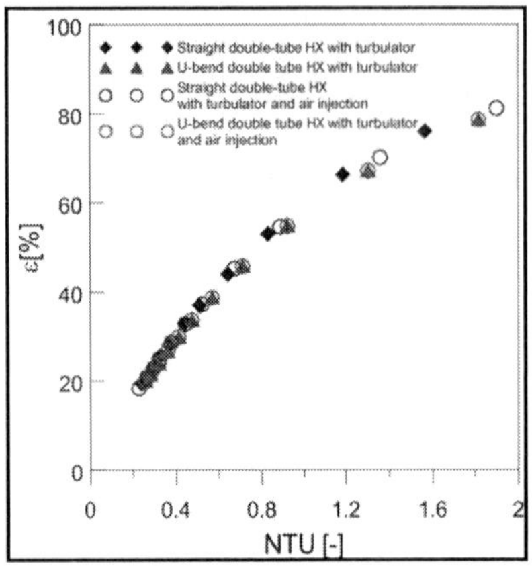

Fig. 9. The dependence of heat transfer efficiency as a function of the NTU for SDTHXT and UBDTHXT in single and two-phase flow regimes.

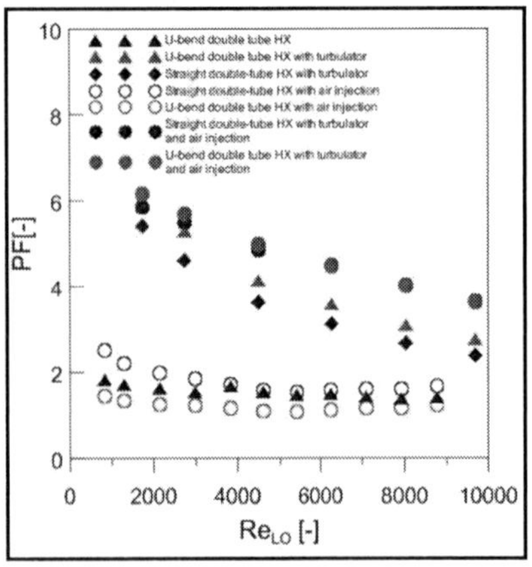

Fig. 10. The dependence of performance factor as a function of Re_{LO} for all heat exchangers configurations.

Both HX configuration with helicoidal turbulator are characterized by more than two times larger values of Nusselt Numbers compare to plain constructions. Outcomes of Garcia et al. [17] correlation for heat exchangers with helicoidal turbulators. The correlation quite good predict results for UBDTHXT but it overestimates results for straight HX with helicoidal turbulator (SDTHXT).Like in the case of plain constructions two-phase flow regime SDTHXT has larger values of Nusselt numbers at a whole range of liquid Reynolds numbers (the same water mass flow rate as in single phase regime). However, the difference in case of U-bend HX construction between results for single-phase and two-phase flow is not significant. The experimental results for the two-phase flow regime were compared with Kim et al. correlations [16]. Kim et al (1999) methodology [16].better predicted experimental results for straight HX but for U-bend HX Kim et al. (2002) [16] offered better fits to experimental data. Again HXs with helicoidal turbulator working at two-phase flow-regime are characterized larger thermal power as well as larger pumping power consumptions. However in two-phase flow regime constructions are characterized by higher NTU values, which ultimately results in a maximum efficiency of 80% for SDTHXT and 78% for UBDTHXT. In order to emphasize the impact of the presented modification of flow geometry in the shell side of double tube HXs, the results were compared with reference data for plain construction of double tube heat exchanger. It should be noted that all heat exchangers configurations have better performance than the plain tube in tube heat exchanger.

There was no significant performance difference found between plain U-bend double tube heat exchanger and SDTHX and UBDTHX working on two-phase flow regime (with air injection). It could be explained by the fact that the air injection increases flow turbulisation and heat transfer but also significantly increase pressure drops. In plain U-bend heat exchanger, there is also heat transfer enhancement mechanism, so-call Dean effect, which slightly increases heat transfer but also doesn't significantly increase pressure drops. This approach could be also used to explain the difference between SDTHXT performance and UBDTHXT. The U-bend heat exchanger with helicoidal turbulator has better performance than the straight heat exchanger with the same turbulator. However, configurations with helicoidal turbulators working in two-phase flow conditions (air-water mixture) had the best performance. It should be emphasized that this combination of active and passive techniques provide larger than 300% performance in the whole flow regime compare to reference plain heat exchanger configuration(see Fig. 10).

5 Conclusions

The article presents passive and active heat transfer enhancement methods in the form of helicoidal turbulator and air bubble injection. Both used heat transfer enhancement techniques provided better heat exchangers performance compare to reference plain double tube heat exchanger. The heat exchangers configurations with helicoidal turbulators working in two-phase flow conditions (air-water mixture) had the best performance. What is also important still such combination of heat transfer enhancement techniques in heat exchangers do not increase significantly relation between heat flux and pumping power. However, it should be noted that for smaller values of Reynolds number (Re_{LO}<3000) it seems that simplest is to use just passive enhancement techniques (e.g. helicoidal turbulator). The experimental results were compared with selected well-known correlation from the open literature. It worth to added that especially in heat exchangers configurations with helicoidal turbulator (single phase flow regime) results of calculations were clearly overestimated compare to experimental data.

Acknowledgements

The authors would like to appreciate funding received from the Research Council of Norway under the FRINATEK Project 231529.

Nomenclature

A – heat transfer area, m^2,
C – thermal capacity, W/K
Cp – specific heat, kJ/(kgK)
D,d – diameter of the annulus, inner tube, m
D_e – equivalent diameter, m
f – friction factor,
G – mass flux, kg/(m^2s)
L – Length of annulus tube, m
$\dot{m}$ – mass flow rate, kg/s
Nu – Nusselt number
NTU – the number of transfer units
$\dot{Q}$ – heat flux, W,
Re – Reynolds number
T ,ΔT– temperature, temperature difference °C,
V – volumetric flow, m^3/s
ε – heat transfer effectiveness
φ – void fraction
m – dynamic viscosity, Pas
ρ – density, kg/m^3

Subscripts

a – annulus
av – average
c – cold, coil wire
h – hot
G – gas
in – inlet
L – liquid
LO – liquid only
max – maximum
min – minimum
out – outlet
R – reference

References

1. Q. Wang, M. Zeng, T. Ma, X. Du, J. Yang, Appl. Energy **135** (2014)
2. R. Andrzejczyk, T. Muszyński, Arch. Thermodyn. **37** (2016)
3. T. Muszynski, S.M. Koziel, Arch. Thermodyn. **37** (2016)
4. D. Taler, P. Ocłoń, Int. J. Therm. Sci. **84** (2014)
5. T. Muszynski, Appl. Therm. Eng. **115** (2017)
6. P. Olszewski, Appl. Therm. Eng. **81** (2015)
7. M. Sheikholeslami, M. Gorji-Bandpy, D.D. Ganji, Renew. Sustain. Energy Rev. **49** (2015)
8. A. Moosavi, M. Abbasalizadeh, H.S. Dizaji, Exp. Therm. Fluid Sci. **78** (2016)
9. R. Andrzejczyk, T. Muszynski, Appl. Therm. Eng. **121** (2017)
10. T. Muszynski, R. Andrzejczyk, C.A. Dorao, Int. J. Refrig. **82** (2017)
11. T. Muszyński, R. Andrzejczyk, C.A. Dorao, Arch. Thermodyn. **38** (2017)
12. R. Andrzejczyk, T. Muszynski, C.A. Dorao, Exp. Therm. Fluid Sci. **83** (2017)
13. D. Chisholm, *Institution of Chemical Engineers* (London, 1983)
14. A.A. Rezaei, M. Ziabasharhagh, Heat Transf. Res. **42** (2013)
15 J. Fernandez-Seara, F.J. Uhía, J. Sieres, A. Campo, Appl. Therm. Eng. **27** (2007)
16. S.-M. Kim, I. Mudawar, Int. J. Heat Mass Transf. **58** (2013)
17 J. Garcia, M.P. Porto, R. Revellin, J. Bonjour, L. Machado, Int. J. Refrig. **73** (2017)

Air flow characteristics of a room with air vortex diffuser

Marek Borowski [1,*], Marek Jaszczur[2], *Daniel* Satoła[2], and *Michał* Karch[1]

[1]AGH University of Science and Technology, Faculty of Mining and Geoengineering, Kraków, Poland
[2]AGH University of Science and Technology, Faculty of Energy and Fuels, Kraków, Poland
[3]Frapol Sp. z o.o., Kraków, Poland

Abstract. The air diffuser is a very important component of any ventilation system, and the comfort level of ventilated space occupants depend among many other factors on properly designed and choice of diffusers. At present a large number of diffusers are produced to meet a different kind of requirements. One of the most efficient and very popular types of the diffuser is the vortex diffuser in which air flow has angular as well as translational velocity components. This paper investigates experimentally airflow characteristics of vortex ceiling diffuser and its effect on airflow in a ventilated space. Two thermal comfort criteria namely: mean age of the air and ventilation effectiveness have been used to predict the comfort zone inside the room. Effect of supply air velocity on the flow field is investigated and hence the on comfort and energy consumption. The results show that significant amount of energy can be saved by using vortex diffuser in reference to the other diffusers type. The values of the velocity decay coefficient were compared for different diameters of vortex diffuser.

1 Introduction

The knowledge and forecasts regarding indoor climate conditions are essential for optimising indoor climate and thermal comfort, as well as for energy efficiency and environmental protection [1]. People spend almost 80% of their time during the day indoors. Therefore, indoor air quality and thermal comfort are two important aspects of the indoor environment that require particular attention.

According ASHARE HVAC [2] design conditions, temperature and relative humidity should be maintained within a range of 20-24°C and 40-60% respectively. Care should also be taken to keep the air clean and control its chemical composition. In addition, indoor air pressure should be maintained, and the air flow in the system is preferred without recirculation [3].

Until now, many researchers have conducted experimental or simulation studies, but not very often simultaneous experimental and simulation studies.

Nielsen [4] described experiments with wall-mounted air supply equipment. In this study, he presented air velocity distributions in the vicinity of a wall. Al-Hamed [5] used computer modelling to simulate room ventilation in order to predict mean air temperatures and air velocities for multiple inlet and outlet locations for air supply to the room. In his considerations, Author took into account the *k-ε* turbulence model. The flow field in this model was verified by experimental measurements made in another study .

Srebric and Chen [6] used a simplified method to describe the flow and thermal conditions for eight different types of diffusers: nozzle diffuser, slot (linear) diffuser, valve diffuser, displacement diffuser, round ceiling diffuser, square ceiling diffuser, vortex diffuser, and grille diffuser.

Zhou and Haghighat [7] developed a simplified method for defining boundary conditions at the inlet of a vortex diffuser. They compared airflow visualisation with (Computational Fluid Dynamics) CFD simulations , which showed the effectiveness of the simplified modelling method for a vortex diffuser. They also presented temperature measurement and air velocity profiles in order to validate the results of CFD simulations. This validation showed that the simplified method can be used to model the airflow from diffusers of this type.

On the other hand, Einberga and Hagstrom [8] discussed the results of modelling of multi-cone air diffusers for industrial spaces from CFD. The Authors compared CFD results with experimental measurements. The results of this study showed that the CFD simulation performed with the standard *k-ε* model can accurately predict the non-isothermal airflow around the diffuser.

Most of the previous indoor airflow studies were carried out in reduced-size spaces. At the same time, numerical simulations were carried out for the reduced scale in order to compare them with the experiments. Due to the fact that the research study was carried out to a scale, there was a lack of experimental data collected under real conditions in order to validate the mathematical models. From literature survey can be seen that the experimental studies conducted so far were carried out randomly on selected diffuser models.

* Corresponding author: borowski@agh.edu.pl

The present study focus on the complex geometry of the diffuser in which three-dimensional (3D) flow occurs [9, 10]. The experimental measurements were performed for 4 diffuser sizes and were made for the air streams flowing through the diffuser from 100 up to 650 m^3/h. The distribution of air velocity in the zone under the diffuser was determined. The study of air diffuser properties and its influence on the indoor airflow distribution was carried out with the use of two approaches: experimental measurements and numerical simulations.

2 Experimental studies

2.1 Experimental set-up

The general layout of the experimental configuration is presented in Figure 1. The measurement is performed in the room, with an artificial ceiling and with the dimensions of AxB=6x6m, in which the diffuser is installed. The diffuser is connected to a ventilation duct, through which the fan supplies air in the required amount – from $100m^3/h$ to $650m^3/h$. Modification in the air flow rate is carried out using the fan with a speed regulator (inverter). The ceiling is installed in a room with the following dimensions: axbxc=7x7x4 m. The studies are carried out on the real scale. Figure 2 presents a photograph of the diffuser installed in the false ceiling.

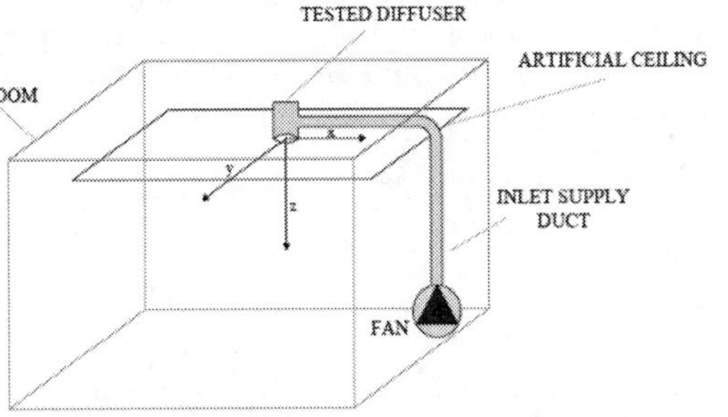

Fig. 1. Schematic diagram of the experimental set-up.

Fig. 2. Diffuser installed in the false ceiling.

The velocity and temperature experimental measurements were made using the AirDistSys 5000 system. This system is used for multi-point measurements of air velocity and temperature. It is used, for example, to assess the thermal conditions and velocity distribution in rooms.

Due to the application of the measuring system of eight SensoAnemo 5100LSF with SF3 probes, measure speed and temperature simultaneously at 8 points with a high level of accuracy. An anemometric transducer measures the airflow velocity in the range from V_w=0.05 to 5.00 m/s. The technical data of the SensoAnemo 5100LSF are as follows:

- velocity sensor type: multidirectional, spherical,
- speed measurement range: 0.05 … 5.00 m/s,
- measurement accuracy: ±0.02 m/s ±1.5%,
- directional sensitivity error above 2m/s: ±2.5%,
- temperature measurement range: -10...+50°C,
- measurement accuracy: ±2°C.

The system also includes the AirDistSys 5000 computer software, which enables the configuration of equipment, measurement of mean air velocity and temperature, the standard deviation of velocity, the intensity of flow turbulence and the degree of draught risk at all analysed measurement points. With the acquisition system , all parameters are recorded and stored on a computer. The data can be recorded at any time and for any length of time.

The measurements were made on a grid of points in the X-Z plane, as shown in Figure 3.

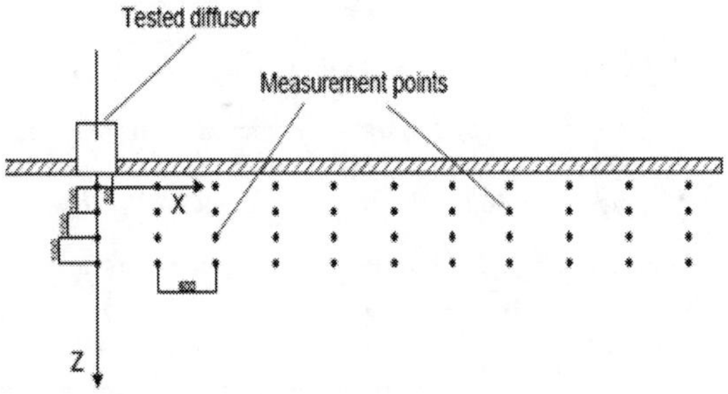

Fig. 3. Measurement points in the X-Z plane.

The measurements were taken in the so-called "lines", i.e. first the measurements were taken on one line for all air streams, and then the thermoanemometers were lowered to the next line along the Z-axis. In the case of the studied diffuser, the first measurement was made at a distance Z=0.05m from the false ceiling. Then, the distance from the ceiling was gradually increased by 0.1 m. The next measurement lines were at the distance of Z=0.15; 0.25; 0.35 m.

2.2 Methodology

A ceiling diffuser, which is available in various sizes, was used for the measurements. The geometry of the diffuser is presented in Figure 4.

Each diffuser was installed in a false ceiling and measured for different air streams. The range of the inlet air velocity of the diffuser ranged from V_m=0.5 to 4 m/s. The test was carried out as follows: after starting the fan and setting the required air flow through the diffuser, the measurement waited until the statistically steady conditions were obtained. Commencing the measurements depends on the degree of temperature stabilisation. The condition is considered steady when the temperature is the same within 2% between two consecutive readings of the measurements. The stationary state is reached after approx. 15 minutes. Then the measurements of air velocity along the determined measurement lines using thermoanemometric probes are started. The duration of the measurement, in which the mean air velocity result was obtained, amounted to 3 minutes. After these measurements, further measurements are taken for the subsequent air streams/air velocities flowing through the diffuser. Then the entire measurement procedure is repeated from the beginning .

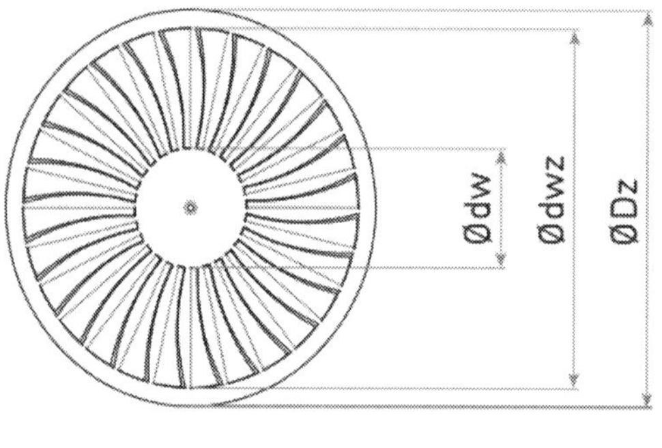

Fig. 4. Simulated diffusers geometry.

The diffusers used in experimental analysis are made in four sizes: 300, 400, 500, 600 mm. Each of them has different dimensions, which are presented in Table 1.

Table 1. List of the characteristic dimensions of diffusers.

Size	*dw*	*dwz*	*Dz*
	mm	mm	mm
300	85	250	300
400	115	350	400
500	150	450	500
600	180	550	600

3 Results and discussion

The measurements for the size 300 diffuser were made for 6 air flow rates V=100, 150, 200, 250, 300 and 350 m^3/h. The measurements were made at eight points with thermoanemometers suspended on a horizontal crossbar from X=0.0 m to X=2.7 m. The measurement points (thermoanemometers) were located at a distance of 0.3 m from each other. The first thermoanemometer was located on the axis of the diffuser X=0.00 m; Z=0.05 m.

Figure 5 shows an example of the velocity distribution for the air flow rates equal to $V=250m^3/h$. It can be infer from the figure that the largest changes in the air flow occur at the distance from the false ceiling in the plane Z=0.05 m. At a larger greater distance from the ceiling, the air velocity is smaller than 0.2 m/s. The analysis shows that the air stream range for the speed of 0.2 m/s equals $L_{0.2}$ = 2.5 m. It should be noted that the diffuser has a significant air stream range.

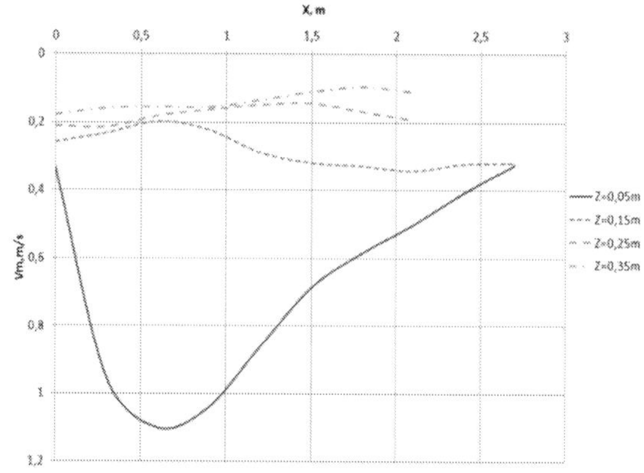

Fig. 5. Velocity distribution for a size 300 diffuser for the air flow rate $V=250m^3/h$.

Due to the fact that the most significant changes take place on the plane Z=0.05m, Figure 6 shows the summary breakdown of the velocity distributions for different air flow rates. This figure shows that the maximum velocity in this plane occurs at a horizontal distance of about 0.75 m and then decreases to disappear at a distance of 1.4 m to 3.5 m, depending on the actual air flow rate.

Figure 7 presents the air velocity profile of the air flowing out of the size 300 diffuser in the plane of Z=0.05 m from the ceiling and for the flow rate equal to $V=100m^3/h$. This figure shows the characteristic shape of the velocity profile from the diffuser.

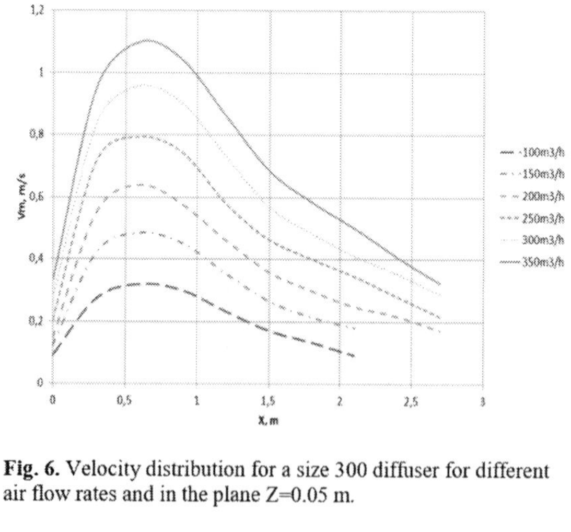

Fig. 6. Velocity distribution for a size 300 diffuser for different air flow rates and in the plane Z=0.05 m.

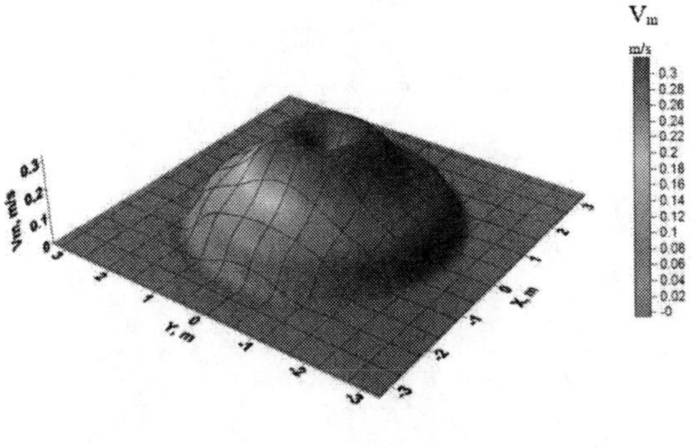

Fig. 7. Velocity profile for a size 300 diffuser in the plane Z=0.05 m for air flow rate V=100 m³/h.

3.1 Description of the air flow from the ceiling diffuser

Figure 8 presents the distribution of velocity from the vortex diffuser with the blades at an angle of 45°. The swirl of the air stream from this diffuser causes the air stream to spread under the ceiling. It seems that the vertical range of the diffuser is low. However, the horizontal range is significant with such a small stream of air and dimensions of the diffuser. This is a signal of the high efficiency of air distribution in the room. One can also observe the swirling motion of the air in the room.

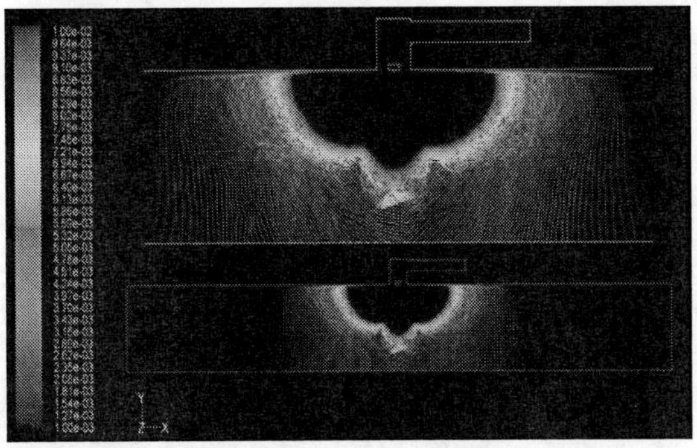

Fig. 8. Vortex diffuser air velocity iso-lines for $V_m = 0.65$ m/s.

3.2 Diffuser velocity decay coefficient

The performance of the jet created by the diffuser can be characterized by a single constant, namely jet velocity decay coefficient *K*. It can be defined as [11]:

$$\frac{V}{V_0} = K\frac{\sqrt{A_0}}{x} \tag{1}$$

The diffuser velocity decay coefficient *K* can be determined by plotted the dimensionless velocity V/V_o versus the dimensionless distance $x/sqrt\ (A_o)$ as shown in Figure 9 where the values of the decay coefficient can be directly determined when the curve intersects the dimensionless velocity at the unity (i.e. $V/V_o=1$). The value of the velocity decay coefficient is 1.6.

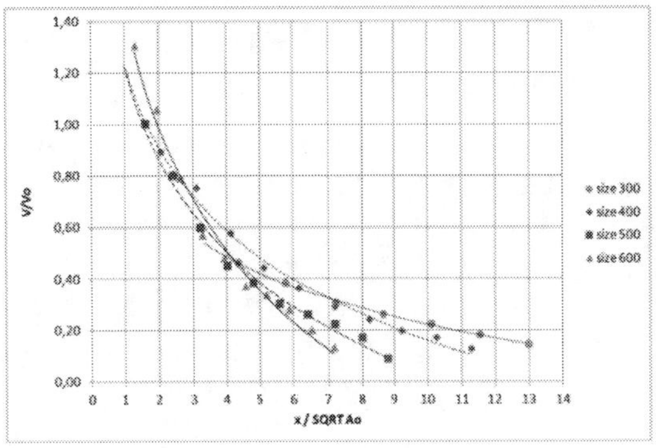

Fig. 9. Dimensionless velocity for different size for vortex diffuser

4 Conclusion

This article presents the results of experimental and numerical studies for different air flow rates and thus different inlet velocities of air into a vortex diffuser.

The velocity vector of the studied vortex diffuser is mainly directed horizontally under the ceiling. The long range is achieved by the swirling movement of the air leaving the diffuser. After the air is distributed under the ceiling, it drops, and the heat gains in the room are observed. In this way, the comfort conditions in the room are effectively ensured. Due to the high horizontal air range, no dead zones are formed in the room. The airflow is also distributed in such a way that the air circulation in the room is reduced. It can be seen that the vortex diffuser is characterised by high efficiency of indoor air distribution.

The buoyancy force has a significant impact on the temperature and velocity field, the diffuser creates a thermal comfort zone with a large radius.

The airflow range for the 300 size vortex diffuser for 0.2 m/s is in the range $L_{0.2}$=1.4-3.5 m. The diffuser has a large range of the air stream.

The maximum velocity under the false ceiling is about 0.75 m horizontally and then decreases to the loss of flow V=0.2 m/s depending on the air stream value.

The pressure coefficient rises sharply for a Reynolds number of less than Re=10,000. An analysis of the dimensionless mean velocity V_m/V_o in relation to the Reynolds Number Re shows that it also rises below 10,000. For small airflows and thus low Reynolds numbers, it is difficult to ensure the correct air stream profile and indoor air distribution and thus the right conditions for thermal comfort.

References

1. Awbi H. B, *Ventilation of Buildings*, second ed., (Spon Press. Publisher: Taylor & Francis 2003)
2. ASHRAE Handbook, Fundamentals, (2008)
3. ASHRAE Standards, Standard 55, (1992)
4. P. V. Nielsen, Energy and Buildings **31**, 179 (2002)
5. Al-Hamed, A. H. *Determining Velocity, Temperature and Occupancy Comfort within a 3-D. ROOM*, (M.Sc. Thesis, King Fahd University, Saudi Arabia 1990)
6. J. Srebric, Q. Chen, HVAC&R Research, **8**, 277 (2002)
7. Zhou L, Haghighat F, *Civil and Environmental Engineering*, Concordia University, Montreal, Canada, (2007)
8. Einberga G, Hagstrom K, *Building and Environment*, **40**, 601 (2005)
9. M. Jaszczur, M. Branny, M. Karch, M. Borowski, Journal of Physics. Conference Series; ISSN 1742-6588, **745**, 032049, 1 (2016)
10. M. Jaszczur, M. Borowski, M. Branny, M. Karch, The study of the velocity field of the air flowing the swirl diffusers using PIV method, Experimental Fluid Mechanics 2016: proceedings of the international conference, Mariánskè Lázně, Czech Republic, Technical University of Liberec, (2016)
11. Mierzwiński S, *Aerodynamics for ventilation*, (in Polish, Publisher: Pol. Śląskiej. Gliwice 2007)

 https://doi.org/10.1051/matecconf/201824002003

An analysis of the innovative exhaust air energy recovery heat exchanger

Marek Borowski[1,*], *Marek* Jaszczur[2], *Daniel* Satoła[3], *Sławosz* Kleszcz[3], and *Michał* Karch[1]

[1]AGH University of Science and Technology, Faculty of Mining and Geoengineering, Kraków, Poland
[2]AGH University of Science and Technology, Faculty of Energy and Fuels, Kraków, Poland
[3]Frapol Sp. z o.o., Kraków, Poland

Abstract. Heating, ventilation and air conditioning systems are responsible for a nearly 50% of total energy consumption in operated buildings. One of the most important parts of the ventilation system is an air handling unit with a heat exchanger for energy recovery which is responsible for effective and efficient energy recovery from exhaust air. Typically heat exchangers are characterised by the producers by heat and humidity recovery efficiency up to 90% and 75% respectively. But these very high values are usually evaluated under laboratory conditions without taking into account a dynamic change of outdoor and indoor air conditions significantly affecting the recovery efficiency. In this paper, results of thermal, humidity and enthalpy recover efficiency of innovative energy recovery exchanger have been presented. The analysed system allows adjustment of the humidity recovery especially useful in the winter period and forefends energy use for an anti-froze system of energy exchanger. Presented result show that analysed innovative system can achieve the value of thermal efficiency recovery higher than 90% and efficiency of humidity recovery about 80%. This is possible because the analysed system is able to work without the use of any primary source energy or other anti-freeze systems. Presented in this research unique solution is able to work without external anti-freeze systems even in extremely adverse outdoor air conditions such as minus 20°C and humidity 100% RH.

1 Introduction

The construction sector is the largest source of energy consumption in the European Union economy, accounting for almost 40% of total energy demand in the EU-28 countries. This is more than transport (32%) or even whole industry sector (26%) [1]. This sector has been classified as one of the key sectors that should meet the 20-20-20 requirement: to reduce greenhouse gas emissions by 20% compared to 1990, to increase energy savings by 20% and increase the share of renewable energy sources by 20% by 2020. Also, the reduction of greenhouse gas emissions from the construction sector by the year 2050 is planned in the range of 88-91% compared to 1990 [2].

To decrease energy consumption in the construction sector in 2010, the European Union (EU) has introduced the directive EPBD [3] on the energy efficiency of buildings, which assumes that from 2019, all public buildings, and then from 2021, all newly constructed buildings, will have to meet the requirements for nearly zero-energy buildings (nZEB) [3]. In Poland, the EPBD directive was implemented in 2014, specifying in the national conditions the definition and requirements for this type of building [4].

It is known that the ventilation, heating, hot water and cooling systems in a typical building account for about 50% of the total energy consumption in the construction sector. Due to the national requirements, these systems have to be designed to meet the criteria for the maximum use of non-renewable primary energy for ventilation, heating/cooling and domestic hot water [5]. Nowadays the growing share of heating loads and ventilation the heat recovery appears as one of the important solutions able to decrease heat losses and create significant energy savings [6-8].

In almost all houses (depends on region) ventilation plays a significant role in the total heat losses. Heat loss due to ventilation is responsible for about 25-55% total heat loss. Taking into account this fact it is recognised that it is practically impossible to achieve a high energy-efficient house without adequately designed mechanical ventilation [9].

The main unit of mechanical ventilation system consists of air-to-air heat exchanger responsible for exhaust air energy (and humidity) recovery. Most manufacturers used of counterflow heat exchangers and declare temperature efficiency values above 90%. However, such high (reported by producers) efficiency results from the fact that it is usually determined in laboratory conditions with minimum values of air flow velocity through the heat exchanger and usually overestimating the results observed during the real operation of the heat exchanger [10]. In fact, during the

* Corresponding author: borowski@agh.edu.pl

year-long operation of the heat exchanger with non-constant parameters (temperature and relative humidity), a significant fall in performance is seen, getting up to 20% of the declared for laboratory measurement efficiency [11]. In addition to that, in locations where the outside temperature in winter falls below minus 5°C, there is a risk of icing on the heat exchanger surface.

The icing on the exchanger surface is caused by the condensation of the incoming from outside air as a result of the temperature fall on the heat exchange surface below the dew point temperature [12]. As a result, the condensate change phase due to the low temperature of the air flowing through the heat exchanger. With the time the ice layer increases significantly reducing the heat exchange surface area. This cause additional resistance of the air flow and leads to an increase in electricity consumption by fans as well as to a meaningful drop in the heat recovery performance. This phenomena in extreme situations can cause the complete destruction of the heat exchanger deactivating the operation of ventilation and air-conditioning systems.

The aim of the study is to evaluate the temperature, humidity and total efficiency of the air handling unit with a new proposed type of high-performance periodic counter flows air-to-air heat exchanger. This work presents the results of studies of the efficiency of recovery of sensible, latent and total heat for a periodic heat exchanger. An analysed heat exchanger significantly gain the energy efficiency of the ventilation as well as comfort air-conditioning system. The proposed system can be also implemented with the recovery of moisture from the used air. The presented solution prevents the heat exchanger icing without increasing its primary energy demand. As a result, very high and regular seasonal performance of heat and humidity recovery, independent of outside air parameters, can be expected.

2 Methodology

The periodic-flow heat exchanger unit consists of a standard counterflow exchanger and a set of opposing air dampers with a short opening/closing time, used to appropriately cyclically modify the direction of air flow through the heat exchanger, allowing the intake air to absorb the moisture from the cold side of the heat exchanger. This solution not only changed flow direction but also protects the exchanger against frosting.

The experimental measurement for the analysed system was done to determine the temperature, humidity and total efficiency of the air handling unit with a new generation of high-performance periodic counterflow air-to-air heat exchanger. The tests were carried out under conditions which, based on experience, were considered to be extremely unfavourable for the possibility of occurrence of the phenomenon of heat exchanger frosting. The details about flow conditions during the study are presented in Table 1.

Table 1. Airflow conditions during the study.

V_{cz} m³/h	T_{cz} °C	ϕ_{cz} %	V_w m³/h	T_w °C	ϕ_w %
670	-20	-70	680	20	-50

The fluid flow and heat transfer experimental measurement of the new type of air-to-air heat exchanger was carried out in the Polish National Research Radom Institute. Two independent calorimetric chambers allowed for precise control of climatic conditions and simulating the desired parameters of the intake air and the parameters of the extract air. The unit, in which the periodic-flow heat exchanger was installed, was placed in the chamber in which the air temperature was maintained at the level of 20.0 +/-1°C, thus reducing possible heat losses through the unit housing.

In the calorimetric chamber which simulated winter conditions, there was a connection of the intake and exhaust air to the unit with the periodic-flow heat exchanger, while in the warm chamber which simulated the conditions inside the ventilated rooms, there was a connection of the extract and supply air to the unit. A diagram of the analysed system is presented in Figure 1.

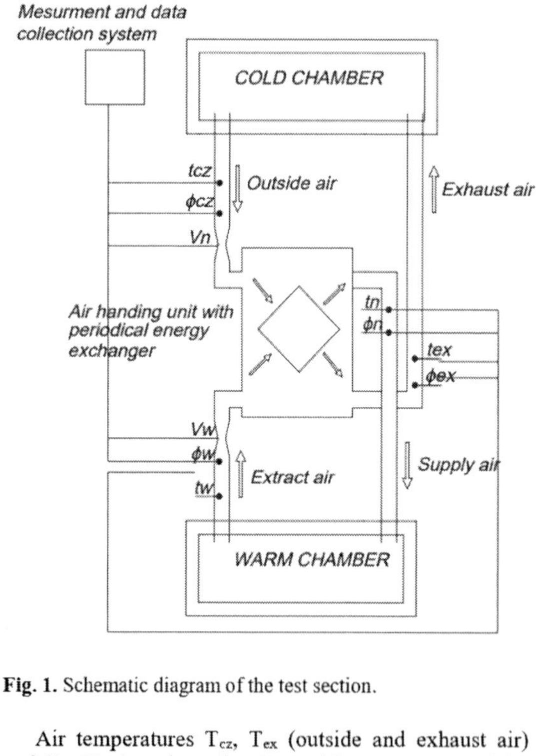

Fig. 1. Schematic diagram of the test section.

Air temperatures T_{cz}, T_{ex} (outside and exhaust air) and T_n, T_w (supply and extract air) were measured by using Geneza GPE-D-A-160-Pt100-klA sensors with 0.1K accuracy. Measuring of air relative humidity ϕ_{cz}, ϕ_{ex}, ϕ_n, ϕ_w were performed by using Introl EE31 transducers with 1% measurement uncertainties. Volume flow rates V_n, V_w (supply and extract air) were measured by using Venturi tube in accordance with PN-81/M-42364 standard.

A data logger (APAR AR207) was used to acquire individual air fluid flow and thermal parameters (air temperatures, volume flow rates, relative humidity). All monitored air thermal parameters and air flow parameters were acquired with a temporal resolution equal to 5 seconds and for a period of 5100 s. Regardless of the configuration, the temperature and humidity of the outside and extract air were stabilized for the period of about 1500 s.

2.1 Air moisture content

The air moisture content is the ratio of the mass of water vapour contained in the air to the mass of dry air. By transforming the gas state equations and using Dalton's law [13], the following equation can be used to determine the air moisture content on the basis of the value of the air temperature and relative humidity:

$$x = 0{,}622 * \frac{\phi * p_{gs}}{p_b - \phi * p_{gs}} \quad (1)$$

where

$$\mathrm{p_{gs}} = 6.1121 * \mathrm{e}^{17{,}502\mathrm{T}/(\mathrm{T}+240.97)} \quad (2)$$

$$\mathrm{p_b} = 1013.25$$

2.2 Air enthalpy

Humid air enthalpy h with moisture content x is the enthalpy of a mixture of 1 kg dry air and x kg water vapour. Assuming that for such a mixture of dry air and the total moisture content in the liquid form at 0°C, the enthalpy equals zero, the following relationship is obtained:

$$h = c_{pg}T + x\left(c''_{pp}T + r_o\right) \quad (3)$$

2.3 Temperature efficiency of the heat exchanger

The temperature efficiency of a heat exchanger is calculated as the ratio of the heat flux recovered by the heat exchanger transferred from the extracted air to the supply air in reference to the total heating power required to heat the outside air to indoor air temperature. The efficiency of the recovery of sensible heat can be determined using the following equation:

$$\eta_t = \frac{(T_n - T_{cz})}{(T_w - T_{cz})} * 100\% \quad (4)$$

2.4 Humidity efficiency of the heat exchanger

The humidity efficiency of a heat exchanger is defined as the ratio of the moisture flux recovered by the heat exchanger transferred from the extracted air to the supply air compared to the total moisture demand that would have to be met in order for the humidity of the outside air to be increased to the level of humidity of indoor air. The efficiency of the recovery of latent heat (the so-called humidity efficiency) can be determined using the following equation:

$$\eta_w = \frac{(x_n - x_{cz})}{(x_w - x_{cz})} * 100\% \quad (5)$$

2.5 Total efficiency of the heat exchanger

The overall efficiency of the heat exchange can be defined as the ratio of the heat flux recovered in the heat exchanger system to the potentially recoverable heat flux. The efficiency of the total heat exchange can be determined from the following equation:

$$\eta_t = \frac{V_n * (h_n - h_{cz})}{V_w * (h_w - h_{cz})} * 100\% \quad (6)$$

3 Results and discussion

Figures 2-5 present the results of the flow rate measurements of supplied air V_n, and extracted air V_w, temperatures T and relative humidity ϕ of the air on each side of the heat exchanger. On the basis of laboratory and experimental test measurements, the interval of rapid air damper position changes was assumed to be 300 s. Changing the position of the dampers results in changing air flow through the exchanger direction, enabling evaporation of the condensate accumulated on the side walls of the exchanger in both operating modes. This condensate is then absorbed by the stream of outside air. The results in Figure 2 shows that by switching the air damper position resulting in a change in the direction of air flow through the heat exchanger and the resistance of the flowing air on both sides of the heat exchanger is changed. This causes a decrease in the extract air flow rate and an increase in the supply air flow rate by a constant value of about 60 m^3/h, which represents 8% of the respective initial rates. The total difference between the supply and extract air balance increases from the initial value of 5% to 15%.

The temperatures of the supply air are presented in Figure 3. It is worth to notice that none of the temperatures did not fall below the value of 18°C during the entire period of the study, with the temperature values for the intake and extracted air stable and amounting to -20°C (+/-0.5°C), and 20°C (+/-0.5°C), respectively. The maximum temperature of the supply air was 18.9°C, and the minimum was 18.2°C, which results in the temperature efficiency of the exchanger in the range of 95-97.5%. The maximum exhaust temperature was -6.2°C and the minimum was -9.1°C. One can infer from the figure that despite critical air conditions, no frost was observed on the heat exchanger surfaces. This was caused by the complete evaporation of the condensate as a result of periodic changes in the direction of air flow through the heat exchanger.

 https://doi.org/10.1051/matecconf/201824002003

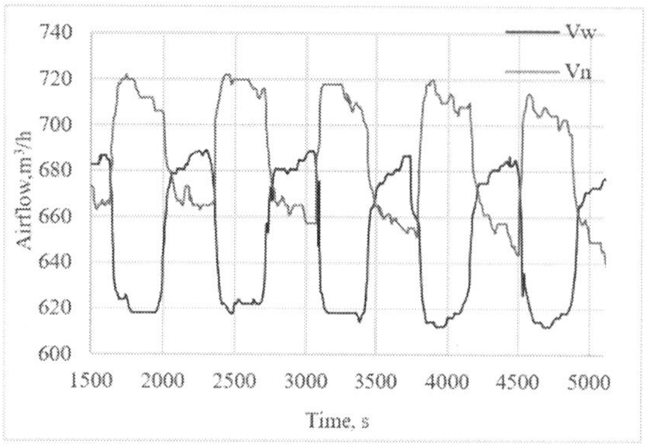

Fig. 2. Supply V_n and extract V_w air volume flow rates.

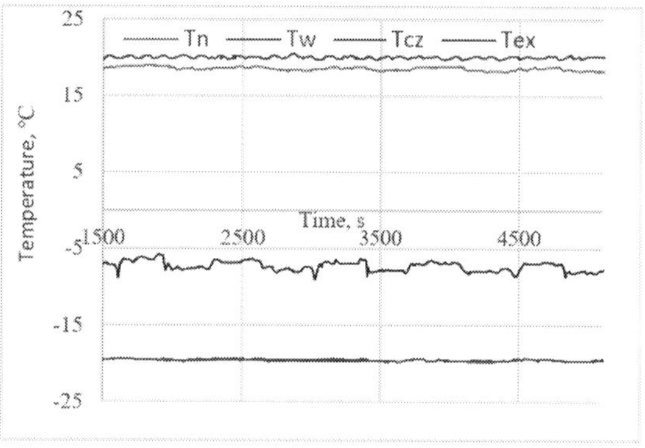

Fig. 3. Outside T_{cz}, supply T_n, extract T_w and exhaust T_{ex} air temperature.

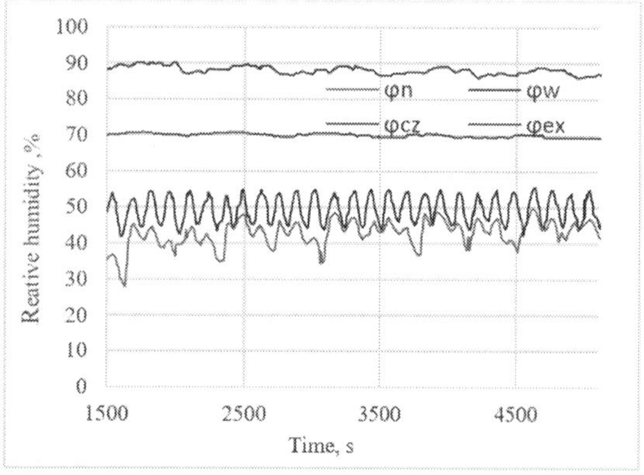

Fig. 4. Outside ϕ_{cz} , supply ϕ_n, extract ϕ_w and exhaust ϕ_{ex} relative humidity.

The humidity of the supply and extract air is presented in Figure 4. During the measurements, humidity changed dynamically between 30%–48% and 44–54% respectively. During the study, the supplied air had the humidity level higher than 30%, which effectively prevented the drying of the ventilated rooms, contributing to maintaining the feeling of thermal comfort for the users of the building.

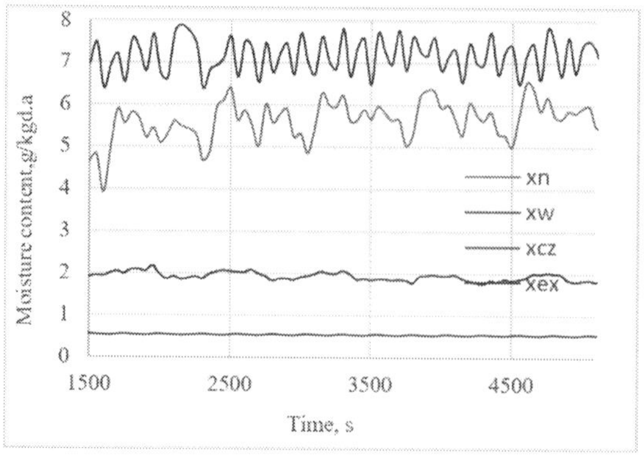

Fig. 5. Outside x_{cz}, supply x_n, extract x_w and exhaust x_{ex} air moisture content.

On the basis of the measurements of temperature and relative humidity of the air flowing through the air-to-air heat exchanger, the moisture content in each of the four air streams was evaluated. All these values, which contribute to the heat and moisture exchange processes, are presented in Figure 5. During the first 150 s of the changes in the working cycles of the quick-acting dampers, the most intensive process of evaporation of the condensate accumulated on the heat exchanger plates is observed, during the following 150 s, the moisture content x_w in the supply air drop as a result of the complete evaporation of moisture located on one side of the heat exchanger.

Table 2. Average temperature and efficiency of the heat exchanger.

V_n m³/h	V_w m³/h	T_{cz} °C	T_n °C	T_w °C	T_{ex} °C	N_t %	Q_t kW
685	651	19.8	18.6	20.1	-7.3	96.3	8.80

Based on the experimental measurement, it was evaluated that the average humidity efficiency is about 79.8%, while the average humidity capacity of the air handling unit with a periodic-flow heat exchanger is 2.96 kW. The detailed results for this case are presented in Tables 2-3. The average overall efficiency of the analysed in this work new heat exchanger type is about 91.4%, which translates into a total capacity of 11.76 kW (see Table 4).

Table 3. Average humidity and efficiency of the heat exchanger.

V_n m³/h	V_w m³/h	x_{cz} x/kg	x_n x/kg	x_w x/kg	x_{ex} x/kg	N_w %	Q_u kW
685	651	0.56	5.82	7.15	1.93	79.8	2.96

Table 4. Average total and efficiency of the heat exchanger.

V_n m³/h	V_w m³/h	h_{cz} kJ/kg	h_n kJ/kg	h_w kJ/kg	h_{ex} kJ/kg	N_c %	Q_c kW
685	651	18.2	33.35	38.2	-2.52	91.4	11.76

4 Conclusions

This work the results of experimental measurements conducted in order to determine the temperature, humidity and total efficiency of an air handling unit equipped with an innovative periodic counter-flow heat exchanger are shown. The analysis indicates that due to the appropriate design of the heat exchanger equipped with a system of air dampers, it is possible to achieve high system performance. The sensible and latent heat recovery at extremely unfavourable simultaneous values, amounting to 96.3% and 79.8% respectively. At the same time an overall heat recovery value is about 91.4% which is unusual for this type of devices. An advantage of the proposed heat exchanger solution is its unique property, which results in the fact that during the experimental measurement, despite low an average air temperature of -7.3°C on the exhaust side, no frost forming has been observed on the heat exchanger side walls or fins. With the proposed system, it is possible to eliminate additional typically implemented in this type of unit the anti-freeze systems that use an electrical heater, which requires high amounts of energy. To achieving the average air temperature of about 18.6°C and the supply air moisture content of 5.82 g/kg, it is unnecessary to use energy-consuming auxiliary heaters and air humidifiers.

Systems with periodic-flow heat exchangers can be recognised as a very good solution to the requirements of nearly zero-energy buildings.

The periodic operation of the heat exchanger should be sought primarily in the optimisation of its design to equate the air resistance on both sides of the exchanger and to minimise the difference in the balance of the supply air and extract air flow during changing the air damper position. It has been recognised that it is essential to carefully evaluate the time of changing the air dumpers positions with relation to the indoor and outside conditions, thus enabling a controlled recovery of moisture from the extract air.

In order to fully implement the proposed solution on a macro scale, it is required to create a numerical model of the air-to-air heat exchanger, which allows for accurate computational fluid dynamics analysis and for the optimisation.

Acknowledgements

The research on the periodic-flow heat exchanger was supported in the Smart Growth Operational Programme 2014-2020 by the Polish National Centre for Research and Development.

References

1. ADEME, Monitoring of energy efficiency trends and policies in the EU, Etude, Rapport, 102 (2015)
2. European Commission, Communication from the commission to European Parliament, the Council, The European Economic and Social Committee of Regions. A Roadmap for moving to a competitive low carbon economy in 2050, (2010)
3. Directive 2012/27/EU of the European Parliament and of the Council of 25 October 2012 on energy efficiency, (2012)
4. Regulation of the Minister of Infrastructure and Construction of November 14, 2017 amending the regulation on technical conditions that should be met by buildings and their location, Dz.U. 2285 (2017).
5. M. Borowski, M. Karch, Chłodnictwo & Klimatyzacja, **4**, 62 (2015)
6. C. Simonson, Energy Build. **37**, 23 (2005)
7. X.P. Liu, J.L. Niu, Appl. Energy **129**, 364 (2014)
8. Y. El Fouih, P. Stabat, P. Rivie`re, P. Hoang, V. Archambault, Energy Build. **54**, 29 (2012)
9. Borowski M, Karch M, Satoła D, *The study of the device "Onyx Experience" intended for the energy-efficient buildings*, Poznań, ISBN: 978-83-61265-09-2, 25 (2017)
10. M. Mijakowski, J. Sowa, P. Narowski, Civil Engineering, **4**, 107 (2010)
11. M. Besler, M. Skrzycki, Rynek Instalacyjny, **3** (2013)
12. D. Seker, H. Karatas, N. Egrican, Int. J. of Refrig. **27**, 367 (2004)
13. Donald H, Jenkins B, *Chemical Thermodynamics at a Glance* (Wiley-Blackwell 2008)

https://doi.org/10.1051/matecconf/201824002004

Evaluation of selected methods of the heat transfer coefficient determination in fin-and-tube cross-flow heat exchangers

Tomasz Bury*, and *Małgorzata* Hanuszkiewicz Drapała

Institute of Thermal Technology, Silesian University of Technology, ul. Konarskiego 22, 44-100 Gliwice, Poland

Abstract. The work is a part of a thermodynamic analysis of a finned cross-flow heat exchanger of the liquid-gas type. The heat transfer coefficients on the liquid and the gas side and the area of the heat transfer are the main parameters describing such a device. The basic problem in computations of such heat exchangers is determination of the coefficient of the heat transfer from the finned surfaces to the gas. The differences in the heat transfer coefficient local values resulting from the non-uniform flow of mediums through the exchanger complicates the analysis additionally. Six Nusselt number relationships are selected as suitable for the considered heat exchanger, and they are used to calculate the heat transfer coefficient for the air temperature ranging from 10°C to 30°C and for the velocity values ranging from 2 m/s to 20 m/s. In the next step, the gas-side heat transfer coefficient is determined by means of numerical simulations using a numerical model of a repetitive fragment of the heat exchanger under consideration. Finally, the Wilson plot method is also used. The work focuses on an analysis of the in-house HEWES code sensitivity to the method of the heat transfer coefficient determination. The authors believe that the analysis may also be useful for the evaluation of different methods of the heat transfer coefficient computation.

NOMENCLATURE

A	-	surface area, m^2
c_p	-	specific heat at constant pressure, J/(kg·K)
h	-	heat transfer coefficient, W/(m^2·K)
k	-	thermal conductivity, W/(m·K)
k	-	overall heat transfer coefficient, W/(m^2·K)
k_l	-	linear heat transfer coefficient, W/(m·K)
L	-	characteristic dimension, m
Nu	-	Nusselt number
Pr	-	Prandtl number
$\dot{q}$	-	heat flux, W/m^2
$\dot{Q}$	-	heat transfer rate, W
Re	-	Reynolds number
St	-	Stanton number
T	-	absolute temperature, K
t	-	temperature, °C
u	-	velocity, m/s
V	-	volume, m^3
$\dot{V}$	-	volumetric flow rate, m^3/s
ρ	-	density, kg/m^3

Subscripts

a	-	air
in	-	inlet
out	-	outlet
w	-	water

Note: Some specific symbols are explained in the text and in Fig. 4.

1 Introduction

The paper presents a part of a thermodynamic analysis of a finned cross-flow heat exchanger of the liquid-gas type. The analysed device, presented in Fig. 1, is a typical car cooler characterized by a non-uniform inflow of air.

Fig. 1. View of the analysed heat exchanger

The measurements were performed on a test stand specially designed and made in the Institute of Thermal Technology of the Silesian University of Technology to enable testing of cross-flow heat exchangers to determine the form of the uneven air flow distribution at the inlet cross-section of the analysed heat exchanger. The test stand diagram is presented in Fig. 2. The stand main element is a thermoanemometric sensor. Figure 3 presents an example air velocity distribution at the analysed heat exchanger inlet. It should be noted that the air inflow has a non-uniform character.

* Corresponding author: tomasz.bury@polsl.pl

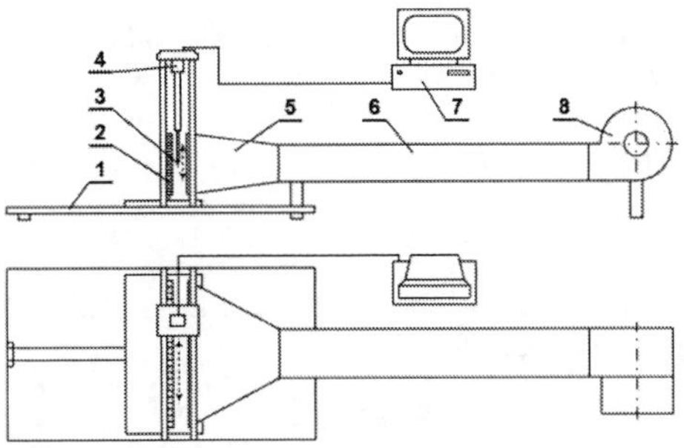

Fig. 2. Test stand – diagram of the air supply module
1 – support plate, 2 – heat exchanger, 3 – thermoanemometric sensor, 4 – measuring probe drive, 5 – diffuser, 6 – air channel, 7 – control computer, 8 – radial fan

The results of the measurements are used to develop a mathematical model of the heat exchanger under consideration. The model utilizes the finite difference method. The final effect of the analyses is a computer code, called HEWES, intended for thermal analyses of the considered type of heat exchangers [1].
It was impossible to validate the code at the initial arrangement of the test stand. For this reason, the test stand was modernized within the current research project and the validation procedure became possible. A hot water supply module, as shown in Fig. 4, was built.

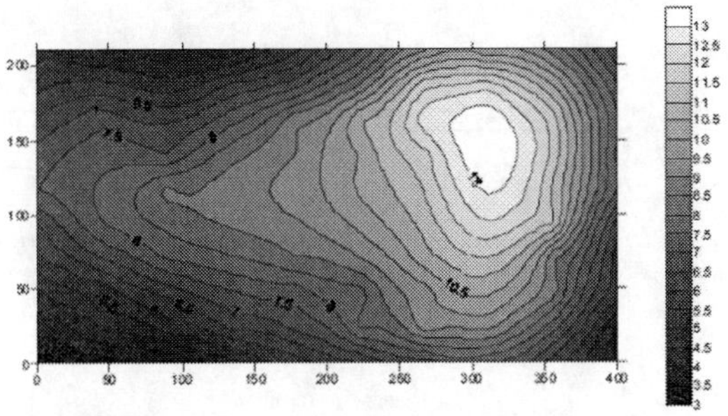

Fig. 3. Example air velocity distribution (in m/s) at the heat exchanger inlet

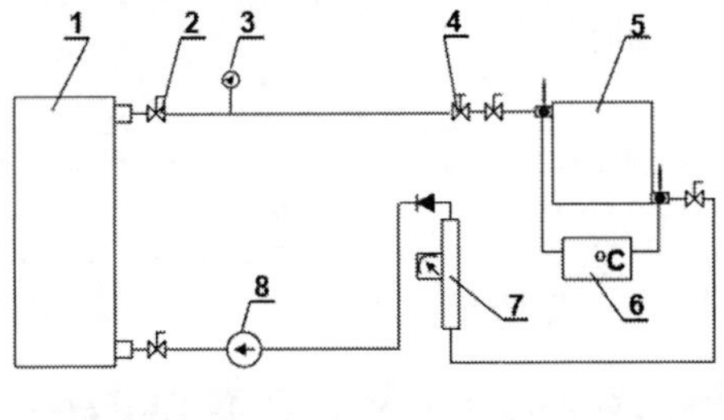

Fig. 4. Test stand – diagram of the hot water supply module
1 – electric boiler, 2 – cut-off valve, 3 – gauge, 4 – regulation valve, 5 – heat exchanger, 6 – temperature measuring system, 7 – flowmeter, 8 – circulation pump

The initial validation procedure showed significant differences between numerical and experimental results [2]. It is suggested that the discrepancies were caused by the method of the heat transfer coefficient determination. The gas-side heat transfer coefficient can be determined through numerical simulations performed for a numerical model of a repetitive fragment of the considered heat exchanger [3] or it can be computed using one of the available Nusselt number correlations.

The heat transfer coefficients on the liquid and on the gas side and the area of the heat transfer are the main parameters describing this type of heat exchangers. Determination of the liquid-side heat transfer coefficient does not pose a big problem. The turbulent flow inside the tubes is a well-recognized issue and most of the existing empirical correlations give comparable results [4]. The basic problem in the computations of such heat exchangers is to calculate the coefficient of the heat transfer from the finned surfaces to the gas. Due to the diversity of designs, it is impossible to formulate a universal correlation for the Nusselt number. It is difficult to select an appropriate relationship for a given geometry It is also difficult to select the characteristic dimensions. The differences in the heat transfer coefficient local values resulting from uneven flows of the mediums through the exchanger complicates the analysis additionally [1].

This paper focuses on an analysis of the in-house HEWES code sensitivity to the method of the heat transfer coefficient determination. The authors believe that the sensitivity analysis may be also useful for the evaluation of different methods of the heat transfer coefficient computation.

2 Heat transfer coefficient computations

2.1 Application of the Nusselt number correlations

A traditional analysis of the convective heat transfer for simple cases is based on the theory of similarity and an analysis of dimensions. The Nusselt number is a dimensionless form of the heat transfer coefficient and it is expressed as:

$$Nu = \frac{h \cdot L}{k} \tag{1}$$

The Nusselt number general form at forced convection is described as follows:

$$Nu = a \cdot Re^{b} \cdot Pr^{c} \tag{2}$$

Constants a, b, and c in Eq. 2 depend on the case. It is very difficult to find an analytical solution for real cases. Therefore, extensive measurements are necessary. A statistical analysis of experimental results makes it possible to determine the values of the constants in Eq. 2 and to formulate a detailed empirical correlation. A large number of such relationships have been worked so far. It should be mentioned, however, that their application is limited to heat exchangers with designs the same as or very similar to those used in the experiments. Six formulae were tested as applicable for the heat exchanger under consideration [1, 4, 7-9]. The diagram presented in Fig. 5 explains the specific symbols used in Eqs. 3–12.

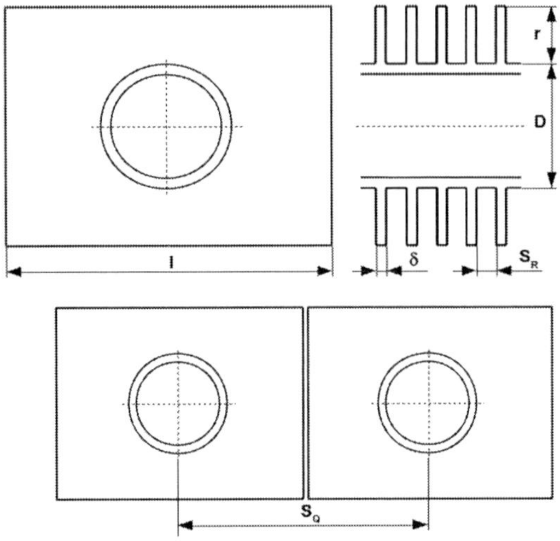

Fig. 4. Tube and fin characteristic dimensions

1. Berman correlation:

$$Nu = 0.225 \cdot Re^{0.633} \quad (3)$$

Equation 3 is valid for gas cooling. While considering the gas heating process, the considered values of the heat transfer coefficient should be by 50% higher.

2. Brigs and Young correlation:

$$Nu = 0.134 \cdot Re^{0.681} Pr^{0.333} \left(\frac{S_R}{r}\right)^{0.2} \left(\frac{S_R}{\delta}\right)^{0.113} \quad (4)$$

3. Schmidt correlation:

$$Nu = 0.3 Re^{0.625} Pr^{0.333} \left(\frac{A_{f+t}}{A_0}\right)^{-0.375} \quad (5)$$

where A_{f+t} is the total surface area of the tube and the fins, and A_0 is the surface area of a smooth, non-finned, tube.

4. Paikert correlation:

$$\mathrm{Nu} = 0.26 Re^{0.6} Pr^{0.333} \left(\frac{A_0}{A_e}\right)^{0.6} \left(\frac{A}{A_{G0}}\right)^{-0.15} \quad (6)$$

where:

$$\frac{A_0}{A_e} = \frac{S_Q(S_R+\delta)}{(S_Q-D)S_R+(S_Q-D-2r)\delta} \quad (7)$$

$$\frac{A}{A_{G0}} = 1 + \frac{2r(r+D+\delta)}{D(S_R+\delta)} \quad (8)$$

The characteristic dimension in Eqs. 3–6 is the tube outer diameter.

5. Norris and Spofford correlation [8]:

$$Nu = 1.0 \cdot Re^{\frac{1}{2}} \cdot Pr^{\frac{1}{3}} \quad (9)$$

The authors of Eq. 9 define the characteristic linear dimension as doubled length of the longitudinal flow path around the fin.

6. Kays and London correlation – the researchers conclude their studies on finned-tube banks using charts presenting the Colburn factor as a function of the Reynolds number [9]:

$$St \cdot Pr^{2/3} = f(Re) \quad (10)$$

The Stanton number is known as the modified Nusselt number and it is defined as:

$$St = \frac{h}{u \cdot \rho \cdot c_p} = \frac{Nu}{Re \cdot Pr} \quad (11)$$

The characteristic dimension in the considered case is the tube bank equivalent hydraulic diameter:

$$d_h = 4l\frac{A_{fr}}{A_{f+t}} \quad (12)$$

where A_{fr} is the flow smallest free path surface area in between the exchanger tubes.

The relationships shown above were used to calculate the heat transfer coefficient for the air velocity values ranging from 2 to 20 m/s and for the air temperatures of 10°C, 20°C and 30°C. The range of the air parameters was established based on experiments. The results are presented in Fig. 6. It can be seen clearly that the results given by the relationships under consideration are very different.

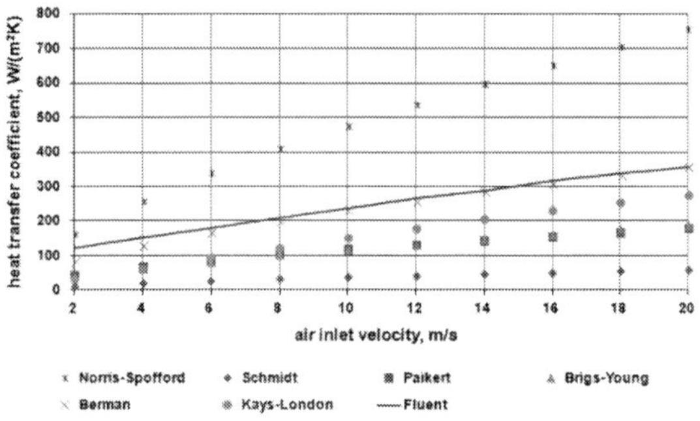

Fig. 6. Heat transfer coefficient calculation results (for the air temperature of 20°C)

The most significant difference is observed for the Norris and Spofford correlation, as well as for the Schmidt correlation. Application of the remaining Nusselt number correlations produced results closer to each other, but the differences still reach up to 200%.

2.2 Numerical simulations using a CFD program

Two geometrical models were made for the numerical computations: one reflected a repetitive element of the considered heat exchanger and the other a repetitive segment. Both models are based on the geometry and dimensions of the tubes and the fins shown in Fig. 7.

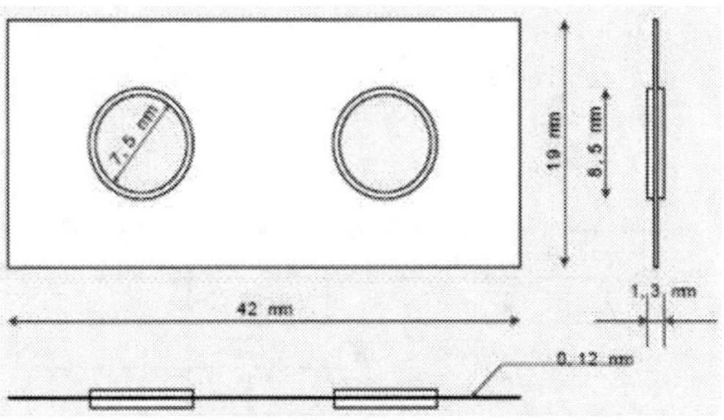

Fig. 7. Geometry and dimensions of the repetitive element

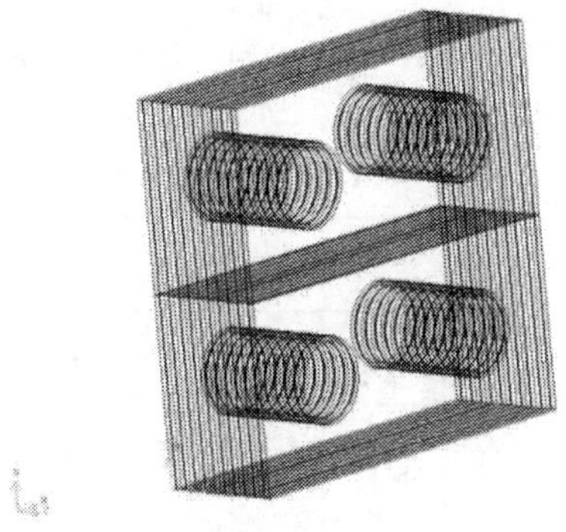

Fig. 8. Geometry of the repetitive segment of the heat exchanger

The simulations were performed using the ANSYS Fluent CFD code. It is assumed that the air inlet is parallel to the x axis of the model. Except the inlet and the outlet surfaces, all the remaining planes are assumed as symmetry planes. At first, the testing computations were performed to select the proper numerical mesh and the turbulence model. The following data were adopted for the computations: the air inlet temperatures of 10°C, 20°C, and 30°C, and the velocity values ranging from 2 m/s to 20 m/s. The Reynolds Stress Model of turbulence was selected for basic computations.

The averaged value of the heat transfer coefficient on the air side is calculated based on the known temperature distribution of the fin and the tube surface areas and the average air temperature and the transferred heat flow rate. This can be written as:

$$h_a = \frac{\sum_{i=1}^{n}(\dot{q}_i A_i)}{\sum_{i=1}^{n} A_i \left[\frac{\sum_{i=1}^{n}(T_i A_i)}{\sum_{i=1}^{n} A_i} - \frac{\sum_{j=1}^{m}(T_{aj} V_j)}{\sum_{j=1}^{m} V_j}\right]} \quad (13)$$

The results, in comparison with the results calculated using the presented Nusselt number correlations, are shown in Fig. 6. It should be noted that the results obtained by means of the numerical model agree very well with the Berman correlation for the Nusselt number.

The values of the convective heat transfer coefficient on the air side obtained from the numerical experiment are higher than those obtained from the majority of the Nusselt number correlations. This may be due to the fact that excellent contact between the fin and the tube in the numerical model and no additional heat resistance resulting, for example, from the fin surface fouling are assumed.

It should also be noted that not all geometrical parameters of the numerical model of the analysed heat exchanger are perfectly in line with the geometry of the devices tested experimentally. The assumed boundary conditions also affect the obtained results. At lower air velocities, the discrepancies between the heat transfer coefficient values obtained from the equations and from the experiment are relatively bigger. This is due to the difficulty of modelling not fully developed turbulent flows.

2.3 Utilization of the Wilson plot

Application of the Wilson method makes it possible to determine the heat transfer coefficient mean value along the heat transfer surface. The term "Wilson plot" is used for various approximation methods of determination of the relationships between the temperature difference and the heat flux for either side of the heat exchanger by measuring the overall temperature difference between the fluids and the heat transfer rate [6].

The gas-side heat transfer coefficient for the considered cooler was determined using the traditional Wilson plot. A set of special experiments was planned and carried out. The water flow rate varied and the air flow parameters were kept constant during the experiments. Assuming that the heat flow rate is proportional to the temperature difference, the heat transfer coefficient has the same value for all the tests.

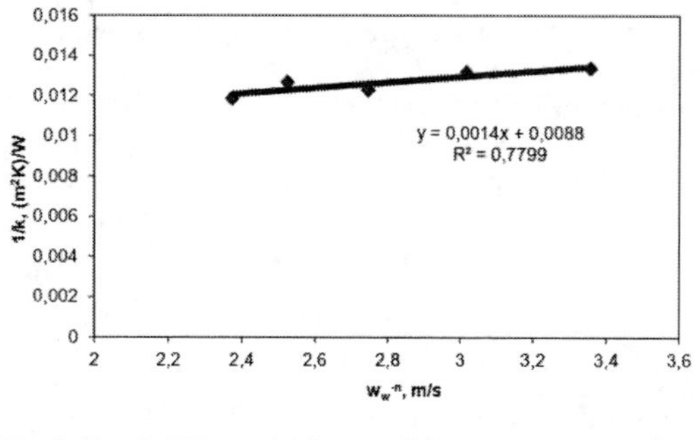

Fig. 9. Simple Wilson plot for one of the measuring series

An example Wilson plot for a selected set of experiments is presented in Fig. 9. According to the Wilson method, the value read out at the intersection of the trend line and the vertical axis is the sum of the tube wall thermal resistance and the thermal resistance on the gas side. Symbol n in the figure is a constant equal to 0.8. The heat transfer coefficient value calculated for the presented data totals 113.7 W/(m^{2}K).

3. Results and discussion

The HEWES code, developed for thermal and hydraulic analyses of fin-and-tube heat exchangers with a non-uniform flow of mediums, was used to evaluate the HTC determination methods. The mathematical model being the basis of the code utilizes the finite difference method. The governing equations are the mass and energy balances written for a repetitive element of the heat exchanger under consideration – this is a single fin with the adjacent section of the tube. The code was developed within the works presented in [1], and then it was validated experimentally [2]. The code was used recently for simulations of other fin-and-tube heat exchangers [10].

The evaluation was based on a comparison of the following parameters:
- total heat transfer rate,
- overall heat transfer coefficient,
- air outlet temperature,
- water outlet temperature.

The total heat flow rate is calculated from the water enthalpy drop:

$$\dot{Q} = \dot{V}_w \rho_w c_{pw} \left(t_{w,in} - t_{w,out}\right) \tag{14}$$

Water density ρ_w is determined for the outlet temperature. The average heat transfer coefficient is computed from the Péclet law:

$$k_l = \frac{\dot{Q}}{L \cdot \Delta t_{ave}} \tag{15}$$

where the average temperature difference is calculated as:

$$\Delta t_{ave} = \varepsilon_{\Delta t} \cdot \Delta t_m \tag{16}$$

The temperature difference correction factor $\varepsilon_{\Delta t}$ is determined based on [7], and Δt_m is the logarithmic mean temperature difference.

The water outlet temperature is measured by a thermocouple, and the HEWES code calculates it as well, so the parameter is found easily.

The HEWES code also computes the average outlet temperature of air, but determination of the temperature experimental value requires averaging the parameter values measured in each measuring field:

$$t_{a,out} = \frac{\sum_i A_i \cdot t_{a,i}}{\sum_i A_i} \tag{17}$$

In equation 17, A_i stands for the measuring field surface area. It should be noted that the thermoanemometric sensor moves along the planes in front of and behind the heat exchanger. The planes are divided into a measuring grid consisting of identical rectangles representing the measuring fields. The air inlet and outlet velocity and temperature are measured in the middle of each measuring field.

The values of relative differences δ are computed for all of the four compared parameters to evaluate the HEWES code accuracy:

$$\delta_X = \left|\frac{X_{exp} - X_{num}}{X_{exp}}\right| \cdot 100\% \tag{18}$$

where X_{exp} is an experimental value of tested parameter, and X_{num} is a computational value.

All the simulations were repeated 8 times to take account of all the methods and formulae for the heat transfer coefficient determination described earlier. The results for two selected experimental cases are presented in Table 1 and Table 2.

Table 1. Evaluation analysis – results for Case 1

Parameter →	$\dot{Q}$	k_l	$t_{a,out}$	$t_{w,out}$
Unit →	kW	W/(mK)	°C	°C
Experiment	13.6	62.9	40.2	54.7
δ	-	-	-	-
Berman	13.0	60.8	39.5	55.5
δ	4.1%	3.4%	1.7%	1.5%
Schmidt	11.1	51.9	37.9	57.4
δ	18.1%	17.5%	5.7%	5.0%
Paikert	11.6	54.2	38.5	56.9
δ	14.5%	13.9%	4.2%	4.1%
Brigs-Young	12.7	59.0	39.0	56.1
δ	6.5%	6.2%	2.9%	2.5%
Norris-Spofford	16.0	73.5	42.4	51.9
δ	17.3%	16.8%	5.5%	5.1%
Kays-London	12.8	59.6	39.4	55.7
δ	5.8%	5.2%	2.1%	1.9%
Wilson	12.4	57.9	38.9	56.3
δ	8.5%	7.9%	3.2%	2.9%
Fluent	14.1	64.9	40.6	54.3
δ	3.5%	3.2%	1.1%	0.8%

Table 2. Evaluation analysis – results for Case 2

Parameter →	$\dot{Q}$	k_l	$t_{a,out}$	$t_{w,out}$
Unit →	kW	W/(mK)	°C	°C
Experiment	7.9	42.2	33.1	45.7
δ	-	-	-	-
Berman	7.6	40.8	32.7	46.2
δ	3.9%	3.2%	1.3%	1.1%
Schmidt	6.5	34.7	31.4	47.9
δ	18.2%	17.7%	5.2%	4.8%
Paikert	6.8	36.6	31.8	47.4
δ	14.2%	13.2%	4.0%	3.8%
Brigs-Young	7.4	39.7	32.3	46.7
δ	6.2%	5.9%	2.5%	2.1%
Norris-Spofford	9.3	49.3	34.8	43.5
δ	17.8%	16.9%	5.2%	4.9%
Kays-London	7.5	40.3	32.5	46.4
δ	5.3%	4.6%	1.8%	1.5%
Wilson	7.3	39.0	32.1	46.8
δ	8.1%	7.6%	2.9%	2.5%
Fluent	8.2	43.5	33.4	45.4
δ	3.3%	3.0%	0.9%	0.7%

Table 1 presents the results obtained from the experiment performed at the maximum volume flow of air (5600 m^3/h), the water volume flow of about 28 dm^3/min and the water inlet temperature of 90°C. The biggest differences were observed while running the HEWES code with the Norris-Spofford and the Schmidt correlations, which give the highest and the lowest values of the heat transfer coefficient on the air side. The simplest Nusselt number correlation proposed by Berman as well as the CFD-based approach for the HTC determination give results closest to the experiment. The results indicate clearly that the HEWES code is sensitive to the HTC determination method. As this parameter concerns the air flow around the finned surface, the second experiment was carried out at the same water flow parameters, but with the minimum air volume flow (about 3000 m^3/h). The results of the comparative analysis for this experiment are listed in Table 2, and they are quite similar to the previous case. The best agreement between experimental and numerical results is found for the CFD model and the Berman correlation.

CONCLUSIONS

The results of the analyses prove that application of the available Nusselt number correlations to determine the HTC on the heat exchanger finned surfaces can produce significantly different results. In this case, there are no unequivocal criteria for the choice of a particular equation. It seems that the typical criterion based on the value of the Reynolds number may be insufficient. It should be added that with complex geometries there are too many variable parameters that can have a significant impact on the fluid flow and the heat transfer, and certainly the cited studies failed to include all these parameters. Moreover, taking account of the fact that many of the commonly used empirical correlations were formulated in the middle of the twentieth century, and considering the technological progress made since then, it can be expected that the possibilities of using them in relation to state-of-the-art devices may be rather limited.
The use of computational fluid dynamics makes it possible to get complete information about the velocity, pressure and temperature fields in the device under consideration. The information can be used to verify the choice of the appropriate empirical relationship, or, in the case of large discrepancies in the obtained results, to identify the heat transfer coefficient for a specific geometry.

The results obtained by applying the calculated HTC values for simulations of experiments by means of the HEWES code indicate that the Berman correlation for the Nusselt number gives results close to experimental ones. The best results were obtained using the CFD-based method of the HTC determination.

This work was realized within the statutory research program of the Faculty of Energy and Environmental Engineering of the Silesian University of Technology.

References

1. R. Piątek, *Thermal analysis of the plate fin-and-tube heat exchanger with a non-uniform inflow of mediums*, (PhD dissertation, Institute of Thermal Technology, Silesian University of Technology, Gliwice 2003)
2. T. Bury, R. Piątek, J. Składzień, Experimental validation of the numerical thermal model of the cross-flow heat exchanger with unequal agent flow, ECOS, Cracow (2008)
3. T. Bury, J. Składzień, The experimental and the numerical analysis of a ribbed heat exchanger with an unequal inlet of the air, Heat Transfer and Renewable Sources of Energy (2006), pp. 419-426
4. E. Kostowski, *Heat transfer* (Publishing House of the Silesian University of Technology, Gliwice 2000)
5. T. Bury, J. Składzień, R. Piątek, 13th Symposium on the Heat and Mass Transfer, pp. 297-304 (2007)
6. J.W. Rose, Experimental Thermal and Fluid Science, **28**, pp. 77-86 (2004),
7. E. Kalinowski, *Heat transfer and heat exchangers*, (Publishing House of the Wrocław University of Technology, 1995)
8. R. H. Norris, and W. A. Spofford, Trans. ASME, **64**, pp. 489-496 (1942)
9. W. M. Kays, and A. L. London, Transactions of ASME, **76**, pp. 387-396 (1954)
10. T. Bury, M. Hanuszkiewicz-Drapała, and K. Widziewicz, Heat Transfer Engineering, **39**,13-14, pp. 1179-1191 (2018)

Numerical analysis and simulation of the heat recovery from wastewater using heat exchanger

Timea Gabor*, *Ancuta-Elena* Tiuc, *Ioana Monica* Sur, and *Iulian Nicolae* Badila

Technical University of Cluj-Napoca, Department of Environmental Engineering and Sustainable Development Entrepreneurship, 103-105 Muncii Avenue, 400641, Cluj-Napoca, Romania

Abstract. The problem of global warming and the reduction of energy consumption have led to an evolutionary progress of research directed towards finding as many solutions as possible to these environmental issues. Firstly, this paper presents the background information on the role of wastewater as a source of heat for the future. Next, the paper includes the analysis elements that define a system for recovering thermal energy from wastewater. The main objective was to identify the parameters that determine the heat transfer. It has started from a conceptual model of the technological system that involves inputs and outputs characterized by technological, physical-chemical, measurable or imposed properties. In the second part this paper presents a numerical model elaborated for the analysis and simulation of the main physical processes, the mass and heat transfer, which underlie the operation of the heat pipe heat exchangers (HPHE). The numerical simulation of heat and mass transfer in the HPHE is computed by using Delphi 7 solver program. This program contained a series of sub-programs for the meshing of the field occupied by the HPHE, another subprogram for solving the meshing equations and the third for post processing. The design of HPHE is the key to provide a heat exchanger system to work proficient as expected. Finally, the result is used to optimize and improving heat recovery systems of the increasing demand for energy efficiency in residential buildings or industry.

1 Introduction

One of the secondary energetic resources that on which a greater accent has been put worldwide are wastewaters (liquid waste). Wastewater from all sources except toilets, enters the sewage system at a relatively high temperature and with an exergy content [1, 2]. Clearly, the heat content of wastewater is a major loss source, so it must be addressed as a solution to reduce energy consumption in both residential and commercial buildings [3]. The potential for domestic wastewater recovery is quite high, approximately 3.5 kWh of energy / person / day, which could be capitalized and used directly in several areas. [2]

The recovery of secondary energetic resources and of liquid waste with high energetic content also led to the development and implementation of various types of heat exchangers and heat recuperators [4, 5]. Heat exchangers are widely used in the industry, transportation, constructions, thermal power plants, heating and air conditioning systems, electronic equipment. In all these applications, improving the efficiency of heat recovery can lead to substantial reduction of costs of production and use. The research is oriented in this direction aim in particular at selecting the work fluid with high thermal conductivity and how to direct the flow towards the surfaces destined for the actual heat transfer [6, 7].

The current paper presents results of research obtained from simulation and numerical analysis of heat recovery from wastewater using a HPHE.

The paper describes the physical mathematical model, establishing the system of governing equations, initial and limits conditions. It also presents the numerical model associated to the mathematical model and numerical methods of discretization when solving the differential system of equations underlying the mathematical model.

Simulation of thermal transfer is conducted for a liquid-liquid type heat exchanger, where heat pipes operate on the principle of phase change of the work fluid. [8, 9]

Worldwide there are a series of papers that deal with theoretical and experimental research of the physical processes associated with the various types of heat exchangers, and more recently a great emphasis is placed on the use of numerical models [10-12].

Numerical models used for the analysis and optimization of mass and heat transfer processes between work fluids in heat exchangers are less costly than experimental research [13]. But it is important to note that these numerical models are improved based on the results of experimental research [14-17].

* Corresponding author: timea.gabor@imadd.utcluj.ro

2 Analysis and modelling

2.1 Description of the experimental equipment

The heat exchanger used for experimental research (Fig. 1) was equipped with 40 heat pipes from *S.C. EnergiQ Indco S.R.L Cluj-Napoca*, with the following characteristics: outside diameter 10 mm; length 1390 mm; working fluid: refrigerant occupied approximately 20% of the total volume of the heat pipe at the vacuum at 10^{-8} Torr.

Fig. 1. The experimental equipment.

The heat exchanger was divided in two compartments (Fig. 2) – the lower one destined for the warm wastewater and the upper one destined for the cold water from the network, which is to be preheated. To these are added the outer wall and partition wall between the two compartments.

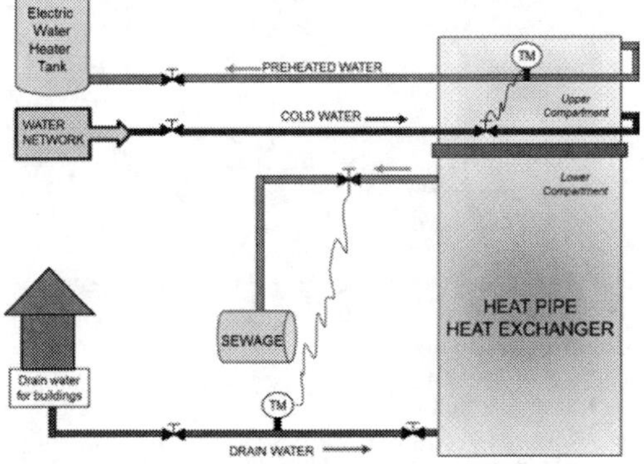

Fig. 2. Schematic diagram of the experimental equipment.

The upper compartment is provided with a horizontal flow divider, thus increasing the time interval in which the main flow of cold water is in contact with the heat pipes. The heat exchanger occupies a range that is approximated by parallelipipedic cells having the *dx*, *dy* and *dz* sides, in number of *im* x *jm* x *km*, where *im*, *jm* and *km* are the number of cells in the direction of coordinate axes *Ox*, *Oy* and *Oz*.

In Fig. 3 is represented a horizontal section parallel with the *xOy* coordinate plane. The 40 heat pipes are represented into two parallel lines. The wall of the heat pipes is approximated also with parallelipipedic cells, with a volume $V_{cell} = dxdydz$.

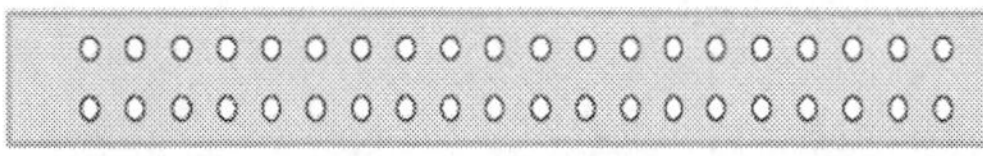

Fig. 3. Horizontal section parallel with the *xOy* coordinate plane.

The flow of wastewater and water from the network is treated as a viscous flow, and the heat transfer at the interface between the heat pipes and water is achieved through conduction. Within the domain occupied by water, the heat transfer is achieved through both conduction and convection.

The structure of the water flow in cavities equipped with heat pipes is complex, mainly because of the occurrence of stationary vortexes above or laterally to the pipes, or of stagnation areas downstream of the pipes. These influence the heat transfer; vortexes contribute to the increase in efficiency of the local heat exchange, while the stagnation areas inhibit the heat exchange. [12, 18]

2.2 Governing equations for the system

In the *xOyz* Cartesian Coordinate System we have the equations of the numerical model, based on fundamental physical principles, expressed by the relationship of conservations laws for movement quantity – Navier Stokes equations (1-3), mass – continuity equation (4) and heat – Fourier equation (5) [19-21]:

$$\frac{\partial \rho u}{\partial t}+\frac{\partial \rho uu}{\partial x}+\frac{\partial \rho vu}{\partial y}+\frac{\partial \rho wu}{\partial z}=-\frac{\partial p}{\partial x}+\frac{\partial}{\partial x}\left(\mu\frac{\partial u}{\partial x}\right)+\frac{\partial}{\partial y}\left(\mu\frac{\partial u}{\partial y}\right)+\frac{\partial}{\partial z}\left(\mu\frac{\partial u}{\partial z}\right) \quad (1)$$

$$\frac{\partial \rho v}{\partial t}+\frac{\partial \rho uv}{\partial x}+\frac{\partial \rho vv}{\partial y}+\frac{\partial \rho wv}{\partial z}=-\frac{\partial p}{\partial y}+\frac{\partial}{\partial x}\left(\mu\frac{\partial v}{\partial x}\right)+\frac{\partial}{\partial y}\left(\mu\frac{\partial v}{\partial y}\right)+\frac{\partial}{\partial z}\left(\mu\frac{\partial v}{\partial z}\right) \quad (2)$$

$$\frac{\partial \rho w}{\partial t}+\frac{\partial \rho uw}{\partial x}+\frac{\partial \rho vw}{\partial y}+\frac{\partial \rho ww}{\partial z}=-\frac{\partial p}{\partial z}-g\rho\beta(T-T_0)+\frac{\partial}{\partial x}\left(\mu\frac{\partial w}{\partial x}\right)+\frac{\partial}{\partial y}\left(\mu\frac{\partial w}{\partial y}\right)+\frac{\partial}{\partial z}\left(\mu\frac{\partial w}{\partial z}\right) \quad (3)$$

$$\frac{\partial \rho}{\partial t}+\frac{\partial \rho u}{\partial x}+\frac{\partial \rho v}{\partial y}+\frac{\partial \rho w}{\partial z}=0 \quad (4)$$

$$\left(\frac{\partial \rho cT}{\partial t}+\frac{\partial \rho cuT}{\partial x}+\frac{\partial \rho cvT}{\partial y}+\frac{\partial \rho cwT}{\partial z}\right)=\frac{\partial}{\partial x}\left(k\frac{\partial T}{\partial x}\right)+\frac{\partial}{\partial y}\left(k\frac{\partial T}{\partial y}\right)+\frac{\partial}{\partial z}\left(k\frac{\partial T}{\partial z}\right) \quad (5)$$

where: *u*, *v* and *w* are components of the velocity vector; μ is dynamic viscosity; *x*, *y* and *z* are independent variables in the *xOyz* system; *T(x,y,z,t)* is temperature; *k* is the thermal conductivity coefficient; T_0 is the reference temperature, in this case the wastewater or cold water input temperature; β is the thermal expansion coefficient; *c* is the specific heat at constant pressure.

For the control the advence of the free surface of the liquid used the equation of the volume fraction (6).

$$\frac{\partial f}{\partial t}+\frac{\partial uf}{\partial x}+\frac{\partial vf}{\partial y}+\frac{\partial wf}{\partial z}=0 \quad (6)$$

where: $f(x,y,z,t)$ is the volume fraction function, if in point (x,y,z) there is liquid it is equal to one and if there is no liquid it is equal to zero. [22, 23]

2.3 Numerical model

During the discretization of the numeric modeling, the differential equations from VOF were used [18, 22, 23]. A horizontal section of the discretization cells applied to the heat pipes is sketched in Fig. 4. Inside the discretization network, cells colored in blue indicate the pipe's wall.

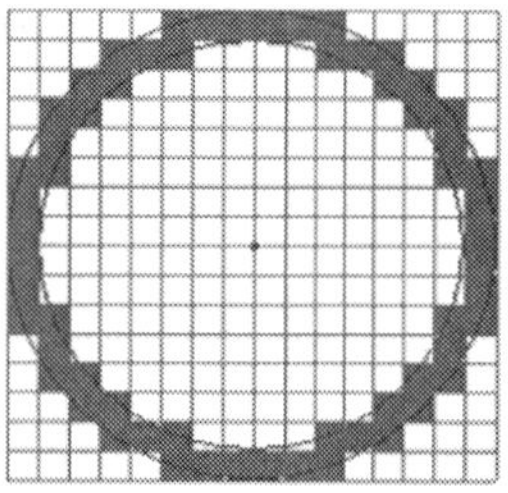

Fig. 4. Horizontal section of a heat pipe, parallel with the xOy coordinate plane.

The current cell with the central point P is presented in Fig. 5. Neighboring cells have centers N, S, W, E, B and T (red dots), and faces of the current cell have centers n, s, w, e, b and t. Thus, in the positive direction of axis Oy noted with E, is the center of the neighboring cell and center of the face in point e. In the negative direction of axis Oy is the center of the neighboring cell in point W, and center of the face in w. So are the analogs in the direction of axes Ox and Oz. For the entire cell, the main characteristic is point P, being treated as a representative point of dimensions ρ, p, T, c, k and μ [19, 20].

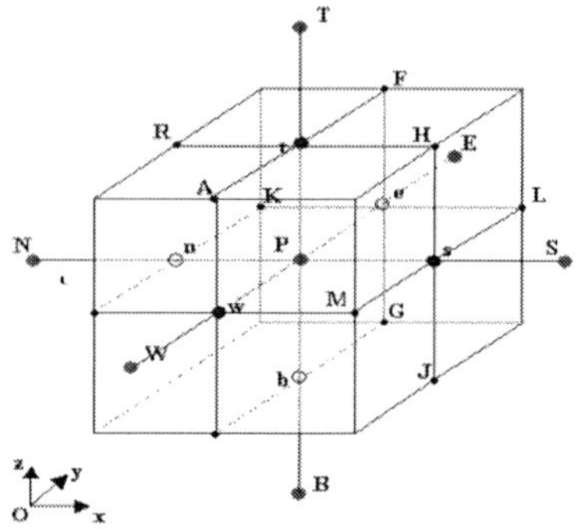

Fig. 5. Current cell P, represented with notations used for face centers and neighboring cells centers.

A parallel section to plane yOz is presented in Fig. 6. It passes through the center of the current P cell and its neighbors, representing the notations and associated grid points; the notations are those used by Patankar [20]. Components of the speed vector are defined in central points of the faces of the current cell, from axis Oy in w and e, and from axis Oz in t and b. It can be observed a cross grid formed by the central points and face centers. The grid points for the v component, by axis Oy, are on the faces parallel with plane yOz. An advantage of the cross grid is the fact that mass flows passing through the faces of the cells can be calculated without interpolation of the speed vector components (Fig. 6).

The parallelepiped centered in point e having the rectangle $tcdb$ as vertical section parallel to yOz, will be the control volume of the v components (i, j, k).

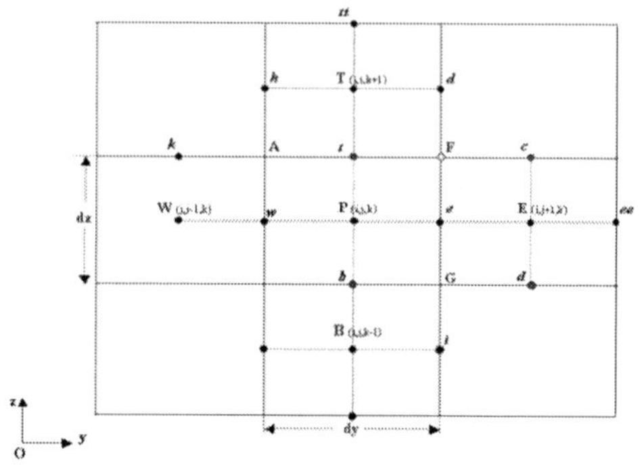

Fig. 6. Vertical section, parallel with plane yOz, through the center of current cell P and its neighbors.

A horizontal section through the center of the current cell P and its neighbors shows the arrangement of the speed components, u and v, in the cross grid, being represented in Fig. 7.

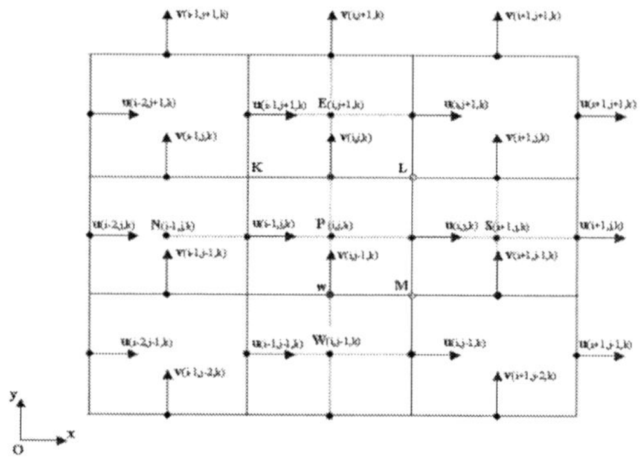

Fig. 7. Horizontal section through the center of the current cell P and its neighbors, indicating the arrangement of u and v components in cross grid.

Due to the fact that network cells can contain various materials, it is required the control of cell content, thus an additional variable was introduced in the form of a tridimensional matrix.

The elements of the variable are noted with $obs[i,j,k]$, $(i,j,k) \in \{1,2,...,im\} \times \{1,2,...jm\} \times \{1,2,...,km\}$, its component for the current cell will have values from one to nine. Their content is further specified:

$obs[i,j,k] = 1$ for cells that contain wastewater in the lower compartment, or cold water from the network in the upper compartment;

$obs[i,j,k] = 2$ for the cells that are partially full with wastewater in the lower compartment, or cold water from the network in the upper compartment;

obs[i,j,k] = *3* for interior cells from the two compartments and that contain air;
obs[i,j,k] = *4* for cells outside the shape that contain air;
obs[i,j,k] = *5* for cells in the supply section of wastewater or cold water;
obs[i,j,k] = *6* for cells in the evacuation section of the wastewater of heated water;
obs[i,j,k] = *7* for cells in the outer wall, divider wall between compartments and flow divider wall in the upper compartment;
obs[i,j,k] = *8* for cells in the heat pipe wall;
obs[i,j,k] = *9* for cells inside the heat pipes.

Next, there was conducted the temporal discretization using an Euler scheme assumed with the time step *dt*. The time step is variable and is determined in such a way that conditions of numeric stability to be satisfied. It integrates the basic equation system on cells in the network in a way that differs by the usual developments in Taylor series that are applied in the classic finite differences methods [21, 24]. The integration of the differential equation with a dependent variable $\Phi \in \{u, v, w, T, p\}$ leads to an equation in the form:

$$a_p\Phi_p = a_0\Phi_0 + a_E\Phi_E + a_w\Phi_w + a_T\Phi_T + a_B\Phi_B + a_S\Phi_S + a_N\Phi_N + b \quad (7)$$

where: $a_0=\rho_0 V_0/dt$, and V_0 and ρ_0 are the volume and density of the cell in point *P* at the previous time step, and *dt* is the dimension of the time step. Coefficients a_E, a_W, a_T, a_B, a_S and a_N contain the sum of advective and diffusive contributions of the neighboring cells (Fig. 5). The diffusive component that was used has the general form $\Gamma f\ A/Dn$ where A is the area of the common face between the current cell and its neighbor, and dn is the distance between the centers of the two cells. The exchange coefficient at the interface between cells, Γf, is calculated as an arithmetic mean for density and viscosity, and as a harmonic mean for temperature values in the neighboring central nodes. When calculating the advective component it's used an "upwind" type interpolation which leads to a coefficient in the form of $\rho A\ max\ (0,\ v)$ where v is the component of the speed vector that is perpendicular on the interface between cells [19, 20].

The system of the mesh equations for the movement quantity conservation (1-3), continuity equation (4) and heat equation (5) are solved in an iterative manner.

The Fourier equation (5) can be solved independently, but the continuity equation (4) and movement quantity equations (1-3) must be solved simultaneously. For the latter, is used the SIMPLE (Semi-Implicit Method for Pressure - Linked Equations) method, which simultaneously solves equations (1-4) using pressure and speed corrections.

Discretization the continuity equation (4) and movement quantity equations (1-3) we obtained (8) equation what is the standard discretization form of the Eq. (2) of movement quantity conservation in the *Oy* axis direction, similarly obtained for *Ox* (9) and *Oz* (10) axis.

$$a_e v_e = a_e^0 v_e^0 + a_{se}v_{se} + a_{ne}v_{ne} + a_{ee}v_{ee} + a_w v_w + a_d v_d + a_i v_i + b_e \quad (8)$$

$$a_s u_s = a_s^0 u_s^0 + a_{ss}u_{ss} + a_n u_n + a_l u_l + a_m u_m + a_h u_h + a_j u_j + b_s \quad (9)$$

$$a_t w_t = a_t^0 w_t^0 + a_{st}w_{st} + a_{nt}w_{nt} + a_c w_c + a_k w_k + a_{tt}w_{tt} + a_b w_b + b_t \quad (10)$$

Following the discretization of the heat conservation equation is resulted eq. (11):

$$a_P T_P = a_P^0 T_P^0 + a_S T_S + a_N T_N + a_E T_E + a_W T_W + a_T T_T + a_B T_B \quad (11)$$

2.4 Stabilization of the speed correction equation

The momentum equations are discretized as follows:

$$\begin{cases} a_s u_s = a_s^0 u_s^0 + a_{ss}u_{ss} + a_n u_n + a_l u_l + a_m u_m + a_h u_h + a_j u_j + b_s \\ a_e v_e = a_e^0 v_e^0 + a_{se}v_{se} + a_{ne}v_{ne} + a_{ee}v_{ee} + a_w v_w + a_d v_d + a_i v_i + b_e \\ a_t w_t = a_t^0 w_t^0 + a_{st}w_{st} + a_{nt}w_{nt} + a_c w_c + a_k w_k + a_{tt}w_{tt} + a_b w_b + b_t \end{cases} \quad (12)$$

Resolution of discretized equations is possible only if the pressure field is given or estimated. If this pressure field is incorrect, the continuity equation cannot be obtained by the speed field given by the momentum equations.

Components of an imperfect speed field (momentum equations with an estimated p^* pressure field) are noted with u^*, v^* and w^* [20]. The following discretized equations had as a solution this "*" speed field, as follows:

$$\begin{cases} a_s u_s^* = a_s^0 u_s^0 + a_{ss}u_{ss}^* + a_n u_n^* + a_l u_l^* + a_m u_m^* + a_h u_h^* + a_j u_j^* + b_s + (p_P^* - p_S^*)dydz \\ a_e v_e^* = a_e^0 v_e^0 + a_{se}v_{se}^* + a_{ne}v_{ne}^* + a_{ee}v_{ee}^* + a_w v_w^* + a_d v_d^* + a_i v_i^* + b_e + (p_P^* - p_E^*)dxdz \\ a_t w_t^* = a_t^0 w_t^0 + a_{st}w_{st}^* + a_{nt}w_{nt}^* + a_c w_c^* + a_k w_k^* + a_{tt}w_{tt}^* + a_b w_b^* + b_t + (p_P^* - p_T^*)dxdy \end{cases} \quad (13)$$

obtained from the discretization equations of the moment, in which the terms containing the pressure forces, and thus the free terms b_s, b_e and b_t, were highlighted, and became $b_s = 0,\ b_e = 0,\ b_t = -g\rho_t\beta(T_t - T_0)V$

The field of stellar speeds must satisfy the continuity equation, in this case p^* must be modified until it does not satisfy this condition. The correct pressure p is obtained with relation [20]:

$$p = p^* + p' \quad (14)$$

where p' is the pressure correction. Modification of the pressure value induces a change in the speed components in the form of:

$$\begin{cases} u = u^* + u' \\ v = v^* + v' \\ w = w^* + w' \end{cases} \quad (15)$$

where u, v and w are the correct speed components, and u', v' and w' are the speed corrections. Subtracting in turn equations (15) from equations (12) we obtained:

$$\begin{cases} a_s u_s' = a_{ss} u_{ss}' + a_n u_n' + a_l u_l' + a_m u_m' + a_h u_h' + a_j u_j' + (p_P' - p_S')A_s \\ a_e v_e' = a_{se} v_{se}' + a_{ne} v_{ne}' + a_{ee} v_{ee}' + a_w v_w' + a_d v_d' + a_i v_i' + (p_P' - p_E')A_e \\ a_t w_t' = a_{st} w_{st}' + a_{nt} w_{nt}' + a_c w_c' + a_k w_k' + a_{tt} w_{tt}' + a_b w_b' + (p_P' - p_T')A_t \end{cases} \quad (16)$$

where relations (14) and (15) where taken into account, and: $A_s=dydz$, $A_e=dxdz$, $A_t=dxdy$

For the calculation of speed corrections obtained with equations (16) we used the SIMPLE procedure. The method is called semi-default because the terms containing speed corrections from the neighboring grid points are omitted, thus the correction speeds became:

$$\begin{cases} a_s u_s' = (p_P' - p_S')A_s \\ a_e v_e' = (p_P' - p_E')A_e \\ a_t w_t' = (p_P' - p_T')A_t \end{cases} \quad (17) \quad \text{or} \quad \begin{cases} u_s' = d_s(p_P' - p_S') \\ v_e' = d_e(p_P' - p_E') \\ w_t' = d_t(p_P' - p_T') \end{cases} \quad (18)$$

where we noted: $d_s = \frac{A_s}{a_s}$, $d_e = \frac{A_e}{a_e}$, $d_t = \frac{A_t}{a_t}$

Correct speeds are given by:

$$\begin{cases} u_s = u_s^* + d_s(p_P' - p_S') \\ v_e = v_e^* + d_e(p_P' - p_E') \\ w_t = w_t^* + d_t(p_P' - p_T') \end{cases} \quad (19)$$

2.5 Establishing the correct pressure equation

The continuity equation (4) was integrated on the control volume centered in point P (Fig. 4), obtaining:

$$(\rho_P - \rho_P^0)\frac{V}{dt} + [(\rho u)_s - (\rho u)_n]dydz + [(\rho v)_e - (\rho v)_w]dxdz + [(\rho w)_t - (\rho w)_b]dxdy = 0 \quad (20)$$

By substituting the corrected speeds, relations (19) in equation (20) we obtain:

$$\begin{aligned} &(\rho_P - \rho_P^0)\frac{V}{dt} + [(\rho u^*)_s + \rho_s d_s(p_P' - p_S') - (\rho u^*)_n - \rho_n d_n(p_N' - p_P')]dydz + \\ &+ [(\rho v^*)_e + \rho_e d_e(p_P' - p_E') - (\rho v^*)_w - \rho_w d_w(p_W' - p_P')]dxd + \\ &+ [(\rho w^*)_t + \rho_t d_t(p_P' - p_T') - (\rho w^*)_b - \rho_b d_b(p_B' - p_P')]dxdy = 0 \end{aligned} \quad (21)$$

and regrouping the terms we have:

$$a_P p_P' = a_S p_S' + a_N p_N' + a_E p_E' + a_W p_W' + a_T p_T' + a_B p_S' + b \quad (22)$$

where we noted:

$$a_S = \rho_s d_s dydz,\ a_N = \rho_n d_n dydz,\ a_E = \rho_e d_e dxdz,\ a_W = \rho_w d_w dxdz,\ a_T = \rho_t d_t dxdy,\ a_B = \rho_b d_b dxdy \quad (23)$$

$$a_P = a_S + a_N + a_E + a_W + a_T + a_B \quad (24)$$

$$b = (\rho_P^0 - \rho_P)\frac{V}{dt} + [(\rho u^*)_n - (\rho u^*)_s]dydz + [(\rho v^*)_w - (\rho v^*)_e]dxdz + [(\rho w^*)_b - (\rho w^*)_t]dxdy \quad (25)$$

It should be noted that, although the notations of the coefficients resemble the notations used in the discretization of the heat conservation equation, here they have a completely different meaning (pressure).

In this context we have seen a lot of interests in the literature about the experimental and theoretical investigation of the heat pipes. [25-28]

3 Results and discussion

For the analysis of the flow and heat transfer phenomena in the upper compartment was used a spatial grid with steps $dx=dy=0.002\ m$ and $dz=0.004\ m$, resulting a number of $im\ x\ jm\ x\ km = 383\ x\ 46\ x\ 78 = 1374204$ computational cells. But observing that the vertical median plane – perpendicular on the wall that contains input and output orifices from the upper compartment – parallel with lateral walls, is a symmetry plane, number of computational cells was reduced in half by replacing $jm=46$ cu $jm=23$, resulting a number of 687102 cells.

Required temperatures: lower wall of upper compartment $t_{inf}=29°C$, wall of heat pipes at constant temperature of $t_{pipe}=32°C$. To take account of heat losses at the interface with the environment, an additional layer of air was added near the lateral faces and the upper face of the compartment, with a constant temperature of $t_{air} = 22°C$.

Thermophysical characteristics of materials that are part of the upper compartment are shown in Table 1.

Table 1. Thermophysical characteristics of materials.

Thermophysical characteristic	Value
Water density, ρ_{water}	977 kg/m^3
Specific heat of water, c_{water}	4180 J/kg°C
Thermal conductivity coefficient of water, k_{water}	0.61 W/m°C
Cinematic water viscosity, μ	0.8 x 10^{-6} m^2/s
Thermal expansion coefficient of water, β	0.000207°C^{-1}
Steel density, ρ_{steel}	7800 kg/m^3
Specific heat of steel, c_{steel}	630 J/kg °C
Thermal conductivity coefficient of steel, k_{steel}	60 W/m °C
Copper density, ρ_{copper}	8920 kg/m^3
Specific heat of copper c_{copper}	390 J/kg °C
Thermal conductivity coefficient of copper, k_{copper}	399 W/m °C
Speed of water input, $v_{w.i.}$	0.27 m/s

Values of the parameters used for the initial conditions are shown in Table 2. Two runs were performed and for each of them the initial condition varied:

Table 2. Initial conditions used for each run.

Run no.	D_0 [m³/h]	$v_{w.i.}$ [m/s]	t_{air} [°C]	t'_2 [°C]	t'_1 [°C]
1.	0.37	0.083	22	12	35
2.	1.18	0.261	27	22	60

The runs were conducted with a variable time step, determined based on the maximum flow speed at the beginning of each cycle.

The flow divider in the upper compartment has the purpose of increasing the heat exchanger's efficiency, because the main flow passes twice between the condensing sides of the heat pipes [11], thus the cold water will accumulate a larger amount of heat – in a given interval of time.

Figure 8 *a)* and *b)* presents the position of the free surface in vertical transversal sections close to the input orifice, at 10 mm from the wall with the input orifice, and in the section that contains the first two heat pipes. We can see the partially filled cells and the formation of swirling currents near the walls, centered in points *A* and *B*. Also, there are swirling currents formed between pipes and lateral walls.

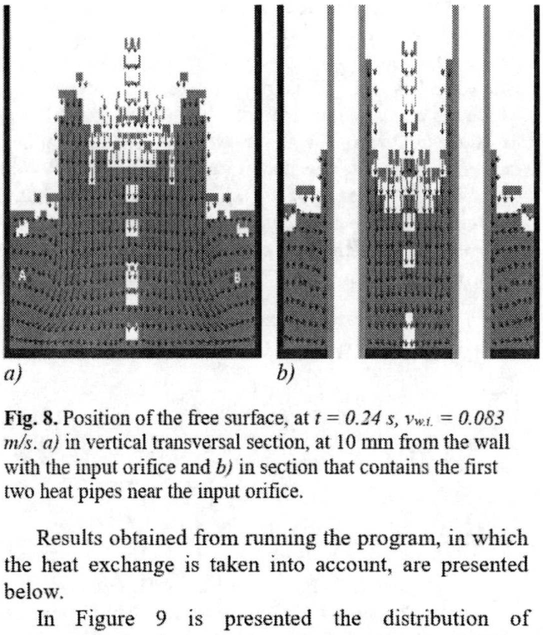

a) *b)*

Fig. 8. Position of the free surface, at $t = 0.24$ s, $v_{w.i.} = 0.083$ m/s. *a)* in vertical transversal section, at 10 mm from the wall with the input orifice and *b)* in section that contains the first two heat pipes near the input orifice.

Results obtained from running the program, in which the heat exchange is taken into account, are presented below.

In Figure 9 is presented the distribution of temperature in the central vertical section at $t=168$ s from the beginning of operation of the heat exchanger and at $t'_1= 35°C$, $t'_2 = 12°C$, $t_{air} = 22°C$, $v_{w.i.} = 0.083$ m/s. Near the lateral walls, the upper wall and in the flow divider plate, the temperature is lower, with values between 13 – 15°C. Near the inferior wall, in contact with the lower compartment and heat pipes, there is a warm layer with temperatures of 25 – 27°C. Inside, there are warmer traces next to the heat pipes. At the left end of the divider plate there is a deformation of these traces, due to a swirling current formed in this phase of operation of the heat exchanger.

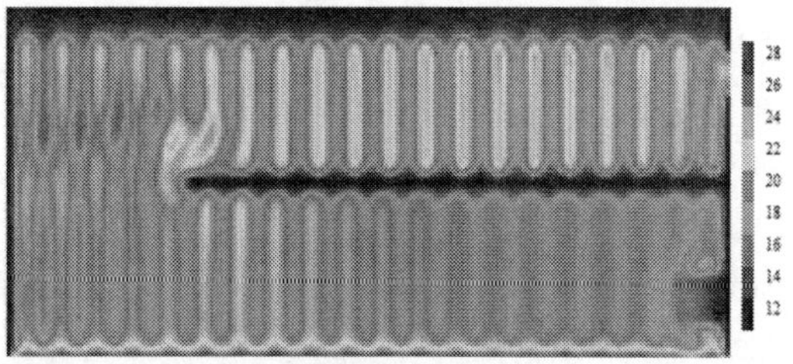

Fig. 9. Distribution of temperature in the central vertical section.

In Figure 10 is shown the intensity of speeds, in horizontal section made a distance of 2 cm from the inferior wall. We can see the main current between the 2 rows of pipes, with maximum speeds in the center of the section, slowing of the flow towards the lateral wall opposed to the input. Speed intensity is lower between two consecutive pipes, but also near the central area of the lateral walls.

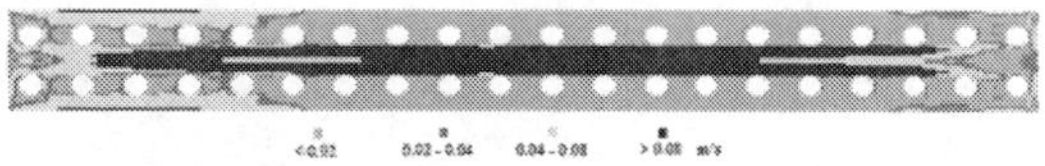

Fig. 10. Intensity of the flow speeds.

Temperature of the preheated water outputting the upper compartment was obtained by a mean of the temperature values in the computational cells, which represent the input sections, namely:

$$T_{output}(t) = \frac{\sum_{i,j,k \in D_{evac}} T[i,j,k]}{N} \tag{26}$$

where: T_{output} is the temperature in the evacuation section; D_{evac} is the domain formed by cells that are part of the evacuation section; t is time that passed since the beginning of operation of the heat exchanger; N is the number of computational cells in the evacuation section.

$T_{ies}(t)$ values were saved every step of time and based on these values were constructed graphs of temperature evolution in the evacuation section. In analogy, was determined the mean pressure in the input sections p_{input} and in the evacuation sections p_{output}, in order to determine the pressure drop between the input orifice and the evacuation orifice:

$$\Delta p = p_{input} - p_{output} \tag{27}$$

In Figure 11 are presented the variation of temperatures of preheated water, at the output orifice from the upper compartment. We can observe that time required to obtain a state of quasi-stationary state depends of the input speed, initial temperature of the cold water and temperature of the environment. In this figure we can observe how the cold water temperature varies depending on time at the output orifice of the upper compartment in the case of the first run.

Due to the heat insulating paint applied on the lateral and upper walls, the heat exchange coefficient between the walls and the environment have a relatively low value of 2 W/m^2 °C. At lower input speeds, the quasi-stationary phase is obtained later. Similarly, at lower environment temperatures, the time required to obtain the quasi-stationary state increases.

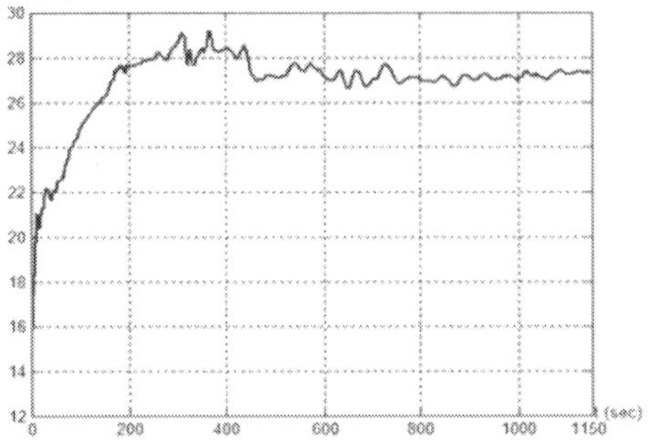

a)

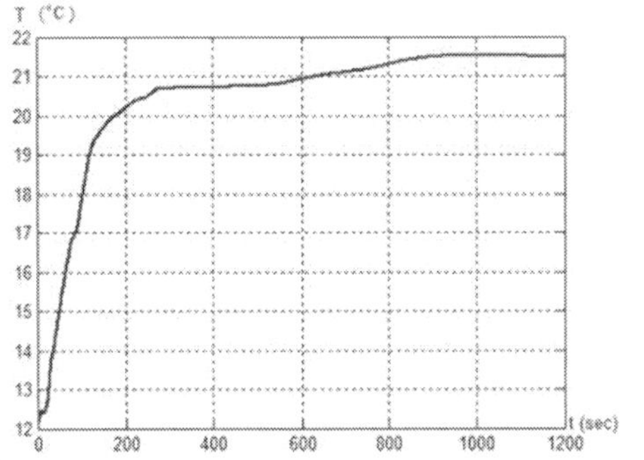

b)

Fig. 11. Variation of temperature of heated water at the output orifice: *a)* first run; *b)* second run.

A centralization of the values obtained for temperature during the two runs, is shown in Table 3. Results obtained for the estimated efficiency are presented in this table based on relation [29].

$$\varepsilon_{estimated} = \frac{t_2'' - t_2'}{t_1' - t_2'} \cdot 100\% \quad (28)$$

where: $\varepsilon_{estimated}$ is the thermal efficiency estimated from the computer simulation; t_2'' is the temperature in the evacuation section; t_2' is the temperature of cold water at the input orifice; t_1' is the temperature of wastewater at the input.

Table 3. Values obtained after running the program.

Run no.	$V_{w.i.}$ [m/s]	t_{air} [°C]	t'$_2$ [°C]	t'$_1$ [°C]	t''$_2$ [°C]	Δp [Pa]	ε_{estim} [%]
1.	0.083	22	12	35	27.6	1931.8	68
2.	0.261	27	22	60	34.11	1962.5	32

We can observe that a maximum efficiency was obtained for run number one, and the lowest efficiency for run number two. The cause of low efficiency for run number two is high input speed, which reduces the contact time of cold water with the heat pipes.

4 Conclusions

The program used in the heat and mass transfer simulation was done in the *Delphi 7* programming language, which is a variant of the Turbo Pascal language. This program contained a series of subprograms for discretization of the domain occupied by the heat exchanger with heat pipes, another subprogram for solving the discretization equations and a third one for post-processing.

Parameters for the flow and heat transfer analysis were chosen to be as close as possible to the real existing conditions, due to the fact that at an experimental level on the laboratory stand we could not use a high enough flow and due to weather conditions (cold water temperature was higher).

The results, obtained by computerized simulation, revealed that the recovered temperature depends greatly on the flow of wastewater and cold water, which validates the experimental results.

Also, the obtained results revealed that the distribution of speed field inside the upper compartment, between the heat pipes, leads to the formation of swirls, where the heat transfer is much lower, due to the way in which the placement of pipes that are in line.

In the context of the above, we can conclude that heat recovery from wastewater in quasi-stationary regime depends on: *the speed of the input cold water, the initial temperature of the cold water and the temperature of the environment*, which was confirmed by the conducted experimental research.

5 Acknowledgements

The results presented in this paper were obtained with the support of the Technical University of Cluj-Napoca. Through the research Contract no. 1994/12.07.2017, CI_2017_IMM_1, Internal Competition CICDI-2017.

References

1. A. Hepbasli, E. Biyik, O. Ekren, H. Gunerhan, M. Araz, J. Energy Conversion and Management **88**, pp. 700–722 (2014)
2. A.R. Mazhar, S. Liu, A. Shukla, J. Energies **11**, 386, pg. 34. (2018)
3. A. Balastov, N. Kachalov, E. Senkiv, O. Kachalova, J. MATEC Web of Conferences **110**, 01008 (2017)
4. M. Omidi, M. Farhadi, M. Jafari, J. Applied Thermal Engineering **110**, pp. 1075–1090 (2017)
5. O. Culha, H. Gunerhan, E. Biyik, O. Ekren A. Hepbasli, J. Energy and Buildings **104**, pp. 215–232 (2015)
6. A. Matveev, D. Zelentsov, A. Louks, J. MATEC Web of Conferences **106**, 06016 (2017)
7. S. Ornelas, S. Sousa, C. Dong, M. Fernelius, M. Hofer, T. Holsclaw, A. Jennison, D. Mai, K. Ninh,

M. Poel, *Mathematical modeling, numerical simulation and statistical optimization of heat pipe design*, (Report of Center for Applied Mathematics, Computation and Statistics Departament of Mathematics, San Jose, CA, pp. 1–37, 2006)
8. M.I. Heredia, J. Siqueiros, J.A. Hernández, D. Juárez-Romero, A. Huicochea, J.G. González-Rodríguez, J. Applied Thermal Engineering **128**, pp. 737–746 (2018)
9. H. Ma, L. Yin, X. Shen, W. Lu, Y. Sun, Y. Zhang, N. Deng, J. Applied Energy **169**, pp. 177–186 (2016)
10. A. Nouri-Borujerdi, M. Layeghi, J. Fluids Engineering, **126**, pp. 442–448 (2004)
11. S. Knudsen, G.L. Morrison, M. Behnia, S. Furbo, J. Solar Energy, **78**, 2, pp. 281–289 (2005)
12. S.F. Tsai, T.W.H. Sheu, J. Computers & Fluids, **27**, 1, pp. 29–46 (1998)
13. D.J. Durrenmatt, O. Wanner, J. Water Research **48**, pp. 548 - 558 (2014)
14. Gupta M, *Numerical Study of Heat Transfer Enhancement in a Plate-Fin Heat Exchanger Using Rectangular Winglet Type Vortex Generator* (Ph.D. Thesis, National Institute of Technology, Kurukshetra, India, 2010)
15. Khan W.A, *Modeling of Fluid Flow and Heat Transfer for Optimization of Pin-Fin Heat Sinks* (Ph.D. Thesis, University of Waterloo, Canada, 2004)
16. Sahiti N, *Thermale and Fluid Dynamic of Pin Fin Heat Transfer Surfaces* (Ph.D. Thesis, Erlangen University, Nürnberg, Germany, 2006)
17. Williamson N.J, *Numerical Modelling of Heat and Mass Transfer and Optimisation of a Natural Draft Wet Cooling Tower* (Ph.D. Thesis, University of Sydney, Australia, 2008)
18. A.A. Ozalp, I.Dincer, International Journal of Thermal Sciences, **49**, pp. 1799–1812 (2010)
19. N.C. Markatos, J. Ironmaking and Steelmaking, **16**, 4, pp. 266–273 (1989)
20. Patankar S.V, *Numerical Heat Transfer and Fluid Flow* (Taylor & Francis, New York, 1980)
21. Olsen N.B, *CFD Algorithms for Hydraulic Engineering* (Norvegian University of Science and Technology, Trondheim, Norway, 2000)
22. C.W. Hirt, B.D, Nichols, J. Computational Physics, **39**, pp. 201–225 (1981)
23. G.Y. Soh, G.H. Yeoh, V. Timchenko, International Journal of Heat and Mass Transfer, **100**, pp. 573–581 (2016)
24. H-J. Laue, H. Kruse, J. Wärmepumpe Aktuell – Informationszentrum Wärmepumpen + Kältetechnik, **8**, 2, pg. 2 (2009)
25. Faghri A, *Heat Pipe Science and Technology* (Taylor & Francis, New York, 1995)
26. X.-S. Yang, M. Karamanoglu, T. Luan, S. Koziel, J. Computational Science **5**, pp. 119–125 (2014)
27. H. Jouhara, S. Almahmoud, A. Chauhan, B. Delpech, G. Bianchi, A. Tassou R. Llera, F. Lago, J.J. Arribas, J. Energy **141**, pp. 1928–1939 (2017)
28. R.T. Dobson, T.M. Harms, *Lumped parameter analysis of closed and open oscillatory heat pipes*, (in: *11th Int. Heat Pipe Conf.* Tokyo, 12–16 September 1999)
29. Proskiw G. 2003, *Design and Analysis of a Residential Greywater Heat Recovery System* (Report Prepared for CANMET Energy Technology Centre, Canada, pg. 4, 5, 52, 2003)

Experimental studies and numerical modelling of heat and mass transfer process in shell-and-tube heat exchangers with compact arrangements of tube bundles

Valery Gorobets[1], *Yurii* Bohdan[2,*], *Viktor* Trokhaniak[1], and *Ievgen* Antypov[1]

[1]Heat and Power Engineering Department, Education and Research Institute of Energetics, Automation and Energy Efficiency, National University of Life and Environmental Sciences of Ukraine, Heroyiv Oborony st., 15, Kyiv, 03041, Ukraine
[2]Ships Power Plant Operation Department, Marine Engineering Faculty, Kherson State Maritime Academy, Ushakova avenue, 20, Kherson, 73000, Ukraine

Abstract. Shall-and-tube heat exchangers based on the bundles with in-line or staggered arrangements have been widely used in industry and power engineering. A large number of theoretical and experimental works are devoted to study of hydrodynamic and heat transfer processes in such bundles. In that, works the basic studies of heat and mass transfer for these bundles are found. However, heat exchangers of this type can have big dimensions and mass. One of the ways to improve the weight and dimensions of the shell-and-tube heat exchangers is to use compact arrangement of tube bundles. A new design of heat exchanger is proposed, in which there are no gaps between adjacent tubes that touch each other. Different geometry of these tube bundles with displacement of adjacent tubes in the direction of transverse to the flow is considered. Numerical modelling and experimental investigations of hydrodynamic, heat and mass transfer processes in such tube bundles has been carried out. The distribution of velocities, temperatures, and pressure in inter-tube channels have been obtained.

1 Introduction

One of the most common equipments that are used in all types of power plants and engines are heat exchangers. Today well known many types of heat exchangers and methods for improving of heat transfer, including those used for engines [1, 2].

It is known that for shell-and-tube heat-exchangers maximal heat transfer in tube bundles is observed in the case of a cross current of heat carriers, i.e. in the case of the transverse flow around such bundles. Results of researches of hydrodynamics and heat exchange of tube bundles (with the tubes of traditional circular cross-section), are described widely enough in the monograph [3]. A significant number of works devoted to the investigation of smooth-tube bundles with various geometry and arrangement in bundles [4, 5]. At the same time, the number of papers devoted to investigations of the processes of hydrodynamics and heat, mass transfer in compact smooth-tube bundles, with minimum intertube intervals, is very insignificantly.

Extensive studies of convective heat transfer of various transversely streamlined bundles of smooth cylindrical tubes with their inline arrangement were carried out by Zhukauskas et al. In that studies, the main laws of the change in heat transfer and hydrodynamics are studied.

Also researches of compact, transversely streamlined bundles of smooth tubes with the inline arrangement were carried out by Aiba [6, 7].

The conducted studies showed that the inline tube bundles with a transverse flow around it have a lower hydraulic resistance in comparison with it staggered arrangement, but they have more low coefficients of heat transfer on the surface. One way to increase the heat transfer coefficient is to use small diameter tubes.

In this paper, the investigation of the processes occurring in heat exchangers is aimed at increasing the efficiency of the compressed smooth-tube bundles with inline configuration and inline shifted configuration with a minimum longitudinal relative step equal to unity. Experimental research, numerical computer modelling and thermohydraulic calculations of processes of heat transfer and hydrodynamics, which occur in channels of compact bundles, are carried out.

2 Research description

2.1 Experimental plant

For realization of investigations the experimental co-generation plant is created with utilizer of waste heat of exhaust gases of new design [8], which installed on the

* Corresponding author: bohdanyurii09@gmail.com

line of exhaust gas channel of high speed reciprocating internal combustion engine (Figure 1).

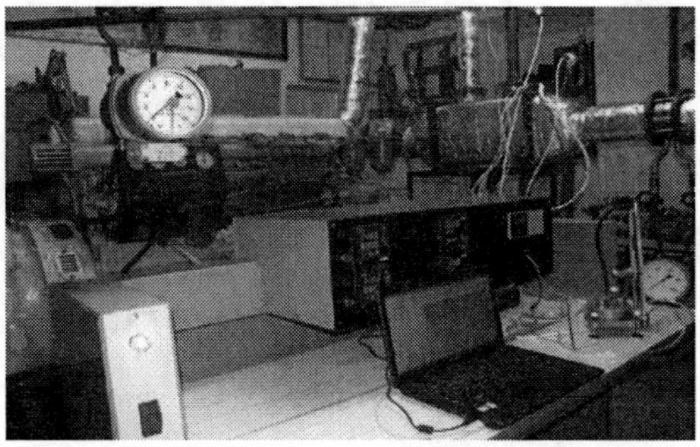

Fig. 1. A general views of the experimental plant.

During realization of experiments collection and treatment of these measurable parameters come true by means of the automated computer complex.

Surface of heating of proposed heat exchanger has next descriptions: diameter and thickness of smooth tubes - $d \times \delta$ = 0.010 × 0.001 m; equivalent diameter of channel between rows of tubes - $D_{chan.eq}$ = 0.00967 m; length of tubes - l = 0.150 m; equivalent diameter of casing - $D_{cas.eq}$ = 0.150 m; total for two heating areas, heat-receiving surface - H = 1.602 m^2.

2.2 Numerical computer modelling

For the co-generation plant on the base of combustion engine with new design waste heat utilizer of exhaust gases the computer modelling of hydrodynamics and heat transfer processes, which occur in channels of investigated tube bundles is conducted. The goal of these calculation is to obtain the local distributions of the field of velocities, temperatures and pressures, and as well as obtaining of dependences of heat transfer characteristics from the dynamic and thermophysical parameters of heat-carriers, including local distributions of coefficients of heat transfer on the circumference of tubes 1-4th rows for the first section of bundle. Geometry of the prospected channel with the compact location of bundles with small diameter tubes is brought in Figure 2, that has next geometrical characteristics: values of transversal and longitudinal step of tubes $S_a \times S_b$ = 0.015×0.010 m; diameter of tubes d = 0.010 m; number of tubes in raw i = 42; the tube bundles consist of j = 9 rows, and the common number of tubes in bundles makes z = 378, thus bundles divide between itself a technological gap, as shown in Figure 2. In Table 1 and Table 2 the geometry parameters of arrangement of tube bundles with inline and shifted compact location of tubes are shown.

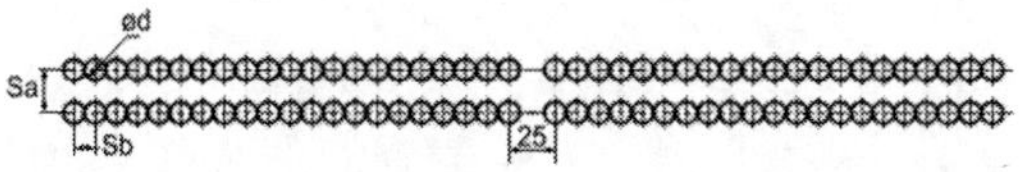

Fig. 2. Geometry of channel with the compact location of tube bundles.

The numerical computer modelling of processes of heat transfer and hydrodynamics in the investigated tube bundles was conducted on the basis of method of eventual elements by means of ANSYS Fluent software. Mathematical model of heat- and mass transfer processes, which take place in the investigated heat exchanger in two-dimensional formulation includes a system of Navier-Stokes equations, equation of energy transfer for convective flows [9] and standard k-ε model of turbulence [10].

Table 1. Geometry parameters of arrangement of tube bundles with the compact location of tubes.

No	Transversal step S_a, m	Longitudinal step S_b, m	Diameter of tube d, m	Number of tubes in raw, i
1	0.013	0.008	0.008	42
2	0.015	0.008	0.008	42
3	0.017	0.008	0.008	42
4	0.015	0.010	0.010	42
5	0.017	0.010	0.010	42
6	0.019	0.010	0.010	42
7	0.017	0.012	0.012	42
8	0.019	0.012	0.012	42
9	0.021	0.012	0.012	42

Table 2. Geometry parameters of arrangement of tube bundles with the shifted compact location of tubes.

No	Transversal step S_a, m	Shifting of tubes K, m	Diameter of tube d, m	Number of tubes in raw, i
1	0.15	0.001	0.010	40
2	0.15	0.002	0.010	40
3	0.15	0.003	0.010	40
4	0.15	0.004	0.010	40
5	0.15	0.005	0.010	40

Numerical calculations in the investigated channels of tube bundles were carried out for 9 constructions (Table 1) with the Reynolds number Re = 7085. As heat carriers are used exhaust gases (hot heat carrier) with a temperature at the inlet of 470 °C, which flow in the channels of the tube bundles. The temperature of the walls of the pipes was assumed constant, but changed for each section as the coolant moved (first section 75 °C, second 34 °C). Similar conditions take place, for example, in multi-pass heat exchangers.

In addition to the inline compact arrangements of tube bundles, numerical simulations were carried out for the more compact arrangements of tube bundle. The geometry parameters of tube bundles with a tube diameter d = 0.010 m is carried out in Table 2, which differ from those considered above by their more compact arrangement with shifting of adjacent tubes by some distance, while five variants of the bundle design are considered, with a shifting of tubes in the transverse direction K from 0.001 to 0.005 m. The calculations were carried out at a Reynolds number of 18,6·10^3. As the heat carriers, the exhaust gases (hot heat carrier) of the internal combustion engine with a temperature of 470 °C at the inlet to the heat exchanger, which flow in the intertubular channels, and the fresh water (cold heat carrier) moving inside pipes with inlet temperature in the heat exchanger equal to 20 °C.

Numerical modelling of heat transfer and hydrodynamics in channels with compact arrangement of tube bundles was performed for bundles that contain 40 tubes in a single row with a diameter of 0.010 m.

2.2 Analysis concept

To evaluate the heat exchange surface, from the energy point of view, the coefficient of thermohydraulic efficiency (M.V. Kirpichev's criterion) is used, which is defined as the ratio of the transferred amount of heat Q through the heat exchange surface to the total power N required for pumping heat carriers through the heat transfer surface on both sides (without taking into account the efficiency of the superchargers and drives) [11].

$$E = \frac{Q}{N} \quad (1)$$

The amount of heat that is taken from the hot or transferred to the cold heat carrier is determined by the formula

$$Q = c_p G \Delta T \quad (2)$$

Power required to pump the heat carrier:

$$N = \frac{\Delta p G}{\rho} \quad (3)$$

3 Results and discussions

The results of the numerical calculations are given for one of the geometries under consideration (d = 0.008 m, Sa = 0.013 m) in Figure 2. The distribution of the velocity field in the channels of the tube bundle is shown in Figure 3, and in Figure 6 shows the distribution of the velocity vectors on a small section of the channel. The value of the average speed of exhaust gas in the narrowest cross section of the channel is 86±2 m/s (Figure 3).

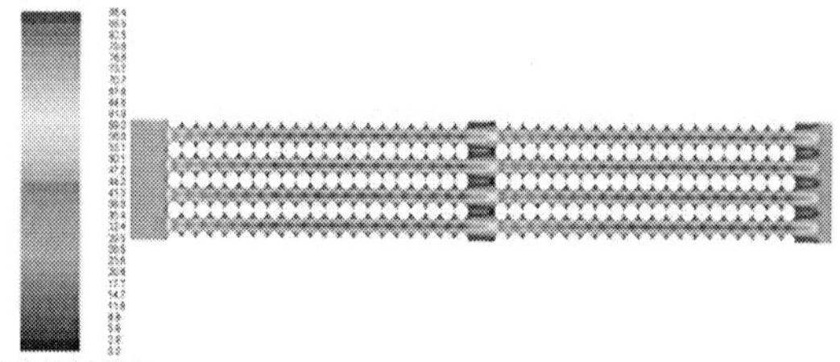

Fig. 3. Velocity of exhaust gases in the channels of tube bundles, m/s (Re = 7085).

Figure 4 shows the temperature distribution in the channels where the exhaust gas temperature at the outlet is 82±3 °C. Figure 5 shows the change in pressure in the flow of exhaust gases flowing along the channels of the tube bundles.

The results of numerical calculations for the channel with a shifting of adjacent tubes of 0.005 m are given in Figures 6-8. The distribution of the velocity of the heat carrier flow in the channel is shown in Figure 6.

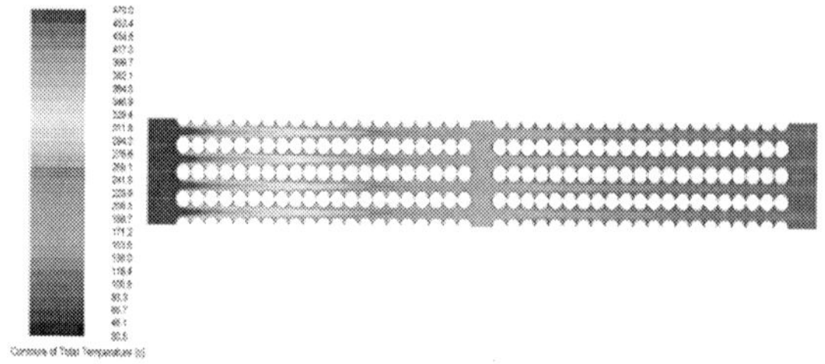

Fig. 4. Change of temperature in the channels of tube bundles, °C (Re = 7085).

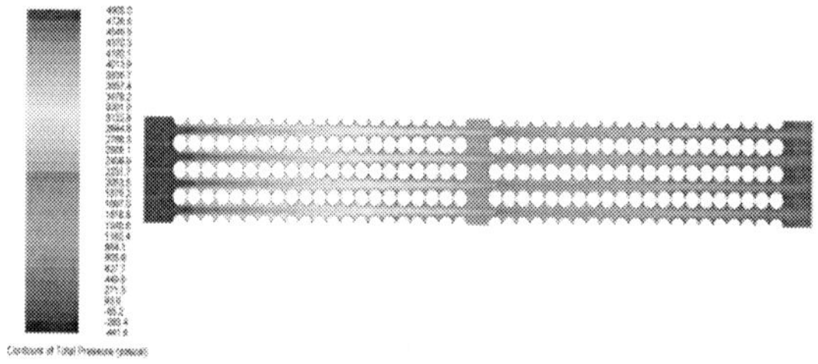

Fig. 5. Change of pressure in the channels of tube bundles, Pa (Re = 7085).

Fig. 6. Velocity of exhaust gases in the channels of tube bundles with shifting, m/s (Re = $18{,}6\cdot10^3$).

Fig. 7. Change of temperature in the channels of tube bundles with shifting, °C (Re = $18{,}6\cdot10^3$).

As we can see in Figure 6, at the upper point of the tube, the boundary layer separates, and at the junction of adjacent tubes, stagnant zones are observed. At separate points of the channel, the velocity of the exhaust gases reaches 55 m/s, and their average velocity in the narrow cross-section of the channel is about 45 m/s, which is greater than in the channels without shifting. Figure 7 shows the temperature distributions in the intertubular channel. The outlet temperature of the cooled gas is 78 °C, which is the lowest temperature among the output temperatures that are achieved in the channels of other structures.

Figure 8 shows the change in the pressure fields in the channel of the investigated structure. From the obtained pressure distributions it follows that the total pressure drop is about 15700 Pa and is unacceptable for heat exchanger intended for the co-generation plant, it is necessary to apply a small shifting of the tubes 0,001÷0,002 m not exceeding the boundary-permissible back pressure of the engine.

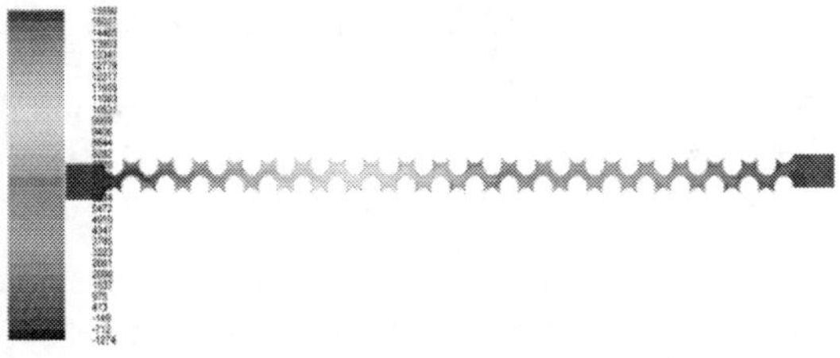

Fig. 8. Change of pressure in the channels of tube bundles with shifting, Pa (Re = $18{,}6 \cdot 10^3$).

In the Figure 9 the dependences of the Nusselt number on the bundles geometry is showed.

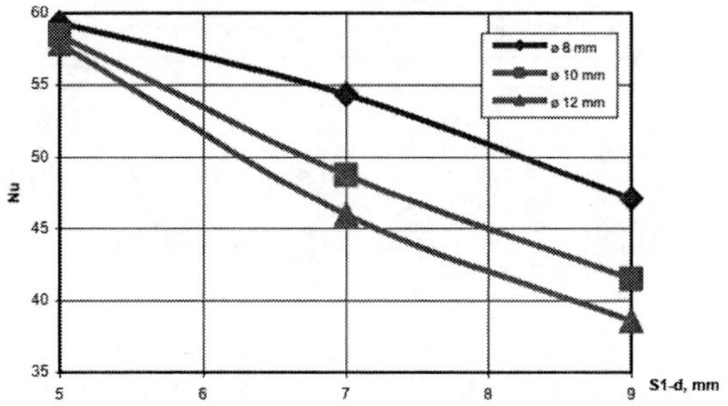

Fig. 9. Dependence of the Nusselt number Nu on the geometry of a tube bundles.

As a result of the analysis of the obtained dependences, it can be concluded that the most effective design for the intensity of heat transfer is a tube bundles with a transverse step of 0.005 m and a diameter of 0.008 m.

Figure 10 shows the dependence that characterizes the pressure loss of the heat exchange surface on the tube bundles of various geometries.

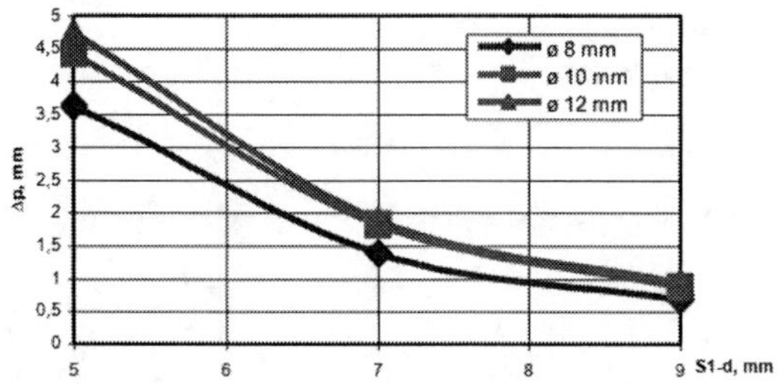

Fig. 10. Dependence of the pressure drop Δp on the geometry of a tube bundles.

As we can see in Figure 11 the thermohydraulic efficiency decreases with decreasing step and diameter of the tubes due to the growth of hydraulic resistances.

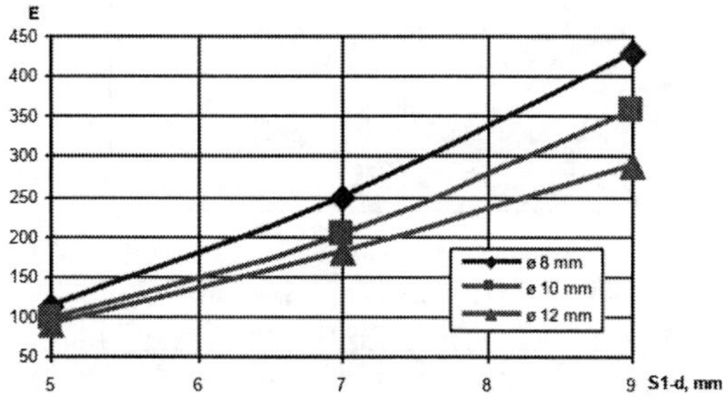

Fig. 11. Dependence of the thermohydraulic efficiency E on the geometry of a tube bundles.

Analyzing the obtained dependences from the point of view of the effect of geometry on the heat and hydrodynamic characteristics of a compact transversely streamlined tube bundle, it can be noted that the best characteristics are tube bundles with a transverse step of 0.013 m and a tube diameter of 0.008 m. In this case, a significant reduction in the temperature of the heat carrier at the outlet of the channel is achieved.

Figure 12 shows the change the Nusselt number, depending on the geometry of the tubes arrangement in the bundles with shifting. The abscissa indicates the displacement values of adjacent tubes equal to 1, 2, 3, 4 and 5 mm, respectively, and the ordinate represents the average values of the Nusselt number over the surface of the tube bundles. As a result of the analysis of the obtained dependence, it can be concluded that the most effective in terms of the intensity of heat exchange is the geometry of the bundle with shifting of the tubes by 5 mm. It should also be noted that when the tubes are shifted by 5 mm, the pressure difference in the channel will be maximum (Figure 13), because of this, when using such a geometry in the heat exchangers, the back pressure on the diesel exhaust exceeds the limit value of 5 kPa and will be to have a significant impact on the engine efficiency. Therefore, in order to intensify the heat transfer, the most suitable small shifting of 1 to 3 mm, which do not cause a significant increase in aerodynamic resistance of the bundle.

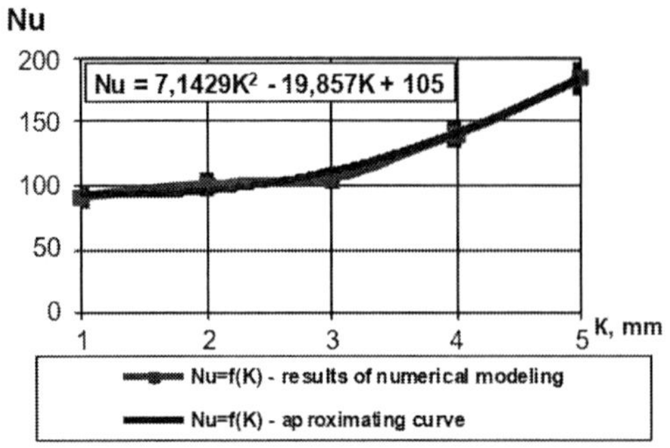

Fig. 12. Dependence of the Nusselt number on the shifting of a tubes in bundles.

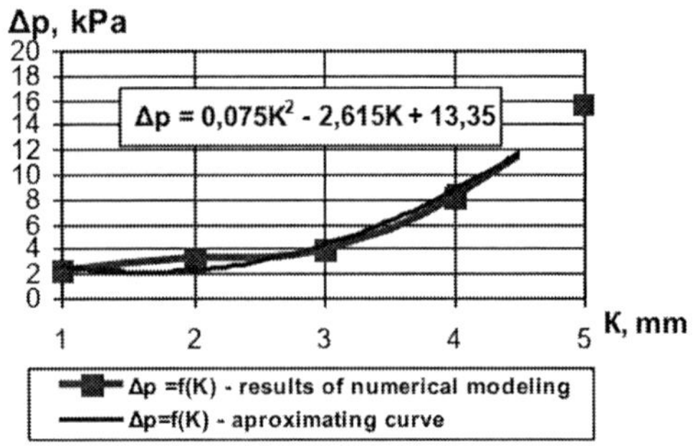

Fig. 13. Dependence of the pressure drop on the on the shifting of a tubes in bundles.

Figure 14 shows the dependence of the thermohydraulic efficiency E (formula 1) on the geometry of tube arrangement in the bundle with shifted tubes. As can we see in Figure 14, the thermohydraulic efficiency in such channels due to the growth of hydraulic resistance decreases with increasing shifting of adjacent tubes.

Analysis of the obtained dependences shows that from the point of view of the intensification of heat transfer processes on the surface of the tube bundles with an insignificant increase in the aerodynamic resistance, the design has better characteristics with a small shifting of 1 to 3 mm (Figures 12-14).

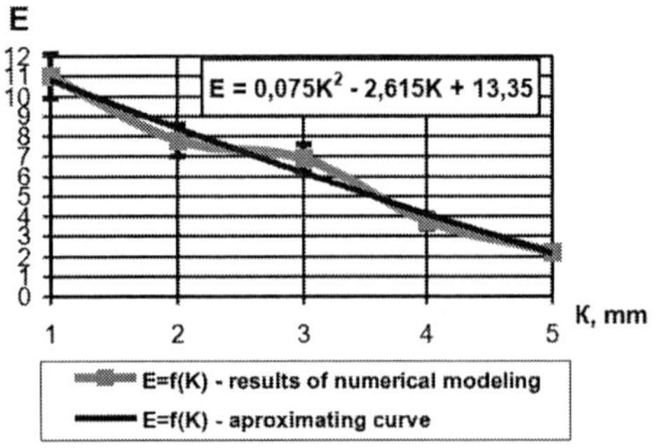

Fig. 14. Dependence of the thermohydraulic efficiency on the shifting of a tubes in bundles.

Intensification of heat transfer in bundle with the compact arrangement of tubes in comparison with an inline bundle is achieved by influencing the heat transfer conditions of next factors: 1) reduction of transverse distance (step) between the nearby rows of tubes, that results in the increase of velocity of external flow at the identical expense of heat carrier in the heat exchangers of the known constructions; 2) creation of longitudinal alternating pressure gradients, which also leads to intensification of heat transfer processes (analog of channels with variable cross-section or "corrugated" channels of type diffuser-confuser).

Results of investigations of the flow and heat transfer in the bundles of new design heat exchanger were estimated by comparing the obtained experimental data with the results of numerical simulation and known experimental data obtained by other researchers. The results of the researches are presented in the form of graphical dependence in Figure 15 for Reynolds numbers in the range Re = 3000 ÷ 6000 and Prandtl number Pr = 0.73.

Figure 15 shows the experimentally determined dependence of Nusselt number averaged over the surface $\overline{Nu}$ on Reynolds number Re.

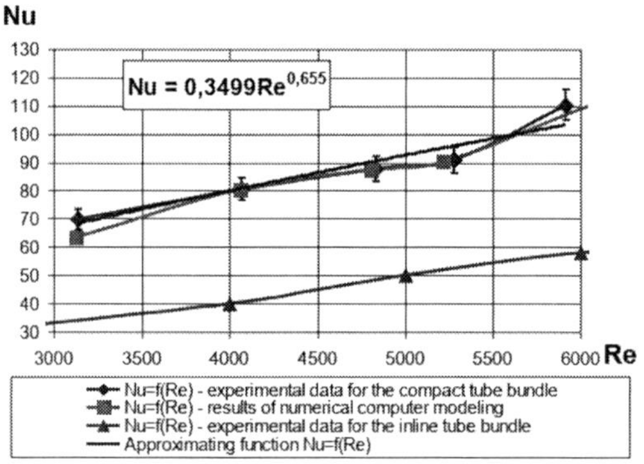

Fig. 15. Dependence of average Nusselt number $\overline{Nu}$ on Reynolds number Re: ♦ – experimental data for the compact tube bundle; ▲ – experimental data for the inline tube bundle; ■ – results of numerical computer modelling.

Figure 15 also shows the experimental dependence $\overline{Nu}$ = f(Re), obtained for the inline tube bundle in the work of Belov and Kudryavtsev [12]. The comparison shows that for the same values of Reynolds number, the intensity of heat transfer for a compact tube bundle is almost in 2 times greater than intensity of heat transfer on the surface of the inline tube bundle.

4 Conclusions

1. Experimental data have been obtained for heat and mass transfer processes in channels of compact tube bundles, which are represented in the form of the dependences of the heat transfer characteristics on the dynamic and thermophysical parameters of heat carriers.
2. A comparative analysis of the thermohydraulic efficiency for channels with different transverse steps of the tube in bundles has been carried out and it is shown that the structures developed are sufficiently effective at

a significant reduction in the mass-dimensions of the heat-exchange surface.

3. Numerical computer modelling of heat and mass transfer processes in channels with transverse flow past extremely compressed tube bundles with inline arrangement and absence of a gap between adjacent tubes in the direction of the heat carrier motion is carried out.

4. The new design of heat exchangers with the compact tube bundles is proposed, which has high efficiency, low aerodynamic and hydraulic resistance, and Nusselt number on the tube surface is approximately in 2 times higher than their values for a bundle with a traditional inline arrangement of tubes.

References

1. G.F. Hewitt, *Heat Exchanger Design Handbook* (Begell House, New York, 2008)
2. J.E. Hesselgreaves, R. Law, D. Reay, *Compact Heat Exchangers: Selection, Design and Operation Hesselgreaves* (Butterworth-Heinemann, United Kingdom, 2016)
3. A. Zhukauskas, R. Ulinskas, *Heat transfer of cross flow tube bundles* (Mokslas, Vilnius, 1986)
4. V.A. Pronin et al. Trans. of the 4rth Russian National Heat Exchange Conf., Moscow, **6**, pp. 285-288, (2006)
5. M.K. Goel, S.N. Gupta, International Journal of Science, Engineering and Technology Research, **3**, no. 4, pp.1166-1172, (2014).
6. S. Aiba. et al., Int. J. Heat Mass Tran., **21**, pp. 311-319, (1980).
7. S. Aiba, J. Heat Transf., **112**, pp. 933-938, (1990).
8. V.G. Gorobets, Yu.O. Bohdan, Heat exchanger. UA Patent 104559, F28D 7/00 filed March 27, 2013, and issued February 10, 2014.
9. H. Schlichting, *Boundary-Layer Theory*, 7th edition, (McGraw-Hill, New York, 1979).
10. ANSYS FLUENT Theory Guide. Release 14. ANSYS, Inc. Southpointe 275 Technology Drive Canonsburg, PA 15317, 2011.
11. M.V. Kirpichev, Izv. ENIN after G.M. Krzhizhanovskogo, **12**, pp.5-9, (1944).
12. I.A. Belov, N.A. Kudryavtsev, *Heat transfer and resistance of tube packs*, (Energoatomizdat, Leningrad , 1987).

Numerical analysis of flow in a modular heat exchanger used in food industrial plants

Wojciech Judt[1,*], and *Jarosław* Bartoszewicz[1]

1 Poznan University of Technology, Chair of Thermal Engineering, Piotrowo 3 St., 60-965 Poznań, Poland

Abstract. In the article is proposed a construction of modular cross-flow heat exchanger. This construction can be used in food industrial plants for example to heating a palm oil for a chips production. The second possibility to make use of that heat exchanger in a food industrial plants will be a heating of a worn out rape oil after a technological process for a improvement of effective utilization in oil boiler. The main advantage of modular heat exchangers is the possibility of simple adjustment of heating power dependent on demand. It is possible by valve system, which can cut off subsequent sections of the heat exchanger from the plant when heat requirement is decreasing. During a numerical simulation was demonstrated, that an exhaust gases flow in a heat exchanger is not homogenous. In numerical analysis was obtained a real flow velocity distribution and temperature distribution for a full geometry of the designed object and for two subsequent section of the heat exchanger.

1 Introduction

The food industry is one of a big beneficiary of heat usage for a production process. The main problem of a production process on food plants is a controlling of a power in installed heat exchangers. During a production process, necessarily is a dosage of an amount of heat delivered. This phenomenon depends on different heat demand for other parts of a fabrication, during a production process.

Authors of a paper analyzed a heat transfer process for a palm oil heating in a food industrial plant for a chips production. For a heating medium in a designed heat exchanger was a waste heat used from a heating boiler, which is used for a steam generation in another process, which is realized in the industrial plant. This heating boiler reaches a nominal power equal to 4 MW. Estimated heating power, which can be recovered from exhaust gases, generated by heating boiler with that power is about a 150 kW. The second possibility of heat exchanger application is heating an used palm oil after a production for a reprocessing process. Heating of used palm oil before a combustion will increase a efficiency of reprocessing. A scheme of a heat recovery plant is shown in figure 1.

Popular construction of heat exchanger for food industrial plants is a plate heat exchangers, but this type of heat exchangers is not popular for heat transfer between gaseous and liquid mediums [1]. The main advantage of a proposed solution of a heat exchanger structure is a modular construction. The second advantage of planned construction is a simplicity of a heat exchanger production. The assumption, that a heat exchanger will be not exposed to a corrosion environment allows making a construction from a carbon steel. It causes, that a production of the heat exchanger will be cheap. The modular structure allows for a heat transfer controlling, between heated and cooled medium.

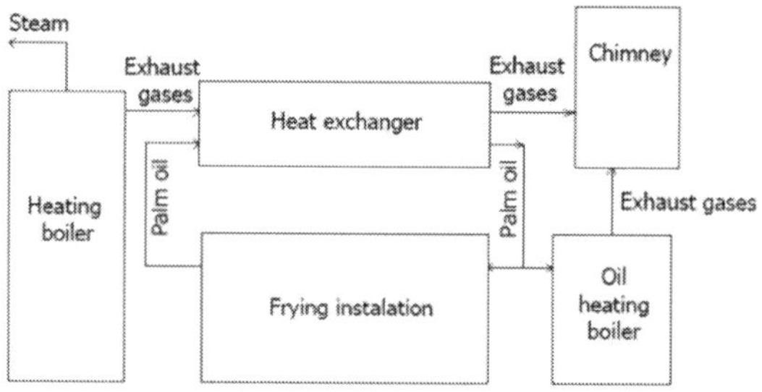

Fig. 1. Scheme of a heat recovery plant.

When a heat demand is lower, then possible is to decreasing an area of heat transfer. Heat power controlling can be controlled by valve set, which can be assembled after every module of the heat exchanger. The flow of a heated medium, which are flowing through pipes can be led out from the heat exchanger and a temperature of exhaust gases in the outlet of the construction can be controlled. This denouement causes that a risk of a heat exchanger damage by corrosion during a liquefaction of acids and water vapour is getting smaller. The phenomenon of a heat exchangers failure during a corrosion environment exposition is commonly present in improperly projected heat exchangers or heating boilers, wherein the coolest parts of heat exchangers is precipitated a liquid water or acids.

The main purpose of the paper is an analysis of exhaust gas real velocity and temperature distribution in

* Corresponding author: wojciech.judt@put.poznan.pl

a designed geometry of the heat exchanger. During a research was prepared a numerical analysis of a distribution of these parameters in the designed heat exchanger. Numerical analysis in a heat exchanger design process is a simple way for an estimation of basic parameters of a fluid flow, without a necessity of a prototype building or realization of experimental research. Numerical analysis is widely present in a research concerning with heat exchangers [2].

2 Analysis and modeling

Authors of article assumed, that a temperature of exhaust gases, which are leaving a heating boiler is around a 260 °C. The minimal temperature of exhaust gases in the heat exchanger is conditioned by an avoiding of corrosion conditions when exhaust gases reach a temperature of water or acids condensation. The minimal temperature of exhaust gases in the outlet of the heat exchanger was assumed to equal a 180 °C. The temperature of palm oil during a production process of a chips frying is equal to 180 °C. For a palm oil returning from a production, line authors assumed, that a temperature at the inlet to the heat exchanger is equal to 130 °C. A counter-current flow of the mediums will be realized through a heat exchanger. The mass flow of exhaust gases results from the amount of exhaust gases, which are generated from a combustion process in a heating boiler. The mass flow of combusted coal was computed from equation 1.

$$\dot{M}_c = \frac{\dot{Q}_b}{\eta_b \cdot W_c} \tag{1}$$

It was assumed that efficiency of combustion process equal 0,8. The low calorific value of combusted coal is equal 23 MJ/kg. The obtained mass flow of combusted coal is equal to 0,2 kg/s. The mass flow of fumes, which was generated from the combustion process was computed from few simple equations [3] describing a quantity of generated mass flow of fumes from solid fuel combustion with combustion air factor equal to 1.8, and known ultimate analysis of combusted coal. The calculated mass flow of exhaust gases is equal to 2.15 kg/s. The mass flow of palm oil is a result of a requisition of oil amount for a production process and can be variable. For a heat exchanger calculation assumed an average oil demand for a nominal production process and it equals to 1.3 kg/s.

The heat exchanger was designed in analytical calculations. After that was prepared a numerical simulation of heat exchanger working in the ANSYS Fluent environment. Numerical analysis was realized with the same thermodynamic parameters of fluids, as in analytical calculations. Authors of paper developed three numerical models. It was a numerical model of heat transfer for a first and second module of designed installation and analysis of two neighboring sections of one designed module of the heat exchanger.

2.1 Analytical calculations

Analytical calculations of the designed heat exchanger are based on the equation 2 which is describing a heat transfer by heat transmission

$$\dot{Q}_e = h \cdot A \cdot \Delta t_{log}. \tag{2}$$

Overall heat transfer coefficient used for calculations is described by equation 3 [9]

$$h = \frac{1}{\frac{d_{ave}}{h_{c\,fumes} \cdot d_{in}} + \frac{g_{pipe}}{k_{pipe}} + \frac{d_{ave}}{h_{c\,oil} \cdot d_{ex}}}. \tag{3}$$

The log mean temperature difference is described by equation 4 [9]

$$\Delta t_{log} = \frac{\frac{t_{fumes,in} - t_{oil,out}}{t_{fumes,out} - t_{oil,in}}}{ln\frac{t_{fumes,in} - t_{oil,out}}{t_{fumes,out} - t_{oil,in}}}. \tag{4}$$

Log mean temperature difference is equal to 59.4 °C. Thermodynamic parameters of fluids, which are exchanging heat in a heat exchanger were implemented for an average temperature of each medium. Parameters, which are needed for a realized calculation are presented in table 1 [4, 5].

Table 1. Thermodynamic parameters of a fluid, which are taking part in a heat transfer process.

Parameter	Fumes [4]	Palm oil [5]
T_{ave} [°C]	215	155
Cp [J/(kg·K)]	1097	2249
ρ [kg/m³]	0,636	825,4
k [W/(m·K)]	0,04	0,16
Pr [-]	0,65	47,7
ν [m²/s]	$3{,}28 \cdot 10^{-5}$	$4{,}18 \cdot 10^{-6}$

Heat transfer coefficient for exhaust gases was calculated by using a Nusselt number defined by Isaczenko in equation 5 [6]

$$Nu = 0.26Re^{0.65} \cdot Pr_f^{0.33} \cdot \left(\frac{Pr_f}{Pr_w}\right)^{0.25}. \tag{5}$$

Heat transfer coefficient for palm oil was calculated by using a Nusselt number defined by Sieder and Tate in equation 6 [6]

$$Nu = 1.86\left(RePr\frac{D}{L}\right)^{0.33} \cdot \left(\frac{\mu_f}{\mu_w}\right)^{0.14}. \tag{6}$$

Results of analytical calculations for both mediums, which are transferring a heat are presented in table 2.

Table 2. Results of analytical calculations of a heat transfer process in a designed heat exchanger.

Parameter	Fumes	Palm oil
Re [-]	13400	400
Nu [-]	86,2	81,4
h_c [W/(m^2·K)]	140	700
h [W/(m^2·K)]	96	
$\dot{Q}$ [kW]	140	

Obtained power of designed heat exchanger during analytical calculations is close to assumed heat waste from exhaust gases, which can be recovered. Area of heat transfer obtained for designed heat exchanger during analytical calculations is equal to 24 m^2.

Installation of the heat exchanger is built from two modules. Palm oil is located inside the pipes of the heat exchanger. Exhaust gases are flowing around the bank of tubes at the right angle to them. Each module has 2,5 meters length and is composed of five sections. After each section of the heat exchanger, flow of oil through pipes can be stopped and can be led out from installation. This solution allows for maneuvering a heat exchanger power. Each part of the heat exchanger is connected serially to next section.

2.2 Numerical analysis of the full geometry of heat exchanger module

A numerical model was implemented as a steady state. It is caused, because the main target of a paper is showing a character of a fluid flow in the designed heat exchanger, in steady state conditions. Numerical analysis was divided into two different models, each for the different module of the designed heat exchanger. Initially, calculations were realized for the first module of heat exchanger construction. After that, results obtained for a first module were implemented to a second model, as a boundary condition. The composition of results from two numerical models shows a whole distribution of flow and heat transfer parameters. Because of a model is a three-dimensional, analysis of a fluid flow requires solving a three equation of a flow velocity and a continuity equation. The turbulence of a flow in a heat exchanger was solved by a k-Epsilon turbulence model implementation.

For a heat transfer modeling, energy equation is needed to be solved. Heat transfer process in a numerical model is realized as an interface between two domains, representing exhaust gas and palm oil location in a heat exchanger volume. Thermal conditions of realized heat transfer are calculated via system coupling. Diaphragm influence for a heat transfer process is implemented to a model in the interface as a steel wall with 2,6 mm thickness. Exterior walls of the heat exchanger are implemented to a model as adiabatic.

2.3 Numerical analysis of two sections of heat exchanger module

Numerical analysis of two neighboring sections was realized for a third and fourth section, counting from a heat exchanger inlet. Two sections of the heat exchanger were analyzed in the detached model because this part is a periodical element of the designed heat exchanger. This analysis can better imitate a character of a fluid flow and heat transfer process, because a grid implemented for this model can be more detailed, as in a simulation for a whole module. Boundary conditions for this model originate from a numerical analysis for the first module of the designed heat exchanger. The boundary condition for the inlet of that module was implemented by a temperature and velocity distribution as a profile function in inlet plane to the modeling section.

3. Results and discussion

Figure 2 shows a temperature distribution for exhaust gases of the first module of designed installation, in a plane located between middle rows of pipes.

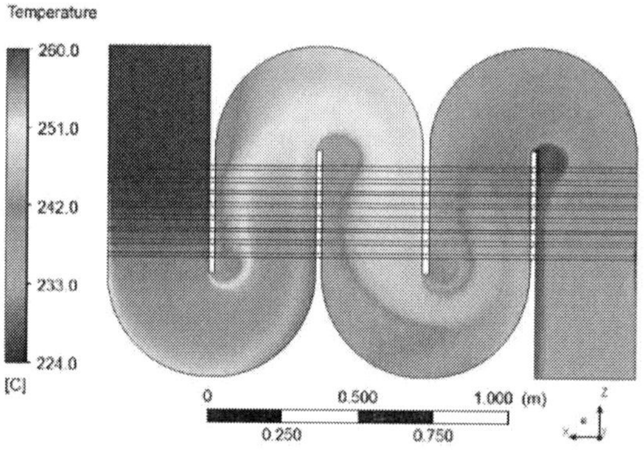

Fig. 2. The temperature distribution of exhaust gases for a first module of the heat exchanger.

Figure 3 shows a temperature distribution for exhaust gas of the second module in a plane located according to the middle row of pipes in the heat exchanger.

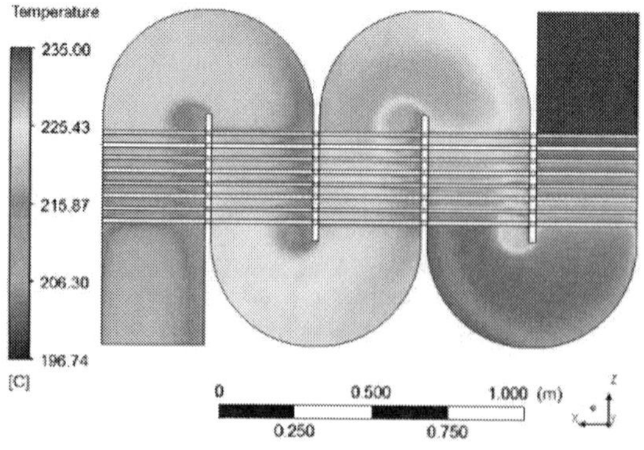

Fig. 3. The temperature distribution of exhaust gases in a second module of the heat exchanger.

Numerical analysis of a temperature distribution for exhaust gases in whole heat exchanger geometry shows, that an average temperature in the outlet of the heat exchanger is equal to 215 °C. It means that a temperature difference between inlet and outlet of the heat exchanger is less than the assumption in analytical calculations. This phenomenon has an influence on a heat exchanger power. Obtained during numerical analysis power of heat exchanger is equal to 106 kW and is to 25% less than a power obtained from analytical calculations. This difference is a result of the accuracy of used equations for a Nusselt number. Increasing a power of heat exchanger requires the extension of a heat transfer area. This operation will increase costs of heat exchanger construction. Also, an extension of heat transfer area can expose a larger area of heat exchanger walls for a cooling under corrosion conditions. The minimal temperature of exhaust gases for designed heat exchanger construction is equal to 196,7 °C. This temperature is less than an average temperature in the outlet of the heat exchanger but is not achieve a temperature of water or acids liquefaction.

Numerical analysis of an exhaust gases flows through a heat exchanger shows, that a flow dissolve is not homogenous. Flow speed is higher for each section of the heat exchanger on a wall located towards an outlet of the heat exchanger. The opposed wall is compassed by fluid with less velocity. This incident causes, that a wall of heat exchanger and area of pipes assembled to these walls are more exposed to an ash deposition. The same aspect occurs for exterior walls of the heat exchanger when a fluid flow direction is changing to the opposite.

On figure 4 is presented a temperature field of exhaust gases for two neighbouring sections of the designed heat exchanger.

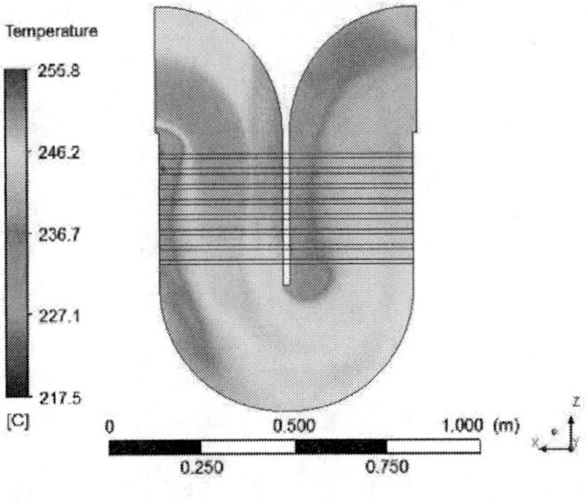

Fig. 4. Temperature distribution in two sections of the first module of the designed heat exchanger.

Realized model shows, that exhaust gases, which are flowing near left walls of each section of heat exchanger have a lower temperature than gases on the opposite side. It means, that a decreasing of temperature to water or acids liquefaction from exhaust gases will first occur on this side of heat exchanger section. It is significant to control a surfaces condition of the heat exchanger in this places during a maintenance. This operation allows preventing a heat exchanger against corrosion.

Figure 5 shows a velocity distribution for exhaust gas of a two neighboring sections.

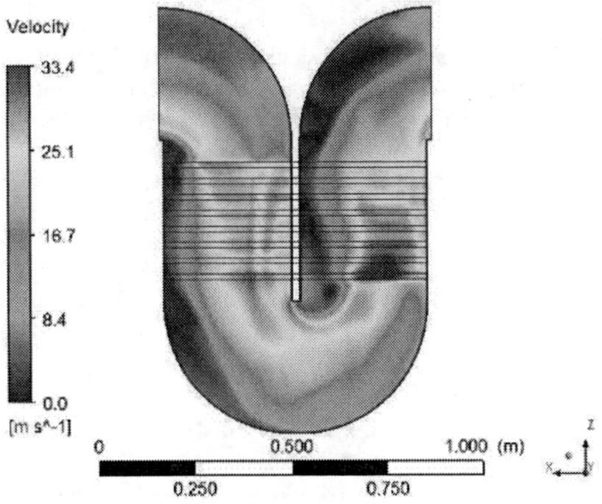

Fig. 5. Velocity distribution in two sections of the first module of the designed heat exchanger.

Velocity distribution of a fluid flow obtained during a numerical analysis in a designed heat exchanger shows, that flow inside sections of the heat exchanger is not identical for a whole volume of arranged construction. A numerical model of a smaller part of heat exchanger shows distinctly locations of detachments of a stream flow and locations of the vortex flow. The flow of exhaust gases inside the heat exchanger has a complex character. Change of direction of fluid flow causes a vortexes generation in specific parts of sections of the heat exchanger. Swirls are generated in the vertexes, where two sections of the heat exchanger are assembled together. The region of flow stagnation covers an area of intensive near left walls of heat exchanger sections. where the center of swirl is located.

The flow of exhaust gases between rows of pipes gives a rise of the velocity of a fluid because a cross-section of the heat exchanger is smaller in this place than in the area of inverting an exhaust gases direction. Numerical analysis shows, that flow of exhaust gases around the first row of pipes generate intense increasing of the velocity of a fluid, but this value is stabilized after flow around next rows of pipes. It is due to decreasing an area of fluid flow stagnation for exhaust gases, after inverting the direction of a flow and a passing an area of vortex generation. Therefore, a velocity distribution of a flow is stabilized.

4 Conclusions

Realized analytical and numerical calculations allowed to show the character of a work for a simple construction of heat exchanger for a food industrial plants. Analytical calculations allowed to predict a design of a heat exchanger geometry. It was done by basic calculations implementation, which is associated with a heat transfer theory. Analytical calculations allow for more intense analysis about a heat transfer process than a simple approach, which is presented in the paper. Information, about more advanced calculations, focused on outlet temperature and dynamics of processes for heat exchangers design can be realized according to papers [7] and [8].

Realized numerical calculations verified analytical calculations and covered the result with acceptable deviation. Numerical analysis showed, that an exhaust gasses flow into subsequent sections of analyzed construction is not homogenous. Realized research showed an area of a flow stagnation and places of vortex generation. During numerical calculations, was showed an area of decreasing a temperature, which is associated with the possibility of corrosion in these places, when a heat exchanger will be incorrectly used.

Numerical calculations are a good solution for a heat transfer analysis not only in heat exchangers modeling. This method will also help for different heat devices modeling, which is popularly used. Paper [9] presents an approach of heat transfer numerical modeling in solid fuel low power heating boilers, which are popularly used in Polish households. Numerical analysis implementation into a designing process also will help for a construction improvement of these heating devices.

Acknowledgements

The research was financed by the Poznan University of Technology financial resources for statutory activity.

References

1. I. Afonso, L. Hes, J. Maia, L. Melo, J. of Food Engineering, **57**, 179-187 (2003)
2. J. Jang, L. Chen, Int. J. Heat Mass Transfer, **40**, 16, 3981-3990 (1996)
3. S. Kruczek, *Heating boilers - constructions and designing*, (Ofic. Wyd. Pol. Wroc., Wrocław, 2001)
4. W. Gogół, *Heat transfer - tables and graphs*, (Wyd. Pol. War., Warszawa, 1976)
5. http://www.lipico.com/technical_references_palm_oil_properties.html, (date of available: 10.01.2018)
6. S. Wiśniewski, T. Wiśniewski, *Heat transfer*, (Wyd. N.-Tech., Warszawa, 2000)
7. D. Taler, *Dynamic of pipes heat exchangers*, (Ucz. Wydaw. Nauk.-Dydakt. AGH, Kraków, 2009)
8. D. Taler, A. Sury, *Mathematical models of cross-flow tube heat exchangers and their application to control of fluid outlet temperature*, (Wydaw. Pol. Krak., Kraków, 2013)
9. R. Urbaniak, J. Bartoszewicz, W. Judt, Heat Trans. Res, **49**, 7, 675-683 (2018)

Prediction or hydraulic performance of shell-and-tube heat exchanger: comparison of 1D and CFD-porous media approaches

Loïck Kalioudjoglou[1,2,*], *Clément* Bonneau[1], *Vincent* Melot[1], *Bruno* Auvity[2], *Christophe* Josset[2], and *Yoann* Merriaux[1]

[1]Naval Group Indret, 44620 La Montagne, France
[2]Laboratoire de Thermique et Energie de Nantes (LTeN) 3 rue Christian Pauc, 44306 Nantes, France

Abstract. The present paper deals with Heat Exchanger sizing methods and offers a comparison between two of them: 1D global method and CFD porous media method. Following Prithiviraj et al. work [1], new developments are based on recent knowledge acquired on porous media, using a coupling strategy of a three-dimensional commercial code with an in-house code library. The distributed hydraulic resistance concept and the numerical model are briefly described and confronted with pressure drop measurements from an experimental E-type STHE setup (shell-and-tube heat exchanger) from the literature. The present paper will put into perspective capabilities and limits of each method with needs for heat exchanger rating. Flow rate repartition is calculated with CFD-porous media using Tinker's current approach. This new analysis provides a complete comparison with 1D global method. It also reveals the major impact of leakage flow rate between baffle and tubes. The numerical estimation of pressure losses, consistent with experimental measurements of Halle et al. [2], implies that our future work will include thermal performance characterization and geometrical optimization.

1 Introduction

In the context of worldwide energy transition, the enhancement of electricity production systems plays a key role. Thus, the optimization of these systems, or of their constitutive components, aims at improving their efficiency, in order to better convert an energy source into electricity. In particular, it covers the complete Rankine cycle and the two-phase heat exchangers, condensers and steam generators. Because of common points with two-phase heat-exchanger flows, a complete understanding of monophasic shell-and-tube heat exchanger fluid dynamic is a preliminary step.

In monophasic STHE, reducing bypass and homogenizing shell side flow contributes to maximizing heat transfer which justify the importance of correct flow predictions. The interdependence between geometrical dimensions, fluid properties and flow involves complex calculations of thermal and hydraulic performances which still remain a challenge.

In order to correctly size heat exchangers, experimental mockup or computational methods may be used: 1D global method, CFD-porous media approach and whole CFD model. The industry sector is increasingly looking for a good compromise between time, cost and accuracy of sizing methods which all have their own benefits and shortcomings.

Experimental mock-up is the most accurate solution but is very costly and slow. Besides, measurements are generally limited to macroscopic measurements.

Numerical models (1D, CFD porous media and whole CFD model) are very cost-saving methods compared to the experimental approach. 1D-global method is the fastest solution with time resolution being a matter of seconds but the unidimensional character of calculation imply sometimes a lack of accuracy. Whole CFD model offers an accurate solution which permit to well understand microscopic local phenomena. However, numerical cost is very important and sometimes models are impossible to use due to a too high mesh number. CFD-porous media is a good compromise between numerical cost and accuracy. It generates a 3D flow field representative of the macroscopic structure within an acceptable time frame and cost.

This paper aims at comparing and criticizing these approaches on an isothermal monophasic shell side water flow case study based on an industrial case introduced by the Argonne National Laboratory [2].

In the tube bundle, baffles are laid-out in order to provide mechanical strength and to mix fluids to improve heat transfers. In practice, clearances between the tube and baffle, and between the baffle and the shell are necessary to assemble the heat exchanger. However, as a result, a leakage occurs through an annular hole. As illustrated in Figure 1 by Tinker [3] the flow could either be transversal or longitudinal to the tube bundle (B-stream), by-pass the bundle (C-stream) or cross the

* Corresponding author: loick.kalioudjoglou@naval-group.com

baffles (A and E streams). Consequently, flow structures are strongly heterogeneous and impact the heat transfer distribution. Therefore, the challenge of numerical computation is to correctly evaluate the effects of different streams induced by the position of the tube bundle and the baffle.

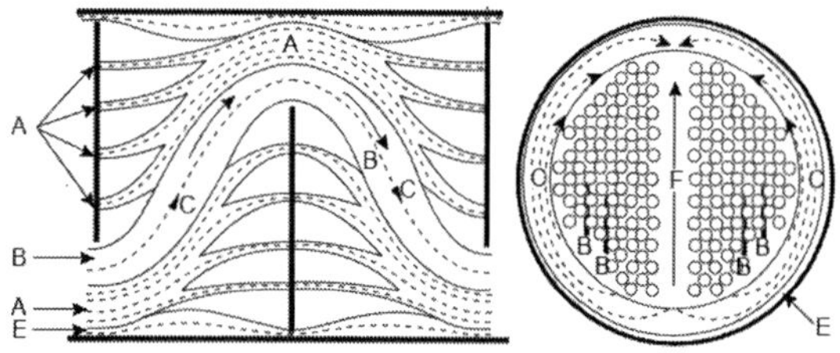

Fig. 1. Flow description through a STHE tube bundle[3].

The challenge of numerical solutions is to correctly estimate the effects of geometry (baffle and geometry layout) on the flow distribution and on pressure losses of shell-and-tube heat exchanger. Therefore, Tinker [3] produced in 1951 a global 1D method which allows to analyze the different current streams aforementioned and to balance the different hydraulic resistance of each one. In this article, a similar iterative global method is used based on Wills and Johnson approach [4].

In 1974, Patankar and Spalding introduced 3D modelling using the first finite difference method [5] with hydraulic resistance distribution model also called porous media model. In this approach, tubes are not simulated in the grid and are replaced by fluid domain considering equivalent volumetric force source terms. This homogenization method permits to know the mean values of velocities and pressures at aimed macroscopic scale and bring a better understanding of local phenomena in STHE with an acceptable calculation cost. Since, Rhodes and Carlucci (1983) [6], Prithiviraj et al. (1998) [1], Vukić et al. (2014) [8], Yang et al. (2014) [7] and Leoni et al. (2017) [9] have applied this approach on finite volume method and improved the accuracy of flow predictions. In this work, a new model, based on a commercial solver is described and compared.

Fig. 2. Test exchanger used by Argonne National Laboratory [2].

More recently, researchers study microscopic flow structure in STHE bundle using CFD simulation applied on real geometries [8], bringing more accuracy on flow prediction. However, these calculations are still applied on small bundle (43 tubes for Heat Exchanger studied by Leoni et al. [8]) because of the limitation of calculation cost.

In this work, a CFD-porous media calculation is solved to better analyze stagnation and recirculation zones and is compared to Halle et al. experiment shown Figure 2. This comparison brings out advantages and shortcomings of all possibilities to sizing heat exchanger.

2 Analysis and modelling

2.1 1D-global Method

A 1-D global method house-code is developed to estimate pressure losses through the shell-side of the heat exchanger with a baffle based on Tinker's stream schematic drawing [3]. With Wills and Johnston method [4], hydraulic resistances are balanced allowing iterative calculation of mass flow rate and pressure losses of each stream.

2.2 CFD-porous media Method

A STHE is composed of a regular solid layout (tube bundle) crossed by a heated fluid. The Patankar's concept [5], also called porous media method consists in physically modelling a tube bundle similar to an anisotropic porous media crossed by the flow. Consequently, the media is represented by a continuous medium of liquid and solid.

Considering this homogenization, it is necessary to use empirical laws to describe macroscopic behavior of fluid through bundle. In a numerical way, this media is only fluid and tubes are not represented in the mesh which contains 4.2 million elements. Navier-Stokes equations are modified including source terms to taking account of their presence.

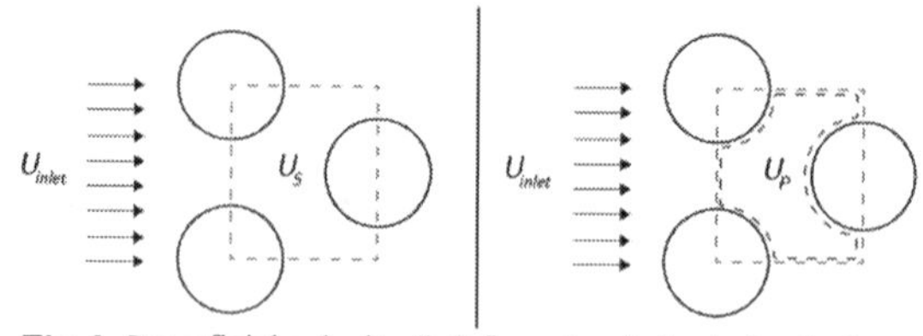

Fig. 3. Superficial velocity (left figure) and physical velocity (right figure) in a tube bundle.

To take account of the tubes presence, fluid porosity is defined as the ratio between fluid volume and total control volume. Superficial velocity is defined as the mean velocity in the porous media Figure 3 illustrated the link between mean physical velocity U_p and mean superficial velocitiy U_s by porosity value.

$$\gamma = \frac{V_{fluid}}{V_{total}} \quad (1)$$

$$U_s = \gamma U_p \quad (2)$$

Since the objective of this work is not to develop a calculation code, this model is applied on the commercial software ANSYS FLUENT® 16.2. The advection-diffusion equation is written in superficial form in order to keep velocity continuity at the interface between free-domain and porous medium. This method requires a volumetric force term source F for momentum conservation equation for modelling an equivalent hydraulic resistance. In the bundle, mass and momentum conservation equations are written:

$$\nabla \cdot (\rho \mathbf{U_s}) = 0 \quad (3)$$

$$\nabla \cdot (\rho \mathbf{U_s} \otimes \mathbf{U_s}) = -\nabla \cdot (\mu_{eff} \nabla \mathbf{U_s}) - \nabla p + \rho \mathbf{g} + \mathbf{F} \quad (4)$$

Hydraulic resistance is expressed in N/m^3 with K the pressure drop coefficients tensor.

$$\mathbf{F} = -\frac{1}{2}\rho \mathbf{K}: \mathbf{U_s}\|\mathbf{U_s}\| \quad (5)$$

Considering Oz the tube axis:

$$\mathbf{K} = \begin{pmatrix} K_{tr} & 0 & 0 \\ 0 & K_{tr} & 0 \\ 0 & 0 & K_{lo} \end{pmatrix} \quad (6)$$

With K_{tr} and K_{lo} transversal and longitudinal pressure drops coefficients. These coefficients are calculated with empirical correlations from literature depending on pitch, diameter, pattern and Reynolds number range. A previous study [10] applied on elementary cases highlights the good agreement of empirical laws used by Chalk River Nuclear Laboratories [6][11] for staggered tube bundle.

The longitudinal coefficient is defined in superficial velocity formulation by [11]:

$$K_{lo} = \frac{4f_{lo}}{D_h \gamma^2} \quad (7)$$

With D_h and Re_h the hydraulic diameter and hydraulic Reynolds number defined by :

$$D_h = \left(\frac{2\sqrt{3}}{\pi}\left(\frac{P}{D_o}\right)^2 - 1\right) \times D_o \quad (8)$$

$$Re_h = \frac{\rho U_p D_h}{\mu} \quad (9)$$

For $Re_h < 2\,250$:

$$f_{lo} = \left(-3.3 + 22.8\frac{P}{D_o}\right) Re_h^{-1} \quad (10)$$

For $2\,250 < Re_h < 25\,000$:

$$f_{lo} = 0.131 Re_h^{-0.294} \quad (11)$$

For $Re_h < 25\,000$:

$$f_{lo} = 0.066 Re_h^{-0.227} \quad (12)$$

The transversal coefficient is defined in superficial velocity formulation by [6]:

$$K_{tr} = \frac{4f_{tr}}{P}\left(\frac{C_t P}{P - D_o}\right)^2 \quad (13)$$

With Re the Reynolds number defined based on the outer diameter :

$$Re = \frac{\rho U_p D_o}{\mu} \quad (14)$$

$Re < 8\,000$:

$$f_{tr} = 0.619 Re^{-0.198} \quad (15)$$

$8\,000 < Re \leq 200\,000$:

$$f_{tr} = 1.156 Re^{-0.2647} \quad (16)$$

The transversal coefficient is slightly modified by a coefficient C_t to take into account the fluid attack angle regarding to the bundle orientation as illustrated in Figure 4.

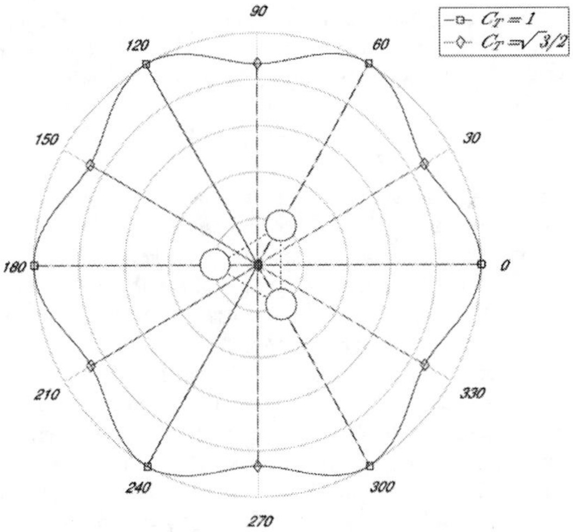

Fig. 4. Variation of coefficient Ct as a function of attack angle.

$$C_t(\theta) = \frac{1 + \frac{\sqrt{3}}{2}}{2} + \frac{1 - \frac{\sqrt{3}}{2}}{2}\cos(6\theta) \quad (17)$$

Indeed, pressure losses are higher for a non-oriented bundle than for 30-degree oriented bundle due to occurrence of preferable pathways noticeable on Figure 5.

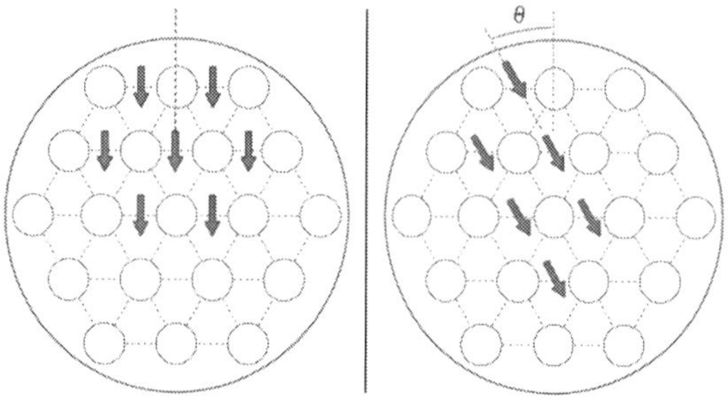

Fig. 5. Non-oriented bundle θ=0° (left figure) and oriented bundle θ=30° (right figure).

Turbulence model for flow in porous media remains a discussed topic in scientific literature. Prithiviraj and Andrews [1] use a $k-\varepsilon$ turbulent model and consider porous source term to modeling generation of turbulence kinetic energy dissipation rate and generation of turbulence kinetic energy due to tubes. Other authors as Chandresis et al. [12] worked to correctly write this source terms for rod bundle flows. Huang et al. [13] compared calculations made with four different turbulent models in the porous medium. The authors conclude that there are no significant differences between the several calculations on the distribution of velocity and temperature fields.

A sensitivity study was conducted and also showed a very low dependence of heat-exchanger pressure-drop on the turbulent model in the porous medium. In this paper, a $k-\omega$ SST Turbulence model is used in annular space around heat exchanger tube bundle. However, turbulence model is not used in porous media. Flow in tube bundle is resolved with a laminar model. Indeed, the K pressure drop coefficient tensor is treated like a catch-all term including turbulence effects. Empirical correlations are based on experimental test results which don't allow isolating tube surface drag term and the Reynolds stress tensor term. Consequently, the use of turbulence model would result in double counting.

Baffles are modelled by a porous media generating leakage rates in tube-to-baffle and shell-to-baffle clearances. Hydraulic resistance is calculated with a pressure drop coefficient tensor using Idel'Cik correlations [14] extended to annular space [15] and constant coefficient $\xi_{in} = 0.5$ et $\xi_{out} = 1$ for specific load losses. This model is both used for tube/baffle and shell/baffle clearance. A sensitivity study depending on baffle's correlation choice has shown a low impact on overall pressure losses.

$$K = \begin{pmatrix} \infty & 0 & 0 \\ 0 & \infty & 0 \\ 0 & 0 & K_{baffle} \end{pmatrix} \quad (18)$$

$$K_{baffle} = \left(\frac{\xi_{in}}{e_{baffle}} + \frac{f_{baffle}}{D_{ttb}} + \frac{\xi_{out}}{e_{baffle}} \right) \gamma^2 \quad (19)$$

3 Results and discussion

The considered E-type STHE is a staggered arranged bundle of 499 plain tubes of Halle's experimental test case referenced with the configuration code F·P·6·14"·30°·29%. Table 1 described geometry dimensions and specification. An isothermal water flow is considered with constant properties calculated for a 293.15K temperature. Volumetric flowrate varies between 0.067 and 0.135 m^3/s.

8 pressure taps are disposed at different positions to quantify pressure losses variation in the heat exchanger. For this configuration, overall pressure drop measurement is related to flowrate by Halle et al. in S.I. units with the expression:

$$\Delta p = 23\,300 \left(\frac{Q}{0.06309} \right)^{1{,}87} \quad (20)$$

Table 1. Dimension and Specification of F·P·6·14" ·30°·29% Heat Exchanger of Argonne National Laboratory [2].

Shell diameter (m)	**0.590**
Tube length (m)	3.580
Tube outer diameter (mm)	19.1
Pitch to diameter ratio (-)	1.25
Tube numbers (-)	499
Baffle numbers (-)	5
Baffle outer diameter (m)	0.587
Baffle thickness (mm)	9.5
Minimum tube-baffle hole diametric clearance D_{ttb} (mm)	0.4
Nozzle diameter (m)	0.337
Baffle cut space as percentage of inside shell diameter (%)	28.9

An overall pressure losses comparison between numerical calculations and experimental measurement is shown on Figure 6 and quantified on Table 2. The numerical evolution of pressure losses is also subject to a volumetric flow rate power law similar to experimental observation. For specified tube-to-baffle clearance $D_{ttb} = 0.4$mm, pressure losses are underestimated by CFD porous media model and overestimated by 1D-global method. These values remain within an acceptable range compared to measurement error which can reach +-5.3% [2]. However, 1D code show significant difference when clearance is not considered. It can be easily noticed that clearance values have a very strong impact on overall pressure drop.

Table 2. Relative deviation between experimental and numerical pressure losses.

	Relative deviation (%)		
Dttb (mm)	0	0.4	0.8
CFD-Porous Media	+18%	-12%	-31%
1D	+125%	+28%	+8%

Leoni et al. [9] has already shown impact of clearance for a small 43-tubes shell and tube heat exchanger with important variations between without-clearance and 0.8mm-clearance configurations. It shows importance of precise specification of clearance of heat exchanger for a correct sizing.

In practice, despite reduction of pressure drop, presence of large clearance isn't desirable because it creates large decrease of heat exchanges.

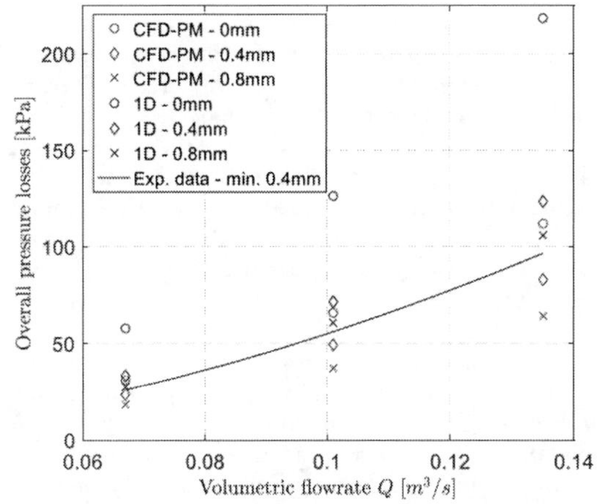

Fig. 6. Comparison of overall pressure losses numerically obtained with experimental measurements.

Figure 7 represents spatial distribution of pressure losses in the STHE. It is obvious that CFD-porous medial model which is three-dimensional allows to record local phenomena like recirculation zones, bypassing or local pressure variation which explain a better agreement compared to 1D-global method.

In addition to pressure losses values, qualitative analysis is a necessary step. An incorrect flow distribution could cause a wrong hydraulic performance prediction. Figure 8 highlights flow structure and local repartition of different streams with streamlines colored by velocity.

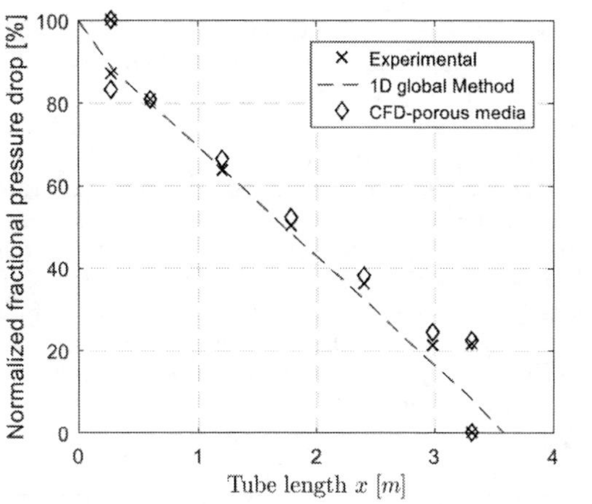

Fig. 7. Fractional distribution of pressure drop averaged and normalized to overall pressure drop for a Volumetric Flow Rate of $Q=100.8m^3/s$.

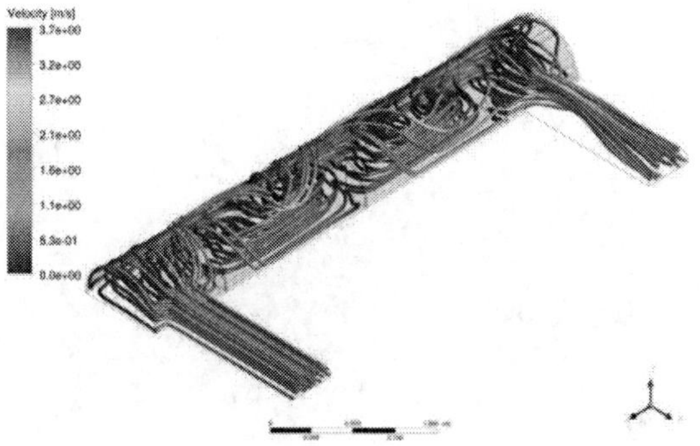

Fig. 8. Streamlines (colored by velocity) in porous media modelled heat exchanger.

Table 3. Tinker's Stream Flow Rate Estimation for a Volumetric Flow Rate of $Q=100.8m^3/s$.

Method	D_{ttb} (mm)	Flow rate (%) A	B	C	E	Ratio Q_B/Q_C
1D	None	0	87.0	13.0	0	6.7
CFD PM	None	0	76.1	23.9	0	3.2
1D	0.4	7.5	74,5	10.9	7.0	6.8
CFD PM	0.4	11.8	64.5	16.1	7.5	4.0
1D	0.8	19.0	65.4	9.4	6.2	7.0
CFD PM	0.8	26.4	53.4	13.9	6.4	3.8
Well designed STHE		9-20	40-70	15-20	6-20	2 - 4.7

It can be shown that the influence of C-stream bypass could decrease strongly the efficiency. A local depression zone can be seen at the outlet. Definitions of plane in post-processing enable numerical calculation of the mass flow rate of each stream. Table 3 compares flow rate repartitions calculated by both methods. These planes are defined on the central crosspass to avoid border effects next to inlet and outlet.

A-stream mass flow rate growth significantly when tube-to-baffle clearance increases. Nevertheless, flow rate ratio Q_B/Q_C does not change with clearance and remains constant for each method. A considerable difference exists between C-stream predictions of 1D method and CFD-porous media method. The latter method forecast a larger fluid bypass. Taborek's feedback [16] identifies in Table 3 ranges of well-designed heat exchanger flow fractions based on industrial size exchangers from the Heat Transfer Research, Inc. data bank. If we only consider results for 0.4mm baffle-to-tube clearances, it can be observed that prediction of ratio Q_B/Q_C with 1D method seem to be slightly superior than experience observations.

Evolution of velocity field plotted on symmetry plane Figure 9 highlights a good distribution of flow : the fluid uses the full depth of the crosspass and the heat exchanger design does not create low-Reynolds recirculation zone.

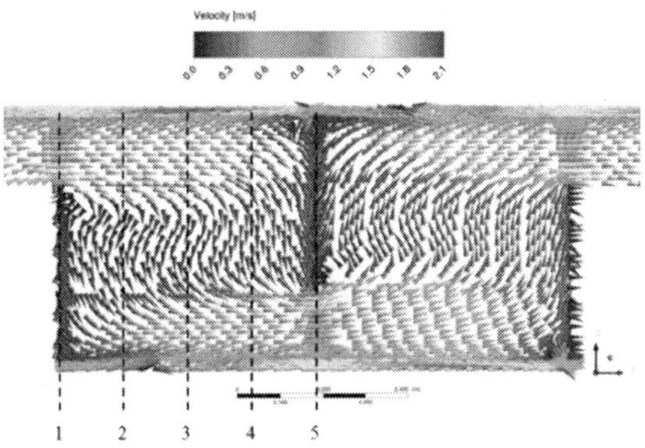

Fig. 9. Velocity field passing through baffle on plane symmetry for a volumetric flow rate of $Q = 100.8\ m^3/s$.

Taborek described in [16] heat exchanger sizing recommendations. He gives optimum zone which relates baffle cut (%), crosspass depth (m) and shell diameter. For the Halle et al. configuration, the optimal baffle cut value recommandate by Taborek is slightly superior : 36% while experimental mock-up consider a 28.9% baffle cut. Based on developed 1D model and CFD-porous media model, a parametric study could verify the quality of these sizing advised and improve geometric design.

In order to examine the impact of porous media on flow structure, five velocity fields between second and third baffle have been exported from simulation. The positions of each plane are indicated on Figure 9.

In plane 1, one portion of fluid crosses the baffle (A-stream and E-stream). The other portion, is distinguished in the baffle window between tube bundle and annular space. The more the space between shell and bundle is significant, the more the bypass flow rate is high [10]. It can be noticed that in plane 5, the flow is completely reversed compared to plane 1 and a similar flow is perceptible.

Between planes 1 and 5, the flow crosses (B-stream) and bypasses (C-stream) the bundle to reach the second window. In contrast to STHE studied by Bonneau [10], the flow is not broken when getting closed to baffle but gradually moves towards the second window and provides a better flow distribution and thus a better thermal efficiency.

4 Conclusions

The application of different sizing tools was set up. A triple confrontation between experimental measurements, 1D global method and CFD-porous media allowed to secure and make reliable their utilization.

It has been demonstrated that CFD-porous media brings an acceptable accuracy lower than 15% for specified value of 0.4mm by experiments [2]. Nevertheless, the CFD-porous media code tends to underpredict pressure losses which involves the use of safety margins for some industrials processes.

1D global method is very fast and allows to consider optimization loops at low calculation costs. However, accuracy is lower than CFD-porous media and overprediction of pressure drops can go up to 30% in comparison to experimental measurements.

It has been shown than not considering the clearance can be damaging for sizing with an overprediction of pressure losses which can reach 18% for CFD-porous media and 125% for 1D global method.

The strong impact of tube-to-baffle clearance has shown the need to correctly specify the clearance for an accurate pressure drop prediction and as a result a correct thermal prediction. Actually, it is difficult to judge the accuracy of numerical code due to clearance value communicated by Halle & al. which is not a nominal but a minimal value.

Calculations confirm designer's experience regarding to bypasss impact, baffle geometry and position influence. The code provides an operating tool for verifying by 3D simulation acquired experiences.

In the future, design optimization study could be considered with sealing strips inserts, bundle geometric adjustments and modifications of positions or shapes of baffles. Besides, design recommendations established by Taborek [16] could be fortified with numerical parametric study.

Finally, a thermal validation study of Shell and Tube Heat Exchanger by CFD porous media would allow to undertake diphasic shell and tube heat exchangers with correct prediction of monophasic cooling and heating regimes.

Acknowledgements

Thanks are due to Mr. S. DITTIERE for the contribution in the calculation step. The authors gratefully acknowledge the supports of Naval Group and Thermal and Energy laboratory of Nantes (LTeN).

Nomenclature

Quantity	Symbol	Coherent SI Unit
Fluid orientation coefficient	C_t	–
Bundle Hydraulic Diameter	D_h	m
Tube to baffle hole diametral clearance	D_{ttb}	m
Outer diameter	D_o	m
Volumetric Term source	$\mathbf{F}$	N/m^3
Longitudinal friction factor	f_{lo}	–
Transversal friction factor	f_{tr}	–
Transversal pressure drop coefficient	K_{tr}	m^{-1}
Longitudinal pressure drop coefficient	K_{lo}	m^{-1}
Baffle pressure drop coefficient	K_{baffle}	m^{-1}
Pitch	P	m
Pressure drop	Δp	Pa
Volumetric flow rate	Q	m^3/s
Physical velocity	U_p	m/s
Superficial velocity	U_s	m/s
Porosity	γ	–
Attack angle	θ	rad
Effective viscosity	μ_{eff}	Pa. s
Singular friction factor	ξ	–

References

1. M. Prithiviraj and M.J. Andrews, Numerical Heat Transfer, **33**, 8, pp. 799-816 (1998)
2. H. Halle, J.H. Chenoweth and M. W. Wambsganss, Journal of heat transfer, **110**, 1, pp. 60-67 (1988)
3. T. Tinker, *Proceedings of the General Discussion on Heat Transfer*, pp. 89-96 (1951)
4. M.J.N. Wills and D. Johnston, *22nd National Heat Transfer Conference*, HTD, **36**, pp. 67-79 (1984)
5. S.V. Patankar and D.B. Spalding, *Heat Exchangers: Design and Theory Source Book* (McGraw-Hill, New York, pp. 155-176, 1974)
6. D.B. Rhodes and L.N. Carlucci, *Proceedings of Canadian Nuclear Society - American Nuclear Society International Conference on Numerical Methods in Engineering,* (Montreal, pp. 935-948, 1983)
7. J. Yang, L. Ma, J. Bock, A.M. Jacobi and W. Liu, Applied Thermal Engineering, **65**, 1-2, pp. 369-383 (2014)
8. M.V. Vukić, G. Vučković, P., Živković, Ž. Stevanović and M. Tomić, Facta Universitatis, Series: Mechanical Engineering, **11**, 2, pp. 169-180 (2014)
9. G.B. Leoni, T.S. Klein and R. De Andrade Medronho, Applied Thermal Engineering, **112**, pp. 497-506 (2017)
10. C. Bonneau, Ph.D. dissertation, Nantes University, France (2017)
11. M.B. Carver, L. N. Carlucci and W. W. R. Inch, No. AECL-7254, Atomic Energy of Canada Ltd. (1981)
12. M. Chandresis, G. Serre and P. Sagaut, International Journal of Heat and Mass Transfer, **49**, 15, pp.2739-2750 (2006)
13. L.Y. Huang, J. X. Wen, T. G. Karayiannis, R. D. Mathews, The PHOENICS Journal of Computations Fluid Dynamics & its applications, **9**, 2, pp. 181-209 (1996)
14. I. E. Idel'Cik, *Collection de la Direction des Etudes et Recherches d'Electricité de France* (Paris, France, 1969)
15. W.R. Van Zyl, J. Dirker and J. P. Meyer, Heat Transfer Engineering, **34**, 13, pp 1112-1123 (2013)
16. J. Taborek, *Heat Exchanger Design Handbook* (1983)

Theoretical modeling and experimental study of auxiliary concrete accumulator for solar heating systems

Jacek Sacharczuk[1,*], *Dawid* Taler[1]

[1] Cracow University of Technology, Faculty of Environmental Engineering, ul. Warszawska 24, 31-155 Kraków, Poland

Abstract. This article presents the issue of the use of the control-volume finite-elements method (CVFEM) to solve transient heat conduction problem in the ceramic or concrete structure of heat storage system. The system can be used as auxiliary storage in solar based domestic hot water (DHW) and heating installations. The storage system consists of modular symmetric components forming parallel air channels. The modular design and symmetry of cross section enables to build a simple numerical model using a coarse mesh of finite volumes. It allows solving the problem using the simple algorithm. Analyzed method of modeling provides a short computation time while maintaining high calculation accuracy.

1 Introduction

The use of the thermal capacity of solid materials for heat storage is an alternative to heat storage systems in water tanks. Despite the lower thermal capacity, the accumulation material can be incorporated into the structure of a building or can be a part of its construction. It creates many possibilities for innovative solutions for heating systems. The idea of heat storage using the thermal capacity of solid materials has been known since ancient times (heating systems based on hypocaust). In solar heating systems, accumulators made of solids are often used in ACPS (active collection - passive storage) systems [1]. In ACPS systems, heat obtained from solar collectors (air, liquid) is transferred directly to heated objects characterized by high thermal capacity. In systems with active charging and discharging, the air flow is forced.

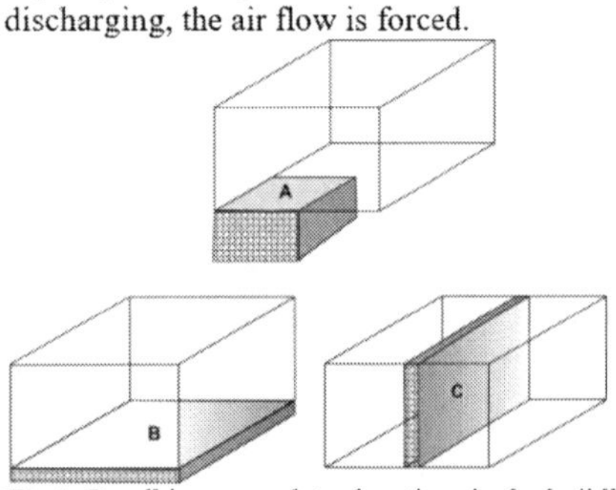

Fig.1. Possible accumulator locations in the building

Adequate selection of geometric parameters of the accumulator, thermal properties of the storage material and convective heat transfer conditions enable heat utilization spread over the day. Ordinary concrete or concrete with improved accumulation properties, chamotte, ceramics, stone and even sand may be used to build the accumulator. The accumulator can be made in the form of a closed space, wall or a horizontal slab. Examples of locations of the heat accumulators in buildings are presented on Fig. 1.

Another group of accumulators consists of accumulators using hollow units with ducted airflow [7-11].

One of the pioneers in that field was James Kachadorian. In 1978 he patented a heating system which used solar energy from the direct gains system or from air collectors and stored it in a gravel-filled or hollowed sub-floor space with air distributing system by natural or forced circulation [8].

Short and Kutscher [1] investigated a hybrid building heating system with a concrete accumulator with a channel structure in the form of a concrete floor slab The air heated in the collector was directed to the air-water heat exchanger, which, as a result, heated domestic hot water.

In the research carried out by YuXiang and Athienitis [12-15], the use of a ventilated ceiling slab (VCS Ventilated Concrete Slab) was described as a heat accumulating element obtained from a solar installation equipped with a BIPV (Building Integrated Photovoltaic) collector integrated with the roof.

2 The concept and the model

2.1 The concept

The concept of the analyzed heating system is based on the simultaneous storage of thermal energy in a water tank and concrete or ceramic accumulator with a channel structure. The heat used to preheat hot water and support the heating of a building is obtained from a traditional solar installation equipped with liquid collectors. Firstly, the water tank is heated, and the surplus of acquired

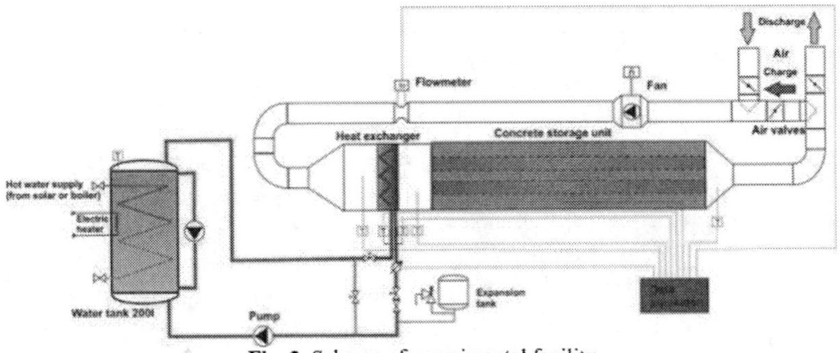

Fig. 2. Scheme of experimental facility

energy is accumulated in the structure of a concrete or ceramic exchanger – an accumulator which is charged and discharged by the air stream (flow in internal channels). During charging, the air circulates in a closed circuit and is heated in a plate-fin heat exchanger. At night, the accumulator is discharged by the stream of air blown directly into the heated space.

2.2 Experimental facility

For the system described above, a model installation was proposed. It consists of an accumulator made of concrete units forming two parallel air channels, air circulation system, two-speed coil plate-fin heat exchanger, and water tank heated by a solar collector. During the experiments, alternative heating with a gas boiler with an electric stabilizing heater was also used. The main element of the installation is a heat accumulator made of 16 concrete blocks, used in a fireplace building technology arranged in two parallel rows. Accumulating material fills a rectangular prism measuring 2240mm x 560mm x 280mm. Inlets and outlets from air ducts were profiled with steel sheet fittings. The accumulating material weights 689.6 kg.

Fig. 3. Experimental facility real view

The heat exchange surface of internal ducts is 2.75m^2, and the hydraulic diameter of a single channel is 178mm. The density of accumulating material is equal ρ=2820kg/m^3, specific heat capacity c=0.94 kJ/kgK and thermal conductivity λ=1.7W/mK.

2.3 Mathematical model

The mathematical model of the heating system based on the assumption that the heat for the storage system was supplied in a stream of water taken from the domestic hot water tank heated by the solar installation. The supply water temperature can, therefore, change according to the scheme of obtaining heat from the solar collectors and disassembling the hot water in the installation. The developed and verified the mathematical model of the plate-fin heat exchanger [17,18] had been used. For modeling transient heat transfer in the accumulator structure, it has been assumed that for the proper description it is sufficient to analyze the temperature distribution in multiple modules limited by the symmetry axes of the cross-section [11]. The concrete blocks used for the verification of the experimental model have a cross-section as presented in Fig. 4 characterized by four symmetry axes. An element corresponding to 1/8 of the cross-section of the fitting can be adopted for modeling the temperature field. In the experimental installation, two parallel rows were made of 8 blocks.

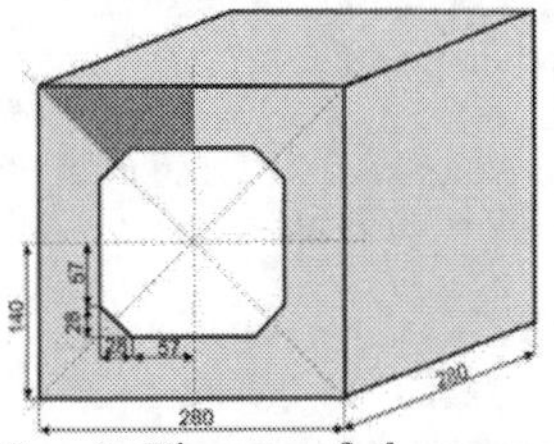

Fig. 4. Diagram of the accumulator model structure and accumulation module geometry

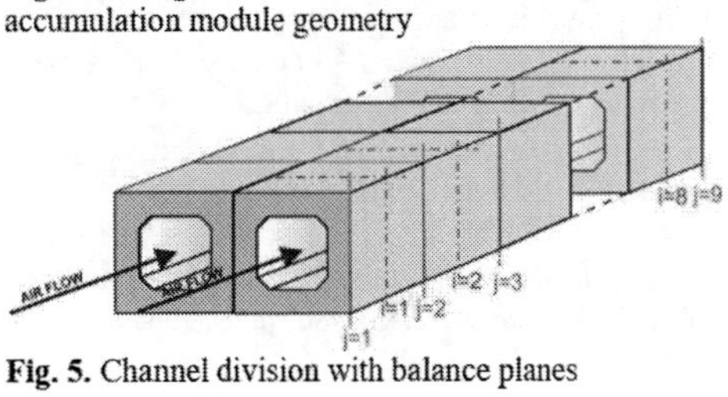

Fig. 5. Channel division with balance planes

It was assumed that the temperature gradient in the material structure in the z-axis direction within individual modules is negligible. The heat conduction equation for steady-state, two-dimensional temperature field in the part of an accumulator cross-section determined by the symmetry planes (Fig. 4) has the following form

$$c(T)\rho(T)\frac{\partial T}{\partial t}=\frac{\partial}{\partial x}\left[\lambda(T)\frac{\partial T}{\partial x}\right]+\frac{\partial}{\partial y}\left[\lambda(T)\frac{\partial T}{\partial y}\right] \quad (1)$$

The outer surface of the air duct is thermally insulated. The convective boundary condition at the inner surface of a duct is as follows:

$$\lambda(T)\frac{\partial T}{\partial n}\bigg|_{S_w}=\alpha\left(T_a-T|_{S_w}\right) \quad (2)$$

The energy conservation equation for the air flowing through the internal channel is given by

$$\rho_a c_p\left(\frac{\partial T_a}{\partial t}+w_a\frac{\partial T_a}{\partial z}\right)=\frac{\alpha(\bar{T}_w-T_a)U_w}{A_a} \quad (3)$$

The initial temperature of concrete units and air are uniform

$$T|_{t=0}=T_a|_{t=0}=T_0 \quad (4)$$

The air temperature at the accumulator inlet $T_{a,in}$ is also known

$$T_a|_{z=0}=T_{a,in}(t) \quad (5)$$

The inlet temperature Ta,in is equal the air temperature downstream a plate fin-and-tube heat exchanger.

The initial-boundary value problem (5) was solved using Control-Volume Finite-Element Method (CVFEM) for determining the temperature in concrete segments and finite difference method (FDM) for determining the air temperature.

The finite difference grid for modeling air flow is defined as follows:

$$z_j=(j-1)\Delta z, j=1,\dots,9 \quad (6)$$

$$z_i=\frac{\Delta z}{2}+(i-1)\Delta z, i=1,\dots,8 \quad (7)$$

where Δz is equal the length of a concrete unit (Figures 4,and,5). The coordinate zi refers to the center of the unit.

To solve the problem of transient heat transfer in the solid accumulator, the following assumptions were adopted:

- the channel was divided into n = 8 segments, with a length corresponding to the length of a single accumulation module.
- the temperature of the storage mass was computed at the cross-section located in the middle of the length of each module (indices i = 1, ...,n)
- the staggered grid was used in the air domain
- section j = 1 corresponds to the inlet plane to the duct (inlet air temperature)
- the cross-sections j, and j+1 correspond to the inlet and outlet of air and coincide with contact planes of the concrete modules
- the cross-section j = n+1 corresponds to the outlet plane from the duct.

Equation (3) was solved using explicit finite difference method. The time derivative in Eq. (3) was approximated by forward difference quotient and space derivative by the backward finite difference

$$\frac{T_{a,j+1}^{k+1}-T_{a,j+1}^{k}}{\Delta t}+w_a\frac{T_{a,j+1}^{k}-T_{a,j}^{k}}{\Delta z}=\frac{\alpha_j^k\left(\bar{T}_{w,i=j}^k-T_{a,i}^k\right)U_w}{A_a\rho_a c_p} \quad (8)$$

The air temperature at the point z_i was assumed to be equal the arithmetic mean of the temperatures $T_{a,j}$ and $T_{a,j+1}$ (Fig. 5)

$$T_{a,i}=\frac{T_{a,j}+T_{p,j+1}}{2} \quad (9)$$

The superscript k+1in Eq. (8) refers to the new time step while k to the old one

$$t_{k+1}=t_k+\Delta t,\ k=0\dots,t_0=0 \quad (10)$$

The time derivative $\frac{\partial T_a}{\partial t}$ in Eq.(3) can be neglected since specific thermal capacity of air $\rho_a c_p$ is very small. The finite difference equation Eq.(8) reduces to

$$w_a\frac{T_{a,j+1}^{k}-T_{a,j}^{k}}{\Delta z}=\frac{\alpha_j^k\left(\bar{T}_{w,i=j}^k-T_{a,i}^k\right)U_w}{A_a\rho_a c_p} \quad (11)$$

Solving Eq. (11) concerning $T_{a,j+1}^k$ gives:

$$T_{a,j+1}^k=\frac{T_{a,j}^k\left(1-\frac{\Delta z\alpha_j^k U_w}{2A_a\rho_a c_p w_a}\right)+\bar{T}_{w,i=j}^k\left(1-\frac{\Delta z\alpha_j^k U_w}{2A_a\rho_a c_p w_a}\right)}{1+\frac{\Delta z\alpha_j^k U_w}{2A_a\rho_a c_p w_a}} \quad (12)$$

$j=1,\dots,9,\ k=0,1,\dots$

Next, the determination of temperature of concrete unit will be discussed.

The control volume finite element method (CVFEM) [18,19] is used to solve the problem of two-dimensional transient heat transfer in the planes of the accumulator cross-section, This method allows modeling the temperature field in a repeatable section of the accumulator section using a triangle mesh based on a relatively small number of nodal points. In the CVFEM method, an energy balance is created for the control area surrounding each node. The control area around the node 1 is formed by one-third of all of the triangles with sides coming out of the joint node 1. The energy balance equation is written for the *1-a-O-c* area, in which the point O is the center of gravity of the triangle *123*, and the points *a* and *c* are respectively the centers of the sides *1-2* and *1-3* (Fig. 6). The surface area of a figure *1-a-O-c* is one-third of a surface area of a triangle *123*.

In the case of a single triangular element *123*, the energy balance equation for the control volume associated with the node *1* can be written as follows:

$$c(T_1)\rho(T_1)\frac{A_{123}}{3}\frac{dT_1}{dt}=\lambda_x(T_0)\frac{y_c-y_a}{2A_{123}}[(y_2-y_3)T_1+(y_3-y_1)T_2+(y_1-y_2)T_3]+\lambda_y(T_0)\frac{x_c-x_a}{2A_{123}}[(x_2-x_3)T_1+(x_3-x_1)T_2+(x_1-x_2)T_3] \quad (13)$$

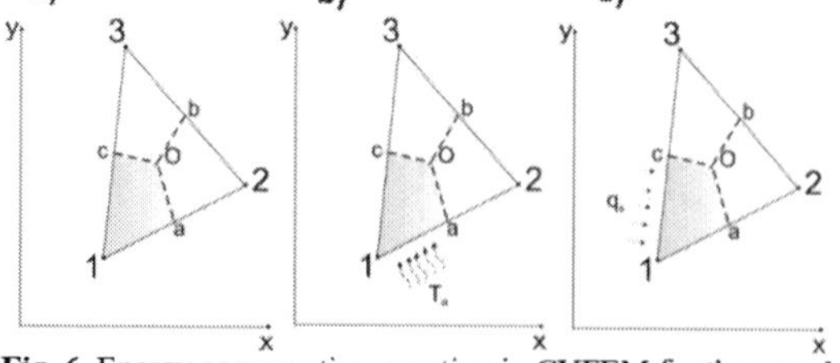

Fig. 6. Energy conservation equation in CVFEM for the area *1-a-c* ; a) inner cell; b) boundary cell with convective boundary condition on the side *1-a*; c) boundary cell with heat flux on the side *1-c*.

If on the edge surface 1-a convection occurs, and the heat flux $\dot{q}_s$ is set on the surface 1-c, Eq. (13) takes the form

$$c(T_1)\rho(T_1)\frac{A_{123}}{3}\frac{dT_1}{dt} = \lambda_x(T_0)\frac{y_c-y_a}{2A_{123}}[(y_2-y_3)T_1+(y_3-y_1)T_2+(y_1-y_2)T_3]+\lambda_y(T_0)\frac{x_c-x_a}{2A_{123}}[(x_2-x_3)T_1+(x_3-x_1)T_2+(x_1-x_2)T_3]+\alpha\left[T_a-\left(\frac{3T_1}{4}+\frac{T_2}{4}\right)\right]s_{1a}+\dot{q}_s s_{1c} \quad (14)$$

Energy balance equations written for all control areas form a system of ordinary differential equations for node temperatures.

In a numerical solution, it is convenient to set a system of balance equations in a matrix form. For perfect insulation of the outer accumulator surface, the equation system is

$$D\cdot A=\frac{1}{\rho c}[(B+C)\cdot T] \quad (15)$$

When the thermal insulation is not perfect, then heat loss occurs on the outer accumulator surface, and equation system has the following form

$$D\cdot A=\frac{1}{\rho c}[(B+C)\cdot T+E\cdot \dot{q}_s] \quad (16)$$

where: D column matrix of temperature-time derivatives at nodes 1 ... *n*, A matrix containing surface areas of elements around nodes, B matrix of coefficients in terms including conduction, C matrix of coefficients regarding convection, E matrix of coefficients defining the length of the element boundary section with the heat flux set on the boundary, T column matrix of temperatures at the nodes and the air temperature in the channel.

For the adopted cross-section with 16 nodes and 6 fictitious nodes resulting from domain symmetry, matrices A and B have the dimensions 16x23, while the matrix T 23x1. The matrix E and the derivative matrix D have the dimension 16x1.

$$\begin{bmatrix}\frac{dT_1}{dt}\\ \frac{dT_2}{dt}\\ \vdots\\ \frac{dT_n}{dt}\end{bmatrix}\cdot[A_1\cdots A_n]=\frac{1}{\rho c}\begin{bmatrix}b_{1,1}\cdots b_{1,m} & b_{1,m+1}\\ b_{2,1}\cdots b_{2,m} & b_{2,m+1}\\ \vdots \quad \ddots \quad \vdots & \vdots\\ b_{n,1}\cdots b_{n,m} & b_{n,m+1}\end{bmatrix}\cdot\begin{bmatrix}T_1\\ T_2\\ \vdots\\ T_m\\ \bar{T}_a\end{bmatrix}+$$

$$\frac{1}{\rho c}\begin{bmatrix}c_{1,1}\cdots c_{1,m} & c_{1,m+1}\\ c_{2,1}\cdots c_{2,m} & c_{2,m+1}\\ \vdots \quad \ddots \quad \vdots & \vdots\\ c_{n,1}\cdots c_{n,m} & c_{n,m+1}\end{bmatrix}\cdot\begin{bmatrix}T_1\\ T_2\\ \vdots\\ T_m\\ \bar{T}_a\end{bmatrix}+\frac{1}{\rho c}\begin{bmatrix}e_1\\ e_2\\ \vdots\\ e_m\end{bmatrix}\cdot\dot{q}_s \quad (17)$$

The matrices A, B, C, and E, can be generated quickly entering the coordinates of individual nodes.

In the case of the analyzed concrete accumulator, the symmetrical fragment of the cross-section was first divided into triangular elements and then 16 control areas were created. Because of the symmetry six fictitious nodes were added (Fig.7).

The derivative of temperature in the normal direction (heat flux) to the symmetry planes *1-2-3-4* and *13-14-16* are equal zero. If the outer surface of a concrete unit was insulated perfectly then, zero heat flux was assumed. In this case, the entire external surface of the analyzed area *1-2-3-4-5-10-11-15-16-14-13* is subject to the boundary condition

$$\lambda\frac{\partial T}{\partial n}=0 \quad (18)$$

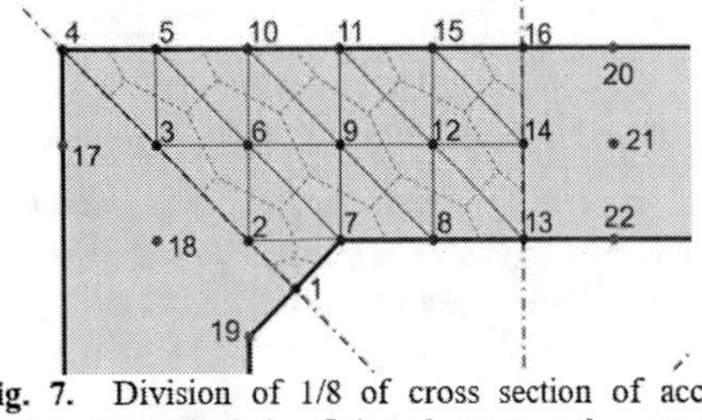

Fig. 7. Division of 1/8 of cross section of accumulating module into triangular finite elements and control volumes around the nodes

Calculations were also carried out for the case of imperfect insulation when there were heat losses from the outer accumulator surface to the environment (nodes *4-5-10-11-15-16*).

The system of ordinary differential equations of the first order for the node temperatures at solid and air domains was solved by the explicit finite difference method. The number of equations for the whole accumulator is 16×(n-1), where n=8 is the number of modules forming the accumulator channel. The use of the explicit method is a significant advantage in online control systems based on the mathematical model of the accumulator. The explicit method does not require iteration or solving a system of algebraic equations at each time step when controlling the air temperature at the accumulator exit.

The computation results are stable if the Fourier and Courant-Friedrichs-Lewy CFL criteria are satisfied:

$$\frac{a\Delta t}{(\Delta x)^2}<\frac{1}{2} \quad \frac{a\Delta t}{(\Delta y)^2}<\frac{1}{2} \quad \frac{w_a\Delta t}{\Delta z}\leq 1 \quad (19)$$

The Gnielinski formula (20) was applied in the analysis of heat exchange in concrete accumulator channels.

$$Nu=\frac{\left(\frac{\xi}{8}\right)(Re-1000)Pr}{1+12.7\left(\frac{\xi}{8}\right)^{\frac{1}{2}}\left(Pr^{\frac{2}{3}}-1\right)}\left[1+\left(\frac{d_h}{L}\right)^{\frac{2}{3}}\right] \quad (20)$$

$0<d_h/L<1,\ 0.6<Pr<1000;\ 4000<Re<10^6;$

For this correlation, the best agreement with the experimental results was obtained.

3 Results and discussion

3.1 Theoretical calculations

At first, the mathematical model developed in the paper was used to analyze the process of heating and cooling the accumulator under test conditions. The constant temperature of 60°C (corresponding to the maximum real temperature obtained from the solar source) was maintained to charge the accumulator. The analysis was performed for three different flow rates 1.1 m/s, 2.2 m/s, and 4.4 m/s (Figs. 8 and 9).

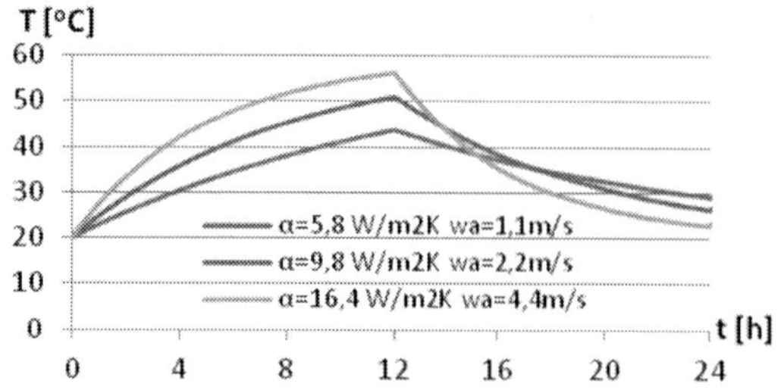

Fig. 8. Mean temperature of accumulator during charge-discharge cycle at different air flow rates

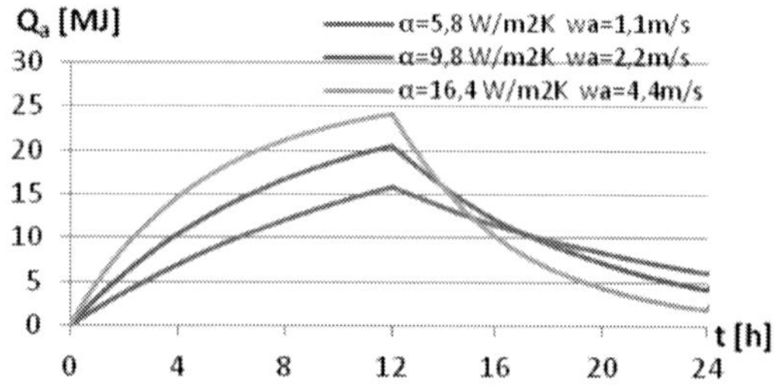

Fig. 9. Energy accumulated during charge-discharge cycle, at different air flow rates

3.2 Experimental verification of the model

Based on the developed mathematical model, Fortran software was written. The software allows to simulate the system operation using input data (calculated or measured) ie. air and water temperatures, air flow velocity upstream the air-cooled heat exchanger. In the tests concerning the full charging and discharging cycle of the accumulator, a comparison between measured heat flow rates on the water and air side of the heat exchanger and the accumulator were done. Also the energy taken from the air and transferred to the concrete accumulator during given period were compared. In addition, the results of temperature measurements at selected points of the solid storage mass were compared with the results of modeling.

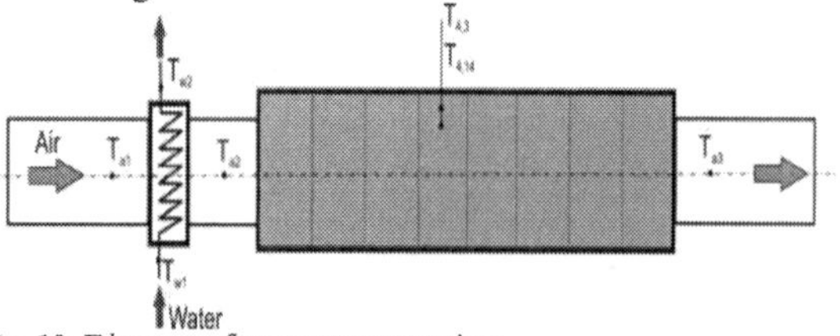

Fig. 10. Diagram of measurement points

To determine the heat flux transferred from the insulation to the environment in the charging period, the heat flux was measured, and compared with computations. The overall heat transfer coefficient between the outer surface of the concrete duct and environment was estimated based on experimental tests. This allowed to take into account heat losses by insulation to the environment in the accumulator mathematical model.

The time variations of air temperatures and accumulated energy obtained by the mathematical model show good compliance with the measurements. A comparison of the temperature of the air, temperature of the concrete accumulator at two points and energy accumulated in the concrete is depicted in Figures 11,12, and 13 for w_a=2.0m/s (Re_a =21600).

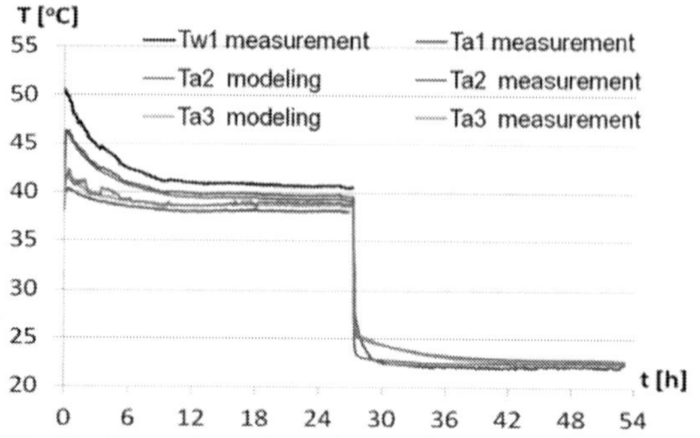

Fig. 11. Comparison of calculated and measured air and water temperatures

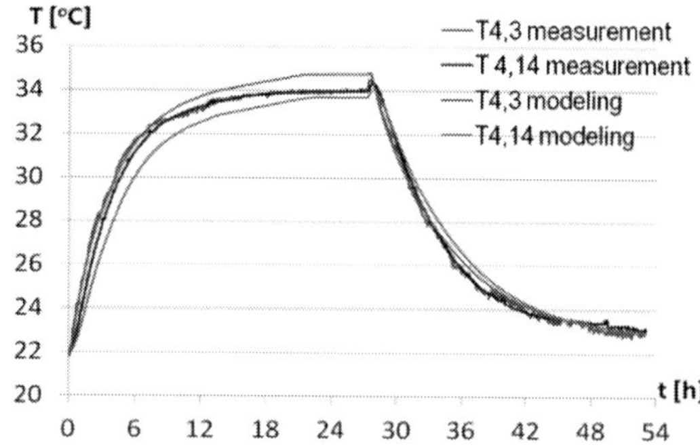

Fig. 12. Comparison of calculated and measured temperatures at selected points of the storage mass; location of points is shown in Fig. 5,7 and 10

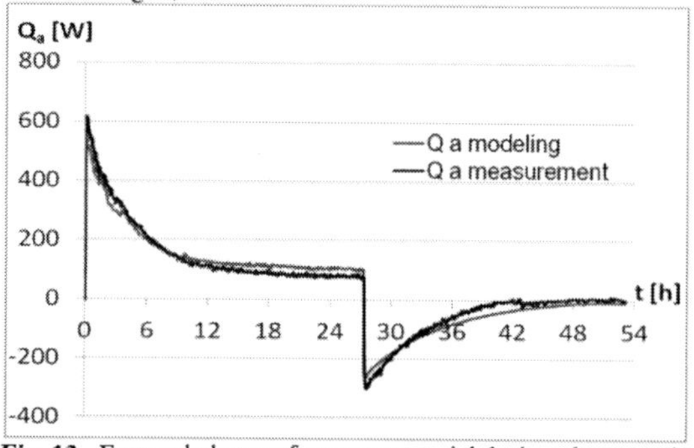

Fig. 13. Energy balance of storage material during charge and discharge, model and the experimental course.

3.3 Conclusions

The proposed method of modeling a heating system with a duct concrete heat accumulator allows simulation of any charging and discharging of the accumulator for any time changes of heating water and air supply parameters. The simple cross-section of the accumulation element adopted for the tests enabled effective laboratory tests and their comparison with numerical simulations. In practice, the analyzed heating system can be used in conditions of a large daily amplitude of ambient air temperature. In the practical implementation of the accumulator, it is recommended to use more extended heat transfer surfaces in the accumulator channels.

NOMENCLATURE

a – thermal diffusivity, m^2/s
A – duct cross-section area, m^2
$A_{i,j,k}$ – finite element surface area, m^2
$b_{i,j}, c_{i,j}, e_i$ – matrix coefficients
c – specific heat of solid material, J/kgK
c_p – specific heat of the air, J/kgK
d – diameter of the tube or air duct, m
L – length of the tube or air duct, m
Nu – Nusselt number
Pr – Prandtl number
$\dot{q}$ – heat flux, W/m^2
Q – heat accumulated in concrete, J
Re – Reynolds number
s – length, m
t – time, s
T – temperature, °C
U_w – duct cross-section perimeter, m
w – airflow velocity, m/s,
x,y,z – Cartesian coordinates, m

Greek symbols

α – heat transfer coefficient, W/m^2K
λ – thermal conductivity, W/mK
ρ – density, kg/m^3
ξ – Darcy-Weisbach friction factor

Subscripts

$1,2,3$ – nodes numbers
a – air, air duct, accumulated
a,b,c – midpoints of triangle sides
h – hydraulic
in – inlet
O – gravity center of the triangle
S_w – inner surface of air duct
w – wetted, water

REFERENCES

1. Short, W., and Kutscher, C.F., Analysis of an Active Charge/Passive Discharge Solar Space conditioning System, American Solar Energy Society, Anaheim (1984)
2. Duffie, J. A., and Beckmann, W. A., *Solar Engineering of Thermal Processes*. 2nd edition, J.Wiley and Sons Inc., New York (1991)
3. Esence, T., and Bruch, A., and Molina, S., and Stutzb, B., and Fourmigué, J.F., A review on experience feedback and numerical modeling of packed-bed thermal energy storage systems, Sol Energy, **153**, pp. 628-654 (2017)
4. Kumar, A., and Man-Hoe, Kim, Solar air-heating system with packed-bed energy-storage systems, Renew Sust Energ Rev, **72**, pp. 215-227 (2017)
5. Singh, H., and Saini, R.P., and Saini, J.S., A review on packed bed solar energy storage systems, Renew Sust Energ Rev, **14**, pp. 1059-1069 (2010)
6. Tyagi,V.V., and Panwar, N.L., and Rahim N.A., and Kothari, Richa, Review on solar air heating system with and without thermal energy storage system, Renew Sust Energ Rev, **16**, pp. 2289– 2303 (2012).
7. Fraisse, G, and Johannes, K, and Trilat-Berdal, V, and Achard, G, The use of a heavy internal wall with a ventilated air gap to store solar energy and improve summer comfort in timber frame houses, Energy and Buildings, **38**, pp. 293-302 (2006)
8. Kachadorian, J., Solar Heated Concrete Slab Building Structure, US patent 4127973 (1978)
9. Pei-Wen Li, Cho Lik Chan, *Thermal energy storage analyses and designs*, Elsevier Academic Press, London (2017)
10. Qing, Li, and Fengwu, Bai, and Bei, Yang, and Yan, Wang, and Mingxu, Han, Dynamic simulations of a honeycomb ceramic thermal energy storage in a solar thermal power plant using air as the heat transfer fluid, Appl Therm Eng, **129**, pp. 636-645 (2018)
11. Sacharczuk, J, and Taler, D, A concrete heat accumulator for use in solar heating systems – a Mathematical Model and Experimental Verification, Archives of Thermodynamics, **35**, pp.281-295 (2014)
12. Yuxiang, C, and Galal, K, Athienitis, A. K, Integrating hollow-core masonry walls and precast slabs into building space heating and cooling, Journal of building engineering, **5**, pp 277-287 (2016)
13. Yuxiang, Chen, and Galal, Khaled, and Athientis, A.K., Modeling Design and Thermal Performance of a BIPV/T System Thermally Coupled with a Ventilated Concrete Slab in a Low Energy Solar House: Part 1 BIPV/T System and House Energy Concept, Sol Energy, **84**, pp 1892-1907 (2010).
14. Yuxiang, Chen, and Athienitis, A. K., and Galal, Khaled, Modeling Design and Thermal Performance of a BIPV/T System Thermally Coupled with a Ventilated Concrete Slab in a Low Energy Solar House: Part 2 Ventilated Concrete Slab BIPV/T System and House Energy Concept, Sol Energy, **84**, pp. 1908-1919 (2010)
15. Yuxiang, Chen, and Galal, Khaled, Athienitis, A. K., Integrating hollow-core masonry walls and precast slabs into building space heating and cooling, Journal of Building Engineering, **5**, pp 277-287 (2016)
16. Fraisse, G., and Johannes, K., and Trilat-Berdal, V., and Achard, G., The use of a heavy internal wall with a ventilated air gap to store solar energy and improve summer comfort in timber frame houses, Energ Buildings, **38**, pp. 293-302 (2006)
17. Taler, D., Energy Mathematical modeling and experimental study of heat transfer in a low-duty air-cooled heat exchanger, Energ Convers Manage, **159**, pp. 232-243 (2018)
18. Taler, D., *Calculations and experimental tests of heat exchangers*, 1st edition, Publishing House of Cracow University of Technology, Cracow (2016) (in Polish)
19. Baliga, B. R, and Atabaki, N, Control-Volume Based Finite-Difference and Finite- Element Methods, in Minkowycz, W. J, and Sparrow, E. M, and Murthy, J. Y, *Handbook of Numerical Heat Transfer*, J. Wiley and Sons Inc., New York (2006)

Numerical and experimental study of flow distribution in tubular space of fin-and-tube heat exchanger with modified inlet manifold

Tomasz Stelmach[1]

[1] Institute of Thermal Power Engineering, Cracow University of Technology, Faculty of Mechanical Engineering

Abstract. This paper presents the experimental and numerical investigation of flow distribution in the tubular space of cross-flow fin-and-tube heat exchanger. The tube bundle with two rows arranged in staggered formation is considered. A modified heat exchanged manifold, with inlet nozzle pipe located asymmetrically is considered. The outlet nozzle pipe is located in the middle of the outlet manifold, with a standard shape. An experimental stand allows one to investigate the volumetric flow rate in heat exchanger tubular space using the ultrasonic flowmeters. Various inlet mass flow rate i.e. 3 m^3/h, 4 m^3/h and 5 m^3/h are considered. The experimental results are compared with CFD simulation performed in ANSYS CFX program using the SSG Reynolds Stress turbulence model. A relatively good agreement is found for tube Re numbers varied from 1800 to 3100. .

1 Introduction

Fin-and-tube heat exchangers are widely used in chemical, petrochemical, refrigeration automotive and energy industry. This kind of heat exchangers operate in cross-flow arrangement of working fluids. In tubular space of heat exchanger flows the liquid with high heat transfer coefficient (water or oil), and in intertubular space flows gas with the low heat transfer coefficient. The extended heat transfer surfaces due to the application of fins (mostly rectangular or circular) allows to transfer a large amount of heat between working fluids. To achieve high thermal and hydraulic efficiency the elliptical tubes are used. The main advantage of elliptical tube is the favorable hydraulic performance compared to circular tube. Therefore, less pumping power is needed for gas flow through the tube bundle [1-4]. The fin-and-tube heat exchangers consists of inlet and outlet manifolds and the tube bundle (forming the tubular space of heat exchanger). The compact shape is achieved thanks to the application of heat exchanger manifolds with small volume [4,5,12]. Therefore, it is needed to ensure the uniform flow distribution from the inlet manifold to heat exchanger tubes, to allow the correct operation of the device [13,14]. If the flow maldistribution occurs, then in case of large heat fluxes transferred, the flow boiling [9,10] may occur. The flow boiling in heat exchanger tubes may intensify the fouling (scaling) processes [8,11], which leads to excessive thermal stresses in tube bundle. The part of tubes can work under the compressive stresses, which may result in tube buckling [12]. In this paper a flow distribution in heat exchanger with a modified shape of the inlet manifold is studied. The experimental stand described in [6,7] is used for experimental investigation of water flow distribution in heat exchanger. Furthermore, the numerical simulations are performed using the SSG Reynolds stress model, to check the agreement between the experimental results and numerical simulation

2 Tested heat exchanger

The studied heat exchanger consists of tube bundle (6) connected with the inlet (5) and the outlet (8) manifolds. The tubes are arranged in staggered formation. The tube length is 2000 mm. The ultrasonic flow meters Sontex Superstatic 749 (7) are installed in each tube of the heat exchanger to capture the volumetric flow rate. The inlet and outlet nozzle pipes are connected with feed water tank (1) through the connecting tubes (4). The pump (2) is used to supply the water to heat exchanger under the specified volumetric flow rate. To control the flow rate the total flow rate meter (3) is used. The flow rate readings from total volumetric flow rate meter are then compared with the sum of individual tube flow rates obtained from the measurements, to check the correctness of tube flow meters indications. The measurements allow one to check if the flow maldistribution exists in heat exchanger, as well as to validate the results of CFD simulations. To ensure the correct inflow of the water from inlet nozzle pipe to heat exchanger inlet manifold, the extended inlet pipe with the length of 100 mm is used.

* Corresponding author: tomasz.stelmach@gkpge.pl

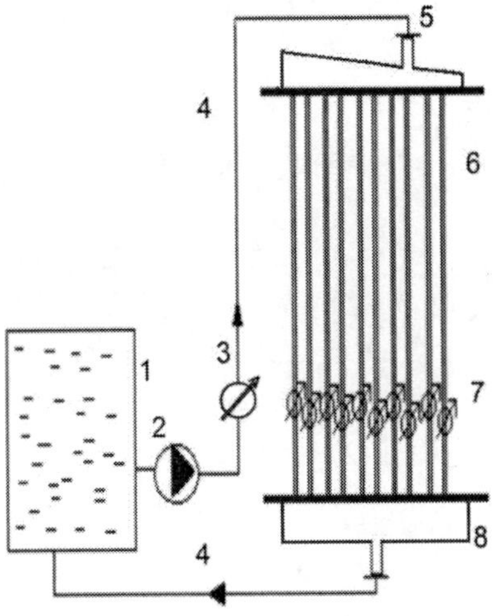

Fig. 1. Scheme of the tested heat exchanger tubular space of fin-and-tube heat exchanger: 1 - water feed tank, 2 - pump, 3 – total volumetric flow rate meter, 4 - connection pipes, 5 - modified inlet manifold with a nozzle pipe, 6 – tube bundle, 7 – single tube flow meter, 8 - outlet manifold and an outlet nozzle pipe.

Fig. 2. Inlet and outlet manifold of the tested heat exchanger.

Figure 2 shows the inlet and outlet manifolds. The inlet manifold is equipped with the nozzle located asymmetrically, the volume of the manifold is extended by using the straight wedges. The outlet manifold is connected with the outlet nozzle pipe located in the middle of the manifold.

3 Volumetric flow rate distribution in heat exchanger tubes

The geometry of the heat exchanger inlet and outlet manifolds is presented in Fig. 3. The nozzle pipe of tested inlet manifold is located in a distance of 195 mm far from the right side of the inlet manifold, and 440 mm from the left side.

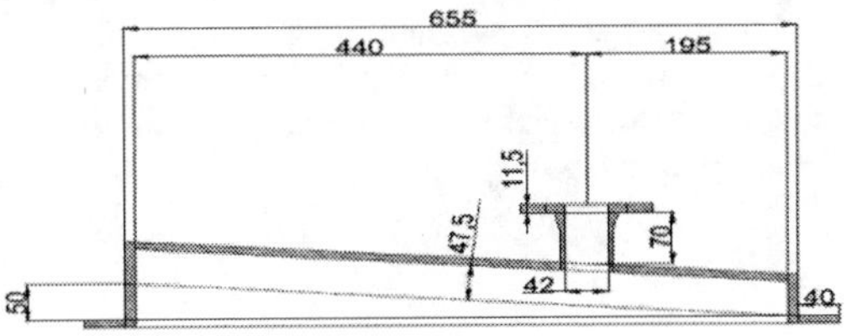

Fig. 3. Scheme of the inlet manifold with main dimensions.

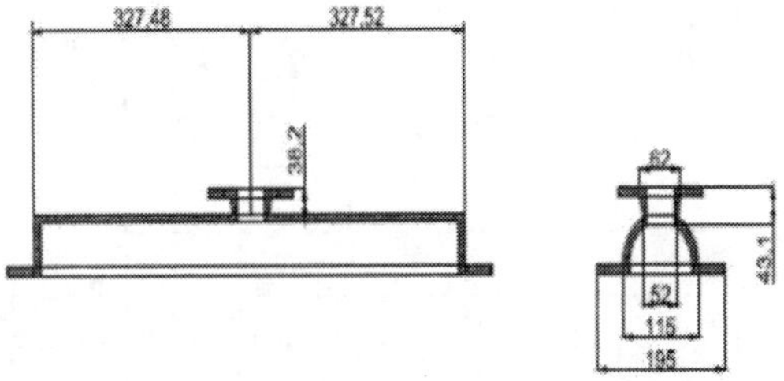

Fig. 4. Outlet manifold geometry.

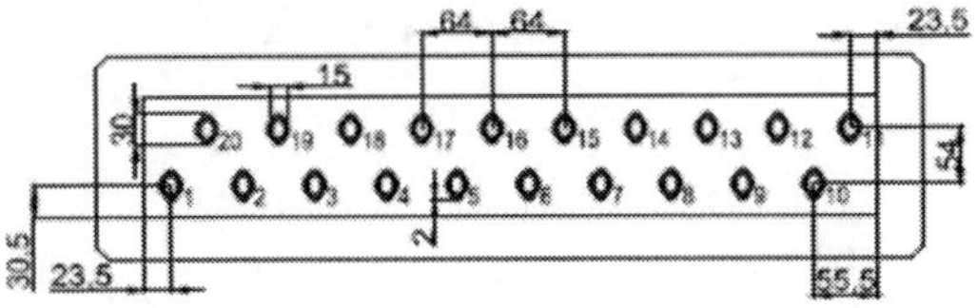

Fig. 5. Tube numbering and sieve plate geometry.

In this study, the measurements are carried out for the specified total volumetric flow rates equal to 3 m^3/h, 4 m^3/h, and 5 m^3/h. That studied volumetric flow range leads to the transitional and turbulent flow regime in heat exchanger tubes. The results are presented for the tubes in the first row of tube bundle 1-10 and the second row of tube bundle 11-20 (Fig. 5). The water temperature is constant and set equal to 20 °C. The readings from flow meters are recorded after the steady-state flow distribution is reached.

Measurement no 1 – total volumetric flow rate of $\dot{V} \approx 3$ [m^3/h]:

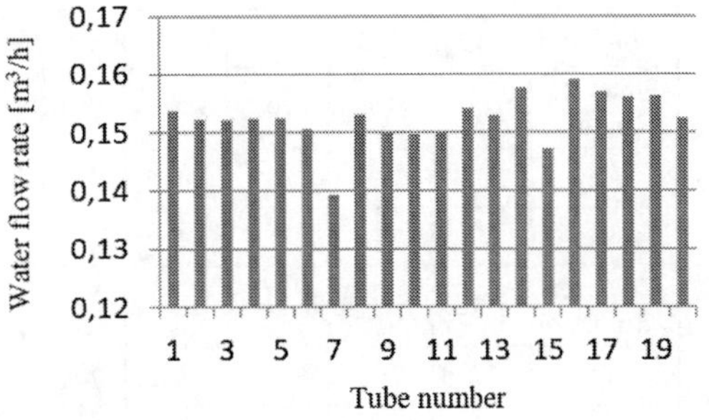

Fig. 6. Volumetric flow rate in heat exchanger tubes for a total flow rate of $\dot{V}$ = 3,0482 [m^3/h].

Table 1. The volumetric flow rate and Reynolds (Re) numbers in heat exchanger tubes for a total flow rate of $\dot{V}$ = 3,0482 [m^3/h].

Tube no.	**1**	**2**	**3**	**4**	**5**
Flow rate [m^3/h]	0,1537	0,1522	0,1521	0,1524	0,1525
Re	1830	1812	1811	1814	1815
Tube no.	**6**	**7**	**8**	**9**	**10**
Flow rate [m^3/h]	0,1506	0,1393	0,1530	0,1500	0,1497
Re	1793	1658	1821	1786	1782

Tube no.	**11**	**12**	**13**	**14**	**15**
Flow rate [m^3/h]	0,1500	0,1541	0,1529	0,1577	0,1472
Re	1786	1835	1820	1877	1752
Tube no.	**16**	**17**	**18**	**19**	**20**
Flow rate [m^3/h]	0,1591	0,1569	0,1560	0,1563	0,1525
Re	1894	1868	1857	1861	1815

Measurement no 2 – total volumetric flow rate of $\dot{V} \approx 4$ [m^3/h]:

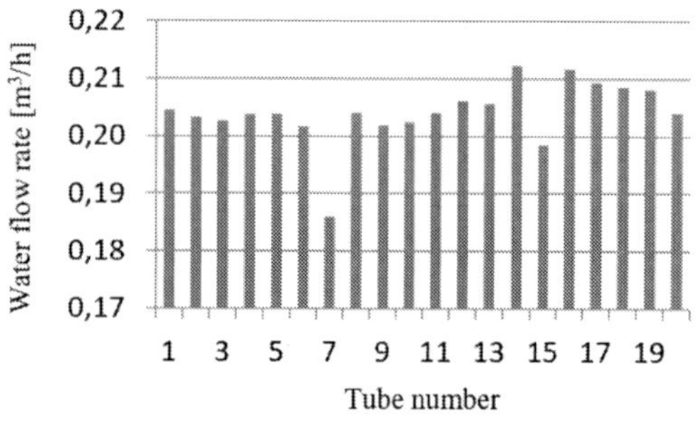

Fig. 7. Volumetric flow rate in heat exchanger tubes for a total flow rate of $\dot{V}$ = 4,0831 [m^3/h].

Table 2. The volumetric flow rate and Reynolds (Re) numbers in heat exchanger tubes for a total flow rate of $\dot{V}$ = 4,0831 [m^3/h].

Tube no.	**1**	**2**	**3**	**4**	**5**
Flow rate [m^3/h]	0,2045	0,2033	0,2027	0,2038	0,2039
Re	2435	2420	2413	2426	2427
Tube no.	**6**	**7**	**8**	**9**	**10**
Flow rate [m^3/h]	0,2017	0,1860	0,2041	0,2019	0,2025
Re	2401	2214	2430	2404	2411
Tube no.	**11**	**12**	**13**	**14**	**15**
Flow rate [m^3/h]	0,2041	0,2062	0,2057	0,2123	0,1985
Re	2430	2455	2449	2527	2363
Tube no.	**16**	**17**	**18**	**19**	**20**
Flow rate [m^3/h]	0,2117	0,2094	0,2086	0,2081	0,2041
Re	2520	2493	2483	2477	2430

Measurement no 3 – total volumetric flow rate of $\dot{V} \approx 5$ [m^3/h]:

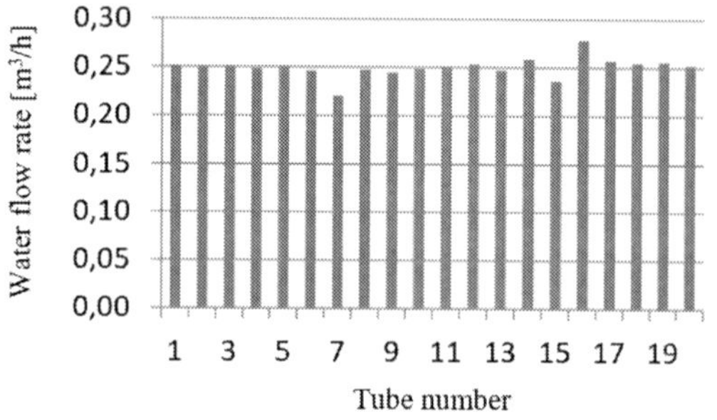

Fig. 8. Volumetric flow rate in heat exchanger tubes for a total flow rate of $\dot{V}$ = 5,0048 [m^3/h].

Table 3. The volumetric flow rate and Reynolds (Re) numbers in heat exchanger tubes for a total flow rate of $\dot{V}$ = 5,0048 [m^3/h].

Tube no.	**1**	**2**	**3**	**4**	**5**
Flow rate [m^3/h]	0,2514	0,2501	0,2506	0,2483	0,2507
Re	2993	2977	2983	2956	2985
Tube no.	**6**	**7**	**8**	**9**	**10**
Flow rate [m^3/h]	0,2465	0,2203	0,2474	0,2446	0,2484
Re	2935	2623	2945	2912	2957
Tube no.	**11**	**12**	**13**	**14**	**15**
Flow rate [m^3/h]	0,2508	0,2540	0,2474	0,2589	0,2362
Re	2986	3024	2945	3082	2812
Tube no.	**16**	**17**	**18**	**19**	**20**
Flow rate [m^3/h]	0,2786	0,2571	0,2548	0,2560	0,2527
Re	3317	3061	3033	3048	3008

4 Fluid flow simulation

CFD simulation are carried out to study the flow distribution in tubular space of the studied heat exchanger. The simulation is performed using the commercial code ANSYS CFX [15]. A geometrical model and a sample mesh is presented in Figure 9. The mesh consists of 1.3 million nodes and 4.47 minion of element. The grid independency study is made and further refinement do not influence the obtained values of volumetric flow rate in heat exchanger tubes.

https://doi.org/10.1051/matecconf/201824002010

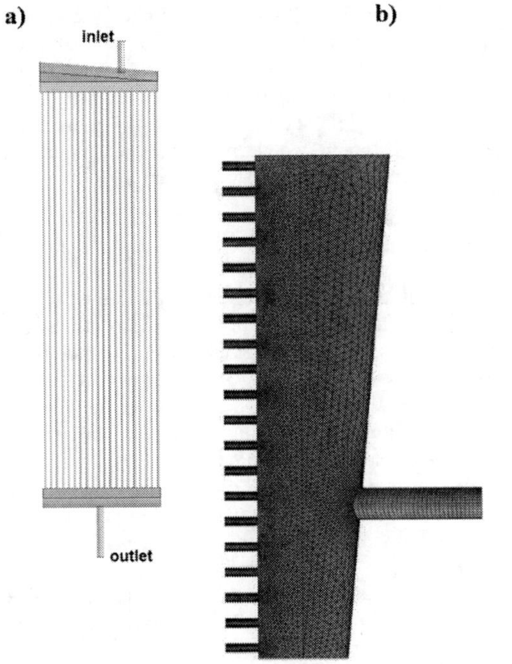

Fig. 9. Geometrical model of the heat exchanger (a) and CFD mesh (b).

As mentioned before the total volumetric flow rate is given based on the total volumetric flow meter indication at the location of inlet nozzle pipe. The inlet boundary condition in CFD simulation is applied at this region. The outflow boundary condition with a given relative pressure valued as 0 is specified at the location of outlet nozzle pipe. The SSG Reynolds Stress Transport turbulence model is employed for simulating the water flow in tubular space and manifolds of heat exchanger. This turbulence model predicts flow separation and secondary flow, therefore it is applied in this simulation. Figure 10 shows the streamline distribution at the location of inlet manifold and within the entire fluid flow domain.

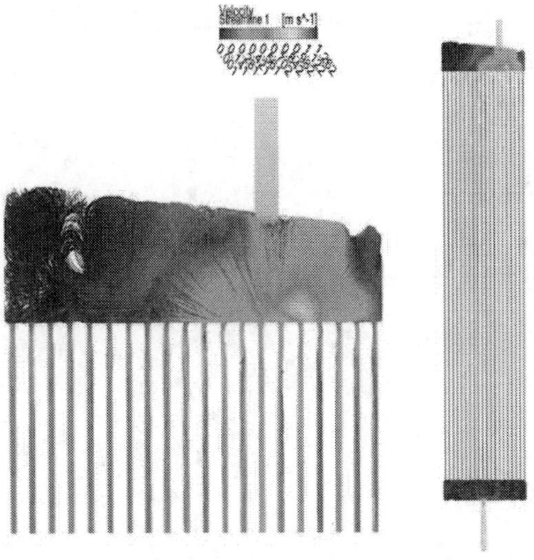

Fig. 10. Velocity streamlines obtained for total volumetric flow rate of $\dot{V}$ = 4,08 [m³/h].

From Figure 10 one can observe the highest flow velocity in tubes located below the inlet nozzle pipe. In other tubes the flow is undisturbed with rater uniform flow rate. The slightly higher value of flow velocity in tubes located downstream the inlet nozzle pipe. The comparison of volumetric flow rates in individual tubes is shown in Figures 11-13.

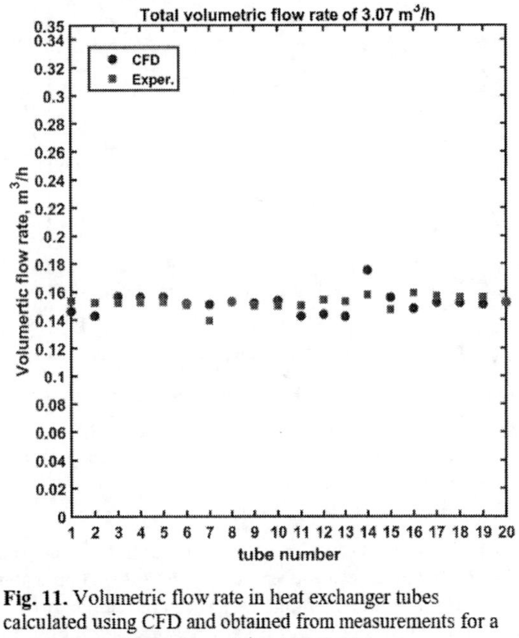

Fig. 11. Volumetric flow rate in heat exchanger tubes calculated using CFD and obtained from measurements for a total volumetric flow rate of $\dot{V}$ = 3,07 [m³/h].

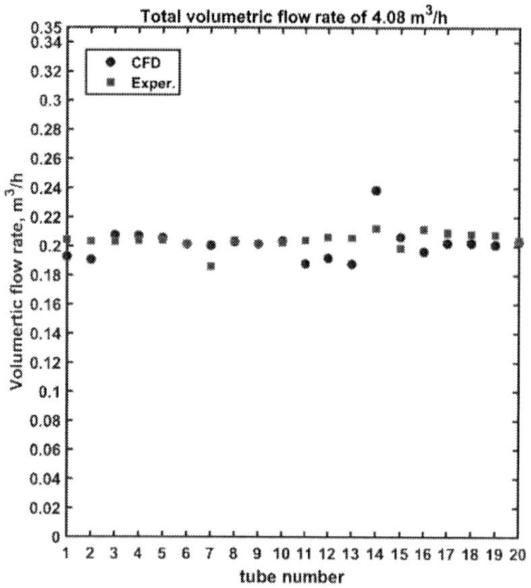

Fig. 12. Volumetric flow rate in heat exchanger tubes calculated using CFD and obtained from measurements for a total volumetric flow rate of $\dot{V}$ = 4,08 [m³/h].

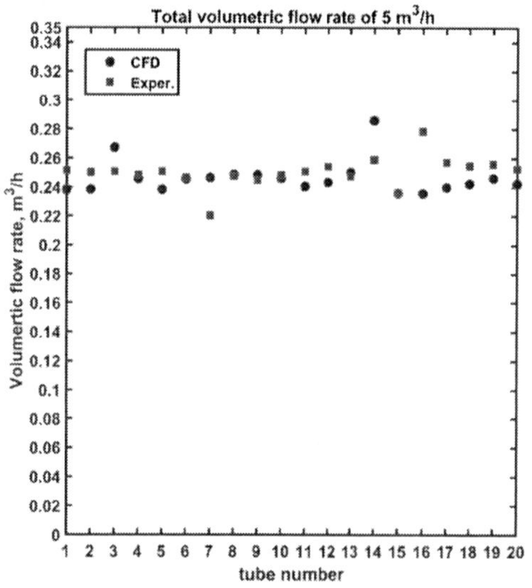

Fig. 13. Volumetric flow rate in heat exchanger tubes calculated using CFD and obtained from measurements for a total volumetric flow rate of $\dot{V}$ = 5 [m³/h].

Figures 11-13 show both the measurements results and CFD simulation results obtained for the case when water flow is laminar and Re number is varied in a range of 1650 to 1900 in heat exchanger tubes (measurement point no. 1, total volumetric flow rate of 3 m^3/h). The CFD simulation results and experimental results are rather close with slight differences observed in a region downstream to inlet nozzle pipe. In this region the applied turbulence model gives the higher value of flow rate in tube no. 14 (located directly below the inlet nozzle pipe), while the lower flow rates are calculated in tubes 12 and 13 nearby the inlet nozzle pipe. The results obtained for measurement point no. 2 are similar. Higher discrepancies between the measurements and CFD simulation are observed for total volumetric flow rate of 5 m^3/h where the measurements indicated the highest flow in tube no 16, while CFD simulation in tube no 14. The maximum percentage difference between measurement results and CFD simulation exceeds 15% in tube no 16.

5 Summary

In this study the CFD simulation are performed to study the flow distribution in a fin-and-tube cross-flow heat exchanger with modified inlet manifold. The tube bundle with staggered arrangement of two rows of tubes is studied. The modification of heat exchanger inlet manifold is done by extending the flow volume through adding the specially designed wedges. The CFD simulations employed SSG Reynolds Stress model that is able to predict the flow separation and secondary flows. The tube Reynolds numbers in a range 1650 – 3300 occurred for the water flow inside heat exchanger tubes. The CFD simulation results (volumetric flow rate in heat exchanger tube) are compared with measurements results. The measurements are made by using the Sontex Superstatic 749 ultrasonic flowmeters installed in each tube of heat exchanger, For Re numbers close to 2000 a good agreement between the measurement results and CFD simulation is found. Higher discrepancies are obtained for tube Re number over 3000, where the CFD simulations are not able to correctly predict the tube location, where the flow is the highest. In this particular case, for total inlet volumetric flow rate of 5 m^3/h the differences between CFD simulation results and measurement results exceeds 15%.

Acknowledgements

The author would like to acknowledge the Polish National Science Center for a financial support. The computational results presented in this paper are the part of the research grant OPUS 6. The project was financed by the Polish National Science Centre awarded based on the decision number DEC-2013/11/B/ST8/00340.

References

1. Łopata S., Ocłoń P. Archives of thermodynamics **31**, 3, pp. 37-53 (2010)
2. Ocłoń P., Łopata S. Technical Transactions. Mechanics **4-M/2011/B**, 7, year 108, pp. 391-400 (2011)
3. Ocłoń P. (PhD Thesis, Kraków ,(2013)
4. Łopata S., Ocłoń P., (Fluid Dynamics, Computational Modeling and Applications, edited by L. Hector Juarez, part 3, chapter **14**, InTech, ISBN: 978-953-51-0052-2, pp. 327 – 356, Croatia 2012)
5. Łopata S., Ocłoń P., Rynek Energii **102**, 5, pp. 112-124 (2012)
6. Łopata, S., Ocłoń, P., Nowak-Ocłoń, M. E3S Web of Conferences, **13**, art. no. 02001 (2017)

7. Ocłoń, P., Łopata, S., Chłosta, K. Procedia Engineering, **157**, pp. 428-435 (2016)
8. Łopata, S., Ocłoń, P. Energy, 92, pp. 100-116. (2015)
9. Ocłoń, P., Łopata, S., Nowak, M. Heat and Mass Transfer/Waerme- und Stoffuebertragung, **51** (4), pp. 553-566. (2015)
10. Ocłoń, P., Nowak, M., Łopata, S. Journal of Thermal Science, **23** (2), art. no. 1003-2169(2014)02-0177-10, pp. 177-186. (2014)
11. Ocłoń, P., Łopata, S., Nowak, M., Benim, A.C. Progress in Computational Fluid Dynamics, **15** (5), pp. 290-306. (2015)
12. Ocłoń, P., Łopata, S. Heat Transfer Engineering, **39** (13-14), pp. 1139-1155. (2018)
13. Taler, D., Ocłoń, P. International Journal of Thermal Sciences, **84**, pp. 309-322. (2014)
14. Taler, D., Ocłoń, P. Chemical Engineering and Processing: Process Intensification, **83**, pp. 1-11. (2014)
15. ANSYS Inc, Ansys CFX Theory Reference Guide, (2016)

Numerical investigation of flow distribution in tubular space of fin-and-tube heat exchanger

Tomasz Stelmach[1]
[1] Institute of Thermal Power Engineering, Cracow University of Technology, Faculty of Mechanical Engineering

Abstract. This paper presents the experimental and numerical investigation of flow distribution in the tubular space of cross-flow fin-and-tube heat exchanger. The tube bundle with two rows arranged in staggered formation is considered. A standard heat exchanged manifold, with inlet nozzle pipe located asymmetrically is considered. The outlet nozzle pipe is located in the middle of the outlet manifold. A developed experimental setup allows one to measure volumetric flow rate in heat exchanger tubes using the ultrasonic flowmeters. The measurement results are then compared with CFD simulation in ANSYS CFX code using the SSG Reynolds Stress turbulence model, and a good agreement is found for tube Re numbers varied from 1800 to 3100.

1 Introduction

Heat exchangers with finned elliptical tubes are widely used in industry, i.e., chemical, petrochemical, energy and automotive industries [1-4]. The main advantage of using those devices is the large heat transfer area in a compact shape. Mostly, the elliptical tubes are used for tubular space of heat exchanger due to the lower pumping power needed from the intertubular space side [13,14]. However, due to the compact shape of fin-and-tube heat exchangers, there may exist a problem of flow maldistribution to heat exchanger tubes [1,5-8]. The reason is the small volume of heat exchanger manifold. The flow maldistribution in heat exchanger tubes may result in high thermal stresses [11,12], as well as flow boiling [9,10] in a region where the flow rate is low. In literature, the flow maldistribution is heat exchangers is not studied extensively. Mostly, the simulation works are performed. However, the number of experimental works is insufficient. Therefore this paper presents an experimental stand for investigation of flow maldistribution in tubular space of fin-and-tube heat exchanger. The experimental results are obtained for various flow rates, set at the inlet tu heat exchanger. The results of experiments are compared with the CFD simulation results to check the correctness of CFD code ANSYS CFX.

2 Experimental setup

The experimental setup consists of the storage tank (1), from which the water is pumped (2) through connection pipes (4) to heat exchanger. The total volumetric flow rate is controlled via the flow meter (3). The inlet manifold (5) is made of steel pipe with an external diameter of 115 mm and a thickness of 6 mm. The inlet nozzle can be positioned in the various location of heat exchanger manifold. In this study, it is assumed that the distance from one side of the manifold to the inlet nozzle pipe axis is 455 mm while 200 mm from the other side. The outlet manifold nozzle pipe is located in the middle of the manifold (8). The tubular space (6) of heat exchanger consists of two rows of elliptical tubes arranged in a staggered formation with 10 tubes per row. The Sontex Superstatic 749 flowmeters (7) are installed in all tubes at the distance of 300 mm before the outlet manifold. The flowmeters allow one to measure the volumetric flow rate in each tube of the heat exchanger. The measurement data is collected through the dedicated data acquisition system.

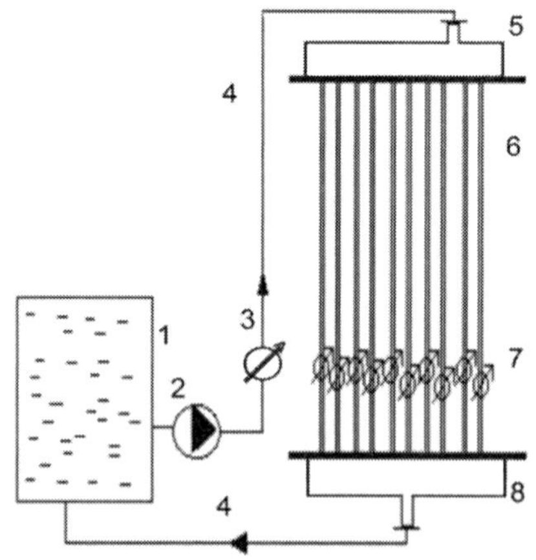

Fig. 1. Scheme of the experimental setup for the investigation of flow conditions in a tubular space of fin-and-tube heat exchanger: 1 - water feed tank, 2 - pump, 3 – flow meter of total volumetric flow rate, 4 - connection pipes, 5 - inlet manifold with a nozzle pipe, 6 – tube bundle, 7 – flow meters for volumetric flow rate in each tube, 8 - outlet manifold with an outlet nozzle pipe.

* Corresponding author: tomasz.stelmach@gkpge.pl

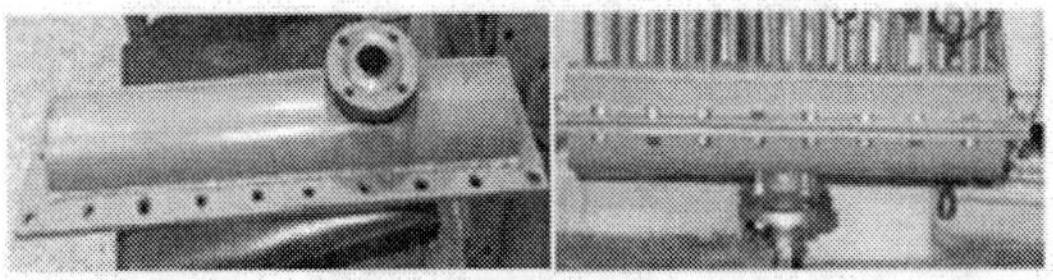

Fig. 2. Inlet and outlet manifold of the tested heat exchanger.

Figure 2 shows the inlet and outlet manifold of the tested heat exchanger. To ensure uniform inflow of the water the straight connecting pipe with the length of 100 mm is installed above and below the inlet and outlet nozzle pipes.

3 Measurements of flow distribution in heat exchanger tubular space

The major geometrical dimensions of the heat exchanger inlet and outlet manifolds are shown in Fig. 3. The nozzle pipe in the tested inlet manifold is located 455 mm from the left side of the inlet manifold, and 200 mm from the right side.

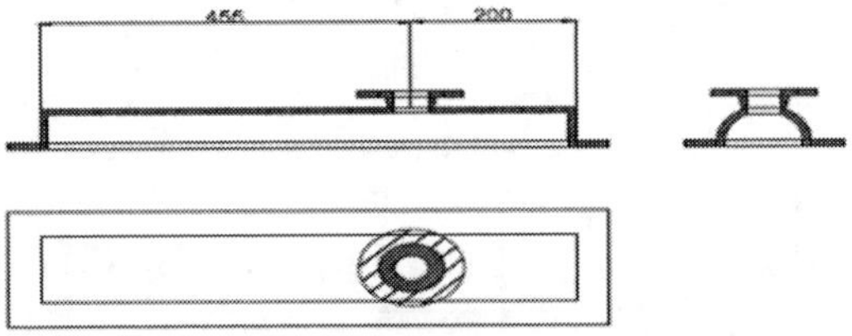

Fig. 3. Dimensions of the inlet manifold.

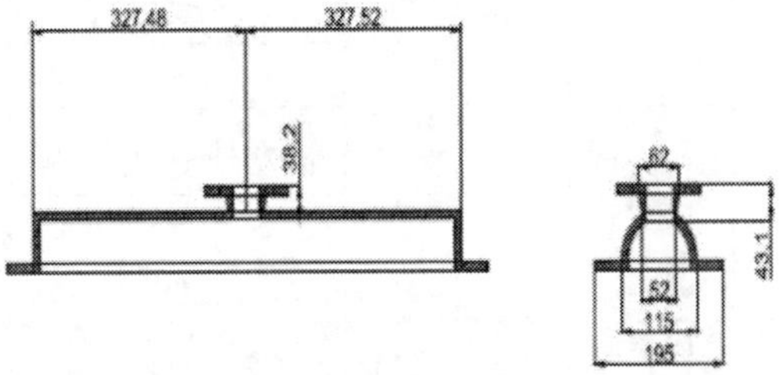

Fig. 4. Dimensions of the outlet manifold.

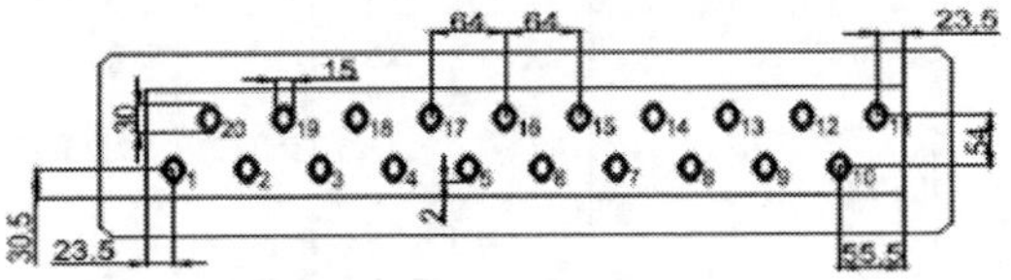

Fig. 5. Scheme of the sieve plate with given tube numbering.

In this study, the measurements are performed for the total volumetric flow rate in a rage from 3 m^3/h to 5 m^3/h. That volumetric flow range corresponds to the transitional and turbulent flow regime. The results are presented for the tubes in the first row of tube bundle 1-10 and the second row of tube bundle 11-20. The water temperature is constant and equal to 20 °C. The readings from the flowmeters were made after the steady-state conditions were achieved.

Measurement no 1 – total volumetric flow rate of $\dot{V} \approx 3$ [m^3/h]:

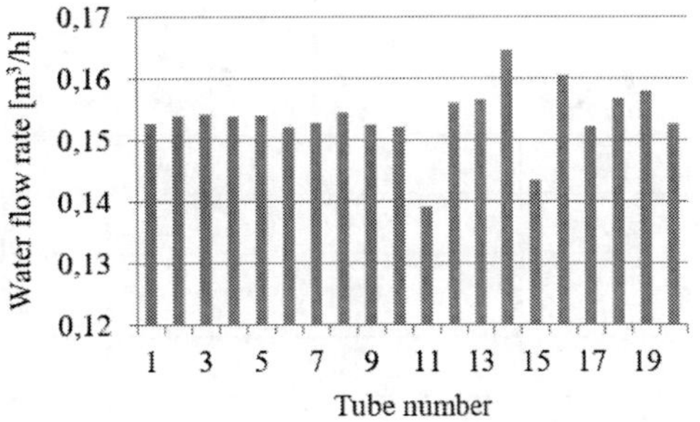

Fig. 6. Volumetric flow rate in heat exchanger tubes for a total flow rate of $\dot{V}$ = 3,0726 [m^3/h].

Table 1. The volumetric flow rate and Reynolds (Re) numbers in heat exchanger tubes for a total flow rate of $\dot{V}$ = 3,0726 [m^3/h].

Tube no.	**1**	**2**	**3**	**4**	**5**
Flow rate [m^3/h]	0,1527	0,1539	0,1542	0,1539	0,1540
Re	1818	1832	1836	1832	1833
Tube no.	**6**	**7**	**8**	**9**	**10**
Flow rate [m^3/h]	0,1521	0,1528	0,1545	0,1525	0,1521
Re	1811	1819	1839	1815	1811
Tube no.	**11**	**12**	**13**	**14**	**15**
Flow rate [m^3/h]	0,1392	0,1560	0,1566	0,1646	0,1435
Re	1657	1857	1864	1960	1708
Tube no.	**16**	**17**	**18**	**19**	**20**
Flow rate [m^3/h]	0,1605	0,1522	0,1568	0,1579	0,1526
Re	1911	1812	1867	1880	1817

Measurement no 2 – total volumetric flow rate of $\dot{V} \approx 3{,}5$ [m^3/h]:

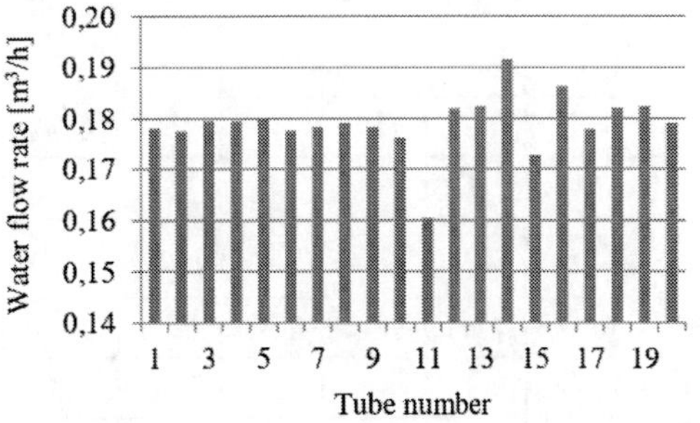

Fig. 7. Volumetric flow rate in heat exchanger tubes for total flow rate of $\dot{V}$ = 3,5811 [m^3/h].

Table 2. The volumetric flow rate and Reynolds (Re) numbers in heat exchanger tubes for a total flow rate of $\dot{V}$ = 3,58 [m³/h].

Tube no.	**1**	**2**	**3**	**4**	**5**
Flow rate [m³/h]	0,1782	0,1775	0,1795	0,1795	0,1799
Re	2121	2113	2137	2137	2142
Tube no.	**6**	**7**	**8**	**9**	**10**
Flow rate [m³/h]	0,1776	0,1783	0,1791	0,1783	0,1762
Re	2114	2123	2132	2123	2098
Tube no.	**11**	**12**	**13**	**14**	**15**
Flow rate [m³/h]	0,1605	0,1819	0,1824	0,1916	0,1728
Re	1911	2165	2171	2281	2057
Tube no.	**16**	**17**	**18**	**19**	**20**
Flow rate [m³/h]	0,1863	0,1779	0,1821	0,1824	0,1791
Re	2218	2118	2168	2171	2132

Measurement no 3 – total volumetric flow rate of $\dot{V} \approx 4$ [m³/h]:

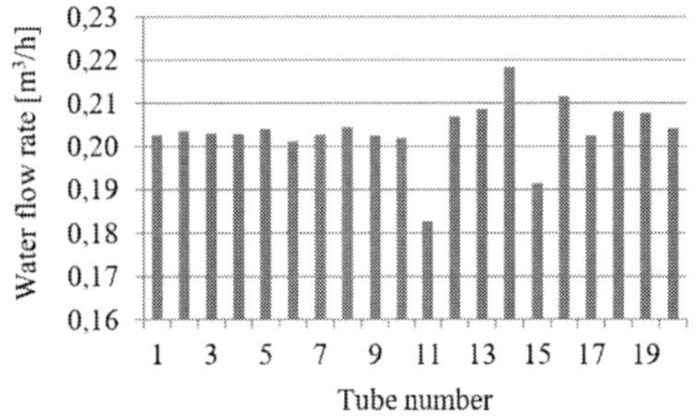

Fig. 8. Volumetric flow rate in heat exchanger tubes for a total flow rate of $\dot{V}$ = 4,0712 [m³/h].

Table 3. The volumetric flow rate and Reynolds (Re) numbers in heat exchanger tubes for a total flow rate of $\dot{V}$ = 4,07 [m³/h].

Tube no.	**1**	**2**	**3**	**4**	**5**
Flow rate [m³/h]	0,2027	0,2035	0,2030	0,2029	0,2041
Re	2413	2423	2417	2415	2430
Tube no.	**6**	**7**	**8**	**9**	**10**
Flow rate [m³/h]	0,2012	0,2028	0,2045	0,2025	0,2019
Re	2395	2414	2435	2411	2404
Tube no.	**11**	**12**	**13**	**14**	**15**
Flow rate [m³/h]	0,1827	0,2069	0,2086	0,2184	0,1915
Re	2175	2463	2483	2600	2280
Tube no.	**16**	**17**	**18**	**19**	**20**
Flow rate [m³/h]	0,2116	0,2025	0,2080	0,2077	0,2042
Re	2519	2411	2476	2473	2431

Measurement no 4 – total volumetric flow rate of $\dot{V} \approx 4,5$ [m³/h]:

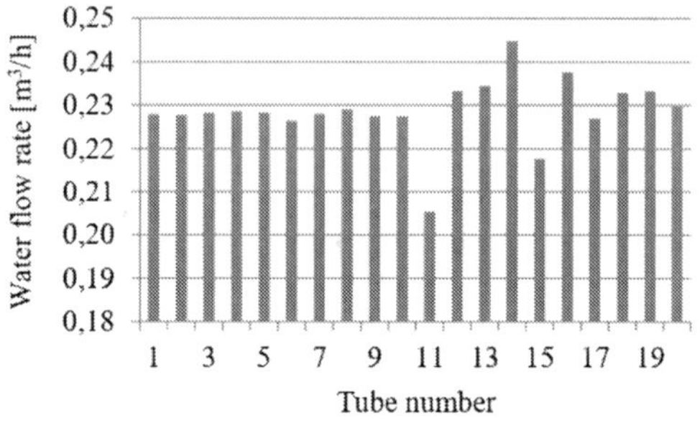

Fig. 9. Volumetric flow rate in heat exchanger tubes for total flow rate of $\dot{V}$ = 4,5760 [m³/h].

Table 4. The volumetric flow rate and Reynolds (Re) numbers in heat exchanger tubes for a total flow rate of $\dot{V}$ = 4,5760 [m³/h].

Tube no.	**1**	**2**	**3**	**4**	**5**
Flow rate [m³/h]	0,2280	0,2277	0,2283	0,2286	0,2284
Re	2714	2711	2718	2721	2719
Tube no.	**6**	**7**	**8**	**9**	**10**
Flow rate [m³/h]	0,2264	0,2281	0,2290	0,2275	0,2275
Re	2695	2715	2726	2708	2708
Tube no.	**11**	**12**	**13**	**14**	**15**
Flow rate [m³/h]	0,2054	0,2333	0,2345	0,2449	0,2178
Re	2445	2777	2792	2915	2593
Tube no.	**16**	**17**	**18**	**19**	**20**
Flow rate [m³/h]	0,2376	0,2270	0,2329	0,2333	0,2298
Re	2829	2702	2773	2777	2736

Measurement no 5 – total volumetric flow rate of $\dot{V} \approx 5$ [m³/h]:

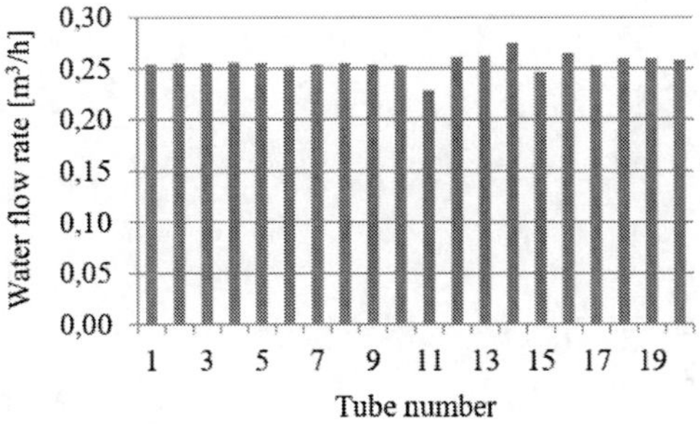

Fig. 10. The volumetric flow rate and Reynolds (Re) numbers in heat exchanger tubes for a total flow rate of $\dot{V}$ = 5,1073 [m³/h].

Table 5. The volumetric flow rate and Reynolds (Re) numbers in heat exchanger tubes for a total flow rate of $\dot{V}$ = 5,1073 [m³/h].

Tube no.	**1**	**2**	**3**	**4**	**5**
Flow rate [m³/h]	0,2541	0,2547	0,2549	0,2556	0,2552
Re	3025	3032	3035	3043	3038
Tube no.	**6**	**7**	**8**	**9**	**10**
Flow rate [m³/h]	0,2518	0,2537	0,2551	0,2536	0,2524
Re	2998	3020	3037	3019	3005
Tube no.	**11**	**12**	**13**	**14**	**15**
Flow rate [m³/h]	0,2284	0,2610	0,2615	0,2745	0,2461
Re	2719	3107	3113	3268	2930
Tube no.	**16**	**17**	**18**	**19**	**20**
Flow rate [m³/h]	0,2648	0,2526	0,2595	0,2596	0,2582
Re	3152	3007	3089	3090	3074

4 CFD simulation

CFD simulation are carried out to study the flow maldistribution in the fin-and-tube heat exchanger. The simulation is performed using the commercial software ANSYS CFX [15]. A sample mesh is shown in Figure 11.

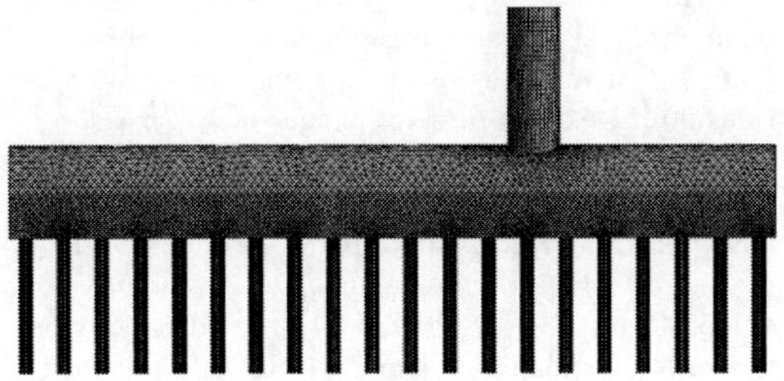

Fig. 11. Mesh used in CFD simulation.

A total volumetric flow rate is specified at the location of inlet nozzle pipe as inlet boundary condition in CFD simulation. At the location of outlet nozzle pipe, the outflow boundary condition with relative pressure set as 0 is given. The SSG Reynolds Stress Transport turbulence model is used to simulate the fluid flow of water in heat exchanger. The model allows to simulate the flow field with flow separation and secondary flows. A sample streamlines obtained for measurement no 1, are shown in Fig. 12.

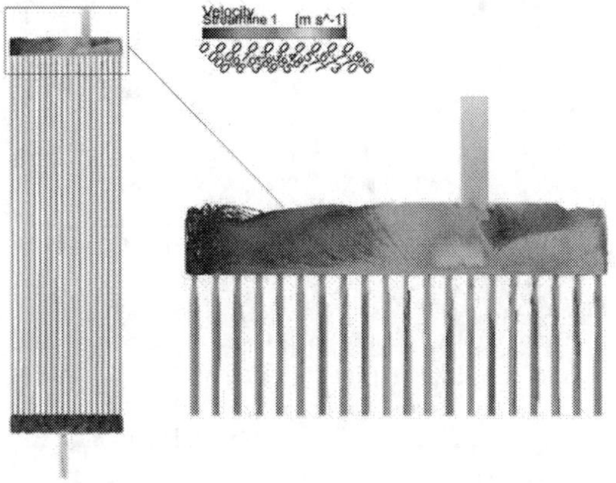

Fig. 12. Velocity streamlines obtained for total volumetric flow rate of $\dot{V}$ = 3,0726 [m³/h].

Figure 12 reveals the slightly higher value of flow velocity in tubes located downstream the inlet nozzle pipe. The comparison of volumetric flow rates in individual tubes is shown in Figures 13-17.

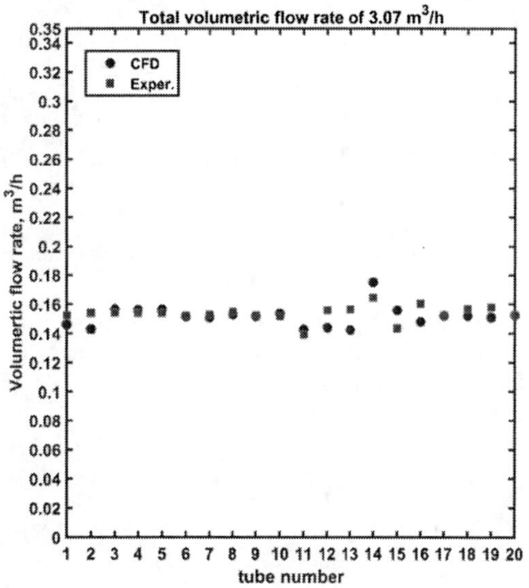

Fig. 13. Volumetric flow rate in heat exchanger tubes calculated using CFD and obtained from measurements for a total volumetric flow rate of $\dot{V}$ = 3,07 [m³/h].

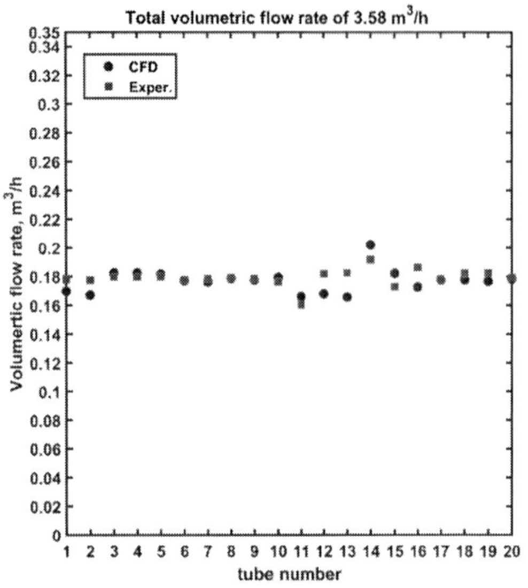

Fig. 14. Volumetric flow rate in heat exchanger tubes calculated using CFD and obtained from measurements for a total volumetric flow rate of $\dot{V}$ = 3,58 [m³/h].

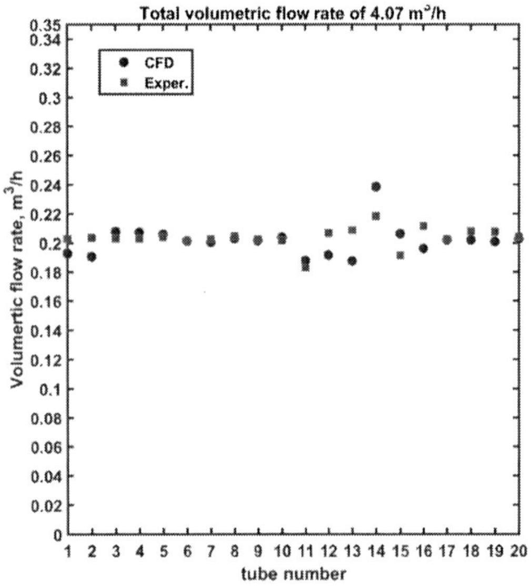

Fig. 15. Volumetric flow rate in heat exchanger tubes calculated using CFD and obtained from measurements for a total volumetric flow rate of $\dot{V}$ = 4,07 [m³/h].

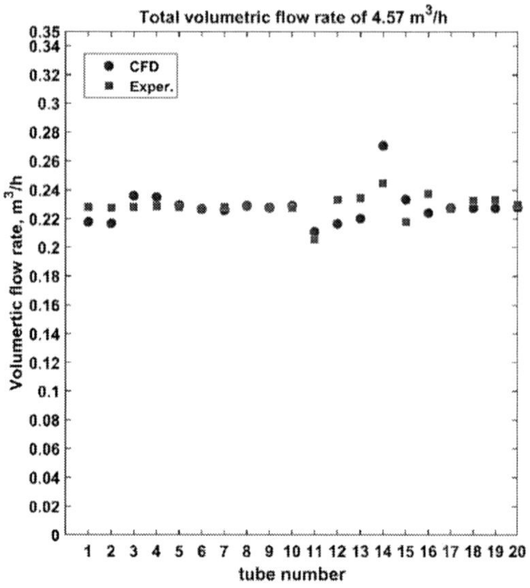

Fig. 16. Volumetric flow rate in heat exchanger tubes calculated using CFD and obtained from measurements for a total volumetric flow rate of $\dot{V}$ = 4,57 [m³/h].

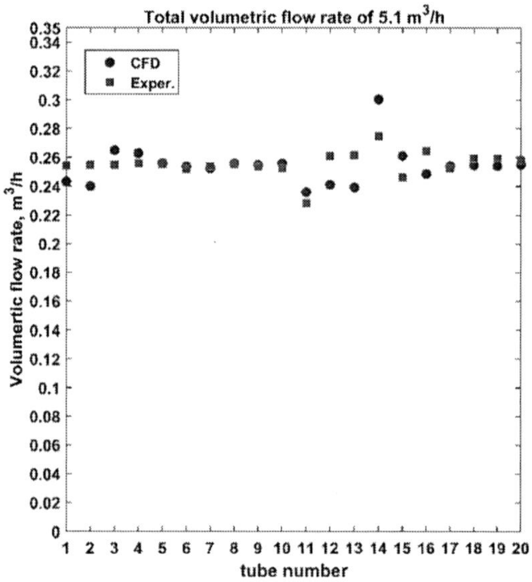

Fig. 17. Volumetric flow rate in heat exchanger tubes calculated using CFD and obtained from measurements for a total volumetric flow rate of $\dot{V}$ = 5,1 [m³/h].

Figure 13 shows the comparison between CFD simulation results and experimental data obtained for laminar flow regime where Re number varies in a range of 1600 to 1800 in heat exchanger tubes (measurement point no. 1). The CFD simulation results and experimental results are in good agreement; slight discrepancies are observed in a region downstream the inlet nozzle pipe, where the CFD simulation predicts the slightly higher value of flow rate in tube no. 14 (located directly below the inlet nozzle pipe), while the lower flow rates are calculated in tubes 12 and 13 nearby the inlet nozzle pipe.

The similar situation occurs for other measurement points (Figs. 14-17), even for turbulent flow (measurement no. 5) where Re number exceeds 3000. Therefore, for tube Re number from 1800-3100, the SSG Reynolds Stress Transport turbulence model provides satisfactory agreement with the measurement data.

5 Conclusions

This paper presents the experimental and numerical investigation of flow distribution in a single pass, double row tubular space of fin-and-tube heat exchanger. The elliptical tube bundle with tubes arranged in staggered formation is examined. The stand installed in the Institute of Thermal Power Engineering of the Cracow University of Technology allows one to measure a volumetric flow rate in all tubes of the heat exchanger. The Sontax Superstatic 749 ultrasonic flow meters are used for flow rate measurements. The measurements were performed for various inlet flow rates, so that the Re number in tubes was examined in a range of 1800 to 3100. In CFD simulation the SSG Reynolds Stress Transport turbulence model was used, and provided satisfactory agreement with the experimental data. The highest difference in measured and predicted flow rate occurs in a tube located downstream the inlet nozzle pipe, where the CFD simulation overpredicts the value of flow rate. In tubes located far from the inlet nozzle pipe, the measurement results and simulation results are in very good agreement.

Acknowledgement

The author would like to acknowledge the Polish National Science Center for a financial support. The computational results presented in this paper are the part of the research grant OPUS 6. The project was financed by the Polish National Science Centre awarded based on the decision number DEC-2013/11/B/ST8/00340.

References

1. Łopata S., Ocłoń P. Archives of thermodynamics **31**, 3, pp. 37-53 (2010)
2. Ocłoń P., Łopata S. Technical Transactions. Mechanics **4-M/2011/B**, 7, year 108, pp. 391-400 (2011)
3. Ocłoń P. (PhD Thesis, Kraków ,(2013)
4. Łopata S., Ocłoń P., (Fluid Dynamics, Computational Modeling and Applications, edited by L. Hector Juarez, part 3, chapter **14**, InTech, ISBN: 978-953-51-0052-2, pp. 327 – 356, Croatia 2012)
5. Łopata S., Ocłoń P., Rynek Energii **102**, 5, pp. 112-124 (2012)
6. Łopata, S., Ocłoń, P., Nowak-Ocłoń, M. E3S Web of Conferences, **13**, art. no. 02001 (2017)
7. Ocłoń, P., Łopata, S., Chłosta, K. Procedia Engineering, **157**, pp. 428-435 (2016)
8. Łopata, S., Ocłoń, P. Energy, 92, pp. 100-116. (2015)
9. Ocłoń, P., Łopata, S., Nowak, M. Heat and Mass Transfer/Waerme- und Stoffuebertragung, **51** (4), pp. 553-566. (2015)
10. Ocłoń, P., Nowak, M., Łopata, S. Journal of Thermal Science, **23** (2), art. no. 1003-2169(2014)02-0177-10, pp. 177-186. (2014)
11. Ocłoń, P., Łopata, S., Nowak, M., Benim, A.C. Progress in Computational Fluid Dynamics, **15** (5), pp. 290-306. (2015)
12. Ocłoń, P., Łopata, S. Heat Transfer Engineering, **39** (13-14), pp. 1139-1155. (2018)
13. Taler, D., Ocłoń, P. International Journal of Thermal Sciences, **84**, pp. 309-322. (2014)
14. Taler, D., Ocłoń, P. Chemical Engineering and Processing: Process Intensification, **83**, pp. 1-11. (2014)
15. ANSYS Inc, Ansys CFX Theory Reference Guide, 2016

Performance of Air-Cooled Heat Exchanger with Laminar, Transitional, and Turbulent Tube Flow

Dawid Taler*
Cracow University of Technology, Faculty of Environmental Engineering, Department of Thermal Processes, Air Protection and Waste Utilization, 31-155 Cracow, Poland

Abstract. Some air-cooled heat exchangers, especially in air conditioning and heating installations, heat pumps, as well as car radiators, work in a wide range of loads when the liquid flow in the tubes can be laminar, transitional or turbulent. In this paper, a semi-empirical and empirical relationship for the Nusselt number on the liquid-side in the transitional and turbulent range was derived. The friction factor in the transition flow range $Re_{w,trb} \le Re_w \le Re_{w,tre}$ was calculated by linear interpolation between the values of the friction factor for $Re_{w,trb} = 2{,}100$ and $Re_{w,tre} = 3{,}000$. Based on experimental data for a car radiator, empirical heat transfer relationships for the air and water-side were found by using the least squares method. The water temperature at the outlet of the heat exchanger was calculated using P-NTU (effectiveness-number of transfer units) method. The heat flow rate from water to air was calculated as a function of the water flow rate to compare it with the experimental results. The theoretical and empirical correlation for the water-side Nusselt number developed in the paper were used when determining the heat flow rate. The calculation results agree very well with the results of the measurements.

1 Introduction

Heat transfer correlations for calculating heat transfer coefficients on both air and water side are needed in the thermal calculations of heat exchangers using engineering methods such as LMTD, ε-NTU, and P-NTU [1-4]. Turbulent flow regime of the fluid in tubes is usually assumed in the design and performance calculations of the plate fin and tube heat exchangers (PFTHEs). Usually, the power-type heat transfer correlations are applied for calculating the water and air-side heat transfer coefficients [1-6]. The most popular heat transfer correlations on the water –side are the well-known relationships of Dittus-Boelter and Sieder and Tate. The Gnielinski [7] relationship is also frequently used. These correlations are appropriate only for high Reynolds numbers when the Re_w is larger than 4000. Recently, Taler [8-9] proposed a new correlation to calculate the Nusselt number for the transitional and turbulent fluid flow inside the tubes at a given tube surface temperature or constant heat flux. In this paper, theoretical and empirical relationships were found. The tube-side heat transfer correlations proposed in [8, 9] were generalized to account that transitional flow begins at $Re_{w,trb}$ and ends at $Re_{w,tre}$. It was assumed in [8, 9] that $Re_{w,trb}$ = 2300 and $Re_{w,tre}$ = 3000. However, the experimental results obtained by Ghajar et al. [10] and Meyer and co-workers [11] showed that the values of $Re_{w,trb}$ and $Re_{w,tre}$ can differ from the values 2300 and 3000, respectively. Air and water-side heat transfer correlations for a PFTHE made of circular tubes were determined by the least squares method using 70 experimental data sets obtained in transitional and turbulent flow regime. To calculate the air and water temperature at the outlet of the heat exchanger, the P-NTU method was used. Empirical water and air side correlations were determined based on 70 measurement series including transitional and turbulent flow regimes. Experimental tests were carried out on the new-design car radiator made of circular tubes. The water-side friction factor was linearly interpolated between ξ_w = $64/Re_w$ for laminar flow at $Re_w = Re_{w,trb}$ and the friction factor for the turbulent flow $\xi_w(Re_{w,trb})$ both in a theoretical and empirical relationship for the Nusselt number. The water-side friction factor in the turbulent flow regime, when the Reynolds number is higher than $Re_{w,trb}$ was determined using explicit formula recently proposed by Taler [12]. The Gnielinski formulas [7] was used to calculate the heat transfer coefficient for the laminar flow of the liquid in heat exchanger tubes. The heat flow rate transferred in the studied heat exchanger from the hot water to the air was calculated by the P-NTU method in which new heat transfer correlations were used. It was shown that computed and experimentally determined heat flow rates are in good agreement.

* Corresponding author: dtaler@pk.edu.pl

2 Performance calculations of the investigated PFTHE

Performance calculations of the investigated PFTHE, i.e., determining the temperature of the water and air at the outlets of the investigated heat exchanger was conducted using the P-NTU method.

a)

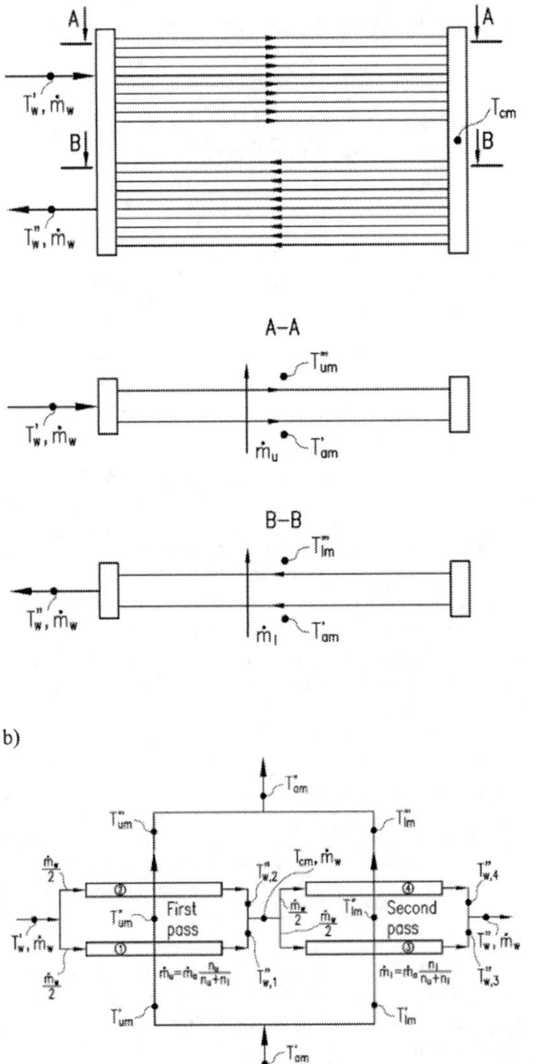

Fig. 1. Flow arrangement of the two-row car radiator with two passes; a) front and top view, b) flow system; 1 – the first tube row in the first (upper) pass, 2 – the second tube row in the first (upper) pass; 3 – the first tube row in the second (lower) pass, 4 – the second tube row in the the second (lower) pass.

Thermo-hydraulic tests of air-cooled plate fin-and-tube heat exchanger were carried out in a laminar, transition and turbulent flow regime of water in the tubes. The tested PFTHE was a car radiator for a spark engine with a displacement volume of 1,600 cm^3. Two-pass car radiator has two rows of round tubes. There are twenty tubes in the first pass, ten in each row, i.e. n_u = 10. The number of tubes in the second pass is smaller and is eighteen, nine in each row, i.e. n_l = 9. The tube arrangement is in-line. Water flows in parallel through the first and second row of tubes in both first and second pass. Between the first and second pass, there is a chamber in which water from the first and second row of tubes is mixed and then flows into the second pass. The flow arrangement of an automobile radiator is shown in Fig. 1. The internal diameter of the tube with the wall thickness δ_t = 0.5 mm is d_{in} = 6.2 mm. The transverse pitch of the tube arrangement is equal to p_1 = 18.5 mm, and the longitudinal pitch is p_2 = 10 mm. The plate fins with a thickness of 0.08 mm are set on tubes with the pitch of 1.5 mm. Dimensions of the heat exchanger are as follows: length L_{ch} = 520 mm, height HL_{ch} = 359 mm, and thickness $W_{ch} = 2p_2$ = 24 mm. The mean water T_{cm} and air temperature T'''_{um} after the first pass and after the second pass T''_w and T'''_{lm} (Fig. 1) were calculated using the P-NTU (Effectiveness - Number of Transfer Units) method.

The mean water T_{cm} and air temperature T'''_{um} after the first pass and after the second pass T''_w and T'''_{lm} (Fig. 1) were calculated using the P-NTU (Effectiveness - Number of Transfer Units) method.

Formulas for calculating the water-side effectiveness P_w and air-side effectiveness P_a were derived based on the mathematical model of the heat exchanger developed by Taler [13].

The numbers of heat transfer units for water NTU$_w$ and air NTU$_a$ are defined in relation to one tube located in the first (upper) or second (lower) pass. The numbers of transfer units are as follows

- first (upper) pass

$$N_{a,u} = \frac{U_{o,u}\,A_o}{\dot{m}_{a,t}\,\bar{c}_{p,a}}, \quad N_{w,u} = \frac{U_{o,u}\,A_o}{\dot{m}^u_{w,t}\,\bar{c}_{p,w}} \tag{1}$$

- second (lower) pass

$$N_{a,l} = \frac{U_{o,l}\,A_o}{\dot{m}_{a,t}\,\bar{c}_{p,a}} \quad N_{w,l} = \frac{U_{o,l}\,A_o}{\dot{m}^l_{w,t}\,\bar{c}_{p,w}} \tag{2}$$

The mass flow rate $\dot{m}_{a,t} = w_0\,p_2\,L_{ch}\,\rho_a\left(T'_{am}\right)$ in Eq. (1) and (2) denotes the air mass flow rate per one pitch p_2 (through the cross-section p_2L_{ch}). Similarly, the water mass flow rates $\dot{m}^u_{w,t}$ and $\dot{m}^l_{w,t}$ per one tube are $\dot{m}^u_{w,t} = \dfrac{\dot{m}_w}{2n_u}$ and $\dot{m}^l_{w,t} = \dfrac{\dot{m}_w}{2n_l}$, respectively.

The numbers of heat transfer units in the first pass for air and water are defined by

$$P_{a,u} = \frac{T'''_{um} - T'_{am}}{T'_w - T'_{am}} \qquad P_{w,u} = \frac{T_{cm} - T'_w}{T'_w - T'_{am}} \tag{3}$$

The following formulas were obtained using the mathematical model of the exchanger presented in the paper [13]

$$P_{a,u}=\left(1-e^{-N_{a,u}}\right)^2\left(\frac{1-e^{-B_u}}{B_u}-e^{-B_u}\right)+ \\ +\left(1-e^{-2N_{a,u}}\right)\frac{1-e^{-B_u}}{B_u} \quad (4)$$

$$P_{w,u}=1-e^{-B_u}-\frac{N_{w,u}}{2N_{a,u}}\left(1-e^{-N_{a,u}}\right)^2 e^{-B_u} \quad (5)$$

where the symbol B_u denotes

$$B_u=\frac{N_{w,u}}{N_{a,u}}\left(1-e^{-N_{a,u}}\right) \quad (6)$$

The water temperature T_{cm} at the outlet from the first pass as well as the average air temperature T'''_{um} after the first pass can be determined using only the formula (5) for water-side effectiveness $P_{w,u}$. The temperature T_{cm} can be obtained from the definition (3)

$$T_{cm}=T'_w-P_{w,u}\left(T'_w-T'_{am}\right) \quad (7)$$

where $P_{w,u}$ is calculated using the relationship (5).

Solving the heat balance equation for the upper pass

$$\dot{m}_w\bar{c}_w\left(T'_w-T_{cm}\right)=\dot{m}_u\bar{c}_{p,a}\left(T'''_{um}-T'_{am}\right) \quad (8)$$

gives the air temperature T'''_{um}

$$T'''_{um}=T'_{am}+\frac{\dot{m}_w\bar{c}_w}{\dot{m}_u\bar{c}_{p,a}}\left(T'_w-T_{cm}\right) \quad (9)$$

The air mass flow rate $\dot{m}_u$ through the upper heat exchanger pass is

$$\dot{m}_u=\dot{m}_a\frac{n_u}{n_u+n_l} \quad (10)$$

With the known temperature T_{cm}, the liquid temperature T''_w at the heat exchanger outlet and the air mean temperature T'''_{lm} behind the second pass can be determined similarly. The air-side and water side-effectiveness of the second pass is defined as

$$P_{a,l}=\frac{T'''_{lm}-T'_{am}}{T_{cm}-T'_{am}} \qquad P_{w,l}=\frac{T_{cm}-T''_w}{T_{cm}-T'_{am}} \quad (11)$$

The relationships for $P_{a,l}$ and $P_{w,l}$ are as follows

$$P_{a,l}=\left(1-e^{-N_{a,l}}\right)^2\left(\frac{1-e^{-B_l}}{B_l}-e^{-B_l}\right)+ \\ +\left(1-e^{-2N_{a,l}}\right)\frac{1-e^{-B_l}}{B_l} \quad (12)$$

$$P_{w,l}=1-e^{-B_l}-\frac{N_{w,l}}{2N_{a,l}}\left(1-e^{-N_{a,l}}\right)^2 e^{-B_l} \quad (13)$$

where the symbol B_l denotes

$$B_l=\frac{N_{w,l}}{N_{a,l}}\left(1-e^{-N_{a,l}}\right) \quad (14)$$

The water temperature T''_w at the outlet from the second pass and the average air temperature T'''_{lm} after the second pass can be determined using the relationships (11) for the air-side effectiveness $P_{a,l}$ and water-side effectiveness $P_{w,l}$. The water temperature T''_w at the outlet from the second pass is

$$T''_w=T_{cm}-P_{w,l}\left(T_{cm}-T'_{am}\right) \quad (15)$$

where $P_{w,l}$ is given by Eq. (13).

Solving the heat balance equation for the second pass

$$\dot{m}_w\bar{c}_w\left(T_{cm}-T''_w\right)=\dot{m}_l\bar{c}_{p,a}\left(T'''_{ul}-T'_{am}\right) \quad (16)$$

gives the mean air temperature T'''_{lm} after the second pass

$$T'''_{lm}=T'_{am}+\frac{\dot{m}_w\bar{c}_w}{\dot{m}_l\bar{c}_{p,a}}\left(T_{cm}-T''_w\right) \quad (17)$$

The air mass flow rate $\dot{m}_l$ through the lower (second) heat exchanger pass is

$$\dot{m}_l=\dot{m}_a\frac{n_l}{n_u+n_l} \quad (18)$$

The mean temperature of the air T''_{am} after the heat exchanger is calculated from the formula

$$T''_{am}=\frac{n_u}{n_u+n_l}T'''_{um}+\frac{n_l}{n_u+n_l}T'''_{lm} \quad (19)$$

If the temperature of the water T''_w at the exit from the heat exchanger is known, then the temperature of the air after the heat exchanger T''_{am} can also be determined from the heat balance equation for the exchanger (Fig. 1)

$$\dot{m}_w\bar{c}_w\left(T'_w-T''_w\right)=\dot{m}_a\bar{c}_{pa}\left(T''_{am}-T'_{am}\right) \quad (20)$$

Solving Eq. (20) for T''_{am} gives

$$T''_{am}=T'_{am}+\frac{\dot{m}_w\bar{c}_w}{\dot{m}_a\bar{c}_{pa}}\left(T'_w-T''_w\right) \quad (21)$$

The presented procedure for calculating the water and air outlet temperature based on the P-NTU method can also be applied to other types of heat exchangers with different flow systems. Many relationships for calculating the effectiveness P as a function of the number of transfer units NTU can be found in the books on the heat exchangers [1-4].

Overall heat transfer coefficients $h_{o,l}$ and $h_{o,u}$ referred to the outer surface area of the bare tube for the lower and upper pass are given by

$$\frac{1}{U_{o,l}}=\frac{1}{h_{w,l}}\frac{r_o}{r_{in}}+\frac{r_o}{k_t}\ln\left(\frac{r_o}{r_{in}}\right)+\frac{1}{h_o} \quad (22)$$

$$\frac{1}{U_{o,u}}=\frac{1}{h_{w,u}}\frac{r_o}{r_{in}}+\frac{r_o}{k_t}\ln\left(\frac{r_o}{r_{in}}\right)+\frac{1}{h_o} \quad (23)$$

The effective heat transfer coefficient h_o taking into account the presence of fins, based on the surface area of the bare tube was determined using the following relationship

$$h_o=\left(\frac{A_{bf}}{A_o}+\frac{A_f}{A_o}\eta_f\right)h_a \quad (24)$$

The continuous fin was divided into 38 rectangular fins due to the symmetry of the temperature field in the continuous fin. The fin efficiency η_f was determined assuming that the air-side heat transfer coefficient h_a is constant. With this assumption, the fin efficiency η_f can be determined using the formula

$$\eta_f=\frac{\bar{T}_{fin}-T_f}{T_b-T_f} \quad (25)$$

where $\overline{T}_{fin}$ is the mean temperature of the fin surface.

The surface temperature of the fin T_{fin} and the average surface temperature of the $\overline{T}_{fin}$ were calculated by the finite element method (FEM) using the software ANSYS v.16. The thermal conductivity of the aluminum fin was k_{fin} = 207 W/(m²·K). The calculations were carried out for the air-side heat transfer coefficient h_a varying from 5 W/(m²·K) to 300 W/(m²·K) with a 5 W/(m²·K) step. The temperature of the fin base T_b was assumed to be 100°C. Temperature distributions on the fin surface for heat transfer coefficients of 5, 50, 100 and 300 W/(m²·K) are depicted in Fig. 2. The analysis of the results presented in Figure 3 shows that the temperature of the fin decreases with the increase in the air-side heat transfer coefficient h_a. The calculated fin efficiencies η_f for different heat transfer coefficients h_a using the ANSYS 16 were approximated using the least squares method.

a)

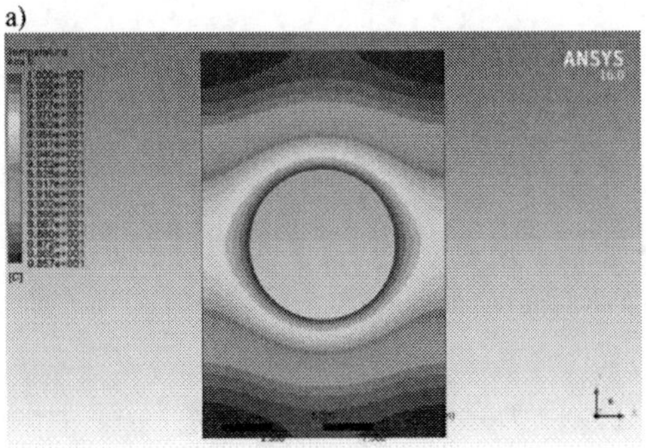

b)

c)

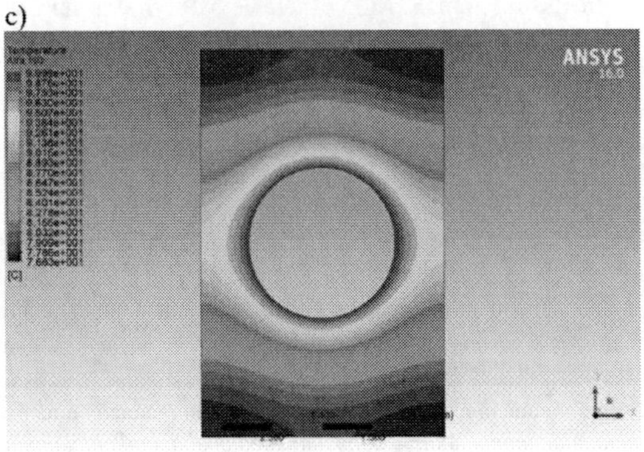

d)

Fig. 2. Temperature distributions on the fin surface for various heat transfer coefficients; a) h_a = 5 W/(m²·K), b) h_a = 50 W/(m²·K), c) h_a = 100 W/(m²·K), d) h_a = 300 W/(m²·K).

The function

$$\eta_f = \frac{0.999882 + 0.0003515\,h_a}{1 + 0.0021342\,h_a} \qquad (26)$$

$$0 \le h_a \le 300 \text{ W/(m}^2\text{K)}$$

When operating the heat exchanger in a wide range of loads, the flow regime inside the tubes can be laminar, transitional or turbulent.

The heat transfer coefficient for the laminar flow on the waterside depends on the assumed boundary condition on the tube surface. It is possible to set a constant tube wall temperature or a constant heat flux. In the case of tubular cross-flow heat exchangers, better compatibility with experimental results was obtained assuming a constant heat flux at the inner surface of the tubes. Fluid temperature variations in the tubular cross-flow heat exchanger are similar to changes in fluid temperatures that occur in counter-flow heat exchangers. The difference between the temperatures of both fluids is almost constant over the length of the entire heat exchanger. It can, therefore, be assumed that the heat flux at the inner surface of the tubes is constant.

2.1 Air-side heat transfer correlation

The Nusselt number on the air side was assumed in the following form

$$\mathrm{Nu}_a = x_1\,\mathrm{Re}_a^{x_2}\,\mathrm{Pr}_a^{1/3} \qquad 225 \le \mathrm{Re}_a \le 560 \qquad (27)$$

where x_1 = 0.5012 and x_2 = 0.4137 are the parameters, which were determined based on the 70 experimental data sets. The hydraulic diameter on the air side of the investigated heat exchanger was d_a = 1.95 mm. The formula proposed by Kays and London was used [1] to calculate the diameter d_a. The air-side Reynolds number was defined as $\mathrm{Re}_a = w_{max}/\nu_a$, where w_{max} is the air velocity in the minimum free flow area. The maximum air velocity w_{max} occurring within the in-line tube rows is

$$w_{max} = \frac{s\,p_1}{\left(s-\delta_f\right)\left(p_1-d_o\right)}\frac{\overline{T}_{am}+273}{T'_{am}+273}\,w_0 \qquad (28)$$

where w_0 is the air velocity upstream of the heat exchanger. The tubes in the radiator were arranged inline. Thus, w_{max} was the air velocity in the narrowest cross-section between two neighboring tubes located in the same row. All physical air properties in Eq. (28) were evaluated at the average temperature $\bar{T}_{am} = \left(T'_{am} + T''_{am}\right)/2$.

2.2 Liquid-side heat transfer correlation

In this section, heat transfer correlations for laminar flow as well as for transition and turbulent flow in tubes used in the mathematical model of the exchanger are presented. The average Nusselt number $\mathrm{Nu}_{m,q}$ for hydraulically and thermally developing laminar fluid flow in a tube with the uniform wall heat flux, which usually occurs in heat exchangers was used. The following formula recommended by the VDI Heat Atlas [7] was applied

$$\mathrm{Nu}_{m,q} = \left[\mathrm{Nu}^3_{m,q,1} + 0.6^3 + \left(\mathrm{Nu}_{m,q,2} - 0.6\right)^3 + \mathrm{Nu}^3_{m,q,3}\right]^{1/3} \quad (29)$$

$$\mathrm{Re}_w \le \mathrm{Re}_{w,trb}$$

The limit value of the Reynolds number $\mathrm{Re}_{w,trb}$, in which the laminar flow ends, and the transitional starts, is usually taken as 2300 [7] or 2100 [14-15]. However, recent experimental studies [10-11] show that the limit value of the Reynolds number $\mathrm{Re}_{w,trb}$ is influenced by the type and shape of the inlet to the tube.
The symbol $\mathrm{Nu}_{m,q,1}$ in Eq. (29) denotes the mean Nusselt number for hydrodynamically and thermally fully developed flow

$$\mathrm{Nu}_{m,q,1} = \frac{48}{11} = 4.364\ ,\ \mathrm{Re}_w \le \mathrm{Re}_{w,trb} \quad (30)$$

The Lévêque solution [16] $\mathrm{Nu}_{m,q,2}$ denotes the mean Nusselt number for hydrodynamically and thermally fully developed flow over the plate with a linear temperature profile in the fluid and constant heat flux at the wall surface

$$\mathrm{Nu}_{m,q,2} = 3^{1/3}\Gamma(2/3)\left(\mathrm{Re}_w \mathrm{Pr}_w \frac{d_{in}}{L}\right)^{1/3} = 1.9530\left(\mathrm{Re}_w \mathrm{Pr}_w \frac{d_{in}}{L}\right)^{1/3},\quad \mathrm{Re}_w \le \mathrm{Re}_{w,trb} \quad (31)$$

Eq. (31) was derived by Lévêque [16] using the local Nusselt number for small values of the parameter $(x/d_{in})/(\mathrm{Re}_w\mathrm{Pr}_w)$, where x is the distance from the tube inlet.

The Nusselt number $\mathrm{Nu}_{m,q,3}$ appearing in the correlation developed for a tube with a constant wall heat flux is given by

$$\mathrm{Nu}_{m,q,3} = 0.924\,\mathrm{Pr}_w^{1/3}\left(\mathrm{Re}_w \frac{d_{in}}{L}\right)^{1/2} \quad (32)$$

$$\mathrm{Re}_w \le \mathrm{Re}_{w,trb}$$

The relationship (32) was obtained by an approximation of Nusselt numbers obtained from a numerical solution of the momentum and energy conservation equations assuming a uniform velocity and temperature profiles at the inlet of the tube.

Taler [8] determined the Nusselt number for the turbulent flow in the tube as a function of Prandtl and Reynolds numbers by integrating the energy conservation equation assuming a radial velocity distribution determined experimentally. Using the Nusselt number values obtained by Taler [8], the correlation for transitional and turbulent flow on the water-side was found by the method of least squares. The friction factor ξ_w was evaluated using the explicit relationship proposed by Taler [12] for turbulent flow regime. In this paper, the relationship proposed by Taler [8] was generalized to the following form

$$\mathrm{Nu}_w = \mathrm{Nu}_{m,q}\left(\mathrm{Re}_{w,trb}\right) + \frac{\frac{\xi_w}{8}\left(\mathrm{Re}_w - \mathrm{Re}_{w,trb}\right)\mathrm{Pr}_w^{1.008}}{1.084 + 12.4\sqrt{\frac{\xi_w}{8}}\left(\mathrm{Pr}_w^{2/3} - 1\right)} \times \left[1 + \left(\frac{d_{in}}{L}\right)^{2/3}\right], \quad (33)$$

$$\mathrm{Re}_{w,trb} \le \mathrm{Re}_w \le 10^6$$

$$0.1 \le \mathrm{Pr}_w \le 1000,\ \frac{d_{in}}{L} \le 1$$

The friction factor in Eq. (36) is given by the following relationships

- turbulent flow regime [12]

$$\xi_w = \left(1.2776\log\mathrm{Re}_w - 0.406\right)^{-2.246},\quad 3{,}000 \le \mathrm{Re}_w \le 10^6 \quad (34)$$

- transitional flow regime

$$\xi_w = 0.030476 + 1.45216\cdot10^{-5}\left(\mathrm{Re}_w - 2100\right),\quad 2.1\cdot10^3 \le \mathrm{Re}_w \le 3\cdot10^3 \quad (35)$$

The formula (35) for the friction factor results from linear interpolation between ξ_w = $64/\mathrm{Re}_w$ for laminar flow at Re_w = 2100 and turbulent friction factor given by Eq.(34) at Re_w = 3000.
The water-side Reynolds number $\mathrm{Re}_w = w_w d_{in}/\nu_w$ is based on the inner diameter of the circular tube.
The inner diameter of the tubes was d_{in} = 6.2 mm. The physical properties of the water were determined at the mean temperature $\bar{T}_w = \left(T'_w + T''_w\right)/2$.

In the relationship (35) there is a Reynolds number $\mathrm{Re}_{w,trb}$, at which the transitional flow begins. It was assumed that $\mathrm{Re}_{w,trb}$ = 2100. The heat transfer correlation (33) can also be used when the transitional flow regime starts at other values of the Reynolds number [10-11]. In heat exchangers, the influence of free convection inside tubes is negligibly small. Natural convection in horizontal pipes occurs due to the temperature difference of the liquid inside the pipe. The temperature of the fluid in the central part of the tube cross-section is different than near its inner surface. The formation of natural convection is favored by the large diameter and length of the pipe and the high heat flux at the inner surface of the pipe. In heat exchangers, there are usually several passes. Between two neighboring passes, there are reversible chambers or bends, in which the fluid is intensively stirred. This, in turn, equalizes the temperature distribution inside the cross-sections of the tubes. At the inlet to each pass, the fluid temperature is almost uniform. The ratio of the length of pipes to their diameters is not very high in heat exchangers. Therefore, there are no conditions to create a large difference in fluid temperature inside the cross-section of the heat exchanger tubes, contributing to the formation of natural convection. For this reason, in the heat transfer correlations (33) and (36) intended for use in heat exchangers, natural convection is omitted.

The water-side correlation $Nu_w = f(Re_w, Pr_w)$ was assumed for the investigated heat exchanger in the form similar to the correlation (36) valid for flow in straight tubes

$$Nu_w = x_3 + \frac{\frac{\xi}{8}\left(Re_w - Re_{w,trb}\right)Pr_w^{1.008}}{1.084 + 12.4\sqrt{\frac{\xi}{8}}\left(Pr_w^{2/3} - 1\right)} \times \left[1 + \left(\frac{d_{in}}{L}\right)^{2/3}\right], \quad (36)$$

$$2100 \le Re_w \le 18000, \quad 2.6 \le Pr_w \le 3.9,$$

$$\frac{d_{in}}{L} = \frac{6.2}{520}$$

The friction factor in the correlation (36) is given by the formulas (34) and (35). The calculations were carried out for $Re_{w,trb} = 2100$. The unknown parameter x_3 in formula (36) corresponds to the average Nusselt number $Nu_{m,q}$ in Eq. (33) which was $Nu_{m,q}\big|_{Re_{w,trb}=2100} = 9.62$ considering that the average Prandtl number for 70 data series was $\overline{Pr}_w = 3.244$. The parameter x_3 = 10.4923 was determined by the least squares method using 70 experimental data sets.

3 Experimental validation of the mathematical model of the heat exchanger at low loads

The heat flow rate $\dot{Q}_{w,calc}$ transferred from the water to air was calculated as follows

$$\dot{Q}_{w,calc} = \dot{V}_w \rho_w\left(T_w'\right)\bar{c}_w\left(T_w' - T_w''\right), \quad (37)$$

where $\dot{V}_w$ - volume flow rate measured at the inlet of the heat exchanger, T_w'-measured temperature at the inlet of the heat exchanger, T_w''- calculated temperature at the outlet of the heat exchanger, $\bar{c}_w$-mean specific heat capacity in the range from T_w'' to T_w'.

The temperature T_w'' was evaluated using the formula (15). The air-side Nusselt Nu_a number was calculated using the correlation (27) and the water-side Nusselt number Nu_w was obtained from the empirical correlation (36) or the theoretical formula (33). If the Reynolds number Re_w is less than 2100, the water-side heat transfer coefficient was calculated using the relationship (29). The measured volumetric flow rate $\dot{V}_w$ at the inlet

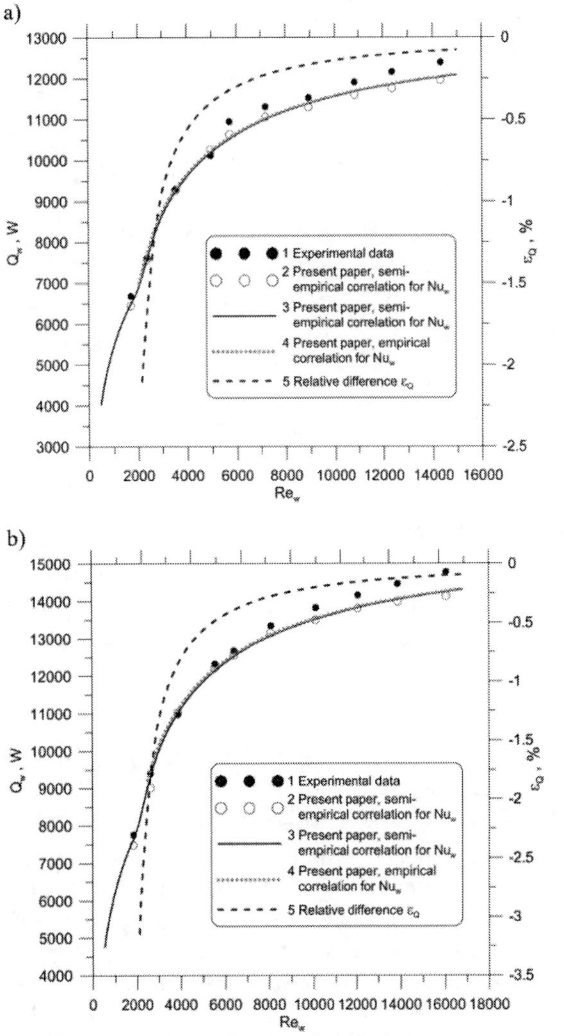

Fig. 3. Comparison of the heat flow rate determined experimentally with the heat flow rate obtained using P-NTU method in the Reynolds number range $500 \le Re_w \le 18000$; a) $\bar{w}_0$ = 2.27 m/s, $\bar{T}_{am}$ = 8.78°C, $\bar{T}_w'$ = 61.70°C; b) $\bar{w}_0$ = 2.27 m/s, $\bar{T}_{am}$ = 8.24°C, $\bar{T}_w'$ = 70.56°C; 1 – heat flow rate $\dot{Q}_{w,exp}$ determined experimentally, 2 – heat flow rate $\dot{Q}_{w,calc}$ calculated using the P-NTU method with theoretical relationship (33) for Nu_w, 3 - heat flow rate $\dot{Q}_3 = \dot{Q}_{w,calc}$ calculated using the P-NTU method and theoretical relationship (33) for Nu_w using the mean values of measured quantities $\bar{w}_0$, $\bar{T}_{am}$, $\bar{T}_w'$, and $Nu_{m,q}\big|_{Re_{w,trb}=2100} = 9.62$, 4 – heat flow rate $\dot{Q}_4 = \dot{Q}_{w,calc}$ calculated using the P-NTU method and empirical relationship (36) for Nu_w using the mean values of measured quantities $\bar{w}_0$, $\bar{T}_{am}$, $\bar{T}_w'$, and $x_3 = 10.49$, 5 - relative difference ε_Q between $\dot{Q}_3$ and $\dot{Q}_4$ $\left(\varepsilon_Q = 100\left(\dot{Q}_3 - \dot{Q}_4\right)/\dot{Q}_3\right)$.

Then the mean values $\bar{w}_0$, $\bar{T}_{am}$, and $\bar{T}_w'$ from the measurement series were calculated and used in the P-NTU method to determine $\dot{Q}_{w,calc}$ as a function of the volumetric flow rate $\dot{V}_w$.

of the heat exchanger, as well as the measured inlet water temperature T'_w and the outlet temperature of the water T''_w, were used for calculating the heat flow rate $\dot{Q}_{w,\exp}$

$$\dot{Q}_{w,\exp} = \dot{V}_w \, \rho_w (T'_w) \bar{c}_w \left[T'_w - (T''_w)^{meas} \right] \quad (38)$$

where: ρ_w - water density, $\bar{c}_w$ - mean specific heat of the water between T''_w and T'_w.

The relative difference ε_Q between the calculated heat flow rate $\dot{Q}_{w,calc}$ and the heat flow rate $\dot{Q}_{w,\exp}$ was determined as follows

$$\varepsilon_Q = \frac{\dot{Q}_{w,calc} - \dot{Q}_{w,\exp}}{\dot{Q}_{w,calc}} \cdot 100\% \quad (39)$$

The comparisons of the calculated and measured heat flow rates are shown in Fig. 3.

The relative difference values ε_Q shown in Fig. 3 were calculated for each measurement series. The heat flow rate $\dot{Q}_{w,calc}$ in Eq. (39) was calculated using the P-NTU method, in which the input data was the air velocity w_0, the volume flow rate $\dot{V}_w$, the air temperature T'_{am} and water temperature T'_w. The heat flow rate $\dot{Q}_{w,calc}$ is indicated in Figure 3 with empty circles (entry no. 2).

The $\dot{Q}_{w,calc}$ values were evaluated using theoretical correlation (33) for Nu_w (solid line in Figure 3, entry no. 3) and empirical correlation (36) (point curve in Figure 3, entry no. 4) and used for determining ε_Q by Eq. (39) (dashed line in Figure 3, entry no. 5).

An inspection of the results presented in Figure 3, shows that the application of the semi-empirical correlation (33) [9] for the water-side Nusselt number in transitional and turbulent flow regime gives very satisfactory agreement between calculated and experimental values. Also, the differences between calculated and experimentally determined heat flow rates in laminar flow range are small (Fig. 3).

Many other experimental tests were carried out for various air velocities and water flow rates. An excellent compatibility between measured and calculated heat flow rates was achieved for all data sets.

Conclusions

New theoretical heat transfer correlation for calculating the water-side Nusselt number in transition and turbulent flow regime was proposed and validated experimentally. Also, empirical heat transfer correlations on the water and air side were parallelly determined. A new formula for the friction factor in the transitional flow regime that is needed to calculate the water-side Nusselt number was proposed. The friction factor in the transitional region $Re_{w,trb} \leq Re_w \leq Re_{w,tre}$ was determined by linear interpolation between the friction factors at $Re_w = Re_{w,trb}$ and $Re_w = Re_{w,tre}$. It should be emphasized that proposed semi-empirical water-side heat transfer correlation may be used for different, experimentally determined values of $Re_{w,trb}$ and $Re_{w,tre}$, not only for $Re_{w,trb} = 2100$ and $Re_{w,tre} = 3000$. The linear interpolation of the friction factor in the transitional flow regime requires the knowledge of $Re_{w,trb}$ and $Re_{w,tre}$. The proposed theoretical and empirical water-side correlations ensure the continuity of the Nusselt number at the transition from laminar to turbulent flow. The heat flow rates calculated using the P-NTU method with new heat transfer correlations agreed very well with the heat flow rates determined experimentally. The new water-side relationships for the friction factor and Nusselt number in transitional and turbulent flow regime proposed in the paper can be used in a wide range of load changes. The liquid flow regime in tubes can vary from laminar through transitional to turbulent, and vice versa. The heat transfer coefficient at the tube inner surface is a continuous function of the Reynolds number without sudden variations during the change of the laminar to transition or transition to the turbulent flow regime. Due to the continuity of variations in the water-side heat transfer coefficient, the fluid temperature at the outlet of the heat exchanger or the heat flow rate transferred from hot to the cold fluid can be adjusted smoothly by changing the rotational speed of the pump, i.e., by changing the volume flow rate of the liquid.

References

1. W.M. Kays, A.L. London, *Compact Heat Exchangers*, 3rd ed. (Krieger, Malabar, Florida, USA, 1998)
2. S. Kakaç, H. Liu, A. Pramuanjaroenkij, *Heat Exchangers. Selection, Rating, and Thermal Design*, 3rd ed. (CRC-Taylor & Francis Group, Boca Raton, USA, 2012)
3. R.K. Shah, D.P. Sekulić, *Fundamentals of Heat Exchanger Design* (Wiley, Hoboken, USA, 2003)
4. T. Kuppan, *Heat Exchanger Design Handbook*, 2nd ed. (CRC-Taylor and Francis Group, Boca Raton, USA, 2013)
5. D. Taler, *Numerical Modeling and Experimental Testing of Heat Exchangers* (Springer International Publishing, Berlin-Heidelberg, 2019)
6. D. Taler, J. Therm. Sci. **26**, 339-348, (2017)
7. V. Gnielinski, *VDI Heat Atlas, Chapter G1* (Springer-Vieweg, Berlin-Heidelberg, 2013)
8. D. Taler, Int. J. Therm. Sci. **108**, 108-122, (2016)
9. D. Taler, Energy Convers. Manage. **159**, 212-243, (2018) 212-243
10. A.J. Ghajar, L.M. Tam,_Exp. Therm. Fluid Sci. **8**, 79-90, (1994)
11. J.P. Meyer, M. Everts, Int. J. Heat Mass Transf. **117**, 1251-1273, (2018)
12. D. Taler, Int. J. Therm. Sci. **105**, 109-122, (2016)
13. D. Taler, Heat Mass Transfer **49**, 1125-1139, (2013)

14. S.W. Churchill, Chem. Eng. **84,** 91-92, (1977)
15. D.C. Rennels, H.M. Hudson, *Pipe Flow. A Practical and Comprehensive Guide* (AIChE-Wiley, Hoboken, 2012)
16. M.A. Lévêque, Ann. Mines, Memoires, Series **12-13**, 201-299, 305-362, 381-415. (1928)

Do ionic liquids replace water or nanofluids to enhance heat transfer in micro-channel systems?

Mushtaq T. Al-Asadi[a,b*], and *Waleed T.* Al-Sallami[a*]

[a]School of Mechanical Engineering, University of Leeds, UK
[b]Refrigeration Department, Eng. Division, Basra Oil Company, Ministry of Oil, Basra, Iraq
*Correspondence author: Email: mushtaq@boc.oil.gov.iq (M. T. Al-Asadi), ml13wtia@leeds.ac.uk (W. T. Al-Sallami).

Abstrct. This study explores the influence of different fluids in a uniform micro-channel geometry on conjugate heat transfer. Nanofluids used here, are SiO_2 and Al_2O_3 nanoparticles with various concentrations up to 4% in water as a base fluid under laminar flow. Nanofluids are compared to an ionic liquid called Butyl Methyl Imidazolium Tetrafluroborate (BMIM BF_4) to examine the influence of heat transfer and fluid flow characteristics. Good agreement was found against previous micro-channel studies, numerical simulations with COMSOL Multiphysics® V5.2a are used at Reynolds number ranging from 50 to1000 with a uniform heat flux of 100 W/cm^2 relevent to cooling systems. Advanced fluids might offer significant potential in combatting the challenges of heat transfer in the technological drive toward lower weight/smaller volume electrical and electronic devices. The results show that the lowest thermal resistance is for nanofluids followed by water, then the ionic liquid. Moreover, Al_2O_3-water offers lower thermal resistance than the SiO_2-water for all given nanoparticle concentrations. It is also found that there is no beneficial using ionic liquids in forced convection systems which work below boiling temperature conditions.

Nomenclature

FEM	Finite Element Method
BMIM **(BF_4)**	Buty lMethyl Imidazolium Tetrafluroborate
k	Thermal conductivity, W/m.K
L	Channel length, μm
P	Pressure, N/m^2
q	Uniform heat flux, W/cm^2
Re	Reynolds number
T	Temperature, K
X	Axial distance, μm
Greek Symbols	
φ	Nanoparticles concentration, %
ρ_f	Density of fluid, kg/m^3
ρ_s	Density of solid particles, kg/m^3
μ	Viscosity, Kg/ms
Sub script	
ave	Average
In	Inlet
Out	Outlet
S	Surface
Θ	Thermal resistance

1 Introduction

Micro-processor cooling systems are gradually becoming a wide area of interest for thermal system researchers due to the rapid development of size shrinking and reduced weight in electronic devices, with the consequence of higher heat generation in limited cross-sectional areas. This brings to fore the need to eliminate the generated heat as it could adversely affect the electronic chip. This can be achieved in a number of ways, using air and water-base cooled heat sinks [1-4] and the use of mini- and micro-channel heat exchangers and/or advanced fluids [1, 5-7].

Micro-channel design is being modified rapidly to meet the recent improvement in electronic chips and to enhance the heat transfer rate [8].

Among different types of micro channel shapes, the rectangular micro-channel was considered to be of ideal geometry based on the numerical investigation by Xia et al. [9].

Heat transfer can as well be enhanced using advanced fluids such as nanofluids which are combinations of a base fluid and nano-sized particles (1-100 nm). Nanoparticles can be from materials such as metallic (Fe, Cu, Ag, Au, Al), carbon or metallic oxide (Fe_3O_4, CuO, Al_2O_3, TiO2, SiC, SiO_2, ZnO). Nanofluids have high thermo-physical properties compared to the base fluid in terms of thermal conductivity [7]. Metallic dioxide nanoparticles are commonly used with water as a base fluid, however, Al_2O_3-water and SiO_2-water nanofluids offer highest heat

*Correspondence author: Email: mushtaq@boc.oil.gov.iq (M. T. Al-Asadi), ml13wtia@leeds.ac.uk (W. T. Al-Sallami)

transfer enhancement among the common nanofluids [2, 7, 10].

Ionic Liuids (ILs) are molten salts which are available at liquid phase under room standard conditions. These salts attracted many researchers due to their promising properties such as their low volatility, exhibit high thermal stability and their physical properties can be controlled easily by changing their chemistry [11]. Thus, ILs have been suggested to be employed in thermal applications for energy storage applications and for many other engineering aspects [11, 12].

In addition, two main points motivate reaearchers to investigate thermo-hydraulic performance of ILs; their high thermal stability which means that they can be used under extreme conditions and the high capability of controlling their physical properties by changing their chemistry. The conventional thermal decomposition of ILs are above 200 ^{0}C and the typical melting temperature is below 0 ^{0}C [12]. Besides, the variation of cation type, anion type, cation chain length and anion chain length can lead to a significant impact on their physical properties [13]. However, due to the lack of data of the thermal and hydraulic properties of ILs, thermal conductivity, specific heat capacity, viscosity and density all togeteher. Only one IL is employed in this study when its properties are obtained from [12, 14].

From the above, the aim of this paper is mainly to suggest using ILs as coolants in various heat exchanging duties under extreme conditions and to compare between one IL with various nanofluids and water when they are employed as coolants for cooling electronic systems.

2 Analysis and Modelling

In this article, a rectangular uniform micro-channel is considered to examine the heat transfer and fluid flow performance using different fluids such as pure water, Al_2O_3 / water, SiO_2 / water and ionic liquid. The details of the modeling are found in [10].

The conjugate heat transfer model and validations against previous studies of the micro-channel is presented as in [1, 5, 6]. Al_2O_3 and SiO_2 nanoparticles are considered to be separated from each other and well distributed in water as a base fluid at all ranges of concentrations to offer homogeneous nanofluid. All required equations are found in [2, 15-18].

The influence of different effects on thermal resistance can be seen in Fig. 1, the effect of nanoparticles concentration up to 4% of the Al_2O_3 and SiO_2 nanoparticles in water with the pure water and ionic liquid in a uniform micro-channel. It is found that thermal resistance decreased as the concentrations of nanoparticles increased due to thermo-physical properties enhancement especially the thermal conductivity of nanofluids.

However, the ionic liquid does not offer a thermal resistance reduction, again because its thermal properties compared to nanofluids. Nanofluids offer low thermal resistance compared to the pure water and ionic liquid, therefore the uniform channel can be presented as an example of heat transfer enhancement using SiO_2 and Al_2O_3 –water compared to water and ionic liquid.

In terms of nanofluids, Fig. 1 shows that Al_2O_3-water offered better heat transfer rate than SiO_2-water at the same concentration. This is because of the high thermal conductivity of the Al_2O_3-water compared to SiO_2-water.

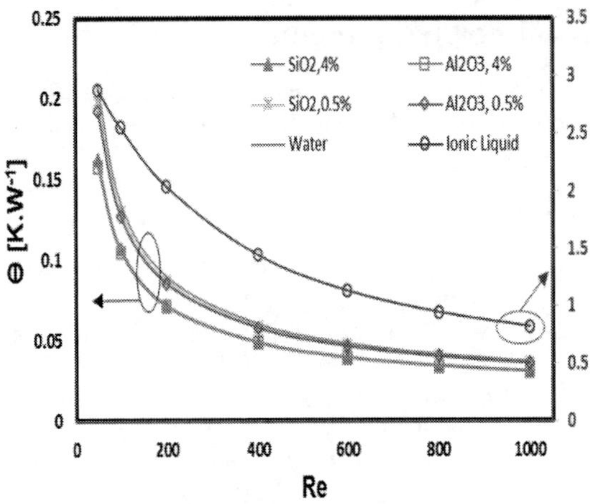

Fig. 1. Influence of different nanoparticles in the range of 0-4%, with ionic lquid and water in water on thermal resistance.

Fig. 2 presents the effect of different fluids on pressure drop. The results showed that the highest pressure penalty is for the ionic liquid, followed by nanofluids with high concentrations, while the lowest pressure drop is for water. Regarding to nanofluids, as to be expected that the pressure drop increased when nanoparticles concentration increased.

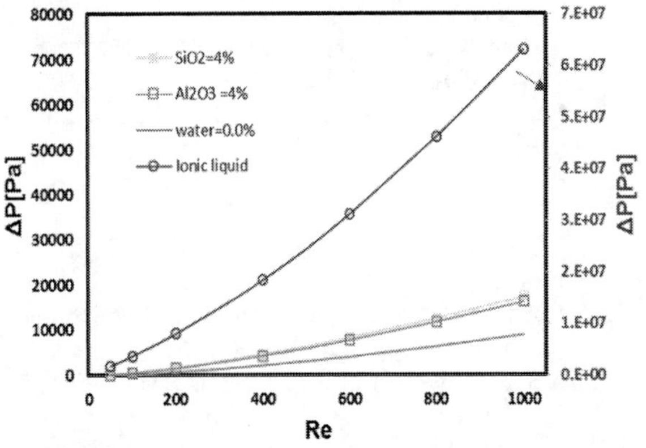

Fig. 2. Impact of different nanoparticles in the range of 0-4% in water with pure water on pressure drop.

3 Conclusions

Thermo-hydraulic comparison has been conducted to examine the influence of different fluids such as water, ionic liquid (IL) and nanofluids. The results reveal that in this particular application and for this particular IL, nanofluids and water offer better heat transfer performance. However, the variation of IL chemistry can lead to a significant enhancement in its thermal properties. Thus, the investigation to find suitable IL for this application will be conducted in the future. Lastly, ILs can be employed as coolants under extreme conditions when the conventional fluids are either boiled or solidified.

Acknowledgements

We thank the Basra Oil Company (BOC) and the Ministry of Oil, Iraq for their sponsorship of Mushtaq T. K. Al-Asadi.

References

1. M.T. Al-Asadi,A. Al-damook, and M.C.T. Wilson. Int. Comm. HMT. **91**, p. 1-10, (2018).
2. , M.T. Al-asadi et al. Int. Comm. HMT. **85**, p. 81-91, (2017).
3. W. Al-Sallami, A. Al-Damook, and H. Thompson. Int. Comm. HMT. **75**, p. 183-191, (2016).
4. W. Al-Sallami, A. Al-Damook, and H. Thompson. Int. J. Thermal Sciences. **121**, p. 266-277, (2017).
5. M.T. Al-Asadi., F.S. Alkasmoul, and M.C.T. Wilson. Int. Comm. HMT. **74**, p. 40-47, (2016).
6. M.T. Al-Asadi, F.S. Alkasmoul, and M.C.T. Wilson. App. Thermal Engg. **130,** p. 571-586, (2017).
7. M.T. Al-Asadi, et al. J. Nanofluids. **6**, p. 987-1007, (2017).
8. Li, L., et al. Int. J. Thermal Sciences. **92,** p. 85-96, (2015).
9. Xia, G.D., et al. Int. JHMT. **80**, p. 439-447,(2015).
10. M.T. Al-Asadi, and M.C.T. Wilson. Int. Conf. Thermal Engg: Theory and Apps, Doha, Qatar February 25-28, (2018).
11. Brennecke, J.F. and E.J. Maginn. AIChE Journal. **47,** 11, p. 2384-2389, (2001).
12. M.E. Van Valkenburg, et al. **425,** 1-2, p. 181-188, (2005).
13. P Wasserscheid. and T. Welton. John Wiley & Sons. (2008).
14. J. Jacquemin, et al.. Green Chemistry. **8,** 2, p. 172-180, (2006).
15. J. Albadr, S. Tayal, and M. Alasadi. Case Studies in Thermal Engg. **1,** 1, p. 38-44, (2013).
16. A.S. Kherbeet, et al. Int. Comm. HMT. **76**, p. 237-244, (2016).
17. A.S. Kherbeet, et al. Int. Comm. HMT. **78**, p. 145-154, (2016).
18. M. T. K. Al-Asadi. PhD thesis, University of Leeds, Whight Rose (2018).

CFD analysis of a one-pass photovoltaic/thermal air system with and without offset strip fins

Moustafa Al-Damook[1,2*], *Darron* Dixon-Hardy[3], *Peter* J. Heggs[3], *Mansour* Al Qubeissi[4], *Khaled* Al-Ghaithi[5], *Patrick E.* Mason[3], *and Joshua* Cottom[3]

[1]School of Mechanical Engineering, University of Leeds, LS2 9JT, Leeds, United Kingdom
[2]Renewable Energy Research Centre, University of Al-Anbar, Iraq
[3]School of Chemical and Process Engineering, University of Leeds, Leeds, United Kingdom
[4]Faculty of Engineering, Environment and Computing, Coventry University, Coventry, United Kingdom
[5]School of computing, Faculty of Engineering, University of Leeds, Leeds LS2 9JT, United Kingdom

Abstract. The focus on managing PV panel temperature has undergone a remarkable development in the last two decades. Specifically, in countries with moderate weather temperature and high insolation, the problem of keeping the PV cell temperature in an optimal range has been managed by use of PV/T collectors. In this work, a single pass PV/T collector using laminar air flow has been assessed. Two PV/T collector designs are utilised, one with and one without offset strip fins. COMSOL Multiphysics v5.3a has been used for the analysis of the thermal and electrical performances. Two assumptions were implemented in order to reduce the computational time from 95 hours to 7 hours, namely ignoring radiative effects between the fins and the wall channels, and representing thin layers as 2D boundaries, whilst ensuring a high level of conformity (4%),. Monocrystalline silicon PV cells were used with a power temperature coefficient of 0.41%. A validation against work in the literature was made, showing a good consistency. The objective of this work is to verify the performance of the air PV/T collector with offset strip fins compared to an unfinned air PV/T collector. The results reveal that the use of offset strip fins has a noticeable impact on both the electrical and thermal efficiencies of the system. In addition, the maximum combined efficiency (η_{Co}) for the finned PV/T system is 84.7% while the unfinned PV/T system is 51.2%.

1 Introduction

The capability of photovoltaic (PV) systems has been developed considerably in recent years to the point that some PVs can absorb more than 75% of the insolation energy and up to 18% of this being converted to electrical power [1]. The remainder of the absorbed energy is released as heat, causing the temperature of the PV cells to rise. Once the temperature increases above standard conditions (25°C and 1000 W m^{-2}), the cell efficiency falls by approximately 0.4–0.65% per degree Celsius [1]. This reduction is defined by the temperature coefficient of the PV cell. For example, for a panel with a temperature coefficient of -0.65%, the panel's maximum power will reduce by 0.65% for every 1°C rise. It is therefore important to implement a means of controlling the temperature rise (i.e. cooling) to keep the panel operating at optimal conditions. If the heat removed from the panel can be usefully exploited, then this provides an additional advantage and the system can be classed as a photovoltaic/thermal system (PV/T). Ideally, PV/T technology should allow the PV cell to operate in a temperature range optimised for electrical efficiency, as well as producing useable heat. The efficiencies of both the PV system and the thermal system can be considered as an overall (hybrid) efficiency.

A number of studies have investigated the effects of extended surfaces on improving PV/T system hybrid efficiency. Extended surfaces can be classed as three main types: traditional fins [2, 3]; interposition of a thin metallic sheet (TMS) [4-8] and obstacles (ribs) [4, 5, 9, 10].

To ensure lower-pressure drop and higher contact area [11-14], a traditional offset strip fin is employed.

The present study numerically examines the use of an offset strip fin design, the performance of which is evaluated by comparison with a bare solar collector design (without fins). An analysis of this kind for a PV/T system with offset fins has not previously been explored.

2 Numerical investigations

Figure 1 (a) illustrates the various layers comprising the PV module and (b) is a schematic view of a single pass PV/T collector without fins. Figure 2 (a) is a 3D view of the offset longitudinal fins arrangement to be attached to the bottom of the PV module and (b) provides the dimensions of these fins. The physical dimensions of the

* Corresponding author: mnmaab@leeds.ac.uk

PV/T collectors are listed in Table 1. The thicknesses and physical properties of the layers in the PV module are detailed in Table 2. Mathematical models of these two collector systems are investigated numerically using COMSOL Multiphysics v5.3a. The model of the one pass air PV/T collector consists of three parts: the PV thermal and electrical absorber plate, the channel ducting, and the back insulation. The second model is similar, but includes the longitudinal offset strip fins attached to the base of the PV module.

Table 1. Specification of PV/T Collectors.

Collector length (L) (mm)	660
Collector width (W) (mm)	340
Symmetry width (W) (mm)	170
Duct depth (δ_{D1}) (mm)	25
Bottom absorber plate thickness (mm)	1
Absorber plate thickness (mm)	0.5
Fin thickness (mm)	0.5
Length of fin (L_f) (mm)	50
Space of fin (S_f) (mm)	20

Table 2. Physical properties of the PV module layers BP 585 [15 ,16].

Layer	t	k	r	cp	e
PV Glass	3	1.8	3000	500	0.84
EVA	0.5	0.35	960	2090	-
PV cells	0.3	148	2330	677	0.7
Tedlar	0.5	0.2	1200	1250	0.87

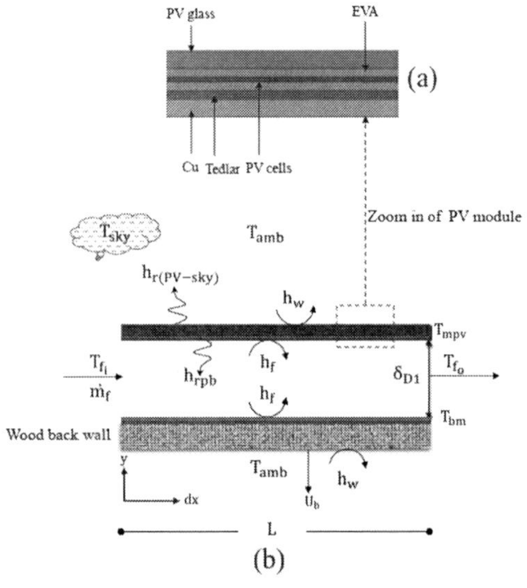

Fig. 1. (a) The thermal schematic of air PV/T collector, and (b) the PV module layers.

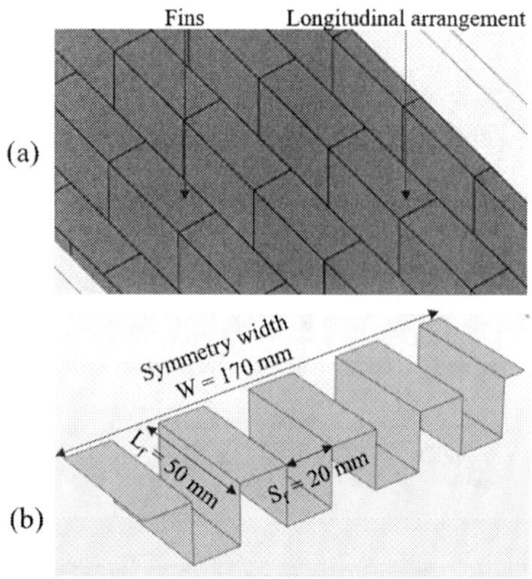

Fig. 2. Offset strip fins longitudinal arrangement: (a) the 3D view of fins array arranged longitudinally and (b) 3D view of one array with fin dimensions.

To simulate realistic and accurate results, a surface-to-surface radiation module was used with an 'External Radiation' node to simulate the incident solar radiation instead of using constant heat flux as illustrated in Eq.(1). Eq.(2) represents the governing equation for modelling the emission and absorption between interfacing surfaces with temperature T into air flow or to the surroundings.

$$q_r = \varepsilon\,(G - E_b) = \varepsilon\,(T_s^4 - T^4) \qquad (1)$$

where ε is the surface emissivity, G is incident solar radiation (W m^{-2}), and E_b is the blackbody hemispherical total emissive power, estimated as:

$$E_b = \sigma\, A_a T^4 \qquad (2)$$

In order to avoid the complexity, the emissivity of PV layers and glass set as constant (independent wavelength and temperature) since the glass is ultra-clear. All thermal properties of the materials coolant (air) are set as temperature dependent.

Eq.(3) is the general governing equation for the 3D conjugate heat transfer system under transient conditions. Assuming steady state conditions, electrical power generation ($\dot{Q}_v$) and a stationary PV panel, Eq.(3) reduces to Eq.(4). The boundary conditions were set as presented in Table 3.

Table 3. Boundary conditions.

Lower wind velocity (m s^{-1})	0.5
Upper wind velocity (m s^{-1})	3
Incident solar radiation (Wm^{-2})	1000
Ambient temperature °C	50
Re	400-2200
Outlet pressure (Nm^{-2})	0
Inlet air temperature °C	50
No slip condition at fins and wall collector (m s^{-1})	$U_{x,y,z}$=0

 https://doi.org/10.1051/matecconf/201824003002

$$\frac{D(\rho\, C_p\, T)}{Dt} = \nabla.(k\nabla T) + \dot{Q}_v \quad (3)$$

$$\nabla.(k\nabla T) + \dot{Q}_v = 0 \quad (4)$$

The extracted power of the PV module generation can be estimated by Eq.(5) [15, 17].

$$P_{module} = I_m V_m = FF\, I_{sc}\, V_{oc} = \eta_{cell}\, A_{cells}\, G\, \alpha_{pv} \quad (5)$$

$$\eta_{cell} = \eta_{ref}\left(1 - \beta_{ref}\left(T_{mpv} - T_{ref}\right)\right) \quad (6)$$

where η_{cell} is the electrical efficiency of PV cell.

The thermal performance can be estimated as [18]:

$$\eta_{th} = \dot{Q}_u / \dot{Q}_s \quad (7)$$

where $\dot{Q}_u$ is the effective heat extraction by PV/T system:

$$\dot{Q}_u = \dot{M}\, Cp\, (T_o - T_i) \quad (8)$$

where $\dot{M}$ is mass flow rate (kg s^{-1}). The overall heat produced by incident solar radiation is given by:

$$\dot{Q}_s = G\, A_a \quad (9)$$

The total collector efficiency (hybrid efficiency) of the PV/T collector can be estimated using Eq(10) [19, 20].

$$\eta_{co} = \eta_{th} + \frac{\eta_{ele}}{C_f} \quad (10)$$

where C_f is the conversion factor of the thermal power plant and lies between 0.29 to 0.4 [19], and is assumed to be 0.36 in our analysis.

Three further assumptions were implemented to reduce the computational time. The first one was ignoring radiative effects between the fins and the wall channels while the second was representing thin layers as 2D boundaries and the last is a symmetry boundary condition in Y flow direction plan. The computational time reduced therefore from 95 hours using a high performance computer (HPC) to 7 hours using a desktop PC.

3 Results and discussion

Numerical investigations were performed using COMSOL Multiphysics v5.3a to assess the influence of the PV panel temperature on the monocrystalline PV. Two approaches were used to ensure the accuracy of results; mesh independence tests and validation against previous literature work.

Table 4. The validation between CFD Model and [21].

PV/T layers	[21] (°C)	CFD (°C)
T_g	30.5	32.3
T_{mpv}	31.3	32.0
T_{Tedlar}	28.8	31.7
T_{fm}	18.4	19.1
T_{fo}	20.5	20.3
T_{ins}	19.7	16.1

The validation with [21] showed good agreement as shown in Table 4.The mesh independence test was performed to enhance the convergence speed and ensure high mesh quality. The number of elements was increased until the solution remained constant, as presented in Table 5. Both thermal efficiency η_{th} and pressure drop Δp (Pa) were studied in this test.

Table 5. Mesh independent test analysis.

Element size	No of elements	η_{th}	Δp (Pa)
Very coarse	114096	65.0	13.3
less coarse	241591	62.8	8.3
Coarse	427075	60.8	5.5
Fine	1201596	57.8	3.3
Very Fine	2097077	57.0	2.8

Figure 3 shows the impact of the PV panel temperature on the PV electrical efficiency for both with and without fins systems for a range of Reynolds numbers (400-2200). The temperature of the PV panel is seen to be inversely proportional to Reynolds number. This is expected because of the convective heat transfer coefficient is directly proportional to the Reynolds number. This decrease in temperature results in a corresponding increase in electrical efficiency as expected by Eq.(6). More importantly at Reynolds numbers between 400-1000, the temperature of unfinned is lower than that of the finned system. As a result, the unfinned system had a better electrical efficiency in that range. However, for Reynolds numbers more than 1000, the temperature of the finned system was markedly lower than that of the unfinned model. This, as can be seen from Figure 3, made the finned system more electrically efficient in this range. This effect is owing to the fin effectiveness at Reynold numbers less than 1000 working as a conductive resistance layer to heat transfer rather than a dissipater of heat. In other words, the ratio of heat transfer rate of the finned PV/T model was lower, compared to the unfinned model at Reynolds number less 1000 under the aforementioned conditions.

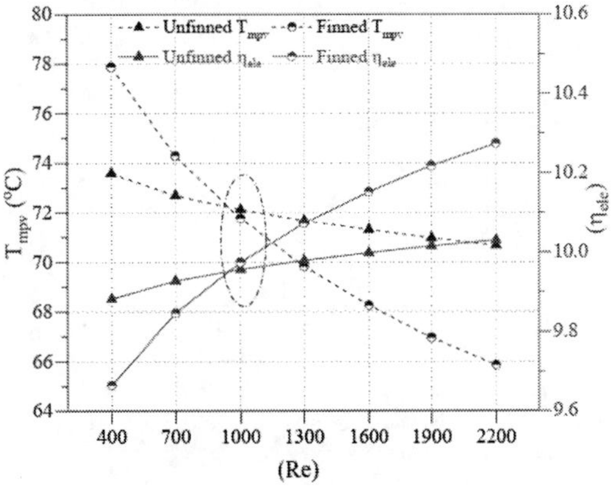

Fig. 3. Effect of PV panel temperature rise on the electrical efficiency under different Reynolds number ranges for finned and unfinned air PV/T collectors.

https://doi.org/10.1051/matecconf/201824003002

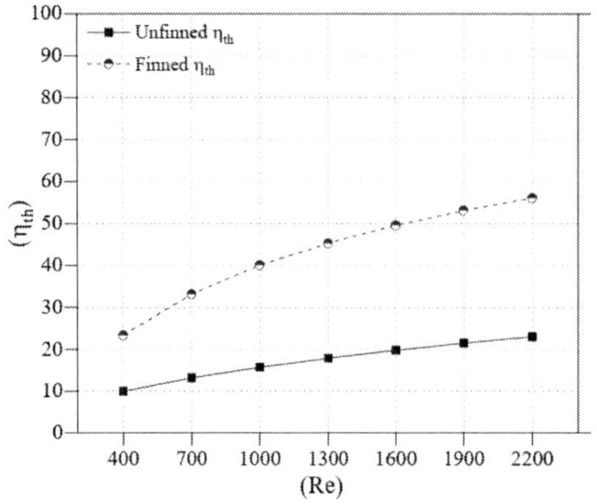

Fig. 4. Effect of Reynolds number on the thermal efficiency for finned and unfinned air PV/T collectors.

The system thermal efficiency enhancements for finned and unfinned air PV/T were calculated using Eq.(7). In Figure 4, a comparison is shown for the thermal efficiencies of the two models with Reynolds number range between 400-2200. The results obtained from this figure revealed that maximum and minimum percentage improvements in thermal efficiency were 143.7% and 134.1%, respectively.

In addition, Figure 5 illustrates this finding for the total system efficiency using Eq.(10), with maximum and minimum percentage enhancements of 66.36% and 34.12%, respectively. The maximum combined efficiency for the finned system was 84.7% in comparison to a maximum of 51.2% for the unfinned system.

This enhancement, however, increased the pressure drop penalty in the system, thus, requiring more pumping power calculated at 2 to 6 times compared with the unfinned PV/T system (see Figure 6).

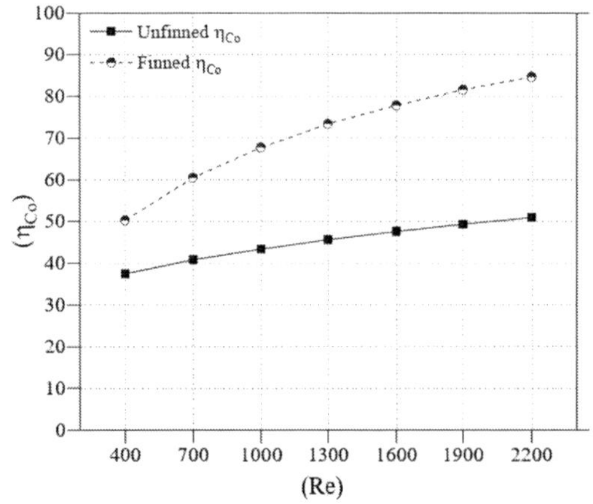

Fig. 5. Effect of Reynolds number on the combined efficiency for finned and unfinned air PV/T collectors.

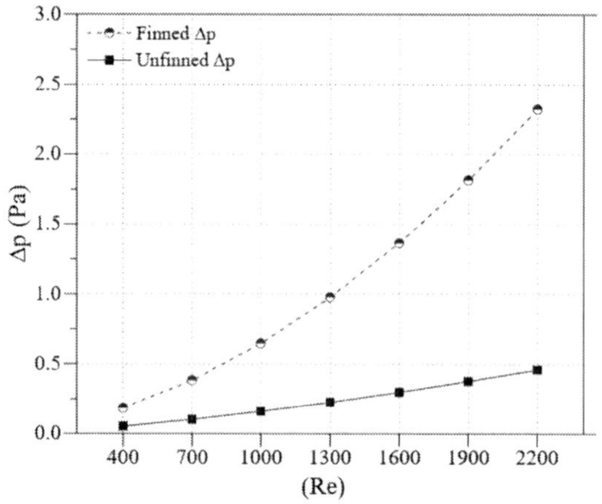

Fig. 6. Effect of Reynolds number on the pressure drop for finned and unfinned air PV/T collectors.

Figure 7 (a) and (b) presents the velocity distribution for the finned model. In (a), the velocity distribution is for a longitudinal YX cut-plane viewpoint at 12.5 mm height above the bottom absorber plate (Y is the dimension in air flow direction, X is the dimension in collector width and Z is the dimension in air depth direction). In (b), the velocity distribution is for several ZX cut-plane viewpoints. The velocity profiles between fins can be seen to have a parabolic profile with 0 velocity at the wall as expected due to the nonslip condition. The entrance length can also be noticed near the inlet as expected after which a developed flow was established.

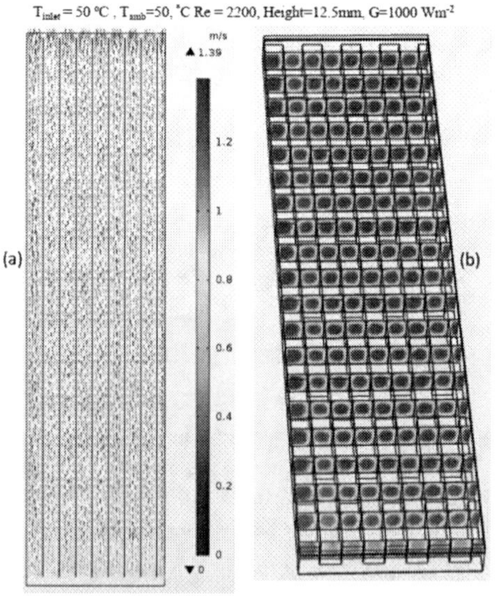

Fig. 7. The velocity profile for a finned air PV/T model, showing (a) XY cut-plane and (b) XZ cut-plane viewpoints.

4 Conclusions

Two air PV/T collectors have been developed using COMSOL Multiphysics v5.3a, namely, a single pass PV/T collector without fins, and a single pass PV/T collector including a longitudinal fin arrangement. The weather conditions employed considered the worst case scenario of 1000 W m^{-2} and 50 °C. For this particular model, the fundamental results revealed that the use of offset strip fins enhanced the thermal efficiency of the PV/T collector and maintained the electrical PV efficiency at an acceptable level. The maximum combined efficiency obtained was 84.7 %, which was very acceptable for an air PV/T collector. Future work will focus on utilising different duct and offset arrangements.

Acknowledgments

The first author is extremely grateful to the Higher Committee for Education Development (HCED), Iraq, and the Renewable Energy Research Centre, University of Anbar, for the their support to the project.

Nomenclature

PV	Photovoltaic
PV/T	Photovoltaic/thermal
t	Thickness (mm)
k	Thermal conductivity (W m^{-1} K^{-1})
h	Heat transfer coefficient (W m^{-2} K^{-1})
Cp	Specific heat capacity of air (J kg^{-1} K^{-1})
T	Temperature °C
CFD	Computational Fluid Dynamics
A	Area (m^2)
P	Power (W)
u	Velocity vector in x direction (m s^{-1})
v	Velocity vector in y direction (m s^{-1})
w	Velocity vector in z direction (m s^{-1})
$\dot{Q}$	Heat rate (W)
G	Incident solar radiation (W m^{-2})
I	Current (A)
V	Voltage (V)
C_f	Conversion factor of the thermal power plant
FF	Fill factor
Greek Symbols	
ρ	Density (kg m^{-3})
β	Power temperature coefficient
ε	Emissivity
η	Efficiency
α	Absorptivity
σ	Stefan-Boltzmann constant 5.67 × 10^{-8} (W m^{-2} K^{-4})
Subscripts	
w	Wind
amb	Ambient weather conditions
s	Sky and solar or surrounding
r	Radiation
mpv	Average PV module temperature
f	Fluid
rpb	Radiation between lower and upper absorbers
o	Out
i	Inlet
A	Aperture
g	Glass
fm	Mean fluid
th	Thermal
Co	Combined
ele	Electrical
ins	Insulation
bm	Bottom Absorber Plate
v	Volume
m	Maximum power point
oc	Open circuit voltage
sc	Short circuit current
ref	Reference

References

1. A. Makki, S. Omer, H. Sabir, Renewable and sustainable energy reviews, **41,** p. 658-684 (2015)
2. M. Tabatabaian, W. Tivy, C. Bibby, Americ. Soci. of Mech. Engin. (2012)
3. R. Kumar, M.A. Rosen, Applied Energy, **88,** 11, 3603-3614 (2011)
4. D. Gupta, S. Solanki, J. Saini, Solar Energy, **51,** 1, p. 31-37 (1993)
5. Y. Tripanagnostopoulos, Solar energy, **81,** 9, p. 1117-1131 (2007)
6. J. Tonui and Y. Tripanagnostopoulos, Renewable energy, **32,** 4, p. 623-637 (2007)
7. J. Tonui, Y. Tripanagnostopoulos, Solar energy, **81,** 4, p. 498-511 (2007)
8. J. Tonui, Y. Tripanagnostopoulos, Solar Energy, **82,** 1, p. 1-12 (2008)
9. J. Zhu, M. Fiebig, N. Mitra, Int. Jour. of Heat and Mass Trans. **38,** 3, p. 495-501 (1995)
10. J. Han, J.S. Park, Int. Jour. of Heat and Mass Trans., **31,** 1, p. 183-195 (1988)
11. L.A. Tagliafico, F. Scarpa, M. De Rosa, Renewable and Sustainable Energy Reviews, **30,** p. 526-537 (2014)
12. M. Yang, et al., Applied Energy, **113,** p. 1349-1362 (2014)
13. S. Youcef-Ali S., J. Desmons, Renewable Energy, **31,** 13, p. 2063-2075 (2006)
14. Handbook, A., *ASHRAE handbook–fundamentals.* (Atlanta, GA, 2009)
15. K. Kant et al., Solar Energy, **134,** p. 147-155 (2016)
16. F. Calise, R.D. Figaj, L. Vanoli, Energies, **10,** 4, p. 491 (2017)
17. S. Dubey, G. Sandhu, Tiwari, Applied Energy, **86,** 5, p. 697-705 (2009)
18. K.E. Amori, M.A. Abd-AlRaheem, Renewable Energy, **63,** p. 402-414 (2014)
19. F. Sarhaddi, et al., Applied Energy, **87,** 7, p. 2328-2339 (2010)
20. N. Aste, C. del Pero, F. Leonforte, Solar Energy, **102,** p. 98-115 (2014)
21. K.E. Amori, H.M.T. Al-Najjar, Applied Energy, **98,** p. 384-395 (2012)

Experimental studies of two-phase liquid metal-gas chain flow with ultrasonic echo pulse method and in the magnetic field of permanent magnets

Artur Andruszkiewicz[1,*], and *Kerstin* Eckert[2]

[1] Wrocław University of Science and Technology, Wybrzeże Stanisława Wyspiańskiego 27, 50-370 Wrocław, Poland
[2] Technische Universitätv Dresden, Faculty of Mechanical Science and Engineering, George-Bähr- Strasse 3c, 01069 Germany

Abstract. Metallurgical processes of refining and degassing liquid metals are related to the blowing-in neutral gasses like for example argon. Effectiveness of the process depends on the flowing gas parameters like: dimensions, velocity, distribution amd gas phase flow zone. These parameters can be changed under the influence of outside magnetic field and formed by it a Lorentz' force. In order to get full information of rising bubbles essential are certain measuring methods. Among many of these the ultrasonic echo pulse method is likely to be the most intensive developed in the past few years. That method enables estimation of gas phase flow parameters in a range from individual bubble flow up to full continuous gas flow at the inlet nozzle. In this paper a measurement principle has been described for an ultrasonic echo method and there also submitted and discussed results of obtained results for diphase flow of GaInSn-argon.

1 Introduction

The two-phase liquid metal-gas flow appears in many technological processes like in metallurgical process of after-furnace steel refining. The process of liquid metal degassing, purification or enrichment are directly related to argon blow-in. A refining efficiency depends on the dimension of rising bubbles, its contact time with metal, the rise zone, and the flow stream of bubbles. A proper recognition of these parameters enables optimization as well as control of metallurgical processes. In order to get a complete knowledge of the rising gas bubbles an experimental studies of two-phase liquid metal-gas flow are essential. Among many methods of investigation two-phase flows recently more and more important becomes the ultrasonic method. For example, the ultrasonic echo pulse method is presently used by TU Dresden for identification of two-phase liquid metal-gas flow parameters in their project studies on "Elektromagnetische Stromungsbeeinflussung in Matallurie, Kristallzuchtung und Elektrochemie" no SFB 609.

The aim of the article is to present selected research results: rise velocity of bubbles, its creation frequency, dimensions and flow zone in the liquid metal using the echo pulse methopd. The method has been elaborated for a container filled with liquid metal Ga^{68} In^{20} Sn^{12} where the argon gas phase was blow-in through a nozzle. Measurements were also done in magnetic field of permanent magnets having magnetic induction of B=300 mT. Magnetic field influences the rise velocity and the final velocity of gas bubbles, the movement trajectory and the zone of gas phase flow as well as the drag coefficient value. The article also compare the gas phase flow parameters in the liquid metal in motion, in the magnetic field and without the magnetic field.

2 Ultrasonic echo pulse method – the measurement principle

Among many methods of two-phase liquid metal-gas flow studies the biggest significance have got the non invasive techniques. Its basic advantages are no interference with the flow, no perturbations of velocity profile distribution and no changes of position distribution and diameter of rising bubbles. A non invasive techniques enable assign the basic two-phase flow parameters practically in a real time. In the studies are used ultrasonic Doppler's methods (UDV) [1-3], tomography techniques: positron-emission PET and roentgen CT [4] as well as methods utilizing eddy current and magnetic sensors [5-6]. The ultrasonic echo pulse method which can be used in identification of two-phase liquid metal-gas flows it is not yet fully developed and propagated comparing to the other above mentioned methods. That method is most widely used in non invasive material investigation - ultrasonic defectoscopy. It is a primary way of material fault detection also giving the localization, dimensions and space orientation of the defects. Presuming a gas phase as flow discontinuity than the echo pulse method can be utilized to estimate characteristic parameters of two-phase liquid metal-gas flow.

The primary work pertaining to estimation of liquid metal-gas flow parameters with utilized echo method

* Corresponding author: artur.andruszkiewicz@pwr.edu.pl

and ultrasonic defectoscope is contributed to Hoffman & Kaiser of TU Dresden. They used two or one ultrasonic heads made by Krautkrammer and sampled signals with digital oscilloscope. The measurement tests done with one ultrasonic head were not difficult (for estimation of volumetric share) but in a case when two heads were used for velocity measurements it was necessary to set an appropriate correlation between ultrasonic pulse frequency and signal recording time [7], and also properly set the signal recording time in order to record all of the flowing gas bubbles. The study results were presented in papers [5-6].

The measuring principle used in echo pulse method, shown on Fig.1, is described in detail in [8]. Short pulses transmitted from ultrasonic head goes through a connection layer (mostly of Vaseline) and front wall of a container than are reflected from gas bubbles moving in liquid and being in the ultrasonic area.

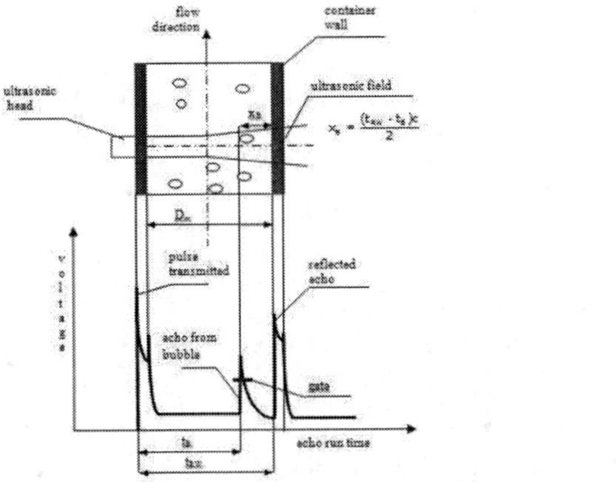

Fig. 1. The measuring principle used in echo pulse method: t_B - run time for an echo reflected from gas bubbles, t_{RW} – echo run time for an echo reflected from back wall of container, c – ultrasonic wave velocity in liquid, x_B – a distance between a bubble and inner part of container back wall.

Ultrasonic pulses reflected from bubbles return as echo back to the ultrasonic head which operate in a transmitter-receiver mode. Part of the transmitted pulses reach the back wall of container and reflected goes back to ultrasonic head. Calculating the time difference between pulses reflected from gas bubbles and container back wall the process characteristic may be presented as a voltage-time curves on ultrasonic defectoscope. It is an A type presentation shown on a Fig.1. The ultrasonic defectoscope records only signals overcrossing a set level, mostly 20% of the maximum screen scale level. For this value the run time for a wave patch from transmitter to reflector and back is calculated. Knowing the wave velocity in the container wall and tested medium a distance from the reflector may be calculated. The upper part of Fig.1. Shows the distance x_B of the bladder from the back wall of the tank. Its determination requires only the knowledge of the velocity of the ultrasonic wave in the medium. The repetition frequency of ultrasonic pulses should be set in such a way that the echo reflected from back wall of the container should be recorded before sending a next pulse.

3 The testing bench – a measurement realization

The test studies were executed on a test bench as shown on Fig.2.

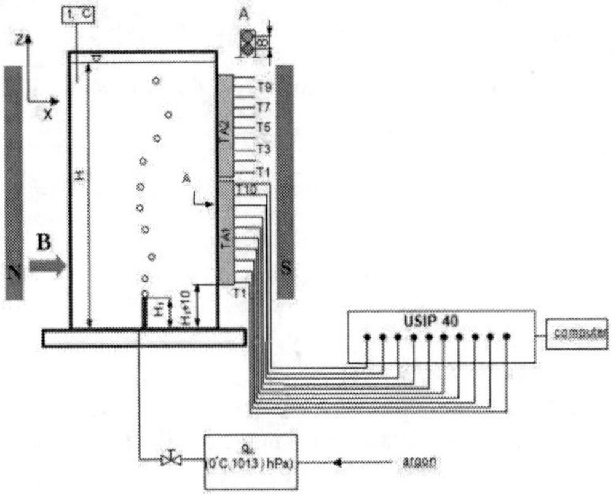

Fig. 2. A testing bench diagram.

A container with a diameter of D_w = 80 mm was filled up with liquid metal GaInSn to the height of H = 210 mm (with density of ρ_{FM} = 6361 kg/m^3, surface tension σ_{FM}= 0.533N/m and temperature t = 20°C). Argon in the form of gas bubbles has been blow-in through a nozzle with a height of 20 mm and aperture diameters of 0.5 mm, 0.9 mm and 1.2 mm, depending on the test number. The volumetric argon gas flow q_v has been measured with thermo-sensor flow meter (Mas-Flo, MKS Instruments) class 1, with ranges: for single rising bubbles 10 sccm, and for a chain bubbles 100 sccm. The ultrasonic head (TA1) was fixed 10 mm above the argon inlet nozzle. The head consisted 10 sensors having diameter of 5mms each with frequency 15MHz, spaced at 8mm (the total distance between sensors has been 72 mm), and operated in transmission-receiving mode. The created ultrasonic zone in liquid metal GaInSn has a characteristic length of local field l_o = 34.3 mm and opening angle of beam $\varphi \approx 2.6°$.

Signals from rising gas bubbles were recorded by means of two measuring heads installed one above the other (TA1, TA2), that enabled to measure bubbles velocity and its changes on the rise path ca. 160 mm. Ultrasonic heads were alternatively connected to a 10 channel defectoscope USIP 40 Box controlled by computer with a software UltraPROOf (by Krautkramer). In the tests only the measurement of echo run time from the rising bubbles were utilized because its very low signal-to-noise ratio that allows the exact estimation of gas bubble contours. The echo overshoot time resolution has been 2.5 ns, that equals to an alteration of bubble position in a liquid metal of ca. 0.004 mm. A pulse repetition frequency was set to a level of 14286 Hz, the highest setting value for 10 ultrasonic sensors, and signal amplification was set in such a way that the amplitude of signals over passed 20% of the level shown on Fig.1. The signal sampling period equals 0.7 ms. The recording period was set at 360 s, according to a principle proved by Sommerlatt [7] that to get representative results a sampling period have to be longer than 2minutes. For such recording time it is possible to record even several thousand rising bubbles and exactly estimate its rise velocity and rise zone, depending on the creation frequency of argon bubbles.

Likewise measurements of rising bubbles in a permanent magnetic field has been done with magnets of average induction B = 300 mT (by Teslameter model 5180). Two magnets 160x100x50 were spaced at 115 mm.

The resultant Lorentz' force in a magnetic field of induction B equals to:

$$\mathbf{F} = \mathbf{j} \times \mathbf{B} = \sigma_{el} (\mathbf{w} \times \mathbf{B}) \times \mathbf{B} \qquad (1)$$

where: **j** – current density, **w** – liquid metal velocity, σ_{el} – electric conductivity.

Drawing no 3 shows the measuring bench container with ultrasonic heads in magnetic field as well as directions and sense of Lorentz' force.

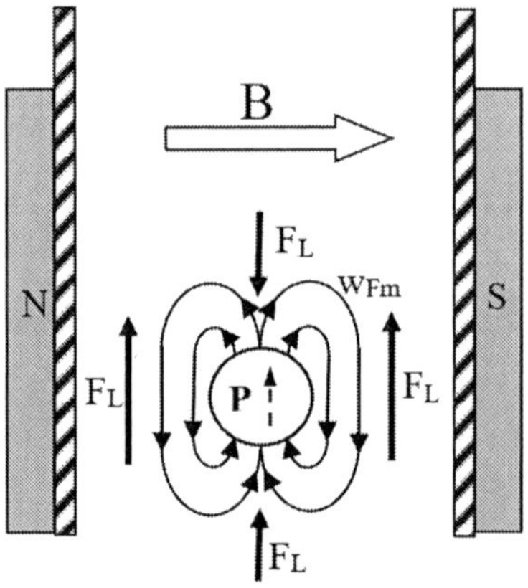

Fig. 3. Testing (measuring) bench in a magnetic field and Lorentz' forces acting on liquid metal in magnetic field: w_{Fm} – liquid metal velocity, P – gas bubble rising up vertically.

4 The results of studies and discussion

4.1 Rise velocity of bubbles

A typical measuring signal of rising bubbles of echo run time method is shown on Fig.4.

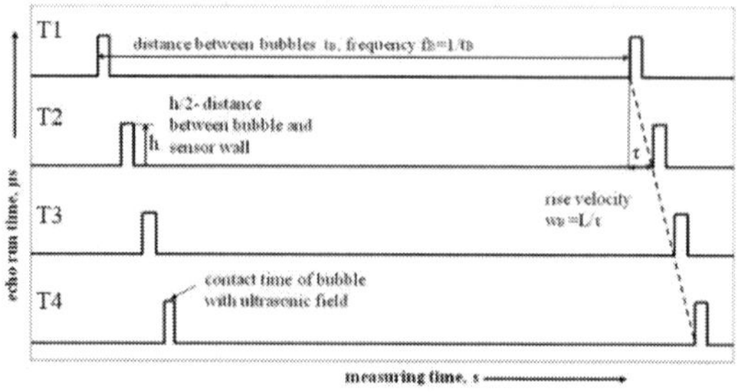

Fig. 4. Measuring signal of echo run time method.

The velocity w_B of the bubble (vertical component of velocity) was calculated by measuring the bubble time τ on the L road between two successive ultrasonic sensors. Fig.5. shows exemplary results of bubble velocity measurements from the height above the argon inlet nozzle for B = 0 mT and B = 300 mT. Fig.6. shows the dependence of average speed on the flow of argon volume calculated for the condition at the inlet nozzle also for B = 0 mT and B = 300 mT.

Starting from the volume flow of argon at the nozzle q_v = 1.8 l/h, the velocity of the bubbles in the magnetic field is lower than the velocity for B = 0 mT. Differences in velocities are increasing with increasing gas flow. The flow value q_v = 1.8 l/h can be taken as the beginning of the influence of Lorentz force on the velocity of rising bubbles.

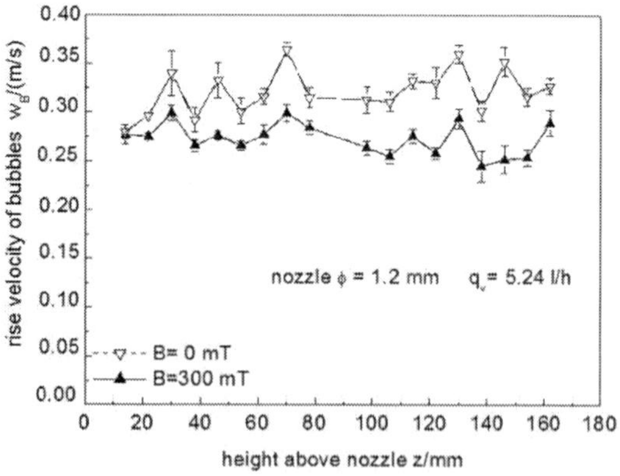

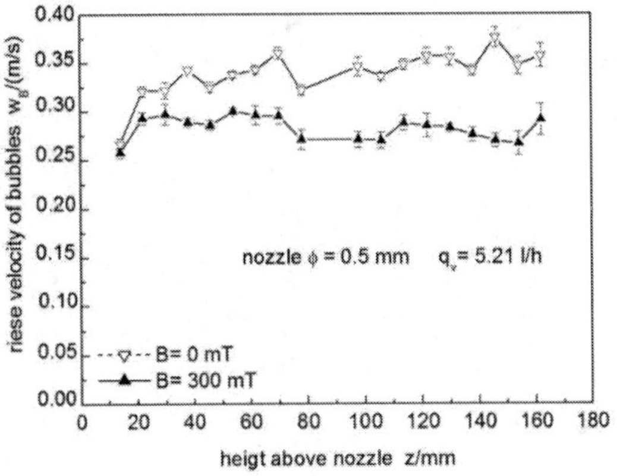

Fig. 5. Rise velocity of bubbles for $B = 0$ mT & $B = 300$ mT.

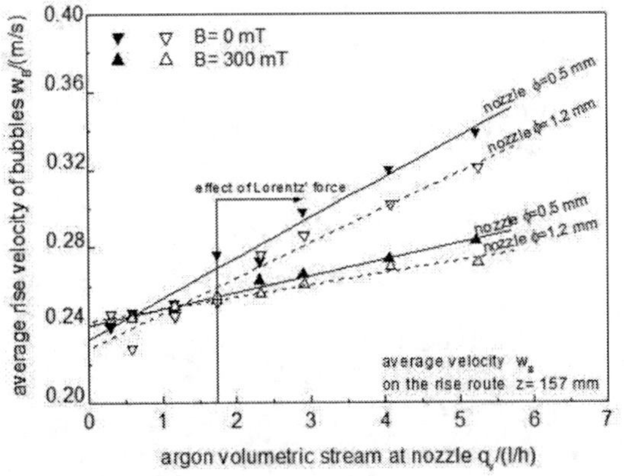

Fig. 6. Rise velocity of bubbles as a function of argon volumetric stream at nozzle.

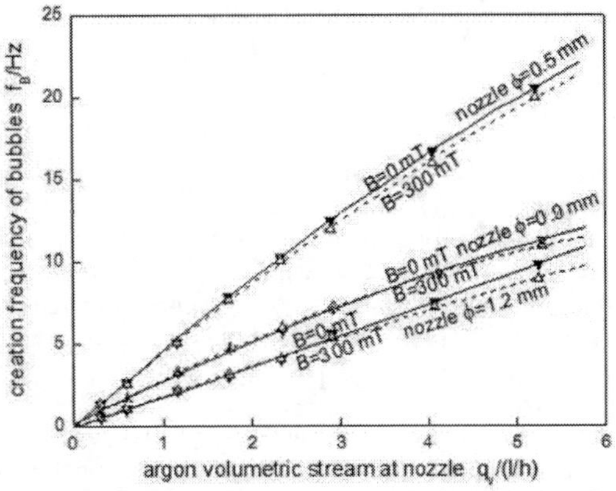

Fig. 7. Creation frequency of argon bubbles as a function of gas volumetric stream at nozzle for $B = 0$ mT & $B = 300$ mT.

The frequency difference between B=0 mT and B = 300 mT exists for argon volumetric flow greater than 6 l/h and nozzles of 0.5 & 1.2 mm. Anyhow they does not exceed 10% and seems to be of measurement errors origin. The diameter of rising bubble d_B is calculated with Mersmann's equate in a form of:

$$d_B = \{3\cdot\sigma\cdot d_o/(g\cdot\rho)+((3\cdot\sigma\cdot d_o/(g\cdot\rho))^2+(15\cdot q_v^2\cdot d_o/g))^{1/2}\}^{1/3} \quad (3)$$

This equate has been delivered on the base of force balance influencing an individual spherical rising gas bubble in motionless liquid. A gas bubbles are introduced into container through a nozzle with a diameter of d_o, and gas stream rises it up q_v.

The above given equation were used by Kaiser [5-6] to estimate diameter of rising gas bubbles in mercury, with gas flow stream q_v = 3.6 l/h. From among many others correlation s it is worth to mention Kantarcia [11] equations for calculating diameter of bubbles moving in barbotage columns.On a Fig.8 is presented relationship of bubble average equivalent diameter d_B on gas volumetric stream for a nozzle q_v. It is calculated by means of:

$$d_B = (6\cdot q_v/(\pi\cdot f_B))^{1/3} \quad (4)$$

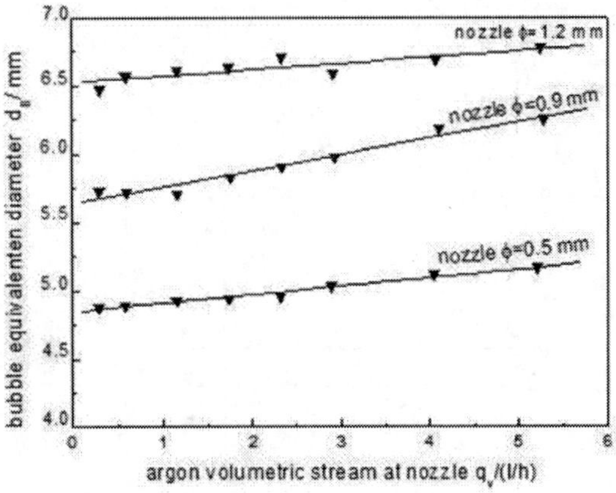

Fig. 8. Dependency of bubble equivalent diameter on volumetric gas stream at nozzle.

4.2 The creation frequency of bubbles and its dimension

The process of gas bubble creation is difficult to describe theoretically and because of it to estimate the frequency and dimension of bubbles empirical approximate equations are used. The basic equation to calculate the frequency of bubble creation is given in a work [9]:

$$f_B = 3.82\cdot10^{-5}\cdot(g^{0.75}\cdot q_v/(d_o^{0.75}\cdot\upsilon^{1.5})) \quad (2)$$

where: ν – kinematic viscosity coefficient, q_v – gas volumetric stream at nozzle.

The authors of study [10] pointed out that utilization of the above mentioned equation for a nozzle of 0.5 mm diameter and water-air flow gives errors of more than 100%. On a Fig. 7 are presented experimental study results showing a relationship of bubble creation frequency as a function of argon stream volume at nozzle. The results were calculated with Fourier quick transform FFT of measured signal. As the studies shown, in practice a magnetic field does not exert an influence on bubble creation frequency changes.

4.3 The gas phase rising zone

Most of works assigned to estimate a zone inside which a gas phase is present are based on analyzes of flow patterns obtained from model studies of bubble movement or gas stream, most often observe in a water filled container. It was proved by many investigators [12-16] who were studied the phenomenon's. From the practice point of view in estimation of gas phase flow zone the most important are computer tomography methods utilizing mainly capacitance or optical sensors [17-18].

Ultrasonic echo pulse method enables as well estimation of a zone inside which exists gas flow phase likewise in nontransparent fluids as in liquid metals. A knowledge of the flow region is helpful in fixing the best disposition of argon inlet nozzles in steel refining processes.

Drawings no's 9 &10 present the argon gas rise bubble distribution in a liquid metal along a route z = 157 mm above inlet nozzle for B = 0 mT & B = 300 mT. On the right side is presented the signal quantity of rise bubbles on each measuring level as well as flow zone width ΔX on some levels. The measurements has shown, that in magnetic field the argon bubbles rises straightforward and the rise zone is even 50% smaller than the rise zone without magnetic field. It is clearly visible on a Fig.10. where the flow of argon bubbles divides into two streams at B = 0 mT and at B =300 mT. The flow goes up nearly straightforward with a maximum flow zone width ΔX = 14 mm. The straight forward movement of bubbles bring forth a greater number of bubbles into the ultrasonic region of measuring sensor as it is salient for regions higher than 82 mm above nozzle. There the number of bubbles found in the ultrasonic field of the sensor is a couple of times greater than in a flow without a magnetic field.

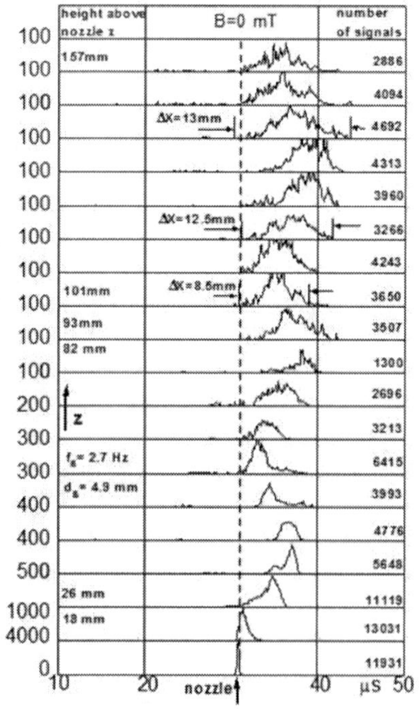

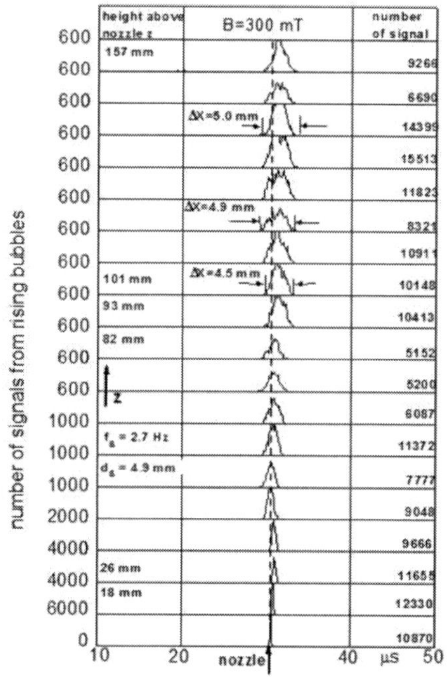

Fig. 9. Rising gas bubble distribution for nozzle ϕ = 0.5 mm and q_V = 0.6 l/h.

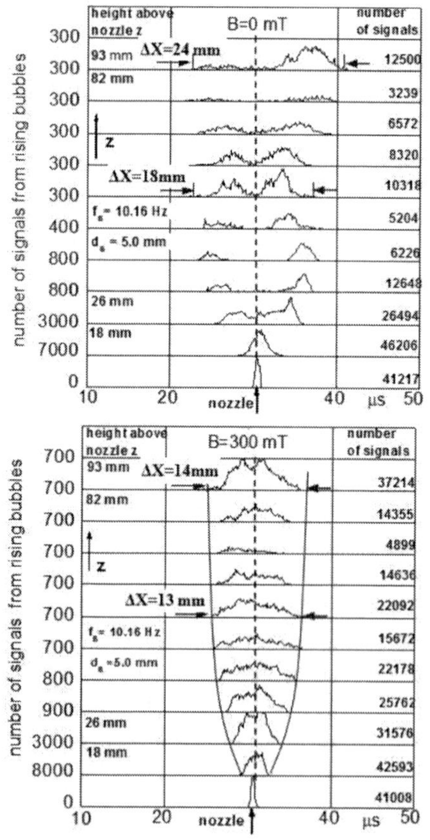

Fig. 10. Rising gas bubble distribution for nozzle ϕ= 0.5 mm and q_V = 2.4 l/h.

5 Summary

The topic of this paper is adaptation of ultrasonic echo pulse method to indentify two-phase liquid metal-gas. Apart from using it in defectoscopy the echo method maybe well used in measurements of characteristic parameters of liquid metal-gas flows. In these studies only a time measurement of echo route from moving bubbles method has been utilized as that method characterize the most favorably signal-to-noise ratio as well as an easy and precise definition of bubble position and by that estimation of the flow zone.

The accomplished measurements and its result analyzes allows to state the following conclusions:
- the Lorentz' force reduces the argon bubbles rise velocity in liquid metal of GaInSn; the influence of Lorentz' force on rise velocity begins with argon flow stream of q_V = 1.8 l/h
- a magnetic field does not change the creation frequency of gas phase bubbles therefore the average equivalent diameter of rising bubble has not change
- a proper values of gas phase creation frequency for different inlet nozzle diameters of gas introduced to liquid metal enables the right estimation of rise bubble dimension and by that the proper sizing of nozzle diameters for a specific processes of reining liquid steel
- a decrease of gas bubble rise velocity in a magnetic field increase the contact time of bubbles with liquid metal and in consequence enable the best utilization of blow-in gas
- the magnetic field influence also the bubbles movement trajectory –practically the bubbles rise up straightforward and its flow zone width is smaller than for the rising without the magnetic field; this effect maybe utilized for suitable disposition and selection of distances between argon inlet nozzles in order to obtain the best distribution of gas phase in processes of refinement and degassing liquid metals without forming dead zones in the refining vat.

References

1. C. Zahng, *Liquid metal driven by gas bubbles in a static magnetic field* (Dissertationt, Fakultät Maschinenwesen der Technische Universität Dresden, Dresden, 2009)
2. C. Zhang, S. Eckert, G. Gerbeth, International Journal of Multiphase Flow **31** (2005)
3. Z.H.Wang, S.D. Wang, X.Meng,M.J.Ni, International Journal of Multiphase Flow **94** (2017)
4. S.Boden, S. Eckert, G. Gerbeth, M. Simonnet, M.Anderhuber, P. Gardin, (*X-ray visualisation of bubble formation and bubble motion in liquid metals*. 6th Int. Conf. on Electromagnetic Proc. of Materials, Dresden, 2009)
5. E. Kaiser, Tm – Technisches Messen, **71** (2004)
6. E. Kaiser, Chemie Ingenieur Technik **76** (2004)
7. H.-D.Sommerlatt, E. Kaiser, Tm – Technisches Messen **71** (2004)
8. A. Andruszkiewicz, K. Eckert, S. Eckert, S. Odenbach, European Physical Journal Special Topics **220** (2013)
9. H. Brauer, *Grundlagen der Einphasen -und Mehrphasenströmungen* (Verlag Sauerländer, Aarau und Frankfurt am Main, 1971)
10. A. Andruszkiewicz, H.-D. Sommerlatt, Chemical and Process Engineering **29** (2008)
11. N. Kantarci, F. Borek, K.O. Ulgen, Process Biochemistry **40** (2005)
12. A.Sokolichin, C. Eingenberger, A. Lapin, A. Lübbert, Chemical Engineering Science **52** (1997),
13. A.Sokolichin, C.Eingenberger, Chemical Engineering Science **49** (1994)
14. Y. Pan, M.P Dudukovic, AICHE **46**, (2000)
15. N.G. Deen, T. Solberg , B.H. Hjertager, Chemical Engineering Science **56** (2001)
16. A. Tomiyama, I. Zun, H. Higaki, Y. Makino, T. Sakaquichi, Nuclear Engineering and Design **175** (1997)
17. 2and International symposium an Process Tomography in Poland 11th -12th September 2001
18. 4and International symposium an Process Tomography in Poland (Computer Engineering Department Technical University Lodz, September 2004)

https://doi.org/10.1051/matecconf/201824003004

Effects of nanoparticles on hydraulic cavitation

Min-rui Chen [1], *Jin-yuan* Qian[1,2,*], *Zan* Wu[2], *Chen* Yang[1], *Zhi-jiang* Jin[1], and *Bengt* Sunden[2]

[1]Institute of Process Equipment, College of Energy Engineering, Zhejiang University, Hangzhou, 310027, PR China
[2]Department of Energy Sciences, Lund University, P. O. Box 118, SE-22100 Lund, Sweden

Abstract. When liquids flowing through a throttling element, such as a perforated plate, the velocity increases and the pressure decreases. If the pressure is below the saturated vapor pressure, the liquid will vaporize into small bubbles, which is called hydraulic cavitation. In fact, vaporization nucleus is another crucial condition for vaporizing. The nanoparticles contained in the nanofluids play a significant role in vaporization of liquids. In this paper, the effects of the nanoparticles on hydraulic cavitation are investigated. Firstly, a geometric model of a pipe channel equipped with a perforated plate is established. Then with different nanoparticle volume fractions and diameters, the nanofluids flowing through the channel is numerically simulated based on a validated numerical method. The operation conditions, such as the temperature and the pressure ratio of inlet to outlet, are the considered variables. As a significant parameter, cavitation numbers under different operation conditions are achieved to investigate the effects of nanoparticles on hydraulic cavitation. Meanwhile, the contours are extracted to research the distribution of bubbles for further investigation. This study is of interests for researchers working on hydraulic cavitation or nanofluids.

1 Introduction

Blocked by a throttling element, the fluid flow velocity increases and the pressure decreases. If the pressure is below the saturated vapor pressure of the liquid, the liquid vaporize to small bubbles, causing hydraulic cavitation. The cavitation may do damage to the hydraulic equipment, on the other hand, cavitation can be used in variety fields, such as sewage treatment, mass transfer and heat transfer promotion. Orifice plates and perforated plates are widely used to induce cavitation, and many scholars have contributed in this filed.

Adopting the tool combined with theoretical, numerical and experimental methods, Ebrahimi *et al.* [1] investigated a high-pressure cavitation flow induced by a thick orifice plate to obtain a critical ratio of downstream pressure to upstream pressure. Rudolf *et al.* [2] conducted experiments to explore dynamics of the cavitation flow downstream orifice plates. Wang *et al.* [3] investigated the cavitation flow and cavitation-induced erosion by mean of experiment. Araoye *et al.* [4] explore the flow through multi-stage restricting orifices by mean of numerical simulation. Zhang *et al.* [5] employed the particle image velocimetry to investigate the cavitation flow characteristics downstream different kinds of triangular multi-orifice plates. He *et al.* [6] investigated the hydrodynamic cavitation in single-hole orifice plate by means of visualized experiment and numerical simulation. Tao *et al.* [7] investigated the sound generation mechanisms of orifice plates. Qian *et al.* [8-9] analyzed the Mach number on multi-stage perforated plates inside a high pressure reducing valve, and put forward a method to reduce the aerodynamic noises. Carpenter *et al.* [10] conducted experiments in detail to explore the effects of geometrical parameters on the production of oil in water emulsion. Kabeel *et al.* [11] investigated the effects of the alumina nanofluid concentration on the characteristics of sharp-edge orifice flow, basing on a model of a single-hole orifice plate. Shaaban [12] optimized the orifice flowmeter for liquid hydrogen and obtained a novel orifice flowmeter. Liu *et al.* [13] applied perforated plate flowmeters to measure the flow rate of cryogenic fluids. Aided by numerical simulation, the characteristics of perforated plates were investigated. Zhang *et al.* [14] investigated how a diffuser orifice plate affects the flow characteristics of air-cooled steam condenser. Qian *et al.* [15] explored the effects of orifice on pressure drop in a pilot-control globe

* Corresponding author: qianjy@zju.edu.cn

valve by mean of experiment and numerical simulation. Hilares *et al.* [16] employed the response surface methodology to investigate the efficacy of hydrodynamic cavitation which was adopted for pretreatment of sugarcane bagasse. Bokhari *et al.* [17] applied the orifice plate to cleaner production of rubber seed oil methyl ester, and conducted a parametric optimization by response surface methodology. Zeng *et al.* [18] applied an orifice plate to a dual-chamber air spring, and analyzed the effects of the shape and number of orifices on the damping characteristics. Karamah *et al.* [19] used a method of ozonation combined with hydrodynamic cavitation produced by orifice plate to disinfect Escherichia coli bacteria.

The effects of nanofluid on hydraulic cavitation are investigated in this paper. The hydraulic equipment used to induce the cavitation is a 5-hole perforated plate. The alumina nanofluid is employed, of which the nanoparticle volume fraction and the particle diameter are considered as variables. Additionally, the investigation is carried out for different operating conditions.

2 Analyses and modeling

In the present research, a pipe channel equipped with a 5-hole perforated plate is adopted as the device inducing hydraulic cavitation. The flow channel is symmetric, so a symmetric model is proposed, as Fig. 1(a), and the arrangement of the 5 holes is depicted in Fig. 1(b).

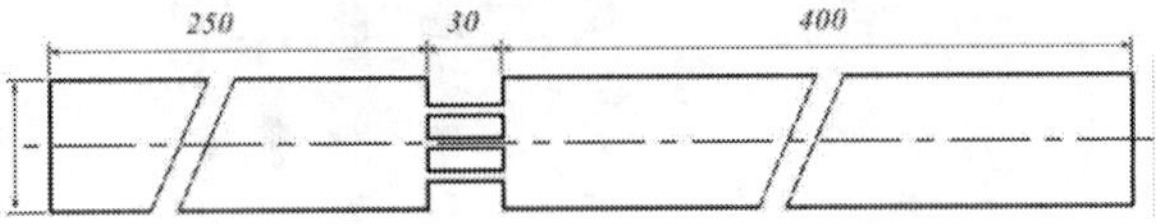

(a) The geometric properties of the model.

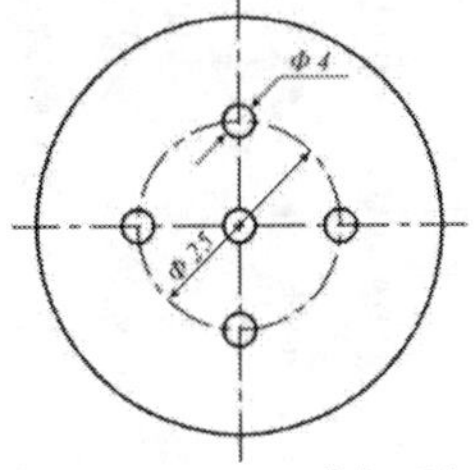

(b) The arrangement of the 5 holes.

Fig. 1. The geometric properties of the model.

The geometric model is discretized by the structured mesh. The elements located at the 5 holes and at the boundary layer have smaller size to promote the simulation accuracy. Mesh independence research has been conducted, and it is found that 1,921,100 elements are sufficient for the accurate numerical simulations.

In this paper, the Realizable k-ε turbulence model, the mixture model and the Schner-Sauer cavitation model are employed. In terms of the boundary conditions, the inlet is set as a pressure inlet and the outlet is set as a pressure outlet. Additionally, no-slip wall boundary is applied for all walls, and the standard wall function is adopted. In order to validate the numerical method, a numerical simulation has been conducted, and the results are compared with a reported experiment, which was carried out by Kim *et al.* [20]. The comparison of experiment and numerical simulation is listed in Table 1. The maximum relative error is 9.2%, which indicates the numerical method is reliable.

Table 1. Comparison of experiment and numerical simulation

Differential pressure (MPa)	Experimental mass flow rate (kg/s)	Numerical simulated mass flow rate (kg/s)	The relative error (%)
0.49	2.76	2.99	8.3
0.44	2.65	2.89	9.2
0.40	2.53	2.73	7.9

3 Results and discussion

In this investigation, the alumina nanoparticle volume fraction and diameter are considered variables. Besides, the numerical simulations are conducted under different operating conditions, that is the inlet and outlet pressure difference and the temperature are changed. The saturated vapor pressure of alumina nanofluid changes with the nanoparticle volume fraction, the particle diameter and the temperature. Tso *et al.* [21] have conducted investigations on nanofluid saturated vapor pressure, and a part of the results are listed in Table 2. The effects of alumina nanoparticle volume fraction and diameter at different operating conditions on cavitation are respectively discussed in the following sections.

Table 2. [21] The saturated vapor pressure of alumina nanofluid for different conditions [unit: Pa]

Particle diameter	Nanoparticle volume fraction	temperature		
		25 °C	30 °C	40 °C
13 nm	0.01%	3,176.6	4,294.8	7,493.3
	0.1%	3,135.8	4,222.7	7,458.9
	0.5%	3,114.1	4,197.7	7,364.2
20 nm	0.01%	3,169.0	4,251.8	7,412.8
	0.1%	3,173.7	4,217.3	7,420.2
	0.5%	3,105.1	4,171.5	7,333.3
80 nm	0.01%	3,138.5	4,220.8	7,395.6
	0.1%	3,092.9	4,200.1	7,376.1
	0.5%	3,057.9	4,137.7	7,259.3

3.1 The contours of vapour volume fractions

Figure 2 shows how the vapor induced by the hydraulic cavitation distributes on the symmetry plane. The nanoparticle volume fraction ranges from 0.01% to 0.5%, the nanoparticle diameter is 20 nm, and the outlet pressure is 0.1 MPa.

 https://doi.org/10.1051/matecconf/201824003004

In Fig. 2(a), the temperature is 30 °C, and the inlet pressure ranges from 0.4 to 1 MPa. It is found that when the inlet pressure is 0.4 MPa, the vapor volume fraction varies, while when the inlet pressure is higher, the vapor volume fraction contours are almost the same. It is indicated that with a lower inlet pressure, the nanoparticle volume fraction affects the cavitation of the nanofluid, while the cavitation is dominated by the inlet pressure with a higher inlet pressure.

In Fig. 2(b), the inlet pressure is 0.4 MPa, and the temperature ranges from 25 to 40 °C. It is found that the vapor volume fraction changes with the nanoparticle volume fraction. For different temperatures, the laws of the vapor volume fraction changing with the nanoparticle volume fraction vary.

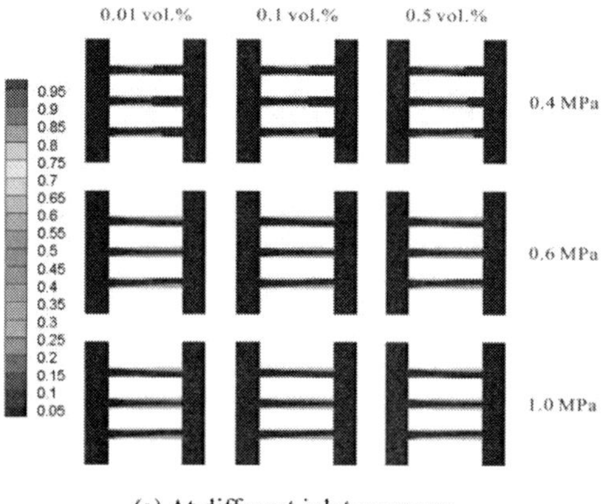

(a) At different inlet pressures.

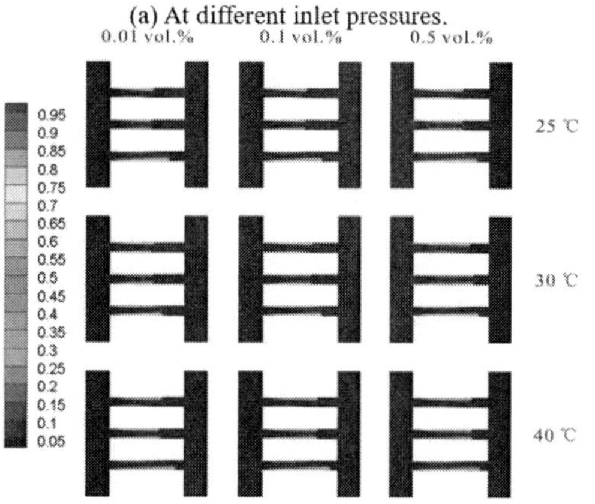

(b) At different inlet temperatures.

Fig. 2. The vapor volume fraction on the symmetric plane for different nanoparticle volume fractions.

Figure 3 shows the vapor distribution on the symmetry plane, of which nanoparticle diameter ranges from 13 to 80 nm, the nanoparticle volume fraction is 0.1%, and the outlet pressure is 0.1 MPa.

In Fig. 3(a), the temperature is 30 °C, and the inlet pressure ranges from 0.4 to 1 MPa. When the inlet pressure is 0.4 MPa, the vapor volume fraction changes with the nanoparticle diameter. When the inlet pressures are 0.6 MPa and 1.0MPa, the distribution of the vapor are almost the same. It indicates that for a low inlet pressure, the nanoparticle diameter impacts on the cavitation, while for a high inlet pressure, the cavitation is mainly dominated by the inlet pressure.

In Fig. 3(b), the inlet pressure is 0.4 MPa, and the temperature ranges from 25 to 40 °C. According to Fig. 3, the vapor volume fraction changes with the nanoparticle diameter. Additionally, the laws of the vapor volume fraction changing with the nanoparticle diameter are different, at different temperatures.

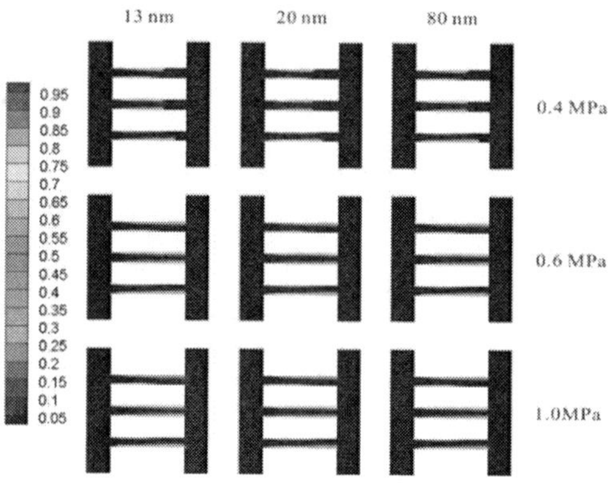

(a) At different inlet pressures.

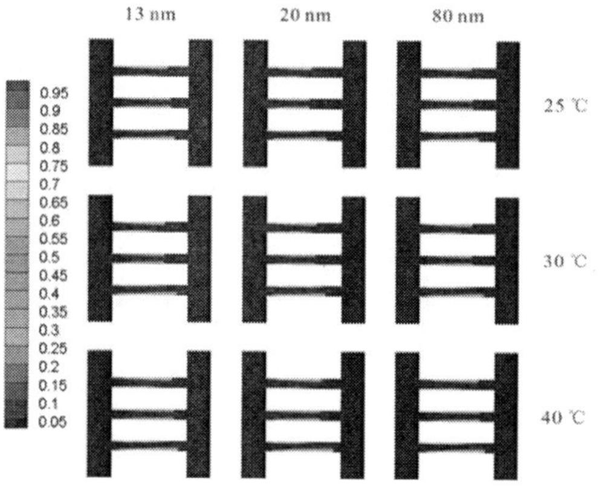

(b) At different temperatures.

Fig. 3. The vapor volume fraction on the symmetric plane for different nanoparticle volume fractions.

3.2 The cavitation number for different conditions

The cavitation number is a parameter used to assess the cavitation level. In fact, a smaller cavitation number means a more serious cavitation. The cavitation number on the cross sections of the perforated plate at different conditions are analysed in this section. The inlet of holes is located at 0 mm, and the outlet of holes is located at 30 mm.

Figure 4 shows the cavitation number with different nanoparticle volume fractions. At a high inlet pressure, the cavitation is mainly dominated by the inlet pressure, so the inlet pressure is set as 0.4 MPa in Fig. 4(a). It can be obtained that, a larger nanoparticle volume fraction leads to a longer section where the cavitation number keeps small. That is to say, for a larger nanoparticle volume fraction, the cavitation will distribute broader. So when the inlet pressure is 0.4 MPa, the cavitation level increases with the nanoparticle volume fraction.

In Fig. 4(b), the curves are grouped by temperature. The section where the cavitation number keeps the minimum is longest when the nanoparticle volume fraction is 0.5%, indicating that the cavitation level is the highest. In terms of the other two nanoparticle volume fractions, the section where the cavitation number keeps the minimum is always longer when the nanoparticle volume fraction is 0.1%, but the differences are very small when the temperatures are 25 °C and 40 °C. That is to say, though a relatively higher nanoparticle volume fraction leads to a more serious cavitation.

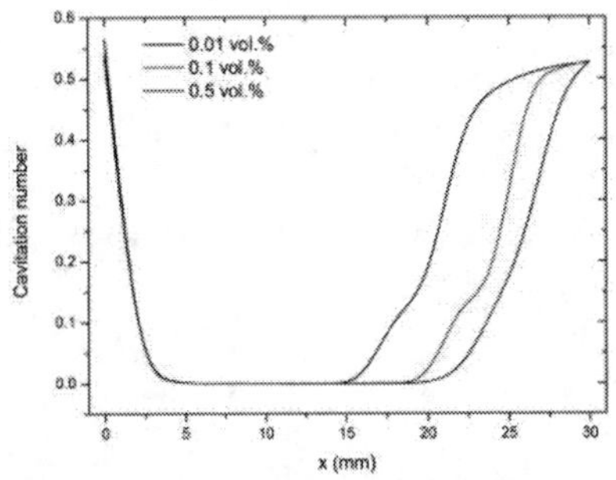

(a) The inlet pressure is 0.4 MPa.

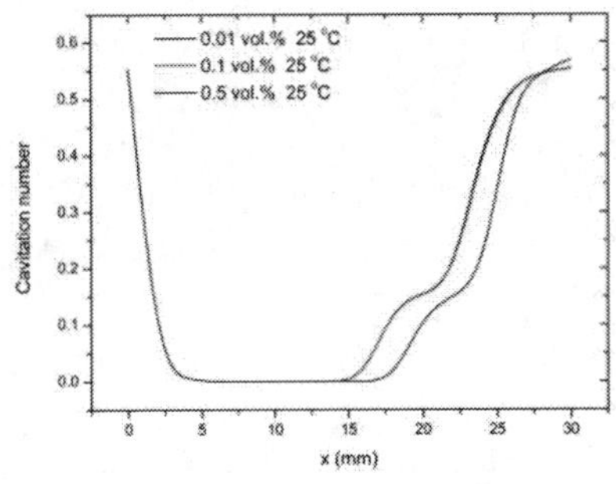

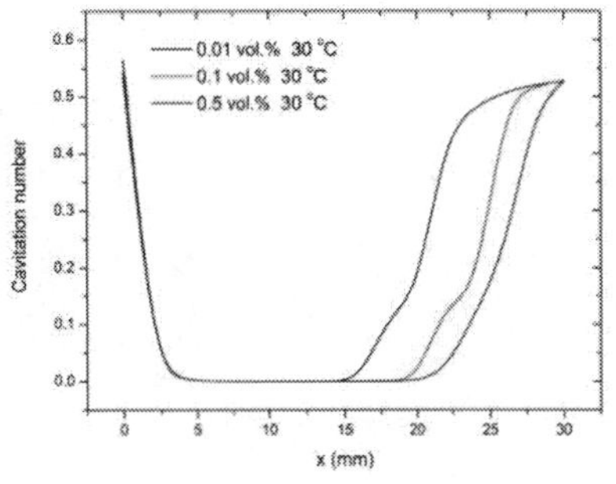

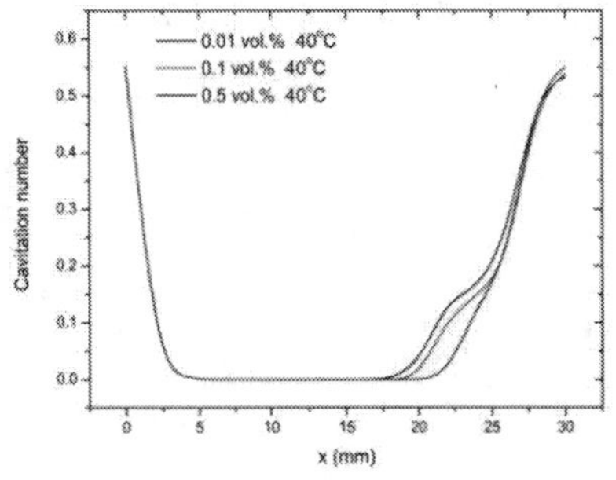

(b) At different temperatures.

Fig. 4. The cavitation number for different nanoparticle volume fractions.

Figure 5 shows the cavitation number with different nanoparticle diameters. Fixing the inlet pressure as 0.4 MPa, Fig. 5(a) depicts the cavitation number. It is found that the three curves are almost overlap, which indicates that when the inlet pressure is 0.4 MPa, the nanoparticle diameter affects the cavitation number insignificantly. Fig. 5(b) shows the cavitation number on cross sections of the perforated plate at different temperatures, and the curves are grouped by temperature. When the temperature is 25 °C, the difference of cavitation number is the largest, while at other two temperatures, the curves are almost coincident. It can be obtained that at a low temperature, the particle diameter affects the cavitation, while at a high temperature the effects are diminishing.

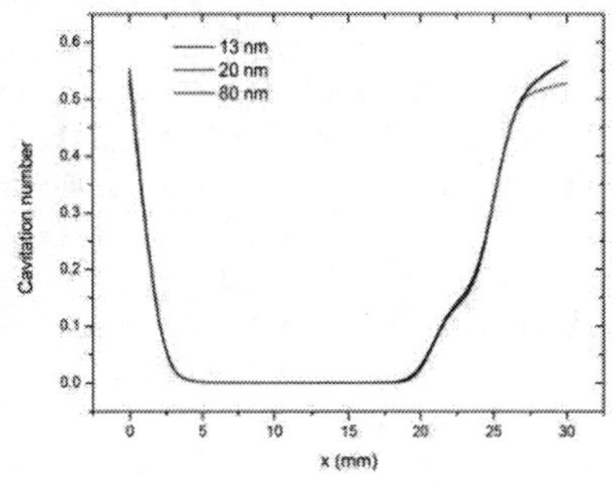

(a) The inlet pressure is 0.4 MPa.

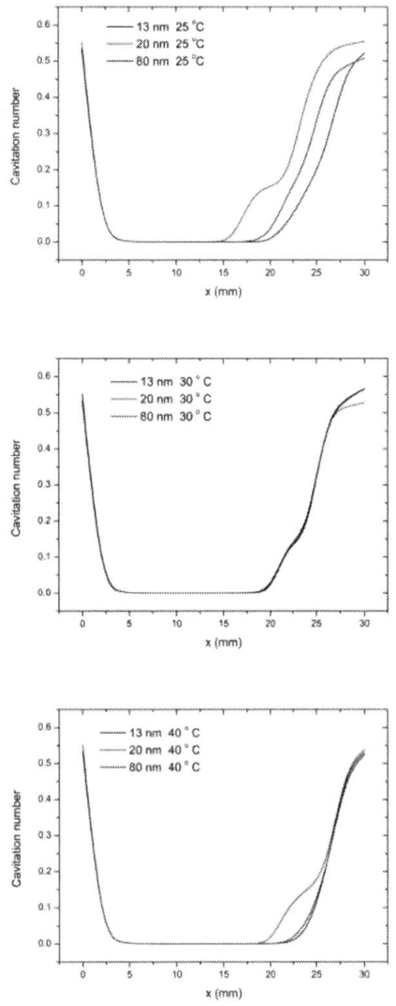

(b) At different temperatures.

Fig. 5. The cavitation number for different nanoparticle diameters.

4 Conclusions

The effects of nanoparticles on hydraulic cavitation at different inlet pressures and different temperatures are investigated by mean of numerical simulation. The nanoparticle volume fraction and diameter are considered as variables. The vapor volume fraction and the cavitation number are analyzed, and the following conclusions are obtained:

For a high inlet pressure, the cavitation is mainly dominated by the inlet pressure. On the other hand, for a low inlet pressure, with the increase of the nanoparticle volume fraction, the cavitation increases.

In the temperature range of this investigation, the cavitation increases with the nanoparticle volume fraction, and when at a low temperature, the nanoparticle diameter affects the cavitation significantly.

This work is supported by the Key project of Natural Science Foundation of Zhejiang Province, China through Grant No. LZ17E050002, and the Fundamental Research Funds for the Central Universities through Grant No. 2018QNA4013.

References

1. B. Ebrahimi, G. He, Y. Tang, M. Franchek, D. Liu, , J. Pickett, F. Springett, D. Franklin, Int. J. Therm. Sci., **114**, pp. 229-240, (2017)
2. P. Rudolf, D. Kubina, J. Kozák, M. Hudec, F. Pochylý, Meeting of Departments of Fluid Mechanics and Thermodynamics (2017)
3. Y. Wang, S. Zhuang, H. Liu, Z. Zhao, M. Dular, J. Wang, J. Mech. Sci. Technol., **31,** 7, pp. 3305-3315, (2017)
4. A. A. Araoye, H. M. Badr, W. H. Ahmed, Ann. Nucl. Energy, **104**, pp. 75-90, (2017)
5. K. Zhang, Z. Dong, W. Zhao, W. Ju, Y. Li, K. Geng, Z. Qin, L. Wang, J. Hydroelectr. Eng., **36**, pp. 56-64, (2017)
6. Z. He, X. Zhang, Y. Chen, C. Ji, J. Jiangsu Univ., Nat. Sci., **38**, pp. 416-422, (2017)
7. F. Tao, P. Joseph, X. Zhang, O. Stalnov, M. Siercke, H. Scheel, J. Acoust. Soc. Am., **142,** 2, pp. 561-572, (2017)
8. J. Y. Qian, M. Zhang, L. N. Lei, F. Chen, L. Chen, L. Wei, Z. Jin, Energy Convers. Manage., **119**, pp. 81-90, (2016)
9. J. Y. Qian, L. Wei, G. R. Zhu, Z. Jin, Energy Convers. Manage., **109**, pp. 86-93, (2016)
10. J. Carpenter, S. George, V. K. Saharan, Chem. Eng. Proc. Process Intensif., **116**, pp. 97-104, (2017)
11. A. E. Kabeel, M. Abdelgaied, Alexandria Eng. J., **55,** 2, pp. 1099-1106, (2016)
12. S. Shaaban, Int. J. Hydrogen Energy, **42**, pp. 14621-14632, (2017)
13. H. F. Liu, H. Tian, H. Chen, T. Jin, K. Tang, J. Zhejiang Univ. A, **17,** 3, pp. 230-239, (2016)
14. X. Zhang, T. Wu, Appl. Therm. Eng., **98**, pp. 179-188, (2016)
15. J. Y. Qian, B. Z. Liu, L. N. Lei, H. Zhang, A. L. Lu, J. K. Wang, Z. J. Jin, Int. J. Hydrogen Energy, **41,** 41, pp. 18562-18570, (2016)
16. H. R. Terãn, G. F. de Almeida, M. A. Ahmed, Technol., **235**, pp. 301-308, (2017)
17. A. Bokhari, F. C. Lai, S. Yusup, J. J. Klemes, M. M. Akbar, M. Kamil, J. Cleaner Prod., **136**, pp. 31-41,

(2016)

18. X. Zeng, L. Zhang, F. Yu, J. Zhou, J. Vibroeng., **19,** 4, pp. 2375-2389, (2017)
19. E. F. Karamah, R. Ghaudenson, F. Amalia, S. Bismo, American Inst. of Phys. Conf. Ser., AIP Conference Proceedings 1904, 020075 (2017); doi: 10.1063/1.5011932
20. 20. C.Y. Tso, C. Chao,Int. J. Heat Mass Transfer, **84**, pp. 931-941, (2015)
21. S. M. Kim, I. C. Bang, Ann. Nucl. Energy, **96**, pp. 12-18 (2016)

https://doi.org/10.1051/matecconf/201824003005

Influence of flow conditions and foam parameters on pressure drop and heat transfer in flow of fluids through channels filled with metal foams

Roman Dyga, *Małgorzata* Płaczek[2,*], and *Stanisław* Witczak

Opole University of Technology, Faculty of Mechanical Engineering, Department of Process Engineering, 5 Mikolajczyka Street, 45-271 Opole, Poland

Abstract. This paper reports the results of research into air, water and oil flow through pipes filled with open-cell metal foams, which form a material with a considerable potential for application in the design of process equipment. This study applied three metal foams with various geometrical parameters. The objective of the experiments involved the measurement of pressure drop within a relatively large range of the variability of the flow conditions. On the basis of data from the literature and analysis of the results of experiments, an assessment was undertaken concerned with the influence of the geometrical parameters of the foams as well as velocities and fluid properties on the flow regime and pressure drop. The results demonstrate that the theory concerned with the fluid flow through porous media has a limited application with regard to metal foams due to the considerable turbulence of the flow through such foams. If flow occurs in other conditions than laminar regime, the permeability of metal foams is relative not only to the geometrical structure of the foams but also depends on the properties and velocity of the fluid. The present study demonstrated that the assessment of the flow regime can apply the modified Reynolds number in which the characteristics dimension is defined on the basis of the parameters accounting for the geometrical foam structure. Three flow regimes were distinguished – laminar flow, transient Forchheimer and transient Froud flow. The ranges corresponding to the occurrence of the particular flow regimes were subsequently determined.

1 Introduction

Open cell metal foams form the type of cellular material that is relatively little known. Such foams consist of thin interwoven ligaments forming cellular materials, whose structure resembles a spatial skeleton comprising relatively large and hollow cells. The adjacent cells are connected by open "windows" – pores - located in the common walls. Such a spatial structure and the considerable porosity of the foams, which usually exceeds 90%, promote fluid flow through the foam skeleton and reduce the cost of fluid transport. The considerable thermal conductivity of the foam skeleton and continuous structure of the skeleton ligaments lead to the decrease of the thermal resistance. The technology of manufacturing open-cell foam provides adequate solutions needed to obtain this material from a variety of metal alloys. As a result, it is possible to utilize foams even in extreme operating conditions, in which high thermal and chemical resistance of material are required.

The above characteristics of open-cell metal foams have led to the identification of these materials as an alternative to more common structural packings applied in process equipment. The literature in the areas lists a number of potential applications of metal foams, such as heat exchangers, heat storage tanks and heat recovery steam generators as well as chemical reactors, including catalytic reactors.

The practical use of open-cell metal foams in the design of process equipment is still at an initial stage, and the insight gained into the phenomena occurring in the spaces occupied by metal foams, including hydrodynamic phenomena, needs further research. The adequate design of the process equipment, in which the fluid is pumped through metal foams needs to involve a process in which the influence of fluid properties and foam parameters on the flow regimes and the value of the pressure drop are understood. The analysis of the results of research carried out in this area for more than a dozen years indicates the existence of many deficiencies and ambiguities, which have encouraged the experimental work and theoretical studies reported here.

2 Hydrodynamics of flow through – State of the art

Fluid flow through metal foams is often described on the basis of the theory of flow through porous media. If this approach is applied, the parameters characterizing the foams include permeability K and inertial coefficient β.

* Corresponding author: m.placzek@po.opole.pl

Various reports in the literature contain a statement regarding the dependence of these quantities on the geometrical foam parameters, such as the foam structure, cell diameter, porosity and the specific surface [1–3]. However, we are not able to conclude unambiguously about the impact of these parameters on the value of permeability. For instance, as reported by Bonnet et al., [4], permeability of foams increases along with the square of the pore diameter, whereas the results gained by Bagci and Dukhan [5] as well as Wang and Guo [6] contain a conclusion that the relation is proportional. The authors of the work in [4] remarked about the effect of fluid properties on the permeability, whereas Kouidri and Madani [7] demonstrated that the permeability and inertial coefficient of foams are also relative to the fluid velocities. Although the results are less straightforward, similar observations were made by Medraj et al. [2] with regard to the transient and turbulent flows. The results found in [5] contain a statement that permeability of foams remains constant only in the conditions of laminar flow. Thus, the ranges corresponding to the occurrence of the particular flow regimes have not been explicitly demonstrated. The authors of papers in this area provide different values of the critical Reynolds number corresponding to the flow regime transitions. The reasons associated with this are related to various quantities which can be applied to define the hydraulic diameter d_h of the foam. A number of authors, including Bagci and Dukhan [5] postulate that the square root of the permeability should play the role of the hydraulic diameter. Other studies adopt various geometrical parameters of foam to represent d_h, such as the cell diameter [8, 9], pore diameter [10] and the specific surface [11].

The calculation of pressure drop $\Delta P/\Delta L$ accompanying flow through metal foams is most commonly implemented by application of the Forchheimer equation,

$$\left(\frac{\Delta P}{\Delta L}\right)_f = \mu_f K^{-1} u_f + \rho_f \beta u_f^2, \qquad (1)$$

which forms a development of the Darcy's law, describing laminar flow through porous media. The second term in the right hand side of Eq. (1) represents the component of the pressure drop resulting from the effect of inertia force exerted on the fluid flow. When inertia is accounted for, Forchheimer equation can be applied not only with regard to the laminar flow but also for transient flow. Along with the increase in the fluid velocity, the pressure drop associated with the inertia of the fluid determine the pressure drop to a much greater degree compared to the conditions when the fluid has a low velocity, and the term expressed as $\rho_f \beta u_f^2$ does not compensate for this increase in a sufficient manner. Hence, the pressure drop determined by the application of the Forchheimer equation can be characterized by a relatively high error if fluid is pumped through the foam with a considerable velocity. The reports in [6, 7, 12] suggest that the Forchheimer equation does not find application when the fluid velocity exceeds 10 m/s.

Pressure drop during flow through metal foams can be derived on the basis of the Ergun equation. It was originally developed for application to flows through a packed bed, but the potential for its use with regard to flows through foams was proposed e.g. by Liu et al. [13] and Beugre et al. [14]. Some authors suggest that the pressure drop accompanying the flow through metal foams on the basis of the correlations where the friction factor plays the measure of the pressure drop. Darcy-Weisbach equation forms an example of this, when it is applied with regard to foams in a number of studies [15–17].

The results of comparisons of the pressure drop derived on the basis of various correlations demonstrate that considerable differences are gained in terms of the calculated values. An opinion provided in [18] contains a suggestion that these discrepancies are due to the lack of the uniform and recognized definitions and approaches applied to determine cell and pore diameters, specific surface, porosity as well as other quantities characteristic for foams.

The majority of the research conducted to this date was realized by application of water or air.

In the own studies presented in this paper, oil was also used - in order to extend the range of viscosity changes of liquids - and the flow was carried out through foams with different geometrical structure.

3 Test stand and research methods

The research was conducted by application of three foams made of aluminum alloys (Fig. 1) – two from AlSi7Mg alloy and one made of Al 6101 alloy. The foams were characterized by the pore density of 20, 30 and 40 PPI (pores per inch). The analyzed foams varied in terms of the porosity ε, cell and pore sizes. The diameter of the pores d_p and cells d_c was derived on the basis of an analysis of microscopic images of the foam skeleton.

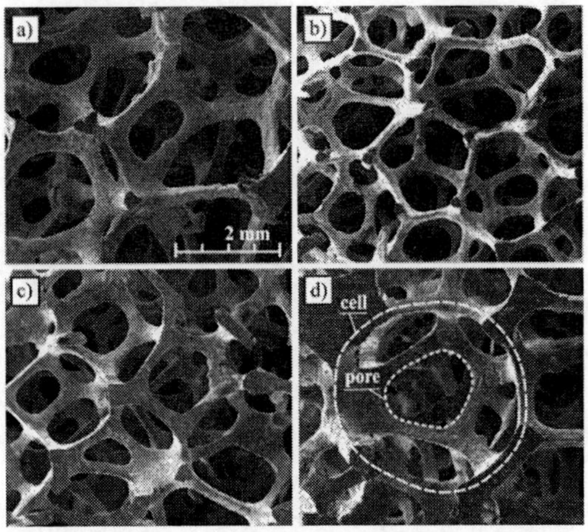

Fig. 1. Microscopic image of foams applied in the research: a) 20 PPI foam, b) 30 PPI foam, c) 40 PPI foam, d) cell and pore.

The fluid applied in the study included air, water and Velol-9Q machine oil. These fluids are distinct in terms of physical properties, including density ρ_{ol} and viscosity μ_{ol}. A summary of the basic parameters of the foams and the impact of temperature on the properties of the Velol-9Q oil is presented in Table 1.

Table 1. Foam parameters and oil properties.

Foam (alloy)	ε, –	$d_c \cdot 10^{-3}$, m	$d_p \cdot 10^{-3}$, m
20 PPI (AlSi7Mg)	0.9336	3.452	1.094
30 PPI (AlSi7Mg)	0.9435	2.255	0.712
40 PPI (Al 6101)	0.9292	2.386	0.824
density of Velol-9Q $\rho_{ol} = -0.669t + 872.76$ viscosity of Velol-9Q $\mu_{ol} = 0.117t^{-0.865}$			

The research was conducted by application of the experimental installation, whose diagram is presented in Fig. 2. Its main component is formed by a horizontal channel completely filled by a metal foam (a separate channel was performed for each foam). The channel was in the form of a pipe with the internal diameter of 0.02 m and comprised three sections – a measurement section with the length of 1.27 m, a section with the length of 0.92 m preceding the measurement section and applied for flow development and an outlet section with the length of 0.42 m located behind the measurement section.

The installation was supplied by compressed air. The liquids were routed into the installation using pumps. The flow rates of the fluids were measured by application of a set of flowmeters with high measurement accuracy. For the case of air, two mass flowmeters were applied for the measurements.

The direct objective throughout the research was to measure the pressure drop. This value was recorded as the difference of the pressure in two points in the channel located at a distance of 1 m from each other. The measurements of the pressure drop were realized by the application of five piezoresistance sensors with the total measurement range of 0–150000 Pa. Overpressure was measured in the channel as there was a need to determine the air density.

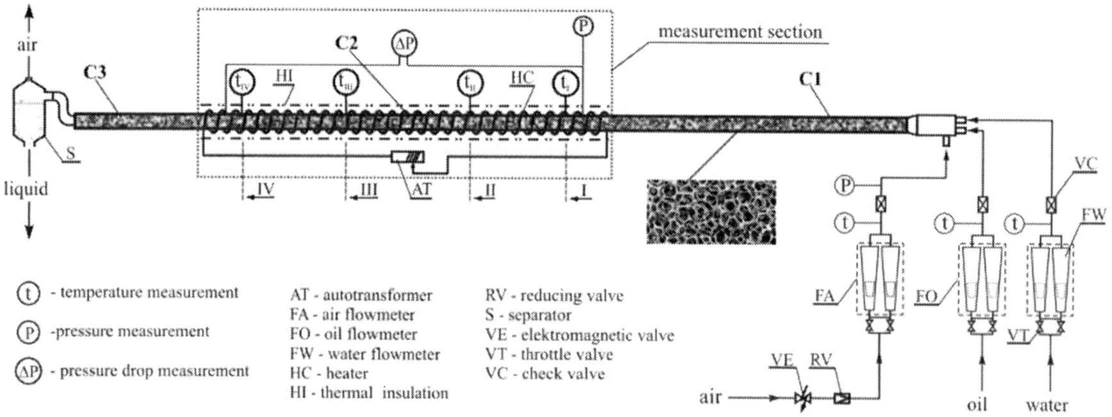

Fig. 2. Scheme of experimental setup.

The measurements were performed within a considerable variability of the flow rates $\dot{V}_f$ so as to provide the results for the laminar and turbulent flows. For all three foams, research was realized in the conditions of the same fluid velocities. For the reason for the diverse foam porosity, the notion of the Darcy velocity was applied u_f,

$$u_f = \dot{V}_f \frac{4}{\pi d^2}, \tag{2}$$

which denotes the mean velocity that is obtained by a fluid passing through an empty measurement channel with diameter d (i.e. not accounting for the presence of the foam).

Table 2. Conditions of the experiment.

fluid, f	u_f, m/s	Re_f^*, –
air, a	0.028–9.88	23–13244
water, w	0.003–0.270	30–4509
oil, ol	0.003–0.167	3–198

*Reynold number Re_f - Eq (3)

4 Results and discussion

During the flow of fluid through pipes filled with metal foams, the greatest pressure drop $(\Delta P/\Delta L)_f$ values were recorded for the case of the 40 PPI foam, whereas the lowest – for the case of the flow through 20 PPI foam. For the purposes of illustration, Fig. 3 contains the

https://doi.org/10.1051/matecconf/201824003005

results of the pressure drop during water flow through all three types of foam.

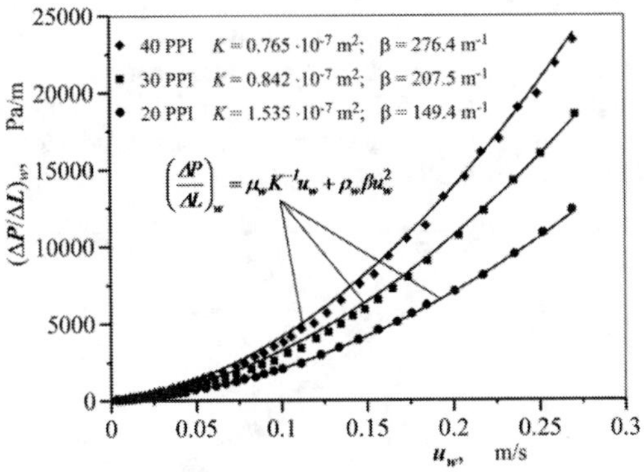

Fig. 3. Pressure drop during water flow.

As the foams applied in the research had similar levels of porosity, the differences in the pressure drop have to be attributed to the differences in the geometrical foam parameters. The 20 PPI foam has the largest cells, which results in the lower pressure drop of the fluid compared to foams with smaller cell diameters. Although the cell diameter in 40 PPI foams is only slightly greater than for the case of 30 PPI foam, the value of the pressure drop during flow through the first of them is considerably greater than for the case of the 30 PPI foams. The reason is likely to be associated with the specific geometrical structure of the 40 PPI foam skeleton. The cells of this foam form a more closed structure due to the occurrence of large nodes at the contact points between the fibers in the foam skeleton. Additionally, there are numerous deformations of the foam skeleton (Fig. 4), which leads to hinders the flow.

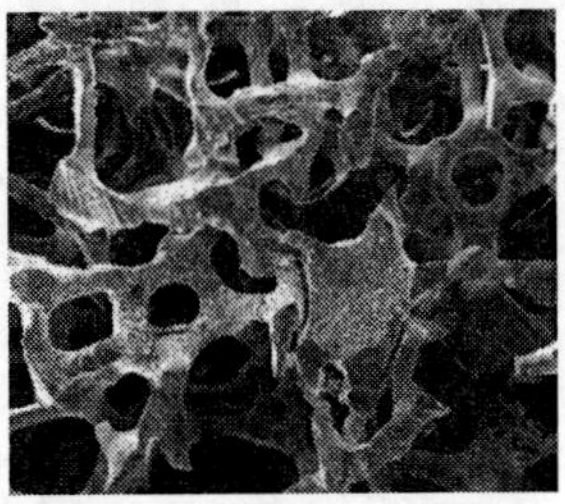

Fig. 4. Deformation of 40 PPI foam skeleton.

The value of the pressure drop is relative to the type and velocity of the fluid pumped through the foams. Under the assumption of the same velocity, the greatest pressure drop values were registered during the flow of oil, whereas the smallest for the air flow, as presented in Fig. 5 (in a logarithmic system due to the considerable variability of the fluid velocities). The pressure drop increase therefore as the viscosity of fluid increase.

The pressure drop also increases along with the increase of the fluid velocities, and the course of the variation of this drop can be described by the application of the Forchheimer equation, as illustrated in Fig. 3.

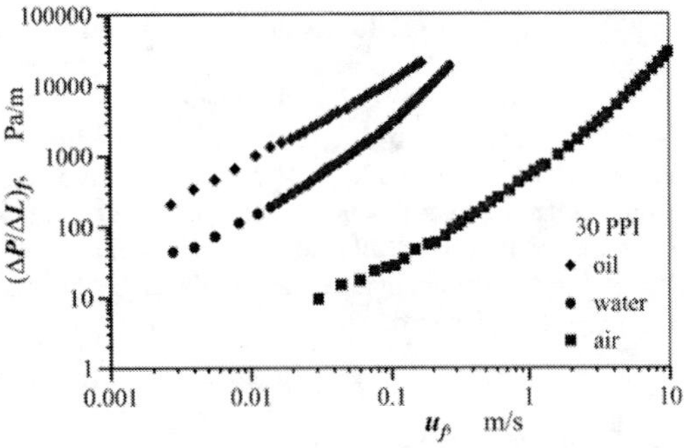

Fig. 5. Comparison of pressure drop during air, water and oil flow through 30 PPI foam.

The permeability K and inertial coefficient β were derived by the application of the measured pressure drop and the least squares method. The values of K and β for the analyzed cases of flows are presented in Fig. 6. The 20 PPI foam is characterized with the considerably greater permeability and lower inertial coefficient compared to the remaining foams. The permeability of the 30 and 40 PPI foams assumes similar values; however, the inertial coefficients of these foams are different. Both permeability and inertial coefficient assume different values for air, water and oil. These quantities are relative not only on the foam structure but also to the type of fluid pumped through the foams. For the case of the permeability, the differences for the particular fluids are in the range of 30%.

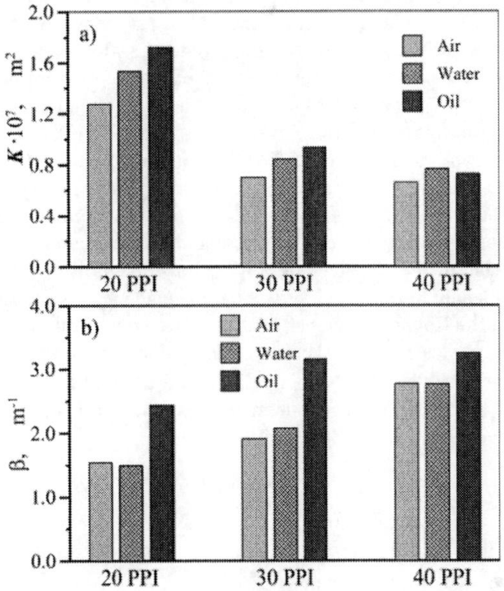

Fig. 6. Comparison of permeability (a) and inertial coefficient (b) of foams for air, water and oil flow.

In addition, the study offered the conclusion that the value of permeability is affected by the fluid velocity as presented in Fig. 7 on the basis of data gained with regard to 20 PPI foam. In a given range of the fluid velocities, permeability assumes a constant value, which is equal to below 0.12 m/s for water, and below 3.5 m/s for the air. For higher fluid velocities, we can note an increase of the value K. For the case of flow through 30 and 40 PPI foams, similar results were gained. For the velocity for which an increase of permeability was recorded, the flow is likely to cease to demonstrate laminar characteristics. For the case of the flow of oil, the increase of permeability does not occur since the oil flow demonstrates laminar characteristics over the entire range of the velocities due to its greater viscosity.

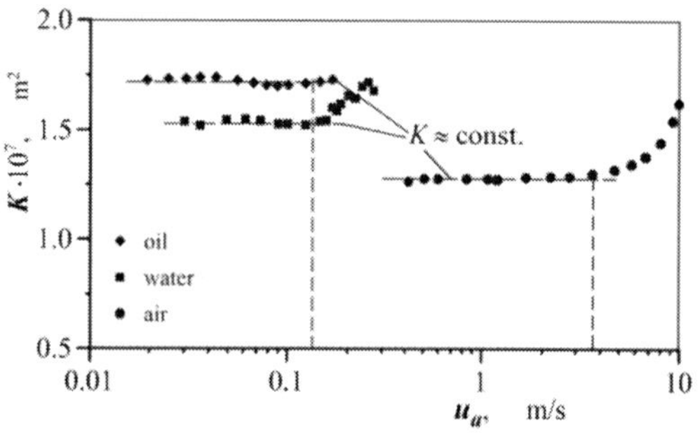

Fig. 7. Influence of fluid velocity on the permeability of 20 PPI foam.

The experimental data confirms the fact reported in the literature that for the case of metal foams, permeability and inertial coefficient do not have the characteristics of material constants. This, in turn, forms an impediment in the application of the Forchheimer equation in the description of the pressure drop through metal foams. In addition, the theory of flow through porous media accounts for the factors associated with the flow turbulence only to a limited degree. It seems that a more universal quantity characterizing the pressure drop through metal foams can be based on an adequately defined friction factor f.

The analysis of the value of the friction factor defined on the basis of the measured pressure drop and Darcy-Weisbach equation (f_{exp} - Fig. 8) confirms the possibility to define the friction factor by means of a relation that is common for various foams and fluids. The key role in this is attributed to an adequate statement of the Reynolds number Re_f. It was demonstrated that the Reynolds number should be derived on the basis of a mean actual fluid velocity (u_f/ε) and accounting for the geometrical foam parameters – i.e. porosity and pore diameter, in accordance with the relation,

$$Re_f = \frac{u_f d_h \rho_f}{\varepsilon \mu_f}, \tag{3}$$

where hydraulic diameter d_h:

$$d_h = \frac{\varepsilon d_p}{1-\varepsilon}. \tag{4}$$

The values of the friction factor presented in Fig. 8 assume similar values for the equivalent flow conditions, which confirm the assumptions adopted for the purposes of calculating the friction factor. Concurrently, the characteristic course of the variations in the friction factor confirms the occurrence of a variety of flow regimes. When the Reynolds number does not exceed the value of around 150, the friction factor decreases linearly (in a dual logarithmic system) along with an increase in Re_f, which demonstrates that the flow is laminar. For the case of oil, laminar flow occupies the entire range of the research that was conducted by application of this liquid.

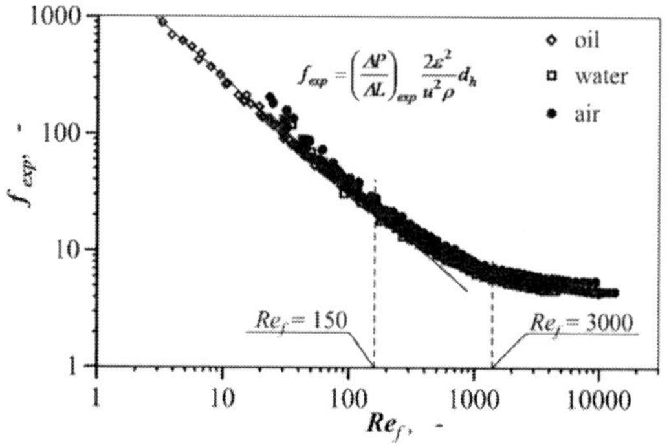

Fig. 8. Value of friction factor for investigated cases of flow.

Along with an increase of the fluid velocity and the Reynolds number corresponding to it, the flow gradually takes on the characteristics of turbulent flow. This is demonstrated by a considerable deviation of the friction factor from a straight line for the case when $Re_f > 150$. For the range of $Re_f \approx 150 - 1300$ the turbulence of the fluid stream are small and the flow has the characteristics of the transient Forchheimer flow. When $Re_f > 1300$, the friction factor demonstrates a tendency to take on a constant value in an asymptotic manner. This indicates that the fluid stream demonstrates a considerable level of instability. The turbulence of the of fluid remains in the range of the transient flow; however, due to the considerable effect of the inertial force, which prevails over the viscosity forces, this type of flow should be considered as a separate regime – called the transient Froud flow.

5 Conclusions

In this study, an assessment of the effect of flow conditions, including geometrical foam parameters, velocities and fluid properties on the flow regimes and pressure drop in channels filled with metal foams was performed on the basis of insight from the literature and results of experimental studies. In addition, the potential for the application of the flow theory through porous media in the description of flow through metal foams.

The following conclusions were derived from the study:

- Pressure drop is considerably relative to the foam porosity, cell size and structure of the cellular skeleton.
- Except for laminar flow, transient flow was observed within a considerable range of the variable fluid flow velocities – Forchheimer flow and Froud flow. The Forchheimer flow is characterized by a

small degree of stream disturbance while the Froud flow shows turbulent flow characteristics.

- A similarity was demonstrated in terms of hydrodynamic parameters during flow of fluids through foams with distinct geometrical structure. The role of a similarity criterion can be taken on by the Reynolds number, in which the hydraulic diameter is defined on the basis of the geometrical foam parameters, i.e. porosity and pore diameter (Eq. (4)).
- Flow is laminar when $Re < 150$. In the range of $Re = 150–1300$, transient Forchheimer flow is recorded, and when $Re > 1300$, we have to do with the Froud flow.
- Permeability and inertial coefficient of metal foams are relative to the geometrical structure of the foam skeleton as well as properties and velocities of the fluids pumped through them. Hence, these quantities cannot be directly derived on the basis of the geometrical foam parameters.
- Influence of flow conditions on permeability forms an impediment in the application of the standard laws with regard to flows through porous media, including the use of Forchheimer law for the description of the flow through metal foams.

References

1. D. Edouard, M. Lacroix, C.P. Huu, F. Luck, Chem. Eng. J. **144**, 299 (2008)
2. M. Medraj, E. Baril, V. Loya, L.-P. Lefebvre, J. Mater. Sci. **42**, 4372 (2007)
3. O. Smorygo, V. Mikutski, A. Marukovich, A. Ilyushchanka, V. Sadykov, A. Smirnova, Acta Mater. **59**, 2669 (2011)
4. J.-P. Bonnet, F. Topin, L. Tadrist, Transp. Porous Media **73**, 233 (2008)
5. Ö. Bagci, N. Dukhan, Int. J. . Heat Mass Transfer **103**, 879 (2016)
6. H. Wang. L. Guo, Chem. Eng. Sci. **155**, 438 (2016)
7. A. Kouidri, B. Madani, Mech. Mater. **99**, 79 (2016)
8. F.E. Teruel. Rizwan-Uddin, Int. J. . Heat . Mass Transfer **52**, 5878 (2009)
9. Z. Wu, C. Caliot, F. Bai, G. Flamant, Z. Wang, J. Zhang, C. Tian, Applied Energy 87, 504 (2010)
10. W. Xu, H. Zhang, Z. Yang, J. Zhang, Chem. Eng. J. **140**, 562 (2008)
11. N. Dukhan, P. Patel, Exp. Therm. Fluid Sci. **32**, 1059 (2008)
12. A. Hamadouche, R. Nebbali, H. Benahmed, A. Kouidri, A. Bousri, Exp. Therm. Fluid Sci. **71**, 86 (2016)
13. J.F. Liu, W.T. Wu, W.C. Chiu, W.H. Hsieh, Exp. Therm. Fluid Sci. **30**, 329 (2006)
14. D. Beugre, S. Calvo, G. Dethier, M. Crine, D. Toyea, P. Marchota, J. Comp. Applied Mathem. **234**, 2128 (2010)
15. R. Dyga, L. Troniewski, Archiv. Thermodynamics **36**, 139 (2015)
16. A.A. Sertkaya, K. Altınısık, K. Dincer, Exp. Therm. Fluid Sci. **36**, 86 (2012)
17. J.-S. Noh, K.B. Lee, C.G. Lee, Int.Comm. Heat Mass Transfer, **33**, 434 (2006)
18. A. Inayat, M. Klumpp, M. Lämmermann, H. Freund, W. Schwieger, Chem. Eng. J. **287**, 704 (2016)
19. W. Azzi, W.L. Roberts, A. Rabiei, Mater. Design, **28**, 569 (2007)
20. S.K. Dhiman, V. Verma, D.P. Rao, M.S. Rao, AIChE J. **51**, 3186 (2005)

Identifying synchronization between flow boiling inside two parallel minichannels using joint recurrence plots

Hubert Grzybowski*, *Iwona* Gruszczyńska, and *Romuald* Mosdorf

Department of Mechanics and Applied Computer Science, Faculty of Mechanical Engineering, Bialystok University of Technology, Wiejska 45 C, 15-351 Bialystok, Poland

Abstract. In this study the flow boiling inside two parallel tubes with a diameter of 1 mm was analysed in order to determine synchronization level between the channels. An experimental setup was built to investigate the pressure and temperature oscillation in parallel minichannels. During the experiments, the two-phase flow patterns were recorded by high speed camera and also the presence of vapour in channel outlet was measured by laser-phototransistor sensor. Various types of two-phase flow instabilities were observed in investigated the system. Experiment was carried out for various heat rate and mass flux. The method of identifying synchronization between flow boiling between parallel channel will be presented on measurement data recorded for heat rate q equal to 50.15 W and the average mass flux $\dot{m}$ equal to 38.8 kg/m^{2}s. The signal was subjected to a nonlinear analysis based on the joint recurrence plot (JRP) method. The JRP method was carried out in order to determine synchronization level between signals from parallel channels. Results of pressure and laser-phototransistor oscillation analysis shows that during flow boiling phase synchronization and phase shift between the channels can be detected using appropriate *RQA* indicators.

Nomenclature

JR	joint recurrence matrix
JRP	joint recurrence plot
l	length of diagonal line
L_{max}	length of the longest diagonal line
m	embedding dimension
$\dot{m}$	mass flux (kg·m^{-2}·s^{-1})
n	embedding dimension
N	number of samples
p	pressure (Pa)
q	heat rate (W)
PS	phase synchronization
$\mathbb{R}$	set of real numbers
R	recurrence matrix
RP	recurrence plot
RQA	recurrence quantification analysis
T	temperature difference (K)
t	time (s)
x	measured value
y	measured value
Greek symbols	
ε	threshold for *RP* computation
Θ	Heaviside function
τ	time delay (samples)
Subscripts	
1,2	number of channel
i, j	indices

1 Introduction

Two-phase flow boiling instabilities can occur in any heat exchanging system equipped in small diameter channels [1]. This makes it difficult to precisely determine the optimal operating conditions of the systems. Some flow oscillations may cause undesirable effects such as mechanical vibrations, burn-out and thermal fatigue. Instabilities are associated with formation of different flow patterns and also leads to formation of various pressure drop and wall temperature oscillations [2]. The oscillations caused by one type of instability are superimposed to another type which may lead to chaotic oscillations [3]. In parallel minichannels also flow maldistribution phenomenon took place [4]. All those effects reduce heat transfer effectiveness. This paper presents a method of identifying synchronization between flow boiling between two parallel minichannels system. The pressure and temperature fluctuations occurring during flow boiling in two parallel minichannels with diameter of 1 mm have been experimentally investigated and analysed. Data acquired for constant heat rate 50.15 W average mass flux $\dot{m}$ equal to 38.8 kg/m^{2}s was taken under consideration. The joint recurrence plot (JRP) method was used to measure synchronization level between signals from each

* Corresponding author: h.grzybowski@pb.edu.pl

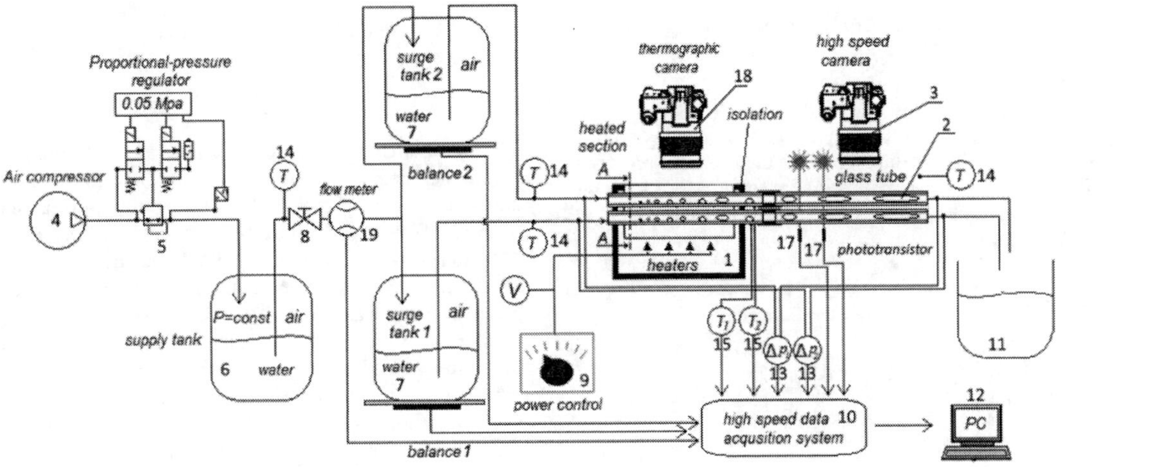

Fig. 1. Scheme of the experimental setup.

channel. Detection of phase synchronization (PS) by means of recurrences was done by calculation of recurrence rate directly from the RP. The $1/L_{max}$ indicator for laser-phototransistor series was used to reveal moments when pressure fluctuations are shifted relatively to each other.

2 Experimental setup

In this study the flow boiling inside two parallel tubes with a diameter of 1 mm was analysed. The experimental procedures were carried out for three different types of liquid supply system. Fig. 1 shows a diagram of the experimental setup in one of three liquid supply configurations. Distilled water (working fluid) has been pushed out of the supply tank (6 – Fig. 1) by means of the compressed air. A proportional pressure regulator (5 – Fig. 1) was used to maintain a constant overpressure of 15 kPa in the supply tank. A ball valve (8 – Fig. 1) was installed between two surge tanks (7 – Fig. 1) and the supply tank (6 – Fig. 1), where it was possible to measure and regulate the liquid flow rate.

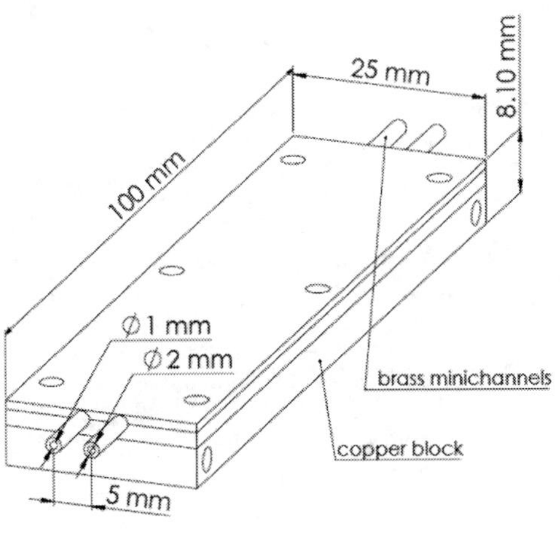

Fig. 2. Schematic diagram of test section.

The water flow starting in the supply tank goes to two separate surge tanks and then goes to a heated section (1 – Fig. 1) consisting of two circular brass minichannels with an inner diameter of 1 mm, an outer diameter of 2 mm and a length of 150 mm. The minichannels were installed in the copper block with dimensions of 25×8.1×100 mm, shown in Fig 2. The copper block was electrically heated and isolated by ceramic fibre wool. The heat was supplied to the pipes on a length of 100 mm. The pressure difference in separate channels was measured using the silicon pressure drop sensors MPX50DP with a response time of 0.001 s. The outlet of brass pipes (length 25 mm) was isolated and the thermocouples (type K with diameter of 0.081 mm and ± 0.8 K accuracy, response time 0.025 s), was placed in the distance of 10 mm from the pipe outlets. The amount of vapour flowing through the glass pipes was measured by laser-phototransistor sensors placed 20 mm downstream from minichannels outlets. During the experiment, the flow patterns inside the glass pipes were recorded using a high speed camera. The mass flux of water was measured by flow meter calculating the mass increment per unit time. All data was collected by a data acquisition system with a sampling frequency of 1 kHz. During the experiment various types of two-phase instabilities were observed. Similar experiment was conducted in our previous work [5] on similar single channel experimental setup.

3 Characteristics of recorded signals

Experimental measurements for parallel channels ware carried out at a constant amount of heat rate generated by heaters 50.15 W. The average mass flux $\dot{m}$ was computed from value recorded by flow meter divided by cross-sectional area of both channels. Thus, the average mass flux was equal to 38.8 kg/m^{2}s and oscillated in range ± 0.5 kg/m2s due to long periodic pressure drop oscillations. Pressure drop was measured by differential pressure drop sensor with a response time of 0.001 s. The content of the minichannel (vapour or liquid) has been qualitatively assessed using the laser-phototransistor sensors (17 –Fig.1). When the channel is filled with water, the sensor generates a high voltage signal (above 3 V). When the bubble or slug occurs inside the minichannel, then the sensor generates a low voltage signal (below 1 V). Above properties of laser-phototransistor sensors allow using to use the signal from sensors for qualitative assessment of the presence of bubbles or slugs in the minichannel. Fig. 3 shows input signal conversion to dash plot which will be used in next figures.

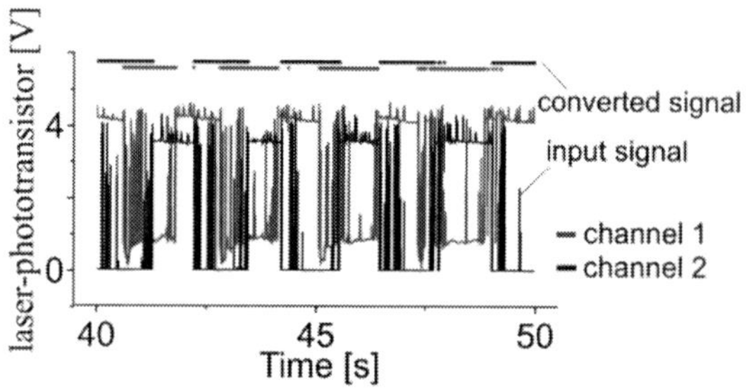

Fig. 3. Laser-phototransistor signal conversion to dash plot.

4 Synchronization analysis method

The times series measured for each channel can be treated as two dynamical system with mutual interaction. The channels can be synchronized by means of thermal or pressure effects. Joint recurrence plot (JRP) is a technique used to visualize recursive states occurring in one dynamic system with happening at the same time in another dynamic system. JRP is an RP extension that refers to only one system. The JRP developed by Romano [6] is associated with the creation of graph which shows times at which a recurrences in one dynamical system occurs in another dynamical system and can be used to detect phase synchronization. Obtained recurrence plots are based on information on the reconstruction of trajectories of attractors from time series using the delay method. This method assumes that the coordinates of the attractor points are calculated based on the order of the samples between which the distance equals of the time delay. Method was performed for the signals obtained from the pressure sensors and laser-phototransistors. However, the JRP and RPs were obtained for selected time windows from all signals containing 5000 samples. Values of time delay, embedding dimension and threshold were chosen for two time series, the gained JRP was the result of the multiplication of RPs (eq. 1,2) according to the formula (eq. 3):

$$\mathbf{R}_{i,j} = \theta(\varepsilon_x - \|x_i - x_j\|,\ x_i \in \mathbb{R}^m,\ i,j-1,\ldots,N, \quad (1)$$

$$\mathbf{R}_{i,j} = \theta(\varepsilon_y - \|y_i - y_j\|,\ y_i \in \mathbb{R}^n,\ i,j-1,\ldots,N, \quad (2)$$

$$\mathbf{JR}_{i,j} = \theta\left(\varepsilon_x - \|x_i - x_j\|\right)\cdot\theta\left(\varepsilon_y - \|y_i - y_j\|\right) \quad (3)$$
$$x_i \in \mathbb{R}^m, y_i \in \mathbb{R}^n, i,j-1\ldots N,$$

where: N is a number of considered states x_i, ε is a threshold distance with m (or n) – dimensional phase space, $\|-\|$ – norm, $\theta(\cdot)$ – Heaviside step function.

The mutual information method [7] was used to determine the time delay for the reconstruction attractors. Time delay for which the mutual information obtained first minimum is the correct value of τ. To determine the proper embedding dimension of the attractors in phase space was chosen the algorithm of the Nearest False Neighbor [8]. The main idea of this method is to see how the number of adjacent points in the immersion space varies with the increase embedding dimension. Proper embedding dimension for attractor reconstruction is defined as the dimension for which the fraction of false points reaches zero. The threshold (ε) can be set as a fixed or selected value to get the overall percentage of recursive points on the recursive graph (ie. to obtain the assumed density graph). In this case, ε was determined by the size of the reconstructed attractor and should be equal to 20% of the attractor diameter in m-dimensional phase space. The threshold value is associated with the number of black dots on the RP. When this value is too large, the RP becomes too black, whereas when it is low, too little information about system dynamics is provided.

Joint recurrence quantification analysis (*JRQA*) is a method for quantitative analysis of characteristic structures of overlapping recurrence graphs of two dynamic systems. *JRQA* is an extension of the *RQA* method, a tool proposed by Zbilut and Webber [9], which provides additional information about the recurrence graph of a single time series. Therefore *JRQA* generates the indicators which used to describe the joint dynamics of two time series. With the 13 *RQA* measures are described and developed by Marwan et al. [10]. For the analysis of this paper *RR* (Recurrence rate) and L_{max} (longest diagonal line) indicators were selected. The number of points in the JRP can be quantified by the recurrence rate, RR, which is a ratio of recurrence points to all possible recurrence points in the RP. It is defined as follows [6]:

$$RR = \frac{1}{N^2}\sum_{i,j=1}^{N} R_{i,j} \quad (4)$$

Second indicator L_{max} is defined as the time in which the state does not change or change very slowly:

$$L_{max} = \max(\{l_i; i = 1,\ldots,N_l\}) \quad (5)$$

The calculations have been made using the CRP toolbox for Matlab® [11].

5 Results and discussion

Fig. 4 presents pressure, temperature and laser-phototransistor time series in comparison with *JRQA* analysis results.

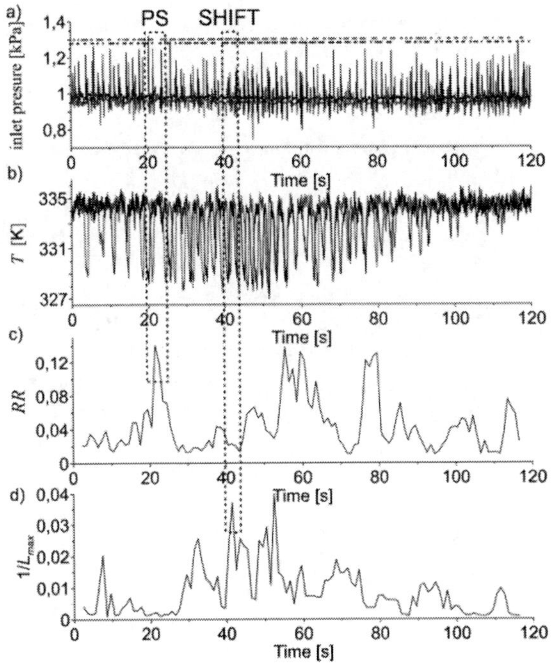

Fig. 4. Inlet pressure and temperature oscillations in comparison with analysis results.

Time series data acquired from first channel was marked in red color and for the second channel was black. The thermocouples time series graphs (Fig. 4a) shows that boiling in each channels causes periodic temperature oscillations, which can be shifted, synchronized or desynchronized. In addition, different flow patterns were observed in each channel (Fig.5).

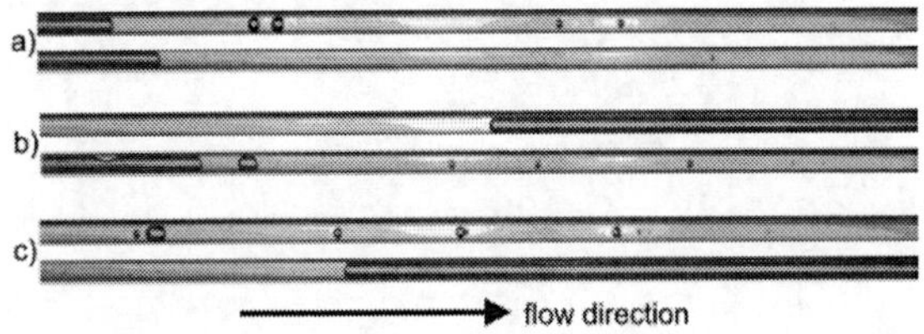

Fig. 5. Flow patterns recorded at the same time in both channels.

Mainly slug or bubbly flows with temporary flow reversal were observed. Due to the chaotic nature of time series associated with flow boiling there is hard to distinguish degree of synchronization between the channels. Therefore, an attempt was made to determine the degree of synchronization based on analysis of pressure and laser-phototransistor time series. The Fig.4c shows RR changes for pressure signals. Detection of phase synchronization (PS) by means of recurrences is done by calculation of recurrence rate directly from the RP. The method can detect PS even in time series which are strongly corrupted by noise [12]. According to the analyzed signals, high values of RR reveals moments when pressure peaks occurs at the same time in both channels. Example of PS ($RR > 0.1$) was marked with dashed line rectangle in Fig.4.

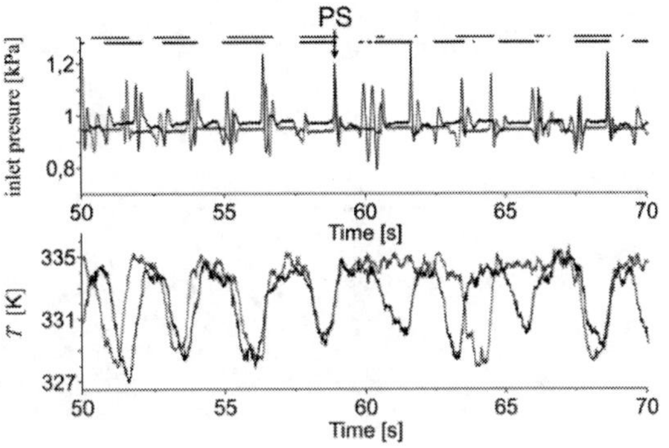

Fig. 6. Inlet pressure and temperature time series for which the moment of phase synchronization can be observed.

For the analyzed series, also the time interval between 58-60 s had the high value of the RR indicator. The part of the signals for which the moment of phase synchronization ($RR > 0.1$) is clearly marked is shown in the Fig.6. It can be also observed form temperature times series that before PS temperature fluctuation also tends to synchronization but after moment of pressure peak synchronization moment there is a temperature loss of synchronization.

Fig.5a presents the flow pattern observed when phase synchronization occurs. At the same time from both channel flows out vapour slugs. Another parameter that was considered during the analysis was L_{max}. This indicator is related with the positive Lyapunov exponents [13], which is a measure of the chaos in system. Rapid changes of this indicator detects chaos to periodic chaos transitions [6]. It was observed that high values of $1/L_{max}$ indicator for laser-phototransistor series reveals moments when pressure fluctuations are shifted relatively to each other or more precisely the moment in which pressure in one channel oscillates and in the second it does not. Two of those moments ($1/L_{max} > 0.02$) were marked with dashed line rectangle in Fig.7. Dependence between pressure oscillation shift and temperature increase can be observed. When pressure oscillations occur in one channel at the same time the temperature increases and then the same phenomenon with a certain shift is repeated in the second channel. Fig.5b and Fig.5c presents the flow pattern observed when phase shift occurs. In Fig.5b vapour slug already flowed out from the upper channel and in the bottom channel is just forming. Next in Fig.5c slugs flowing in bottom channel can be observed.

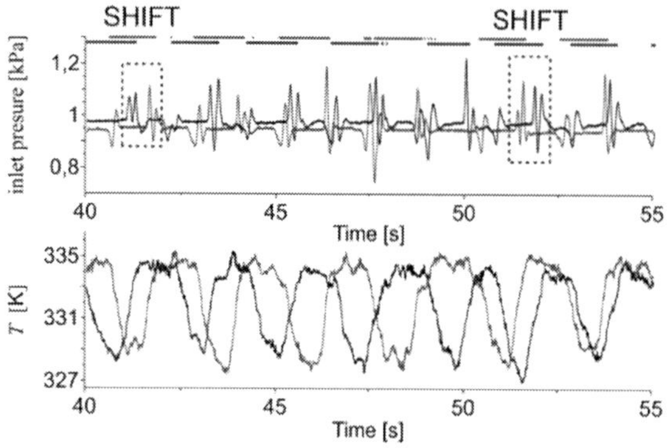

Fig. 7. Inlet pressure and temperature time series for which the moment of phase shift can be observed.

6 Conclusions

In this study the flow boiling inside two parallel tubes with a diameter of 1 mm was analysed in order to determine synchronization level between the channels. Due to the chaotic nature of time series associated with flow boiling it is hard to distinguish degree of synchronization between the channels. The JRP method was used to detect phase synchronization and phase shift between signals from each channel. Recurrence rate indicator was used in order to detect moment of phase synchronization and $1/L_{max}$ indicator was used to detect phase shift between the signals. It was observed that before PS temperature fluctuation also tends to synchronization but after the pressure peak synchronization moment there is a temperature loss of synchronization. Also dependence between pressure oscillation shift and temperature increases in both channel was observed. The recordings from high speed camera confirmed the areas detected by JRP analysis.

Acknowledgements

The study has been accomplished under the research project UMO-2015/17/N/ST8/03079 financed by the Polish National Science Centre.

References

1. T. G. Karayiannis, M. M. Mahmoud, *Appl. Therm. Eng.* **115**, 1372–1397 (2017)
2. L. C. Ruspini, C. P. Marcel, A. Clausse, *Int. J. Heat Mass Transf.* **71**, 521–548 (2014)
3. Chin-Jang Chang, R. T. Lahey, *Nucl. Eng. Des.* **167**, 307–334 (1997)
4. E. R. Dario, L. Tadrist, J. L. G. Oliveira, J. C. Passos, *Appl. Therm. Eng.* **91**, 924–937 (2015)
5. H. Grzybowski, R. Mosdorf, *Int. J. Heat Mass Transf.* **73**, 500–510 (2014)
6. N. Marwan, M. Carmen Romano, M. Thiel, J. Kurths, *Phys. Rep.* **438**, 237–329 (2007)
7. M. S. Roulston, *Phys. Nonlinear Phenom.* **125**, 285–294 (1999)
8. M. B. Kennel, R. Brown, H. D. I. Abarbanel, *Phys. Rev. A* **45**, 3403–3411 (1992)
9. J. P. Zbilut, A. Giuliani, C. L. Webber, *Phys. Lett. A* **246**, 122–128 (1998)
10. N. Marwan, N. Wessel, U. Meyerfeldt, A. Schirdewan, J. Kurths, *Phys Rev E* **66**, 026702 (2002)
11. N. Marwan, Cross Recurrence Plot Toolbox for MATLAB®, Ver. 5.21 (R31c), http://tocsy.pik-potsdam.de/CRPtoolbox/, accessed 2017-01-02
12. M. C. Romano, M. Thiel, J. Kurths, I. Z. Kiss, J. L. Hudson, *EPL Europhys. Lett.* **71**, 466 (2005)
13. L. L. Trulla, A. Giuliani, J. P. Zbilut, C. L. Webber, R*Phys. Lett. A* **223**, 255–260 (1996).

Possibility of the drag reduction effect application to reduce the cost of transport energy in water systems

Zbigniew Matras[1], and *Bartosz* Kopiczak[1,*]

[1] Institute of Thermal and Process Engineering, Cracow University of Technology, Cracow, Poland

Abstract. The main aim of this paper is to present possible drag reduction effect application to reduce the energy costs in water transport systems. Results obtained in laboratory scale experiment present possibility to enhance the drag reduction effect in pipe flow by simultaneous addition to the transported water small amount of high molecular polymers and surfactants. The hypothetic mechanism of drag reduction by polymer-micellar aggregates is presented. Qualitative analysis of polymer-micellar additives influence on shape and character of flow resistance curves is performed. Complex polymer-micellar solution flow resistance curves are compared with appropriate single additive polymer or surfactant one.

1 Introduction

The most commonly used drag reduction (DR) agents in scientific research as well as in engineering applications are high molecular weight polymers and surfactants. Adding even a small amount of these substances to the transported solvent induces significant and extremely advantageous, from the energetic point of view, reductions of pressure losses caused by fluid internal friction in the turbulent range of flow. This results in a significant increase in the flow rate without the necessity of increasing the power demand, or vice versa – to reduce the power demand while maintaining a constant flow rate [1-7]. Therefore, it provides large potential possibilities for application of this effect in different industry branches, particularly in oil industry [3, 8], or in heating [9], fire-fighting [10], transport of slurries, sludge and brines [2, 11], while increasing the efficiency of sewerage and anti-flood systems in time of heavy rainfall [12]. The drag reduction effect caused by polymers was utilized in designing an industrial installation for crude oil transportation in Norway, USA and India [2, 3, 8].

The new internal structure, which occurs when special additives are introduced into the solution, was perceived to be the cause of the drag reduction effect.
Adding a small amount of a polymer to the water causes a change of the internal structure of the solution creating so called macromolecules. Due to the characteristic chain structure of polymers in solutions, macromolecules form different conformations.

The drag reduction effect caused by small amounts of surfactant addition is particularly interesting from a cognitive point of view, as well as its utilitarian character [1, 2, 13-19]. Surfactants added to the water solvent create a low concentrated solution. When the surfactant concentration exceeds a certain critical level of concentration, individual surfactant molecules start to organize into new structures called micelles. The drag reduction effect in the flow of such a micellar solution takes place only when cylindrical/threadlike micelles are present in the solution. Reformation of spherical micelles into threadlike micelles is usually induced by adding a small amount of electrolytes (e.g. salts or alcohols) into the surfactant solution [14-17].

At no motion condition, when the fluid is at rest, the above mentioned structures are chaotic. Only during fluid flow shearing, both macromolecules and micelles start to arrange in a characteristic orientation, in accordance with the principle of minimum resistance. Solution concentrations are very small. In case of high molecular polymers it is order of magnitude of several or several dozen of ppm, and for surfactants it is order of tenths of mM/dm3.

The effect of drag reduction of flow by high molecular polymers or surfactants has also some limitations resulting from the properties of additives. The main disadvantage of high molecular polymers is their susceptibility to mechanical and thermal degradation. In case of surfactant solutions it is a certain critical Reynolds number. When exceeded, micelles orientation is lost, and the drag reduction effect collapses. The number depends not only on the pipe's diameter, but also on the solution concentration.

An important distinguishing factor for the phenomenon of drag reduction caused by surfactant additives comparing to drag reduction effect induced by high molecular polymers is the reversibility of solution degradation [1, 15]. In case where the value of Re number falls below critical Re value, the drag reduction effect occurs again. Moreover, the drag reduction effect in surfactant solution is not weakened in time of long-term liquid pumping comparing to the polymer solution.

* Corresponding author: bartosz.kopiczak@mech.pk.edu.pl

This is a great advantage for the transport of liquids over long distances.

The new and poorly recognized effect is phenomenon of fluid flow drag reduction by co-addition into the solvent both the high molecular polymer and the surfactant with salt. Mainly internal structure formation and chemical reaction process in polymer-micellar solutions are highlighted in the few published works related with this subject [18-20]. First attempts of drag reduction effect experimental examination had been performed, confirming that simultaneous introduction into the solvent polymer and surfactant additives combine positive features of their purely polymer and micellar analogues providing additional extension of drag reduction zones [15, 16, 21]. The pipeline drag reduction measurements reveal a considerable synergistic effect, that is, the mixed polymer–surfactant aggregates gives a significantly higher drag reduction comparing to their pure polymer or pure surfactant equivalents.

The aim of the work is experimental confirmation of the thesis that co-addition to the solvent both surfactants and high molecular polymer causes intensification of drag reduction effect and significant extension of drag reduction zones, comparing to the drag reduction effect obtained by application of pure polymer or pure surfactant additives. Possible drag reduction effect application to reduce the costs of transport energy in water transport systems will be proposed.

2 Laboratory set-up and DR additives

Measurements were performed using modern capillary-pipe rheometer, designed and constructed in the Division of Fluid Mechanics laboratory at the Cracow University of Technology [22]. Additionally rheological characteristics of the analysed solution have been assigned by the use of the Anton Paar rotational rheometer MRC 301 with the plate-cone measuring system. The following drag reduction agents were used for experimental analysis: the non-ionic polymer – poly(ethylene oxide) – [CH2 CH2 O]n (PEO) with high molecular weight Mv = 8·106 g/mol and the cationic surfactant – cetyltrimetyl ammonium bromide – [CH3(CH2)and5N(CH3)3]+Br~ (CTAB).
In order to lower the CMC value, salt sodium salicylate C7H5NaO3 (NaSal) was used. Distilled water was used as the solvent.

Adiabatic steady flow of homogenous solutions were examined in 4 different straight pipes with diameters between 10[mm] and 21[mm], in temperature equal to 25°C.

3 Experimental analysis and discussion

In order to identify rheological characteristics of analysed solution, flow curves have been drawn for each of them in form of relationship described by the equation (1):

$$\tau = f(\dot{\gamma}) \tag{1}$$

where
τ – the shear stress,
$\dot{\gamma}$ – the shear rate.

Interpretation of experimental results presented in form of flow curves indicates, that analysed solutions can be successfully approximated with the Ostwald de Waele power-law fluid model. Representative rheological characteristics are illustrated in figure 1.

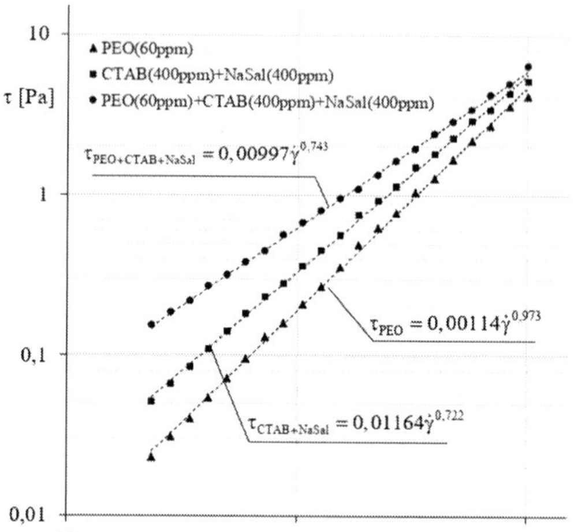

Fig. 1. Representative flow curves of polymer, surfactant and polymer-surfactant solutions.

In order to perform a better interpretation of the effect of the simultaneous addition of the polymer, the surfactant and salt, comparing to the adequate addition of pure polymer or pure surfactant, measurement data was presented in the modified system of "pseudorheostable" numbers [Re_M, c_{fM}], and described by formulas (2) and (3) [7]:

$$c_{fM} = \frac{\frac{D\Delta p}{4L}}{\frac{\rho u_m^2}{2}} \left[\frac{2(n+1)}{3n+1} \right]^{2,5} \tag{2}$$

$$Re_M = \frac{D^n u_m^{2-n} \rho}{K\left(\frac{3n+1}{4n}\right)^n 8^{n-1}} \left[\frac{2(n+1)}{3n+1} \right]^{-2,5} \tag{3}$$

where:
D – the pipe diameter,
ρ – the fluid density,
n, K – the flow index and the consistency constant of power-law fluid model.

The flow resistance curves of rheostable (purely viscous) non-Newtonian fluids in such defined dimensionless numbers system are transformed to a single curve – in the whole range of the modified Reynolds number (3) – identical to the classical Newtonian curve described in the laminar range by Fanning equation and in the turbulent flow by Blasius formula. The choice of such a coordinate system was dictated additionally by the fact that it facilitates the identification and description of the characteristic drag reduction flow zones. In this modified system of pseudorheostable numbers [Re_M, c_{fM}] each deviation of the experimental flow resistance curve which indicates abnormal drag reduction from pseudorheostable Blasius curve, allows the identification of specific additives' influence (polymers or/and surfactants with salt) on a range of analyses of the drag reduction effect.

Representative flow resistance curves of the analysed polymer-micellar solution and the corresponding purely polymer solution and purely micellar solution are presented in figure 2.

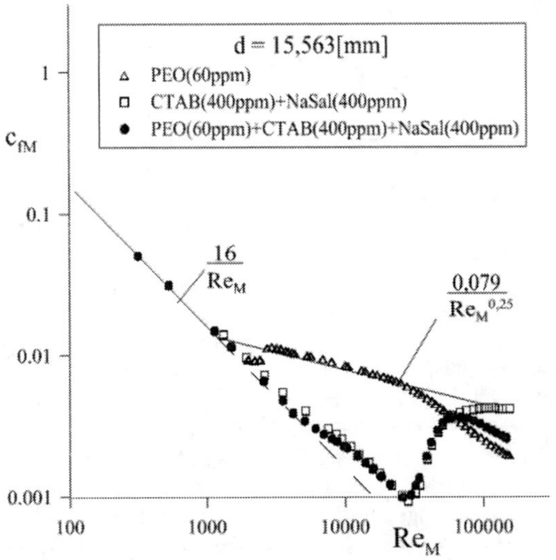

Fig. 2. Representative experimental flow resistance curves of polymer, surfactant and polymer-surfactant solutions.

The analysis of presented flow resistance curves indicates that in the laminar range of flow simultaneous addition of polymer and surfactant to the solvent causes inconsiderable increase of flow resistance and significant extension of the stable transitional zone of flow. The existence of the third significantly extended reduction zone is also observed, in which the viscoelastic properties of the solution are the dominant factor. It is particularly well illustrated by figure 3.

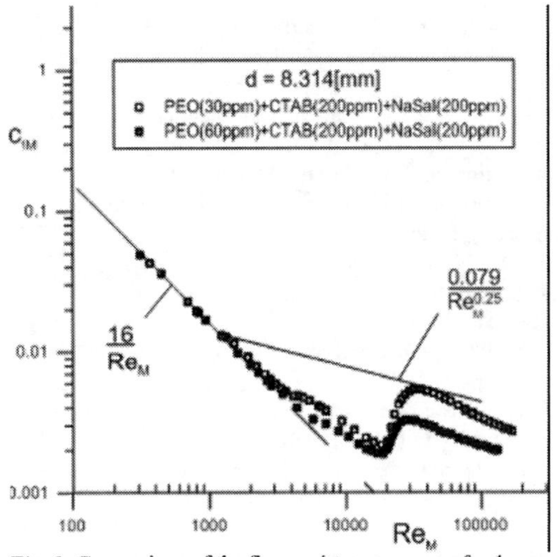

Fig. 3. Comparison of the flow resistance curves of polymer-surfactant water solutions for different polymer concentration.

In comparison to a purely micellar solution, the collapse of drag reduction is normally observed in this region. It should also be noted that addition to the micellar solution of even small amounts of high molecular weight polymer (about 10-100[ppm]) causes a reduction of the non-Newtonian properties of the solution.

The results of drag reduction measurements analysis indicate the effect of pipe diameter influence on drag reduction efficiency. Increasing the pipe diameter d results in clear extension of the stable transitional zone towards higher values of the Reynolds number. Moreover, decreasing the pipe diameter value d results in an increase of the drag reduction effect in the third additional turbulent range of flow – figure 4.

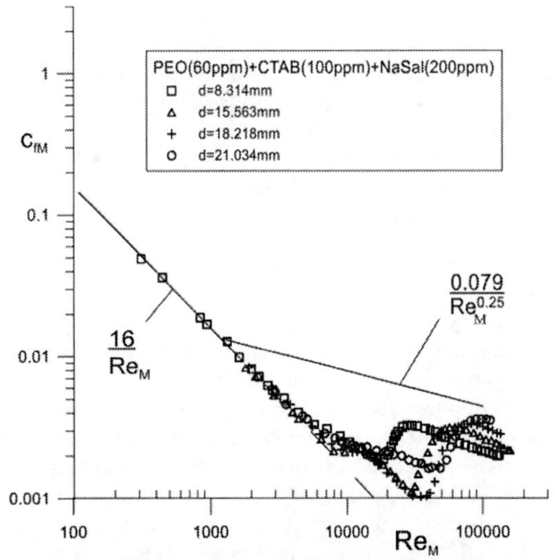

Fig. 4. Representative experimental flow resistance curves of polymer, surfactant and polymer-surfactant solutions.

Simultaneous introduction of small amounts of polymer and surfactant additives to the solvent provides initiation of micellarization process at much lower concentration, comparing critical micelles concentrations (CMC). This concentration is called the critical aggregation concentration (CAC) [15, 20].

In case of an ionic surfactant mixed with a counter charged polyelectrolyte only a small part of the polyelectrolyte is adsorbed by the surfaces of the micelles. Furthermore, CAC has a lower order of magnitude than the original CMC due to the facts electrostatic or hydrophobic interaction occurs between the polymer and the surface of the micelles. Polymer can trigger a certain amount of concentrated counter-ions, as they combine with the micelles.

In the initial state, there is a long polymer chain with a great number of concentrated counter-ions. The final state of the mixture has single threadlike micelles with a part of polymer macromolecule chain coiled around rigid micelles. According to [15, 20, 21], these molecules form the so-called aggregates.

Such newly created polymer-micellar solution can be characterized by a lower susceptibility to mechanical degradation during flow or its degradation can be almost invisible. Schematically illustrated aggregates (figure 5) subjected to the shear stress take orientations consistent with the aforementioned principle of minimum resistance. With increasing value of the Reynolds number, internal friction forces recoil, stretch and extend the aggregates leading to the laminarization of the initial phase of the turbulent flow.

Therefore, it may be hypothesized that the rigid rodlike micelles, which create the core of the aggregates, are responsible for reducing the flow resistance in the extended transitional zone between the laminar and turbulent flow.

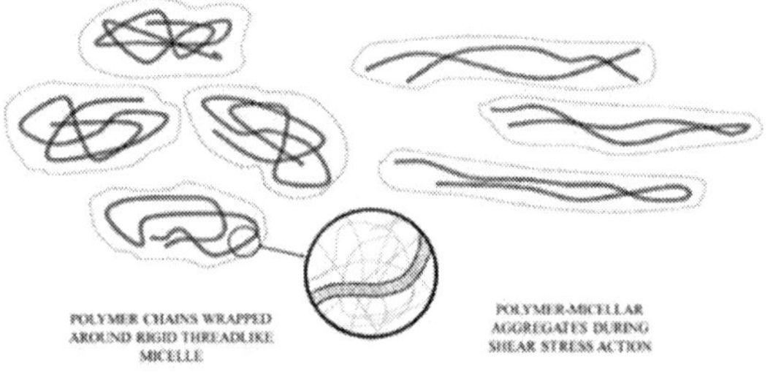

Fig. 5. Polymer-micellar aggregates.

The aggregates and micelles are responsible for transmission of internal friction in the liquid. The value of the critical Reynolds number for which the transition to the turbulent zone is observed is greater for polymer-micellar solutions. This means that the stable transition zone is extended. The reason for such behaviour can be the partial disintegration of aggregates to original forms, i.e. micelles (formed from the surfactant) and macromolecules (formed from the polymer) due to a significant increase of the shear rate. From this moment, both micelles and macromolecules interact separately on the transported solution causing a further drag reduction effect. Passing further in the turbulent range of flow micelles lose their orientation and no longer have a major impact on the drag reduction. A key role is played in this zone by the polymer. Not having undergone an earlier degradation, the polymer macromolecules still cause the low reduction.

In drag reduction caused by the use of polymer-surfactant solution, one cannot talk about the so-called collapse of the drag reduction. It occurs permanently over a wide range of Reynolds numbers. In the turbulent zone polymer macromolecules undergo a certain mechanical degradation. Decreasing the shear rate leads to the reconstruction of the solution's internal structure. As a result of electrostatic interaction, the recreated micelles are combined with the polymer chains by coiling around them. These chains are much shorter and such newly created aggregates do not have the same rheological properties as the original ones. This results in a slight increase of the flow resistance in comparison with a freshly prepared solution.

4 Conclusions

The experimental analysis of the influence of simultaneous addition of polymer and surfactant additives to a water solvent indicates that in the laminar range of flow the simultaneous addition of a polymer and a surfactant to the solvent causes an inconsiderable increase of the flow resistance comparing to the pure solvent's flow resistance. A significant extension of the stable transitional zone of flow has been observed, however. The analysis of experimental flow resistance curves (fig.2-4) allows to observe the existence of the third significantly extended reduction zone in the turbulent range of flow. In this zone the viscoelastic properties of the solution are the dominant factor. It is found that in this zone the drag reduction is influenced by the polymer concentration in the solvent. The higher concentration of polymer in the solution is applied, the greater drag reduction in the flow is obtained.

Simultaneous addition of a polymer and a surfactant with salt significantly reduces the mechanical degradation of the polymer-micellar solution's internal structure.

The comparison analysis indicates that polymer-micellar solutions combine and intensify the positive features of their pure polymer and micellar equivalents. Moreover, they provide efficient reduction of flow resistance in a wider range of the Reynolds number.

Significant increase in the flow rate without the necessity of increasing the power demand, or vice versa – to reduce the power demand while maintaining a constant flow rate, which can be obtained by co-addition of polymer and surfactant to water solvent provides large potential possibilities for application of this effect in different industry branches. There are already know applications of pure surfactant solution in heating [9]. As it is expected, the drag reduction effect is accompanied by the effect of reduction of heat transfer abilities. It is anticipated from the analogues relationship between heat transport and momentum transport. This fact can be beneficial to some particular practice use of drag reduction solutions. In case of central heating systems and cooling systems it can greatly reduce the costs of heat insulation in pipelines – especial when water is transported at long distances. Mechanical and thermal degradation of polymer will be beneficial in places where heat exchangers are expected to supply or remove heat.

It must be pointed out, that currently, parallel to the experimental studies on the drag reduction effect, research on a local intensification of the heat exchange are conducted in the areas in which they are expected to occur.

Increase of the flow rate in third significantly extended turbulent zone of flow can be extremely beneficial when increase of the efficiency of the water transport system is required.

It can be applied in fire-fighting or to increase the efficiency of sewerage and anti-flood systems, or in transport of slurries, sludge and brines, when transported medium is going to be subjected to further filtration or other chemical processes. By the application of drag reduction effect pumping cost can be reduced significantly in such situations.

In case of anti-flood systems application biodegradable polymers should be used. High uncertainty regarding the toxicity is observed for cationic surfactants which are investigated in this paper. However, for such propose more toxic ionic surfactants can be replaced with non-ionic, biodegradable surfactants for which drag reduction effect is also observed.

References

1. Y. Wang, B. Yu, J. L. Zakin, H. Shi, *Advance in Mech. Eng.*, **10**, 2011
2. J. L. Zakin, W. Ge, *Polymer Physics: From Suspensions to Nanocomposites and Beyond*, John Wiley & Sons, Inc., Hoboken, NJ, USA, 89-127, 2010.
3. C. M. White, M. G. Mungal, Annu. *Rev. Fluid Mech.*, **40**, 235-256, 2008.
4. W. Borostow, 2008, J. Ind. Eng. Chem., **14**, 409-416.
5. C. Shu-Peng, *J. Hydrodynam.*, **24**, 2, 202-206, 2010.
6. S. Tamano, M. Ito, K. Kato, K. Yokota, *Phys. Fluids*, **22**, 5, 055102, 2010.
7. Z. Matras, *Cracow University of Technology, Monografia* **29**, 1984.
8. T. Dujmovich, A. Gallegos, *Offshore*, **65**, no. 12, 1–4, 2005.
9. J. F. Motier, L. C. Chou, N. S. Kommareddi, *Proceedings of the ASME Fluids Engineering Division Summer Meeting*, San Diego, Calif, USA, 1996.
10. R. C. R. Figueredo, E. Sabadini, *Colloid. Surface. A*, **215**, no. 1–3, 77–86, 2003.
11. J. Golda, *Chem. Eng. Commun.*, **43**, no. 1–3, 53–67, 1986.
12. G. Dembek, H. W. Bewersdorff, *GWF, Wasser/Abwasser*, **122**, no. 9, pp. 392–395, 1981.
13. N. A. Tuan, H. Mizunuma, *J. Nonnewton. Fluid. Mech.*, **198**, Pages 71-77, ISSN 0377-0257, 2013.
14. G. Aguilar, K. Gasljevic, E. F. Matthys, *Int. J. Heat Mass Transfer.*, **44**, 15, 2835-2843, 2001.
15. Z. Matras, T. Malcher, B. Gzyl-Malcher, *Thin Solids Films*, **516**, 8848–8851, 2008.
16. Z. Matras, J. Głód, T. Malcher, *Applied Mechanics and Engineering*, **4**, 441, 1999, 441-446, 1999.
17. B. Bębenek, *Chem. and Proc. Engineering*, **2**, No 179, 1994.
18. A. Knop, R. Colby, *ACS J. Surf. Colloids*, **15**, (1) 58-65, 1999.
19. J. Myska, *AIChE Journal*, **44**, No 6., p.1467, 1998.

The influence of the numerical solver selection on the nozzle impulse flow simulation results

Przemysław Młynarczyk[1]

[1] Cracow University of Technology, Institute of Power and Process Engineering, Mechanical Faculty 31-864 Krakow, al. Jana Pawla II 37 Poland

Abstract. Numerical simulations are currently used for different applications in a various fields of science. Certain solutions are not as obvious as the others while the results can give very valuable conclusions. Computational Fluid Dynamics (CFD) is one of the tools that can be used to solve different problems related with the mass and heat transfer. Nowadays it is already known that the impulse flow simulation allows to determine pressure pulsation attenuation parameters by a given geometry. However, the nozzle shape optimization method strongly depends on the numerical results obtained from the impulse flow simulation. In commercial CFD software Ansys-Fluent the obtained results depends strongly on the chosen numerical methods, especially the spatial discretization method. This is the reason to use other software as a benchmark. Alternative software FlowVision was used to perform the impulse flow simulation for the same geometries to compare the results. As there is a different problem definition in both systems the calculations, accuracy and results differ from each other. The paper describes the numerical differences between solvers. Article contains discussion about obtained results and includes hints how to avoid mistakes when user change software, especially in solving unusual CFD problems..

1 Introduction

Numerical calculations used to simulate reality in the modern world are a typical engineering tool used to improve design process. Users often don't have knowledge about the calculations performed by the software. In the case of well-known problems, the experienced user is able to determine the value of the obtained results. Interpretation problems increase when a new problem is solved using numerical methods.

Using CFD software numerical calculations can give, for the same problem, different results which depends strongly on the solver algorithms. Most common problems described in the literature are related with the appropriate selection of the closing models. However, this is a matter related with the simulated phenomena physics and is not related to the calculation method. Numerical problems are related with the solution methods such as a spatial discretization schemes, time integration schemes etc. In the paper [1] the differences between implicit and explicit time integration schemes are described for the turbulent incompressible flows. Influence of the spatial discretization methods on the numerical results for different calculations are described in [2,3].

Verification of newly introduced CFD solvers using the established commercial software is very desirable and the results of such comparisons are regularly presented at international conferences [4]. Modelling the impulse flow, damping is not only the effect of calculated physical phenomena but is also a numerical dissipation result. Authors in [5] describes the interpretation problems resulting from numerical dissipation. This paper presents impulse flow damping comparison for different shapes using two solvers with the same flow parameters, boundary conditions and additional models (turbulence, wall functions etc.).

2 Solvers

In the both analysed systems different solvers can be used. In the Ansys-Fluent software there are two main solvers: Pressure- and Density- Based. FlowVision have three different approaches from which the preferred is the AST "implicit-new" solver. This approach also define the time and spatial discretization schemes. In this paper the impulse flow simulations for two solvers are compared: Ansys-Fluent pressure-based solver and FlowVision AST "implicit new" solver. Algorithms of selected solvers are presented in the figures 1. and 2. Linearization of the governing equations using both softwares use the implicit approach. In the FlowVision the "implicit" solver developed by Aksenov [6] is used.

2.1 Ansys-Fluent

In the Ansys-Fluent pressure-based segregated solver the individual governing equations for the solution variables are solved one after another [7]. This approach makes

* Corresponding author: pmlynarczyk@pk.edu..pl

this solver memory-efficient, although achieving the convergence is more time-consuming. Because the impulse flow is not a complicated state of flow the SIMPLE pressure-velocity coupling algorithm is selected. Calculating the variables the QUICK spatial discretization scheme is used for all values. In the literature it is described that QUICK scheme gives most satisfying results for the nozzle impulse flow damping [2].

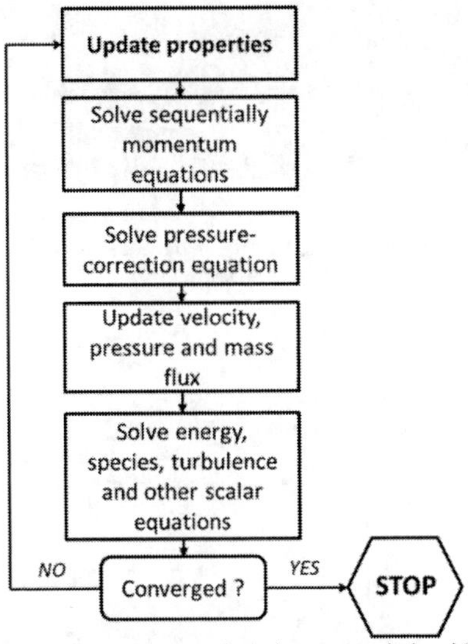

Fig. 1. Ansys-Fluent pressure-based segregated algorithm [7].

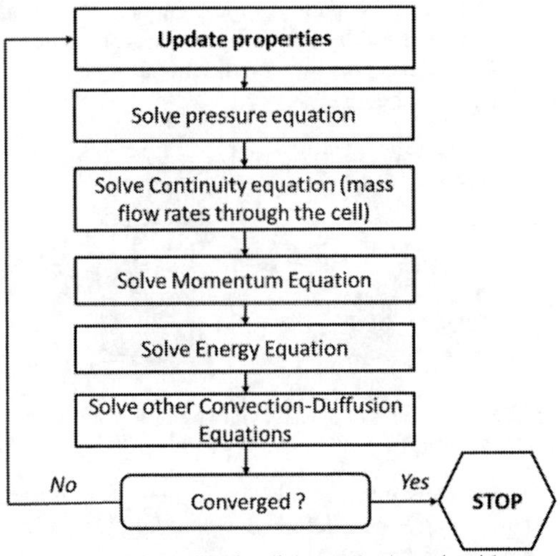

Fig. 2. Flow Vision AST "implicit-new" solver algorithm.

The general mass (1), momentum (2) and energy (3) conservation equations are solved in the forms [7]:

$$\frac{\partial p}{\partial t} + \nabla \cdot (\rho \vec{w}) = S_m \quad (1)$$

$$\frac{\partial}{\partial t}(\rho \vec{w}) + \nabla(\rho \vec{w}\vec{w}) = -\nabla p + \nabla \cdot (\bar{\bar{\tau}}) + \rho \vec{g} + \vec{F} \quad (2)$$

$$\frac{\partial}{\partial t}(\rho E) + \nabla \cdot \left(\vec{w}(\rho E + p)\right) = \nabla \cdot \left(k_{eff} \nabla T + (\bar{\bar{\tau}}_{eff} \cdot \vec{w})\right) + S_h \quad (3)$$

where Sm is an additional mass source, S_h is an additional volumetric heat source and F is an additional force defined by the user.

2.2 Flow Vision

The FlowVision AST "implicit-new" solver use velocity-pressure split method, where the Navier-Stokes equations are solved using conservative velocities at a given time step. Variables like density, pressure, temperature or velocity are stored in the cell centers, although the mass flow rates are determined at the cell faces. This is a new type of the algebraic solver, it is a combination of the Aggregation AMG, selective AMG and RParFBBS solvers [8]. The P-V split approach is widely described in the paper [6]. More details of the general solver algorithm and the discretization schemes are the company industrial secret. Continuity equation in the FlowVision takes the same form as in the Fluent software. However, momentum (4) and energy (5) equations are treated in a slightly different way [8]. General momentum equation takes the form:

$$\frac{\partial}{\partial t}(\rho \vec{w}) + \nabla(\rho \vec{w}\vec{w}) = -\nabla p + \nabla \cdot (\bar{\bar{\tau}}) - D\vec{w} - \widehat{D}\vec{w} + F + F_{el} + F_{user} \quad (4)$$

As can be seen there are additional expressions which include to the equation isotropic (D) and anisotropic ($\widehat{D}$) resistance. Additional forces are: F_{el}- force exerted by the electrostatic field, F_{user}- additional user defined volume force and F which is calculated using equation (5) :

$$F = \left(\rho - \rho_{hyd}\right)g - \rho(2\omega \times V + \omega \times \omega \times r) \quad (5)$$

Using FlowVision user can choose one of three energy equations implemented in the software :

- equation for total enthalpy,
- equation for thermodynamic enthalpy,
- simplified equation for thermodynamic enthalpy, in which the convective term is absent.

For slow gas flows it's recommended to use the simpler energy equation (6), which is formulated via the thermodynamic enthalpy [8]:

$$\frac{\partial}{\partial t}(\rho h) + \nabla(\rho \vec{w} h) = \frac{\partial p}{\partial t} + \mathrm{w} \cdot \nabla \mathrm{p} - \nabla \cdot J_q + Q_{vis,L} + Q_{rad} + Q_{user} \quad (6)$$

where J_q is a specific heat flux and Q letter variables are additional volume source of energy due to viscous dissipation (Q_{vis}), radiation (Q_{rad}) and user sources (Q_{user}).

3 Numerical Model

Numerical analysis of impulse flow were performed for three different geometries. As a basic shape the straight pipe is selected, the Venturi orifice represents geometry with linear neck-shape and as a most complicated geometry the pipe with hyperboloid shape neck is chosen. Geometries are defined as a surface models for the axisymmetric Fluent calculations as it is presented in the figure 3. In the FlowVision software the STL geometry file must be imported into the program because there is no geometry design system. To perform axisymmetric calculations model must be 3D element created from profile turned by, at last, two degrees as it is presented in the fig. 4.

Fig. 3. Investigated geometries designed in the Ansys Design Modeler: A- empty pipe, B- Venturi nozzle, C- hyperboloidal nozzle.

Fig. 4. Geometry model turned by 2 degrees in the Flow Vision software.

Investigated geometries for both simulations are designed in the ANSYS software.

3.1 Mesh generation

Mesh generation conducted in the Ansys-Fluent is well known process. User can define global parameters for the model discretization or can define local grid parameters. Software generate grid that meet all defined conditions. For QUICK discretization scheme the grid have to be Cartesian or equivalent (the subsequent cells must be specified). Mesh detail for the Venturi orifice is presented in the fig. 5.

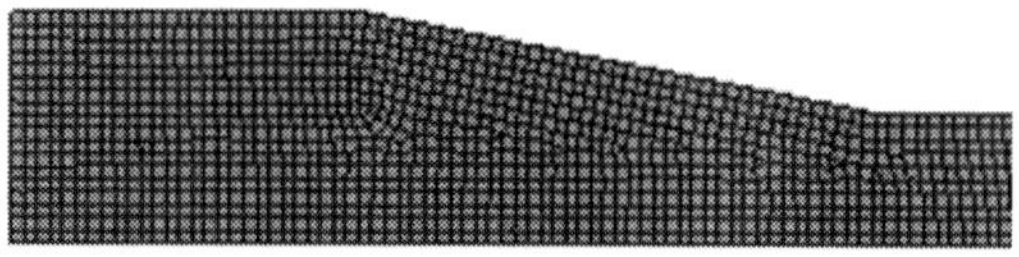

Fig. 5. Venturi orifice grid detail. Grid performed in the Ansys mesher module.

Mesh generation philosophy in the FlowVision software is very different than in a classical mesh systems. Program calculate the dimensions of the computational domain and creates box-shaped catesian grid around it. Later, only this part of the initial grid, which includes the model geometry is calculated. For the pipe element with hyperboloidal neck the initial grid detail is presented in the fig. 6. For the initial grid user can define only number of cells in each direction. In presented case 2 cells are defined in the depth direction. FlowVision mesh automatic adaptation module is very sophisticated and allows to create very detailed grid without user experience. Simplifying: the grid adaptation is a consequence of the cell splitting and merging. In automatic mesh refinement the cell is divided into four smallest cells and the process is conducted due the appropriate dimension of the smallest cell is obtained as it is presented in the figure 7 [8].

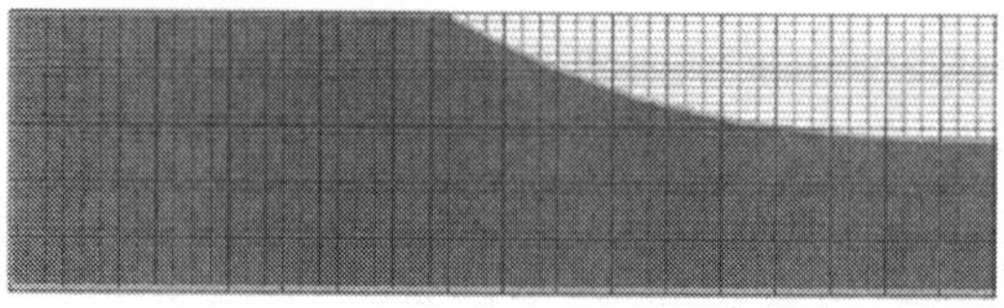

Fig. 6. Initial grid created on the hyperboloidal geometry imported into the Flow Vision software.

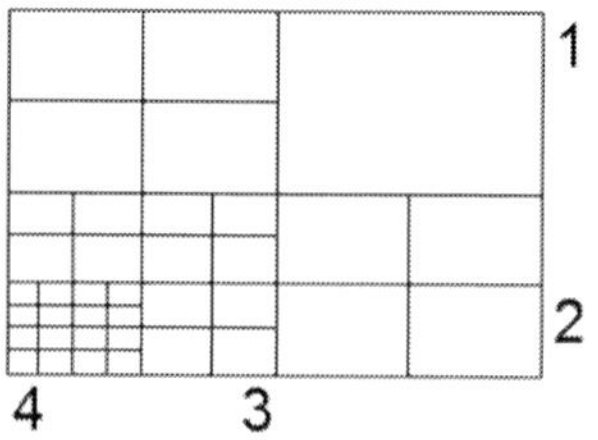

Fig. 7. Levels of splitting the computational grid.

As the model geometry of the computational domain is a set of surfaces the Cartesian initial grid is sliced by the model facets. This increase the number of cells as it is presented in the figure 8.

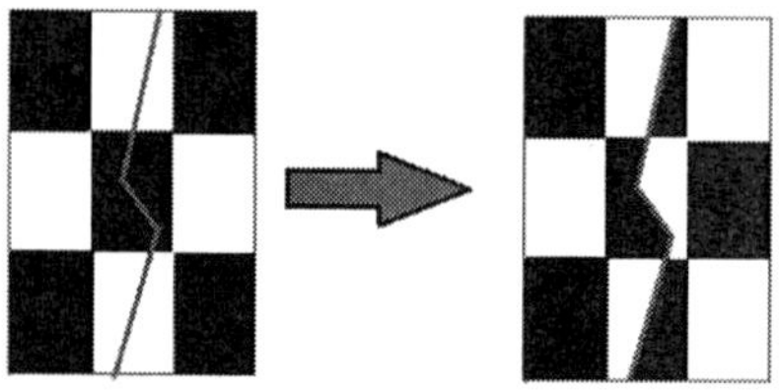

Fig. 8. Splitting cells of the initial grid by surface of the geometry model.

Sliced cells must be adapted as they have different shapes. Adaptations define the splitting and merging parameters of those cells:
- that cell belong to the specified object volume,
- that cell is adjacent to the specified object surface,
- that cell is adjacent to a surface that has certain boundary conditions defined.

Adaptation also impacts the surface or the volume of the object. The grid refinement is performed mainly by two types of adaptation. Basic types are splitting and merging, but program can use also additional adaptation types. One of the most important is the "adaptation to solution" which allows to adapt the grid in the object area where big gradients of some variable occurs. However the program has a limitation, especially for axisymmetic flow- the computational domain and the initial grid should be oriented parallel. This could be a problem if the model is not parallel to the coordinate system of the FlowVision workspace.

3.2 Boundary Conditions

Nozzle impulse flow in the numerical calculations is defined as a one-time step excitation with known value. In this case, the model consist of 4 boundary conditions for each software as it is presented in the fig. 9. Main difference between boundary conditions using Fluent or FlowVision is that in Fluent the axisymmetric calculations are performed on a "flat" model with one boundary defined as an axis and using the FlowVision there is a 3D turned model and the axisymmetric case is obrained by defining the symmetry BC on the side surfaces.

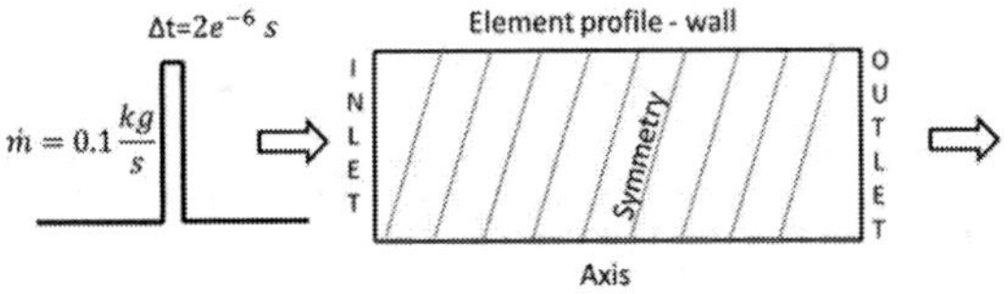

Fig. 9. Boundary conditions.

Differences in the boundary conditions in the two compared softwares are neglible so they are not described in detail in the paper.

3.3 Turbulence model

Numerical simulations are performed using basic k-ε turbulence model. Ansys Fluent has three k-ε models:

- Standard,
- RNG,
- Realizable.

FlowVision software have implemented four models named:

- KES (Standard),
- KEAKN (Abe, Kondoh, Nagano),
- KEFV (FlowVision),
- KEQ (Quadratic),

where the KES model is the default turbulence model used in this software. In the presented simulations the standard k-ε model, which is a default turbulence model in both system, is used. Software developers describes this models with some differences. As the paper have to concentrate on the grid and solver influence on the results, due to lack of space, the turbulence models are not investigated in details.

3.4 Wall functions

For some calculations proper choose of the wall functions model have significant influence on the final result. In the presented case the same wall functions models are defined. In the FlowVision two models WFFV and WFS can be chosen for the k-ε turbulence model. In the Ansys-Fluent software there is a more wall functions calculations methods defined. Except the standard method software have defined four more wall function calculation methods. In the presented investigations the standard wall functions in both softwares are used. In the FlowVision standard wall functions model is presented by the WFS model.

4 Results

In the presented investigations the working fluid is air modelled as an ideal gas. The environmental pressure is equal to 1 atmosphere and the fluid temperature is defined as a 300 Kelvin. In the presented case 20000 time steps were performed which gives the simulation time equal to 0,04 s. Calculations performed in both systems takes a similar time on the default calculation speed-up settings. As a simulation results the spatially averaged value of the mass flow on the outlet is investigated. For the impulse flow simulation such a result is a damped, oscillating function. In the figures 10., 11. and 12. the comparison between signals obtained in Ansys-Fluent and FlowVision software for three different shapes are presented.

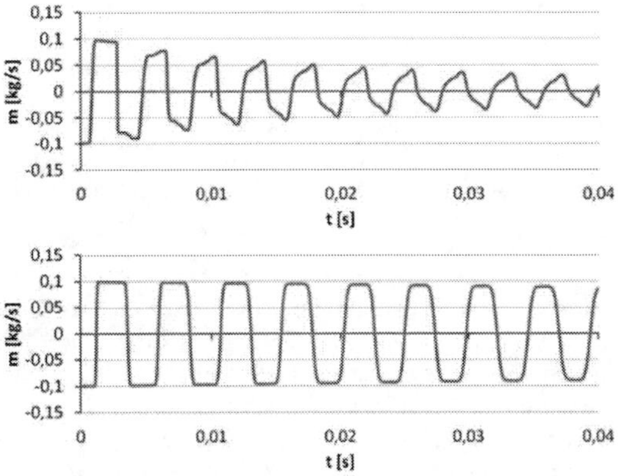

Fig. 10. Signal obtained for the empty pipe impulse flow simulation using the Ansys-Fluent (red) and FlowVision (blue) software.

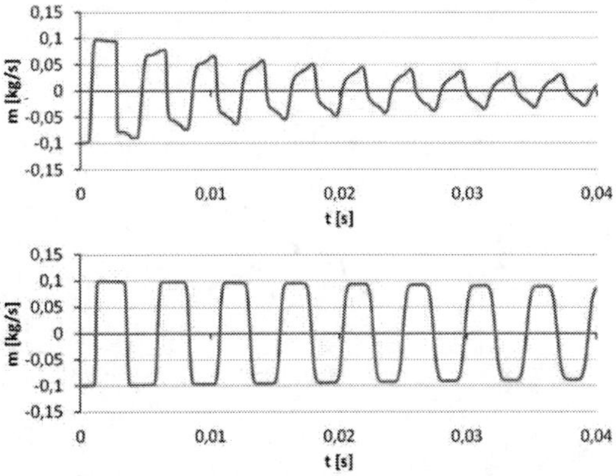

Fig. 11. Signal obtained for the Venturi orifice impulse flow simulation using the Ansys-Fluent (red) and FlowVision (blue) software.

As it can be seen in the figures the flow rate signal is damped in a different way in both software's. Ansys-Fluent software suppress the impulse flow much more than FlowVision software and this applies to all nozzles. In this case, can be concluded that the geometry shape and mesh definition don't have a significant influence on the flow damping. Analysing the graphs can be seen that the amplitudes of the forthcoming peaks are constantly falling in the Ansys-Fluent calculations and, that have rather pulsation-characteristic using FlowVision software. It is also visible that the flow is more intensive in the Ansys-Fluent software as the obtained signal have much more sharp peaks.

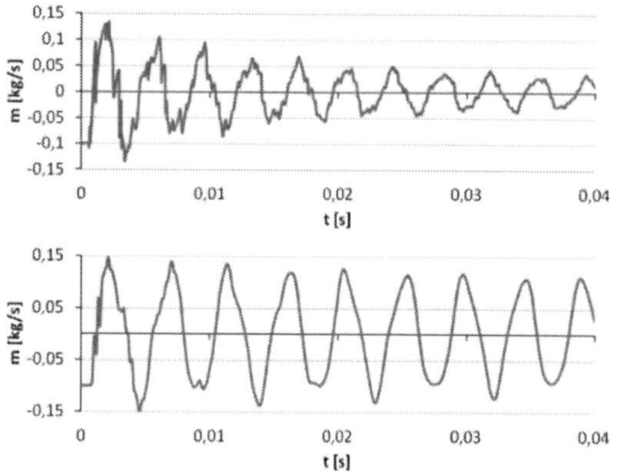

Fig. 12. Signal obtained for the twin hyperboloidal nozzle impulse flow simulation using the Ansys-Fluent (red) and FlowVision (blue) software.

As the pipe lengths, mass flow impulse value, fluid parameters: temperature and pressure are the same, very interesting is that the signal period is different in both systems. It means that the flow speed is different. It is possible that it's influenced by differences in the outflow boundary, minor gas parameter differences due to the accuracy, cell size differences, numerical dissipation and/or the model viscosity importance, which seems to be higher in the Ansys-Fluent software. In the figure 13., the detailed first amplitude of the twin hyperboloid nozzle flow signal for both investigations is presented.

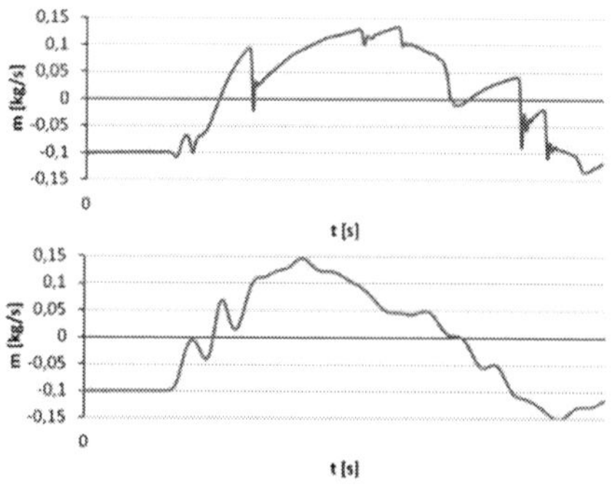

Fig. 13. Detailed first aplitude of the twin hyperboloid nozzle flow signal using the Ansys-Fluent (red) and FlowVision (blue) software.

In the figures can be seen global similarities and local differences. For both softwares the maximum value is preceded by the function oscillation which consist of two peaks. In the FlowVision software the first mass flow rate value peak is bigger and the second one is smaller. The graph is smooth and free of sudden value drops, while in the Ansys-Fluent results can be seen this substantial behavior. Another difference which is not clearly visible in this case, is that the mass flow rate value rises faster for the FlowVision calculations, and falls slower, when in the Ansys-Fluent software is exactly the opposite. This is more visible for the empty pipe and Venturi orifice comparison of the first amplitude presented in the figures 14. and 15.

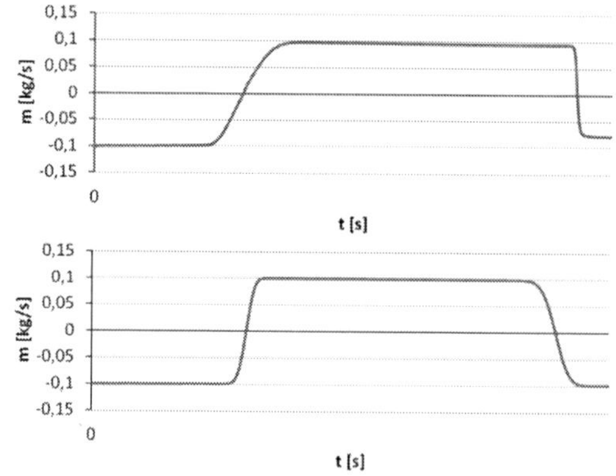

Fig. 14. Detailed first aplitude of the empty pipe flow signal using the Ansys-Fluent (red) and FlowVision (blue) software.

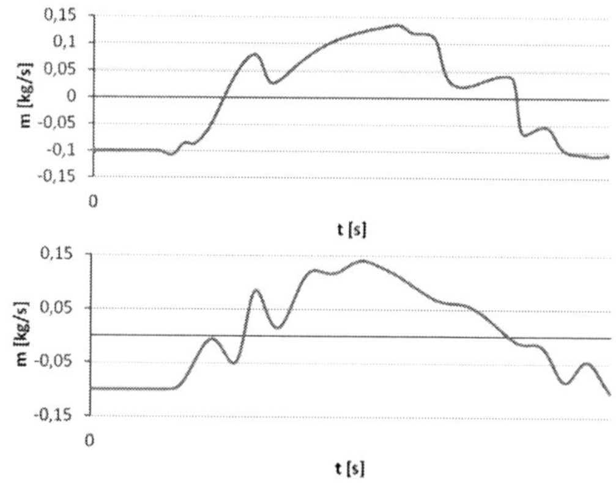

Fig. 15. Detailed first aplitude of the Venturi orifice flow signal using the Ansys-Fluent (red) and FlowVision (blue) software.

To get the answer how and why presented differences exist more detailed investigations of the numerical model are needed. In order to evaluate some differences between the obtained results, the pressure distribution maps study can be performed as a preliminary calculations assessment. In the figures 16.-18. the comparison of the pressure contours for investigated elements are presented. As there is no need to make quantitative comparison the pressure values are

 https://doi.org/10.1051/matecconf/201824003008

not presented to obtain better quality of the figures. Qualitative comparison allows to asses how much the pressure distribution differs in both cases and how it can affect the results. In the figures red colour represents higher pressure and blue colour represents lower pressure.

Fig. 16. Empty pipe flow pressure contours after 1000 time steps (0,002 sec) A-Ansys-Fluent, B- FlowVision.

For the empty pipe flow it's very significant that the under- and over- pressure occurs on different sides of the moving disturbance. In the figure 16. the disturbance moves backward after it is reflected from the outlet and it is moving into the left side. It seems that in the Ansys-Fluent the gas in expanded behind the disturbance when in the FlowVision gas is sucked thorught the outlet to follow disturbance. This is another indication that, to better understand the differences in obtained results, special attention should be paid to the outflow boundary condition. For complicated shapes the flow is not as simply as for empty pipe which means that analysis of the pressure contours cannot give much more information. In the figs. 17. and 18. Can be seen that the distribution of high and low pressure is in some places more and in another places less similar.

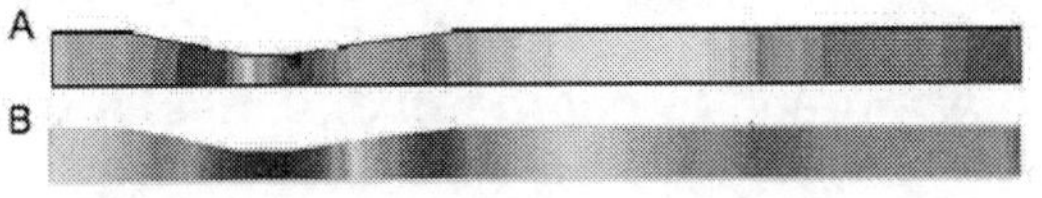

Fig. 17. Venturi orifice pressure contours after 1000 time steps (0,002 sec) A-Ansys-Fluent, B- FlowVision.

Fig. 18. Twin hyperboloidal nozzle pressure contours after 1000 time steps (0,002 sec) A-Ansys-Fluent, B- FlowVision.

Due to lack of space in this paper the FFT analysis is not presented but it should be known that for the impulse excitation (dirac delta in a mathematic sense) the Fast Fourier Transform analysis has a fundamental meaning.

4 Conclusions and discussion

Using the impulse flow simulations as a tool to estimate the damping parameters of the different nozzles shapes requires a suitably prepared numerical model. Previous studies have shown that the impulse damping in the numerical simulations using Ansys-Fluent software overestimate the damping in the empty pipe and underestimate the damping in other nozzles. This attempt, to use another software for such a non-standard simulations, allows to be aware how much the results depends on the solver, and minor differences in the calculation algorithms. This paper presents differences and similarities between Ansys-Fluent and FlowVision software without assessment, which software is better, because in this case such an opinion is not possible.

Presented FlowVision software can be distinguished by a very interesting and effective grid generation philosophy which is often a weakest element in the other software. One of the conclusions of this paper is also that using commercial software, operator is not able to determine in 100% how the problem is resolved, especially if the problem is unusual. The author also want to leave the reader with an assessment whether one or another software will be better to calculate his specific problem and what is important, to pay attention to, when changing the calculation software.

References

1. G. Noventa, F. Massa, F. Bassi, A. Colombo, N. Franchina, A. Ghidoni, Computers and Fluids, **139**, pp. 248-260 (2016)
2. P. Cyklis, P. Młynarczyk, Procedia Engineering, **157**, pp. 396-403 (2016)
3. X. Ch. Mi, A. J. Higgins, S. Goroshin, J. M. Bergthorson, *Proceedings of the Combustion Institute.* **36**, pp. 2359-2366 (2017)
4. J. A. Keep, S. Vitale, M. Pini, M. Burigana., Energy Procedia, **129**, pp. 1071-1077 (2017)
5. F. Cadieux, S. Guangrui, J.A. Domaradzki, Computers and Fluids, **154**, pp. 256-272 (2017)
6. A.A. Aksenov, S.V. Zhluktov, D. V. Savitskiy, G. Y. Bartenev, V. I. Pokhilko, Journal of Physics: Conference Series, **653** (2015)
7. ANSYS theory guide v15.0
8. FlowVision User's guide v3.09.05

Meshing strategy for hybrid tet-hex PHES layered bed heat storage

Jędrzej Mosiężny[1]*, and *Bartosz* Ziegler[1]

[1]Poznań University of Technology, Chair of Thermal Engineering, Piotrowo 3, 60-965 Poznań, Poland

Abstract. Article describes a meshing strategy for efficient computation of generic layered heat storage device for pumped heat energy storage (PHES). Mesh for structural solid body and fluid boundary layer is with unstructured flow field is presented. A rationale for geometry choice, mesh topology and different types of mesh topologies considered in this study. The effect of coarse and fine mesh on computational time and accuracy of results for transient analysis of coupled fluid and heat flow is described.

1 Nomenclature

TES – Thermal Energy Storage
ES – Energy Storage
FES – Flywheel energy storage
CAES – Compressed air energy storage
PS – Power Storage
PSH – Pumped Storage Hydroelectricity
PHES – Pumped Heat Energy Storage
RE – Renewable Energy

2 Background

Rapid growth of global electrical power demand next to the climate change issue [1] is influenced by introduction of new environmental regulations concerning emission reductions of carbon dioxide [2]. This goal is realized by fast increase of renewable energy (RE) technologies share to energy mix of many countries, mostly solar and wind power. Utilization of these sources depends on weather conditions and potential of power generation is highly variable. Unproductive periods without wind and sunless days are interlaced with the peaks of electricity production exceeding the current demand. This surplus should be stored or otherwise is wasted, what causes rise of production costs as well as premature network wear.

Literature analysis has indicated that TES offers relatively high energy density and much lower price per kWh of built storage comparing to very high energy density batteries. Classical TES approach allows to obtain efficiencies up to 60%. However such high value is very difficult to achieve and in case of low short term and long term energy prices fluctuations this value does not provide sufficient cost effectiveness. Therefore development of this technology is focused on investigations of round-trip efficiency (the ratio of energy put in to energy retrieved from storage) increasing concepts. Pumped heat energy storage is one of the most promising TES derived approach which has the potential to meet this demand [3]. PHES concept is based on heat pump principle of operation during loading and turbomachinery devices during discharging of hot side storage tank.

Small number of papers describe existing prototypes of PHES in the context of heat transfer and storage components of the system [4]. This paper concerns the mesh preparation for conjugate heat transfer CFD calculations of fluid and heat flow in fixed bed heat storage reservoir.

3 Approach to geometry and meshing

3.1 Preparation of the geometry

The study is performed on a segment of generic layered thermal storage as presented in Fig 1.

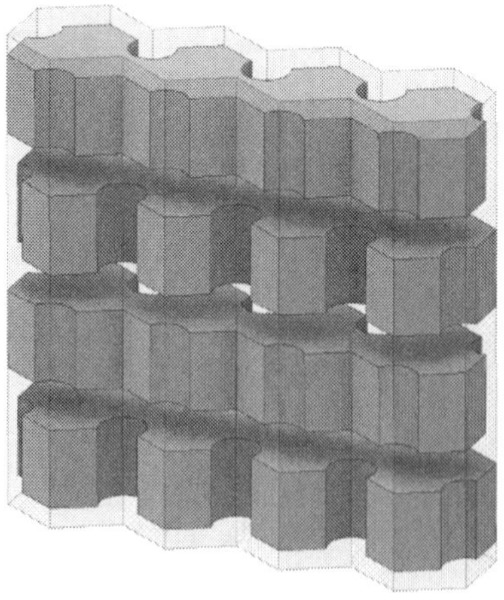

Fig. 1. Segment of layered thermal storage.

* Corresponding author: jedrzej.mosiezny@put.poznan.pl

The CAD geometry is composed of two solid body (in context of CAD modelling) regions representing layers of storage and fluid region respectively.

Individual holes in the discussed geometry are arranged in a triangular/hexagonal pattern and are shifted in relation to upper and lower level. Hole diameter is 40mm, distance between hole centrelines is 100mm, layer thickness is 50mm and distance between layers is 25mm. Geometry does not represent any particular storage design (Fig. 2).

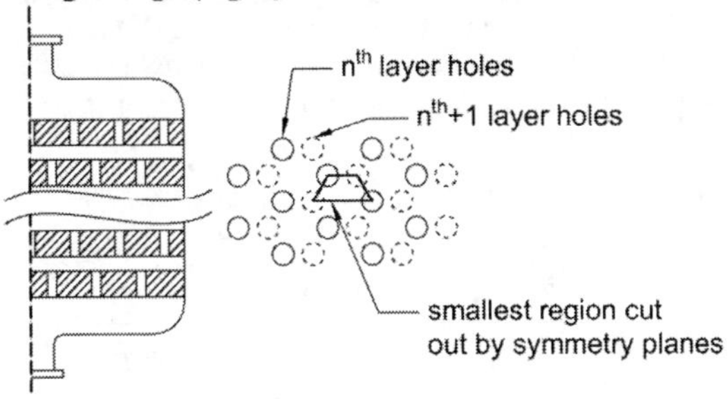

Fig. 2. Hole pattern overlapping.

It was found that the flow phenomena occurring in such entity are mostly in the direction of flow with minimal activity in radial, with regard to hole centrelines, direction. Therefore presented generic geometry was then sliced to obtain smallest possible volume that represents a single "cell" of the storage (Fig 3.). Side walls will be marked as symmetry boundary condition.

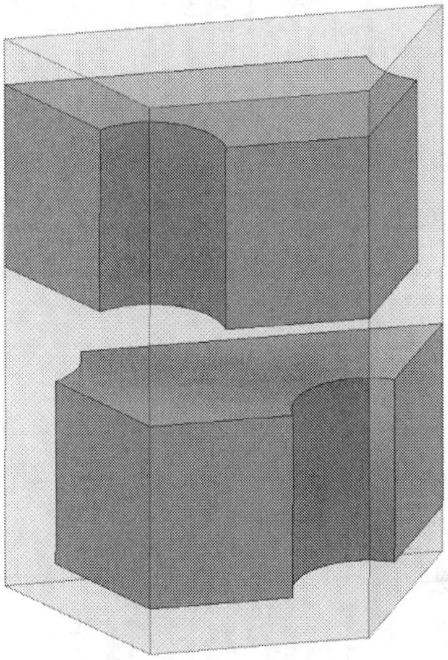

Fig. 3. Basic Geometry cell for CFD analysis.

3.2 Meshing approach and topology

Mesh was created in Pointwise V18.0R4 meshing software.

Based on previous experiences of the research team, a structural, hexahedral mesh was intended for this study. Although not very complex, the geometry does not mesh with fully hexagonal meshes without interfaces, therefore a following approach was made.

A topology for solid body (dark grey in Fig. 2) was created on the face of the solid body perpendicular to the flow direction (Fig 4). As the software used allows for "patch independent" meshing, that is: mesh information can be independent from the geometry, both solid and fluid regions were meshed during this operation.

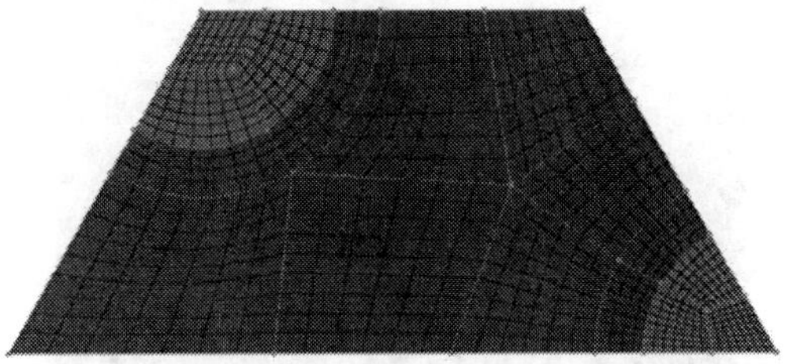

Fig. 4. Coarse mesh topology for solid region (dark grey) and fluid region (light gray)

The presented topology was then extruded in the direction of the flow to create the volume mesh for the solid body region and flowpath. Upper and lower surfaces were then extruded again to create a flow boundary layer. Such approach resulted in a typical h-topology of the fluid boundary layer. Final solid region with fluid boundary region was mirrored and translated to form the second solid.

Remaining fluid volume was then filled with unstructured, isotropic tetrahedra with pyramid transitioning layer (Fig 5).

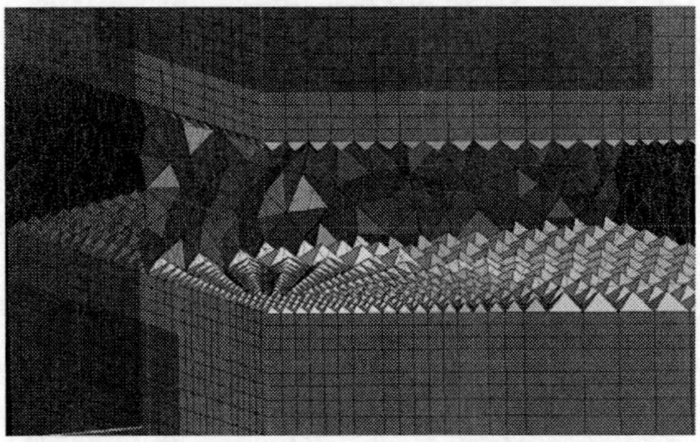

Fig. 5. Overview of mesh elements in coarse mesh; dark grey hexa: solid, light grey hexa: fluid BL, light grey pyramids: transition layer, dark grey tet: flowfield.

Such basic geometry was then copied to form 32 solid domain layers (or 16 pairs with shifted hole patten) with one fluid domain around them.

One of the early approaches for meshing was to create an o-grid topology in the fluid boundary. Although the mesh quality in the fluid boundary was more than acceptable, the rounded section of the o-grid made it impossible to create a high quality pyramid transition layer and transition to tetrahedra.

Final mesh consists a 4 meter long cut from layered bed heat storage device, including 30cm inlet and outlet regions. Such geometry represents the discussed device in 1:1 scale. Results on heat flux, heat capacity and mass flow are easily scalable for a real device with given area of thermal storage cross-section.

Fig. 6. Overview of fine mesh on Basic Geometry cell

3.3 Mesh sizing and quality

Initial meshes considered for this study were created for high resolution fluid boundary layer with y+ parameter calculated to value of 5, based on reference velocity and hydraulic diameter of the bed holes. Real value of y+ based on flow-field values is in order of magnitude of 2, ranging up to 4 where velocity magnitude increases locally. Full 32 layers of fine mesh result in around 14.2 million elements. Estimated physical flow time for completing a single charge/discharge cycle was 6 hours. Completing this physical flow on a 120 core HPC cluster was estimated to be completed in 32 years for a single charge/discharge cycle. Although the quality of the results was superior to the coarse grid, the computational wall-time made the calculation inherently impractical.

For coarse mesh element count was reduced by half in each direction of the mesh for both structured and unstructured domains. This resulted in a mesh with y+ parameter in order of magnitude of 50. Cell count is also drastically reduced to 2 million cells. Real y+ values calculated from flow data range from 13 to 23 where local velocity magnitude is increased. Using a coarsened mesh resulted in achieving relatively practical wall-time with three charge/discharge cycles completed in 21 days.

Mesh quality is comparable for both meshes. Quality figures for full 32 layer mesh are presented in figures 7 thru 9.

Table 1. Element count per basic geometry.

Element type	Fine grid	Coarse grid
Hexa	439 264	47 774
Tet	398 443	74 133
Pyramid	11 872	2 984
Total	849 579	124 861

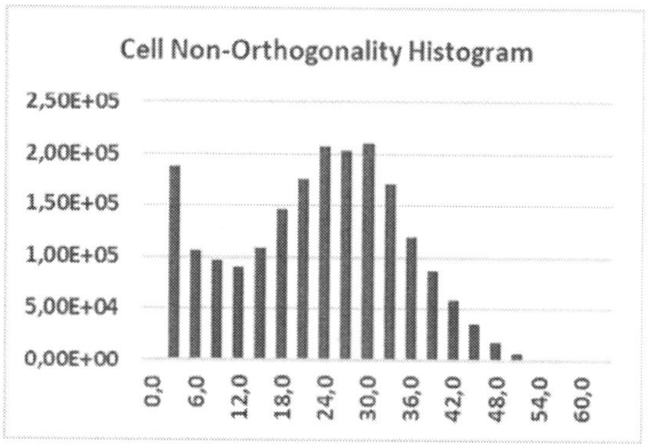

Fig. 7. Coarse mesh cell non-orthogonality histogram. Lower values – better mesh

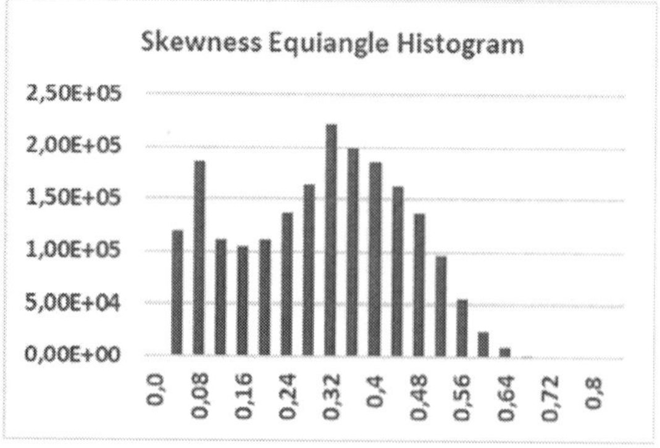

Fig. 8. Coarse mesh cell skewness equiangle histogram. Lower values – better mesh

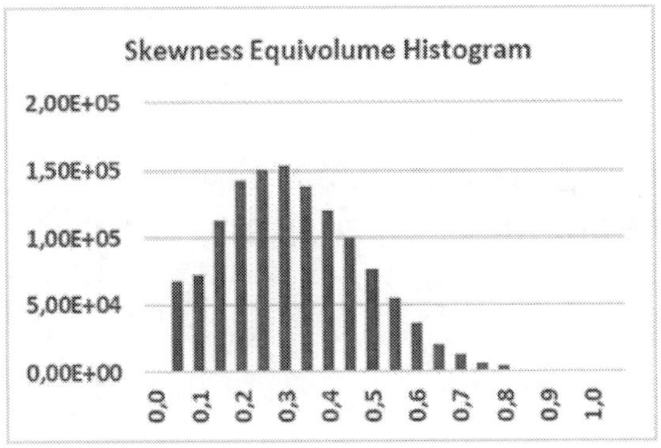

Fig. 9. Coarse mesh cell equivolume histogram. Lower values – better mesh

4 Results and discussion

Mesh sizing in the boundary region for conjugate heat transfer require y+ wall sizing in range of 1. This wall sizing ensures accurate resolution of temperature field and resulting heat flux.

Mesh influence ne the results was observed for both meshes. Heat flux towards the solid body domains was measured during first 50 seconds of flowtime. Although total heat flux towards the solid body domains was comparable for two meshes, with difference of 0,3% lower heat flux on the coarser mesh, comparison of the first 5 layers showed underestimation of heat flux in range of 10% for coarser mesh in comparison to fine mesh. As seen on fig 10. fluid temperature drop is more observable on fine mesh in comparison to coarse mesh.

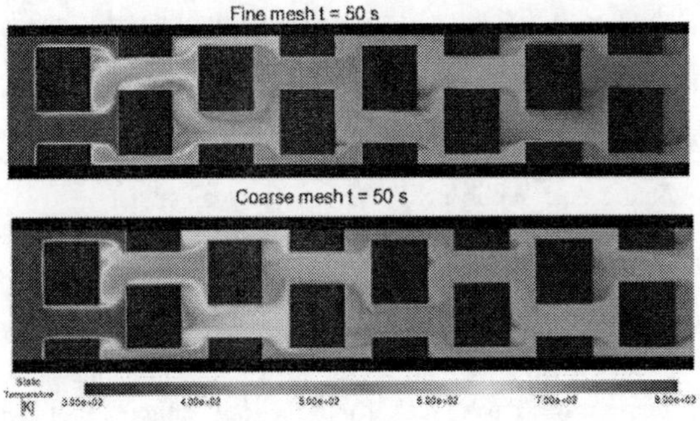

Fig. 10. Overview of fine mesh on Basic Geometry cell

Coarser mesh resulted in better convergence and stabilisation of the calculations. Some of the small scale phenomena were filtered out resulting in more uniform flow field. More details on the numerical approach and results of the calculations are presented in [5].

4 Conclusions

It is concluded, that for best quality results, mesh sizing must retain y+ sizing in order of magnitude of 1 for best conjugate heat transfer resolution. However for large scale computations, even implementing a section of the storage requires very large computational resources.

It is suggested that for large scale computations, one should perform a detailed analysis of a single heat storage cell (such as the basic geometry, fig. 3) in order to obtain values of heat capacity of such cell per unit volume and unit area, followed by the implementation of these values in properties of porous material. Further work is described in [6]

This research was supported in part by PLGrid Infrastructure.

References

1. R. A. Betts, C. D. Jones, J. R. Knight, R. F. Keeling, J. J Kennedy, Nature Climate Change, Nature Research, **6(9)**, 806. (2016)
2. European Commission, Roadmap 2050. Policy. (2012)
3. J. D. McTigue, A. J. White, C. N. Markides, Applied Energy, **137**, 800 (2015)
4. A. Smallbone, V. Jülch, R. Wardle, A. P. Roskilly, Energy Conversion and Management, **152**, 221. (2017)
5. B. Ziegler, J. Mosiężny, P. Czyżewski, Int. J. Numerical Methods for Heat & Fluid Flow. (to be published)
6. J. Mosiężny, B. Ziegler, Int. J. Numerical Methods for Heat & Fluid Flow. (to be published)

Experimental validation of pressure drop models during flow boiling of R134a – effect of flow acceleration and entrainment

Tomasz Muszynski[1,*], *Rafal* Andrzejczyk[1], and *Carlos* Dorao[2]

[1] Gdansk University of Technology, Gdansk, Poland
[2]NTNU, Trondheim, Norway

Abstract. A crucial step to assure proficient work of power and process apparatus is their proper design. A wide array of those devices operates within boiling or condensation of the working fluid to benefit from high heat transfer rates. Two-phase flows are associated with high heat transfer coefficients because of the latent heat of evaporation and high turbulence level between the liquid and the solid surface. Predicting heat transfer coefficient and pressure drop is a challenging task, and has been pursued by researchers for decades. In the case of diabatic flows, the total pressure drop is due to the change in kinetic and potential energy. The article presents detailed boiling pressure drops data for R134a at a saturation temperature of 19.4°C. Study cases have been set for a mass flux varying from 300 to 500 kg/m^{2}s. Presented data along with the data reduction procedure was used to obtain the momentum pressure drop values during flow boiling. The study focuses on experimental values of momentum pressure drop component and its prediction based on various void fraction models and entrainment effects.

1 Introduction

Rising energy costs and environmental considerations are driving industry for renewable energy use [1], major gains can be made by increasing the efficiency of energy utilization, i.e. recovering low-grade waste heat [2], improving process efficiency [3], or utilizing thermal energy [4].

An essential step to assure efficient work of power and process apparatus is their proper design [5,6]. A wide range of those devices operates within boiling or condensation of the working fluid to take advantage of high heat transfer coefficients [7]. Two-phase flows are associated with large rates of heat transfer because of the latent heat of evaporation and enhancement of the turbulence level between the liquid and the solid surface [8,9]. High heat fluxes can also be obtained by means of numerous enhancement technologies [10–12]. Predicting heat transfer coefficient and pressure drop is a challenging task, and has been pursued by various research for decades. In the case of diabatic flows, it has been shown that total pressure drop is the summation of acceleration, static and frictional pressure drop [13].

The frictional pressure drop of two-phase flow is a function of fluid properties and significantly depends on mass and volume fractions of the mixture. Past research aimed at predicting the frictional pressure drop, includes a large number of studies that rely on either the Homogeneous Equilibrium Model (HEM) or semi-empirical correlations [14].

The research by Revellin and Haberschill [15] focused on experimental frictional pressure drop predictions, compared to 485 database points for smooth tubes from the literature. It was reported that the Quibén [16] displayed the best consistency with experimental data points. The 74% of the data were predicted by the latter model within a 30% error band. Furthermore, model predicts a maximum pressure drop before the annular-to-dryout transition, i.e. as usually in the annular flow regime. The frictional pressure drop of R-134a investigations in 5mm diameter smooth pipes was presented by Andrzejczyk et al. [17]. The study showed that updated Quiben [16] model by Thome [13] and Zhang and Webb [18] obtain the best consistency with experiments with a mean absolute deviation (MAD) under 20%. It was also noted that both models predict well peak pressure drop.

Garcia [19] presented an experimental study on R-407C two-phase flow in a smooth copper tube, alternating from 4.5 to 8.0 mm inner diameter, evaporation pressures (10 and 12 bar), mass flux (from 180 kg/(m^{2}s) to 500 kg/(m^{2}s)). The effects of vapor qualities, diameters, mass fluxes and saturation pressures on the two-phase flow pressure drop were studied. The experimental data were compared with 18 most common correlations. Results showed that Cicchitti et al. [20] and Sun and Mishima [21] correlations provide the best results.

The article presents detailed boiling pressure drops data for R134a at a saturation temperature of 19.4°C with inlet subcooling of 7K and 14 K. Study cases have been set for a mass flux varying from 300 to 500 kg/(m^{2}s). Presented data along with data reduction procedure was

* Corresponding author: Tomasz.Muszynski@pg.edu.pl

used to obtain the momentum pressure drop values during flow boiling.

The study focuses on experimental values of momentum pressure drop component and its prediction based on various void fraction models and entrainment effects. The primary objectives of the present study are to:

- Provide an experimental database including multiple data points in transition and dryout flow regions for two-phase total pressure drop.
- Provide an experimental procedure for obtaining acceleration pressure drop in a horizontal channel, and recognize its effect on values of total pressure drop in a horizontal tube.
- Conduct a systematic assessment of the entrainment on total pressure drop.

Experimental Setup

In order to obtain experimental values of momentum pressure drop component, the experimental facility is an R134a loop consisting of a reservoir of refrigerant, a pump, a conditioner, a flow meter, a heated test section, a visualization glass, an adiabatic section and a condenser. The loop is schematically represented in figure 1.

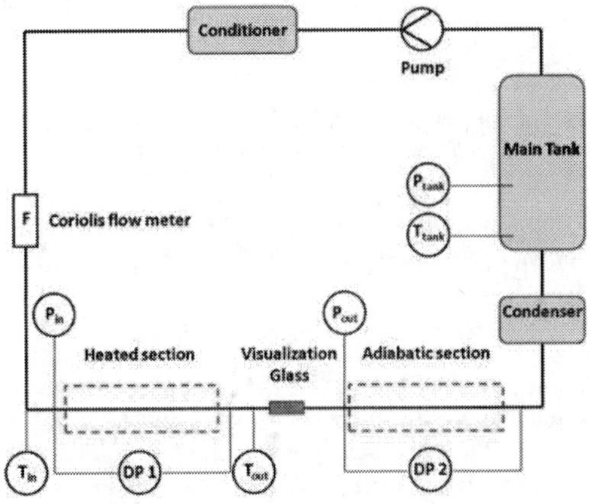

Fig. 1. Schematic of the test facility.

The fluid pressure in the loop is set by controlling the temperature in the main tank where the refrigerant is at saturation conditions. The fluid is circulated by a magnetically coupled gear pump. The conditioner is a shell and tube heat exchanger with glycol in the shell side which is used for adjusting the R134a inlet temperature. The glycol temperature is controlled by the air-cooled chiller. The refrigerant flow rate is measured by a Coriolis type mass flow meter. The heated section is made of stainless steel, as a seamless pipe. The tube is electrically heated by Joule effect with the use of a low voltage DC power supply. The section is thermally insulated with a thick layer of mineral wool, thus thermal losses are neglected. The section is 2m long, with 5mm and 8mm internal and external diameter respectively.

Nine galvanically separated thermocouples are distributed along the tube wall surface, additional two are inside the tube in order to measure the local fluid temperature. The adiabatic test section is a thermally insulated 1m long stainless steel pipe with an inner diameter of 5mm and 8mm outer diameter. The piping system at the supply and outlet of the sections is made with similar pipe dimensions, thus compression and expansions effects are eliminated.

The measurements data are acquired by National Instruments CompactRIO data acquisition system. The signals from measuring devices were processed with the aid of the LabVIEW application. The temperatures, absolute pressures, pressure differences and mass flow rates were acquired at a frequency of 2 Hz. For every experimental point - about 100 data points were acquired. Additionally, in order to verify the steady-state conditions, data points were doubled after 15 minutes. Two-phase total pressure drop was directly measured with a differential pressure transducer, same type differential pressure transducer was used to measure adiabatic pressure drop at the second test section. The heat flux was determined according to eq.:

$$\dot{q} = \frac{\dot{Q}_{el}}{A_w} \tag{1}$$

The absolute pressure at the inlet and outlet of the heated section was also recorded and was used for obtaining the saturation temperature T_{sat} of the fluid based on the equilibrium thermodynamic properties using NIST REFPROP fluid database [22]. The fluid outlet quality (x_{out}) was determined from mass and energy conservation equation (2).

$$x_{out} = \frac{\dot{q}\pi D_i L - \dot{m} c_{pl} \Delta T_{sub}}{\dot{m} h_{lv}} \tag{2}$$

where: h_{lv} – is the latent heat of evaporation, L – is the length of the heated section, $\dot{m}$ – mass flow rate of the refrigerant, $\dot{q}$ – heat flux density, c_{pl} – specific heat of liquid phase, ΔT_{sub} – is the inlet liquid subcooling.

The local heat flux as the function of generated Joule heat is assumed to be constant during the evaporation process along the length of the tube. In order to determine the reliability of the experimental results, an uncertainty analysis was conducted on all measured quantities as well as the quantities calculated from the measurement results. Uncertainties were estimated according to the standard procedures described by NIST [23].

Results

The total pressure drop of fluid Δp_{total} is the summation of acceleration, static and frictional pressure drop according to eq (3) if acceleration and static pressure drop are present.

$$\Delta p_{total} = \Delta p_{mom} + \Delta p_{frict} + \Delta p_{grav} \quad (3)$$

where Δp_{grav} is the elevation head pressure drop and can be omitted in case of a horizontal tube, Δp_{mom} is the momentum pressure drop created by the acceleration of the flow in a heating/cooling process, and Δp_{frict} is the frictional pressure drop of two-phase flow. Most commonly [24–27] the momentum pressure is expressed by means of eq. (4).

$$\Delta P_{mom} = G^2 \left\{ \left[\frac{x^2}{\rho_g \varepsilon} + \frac{(1-x)^2}{\rho_l (1-\varepsilon)} \right]_{out} - \left[\frac{x^2}{\rho_g \varepsilon} + \frac{(1-x)^2}{\rho_l (1-\varepsilon)} \right]_{in} \right\} \quad (4)$$

Differences between void fraction correlations are minor, therefore in the further analysis, only three formulas were used. The author's previous studies [17,28,29] have shown that the predictions given by various methods differ significantly, therefore a modification of Thome [13] approach was proposed. The modification proposes to calculate the friction pressure drop at the interface of the liquid-vapor mixture based on modified velocity due to entrainment.

The velocity is calculated based on the density of the vapor-droplet mixture at the core of the annular flow. It is calculated based on a homogeneous flow for the core flow as follows

$$\rho_c = (1-\varepsilon_C)\rho_L + \varepsilon_C \rho_V \quad (5)$$

where ε_c is the void fraction of the core flow:

$$\varepsilon_C = \frac{\varepsilon}{\varepsilon + E(1-\varepsilon)} \quad (6)$$

The liquid entrainment fraction correlation by Tibiriçá [30] used in this study:

$$E = \left(1 + 10^{6.8592} \left(\frac{\rho_V}{\rho_L} \right)^{0.6267} We_{jv}^{-1.1641} Re_L^{-0.3988} \right)^{-1} \quad (7)$$

where the superficial gas velocity Weber number is defined as:

$$We_{jv} = \frac{\rho_v \cdot j_v^2 \cdot D}{\sigma} \quad (8)$$

The superficial gas velocity is calculated as:

$$j_v = \frac{x \cdot G}{\rho_v} \quad (9)$$

and liquid Reynolds number :

$$Re_L = \frac{G \cdot D}{\mu_L} \quad (10).$$

The details on calculation of the friction factor during the annular flow are presented in detail in [13].

The obtained experimental pressure drop data at the were compared with the Darcy-Weisbach correlation, assuming the friction coefficient according to the Haaland [31] and Blasius equation, with pipe internal roughness height given by supplier lower than 0.03mm.

Adiabatic pressure drops as indicated in the introduction can be measured after preparation of vapor-liquid mixture in heated section. Varying the inlet subcooling temperature of working fluid will influence the amount of heat necessary to obtain same vapor quality. Figure 2 shows the experimental two-phase pressure drop of the 5 mm tube as a function of mass flux and heat flux of R134a at different subcooling temperatures, at a saturation temperature of 19.4 °C.

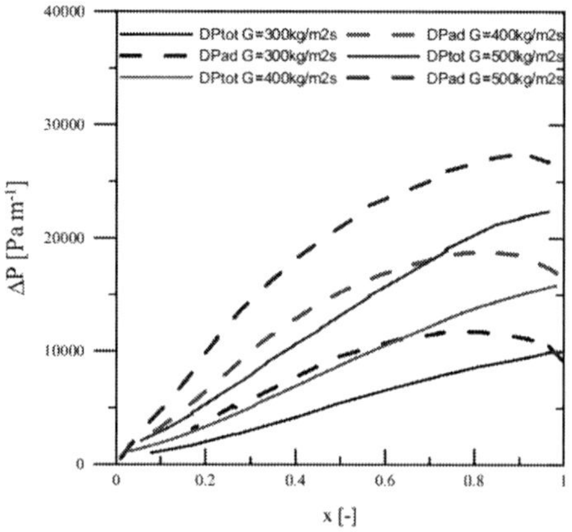

Fig. 2. Experimental values of the total and adiabatic pressure drop of R134a versus exit vapor quality of the 5 mm tube at Tsat=19.4 °C.

Figures 3 to 5 shows the experimental adiabatic frictional pressure drops, plotted versus values predicted with selected correlations. Pressure drop models used in this comparison are presented in detail in the introduction section i.e. Zhang and Webb [18] and Thome et al. [13]. The pressure drop increases exponentially with increasing mass flux. It can also be seen that the two-phase pressure drop is increasing for higher exit vapor quality with a maximum around vapor qualities around x=0.8-0.9, as reported in the literature.

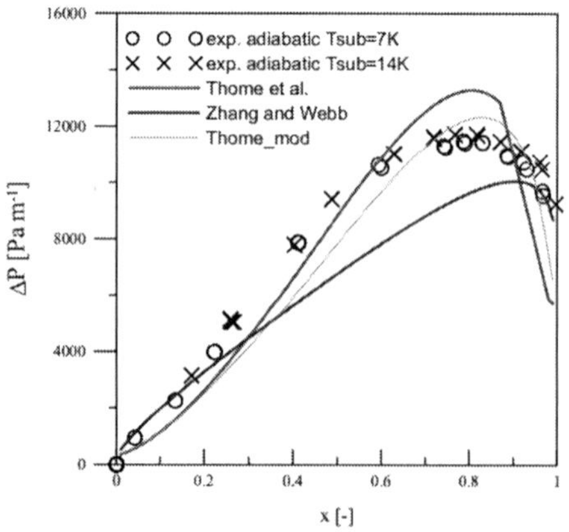

Fig. 3. Exp. values of the adiabatic frictional pressure drop of R134a versus predictions for G=300 kg/(m²s).

https://doi.org/10.1051/matecconf/201824003010

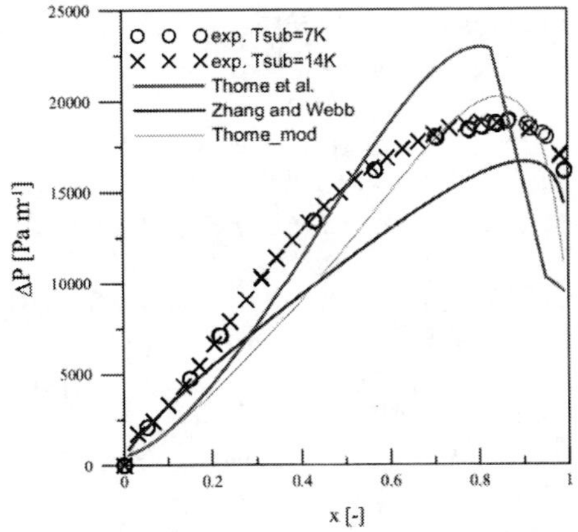

Fig. 4. Exp. values of the adiabatic frictional pressure drop of R134a versus predictions for G=400 kg/(m²s).

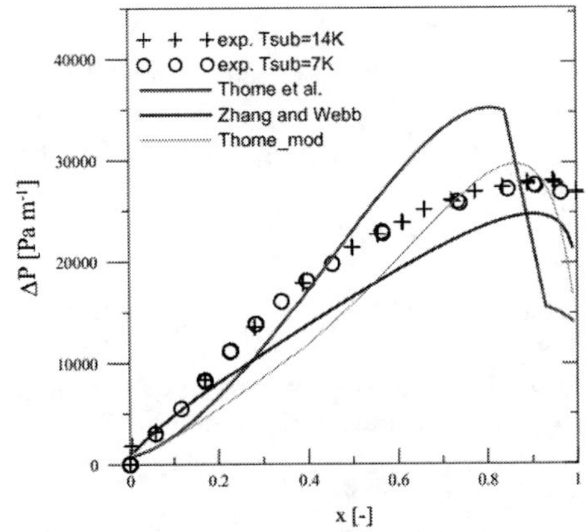

Fig. 5. Exp. values of the adiabatic frictional pressure drop of R134a versus predictions for G=500 kg/(m²s).

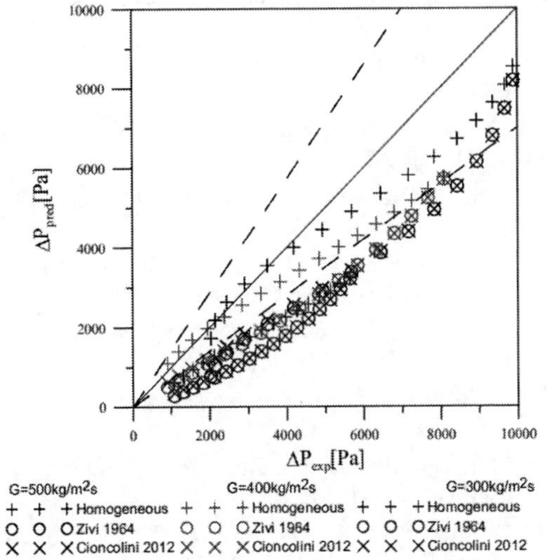

Fig. 6. Exp. values of momentum pressure drop of R134a versus predictions.

The proposed model modification with entrainment effect allows to better predict the peak pressure drop values. As can be clearly seen proposed modification provides better data representation for selected flow range. As the mass flux is lower thus the lowers the entrainment effect thus the modification in case of G=300kg/m²s is closer to original model [13].

The pressure drop component due to acceleration varies with outlet vapor quality and void fraction. Literature models of a void fraction are created with the assumption that the same flow regimes are obtained at the same local liquid and vapor flow rates. In order to calculate the momentum pressure drop from adiabatic and diabatic experiments, a piecewise decomposition of the flow must be performed. Due to the limited data points, polynomial interpolation of the data is necessary as described in [16,29]. Electric heating of the channel allows us to assume constant heat flux density along the heated channel length, therefore a linear change in local thermodynamic vapor quality of the flow can be assumed. Therefore the pressure drop along the heated channel can be recreated by using friction pressure drop values at "local" vapor qualities at heated channel length. Rearranging eq.(3) the momentum pressure drop can be calculated by subtracting frictional pressure drop from measured values of total pressure drop in heated section. This operation for an infinite number of points can be expressed as an integral in form of eq.(11). The accuracy of this method highly depends on polynomial interpolation of the data. For presented data, the best fit error was below 1%, which corresponds to total momentum pressure drop of 1.15%

$$\Delta P_{ad,decomposition} = \int_0^L \frac{d}{dz}\Delta P_{ad}(x)dz \quad (11)$$

From the analysis of data presented in figure 6, a good consistency with the homogeneous model is visible for all of the cases. Predictions given by homogeneous model compared to other correlations in that flow range differ as much as 50% corresponding up to 1kPa, for vapor quality range 0.2 to 0.8. As can be seen, all of the void fraction correlations under predict the experimentally obtained momentum pressure drop. The influence of various void fraction correlations is more pronounced in higher velocities.

Conclusions

As the first step of this work, an experimental study was undertaken in order to obtain total and frictional pressure drop values during flow boiling of R134a in a smooth horizontal tube, at a saturation temperature of 19.4°C with inlet subcooling of 7 and 14K.

The experimental campaign acquired over 300 experimental data points. Based on gathered experimental data, momentum pressure drop was obtained, from a comparison of the adiabatic frictional pressure drop and total pressure drop in diabatic section.

The flow range was chosen to represent an annular flow pattern.

Verification of the momentum pressure drop predictions showed that all void fraction correlations predict experimental data fits in the range of ±30%. For experimental flow range, the homogeneous model yields the best consistency with experimental data.

The proposed modification of the annular flow model allowed to obtain better peak pressure drop prediction.

The authors would like to appreciate funding received from the Research Council of Norway under the FRINATEK Project 231529.

Nomenclature

c_{pl}	specific heat, [kJ/(kgK)]
D	diameter, [m]
E	liquid entrainment fraction, [-]
f	friction factor, [-]
Ft	Froude rate, [-]
G	mass flux, [kg/(m^{2}s)]
g	acceleration due to gravity [m/s^2]
h	enthalpy [kJ/kg]
j	superficial velocity [m/s]
L	length [m]
m	mass flow rate [kg/s]
ΔP	pressure drop [Pa]
$\dot{Q}$	heat flux [kW]
$\dot{q}$	heat flux density[kW/m^2]
Re	Reynolds number[-]
We	Weber number [-]
x	quality [-]
Greek symbols	
ε	void fraction [-]
μ	viscosity [Pas]
ρ	density [kg/m^3]
Superscripts	
ad	adiabatic
c	core
fric	frictional
h	heated
in	inlet
l	liquid
mom	momentum
mix	mixture
out	outlet
sat	saturation
sub	subcooling
static	gravitational
tot	total
v	vapor
w	wall

References

1. B. Igliński, G. Piechota, A. Iglińska, M. Cichosz, R. Buczkowski, Clean Technol. Environ. Policy. **18**, 45–61 (2016)
2. T. Muszynski, Appl. Therm. Eng. **115**, 782–792 (2017)
3. P. Ostrowski, M. Pronobis, L. Remiorz, Appl. Therm. Eng. **84**, 390–398 (2015)
4. G. Besagni, R. Mereu, G. Di Leo, F. Inzoli, Int. J. Refrig. **58**, 154–171 (2015)
5. D. Taler, P. Ocłoń, Chem. Eng. Process. Process Intensif. **83**, 1–11 (2014)
6. T. Muszynski, S.M. Koziel, Arch. Thermodyn. **37**, 45–62 (2016)
7. R. Andrzejczyk, T. Muszynski, Arch. Thermodyn. **38**, 3–21 (2017)
8. T. Bohdal, H. Charun, M. Sikora, Int. J. Refrig. **59**, 210–223 (2015)
9. T. Muszynski, D. Mikielewicz, Appl. Therm. Eng. **93**, 1403–1409 (2016)
10. R. Andrzejczyk, T. Muszyński, Arch. Thermodyn. **37**, 137–159 (2016),
11. R. Andrzejczyk, T. Muszynski, Appl. Therm. Eng. **121**, 115–125 (2017)
12. T. Muszynski, R. Andrzejczyk, Appl. Therm. Eng. **93**, 1360–1366 (2016)
13. J.R. Thome, *Engineering data book III* (Wolver. Tube Inc. 2010)
14. [14] Y. Xu, X. Fang, Appl. Therm. Eng. **64**, 242–251 (2014)
15. R. Revellin, P. Haberschill, Int. J. Refrig. **32**, 487–497 (2009)
16. J.M. Quiben, J.R. Thome, Int. J. Heat Fluid Flow. **28**, 1060–1072 (2007)
17. R. Andrzejczyk, T. Muszynski, C. Alberto Dorao, Exp. Therm. Fluid Sci. **83**, 78–87 (2017)
18. [18] M. Zhang, R.L. Webb, Exp. Therm. Fluid Sci. **25**, 131–139 (2001)
19. J. Garcia, M.P. Porto, R. Revellin, J. Bonjour, L. Machado, Int. J. Refrig. **73**, 163–174 (2017)
20. A. Cicchitti, C. Lombardi, M. Silvestri, Energ. Nucl. **7**, 417–425 (1960)
21. L. Sun, K. Mishima, Int. J. Multiph. Flow. **35**, 47–54 (2009)
22. [22] E.W. Lemmon, M.L. Huber, M.O. McLinden, *Phys. Chem. Prop.* (NIST 2010)
23. B.N. Taylor, C.E. Kuyatt, NIST Tech. Note. **1297** 20 (1994)
24. Y. Xu, X. Fang, D. Li, G. Li, Y. Yuan, A. Xu, Int. J. Heat Mass Transf. **98**, 150–163 (2016)
25. K. Aroonrat, S. Wongwises, Int. J. Heat Mass Transf. **106**, 437–448 (2017)
26. R. Revellin, *Experimental two-phase fluid flow in microchannel* (EPFL, 2005)
27. [27] R. Charnay, R. Revellin, J. Bonjour, Int. J. Refrig. **54**, 98–125 (2015)
28. T. Muszyński, R. Andrzejczyk, C.A. Dorao, Arch. Thermodyn. **38**, 101–118 (2017)
29. T. Muszynski, R. Andrzejczyk, C.A. Dorao, Int. J. Refrig. **82**, 163–173 (2017)
30. C.B. Tibiriçá, D.M. Rocha, I.L.S. Sueth, G. Bochio, G.K.K. Shimizu, M.C. Barbosa, S. dos S. Ferreira, Appl. Therm. Eng. **126**, 774–795 (2017)
31. S.E. Haaland, J. Fluids Eng. **105**, 89–90 (1983)

https://doi.org/10.1051/matecconf/201824003011

Computational analysis of a cross flow turbine performance

Vanessa Ruiz Gómez [1*], *Edison A. Palacio* Higuita[1], and *Aldo Germán Benavides* Morán[1]

[1]Mechanical Engineering Department, Facultad de Minas, Universidad Nacional de Colombia, sede Medellín, Colombia

Abstract. In the electrical energy generation context in Colombia, the water resources represent the 64% of the potential generated according to UPME in the 2015 year; becoming into a solution to the growing energy demand and to the supply of energy in non-interconnected zones. The cross-flow turbines as Michell-Banki type, become an efficient and economically attractive choice. This paper shows the fluid-dynamic performance of a laboratory's model turbine under several operating conditions. The development of this analysis is supported by the results of experimental tests, uses the computational fluid dynamics as a tool for modelling, estimate, and analyse the turbine behaviour under different operating conditions, with ANSYS-Fluent software; the computational model considers the most important geometric aspects of the turbine and the opening percentage effect of the guide blade. The water flow through the rotor is approach through a turbulence model as $\kappa - \varepsilon$ type. The numerical study results agree satisfactorily with the turbine performance observed in the laboratory.

1 Introduction

In the electrical energy generation context in Colombia, the water resources are highly exploited, these provides about the 64% of potential generated for national interconnected system [5], besides represents a solution to growing energetic demand and its supply in non-interconnected zones.

Because of this is necessary to implement means that make up for the demand in an economical and efficient way, Michell – Banki turbine has been an alternative used for it, and this paper shows the performance evaluation for a laboratory model under different operating conditions.

Functioning of this turbomachinery type lies in move rotor due to blades interaction with fluid flow and pressure difference generated here, the fluid gets in by a nozzle where is valve that regulates the flow according to the angular velocity needed by the electric generator, it is a coupled to the rotor shaft.

Computational fluid dynamics has been used to determine the behaviour of different models [4,6,7], propose enhancements in design according to most influential parameters in its performance, like dimensions, number of blades and turbine's valve opening percentage, and find the optimal conditions for the fluid in the system. Development of this paper is based in previous researching results and it focuses on analyse, under different states, the model operating conditions, figure 1, which main features are included in table 1, where α is the angle between the circumferences that describes blades trajectory, and λ angle defines the opening circumference arc towards the rotor.

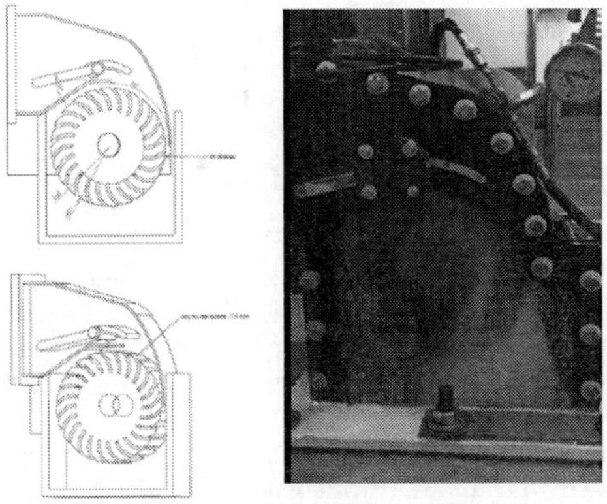

Fig. 1. CAD (left) y turbine model (right).

Table 1. Model geometric features.

Rotor internal diameter	0,122 m	α	33°
Rotor external diameter	0,194 m	λ	90°
Blades number	24	Rotor width	0,115 m

2 Methods

Previous developed analyzes around this turbine types [3], allow to compare and establish the most important geometrical parameters in the system, like the nozzle curvature, profile and number of blades, opening valve position, rotor size, well, affect directly to fluid behavior and relate operating features, like flow, flow direction and smoothness, and distribution of pressure and velocity.

Obtained experimental results, as well as advices from previous studies, offer reference points to prove

* Corresponding author: varuizgo@unal.edu.co

computational simulations completely and to characterize turbine functioning, comparing pressure drop through this, with the available energy that is in the flow initially. For that, equation (1) is used, where T is the rotor torque at a certain angular velocity, ω, Q is the flow, ρ is density, g is gravity acceleration and H is hydraulic head; to know the efficiency some measurements were made at different operating conditions.

$$\eta_t = \frac{T \cdot \omega}{\rho g Q H} \tag{1}$$

With the objective to model the flow behavior in the turbine in a computational and suitable way, turbulence model κ - ε realizable is used, this model is based on Navier-Stokes equations for mean velocity field (RANS), it looks for solve Reynolds stress under Eddy's viscosity linear model or turbulent viscosity. This model includes two transport equations more, one of this for turbulent kinetic energy, κ, and other for its dissipation rate, ε. This model presents a realizability constrain that imposes a limit on turbulent viscosity. This is a suitable model for flows with high curvature in its streamlines or flow in rotation.

Now RANS (2) and transport equations (3) y (4) are presented, where p is fluid density, it is taken as constant property, is mean velocity, is fluid velocity fluctuation and term represents Reynolds stress.

$$\rho\left[\frac{\partial U_i}{\partial t} + U_j\frac{\partial u_i}{\partial x_j}\right] = -\frac{\partial p}{\partial x_i} + \frac{\partial}{\partial x_j}\left[T_{ij} - \rho\overline{u_i u_j}\right] \tag{2}$$

In equation (2) the first term to the left represents mean velocity local change rate, second term is convective acceleration, first term to the right is pressure gradient in flow direction, next term represents viscous stress gradient due to fluid properties and, last is a term known as Reynolds stress that represents momentum transfer due to flow fluctuations.

$$\frac{\partial k}{\partial t} + U_j\frac{\partial k}{\partial x_j} = \frac{\partial}{\partial x_j}\left[\left(v + \frac{v_t}{\sigma_k}\right)\frac{\partial k}{\partial x_j}\right] + \mathcal{P}^k - \epsilon \tag{3}$$

$$\frac{\partial \epsilon}{\partial t} + U_j\frac{\partial \epsilon}{\partial x_j} = \frac{\partial}{\partial x_j}\left[\left(v + \frac{v_t}{\sigma_\epsilon}\right)\frac{\partial \epsilon}{\partial x_j}\right] + \frac{\epsilon}{k}\left(c_{\epsilon 1}\mathcal{P}^k - c_{\epsilon 1}\epsilon\right) \tag{4}$$

Transport equations right side (3) and (4) for κ – ε respectively, contain corresponding advective terms, local change rate, in mentioned order for each variable, turbulent kinetic energy and its dissipation rate. Left side of both equations involves transport, production and dissipation terms, respectively.

3 Results

Fluid behavior was analyzed inside turbine in steady and transient state, model was simplified by a bidimensional geometry and different operating conditions were considerate, moreover a measurements set were made on laboratory model to compare results and know turbine performance.

3.1 Steady state results

Three different angular velocities were considered in steady state: 200, 220, 250 RPM for each flow supposed: 50 and 91 L/seg. In the last case efficiency drastically drops due to a high flow and low angular velocities for the rotor while at 50 L/seg flow, turbine can work with low angular velocities and still gets good efficiencies at certain pressure drops, in fact, is found that better turbine efficiency is 84%, for evaluated cases, at 50 L/seg and pressure drop of 70 kPa. Some results are reported in table 2.

Table 2. Numerical results for 50 L/s and some angular velocities.

Flow [L/s]	angular velocity [rpm]	Efficiency	Pressure drop [kPa]
50	250	84%	70
50	220	47%	68
50	200	35%	65

Figure 2 shows that fluid velocity decreases just a bit when it hits a blade for first time, and nozzle opening arc allows to fluid get it to the rotor through six blades at the same time. Small space at the right side between turbine wall and rotor, causes some acceleration in outgoing fluid, while in practice it was observed that there are some recirculation at the left side.

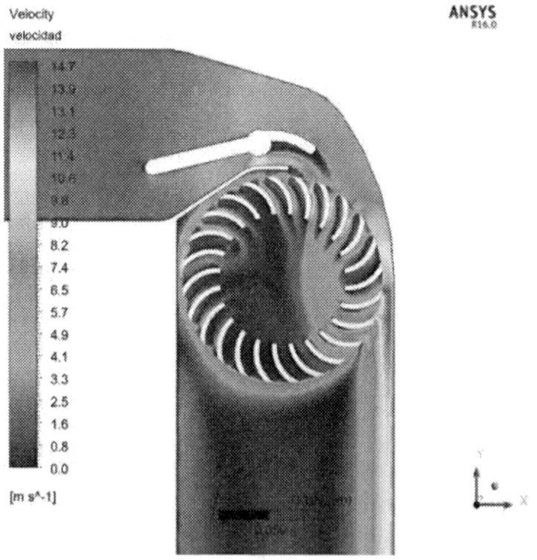

Fig. 2. Velocity contours at 50 L/s and 250 rpm.

Turbulent kinetic energy gets big magnitudes in flow zones that are below director blade bearing surface, and it is significantly biggest between blades zones at rotor entrance, like figure 3 shows. It could infer that in this zones turbulence is generated in the flow due to high velocity gradients, in other flow zones turbulent kinetic energy has a small magnitude and is uniform because streamlines don't have big fluctuations.

Table 3. Numerical results in transient state ad different operating condition.

Flow [L/s]	Angular velocity [rpm]	Pressure drop [kPa]	Power [kW]	Efficiency
28	600	31,5	1,2	30 %
32	956	47	1,5	44%
33	510	35	1,8	33%
40	1250	76	3,3	71%
43	1329	82	4	76%

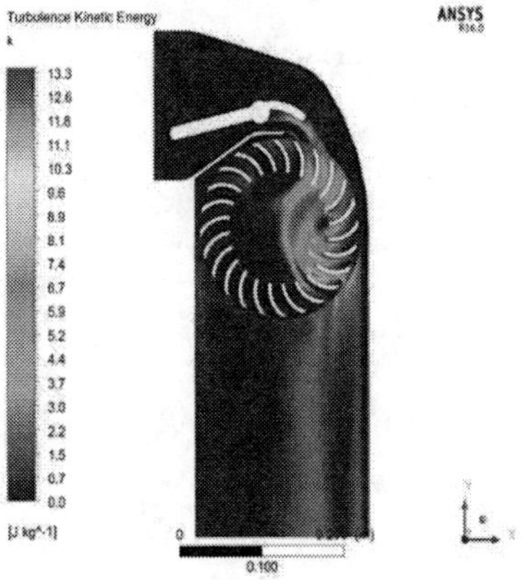

Fig. 3. Turbulent kinetic energy contours at 50 L/s and 250 rpm.

3.2 Transient state results

To determine turbomachinery behavior is important to include time in developed analysis to get more accurate results, because of this, turbulence model $\kappa - \varepsilon$ realizable, is used in a transient state. In table 3, different flows and angular velocities values are summarized. Until a certain angular velocity, increasing flow takes to increasing pressure drop and efficiency, but in other hand, when angular velocity is too low for determined flow, generated power will be smaller, and thus, turbine efficiency too. It is pointing out that turbine is working below its optimal operating point.

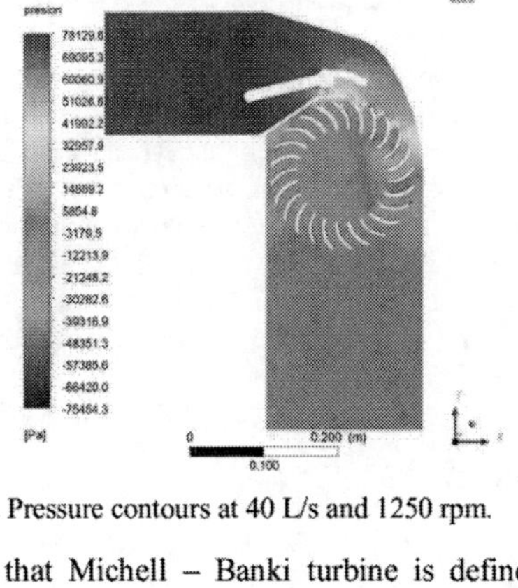

Fig. 4. Pressure contours at 40 L/s and 1250 rpm.

Given that Michell – Banki turbine is define as a reaction turbine, there is a significantly pressure drop through rotor as can see in figure 4. Worth to advise that energetic losses from turbine entrance to crossing to rotor, are small, valve opening percentage, at turbine entrance, make that fluid pressure decreases, in a small zone there. While the fluid that gets to go over the valve, arrives with a bigger pressure compared with the flow below the valve.

Moreover, it can see with colors scale help, how pressure drops through rotor and after first fluid path through blades. In other hand, there is a pressure difference between blade faces to propel its movement, this difference can be seen in external blades tip.

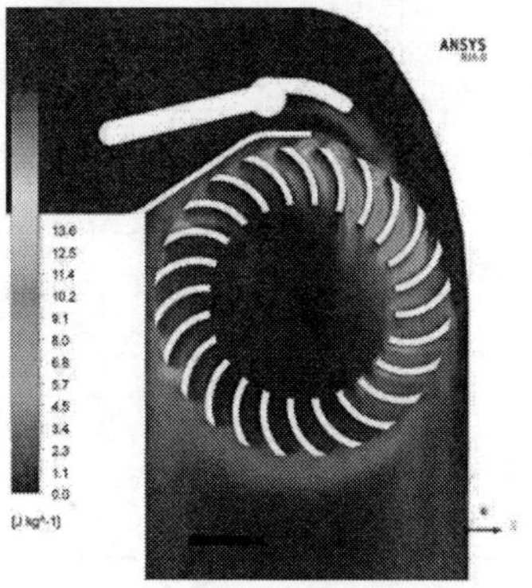

Fig. 5. Turbulent kinetic energy contours at 40 L/s and 1250 rpm.

Comparing turbulent kinetic energy contour in steady and transient state, we can see at first case energy dissipation along streamline through rotor, while in transient case dissipation occurs intensity around blades, like figure 5 shows. A possible cause for this, is increase in viscous effects with rotor movement.

After pass by the valve opening that regulates flow through rotor, fluid increases its velocity significantly and keeps high value when cross rotor and finally, increases again by its second crossing through blades, as can be seen in figure 6, in this way, energy is pulled out from fluid in two different points.

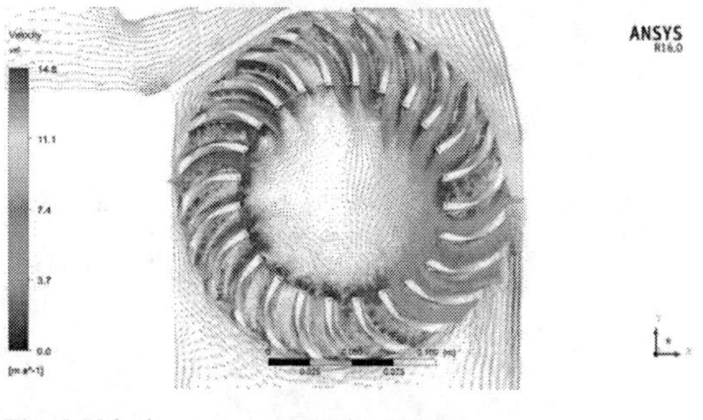

Fig. 6. Velocity vectors at 40 L/s and 1250 rpm.

3.3 Experimental results

Aiming to characterize real turbine behavior at different opening valve percentage, flow, pressure and angular velocity measurements were carried out, applying four different force values to rotor through a pneumatic brake, like are summarized en table 4. Finding that at more braking force, more flow is necessary to make up for required power by electric generator shaft. At the same time there is a pressure decrease upstream, making to decrease turbine efficiency, this behavior trend is shown in graph 7.

Table 4. Model laboratory behavior for different positions of opening valve.

Valve opening	Pressure drop [kPa]	Brake force [N]	Efficiency
Totally open	54,4	0	51%
	53,1	8,9	49%
	52,4	13,4	49%
	49,6	26,7	46%
	48,2	84,6	45%
Open 3/4	58,6	0	54%
	57,9	4,4	54%
	57,2	8,9	53%
	56,5	17,8	53%
	48,2	66,8	45%
Open 1/2	63,4	0	59%
	62,7	13,4	58%
	61,3	17,8	57%
	59,9	49	56%
	56,5	155,8	53%

A trend observed as in numerical as in experimental result is, efficiency increases proportionally to flow, for studied values, moreover, is found that efficiency is smaller for experimental data than for numerical, for the same flow, as can be seen in figure 8. This is due to non-accurate measurements, but in practice energetic losses in flow are bigger in experimental tests than in numerical simulations.

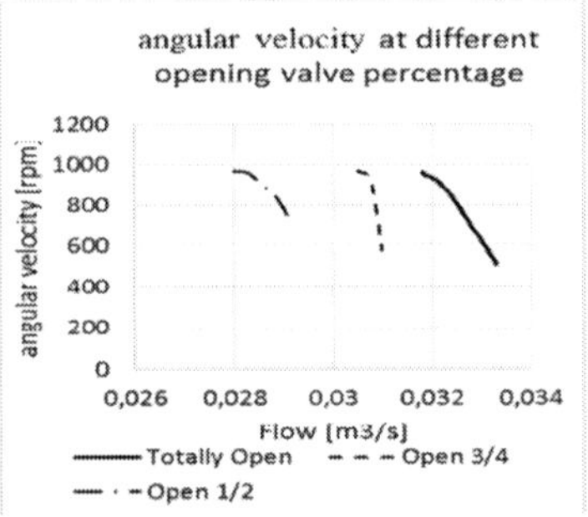

Fig. 7. Experimental results for angular velocity against flow for different entrance valve openings.

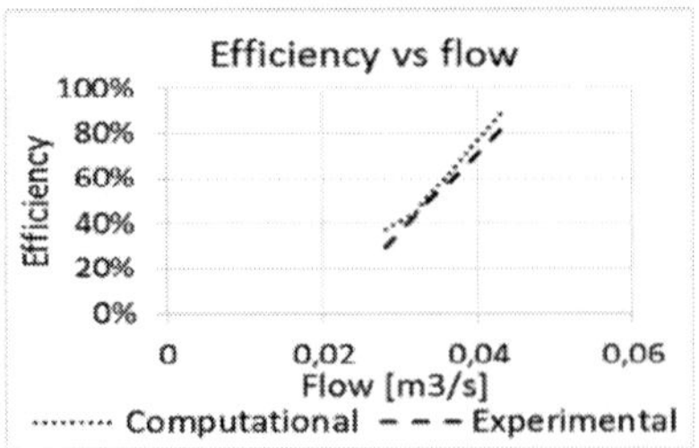

Fig. 8. Efficiency for experimental and computational results.

4 Conclusions

Michell–Banki turbine has a simple and economical design, moreover it can reach high efficiencies since it operates under suitable conditions, like was found in performed simulations, in this case maximum efficiency was 84% for 50 L/s, 250 rpm y 9 m of hydraulic head, like operating conditions; this make interesting to go in depth in studies around this model and implement them in electric energy generation sector to make the most of water resources that are close to non-interconnected zones.

Computational methods in the study of this turbines type allows to know model functioning and performance, under different operating conditions, which allows to identify improvement points on hydrodynamic and structural design and allows to simulate conditions that in practice are no easy to achieve, and although is possibly get very accurate results, it has high effort, time and cost computational.

Simulations in steady state were made under the supposition that the flow keeps fixed while angular velocity changes, to know what the effects could be of reach certain velocity in flow but, is known that in experimental case angular velocity can't change without flow changes too.

In velocity contours observed for different simulations settings, flow is affected by area reduction due to a small space between turbine casing and rotor, increasing its velocity, allowing a possible appearance for drag force in blades face that is in contact with flow that doesn't cross the rotor. Furthermore, in velocity contours it can be seen that in transient state is possible corroborate wall effect on flow, in other hand, in practice was seen that this zone is highly susceptible to flow recirculation and air bubble stack, so it is recommended enlarge this width and the walls towards discharge.

Although, experimental results serves like a reference point to compare computational results and establish a trend in turbine behavior, these aren't determining enough so, there are several effects that takes part negatively on reliability and accuracy, like measurement instrument calibration, leaks in system of tubes upstream to turbine, flow losses that are not in contact with rotor, and flow recirculation present by geometric features on turbine discharge position.

Is necessary experimental data measurement and analysis for different turbine operating conditions, make graphs and curves that characterize its performance, to compare with numerical results. Computational fluid dynamics results are a close relationship with experimental data trend, nevertheless results are a little different, no more than 15%, due to all events without considering in simulations as flow irreversibleness and mistakes in measurements.

References

1. Gómez et al., *La turbina Mochell-Banki y su presencia en Colombia* (Avances en recursos hidráulicos, 2008)

https://doi.org/10.1051/matecconf/201824003011

2. Paris et al., *Construction and performance evaluation of Michell-Banki turbine prototype* (2013)
3. Sammartano et al., *Banki-Michell optimal design by computational fluid dynamics testing and hydrodynamic analysis* (2013)
4. Sammartano et al., *Experimental study of cross-flow micro-turbine for aqueduct energy recovery* (2014)
5. UPME, *Atlas potencial hidroenergético de Colombia* (pp. 25-36, 2015)
6. Choi et al., *Performance improvement of a cross-flow hydro turbine by air layer effect* (2010)
7. Velasquez et al., *Análisis de la influencia del porcntaje de apertura del álabe director sobre la potencia hidráilica de un picogenerador de flujo cruzado mediante un estudio númerico en software CFD* (2013)

Analysis of flow through channel with mounted blades

Łukasz Semkło[1],* and *Łukasz Gierz*[2]

[1]Poznan University of Technology, Chair of Thermal Engineering, Poznań, Poland
[2]University of Technology, Institute of Machines and Motor Vehicles, Poznań, Poland

Abstract. Numerical analysis of parts of the installation for transporting the mixture. The mixture consists of sand, stones and water. The analysis concerns the possibility of reducing the flow velocity of the mixture in the installation by means of installing the blades. The article presents an analysis of 10 blades that have been simulated. Speed distributions are presented after passing through the blades and determined which slows the speed of the mixture to the best possible extent.

1 Introduction

The suction dredgers are devices used to extract the aggregate from the bottom of the tank. The continuous increase in aggregate extraction in Poland requires manufacturers of dredgers to improve these machines. For the first 6 years since Poland joined the EU, aggregate production increased by over 100%. Despite such a big increase, Poland produces less aggregate than the European average indicates. Our country is only 19th in this respect behind the European powers such as Norway, Austria, Finland, Denmark, Sweden, the Netherlands and others. Forecasts for the coming years predict that the production of natural gravel aggregates will stabilize and slightly increase by launching new sites of exploitation and possible legalization of the places of their previous extraction without concessions. The growing cost of transport plays a special role and will be appreciated [1,2,3].

The aim of the research was to develop a steering structure for the mixture of sand and gravel in the aspect of reducing the speed of moving particles. Optimizing the shape of the proposed blade should contribute to increasing the life of the rotor. The simulation tests carried out allowed to determine the most favorable shape of the working body. For the needs of these studies, the concepts of steering wheels of various shapes have been proposed and modeled. Each of these blades was placed 100 mm from the suction port of the pump. By concealing the various shapes of the steering wheels, it was assumed that they would constitute an obstacle to the transported mixture. Lower speed or lower impact force in the central part of the rotor will probably contribute to increasing its durability by changing the trajectory of the flow of a mixture of sand and gravel (previously you wrote gravel and water.

Figure 1 prezent drigger and figure 2 shown element in engine room wich was model.

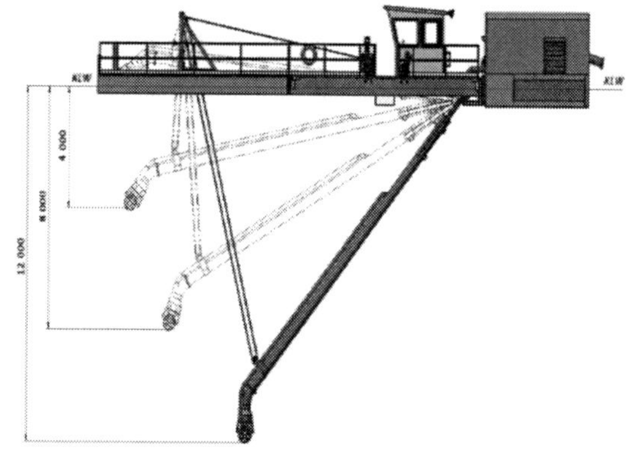

Fig. 1. Żwir-kop model ZK 10/8 300/250 dredger [4].

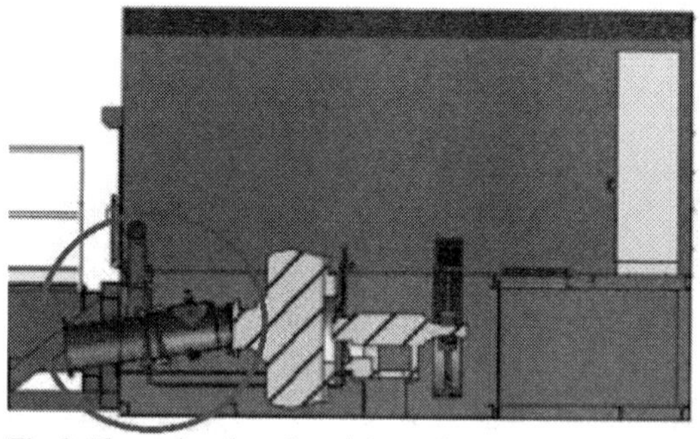

Fig. 2. The analyzed section of the engine room [4].

The Finite Volume Method was used for the research, which differs from the FEM method in that it is based on the analysis of the field variable behavior inside the grid cells, and not in the grid nodes. The FVM method consists in estimating the integral inside the cells of the grid or their mean values of the integral is divided by the cell volume. In subsequent time stages, by using the stream function, which flows through the volume

* Corresponding author: lukasz.semklo@put.poznan.pl

elements of the model, field variables are updated. The space of the tested pipe was described by means of a grid, i.e. small sub-areas in the shape of tetrahedrons. This process is called discretization. The simulation was performed in the Ansys Fluent 16.0 program. The methods used in this program belong to the numerical fluid mechanics (CFD - Computational Fluid Dynamics) involving the flow of liquids, gases and related phenomena. [5, 6].

2 Analysis and modelling

For the purpose of this research, it is necessary to determine the parameters of a mixture of gravel and aggregate. First, the analysis of the grain size of the soil was made and then the specific density was determined in the next order. A detailed description of the methods used and a summary of the results obtained are described below. To determine the granulation or division into individual fractions, a shaker was used along with sieves of various mesh sizes. According to the Polish standard [9], we distinguish the following five factions:
- stony with grains d> 40mm;
- pebble with grains d = 40 ÷ 2mm;
- sand with grains d = 2 ÷ 0.05 mm;
- dust with grains d = 0.05 ÷ 0.002 mm;
- iowa with grains d <0.002mm. [7]

Analysis of the percentage composition of individual fractions of the mixture of sand and aggregate was carried out in laboratory conditions. For these tests, a representative sample of 696.91 g was taken from the bottom of one of the dumped water reservoirs, which was then dried at 90oC to 0.8% humidity.

2.1 Numerical analysis

The developed models of both the suction pipe and the steering wheel were imported into the Ansys program, in which they underwent discretization of the three-dimensional space of the fluid (water with sand) using FVM. The next stage of the research was to determine the inlet and outlet connections so that the flow direction is known. After completing the grid creation process, the virtual model was moved to the simulation program. Before starting essential research, it was necessary to establish and task boundary conditions that are classified as follows:

- o the aggregate expenditure at the level of 500 t / h based on data from the producer;
- o specific gravity of aggregate 3.7 g / cm3 - which was determined on the basis of completed tests;
- o the speed at which the 5m / s spoil flows, based on data from the manufacturer;
- o the share of the quartz mineral fraction in the mixture with water was about 15%
- o in order to simplify the research fraction sizes were adopted: 0.05m;
- o length of the modelled pipe: 2 m;
- o diameter of the pipe: 250 mm.

Figure 3 presents geometry of the pipe. Figures from 4 to 13 shows the geometry of blades.:

- o the aggregate expenditure at the level of 500 t / h based on data from the producer;
- o specific gravity of aggregate 3.7 g / cm3 - which was determined on the basis of completed tests;
- o the speed at which the 5m / s spoil flows, based on data from the manufacturer;
- o the share of the quartz mineral fraction in the mixture with water was about 15%
- o in order to simplify the research fraction sizes were adopted: 0.05m;
- o length of the modeled pipe: 2 m;
- o diameter of the pipe: 250 mm.

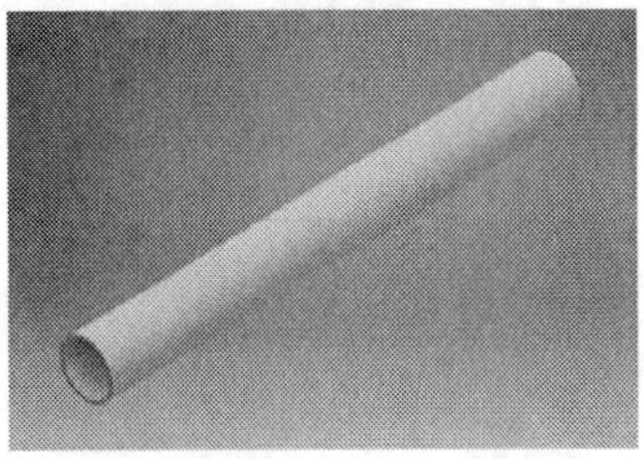

Fig. 3. The analyzed section of the engine room [4].

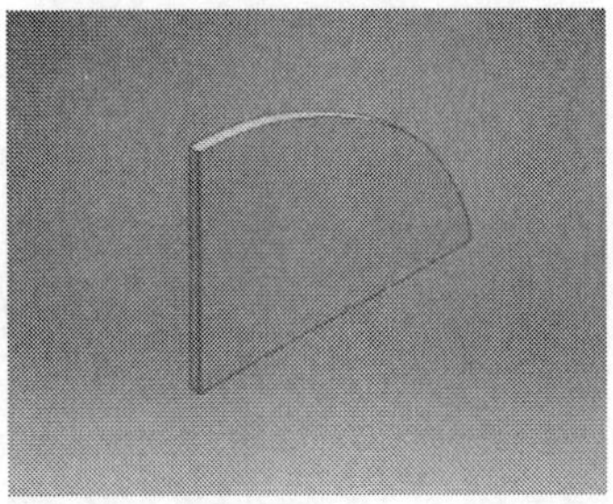

Fig. 4. The analyzed section of the engine room [4].

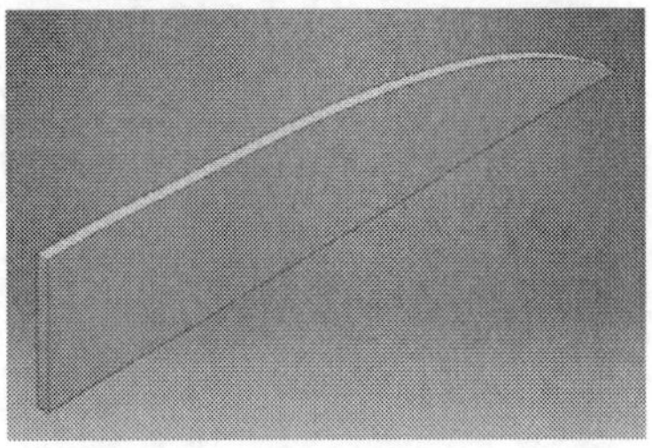

Fig. 5. The analyzed section of the engine room [4].

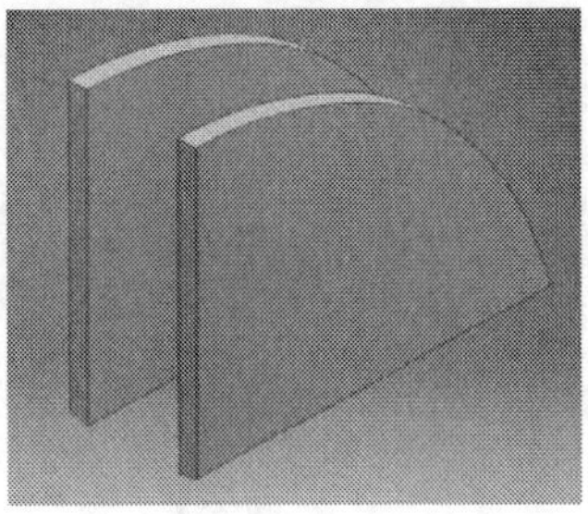

Fig. 6. The analyzed section of the engine room [4].

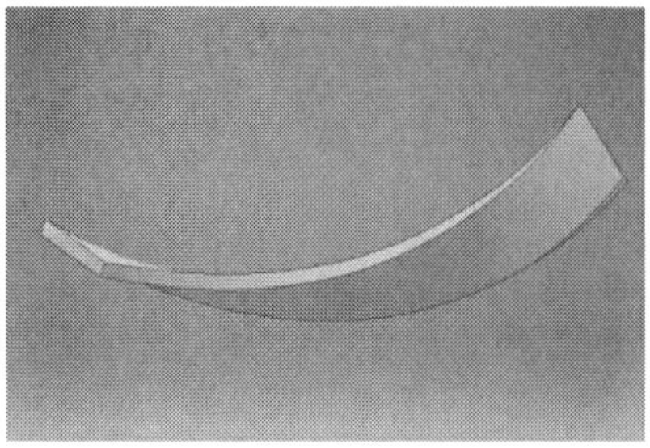

Fig. 7. The analyzed section of the engine room [4].

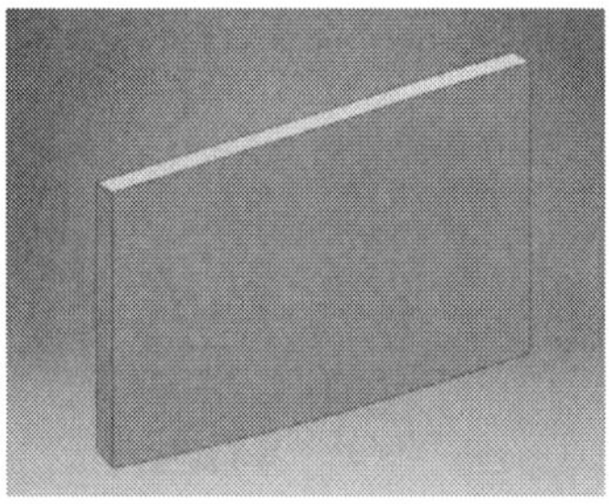

Fig. 8. The analyzed section of the engine room [4].

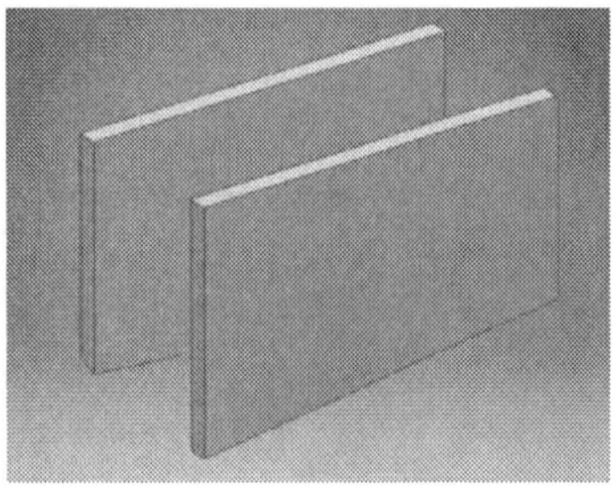

Fig. 9. The analyzed section of the engine room [4].

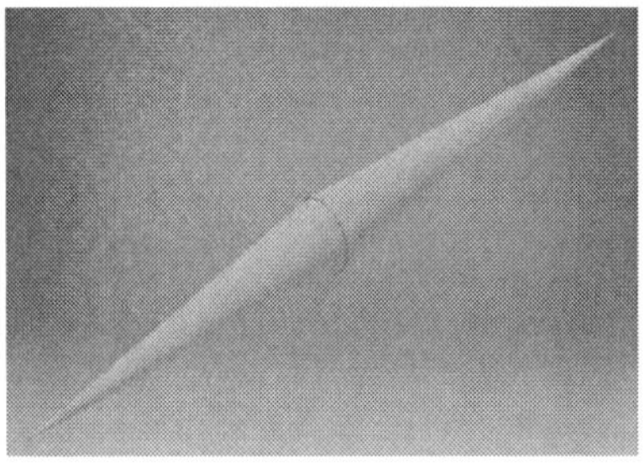

Fig. 10. The analyzed section of the engine room [4].

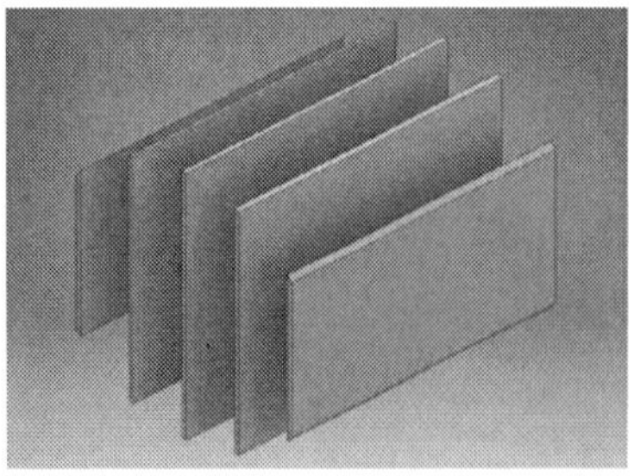

Fig. 11. The analyzed section of the engine room [4].

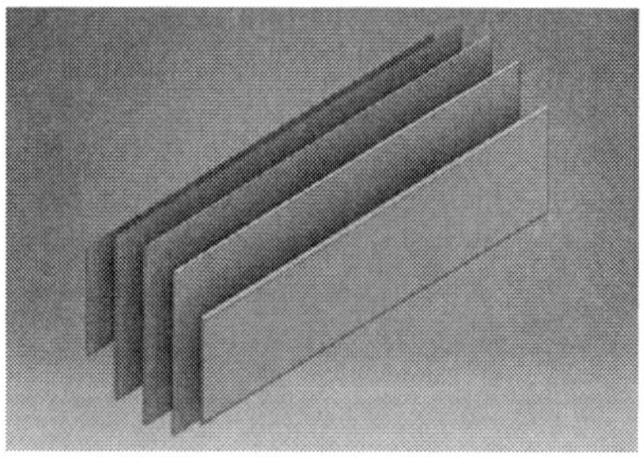

Fig. 12. The analyzed section of the engine room [4].

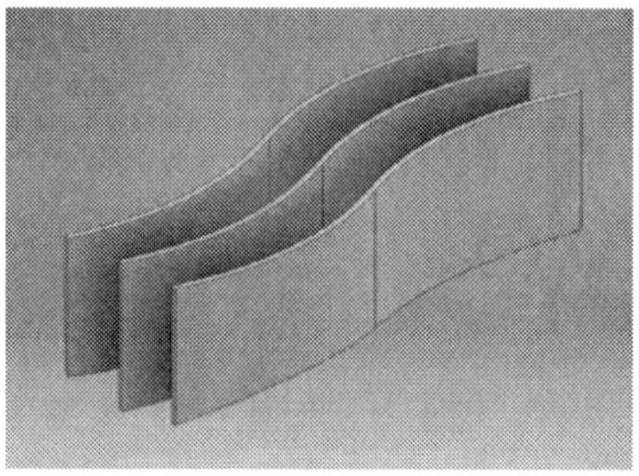

Fig. 13. The analyzed section of the engine room [4].

3 Results

As part of the simulation tests, the results of the trajectories of the moving particles before and after the collision with the hearth were obtained. The obtained test results do not take into account the direct collision between the molecules, because the overriding goal was to observe the reaction between the element controlling the particles of gravel and aggregate mixture.

The carried out studies of the flow of spoil in the suction dredger pipeline with the modelled vanes placed inside, allowed to analyze the velocity and trajectory of turbulence with which the tested mixture of water and aggregate hits the pump impeller. The tests were carried out for the pipeline without the blades and with all ten proposed blades. During the calculation, the mixture was given an initial speed of 5 m/s.

The flow without the working body is not disturbed and is shaped as it would be expected. At the outlet we notice a slight acceleration. The results are presented in Figures 14 – 34.

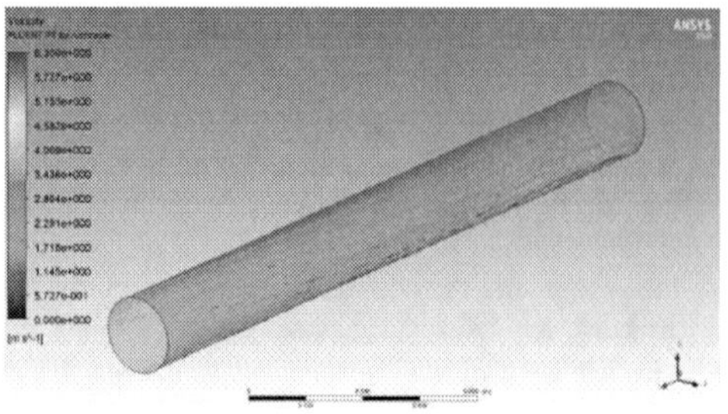

Fig. 14. Geometry without blade velocity distribution in the tube.

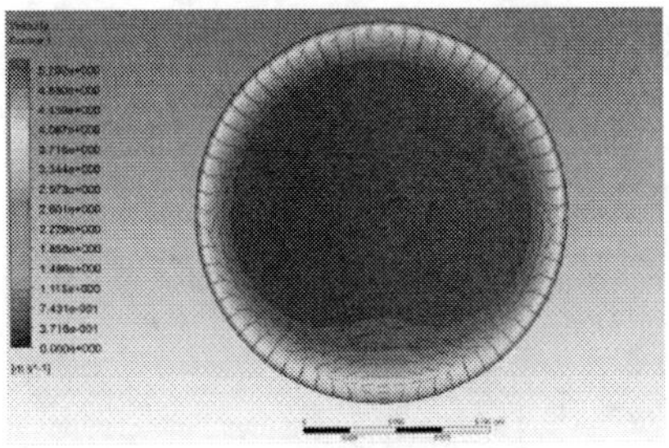

Fig. 15. Geometry without blade velocity distribution at the inlet of the tube.

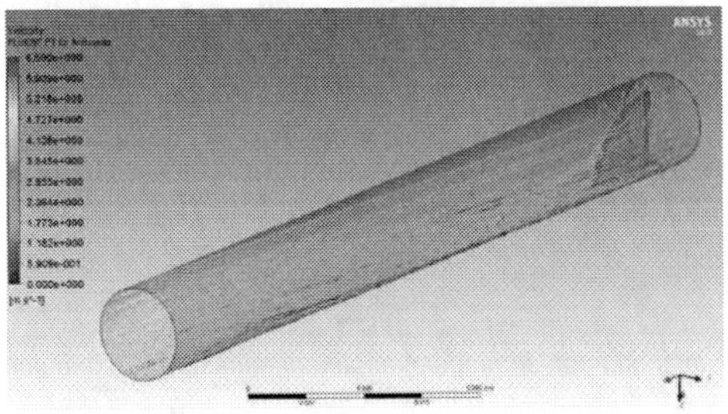

Fig. 16. Blade no. 1 velocity distribution in the tube.

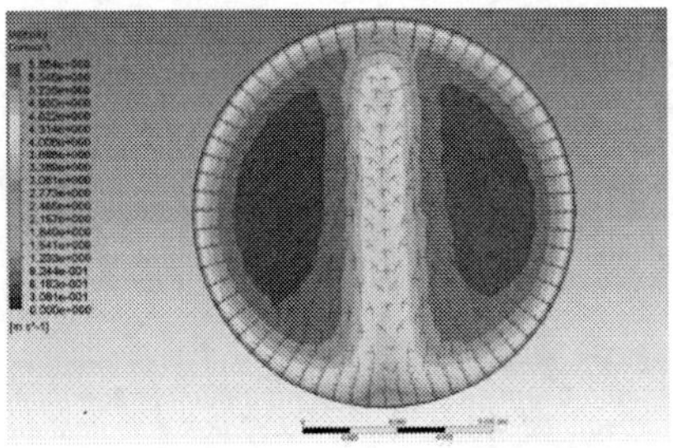

Fig. 17. Blade no. 1 velocity distribution at the inlet of the tube.

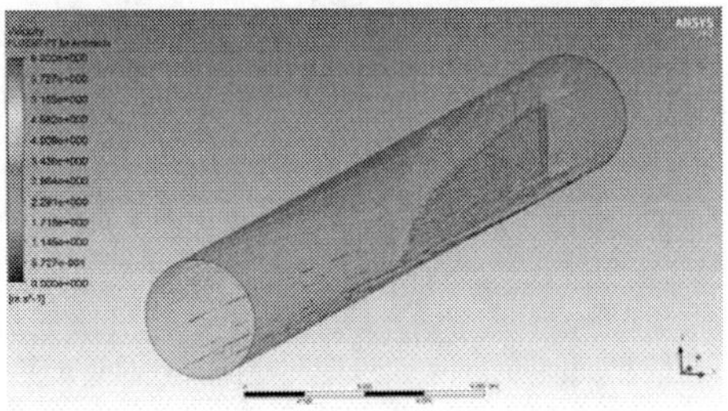

Fig. 18. Blade no. 2 velocity distribution in the tube.

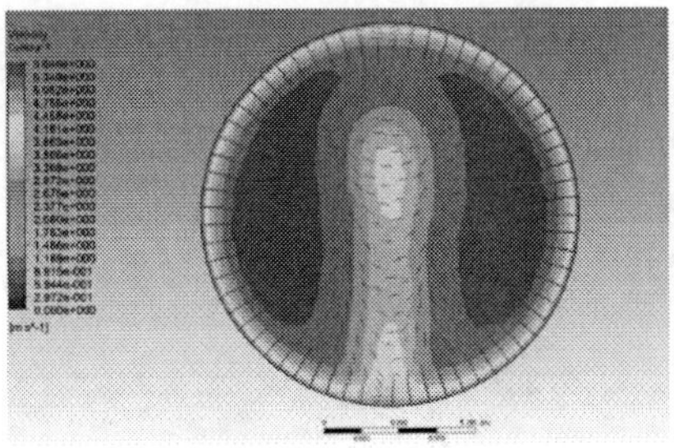

Fig.19. Blade no. 2 velocity distribution at the inlet of the tube.

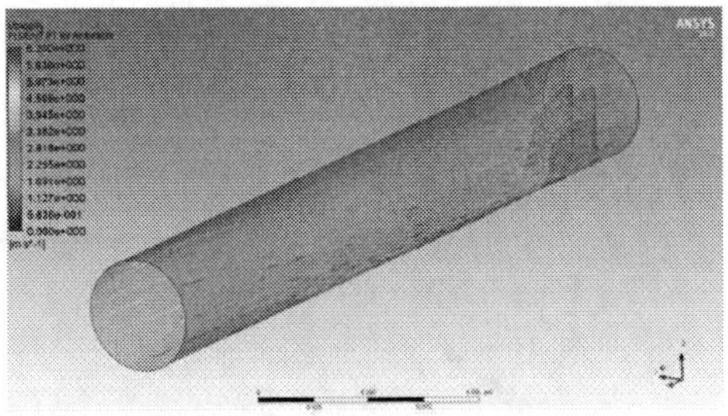

Fig.20. Blade no. 3 velocity distribution in the tube.

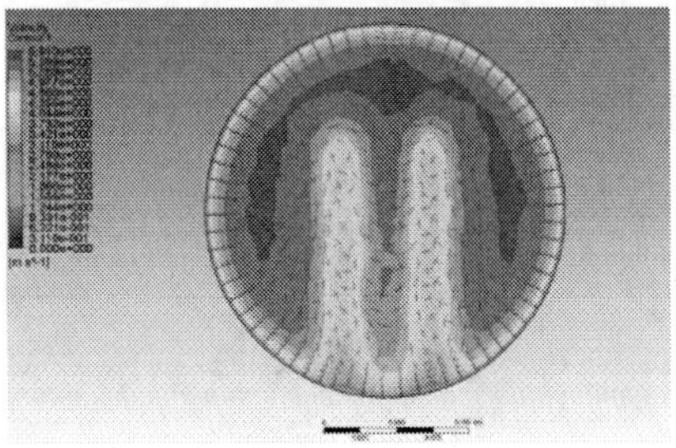

Fig. 21. Blade no. 3 velocity distribution at the inlet of the tube.

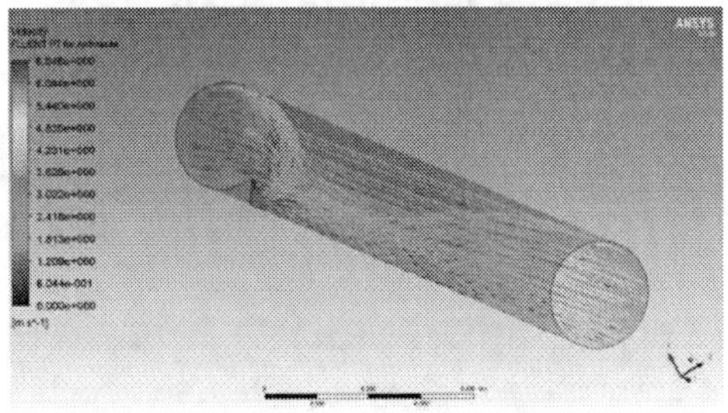

Fig. 22. Blade no. 4 velocity distribution in the tube.

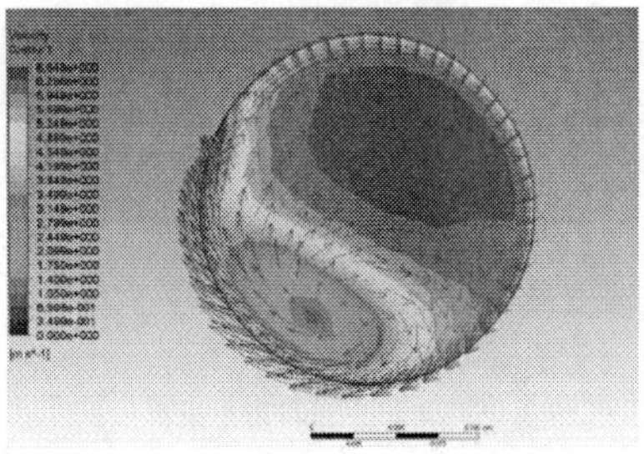

Fig. 23. Blade no. 4 velocity distribution at the inlet of the tube.

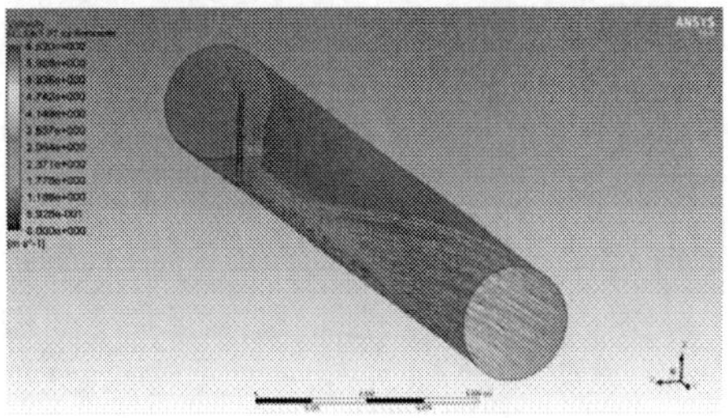

Fig. 24. Blade no. 5 velocity distribution in the tube.

https://doi.org/10.1051/matecconf/201824003012

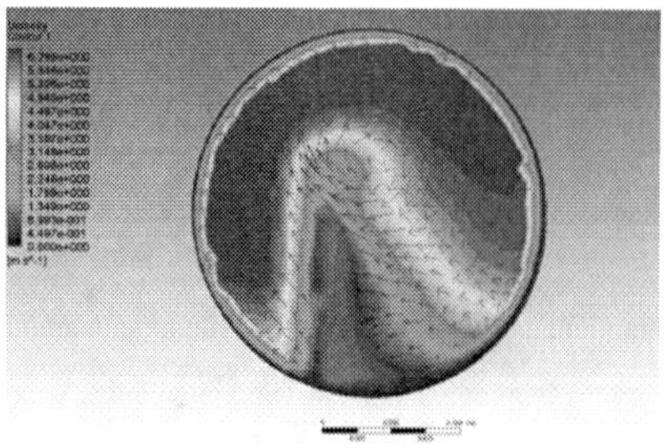

Fig. 25. Blade no. 5 velocity distribution at the inlet of the tube.

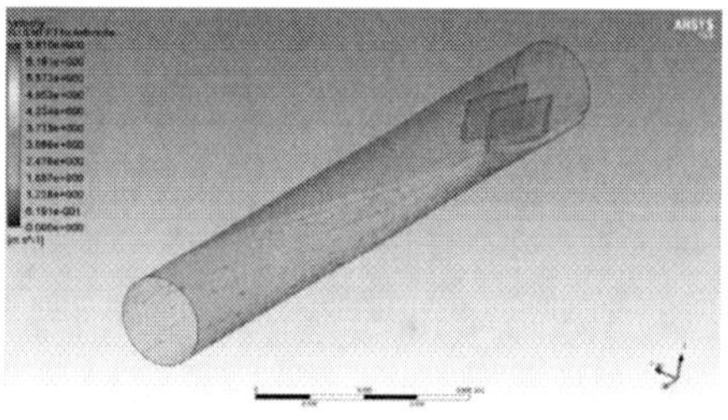

Fig. 26. Blade no. 6 velocity distribution in the tube.

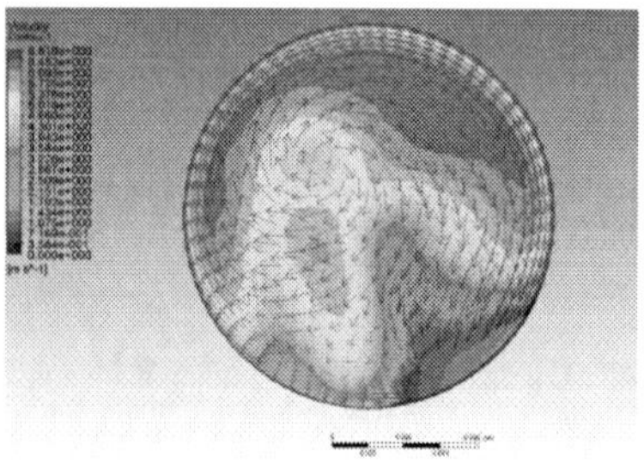

Fig. 27. Blade no. 6 velocity distribution at the inlet of the tube.

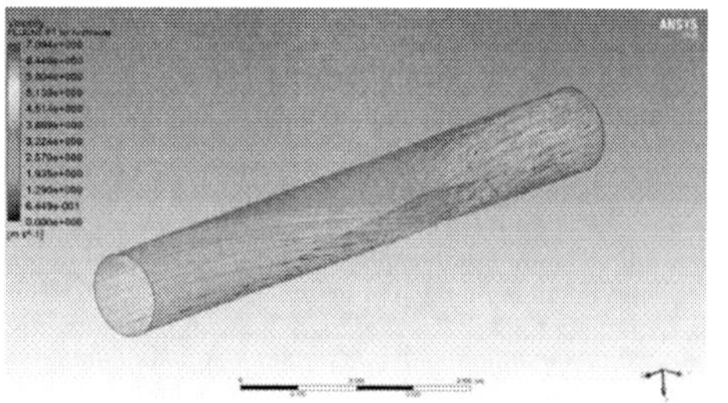

Fig. 28. Blade no. 7 velocity distribution in the tube.

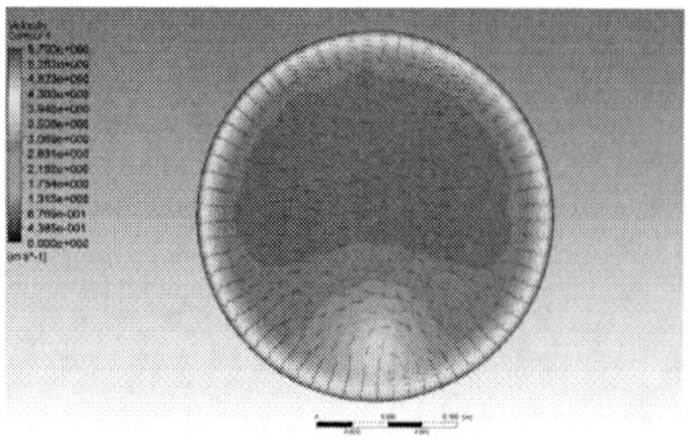

Fig. 29. Blade no. 7 velocity distribution at the inlet of the tube.

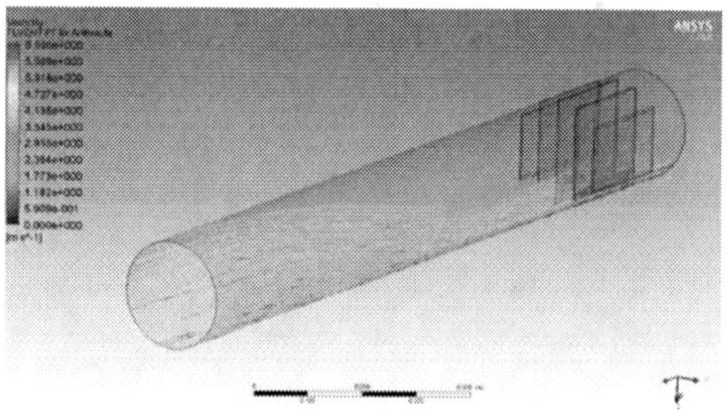

Fig. 30. Blade no. 8 velocity distribution in the tube.

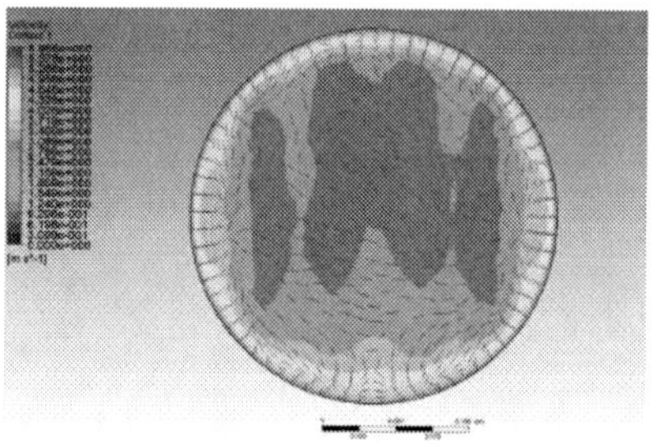

Fig. 31. Blade no. 8 velocity distribution at the inlet of the tube.

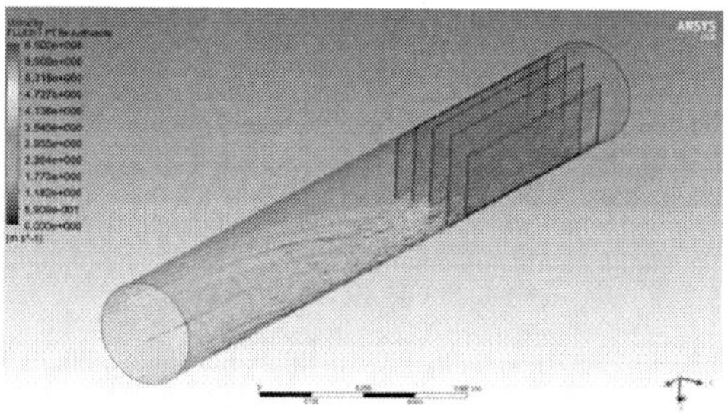

Fig. 32. Blade no. 9 velocity distribution in the tube.

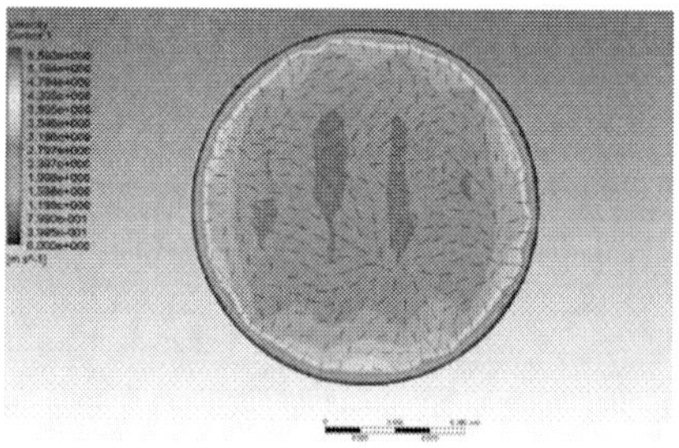

Fig. 33. Blade no. 9 velocity distribution at the inlet of the tube.

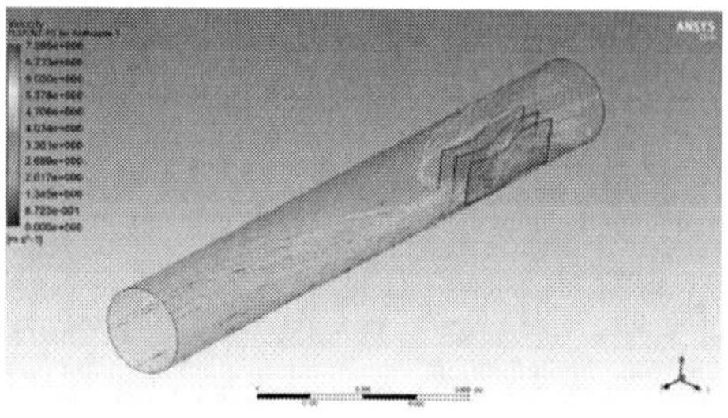

Fig. 34. Blade no. 10 velocity distribution in the tube.

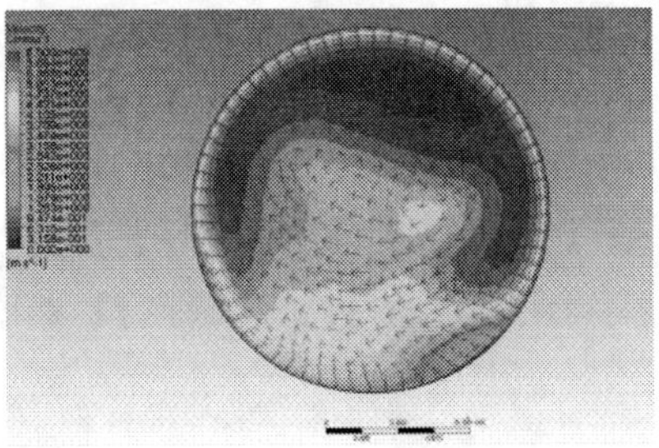

Fig. 35. Blade no. 10 velocity distribution at the inlet of the tube.

3 Summary

As a result of the conducted simulations, it can be concluded that the proposed shapes of blades numbered 7, 8 and 9 best meet the assumed criteria. Model 8 gives optimistic results of the average speed of solids. In addition, the particle paths after impact with the control element have a favourable orientation.

The calculations did not take into account the collisions between molecules, therefore during the simulation of grains, which theoretically should hit each other, penetrate. Therefore, the speed results obtained after using the working organs may differ from the actual ones. In fact, after hitting the grain with a grain or other solid, both elements should lose speed. The only way to assess the behaviour of particles after collisions was to analyze the trajectory of particle movement.

References

1. S.F. Kim, F. Low, J. Film Cooling, **122**, pp. 21-26, 1997
2. H. Schlichting, *Boundary-Layer Theory* (7th edition, McGraw-Hill, New York, 1979)
3. A. Bęben, *Maszyny i urządzenia do wydobywania kopalin pospolitych bez użycia materiałów wybuchowych* (Wydawnictwo AGH, Kraków, 2008)
4. W. Kozioł, Ł. Machniak, Surowce i Maszyny Budowlane, nr 3 (2011)
5. http://www.imsdredge.com
6. M. Dudek, *Dokumentacja techniczno ruchowa pogłębiara ssąca (refuler)* (2016)
7. M. Kmiotek, Scientific Bulletin of Chelm, no. 1 (2008)
8. H. Kudela, *Laboratorium numerycznej mechaniki płynów* (fluid.itcmp.pwr.wroc.pl/~znmp/dydaktyka/wzzmp/WSTEP_LAB_NUM.pdf)
9. PN-86/B-02480

Influence of crystal growth kinetics on prediction of macro-segregation by micro-macroscopic simulation of binary alloy solidification

Mirosław Seredyński[1,*], and *Jerzy* Banaszek[1]

[1] Institute of Heat Engineering, Faculty of Power and Aeronautical Engineering, Warsaw University of Technology, Nowowiejska 21/25, 00-665 Warsaw, Poland

Abstract. Using the micro-macroscopic computer simulation model of binary alloy solidification, based on a single domain control-volume calculations, and involving the tracking of columnar dendrite tips envelope for distinguishing different grain structures within the two-phase liquid-solid region, the paper reports the outcomes of the detailed analysis addressing the question of the influence of enhanced KGT crystal growth models on the predicted macro-segregation pattern in the Al-4wt.%Cu alloy cast. In particular, the analysis concerns the impact of a proper selection of the stability constant for the KGT model (based on a crystal-melt surface energy anisotropy strength and linear scaling law of the marginal stability theory) on the predicted dendrite tip under-cooling, volumetric solid fraction and actual solute concentration along the front of columnar dendrite tips. Differences in the resulting evolution of both: the under-cooled melt region within the mushy zone and the solute concentration fields calculated for the considered various kinetics of a grain growth are also reported and discussed.

1 Introduction

Complex fluid flow, heat and mass transfer processes, accompanying solidification of metal alloys at the scale of individual crystals and in a whole cast, are responsible for the developing micro and macroscopic grain structures, and thus for mechanical and thermo-physical properties of a final manufactured product. Macro-segregation, i.e. a compositional heterogeneity, is a very common and serious casting defect arising due to the relative motion between the liquid and solid phases. Since this compositional heterogeneity is a large length scale phenomenon, it cannot be removed by post-casting diffusion mechanisms and homogenization heat treatment.

Tracing macro-segregation formation experimentally during the evolving process of solidification is extremely difficult; so far its ultimate picture has only been detected by post-mortem examination of the final cast. Therefore, a computationally efficient and reliable computer simulation becomes nowadays a key powerful engineering tool for designing and control of casting processes.

Such simulation involves multi-scale modelling of alloy solidification across the whole spectrum of length and time scales - including the levels of individual dendrite arms (micro-scale), a single crystal (meso-scale) and transport processes in a whole cast (macro-scale). Since a direct numerical simulation of the solidification process in its full-length and time scales for a real whole cast is still very limited due to the lack of adequate computational resources, the idea of micro-macroscopic modelling has been developed. Key calculations are performed by solving average macroscopic transport equations, and information about the developing micro-structures within the two-phase region, called a mushy zone, is transferred to the macroscopic model through the effective properties of the two-phase region or/and by effective coupling of direct calculations in both these scales.

Macroscopic equations of mass, momentum, energy and species conservation can be obtained by using the classical theory of mixture (e.g. [1]) or the volume averaging (e.g. [2]). Nowadays, the most widely used formulation here is based on a fixed grid/single-domain approach and the enthalpy-porosity model, where the total enthalpy concept is applied to account for a latent heat effect, and the momentum transport within columnar mushy zone is simulated by porous medium models (e.g. [3]). Since two different dendritic structures develop within the mushy zone, i.e. columnar and equiaxed ones, it is crucial to distinguish them and then to apply different models of the momentum transport in each of them, i.e. a porous medium model or a slurry one. A common approach is based on the *coherency point* model [4,5], where these distinct grain structures are identified by the inspection of the solid volume fraction; when its local value exceeds some previously specified critical packing value, the region is treated as a stationary columnar mush, otherwise as an equiaxed zone. Banaszek and Seredynski [6] have contested, however, the assumption that this critical value, usually

* Corresponding author: msered@itc.pw.edu.pl

taken from experiments as equal to the solid fraction at the dendrite coherency point [5] remains constant during progressing solidification. And, in [6] they have proposed distinguishing the developing columnar and equiaxed regions through direct tracking an imaginary envelope encompassing columnar primary dendrite tips on a fixed control volume grid.

The idea of tracing the locus of columnar primary dendrite tips moving according to an assumed kinetics of crystal growth, originated by Browne and Hunt [7] for binary alloy solidification in microgravity conditions, has been further developed for alloy solidification driven by thermal and thermo-solutal natural convection to predict macro-segregation patterns on regular and non-structural control volume grids [8,9], and for solidification in semi-transparent materials [10].

In most micro-macroscopic calculations of alloy solidification a proper choice of the crystal growth kinetics is essential. Indeed, the dendrite tip kinetics is inherent part of computer simulations based on the above-mentioned front tracking approach, on the cellular automaton coupled with the finite element method [11], in the recently developed five-phase model [12] for deducing the growth of mesoscopic envelope linking active dendrite branches, and in others.

Recently, Rebow and Browne [13] extended the most popular constrained (columnar) dendritic growth model of Kurz-Giovanola-Trivedi (KGT) [14] for aluminium alloys by new estimations of the dendrite tip stability parameter, based on measured values of their crystal-melt surface energy anisotropy strength and a simple linear scaling law of the microscopic solvability theory (MST). The influence of such an augmented 2D and 3D KGT models on the columnar mush – undercooled liquid interface position, on the solutal undercooling and the average temperature gradient at this interface, on the relative tendency to form an equiaxed zone and on the columnar to equiaxed transition (CET) phenomenon was, recently, analysed by Seredynski et al. [15] for two selected examples of Al-4wt%Cu alloys solidification driven only by conduction.

Considering the above, the paper presents some new results of the extended analysis involving the influence of a chosen KGT dendrite tip kinetics (with different dendrite tip stability parameters) on the predicted development of columnar and equiaxed grain regions within the mushy zone, macroscopic fields of temperature and species concentration, and thus on the numerically obtained macro-segregation patterns, for the Al-4wt%Cu alloy solidification driven by both molecular diffusion and thermo-solutal convection. The issue of the role of thermal and solutal buoyancy forces in the process development, when compared to a pure diffusion case, is also addressed. For these purposes, the original authors' micro-macroscopic computer simulation model is used, which is based on single domain control-volume calculations involving the tracking of columnar dendrite tips envelope (for distinguishing different grain structures within the mushy zone).

2 Computational model

Progress of the binary alloy solidification is described by a single set of mass, momentum, energy and solute conservation equations based on the mixture theory approach [1]:

$$\nabla \cdot \mathbf{V} = 0 \tag{1}$$

$$\begin{aligned} \frac{\partial(\rho \mathbf{V})}{\partial t} + \nabla \cdot (\rho \mathbf{V}\mathbf{V}) &= \nabla \cdot (\mu_l \nabla \mathbf{V}) \\ &- \nabla \cdot (\mu_l f_s \nabla \mathbf{V}_s) + \nabla \cdot (\mu_s f_s \nabla \mathbf{V}_s) \\ &- \nabla \cdot \left[\rho \frac{f_s}{f_l} (\mathbf{V} - \mathbf{V}_s)(\mathbf{V} - \mathbf{V}_s) \right] \\ &+ \rho \mathbf{B} - (1 - g_s \tilde{V}) \nabla p - \mu_l g_l \mathbf{K}^{-1} \mathbf{V} \tilde{V} \end{aligned} \tag{2}$$

$$\begin{aligned} \frac{\partial(\rho c T)}{\partial t} + \nabla \cdot (\rho c \mathbf{V} T) &= \nabla \cdot (k \nabla T) \\ &+ \frac{\partial (f_s \rho L_h)}{\partial t} - \nabla \cdot (f_l \rho L_h \mathbf{V}_l) \end{aligned} \tag{3}$$

$$\begin{aligned} \frac{\partial(\rho C)}{\partial t} + \nabla \cdot (\rho \mathbf{V} C) &= \nabla \cdot (D \nabla C) \\ &+ \nabla \cdot (f_l D_l \nabla (C_l - C)) \\ &+ \nabla \cdot (f_s D_s \nabla (C_s - C)) \\ &- \nabla \cdot (\rho (\mathbf{V} - \mathbf{V}_s)(C_l - C)) \end{aligned} \tag{4}$$

where V, p, T, C are, respectively, the velocity vector, pressure, temperature and solute concentration of the solid–liquid mixture. Eqs (1)-(4) are supplemented with the following assumptions and the resulting closure relations.

All material properties, except for the solutal diffusivity, are temperature independent and the same for both phases. The assumption of even phase densities implies the equality of the volumetric and mass phase fractions, i.e. $g_i = f_i$ for $i = s,l$.

Momentum equation, eq. (2), includes a special source term with the switching function, $\tilde{V}$, to mimic distinct fluid flow resistances in both porous and slurry regions developing in the mushy zone.

The volumetric force term $\rho\mathbf{B}$ of eq. (2) accounts for thermal and solutal buoyancy forces, and according to the Boussinesq's model assumption reads:

$$\begin{aligned} \rho \mathbf{B} &= g_l \rho_{l,ref} \mathbf{g} \left[\beta_{T,l} (T - T_{ref}) + \beta_{C,l} (C_l - C_{l,ref}) \right] \\ &+ g_s \rho_{s,ref} \mathbf{g} \left[\beta_{T,s} (T - T_{ref}) + \beta_{C,s} (C_s - C_{s,ref}) \right] \end{aligned}$$

where the reference densities for the solid and liquid phases are assumed equal.

The model of micro-segregation obeys the *lever rule* law, so the solute concentrations in both phases are uniform and equal to C_l and C_s, respectively. Due to distinct solute solubility in both phases, and in virtue of the equilibrium assumption, they are related by $C_s = k_p C_l$, where k_p stands for the partition coefficient. The concentration of solute in the mixture, C, is an average one of the two phases concentrations, and thus equal to $C = f_s C_s + f_l C_l$.

The porous zone of columnar grains is considered to be isotropic, and the Carman-Kozeny law is used: $K = \lambda_2^2 (1-g_s)^3 / (180 g_s^2)$, where λ_2 is a secondary dendrite arm spacing, taken as a time- and spatially-independent microstructure parameter.

The effective solid viscosity, μ_s, is related to the effective mixture viscosity, μ_m, with the relation, $\mu_s = [\mu_m - (1-g_s)\mu_l]/g_s$. The mixture viscosity is given by the commonly used rheological model, $\mu_m = \mu_l (1 - g_s/g_{s,coh})^{-2.5 g_{s,coh}}$, where $g_{s,coh}$ is the coherency solid fraction, assumed as equal to 0.3.

The velocity V is an average of the solid and liquid phase velocities $\mathbf{V}_s$ and $\mathbf{V}_l$, respectively, namely $\mathbf{V} = f_l \mathbf{V}_l + f_s \mathbf{V}_s$. In the slurry region, in front of the columnar dendrite tip surface, velocities of both phases are assumed to be equal, $\mathbf{V}_l = \mathbf{V}_s$. In the porous medium, velocity of the solid phase is equal to 0. Close to the moving front of columnar dendrite tips, the phase velocities are smoothed with the switching function, $\tilde{V}$, taking $\mathbf{V}_l = \mathbf{V}/(1 - f_s \tilde{V})$ and $\mathbf{V}_s = (1 - \tilde{V})\mathbf{V}_l$.

The conservation equations, eq. (1)-(4), are discretized on non-orthogonal, triangular and collocated control volume grid [16]. All variables are stored in the centres of the computational cells. The SIMPLE scheme [17] is used to solve coupled mass and momentum equations, whereas all transient terms are integrated with the fully implicit Euler scheme [17]. The second order upwind scheme is used to discretize the diffusive terms, and Rhie and Chow scheme [18] is employed to evaluate mass fluxes across the faces of control volumes.

3 Front tracking procedure

The envelope of columnar dendrite tips is a virtual surface, represented in the model as the loop of linear segments with mass-less markers at their ends (Fig. 1). The undercooling is determined at each marker after each time step and it is moved towards the undercooled liquid along the direction normal to the front. The actual velocity of the i-th marker, $\mathbf{V}_{t,i}$, is equal to those predicted with dendrite growth kinetics model and expressed as a function of the local undercooling, $\Delta T = T_m + m_l C - T$, where C and T are, respectively, the mixture concentration and the temperature interpolated at the actual marker. A new position of the marker is determined as $\mathbf{x}_i^n = \mathbf{x}_i^0 + \mathbf{V}_{t,i}\Delta t$ (see [8]-[11]).

At the beginning of the next following time step, the *switching function* value is determined, based on the actual positions of markers (red points in Fig. 1) and the control volume grid nodes lying behind the front (green points in Fig. 1). Additionally, cross-sections of the front and the control volume faces are found (temporary markers - blue points in Fig. 1). The switching function is then calculated as the ratio of the shaded region in Fig. 1 to the volume of the actual control volume.
The proposed model has been validated by comparing its predictions with the available in the literature experimental measurements [19] and verified with the other results of computer simulation (e.g. [20]) in [9] and [15].

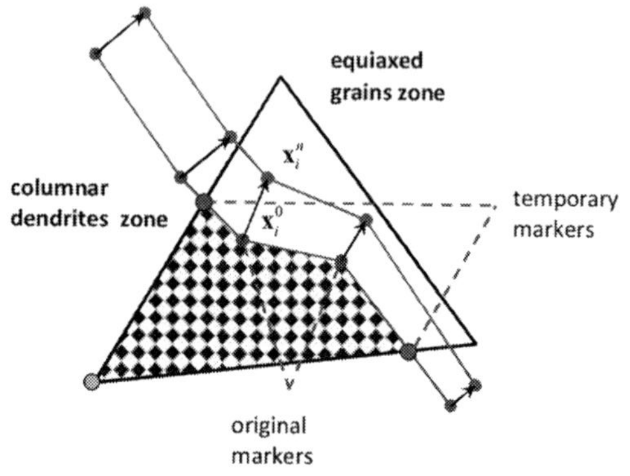

Fig. 1. Graphical representation of the front and the *switching function* evaluation in a control volume

A dendritic growth model is a crucial issue in micro-macroscopic computer simulation of solidification because it decides on the predicted development of grain microstructure, which in turn highly influences the mechanical properties of a final casting product. The accuracy of the commonly used models of dendrite growth relies on the dendrite tip stability parameter. It is a common practice to use its constant value $\sigma^* = 0.025$ for all alloys, which comes from the marginal stability theory (e.g. [14]). However, it was shown recently, by using 3D simulation based on the sophisticated phase field simulation [21,22] or on the theoretical analysis founded on the linear scaling law of the microscopic solvability theory [13], that this value is too small and more than twice in the case of Al-Cu alloys. Very recently, it has been presented (see [15]) that for the Al-4wt%Cu alloy a chosen value of the dendrite tip stability parameter significantly influences the growth velocity, which is the function of under-cooling (Fig.2). Moreover, it also has an impact on the average temperature gradient at the columnar mush/under-cooled liquid interface, the relative tendency for forming an equiaxed zone and on the development of the columnar to equiaxed transition phenomenon. However, this analysis has been restricted to only diffusive transport, i.e. thermo-solutal convection has been neglected.
In the present paper, there is presented the extended study - involving much more challenging and real case of solidification in the terrestrial conditions, where thermal and solutal natural convection plays a critical role in the developing macroscopic heterogeneity of a species concentration. Two distinct KGT kinetics models for a columnar dendrite growth are compared. They differ in the dendrite tip stability parameter; its two different values are taken in the analysis, namely the

commonly used one, i.e. σ^*=0.025, and the over twice higher proposed in [15], i.e.: σ^*=0.058.
The relationship between the dendrite tip velocity V_t and the solutal tip under-cooling ΔT (shown in Fig. 2) is assumed as $V_t = A(\Delta T)^n$, where n = 3.1 and A [$ms^{-1}K^{-n}$] takes values of A = 0.00096 and A = 0.0022 for σ^*=0.025 and σ^*=0.058, respectively.

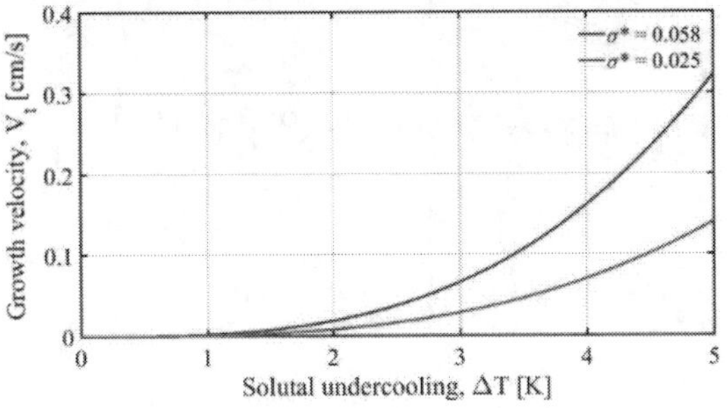

Fig. 2. Influence of the dendrite tip stability parameter on the dendrite tip velocity as a function of solutal undercooling.

Growth velocity predicted by the KGT model with σ^*=0.058 is more than twice higher than the one based on σ^*=0.025 in the considered range of undercooling (Fig.2).

4 Problem setup

To study the impact of the dendrite tip kinetics on the predicted formation of the undercooled slurry zone and the macro-segregation development, the above described computational model has been applied to simulate solidification of the Al-4wt%Cu alloy in a square casting mould of 18cm x 18cm, similar to the one used in [8]. The domain geometry, initial and boundary conditions are shown in Fig.3, whereas thermo-physical properties of the alloy are collected in Table 1.
All walls of the mould are convectively cooled with the same rate. Half of the domain (due to symmetry) is discretized with the uniform triangular grid of 8520 control volumes, and the fully implicit Euler scheme is used with time steps varying from 0.001s to 0.01s.

5 Results and discussion

Solute concentration, determined with the two considered models, is compared in Fig. 4 for two different times of the cooling and solidification process, i.e. after 150s and for completely solidified alloy after 450s.

Differences in the predicted macro-segregation pattern increase in time, but they are not pronounced. The differences of the solute concentration distribution in the completely solidified alloy are shown in Fig. 5. Positive values denote the zones, where macro-segregation calculated for σ^*=0.025 is higher than for σ^*=0.058.
Maps of the volumetric solid fraction (Fig. 6) show the similar pattern - the grains are transported with the bulk fluid downwards, and the deposit of equiaxed grains develops in the bottom of the undercooled liquid zone.
In Fig.7 temporal shapes of the envelope separating the columnar mush and undercooled/equiaxed region are compared for both dendrite tip kinetics models. The difference in the velocity – undercooling relation, shown in Fig.2, makes the averaged columnar dendrite tip growth rate predicted for σ^*=0.058 clearly higher than for σ^*=0.025. So, the size of the undercooled zone, where equiaxed grains can nucleate and grow, is smaller for the former case in the whole solidification process.

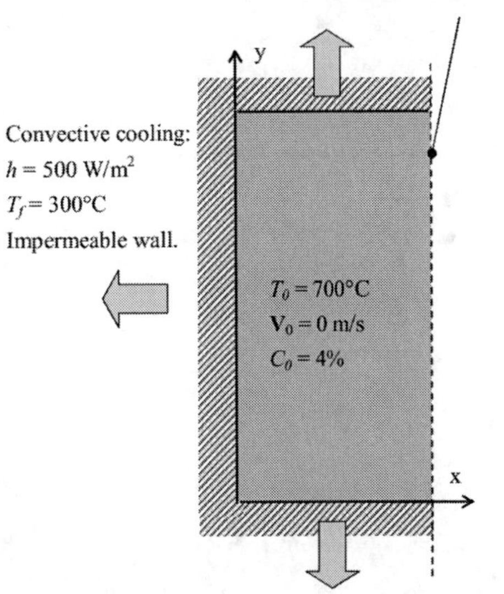

Fig. 3. Domain geometry, initial and boundary conditions for the problem analysed.

Table 1. Material properties of Al-4wt.% Cu

	Units	Value
Density, ρ	kg/m^3	2600
Specific heat, c	J/(kgK)	1000
Thermal conductivity, k	W/(mK)	140
Latent heat of fusion, L	kJ/kg	390
Viscosity of liquid, μ_l	kg/(ms)	$1.3 \cdot 10^{-3}$
Thermal expansion coeff. of liquid, $\beta_{T,l}$	1/K	$1.17 \cdot 10^{-4}$
Solutal expansion coeff. of liquid, $\beta_{C,l}$	1/(wt%)	-0.0073
Thermal expansion coeff. of solid, $\beta_{T,s}$	1/K	$2.25 \cdot 10^{-5}$
Solutal expansion coeff. of solid, $\beta_{C,s}$	1/(wt%)	-0.0087
Melting temperature of pure Al, T_M	°C	660.35
Eutectic temperature, T_E	°C	580
Partition coefficient, k_p	-	0.17
Liquidus slope, m_l	K/(wt%)	-3.403
Solutal diffusion coefficient in liquid, D_l	m^2/s	$5.2 \cdot 10^{-9}$
Solutal diffusion coefficient in solid, D_s	m^2/s	$5.2 \cdot 10^{-11}$
Gibbs-Thompson coefficient, Γ	mK	$2.41 \cdot 10^{-7}$
Secondary dendrite arm spacing, λ_2	µm	100

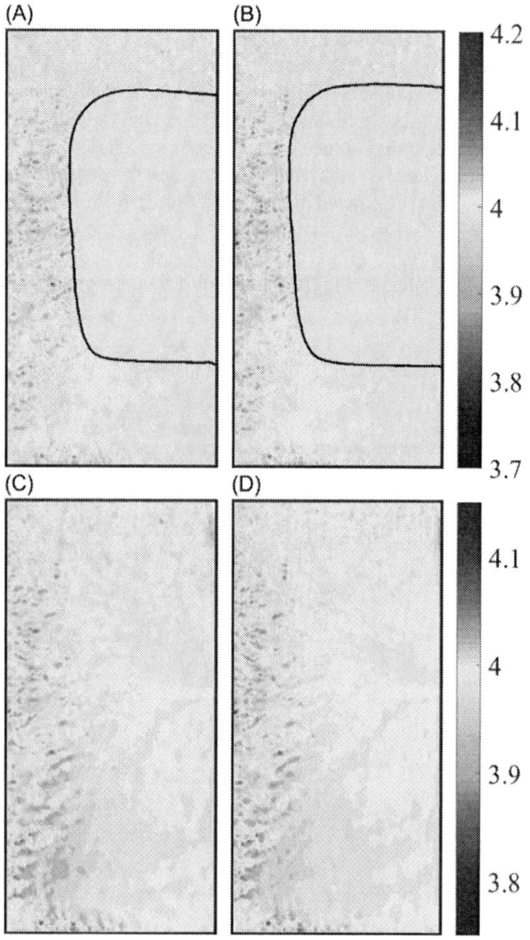

Fig. 4. Solute concentration predicted for σ*=0.058 (the left column) and σ*=0.025 (the right column); after 150 s. (A and B) and 450 s. (C and D).

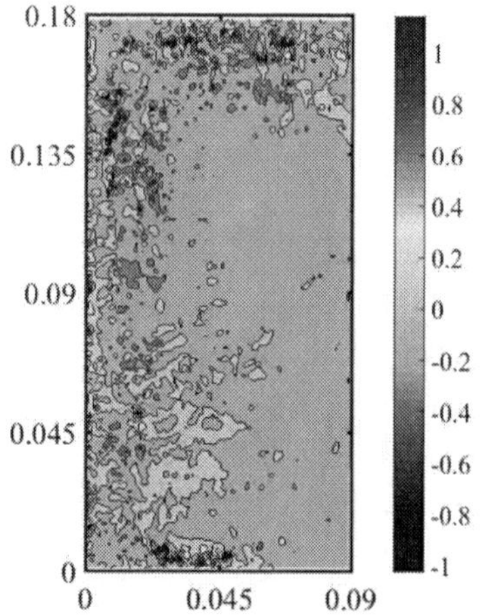

Fig. 5. Relative differences in the predicted macro-segregation patterns in the completely solidified alloy.

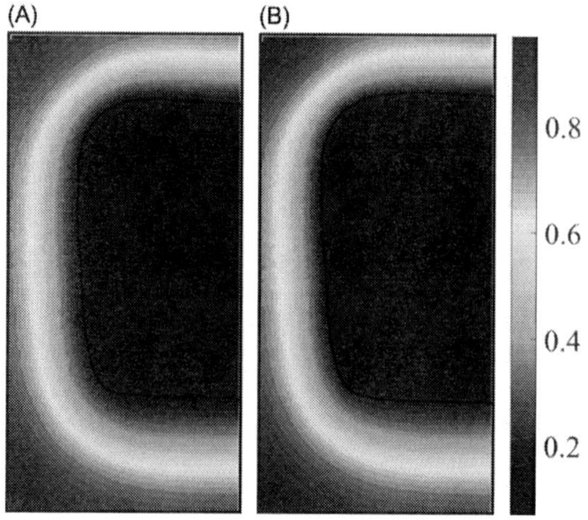

Fig. 6. Volumetric fraction of the solid phase after 150 s. predicted for (a) σ^*=0.058 and (b) σ^*=0.025.

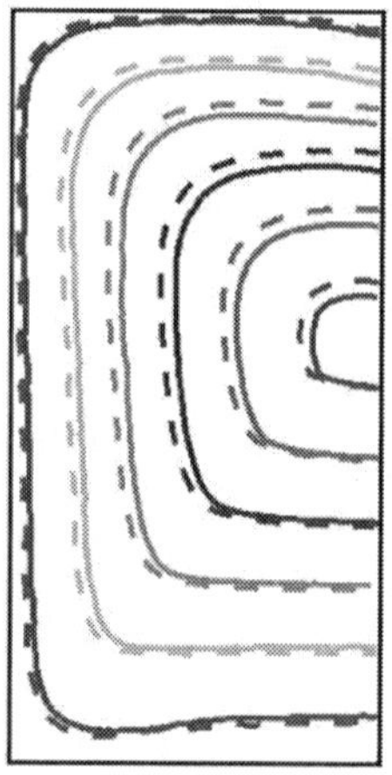

Fig. 7. Position of the front determined at different times of solidification: 50s. –blue, 100s. – yellow, 150s. – green, 200s. – marmot, 250 s. – orange and 300 s. – violet; a continuous line for σ^*=0.058 and a dashed line for σ^*=0.025.

Temporary variations of the averaged under-cooling along the columnar dendrite tips envelope are presented in Fig. 8. They are similar for both analysed dendrite kinetics models. After the short time period of nucleation at the cooled walls, the under-cooling predicted by both models stabilizes nearly at the same moment. Then, it slightly decreases, attains its minimum at about 100 s of the process, and finally slowly increases. However, the under-cooling predicted with σ^*=0.058 is lower in the whole analysed time.

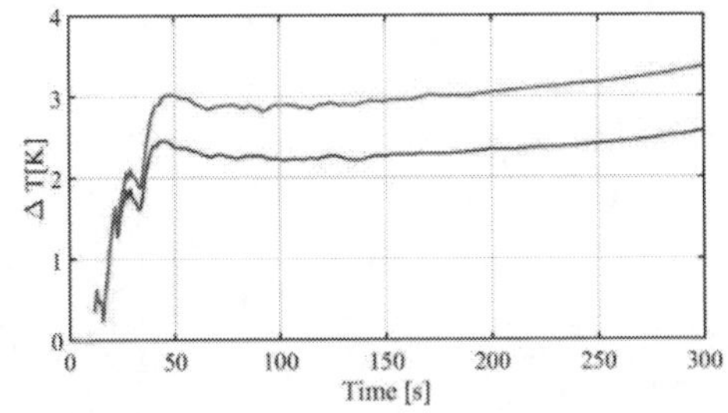

Fig. 8. Temporal changes of the averaged undercooling, for σ^*=0.058 – blue line, and for σ^*=0.025 – red line.

The predictor of the equiaxed solidification and the CET development, the *equiaxed index* originally proposed by Browne [23], has been further used to study the impact of the dendrite tip kinetics on the developing grain structures in the analysed Al-4wt%Cu solidification, in the presence of the thermo-solutal convection. This index is defined as the sum of products of local under-cooling and the volume of computational cells gathered within the under-cooled zone, the domain bounded by the *liquidus line* and the front of columnar dendrite tips. Its time changes predicted by both the analysed dendrite tip kinetics are presented in Fig.9. The higher under-cooling at the dendrite front and the larger under-cooled zone predicted for σ^*=0.025 result in higher values of the index, what means a better promoting the equiaxed solidification.

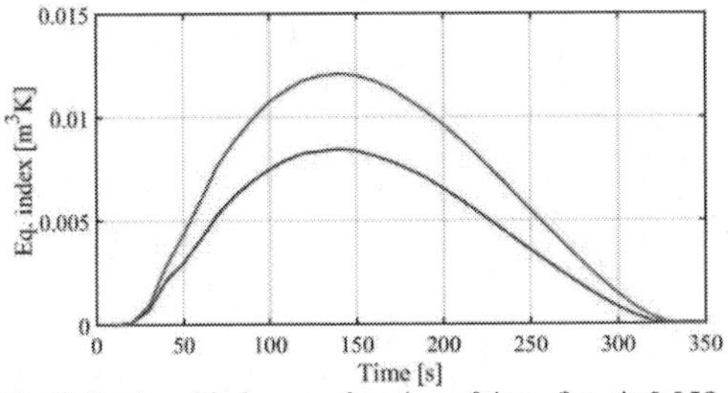

Fig. 9. Equiaxed index as a function of time, for σ^*=0.058 – blue line, and for σ^*=0.025 – red line.

6 Conclusions

The presented analysis shows rather a weak impact of the dendrite tip stability constant in the KGT kinetics model on the numerically predicted overall macro-segregation pictures in the analysed case of the Al-4wt%Cu alloy solidification. One of the reasons for this may be a low mass fraction of copper in the analysed alloy, much smaller that its eutectic composition. This gives rather weak heterogeneity of the species concentration itself, comparing to alloys reach in Cu. Moreover, in the presented calculations the equilibrium solidification was assumed and the simple model of micro-segregation, i.e. the lever-rule, used. More enhanced simulation models are needed for more precise study of the issue of the role of the dendrite tip kinetics on the accuracy of computer simulation of a multi-scale process of alloy solidification.

However, the carried out research showed visible influence of the dendrite tip kinetics on the local undercooling, temporal position of the interface between regions of different dendritic structures evolving within the mushy zone and on the potential for the development of equiaxed grain structures. Moreover, the comparison of the obtained results with those presented in [15] for Al-4wt%Cu solidification driven only by diffusion reveals that the presence of thermo-solutal natural convection visibly diminishes the influence of the dendrite tip kinetics model on the numerically predicted tendency for the development of equiaxed grain structures.

We are extremely grateful to Dr Marek Rebow, from the College of Engineering & Built Environment, Dublin Institute of Technology Ireland, for sharing with us His knowledge and expertise in modelling the dendrite tip kinetics and the CET phenomenon.

References

1. J. Ni, F.P. Incropera, Int. J. Heat Mass Transfer **38**, 1271 (1997)
2. J. Ni, C. Beckermann, Metall. Mater. Trans. B **22B**, 349 (1991)
3. C.R. Swaminathan, V.R. Voller, Int. J. Heat Mass Transfer **40**, 2859 (1997)
4. J. Ni, F.P. Incropera, Int. J. Heat Mass Transfer **38**, 1285 (1995)
5. O.J. Ilegbusi, M.D. Mat, Mater. Sci. Eng., A **247A**, 135 (1998)
6. J. Banaszek, M. Seredyński, Int. J. Heat Mass Transfer **55**, 4334 (2012)
7. D. J. Browne, J.D. Hunt, Numer. Heat Transfer, Part B **45B**, 395 (2004).
8. J. Banaszek, D.J. Browne, Mater. Trans., JIM **46**, 1378 (2005)
9. M. Seredyński, J. Banaszek, J. Heat Transfer **132**, 102301 (2010)
10. P. Łapka, P. Furmański, J. Heat Transfer **132**, 023504 (2010)
11. Ch.-A. Gandin, J. Desbiolles, M. Rappaz, Ph. Thevoz, Metall. Mater. Trans. A **30A**, 3153 (1999)
12. M. Ahmadein, M. Wu, A. Ludwig, J. Cryst. Growth **417**, 65 (2015)
13. M. Rebow, D.J. Browne, Scr. Mater. **56**, 481 (2017)
14. W. Kurz, B. Giovanola, R. Trivedi, Acta Metall. **34**, 823 (1986)
15. M. Seredyński, M. Rebow, J. Banaszek, J. Heat Mass Transfer (2017) https://doi.org/10.1007/s00231-017-2070-z
16. H.K. Versteeg, W. Malalasekera, *An introduction to computational fluid dynamics. The finite volume method* (Pearson Prentice House, 2007)
17. S.V. Patankar, *Numerical heat transfer and fluid flow* (Mc Graw-Hill, 1980)

18. C.M. Rhie, W.L. Chow, AIAA J. **21**, 1525 (1983)
19. D.J. Hebditch, J.D. Hunt, Metall. Trans. **5**, 1557 (1974)
20. P. Roux, B. Goyeau, D. Gobin, F. Fichot, M. Quinard, Int. J. Heat Mass Transfer **49**, 4496 (2006)
21. K. Oguchi, T. Suzuki, Mater. Trans. JIM **48**, 2280 (2007)
22. A. Choudhury, K. Reuther, E. Wesner, A. August, B. Nestler, M. Rettenmayr, Comput. Mater. Sci **55**, 263 (2012)
23. D. J. Browne, ISIJ Int., **45**, 37 (2005)

Transient free convection in an inclined square porous cavity filled with a nanofluid using LTNE and Buongiorno's models

Mikhail Sheremet[1], and *Ioan* Pop[2,*]

[1]Laboratory on Convective Heat and Mass Transfer, Tomsk State University, 36 Lenin Avenue, 634050 Tomsk, Russia
[2]Department of Mathematics, Babeş-Bolyai University, 400084 Cluj-Napoca, Romania

Abstract. The combined effect of Brownian diffusion, thermophoresis and cavity inclination angle on natural convective heat transfer in an inclined porous enclosure has been studied numerically. Fluid containing nanoparticles of low concentration circulates inside the cavity under the effect of the buoyancy force. Governing equations with corresponding boundary conditions formulated using the non-dimensional stream function and vorticity variables have been solved by the finite difference method. An influence of the cavity inclination angle, Darcy and Nield numbers on nanofluid flow and heat transfer has been investigated. It has been found that high Nield numbers illustrate more equilibrium temperature distribution inside the porous cavity.

1 Introduction

Fluid flow in porous media has been an area of intensive investigation for the last decades. The growing emphasis on effective granular and fibrous insulation systems, for the successful containment of the transport of radio-nuclide from deposits of nuclear waste materials, has stimulated various studies in fluid saturated porous media and many results were obtained for the forced and convective flow in the fundamental geometries of internal (cavities, annulus, etc.) and external (over surfaces) flows (Nield and Bejan [1]). Comprehensive reviews of heat transfer mechanisms in geothermal systems has been first presented by Cheng [2, 3], and Bejan [4] with emphasis on its applications in geothermal and energy systems research. Since then a large number of practical applications, both industrial and environmental, have caused a rapid extension of the research, and a substantial number of articles, which relate the boundary-layer flow past surfaces of various flow configuration models, have been published. Nield and Bejan [1], Ingham and Pop [5–7], Pop and Ingham [8], Vafai [9, 10] and Bejan et al. [11] identified many applications that highlight the directions where further theoretical and experimental developments and investigations are required.

The aim of the present study is to analyze numerically transient free convection in an inclined square porous cavity filled with a water-based nanofluid using LTNE and Buongiorno's models. The partial differential equations with corresponding initial and boundary conditions have been solved using a developed in-house computational fluid dynamics code on the basis of the finite difference method of the second order accuracy.

2 Analysis and modelling

A two-dimensional inclined square porous cavity with sides of length L, filled with a water-based nanofluid is considered (Fig. 1). The bottom horizontal wall has a local zone of constant hot temperature (T_h) with length d, while the remainder of this wall is adiabatic. The left and right vertical walls are maintained at constant cold temperature (T_c), where $T_h > T_c$, while the top horizontal wall is adiabatic. The following assumptions are made in the study:

(a) The Darcy model and the Boussinesq approximation are applicable.

(b) Nanofluid and the solid matrix of the porous medium are not in local thermodynamic equilibrium.

(c) The properties of the nanofluid and the porous medium are homogeneous and isotropic with permeability K.

(d) The base fluid physical properties are constant except for its density.

(e) The boundaries of the cavity are assumed to be impermeable.

(f) The local thermal equilibrium model is also employed to account the temperature difference between the base fluid and nanoparticles.

(g) Nanoparticles are well suspended in the nanofluid by using either surfactant or surface charge technology that prevents the nanoparticles from agglomeration and deposition on the porous matrix.

The equations governing the unsteady Darcy free convection flow of the nanofluid using the mathematical nanofluid model proposed by Bungiorno [12, 13] and with the above mentioned assumptions can be written in the form (see Nield and Bejan [1]), including the effects

* Corresponding author: popm.ioan@yahoo.co.uk, ipop@math.ubbcluj.ro

of porous matrix, Brownian diffusion, thermophoresis and heat transfer between porous matrix and nanofluid, while nanoparticles and the base fluid are in the local thermal equilibruim.

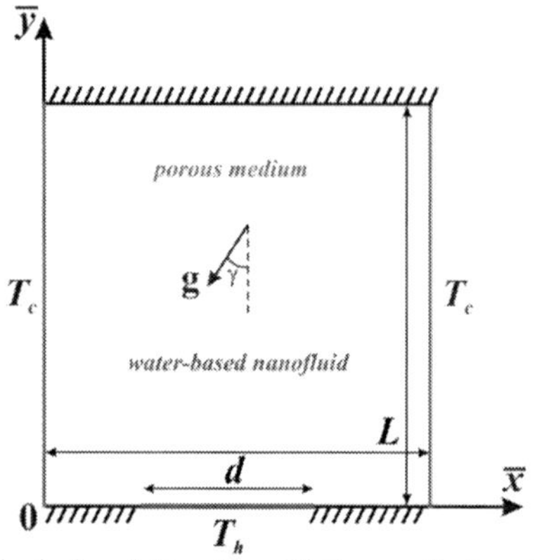

Fig. 1. Physical model together with the coordinate system.

Taking into account the dimensionless stream function and vorticity variables as well as above-mentioned approaches, the governing equations can be formulated in the following form [12]:

$$\frac{\partial^2\psi}{\partial x^2}+\frac{\partial^2\psi}{\partial y^2}=Da\sqrt{\frac{Ra}{Pr}}\left\{\frac{\partial\theta_f}{\partial y}sin(\gamma)-\frac{\partial\theta_f}{\partial x}cos(\gamma)\right\}+Da\sqrt{\frac{Ra}{Pr}}Nr\left\{\frac{\partial\phi}{\partial x}cos(\gamma)-\frac{\partial\phi}{\partial y}sin(\gamma)\right\} \quad (1)$$

$$\varepsilon\frac{\partial\theta_f}{\partial\tau}+\frac{\partial\psi}{\partial y}\frac{\partial\theta_f}{\partial x}-\frac{\partial\psi}{\partial x}\frac{\partial\theta_f}{\partial y}=\frac{\varepsilon}{\sqrt{Ra\cdot Pr}}\left(\frac{\partial^2\theta_f}{\partial x^2}+\frac{\partial^2\theta_f}{\partial y^2}\right)+\frac{\varepsilon\cdot Nb}{\sqrt{Ra\cdot Pr}}\left(\frac{\partial\phi}{\partial x}\frac{\partial\theta_f}{\partial x}+\frac{\partial\phi}{\partial y}\frac{\partial\theta_f}{\partial y}\right)+\frac{\varepsilon\cdot Nt}{\sqrt{Ra\cdot Pr}}\left[\left(\frac{\partial\theta_f}{\partial x}\right)^2+\left(\frac{\partial\theta_f}{\partial y}\right)^2\right]+\frac{H}{\sqrt{Ra\cdot Pr}}\left(\theta_s-\theta_f\right) \quad (2)$$

$$\frac{\partial\theta_s}{\partial\tau}=\frac{\alpha_s/\alpha_f}{\sqrt{Ra\cdot Pr}}\left(\frac{\partial^2\theta_s}{\partial x^2}+\frac{\partial^2\theta_s}{\partial y^2}\right)+\frac{H\xi}{\sqrt{Ra\cdot Pr}}\left(\theta_f-\theta_s\right) \quad (3)$$

$$\varepsilon\frac{\partial\phi}{\partial\tau}+\frac{\partial\psi}{\partial y}\frac{\partial\phi}{\partial x}-\frac{\partial\psi}{\partial x}\frac{\partial\phi}{\partial y}=\frac{\varepsilon}{Le\sqrt{Ra\cdot Pr}}\left(\frac{\partial^2\phi}{\partial x^2}+\frac{\partial^2\phi}{\partial y^2}\right)+\frac{\varepsilon}{Le\sqrt{Ra\cdot Pr}}\frac{Nt}{Nb}\left(\frac{\partial^2\theta_f}{\partial x^2}+\frac{\partial^2\theta_f}{\partial y^2}\right) \quad (4)$$

Following the geometry in Fig. 1, we assume that the initial and boundary conditions for these equations are

$\psi=0,\ \theta_f=\theta_s=0$ at $\tau=0$

$\psi=0$ on the walls of the cavity

$\theta_f=\theta_s=0,\ Nb\,\partial\phi/\partial x+Nt\,\partial\theta_f/\partial x=0$ at $x=0,1$

$\frac{\partial\theta_f}{\partial y}=\frac{\partial\theta_s}{\partial y}=\frac{\partial\phi}{\partial y}=0$ at $y=0$ outside the heater

$\theta_f=\theta_s=1,\ Nb\frac{\partial\phi}{\partial y}+Nt\frac{\partial\theta_f}{\partial y}=0$ at $y=0$ at the heater

$\frac{\partial\theta_f}{\partial y}=\frac{\partial\theta_s}{\partial y}=\frac{\partial\phi}{\partial y}=0$ at $y=1$

Here Ra is the Rayleigh number, Nr is the buoyancy ratio parameter, Nb is the Brownian motion parameter, Nt is the thermophoresis parameter and Le is the Lewis number, which are defined as

$$Ra=\frac{\rho_f g\beta\Delta T\left(1-C_0\right)L^3}{\mu\alpha_f},\ Nr=\frac{\left(\rho_p-\rho_f\right)C_0}{\left(1-C_0\right)\rho_f\beta\Delta T},$$
$$Pr=\frac{\mu}{\rho_f\alpha_f},\ Da=\frac{K}{L^2},\ Nb=\frac{\sigma D_B C_0}{\alpha_f},\ Le=\frac{\alpha_f}{D_B}, \quad (5)$$
$$Nt=\frac{\sigma D_T\Delta T}{\alpha_f T_c},\ H=\frac{L^2h}{k_f},\ \xi=\frac{\left(\rho c\right)_f}{\left(\rho c\right)_s\left(1-\varepsilon\right)}$$

The physical quantities of interest are the local fluid and solid Nusselt numbers Nu_f and Nu_s evaluated at the heated part of the bottom wall, which are defined, respectively, by

$$Nu_f=-\left.\frac{\partial\theta_f}{\partial y}\right|_{y=0},\ Nu_s=-\left.\frac{\partial\theta_s}{\partial y}\right|_{y=0} \quad (6)$$

and the average Nusselt numbers $\overline{Nu_f}$ and $\overline{Nu_s}$ evaluated at the heated part of the bottom wall, which are given by

$$\overline{Nu_f}=\frac{1}{D}\int_{(1-D)/2}^{(1+D)/2}Nu_f\,dx,\ \overline{Nu_s}=\frac{1}{D}\int_{(1-D)/2}^{(1+D)/2}Nu_s\,dx \quad (7)$$

Equations (1)–(4) with corresponding initial and boundary conditions were solved using an in-house computational code developed by means of the finite difference method of the second order accuracy [12, 14, 15]. The developed model and numerical method have been verified comprehensively using the numerical and experimental data of other authors [12, 14–17].

The grid independency test has been performed for the present formulation (see Fig. 1) at $Ra = 10^5$, $Pr = 6.82$, $Da = 10^{-3}$, $\varepsilon = 0.5$, $Nr = 5$, $Nb = 10^{-6}$, $Nt = 10^{-6}$, $Le = 1000$, $\alpha_s/\alpha_f = 649.7$, $\xi = 3.4$, $H = 10$. During analysis the porous medium is the aluminum foam. Four uniform grids are tested. Table 1 presents an influence of the grid elements on the average Nusselt numbers.

Table 1. Effect of mesh on the average Nusselt numbers.

Uniform mesh	$\overline{Nu_f}$	$\overline{Nu_s}$
200×200	4.63	1.88
300×300	5.77	1.89
400×400	5.94	1.90

Using the results presented in Table 1, the uniform grid of 300×300 points has been chosen for the present study.

3 Results and Discussion

Analysis has been carried out for the following values of the governing parameters: $Ra = 10^5$, $Pr = 6.82$, $Da = 10^{-5}$–10^{-3}, $\varepsilon = 0.5$, $H = 10$–1000, $Nr = 5$, $Nb = 10^{-6}$, $Nt = 10^{-6}$, $Le = 1000$, $\alpha_s/\alpha_f = 649.7$, $\xi = 3.4$, $\gamma = 0$–$\pi/2$. Particular efforts have been focused on the effects of the Darcy number, Nield number and cavity inclination angle on streamlines, isotherms, isoconcentrations and average Nusselt numbers for the nanofluid and solid matrix.

Figure 2 shows streamlines, isotherms for nanofluid and porous matrix and isolines of nanoparticles concentrations inside the cavity for $Da = 10^{-3}$, $\gamma = 0$ and different values of the Nield number.

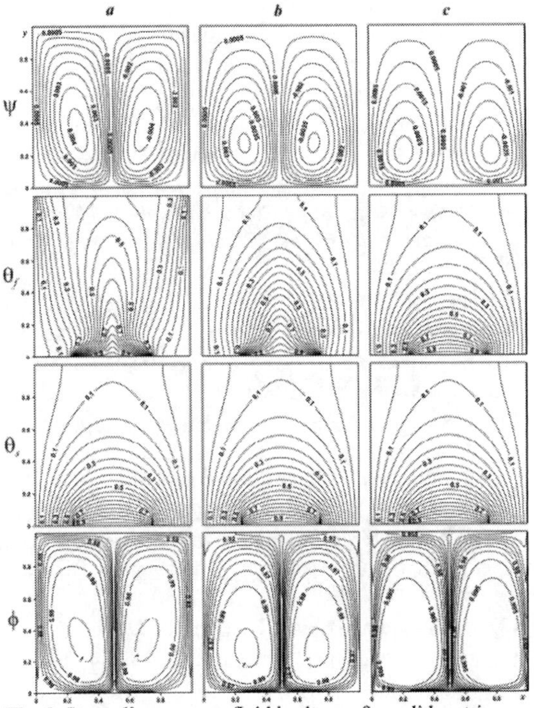

Fig. 2. Streamlines ψ, nanofluid isotherms θ_{nf}, solid matrix isotherms θ_s and isolines of nanoparticles volume fraction ϕ at $Da = 10^{-3}$, $\gamma = 0$ for: $H = 10 - a$, $H = 100 - b$, $H = 1000 - c$.

It should be noted that the Nield number characterizes the interface heat transfer between the nanofluid/solid–matrix phases. Low values of the Nield number illustrate weak thermal interaction between the fluid phase and solid matrix and as a result, one can find more intensive circulation inside the cavity with a stable thermal plume formed over the local heater ($D = d/L = 0.5$). Regardless of the Nield number values two convective cells are appeared within the cavity with clockwise and counter-clockwise motions. It is worth noting that intensive circulation is characterized by upper position of the convective cells cores, while an attenuation of the convective flow leads to a reduction of the convective cells cores up to the bottom wall. Growth of the Nield number illustrates more intensive thermal interaction between the fluid phase and solid matrix that results in a domination of the heat conduction (isotherms of the nanofluid become similar to the solid matrix isotherms for high values of H). Convective fluid rate decreases with the Nield number and distributions of nanoparticles become more homogeneous with low differences in ϕ inside the cavity, while high differences in ϕ can be found near the walls.

Figure 3 demonstrates the average Nusselt numbers at the heater for the cases of the fluid phase (*a*) and solid matrix (*b*) in dependences on the dimensionless time and the Nield number. It should be noted that $\tau = 300$ can be considered as a steady state, while for $H = 10$ one can find small changes in $\overline{Nu_f}$ with τ. As has been mentioned above, low values of H reflect weak interaction between fluid and solid and as a result $\overline{Nu_f}$ is high, while a rise of the Nield number leads to the reduction of the heat transfer rate inside the nanofluid. At the same time, $\overline{Nu_s}$ decreases also with H. Further growth of the Nield number can lead to the equality of the average Nusselt numbers for the fluid phase and solid matrix.

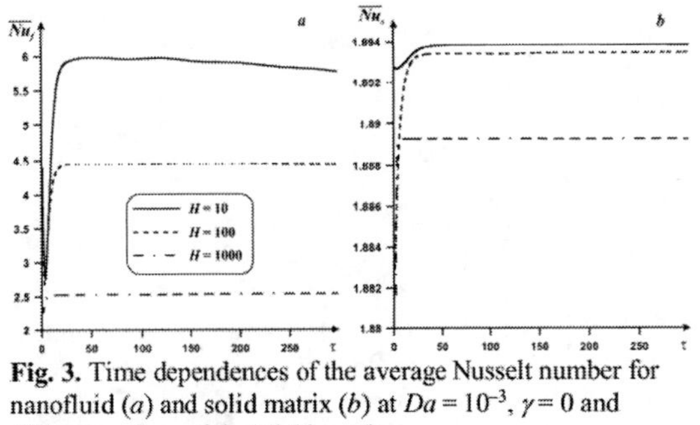

Fig. 3. Time dependences of the average Nusselt number for nanofluid (*a*) and solid matrix (*b*) at $Da = 10^{-3}$, $\gamma = 0$ and different values of the Nield number.

Effects of the cavity inclination angle, Darcy and Nield numbers on the average Nusselt numbers for the fluid phase and solid matrix are shown in Figs. 4 and 5.

Figure 4 illustrates an influence of the cavity inclination angle and the Darcy number on the heat transfer rate in the nanofluid and solid matrix. Low values of the Darcy number characterize a porous medium with low permeability, therefore a growth of Da leads to the heat transfer enhancement inside the fluid phase. At the same time, the average Nusselt number inside the solid matrix has non-linear behavior with Da. A growth of the cavity inclination angle between 0 and $\pi/2$ results in the heat transfer rate reduction inside the fluid phase for high values of the Darcy number, while for low values of Da a raise of γ leads to weak changes of $\overline{Nu_f}$. It is interesting to note that $\overline{Nu_s}$ diminishes with γ for $H = 10$, while for $H = 100$ and $H = 1000$ $\overline{Nu_s}$ increases with the cavity inclination angle.

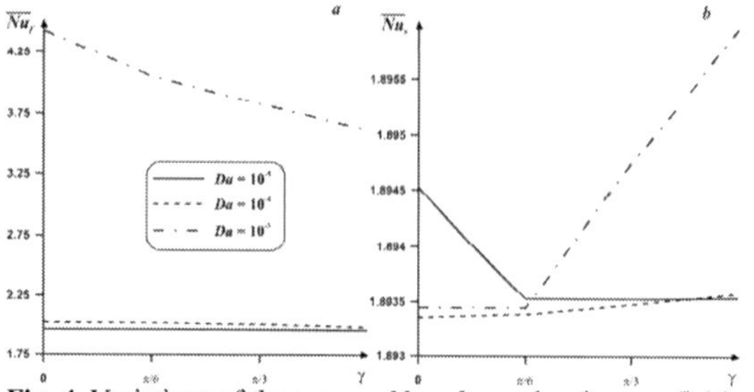

Fig. 4. Variations of the average Nusselt number for nanofluid (*a*) and solid matrix (*b*) for $H = 100$ and different values of the cavity inclination angle and Darcy number.

Figure 5 demonstrates variations of the average Nusselt numbers with the cavity inclination angle and the Nield number for $Da = 10^{-4}$. It should be noted that for $H = 10$ a growth of γ from 0 till $\pi/3$ leads to the heat transfer enhancement inside the nanofluid, while for other values of H $\overline{Nu_f}$ decreases with γ. A raise of the Nield number leads to the average Nusselt number increase in the nanofluid phase, while $\overline{Nu_s}$ decreases with H. Such behavior with the Nield number can be exlained by the considered conduction regime for $Da = 10^{-4}$.

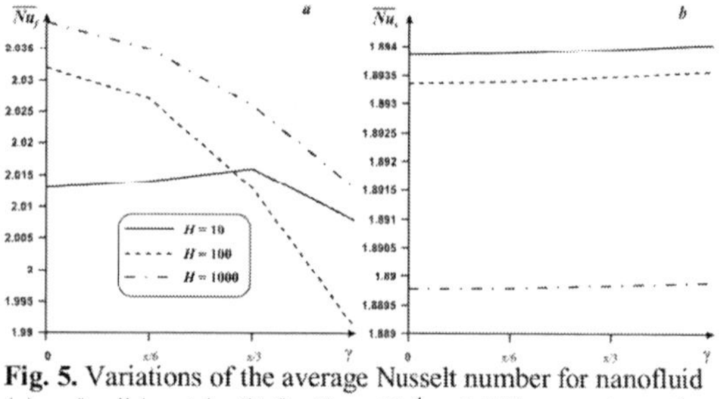

Fig. 5. Variations of the average Nusselt number for nanofluid (*a*) and solid matrix (*b*) for $Da = 10^{-4}$ and different values of the cavity inclination angle and Nield number.

4 Conclusions

Natural convection of nanofluid inside an inclined porous cavity with a local heater on the bottom wall has been studied numerically using the two-phase nanofluid model and local thermal non-equilibrium approach. Particular efforts have been focused on the effects of the cavity inclination angle, Darcy number and Nield number on flow field, temperature and nanoparticles volume fraction distributions as well as average Nusselt numbers at the local heater inside the nanofluid phase and solid matrix. It has been found that for high values of the Darcy number (convective flow and heat transfer regime) a growth of the Nield number leads to the reduction of the heat transfer rate inside the fluid phase. At the same time, $\overline{Nu_f}$ decreases with the cavity inclination angle also for high Da.

Acknowledgements

This work of the first author was supported by the Grants Council under the President of the Russian Federation (Project No. MD-2819.2017.8). The work of the second author was supported from the grant PN-III-P4-ID-PCE-2016-0036, UEFISCDI, Romania.

References

1. D.A. Nield, A. Bejan, *Convection in Porous Media* (4th edition, Springer, New York, 2013).
2. P. Cheng, Adv. Heat Transfer **14** (1978).
3. P. Cheng, In: S. Kakaç, W. Aung, R. Viskanta (eds.) *Natural Convection: Fundamentals and Applications*. (Hemisphere, Washington, DC, 1985).
4. A. Bejan, In: S. Kakaç, R.K. Shah, W. Aung (eds.), *Handbook of Single-Phase Convective Heat Transfer*, Chapter 16 (Wiley, New York, 1987).
5. D.B. Ingham, I. Pop, *Transport Phenomena in Porous Media* (Vol. I, Pergamon, Oxford, 1998).
6. D.B. Ingham, I. Pop, *Transport Phenomena in Porous Media* (Vol. II, Pergamon, Oxford, 2002).
7. D.B. Ingham, I. Pop, *Transport Phenomena in Porous Media* (Vol. III, Elsevier, Oxford, 2005).
8. I. Pop, D.B. Ingham, *Convective Heat Transfer: Mathematical and Computational Modelling of Viscous Fluids and Porous Media* (Pergamon, Oxford, 2001).
9. K. Vafai, *Handbook of Porous Media* (Marcel Dekker, New York, 2000).
10. K. Vafai, *Handbook of Porous Media* (Taylor & Francis, New York, 2005).
11. A. Bejan, I. Dincer, S. Lorente, A.F. Miguel, A.H. Reis, *Porous and Complex Flow Structures in Modern Technologies* (Springer, New York, 2004).
12. A. Shenoy, M. Sheremet, I. Pop, *Convective Flow and Heat Transfer from Wavy Surfaces: Viscous Fluids, Porous Media and Nanofluids* (CRC Press, New York, 2016).
13. J. Buongiorno, J. Heat Transfer **128** (2006).
14. M.A. Sheremet, I. Pop, Int. J. Heat Mass Transfer **79** (2014).
15. M.A. Sheremet, I. Pop, Transport in Porous Media **105** (2014).
16. M.Z. Saghir, A. Ahadi, A. Mohamad, S. Srinivasan, Int. J. Thermal Sciences **109** (2016).
17. M.Z. Saghir, A. Mohamad, Int. J. Numerical Methods Heat Fluid Flow **28** (2018).

Effect of viscous dissipation in two-dimensional Stokes flow between rotating cylinders – preliminary numerical investigation

Tomasz Janusz Teleszewski[1,*]

[1]Department of HVAC Engineering, Faculty of Civil and Environmental Engineering, Bialystok University of Technology, Wiejska 45A, 15-351 Bialystok, Poland

Abstract. This paper introduces the modelling of two-dimensional laminar flow of Newtonian fluid in a system of concentric or eccentric rotating cylinders with regards to viscous dissipation. Viscous dissipation is a main part, where the viscosity is large for example in oils. The dependence of the Nusselt number on the ratio of the radius of the inner cylinder to the radius of the external cylinder for the selected distance of the cylinder axes was investigated. In order to determine the velocity fields and the temperature distribution, the boundary element method was used. The results of the calculations were presented in the form of diagrams.

1 Introduction

The problem of mathematical modelling of viscous fluid movement in eccentric rotating cylinder systems (Fig. 1) is interesting in a scientific aspect as a peculiar form of Taylor-Couette flows [1], and extremely important from an engineering point of view in lubrication theory [2] and sliding designs of hydrodynamic radial bearings [2], liquid mixers with different physical properties [3], rotational viscometers [4].

Although the flow of viscous fluid in the space between cylinders (two, or more generally many) is a three-dimensional flow, assuming a considerable length of cylinders, a sufficient approximation is the adoption of the two-dimensional flow model described in the Stokes equations and continuity for a viscous, incompressible fluid.

The viscosity dissipation in the flows between the rotating cylinders is usually neglected. Flows between rotating cylinders are solved by analytical [1, 5, 6] or numerical [7-11] methods. Considerations of viscous dissipation for flows with a small Reynolds number usually relate to flows in microchannels [12-14]. Examples of the application of the boundary element method for solving the viscous dissipation effect for laminar flow in straight ducts were presented in [15, 16]. Examples of viscous dissipation solutions in flows between rotating cylinders can be found in the literature [17-19].

The aim of this study is to preliminary investigate the effect of viscous dissipation in Stokes flows of Newtonian fluid between rotating cylinders. The physical model of the discussed problem is presented in Figure 1. The main geometric parameter is the ratio of the radius of the inner cylinder to the radius of the external cylinder r_1/r_2 and the distance between the centres of the cylinders ε.

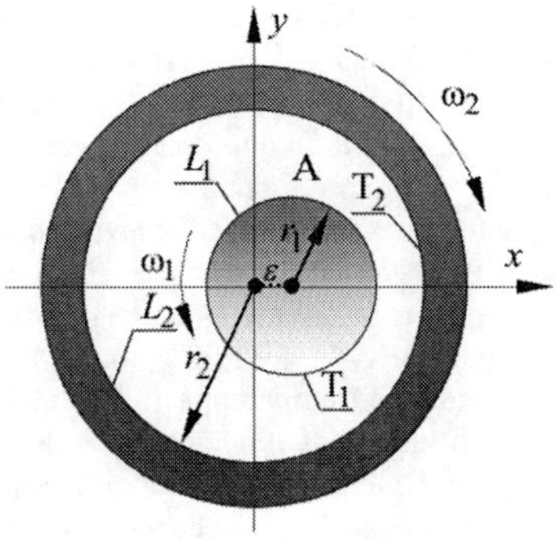

Fig. 1. Flow between two eccentric rotating cylinders in two-dimensional problem.

2 Mathematical formulation

The presented problem of the effect of viscous dissipation in two-dimensional Stokes flow between rotating cylinders is described by the momentum (1) and the conservation of energy equation (2) with consideration of the viscous dissipation term $\Phi(x,y)$ (3):

$$\frac{\partial^2 u_x}{\partial x^2}+\frac{\partial^2 u_x}{\partial y^2}=\frac{1}{\mu}\frac{\partial p}{\partial x},\qquad \frac{\partial^2 u_y}{\partial x^2}+\frac{\partial^2 u_y}{\partial y^2}=\frac{1}{\mu}\frac{\partial p}{\partial y}, \tag{1}$$

$$k\left(\frac{\partial^2 T}{\partial x^2}+\frac{\partial^2 T}{\partial y^2}\right)=\Phi(x,y) \tag{2}$$

* Corresponding author: t.teleszewski@pb.edu.pl

$$\Phi(x,y) = -\mu\left[2\left(\frac{\partial u_x}{\partial x}\right)^2 + 2\left(\frac{\partial u_y}{\partial y}\right)^2 + \left(\frac{\partial u_y}{\partial x} + \frac{\partial u_x}{\partial y}\right)^2\right] \quad (3)$$

where u_z and u_y are a velocity component, p is the pressure, μ is the dynamic viscosity, T is the temperature and k is the thermal conductivity.

The differential equation 2 includes the temperature condition (Dirichlet condition) and the heat flux density condition (Neumann condition). In this study, energy equation (2) is subjected to the following boundary conditions for the cylinder inner wall and outer wall respectively:

$$T_1 = T_2 = \text{const., at } L_1 \text{ and } L_2 \quad (4)$$

It should be noted here that the determined issue, the main role is played by viscous dissipation and the walls of the rotating cylinders are maintain at a constant temperature.

3 Results and analysis

In order to perform the simulation, the author's calculative program was written in Fortran language, in which the calculations are carried out in two stages. First, the energy dissipation function from equation (1) is determined, which is the condition for equation (2). In the second part of the program, the equation (2) is solved to determine the temperature field. The calculations were made using the boundary element method (BEM) for 1000 linear elements on the boundary $L=L_1+L_2$. and 80400 triangular cells in area A. In order to verify the method, the values of temperature fields in area A were compared with the known analytical solution of the temperature field including energy dissipation for the Stokes flow between two concentrically rotating cylinders [1]. For the above assumed number of elements on the L boundary and area A, the maximum relative differences in temperature did not exceed 0.001%.

The Nusselt number is equal to

$$\mathrm{Nu}_{vd} = \frac{q_{tot} r_1}{k\left(T_w - T_m\right)} \quad (5)$$

where the mean temperature is defined as

$$T_m = \frac{1}{A}\iint_A T(x,y)dA \quad (6)$$

where T_w is the temperature at the boundary L equal to $T_w=T_1=T_2$.

Figure 2 shows of the graphical results of calculations of Nusselt numbers as a function of the geometrical parameter r_1/r_2 with ε=0.1. As the ratio between the radius of the inner cylinder and the radius of the outer cylinder increases, the number of Nusselt increases.

The dimensionless velocity, temperature distribution and dimensionless dissipation function in area A is obtained from the following formulas, respectively

$$\vartheta = \frac{u}{u_m} \quad (7)$$

$$\theta = \frac{T}{T_m} \quad (8)$$

$$\varphi = \frac{\Phi}{\Phi_m} \quad (9)$$

where, the mean velocity and mean dissipation function is defined as

$$u_m = \frac{1}{A}\iint_A u_z(x,y)dA \quad (10)$$

$$\Phi_m = \frac{1}{A}\iint_A \Phi\ (x,y)dA \quad (11)$$

Figures 3a-c respectively show the velocity field with streamlines, dimensionless temperature with heatlines and viscous dissipation function $\Phi(x,y)$ for ε=0.1 and r_1/r_2=0.1.

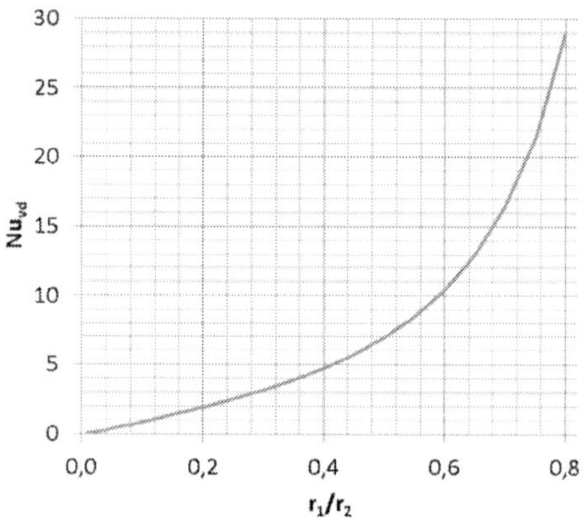

Fig. 2. The influence of Nu_{vd}, on the r_1/r_2 for ε=0.1.

The results of calculations of velocity fields and streamlines (Fig. 3a) are consistent with the results of calculations from work [6]. A characteristic feature of a given geometry (r_1/r_2=0.885) is the creation of a single eddy. The trends of the temperature distribution (Fig. 3b) are consistent with the results of calculations in [17], where the highest temperature increase is located at the narrowest point between the rotating cylinders. In the case of the viscous dissipation function (Fig. 3c) for the given geometry, the maximum value $\Phi(x,y)$ is also at the narrowest point between the rotating cylinders.

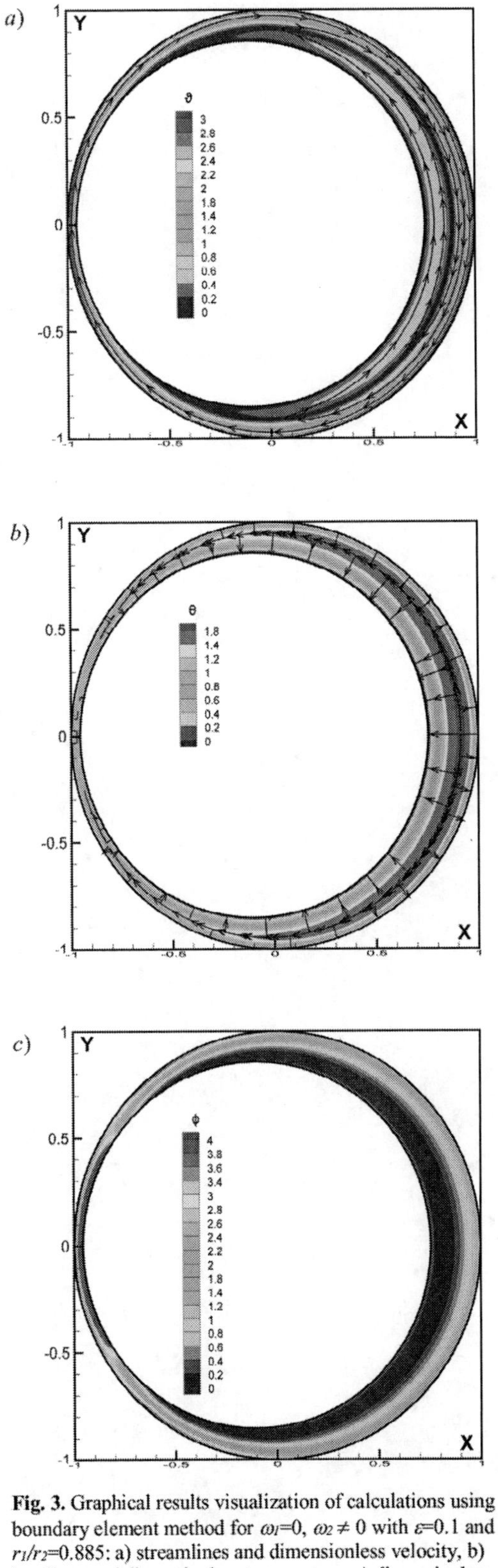

Fig. 3. Graphical results visualization of calculations using the boundary element method for ω_1=0, $\omega_2 \neq 0$ with ε=0.1 and r_1/r_2=0.885: a) streamlines and dimensionless velocity, b) heatlines and dimensionless temperature, c) dimensionless dissipation function.

4 Conclusions

The above study shows the usefulness of the boundary element method for solving the problems described in Stokes equations, in the analysed case for the determination of a two-dimensional viscous fluid flow including viscous dissipation in a system of concentric or eccentric rotating cylinders at different angular velocities, reduced to the analysis of the flow of a viscous fluid in a plane perpendicular to the cylinder axis. The paper presents the results of calculations using the boundary element method for example with a given geometry. The dependence of the Nusselt number increases with the increase of the radius of the inner cylinder relative to the radius of the outer cylinder.

Acknowledgements

This work has been carried out in the Department of HVAC Engineering, Faculty of Civil and Environmental Engineering, Bialystok University of Technology, Bialystok, Poland. This scientific project was financed within the framework of science research S/WBiIS funds at Bialystok University of Technology.

References

1. F.M. White, *Viscous Fluid Flow* (McGraw-Hill, 2005)
2. M.M. Khonsari, E.R. Booser, *Applied Tribology: Bearing Design and Lubrication* (John Wiley & Sons Ltd., 2008)
3. D.G. Thomas, AIChE. J. **10**(4), 517 (2009)
4. L. Bin, W. Yu, Procedia Engineering **29**, 3513
5. G.I. Taylor, Phil. Trans. Royal Society **223**, 289 (1923)
6. B.Y. Ballal, R.S. Rivlin, Arch. Rational Mech. Anal. **62**, 237 (1976)
7. A. San Andres, A.Z. Szeri, J. Appl. Mech. **51**(12), 869 (1984)
8. D.L. Young, C.L. Chiu, C.M. Fan, C.C. Tsai, Y.C. Lin, Eur. J. Mec. B-Fluids **25**, 877 (2006)
9. P. Zhang, X. Zhang, Mathematical Problems in Engineering, Article ID 451546 (2014)
10. S. Meena, P. Kandaswamy, L. Debnath, Int. J. Math. Math. Sci. **25**(2), 93 (2001)
11. L.M. de Socio, L. Marino, Eur. J. Mec. B-Fluids **22**, 85 (2003)
12. G.L. Morini, M. Spiga, J. Heat Transfer **121**(4), 1083 (1999)
13. M. Avci, O.A. Aydin, C.R. Mecanique 324, 164 (2006)
14. G.L. Morini, M. Spiga, J. Heat Transfer, **129**(3), 308 (2007)
15. T.J. Teleszewski, S.A. Sorko, Procedia Engineering **157**, 171 (2016)
16. T.J. Teleszewski, S.A. Sorko, Journal of Numerical Methods for Heat & Fluid Flow, **28**(1), 220 (2018)

17. J.H.C. Araujo, V. Ruas, A.S. Vargas, Int. J. Numer. Methods Fluids **11**, 849 (1990)
18. N. Ashrafi, A. Hazbavi, Arch. Appl. Mech. **83**(12), 1739 (2013)
19. A. Hazbavi, ASME. J. Thermal Sci. Eng. Appl., **8**(2), 021020-021020-9 2016

https://doi.org/10.1051/matecconf/201824003016

Viscous dissipation effect on laminar forced convection in a circular duct with adiabatic walls – preliminary numerical investigation

Tomasz Janusz Teleszewski[1,*]

[1]Department of HVAC Engineering, Faculty of Civil and Environmental Engineering, Bialystok University of Technology, Wiejska 45A, 15-351 Bialystok, Poland

Abstract. In this work, an analysis of laminar forced convection in a pipe with heated and adiabatic walls for a Newtonian fluid with constant properties is performed by taking the viscous dissipation into account when the axial heat conduction in the fluid is neglected. The Nusselt number versus the Brinkmann, which is based on the total wall heat flux density, have been investigated. In order to determine the temperature field, an analytical solution describing the velocity field in the pipe was used, while the energy equation was determined by the boundary element method (BEM). The results of the calculations of Nusselt numbers as a function of the Brinkman number for different thermal insulation heights to the diameter of the circular duct were presented in the form of diagrams.

1 Introduction

Forced convection in ducts is fundamental in heat transfer in many different systems, such as solar energy collection, nuclear heating, microflow sensors, heat exchangers and cooling of electronic devices. Research on issues related to convection forced by various channels showed that the effect of viscous dissipation cannot be neglected in the case of a number of applications, such as flows in microchannels and micro exchangers. The effect of viscous dissipation in the laminar flow of forced convection through channels is the subject of many studies. The first studies on the effect of viscous dissipation of a Newtonian fluid flow through circular tubes were made by Brinkman [1]. Numerical studies on the effect of viscous dissipation in the laminar convection of a forced Newtonian fluid in circular ducts for various boundary conditions are widely described in the literature for different boundary conditions, e.g. for a constant wall temperature [2-4] on the duct wall, for a constant heat flux [3-5] and for a periodic wall temperature or wall heat flux [6-10]. Investigations on laminar forced convection in a non-circular duct, for a Newtonian fluid at constant properties, by taking into account the viscous dissipation effect, have been presented in the literature [2, 11-17]. Several studies have focused specifically on the viscous dissipation effects for laminar forced convection of power-law fluids [18-22] and a Bingham fluid [23-28] under different thermal boundary conditions in circular ducts.

In the construction of micro-heat exchangers, adiabatic walls are often built-in. Examples of laminar flow analysis in forced convection through non-circular ducts with adiabatic walls are presented in the literature [11, 12, 29-32]. In the literature, no studies on the effect of viscous dissipation in laminar convection forced by circular pipes with adiabatic walls were found. The aim of the publication is to preliminarily study the effect of viscous dissipation through a circular tube where the height of thermal insulation s (adiabatic wall) is described by the dependence (Fig. 1):

$$\sigma_{(s/D)} = \frac{s}{D} \qquad (1)$$

where D is the diameter of the pipe.

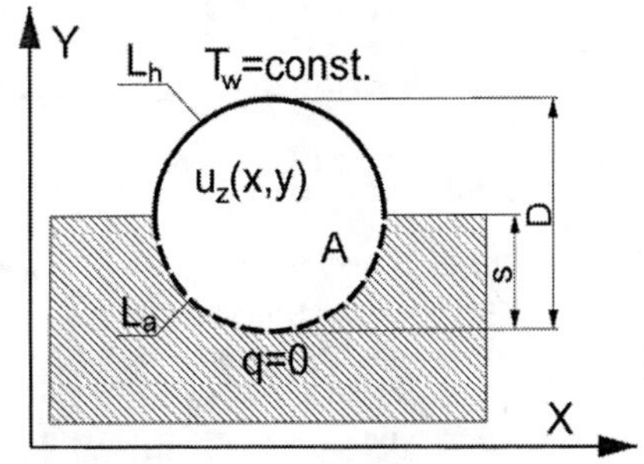

Fig. 1. Geometry of physical problem and boundary conditions.

2 Mathematical formulation

The presented problem of laminar forced convection in laminar flow through any duct with viscous dissipation is described by the continuity (2), momentum (3) and the

* Corresponding author: t.teleszewski@pb.edu.pl

conservation of the energy equation (4) with the consideration of the viscous dissipation term Φ (5):

$$\frac{\partial u_z}{\partial z} = 0 \tag{2}$$

$$\mu\left(\frac{\partial^2 u_z}{\partial x^2}+\frac{\partial^2 u_z}{\partial y^2}\right)=\frac{\partial p}{\partial z}, \qquad \frac{\partial p}{\partial x}=\frac{\partial p}{\partial y}=0 \tag{3}$$

$$k\left(\frac{\partial^2 T}{\partial x^2}+\frac{\partial^2 T}{\partial y^2}\right)=\rho c_p \frac{\partial T}{\partial z} u_z + \Phi(x,y) \tag{4}$$

$$\Phi(x,y)=\mu\left[\left(\frac{\partial u_z}{\partial x}\right)^2+\left(\frac{\partial u_z}{\partial y}\right)^2\right] \tag{5}$$

where u_z is the axial velocity, p is the pressure, μ is the dynamic viscosity, T is the temperature, k is the thermal conductivity, c_p is the specific heat and ρ is the density. The solution of equation (3) is a well-known velocity profile for fully developed circular duct flow in cylindrical coordinates:

$$u_z(r)=\frac{dp}{dz}\frac{1}{4\mu}\left(r_o^2-r^2\right) \tag{6}$$

where $r_0=D/2$ is the radius of the pipe, r is the cylindrical coordinates and dp/dz is the pressure gradient.

The axial gradient dT/dz is independent of x and y. The differential equation (4) includes the temperature condition and the heat flux density condition, and it neglects the axial heat conduction in the fluid. In this work, equation (4) is subjected to the following boundary conditions for the adiabatic wall (L_a) and heat wall (L_h), respectively:

$$q=-k\frac{\partial T}{\partial n}=0, \text{ at } L_a \tag{7}$$

$$T=\text{const., at } L_h \tag{8}$$

3 Results and analysis

The field of velocity $u_z(x,y)$ was determined analytically from equation (6), while the temperature field $T(x,y)$ was determined by the boundary element method for 5000 linear boundary elements. The description of the solution of the energy equation using the boundary element method (BEM) can be found in [17, 33].

The representation of the viscous heating effect is the Brinkman number, which is based on the wall heat flux density, is given by

$$\text{Br}_q=\frac{\mu u_m^2}{D\, q_{tot}} \tag{9}$$

where D is the diameter of the pipe and the mean velocity is defined as

$$u_m=\frac{1}{A}\iint_A u_z(x,y)dA \tag{10}$$

where A is the cross-section area of the pipe and the total wall heat flux density is defined as

$$q_{tot}=\frac{1}{L_h}\int_{L_h} q dL_h \tag{11}$$

The Nusselt number is equal to

$$\text{Nu}=\frac{q_{tot}D}{k\left(T_w-T_b\right)} \tag{12}$$

where T_w is the wall temperature on the boundary of L_h, while T_b is the value of the fluid bulk temperature, which is described by dependence

$$T_b=\frac{1}{Au_m}\iint_A u_z(x,y)T(x,y)dA \tag{13}$$

Figure 2 shows examples of the graphical results of calculations of Nusselt numbers as a function of the Brinkman number for $\sigma_{(s/D)}=0$ and $\sigma_{(s/D)}=0.7$. The trends and shape of the function of the Nusselt number dependence on the number of Brinkman for various geometric parameters $\sigma_{(s/D)}$ are similar to those of Nu=f(Br) presented in [3, 4, 8-17, 30]. The dependencies Nu=f(Br) are characterised by the creation of singular points in which the temperature difference between the wall temperature and the fluid bulk temperature ($\Delta T=T_w-T_b$) changes its sign from positive to negative and the heat flux supplied by the wall is balanced by viscous dissipation.

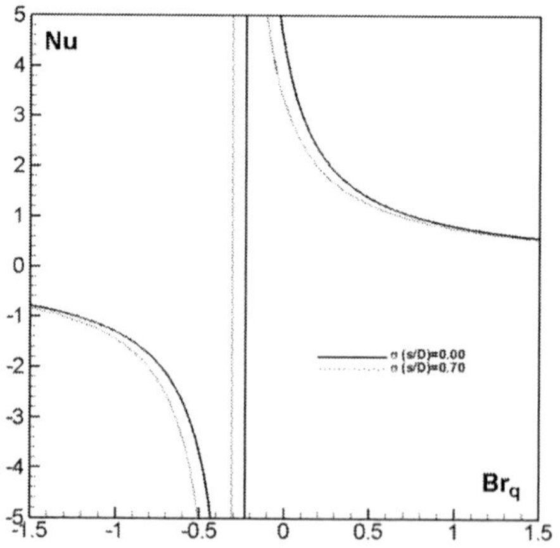

Fig. 2. The influence of Br_q, on the Nu with $\sigma_{(s/D)}$.

The dimensionless temperature distribution in the cross-section of the circular duct is obtained from the following formula

$$\theta=\frac{T}{|T_b|} \tag{14}$$

https://doi.org/10.1051/matecconf/201824003016

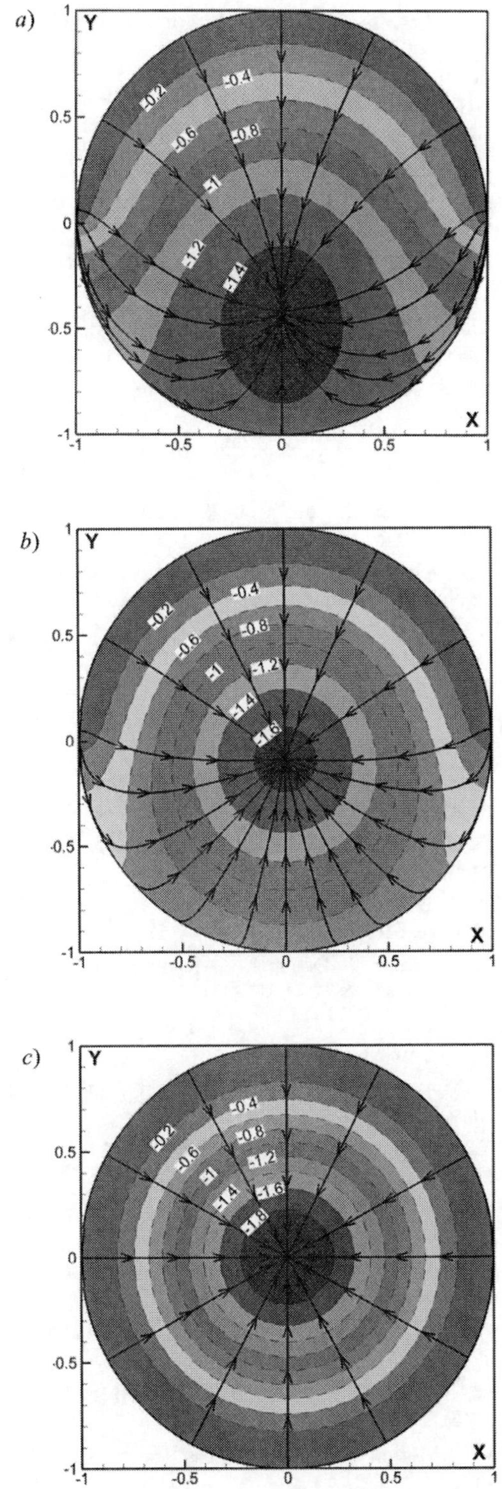

Fig. 3. Heatlines and dimensionless temperature θ obtained from the BEM solution in a circular duct with an adiabatic side ($\sigma_{(s/D)}$=0.5), for different values of Brinkman (Br_q) and Nusselt number (Nu): (a) Br_q=0.03, Nu=3.18; (b) Br_q=0.31, Nu=1.68; (c) $Br_q=\pm\infty$, Nu=0.

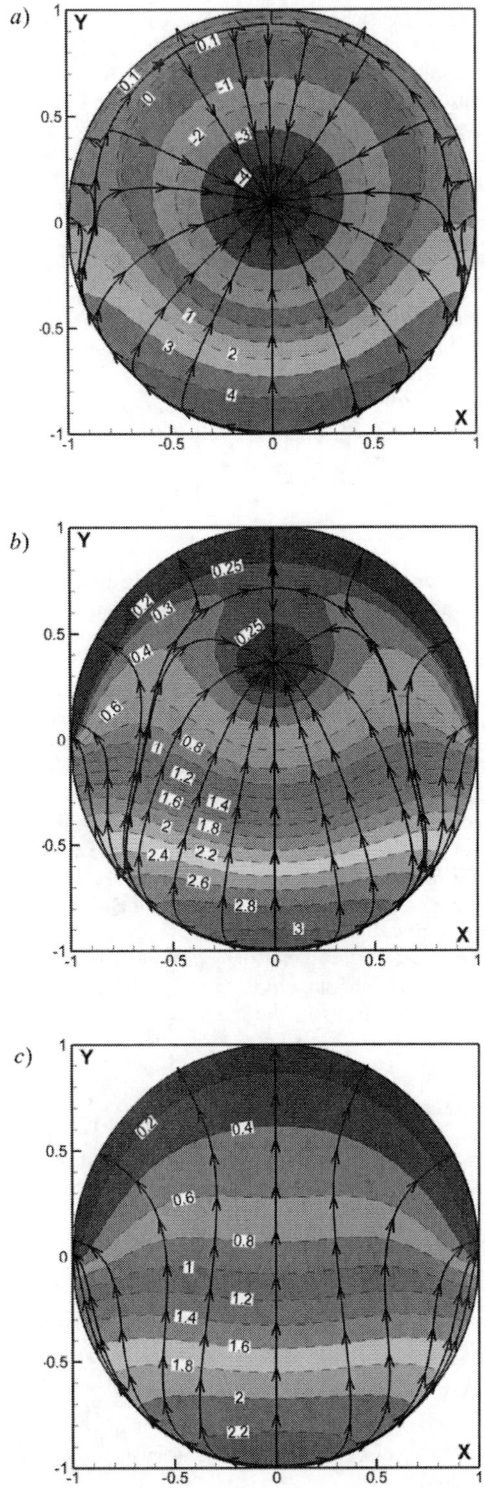

Fig. 4. Heatlines and dimensionless temperature θ obtained from the BEM solution in a circular duct with an adiabatic side ($\sigma_{(s/D)}$=0.5), for different values of Brinkman (Br_q) and Nusselt number (Nu): (a) Br_q=-0.38, Nu=-10.61; (b) Br_q=-0.14, Nu=6.98; (c) Br_q=-0.1, Nu=5.19.

The dimensionless distribution and the heatlines through the circular tube for laminar forced convection in a circular duct by taking into account the effect of viscous dissipation for $\sigma_{(s/D)}$=0.5 are shown in Figs. 3a-c and 4a-c. The point of minimum temperature departs from the adiabatic side and vanishes on the heated perimeter when the flux production by viscous forces increases. As is well known, as the rate of shear is stronger near the wall region, the thermal energy generated due to viscous dissipation is more pronounced near the wall region compared to the central region of the circular tube. Fig. 3c refers to the case when the Nusselt number is equal to zero and the wall heat flux is equal to the wall heat flux generated by viscous dissipation ($Br_q \to \pm\infty$); it is interesting to note that the minimum point of temperature is located is in the centre of a circular pipe.

4 Conclusions

The effect of viscous dissipation of laminar forced convection in circular ducts with heated and adiabatic walls, under the assumption that the axial conduction in the fluid can be neglected, has been studied. As the main geometric parameter, the ratio of the thermal insulation height to the pipe diameter was adopted. In order to determine the temperature fields in the cross-section of the circular pipe, the analytical equation of the velocity field in the pipe was used, and the energy equation was solved with the boundary element method. The shapes of the determined variation of Nusselt's number from the number of Brinkman for different insulation height ratios to the pipe diameter are similar to each other. The points of singularities in the dependences of the Nusselt number in the function of the Brinkman number are also observed. The publication presents examples of solutions of temperature fields in the cross-section of a circular pipe.

Acknowledgements

This scientific project was financed within the framework of science research S/WBiIS funds at the Bialystok University of Technology.

References

1. H.C. Brinkman, Appl. Sci. Res. **2**, 120 (1951)
2. V.P. Tyagi, J. Heat Transfer **88**(2), 161 (1996)
3. O.A. Aydin, Energ. Conver. Manage. **46**, 757 (2005)
4. O.A. Aydin, Energ. Conver. Manage. **46**, 3091 (2005)
5. E. Zanchini, Int. J. Heat Mass Transfer **40**(1), 169 (1997)
6. C-J. Hsu, A.I.Ch.E. Journal **11**(4), 690 (1965)
7. J.N.N. Quaresma, R.M. Cotta, Int. Commun. Heat Mass **21**(5), 729 (1994)
8. D-Y Lee, S-J Park, S.T. Ro, Int. J. Heat Mass Transfer **38**(14), 2529 (1995)
9. A. Barletta, E. Zanchini, Heat and Mass Transfer **31**(1), 41 (1995)
10. A. Barletta, E. Rossi di Schio, Heat and Mass Transfer **35**(1), 9 (1999)
11. G.L. Morini, M. Spiga, J. Heat Transfer **121**(4), 1083 (1999)
12. G.L. Morini, M. Spiga, J. Heat Transfer, **129**(3), 308 (2007)
13. G.L. Morini, M. Spiga, P. Tartarini, Int. Comm. Heat Mass Transfer **25**(4), 551 (1998)
14. M. Avci, O.A. Aydin, C.R. Mecanique 324, 164 (2006)
15. T.J. Teleszewski, S. A. Sorko, Procedia Engineering **157**, 171 (2016)
16. T.J. Teleszewski, IOP Conf. Ser.: Mater. Sci. Eng. **269**, 012038 (2017)
17. T.J. Teleszewski, S.A. Sorko, International Journal of Numerical Methods for Heat & Fluid Flow **28**(1), 220 (2018)
18. A. Barletta, Int. J. Heat Mass Transf. **40**(1), 15 (1982)
19. A. Lawal, A.S. Mujumdar, Wärme - und Stoffübertragung **27**(7), 437 (1992)
20. T. Boaca, I. Boaca, Trans. on Comput. Sci. I, LNCS **4750**, 139 (2008)
21. A.M. Dehkordi, A.A. Mohammadi, Energ. Conver. Manage. **50**(4), 1062 (2008)
22. A.M. Dehkordi, A.A. Mohammadi, Energ. Conver. Manage. **51**(5), 1065 (2010)
23. G.C. Vradis, J. Dougher, S. Kumar, Int. J. Heat Mass Transf. **36**(3), 543 (1993)
24. T. Min, J.Y. Yoo, H. Choi, Int. J. Heat Mass Transf. **40**(13), 3025 (1997)
25. T. Min, H.G. Choi, J.Y. Yoo, H. Choi, Int. J. Heat Mass Transf. **40**(15), 3689 (1997)
26. T. Min, J.Y. Yoo, J. Heat Transfer **121**(3), 556 (1999)
27. R. Khatyr, D. Ouldhadda, A. Il Idrissi, Comptes Rendus Mecanique **330**(2), 69 (2002)
28. R. Khatyr, D. Ouldhadda, A. Il Idrissi, Int. J. Heat Mass Transf. **46**(4), 589 (2003)
29. F. W. Schmidt, M.E. Newell, Int. J. Heat Mass Transf. **10**, 1121 (1967)
30. M. Spiga, G.L. Morini, Int. J. Heat Mass Transf. **39**(6), 1165 (1996)
31. G.L. Morini, Int. J. Heat Mass Transf. **43**(5),741 (2000)
32. V.V. Dharaiya, S.G. Kandlikar, J. Heat Transfer **134**(2) (2012)
33. C.A. Brebbia, J.C.F. Telles, L.C. Wrobel, Boundary Element Techniques-Theory and Applications in Engineering (1984)

Prediction and measurement of the aerodynamic performance of a wind turbine

Ali Cemal Benim*, *Michael* Diederich, and *Fethi* Gül

Center of Flow Simulation (CFS), Department of Mechanical and Process Engineering, Düsseldorf University of Applied Sciences, Münsterstr. 156, D-40476 Düsseldorf, Germany

Abstract. Aerodynamic behavior of a small wind turbine is analyzed, both experimentally and numerically. Mainly, an unsteady three-dimensional formulation is adopted, where the flow turbulence is modelled by an Improved Delayed Detached Eddy Simulation framework, using the four-equation transitional Shear Stress Transport model, as the turbulence model. A quite good agreement between the measurements and calculations is observed.

1 Introduction

Wind energy is increasingly occupying an important position in power generation from the renewable energy resources.

Apart from the large-sized wind turbines with large capacities (MW range), which can be installed on-shore or off-shore, small-sized wind turbines with relatively small capacities (approximately 10 kW and smaller) are being increasingly installed in urban regions, for a much wider and more thorough usage of the wind energy.

Small wind turbines are also significant for decentralized power generation and power autarchy. A serious problem pertaining the usage of small wind turbines in urban regions is noise generation, which is also an issue in the project in which the present study is imbedded.

Unsteady, three-dimensional calculations, along with experiments are conducted for an available small, horizontal-axis wind turbine power plant.

Flow turbulence is modelled by applying high-resolution methods [1-3] that will be discussed in more detail below.

Amongst the considerable number of the previous analyses on the aerodynamics of the wind turbines, there is a large number of investigations that deal with the aerodynamic optimization [4].

However, the larger number of those investigations adopt rather simplified approaches, such as the Blade Element Momentum (BEM) methodology for predicting the flow field.

Investigations that apply a CFD approach [5-7] are rather smaller in number, where comparably simpler, i.e. one-equation turbulence models were used. In the present work, a more accurate turbulence modelling is applied.

2 The investigated wind power plant

The investigated wind power plant is a small, horizontal axis wind turbine with three rotor blades and a nominal electrical power of 12 kW. The wind power plant is installed near Husum, in northern Germany and in operation. The pole has the height of approximately 30 m. The nominal rotor speed is 2 revolutions within a second. The wind speed and the electrical power output of the power plant has been continuously monitored.

3 Modelling approach

The CFD analyses are conducted within the framework of the finite volume method based general-purpose CFD software ANSYS Fluent [8].

The Navier-Stokes equations, which are coupled with the equation of continuity are numerically solved, coupled with the equations of the turbulence model, as will be described below. For the material properties, constant values are used that correspond to those of air at air ambient conditions. It is assumed that the flow is incompressible. This can be deemed as realistic at the occurring velocities, which lead to maximum Mach numbers lower than 0.2.

A three-dimensional modelling is applied. Here, the problem is defined in a rotating frame of reference, which is a typical procedure in the mathematical modelling of flows in turbomachinery [9-11].

In the three-dimensional, unsteady CFD calculations, a recent version of the DES, i.e. the Improved Delayed Detached Eddy Simulation (IDDES) [8,12] is applied to model the turbulence, which is a hybrid RANS-LES approach.

As the RANS complementary, the four-equation transitional SST model [8,13] is adopted to consider the transitional effects in the boundary layer, which was also previously used successfully in the prediction of

* Corresponding author: alicemal@prof-benim.com

different categories of problems [14]. In preliminary calculations, which were also performed as part of grid independence studies, steady-state (RANS) modelling is also applied. Here, the two-equation SST [8,15] model is adopted as the turbulence model, which has also previously been successfully used in a wide range of wall driven turbulent shear flows [16-20].

In the unsteady calculations, a second-order accurate backward differencing scheme is adopted for the time discretization.

The time step sizes are chosen in such a way that the cell Courant numbers remain smaller than unity. For the discretization in space, the central and second-order upwind schemes are used in a combined manner depending on the local turbulence formulation within the framework of the hybrid (LES/RANS) turbulence modelling approach. The velocity-pressure coupling is formulated by the PISO algorithm [8].

4 Results

In the computational modelling, the effect of the pole and the ground are omitted, since the associated additional computational effort would explode available framework.

At the outlet boundary, a constant static pressure together with vanishing normal gradients for the convectively transported variables are prescribed as boundary condition.

The boundary condition at the inlet boundary is defined to be, a prescribed uniform velocity field parallel to the turbine axis is prescribed together with a uniform turbulence intensity of 4% and a length scale equal to 25% of average chord length of the turbine blade. As the value of the constant wind velocity at the inlet boundary, 9.5 m/s is prescribed (in stationary frame of references).

A formal grid independence study for the unsteady, three dimensional calculations using the IDDES turbulence model has not been performed within the present investigation. Grid independence studies were conducted as a preliminary study, using RANS models, in steady-state, which have led to design of the final three-dimensional grid, as also guided by the previous applications [9-11].

The computational grid is generated by an unstructured methodology, using tetrahedral cells. However, along the solid walls, the local grid is generated by using prism layers, for a better resolution of the boundary layer.

For arriving at an favorable distribution of the grid cells, the fact has been taken into account that the power generation takes place at rather high radial positions of the blade, where the local velocities are comparably high, whereas the radially lower parts of the blade do not considerably contribute to the power generation, in comparison, but are responsible, rather, for the stability of the blade construction.

Based on this consideration, the radially upper part of the blade is discretized by a finer grid resolution compared to its lower part. The finally applied computational grid consisted of approximately 6 million nodes. A view of the surface grid near the tip region is displayed in Fig. 1.

The predicted distribution of the time-averaged static pressure field around the blade, as well as the "pseudo streamlines" of the time-averaged velocity field in the rotating frame of reference are illustrated in Figure 2 for a constant-radius cross section along the height of the blade.

Please note that the displayed streamlines in Figure 2 are not the "real" streamlines. They are based on the two-dimensional velocity field that is obtained by the projection of the predicted three-dimensional velocity field onto the displayed plane of the considered blade section. Thus, they don't deliver the exact information on the flow field, as, for example, the mass flow rate conservation between neighboring streamlines is not guaranteed. Still, the flow structure can reasonably be visualized by means of the pseudo streamlines, where a local recirculation zone on the suction surface, near the trailing edge, can be observed.

The calculated aerodynamic efficiency for the conversion of the accessible wind power to the power on the shaft ($\eta_{a,pred}$), as well as these measured efficiency for the conversion of the accessible wind power to the electrical power ($\eta_{e,meas}$) are given in Table 1.

Fig. 1. A view of the surface grid near the tip.

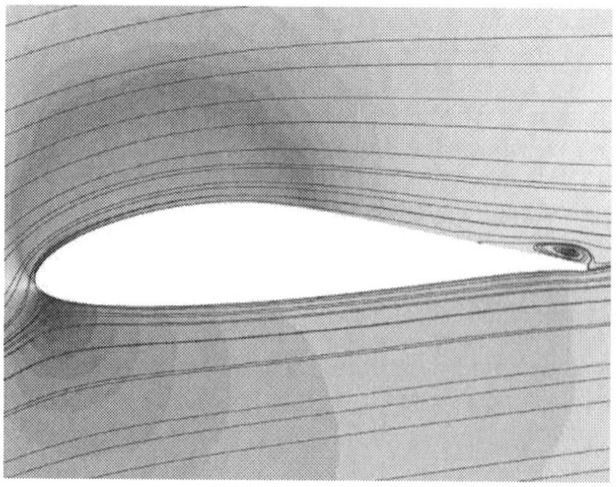

Fig. 2. Distribution of the time-averaged static pressure field and "pseudo streamlines" at a constant-radius section.

Table 1. Efficiencies.

$\eta_{a,pred}$	$\eta_{e,meas}$
0.25	0.21

One can see that the predicted aerodynamic efficiency by the present computational procedure is in a rather reasonable relationship with the measured efficiency of the wind turbine plant for the conversion of the wind energy to the electrical power (Table 1).

5 Conclusions

The aerodynamics of a small wind turbine is analyzed by means of experiments and numerical predictions. In the latter, an unsteady 3D formulation is used, modelling the flow turbulence by an IDDES approach in combination with a transitional SST model. A good agreement between the experiments and calculations is observed.

We gratefully acknowledge the financial support by the German Federal Ministry of Economy of Technology for this research project.

References

1. A. C. Benim, A. Nahavandi, K. J. Syed, Progress in Computational Fluid Dynamics, **5**, 8, pp.444-454 (2005)
2. A. C. Benim, M. P. Escudier, A. Nahavandi, A. K. Nickson, K. J. Syed, F. Joos, International Journal of Numerical Methods for Heat & Fluid Flow, **20,** 3, pp.348-370 (2010)
3. A. C. Benim, H. Chattopadhyay, A. Nahavandi, International Journal of Thermal Sciences, **50**, 10, pp. 1973-1983 (2011)
4. A. Chehouri, R. Younes, A. Ilinca, J. Perron, Appl. Engy **142**, pp.361-388 (2015)
5. X. M. Chen, R. Agarwal, J. Aircraft, **49**, pp.622–629 (2012)
6. A. F. P. Ribeiro, A. M. Awruch, H. M., Gomes, Applied Math. Modelling, **36**, pp.4898–4907 (2012)
7. Y. P. Ju, C. H. Zhang, J. Power, **226**, pp.245–261 (2012)
8. www.ansys.com
9. A. C. Benim, M. Geiger, S. Doehler, M. Schoenenberger, H. Roemer, *Modelling the flow in the exhaust hood of steam turbines under consideration of turbine-exhaust hood interaction*, in: Proceed. 1st European. Conf. Turbomachinery - Fluid Dynamic and Thermodynamic Aspects: Computational. Methods, Erlangen, Germany, March 1-3, 1995, Book Series: VDI Ber., Vol.1185, pp.343-357 (VDI Verlag, Duesseldorf, 1995)
10. A. C. Benim, D. Brillert, M. Cagan, ASME Paper No. GT2004-54151, **4**, pp. 453-460, (2004)
11. A. C. Benim. M. Diederich, B. Pfeiffelmann, Computation, **6,** 2, 34 (2018)
12. M. S. Gritskevich, A. V. Garbaruk, J. Schutze, F. R. Menter, Flow, Turbulence, Combustion, **88**, pp.431-449 (2012)
13. F. Menter, T. Esch, S. Kubacki, "Transition modelling based on local variables", *Engineering Turbulence Modeling and Experiments 5*, W. Rodi, N. Fueye, Eds., pp.555-564 (Elsevier, Amsterdam 2002) pp.555-564.
14. P. Oclon, S. Lopata, M. Nowak, A. C. Benim, Progress in Computational Fluid Dynamics, **15**, 5, pp.290-306 (2015)
15. F. R. Menter, AIAA Journal, **32**, pp. 1598-1605 (1994)
16. A. C. Benim, A. Nahavandi, "A computational analysis of turbulent swirling flows", in: K. Hanjalic, Y. Nagano. M. J. Tummers (Eds.) *Turbulence, Heat and Mass Transfer*, vol. 4, pp.715-722 (Begell House, Danbury, Connecticut, 2003)
17. A. C. Benim, M. Cagan, D. Gunes, International Journal of Thermal Sciences, **43,** 8, pp.725-732 (2004)
18. A. C. Benim, M. Cagan, A. Nahavandi, E. Pasqualotto, *RANS predictions of turbulent flow past a circular cylinder over the critical regime*, (Proc. 5th IASME/WSEAS International Conference on Fluid Mechanics and Aerodynamics, Athens, Greece, pp.232-237, 2007)
19. A. C. Benim, K. Ozkan, M. Cagan, D. Gunes, International Journal of Numerical Methods for Heat & Fluid Flow, **17**, 3, pp.284-301 (2007)
20. A. Assmann, A. C. Benim, F. Gül, P. Lux, P. Akhyari, U. Boeken, F. Joos, P. Feindt, A. Lichtenberg, Journal of Biomechanics, **45,** 1, pp.156-163 (2012)

Numerical study of an asymmetrically heated rectangular duct with suspended cylinders

Gaurav Kumar Chhaparwal, *Ankur* Srivastava*, and *Ram* Dayal

Manipal University Jaipur, India 303007

Abstract. An asymmetrically heated (only one side of the duct is heated) solar air heater duct is numerically investigated using openFOAM to study the effect of passive turbulators (suspended cylinders) installed within the boundary layer close to the absorber plate to enhance heat transfer from it. Vortex flow is created behind the suspended cylinders, which disturbs the boundary layer and causes mixing of hot stream lines near the surface with relatively cold stream away from the absorber plate. Diameter of the suspended cylinders and its distance from the absorber plate is varied to study its effect on wall temperature of absorber plate in the turbulent flow regime.

Nomenclature

f	Friction factor
Nu	Nusselt number
W[mm]	Width of the rectangular duct
H[mm]	Height of the rectangular duct
L[mm]	Length of the rectangular duct
AR	Aspect ratio (W/H)
D[mm]	Diameter of the cylinder
S[mm]	Transverse distance of cylinder
P[mm]	Pitch of two the cylinders
δ[mm]	Boundary layer thickness
U^+	Dimensionless free stream velocity
U[m/s]	Free stream velocity
U_{ref} [m/s]	Reference velocity
y^+	Dimensionless transverse distance
Δy[mm]	Distance of the first node from the wall
T[K]	Temperature of the absorber plate
x[mm]	Distance from the leading edge
t[sec]	Real time during simulation
Re_x	Reynolds number at distance x
ω	Turbulent Frequency
ρ	Density of the air
U_j	Free stream velocity in y direction
μ	Dynamic viscosity
μ_t	Turbulent dynamic viscosity
υ	Kinematic viscosity
ε	Turbulent eddy dissipation
P_k	Production rate of turbulence

1 Introduction

A solar air heater is a device which uses solar energy to heat up air that is made to pass through its duct. This hot air can be used for various applications such as; cloth dryer, crop dryer, packaging product, room heating etc. Aim of the present study is to improve its poor Thermal-Hydraulic Performance Parameter (THPP).

Absorber plate of the solar air heater absorbs solar radiation and gets heated up. Air blower is used to blow air through a duct, which is covered by absorber plate. Rectangular duct is perfectly insulated except on absorber plate part. Flowing air takes up heat from absorber plate and gets heated mostly though convection mode of heat transfer. But the THPP of such solar air heater is low. This is because convective heat transfer coefficient between absorber plate and air is very low. Many researchers have used different artificial roughness such as: V shape continues inclined and straight ribs, broken ribs, wire mesh, circular ribs etc. to enhance heat transfer. After comprehensive literature review it has been found that V shape rib is most efficient among all the other type of ribs. All these ribs are engraved on the surface of absorber plate. None of the researchers have used suspended cylinders installed at some distance away from the absorber plate. When air strikes on these cylinders, vortex flow is created behind the face of the cylinders. This vortex flow causes air in the vicinity of absorber plate (which has maximum air temperature) to get mix with relatively cold air in the middle of duct.

The different kinds of roughness, used by researchers have been shown in table 2 (see at the end of the paper).

* Corresponding author: ankur.srivastava@jaipur.manipal.edu

Table 2 is represented in graphical form in the fig 1, fig 2 and fig 3 which shows the variation of friction factor, Nusselt number and THPP at the different combination of e/D, P/e ratios of the different artificial roughness. The behavior of the fluid flow by varying the certain parameters of the ribs or the geometry is not fixed. CFD analysis can be helpful in the predictions of the various parameters of the fluid by changing the geometries very accurately, which is the motive of this project.

Experimental study is common in the field of solar air heater, only few numerical studies have been done so far, to analyze heat transfer enhancement in a solar air heater. The objective of the proposed study is to perform a series of numerical simulations using computational fluid dynamics (CFD) tools to arrive at the optimum configuration for maximum THPP of the solar air heater.

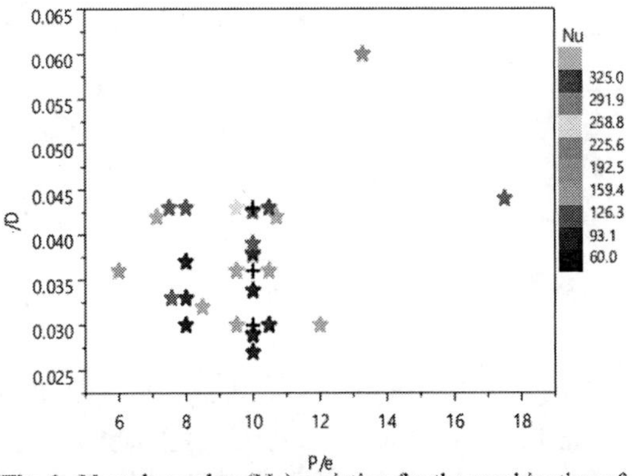

Fig. 1. Nusselt number (Nu) variation for the combination of 'e/D' and 'P/e' of the roughness.

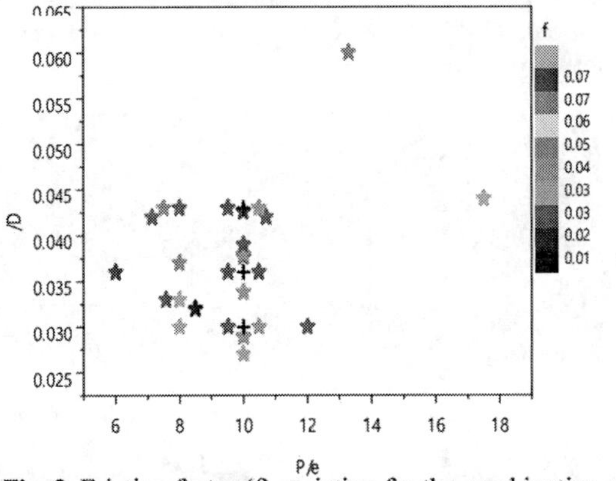

Fig. 2. Friction factor (f) variation for the combination of 'e/D' and 'P/e' parameters of the roughness.

$$THPP = \frac{\left[\frac{Nu\ no\ for\ roughned\ duct}{Nu\ no\ for\ smooth\ duct}\right]}{\left[\frac{f\ for\ roughned\ duct}{f\ for\ smooth\ duct}\right]^{\frac{1}{3}}} \quad (1)$$

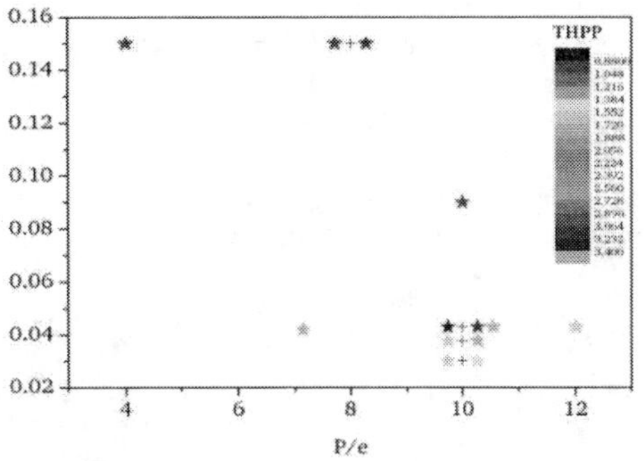

Fig. 3. THPP variation for the combination of 'e/D' and 'P/e' parameters of the roughness.

2 Modeling

2.1. Mathematical modeling

Three turbulent models; k-ω, k-ε and SST were used and results were compared subsequently. And it is found that the k-ω model gives the suitable results for our computational domain. It solves two transport equations, one for the turbulent kinetic energy k, and another for the turbulent frequency ω. The stress tensor is computed from the eddy-viscosity concept.
k-equation is given by

$$\frac{\delta \rho k}{\delta t} + \frac{\delta}{\delta x_j}(\rho U_j k) = \frac{\delta}{\delta x_j}\left[\left(\mu + \frac{\mu_t}{\sigma_k}\right)\frac{\delta k}{\delta x_j}\right] + P_k - \beta' \rho \varepsilon \omega + P_{kb} \quad (2)$$

ω-equation is given by

$$\frac{\delta \rho \omega}{\delta t} + \frac{\delta}{\delta x_j}(\rho U_j \omega) = \frac{\delta}{\delta x_j}\left[\left(\mu + \frac{\mu_t}{\sigma_\omega}\right)\frac{\delta \omega}{\delta x_j}\right] + \alpha \frac{\omega}{k} P_k - \beta' \rho \varepsilon \omega^2 + P_{\omega b} \quad (3)$$

In addition to the independent variables, the density ρ, and the velocity vector U, are treated as known quantities from the Navier-Stokes method. k is the production rate of turbulence. The model constants are given by: $\beta' = 0.09$, $\beta = 0.075$, $\alpha = 5/9$, $\sigma_k = 2$, $\sigma_\omega = 2$, $k = 0.00135$, $\varepsilon = 0.2278125$, $\omega = 1875$, $\mu_t/\mu = 1$

2.2. Physical model

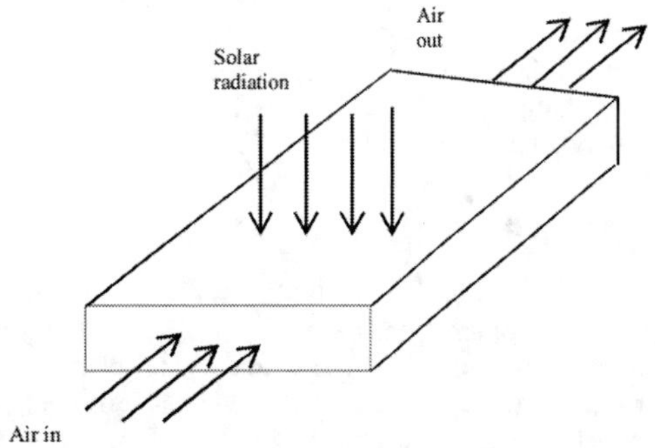

Fig. 4. Schematic diagram of a solar air heater duct.

Fig. 4 shows the schematic diagram of a rectangular duct of a solar air heater to absorb solar radiation and transfer this solar energy to incoming air Solar air heater duct is rectangular in shape. It has four surfaces; two lateral surfaces, one upper surface which is known as absorber plate and one lower surface which is called back plate. Width W and height b both are important design parameters of a solar air heater and decide the pressure loss and the heat transfer in the duct. Reducing height increases heat transfer but pressure loss too in the duct and vice versa.

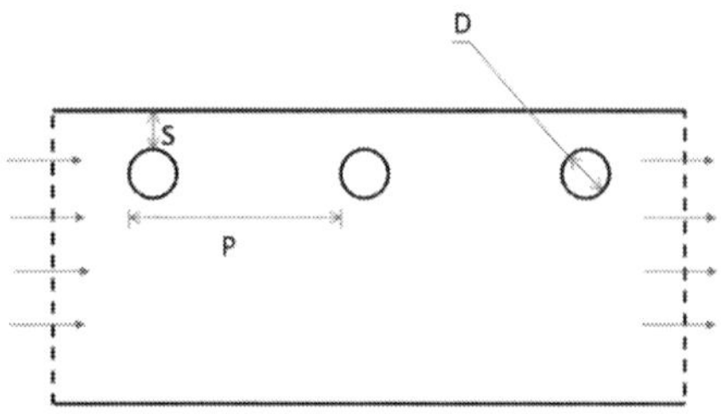

Fig. 5. Cross section diagram of a solar air heater duct along the length.

Fig. 5 shows the cross section of the rectangular duct, but here suspended cylinders are installed in the same duct to exploit the von-Karman effect of vortex flow behind these cylinders. Apart from width W, height b, we have, three new important parameters; those are needed to be studied. These parameters are; distance of cylinder from the absorber surface S, distance between centres of any two cylinders, i.e. pitch P and diameter of the cylinder D.

2.3 Numerical model

2.3.1 Design of experiment

We limited our study for two parameters only; S and D. All possible different cases, in terms of different combinations of S and D are taken for the study. Preliminary results rejected unfruitful cases and we finally left with only four cases as shown in table 1. Reference for the variation in D and S is taken from the value of boundary layer thickness δ = 4.8 mm, at 200 mm from the leading edge, that can be given by

$$\frac{\delta}{x} = 0.382\, Re_x^{-1/5} \tag{4}$$

In numerical modeling, 3D domain is avoided due to increase in number of mesh elements that will increase the computing time. 2D geometry and its meshing is done in such a way that it resembles the 3D domain, and the results obtained are within the acceptable limits.

2.3.2 Geometry

Geometry is modeled in ANSYS workbench as shown in fig. 6. A 2D domain consists of a circle which is representing cylinder in a plane, whose diameter and distance from the surface vary case to case; therefore four different cases are modelled as given in table 1.

Table 1. Various possible as per different S and D.

Case	Diameter (D) (mm)	Distance from wall (S) (mm)
Case 1	5.0	5.0
Case 2	5.0	2.5
Case 3	2.5	5.0
Case 4	2.5	2.5

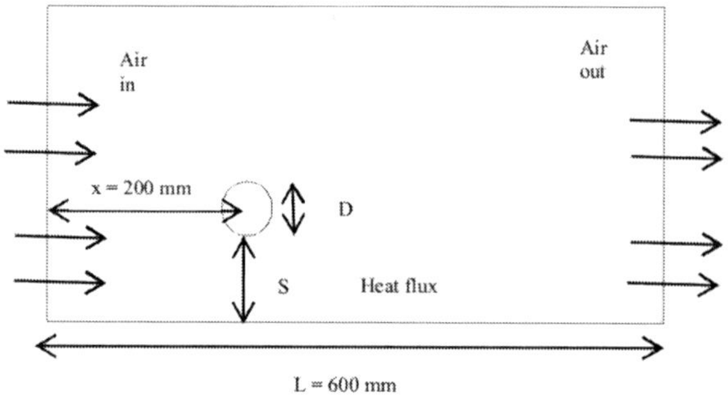

Fig. 6. Schematic diagram of computational domain.

2.3.3 Meshing

Geometry is then imported in ICEM CFD meshing software. Here mesh is generated; mesh elements near the cylinder are kept high to observe the von Karman effect of the vortex flow behind the cylinder. We have taken, $y^+ = 1$, for which $\Delta y = 4 \times 10^{-6}$ mm. Fig 7 shows the overall mesh of the configuration. Fig 8 and fig 9 show the enlarge view meshing near the cylinder boundary.

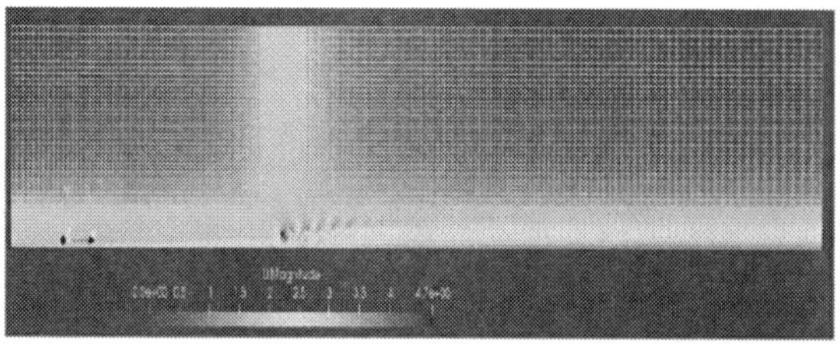

Fig. 7. Overview of the mesh along the length of the flat plate.

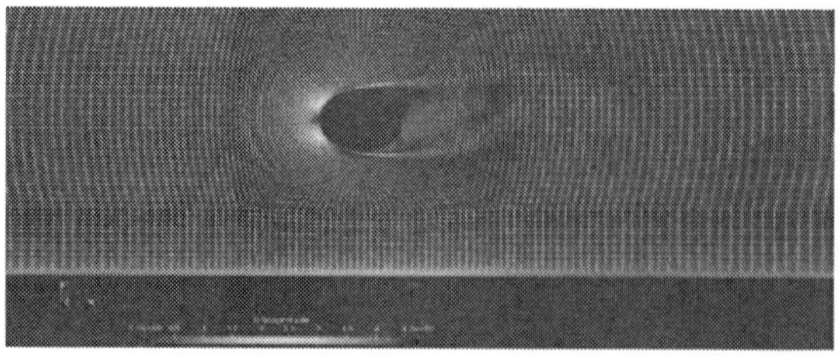

Fig. 8. Mesh shape around the suspended cylinder.

3 Result analyses

Mesh created in ICEM CFD is imported in openFOAM. Boundary conditions are applied and problem domain is solved; first for the mesh independency test then for the validation of results and finally results are obtained and plotted with the help of PARAVIEW.

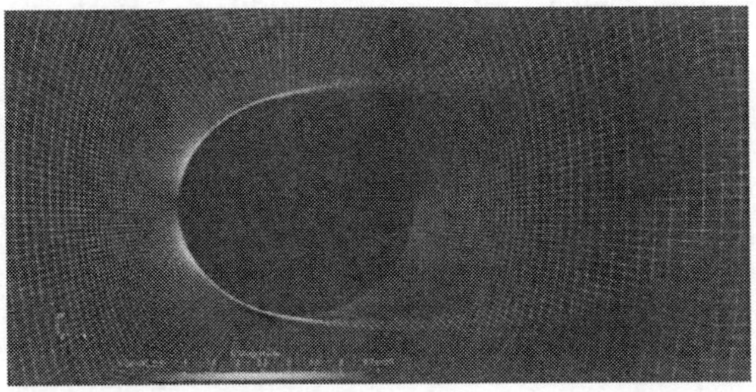

Fig. 9. Enlarged view of mesh around the suspended cylinder.

3.1 Boundary condition

2D domain can be treated as a flat plate over which a cylinder is placed 200 mm from the leading edge. Inlet: turbulent flow is observed after distance of 100 mm from the leading edge for the critical Reynolds number 5×10^5. Corresponding to this Reynolds number, the input velocity of the air flow is kept 3.0 m/s. Temperature of the inlet air is 300 K, as room temperature. Outlet: Zero static pressure is assumed, to get fully developed flow. Absorber plate: A constant heat flux of 1000 W/m^2 is applied on the absorber plate. Our aim is to diffuse this heat in the fluid domain to the maximum possible extent. It is assumed that suspended cylinder over the absorber plate doesn't participate in any mode of heat transfer; its only purpose is to create vortex flow.

3.2 Validation

For the baseline geometry (without suspended cylinders) and geometry with suspended cylinder, validation of numerical results is conducted. In validation, numerical results are compared with the standard results for U^+ and y^+ value as shown in the fig 10 and fig 11. Simulation results performed in OpenFOAM matches satisfactorily with the standard/benchmark results.

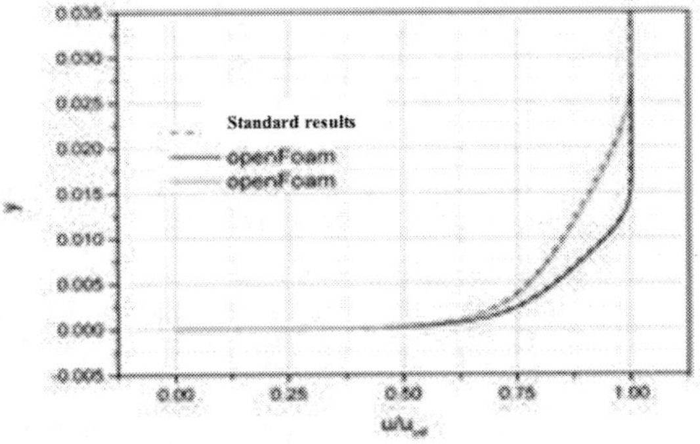

Fig. 10. Comparison of the benchmark result with the OpenFOAM results in terms of variation of U^+ value with logarithmic y^+ value.

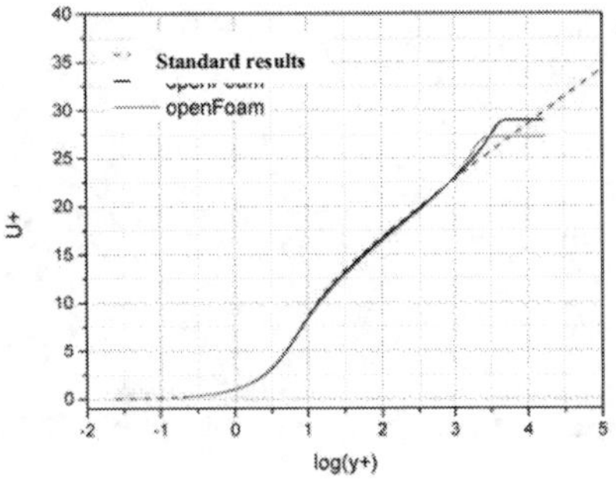

Fig. 11. Comparison of the benchmark result with the OpenFOAM results in terms of variation of y value with free stream velocity.

3.3 Comparison of different cases

Different cases are compared with baseline results. Geometry with suspended cylinder showed heat transfer increment over the baseline result.

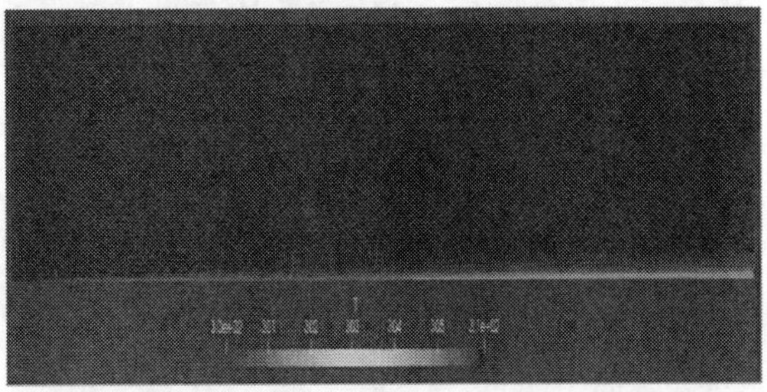

Fig. 12 Temperature profile along the length of the domain for the case 2.

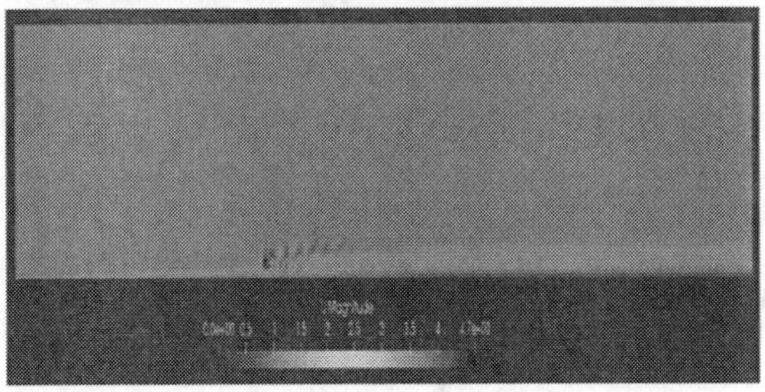

Fig. 13. Velocity profile along the length of the domain for the case 2.

Fig 12 and fig 13 show the pictorial results of change in the velocity and temperature profile, after inclusion of the suspended cylinder for the case 2. It can be clearly observed with the help of contour plot of velocity and temperature profile that due to presence of cylinder, mixing of hot air near the heated surface with air of relatively lower temperature takes place. Hence the wall temperature of the heated plate decreases behind the cylinders. Similarly, for the case 4, as shown in fig 14 and fig 15, velocity profile and temperature profile show the similar pattern after and before cylinder impediment. But in case 4, the reduction in wall temperature is slightly less the case 2. Velocity contour in fig 15 shows turbulence for the case 4 than case 2, similarly darker temperature contour in fig 2 suggest more heat supplied to the streamlines away from the wall/surface.

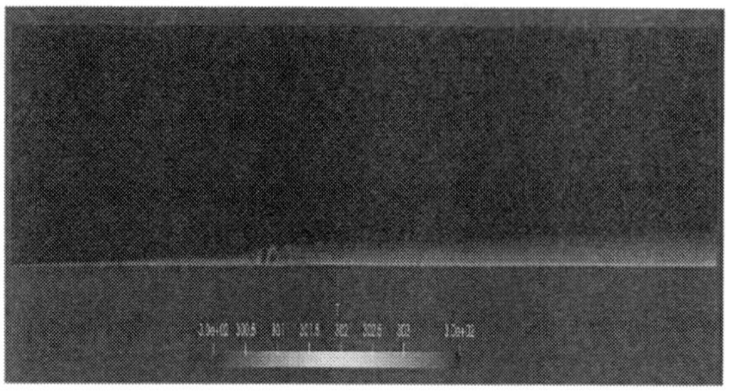

Fig. 14. Temperature profile along the length of the domain for the case 4.

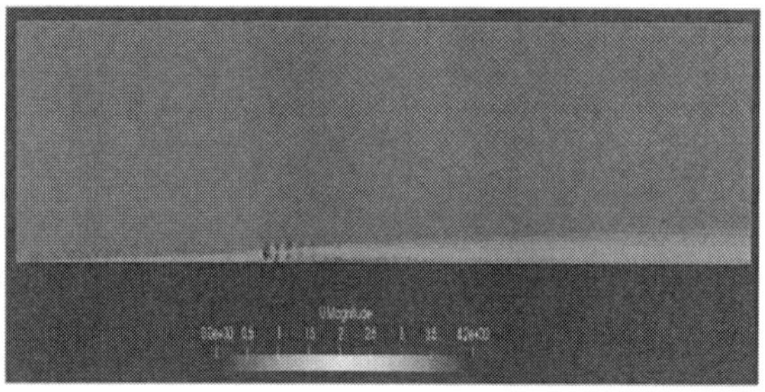

Fig. 15. Velocity profile along the length of the domain for the case 4.

Fig 16 shows the graphical variation of temperature with time along the length of the duct of the baseline geometry. It shows the transient nature of temperature variation along the along the duct, hence all cases are studied after steady state is achieved. Fig 17 shows the variation of the wall temperature along the length of the duct for different cases, based on positioning of suspended cylinders and change in diameter. Temperature increases as we move from the leading edge of the geometry. But this increment in temperature is more for the baseline geometry than the different cases of suspended cylinder. Temperature increment curve for the baseline geometry is smooth. For the different cases of cylinders, temperature increment curve has abrupt change in slope around the cylinder at a distance of 0.2 m from the leading edge. Temperature decreases for a moment then it increases again after passing the vortex flow behind the cylinder. But this maximum temperature reached is less then maximum temperature of baseline geometry. Case 4 has minimum rise in the temperature as compared to the other geometries.

In case 4 we have D = 2.5 mm and S = 2.5 mm, both values comes within the boundary layer thickness at that point i.e. δ = 4.8 mm, hence results are favourable for these values.

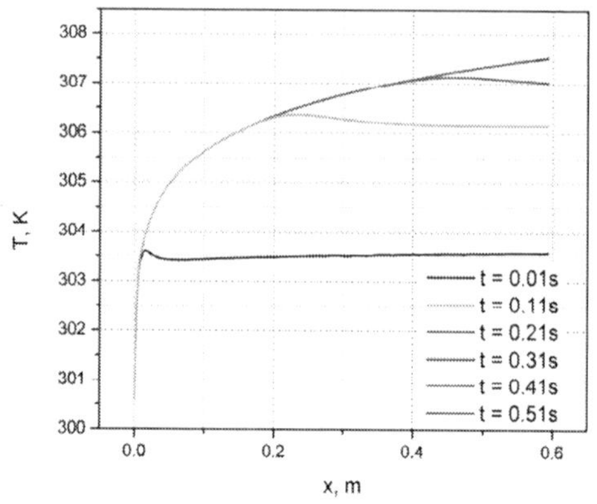

Fig. 16. Temperature variation with time along the length of the SAH duct.

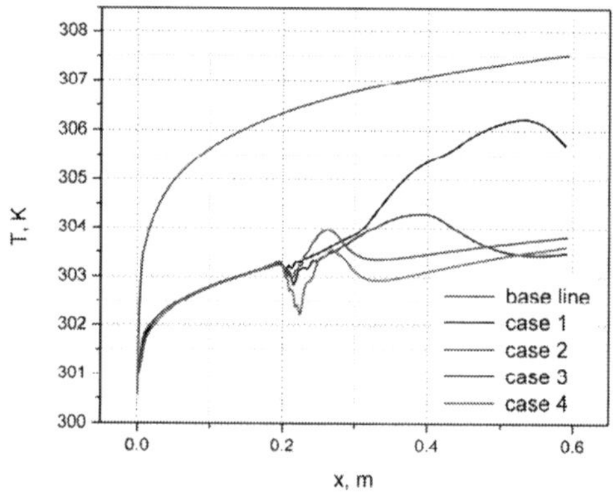

Fig. 17. Temperature variation along the length of SAH duct for different cases.

4 Conclusions

Following points can be concluded from the above observation and discussion of this numerical study;

1. Periodic vortex shedding behind the cylinder induces mixing of hot and cold stream, thereby enhancing heat transfer from the heated wall.
2. Positioning of cylindrical turbulator within the boundary layer causes higher heat transfer compared to the case when it is positioned beyond boundary layer thickness.
3. In case of a rectangular duct, it is recommended to place the cylindrical turbulators at adistance beyond entrance length, in order to have higher heat transfer from the heated wall

Acknowledgements

We are very thankful to Manipal University Jaipur for providing financial assistance seed grant name MUJ/REGR/1403/11 for this research and lab facility to carry out simulation work.

References

1. Kumar S, Sharad R.P., Renewable Energy **34** (2009) 1285–1291
2. Bhagoria J.L., Saini J.S., Solanki S.C., Renewable Energy **25** (2002) 341–369
3. Bopche Santosh B., Tandale Madhukar S., International Journal of Heat and Mass Transfer **52** (2009) 2834–2848.
4. Tanda G, International Journal of Heat and Mass Transfer **47** (2004) 229–243
5. Karwa R., Solanki S.C.,International Journal of Heat and Mass Transfer **42**(1999) 1597-1665
6. Hans V.S., Saini R.P., Saini R.C., Solar Energy **84** (2010) 898–911
7. Jaurker A.R., Saini J.S., Gandhi B.K., Solar Energy **80** (2006) 895–907
8. Karwa R., Solanki S.C., Saini J.S., Energy **26** (2001) 161–176
9. Kumar A., Energy Procedia **57** (2014) 2102 – 2111
10. Patnaik A., Saini R.P., Singal S.K., Renewable Energy **34** (2009) 2914–2922
11. Schlichting, H., 1979, Boundary-Layer, Theory, 7th edition, McGraw-Hill, New York.

Table 2. Summary of different artificial roughness used by researchers with details of important parameters

S. No	Geometry of the rib	p/e	e/D	St/St_s	Nu/Nu_s	Re	f/f_s	THPP	Name of the author
1.	Rib- Groove	6	0.0363	---	2.667	20000	3.333	1.785	A.R.Jaurket et al.
2.	Multiple V-ribs facing downward	12	0.043	---	4.629	20000	27.38	1.56	V.S.Hans et al.
3.	Transverse Continuous ribs	8	0.15	---	2.1	40000	12.1428	0.914	G.Tanda et al.
4.	Transverse broken ribs	4	0.15	---	2.5	40000	18.57	0.944	G.Tanda et al.
5.	Broken v-ribs	8	0.15	---	2.02	40000	10.71	0.916	G.Tanda et al.
6.	V-shaped continuous	10	0.03	---	1.579	13500	1.381	1.418	Aharwal and lanjewar et al.
7.	V-shaped with gap	10	0.03	---	1.842	13500	1.534	1.601	Aharwal and lanjewar et al.
8.	Inverted u-shaped ribs	6.67	0.0398	--	2.917	18000	3.833	1.864	Bopche and Tandale et al.
9.	Transverse wedge ribs	7.57	0.033	--	2.3	18000	3	1.596	Bhagoria et al.
10.	Chamfered repeated ribs	4.58	0.0321	2	---	17000	2.625	1.45	Karwa et al.
11.	Arc-shaped ribs	10	0.0426	--	1.474	18000	1.744	1.225	Kumar and saini et al.
12.	Dimple shaped ribs	10	0.0289	--	2.57	11500	4	1.62	Saini and verma at el.
13.	Inclined continuous with gap	10	0.0377	--	2.58	18000	2.9	1.809	K.R.Aharwal et al.
14.	Inclined continuous	10	0.0377	--	2.3	18000	2.5	1.695	K.R.Aharwal et al.
15.	V-shaped rib	10	0.043	--	2	20000	3	1.387	A.Kumar et al.
16.	Multi V-shaped rib	10	0.043	--	5	20000	4	3.15	A.Kumar et al.
17.	Multi V-shaped with gap	10	0.043	--	5.5	20000	4.25	3.395	A.Kumar et al.
18.	Transverse rectangular ribs	10	0.09	--	2.12	40000	14	0.88	G.Tanda et al.
19.	Transverse square ribs	7.14	0.042	--	2.78	18000	3.557	1.824	A.S.Yadav et al.
21.	Equilateral triangular rib	14.29	0.021	--	2.13	18000	2.357	1.565	A.Yadav and Bhagoria et al.
22.	Equilateral triangular rib	10	0.03	--	2.766	18000	2.714	1.983	A.Yadav and Bhagoria et al.
23.	Equilateral triangular rib	7.14	0.042	--	3.276	18000	3.286	2.204	A.Yadav and Bhagoria et al.
20.	Chamfered repeated ribs	4.58	0.0441	--	2.2	11000	3.8	1.6021	R.Karwa et al.

Temporal load resolution impact on PV/grid system energy flows

Marek Jaszczur[1,*], *Qusay* Hassan[1,2], and *Janusz* Teneta[3]

[1]AGH University of Science and Technology, Faculty of Energy and Fuels, Krakow, Poland
[2]University of Diyala, Mechanical Engineering Department, Diyala, Iraq
[3]AGH University of Science and Technology, Faculty of Electrical Engineering, Krakow, Poland

Abstract. In this paper, an investigation of the electrical load temporal resolution on the PV/Grid energy system flows, and self-consumption is done in order to determine the optimum parameters for modelling and simulation. The analysed PV/Grid power systems include a photovoltaic system with the nominal power of $P_{max@STC}$=1.5, 2.5, 3.5 kW without storage unit connected to the grid. The results show that the temporal load resolution may have a high impact on energy flows as well as can be a critical issue for the system analysis accuracy even for the single household. It has been found that the load temporal resolution for energy consumption of 1-min yields reliable results, while data resolutions of 5 and 15 min are still sufficient, however, in that case, the daily electrical energy flows and in consequence energy self-consumption estimation error for selected days may exceed 15%. Acquisition time step longer than 15-minutes may increase error above 20% and from the designer's point of view should not be used. The high and low temporal resolution experimental data of the electricity consumption (load) for a household are available in digital form on the author's website http://home.agh.edu.pl/jaszczur.

1 Introduction

The Sun provides energy for all parts of the Earth, and this type of energy can be considered as a one of the most promising among renewable energy resources, due to free availability, clean and sustainable [1]. However, many parts of the World are still struggling to meet their demand. The photovoltaic (PV) module is capable of harnessing and converting energy from the Sun directly into most useful electrical energy [2]. At present, the efficiency of PV systems has not achieved the satisfying level, due to influenced by a large number of technical factors (cell efficiency, optical efficiency). PV system performance also depends on a large number of environmental factors as: air temperature, wind speed, air pollution, the angle of incident irradiation, solar radiation spectrum, snow and shadowing, and scientists are still working to gain the performance of the PV systems [3]. In recent years, a large number of countries have shown a growing interest in the installation of solar energy based systems on the roofs of the household/building. Notably, on-grid PV systems received the most attention for overcoming the intermittent nature of renewable sources, the issue of over-scaling and enhancing supply reliability [4,5]. Modelling and optimisation of such systems are the key issue and subject to specific constraints that are not only of a technical point of view. From contributors in the energy system, optimisation can see as obtaining the best result of committed resources, including both energy sources and economic factor, under some constraints derived from the number of specific objectives. The electrical energy flows in PV/Grid power systems depends on many system parameters, e.g., system size, system type, solar radiation and load (energy consumption) profile. For analysing renewable energy system connected to the grid, the electrical energy flows from/to the grid have to be investigated. However, all these types of investigation should rely on reliable data for a load of residential electricity consumption. The electrical energy flows in the PV real system depend the on the actual data only, but for the system analysis, it may also depend on the analysis time step and temporal resolution of the input data. Usually, due to technical limitation and storage units, it is not practical to use high temporal resolutions data in the order of seconds for solar radiation or energy consumption, however, for electric digital energy meter, at least one-second sampling rate is used whereas for analogue electrical meter fully continuous sampling rate is implemented. For this reason, it is an important issue to get detailed knowledge about the influence of the temporal load resolution on the energy flow in such systems when analysed.

In the literature, a few studies have been reported modelling PV/grid energy systems with an indirect focus on load temporal resolution. Osawa et al. [6] investigate improving PV self-consumption for charging the vehicle. The simulation was done for the one-year time span, resulting with 3.4 kW PV system and the result shows that self-consumption reach up to 79% with optimised charge/discharge control and utilisation of the electric vehicle battery. For a single-family house with two parents and one child, 10.8 kW PV system was studied

* Corresponding author: jaszczur@agh.edu.pl

by Vrettos et al. [7] resulting in the self-consumption percentage increase around 17% by using optimising algorithms for batteries charging. In another work Femia and Zamboni [8] show the results of self-consumption analysis and obtain value of about 16% without storage and without load management, 31% without storage and with load management, 33% with storage and without load management and 48% with storage and with load management for a flat with two people. Similar work was done by Cagigal et al. [9] for two weeks span and 5.55 kW PV system with 5.4 kWh battery storage. The results showed that the PV self-consumption could be improved by around 33% with storage and without load management, 42% with storage and with load management. From presented examples and as written by Petrollese and Cocco [10] there are two main effective ways for increased PV self-consumption: the energy storage and the load management. In the literature [11-20] a large number of studies have been done using battery storage for increase self-consumption. In [21-24] management shifting of flexible electric loads in households have been analysed to optimise the utilisation of the PV production. Two methods of consideration provided to enhance self-consumption are described in [25–28]. In a study performed by Denholm & Margolis [29] authors conclude that by using electrical storage systems and load shifting it is possible to maximise PV energy self-consumption. At present, the PV grid-connected systems offer constructive solutions to improve the grid and services at the same time; the grid network should be designed to respond for this shape by using a modern updated solution such as a storage system using, controlling consumption or adapting generation.

The effects of the PV size on self-consumption on the aggregated consumption of electrical grid have been studied by Calpa et al. [30]. The authors reported that the PV self-consumption could facilitate the overall consumption of the system. On the other hand, direct use of the energy from the PV system can increase the overall consumption volatility of high penetration levels. For the large-scale PV power plant, the self-consumption mechanism has been analysed by Şenol et al. [31]. Authors concluded a need for installing large scale PV plant with high self-consumption and with optimal power capacity by considering technical and economic aspects. In the research done by Castillo-Cagigal et al. [32], the PV self-consumption was evaluated using artificial intelligence techniques. Optimal operating strategies of the PV system analysis have been the subject of extensive research on the PV/Grid systems.

Available in the literature studies use different modelling approaches and strategies however in most of the studies a single arbitrarily selected input temporal resolution or simulation time step are used for analysis. In other words energy consumption (load) and solar radiation data, temporal resolution was used without any justification. It is clear that obtaining temporal resolution data in the order of seconds for solar radiation and load is complicated and maybe unpractical and for this reason, 15 or 60 min data are usually used. However, input data temporal resolution and time step may generate significant analysis errors in any analysis of this type. For this reason, this work aims to determine the optimum temporal resolution for electrical load and solar radiation measurement.

2 Methodology

In order to determine the impact of the electrical load and solar radiation temporal resolution on the energy flows as well as PV system self-consumption, detailed system analysis were performed for a single household and three different PV systems with nominal power equal to 1.5, 2.5 and 3.5 kW_p. The PV system is equipped with a single-phase inverter. Polycrystalline silicon PV panels with 13% efficiency at STC condition we used. The system under consideration has the following orientation: tilted angle β=35° and azimuth angle γ=20° (south), and the system has been located in Kraków city (50.066092N, 19.921835E), Poland.

2.1 Load demand

The experimental measurements of the electrical AC load for a household in Krakow, Poland was performed using energy meter (OWI 2.8:1096) with transmitter (CM180) for the period 01.01.2017-31.12.2017. The considered household is flat with area of $70m^2$, has three rooms and is occupied by two adults. The heating system, hot water, and the cooker are supplied with natural gas. The yearly electrical energy consumption was 1.815MWh in 2017.

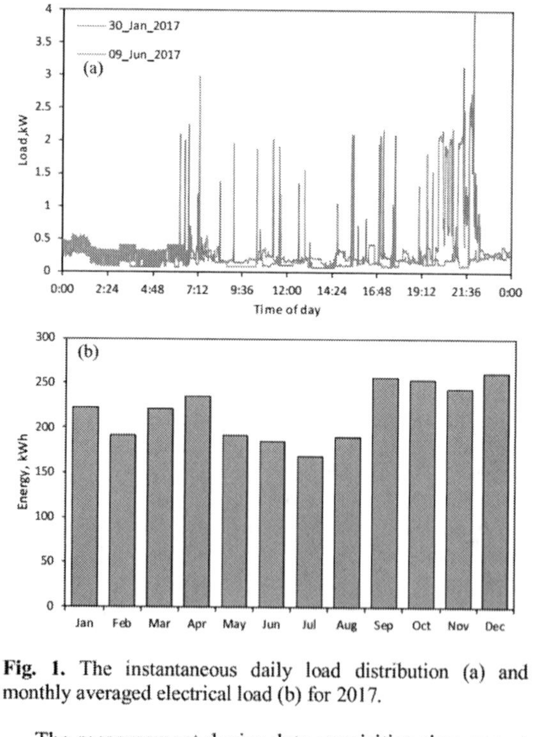

Fig. 1. The instantaneous daily load distribution (a) and monthly averaged electrical load (b) for 2017.

The measurement device data acquisition time was at least 1-minute (typically 10-60 sec) and depends on the electrical power fluctuation.

However, for the present study, an original non-uniform time step measurement has been re-sampled by means averaging procedure to the 60 seconds temporal resolution. Base on default 60 seconds resolution, four different temporal resolutions for the load has been created equal to: 5, 15, 30 and 60 minutes. The daily load distribution and monthly average energy consumption from experimental measurements are presented in Figure 1(a) and 1(b) respectively.

2.2 Solar irradiation

In order to performed on gridPV system analysis, require the global (GHI) and diffused components (DHI) of solar radiation were measured by pyranometers (Kipp & Zonen CM 21). Measurements were obtained every second, then averaged over 5 min, and recorded by a Combilog 1200 datalogger (Theodor Friedrichs). These measurements have been recorded over the entire year of 2017. The daily distribution and monthly averaged solar radiation are shown in Figure 2(a)-(b) respectively. Results show that the monthly averaged solar radiation and daily solar radiation distribution vary significantly by the month.

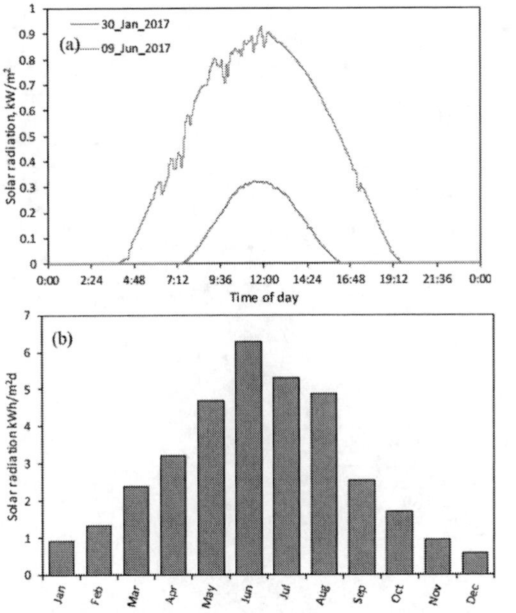

Fig. 2. Daily (a) and monthly averaged (b) total solar radiation distribution.

3 Mathematical modelling

3.1 Power of PV

The PV panel electrical power depends on the incident solar radiation G_T (global component) and can be calculated according to the following formula [3]:

$$P_{PV} = C_{PV} D_{PV} \left[1+\alpha_P \left(T_c - T_{c,STC}\right)\right]\left(\frac{G_T}{G_{T,STC}}\right) \quad (1)$$

where P_{PV} is the photovoltaic panel generated power, C_{PV} is the PV panel rated capacity, D_{PV} is the PV derating factor, α_P is the PV thermal coefficient of power, T_c is the PV module temperature, $G_{T,STC}$ and $T_{c,STC}$ are the solar irradiation and panel temperature at STC conditions. In this work the effect of the module temperature on the PV power not taken into account.

3.2 Load balance equation

The instantaneous load balance equation ensures that the electricity demand of the household is satisfied at all times from the photovoltaic system or from the grid system, as follows:

$$P_{Laod,t} = P_{From\ grid,t} + P_{PV,t} \quad (2)$$

where $P_{Load,t}$ is the instantaneous electrical load demand, $P_{From\ grid,t}$ instantaneous power taken from the grid and $P_{PV,t}$ instantaneous PV generated power. The load balance equation, describes the power flows between the system components, and can be calculated with the following constraints:

$$P_{Laod,t} = \begin{cases} P_{From\ grid,t} & for\ \ P_{PV,t} = 0 \quad ; P_{To\ grid,t} = 0 \\ P_{From\ grid,t} + P_{PV,t} & for\ \ P_{PV,t} < P_{Load,t} ; P_{To\ grid,t} = 0 \\ P_{PV,t} & for\ \ P_{PV,t} \geq P_{Load,t} \ ; P_{To\ grid,t} \geq 0 \end{cases} \quad (3)$$

3.3 PV self-consumption

The outline of the power profiles of on-site PV power generation and consumption can divide into 3 components: the electricity generation, the total net electricity demand, and the absolute self-consumption [33]. The self-consumption is the electricity generated by the photovoltaic system that is consumed directly by the household. Self-consumption (SC) and self-sufficiency (SS) are defined as follows:

$$SC = E_{PV,consumed} \,/\, E_{PV,generated} \quad (4)$$

$$SS = E_{PV,consumed} \,/\, E_{total} \quad (5)$$

where $E_{PV,\ consumed}$ is the PV energy consumed directly by the household, $E_{PV,\ generated}$ photovoltaic system energy generation and E_{total} household energy demand. The analysed here on-grid PV system was calculated with time step Δt=1 min as follows:

$$\text{Energy} = \sum_{t=1}^{t=n} E_{From\ grid,t} - \sum_{t=1}^{t=n} E_{To\ grid,t} \quad (6)$$

where $E_{From\ grid}$, $E_{To\ grid}$ and n are the energy from the grid, energy fed to the grid, and n is equal to the number time steps respectively.

4 Results

All performed analysis is based on experimental measurement for load and solar irradiation. Figures 3(a)-3(d) and Figures 4(a)-3(d) shows the load and the PV system instantaneous power for the selected sunny and cloudy days respectively and the load input data temporal resolution equal to 1, 15, 30 and 60 min respectively. For the solar irradiation fixed temporal resolution equal to 5 minutes was considered. The common area under the curve of the solar radiation and load represents self-consumption. One may easy inferred from results the impact of the different load temporal resolution on the energy-self consumption (common area). However, one should notice that changes in self-consumption are apparent created by the input data resolution and the real system self-consumption is fixed.

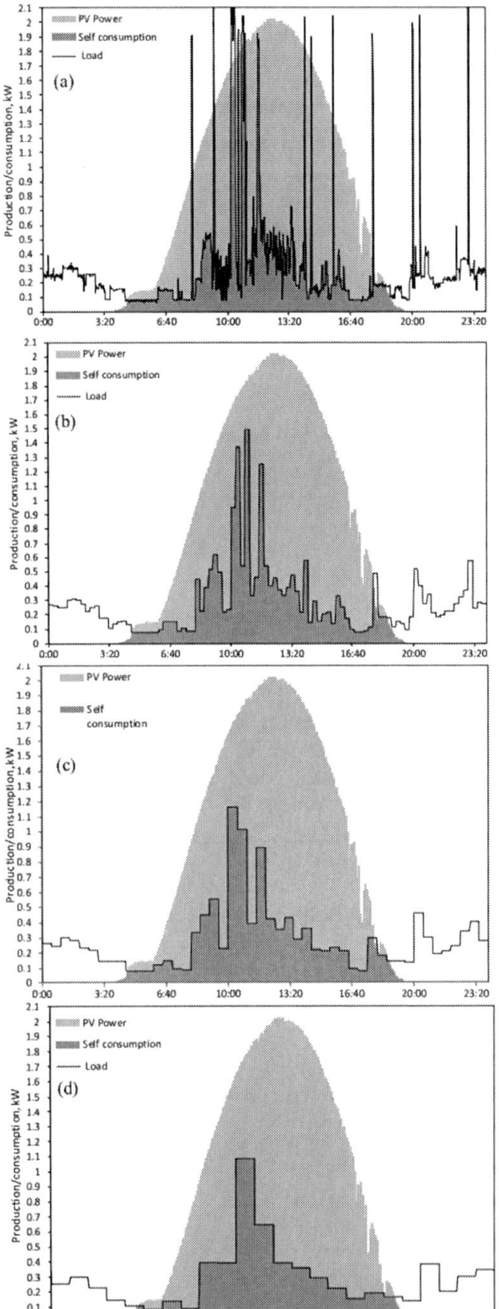

Fig. 3. Load, PV power and self-consumption for 2.5 kW_p PV system for a sunny day (03.06.2017) and load temporal resolution: 1min(a), 15min(b), 30min(c), 60min(d).

The total energy consumption for the selected sunny day is 6.98 kWh, while the energy supplied by the 2.5 kW_p PV system is 16.3 kWh. The real daily energy consumption is temporal resolution independent and constant by definition; however, the energy distribution due to averaging may vary significantly for different resolutions. Particularly, the maximum power peaks become much lower and wider.

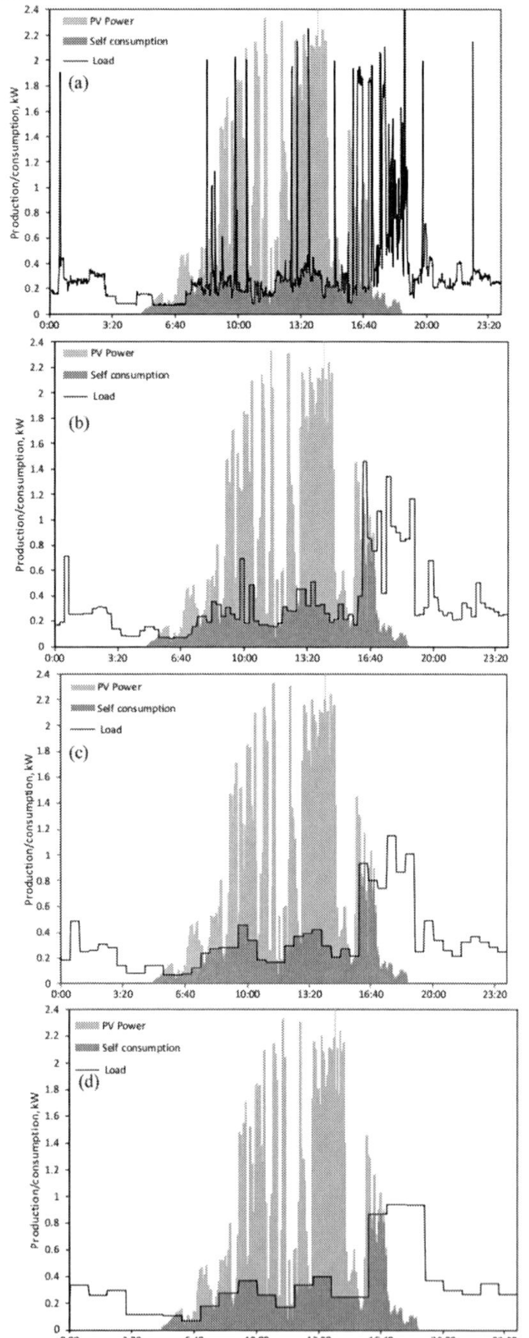

Fig. 4. Load, PV power and self-consumption for 2.5 kW_p PV system for a partly cloudy day (05.06.2017) and load temporal resolution: 1min(a), 15min(b), 30min(c), 60min(d).

The daily energy consumption for the selected cloudy day is 8.08 kWh, while the total energy generated by the PV system is 10.47 kWh.

The energy flows obtained from the grid, fed to the grid and PV self-consumed for different load temporal resolutions for presented sunny and cloudy days are shown in Tables 1 and 2 respectively.

Table 1. Daily energy from/to the grid and self-consumption for 2.5 kW_p PV system on presented sunny day.

Resolution of the load (min)	Energy taken from the grid (kWh)	Energy fed to the grid (kWh)	PV self-consumption (kWh)
1	2.75	10.47	4.22
5	2.61	10.33	4.37
15	2.42	10.14	4.55
30	2.38	10.1	4.59
60	2.36	10.08	4.61

Table 2. Daily energy from/to the grid and self-consumption for 2.5 kW_p PV system on presented cloudy day.

Resolution of the load (min)	Energy taken from the grid (kWh)	Energy fed to the grid (kWh)	PV self-consumption (kWh)
1	5.26	6.58	2.82
5	5.08	6.4	3.00
15	4.83	6.15	3.26
30	4.71	6.03	3.37
60	4.69	6.01	3.39

The different load temporal resolutions affect the apparent energy flow in the PV system. The errors related to the apparent flows can be calculated in reference to the high-resolution results (1 minute) using SC at two different resolutions.

$$\text{Absolute errors} = SC_{2\ resolution} - SC_{1\ resolution} \quad (7)$$

$$\text{Relative errors} = SC_{2\ resolution} - SC_{1\ resolution} / SC_{1\ resolution} \quad (8)$$

The absolute and relative errors for PV self-consumption caused by the different load temporal resolutions for presented sunny day and 2.5 kW_p PV system are shown in Figure 5(a) and (b) respectively.

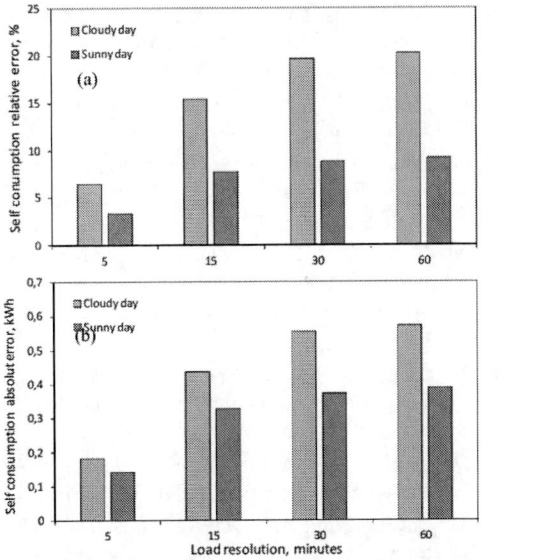

Fig. 5. The self-consumption absolute (a) and relative errors (b) vs. temporal load resolution for cloudy and sunny days.

The yearly energy flows, from/to grid and self-consumption at different load resolution for three different PV systems 1.5, 2.5 and 5.3 kW_p and for the year 2017 are shown in Tables 3, 4 and 5 respectively.

Table 3. Energy from/to the grid and self-consumption for 1.5 kW_p PV system (whole year 2017).

Resolution of the load (min)	Energy taken from the grid (kWh)	Energy fed to the grid (kWh)	PV self-consumption (kWh)
1	2091.80	796.03	528.26
5	2069.19	772.87	551.43
15	2048.42	752.02	572.28
30	2033.00	736.60	587.70
60	2016.73	720.33	603.97

Table 4. Energy from/to the grid and self-consumption for 2.5 kW_p PV system (whole year 2017).

Resolution of the load (min)	Energy taken from the grid (kWh)	Energy fed to the grid (kWh)	PV self-consumption (kWh)
1	1977.96	1563.61	642.10
5	1951.32	1536.41	669.30
15	1926.92	1511.93	693.78
30	1908.52	1493.53	712.18
60	1893.42	1478.43	727.28

Table 5. Energy from/to the grid and self-consumption for 3.5 kW_p PV system (whole year 2017).

Resolution of the load (min)	Energy taken from the grid (kWh)	Energy fed to the grid (kWh)	PV self-consumption (kWh)
1	1902.86	2372.84	717.19
5	1876.98	2346.40	743.65
15	1852.14	2321.49	768.55
30	1835.20	2304.55	785.49
60	1821.69	2291.03	799.01

The absolute and relative errors for on-grid PV system self-consumption due to different load temporal resolutions for three PV system sizes are shown in Figure 6(a) and Figure 6(b) respectively.

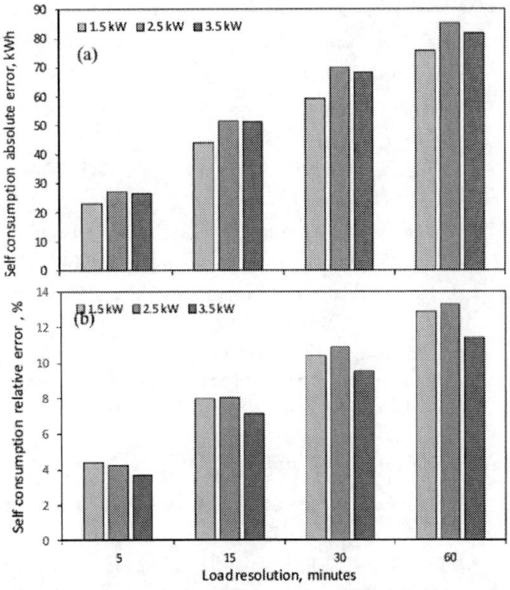

Fig. 6. The yearly self-consumption absolute (a) and relative errors (b) due to load temporal resolution.

5 Conclusions

The study investigated electrical load temporal resolution, and the photovoltaic system size (1.5, 2.5 and 3.5 kW_p) effect on the on-grid PV system energy flows. The results showed that power also flows might differ significantly if the load temporal resolution is too high. The relative and absolute errors of the analysis increase with increasing load temporal resolution. The energy self-consumption underestimation for presented cloudy day for the 2.5kW_p PV and the resolutions 5, 15, 30 and 60 minutes is 0.18, 0.44, 0.55, 0.57 kWh while for the presented sunny day is 0.15, 0.32, 0.37, 0.39 kWh (for sunny day) respectively. The relative error recorded is 6.4, 15.6, 19.5, 20.2% for the cloudy day, and 3.6, 7.8, 8.8,9.2% for sunny day respectively.

The relative yearly error for self-consumption estimation due to load temporal resolution 60 min depends on system size and for the 1.5 kW_p PV is 14.33% while for the 2.5, and 3.5 kW_p PV are 13.3 % and 11.4 % respectively. It has been found that the load temporal resolution of 1-min yields reliable results, while data lower resolutions of 5 and 15 min are still sufficient, however, lower than 15-minutes resolution may increase daily error above 20% and should not be used by system designers, computer software and system analysis tools.

The high and low temporal resolution experimental data of the electricity consumption (load) for a household are available in digital form on the author's website http://home.agh.edu.pl/jaszczur.

Acknowledgements
The present work supported by the Polish Ministry of Science (Grant AGH 15.11.210.436 and 11.11.210.312).

References

1. H. Gujba, Y. Mulugetta, A. Azapagic, Energy Policy **39**, 968-980 (2011)
2. S. Dubey, J.N. Sarvaiya, B. Seshadri, Energy Procedia **33**, 311-321 (2013)
3. Q. Hassan, M. Jaszczur, E. Przenzak, E3S Web of Conferences **14**, 01028, (2017)
4. A. Franco, M. Shaker, D. Kalubi, S. Hostettler, Sustainable Energy Tech. and Assessments (2017)
5. I.P. Panapakidis, D.N. Sarafianos, M.C. Alexiadis, Rene. and Sust. Ene. Rev. **16**, 551-563 (2012)
6. M. Osawa, K. Yoshimi, D. Yamashita, R. Yokoyama, T. Masuda, H. Kondou, T. Hirota, (*ISGT Asia, IEEE*, 1-6, 2012)
7. E. Vrettos, A. Witzig, R. Kurmann, S. Koch, G. Andersson, (*EU PVSEC*, 2013)
8. N. Femia, D. Toledo, W. Zamboni, (*IECON 39th IEEE* , 6800-6805, 2013)
9. M. Castillo, E. Matallanas, D. Masa, E. Caamaño, A. Gutiérrez, F. Monasterio, J. Jiménez, (*GeDELOS-PV system,* 2010)
10. M. Petrollese, G. Cau, D. Cocco, Applied Energy, **212**, 746-758 (2018)
11. J. Widén, E. Wäckelgård, D. Lund, Solar Energy **83**, 1953-1966 (2009)
12. J. Li, M.A. Danzer, Journal of Power Sources **258**, 365-373 (2014)
13. Q. Hassan, M. Jaszczur, M. Mohamed, K. Styszko, K. Szramowiat, J. Gołaś, E3S Web of Conferences **10**, 00093 (2016)
14. R. Kanamori, T. Ito, N. Mizutani, (*SOCA, IEEE*, 1-5, 2011)
15. G. Mulder, D. Six, B. Claessens, T. Broes, N. Omar, J. Van Mierlo, Applied Energy **111**, 1126-1135 (2013)
16. B. Ceran, Q. Hassan, M. Jaszczur, K. Sroka, E3S Web of Conferences **14**, 01020 (2017)
17. A. Bayod-Rujula, E. Haro-Larrode, A. Martinez-Gracia, Solar Energy **98**, 582-591(2013)
18. L. Nge, M. Midtgard, L. Norum, (*PowerTech, IEEE* Trondheim, 1-6, 2011)
19. B. Song, Y. Jung, H. Ji, C. Choi, W. Lee, Y. Won, (*Asia, (ICPE & ECCE), IEEE*, 2625-2630, 2011)
20. E. Waffenschmidt, Energy Procedia **46**, 88-97 (2014)
21. J. Munkhammar, J. Widén, WREF and CRES, 1858-1865 (2012)
22. M. Chaabene, M.B. Ammar, A. Elhajjaji, Applied Energy **84**, 992-1001 (2007)
23. Q. Hassan, M. Jaszczur, E. Przenzak, J. Abdulateef, (*Contemporary Problems of Power Engineering and Environmental Protection* **33**, 2016)
24. Y. Zong, L. Mihet-Popa, D. Kullmann, A. Thavlov, O. Gehrke, W. Bindner, (*ISGT, IEEE*, 1-8, 2012)
25. E. Matallanas, M. Castillo, A. Gutiérrez, F. Monasterio, E. Caamaño, D. Masa, J. Jiménez, Applied Energy **91**, 90-97 (2012)
26. C. Clastres, T.H. Pham, F. Wurtz, S. Bacha, Energy **35**, 55-64 (2010)
27. A. Barbato, A. Capone, G. Carello, M. Delfanti, M. Merlo, A. Zaminga, (*SmartGridComm, IEEE,* 345-350, 2011)
28. M. Jaszczur, Q. Hassan, J. Teneta, J. Abdulateef, Journal of Thermal Science, (2018)
29. P. Denholm, M. Margolis, Energy Policy **35**, 4424-4433 (2007)
30. M. Calpa, M. Castillo-Cagigal, E. Matallanas, E. Caamaño-Martín, A. Gutiérrez, Procedia Computer Science **83**, 816-823 (2016)
31. M. Şenol, S. Abbasoğlu, O. Kükrer, A.A. Babatunde, Solar Energy **132**, 518-537 (2016)
32. M. Castillo-Cagigal, E. Maallanas, F. Monasterio-Huelin, E.C. Martín, A. Gutiérrez, IEEE Transactions on Industrial Informatics **12**, 941-951 (2016)
33. R. Luthander, J. Widén, D. Nilsson, J. Palm, Applied Energy **142**, 80-94 (2015)

https://doi.org/10.1051/matecconf/201824004004

An analysis of temperature distribution in solar photovoltaic module under various environmental conditions

Marek Jaszczur[1,*] *Qusay* Hassan[1,4], *Janusz* Teneta[2], *Ewelina* Majewska[1], *Marcin* Zych[3]

[1]AGH University of Science and Technology, Faculty of Energy and Fuels, Poland
[2]AGH University of Science and Technology, Faculty of Electrical Engineering, Automatics, Poland
[3]AGH University of Science and Technology, Faculty of Geology, Geophysics and Environmental Protection, Poland
[4]University of Diyala, Department of Mechanical engineering, Diayal, Iraq

Abstract. The operating temperature of the photovoltaic module is an important issue because it is directly linked with system efficiency. The objective of this work is to evaluate temperature distribution in the photovoltaic module under different environmental conditions. The results shown that photovoltaic module operating temperature depends not only on the ambient temperature or solar radiation dependent but also depends on wind speed and wind direction. It is presented that the mounting conditions which are not taken into consideration by most of the literature models also play a significant role in heat transfer. Depends on mounting type an increase in module operating temperature in the range 10-15°C was observed which cause further PV system efficiency decrease of about 3.8-6.5 %.

1 Introduction

Due to the international regulations regarding global warming and increasing CO_2 emissions, the electrical energy production has to take into account the environmental aspect. The care for the environment is required because the Earth pollutions significantly influence the quality and the length of life for all living organisms. The additional problem is that fossil fuels are non-renewable and one day will be depleted. On the other hand, rapid growth in renewable energy systems has reached in 2016 the faster growing with global investments than traditional systems [1]. One of the widely used renewable energy systems which provide the direct method of converting solar energy into electrical energy is the photovoltaic system. Despite the significant progress that has been already made, the increase in electrical efficiency and the reduction of the production cost are to be still important goals. The performance of the photovoltaic system depends on a large number of factors such as module specifications, mounting scheme, module inclination, solar radiation, wind speed, ambient temperature, pollution or shadowing. The energy conversion efficiency in the photovoltaic module has high depends on the module temperature. The weak cooling mounting, high ambient temperature and high solar irradiation have a negative influence on the module energy conversion efficiency [2]. In the critical situation of inverter overheat the system will radically decrease the generating power or even shut down the system. The photovoltaic cell efficiency designed under Standard Test Conditions STC and varied depending on the cell type (polycrystalline silicon 16.0-17%, monocrystalline silicon up to 20% and thin film amorphous silicon about 7.5%) [3] and the cell operating temperature. With increasing the photovoltaic temperature, the current increase but the voltage decreases and final effective electrical power decrease [4]. The module efficiency is usually defined at the Standard Test Conditions STC (temperature 25 °C, solar radiation 1000W/m^2, wind speed 1 m/s).

There are several studies investigated the influence of temperature and thermal effect on photovoltaic energy and efficiency. The review concentrating on model validation, i.e. on temperature from measured and available models has been prepared by Santiago et al. [5]. The study analysed twenty models for predicting module operating temperature with different complexity. Perović et al. [6] have presented a model for estimation PV module temperature used energy balance equation. Authors validated the model with experiments and compared with two other models using different methodology. The results illustrated good agreement with the experiments measured data. Estimation for the heat transfer coefficient for photovoltaic module based on the empirical or theoretical relationship has been prepared by Kendoush [7]. The study was conducted using theoretical analysis for the fluid flowing across the flat plate in order to determine mass and heat flow. Zondag et al. [8] analysed the thermal performance of hybrid collectors with a-Si photovoltaic cells. The results showed that the annual degradation in the electrical efficiency is 7.6% due to temperature effect. By using experimental measurements, Hassan et al. [9] analysed the temperature effect on the performance of the photovoltaic system and energy production. The results of the analysis confirmed that the operating temperature of a module plays a vital

* Corresponding author: jaszczur@agh.edu.pl

role in the conversion process and the electrical efficiency and power output are significantly influenced. Ceylan et al. [10], analysed an effect of ambient temperature on the photovoltaic module temperature. Their results developed linear and nonlinear models for predicting module temperature effect on the output power and efficiency drop. They concluded that the non-linear model is more efficient than linear. Comprehensive analysis of PV modules by in-situ experimental measurement performed with 3-D numerical simulation in order to model heat transfer from PV module have been by Jaszczur et al. [11]. The results allow developing a new model for PV panel evaluation which includes all vital environmental factors (ambient temperature, incident solar, wind speed) as well as geometrical conditions or mounting conditions (using Ross coefficient). Hassan et al. [12] used energy balance equations and NOCT to prepare the new model for predicting PV module temperature. Their results showed that by using this model the yearly energy losses due to temperature, effects could be about 23% for 1kW PV system. An accurate determination of the PV module operating temperature is an important issue. Koehl et al. [13] validate two models for prediction module temperature in different climatic regions and getting accurate results for PV module temperature evaluation.
In this research, an analysis of temperature distribution in the photovoltaic module under various environmental conditions was performed. The real system analysis was base on locally measured weather conditions and verify using in-situ experimental measurement and CFD computer modelling.

2 Experimental and numerical set-up

The experimental set-up consists of two BrukBet BEP260W polycrystalline photovoltaic modules (temperature coefficient of power -0.40 %/°C and 1.62 m^2 of surface area, white back sheet) grid-connected, and two BrukBet BEP260W like modules without PV cells (first with white and second with a black back sheet) presented in Fig. 1(a). An analysed modules were adjusted at tilt angle β=35° and azimuth γ=20° West and located at the roof of C3 building, AGH University of Science and Technology in the city centre of Kraków, Poland (50.066354N, 19.918191E).

The acquisition system for temperature and solar radiation include 4 Advantech ADAM4018 DAQ modules that provide programmable input ranges on all channels, computer unit with data acquisition software and storage. As temperature sensors, eight T-type thermocouples were used (four thermocouples per PV modules) located in the recommended by standard point - see Fig.1(b). The signals from thermocouples were acquired every second, and the average value was recorded every 60 seconds.
The ambient temperature, wind speed and solar irradiation were obtained from the local weather station. The signals from weather station sensors were acquired every second, and the average value was recorded for analysis every 5 minutes.

Fig.1. An experimental set-up with the mounting system (a) and thermocouples arrangement (b).

3-D model of the photovoltaic module with the mounting system has been created in order to analyse fluid flow and temperature distribution. As a numerical solver Fluent v.18 was used. The geometry under consideration and key heat transfer components are shown in Fig. 2. The module size is 1.654×0.989×0.04m and glass thickness is 5mm. Instead of the multilayer system, it is assumed the modules one layer with an averaged thermal properties. Computational domain size was 4.2×2.8×2.2m, and has been filled with hex-dominant mesh elements, while in order to solve for boundary layer prismatic elements were adjusted. Additionally for the region with the high turbulence intensity finer mesh has been used. Based on the preliminary test analysis mesh with about 500,000 elements has been found to give a mesh-independent solution.

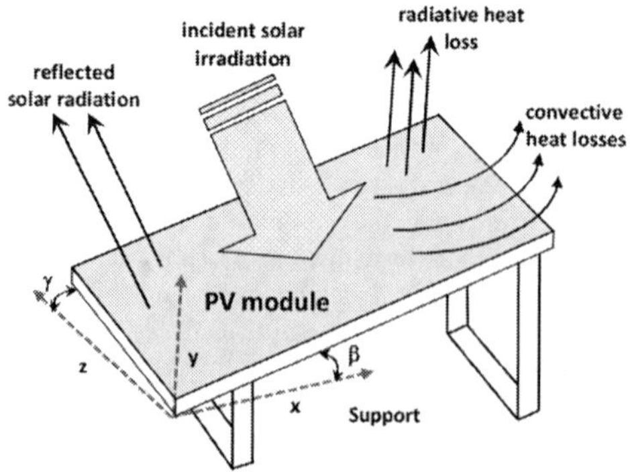

Fig.2. The geometry and the key heat transfer components.

It assumed that that fluid is Newtonian with constant fluid and solid properties presented in Table 1.

Table 1. Fluid and solid properties.

Material name	Specific heat J/kgK	Density kg/m^3	Viscosity/ Emissivity kg/ms / -	Thermal conductivity W/mK
Air	1006.43	1.225	$1.78 \cdot 10^{-5}$	0.0242
Panel	700	2700	0.92	197
Soil	1600	1600	0.95	2

The considerations for the mathematical model eq. (1)-(3) are: steady state negligible viscosity dissipation and the use of the Boussinesq approximation. The time-averaged equations for continuity, momentum and energy can be written as follows:

$$\frac{\partial \overline{U}_i}{\partial x_i} = 0, \tag{1}$$

$$\overline{U}_j \frac{\partial \overline{U}_i}{\partial x_j} = -\frac{1}{\rho}\frac{\partial \overline{P}}{\partial x_i} + \nu \frac{\partial^2 \overline{U}_i}{\partial x_j^2} - \frac{\partial}{\partial x_j}\left(\overline{u_i u_j}\right), \tag{2}$$

$$\overline{U}_j \frac{\partial \overline{T}}{\partial x_j} = \kappa \frac{\partial^2 \overline{T}}{\partial x_j^2} - \frac{\partial}{\partial x_j}\left(\overline{u_j \theta}\right), \tag{3}$$

where x_i, x_j are the Cartesian coordinates of the system, $\overline{U}, \overline{P}, \overline{T}$ are the mean velocity, dynamic pressure and temperature; ν, κ are fluid viscosity and thermal diffusivity. Turbulent a semi-empirical 2-equation eddy viscosity realizable k-ε model was used. This model is based on the Boussinesq hypothesis, and Reynolds stresses are expressed regarding mean velocity gradients and that the turbulent eddy viscosity is related to the turbulent kinetic energy and the dissipation rate of turbulent kinetic energy as follows:

$$-\rho \overline{u_i u_j} = \tau_{ij} = \mu_T \left(\frac{\partial \overline{U}_i}{\partial x_j} + \frac{\partial \overline{U}_j}{\partial x_i}\right) = 2\mu_T \cdot S_{ij}, \tag{4}$$

$$\mu_T = \rho \cdot C_\mu \frac{k^2}{\varepsilon}; \; \overline{u_i \theta} = -\left(\kappa + \frac{\nu_T}{\mathrm{Pr}_T}\right)\frac{\partial \overline{T}}{\partial x_i}. \tag{5}$$

Turbulence model Realizable k-ε posses an improved formulation for the turbulent viscosity. The equation for the turbulent kinetic energy and the dissipation rate are as follows:

$$\frac{\partial}{\partial x_i}\left(\overline{U}_i \cdot k\right) = \frac{\partial}{\partial x_i}\left(\frac{\nu_{eff}}{\sigma_k}\frac{\partial k}{\partial x_i}\right) + P_k - \varepsilon, \tag{6}$$

$$\frac{\partial}{\partial x_i}\left(\overline{U}_i \cdot \varepsilon\right) = \frac{\partial}{\partial x_i}\left(\frac{\nu_{eff}}{\sigma_\varepsilon}\frac{\partial \varepsilon}{\partial x_i}\right) + C_1 \cdot S \cdot \varepsilon - \frac{C_2 \cdot \varepsilon^2}{k + \sqrt{\nu \cdot \varepsilon}}, \tag{7}$$

where k and ε are turbulence kinetic energy and dissipation rate, P_k is the turbulent kinetic energy production $\mu_{eff} = \mu + \mu_T$ and $\nu_{eff} = \mu_{eff} / \rho$ are turbulent effective viscosity. The model constants are:

$C_1 = \max\left[0.43, \frac{\eta}{\eta+5}\right]$; $\eta = S\frac{k}{\varepsilon}$; C_{e1}=1.44; σ_k =1.0; σ_ε = 1.2

$$S = \sqrt{2S_{ij}S_{ij}}; \qquad C_\mu = \frac{1}{A_0 + A_S \frac{k\sqrt{S_{ij}S_{ij} + \tilde{\Omega}_{ij}\tilde{\Omega}_{ij}}}{\varepsilon}};$$

$$S_{ij} = \frac{1}{2}\left(\frac{\partial \overline{U}_j}{\partial x_i} + \frac{\partial \overline{U}_i}{\partial x_j}\right); \; \tilde{\Omega}_{ij} = \Omega_{ij} - 2\varepsilon_{ijk}\omega_k;$$

$$\Omega_{ij} = \overline{\Omega}_{ij} - 2\varepsilon_{ijk}\omega_k; \mathrm{A}_0 = 4.04; \; \mathrm{A}_S = \sqrt{6}\cos\phi;$$

$$\phi = \frac{1}{3}\cos^{-1}\left(\sqrt{6}W\right); \; W = \frac{S_{ij}S_{jk}S_{ki}}{\tilde{S}^3}.$$

The heat transfer model includes solar model and radiation heat transfer, model. For the radiative heat transfer Surface-to-Surface (S2S) radiation model is used:

$$q_{out,k} = \varepsilon_k \sigma T_k^4 + \rho_k q_{in,k}, \tag{8}$$

$$A_k q_{in,k} = \sum_{j=1}^{N} A_j \, q_{out,k} F_{jk}, \tag{9}$$

where $q_{out,k}$ and $q_{in,k}$ are energy flux leaving the surface and energy flux incident on the surface from surroundings; A_k is an area of k-surface, F_{jk} is view factor between k-surface and j-surface. Average air velocity and turbulence intensity profiles for roughness length 0.03 m are employed as an inlet boundary conditions. At the domain exit, a uniform zero gauge pressure boundary condition is used. At the side walls pressure outlet boundary condition were applied. The bottom wall has been treated as a rough wall with no-slip boundary condition. At the inlet of the domain the air temperature T_a equal to the values obtained from measurements have been applied. Solar radiation is applied to the thermal radiation model as an incident solar radiation in the range 100-1000W/m^2 depend on the case.

3 Results and discussion

The incident global and diffuse solar radiation at horizontal surface β=0° and tilted at β=35° (azimuth γ=20°) and wind speed and ambient temperature are shown in Fig. 3 for selected sunny day 07.06.2018.

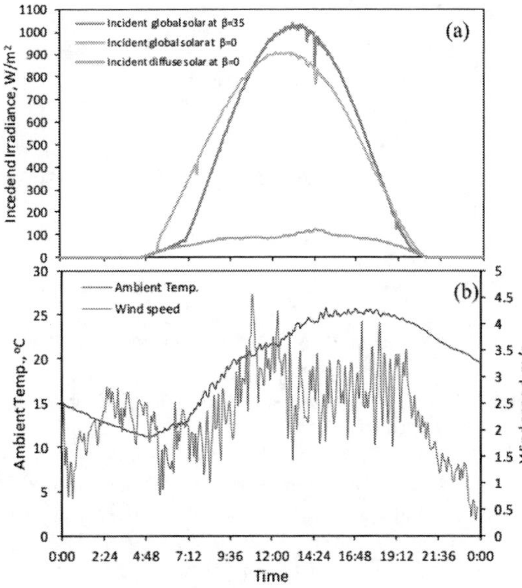

Fig. 3. The solar irradiation (a), wind speed and ambient temperature (b).

One may infer from results that the incident solar radiation, at β =35° has lower magnitude and the signal is a delay in the morning hours due to pyranometer azimuth and inclination.

Figure 4(a) shows the temperature of polycrystalline panels with tilted angle β =35° for selected sunny day 07.06.2018. The maximum temperatures recorded during the day is 50.9°C. It can be seen that at different points of the panel (T1-T4) temperature during high solar irradiation may vary significantly as a result of different heat transfer from various part of the photovoltaic module (top, side or corners). In Figure 4(b)-(c) temperature from the modules with black and white back sheet (and without silicon cells) are presented for β = 35°. Much higher temperature raising up to equal to 57.5°C is observed for a panel with a black back sheet while for the panel with a white back sheet the maximum temperature 34.4°C was recorded during the day.

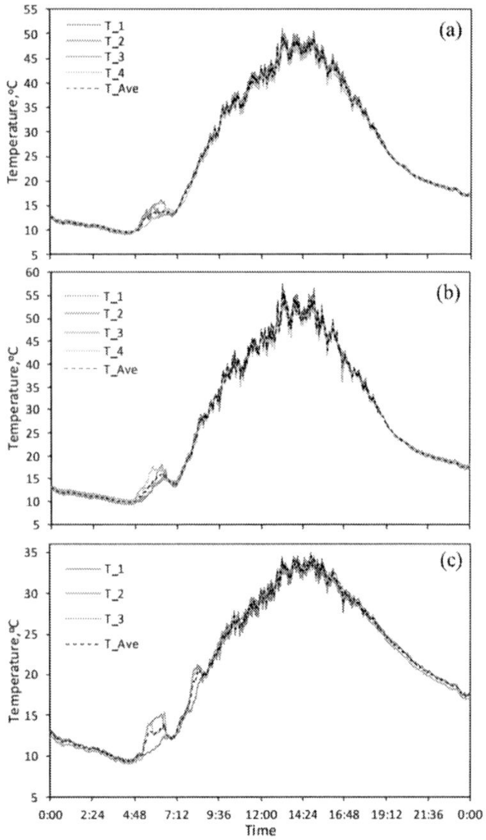

Fig. 4. Module temperature: with Si-cells (a), black back sheet (b), white back sheet (c) and for β = 35°.

In order to evaluate temperature distribution at module surface, Computational Fluid Dynamics (CFD) analysis was performed for the panels with titled angles β = 0 , 15° and 35° and environmental conditions, solar radiation G_T = 470 W/m^2, wind speed V_w = 3.1 m/s and ambient temperature T_a=8 °C. In Figure 5(a)-4(d) the flow field around the module is shown. Except the configuration presented in Figure 5(b) wind direction is opposite to the X-axis direction (V_w = -3.1 m/s) while in Figure 5(b) air flow has the same magnitude but the flow is from the module left side. In Figure 5(d) module tilt angle is β = 0°, and the space under the module is closed which significantly reduces heat transfer from the bottom surface.

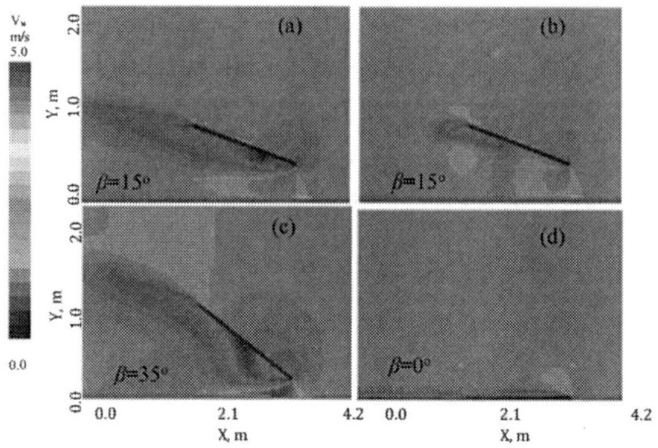

Fig.5. Velocity field around module for different tilt angle: β=15° (a), β=15° (b), β=35° (c), β=0° (d) and air flow direction

The temperature field for the same conditions as presented in Figures 5 are presented in Figures 6. One can see from this figures that temperature distribution for similar weather parameters differs significantly. The highest temperature has been obtained for module with tilt angle β = 0° when the cooling conditions are the worst cooling while the lowest temperature of the PV module is seen for the tilt angle equal to 15° when the of air flow is from the module (left) side - see Fig. 6(b).

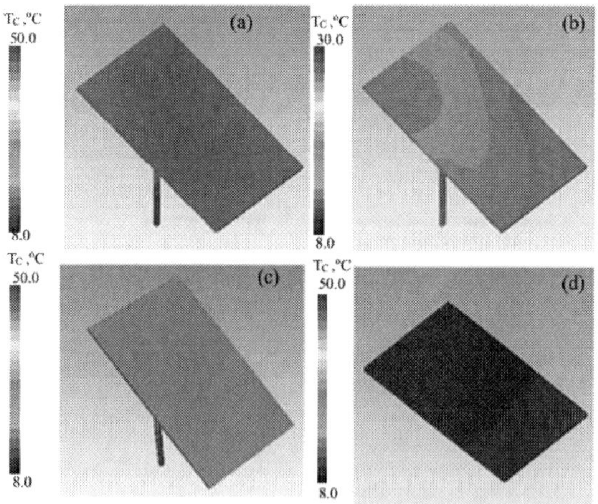

Fig. 6. Temperature at the module surface for different tilt angle: β=15° (a), β=15° (b), β=35° (c), β=0° (d) and air flow direction.

The essential parameters which have considerable influence on the module operating temperature are the solar irradiation, ambient temperature T_a and wind speed V_w The impact of above parameters on the mean module operating temperature are shown in Figures 7 together with experimental measurement. It can be seen that for low airspeed the module temperature becomes very high which effect PV module performance.

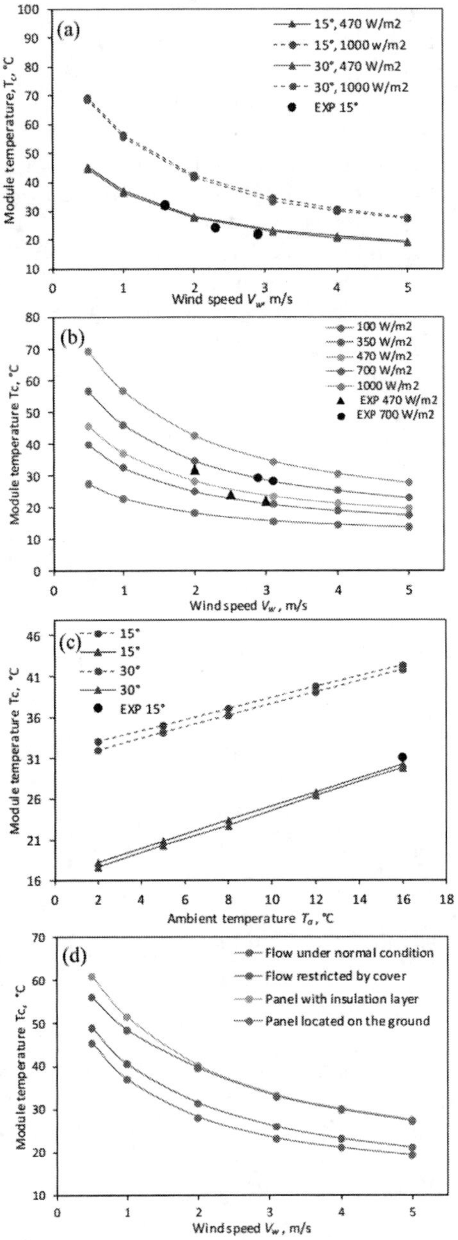

Fig. 7. The average module temperature T_c vs.: wind speed V_w (a-b), ambient temperature T_a, (c) mounting configuration (d).

One of the important parameters is related to the module mounting conditions. Typically module top surface is well cooled. However, the bottom module surface heat transfer can differ depending on mounting configuration significant. For the conditions which allow for air flow from two sides the module temperature become the lowest. When the flow below module is restricted the temperature, an increase about 2-4°C is observed. On the other hand when the module is mounted horizontally, or the module bottom surface is insulated (wall integrated configuration) poor heat transfer causes additional module temperature increase about 10.0°C.

Using numerical and experimental data correlation for module temperature T_c can be created which takes into account all important phenomena parameters: solar irradiation, ambient temperature, wind speed and the mounting conditions. The proposed correlation uses four parameters responsible for all crucial phenomena. In the literature, several relationships have been advised however most of them use two or three parameters only. In the present work modification of the following relationship will be used [14]:

$$T_C = T_a + kG_T. \tag{10}$$

In this very popular and straightforward linear expression, no wind effect is taken into account, but the mounting condition does. eq. (10) depends on the one k, which is known as the Ross coefficient [15]. In the literature models which take into account wind speed are proposed however without mounting configuration consideration [16]:

$$T_C = 0.943T_a + 0.028G_T - 1.528V_w + 4.3. \tag{11}$$

In this work the extended formula of eq.(10) but taking into account air velocity is proposed in the as follows:

$$T_C = T_a + kG_T + e^{(3.75-0.75V_w)}. \tag{12}$$

Proposed correlation takes into account ambient temperature, solar irradiation, wind speed as well as a mounting configuration including PV modules arrangement. The model correlation eq.(12) accuracies have been validated using: root mean squire error (RMSE), the mean bias error (MBE) and the coefficient of determination (R^2):

$$RMSE = \sqrt{\frac{1}{N}\sum_{i=1}^{N}\left(T_{cal,i} - T_{exp,i}\right)^2}, \tag{13}$$

$$MBE = \frac{1}{N}\sum_{i=1}^{N}\left(T_{cal,i} - T_{exp,i}\right), \tag{14}$$

$$R^2 = \frac{\sum_{i=1}^{N}\left(T_{cal,i} - \bar{T}_{exp}\right)^2}{\sum_{i=1}^{N}\left(T_{exp,i} - \bar{T}_{exp}\right)^2}, \tag{15}$$

where N indicates a number of measurements, while $T_{cal,i}$ and $T_{exp,i}$ are obtained from the model equation or measured respectively. Table 2 shows the proposed model's accuracy indicators. As can be seen, all indicators show very good accuracy of the proposed model.

Table 2. The models accuracies statistical indicators.

Model	MBE	RMS	R^2
Proposed model eq. (12)	0.72	2.71	0.93

4 Conclusions

The operating temperature of the PV module panel is an important issue because it is directly linked with photovoltaic system efficiency. In the present work, the temperature distribution in the PV module is analysed using computer modelling and experimental measurement. Presented results show large temperature variations, and measured temperature for the white back sheet panel was more than 20°C lower than for other configurations which translate into at least an 8% decrease in module performance.

The analysis shows that module operating temperature depends not only on ambient temperature and solar irradiation what is taken into account by most of the models but also depends on wind speed and wind direction. It has shown that the mounting conditions also play a significant role and may cause an additional module temperature increase of about 10°C which to turn into further module efficiency decrease about (3.8-6.5) % for mono and polycrystalline modules (assuming (-0.38 - 0.42) %/°C).

The results enable the development of a model for the photovoltaic module operating temperature. The proposed model is very much sought after by the system designers, computer software and system malfunction detection tools and takes into account all basic phenomena parameters such as solar irradiation, ambient temperature, wind speed and the mounting conditions.

Acknowledgement

The photovoltaic installation was funded by the National Fund for Environmental Protection and Water Management, Grant 139/2015/Wn06/OA-is-ku/D. This work was partially financed by AGH UST Rector Grant 35/2018 and AGH University Grant No.11.11.210.312. The authors would like to thank MESCO Sp. z o.o. company to provide ANSYS software used in this computer simulation.

References

1. H. E. Adib, F. Murdock, A. Appavou, B. Brown, A. Epp, A. Leidreiter, C. Lins et al. Renewables 2016 Global Status Report. Global Status Report (2016)
2. R. Anis, M. Abdul-Sadek Nour, Energy Conversion and Management, **36**, 1107 (1995)
3. S.A. Kalogirou, R. Agathokleous, G., Energy, **51**, 439 (2013)
4. C.W.A. Baltus, J.A. Eikelboom, R.J.C. van Zolingen, R. J. C., 14th European Photovoltaic Solar Energy Conference, Barcelona, 30.06-4.07 (1997)
5. I. Santiago, D. Trillo-Montero, I. M. Moreno-Garcia, V. Pallarés-López, J. J. Luna-Rodríguez, Renewable and Sustainable Energy Reviews, **90**, 70 (2018)
6. B. Perović, D. Klimenta, M. Jevtić, M. Milovanović, Zbornik Međunarodne konferencije o obnovljivim izvorima električne energije–MKOIEE, **5**, 89 (2017)
7. A. A. Kendoush, International Journal of Thermal Sciences, **48**, 188 (2009)
8. H. A. Zondag, D. W. De Vries, W. G. J. van Helden, R. J. C. van Zolingen, A. A. Van Steenhoven, Solar energy ,**74**, 253 (2003)
9. Q. Hassan, M. Jaszczur, E. Przenzak, E3S Web of Conferences, EDP Sciences, 01028 (2017)
10. I. Ceylan, O. Erkaymaz, E. Gedik, A. E. Gürel, Case Studies in Thermal Engineering, **3**, 11 (2017)
11. M. Jaszczur, J. Teneta, K. Styszko, Q. Hassan, P. Burzyńska, E. Marcinek, N. Łopian, Environmental Science and Pollution Research, 1-16. (2018)
12. Q. Hassan, M. Jaszczur, E. Przenzak, J. Abdulateef, Contemporary Problems of Power Engineering and Environmental Protection, 33 (2016)
13. M. Koehl, M. Heck, S. Wiesmeier, J. Wirth, Solar Energy Materials and Solar Cells, **95**, 1638 (2011)
14. X. Ju, A. Vossier, Z. Wang, A. Dollet, G. Flamant, Solar Energy, **93**, 80 (2013)
15. J. Zhou, Q. Yi, Y. Wang, Z. Ye, Solar Energy, **111**, 97 (2015)
16. R. Chenni, M. Makhlouf, T. Kerbache, A. Bouzid, A., Energy, **32**, 1724 (2007)

 https://doi.org/10.1051/matecconf/201824004005

Study of dust deposition and temperature impact on solar photovoltaic module

Marek Jaszczur[1,*], *Qusay* Hassan[1], *Janusz* Teneta[2], *Katarzyna* Styszko[1], *Weronika* Nawrot[1], and *Robert* Hanus[3]

[1]AGH University of Science and Technology, Faculty of Energy and Fuels, Poland
[2]AGH University of Science and Technology, Faculty of Electrical Engineering, Automatics, Poland
[3]Rzeszów University of Technology, Faculty of Electrical and Computer Engineering, 35-959 Rzeszów, Poland

Abstract. This work investigates the effect of the dust deposition and temperature on the photovoltaic module working under varying environmental conditions. The impact of the module temperature caused by the dust deposition, as well as direct impact caused by the dust particle absorption, scattering and reflectance on module performance, has been analysed. Presented results showed that dust deposition even in desert free country Poland influences the module temperature and power output significant. The estimated join effect of dust deposition and temperature increase, on power loss after three-four days of natural dust exposure in the city environment, was about 0.32 and 0.37 point of percentage for the selected sunny and cloudy days while evaluated PV system efficiency was in the range 13.37-14.14%.

1 Introduction

Observed since 2011 the World energy demand growth rate is about 1.8% per year [1]. This continuous growth deteriorates the Earth ecological system and causes global warming phenomenon as well as rise CO_2 emissions rate. The factors mentioned above motivate researchers to find alternative sources for conventional energy, in which the environmental aspect is taken into consideration. On the other hand the year 2016 has been recorded as the highest growth in the installation with the use of renewable energy [2].

Solar energy has been concerned as one of the most promising renewable energy resources with the World solar photovoltaic (PV) systems power cumulative capacity 301473 MW (in 2016) [3]. It is known that the performance of photovoltaic modules can be influenced by a large number of environmental and climatic factors such as incident irradiation, solar irradiation spectrum, temperature, wind speed, air pollution, tilt angle, snow and shadowing etc. [4].

One of the parameters that influence the energy conversion in PV modules, which can be a significant in many regions, is dust deposition. The dust particles deposited on the surface directly affect the transmittance of the solar collector, and photovoltaic modules lead to a substantial decrease in its performance [5]. There are number of studies in the literature which take into account the impact of the accumulative dust and environment conditions on the PV efficiency and energy production. Jamil et al. [6] presented a review of the research studies focusing on the effects of the environmental variables such as accumulative dust, wind speed, ambient temperature and humidity in addition to module tilt angle and orientation on the dust deposition on the PV modules. Mekhilef et al. [7] have done a similar review for effect of the dust, wind velocity, and relative humidity on the PV modules.

An experimental study carried out by Hassan et al. [8] was conducted under various environmental conditions in order to estimate PV modules energy conversion loss due to two independent effects: module operating temperature and natural dust deposition. A proposed new correlation allows to calculate the efficiency loss of the PV module caused by dust deposition, in addition to that authors obtained the module efficiency loss 4.6% and 5.8% at module temperature 30.9°C and 37.1°C respectively. Sarver et al. [9] prepared a review for the dust effect on PV performance for the different geographical region. Authors conclude that the duration of the dust deposition shows considerable variations depend on exposure time.

The module energy yield losses caused by dust deposition were studied by Sayyah et al. [10]. In that work authors concluded that the PV modules operate in different regions of the world, may require the additional cleaning processes including manual and automatic for assessing the most effective method for dust removing. The dust deposition accumulated on the PV modules surface depends upon two main factors: the local environmental conditions of the site (i.e. climate) and location with the orientation of the modules [11]. El-Shobokshy and Hussein [12] illustrate the effect of the chemical and physical properties of the dust particles on the PV modules performance. In their investigation authors used artificial dust: limestone, carbon and cement with different diameters of 5, 10, 50, 60, and 80

* Corresponding author: jaszczur@agh.edu.pl

https://doi.org/10.1051/matecconf/201824004005

μm. The particles were deposited on the PV module front glass surface with a mass density of about 25 g/m^2. Authors conclude that the finer particles have a more degrading effect on cell performance than coarse particles.

The natural dust deposition on the PV modules front cover glass was experimentally analysed in [13] by means of several PV modules during exposure periods from 1 day up to 1 week. The results showed that dust deposition intensity on the module surface ranged from 7.5 to 42.1 mg/m^2 for exposure periods of one day while the measured dust deposition for one week ranged from 25.8 to 277.0 mg/m^2. The results also showed that the intensity and volume of precipitation in addition to the air humidity has a high impact on the dust particle deposition.

Kaldellis and Kokala [14] evaluated the effect of the air pollution on the photovoltaic module energy yield. The analysis was done based on the performance of five pairs of PV modules located on the building roof and operated in the urban environment of the city of Athens. The reported results showed that even a small amount of dust deposition about 1 g/m^2 could lead to energy losses corresponding almost 4% on an annual basis. Cabanillas and Munguia [15] in 20 days outdoor studies performed in city Hermosillo, Mexico investigate the dust effect on the PV module generated power. The authors showed that the dust deposition decreased the maximum power of a-Si, mono-Si, and poly-Si PV modules by 14%, 8.5%, and 5.2%, respectively. The type of glass used on the front surface of the PV modules has a significant impact on the amount of accumulated dust.

Using the textured glass, module performance will increase due to a decrease in reflection losses at the same time this reduces the rate of dust accumulation [16]. The comparison of two different textured glass cover for the monocrystalline modules with 30° tilted angle performed in order to the evaluated rate of accumulative dust was done by Piliougine et al. [17]. For the one year exposition, results show some difference in dust soiling between the textured glass and nontextured glass surfaces. In a study done in three different sites in the desert southern California, US by Caron and Littmann [18] investigate also PV modules energy losses due to the accumulative dust. The obtained maximum losses recorded with a fewer rainfall period were about 8.6%. Based on the experimental work for PV energy losses due to soiling for three PV systems Kimber [19] validate a mathematical model for prediction energy losses.

It has been seen from the literature review that the dust deposition can significantly influence PV module performance. However, the studies were performed mainly for artificial dust or in the very dusty region. Most of the investigations presented in the literature were conducted in arid and semi-arid areas.

Only a few studies deal with natural dust. The key point which may be drawn from a review is that there is no theoretical or empirical model for PV module efficiency loss due to dust which is applicable to all the types of pollutants and different types of PV. Additionally, the models represent specific and particular cases of a narrow range of functionality. On the other hand, none of the work is taking into consideration at the same time effect of dust deposition and operating temperature and still it is not clear if deposited dust increases or decrease PV module operating temperature. For this reason, the objective of this work is to study - based on the experimental measurement - the impact of the temperature and natural dust deposition on the PV modules performance.

2 Experimental and numerical set-up

The experimental measurement for the dust deposition was conducted using two types of photovoltaic modules; the first type (11 pcs.) was polycrystalline BrukBet BEP260W module (temperature coefficient of power -0.40 %/°C and 1.62 m^2 of surface area, white colour back sheet), the nominal module power is equal to 260W, and orientated at tilt angle β=35°. The second type (2 pcs.) was a polycrystalline Sharp ND-RJ260 photovoltaic module (temperature coefficient of power -0.42 %/°C and 1.6 m^2 of surface area, white colour back sheet orientated at tilt angle β=15°). The azimuth for all PV modules was the same γ=20° West and all modules were located in building C3 AGH University of Science and Technology in the city centre of Kraków, Poland (50.066169 N, 19.921547 E).

Fig. 1. The polycrystalline Sharp ND-RJ260 PV modules: front view with collected dust sample (a) and backside view with thermocouples arrangement (b).

The experimental set-up for the efficiency analysis consists of two described above Sharp ND-RJ260 photovoltaic modules connected to the Envertech EVT500 microinverter. The analysed modules electrical

 https://doi.org/10.1051/matecconf/201824004005

parameters (AC and DC) were acquired with 3-minutes temporal resolution using EnverBridge. The solar irradiation (global horizontal - GHI and diffuse horizontal – DHI components), ambient temperature Ta and other environmental factors were obtained from the local sensors (Pyranometers, Thermometer), at location AGH University of Science and Technology in the city centre of Kraków, Poland.

The main hardware components of the data logging system for temperature and solar irradiation measurement include four Advantech ADAM4018, 16-bit, an 8-channel analogue input module that provides programmable input ranges on all channels, RS485 transceiver, computer unit with data acquisition software and storage. As a temperature sensors laboratory, calibrated (using certified 100-TP thermometer accuracy ±0.02°C) T-type thermocouples (Copper-Constantan) were used. Four thermocouples were attached to each analysed PV modules backsheet in the recommended by standard arrangement (see Fig. 1b).

The signals from all temperature sensors were acquired every second, and the average value was recorded every 60 seconds.For the present analysis of dust mass deposition, the naturally deposited dust was collected from all modules at morning hours at about 9 a.m. with different daily schedule in order to get different exposure period from 1 day up 7 days while for PV module performance analysis the dust from polycrystalline Sharp ND-RJ260 photovoltaic two modules were used olny. The right Sharp module was cleaned at the afternoons hours at 3:00 p.m., while the left (reference see Fig. 1a) module was cleaned at the morning hours at 9:00 a.m.

The deposited dust particles were removed using deionised water and a scratcher with suction pump [13]. The water with dust was collected in plastic containers, followed by freeze-drying (Liophilizator Alpha 1-4 LD). Dry mass of deposited dust was evaluated with laboratory class balance (OHAUS Discovery DV215CD). The PV module instantaneous power, P_{PV} depends on the solar irradiance incident on the PV surface G_T, but it also depends on several additional factors such as dust and soiling, losses, shading, age, temperature and can be calculated from the following equation [20]:

$$P_{PV} = \eta A_c G_T\left[1 + \alpha_P\left(T_c - T_{c,STC}\right)\right] \quad (1)$$

where η is the PV module efficiency, G_T is the incident solar irradiation, α_P is the PV thermal coefficient of power, Tc is the PV module operating temperature, $T_{c,STC}$ is the cell temperature at Standard Test Condition and A_C is the module area. When incident solar irradiation G_T, module operating temperature T_c and module power P_{PV} are known from the measurement than eq. (1) can be used to calculate the PV module efficiency η taking into account temperature effect or not as follows:

$$\eta_p = \frac{P_{PV}}{A_c G_T\left[1+\alpha_P\left(T_c - T_{c,STC}\right)\right]} \quad (2)$$

$$\eta = \frac{P_{PV}}{A_c \cdot G_T} \quad (3)$$

where η_p is the PV module efficiency taking into account PV module operating temperature eq. (2) and η is the PV efficiency without temperature consideration eq. (3)

3 Results and discussion

The dust samples were collected from all modules using the different exposure period from one day up to seven days for three consecutive weeks and were the subject of analysis. Dust deposition for modules with title angle β=15°, β=35° are presented in Figures 2. Smaller title angle promotes substantial dust particle deposition.

Figure 3 shows the instantaneous solar irradiation - global horizontal irradiation (GHI) and diffuse horizontal irradiation (DHI) components for two selected days: cloudy (04.06.2018) and sunny (07.06.2018) respectively. For the sunny day (Fig. 3a), dust particles were deposited for four days while for the cloudy day (Fig. 3b) exposition time was three days.

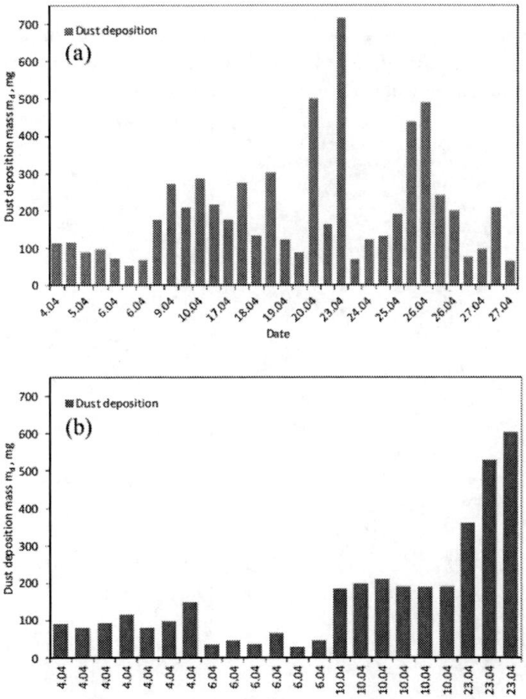

Fig. 2. Dust deposition on the PV module surface with a titled angle: β=15° (a) and β=35° (b).

https://doi.org/10.1051/matecconf/201824004005

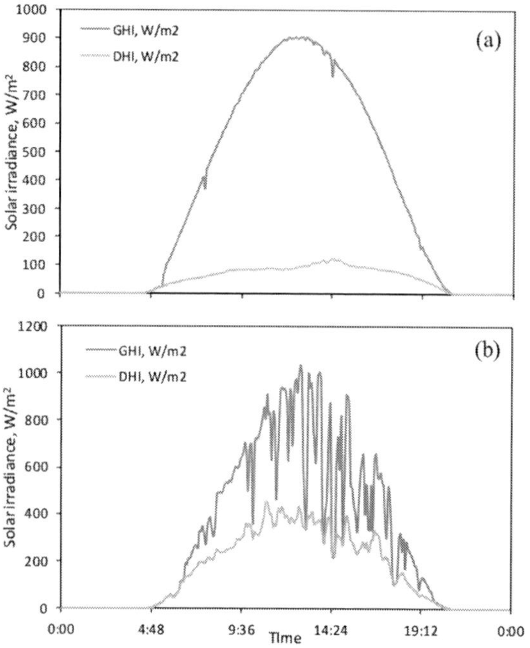

Fig. 3. Instantaneous solar irradiation components for: sunny (a) and cloudy (b) days.

Figure 4 show the instantaneous PV module temperature T_c for clean N1 and dusty N2 module and for two selected days: cloudy and sunny respectively.

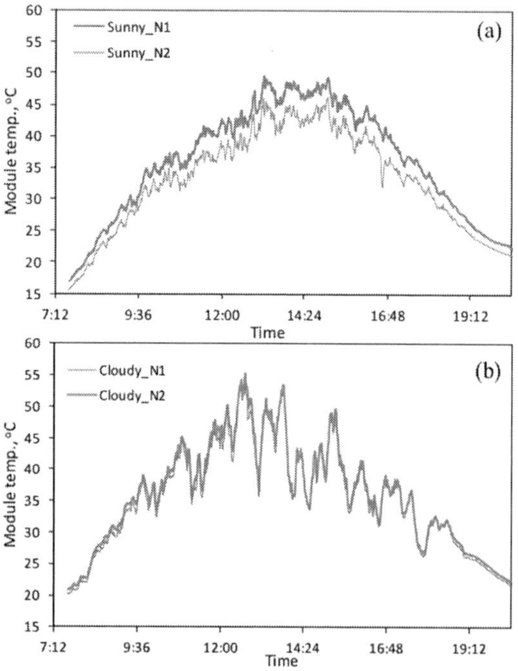

Fig. 4. PV module temperature T_c for sunny (a) and cloudy (b) days.

In order to calculate PV module efficiency incident solar irradiation and electrical power from the PV system for a selected period is required. The PV system 10 minutes averaged AC electrical power is presented in Figures 5 for the cloudy (04.06.2018) and sunny (07.06.2018) days. The results show that the power generated by the system with the clean module is slightly higher than from unclean one.

In Figure 6 module efficiency η is presented calculated with eq. (3). One may infer from this figures that the clean PV module efficiency η is higher than from unclean one. This is observed for sunny (Fig. 6a) as well as (Fig. 6b) cloudy days. However, it is also worth to notice that for the sunny day almost continuous efficiency decreases during the day was observed. This is better seen when PV system power losses (absolute value in percentage points) or percentage power loss due to dust deposition and temperature for cloudy and sunny days respectively are calculated from as follows:

$$\Delta_{\eta_{loss}} = \eta_c - \eta_d \tag{4}$$

$$\delta_{\eta_{loss}} = \frac{\eta_c - \eta_d}{\eta_c} \cdot 100\% \tag{5}$$

where η_c, η_d represent the efficiency of the clean and dusty modules calculated using the eq. (2) or eq. (3) for the clean or dusty modules data. In Figure 7 photovoltaic module power losses are presented. For sunny day an increase in the power losses is observed up to 2 p.m. For cloudy day continuous increase is seen, however, at 11:00 a.m. negative efficiency loss is observed.

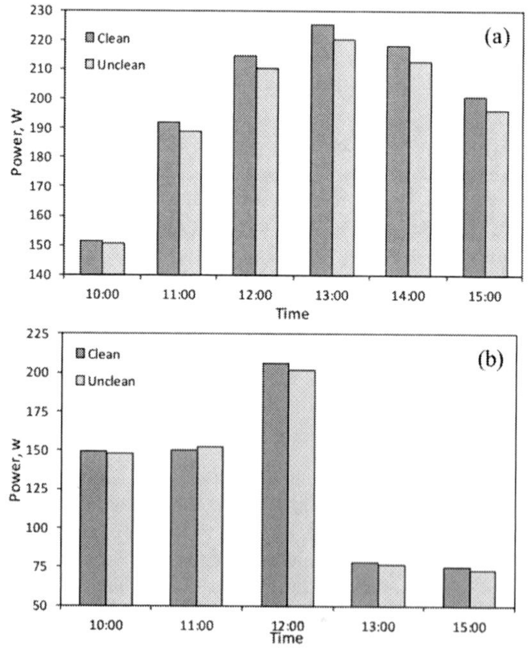

Fig. 5. Photovoltaic module 10-minutes averaged electrical power for: sunny (a) and cloudy (b) days.

The effect is due to very high solar irradiation fluctuation which creates difficulties for accurate

https://doi.org/10.1051/matecconf/201824004005

maximum power point (MPPT) measurement simultaneously at two different analysed photovoltaic modules. The observed maximum efficiency losses $\Delta_{\eta_{loss}}$ are about 0.32 and 0.37 point of percentage for selected sunny and cloudy day respectively.

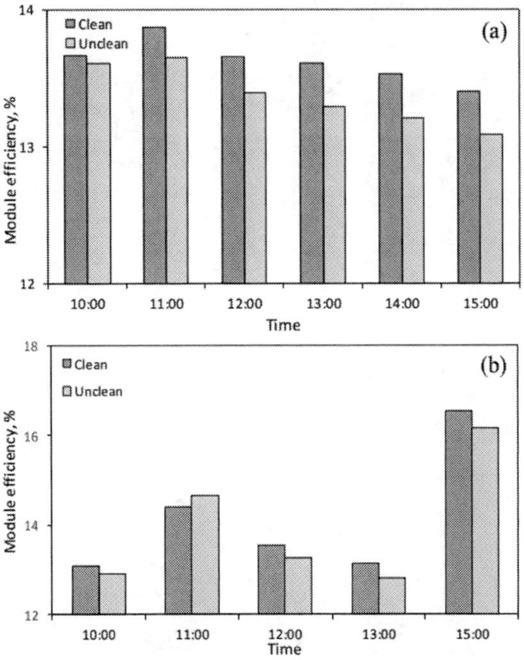

Fig. 6. Photovoltaic module 10-minutes averaged efficiency at a selected time and for: sunny (a) and cloudy (b) days.

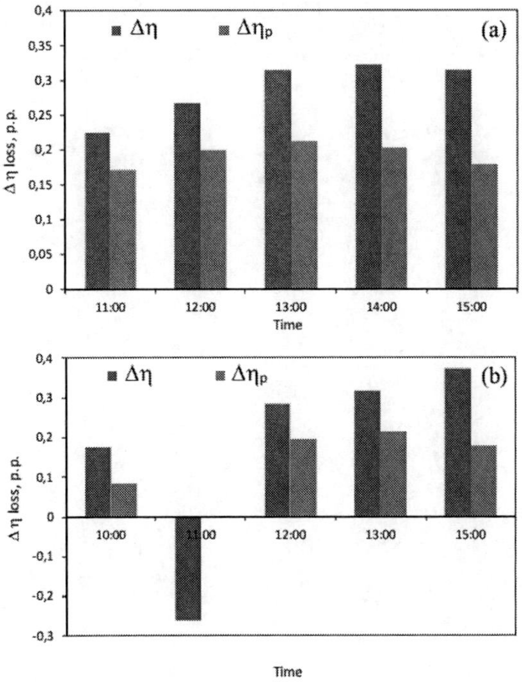

Fig. 7. Photovoltaic module efficiency loss (in percentage points) at a selected time and for: sunny (a) and cloudy (b) days

In Figure 7 two different efficiency losses are presented. The first one is the total efficiency loss $\Delta\eta$ calculated with eq. (3,4) and it is related to dust deposition and to PV module operating temperature difference (in reference to the clean module) caused by the dust deposition.

While the second efficiency loss $\Delta\eta_p$ is the efficiency loss which takes into account the only direct dust deposition effect excluding temperature increase due to dust at the module surface. In that case, the efficiency of the clean and the dusty modules are calculated from the eq. (2,4) and takes into account the clean and unclean module operating temperature T_c.

One may infer from Figures 7 that efficiency loss excluding dust-temperature related phenomena is 24-43% lower. This confirmed that direct dust deposition effect which causes reduction of solar irradiation intensity that can reach the photovoltaic module surfaces has a dominant effect. In another word absorption, scattering and reflectance of the incident solar radiation due to particle layer at the PV module from surface plays a more important role than dust related indirect thermal effect (module temperature increase due to dust particle deposition), however, the secondary effect for the presented cases is not negligible.

4 Conclusions

This study investigates the impact of dust deposition and the operating temperature on the photovoltaic module performance working under varying environmental conditions.

The focus is on the evaluation of the PV modules energy conversion loss due to two effects: one related to absorption, scattering and reflectance of the incident solar radiation due to dust layer at the PV module, and the second one related to the indirect thermal effect - module temperature change due to dust particle deposition.

In the experimental study conducted under variable environmental conditions, it was seen that the PV module power output is influenced by the dust deposited particles and the effect is non-negligible even for as short as three-four days of natural dust deposition in desert free country Poland.

The maximum observed efficiency loss equal to 0.32 and 0.37 point of percentage (which results in the power losses of about 2.39% and 2.26% see eq. (5)) and was observed for presented sunny and cloudy days. Results shows that efficiency loss excluding dust-temperature related phenomena is 24-43% lower which confirmed that direct dust deposition effect has a dominant effect.

On the basis of the presented analysis, one may state that the particles which exist in the air of urban areas negatively affect the PV module efficiency and the level of power losses is non-negligible. The efficiency loss gradually increased with the dust deposited mass which was presented in another authors work [5] together with the mathematical model of this phenomena.

Acknowledgements

The photovoltaic installation was funded by the National Fund for Environmental Protection and Water Management, Grant 139/2015/Wn06/OA-is-ku/D. This work was partially financed by AGH University Grant No.11.11.210.312 and No. 15.11.210.436.

References

1. L. Sawin at al. Renewables, Global Status Report, (2017)
2. Renewable energy Mid-Term market report, https://www.iea.org, (2016)
3. British Petroleum Statistical Review of World Energy, http://www.bp.com/statisticalreview, (2017)
4. Q. Hassan, M. Jaszczur, E. Przenzak, J. Abdulateef, *Contemporary Problems of Power Engineering and Environmental Protection*, **33**. (2016)
5. M. Jaszczur, J. Teneta, K. Styszko, Q. Hassan, P. Burzyńska, E. Marcinek, N. Łopian, *Environmental Science and Pollution Research, 1-16.* (2018)
6. J. Jamil, A. Rahman, S. Shaari, *Renewable and Sustainable Energy Reviews* **67** (2017)
7. S. Mekhilef, R. Saidur, M. Kamalisarvestani, *Renewable and Sustainable Energy Reviews*. **16** (2012)
8. Q. Hassan, M. Jaszczur, K. Styszko, J. Teneta, M. Dudek, *Thermal Science Journal* (2018)
9. T. Sarver, A. Qaraghuli, L. A. Kazmerski, *Renewable and Sustainable Energy Reviews*. **22**, (2013)
10. A. Sayyah, N. Horenstein, K. Mazumder, *Solar Energy* **107**, 576 (2014)
11. M. Mani, R. Pillai, *Renewable and Sustainable Energy Reviews* **14**,3124 (2010)
12. S. El-Shobokshy, M. Hussein,.*Solar energy* **51**, 505 (1993)
13. K. Styszko, M. Jaszczur, J. Teneta, Q. Hassan, P. Burzyńska, E. Marcinek, L. Samek, *Environmental Science and Pollution Research*, 1 (2018)
14. K. Kaldellis, A. Kokala, *Energy,* **35,** 4862 (2010)
15. E. Cabanillas, H. Munguía, *Journal of Renewable and Sustainable Energy*, **3,** 043114 (2011)
16. A. Nositschka, D. Neumann, O. Prast, F. Gromball, *In 22nd European Photovoltaic Solar Energy Conference* (2007)
17. M. Piliougine, J. Carretero, M. Sidrach, D. Montiel, P. Sánchez-Friera, *In Proceedings of 23rd European Solar Energy Conference* (2008)
18. R. Caron, B. Littmann, *IEEE Journal of Photovoltaics,* **3,** (2013)
19. A. Kimber, *In Proceedings 22nd European Photovoltaic Solar Energy Conference* (2007)
20. Q. Hassan, M. Jaszczur, E. Przenzak, *In E3S Web of Conferences (Vol. 14, p. 01028). EDP Sciences,* **14,** 01028 (2017)

Tilt optimization of a double-glazed air solar collector prototype

Hocine Mzad[1,*], *Abdessalam* Otmani[1], *Abdallah* Haouam[1], *Stanisław* Łopata[2], and *Paweł* Ocłoń[2]

[1]Department of Mechanical Engineering, Badji Mokhtar University of Annaba, P.O. Box 12, DZ-23000, Annaba, Algeria
[2]Institute of Thermal Power Engineering, Cracow University of Technology, Poland

Abstract. To understand usage of solar air collector there are factors that have influence on the performance. The goal of this paper is to investigate a solar air collector prototype in order to perform a performance optimization using the possible minimum of equations. The thermal efficiency, fluid outlet temperature, heat increase and heat losses of the collector are calculated depending on the collector geometry, fluid properties, fluid inlet temperature, air flow rate, solar insulation and ambient temperature. The calculations were performed using actual data of temperature, wind speed and the global-horizontal radiation H_h over the year 2011 at the international airport of Annaba city. Comparison of results reveals that optimal efficacy is obtained for south facing panels with an inclination angle β comprised in the interval [15°, 35°]. These conditions provide a useful energy exceeding 4300 W and an efficiency of about 51 %.

Keywords: Global irradiation, Panel tilt, Measured data, Optimization, Efficiency.

1 Introduction

The main advantages of solar air collectors are: the working fluid does not freeze or boil, they have noiseless, safely and low cost operation, they do not produce any kind of wastes and have long life cycle [1]. The solar air collectors have the following disadvantages: low density, low thermal capacity and small heat conductivity of air which lead to low thermal efficiency, high cost of production and installation and non-uniform flow [2].

The utilizability concept was originally developed by Klein and Beckman [3] for estimating the performance of flat-plate solar collectors over long time periods. In this application, low critical radiation levels are desirable in that they result in high utilizability values and thus efficient collection of solar energy.

The effect of parallel barriers on the collector efficiency of flat-plate solar air heaters has been investigated theoretically and experimentally. The theoretical predictions agree reasonably well with experimental results, Yeh and Lin [4].

In the same way, Abene et al. [5] conducted an experimental study, which consists of a solar energy simulation; they have sought to improve the efficiency temperature rise couple of the flat plate solar collector by considering several types of obstacles disposed in rows in the dynamic air vein of the flat collector.

Mzad [6] presents an approach to determine thermal parameters such as fluid temperature and efficiency of a double-glazed solar collector tested in north-western part of Algeria based on meteorological parameters collected during the year 2006.

Transient simulation program is used by Erbs et al. [7] to calculate the annual performance of solar energy systems using several correlations. A seasonally dependent daily diffuse correlation is developed from the data, and this daily relationship is used to derive a correlation for the monthly average diffuse fraction.

Comparisons between the optimal tilt angle of a collector based on the predicted monthly diffuse radiation and that based on the actual monthly diffuse radiation carried out by Tang and Wu [8], showed that the method in this paper gives a good estimation of the optimal tilt angle, except for the areas with a lower clearness index.

Ahmad and Tiwari [9] examines the theoretical aspects of choosing a tilt angle for the solar flat plate collectors used at ten different stations in the world and makes recommendations on how the collected energy can be increased by varying the tilt angle.

In the present work a mathematical procedure of calculation was performed. Based on analytical

*Corresponding author: h_mzad@yahoo.fr

formulations and empirical correlations using measured data, the study of a solar air collector was conducted to determine its performances under specific conditions. In this study we try to estimate the useful energy and the efficiency in terms of tilts and orientations of the tested solar air panel in order to achieve best desired results.

2 Data and solar air heater specifications

Weather is one of the meteorological data that is rich by important knowledge. The considered site is located in the North-Eastern corner of Algeria at 36° 54' North latitude and 7° 45' East longitude. The city benefits from a Mediterranean climate known for its long dry hot summers.

The wind speed hitting the solar collector has a significant influence on the estimated energy delivered by the solar air panel (Fig. 1).

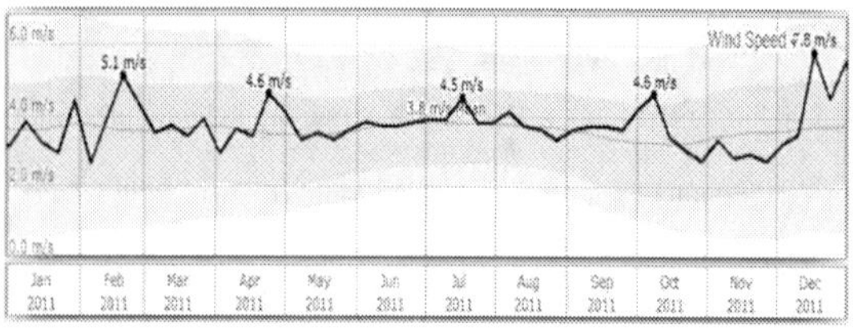

Fig. 1. Yearly average daily wind speed.

In the present studied case, the measured temperatures and solar radiation (Tab. 1) recorded over the year period of 2011 for the city of Annaba were obtained from the Algerian national meteorological office database. The monthly average amount of the total solar radiation incident on a horizontal surface for a given month is evaluated as the numerical average of 3-hourly values for the given month. The primary disadvantage of solar air heaters is the need for handling relatively large volumes of air with low thermal capacity as working fluid. They utilize both beam and diffuse solar radiation for temperatures below 60 °C, and do not require sun tracking. The direct use of air as the working substance reduces the number of required system components. Specifications of the tested solar air collector are summarized in table 2.

Table 1. Climate data for Annaba site, year 2011.

Months	$T_{daily\ mean}$ [°C]	H_h [MJ/m²/day]
Jan/Feb	11.8	2.6
Mar.	13.4	4.53
Apr.	15.2	5.51
May	18.4	6.69
Jun/Jul	23.4	7.6
Aug.	25.7	7.11
Sep.	23.6	5.28
Oct.	20.4	3.93
Nov/Dec	14.4	2.3

In order to minimize the heat losses, a double glazing with the thickness of 3.5 mm were recommended for the flat plate solar air collector under consideration. The absorber made by steel plate with an area of (2×1) m² and thickness of 0.5 mm was covered by black paint (Fig. 2).

Table 2. Solar air heater characteristics.

Parameter	Value
Absorber emissivity, ε_a	0.1
Absorber mean temperature, $T_{p,m}$	50 °C
Air-mass flow rate, $\dot{m}$	0.044 kg/s
Air specific heat, C_p	1005 J/(kg·K)
Albedo of ground, ρ	0.2
Ambient temperature, T_a	28 °C
Collecting area, A_C	2 m²
Collector efficiency, F'	0.75
Collector side area, A_e	0.4 m²
Glass emissivity, ε_g	0.88
Insulator thermal conductivity, k	0.025 W/(m·K)
Inlet air temperature, $T_{f,i}$	11÷26 °C
Optical efficiency, η_o	0.8
Stephan-Boltzmann constant, σ	5.7×10^{-8} W/(m²·K⁴)
Thickness of bottom insulation, L	0.06 m
Thickness of side insulation, L'	0.03 m

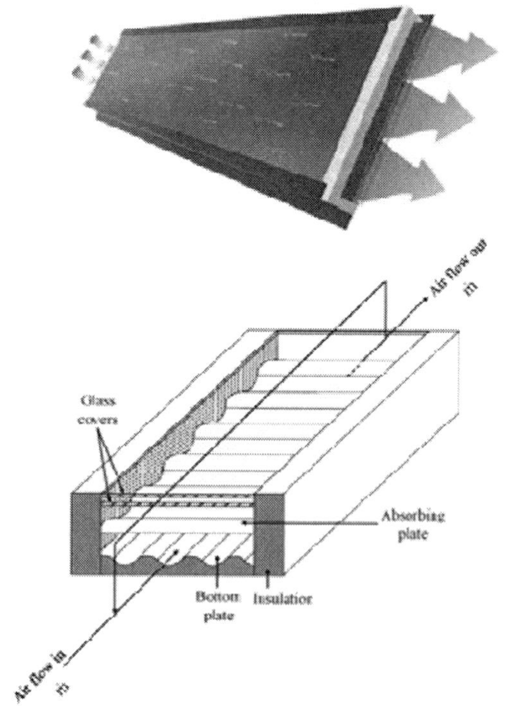

Fig. 2. Air solar collector description.

3 Overriding equations

The efficiency of a solar collector is defined as the ratio of the amount of useful heat collected to the total amount of solar radiation striking the collector surface during any period of time

$$\eta_C = \frac{Q_U}{A_C I_T} \quad (1)$$

In this section various relations that are required in order to determine the useful energy collected and the interaction of the various constructional parameters on the performance of a collector are presented.
The useful energy collected from a collector can be obtained from the following formula

$$Q_U = A_C\left[I_T\eta_O - U_G\left(T_{P,m} - T_a\right)\right] = \dot{m}C_P\left(T_{f,o} - T_{f,i}\right) \quad (2)$$

Eq. (2) can be modified by substituting inlet fluid temperature for the average plate temperature, if a suitable correction factor is included. The resulting equation is

$$Q_U = A_C F_R\left[I_T\eta_O - U_G\left(T_{f,i} - t_a\right)\right] \quad (3)$$

The correction factor F_R, can be considered as the ratio of the heat actually delivered to that delivered if the collector plate were at uniform temperature equal to that of the entering fluid. F_R is affected only by the solar collector characteristics, the fluid type and the fluid flow rate through the collector, Duffie and Beckman [10]

$$F_R = \frac{\dot{m}C_P}{A_C U_G}\left[1 - \exp\left(-\frac{A_C U_G F'}{\dot{m}C_P}\right)\right] \quad (4)$$

where F' is the collector efficiency.

The overall heat loss coefficient of the collector (U_G) is the sum of the top, bottom and edges losses coefficients. Following the basic procedure of Hottel and Woertz [11], Klein [12] developed an empirical equation for the collector forwards heat loss coefficient as

$$U_t = \frac{1}{\dfrac{N_g}{\dfrac{C}{T_{P,m}}\left[\dfrac{T_{p,m} - T_a}{N_g + f}\right]^e} + \dfrac{1}{h_w}} + \frac{\sigma\left(T_{p,m} + T_a\right)\left(T_{p,m}^2 + T_a^2\right)}{\dfrac{1}{\varepsilon_a + 0.00591 N_g h_w} + \dfrac{2N_g + f - 1 + 0.133\varepsilon_a}{\varepsilon_g} - N_g} \quad (5)$$

Parameters in the above equation are correlated as follows

$$f = \left(1 + 0.089h_w - 0.1166h_w\varepsilon_a\right)\left(1 + 0.07866N_g\right) \quad (5.1)$$

$$C = 520\left(1 - 0.00005\beta^2\right) \quad (5.2)$$

$$e = 0.43\left(1 - \frac{100}{T_{p,m}}\right) \quad (5.3)$$

A convective heat transfer coefficient, McAdams [13], for the air flowing over the outside surface of the glass cover depends primarily on the wind speed (Fig. 1)

$$h_w = 5.7 + 3.8\,V_w \quad (5.4)$$

The collector backwards heat loss coefficient for a unit surface is

$$U_b = \frac{k}{L} \quad (6)$$

The edges heat loss coefficient is given by

$$U_e = \frac{k}{L'}\left(\frac{A_e}{A_c}\right) \quad (7)$$

On the other hand, the total radiation on a tilted surface per hour is calculated using the following equations from Duffie and Beckman [10]

$$I_T = I_b R_b + I_d\left(\frac{1 + \cos\beta}{2}\right) + I_h\rho\left(\frac{1 - \cos\beta}{2}\right) \quad (8)$$

where R_b is the dimensionless ratio of solar beam radiation on a tilted surface to that on a horizontal surface

$$R_b = \frac{\cos(\varphi - \beta)\cos\delta\cos\omega + \sin(\varphi - \beta)\sin\delta}{\cos\varphi\cos\delta\cos\omega + \sin\varphi\sin\delta} \quad ...(9)$$

The solar declination according to the Julian day n of the year is

$$\delta = 23.45\sin\left(\frac{360(284 + n)}{365}\right) \quad (10)$$

The hourly angle from the local meridian (ω) varies at 15 degrees per hour, negative in the morning positive the afternoon.

According to Liu and Jordan [14], the hourly diffuse radiation

$$I_d = \frac{\pi}{24}\left[\frac{\cos\omega - \cos\omega_s}{\sin\omega_s - \left(\dfrac{\pi\omega_s}{180}\right)\cos\omega_s}\right]H_d \quad (11)$$

The hourly total solar radiation falling on a horizontal surface [15] results from the sum of hourly beam and diffuse radiations

$$I_h = I_b + I_d$$

$$= \frac{\pi(a + b\cos\omega)}{24}\left[\frac{\cos\omega - \cos\omega_s}{\sin\omega_s - \left(\frac{\pi\omega_s}{180}\right)\cos\omega_s}\right]H_h \quad (12)$$

a and b are empirical functions, expressed as

$$a = 0.4090 + 0.5016\sin\left(\omega_s - \frac{\pi}{3}\right) \quad (12.1)$$

$$b = 0.6609 - 0.4767\sin\left(\omega_s - \frac{\pi}{3}\right) \quad (12.2)$$

The world radiation centre assumes for the solar constant I_o the value of 1367 W/m^2 with an uncertainty of the order of 1 %. The equation below of the average sun irradiation outside the earth's atmosphere, is expressed as

$$I_{on} = I_o\left[1 + 0.033\cos\left(\frac{360n}{365}\right)\right] \quad (13)$$

The correlation of Erbs et al. [7] is used to evaluate the daily diffuse fraction

$$\frac{H_d}{H_h} = 1.0045 + 0.04349K_t - 3.5227K_t^2 + 2.6313K_t^3 \quad (14)$$

K_t is an hourly clearness index and can be defined as the ratio of the total solar radiation to the extraterrestrial radiation for that specific hour

$$K_t = \frac{H_h}{H_o} = \frac{H_b + H_d}{H_o} \quad (15)$$

where:

$$H_o = \frac{86400}{\pi} I_{on}\cos\varphi\cos\delta\left(\sin\omega_s - \frac{\pi\omega_s}{180}\cos\omega_s\right) \quad (16)$$

ω_s is the sunset hourly angle, given as

$$\omega_s = arc\cos(-\tan\varphi\tan\delta) \quad (17)$$

K_t is estimated for the monthly typical day (Fig. 3). The day-type of the month is the day for which the declination is closest to the average declination of this month.

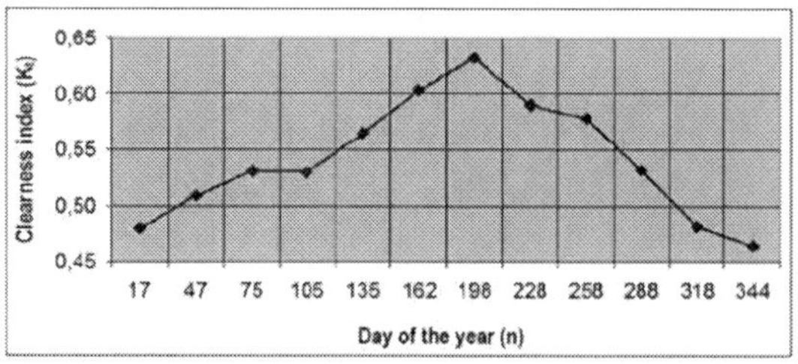

Fig. 3. Clearness index year evolution for typical days.

4 Results and analysis

A wide collector tilt range was used in function of different south orientations. The Following graphs were drawn by using average values of radiation for each month.

The comparison shown in figure 4 between daily solar radiation on earth and extraterrestrial radiation on a horizontal surface, confirm that the solar radiation before entering earth atmosphere is integral, this correspond to the highest curve. Thereafter, the attenuation occurring is due to reflexion, diffusion and absorption by gases and particles present in earth atmosphere. Sometimes, the diminution reaches 50 % of the total value of extraterrestrial radiation.

Curves plotted in figure 5 are similar to above ones, they illustrate the hourly total solar radiation and its components diffuse and beam, on a horizontal surface. Peaks are noticed during summer for total and beam radiation, however diffuse radiation is constant almost the year.

In order to calculate the collector efficiency, a heat balance is used in which the absorbed solar-ray needs to be calculated. Detailled behaviour of the ratio of beam radiation on the tilted surface to that on horizontal surface (R_b) is studied in figure 6, where it is clearly shown the same profile evolution of R_b function. During winter it is very high as the sun is low on the sky. It can cause high results during winter periods.

Yearly useful power evolution is presented on figure 7 where a similarity of curves profile can be seen. Tilts between 15° and 30° provide the best useful power, the maximum is reached in August with a values exceeding 4300 Watt when the air solar collector is facing south. It is clearly visible for tilts upper to 45° the decrease in the intensity of the collector delivered power. However, tilting angles from 30° to 45° are even acceptable for lower azimuth angles. Globally, the overall power decreases when the azimuth angle (γ) increases, modifying the curves by increasing the gap between them.

The figure 8 shows how change the efficiency across each month. This efficiency is higher during the summer months and lower during the winter months because during the summer the energy that arrives to the solar collector is higher than winter. In summer the intensity of the energy that arrives to the solar collector is higher and longer and this is the most important reason of this difference. During summer the energy transferred to air-flow is bigger as the air is at low temperature and gains much heat from the sun. The sun radiation is more stable, so the efficiency is similar at this time of year. Regardless of the orientation, it is noticeable that for $\gamma = 0°$ and 22.5° the collector efficiency is approximately constant over a long period, from March to October.

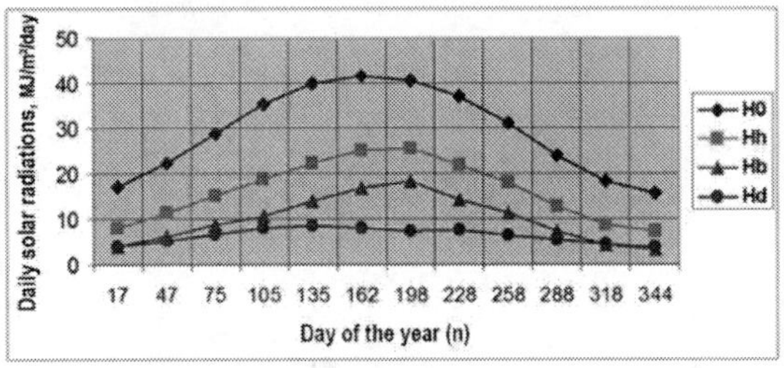

Fig. 4. Daily solar radiations for typical days.

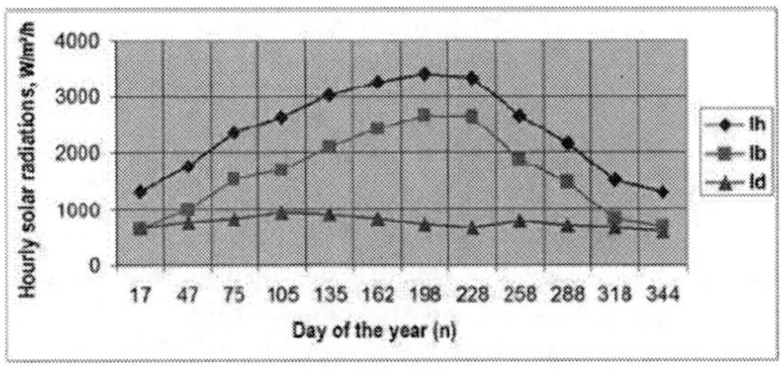

Fig. 5. Hourly solar radiations estimation.

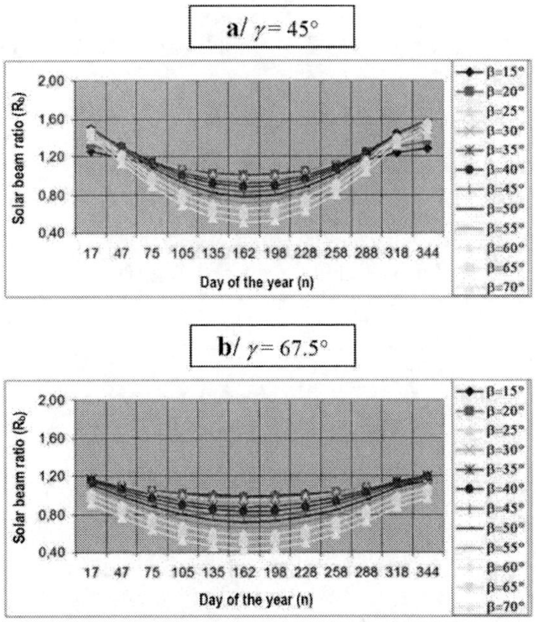

Fig. 6. Ratios of beam radiation on a tilted surface to that on a horizontal surface (a, b).

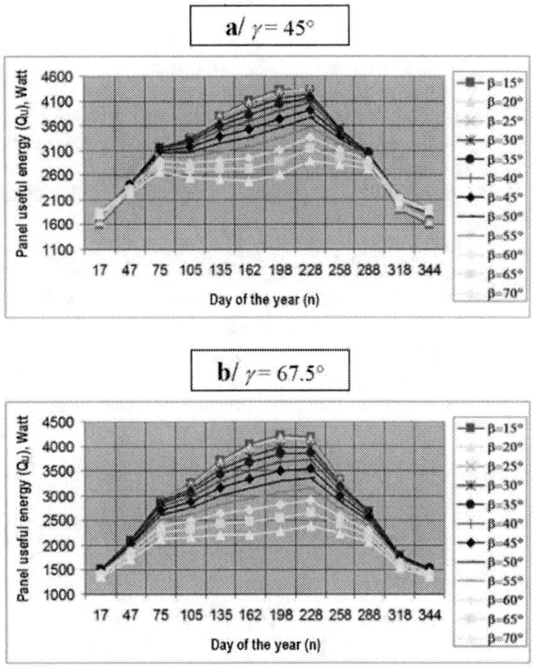

Fig. 7. Yearly air solar heater useful power (a, b).

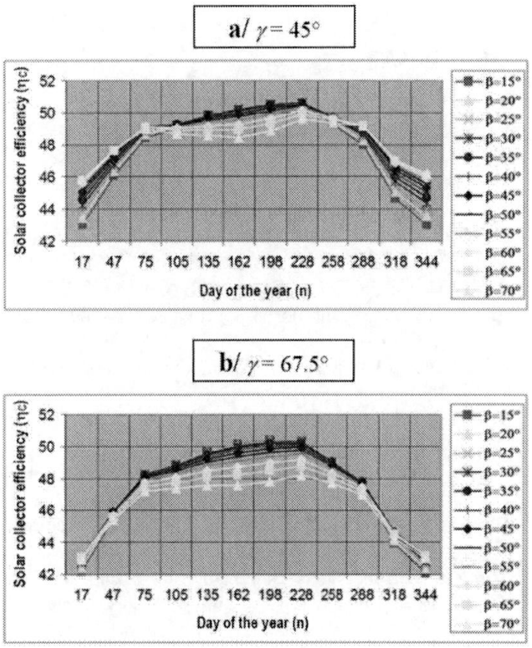

Fig. 8. Yearly collector efficiency (a, b).

5 Conclusions

Knowing the measured global-horizontal radiation H_h, a numerical calculation was performed to study and analyze the influence of tilt and orientation on the thermal performance of a solar air flat plate collector with a double glass cover.

Solar collectors need to be inclined at the optimum angle to maximize the receiving energy. The purpose of the paper was the optimisation of the solar collector position within a wide range of tilts compared with a given orientation.

The yearly clearness index evolution seems to be a parabolic function with the maximum point recorded in July. Obviously, K_t goes down during months when weather is bad. Increasing the tilt has the effect of decreasing the useful energy absorbed, however lower tilts are recommended if solar collectors cannot be oriented south.

However, the system designer may not necessarily be interested in an annually optimal solar collector orientation, but rather, an optimal orientation at a specific time period where there is an energy production deficit. This study can be expanded to determine optimal collector orientations for specific time periods of the day and season.

References

1. A.S. Abdullah, M.K. Bassiouny, Energy Conversion and Management **88**, 88-95 (2014)
2. H.F. Öztop, M.M. Rahman, A. Ahsan, M. Hasanuzzaman, R. Saidur, K. Al-Salem, N.A. Rahim, Int. J. Heat Mass Transfer **55**, 1844-1854 (2012)
3. S.A. Klein, W.A. Beckman, J. Solar Energy Eng. **106**, 393-402 (1984)
4. H.M. Yeh, T.T. Lin, Energy, **21**, 6, 435-443 (1996)
5. A. Abene, V. Dubois, M. Le Ray, A. Ouagued, J. Food Eng. **65**, 15-22 (2004)
6. H. Mzad, Archives of Thermodynamics **29**, 1, 71-86 (2008)
7. D.G. Erbs, S.A. Klein, J.A. Duffie, Solar Energy **28**, 4, 293-302 (1982)
8. R. Tang, T. Wu, Applied Energy **79**, 3, 239-248 (2004)
9. M.J. Ahmad, G.N. Tiwari, The Open Renewable Energy J. **2**, 19-24 (2009)
10. J.A. Duffie, W.A. Beckman, *Solar Engineering of Thermal Processes* (2nd edition, Wiley, New York, 1991)
11. H.C. Hottel, B.B. Woertz, Transactions of the ASME **64**, 64-91 (1942)
12. S.A. Klein, Solar Energy **17**, 79-80 (1975)
13. W.H. McAdams, *Heat Transmission* (3rd edition, McGraw-Hill, New York, 1954)
14. B.Y.H. Liu, R.C. Jordan, Solar Energy **7**, 2, 53-74 (1963)
15. M. Collares-Pereira, A. Rabl, Solar Energy **22**, 155-164 (1979)

Energy saving and optimum insulation thicknesses in building upgrading

Mateusz Orzechowski[1], and *Tadeusz* Orzechowski[2,*]

[1]Statistical Office in Kielce, Wróblewskiego 2, 25-369 Kielce,Poland
[2]Kielce University of Technology, Al. Tysiąclecia PP 7, 25-314 Kielce, Poland

Abstract. Heat pumps are classified as ecologically friendly heat-generating devices and are one of the most suitable solutions for improving energy efficiency in buildings. Systems based on such solutions replace conventional sources very well. Household sector is one of the major energy consumers, where energy is used mainly for space heating and providing hot water. To increase renewable energy consumption levels, it is necessary to take decisive actions aimed at energy modernization of existing buildings. Particular emphasis must be put on the use of renewable energy sources. A wide selection of insulation materials is available on the market, however proper technologies are required to apply those materials, especially for historical buildings. Therefore, while planning investment, a question must be answered what type of material should be used, and what costs that will generate. The paper presents simple methodology for calculating the optimal insulation thickness so that the costs of the thermo-modernization would be covered by savings in heating expenses, when bank loan for the project is available at fixed interest rate.

1 Introduction

In the 21st century, communities have to face different threats that result from a carefree attitude to the use of natural resources in the last century. To a far extent that refers to over-exploitation of conventional energy sources, which directly leads to the natural environment deterioration. In particular, that refers to large urban agglomerations with dense built-up environment, in which the limit values of air pollutant levels are periodically exceeded. The situation occurs clearly more frequently during the heating period, especially in the areas where heat for buildings is predominantly provided by local facilities. The solutions to the problem involve boiler modernisation and advanced combustion processes [1,2], and also a reduction in energy consumption due to thermal upgrading. The best results are obtained when both methods are combined [3].

Heat pumps are classified as ecologically friendly heat-generating devices and are one of the most suitable solutions for improving energy efficiency in buildings [4]. Heating systems based on such solutions replace conventional schemes, such as electric, coal, gas and oil boilers, etc. for the generation of heat and domestic hot water. As a result, the heat pump market has been growing continually. According to [5], the highest sales of new devices per 1000 households in the years 2007-2016 were recorded in Norway, Finland and Greece. Poland took the sixteenth place, after the Czech Republic, and in front of Slovakia. The devices are mostly compressor heat pumps, half of which are air-to-water pumps. Other types, e.g. absorption or gas heat pumps are very rarely found.

In Poland, heat is mainly used for space heating purposes, domestic hot water and air conditioning. This represents more than 31% of total energy consumption. With such a high percentage of energy consumption, a reduction in the share of conventional heat production methods has become a priority for the coming years. Additionally, air-to-water type heat pumps are particularly suitable as a replacement of heating devices in energy upgrading of buildings [6]. They are easy to install and integrate into existing heating systems. External collectors or boreholes are not required [7]. Air is a good source of renewable energy and it is also possible to combine this solution with solar collectors, which offers one of the most effective options [8]. Air-source heat pumps work well enough under severe climatic conditions [9]. However, in order to optimally set their parameters, it is necessary to take into account locally prevailing conditions [10], including the parameters of supply and the heating medium return [11]. The investigations indicate that when selecting a device, it is also necessary to consider the impact of the working medium type on the coefficient of performance (*COP*) [12].

According to the Energy Market Agency forecasts, ultimate energy consumption in the residential sector should be reduced by 3.8% in the next 15 years. To achieve those levels, it is necessary to take decisive actions aimed at energy modernization of existing buildings. That requires proper selection of optimal

* Corresponding author: todek@tu.kielce.pl

solutions, especially in the case of old buildings, which being under supervision of conservation officers cannot be thermally upgraded using typical technologies that involve the external walls insulation retrofit by over-cladding.

Old buildings, which are in various condition, were constructed using different technologies and materials used at the times a given structure was built. Every time the building envelope upgrading is planned, it is necessary to check the chemical composition, or conduct an analysis of the materials that will be used [12] with respect to the various material technologies applied [14] and its potential contribution to global warming [15]. The installation technology is also important, which was discussed, on the example of insulating lime-perlite render, in study [16]. Overall heat losses can be also reduced by means of wall orientation [17] as well as correct operation of ventilation and air conditioning, especially if passive systems are used [18].

Historical buildings can have different use: residential, religious, museums, theatres, libraries educational institutions and palaces [19]. For each type, it is necessary to select an appropriate heating scheme that would provide desirable indoor climatic conditions [20], and account for the thermal properties of the material used for thermal upgrading [21]. Heat pumps offer a good option as they rely on heat sources that are environmentally friendly because they reduce greenhouse gas emissions [22]. A heat pump should be selected depending on local climate, settings and building characteristics [23].

It poses a real challenge to find technical solutions that would ensure energy savings, and at the same time, allow the building users undertake a typical range of activities, the more so the other issue can only undergo qualitative assessment [24]. A wide selection of insulation materials is available on the market, however proper technologies are required to apply those materials. Therefore, while planning investment, a question must be answered what type of material should be used, and what costs that will generate. The outcome of selection, in other words, which solution will be implemented, depends on the energy source quality. The paper discusses simulation investigations into heat demand when the building is heated by an air-source heat pump.

2 Climatic conditions

All simulations of the supply of heat or cold to the buildings rely on the information on climatic conditions. For the sake of the design, rough estimations given in relevant standards, guidelines and recommendations are sufficient. As regards climatic conditions, in accordance with [25], five climatic zones are differentiated. In the zones, the design lowest winter external temperature ranges from -16°C for the mildest zone located at the seaside, to -24°C for the coldest mountainous areas. Similarly, annual average temperature for those two zones is +7.7 °C and 5.5 °C, respectively. For the city of Kielce, the respective temperature values are -20 and 7.6. The calculations conducted in this way are useless while seeking optimal technical and economic solutions. That is a concern of particular importance when calculating the heating performance of the device that directly captures energy from the environment, the temperature of which depends, indirectly or directly, on ambient air. Heat pumps operate in this way. The ones for which the low temperature source are near-surface layers of the ground, and which harvest energy using a horizontal heat exchanger, depend on external conditions that vary over an approximately weekly cycle. As regards air-to-water heat pumps, their output directly depends on the momentary values of ambient air temperature. Then, if available, climatic data collected at the required frequency should be used.

In Poland, a high diversity in weather conditions is observed. The division into climatic zones acc. [26] does not account for local conditions, related, e.g. to the height of the location above sea level of a site. Simulation of the operation of the heat pump concerns a device operating in a building located in Kielce (20°37'E 50°53'N), approx. 300 m asl. The distribution of mean temperatures for all the months of the year is shown in Fig. 1. Weather conditions undergo changes that are statistical in character. Figure 1 also identifies the ranges of temperature variation.

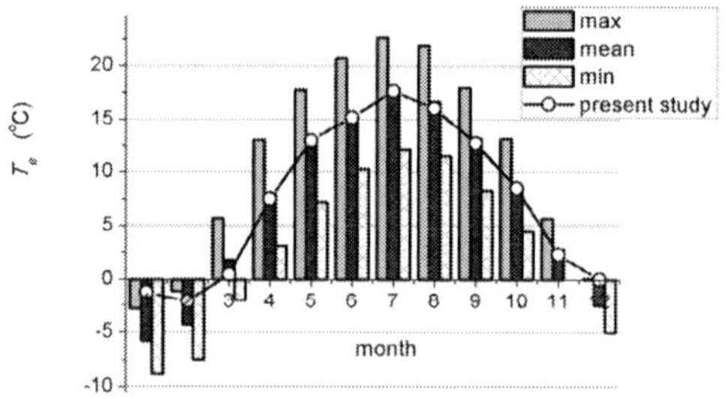

Fig. 1. Mean, minimal and maximal temperatures in Kielce in individual months of the year [26].

Meteorological data quoted in various sources depend on the site and time of measurement taking. In further computations, an hourly temperature distribution was adopted in accordance with [27]. The temperature data were obtained from the meteorological station in Suków, located closest to Kielce. The mean values were calculated for all the months. They do not differ much from the data shown in Fig. 1. Hourly counts of temperature with 1°C increment are shown in Fig. 2. As can be seen, that really low temperatures hardly ever occur. The temperatures below -10°C are recorded for approx. 3% of the heating season. Additionally, temperature measurements taken at high sampling frequency indicate that such cold conditions do not last all the day long. Temperature grows during daylight hours, consequently the demand for space heating of the building is correspondingly lower. The information of that kind should be taken into account while seeking the optimal heat pump for a given building.

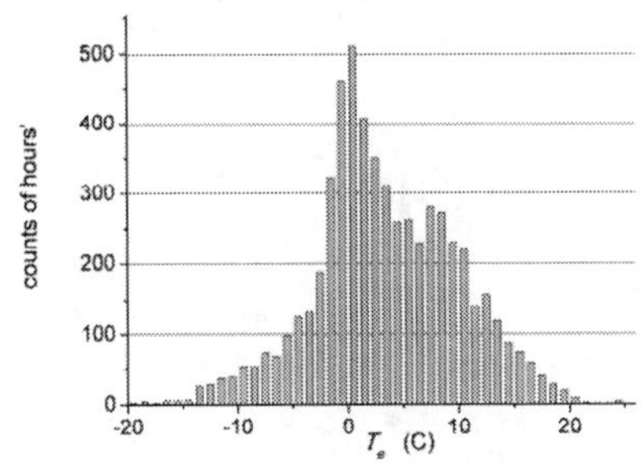

Fig. 2. Hourly counts of temperatures in Kielce [27].

3 Heat pomp sizing and insulation material options

Air-source heat pumps do not occupy much space, consequently they can be used in cities with dense built environment. Depending on power demand, a single- or multi-unit installations can be used. For moderate demand, single units are employed, which can be one- or two-compressor devices. The latter usually have higher output compared with the one-compressor pumps. The costs of purchase and later servicing of two-compressor heat pump is lower than the costs for two separate compressors. Exemplary characteristics of such a heat pump are shown in Fig. 3.

When trying to balance the thermal upgrading costs and benefits, the selection of materials options should be considered. That particularly refers to heritage buildings, in which all the works must be supervised by conservation officers. In a majority of cases, alterations to the building facades will not be approved. Then, the only option available, aimed at energy efficiency improvements and reduction in the heating costs, is to install insulation on the interior.

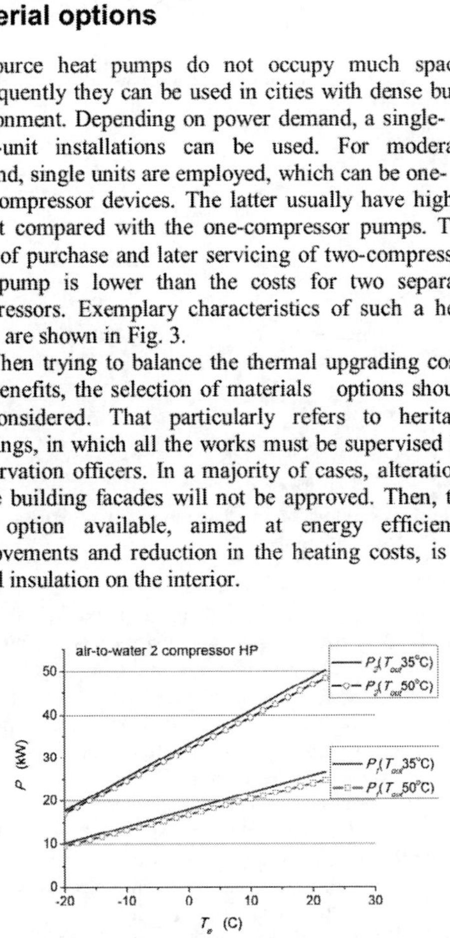

Fig. 3. Characteristic curves of two-compressor air-to-water heat pump.

The project must be preceded by moisture analysis of the envelope and space dividing elements, which should be carried out before and after insulation placement. That especially refers to materials such as Styrofoam or mineral wool. Thermally insulating plasters of various types, or cellular concrete blocks provide a safer solution in this respect. The thermal conductivity of a typical cement-lime plaster is approx. 0.82 W/mK. The need to thermally upgrade heritage buildings contributed to the production of plasters that have good insulation properties, and additionally, sufficient moisture permeability. Many products are available, the most popular ones are mortars using perlite as the main ingredient. The cheapest option is the preparation of the material on the site, the resulting product has the thermal conductivity coefficient that is a few times lower in comparison to the typical cement-lime plaster. Another option is the use of prefabricated lightweight cellular concrete blocks. The major drawback of this technology is a relatively large minimal thickness of the insulation. Additionally, it is also necessary to apply a finishing coat. As a result, the project costs increase. Another option is rigid polyurethane foam that is finished on one side with gypsum wallboard. That, however, is the most expensive technology, therefore less frequently used.

Other insulation materials to be installed on the building interior surface are also available on the market. The choice results from technical and economic analysis that needs to be carried out on individual basis, or from the investor's preferences. In this paper, the analysis aimed at finding the optimal solution was carried out using the example of a thermally insulating plaster based on perlite. It was assumed that the building envelope, 0.4 m in total thickness, was made of brick and the envelope thermal resistance is 0.506 m^{2}K/W. The overall heat load was assumed to be equal to 36 kW (including 3 kW for domestic hot water generation). As a result of the modernisation of the heating system, the load requirement is to be satisfied by the air-source heat pump, the characteristics of which are presented in Fig. 3. It is a two-compressor device, with the first compressor operating at moderate external temperatures, while the second one starts functioning when the temperatures drop. Thermal power demand is illustrated, in a simplified manner, in Fig. 4.

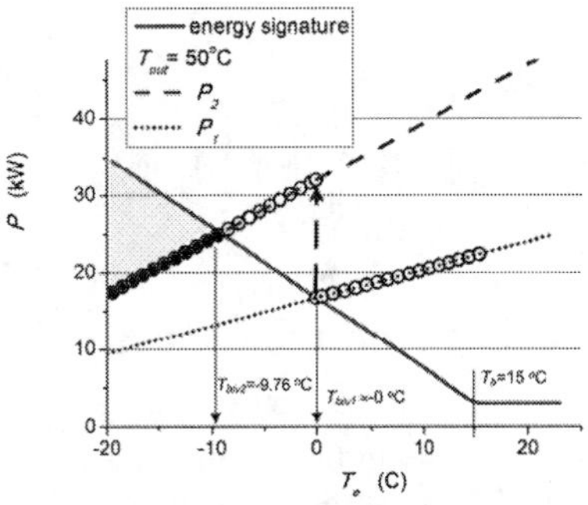

Fig. 4. Heat pump contribution to the building energy signature.

It was assumed that the object space heating starts when the external temperature drops below the base temperature T_b=15 °C. The heating curve intersects the pump characteristics curve at two points. The first point specifies the maximum power that can be utilised when the heat pump operates using one compressor. At the external temperature lower than T_{biv1}, the pump starts the operation mode with two compressors activated. It is possible to fully meet power demand only until the ambient air temperature is above the bivalence point, i.e. $T_e > T_{biv2}$. Then, the heat pump must be supplemented by an additional heat source, e.g. a boiler. In the paper, a monovalent mode with a backup electric heater was assumed. For a vast range of external temperatures, the heat pump operates generating surplus-to-demand power. That makes it necessary to frequently switch the pump on and off, which results in decreasing the actual value of the coefficient of performance. The manufactures of heat pumps do not always specify this value. Coefficient of performance of heat pump under partial load conditions COP_{bin} at temperature T_j may be calculated in accordance with [28]:

$$COP_{bin}(T_j) = COP\frac{P_h}{CcP_h + (1 - Cc)P_d} \qquad (1)$$

where: $P_h = P_h(T_j)$ – heat load of the building at temperature T_j, $P_d = P_d(T_j)$ – power output of the heat pump at temperature T_j,

Cc (*i*) in Eq. (1) is the heat pump capacity ratio, namely the ratio between the thermal power delivered by the heat pump and the maximum power which the heat pump could deliver at the *i*th bin conditions.

The values of this coefficient for the individual ranges of the heat pump operation are shown in Fig. 5.

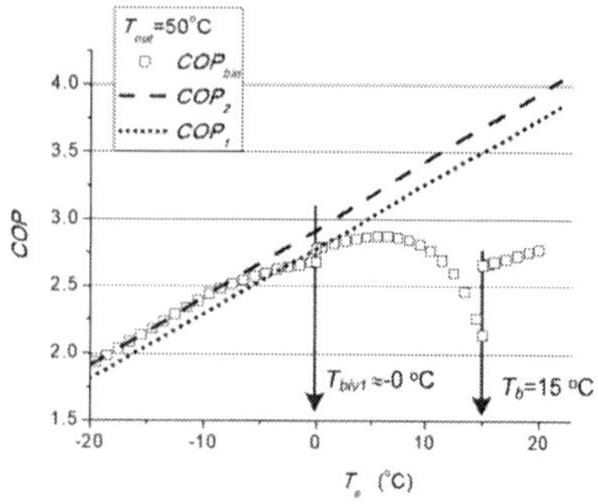

Fig. 5. Coefficient of performance of heat pump under partial load conditions COP_{bin} and full load for one- and two-compressor mode of operation.

4 Optimal insulation thickness

When seeking optimal solutions for a thermal upgrading project, it is necessary to balance overall costs $C_{A,N}$ that include space heating costs, insulation placement costs and the costs of necessary materials. The following notation can be used: f_1 and f_2=1-f_1 represent the share of the building opaque part and the part of it (e.g. windows) that is not thermally upgraded, respectively, U_1 and U_2 denote respective overall heat transfer coefficients, R_w stands for thermal resistances of subsequent layers of the envelope element, d_{ins} and λ_{ins} refer to thickness and thermal conductivity of the insulation material. The total costs $C_{A,N}$ in reference to 1 m² can be calculated from the following formula [29]:

$$C_{A,N} = \left(\frac{f_1}{R_w + d_{ins}/\lambda_{ins}} + U_2 f_2\right) NC_{Em} * $$
$$* \sum_j (T_i - T_j)\frac{\tau_j}{COP_{bin}(T_j)} + NC_{Np}(d_{ins}C_{ins} + C_d) \qquad (2)$$

If the thermo-modernisation investment is carried out with the funds obtained through the bank loan at fixed interest rate i_r at all *N* years, coefficient describing the cost of credit C_{Np} is:

$$C_{Np} = i_r \frac{\left(1 + \frac{i_r}{12}\right)^{12N}}{\left(1 + \frac{i_r}{12}\right)^{12N} - 1} \qquad (3)$$

Figure 6 shows a change in heating costs, insulation costs and overall costs depending on the insulation thickness.

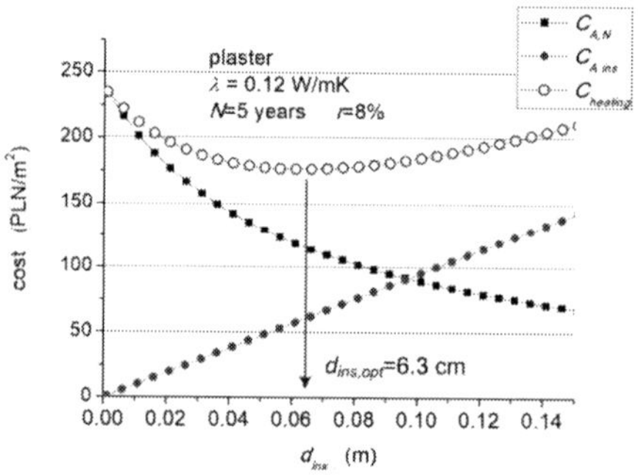

Fig. 6. Costs of plaster insulation for five-year term of loan and at 8 % fixed interest rate.

The computations were made for thermally insulating mortar with thermal conductivity of λ=0.12 W/mK, as declared by the manufacturer. Plaster is applied in coats of up to 3 cm in thickness. For multi-coat insulation, the placement costs increase proportionally. A fixed interest rate i_r= 8% and 5-year loan repayment term were assumed. Ambient air is a low temperature heat source for the compressor heat pump. The hourly distribution of mean thermal conditions of the ambient air is shown in Fig. 2. In the case under consideration, the optimum insulation thickness is d_{ins_opt}=6.3 cm, and the insulation should be placed in two approx. 3cm thick coats.

The investor's decision is to choose the loan period for the investment, which - as shown in Fig. 7 - affects the optimal insulation thickness. Along with the extension of the repayment period, the initial costs also increase. It also causes a corresponding change in the energy demand. For data from Fig. 7, energy consumption compared to the state before thermo-modernization decreases by approximately 38%, 51% and 58% at respectively 3, 5 and 7 years of loan repayment terms.

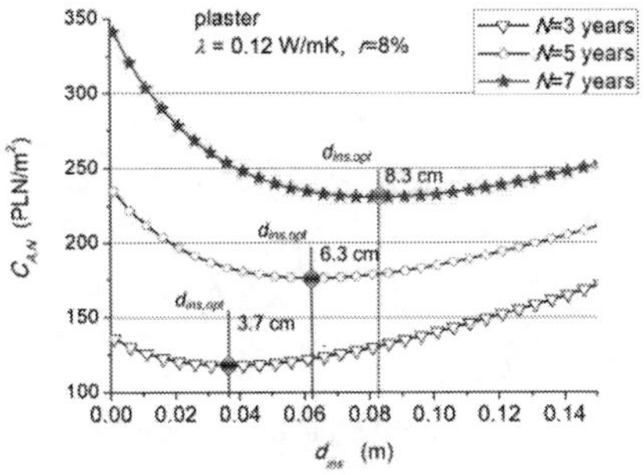

Fig. 7. Change of the optimal insulation thickness at different terms of loan with a fixed interest rate.

5 Conclusions

From equation 2, the insulation optimal thickness can be calculated. The value is equal to approx. 6.3 cm. The optimal thickness is the one for which the overall costs will be minimal [21]. In the case presented in Fig 6, a relatively low increase in costs can be seen in the proximity of d_{ins_opt}. In the plaster thickness range from 1.7 cm to 10 cm, the increase is not higher than 5 %, which produces a rational change in the thermal performance of the envelope elements (walls). On the basis of such information, the investor can make a decision on increasing the thickness of the plaster applied, unless limits are put on, e.g. floor area reduction.

In addition to the costs of space heating and insulation materials, the costs of the heating device replacement must also be taken into account in the thermal upgrading project. The price of the heating pumps available on the market is related to their output. The heat pump under consideration in this study has the output of 32 kW when it is supplied with air at the temperature of T_e=0°C, and the bivalence point is T_{biv}=9.2°C. After thermo-modernization, the heating power needed drops. For the building of concern, it is possible to select a one-compressor heat pump having such characteristics so that the bivalence point would remain unchanged. The required output should be about 20 kW at the temperature of T_e=0°C, the purchase price will be about 30% lower. Assuming a fixed interest rate of 8%, net present value (NPV) can be calculated. It has a positive value after 18 years or 12 years, for two- and one-compressor heat pump, respectively. That indicates when designing thermal modernisation of buildings with thermal upgrading chosen on individual basis, it is also necessary to account for the ultimate parameters of the heating device.

References

1. D. Król, S. Poskrobko, Energy & Fuels, **31**(4) (2017)
2. D. Król, S. Poskrobko, Energy, **116** (2016)
3. T. Orzechowski, M. Orzechowski, E3S Web Conferences, **14** (2017)
4. M. Ali, V. Vukovic, N.A. Sheikh, H.M. Ali, Energy Energy Convers. Manage. **97** (2015)
5. *http://hpc2017.org/wp-content/uploads /2017/06/k211.pdf.* (Last accessed Feb. 2018)
6. V. Bianco, F. Scarpa, L.A. Tagliafico, Appl. Therm. Eng. **114** (2017)
7. W. Wu, W. Shi, B. Wang, L Xianting, Energy Convers. Manage. **98** (2015)
8. W. Deng, J. Yu, Energy Convers. Manage. **120** (2016)
9. Y. Zhang, Q. Ma, B. Li, X. Fan, Z. Fu, Energy Build. **138** (2017)
10. M. Dongellini, C. Naldi, G.L. Morini, Appl. Therm. Eng. **114** (2017)
11. M. Maivel, J. Kurnitski, Energy Build. **94** (2015)
12. L. Yang, H. Yuan, J-W. Peng, C-L. Zhang, Renewable Energy, **87** (2016)
13. Z. Owsiak, J. Zapała-Sławeta, P. Czapik. Bull. PAS: Tech. Sci. **63**(1) (2015)
14. R. Dachowski, P. Kostrzewa, E3S Web Conferences, **10** (2017)
15. C. Hill, A. Norton, J. Dibdiakova, Energy Build. **162** (2018)
16. Y. Govaerts, R. Hayen, M. Bouw, A. Verdonck, W. Meulebroeck, S. Mertens, Y. Grégoire, Constr. Build. Mater. **159** (2018)
17. I. Csáky, F. Kalmár, J. Build. Phys. **39**(2) (2015)
18. A. Rackes, A.P. Melo, R. Lamberts, Appl. Energy, **174** (2016)
19. A. Martínez-Molina, I. Tort-Ausina, S. Cho, J-L. Vivancos, Renew. Sust. Energ. Rev. **61** (2016)
20. S. Bichlmair, S. Raffler, S. Kilian, Energy Build. **95** (2015)
21. R. Walker. S. Pavía, Build. Environ. **94** (2015)
22. R. Dylewski, J. Adamczyk, Build. Environ. **46** (2011)
23. C. Naldia, G.L. Morinia, E. Zanchini, Sust. Cities Soc. **12** (2014)
24. K. Fabbri, J. Cult. Heritage, **14S** (2013)
25. Polish standard PN-EN **12831** (2006)
26. Local Data Bank Warsaw: *www.stat.gov.pl.* (Last accessed October 2017)
27. *www.mib.gov.pl.* (Last accessed Oct. 2017)
28. Standard EN **14825** (2013)
29. T. Orzechowski, M. Orzechowski, J. Build. Phys. **1** (2017)

The influence of temperature on algal biomass growth for biogas production

Monika Pawlita-Posmyk[1,*], *Małgorzata* Wzorek[1], and *Małgorzata* Płaczek[1]

[1]Department of Process Engineering, Faculty of Mechanical Engineering, Opole University of Technology, Opole, Poland

Abstract. The use of algae is gaining popularity in recent years. Algae can form a suitable substrate for biogas production because they are low in cellulose and have low lignin content. The paper contains a summary with the list of the algae species applied for the biogas production process to this date, coupled with the description of the characteristic conditions in which the process is conducted. The key importance factors during the production of biogas include, among others, grade of algae species used, temperature of the process, mode in which the material is supplied into the tank, mixing method, use of co-substrate, etc. This paper reports the results of the study involving the effect of various temperature conditions on the characteristics of the growth rate of algal biomass. The research was performed for the *Chlorella vulgaris BA-0167* strain in the temperature range corresponding to the mesophilic fermentation processes in anaerobic conditions.

1 Introduction

Global demand for energy from renewable sources is constantly growing. What shows a promise of development is the production of energy from the biomass of algae, which allow fulfilment of energy-supply needs and limit CO_2 emission at the same time. Moreover, microalgae grow at exceptional pace (100 times faster than land plants) and some strains of microalgae, including *Chlorella vulgaris*, can accumulate large amounts of lipids, which can be transformed into biodiesel [1]. The recent years have witnessed a significant increase in the use of algae-derived biomass in various industry branches, e.g. in medicine, pharmacy, cosmetology, food industry and agriculture [2]. Algae are also used in the sewage treatment process [3,4]. Results of studies indicate that it is possible to use algae-derived biomass as a source of organic matter in production of biofuels, including biogas (Fig. 1) [5,6].

Biogas can be produced from agricultural waste, solid municipal waste, sewage, waste from wood processing processes, industrial remains and animal waste. The composition of biogas depends on the applied substrate. Its main ingredients are CH_4 (40–70%) and CO_2 (15–60%), other chemical compounds (H_2O, N_2, O_2, H_2, NH_3, H_2S) occur in low amounts [7]. Among factors regarded as key in production of biogas using algae, there are for instance the types of used algae, process temperature, way of reactor feeding, mixing manner or the presence of a co-substrate.

Table 1 shows literature data [8-14] concerning the use of various algae species in production of biogas along with the description of conditions under which the process was conducted.

Dębowski et al. [5] point to factors which can considerably limit the effectiveness of methane fermentation and, in turn, contribute to decrease in or full inhibition of biogas production. Those factors include the presence of cellular wall, which is resistant to degradation under anaerobic conditions due to the presence of cellulose or hemicellulose compounds, presence of strains of microorganisms producing compounds toxic for anaerobic bacteria and an unfavourable C:N relation in the biomass subjected to fermentation.

The aim of the study is to determine the effect of temperature on the effectiveness of production of biomass from microalgae *Chlorella vulgaris BA-0167*. Taking into consideration that *Chlorella vulgaris* is treated as a potential source of biomass for energy supply purposes, including biogas production, the possibility to cultivate algae within temperatures adequate for anaerobic mesophilic fermentation, i.e. 33°C–40°C was examined.

2 General characteristic of microalgal biomass as a raw material for biogas production

Algae form a source of valuable substances, i.e. proteins, lipids, hydrocarbons, vitamins and microelements. Water forms the principal component of the biomass, as it is 75–90% of the harvested mass before processing. The remaining ratio of the content is made by mineral salts and hydrocarbons (30–50%) and lipids (7–15%) of the dry mass [3,15].

Algae biomass productivity in laboratory scale depends on many factors, e.g. on technical conditions under which algae grow, including design of photobioreactor, availability of food ingredients, light intensity, CO_2 concentration and pH level.

* Corresponding author: mpawlita@wp.pl

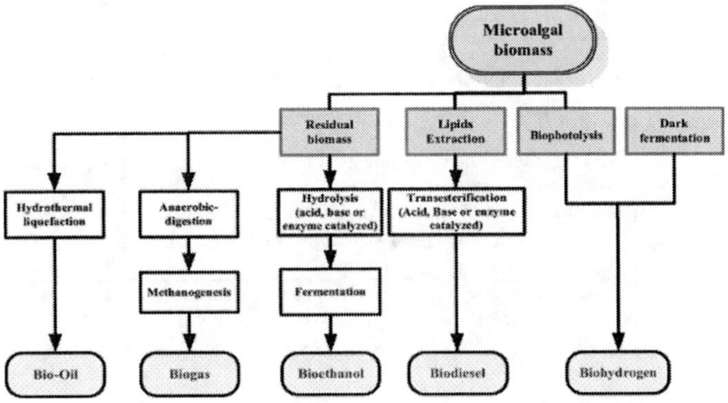

Fig. 1. Processes and products obtained from algae-derived biomass [6].

Mixing is another important aspect of algae cultivation as it guarantees maintenance of homogenous conditions in the entire culture, i.e. it protects against temperature gradient in the tank, guaranteeing, in turn, uniform access to nutrients and light [3,16]. The optimal temperature for algae growth should oscillate within 20 to 30°C [17-21] and according to the authors [22] it should remain within the range of 15 to 26°C. The authors of the paper [22] indicate that the maximum algae growth occurs at 23 °C and the minimum one – at 7 °C. Temperature below 16 °C contributes to retardation of algae cell growth and temperature higher than 35 °C is regarded as lethal point [21]. Therefore, in order to achieve the maximum algae productivity, it is recommended to hold a culture at the temperature corresponding to the optimum for a given species [18,19, 21]. Example of optimal temperature and favourable pH value for the growth of selected algae strains is presented in Table 2. The authors of study [28] stated that algae strains cultivated at temperatures 35°C and higher show a high level of photoinhibition and in this conditions the algae cells can start to decay. In turn, mechanical mixing may contribute to destruction of cells due to occurrence of shear forces [18,19]. The use of natural or artificial light is necessary to provide correct conditions for algae growth. Artificial lighting in the photobioreactor considerably increases the biomassproduction cost, but can also provide higher productivity of culture [17,19].

Chisti [29] indicates that in order to obtain high algae growth, the following mineral ingredients are to be included in the nutrient medium: nitrogen (N), phosphorus (P), iron (Fe) and silicon (Si) in some cases. In turn, Voloshin et al. [30] state that for the efficient biomass increase of the desirable content, it should contain sufficient sources not only of carbon, but also of various necessary mineral elements: nitrogen and phosphorus in significant amounts, and magnesium, calcium, chlorine, silicon, manganese, iron in smaller amounts.

Algae cultures can be cultivated in various types of photobioreactors as apparatuses where it is possible to create proper conditions for cultivation and, at the same time, obtaining of high effectiveness of algae-derived biomass production [3].

Table 2. Example of optimal temperature conditions for algae cultivation.

Algae species	Optimal conditions		Ref.
	temperature	pH	
Chlorella thermophila (HTA1-65)	33°C	n.d.	[1]
Chlorella vulgaris	25–28°C	n.d.	
Chlorella mirabilis	10.1–20.5°C	n.d.	[23]
Chlorella pyrenoidosa	33°C	9.0	[24]
Chlorella sp.	37.5°C	7.0	[25]
Chlorella marina	37.5°C	8.0	
Chlorella vulgaris	27°C	n.d.	[26]
Chlorella	25–30°C	n.d.	[27]

Table 1. Example conditions for biogas production process.

Algae species	Co-substrates	Operating conditions	Biogas/methane production	Ref.
Isochrysin galbana, Selenastrum capricornutum	sewage sludge	**Reactor:** 2 batch reactors (120 mL each), reactor feed: 100% sludge and 25%, 50%, 75% and 100% of the sludge was replaced with microalgae respectively, **Temperature:** 33°C; 50°C, **Duration:** 35 days (33°C), 20 days (50°C), **Conclusions:** during co-digestion with the higher microalgae content was the lower the biogas production	for *Isochrysin galbana* - 33°C - 338±3 mL_{CH4}/ $g_{volatile\ solids\ (VS)}$ - 50°C - 219±10 mL_{CH4}/g_{VS} for *Selenastrum capricornutum* - 33°C - 209±5 mL_{CH4}/g_{VS} - 50°C 152±6 mL_{CH4}/g_{VS} for sewage sludge separately: - 33°C 347±9 mL_{CH4}/g_{VS} - 50°C 464±4 mL_{CH4}/ g_{VS}	[8]

* Corresponding author: mpawlita@wp.pl

Table 1 cont.

macro-algae: Cladophora glomerata (CG), Chara fragilis (CF), Spirogyra neglecta (SN)	-	**Reactor:** batch reactors with a total volume of 5 L and a working volume of 3 L (in the bioreactor substrate amounted to 3 L), **Temperature:** 35 ± 1°C, **Mixing:** mechanical shaking 10-15 min/day, **Duration:** 15 days, **Conclusions:** the amount of biogas and methane, indicates that the most suitable algae for the production of energy are *SN* and *CG*	for *Spirogyra neglecta*: biogas: 0.23 m^3/m^3 day, 46.5% CH_4, for *Cladophora glomerata*: biogas: 0.20 $m^3/(m^3day)$, 51.4 %CH_4, for *Chara fragilis* : biogas: 0.12 $m^3/(m^3day)$, 36.9% CH_4	[9]
Chlorella vulgaris (pure strain 211/11B)	-	**Reactor:** double wall glass reactor (1L effective volume) coupled to photobioreactor with an settling facility to concentrate microalgae before feeding reactor, **Temperature:** 35°C, **Mixing:** magnetic stirrer (continuous), **Duration:** 16 and 28 days HRT, **Conclusions:** carbon and nitrogen fractions of the microalgae showed different responses towards anaerobic degradation; fast nitrogen release is liable to limit intensification capacities of the process	-16 days - 110 mL_{CH4}/g_{CODin}; 29% CH_4 in gas, -28 days - 180 mL_{CH4}/g_{CODin}; 49% CH_4 in gas	[10]
Scenedesmus sp. and *Chlorella sp.*	waste paper	**Reactor:** volume 5.25 L with gas storage tank 7.52 L; working volume 4 L; semi-continuous feeding; fed 1 time per day following the removal of the same volume of effluent, **Temperature:** 35±1°C, **Mixing:** manual mixing by use of permanently installed mixing rod, **Duration:** 10 days HRT, **Conclusions:** the optimized C/N range for the co-digestion was 20–25/1; increase in cellulase activity helpful in the biodegradation of algal sludge	- for algal sludge: 573±28 CH_4 ml/(Lday), - for 25% waste paper + algal sludge - 968 ± 73 CH_4 mL/(Lday), - for 50% waste paper + algal sludge - 1170±75 CH_4 mL/(Lday), - for 75% waste paper + algal sludge 317±114 CH_4 ml/(Lday), - for 100% waste paper - 452±36 CH_4 mL/(Lday)	[11]
Chlorella vulgaris	-	**Reactor:** batch reactors 0.120 L; glass bottles closed with butyl rubber seals and aluminium caps; working volume 0.070 L, 42% head space for gas production, high pressure thermal pre-treatment 140°C (3 bar) and 160°C (6 bar) and 180°C (10 bar) **Temperature:** 35°C, **Mixing:** orbital shaker, **Duration:** 29 days, **Conclusions:** high pressure thermal pretreatment resulted anaerobic biodegradability of *C. vulgaris*; highest anaerobic biodegradability (74%) was provided by the biomass subjected to 160°C, while higher temperatures did not enhance methane yield	- average methane content in the biogas - 66–68%, - raw biomass gave methane yield of 156.4 mL_{CH4}/g_{CODin}	[12]
Chlorella sp.	waste activated sludge	**Reactor:** batch reactor (150 mL), Reactor feed: 10% of sludge by volume; alge: 0%, 4%, 11%, and 41%, and 100%, **Temperature:** 37°C, **Mixing:** multi-magnetic stirrers, **Duration:** 45 days, **Conclusions:** mesophilic anaerobic co-digestion was a suitable approach to increase the biogas yield	- the methane yield of pure algae digestion 123 $mLCH_4/gVS_{fed}$ on 45 day	[13]
Chlorella vulgaris	-	**Reactor:** serum bottles; thermal pre-treatment 120°C (for 20 and 40 min) and adding 4 M sulphuric acid to achieved pH 2 for acid and adding 4 M sodium hydroxide to get pH 10 for the alkali treatment, **Temperature:** 35°C, **Mixing:** shaker (120 rpm), **Duration:** 30 days, **Conclusions:** the pretreatment improved methane production, thermal treatment at 120°C for 40 min led to the highest anaerobic biodegradability; this pretreatment doubled methane production compared to that of the raw microalgae	- not treated 138.9 $mLCH_4/gCOD_{in}$ (69.9% CH_4) -thermal 20 min 180.3 mL $CH_4/gCOD_{in}$ (67.3% CH_4); 40 min 267.7 (68.3% CH_4) -20 min.+acid 221.8 $mLCH_4/g$ COD_{in} (70.6 % CH_4) -40 min.+acid 228.8 $mLCH_4/g$ COD_{in} (66.3 % CH_4) -20 min.+alkali 237.9mL CH_4/g COD_{in} (68.8 % CH_4) -40 min.+alkali 240.6 (70.1 % CH_4)	[14]

3 Methodology

Chlorella vulgaris BA-0167 algae were cultivated in four glass tanks with capacity of 2 dm^3 each. The test stand is presented in Fig. 2.

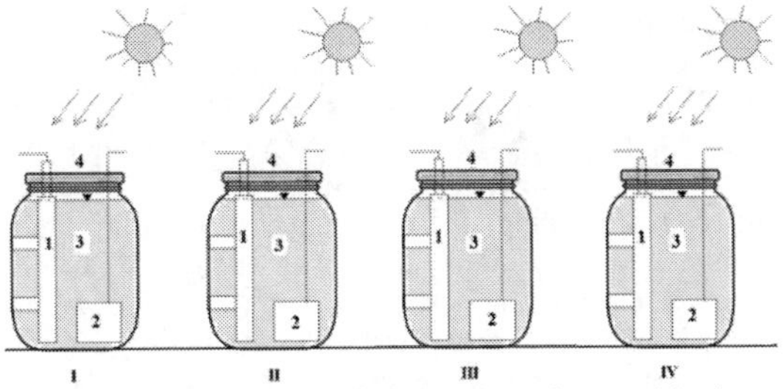

Fig. 2. Experimental stand for *Chlorella vulgaris* cultivation. 1-heater with a thermostat, 2 - circulation pump, 3 - tank with culture medium, 4 - semipermeable barrier for gases.

Four glass tanks were inoculated with *Chlorella vulgaris BA-0167* strain (Fig. 3.) The inoculum was cultivated at the surroundings temperature of 22-24 °C on a standard f/2 medium, in 500 mL Erlenmeyer flasks. For protection against contamination, the flasks were sterilised in the autoclave at 121°C for 20 minutes. The strain was obtained from the CCBA (Culture Collection of Baltic Algae, Institute of Oceanography, University of Gdańsk). To initiate cultivation of algae under laboratory conditions, the same volume equal 0.2 dm^3 of inoculum was used, which was put in each tank and filled with demineralised water to the working volume of 2 dm^3.
Heaters with thermostat (1) were put in the culture tanks (Fig. 2), which allowed maintaining the same set temperature throughout algae cultivation. The literature reports various ranges of changes in the temperature for anaerobic mesophilic fermentation. According to Jankowska et al. [31], the range of temperature should be within 30°C-38°C. On the other hand Santosh et al. [32] indicates the temperature range of 30-40°C and Weiland [33] 35-42°C. In the absence of the temperature range unequivocally determined for mesophilic fermentation, the experiment was conducted at various temperatures, i.e. 33°C (1st tank), 35°C (2nd tank), 38°C (3rd tank) and 40°C (4th tank).

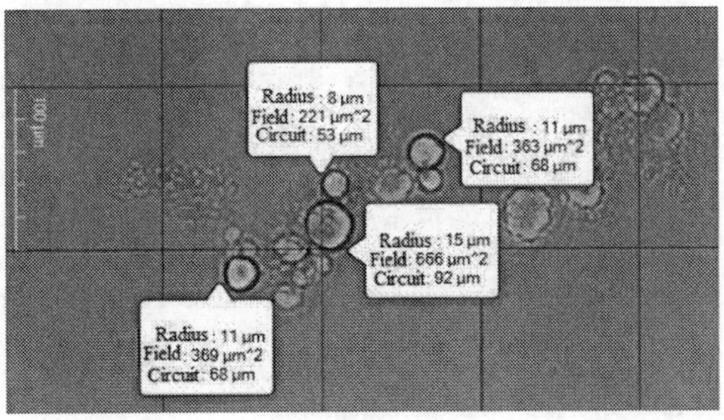

Fig. 3. Microscopic view of *Chlorella vulgaris BA-0167* (typical size of single cell is 4 µm, in the picture visible clusters of cells after a few days of cultivation).

Algae are cultivated in natural light. Algae suspension was mixed on a stream basis in a full immersion circulation pump (2). The pump worked in 15-minutes cycles. Thanks to a time programmer unit it was possible to control the operation of the pump, which guaranteed proper circulation of the culture medium and reduced the risk of photoinhibition. Before the experiment was started, the algae culture was supplied with commercially available nutrient medium, i.e. NPK fertiliser solution with macroelements, in the amount of 1 mL per tank. Algae were cultivated in batch mode for 10 days, (after this time it was observed the reduction in biomass concentration).

The algae was cultivated until the algae biomass concentration decreased in all tanks, i.e. for 10 days. Every day, pH and temperature were taken with the Elmetron CP-401 pH-meter, which made it possible to measure both those values.

The effectiveness of algae biomass growth was determined with the Thermo Scientific Orion Aquamate 8000 UV-Vis spectrophotometer. Each time, a 3500 µL algae sample was collected with automatic pipette Sartorius from a culture tank and its optical density was determined. The measurement of absorbance of algae suspension at 684.0 nm wave length allowed monitoring of algae biomass growth. A developed analytical curve defining the change in absorbance in relation to algae concentration directly allowed reading algae concentration in mg/L from spectrophotometer. An algae sample was always observed microscopically to monitor the condition of a culture (e.g. quantity and size of algae cells) with the Inverted microscope Motic AE 2000-T (under 40x magnification) equipped with digital camera Moticam 5+ and image analysis software (Fig. 3).

4 Results and discussion

A significant effect on the efficiency of biogas production and algae cultivation is mixing, temperature and pH. The applied mixing allowed, for instance, transport of CO_2, removal of O_2 generated in photosynthesis from a tank, favourable access to light and uniform distribution of nutrients as well as heat throughout the reactor. The pH values measured in the culture tanks were changing from 6.88 to 8.07. The pH range in own studies corresponded to the conditions of growth of most algae on culture media, i.e. pH of 7-9 [17,21]. Moreover, that range corresponded with the optimal conditions for biogas production, i.e. pH 6.5 – 8.0 [34]. The oscillation of temperature in algae culture did not exceed ±1°C. Fig. 4 shows a change in the biomass concentration over time.

The highest increase in biomass concentration, almost twofold, was achieved in a culture tank at 33°C, where the biomass concentration increase from 0.176 to 0.331 mg/dm^3, and after day 8 of cultivation, there was a slight decrease in the algae biomass concentration up to 0.326 mg/dm^3. Experimental data indicate that after a relatively short period, the rate of algae biomass gain at 33°C stabilised. At 35°C, there was a minimum increase

* Corresponding author: mpawlita@wp.pl

in the algae biomass concentration only to day 3 of the cultivation, i.e. from 0.176 mg/dm^3 to 0.237 mg/dm^3, and subsequently a concentration decreased to 0.172 mg/dm^3. In turn, in the case of algae cultivation at 38°C and 40°C, the algae concentration was constantly decreasing from 0.176 mg/dm^3 to 0.072 mg/dm^3 and 0.054 mg/dm^3, respectively.

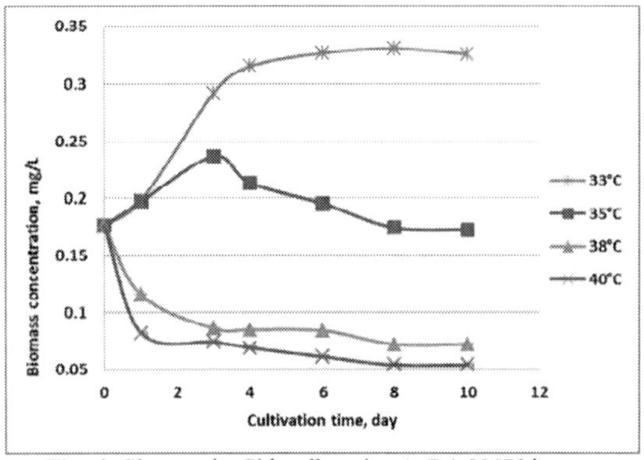

Fig. 4. Changes in *Chlorella vulgaris BA-0167* biomass concentration over time.

The algae biomass concentration decrease observed in the experiments at 35°C is also confirmed by studies conducted by Fernandez et al. [28], who noted similar behaviour of microalgae *Phaeodactylum tricornutum* in a spiral reactor. The authors of [28] indicated for instance that temperature and dissolved oxygen are factors limiting algae growth. Dziosa and Makowska [21], who investigated growth of biomass of fresh-water microalgae cultivated in laboratory bioreactors, also did not note any increase in biomass productivity upon temperature increase up to 30°C, only an increase in cultivation costs. In turn, Converti et al. [35] state that the rate of growth of *Chlorella vulgaris* at 35°C decrease by 17% compared to the productivity observed at 30°C. At 38 °C, microalgae stopped growing and cells started to decay. As the temperature increased to 30°C, the rate of cell growth increased; on the other hand, as the temperature increased from 30 to 35-°C [36,37], the rate went down.

Present studies showed decrease in algae concentration (Fig. 4), which most probably results from the nutrient medium losing its nutrients (batch culture, no medium feeding) and the application of unfavourable temperature conditions. Ma et al. [1] stated that the optimal temperature conditions for *Chlorella vulgaris* are 25° - 28°C; on the other hand, Leyva et al. [26] state that it is 27°C. These temperatures are lower than the temperature applied in algae cultivation system examined in present study. During cultivation, algae grouped into clusters, or floccules (Fig. 5). On walls of photobioreactors, biofilm was formed, which indicates that the algae needed a higher access to light and/or the applied mixing method was insufficient. The algae culture was exposed to sunlight of variable intensity. Photoinhibition was observed (visible in bleaching cells in Fig. 6), which contributed to algae death (Fig. 7).

An analysis of results obtained from studies on change in algae concentration over time at various cultivation temperatures (Fig. 4) allows stating that algae cultivation at the set temperatures of 33 °C, 35 °C, 38 °C and 40 °C does not allow high algae biomass gains.

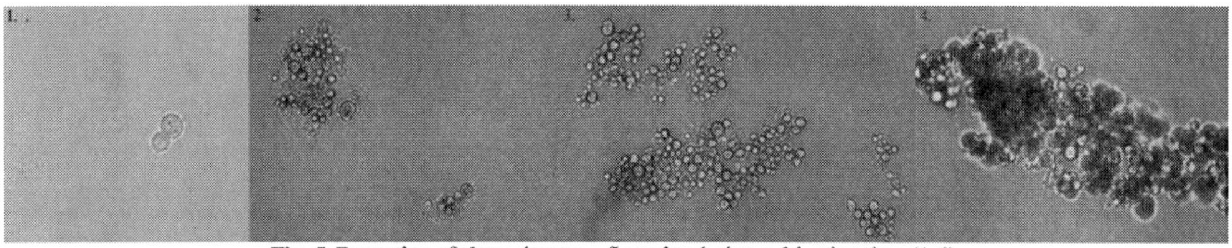

Fig. 5. Formation of algae clusters – floccules during cultivation time (1-4).

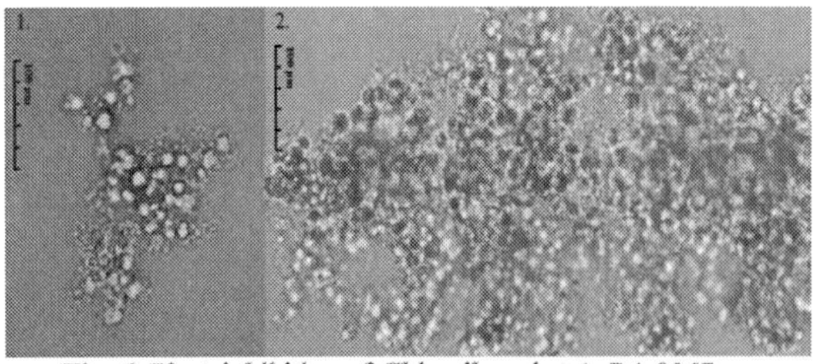

Fig. 6. Photoinhibition of *Chlorella vulgaris BA-0167*.

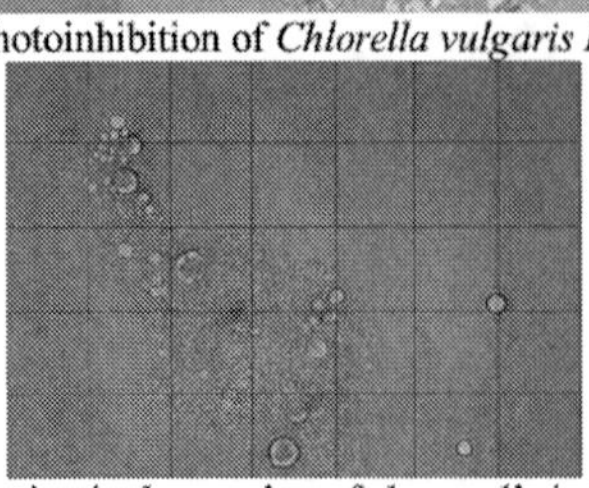

Fig. 7. Reduction in the number of algae cells in the medium.

Experimental test of culture of algae is presented in Fig. 8.

The authors of papers [8-14,31] evaluated the usefulness of various algae species for biogas production by way of fermentation or co-fermentation using various substrates and determining diverse conditions of the process. Montingelli et al. [38] presented the process of biogas production at various temperatures, using such algae species as *Ulva lactuca* (50°C, 53°C), *Laminaria hyperborea* (35°C), *Laminaria sp.* (50°C), *Ascophyllum nodosum* (35°C), *Chlorella vulgaris* (35°C), *Chlorella sp.* (25°C), etc. For example the methane potentials of cyanobacteria and *Chlorella* have been investigated in eight different lab scale reactors (25°C, three-day HRT). It was stated that the highest methane production rates of *Chlorella* were 100±25 mL/(L·d), at loading rate of 7g VS [38,39]. The results of experimental research indicate [8-14] that it is possible to use various algae species as substrates or co-substrates in the process of biogas production and depending on the conditions under which the process is conducted and the temperature optimum characteristic for individual algae species, one should expect diverse amounts of produced biogas.

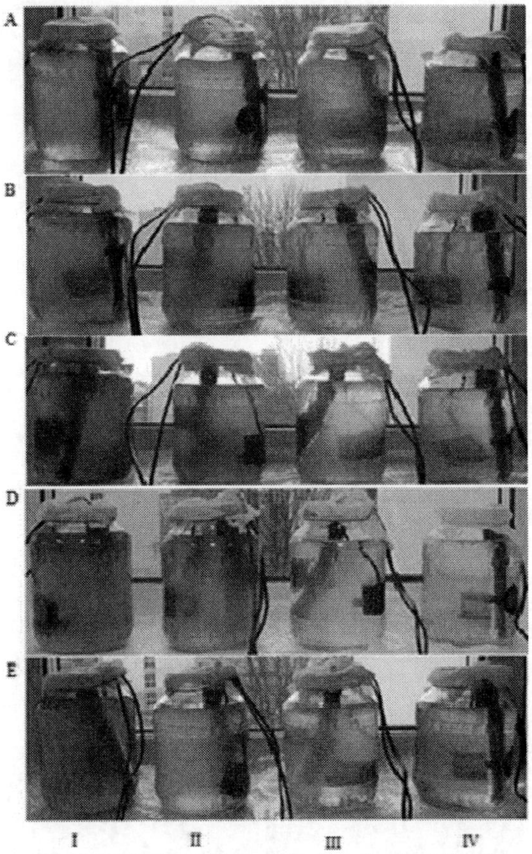

Fig. 8. Cultivation of algae *Chlorella vulgaris BA-0167*
A - start of cultivation (day 0), B - 1st day of cultivation, C- 3rd day of cultivation, D - 8th day of cultivation, E - end of experiment (10th day of cultivation).

5 Conclusions

Mostly due to high rate of growth (photosynthesis), ability to accumulate in cells of lipids and other organic compounds as a result of conversion of CO_2 into biomass and low nutritional requirements (water, light, CO_2, N, P, etc.), algae are treated as a potential substrate or co-substrate used in the production of biofuels, including biogas. Depending on the species of algae being fermented, the amount of produced biogas can differ considerably. Among others, very important factor influencing the effectiveness of algae biomass productivity under conditions favourable for anaerobic fermentation is the temperature of the process. In present studies under conditions of anaerobic mesophilic fermentation conducted at 35°C, 38°C and 40°C, the productivity of the biomass of algae *Chlorella vulgaris BA-0167* was decreasing virtually from the first day of cultivation (at 38°C and 40°C, respectively), which indicates decay of algae cells under those conditions. It was also found that only the temperature of 33°C allows a nearly twofold increase in biomass concentration.

The research confirms the necessity to modify the conditions of cultivation, because of limited algae biomass growth. The changes concern such aspects as increase in the frequency of nutrient medium supply, use of artificial light source and constant magnetic mixing. Expanding the experimental data base in this manner will allow to determine whether the improvement in culture conditions influences an effective growth of algae under the assumed cultivation conditions in terms of temperatures favourable for anaerobic mesophilic fermentation, i.e. 33°C - 40°C.

References

1. S. Ma, B. Han, V.A.R. Huss, X. Hu, X. Sun, J. Zhang, Hydrobiologia **760**, 1 (2015)
2. C. Posten, G. Shaub, J. Biotechnol. **142** (2009)
3. M. Płaczek, A. Patyna, S. Witczak, E3S Web of Conferences **19** (2017)
4. M. Pawlita-Posmyk, M. Wzorek, Chemik **70**, 10 (2016)
5. M. Dębowski, M. Zieliński, A. Grala, M. Dudek, Renew. Sust. Energ. Rev. **27** (2013)
6. S.H. Shah, I.A. Raja, M. Rizwan, N. Rashid, Q. Mahmood, F.A. Shah, A. Pervez, Renew. Sust. Energ. Rev. **81** (2018)
7. M. Pawlita-Posmyk, M. Wzorek, E3S Web of Conferences **19** (2017)
8. M. P. Caporgno, R. Trobajo, N. Caiola, C. Ibáñez, A. Fabregat, C. Bengoa, Renew. Energ. **75** (2015)
9. P. Baltrėnas, A. Misevičius, J. Environ. Health Sci. Eng. **13**, 18 (2015)
10. M. Ras, L. Lardon, S. Bruno, N. Bernet, J.P. Steyer, Bioresour. Technol. 102, 1 (2011)
11. H.W. Yen, D.E. Brune, Bioresour. Technol. **98** (2007)
12. L. Mendez, A. Mahdy, M. Demuez, M. Ballesteros, C. González-Fernández, Fuel **117** (2014)
13. M. Wang, A.K. Sahu, B. Rusten, C. Park, Bioresour. Technol. **142** (2013)
14. L. Mendez, A. Mahdy, R.A. Timmers, M. Ballesteros, Bioresour. Technol. **149** (2013)
15. D. Kępska, Ł. Olejnik, Chemik **68**, 11 (2014)
16. Q. Béchet, A. Shilton, B. Guieysse, Biotechnol. Adv. **31** (2013)
17. M. Frąc, M. S. Jezierska-Tys, J. Tys, Acta Agrophysica **13**, 3 (2009)
18. Y. Chisti, Biotechnol. Adv. **25** (2007)
19. M. Frąc, S. Jezierska-Tys, J. Tys, Afr. J. Biotechnol. **9**, 54 (2010)
20. W. Kozieł, T. Włodarczyk, Acta Agroph. **17**, 1 (2011)
21. K. Dziosa, M. Makowska, Inż. Ap. Chem. **54**, 4 (2015)

22. A. Kumar, S. Ergas, X. Yuan, A. Sahu, Q. Zhang, J. Dewulf, F. X. Malcata, Trends in Biotechnology **28** (2010)
23. S.P. Shukla, J. Kviderová, J. Tříska, J. Elster, Front Microbiol. **4** (2013)
24. S. Cao, X. Zhou, W. Jin, F. Wang, R. Tu, S. Han, H. Chen, C. Chen, G.J. Xie, F. Ma, Bioresour. Technol. **244**, 2 (2017)
25. Z. Ouyang, X. Wen, Y. Geng, H. Mei, H. Hu, G. Zhang, Y. Li, J. Wuhan Bot. Res. **28**, 1 (2010)
26. L.A. Leyva, Y. Bashan, A. Mendoza, L.E. de-Bashan, Naturwissenschaften **101**, 10 (2014)
27. H.P. Kusumaningrum, M. Zainuri, Int. J. Mar. Aquat. Res. Conserv. Co-Exist. **1**,1 (2014)
28. F.G. Acién Fernández, D.O. Hall, E. Cañizares Guerrero, K.K. Rao, E. Molina Grima, J. Biotechnol. **103** (2003)
29. Y. Chisti, Biotechnol. Adv. **25** (2007)
30. R.A. Voloshin, M.V. Rodionova, S.K. Zharmukhamedov, T.N. Veziroglu, S.I. Allakherdiev, Int. J. Hydrogen Energ. **41**, 39 (2016)
31. E. Jankowska, A.K. Sahu, P. Oleskowicz-Popiel, Renew. Sust. Energ. Rev. **75** (2017)
32. Y. Santosh, T.R. Sreekrishnan, S. Kohli, V. Rana, Bioresour. Technol. **95**, 1 (2004).
33. P. Weiland, Appl. Microbiol. Biotechnol. **85**, 4 (2010)
34. S. O. Dahunsi, S. Oranusi, V.E. Efeovbokhan, J. Clean. Prod. **156** (2017)
35. A. Converti, A. A. Casazza, E. Y. Ortiz, P. Perego, M. D. Borghi, Chem. Eng. Process. **48** (2009)
36. K.O. Cassidy, Theses and Dissertations-Biosystems and Agricultural Engineering, 2011
37. S. Daliry, A. Hallajisani, J.M. Roshandeh, H. Nouri, A. Golzary, Global J. Environ. Sci. Manage **3**, 2 (2017)
38. M.E. Montingelli, S. Tedesco, A. G. Olabi, Renew. Sust. Energ. Rev. **43** (2015)
39. A. O. Jegede, Int. J. Agric. & Biol. Eng. **5**, 3 (2012)

A computational investigation of the hydrodynamics of the Badush dam in northern Iraq

Younis Saida Saeedrashed, and *Ali Cemal* Benim*

Center of Flow Simulation (CFS), Department of Mechanical and Process Engineering, Düsseldorf University of Applied Sciences, Münsterstr. 156, D-40476 Düsseldorf, Germany

Abstract. A computational analysis of the hydrodynamics of the Badush dam in Iraq is presented, which is planned to be reconstructed as a repulse dam, to prevent the Mosul city, in case of a failure of the Mosul dam. Computational Fluid Dynamics (CFD) is applied in combination with Geometric Information System (GIS) and Digital Elevation Model (DEM). In the first part of the study, a hydrologic study of a possible Mosul dam failure is performed, predicting the important parameters for a possible flooding of Mosul city. Here, a two-dimensional, depth-averaged shallow water equations are used to formulate the flow. Based on GIS and DEM, the required reservoir size and the water level of the Badush dam are predicted, for its acting as a repulse dam. Subsequently, a computational model of the reconstructed Badush dam is developed, combining the proposed construction with the local geographic topology to achieve a perfect fit. Finally, the water flow through the bottom outlets and stilling basin of the proposed dam is calculated by an unsteady, three-dimensional CFD analysis of the turbulent, free-surface flow. The CFD model is validated by comparing the predictions with measurements obtained on a physical model, where a quite satisfactory agreement is observed.

1 Introduction

In eighties, Mosul dam, which is a large multi-purpose dam located on the Tigris river at about 60 km Northwest of the Mosul city was constructed. Its total storage at maximum operation pool is 11.1 billion cubic meters and the maximum discharge of spillway at maximum pool is 13,000 m^3/s. It is constructed on a foundation of soluble soils that are continuously dissolving, resulting in the formation of underground cavities and voids that place the dam under some continuing risk and thus requires a continuous grouting program to mitigate. The most significant feature regarding the dam foundation is the presence of anhydrite and gypsum bedding and the perpetual grouting program undertaken.

Since 1989, Mosul dam is considered as a floating dam because of the above-mentioned problems at its foundation. Thus, the dam is continuously under an intensive monitoring by the local and international expertise. It is obvious that a failure of the Mosul dam would have severe consequences.

For the case of a failure of the Mosul dam, it is of utmost importance to have a repulse dam between the Mosul dam and the Mosul city that can absorb the flooding water and protect the city from flooding. Such a dam should have the capability of storing a huge amount of water that would be released by the breaking Mosul dam in its upstream. As already identified by the previous studies, the already existing, partially constructed (less than 50%) Badush dam, which is situated 40 km downstream the Mosul dam (16 km northwest of Mosul) has the potential of acting as a repulse dam, provided that it is reconstructed for this purpose. The Badush dam currently has a design flood level of 250 m.a.s.l. corresponding to a storage capacity of 0.39×10^9 m^3, which is too small. A reconstruction of the dam to a larger capacity to take the role of a repulse dam was already planned and various investigations for its reconstruction were performed [1,2]. However, a reconstruction of the dam could not be performed until now.

In the present work, the hydrodynamics of the bottom outlets and the stilling basin of the proposed construction [2] are computationally investigated. The present goal of the analysis is the validation of the computational models by comparisons of the predictions with the detailed measurements performed on the downscaled physical model [2]. A further goal, for the future studies, is to contribute to the improvements of the design, applying the validated computational models.

2 Modelling

The present work is a part of a study with a broader perspective, which also addresses the Mosul dam brake and the resulting flood scenarios with an accurate consideration of the terrain topology.

Remote sensing and Geographic Information Systems (GIS) are being effectively used as important tools to determine the quantitative description of

* Corresponding author: alicemal@prof-benim.com

morphometry of a basin [3], which provide an integrated platform for using Digital Elevation Model (DEM) data. In the present work, the package ArcGIS 10.5 [4] is used as the GIS software tool in order to process 10 m resolution of DEM for extracting related hydrological feature classes and information of the regions under study.

Although the simulation of flood in case of Mosul dam break is not the main scope of the present paper, the modelling framework of the flood simulation shall shortly be outlined here: For the flood simulation, a two-dimensional approach is used, based on a depth-averaged shallow water formulation. This is done within the framework of the hydrologic prediction software HEC-RAS 5.0.3 [5], while adapting commonly employed [6] Manning's roughness coefficients [7].

The unsteady, three-dimensional flow simulations for the Badush dam are performed by the general-purpose finite-difference/finite-volume based CFD code FLOW-3D [8], which uses a special technique, based on open area and volume fractions, for modelling complex geometries on a rectangular grid. The prevailing flow is a two-phase flow of water and air. Unlike entrained two-phase flows, where, e.g. one phase is dispersed in the other [9], the air and water phases are essentially separated by an interface (free-surface). In such cases, the rather state-of-the art approach would be to define a single velocity field, which is governed by the Navier-Stokes equations in the water and air regions, both, based on the local material properties in the corresponding domains, respectively, along with an adequate interface treatment.

Besides this type of modelling, which belongs to the category of volume-of-fluid (vof) [10] model, a simpler model, the so-called "one-fluid" model, is additionally provided in FLOW-3D [8] as a reasonable and efficient one, adequate for the present class of flows, which leads to a much sharper representation of the free-surface [8]. Here, the governing equations are solved only in the liquid domain, but not in the gas one. Although the governing equations are not solved for the gas (in our case air), the model has the capability of taking air entrainment (e.g. by turbulent mixing) empirically into account [11].

In turbulence modelling, advanced, high-resolution approaches such as Large Eddy Simulation (LES) [12] or Detached Eddy Simulation (DES) are being increasingly used in different classes of flow problems [13-15]. However, in the present work, which presents an initial effort in the field, a RANS (Reynolds Averaged Numerical Simulation) [16] approach is adopted as the turbulence modelling framework, which is being applied in the majority of engineering problems in a large spectrum of applications [17-22]. As turbulence model, the RNG (ReNormalization Group) k-ε model [23] is used, which is amended by the so-called wall-functions [8,16].

The velocity-pressure coupling is treated by a pressure-correction method [8], along with a staggered-grid arrangement for the velocity and pressure nodes on a structured grid. The convective terms are discretized by a second-order accurate monotonicity-preserving upwind scheme [24]. In time, a second-order implicit scheme is used [8].

3 Mosul dam break

Although this is not the main topic of the present paper, this issue is shortly addressed, as it provides the main motivation for the studies of the Badush dam, and also dictates the boundary conditions for the modelling of the Badush dam.

DEM data with 10 m resolution is used. Landsat satellite image with 15 m resolution is used for land use/land cover (LULC) classification. For the Manning friction coefficients, the following values are used: 0.03 for Tigris river channel, 0.04 for vegetation and river banks, 0.01 for Mosul city and for 0.05 soil texture.

It is assumed that the process of the flooding will start at elevation 330 m.a.s.l. as a worst scenario. The distance between the Mosul dam to the Mosul city is considered to be 69,000 m. The approximate total area of Mosul city is 209.5 km^2, the elevation of Mosul city is about 228 m.a.s.l. and elevation of the Tigris river bed is 210 m.a.s.l..

The two-dimensional domain is meshed by an equidistant mesh. As a result of grid-independence study, a grid with 500x500 cells is used.

The predicted flooding situation after the reach of the flood wave to Mosul city is displayed in Figure 1, where the position of the Badush dam intended as the repulse dam is also indicated.

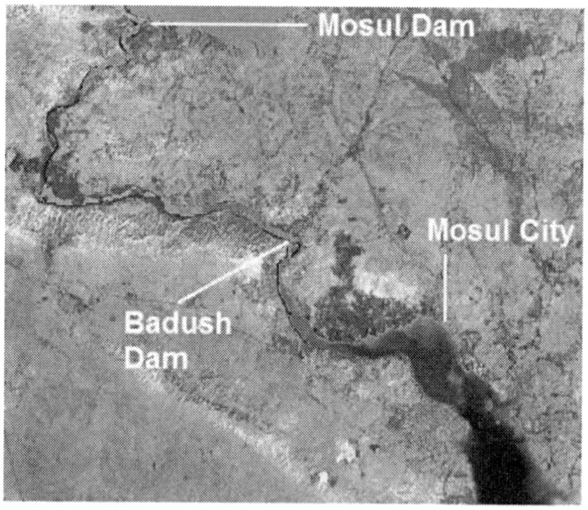

Fig. 1. Predicted flooding after Mosul dam break.

Table 1 summarizes some of the important data predicted for Mosul city for a possible Mosul dam failure.

According to the prediction, the flood wave would reach Mosul city approximately in two hours after the Mosul dam failure, and the maximum flood wave height of approximately 25 meters would be reached in the city in approximately 8 hours.

 https://doi.org/10.1051/matecconf/201824004009

Table 1. Predicted parameters for Mosul city after dam break.

Time of wave arrival to Mosul city	1.9 h
Time to max. wave height in Mosul city	8.0 h
Average flood velocity in Mosul city	3.9 m/s
Average flood depth in Mosul city	36.1 m
Maximum flood depth in Mosul city	44.7 m
Maximum wave height in Mosul city	24.4 m
Inundated area in Mosul city	58%

4 Badush dam

Badush dam should have the capability of storing a huge amount of water, as the water volume kept by the Mosul dam is about 11.1 billion cubic meters. With the help DEM, we have determined that a reconstruction of Badush dam to a flood level up to 307 m.a.s.l. would enable a quite large reservoir of $9.8x10^9$ m^3, which is sufficiently close to the required capacity. The corresponding reservoir area, which amounts in $357.69x10^6$ m^2 is indicated in Figure 2, based on a DEM with a resolution of 10m.

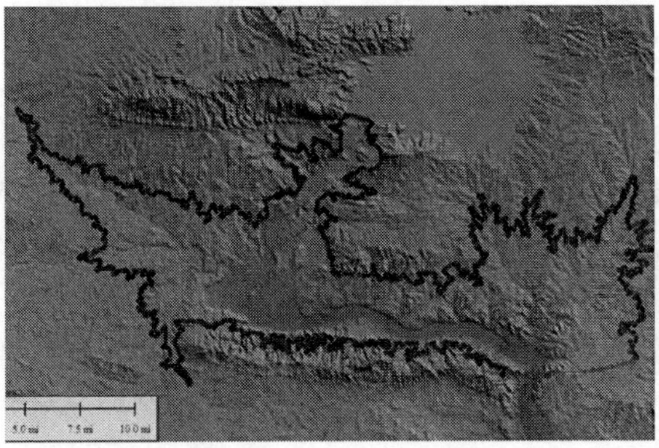

Fig. 2. Suggested Badush dam reservoir area (DEM with 10m resolution).

The important parameters of the intended Badush dam re-construction are summarized in Table 2.

Table 2. Summary of Badush dam specifications.

Dam crest level	312 masl
Spillway crest level	300 masl
Maximum height	102 m
Maximum width	248 m
Number of bottom outlets	8
Abutments crest level	312 masl
Maximum height of abutments	92 m
Total length (Concrete dam + Abutment dam)	3,686 m
Max. reservoir level (Mosul dam break)	307 masl
Total capacity	10^9 m^3
Maximum spillway capacity	4000 m^3/s
Maximum bottom outlets capacity	8000 m^3/s
Design flood during operation	8000 m^3/s

A virtual model of the Badush dam integrated in the topologic landscape is developed. This model relies, essentially, on the previously proposed construction [1,2]. A novelty of the present model is in the advanced methodology applied to fit the dam structure to the topology utilizing DEM, whereas surveyance based topologic data was used in the previous studies as the basis [1,2].

To this purpose, both concrete part and earthfill part of the dam structure which are saved in STL format are investigated to ensure that the dam structures are locating on the exact location that should be. The exact location of the large scaled dam was already determined as ground control points (GCPs) by locally measured data which have been collected via the process of surveying using both GPS and total station instruments. During preprocessing of the collected data, the results showed that the dam structures are coming in fit with its location on the DEM, exactly as the case in the reality; this insured that there is a perfect compatibility between designed dam structure and the DEM.

The central parts of the created virtual dam model is depicted in Figure 3. In the figure, the bottom outlets and the stilling basin are indicated which are of main concern in the present study, as well as the spillways, which play a role especially for high values of the water level (e.g. for a situation that would occur in case of Mosul dam failure). Left to the block containing the bottom outlets, stilling basin and the spillways, the hydropower unit can be seen.

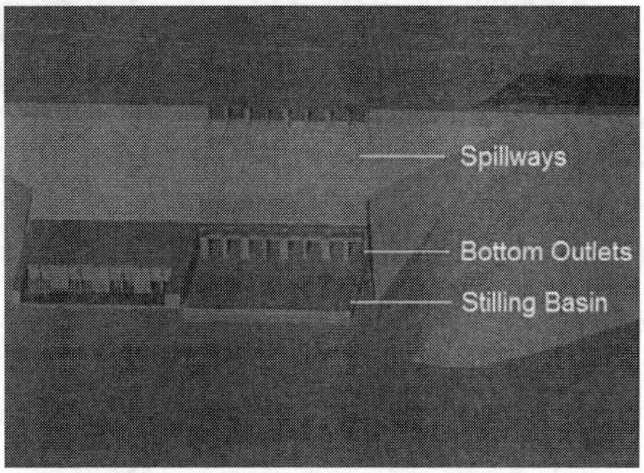

Fig. 3. DEM based created virtual model of Badush dam (detail view).

The number of bottom outlets, which is planned to be eight in total in the considered design (Table 2), and their opening scenarios are important for preventing possible harmful effects on the downstream side of the dam.

It is necessary to define the number of bottom outlets that can be opened as a function of the headwater elevation. For headwater equal to or lower than the normal water level (NWL) of 250 m.a.s.l (meters above sea level), all the eight can be opened (Case 1), while for the headwater attaining a level higher than the NWL (exceptional, but critical operation in the case of Mosul dam failure) not all of the bottom outlets can be totally

https://doi.org/10.1051/matecconf/201824004009

opened at the same time (Case 2). In that case, bottom outlets need to be operated possibly in different combinations with the spillways, depending on the prevailing water level in the reservoir.

The main role of the stilling basin is energy dissipation including the spillway and low level outlets operation, to minimize any undesirable effects in the downstream including scouring. The dissipation of the kinetic energy in the stilling basin can be affected by the basic layout of the stilling basin, with respect to the prescribed lengths and elevations of the apron part and sill part, which may be designed to trigger a hydraulic jump, as well as the introduced artificial structures, roughness elements such as rows and blocks. The definition of the solution domain is depicted in Figure 4.

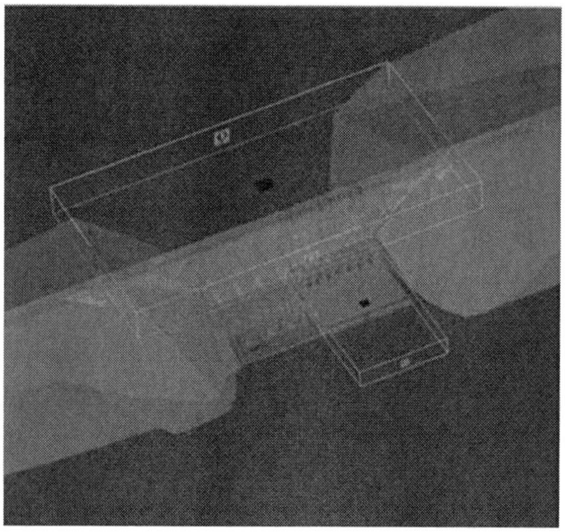

Fig. 4. Solution domain.

On the free lateral boundaries of the solution domain, symmetry conditions are applied. The boundary surfaces that enclose the domain on the top are also prescribed as symmetry boundaries. At the downstream, a pressure outlet boundary is applied, which is further adjusted for local flow being sub- or supercritical [8]. At the inlet a pressure distribution corresponding to the hydrostatic pressure field is prescribed. The remaining boundaries are solid walls, along which no-slip conditions are apply that are augmented by wall-functions in conjunction with the applied two-equation turbulence model [8].

For achieving an adequate grid resolution, a grid independence study is performed. A grid with approx. 2 million cells is found to deliver sufficient grid independence with respect to the monitored value, the flow rate, which is, then, used in the subsequent calculations. A view of the used computational grid is presented in Figure 5. As can be seen in Figure 5, a structured, rectangular grid, consisting of two main blocks is used. The curved/irregular boundaries that do not follow the straight grid lines, i.e. the solid walls of the dam structure and the ground, are approximated by fractional areas and fractional volumes of the cells on interface surfaces, by the feature FAVOR

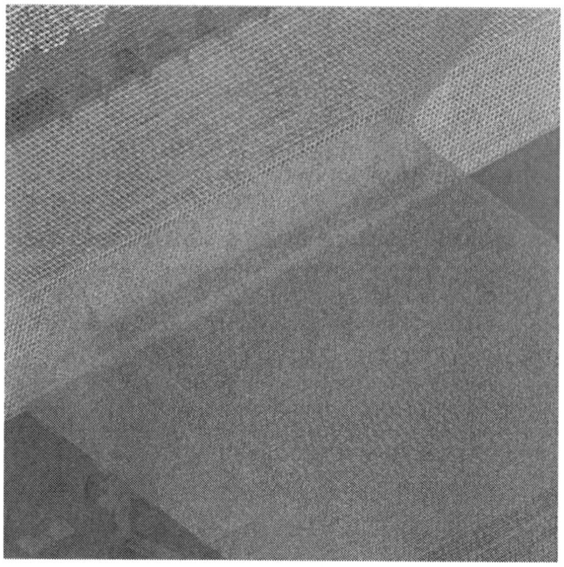

Fig. 5. A view of the computational grid.

of FLOW-3D [8].

The shape of the free-surface at an instant of time is presented in Figure 6, where the color represents the velocity magnitude (m/s) in Fig. 6a, and the gage static pressure (Pa) in Fig. 6b.

The predicted fields of velocity magnitude (m/s), the velocity component in the horizontal direction, the gage static pressure (Pa), the fluid depth (m) as well as the Froude number based on the depth-average velocity are displayed in Figure 7, for an instant of time, in a plane extending in the main flow direction through the middle of a bottom outlet.

As one can see in the figure, the flow abruptly accelerates at the bottom outlet, filling up the initial part, but forming a free surface in the downstream. The thinning of the water layer in further downstream with a corresponding increase of the water velocity indicate a supercritical flow state, which is transformed into a subcritical one by a hydraulic jump in the stilling basin (Fig. 7).

Inspecting the horizontal velocity component (which is defined to be negative in the main flow direction, i.e. form right to left) recirculation patterns can be detected. Just downstream the hydraulic jump, a rather strong recirculation via a reverse flow on the surface is observed (Fig. 7b). Another major recirculation zone is observed downstream the sill.

Measurements were performed on a 1:60 physical model of the dam, which are documented in [2]. For the water level of 250 m.a.s.l., the measured (and upscaled) value of the discharge (Q), the depth-averaged velocity magnitudes at the apron (V_{APRON}) and sill (V_{SILL}), as well as the water depths at the apron (h_{APRON}) and sill (h_{SILL}) of the stilling basin are compared with the predicted value in Table 3.

One can see that the predictions show a reasonable agreement with the available experimental data obtained on a 1:60 scale physical model [2].

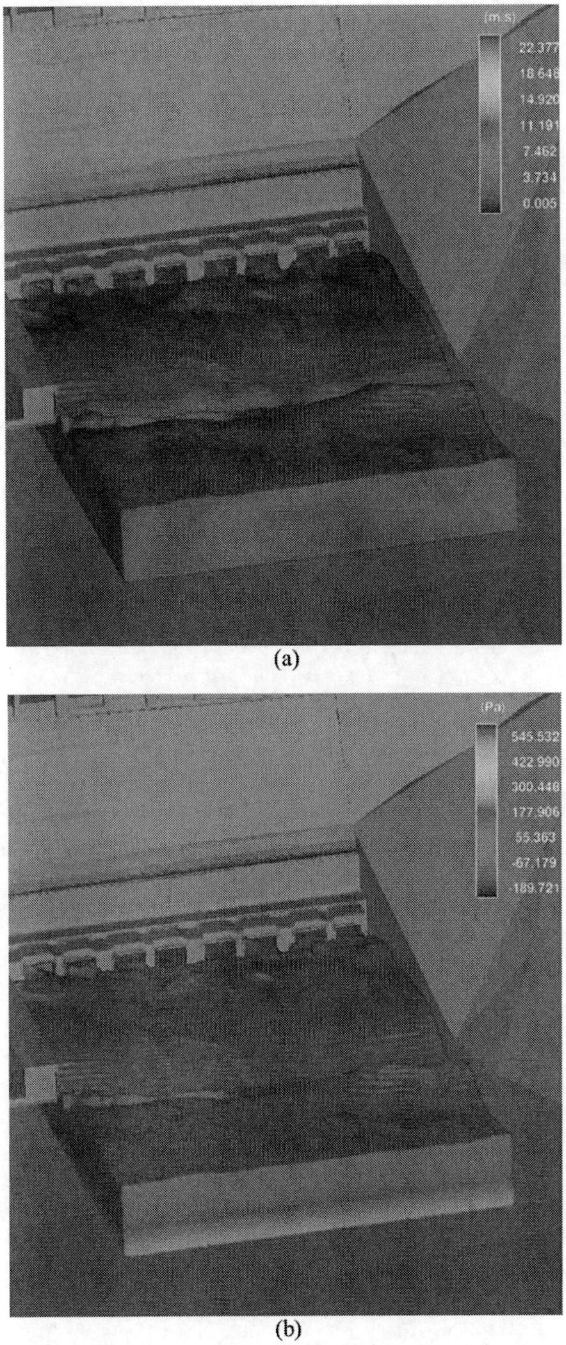

Fig. 6. Predicted instantaneous free surface colored by (a) velocity magnitude (m/s), (b) gage static pressure (Pa).

Table 3. Summary of Badush dam specifications. Comparison of the predicted and measured [2] values.

	EXP	PRED
Q (m^3/s)	8200	8323
V_{APRON} (m/s)	6.0	5.9
V_{SILL} (m/s)	4.7	5.4
h_{APRON} (m)	17.9	16.2
h_{SILL} (m)	13.4	13.8

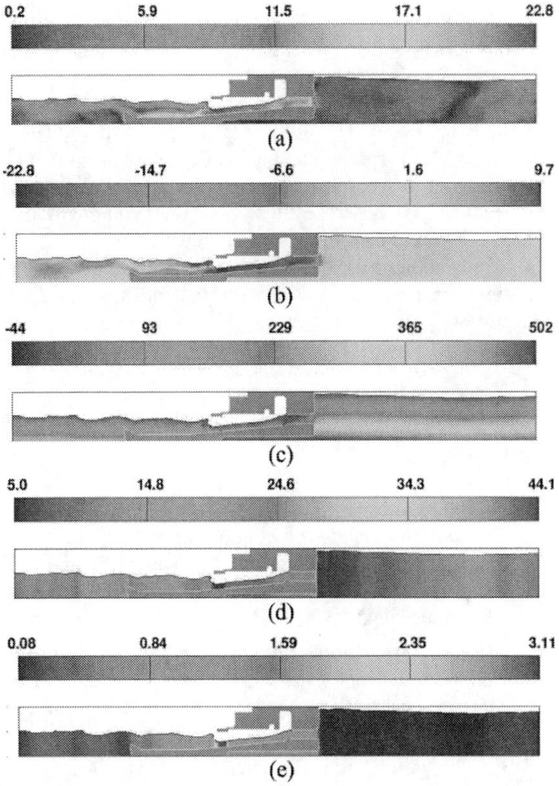

Fig. 7. Predictions in plane extending in main flow direction, through a bottom outlet, at an instant of time, (a) velocity magnitude (m/s), (b) velocity component in horizontal direction (m/s) (defined negative in the main flow direction, i.e. from right to left), (c) gage static pressure (Pa), (d) surface elevation (m), (e) Froude number.

5 Conclusions

In the present study, the hydrodynamics characteristics of the Badush dam in northern Iraq is computationally investigated, the reconstruction of which as a repulse dam is being planned to protect the Mosul city from flooding, in case of a possible failure of the Mosul dam. Different computational tools such as CFD, GIS and DEM are used in combination. The study is performed in two stages. In the first stage, a hydrologic study of a probable Mosul dam break is performed, computationally, based on a two-dimensional, depth-averaged shallow water formulation. As result, besides the important parameters for the flooding of the Mosul city, the required reservoir size and water level of Badush dam are predicted for its reconstruction as a repulse dam.

Subsequently, a virtual model of the reconstructed Badush dam is developed, combining a previously proposed construction with the actual local geographic topology to achieve a perfect fit. The water flow through the bottom outlets and stilling basin of the proposed Badush dam are calculated by an unsteady, three-dimensional CFD analysis of the turbulent, free-surface flow of water, for a given water level. A specialized version of the volume of fluid model is used to treat the two-phase flow that solves the transport equations in the water domain only, and the turbulence is modeled by the renormalization group theory version of the k-ε turbulence model. The applied CFD model is validated by comparing the predictions with the measurements obtained on a 1:60 physical model. A quite satisfactory agreement is observed between the computational predictions and the experimental analysis.

Application of the developed and validated model for design optimizations of the Badush dam is planned as future work.

We gratefully acknowledge the financial support by the Alexander von Humboldt Foundation.

References

1. J. Muskatirovic, *Badush dam project: hydraulic model tests, simulation of flood waves* (Report of the Institute for Development of Water Resources, Hydraulic Laboratory, Jarislav Cerni Institute, Belgrade, Jugoslavia, 1988)
2. Consultants, *Badush dam project, Phase B: Design adjustment and final design, Vol. 1: Hydraulic design verification* (Ministry of Water Resources, Republic of Iraq, 2009)
3. K. Vaidya, R. Chauhan, Int. J. Geomatics Geosciences, **3**, pp.464-473 (2013)
4. ArcGIS, 2017, www.esri.com/argis
5. HEC-RAS, 2017, http://www.hec.usace.army.mil/software/hec-ras/
6. P.B. Bedient, W.C. Huber, B.E. Vieux, *Hydrology and Floodplain Analysis* (Pearson, London, 2008).
7. H. Dorn, M. Vetter, B. Hofle, Rem. Sensing, **6**, pp.1739-1759 (2014)
8. FLOW-3D, 2017, www.flow3d.com
9. A.C. Benim, B. Epple, B. Krohmer, Progress in Computational Fluid Dynamics, **5**, 6, pp.345-361 (2005)
10. C.W. Hirt, B.D. Nichols, J. Computational Physics, **39**, pp.201-225 (1981)
11. C.W. Hirt, *Modeling turbulent entrainment of air at a free surface* (Technical Note FSI-03-TN61-R, Flow Science Inc., Santa Fe, NM, USA, 2012)
12. P. Sagaut, *Large Eddy Simulation for Incompressible Flows – An Introduction*, 3rd ed. (Springer, Berlin, 2006)
13. A.C. Benim, A. Nahavandi, K.J. Syed, Progress in Computational Fluid Dynamics, **5**, 8, pp.444-454 (2005)
14. A.C. Benim, M.P. Escudier, A. Nahavandi, A.K. Nickson, K.J. Syed, F. Joos, International Journal of Numerical Methods for Heat and Fluid Flow, **20**, 3, pp.348-370 (2010)
15. A.C. Benim, H. Chattopadhyay, A. Nahavandi, International Journal of Thermal Sciences, **50**, 10, pp. 1973-1983 (2011)
16. P.A. Durbin, B.A.P. Reif, *Statistical Theory and Modeling for Turbulent Flows*, 2nd ed. (Wiley, Chichester, 2011)
17. A.C. Benim, M. Geiger, S. Doehler, M. Schoenenberger, H. Roemer, *"Modelling the flow in the exhaust hood of steam turbines under consideration of turbine-exhaust hood interaction"*, Book Series: VDI Berichte, Vol. 1185, pp.343-357 (VDI Verlag, Düsseldorf, 1995)
18. A.C. Benim, A. Nahavandi, *A computational analysis of turbulent swirling flows*, in: K. Hanjalic, Y. Nagano. M. J. Tummers (Eds.) *Turbulence, Heat and Mass Transfer*, vol. 4, pp.715-722 (Begell House, Danbury, Connecticut, 2003)
19. A.C. Benim, D. Brillert, M. Cagan, *Investigation into the computational analysis of direct-transfer pre-swirl systems for gas turbine cooling*, ASME Paper No. GT2004-54151, in: Proc. ASME Turbo Expo 2004,Vienna, Austria, June 14-17, 2004, Vol. 4, pp. 453-460, 8 pages, doi: 10.1115/GT2004-54151 (2004)
20. A.C. Benim, M. Cagan, A. Nahavandi, E. Pasqualotto, *RANS predictions of turbulent flow past a circular cylinder over the critical regime*, Proc. 5th IASME/WSEAS International Conference on Fluid Mechanics and Aerodynamics, Athens, Greece, August 25-27, 2007, pp.232-237 (2007)
21. H. Chattopadhyay, A.C. Benim, Journal of Heat Transfer – Transactions of the ASME, **133**, 10, Article Nr.: 104502, 5 pages (2011)
22. A. Assmann, A.C. Benim, F. Gül, P. Lux, P. Akhyari, U. Boeken, F. Joos, P. Feindt, A. Lichtenberg, Journal of Biomechanics, **45**, 1, pp. 156-163 (2012)
23. V. Yakhot, L.M. Smith, J. Scientific Comput., **7**, pp.35-61 (1992)
24. B. van Leer, J. Comput. Phys., **23**, pp.276-299 (1977)

A numerical analysis of the thermohydraulics of an EGS project in Turkey

Ali Cemal Benim[1,*], *Aydin* Cicek[2], and *Arif Mert* Eker[3]

[1]Center of Flow Simulation (CFS), Department of Mechanical and Process Engineering, Düsseldorf University of Applied Sciences, Münsterstr. 156, D-40476 Düsseldorf, Germany
[2]Department of Energy Raw Materials and Survey, General Directorate of Mineral Research and Exploration of Turkey
[3]HD Energy, Inc., Turkey

Abstract. A numerical study of the thermohydraulics of an enhanced geothermal system project in Turkey is presented. The solid structures are modelled as porous media, using the numerically determined hydraulic fracturing data of other authors. The influence of several numerical modelling aspects such as the domain size, grid resolution, temporal resolution as well as the discretization scheme are investigated and assessed to obtain highly accurate numerical solutions under the applied modelling assumptions. Using the suggested mathematical and numerical model, different production scenarios are investigated.

1 Introduction

Geothermal energy is becoming increasingly important in the energy research due to its large capacity and independence on weather conditions, unlike the solar and wind energy. Turkey is positioned as the seventh richest country in the world in its potential on geothermal energy. The estimated total geothermal energy potential of Turkey is about 60 GW [1]. Currently, there are approximately forty installed geothermal power plants in Turkey, and their total installed capacity is approximately 1100 MW_e. The existing, conventional geothermal energy systems are based on natural cracks and reservoirs that are water-dominated. In cases, where the natural cracks and pores do not permit an economic operation, the permeability can be improved by hydraulic fracturing of the rock structures. Such systems are designated as Enhanced Geothermal Systems (EGS). Indeed, applying hydraulic fracturing, i.e. by EGS, the geothermal energy can be gained without a need for a natural water reserve. In Turkey, in a region near Dikili of the Izmir province, with considerable geothermal properties such as a geothermal gradient of approximately 7K/100m, an EGS project is planned. Hou et al. [2] published an investigation on several aspects of this project, including geology and geochemistry.

In the present analysis, a further numerical investigation of the thermohydraulics of this EGS project is presented, based on the boundary conditions and the numerically obtained hydraulic fracturing data published by Hou et al. [2]. The presently applied modelling is similar to that used by Hou et al. [2], which was performed by the software FLAC3D [3], which is a code specialized on geotechnical analysis. However, in [2] essentially no details on the mathematical and numerical modelling aspect were provided. In the present analysis, modelling details are provided and discussed. For guaranteeing high numerical accuracy, the influences of the domain size, grid resolution, temporal resolution as well as the discretization schemes on the numerical prediction are investigated. Thus, an accurate numerical model of the problem within the framework of the applied mathematical model is reached. The developed modeling strategy is, then, used to predict different production scenarios.

2 The considered EGS configuration

A triplet system with a single injection well, and two production wells are investigated in the current analysis. The configuration of the wells with the considered hexagonal region of the rock structures with its depth, thickness and width are is in Figure 1.

The considered geometrical arrangement has two symmetry planes. These symmetry planes are considered in the modelling. Thus, a ¼ th of the region shown in Figure 1 is defined to be the solution domain. The latter, which is based on the definitions provided in [2] is illustrated in Figure 2, where the different rock zones, dimensions (numbers indicate length in meters) as well as boundary conditions are also indicated.

The considered block of the rocks starts at a depth of 1710 m measured from the ground surface and reaches a depth of 3010 m. Each of the zones denoted as "upper" and "lower" has a depth of 150 m. The fractured zone, which is designated as "frac" is positioned between the "upper" and "lower" zones and has the dimensions of 1200x33x1000 (all dimensions in meters). The solid

* Corresponding author: alicemal@prof-benim.com

 https://doi.org/10.1051/matecconf/201824005001

structure around the fractured zone is designated as "rock". The depths of the injection and extraction ports, measured from the upper border of the "frac" zone, are assumed to be the same and denoted by H. The separation of the injection and production wells is denoted by L. The extensions of the "rock" region in x and y directions are indicated by X and Y.

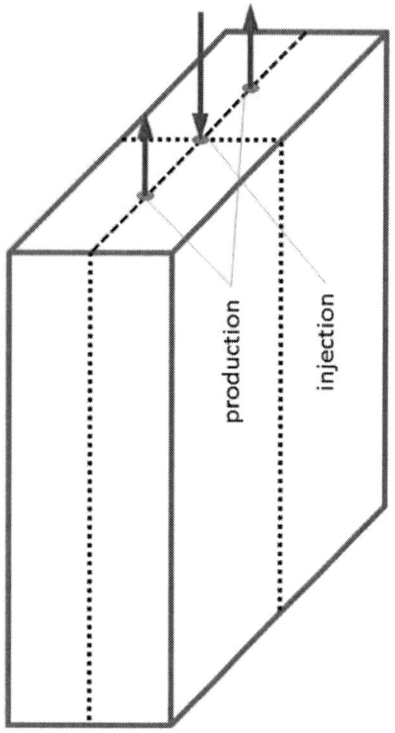

Fig. 1. Arrangement of injection and production wells.

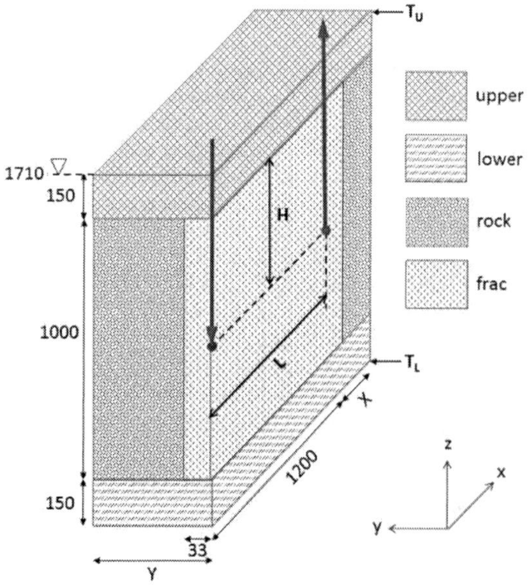

Fig. 2. Solution domain, rock zones, dimensions (numbers indicate length in meters), boundary conditions.

3 An outline of the modelling

The computational investigation is conducted within the framework of the general purpose Finite Volume Method [4] based Computational Fluid Dynamics software ANSYS Fluent 18.0 [5]. The rock regions are modelled as porous regions. Since the time scales of the process is quite long, thermal equilibrium between fluid and solid zones is assumed, which means that a single energy equation is solved [4,5] accounting for the heat transfer by conduction and convection, while the mechanical energies are neglected. Apparently, the radiative heat transfer [6] does not play a role in the present problem.

The introduced fluid is water at moderate temperatures. The water pressures that are encountered are expectedly quite high, also due to the hydrostatic accumulation of the pressure. A boiling of the injected water in the low, hot regions of the rock, followed by a condensation in the upper rock layers before, or during the production is principally possible, as it can happen in some EGS systems. However, in the present case, a study of the prevailing temperatures and pressures together with the properties of water has implied that it is unlikely that the liquid water would undergoes a phase change. Thus, the fluid considered in the system is modelled to be the liquid, incompressible water. For the fluid, i.e. liquid water, and solids, constant material properties are assumed, corresponding approximately to the prevailing thermodynamic conditions, on the average. The material properties of solids are assumed to be the same for all rock zones.

The material properties used in the calculations are given as follows: Fluid viscosity: 0.00019 Pa.s, fluid density: 940 kg/m^3, fluid specific heat capacity: 4176 J/kgK, fluid thermal conductivity: 0.717 W/mK, solid density: 2670 kg/m^3, solid specific heat capacity: 965 J/kgK, solid thermal conductivity 2.83 W/mK.

Turbulence modelling in porous media is a challenge [7]. In the current formulation, the turbulence effects are omitted as a first approximation. This neglect may at least partially be reasoned by the expectedly quite low superficial, and also physical, velocities in most parts of the solution domain. Thus, unlike the majority of engineering heat transfer applications [8-17], no turbulence model is applied. The influence of the porous solid region on the hydrodynamics of the fluid is considered by a sink of momentum in the Navier-Stokes equations that corresponds to a drop of static pressure according to law of Darcy [18]. The inertial resistance is omitted based on the above arguments and lack of data.

For the parameters that describe the rock zones, the hydraulic fracturing data provided by Hu et al. [2] is used as guidance in the current investigation. The currently applied rock zones' porosities and permeabilities (perm) [2] are provided in Table 1 (permeabilities in m^2). As one can see in the table, the porosities and permeabilities are, in general, quite small. Different from the remaining rock regions, the permeability of the fractured region is non-isotropic (Table 1). In the fractured region, the permeability is several orders of magnitude greater in the main plane of fracturing, i.e. in the x-z plane of Fig. 2), in comparison

Table 1. The porosities and permeabilities (m^2) of the rock zones.

	upper	frac	rock	lower
porosity	0.013	0.02	0.02	0.02
perm (x)	$4x10^{-18}$	$1x10^{-13}$	$4x10^{-18}$	$4x10^{-18}$
perm (y)	$4x10^{-18}$	$4x10^{-18}$	$4x10^{-18}$	$4x10^{-18}$
perm (z)	$4x10^{-18}$	$4x10^{-14}$	$4x10^{-18}$	$4x10^{-18}$

to the perpendicular direction (y, Fig.2). The four vertical boundaries of the solution domain (Fig. 2) are prescribed to be symmetry boundaries. The two horizontal boundaries are defined as isothermal walls as indicated in Figure 2 (T_U: upper wall temp., T_L: lower wall temp.).

According to the available information [2], the determined upper and lower wall temperatures have been: T_U=137°C, T_L=240°C. In the current, time dependent problem, initial conditions are also needed, together with the boundary conditions. As the initial condition, a stationary fluid with a linearly varying temperature distribution between the upper (T_U) and lower (T_L) isothermal planes (Fig. 2) is assumed.

Between the diameters of the boreholes and the size of the solution domain, there is a very large difference of scales, which is very difficult to resolve by a computational mesh, if a modelling of the pipes coupled with the rest of the solution domain is aimed at.

Therefore, currently, it is found convenient not to calculate the flow in the pipes and the heat transfer between the fluid in the pipe and the solid structures. This may be considered as a reasonable approach for the present purpose, since the diameters of the pipes are very small (approximately 0.2 m) in comparison to the size of the domain and the pipes are usually coated by cement with a low thermal conductivity. Consequently, the modelled outer walls of the pipes are assumed to be adiabatic walls, and the inlet and outlet boundaries of the solution domain are positioned at the location (Fig. 2) of the outlet surface of the injection pipe, and the inlet surface of the production pipe, respectively.

Moreover, a fine resolution of the processes in the near-fields of the inlet and exit planes of the pipes is not attempted, either. The prescribed inlet and outlet boundaries are planes with much greater dimensions (20 m x 20 m) in comparison to the cross-sectional areas of the real pipes. Since the low porosity and permeability (Table 1) tend to produce a fast dispersion of the flow, the currently adopted modelling may be considered to be convenient for the present purpose. At the inlet plane, a homogeneous inlet velocity normal to the plane, and a constant inlet temperature are prescribed as boundary conditions, while a uniform static pressure and the zero-gradient condition for the velocities are applied at the outlet. The inlet temperature is prescribed to be 60°C.

In the numerical formulation, for the time discretization, a backward differencing scheme [4,5] that is second order accurate is mainly used. A first order scheme is also applied, for comparison. For the spatial discretization, the formally third order accurate QUICK [19] discretization scheme is mainly used for the convective terms. For the purpose of comparison, the second order upwind [20], power law [21], and first order upwind schemes [21] are also applied for the convective terms. The velocity-pressure coupling is treated by the PISO scheme [22].

Within the iterative procedure of a time step, the components of the velocity vector are under-relaxed by an under-relaxation coefficient of 0.5. No under-relaxation is applied to pressure and temperature. As the convergence criteria within each time step, the threshold values for the normalized residuals are fixed to 10^{-12} for the energy equation and to 10^{-6} for the continuity and momentum equations. The applied backward time differencing scheme is, essentially, unconditionally stable. Nevertheless, the time-step sizes (Δt) are prescribed as such that the cell Courant numbers (Co) [4] remain below unity, as it would be required for a convection problem by an explicit scheme.

4 Numerical accuracy studies

4.1 Domain size

In the current time-dependent problem, the temperature front, which propagates in time, shall not interact with the spatial boundaries, i.e. be falsified by the prescribed boundary conditions, within the considered time domain. It has been aimed to analyze a time period of twenty years. The vertical boundary locations with the prescribed temperatures (Fig, 2), were acquired form [2]. Based on this definition [2] it can be assumed that the locations of the boundaries, each with a distance of 150 m from the fractured region allow sufficient room. This can similarly be assumed for the dimensions of the domain in the remaining directions (X, Y, Fig. 2).

Nevertheless, a course estimation was carried out, based on the analytical solution of a one-dimensional, transient heat conduction problem in a semi-infinite domain [23]. This implied a penetration depth of approximately 90 m, supporting the convenience of the assumed distance of 150 m. Initial numerical predictions were also carried out with varying X and Y (Fig. 2), which also supported the convenience of the above-mentioned dimensions. Therefore, for the main calculations the domain size is prescribed by the dimensions X=150m and Y=200m (Fig. 2).

4.2 Grid independence study

The solution domain is discretized by essentially equidistant, orthogonal, block-structured, hexahedral grids. For ensuring grid independence three different grid sizes are investigated. The total number of grid nodes of these three grids are given as: The Coarse Grid: 54,000, The Medium Grid: 185,000, The Fine Grid: 410,000. The predictions are obtained for H = 590 m, for the same injection rate, using the QUICK discretization scheme, the second-order backward time differencing and the same time-step size for all three calculations.

The predicted variations of the temperature by the three grids along a horizontal line between the production and extraction ports, for a moment 90 days after the injection start are presented in Figure 3. As can be seen, the predictions of Medium and Fine grids are

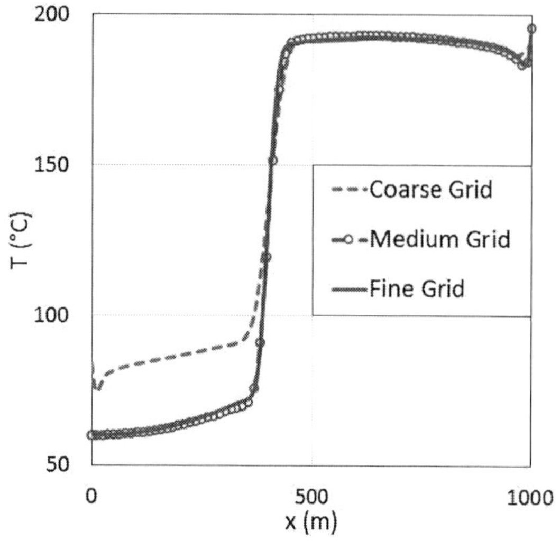

Fig. 3. Temperature profile between injection and extraction ports after 90 days: Grid independence.

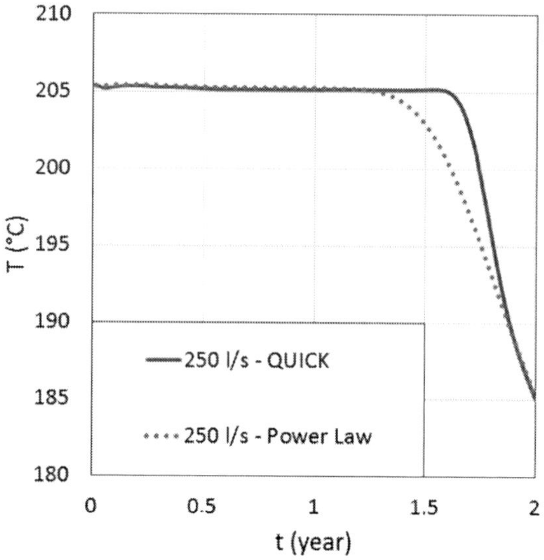

Fig. 4. Temporal variation of production temperature for a period of two years: Spatial discretization schemes.

nearly identical. Therefore, for the Medium grid with a total number of grid nodes 185,000 is taken as the basis for further analysis.

4.3 Effect of spatial discretization schemes

The predictions are obtained applying the same grid (Medium), for H = 590 m, applying the QUICK, second order upwind, power law and first order upwind schemes, applying the second order backward differencing in time, as well as the same time-step size.

The temporal variations of the production temperature, i.e. the temperature at the exit of the solution domain, predicted for an injection volume flow rate of 250 l/s by the QUICK and power law schemes, are presented in Figure 4, for a period of two years.

The predictions of the second order upwind and the first order upwind schemes are not indicated separately in Figure 4, since they are practically identical with those of the QUICK and power law schemes.

As can be seen, the production temperature is constant for a certain period of time and starts to decay after a production time of approximately 1.5 years. One can also see that the different discretization schemes agree with each other quite well, except for the transition period around 1.5 – 2 years, where the power law scheme (as well as the first order upwind scheme) predicts an earlier decline of the production temperature compared to the QUICK scheme (as well as the second order upwind scheme). In the present, main predictions, the principally more accurate QUICK scheme is used.

4.4 Effect of temporal discretization

Mainly, a second order backward time differencing is used, along with a time step (Δt) that ensures a cell Courant number (Co) not exceeding the unity. Because the current problem is dominated by the conduction heat transfer, especially in the non-fractured parts, the stability condition for sole conduction may also be checked, which requires the cell Fourier number (Fo) to remain below ½. A comparison of the occurring Co and Fo indicates that the ratio of Co to Fo is approximately 20 and, the satisfaction of the condition Co < 1 would directly satisfy Fo < ½ tenfold. For the Medium grid and a volume flow rate of 250 l/s, a time step size of Δt=6h, sufficiently fulfills the latter condition. For testing the accuracy of the temporal resolution, a computation is carried out using a time-step size, which is halved (Δt=3h). In the computations, the QUICK scheme is applied for the spatial discretization. All predictions are observed to be essentially identical, confirming the applied strategy in temporal discretization.

5 Results

Using the described numerical model (X=150m, Y=200m, QUICK and Power Law schemes, second order backward time discretization, Δt ensuring Co < 1), the case with H = 590 m, L = 1000 m is analyzed for a period of 20 years, for a volume flow rate of 250 l/s. The contours of predicted temperature contours in plane y-z through the injection wells (Fig. 2) 20 years after production are presented in Figure 5, as predicted by the QUICK and Power Law schemes.

In this plane, with low permeability the conductive heat transfer is dominating. One can see that the temperature front has a rather low penetration in the y direction and remains quite confined to the fractured zone, as estimated before (Section 4.1). It is interesting to note that the results obtained by the QUICK (Fig. 5a) and Power Law (Fig. 5b) scheme for the temperature distribution in this plane, after 20 years are very similar.

The predicted temperature contours in plane x-z through the wells (Fig. 2) 20 years after production are presented in Figure 6. One can see that although the low temperature front has reached the production well in

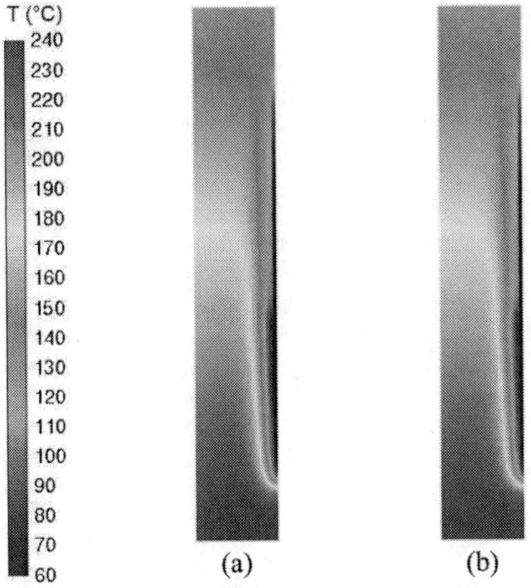

Fig. 5. Temperature distribution in y-z symmetry plane (Fig. 2) through injection well for volume flow rate 250 l/s after 20 years of production (a) QUICK, (b) Power Law.

20 years, the production temperature is still above the injection temperature, due to the mixing of the warmer water, which is heated up by the hot rocks that surround the fractured zone and convectively transported to the production well by a recirculating flow behind the production well (Fig. 6). Comparing the predictions provided by the QUICK (Fig. 6a) and Power Law (Fig. 6b) schemes, once can see that both discretization schemes deliver very similar predictions for the temperature distribution in this plane, after 20 years of production. However, some differences can still be observed, especially in the vicinity of the production well, where it can be observed that the QUICK scheme (Fig. 6a) predicts a slightly higher temperature level for the recirculating flow into the production port, compared to the Power Law scheme (Fig. 6b).

The influence of the injection rate is investigated by considering four values of the water volume flow rates, namely 250 l/s, 200 l/s, 150 l/s and 100 l/s. In all these calculations, the QUICK scheme is used. The temporal variations of the production temperature, i.e. the static temperature at the exit of the solution domain (i.e. inlet of the production well) for a period of 20 years, predicted for injection water volume flow rates of 250 l/s, 200 l/s, 150 l/s and 100 l/s are presented in Figure 7. At the start of the process, the production temperature is only a little above 200 °C and has the same value for all cases. As a common tendency for all cases, it can be seen that the production temperature stays constant and then declines in time. The period before the moment of decline increases with decreasing water volume flow rate. For a water volume flow rate of 250 l/s and 100 l/s this moment is reached after approximately 1.5 and approximately 4 years, respectively, with values in between for the remaining water volume flow rates.

One can observe that the production temperature decays more quickly with increasing water volume flow rate, as it would be expected. Because the gained thermal

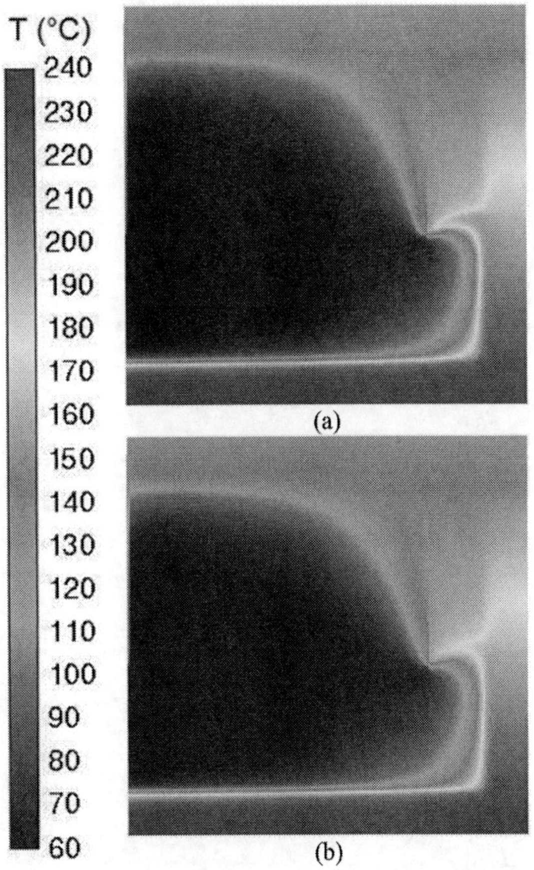

Fig. 6. Temperature distribution in x-z symmetry plane (Fig. 2) through wells for volume flow rate 250 l/s after 20 of production (a) QUICK, (b) Power Law.

energy is aimed to be used for power generation, the production temperature should be high enough. The temperature after a production period of 20 years is near and a little below 110°C for the cases with water volume flow rates 200 l/s and 250 l/s, respectively. For the cases with water volume flow rates of 150 l/s and 100 l/s, the production temperatures predicted after 20 years are substantially higher, i.e. approximately 120°C, and 140 °C, respectively. The favorable injection rate should be defined by accounting for the geothermal capacity, the temperature draw-down and the economic aspects. The current information implies that the preferred injection rate should be within the range 100 l/s – 150 l/s.

Conclusions

A numerical investigation of an intended EGS project in Turkey is carried out. Special attention is devoted to the issues pertaining to the numerical accuracy. For a given configuration, the velocity, temperature and pressure fields are predicted for 20 years, for different injection rates. The predictions show that the injection rates 150 l/s and 100 l/s are adequate, since they can ensure a sufficiently high production temperature level over a period of 20 years. The investigation will be extended in

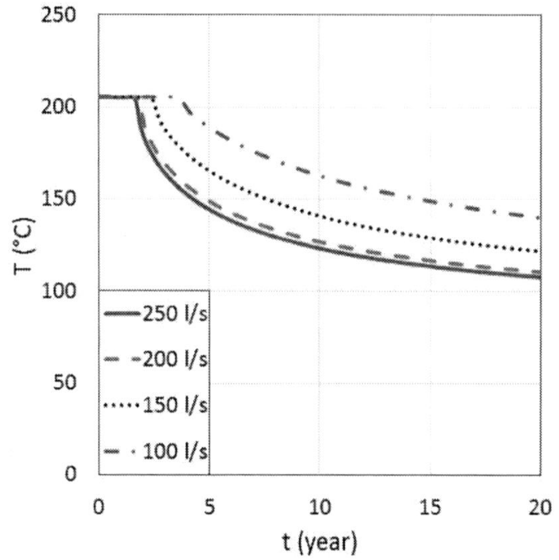

Fig. 7. Variation of production temperature in time for different water volume flow rates.

the future work. This includes the investigation of the possibilities of accounting for local inertial and turbulence effects in modelling the porous media. On the other hand, alternative concepts for the configuration of the injection and production wells, as well as for intensifying the heat exchange between the working fluid and the hot rocks will be investigated.

References

1. M. Parlaktuna, O. Mertoglu, S. Simsek, H. Paksoy, N. Basarir, *Geothermal country update report of Turkey* (European Geothermal Congress Pisa, Italy, 2013)
2. Z. Hou, et al., Environ.Earth Sci.,**73**, pp.6747-6767 (2015)
3. https://www.itascacg.com/software/flac3d
4. H.K. Versteeg, W. Malalasekera, *An Introduction to Computational Fluid Dynamics – The Finite Volume Method, 2nd ed* (Pearson, Harlow, 2007)
5. www.ansys.com
6. A.C. Benim, Computer Methods in Applied Mechanics and Engineering, **67**, 1, pp. 1-14 (1988)
7. N.F. Jouybari, T.S. Lundstrom, J.G.I. Hellstrom, M. Maerefat, M.E. Nimvari, J. Porous Media, **19**, pp.497-513 (2016)
8. A.C. Benim, A. Nahavandi, *A computational analysis of turbulent swirling flows*", in: K. Hanjalic, Y. Nagano. M. J. Tummers (Eds.) *Turbulence, Heat and Mass Transfer*, vol. 4, pp.715-722 (Begell House, Danbury, Connecticut, 2003)
9. A.C. Benim, M. Cagan, D. Gunes, International Journal of Thermal Sciences, **43**, 8, pp.725-732 (2004)
10. A.C. Benim, D. Brillert, M. Cagan, ASME Paper No. GT2004-54151, **4**, pp. 453-460, (2004)
11. A.C. Benim, K. Ozkan, M. Cagan, D. Gunes, International Journal of Numerical Methods for Heat & Fluid Flow, **17**, 3, pp.284-301 (2007)
12. A.C. Benim, H. Chattopadhyay, A. Nahavandi, International Journal of Thermal Sciences, **50**, 10, pp.1973-1983 (2011)
13. H. Chattopadhyay, A.C. Benim, Journal of Heat Transfer – Transactions of the ASME, **133**, 10, Article Nr.: 104502, 5 pages, (2011)
14. I. Taymaz, E. Aslan, A.C. Benim, Thermal Science, **19**, 2, pp.537-547 (2015)
15. P. Oclon, S. Lopata, M. Nowak, A.C. Benim, Progress in Comput. Fluid Dynamics, **15**, 5, pp.290-306 (2015)
16. A.C. Benim, A. Nahavandi, K.J. Syed, Progress in Computational Fluid Dynamics, **5**, 8, pp.444-454 (2005)
17. A.C. Benim, M. Cagan, A. Nahavandi, E. Pasqualotto, *RANS predictions of turbulent flow past a circular cylinder over the critical regime*, (Proc. 5th IASME/WSEAS International Conference on Fluid Mechanics and Aerodynamics, Athens, Greece, pp.232-237, 2007)
18. S. Whitaker, Transport in Porous Media, **1**, pp.3-25 (1986)
19. B.P. Leonard, Comput. Meths. Appl. Mechanics and Eng., **8**, pp.59-98 (1979)
20. T.J. Barth, D.C. Jespersen, AIAA-89-0366 (1989)
21. S.V. Patankar, *Numerical Heat Transfer and Fluid Flow* (Hemisphere, New York, 1980)
22. R.I. Issa, Journal of Computational Physics, **62**, pp.66-82 (1986)
23. J.P. Holman, *Heat Transfer, 10th edition*, (McGraw-Hill, New York, 2010)

Simulation of water turbine integrated with electrical generator

Dariusz Borkowski [1,*], *Michał* Węgiel [1], *Paweł* Ocłoń [1], and *Tomasz* Węgiel [1]

[1] Cracow University of Technology, Warszawska St. 24, 31-155 Cracow, Poland

Abstract. The paper presents the analysis of the small hydropower plant working at variable speed. The hydro-set that consists of the guide vanes and propeller turbine integrated with the permanent magnet synchronous generator is simulated by using Computational Fluid Dynamics (CFD) in Ansys Fluent v18.0. The k-ε and k-ω SST models, as well as the one-equation Spalart-Allmaras (SA) model, were tested. The comparison showed the significant divergence of the calculation results. The turbulence model selection influences the average value of the power and also the speed for which the power is maximal.

1 Introduction

Small Hydropower Plants (SHPs) are objects up to 10MW. The SHPs are mainly the run-of-the-river schemes, where the water storage is limited or impossible. Therefore, the operating parameters are influenced by the actual hydrological conditions. In order to improve the object profitability, the operating range of the turbine discharge needs to be maximal and the average efficiency of the energy conversion have to be possible high [1]. The standard turbine solutions implementing double regulation (e.g. Kaplan turbine) are relatively expensive and can be replaced by the simple propeller turbine working at variable speed [2]. The variable speed operation is provided by the power electronic converter (PEC) that matches electrical energy of the generator with power system requirements [3]. In order to improve the overall system efficiency, the permanent magnet synchronous generator (PMSG) can be used. The PMSG allows to eliminate the mechanical transmission gear by construction with high pair pole numbers, what decreases the generator nominal speed. Furthermore, the simple propeller turbine and flexible generator construction allows to integrate these two elements into one unit, what simplify the hydro-set construction [4].

The hydro-set containing propeller turbine integrated with electrical generator is not well investigated yet. In order to support the design process and determine the operating conditions the detailed modeling is required. The computational fluid dynamics (CFD) technique seems to be the most suitable taking into account qualitive and quantitive results and analysis of the fluid flow machines of complex geometries [5]. However, it requires a properly defined calculation model, which includes: precise geometry, proper grid generation, suitable turbulence model and correct boundary conditions [6].

The paper presents the CFD analysis using three turbulence models. First is the k-ε realizable model providing good performance for flows with rotation and adverse pressure gradients influencing the boundary layers [7, 8]. The second analyzed model is k-ω SST (Shear-Stress Transport), which combines standard k-ω model near the wall and k-ε model in the outer portion of the boundary layer. It gives the accurate results for flows with adverse gradient pressure (flows with separation), therefore it is reliable for turbines [9,10]. Finally, the one-equation Spalart-Allmaras turbulent model is applied. This is a robust model suitable for mildly complex flow under pressure gradient (airfoils, wings). By applying the Wall Treatment technique, it is also applicable in fluid flow machines with low grid resolution of boundary layer [11]. The hydro-set model is prepared and analyzed in Ansys Fluent v18.0 [12].

2 Numerical modelling

2.1. Geometry description

The analyzed scheme is the run-of-the-river SHP, which is situated in a bypass of the overfall weir. The cross-section of the SHP is presented in fig. 1. The SHP contains an axial-flow propeller turbine integrated with the permanent magnet synchronous generator (PMSG) [13]. The turbine is located in the generator rotor. The permanents magnets are mounted outside the rotor ring. The hydro-set contains also the guide vanes that controls the water volume flowing through the turbine (fig. 2). The fundamental and operating parameters of the modelled turbine are given in Table 1 [14].

* Corresponding author: dborkowski@pk.edu.pl

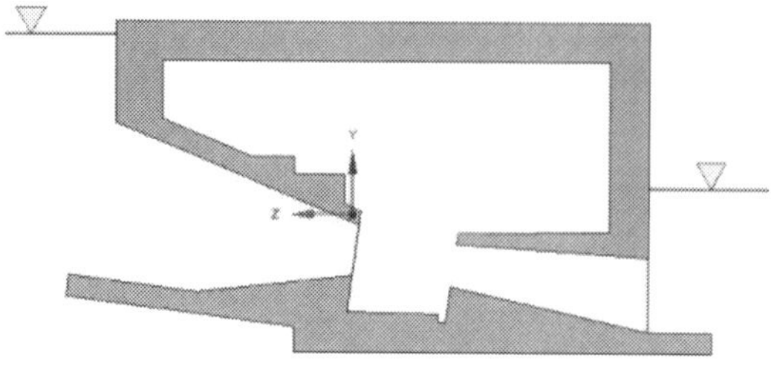

Fig. 1. SHP cross-section.

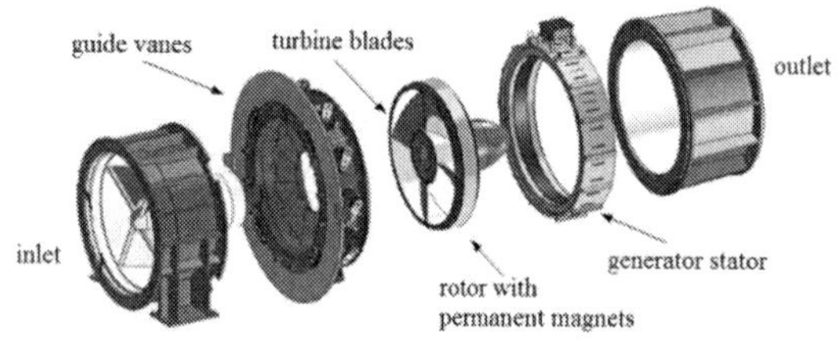

Fig. 2. Main elements of hydro-set.

Table 1. Setting Word's margins.

Parameter (Unit)	**Value**
Net head (m)	2.75
Discharge (m3/s)	3
Rotational speed (rpm)	300
Output power (kW)	75
Runner diameter (m)	0.95
Number of runner blades	4
Number of guide vanes	12

2.2. Mesh grid generation

The computational mesh was prepared in ANYS Meshing application based on the CAD model (Fig. 3.) for non-conformal components: the inlet pipe, hydroset tube that includes the guide vanes and the draft tube. The components were selected considering structure details in order to optimize the mesh. Additionally, to eliminate the computational complexity of the remeshing function the turbine rotor was prepared as a separate component. The curvature size function with a curvature normal angle equals 20° and maximal mesh element of 30 mm was applied. Additionally, due to the significant geometry complexity of the guide vanes and rotor the element size was reduced to 20 mm. The resulted mesh model is presented in Fig. 4.

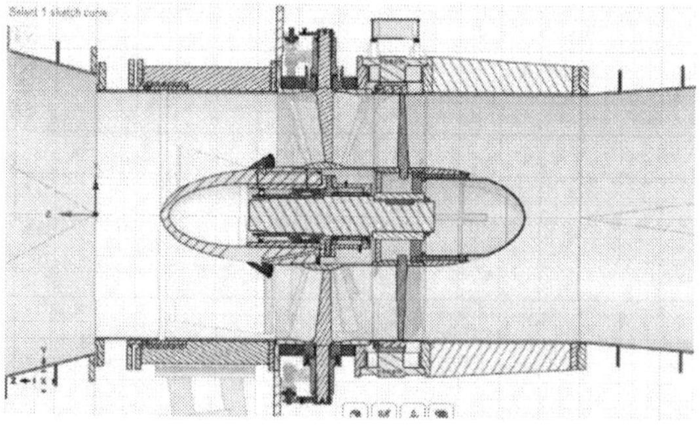

Fig. 3. SHP cross-section.

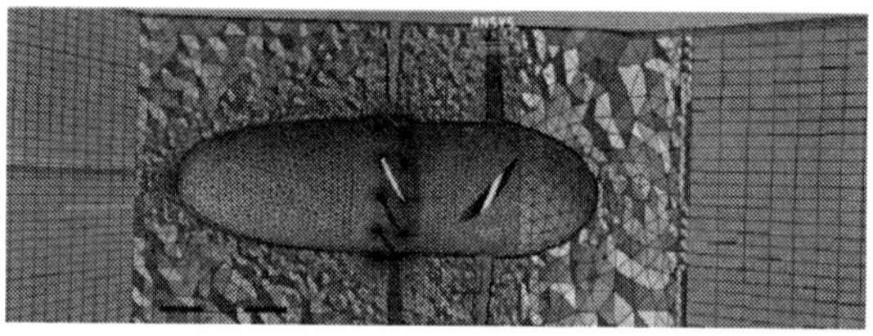

Fig. 4. SHP cross-section.

The generated mesh consists of about 1.2 million elements. The mesh quality was assessed using orthogonal quality and skewness parameters. The average value of the orthogonal quality equals 0,745 while the average skewness is 0,266. Both parameters satisfy the recommended values (orthogonal quality > 0.005, skewness < 0.98) for fluid flow calculations [1].

The calculations were performed for the three different models: Spalart-Allmaras, k-ε realizable and k-ω SST. The model convergence was assessed by the residuals, which measure the discretized equation solution inaccuracy [15]. The obtained residual curves show that the solutions of all models are converging to the exact solution after 50-100 iterations. The fastest stabilization provides the Spalart-Allmaras model, however some variations are presented in the obtained results.

2.3. Boundary conditions and solver settings

At the location of the inlet boundary condition the total pressure (152600 Pa) is given, which is the sum of the static pressure and dynamic pressure. The static pressure (149885 Pa) is defined by the hydrostatic head at the inlet (4,95 m), while the dynamic pressure (2715 Pa) results from the velocity of the fluid at the inlet (2,33 m/s). At the outlet boundary the backflow pressure (125360 Pa) caused by the hydrostatic head at the outlet (2,45 m), is specified (Fig. 5).

The pressure based solver allows to perform calculations of the fluid flow problems using separate or split equations. In case of the transient fluid flow, complex mesh grid or large time steps, the more accurate results are obtained by using the coupled algorithm. The most important solver parameters are listed in Table 2.

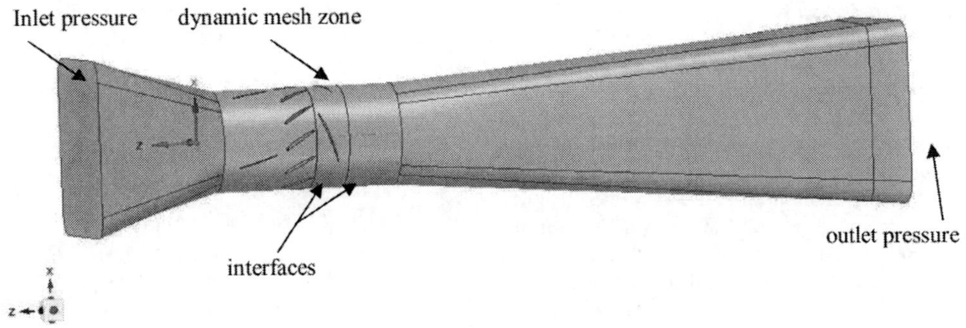

Fig. 5. Boundary conditions scheme.

Table 2. Solver settings and model relaxation factors.

Solver Parameter	**Value**
Pressure-Velocity Coupling	Coupled
Discretization of Gradient	Least Squares Cell Based
Discretization of Pressure	Presto
Discretization of Momentum	Second order Upwind
Discretization of Turbulent Viscosity	Second order Upwind
Transient Formulation	Second Order Implicit
Flow Courant number	100
Momentum and Pressure Explicit Relaxation Factor	0.75
Density and Body Forces Under-Relaxation Factors	1

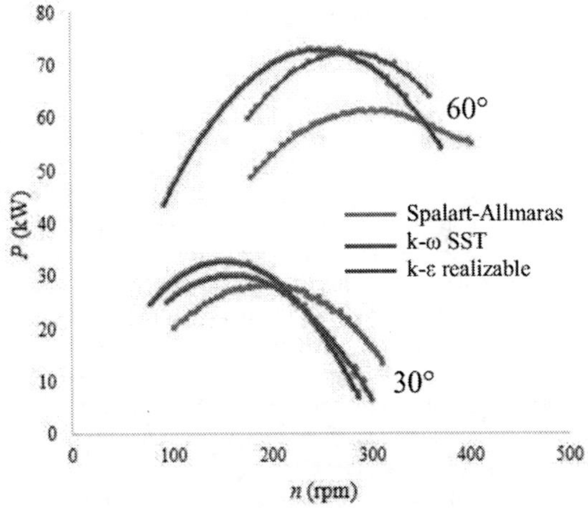

Fig. 6. Power as a function of turbine speed.

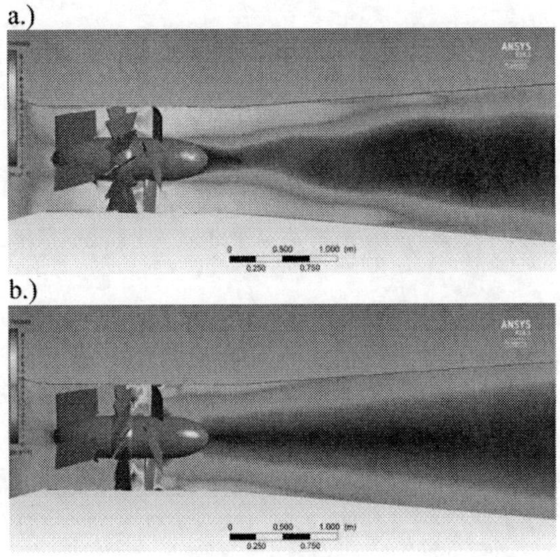

Fig. 7. Distribution of the axial velocity in the hydro-set for 60° (a.) and 30° (b.) of the guide vane angle.

3 Results and discussion

The calculation results of the CFD model for the three turbulence models are presented in Figure 6 in the form of power as function of turbine speed. Two values of the guide vanes angles are analyzed: 30° and 60°, where 0° corresponds to the close guide vanes.

Presented curves indicate the significant divergence of the calculation results. The turbulence model selection influences the average value of the power and also the speed for which the power is maximal.

The axial velocity distribution of hydro-set for both guide vane angles (30° and 60°) are presented (Fig. 7). It can be seen that velocity increases with the guide vane angle increase, which is due to increases in discharge. It is seen that the flow is smooth for nominal guide vane angle (60°), whereas there is significant velocity gradient at guide vanes for lower angle (30°).

* Corresponding author: dborkowski@pk.edu.pl

4 Conclusions

This paper determines the mechanical characteristics of a propeller turbine integrated with an electrical generator, operating at a variable rotational speed in a small hydropower plant, using the ANSYS Fluent software. For this purpose, a geometric model of the fluid domain was prepared. The numerical mesh was created that met the requirements of flow problems, and then using the available models of turbulence and calculation methods the characteristics are determined. The k-ε and k-ω SST models, as well as the and one-equation Spalart-Allmaras (SA) model, were tested. The SA model is preferred in the fluid flow calculations of flow machinery when the boundary layer mesh resolution is low. An additional advantage of using SA model is the shortened calculation time, which is especially crucial for variable time-dependent flows and large mesh models.

The simulations proved that the modeling methodology in the ANSYS FLUENT software can be successfully used in the process of designing and optimizing rotating machines, and in particular the facilities for which the prototyping is done. This mainly applies to prototype objects, to which practically every new hydropower plant can be calculated. At the stage of turbine design due to high acceleration resulting from the application of numerical methods, it is possible to analyze much more design variants at the same time and select the best option. Simultaneously, such a solution also brings financial benefits because in the numerical analysis only the computer model is changed and optimized, so there is no need to produce expensive prototypes every time the model changes.

References

1. G. A. Aggidis, E. Luchinskaya, R. Rothschild, D. C. Howard, Renewable Energy, **35**, pp. 2632–2638 (2010)
2. G. Ardizzon, G. Cavazzini, G. Pavesi, Renew. Sustain. Energy Rev., **31**, pp. 746–761 (2014)
3. J. Fraile-Ardanuy, J. R. Wilhelmi, J. J. Fraile-Mora, J. I. Pérez, IEEE Trans. Energy Convers., **21**, pp. 569–574 (2006)
4. D. Borkowski, T. Wegiel, IEEE Trans. Energy Convers., **28**, pp. 452–459 (2013)
5. J. H. Ferziger, M. Peric, *Computational Methods for Fluid Dynamics*, Springer Verlag, Heidelberg (2002)
6. C. Hirsch, *Numerical Computation of Internal and External Flows: The Fundamentals of Computational Fluid Dynamics*, 2nd Edition, Butterworth-Heinemann (2007)
7. T. C. Vu, M. Koller, M. Gauthier, C. Deschênes, , 25th IAHR Symposium on Hydraulic Machinery and Systems, Timisoara, Romania (2010)
8. A. Lipej, C. Poloni, Journal of Hydraulic Research, **38**, pp. 73-79 (2000)
9. F. R. Menter, AIAA Journal, **32**, pp. 1598-1605 (1994)
10. V. Prasad, K. S. Sayann, P. Krishnamachar, Indian Journal of Engineering & Material Sciences. **16**, pp. 229-236 (2009)
11. M. Kaniecki, Z. Krzemianowski, M. Banaszek, Transactions of The Institute of Fluid-Flow Machinery. **123**, pp. 71–84 (2011)
12. ANSYS Inc., *ANSYS Fluent Theory Guide, Canonsburg*, (PA: release 18.0, 2016)
13. D. Borkowski, IEEE Transactions on Energy Conversion, **32**, pp. 1090-1098 (2017)
14. T. Węgiel, D. Borkowski, D. Liszka, International Conference on the Sustainable Energy and Environment Development (SEED) (2016) doi: 10.1051/e3sconf/20161000100
15. H. H. Peng, W. Qiu, S. Ni, Ocean Engineerings, pp. 304-317 (2013)

A model of the novel concept liquid piston engine sourced by waste heat

Wojciech Bujalski[1], *Kamil* Futyma[2], *Jarosław* Milewski[3] and *Arkadiusz* Szczęśniak*[4]

[1]Warsaw University of Technology, Institute of Heat Engineering, Nowowiejska 21/25, 00-665 Warszawa
[2]Warsaw University of Technology, Institute of Heat Engineering, Nowowiejska 21/25, 00-665 Warszawa
[3]Warsaw University of Technology, Institute of Heat Engineering, Nowowiejska 21/25, 00-665 Warszawa
[4]Warsaw University of Technology, Institute of Heat Engineering, Nowowiejska 21/25, 00-665 Warszawa

Abstract. This paper describes the model of the novel concept liquid piston engine, which is designed to convert low-grade waste heat into electricity. The proposed dynamic oriented model is implemented in Aspen Hysys that enables simulations dynamic simulation of various working agents. The simulation results were verified with experimental data obtained from the research installation. The proposed model demonstrated relatively small discrepancies with respect to experimental research, hence it could be used as a tool for research on optimization of an innovative power plant operation, i.e. various working agents, various operating pressures.

1 Introduction

Nowadays, energy becomes more and more expensive what drives a research on new innovative power systems e.g system based on fuel cells [1], electrolysis combined with turbines [2, 3], compressed air energy storage [4,5] etc). One of the promising sources of power is an utilization of low-grade waste heat, which is being released into the environment. In addition, the utilization of waste heat could lead to reduction of fossil fuel consumption and alleviating environmental problems.

The promising concept for waste recovery seems to be an liquid piston engine, which is recognized as a type of Stirling engine [6]. The concept of liquid piston engine was firstly reported in the available literature in 1971 and denoted as Fluidyne engine [7]. Together with the basic system layout, the several modifications were proposed [8-10]. Slavin et al. [8] proposed a new scheme of engine that is characterized by use of offset heat exchangers instead of the usual heater and cooler. The system layout was firstly reported by [9]. In addition, Author [8] developed numerical model and simulated engine working cycle to define characteristic parameters for system design. Author concluded that the energy conversion based on the proposed principle is a promising area, however he [8] indicated that the realization of the prosed scheme is debatable due to difficulties related to its construction such as requirements for valves related to short actuation period, the realization of components sealing such as hydro-turbine.

We proposed the novel concept of fluid piston engine called "TURMES" [11], where the vapor of low-boiling fluids expands inside the tank and forces a flow of liquid to ensure a continuous liquid flow through the expander – see Fig 1.

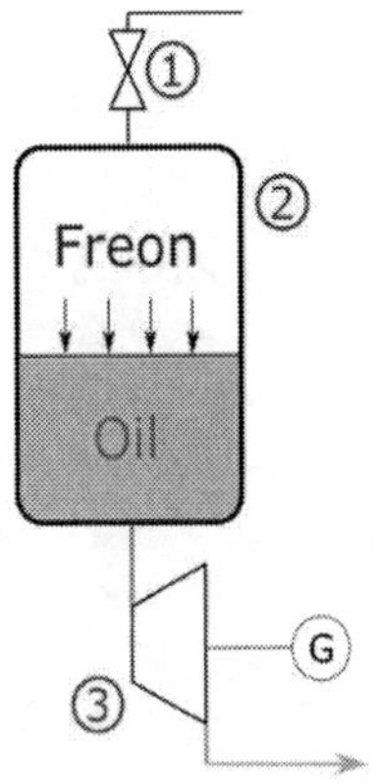

Fig. 1. General idea of the proposed concept TURMES; 1 – valve, 2 – pressurized tank, 3 – oil expander.

The proposed system was constructed in lab-scale (1 kW power installation) and used for model validation. The prelimary study [12-14] revealed that the most suitable pair of working agents with respect to operating conditions are vapor of R-245fa and oil: Ergolid A – selected parameters of working agents are displayed in table 1.

* Corresponding author: Arkadiusz.szczesniak@itc.pw.edu.pl

https://doi.org/10.1051/matecconf/201824005003

Table 1. Setting Word's margins.

Property	Value
Freon	
Trade name	R-245fa
CAS Number	460-73-1
Liquid working agent	
Trade name	Ergolid A (-35°)
CAS Number	107-21-1
% concentration of glycol	48
Type of glycol	ethylene

2 Working principle of the proposed concept of liquid piston engine

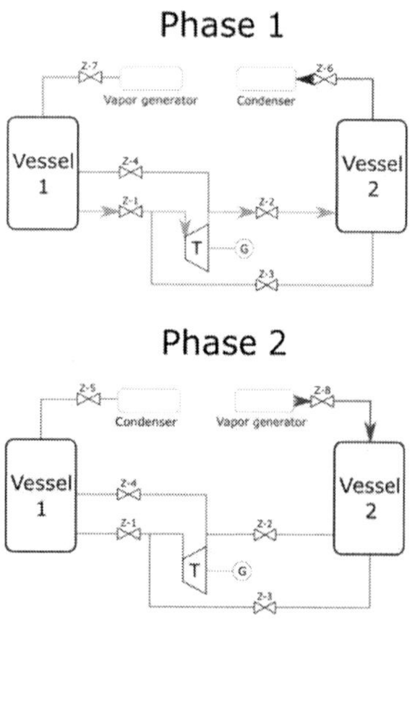

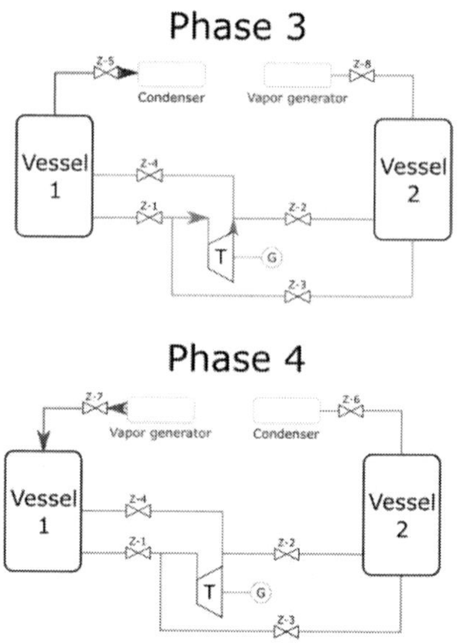

Fig 2: Simplified diagram the innovating power plant equipped with liquid turbine and hydraulic-gas generator of working agent's flow, sources by waste heat

The working principle of novel concept liquid piston engine is guided by opening and closing valves that are used to control the process of oil and liquid motion. In general, the four operation modes are distinguished at liquid piston operation (see Fig. 2):

- Phase 1 - freon expands inside the vessel 1 and forces the flow of Ergolid, which drives a turbine. Valve Z1 and Z2 are opened. Valve Z3 and Z4 are closed
- Phase 2 - when the pressure inside the vessel 1 drops to a required value and oil level in the vessel 1 reaches the minimum level, the vessel 2 is loaded by a live vapor from condenser (valve Z-8 is open, while valve Z8 and Z9 are closed. Pressure inside the vessel 2 rises up to designed value
- Phase 3 - vessel 2 starts to work. Valve Z3 is opened and liquid agent (Ergolid) flows from vessel 2 to vessel 1 and drives a turbine.
- Phase 4 - vapor loading to vessel 1 - phase very similar to stage 2

3 Dynamic oriented model of liquid piston engine

The basic thermodynamic assumption for modeling a working agents behavior is the proper selection of the gas model. In the examined case, modeling the work of the system using the ideal gas model would be a huge mistake. Modeling the operation of the system required the use of a real gas model. Among the many available models [15], the Peng-Robinson model [16] was selected, which is characterized by a very good mapping of the transformation of organic compounds, especially

https://doi.org/10.1051/matecconf/201824005003

in the area of phase transformations. This model, being a semi-empirical modification of the Van der Waals equation, is described by the basic dependence:

$$p = \frac{R \cdot T}{V_m - b} - \frac{a \cdot \alpha}{V_m^2 + 2b \cdot V_m - b^2} \qquad (1)$$

Where,
coefficient which defines the intermolecular interactions:

$$a = \frac{0.4572 \cdot R^2 \cdot T_c^2}{p_c} \qquad (2)$$

Coefficient which defines the volume of gas molecules:

$$b = \frac{0.0778 \cdot R \cdot T_c}{p_c} \qquad (3)$$

$$\alpha = (1 + (0.37464 + 1.54226 \cdot \omega - 0.269 \qquad (4)$$

$$T_r = \frac{T}{T_c} \qquad (5)$$

In general, the liquid piston engine consist of four main elements, which operation have to be simulated i.e.

- **Vessels (1 & 2)** Vessels are used for the working agent expansion to force the flow of liquid through the turbine (see Fig 2). The key thermodynamic assumption is an adiabatic efficiency of expansion process that occurs inside the vessels. The dynamic model assumes the adiabatic efficiency is equal to 99%. Due to the fact that adiabatic efficiency of expansion process in the vessel is greater that expansion in turbine (the adiabatic efficiency in turbine is within the range 80%..90%), the assumption seems to be resonable. Hence, it was assumed that heat transfer could be neglected. Vessels are simulated by the available in Aspen Hysys component "vessel" [15]. It requires to provide vessel geometry, the method of heat losses and definition of position of inlet and outlet port. The amount of liquid phase (Ergolid A) in the tank is determined by the height of the liquid column in % of the tank height. The geometric parameters used for simulations are shown in Table 2. It was assumed that there is no heat transfer between vessel and surrounding
- **Expander** – power generated by an oil expander is calculated based on the formula

$$W = \frac{(p_1 - p_2) \cdot \dot{m}}{\rho} \qquad (6)$$

where: p—pressure, bar; 1, 2—indices which denotes inlet and outlet, respectively; m—mass flow rate, kg/s; ρ —density, kg/m^3.
The characteristic curves of the simulated oil expander are shown in Fig. 3

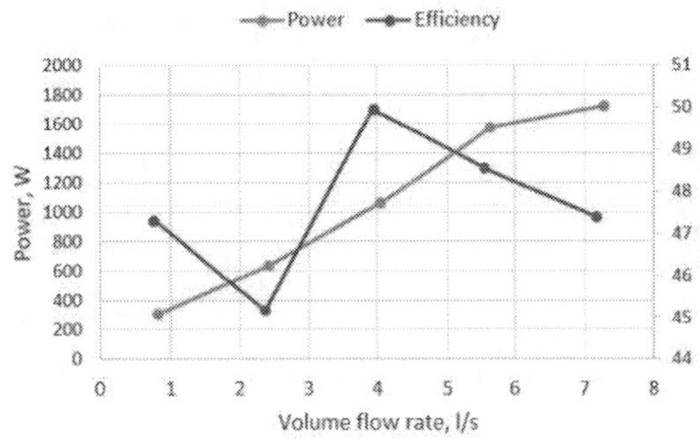

Fig 3: Characteristic curves of an expander operating on Ergolid

- **Valves** Valves are responsible for controlling the process of gas/liquid motion, hence the proper simulation of valves is a key element for accurate modeling of system operation. Inside the valve, the isenthalpic throttling process occurs in order to obtain a designed flow rate or pressure difference.

Table 2 Vessels parameters used in Hysys

Property	Value
Vessel orientation	Vertical cylinder*
Geometry	Flat cylinder*
Volume, m3	4e-2
Diameter, m	0,3238
Height, m	0,4857
Location of nozzles for gas inlet/outlet	Top of the vessel
Location of nozzles for oil inlet/outlet	Vessel bottom
Volume occupied by a Ergolid at the initial phase of expansion process, %	95

3.1 Model validation

The presented model was validated with experimental data. The comparison of experimental data against the model is presented in Figures 4 – 5.

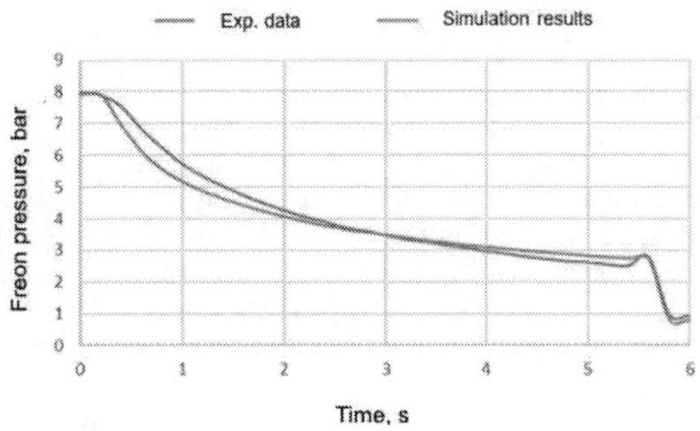

https://doi.org/10.1051/matecconf/201824005003

Fig 4: Comparison of simulated freon pressure and experimental data

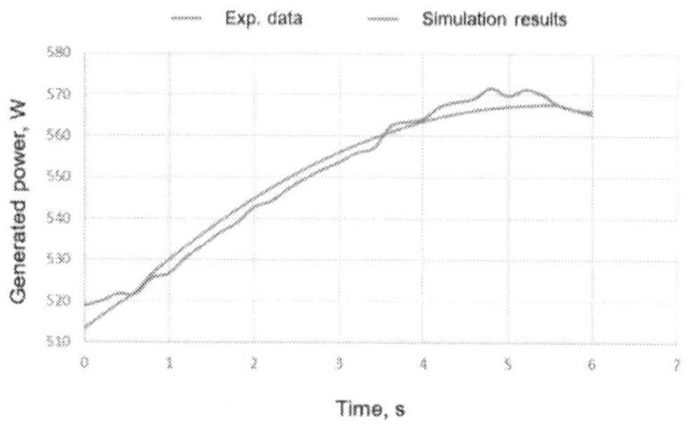

Fig 5: Comparison of experimental data against the model prediction

Figure 5 shows a comparison of measured power and model predictions. At t = 0s (the beginning of the cycle), the power is greater than 0 due to flywheel (mass of flywheel was 20,84 kg and moment of inertia was equal to 0,783 kgm2), which is integrated with the expander. Model predicts power 511 W at time t=0 s, while in the examined installation generated 520 W. The maximum simulated power is 568 W, while the real is slightly greater than 570 W. The discrepancy is ca. 2%.

4 Results and discussion

Nowadays, mathematical modeling is recognized as a basic method for analyzing novel concepts and most often a zero-dimensional approach is used for research on understanding novel cycles, hence Authors developed a fully-physical dynamic oriented model of a liquid piston engine that can aid in the evaluation of sensivity of numerous parameters for current operating conditions. A proposed model was implemented in Aspen Hysys modeling software and validated against experimental data. The comparison of experimental data against the simulation shows little discrepancy (ca. 2%). It proves that the proposed model could be not only used a tool for better understanding the system operation but also as a tool for system optimization.
The preliminary simulations shows that the low grade waste heat conversion into electricity based on liquid piston engine is a promising area, however the detailed study are required to optimize the system operation with respect to operating conditions.

Acknowledgments

The authors would like to gratefully acknowledge the support from the the European Regional Development Fund under the Smart Growth Operational Programme 2014-2020 through the project "The innovative power plant equipped with liquid turbine and hydraulic-gas generator of working agent's flow, sourced by waste heat".

References

1. L. Donghyeon, C., Yujin and R., Jun-Hyung and L., In-Beum,, INTERNATIONAL JOURNAL OF HYDROGEN ENERGY (2017), 25518-25530.
2. J. Kotowicz and L Bartela and K Dubiel-Jurgas, ARCHIVES OF THERMODYNAMICS (2017), 65-87.
3. J. Kupecki, K Motylinski, M. Skrzypkiewicz, M Wierzbicki, and Y Naumovich, ARCHIVES OF THERMODYNAMICS (2017), 53-63.
4. L Szablowski, P Krawczyk, K Badyda, S Karellas, E Kakaras, & W Bujalski (2017). Energy, 138, 12-18.
5. Milewski, J., Badyda, K., & Szabłowski, Ł. (2016). Journal of Power Technologies, 96(4), 245-260.
6. Reader, GT and Hooper, C,.: Stirling Engines, London: E. & FN Spon (1983).
7. West, C,: The Fluidyne heat engine", Electronics and Applied Physics Division, Atomic Energy Research Establishment (1971).
8. Slavin, VS and Bakos, GC and Finnikov, KA,, Applied energy Vol 86, Dec 2009, pp. 1162--1169.
9. Bakos, G and Slavin, VS and Finnikov, KA,., Patent of Russian Federation (2007).
10. S Narayan, A Gupta, and R Ranjeet Mechanical Testing and Diagnosis 5.2 (2015): 12.
11. A RUDOWICZ, Zbigniew; WASKIEWICZ, Jan; WOLOSZYNSKI, Wlodzimierz. Method and a system for driving a turbine. U.S. Patent Application No 15/114,855, 2016.
12. Bujalski, W., Futyma, K., Grzebielec, A., Milewski, J., Kucowski, J., Rudowicz, Z., Rusowicz, A. Analiza czynników niskowrzących do rozprężania w cylindrze z tłokiem hydraulicznym. Chłodnictwo: organ Naczelnej Organizacji Technicznej, 51, (2016)..
13. J Kajurek, A Grzebielec, W Bujalski, K Futyma, A Szczęśniak, A and Z Rudowicz,.: Wpływ właściwości czynnika roboczego na pracę organicznego obiegu Rankine'a, Rynek Energii (2017).
14. J Kajurek, A Rusowicz, A Grzebielec, W Bujalski, K Futyma, and Z Rudowicz,.: Wybór czynników chłodniczych dla zmodyfikowanego obiegu Rankine'a, Wydawnictwo Instytutu Techniki Cieplnej (2017).
15. HYSYS, ASPEN. 7.3 Manual. ASPEN TECH.

Time, s
Time, s

https://doi.org/10.1051/matecconf/201824005003

16. Peng D-Y, Robinson DB. A new two-constant equation of state, Industrial & Engineering Chemistry Fundamentals 1976;15(1):59-64.

About the shortwave multiple reflections in an urban street canyon building related to three different European climates

Chiara Colucci[1,*], *Luca* Mauri[2], and *Andrea* Vallati[1]

[1]Department of DIAEE, Sapienza University of Rome, Via Eudossiana 18, 00184 Rome, Italy
[2]Department of Mechanical Engineering, University of Roma TRE, Via della Vasca Navale 79, 00146 Rome, Italy

Abstract. Energy exchanges between buildings is affected by urban fabric. As a matter of fact, heat exchange between adjacent buildings is due to convective and radiative heat flows. The main parameters which influence these heat exchange mechanisms are due to climate conditions such as air temperatures/humidity, wind speed/direction and solar irradiance. Most building energy simulations are done on an independent single building with typical meteorological year (TMY). These TMY meteorological data cannot represent the state of the urban microclimate and rather ignores the microclimate influence on buildings adjacent to street canyons. However, solar radiation shading and reflection of the environment within the street canyons are important factors affecting the energy consumption of buildings. In this work, a building energy simulation tool is used to study the impact of multiple shortwave inter-reflections in an urban environment. A street canyon model validated in a previous work was modeled in TRNSYS in order to simulate the effects of the urban radiative trapping. An urban canyon with aspect ratio H/W=1 was chosen, with South-North orientation, with transparent/opaque surfaces ratio Atr/Aop=0.5 and 4 values (0.2, 0.4, 0.6, 0.8) of reflectance factor of the envelope surfaces. The goal is to characterize how solar absorption influence the urban energy requirements. The analysis was conducted for 3 cities in different climatic zones: Rome, Palermo and Krakow.

1 Introduction

Energy saving is an important priority in developed countries where a substantial part of energy consumption is represented by buildings. In recent years, many efforts have been made worldwide to improve buildings' energy efficiency and many political decisions have been discussed on this subject [1]. In the EU region, buildings account for about 40% of total energy consumption. For this reason, the "Building Energy Efficiency Directive" has been introduced [2] to prescribe requirements for either new and existing buildings. Building energy efficiency can be improved by implementing both active or passive energy efficient strategies. Improvements to heating, ventilation and air conditioning (HVAC) systems, electrical lighting, etc. can be classified as active strategies, whereas, improvements to building envelope elements can be classified under passive strategies [3-7]. The climatic characterization of an urban environment can present considerable difficulties depending on the topographic configuration of the urban environment, including narrow streets surrounded by buildings, squares, green areas and highly congested areas. Cities are also heat and pollution anthropogenic sources, as demonstrated by several studies that have attributed to urbanization radical changes in the radiative, thermodynamic and aerodynamic characteristics of the surface compared to those of the surrounding rural areas [8,9]. More in detail, meteorological parameters are modified such as: surface temperature, humidity, precipitation, fog, air quality, height of the mixing layer, wind profile. Many studies have also shown that building walls and road surfaces, acting as reflecting elements for solar radiation, cause radiative component bouncing and the accentuation of the overall effects on walls, ground and interposed air volume. Poor ventilation conditions caused by obstacles (e.g. buildings) help to generate and keep such stable and overheated air mass (named "thermal trap"). This mechanism boosts urban texture heat storage capacity in comparison to non-urban surroundings to an extend that depends on the different use of soil and buildings' geometry and materials. The whole phenomenon definitely builds the basis for a pollution increase and summer discomfort. To fully address this topic, it must be considered that a building is a system whose borders exchanges thermal energy with the surrounding environment. It means that urban areas conformation influences local microclimate, introducing differences in the boundary conditions compared to those occurring outside of the city, in rural areas. In this regard it is known, as confirmed by several studies, that the urban heat island (UHI) phenomenon plays an important role for what concerns the air temperature increase in the urban environment [10] and thus an alteration in buildings energy use [11, 12]. In other words, in urban canyons solar radiation affects buildings facades and tarmac and it is trapped in the form of multiple reflections. Of course, the same phenomenon (known as "canyon effect") does not

* Corresponding author: chiara.colucci@uniroma1.it

Nomenclature

α	Absorptivity coefficient
ε	Thermal emissivity
ρ	Reflectance coefficient
g	Window's solar factor
A_{tr}	Trasparent surfaces
A_{op}	Opaque surfaces
H	Henght of building
W	Width of building

occur in non-urban areas. Therefore, in cities, at night, air cooling inside the canyons is much slower than in outer space: infrared energy radiated towards the environment by canyons surfaces are mostly captured and reflected once again by the surrounding buildings rather than thin out in the free space. Various studies have been carried out on the intensity of the heat island for many cities. A study on energy consumption trends in Taiwan showed that energy consumption in urban areas is 81% higher than in rural areas [13].

In short words, the mutual interaction between buildings and urban environment generate the aggravation of operative conditions. For this reason, a holistic approach in the definitions of design choices for new and existing buildings is required to reach a smart planning for modern cities [14,15]. Building energy simulation codes (BES) are often used to estimate and predict the energy consumption of a building and the outdoor thermal conditions. Most energy simulations are performed considering a single isolated building [16] using a typical meteorological year (TMY) which do not account of the urban microclimate, since they were obtained from weather stations located in most cases in rural or suburban areas. Therefore, it is necessary to impose conditions that modify the TMY with the real conditions of the microclimate [17]. Recently, some researchers have begun using BES to study the space cooling demands of buildings, modeling their urban context more accurately, especially for what concerns radiations exchange [18].
The objective of this study is to evaluate how shortwave multiple reflections affect thermal energy demand (cooling and heating) of a street canyon building located in three different climatic zones: Rome, Palermo and Krakow. For this purpose, TRNSYS [19] has been exploited since its radiation data module considers inter-reflections between buildings, based on Gebhart factors formulation [20, 21], when a "false" zone method is adopted. More specifically, by modelling the street canyon as an interior thermal zone, TRNSYS applies the detailed radiation module calculating the multiple reflections both in the shortwave and in the longwave field. An urban canyon with aspect ratios H/W=1, orientations N-S, trasparent/opaque surfaces ratios A_{tr}/A_{op}=0.5 and 0.6 of solar absorption factor of the envelope surfaces. The total absorbed solar radiation on the external walls of the street canyon building were evaluated, such as the seasonal absorbed radiation. Finally, the influence on building thermal energy demands of multiple shortwave inter-reflections was evaluated comparing the results with a reference case with no inter-reflections. The results of study may contribute to rectify, by corrective factors, the data of predicted consumptions obtained by simulation tools that doesn't account of multiple reflections but only of shadowing factors. Considering that increases in cooling demand up to 54% were found in Krakow, our numerical model underlines the importance of considering the multiple reflections inside urban environment for an accurate evaluation of thermal energy demands of buildings inserted in their actual context. The details about the conditions imposed for the numerical model in this work to account only of the shortwave multiple reflections are reported in Section 2.

2 Methodology

In this document, the dependence of the shortwave multiple reflections on the heating and cooling demand of a three-floor building for residential use was analyzed for three cities in different climatic zones: Rome, Palermo and Krakow. Simulations were performed for a building considered in an urban context, adjacent on both main long sides to street canyons. The BES software used to perform the numerical simulations in this work is TRNSYS 17.0, a transient 3D multi-zone building code able to simulate dynamically the energy behavior of a building with time steps of less than one hour.

2.1 Numerical model

In TRNSYS, conduction heat flow through enclosure components is modeled using the 1D transfer functions. About convection, constant values of the CHTC convection factor were chosen according to the standards (CHTC equal to 17.8 W/m²K, while it is equal to 3 W/m²K for indoor surfaces). For solar irradiation, however, there is a model of radiation that considers multiple reflections only for internal thermal zones. This fact implies that, in case of a street canyon building modeled as an outdoor environment, the radiative model does not consider either the shortwave or the longwave inter-reflections between the façades of street canyon buildings. Since the final aim of this work is to investigate the influence of radiative trapping effect of the street canyon environment on the energy demand of a building, the space between the buildings was modeled as an interior thermal "false" zone with virtual border roof and walls.

As for the characteristics of the building studied, the length is 110.5 m, since one of the purposes was to represent a long building with similar adjacent apartments but also to minimize the effects of the boundary conditions on the short sides of the building. Both the height (H) and the width (W) of the building are 13.5 m. The short side walls are modeled adiabatic assuming they are continuous, thus modeling a long row of identical thermal zones. The transparent surface on the long façades of buildings is 50% of the total surface. The H/W (aspect ratio) is 1 and the orientation of the buildings is S-N, so the long façades facing the canyon are oriented E-W. The short side walls are modeled adiabatically assuming they are continuous, thus modeling a long row of identical thermal zones.

The street canyon building under analysis is surrounded by two other rows of buildings, geometrically identical to the studied building (Fig. 1). More details of the numerical model are reported in a previous work where a validation process has been performed [22]. Building's thermal energy demand for both heating and cooling is a function of the temperatures of its facades. These temperatures are influenced by both shortwave and longwave radiative exchange (in addition to convection and conduction).

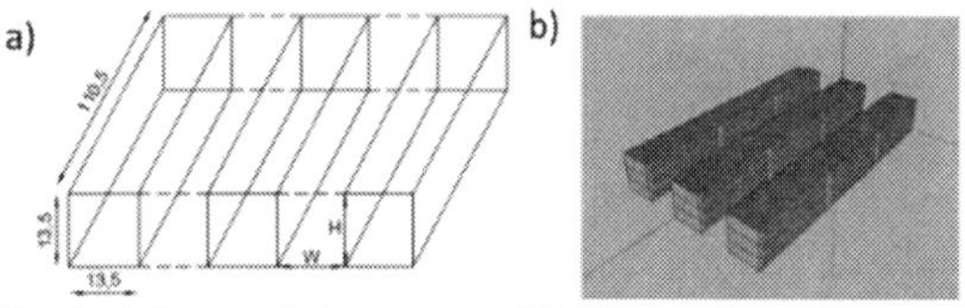

Fig. 1. a) Geometrical features and b) 3D overview of the street canyon.

However, the objective of this work is to evaluate the influence of the shortwave reflections on the thermal energy demand of the building only. For this reason, the longwave radiative exchange was deactivated by imposing the emissivity equal to 0. By setting the model in this way, the longwave radiation losses to sky and to other surfaces has been disabled and the "false" zone method gives only the possibility to investigate the influence of shortwave inter-reflection as a comparison with a model without inter-reflection. The inter-reflection free model is such as the building envelope elements under study don't receive any reflections by surrounding surfaces. It means all the street canyon surfaces around the building under study have absorbance factor equal to 1: this model is named City_OFF. The model named City_ON is the model in which the shortwave multiple reflections are present. In this case, the value of the reflectance factor of the street canyon pavement is constant (ρ=0.2). The simulations were carried out with the variation of the reflectance factors of the street canyon buildings façades change from ρ=0.2 up to ρ=0.8 as shown in Figure 2.

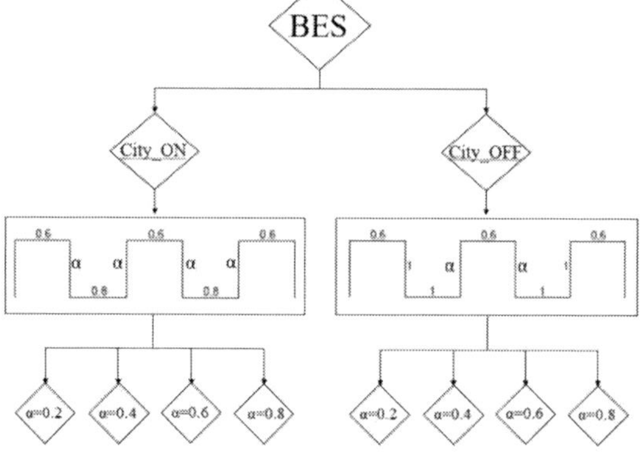

Fig. 2. Simulations scheme.

2.2 Building energy simulation model

The 3-floor central building is being analysed. The characteristics of the opaque and transparent elements of the envelope of all buildings are shown in Table 1.

In Tab. 1, g indicates windows solar factor. Internal gains due to the lighting system, electrical devices, users and occupancy schedules are set according to residential use as indicated in [23]. The gains reported in Table 2 are related to the total floor area. The presence of the users is considered for 24 hours per day. The usage schedule for the electrical devices is set "on" from 8.00 to 24.00 while for lights is set "on" from 17.00 to 24.00.

Table 1. Main features of construction envelope.

Envelope element	Materials	Thickness (m)	U (W/m²K)	g
Walls	Plaster, brick, insulation	0.44	0.36	
Roof	Plaster, concrete, screed, insulation	0.37	0.32	
Pavement	Ceramics, concrete, isolation, plaster	0.54	0.34	
Windows	Glass, wood	-	1.40	0.6

Table 2. Specific values of internal gains.

	Persons (W/m²)	Devices (W/m²)	Lights (W/m²)
Radiative	1.51	0.35	3.5
Convective	3.01	1.05	1.5

* Corresponding author: chiara.colucci@uniroma1.it

The daily natural ventilation rate is 0.3 vol/h while the envelope infiltration rate is assumed to be 0.2 vol/h while during night-time only the infiltration rate is imposed.
The weather files distributed with TRNSYS 17 are organized in directories according to the datasources. Currently there are two data sources:

- The US-TMY2 data sets from NREL (redistributed with NREL's permission). The dataset includes 237 Locations in the US (contiguous states + Alaska and Hawaii), and 2 files for Puerto-Rico and Guam.
- Selected worldwide stations from Meteonorm (distributed under license from Meteotest).

Weather data for the city of Rome, Palermo and Krakow are used as input for the BES. The heating period is imposed to be from 1st November to 15th April and the cooling period is not imposed by laws. Thus, the space cooling and space heating demands are determined considering the room air temperatures controlled to be at 20 °C during the winter and 26 °C during summer (considering as cooling period the days in which the internal temperature exceeds 26 °C).

3 Numerical results

The results show: i) the total amount of the solar energy absorbed with respect to the City_ON and the City_OFF cases in both summer and winter. Therefore, the percentage variation between the two conditions was obtained; ii) the standardized and absolute cooling and heating demands of the building in the City_ON and the City_OFF models. Measurements were performed in the cities of Rome, Palermo and Krakow. It is important to underline that the absorbed radiation values obtained are specific hourly average values and they refer only to the shortwave radiation field.

3.1 Analysis of absorbed solar radiation

In Table 3, on the left side, total absorbed solar radiation for each season is shown. These values are used to determine the percentage variations between the City_ON and City_OFF cases, reported on the right part of the same table. Such difference, is defined as follows:

$$\Delta Sol.Rad_{\%} = \left(\frac{\text{City}_{\text{ON}} - \text{City}_{\text{OFF}}}{\text{City}_{\text{OFF}}}\right) \times 100$$

In general, observing Tab. 3 it can be stated that:

1. As the solar absorption factor α increases, no matter what the season is, the radiation absorbed by the walls of the building increases, both for the City_ON and the City_OFF cases. Accordingly, with the increase of α, and the consequent reduction of the reflectance coefficient ρ=1-α, the reflected solar radiation is smaller, hence the contribution of the inter-reflections is also reduced;
2. The East and West facades of the street canyon building undergo the same increase in radiation absorption, due to their symmetrical exposure to the daily solar path;
3. With the same ρ value, the total solar radiation absorbed by the facades is greater in Palermo than in Rome and Krakow, especially in the summer.

Consequently, the percentage of the inter-reflections is also higher in Palermo.

Table 3. Values of absorbed solar energy (in MJ/m²) in Inter_ON and Inter_OFF and Percentage variation of absorbed solar energy between Inter_ON and Inter_OFF cases for heating and cooling season.

		Building Facade											
		Tot Rad [MJ/m²]								% Rad			
		Heating Season				Cooling Season				Heating Season		Cooling Season	
		West		East		West		East					
		ON	OFF	ON	OFF	ON	OFF	ON	OFF	West	East	West	East
Rome	ρ=0.8	159.8	123.9	158.6	122.6	241.9	185.9	240.5	184.3	28.9%	29.4%	30.1%	30.5%
	ρ =0.6	303.2	247.4	300.9	244.8	459.0	371.3	274.8	272.7	22.5%	22.9%	23.6%	23.9%
	ρ =0.4	432.8	370.6	429.2	366.8	655.1	556.1	650.8	551.5	16.8%	17.0%	17.8%	18.0%
	ρ =0.2	550.6	493.5	545.6	488.3	833.2	740.5	827.2	734.3	11.6%	11.7%	12.5%	12.7%
Palermo	ρ=0.8	170.7	132.3	167.6	128.7	250.1	192.7	244.3	185.9	29.0%	30.2%	29.8%	31.4%
	ρ =0.6	324.2	264.3	317.6	257.0	475.1	384.9	462.9	371.2	22.7%	23.6%	23.4%	24.7%
	ρ =0.4	463.0	395.9	452.9	384.9	678.7	576.5	659.7	556.1	17.0%	17.6%	17.7%	18.6%
	ρ =0.2	589.3	527.1	575.3	512.5	863.9	767.6	837.7	740.4	11.8%	12.2%	12.5%	13.1%
Krakow	ρ=0.8	88.5	69.3	85.1	65.3	177.8	137.4	174.8	133.9	27.7%	30.2%	29.4%	30.5%
	ρ =0.6	168.2	138.4	161.1	130.5	337.7	274.5	331.4	267.5	21.5%	23.5%	23.0%	23.9%
	ρ =0.4	240.5	207.3	229.5	195.4	482.2	411.2	472.5	400.7	16.0%	17.4%	17.3%	17.9%
	ρ =0.2	306.4	276.1	291.2	260.2	613.7	547.5	600.2	533.5	11.0%	11.9%	12.1%	12.5%

* Corresponding author: chiara.colucci@uniroma1.it

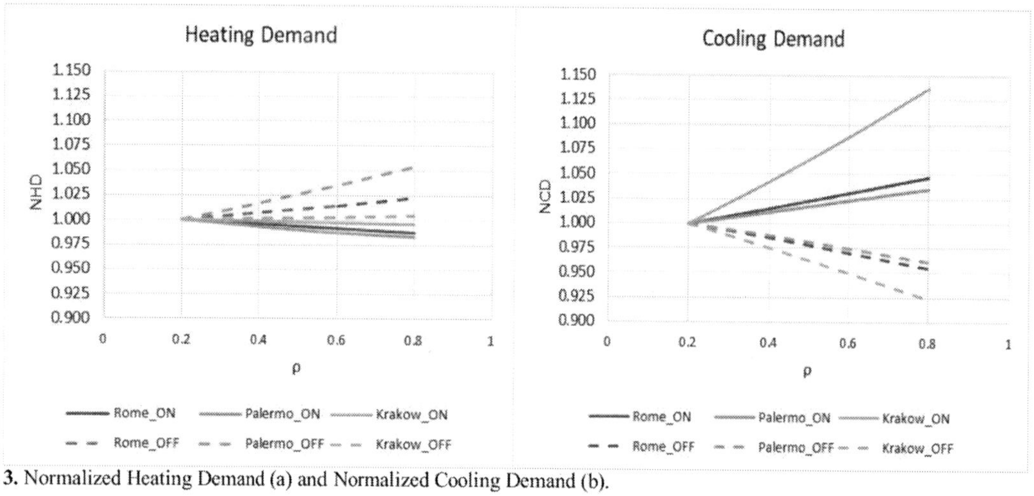

Fig. 3. Normalized Heating Demand (a) and Normalized Cooling Demand (b).

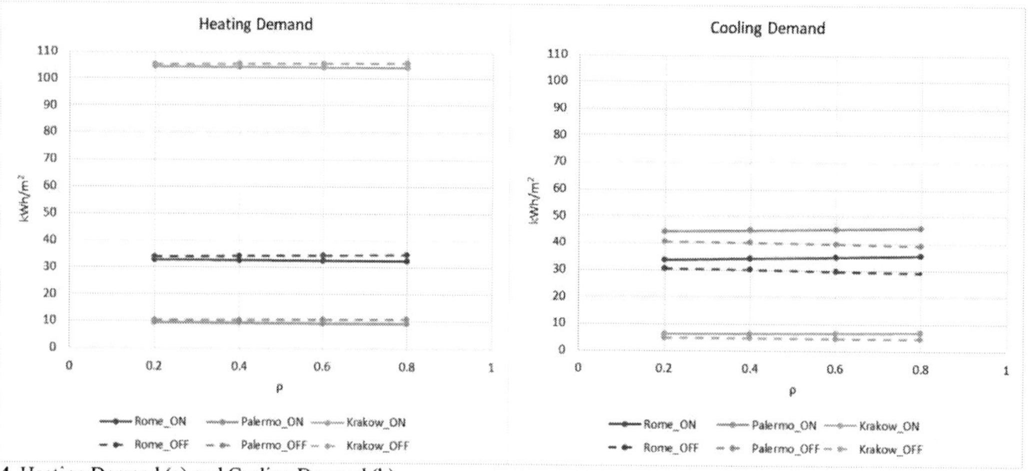

Fig. 4. Heating Demand (a) and Cooling Demand (b).

3.2 Building thermal energy demand

The work final purpose is to evaluate the impact of multiple canyon inter-reflections on the energy demands of buildings in Rome, Palermo and Krakow. The following figures show the thermal energy demand in both the City_ON and City_OFF cases while the solar reflectance coefficient ρ and the city vary. The results show how the contribution of multiple inter-reflections affects heating and cooling demands when coefficient ρ changes. For a better understanding of the energy demand variation as ρ changes, in Fig. 3 results normalized with respect to the ρ = 0.2 case are shown for both City_ON and City_OFF models.

Fig. 3 (a) shows, for the winter case, an increase in the heating demand accordingly to the reflectance coefficient ρ growth for the City_OFF model (the less solar radiation absorbed by the building, the greater the heating demand). An opposite trend is shown for the City_ON case. This inversion of tendency is caused by inter-reflections. As seen in Table 3, as ρ increases, inter-reflections become greater and this leads to a decrease for heating demand. In general, the normalized heating demand range between 1.05 and 0.98.

The opposite situation can be seen in Figure 3 (b), which shows cooling demand for the summer period. In this case, as ρ increases, cooling becomes more demanding for the City_ON case and less demanding in the City_OFF case (see Tab. 3). In general, normalized cooling demand ranges between 1.14 and 0.92. Figure 4 makes clear how inter-reflections affect building energy demand by a different amount for each city. In Fig. 4 (a), as ρ varies, inter-reflections impact on heating demand ranges from -7.1% (ρ=0.8) up to -3.8% (ρ=0.2) for the city of Rome, from -14.3% (ρ=0.8) up to -8.1% (ρ=0.2) for the city of Palermo, from -1.8% (ρ=0.8) up to -1% (ρ=0.2) for the city of Krakow.

Cooling demand for the summer season is shown in Fig. 4 (b): in Rome it increases from 10.5% (ρ=0.2) up to 21.1% (ρ=0.8), from 8.7% (ρ=0.2) up to 17.1% (ρ=0.8) for Palermo and from 25.1% (ρ=0.2) up to 54% (ρ=0.8) in the city of Krakow. The last percentage value is due to the fact that Krakow's standard cooling load is approximately just 6 kWh/m^2.

Considering both the two figures, for the same city and ρ value, percentage difference in demand between City_ON and City_OFF models is lower in the winter than in the summer (7,1% compared to 21,1%). The reason for this is that in the winter temperature difference between the inside and the outside of the building is greater than in the summer. Therefore, when defining building's thermal load, it has a major influence than inter-reflections.

4 Conclusions

For this work, a virtual residential building flanked by two other identical building has been simulated in TRNSYS using the "false zone" method developed and validated in a previous work. In this way, the model allows to account of the multiple radiative reflections between facing surfaces by calculating the values of the Gebhart factors. Measurements were carried out considering three different cities, each of them representative of different climatic zones: Krakow, Rome and Palermo.

Since the final scope was a comparison with respect to a model without inter-reflections, two similar models were set up. In the model without inter-reflection, named City_OFF, the building envelope elements under study doesn't receive any reflections by surrounding surfaces for a specific setting of albedo values. The model named City_ON is the one in which the shortwave multiple reflections are present. The transparent surface on the long façades of buildings is 50% of the total surface. The H/W (aspect ratio) is 1 and the orientation of the buildings is South-North, so the long façades facing the canyon are oriented East-West.

By comparing the results of the two models (City_ON and City_OFF), an increase in the heating demand for the winter case is shown, accordingly to the reflectance coefficient ρ growth for the City_OFF model. An opposite trend is shown for the City_ON case. In the summer, instead, as ρ increases, cooling demand increases for the City_ON case, whilst it decreases in the City_OFF case. This due to inter-reflections. At a deeper lever, having fixed ρ = 0.8 in order to analyze the phenomenon at his greatest effect, simulations show an increase in cooling demand up to 54% and a decrease in heating demand up to 1.8% in the city of Krakow. The same variations are +21.1% and -7.1% for the city of Rome and +17.1% and -14.3% for the city of Palermo. For what has just been said, the analysis shows the extend of the error in thermal energy demand evaluation that occurs if the inter-reflection effect isn't taken on account. In general, during the summer the radiative trap phenomenon is greater than in the winter. This led to a bigger growth of cooling demand rather than heating demand. The reason for this is that in the winter temperature difference between the inside and the outside of the building is greater than in the summer. Therefore, when defining building's thermal load, it has a major influence than inter-reflections.

Given the results shown, this work suggests to consider as driving parameters to define thermal energy demand of new buildings also their interactions with the surrounding environment. This in order to take on account every possible solution to optimize the energy performance of the building itself.

References

1. R. Pacheco, J. Ordóñez, G. Martínez, Renew. Sustain. Energy Rev., **16** (2012)
2. EPBD 2012 - Directive 2012/27 / EU of the European Parliament and of the Council of 25 October 2012 on energy efficiency
3. R. De Lieto Vollaro, F. Faga, A. Tallini, L. Cedola, A. Vallati, Energy Procedia, **82** (2015)
4. A. Tallini, A. Vallati, L. Cedola, Energy Procedia, **82** (2015)
5. A. Vallati, S. Grignaffini, M. Romagna, L. Mauri, EEEIC 2016 - International Conference on Environment and Electrical Engineering 7555731 (2016)
6. A. Vallati, S. Grignaffini, M. Romagna, Sustainability (Switzerland), **7** (2015)
7. P. Ocłoń, M. Bittelli, P. Cisek, E. Kroener, M. Pilarczyk, D. Taler, R.V. Rao, A. Vallati, Applied Thermal Engineering, **108** (2016)
8. M. Ignatius, N.H. Wong, S. K. Jusuf, Sustainable Cities and Society, **19** (2015)
9. E. Erell, D. Pearlmutter, D. Boneh, P.B. Kutiel, Urban Climate, **10** (2014)
10. G. Evola, A. Gagliano, A. Fichera, L. Marletta, F. Martinico, F. Nocera, A. Pagano, Energy Procedia, **134** (2017)
11. M. Palme, L. Inostroza, G. Villacreses, A. Lobato-Cordero, C. Carrasco, Energy and Buildings, **145** (2017)
12. A.D.L. Vollaro, G. Galli, A. Vallati, R. Romagnoli, Journal of Physics: Conference Series 655(1),012056 (2015)
13. R.-L. Hwang, C.-Y. Lin, K.-T. Huang, Energy and Buildings, **152** (2017)
14. G. Battista, E. Carnielo, L. Evangelisti, M. Frascarolo, R. De Lieto Vollaro, Sustainability, **7** (2015)
15. G. Battisti, E. Carnielo, R. De Lieto Vollaro, Energy and Buildings, **133** (2016)
16. F. Calcerano, L. Martinelli, Energy and Buildings, **112** (2016)
17. J. Allegrini, V. Dorer, J. Carmeliet, Energy and Buildings, **55** (2012)
18. J. Allegrini, V. Dorer, J. Carmeliet, Energy and Buildings, **127** (2016)
19. S.-A. Klein, TRNSYS 17: A Transient System Simulation Program. Solar Energy Laboratory, University of Wisconsin, Madison, USA (2010)
20. B. Gebhart, International Journal of Heat and Mass Transfer, **3** (1961)
21. B. Gebhart, Heat Transfer. 2nd. New York: McGraw-Hill (1971)

22. A. Vallati, L. Mauri, C. Colucci, P. Ocłoń, Energy and Buildings, **149** (2017)
23. 11300-1, UNI TS. Energy Performance of Buildings—Part 1. Calculation of Energy Use for Space Heating and Cooling. 1st ed. Milano: UNI, (2014)

Thermal management of machine compartment in a built-in refrigerator

Milind Devle[1,*], *Ankur* Garg[1], and *Darci* Cavali[2]

[1]Global Technology and Engineering Center, Whirlpool of India, Pune, 411014, India
[2]Whirlpool Corporation, Michigan, 49022, USA

Abstract. In general a multi-door refrigerator machine compartment comprises of fan, condenser, compressor, control box, drain tray, and drain tubes. The performance of machine compartment depends upon the efficiency of heat extraction or heat exchange from heat generating components such as condenser and compressor. The efficiency of heat exchange can be improved by addressing two major factors, namely (1) Air bypass and (2) Hot air recirculation. The hot air recirculation in the machine compartment for built-in multi-door refrigerator configuration is the focus of this study. The results from Computational Fluid Dynamics (CFD) simulations show that efficiency of heat exchange for built-in application is lower than that for free-standing configuration. Recirculation of hot air and reduction in airflow are the two major factors which contribute towards the variation in machine compartment performance. The CFD simulations were coupled with Partial Factorial Design of Experiment (DoE) approach to systematically investigate the effect of variables such as (a) side gap and top gap between kitchen cabinetry and the refrigerator, (b) the baffle/flap (i.e. back and bottom of machine compartment) on the performance effectiveness of machine compartment. The results of the simulation provided critical design improvement directions resulting in performance improvement. Furthermore, the CFD simulation results were also compared to test data and the results compared favourably.

1 Introduction

Refrigerator machine compartment comprises of compressor, condenser, fan and many other module sets [1-2]. The components and their arrangement must be designed to improve the cooling, structural and noise performances of the refrigerator. A typical multi-door refrigerator machine compartment is as shown in Fig. 1.
The overall performance of a refrigerator depends on the compressor and condenser heat dissipation or exchange from the system during operation [3]. In addition, the performance of machine compartment depends on the refrigerator installation configuration (i.e. free-standing or built-in configuration). In general, the depth of a free-standing multi-door refrigerator is approximately double that of built-in multi-door refrigerator. When a free-standing multi-door refrigerator is installed as a built-in configuration, the multi-door refrigerator typically does not perform well, resulting in compressor tripping due to heat-up of compressor. To provide installation flexibility and a greater customer choice, it is important that the machine compartment in a multi-door refrigerator be designed such that the thermal loading of the machine compartment remains within acceptable limits for both free-standing and built-in configurations.

Computational Fluid Dynamics is a well known tool for understanding the flow and thermal behaviour [4-5]. In this study, Computational Fluid Dynamics (CFD) simulation has been used to investigate the built-in refrigerator thermal dynamics. Visualization of temperature rise inside the machine compartment is critical to understand the thermal loading that exists inside the refrigerator machine compartment.

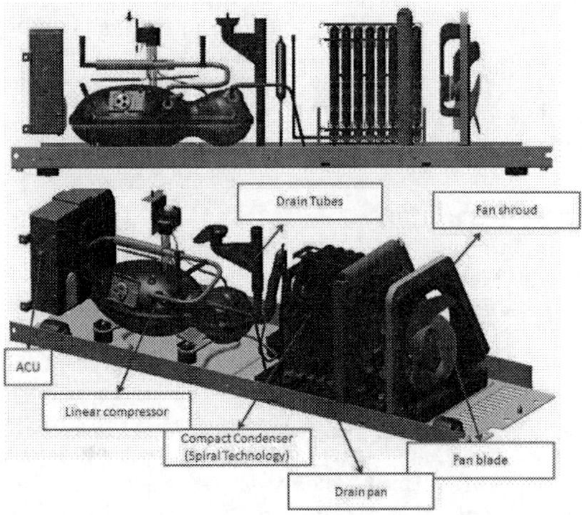

Fig. 1. Machine compartment architecture with spiral condenser and linear compressor.

The current study was conducted in two phases. For phase-1, CFD simulation was used to investigate the

* Corresponding author: milind_devle@whirlpool.com

performance of multi-door refrigerator for the two installation configuration. The results from these simulations were analyzed to identify the root cause behind the thermal overloading of the machine compartment, particular the compressor. For phase-2, CFD simulations were coupled with a Partial factorial Design of Experiments (DoE) to assess the impact of various parameters on the thermal performance of the machine compartment.

2 Modelling methodology

2.1 Description of multidoor refrigerator

Generally, free-standing and built-in refrigerators are multi-door, in which evaporators are placed in compartments which are covered so as to have a unique aesthetic look. Fan allows the warm air to flow over the evaporator from compartments to lower the temperature and dehumidify it. The remaining components of cooling systems such as condenser and compressor are placed at back and bottom side of the refrigerator compartment called as machine compartment. The machine compartment also consists of the fan for forced cooling. In built-in, the multi-door refrigerator is positioned inside kitchen cabinetry. The warm air passage in built-in refrigerators is important to model as it affects the performance and energy efficiency.

2.2 Assumptions

The following assumptions are made in this study:
• Convective temperature profiles are considered for the freezer and refrigerator compartments,
• Constant temperature profiles are considered for condenser and compressor,
• Steady state condition is assumed for simulation,
• Boussinesq assumption is used to model the buoyancy effects,
• The flow is assumed to be turbulent and Spalart Allmaras turbulence model is used,
• Fluid flow is assumed to be incompressible.

2.3 Fluid or CFD model

Fluid modeling was carried out in Creo 3.0 using the master assembly of multi-door refrigerator. The freezer and refrigerator compartment fluids are modelled as void, but the refrigerator structure that are body and door are modelled as solid structure. The compressor and condenser are modelled as void and the fan is modelled as rotating fluid domain considering Moving Reference Frame (MRF) methodology. The base plate and back plate are modelled as solid domains whereas the machine compartment air is modelled as fluid domain. Surrounding air is assumed as ambient fluid domain. For built-in configuration, the kitchen cabinetry is also modelled. Fig 2 shows the CFD model for free-standing and built-in configurations.

The fluid model is meshed using Fluid-Nexus version 1.5.4 software. The material properties used in this study are summarized in Table 1. The boundary conditions used in this study are tabulated in Table 2. The commercial CFD solver, Acusolve version 13 is used for the CFD simulations and the solver settings used in this study are summarized in Table 3.

Free-standing Built-in

Fig. 2A. Typical CFD model for free-standing and built-in.

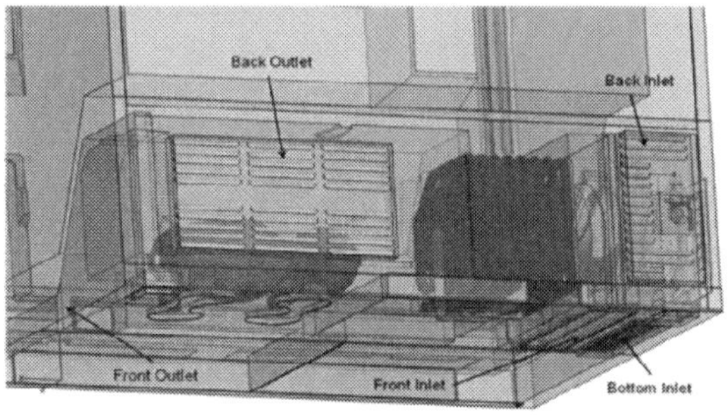

Fig. 2B. Locations where the simulation data extracted.

Fig. 2. Fluid/ CFD model.

Table 1. Material properties.

	Plastic	**Al**	**Foam**
Density, ρ (kg/m3)	1021.5	2658	32
Sp. Heat, Cp (J/(kg-K))	960	998	500
Conductivity, k (W/(m-K))	0.16	235	0.018
	Gasket	**Steel**	**Air**
Density, ρ (kg/m3)	20	7850	1.185
Sp. Heat, Cp (J/(kg-K))	960	449	1005
Conductivity, k (W/(m-K))	0.06	72.8	0.0233
Viscosity, μ (kg-m/s^2)	--	--	1.6E-05

Table 2. Boundary condition details considered for simulation.

Location	Boundary condition type
Freezer compartment	Convective wall
Refrigerator Compartment	Convective wall
Ambient domain	Convective wall
Condenser coils	Weighted average const. temp

Table 3. Acusolve Solver Settings.

	Mode	steady-state
Equation	Flow	Navier Stoke
	Temperature	Advection Diffusion
	Radiation	Enclosure
	Species	None
	Turbulence	Spalart Allmaras
	Mesh	Eulerian
Solution	num_krylov_vectors	70
	Relaxation	0.3

3 Result and discussion

3.1 Phase 1

To identify the performance variation of same multi-door refrigerator between two different installing conditions through Computational Fluid Dynamics (CFD) simulation, parameters like the flow rate, the heat exchange from condenser-compressor and the air temperature before fan inlet data are extracted. The data are summarized in Tables 4 and 5. For a better insight, the temperature contour on a plane passing through the machine compartment and temperature streamline are captured as shown in Fig 3 and 4.

Through the CFD simulation of Phase1, it is observed that, the total flow rate for built-in configuration is reduced by 8% as compared to free-standing, as the flow passage around the refrigerator in built-in got reduced. The total heat exchange for built-in configuration is reduced by 45% as compared to free-standing configuration, which can be attributed to high inlet temperature (as hot air recirculation observed) and less flow rate. The compressor heat exchange seems to be similar in both situations as compared to condenser since compressor is always surrounded by hot air.

3.2 Phase 2

Based on the Phase1 results, hot air recirculation was identified as the major contributor for the refrigerator performance reduction. To increase the performance of built-in refrigerator, the partial factorial Design of Experiment (DOE) is considered with following performance affecting variables with the ranges as shown in Table 6.

Table 4. Normalized flow rate details.

Normalized Flow Rate [cfm]						
	Total	Back Inlet	Bottom Inlet	Front Inlet	Back Outlet	Front Outlet
Free Standing	100.0	55.77	31.03	14.2	79.02	20.98
Built-in	91.84	48.58	25.43	18.3	70.35	21.49

Table 5. Normalized heat exchange and average temperature of plane before fan inlet.

Normalized Heat Exchange [W]				Normalized Avg. Temp @ fan front plane [°F]
	Condenser	Compressor	Total	
Free Standing	72.04	27.96	100.00	100.00
Built-in	33.16	22.16	55.34	107.94

Free standing Built-in

Fig. 3. Temperature contour on plane passing through the machine compartment.

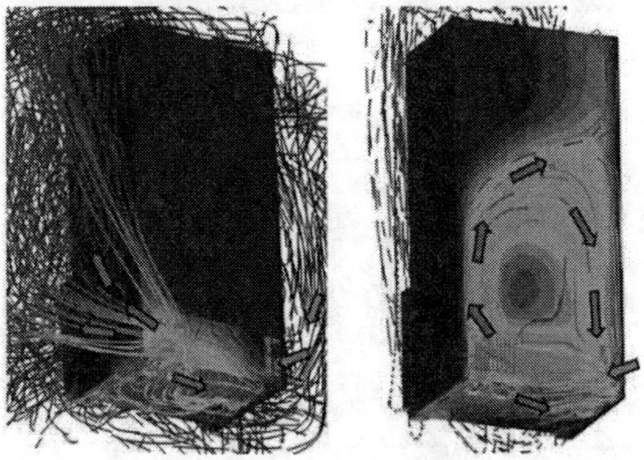

Free standing Built-in

Fig. 4. Temperature contour with streamline.

https://doi.org/10.1051/matecconf/201824005005

Table 6. DoE variables.

X's	Factors	-	+
X1	Back flap	-	+
X2	Side restriction	-	+
X3	Top restriction	-	+
X4	Bottom flap	-	+

Y's	Factors
Y1	Airflow over condenser fan (cfm)
Y2	Overall heat exchange (W)
Y3	Average temperature at compressor (°F)

As per the partial factorial DOE, eight cases were considered for simulation. Case details are summarized as in Table 7. Flowrate, heat exchange from compressor-condenser and the temperature at fan inlet results are tabulated in Tables 8 and 9.

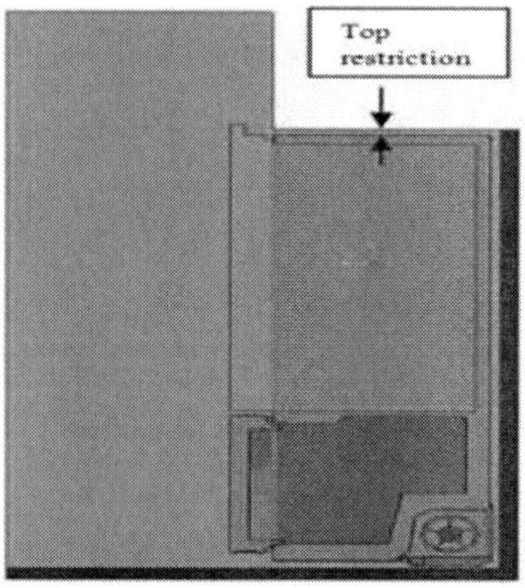

Fig. 5A. Representing top restriction.

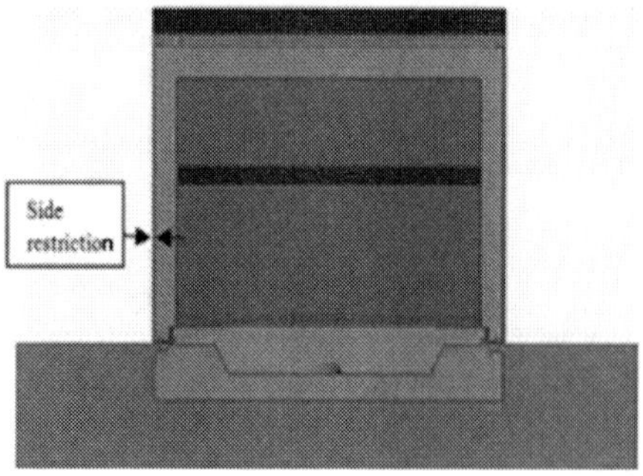

Fig. 5B. Representing side restriction.

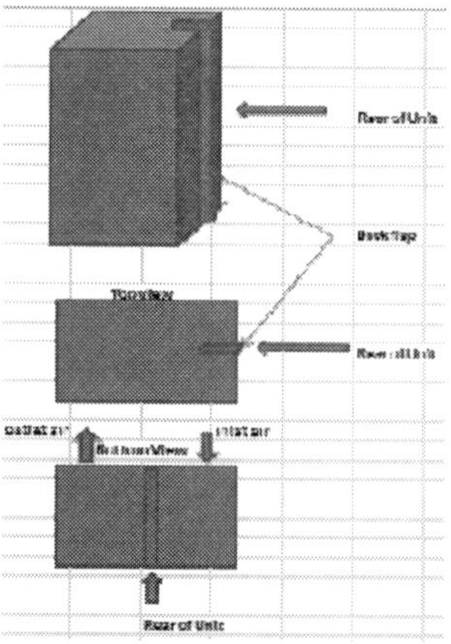

Fig. 5C. Representing back and bottom flaps.

Fig. 5. DOE variable schematic representation.

Table 7. DoE case details.

	Back flap	Side restrictions	Top restrictions	Bottom flap
Run 1	-	+	+	-
Run 2	-	+	-	+
Run 3	-	-	+	+
Run 4	-	-	-	-
Run 5	+	+	+	+
Run 6	+	+	-	-
Run 7	+	-	+	-
Run 8	+	-	-	+

Table 8. Partial DOE flowrate details in cfm (change over the Phase 1 built-in configuration).

	Total	back inlet	bottom inlet	front inlet	back outlet	front outlet
Run 1	1.04	1.52	-0.87	0.24	1.88	-1.21
Run 2	0.79	2.51	-1.02	-0.85	2.12	-1.78
Run 3	0.01	2.57	-1.33	-1.38	1.67	-1.87
Run 4	0.43	1.07	-0.91	0.12	1.58	-1.29
Run 5	-5.18	-2.12	-1.68	-1.54	-3.68	-1.82
Run 6	-1.91	-2.23	-0.15	0.31	-1.67	-0.77
Run 7	-2.24	-1.80	-0.89	0.30	-1.57	-1.00
Run 8	-5.13	-1.55	-2.43	-1.30	-4.01	-1.69

Table 9. Partial DOE heat exchange and average temperature of plane before fan inlet (change over the Phase 1 built-in).

	Heat Exchanged [W]			Avg. Temp @ fan front plane [°F]
	Total	Condenser	Compressor	
Run 1	-1.38	-0.06	-1.31	-0.10
Run 2	6.94	6.82	0.12	-2.07
Run 3	4.84	5.34	-0.49	-1.99
Run 4	1.28	1.90	-0.61	-0.65
Run 5	20.39	20.02	0.37	-8.67
Run 6	10.45	10.61	-0.15	-4.45
Run 7	10.22	10.14	0.08	-4.07
Run 8	19.11	19.01	0.11	-8.38

3.3 Parametric study

CFD data from Tables 8 and 9 are evaluated using the statistical JMP software to find the effectiveness of each variable against the Y's mentioned in Table 6. It is observed that, the total airflow varies with delta variation of 6.22 cfm, total heat exchange varies with delta variation of 20.47 W and average temperature at fan front plane varies with delta variation of 8.57 °F. Also, the fit model analysis for each variable against Y's are also plotted.

3.3.1 Effect on total airflow

The total airflow fit model analysis is shown in Fig 6. The major variation in the total airflow is coming from back flap and the bottom flap. Total airflow of machine compartment got reduced by 11% and 7% because of back flap and bottom flap. Top and side restrictions are having negligible effect on total airflow.

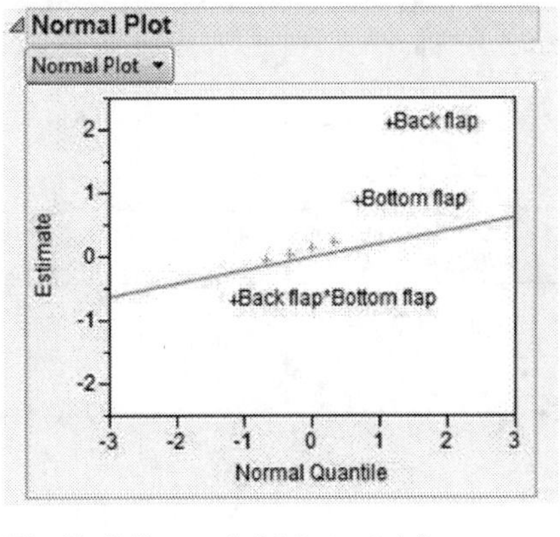

Fig. 6A. JMP normal plot for total airflow.

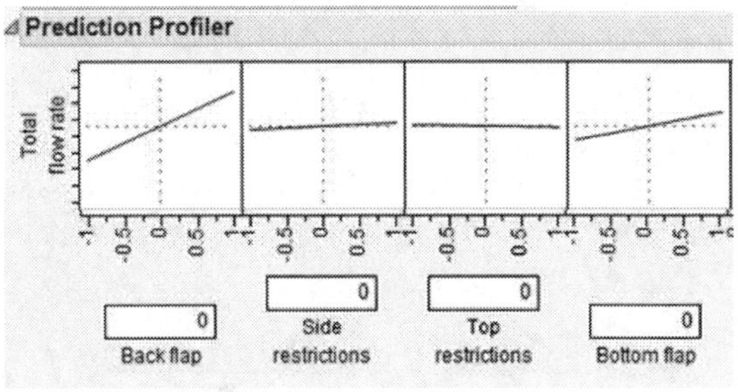

Fig. 6B. Fit model analysis for total airflow.

Fig. 6. Fit model analysis for total airflow.

3.3.2 Effect on total heat exchange

The total heat exchange fit model analysis is shown in Fig 7. The major variation in the total heat exchange is coming from back flap and the bottom flap. The total heat exchange in machine compartment is increased by 35% and 20% because of the back and bottom flap respectively. Top and side restrictions are having negligible effect on total heat exchange.

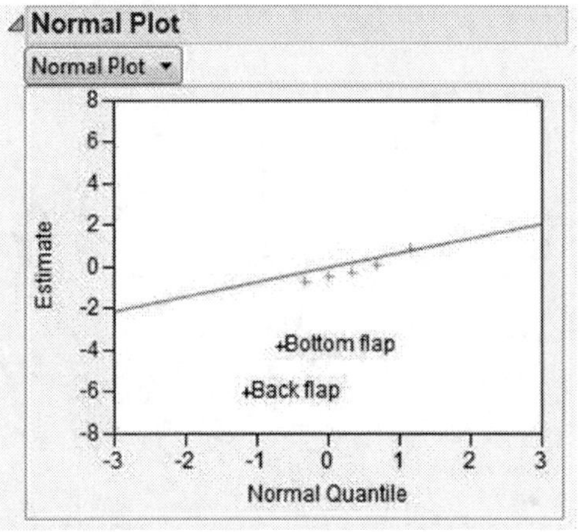

Fig. 7A. JMP normal plot for total heat exchange.

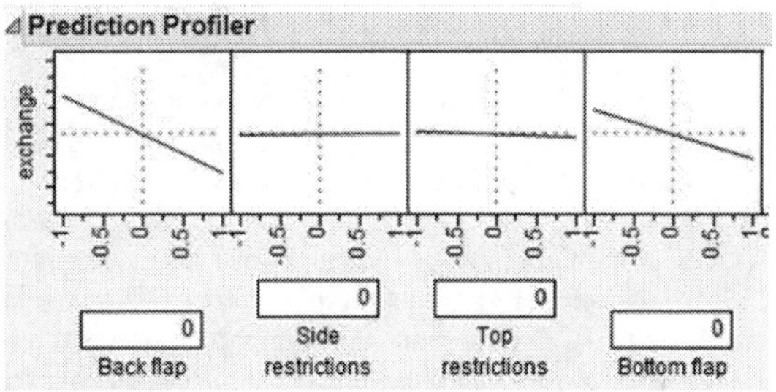

Fig. 7B. Prediction profile of each variable for total heat exchange.

Fig. 7. Fit model analysis for Total heat exchanged.

3.3.3 Effect on average temperature at fan front plane

The average plane temperature fit model analysis as shown in Fig 8. The major variation in the average temperature at fan front plane is coming from back flap and the bottom flap. The average temperature at fan front plane is reduced by 5% and 3% because of the back and bottom flap respectively. Top and side restrictions are having negligible effect on average temperature at fan front plane.

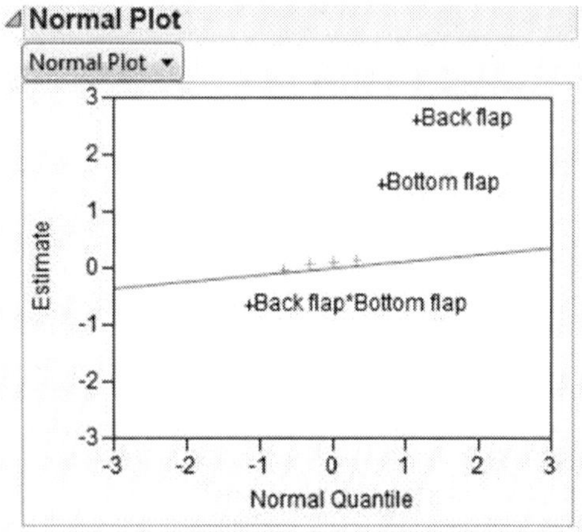

Fig. 8A. JMP normal plot for total average temperature at plane before fan inlet.

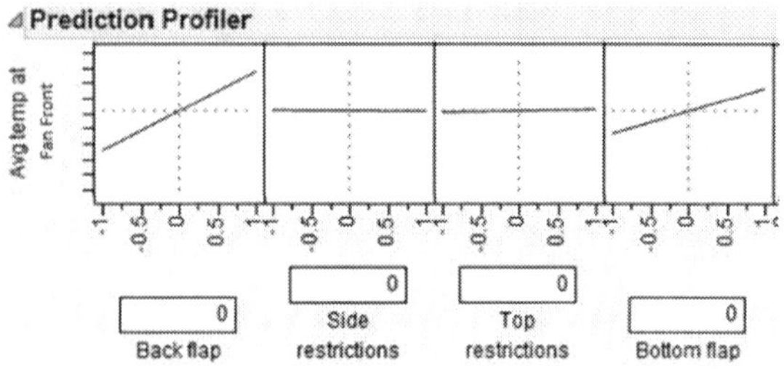

Fig. 8B. Prediction profile of each variable average temperature at plane before fan inlet.

Fig. 8. Fit model analysis for average temperature of plane before fan inlet.

3.4 1D energy simulation

1D energy simulation with and without flaps were performed. The data of 1D simulation is summarized in Fig 9. Through 1D simulation data it is observed that the refrigerator in built in application with current condition (i.e without flaps) is not recommended, as it generates 75% of hot air recirculation at inlet of machine compartment. Due to this, the compressor temperature increases over the operating limit. The proposed changes reduces the hot air recirculation from the original 75% down to 26%, which provides 20% energy benefit and allows the system to operate back within the allowable range of the compressor.

3.5 Field test

1D From overall simulation study it was observed that, the built-in refrigerator with flaps can prove to be an efficient option for improved performance. Thus, testing was performed on built-in multi-door refrigerator with bottom and back flaps using field condition where operating conditions are cyclic in nature. The test data is summarized in Table 10. From the testing, refrigerator with flaps shows following improvements:

- Average freeze compartment (FC) temperature colder by 0.4 °F
- Average refrigerator compartment (RC) temperature colder by 2 °F
- Condenser in temperature: colder by 14 °F
- Condenser out temperature: colder by 7 °F
- Compressor temperature is less than the operating limit so tripping would be avoided
- Compressor run time: 2.5% less
- Energy consumption: 4% improvement

Product characteristic	Unit	Difference
RC temp	[F]	-0.4
PN temp	[F]	-0.2
FC temp	[F]	-0.3
IB temp	[F]	-0.3
FC evap temp	[F]	-1.2
RC evap temp	[F]	-3.7
FC cond temp	[F]	-21.3
RC cond temp	[F]	-33.0
Deck air flow	[CFM]	-6.1
FC power	[W]	-4.9
RC power	[W]	-17.7
FC run time	[%]	-5.7%
RC run time	[%]	-6.0%
Total run time	[%]	-11.7%
Test energy	[kWh/day]	-22.4%
Total energy consumption	[kWh/day]	-19.8%

Fig. 9. 1D energy simulation data: Difference between without and with flaps.

Table 10. Normalized test data with and without flaps.

Case	Avg. FC Temp [°F]	Avg. RC Temp [°F]	Condenser In Temp [°F]	Condenser Out Temp [°F]
Refrigerator with flaps	0	31.8	150.84	102.56
Refrigerator without flaps	0.4	33.7	164.84	109.4

Case	Energy Consumption [kWh/day]	Compressor On Time [%]
Refrigerator with flaps	0.96	0.98
Refrigerator without flaps	1.00	1.00

4 Conclusions

The study compares the performance of the multi-door machine compartment in free-standing and built-in configurations. The study includes the parametric study which proved to be an effective tool to show the effect of various parameters on performance of machine compartment in built-in situation.

The results from the CFD show that the performance of machine compartment in built-in configuration is negatively impacted and there is need to study this aspect. From the overall DOE analysis it is observed that, the back flap and the bottom flap have a major impact on the amount of heat exchange and on the temperature at the fan front. Due to flaps, the efficiency of built-in refrigerator machine compartment is increased by around 35% thereby increasing almost 50% of fresh air intake at fan inlet. Also, the condenser out temperature is reduced by around 20%. Thus, the proposed flaps in built-in refrigerator provide to be efficient since the operating condition seems to be in controlled working condition of compressor and temperature overshoot could be avoided to a large extent.

References

1. N. Z. Adkane, S. V. Borkar, and R. D. Bokde, IRJET, **4**, 6 (2017)
2. P. Tiwari, P. Pandey, and Dr. S. P. S. Rajput, IRJET, **4**, 5 (2017)
3. J. R. Barbosa, Jr. Christian, J. L. Hermes, and C. Melo, J. Braz. Soc. Mech. Sci. & Eng, **32**, 445-453 (2010).
4. S Logeshwaran, JCHPS, **9**, 4 (2016).
5. M. E. Haque, R. A. Bakar, G. L. Ming, and M. Shakaib, JEAS, **12**, 3054-3060 (2017).

Shape and operation optimization of a thick-walled power boiler component

Piotr Duda[1,*], and *Mirosław W.* Mrzygłód[2]

[1]Institute of Thermal and Process Engineering, Cracow University of Technology, Al. Jana Pawła II 37, 31-864 Kraków, Poland
[2] Department of Mechanics and Machine Design, Opole University of Technology, Mikolajczyka 5, 45-271 Opole, Poland

Abstract. The aim of this article is the proposition of a two-stage optimization of power boiler components subjected to transient thermo-mechanical loads. As the basis of transient load regime, the TRD 301 regulations are used. In the first stage of an investigation, the genetic algorithms are used to determine the optimal shape of the structure. As optimization constraints, the maximum total stresses in the whole cycle are assumed. In the next stage, the method for the start-up and shut-down operation optimization is shown. The problem is defined to minimize the time of start-up and shut-down processes with allowable stress constraints according to TRD regulations. These new methods are of high practical significance and can be implemented throughout the industry, wherever heating and cooling processes take place. The obtained optimization results translate directly into the power unit energy effectiveness because the shut-down and start-up losses are reduced.

1 Introduction

A challenge of the XXI century is providing abundant and affordable electricity to meet the need of a growing global population, while lowering and controlling CO2 emissions. Both of these goals can be met through the development of highly efficient technologies of power units. During exploitation of power boiler components, high thermal stresses occur. The cyclic character of such stresses causes low-cyclic fatigue, which may lead to fracture. Manufacturers of such equipment advise keeping within prescribed limits for maximum heating and cooling rates. A long start-up and shut-down operation time of power boilers causes big losses in the electricity generation. On the other hand, rapid transient operations of power boilers cause high temperature differences and thermal stresses in the construction elements [1]. Methods to find optimum operating parameters, that can assure safe heating and cooling processes can be found in [2,3,4]. Reducing the thickness of the walls of boiler components while maintaining the level of stress during the cycle within established limits can bring additionally significant economic benefits. Furthermore, it can improve the thermo-mechanical characteristics of the device.

The aim of this article will be the proposition of a method for lightweight optimization of power boiler components subjected to transient thermo-mechanical loads according to TRD 301 regulations [5]. As optimization constraints, the maximum total stresses are assumed. Next, the method for the start-up and shut-down operation optimization is shown. The problem is defined to minimize the time of start-up and shut-down processes with stress constraints.

Examples of cyclic load thermo-mechanical structural optimization are rarely published. Analytical expressions for shape design sensitivities in the optimization of engine exhaust manifold with respect to its geometry were presented in the work of Yang and Poe [6]. The manifold was formulated as a linear heat transfer and thermo-elasticity problem in a variational form. An example of design and optimization of high-temperature engine components was presented in the paper of Tappeta et al. [7]. These authors used multiple analysis codes and response surface methodology for the design and optimization of large-scale real-life engineering systems.

The majority of available examples of the application of modern structural optimization of power block devices refer to the steady state. Wen et al. [8] present an example of structural optimization of cellular cores of metallic sandwiches being subjected to laminar forced convection at fixed pumping power. Yoon et al. [9] use isogeometric analysis to solve the heat conduction equation and compute shape design sensitivity coefficients. Xie et al. [10] employ a globally convergent method of moving asymptotes (GCMMA) in the lightweight optimization of a corrugated sandwich panel using heat transfer and thermo-mechanical analyses. Xiao and Huang [11] combine particle swarm optimization (PSO) with the finite element method (FEM) to find the optimal design for achieving rapid and uniform mold heating system in RTCM molds. Nowak and Rusin [12] present the problem of shape optimisation of selected areas of the rotor of the high

* Corresponding author: pduda@mech.pk.edu.pl

pressure part of an ultrasupercritical steam turbine together with the optimisation of the turbine startup method.

In this article, the genetic algorithms will be employed to determine the optimal dimensions of a thick-walled boiler component. Next, the optimization of the start-up and shut-down operations is proposed by the use of the successive quadratic programming algorithm [13].

2 Problem formulation

The heat transfer in a thick-walled boiler component is controlled by heat conduction. The transient heat conduction equation in three-dimensional Cartesian coordinates reads:

$$\frac{\partial}{\partial x}\left(k\frac{\partial T}{\partial x}\right)+\frac{\partial}{\partial y}\left(k\frac{\partial T}{\partial y}\right)+\frac{\partial}{\partial z}\left(k\frac{\partial T}{\partial z}\right)=\rho c\frac{\partial T}{\partial t} \qquad (1)$$

where k is the thermal conductivity, T is the temperature, ρ is the density, c is the specific heat capacity, and t is the time instant.

The highest stresses in the component should be limited to allowable values. This restriction can be captured by a simple structural analysis based on elementary equations of thermoelasticity.

$$\sigma_{ij,j}+f_i=0 \qquad (2)$$

$$\varepsilon_{ij}=\frac{1}{2}\left(u_{i,j}+u_{j,i}\right) \qquad (3)$$

$$\sigma_{ij}=2G\varepsilon_{ij}+\lambda\varepsilon_{kk}\delta_{ij}+\beta T\delta_{ij} \qquad (4)$$

The equation (2) is the force equilibrium equation, (3) is the geometric equation, and (4) is the constitutive equation. The variable σ is the stress tensor, **f** is the body forces vector, **u** is the displacement vector, **ε** is the strain tensor, G and λ are the Lamé constants, β is the thermal expansion coefficient and δ_{ij} is the Kronecker delta.

Set of equations (1-4) should be satisfied throughout the boiler component volume, which can be designed as V. Transient heat conduction problems are initial-boundary problems and appropriate initial and boundary conditions should be defined. Initial conditions, are temperature values of a component at its initial moment. There are four basic types of boundary conditions, but usually two of them are used. One part of the component is insulated and 2nd order boundary conditions simplifies to the form

$$\left.\frac{\partial T}{\partial n}\right|_A=0 \qquad (5)$$

where n is the external normal direction to the component edge surface A. The second part of boundary is in contact with fluid and the boundary condition of 3rd order can be written

$$-k\left.\frac{\partial T}{\partial n}\right|_A=h\left[T(x,y,z,t)\big|_A-T_f\right] \qquad (6)$$

where h is heat transfer coefficient and T_f is fluid temperature. In general both h and T_f can vary in time and space. Equations of thermoelasticity (2-4) should be accompanied by boundary conditions defined by displacement or force. Finite element method will be used for discretization of the above equations.

2.1 Shape optimization

The shape of a thick-walled boiler component determine the level of stress that arise at different phases of its operation. Reducing the thickness of the component will result in an increase in stresses caused by internal pressure but will reduce thermal stresses. So the thickness of the walls can be reduced without increasing total stresses. The advantage of the component with thin walls is its lower weight, volume, and it can be faster heated or cooled.

In the optimization process, the design space search is performed using evolutionary algorithms and a parametric FE model (see Fig. 1). As the tool of the investigation, the Evolutionary Optimization System (EOS) software has been chosen [14, 15]. The EOS is designed to solve single and multicriteria optimization problems for nonlinear programming problems. The single criterion optimization thermo-mechanical problems can be formulated as follows:

- find the vector of decision variables (e.g. dimensions of a thick-walled boiler component)

$$\mathbf{y}_k=[y_1,y_2,\ldots,y_K] \qquad (7)$$

which will satisfy the J=2 inequality constraints:

- for the heating operation

$$g_1(\mathbf{x})=S_{3\min}(\mathbf{x})-\breve{\sigma}_a\geq 0 \qquad (8)$$

- for the cooling operation

$$g_2(\mathbf{x})=\hat{\sigma}_a-S_{1\max}(\mathbf{x})\geq 0 \qquad (9)$$

where $g_j(\mathbf{x})$, j=1,2 are the functions of constrains; $S_{3\min}$ denotes the lowest third principal stress in component during the heating process, $S_{1\max}$ is the highest first principal stress in component during the cooling process, $\breve{\sigma}_a$ and $\hat{\sigma}_a$ denote allowable stress for the heating and cooling operations. Both values $S_{3\min}$ and $\breve{\sigma}_a$ are compressive stresses and are negative. Tensile stresses $S_{1\max}$ and $\hat{\sigma}_a$ are positive.

For the objective function, the volume of the structure has been chosen. The algorithm can be adjusted to the investigated object using the following control parameters [14]: population size PJ, parameter for the operation of crossing p_c, parameter for the operation of mutation p_m, number of generations L_g.

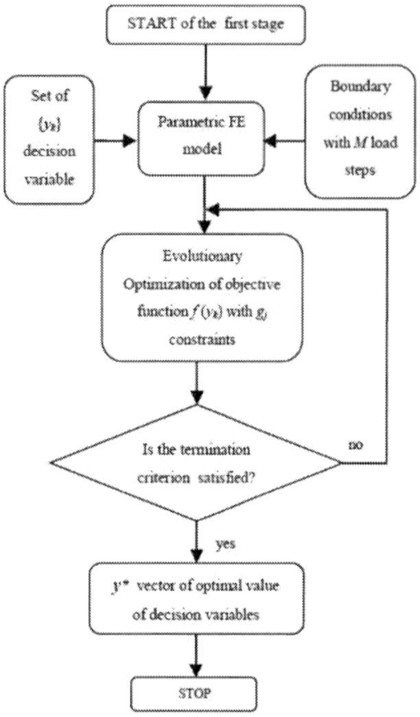

Fig. 1. Proposed parametric GA optimization methodology.

2.2 Optimization of start-up and shut-down operations

The aim of this optimization is to find the fluid transient temperature $T_f(t)$, which heats or cools a thick-walled boiler component in such a way that the total time of heating and cooling operation is the lowest and maximum tensile and compressive total stresses in component do not exceed the allowable values during entire process.

Set of equations (1-4) should be satisfied throughout the boiler component volume, which can be designed as V. Transient heat conduction problems are initial-boundary problems and appropriate initial and boundary conditions should be defined. Initial conditions, are temperature values of a component at its initial moment. There are four basic types of boundary conditions, but usually two of them are used. One part of the component is insulated and 2nd order boundary conditions simplifies to the form

The fluid temperature $T_f(t)$ is assumed as a piecewise linear function. It consists of initial fluid temperature T_{f_1}, temperature change rate v_{T_1}, second fluid temperature T_{f_2} and final temperature change rate v_{T_2}. The second fluid temperature change occurs immediately after obtaining the superheated steam. During the heating process, fluid enters the component of initial temperature T_0. For the cooling process, fluid enters a thick-walled boiler component of working temperature T_w. The proposed course of the fluid temperature is presented in Fig. 2.

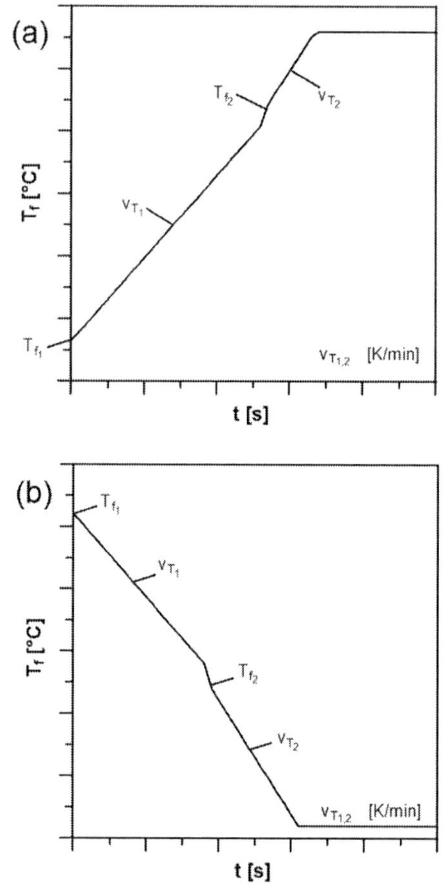

Fig. 2. The chosen function of fluid temperature during heating (a) and cooling operation (b).

A nonlinear optimization problem with inequality constraints is formulated. The aim is to find the best vector $\mathbf{x} = \{T_{f_1}, v_{T_1}, T_{f_2}, v_{T_2}\}$. The optimum vector $\mathbf{x}$ should minimize the objective function S, which is defined as a total time of heating or cooling operation t_{tot}

$$S(\mathbf{x}) = t_{tot} \tag{10}$$

while limiting inequality constraints (8-9). The design vector $\mathbf{x}$ changes in the range

$$\mathbf{x}_l \le \mathbf{x} \le \mathbf{x}_u \tag{11}$$

where subscripts l and u denote lower and upper bounds of searched parameters, respectively. A general nonlinear programming problem is solved using the successive quadratic programming algorithm [13].

The total stress can be calculated adding the influence of inner pressure. When fluid is the pressurized water or superheated steam, the pressure value p_f is independent of temperature. In the case of saturated water, steam-water mixture or saturated steam, p_f is the saturation pressure p_s defined as the function of fluid temperature $T_f(t)$ in °C.

 https://doi.org/10.1051/matecconf/201824005006

3 Numerical example

Shape and operation optimization will be presented for a T-pipe, which is the part of the fresh and preheated steam pipeline in the 360 MW power plant. It was designed for the pressure p_w=18 MPa and steam temperature T_w=540°C. It weighs 1378 kg. The geometry of the T-pipe is presented in Fig. 3. The T-pipe is made of alloy steel 14 MoV63 (13 HMF) [16].

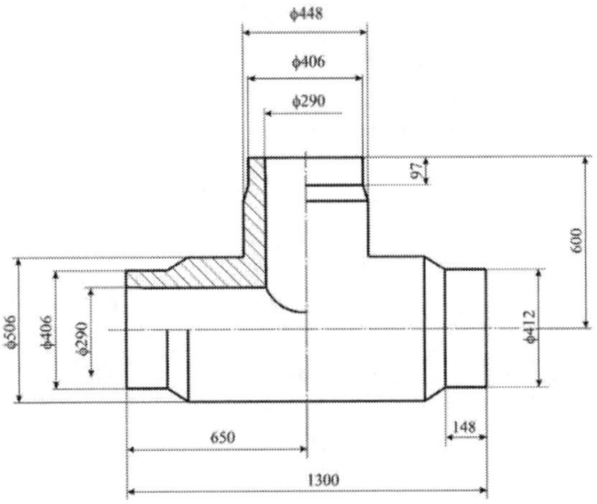

Fig. 3. Geometry of the T-pipe.

TRD301 code [5] can regulate the method of heating or cooling the chosen thick-walled boiler component. It assumes a quasi-steady state and one-dimensional temperature distribution through the wall of construction element. For the chosen T-pipe, the following stress concentration coefficients are considered: $\alpha_p = 3.2$, $\alpha_T = 2$. The calculated allowable stresses for heating and cooling operations according to TRD301 are $\breve{\sigma}_a = -214.4\,\text{MPa}$, $\hat{\sigma}_a = 303\,\text{MPa}$. The allowable rates of temperature change at the beginning and end of the process are obtained based on allowable stresses: $v_{T1(TRD)} = 2.119\,\text{K/min}, v_{T2(TRD)} = 2.993\,\text{K/min}$ - for heating and $v_{T1(TRD)} = -2.119\,\text{K/min}, v_{T2(TRD)} =$ $= -2.993\,\text{K/min}$ - for cooling.

Once the pressure at the beginning of the heating process equals 0 MPa and the pressure at the end of the heating process equals p_w, the fluid temperature can be calculated from the equation:

$$\frac{dT_f}{dt} = v_{T1(TRD)} + \frac{v_{T2(TRD)} - v_{T1(TRD)}}{p_w} p_f(T_f) \tag{12}$$

In a similar way, the fluid temperature for the cooling process can be determined. The calculated fluid temperature and pressure for heating and cooling operation are presented in Fig. 4.

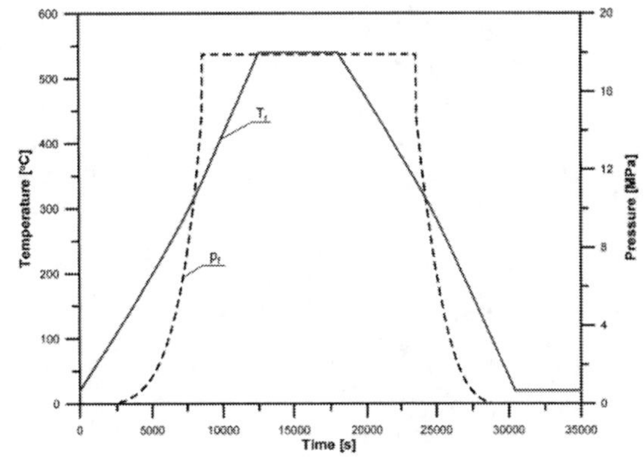

Fig. 4. The calculated medium temperature and medium pressure according to TRD 301.

Fig. 5. T-pipe division into finite elements.

One quarter of the T-pipe was divided into 20-node finite brick elements based on the symmetry conditions. This is shown in Fig. 5. Temperature and stress distribution in the T pipe is calculated using ANSYS software [17]. Temperature distribution in time and space is calculated assuming that the heat transfer coefficient on the inner surfaces of the T-pipe equals 1000 W/(m^{2}K) [18]. Next, stress analyses are carried out. The stress model is constrained to ensure symmetry conditions. It allows for the possibility of unrestricted lengthening in the direction of the horizontal and vertical pipe. Additionally, the surfaces connecting the T-pipe with the pipeline must be planes and the calculated temperature distribution in space and time is used as the body load in the stress model.

The maximum value of equivalent stress and the first principal stress caused by pressure and temperature occurs at about 23100 s and equals $\sigma_{HMH\,max}$=246.8 MPa and $S_{1\,max}$=229.2 MPa respectively. The maximum compressive stress can be estimated by the minimum value of the third principal stress. The minimum value of the third principal stress occurs at about 10300 s and equals $S_{3\,min}$=-139.1 MPa. Throughout the entire cycle the allowable stresses $\breve{\sigma}_a = -214.4\,\text{MPa}$, $\hat{\sigma}_a = 303\,\text{MPa}$ are not exceeded. Principal stresses $S_{1\,max}$ and $S_{3\,min}$ will be adopted for the bounds of constraints during the T-pipe shape optimization.

3.1 Shape optimization

The optimization problem is defined to minimize the volume (mass) of the structure assuming for the following allowable stresses: $\breve{\sigma}_a = -139.1\,\text{MPa}$, $\hat{\sigma}_a = 229.2\,\text{MPa}$.

In the optimization framework, the solution is searched for in 4-dimensional design variable space (2 radii and 2 lengths of the T-pipe reinforcements – see Fig. 6).

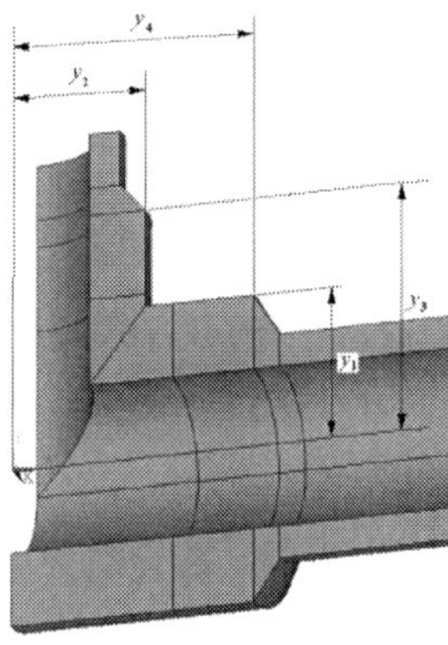

Fig. 6. The 4-dimensional design variable space of GA shape optimization.

For decision variables, the following range of variation is proposed: y_1 = {210 mm, 253 mm}, y_2 = {205 mm, 224 mm}, y_3= {380 mm, 680 mm}, y_4 = {360 mm, 590 mm}.

The parametric FE model is subjected to GA optimization by Evolutionary Optimization System (EOS) software [14]. Moreover, the following steering parameters of EOS are chosen: population size $PJ = 50$, number of generations $N = 25$, probability of crossover p_c= 0.7, probability of mutation p_m= 0.06. The EOS software is connected with the FE program ANSYS for performing batch processing calculations of the objective function value and stress constraints. The best solution **y*** (y_1* = 237.59 mm, y_2* = 215.75 mm, y_3* = 548.81 mm, y_4* = 366.84 mm) is about 10% lighter than the original one.

Fig. 7 presents $S_{1\,max}$ - maximum values of the first principal stresses in the whole T-pipe over the load time history for the original and optimal shape. Fig. 8 shows $S_{3\,min}$ - minimum values of the third principal stresses in the T-pipe for the both geometries. Modifying the shape does not result in an increase in the maximum stress values in time and space. The highest recorded first principal stress for the optimal shape equals 226.9 MPa and is lower than maximum value for the original shape, which equals 229.2 MPa. Additionally, the value of $S_{1\,max}$ for modified geometry is lower than for the original one during almost the whole heating and cooling process (0 s - 13300 s and 22800 s - 35000 s). The maximum compressive stresses in the modified geometry are smaller than in the original T-pipe during the whole cycle. Lower stress values recorded during transient processes make it possible to speed up these processes and reduce the shut-down and start-up losses.

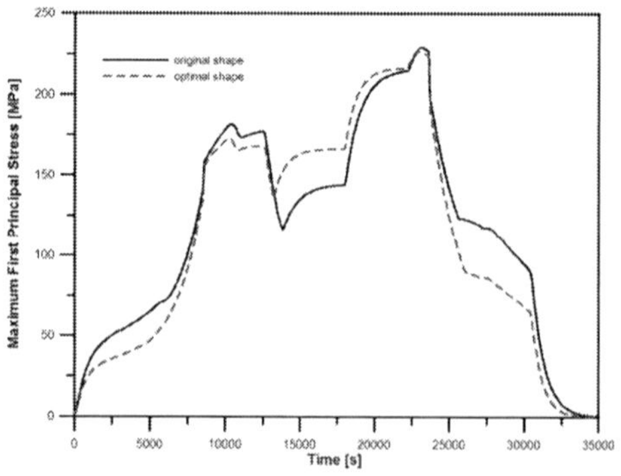

Fig. 7. Maximum first principal stress in the whole structure versus time.

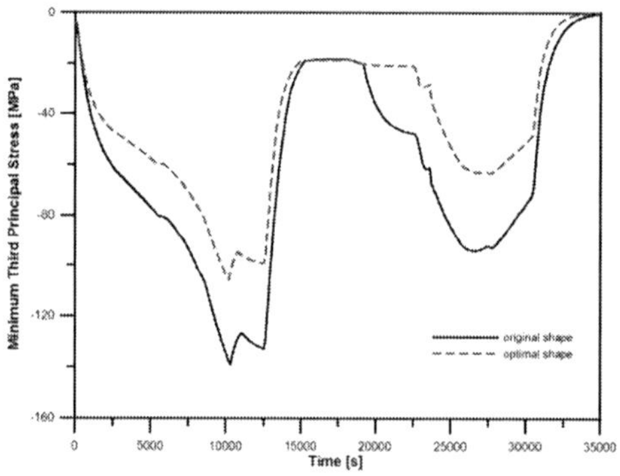

Fig. 8. Minimum third principal stress in the whole structure versus time.

3.2 Optimization of start-up and shut-down operations

The heating process optimization is carried out for the T-pipe with the modified shape, which was obtained in the previous Section. The optimization problem is to minimize the start-up process time with the constraint $\breve{\sigma}_a = -214.4\,\text{MPa}$. Vector of searched parameters **x** consists of { T_{f_1}, v_{T_1}, T_{f_2}, v_{T_2} }. For the lower and upper bounds of the searched-for parameters, in vector **x**, the values of {0°C, 0K/min, 0°C, 0K/min} and {250°C, 40K/min, 540°C, 40K/min} are selected. The starting vector is assumed as {100°C, 4K/min, 300°C, 4K/min}. Using the proposed optimization method, which is presented in Section 2.2, the following parameters are found {147.1°C, 7.01K/min, 359.97°C, 6.27K/min}. Start-up time is reduced from 12600 s to 3622 s as can be seen in Fig.9. Fig. 10 presents $S_{1\,max}$ - maximum values of the first principal stresses and $S_{3\,min}$ - minimum values of the third principal stresses in the whole T-pipe over the start-up load history. The start-up operation based on the proposed optimization method causes higher stresses than the heating according to TRD301. However, all stresses are limited to the allowable values.

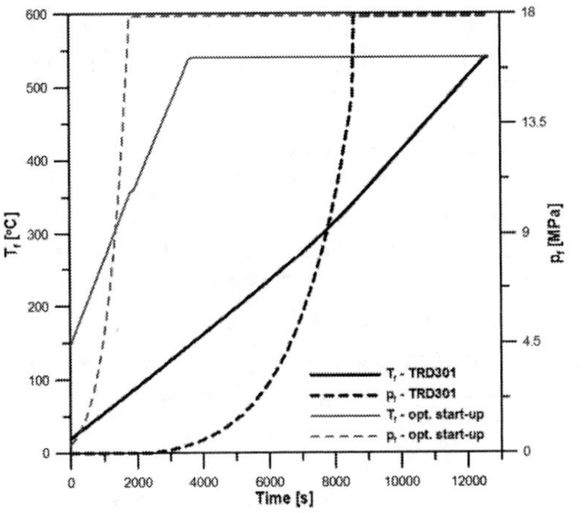

Fig. 9. Start-up temperature and pressure transients based on TRD301 regulations and optimization method.

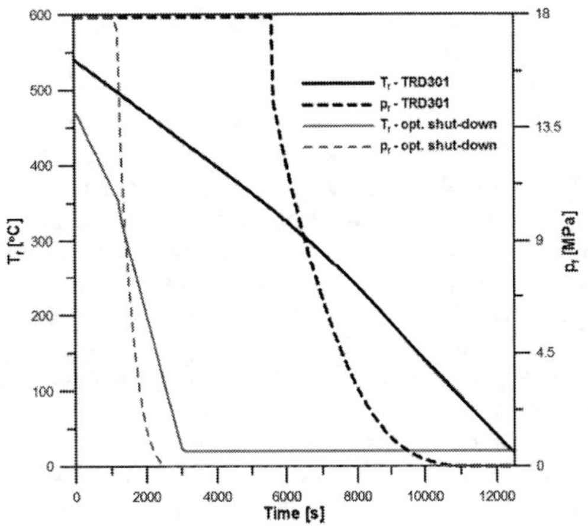

Fig. 11. Shut-down temperature and pressure transients based on TRD301 regulations and optimization method.

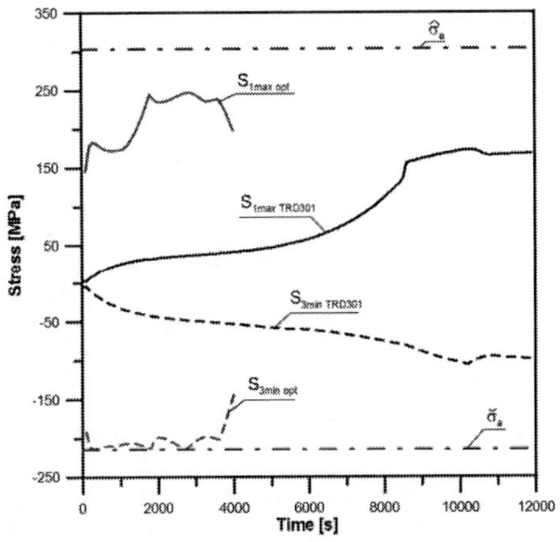

Fig. 10. Maximum first and minimum third principal stresses in the whole structure during start-up operation based on TRD301 regulations and optimization method.

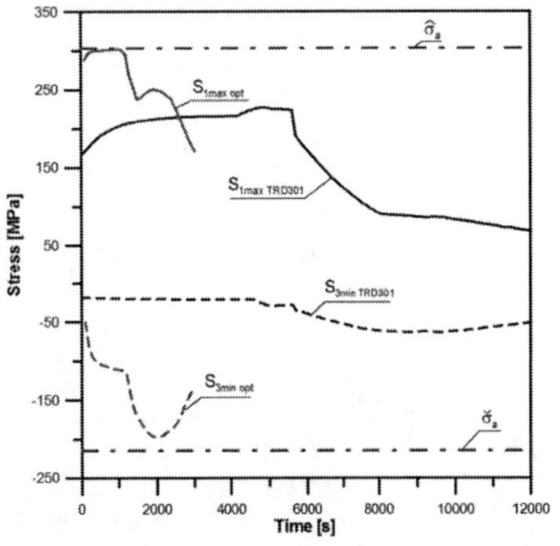

Fig. 12. Maximum first and minimum third principal stresses in the whole structure during shut-down operation based on TRD301 regulations and optimization method.

Next, the cooling process optimization is carried out. The shut-down process time is minimized with the constraint $\hat{\sigma}_a = 303$ MPa . Vector of searched parameters **x** consists of { T_{f_1} , v_{T_1} , T_{f_2} , v_{T_2} }. For the lower and upper bounds of the searched-for parameters the values of {0°C, -40K/min, 0°C, -40K/min} and {600°C, 0K/min, 600°C, 0K/min} are selected. The starting vector is assumed as {400°C, -4K/min, 300°C, -4K/min}. Using the proposed optimization method the following parameters are found {472.2°C, -5.91K/min, 324.2°C, -10.6K/min}. Shut-down time is reduced from 12450 s to 3022 s as can be seen in Fig.11. The shut-down operation based on the proposed optimization method causes higher stresses than the cooling by TRD301, but all stresses are limited to the allowable values as can be seen in Fig. 12.

4 Conclusions

The work presents a two-stage methodology and a successful application of structural optimization of a T-pipe under transient thermal and mechanical loads. Genetic algorithms are applied in the first stage for the component shape modification. The optimization problem is defined to minimize the volume (mass) of the structure with stress constraints. The obtained geometry is about 10% lighter than the original one because the modified T-pipe wall thickness is smaller. This result could be considered as significant if bearing in mind that this structure has been used for many years and was probably already optimized by the designers. The maximum tensile and compressive stresses in the modified geometry are smaller than in the original T-pipe during the whole cycle. Additionally, lower total stress values are recorded during heating and cooling processes. Therefore these transient processes can be accelerated and the shut-down and start-up losses can be reduced. In the second stage, the optimization of the modified T-pipe start-up and shut-down operations is conducted. The problem is defined to minimize the time of start-up and shut-down processes with stress constraints. The time of the start-up and shut-down processes is reduced from 12600 s to 3622 s and from 12450 s to 3022 s respectively. The total stresses, which are recorded during these operations, do not exceed the allowable values. Transient heating and cooling operations can be accelerated without decreasing the longevity of the T-pipe.

The proposed methods can be applied throughout the industry, wherever heating and cooling processes take place. The techniques used to find the optimum layout of the structure, optimum start-up and shut-down transients are fast and do not require a high financial outlay. The presented methods can be used for supercritical power units. The obtained optimization results translate directly into the power unit energy effectiveness, because the shut-down and start-up losses are reduced.

This research was financed by the National Science Centre, Poland, UMO-2015/19/B/ST8/00958

References

1. T.S. Kim, D.K. Lee, S.T. Ro, Applied Thermal Engineering **20**, 11, 977-992 (2000)
2. P. Duda, R. Dwornicka, Structural and Multidisciplinary Optimization **40**, 529-535 (2009)
3. P. Duda, D. Rząsa, International Journal of Energy Research **36** 703–709 (2012)
4. K. Krüger, M. Rode, R. Franke, Energy **29**, 12–15, 2239–2251 (2004)
5. TRD 301, *Technische Regeln für Dampfkessel*, Carl Heymans Verlag, Köln und Beuth-Verlag, Berlin (2001)
6. R.J. Yang, S.C. Poe, Structural Optimization **5** 233-239 (1993)
7. R.V. Tappeta, S. Nagendra, J.E. Renaud, Structural Optimization **18** 134-145 (1999)
8. T. Wen et al., International Journal of Heat and Mass Transfer **50** 2590–2604 (2007)
9. M. Yoon, S.-H. Ha, S. Cho, International Jour-nal of Heat and Mass Transfer **62** 272-285 (2013)
10. G.N. Xie, Q. Wang, Applied Thermal Engineering **59** 425–434 (2013)
11. C.L. Xiao, H.X. Huang, Applied Thermal Engineering **64** 462–470 (2014)
12. G. Nowak, A. Rusin, Energy Conversion and Management **74** 417–425 (2013)
13. K. Schittkowski, Annals of Operations Research **5** Issue 1-4, 485-500 (1986)
14. A. Osyczka et al., *Evolutionary optimization system (EOS) for design automation.* In: Burczynski T, Osyczka A (eds) Evolutionary methods in mechanics. Kluwer, Dordrecht (2004)
15. M. Mrzyglod, A. Osyczka, *Parallel Computing for Design Optimization with Computationally Expensive Functions us-ing Evolutionary Algorithms* (in B.H.V. Topping, (Editor), Proceedings of the Eighth International Conference on the Application of Artificial Intelligence to Civil, Structural and Environmental Engineering, Civil-Comp Press, Stirlingshire, UK, Paper 26, 2005)
16. F. Richter, *Physikalische Eigenschaften von Stählen und ihre Temperaturabhängigkeit.* Mannesmann Forschungberichte 930, Düsseldorf (1983)
17. ANSYS Theory Guide, Release 14, (2013)
18. J. Taler, P. Duda, *Solving direct and inverse heat conduction problems*, Springer-Verlag, Berlin, (2006)

New start-up curves for a 200 MW steam boiler with natural circulation

Piotr Dzierwa[1]

[1]Institute of Thermal Power Engineering, Cracow University of Technology, Cracow, Poland

Abstract. The paper shows the method of determining the start-up curves of a steam boiler, which enables its faster start-up. Faster temperature changes cause thermal stress, especially in thick-walled boiler elements. Critical elements that limit the rate of change of working fluid units include drums, outlet chamber, injection chamber, fresh steam superheater of steam coolers, tees and valve casings, rotor and turbine casing. The walls of these elements have the thickest walls due to the high pressure values occurring in them during operation. Allowable changes in temperature and pressure during start-up and shut-down can be determined by the European Standard EN 12952-3. The article shows how the start-up curves of a steam boiler can be determined using the European Standard and the method proposed by the author. Calculation results show that the starting time of the analyzed block can be significantly shortened.

1 Introduction

Due to the increasing share of wind farms in electricity generation, a need arises to speed up the start-up process in conventional thermal power plants. This results from great variations in the amount of power supplied to the power grid by wind turbines [3-6]. During start-up and shutdown, as well as at changes in loads, considerable thermal stresses arise in the power boiler thick-walled pressure components operating at high temperatures. The critical components that limit the rate of changes in the working fluid parameters include the natural circulation boiler drums, water separators in once-through boilers, chambers of steam injection coolers, T-pipes and valve housings [3]. Considering high-pressure values and large diameters, the walls of these elements are much thicker compared to the boiler other components operating under the same pressure. Considerable differences in temperature occur in very thick walls. The life of the elements may be reduced dramatically if heating and cooling rates are too high. The biggest stresses and strains usually occur at the edges of openings, where, after a longer period of operation, cracks may develop due to low-cycle fatigue. The stresses and strains determine allowable heating and cooling rates for pressure components. Start-up and shutdown should be performed in a manner that ensures that :

- the selected rate of heating (cooling) the boiler components from the initial state to the design value should ensure that allowable stresses and temperatures are not exceeded,
- energy losses are as small as possible,
- the process is safe for staff and equipment.

The regulations concerning the optimum performance of the power unit start-up and shutdown processes have not been developed in Poland yet. For most boilers, allowable rates are specified in the operating instructions prepared by the manufacturer. As a rule, they are constant values which are independent of the nature of any time-dependent changes in pressure in a thick-walled component. The methods that have already been worked out to specify the optimum procedure for carrying out the heating or cooling process concern flat or cylindrical walls. In the boiler technology, the problem that has to be faced most often is a weakened cylindrical or spherical component. The application of optimization methods developed for components with simple shapes is burdened with too big errors. Such methods may only be used to estimate the limit values of start-up or shutdown parameters. The allowable rates of the element heating and cooling determined according to the German regulations concerning boilers (TRD-301) or to the European standard EN 12952-3 is based on the assumption of a quasi-stationary temperature field in the wall of a structural component [1, 2]. This is difficult to achieve in practice as this state settles in the pressure element wall only after it has been heated at a constant rate for 20-30 minutes. Due to high rates of changes in steam temperatures in time, the quasi-stationary temperature field is very seldom to occur [7, 8]. Moreover, the element heating and cooling times determined based on the TRD-301 regulations or the EN 12952-3 standard can cause excessive stress on the edges of the holes. This, in turn, leads to the shorter service life of pressure components of the boiler. The German boiler regulations TRD-301 make it possible to determine allowable

* Corresponding author: pdzierwa@pk.edu.pl

cooling and heating rates for the boiler pressure elements for several selected shapes of structural components. The criterial point is located on the hole internal surface in the plane passing through the pressure element longitudinal axis and the opening axis. However, calculations performed using the Finite Element Method indicate that as heating proceeds, the biggest total stresses may also occur on the hole edge in the plane perpendicular to the pressure component axis and passing through the hole edge axis. Another drawback of the TRD-301 regulations is the assumption of a constant thermal stress concentration coefficient $\alpha_T = 2$, i.e. like in the case of an infinite plate with a hole on the thickness along which temperature drops linearly. The fields of both temperature and stresses in the area of holes are complicated due to the occurrence of connecting pipes. It also turns out that for a three-dimensional temperature field the thermal stress concentration coefficient in a quasi-steady state depends on the value of the heat transfer coefficient h on the external surface [16]. The TRD-301 regulations do not take this fact into account at all. Although the European standard EN 12952-3 takes into account the dependence of the thermal stress concentration coefficient α_T on the heat transfer coefficient h, the quasi-steady temperature field is still adopted in this standard. For components with complicated shapes, if pressure-related stresses are to be calculated, stress concentration coefficients are used making it possible to find the maximum value of these stresses in the concentration zones. In real conditions, the value of the stresses can be calculated using the Finite Element Method (FEM) structural analysis. It often turns out that the values of stress concentration coefficients taken from the TRD-301 or literature [4] deviate considerably from those determined using the FEM. Due to the transient nature of temperature changes, it is very important to monitor stress conditions in thick-walled walls of criteria elements [9-11]. This requires precise determination of transient temperature changes of the working medium [12-15].

The above-mentioned limitations of the TRD-301 regulations and the EN 12952-3 standard make it impossible to calculate allowable rates of heating and cooling the boiler thick-walled elements at which real stresses on the hole edge calculated using the FEM do not exceed prescribed allowable stress values. Therefore, an own method of determining the optimum changes in the fluid temperatures during the heating and cooling of pressure elements was developed at which the total of thermal and pressure-related stresses on the hole edge does not exceed the allowable values. In this paper, the start-up pressure and temperature curves were determined for the critical thick-walled components – the drums and outlet headers – using a standard EN 12952-3 and the method developed in papers. The objective of an extensive structural analysis of boiler pressure components was to improve the flexibility of the power unit.

2 Start-up curves of OP-650-012 boiler

The OP-650-012 boiler is a steam boiler fired with pulverized hard coal. Boiler achieves design electrical power 210 MWe. When designing the OP-650 boilers, the possibility of fast start-ups from cold, warm or hot states was not taken into account. The current operating manual of this boiler allows for heating with a constant rate of 1K/min for the fluid temperature in the boiler drum, no matter what the pressure in the boiler is. This results in a significant extension of the start-up time. The time changes of generator active power, superheater steam mass flow, the pressure in the boiler drum, the pressure in the superheater, during the start-up of the OP-650-012 from the cold state are shown in Fig. 1.

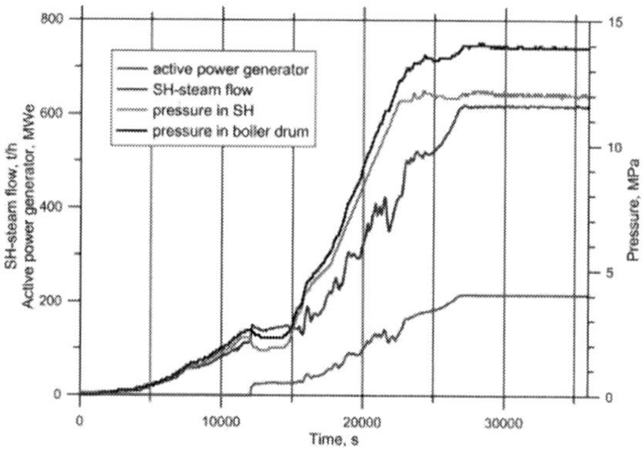

Fig. 1. Start-up curves of the OP-650-012 boiler from the cold state.

It can be seen from Fig. 1 that the start-up time of the OP-650-012 boiler is about 7.5 hours. The start-up duration was calculated as the time interval between the pressure rise in the drum and the moment when the drum pressure reached the nominal value.

3 A new method of determination of the optimum steam curves during the drum heating

The permissible stresses values during start-up and shutdown of the OP-650-012 boiler have been determined from the standard EN 12952-3. First, the allowable rates of temperature changes during the start-up of the boiler drum were determined. Next, the steam temperature curve $T_f(t)$ was found from the solution of the following differential equation

$$\frac{dT_f}{dt} = f(p) \tag{1}$$

with

$$f(p) = \frac{p_2 v_{T1} - p_1 v_{T2}}{p_2 - p_1} + \frac{v_{T2} - v_{T1}}{p_2 - p_1} p \tag{2}$$

Pressure p depends on the temperature in the case of a drum containing water with the saturation temperature. The gauge pressure p_1 at the beginning of the start-up process was 0 MPa, while at the end of the process it was p_2 = 15.95 MPa. The pressure in the outlet header is by about 15% lower than in the boiler drum. The Eq. (1) has been solved with the Runge-Kutta method. Figure 3 and Table 1 shows the allowable heating and cooling rates v_T for the boiler drum obtained by using the standard EN 12952-3. The drum is made of low alloy steel, grade 15NCuMNb (K32Nb). The chemical composition of the steel is as follows: C ≤ 0.17%, 0.25 ≤ Si ≤ 0.5%, 0.8 ≤ Mn ≤ 1.2%, 0.035 % ≤ P, 0.035 % ≤ S, 0.5 ≤ Cu ≤ 0.8%, 0.25 ≤ Mo ≤ 0.5%, 1.0 ≤ Ni ≤ 1.3%.

Table 1. The allowable rates of fluid temperature changes for a drum of OP-650-012 boiler.

	Start-up		**Shutdown**	
	Beginning	**End**	**Beginning**	**End**
Boiler drum				
p, MPa	0	15.95	15.95	0
v_T, K/min	4.52	13.5	-3.91	-12.91
ΔT, K	-26.41	-79.82	22.88	76.3
Outlet header				
p, MPa	0	13.65	13.65	0
v_T, K/min	3.51	4.72	-3.51	-4.72
ΔT, K	-33.84	-45.33	33.84	45.33

The outlet header is made of boiler steel, grade 13CrMo44. The chemical composition of the steel is as follows: 0.1 C ≤ 0.18%, 0.15 ≤ Si ≤ 0.35%, 0.4 ≤ Mn ≤ 0.7%, 0.04 % ≤ P, 0.04 % ≤ S, 0.7 ≤ Cr ≤ 1.0%, 0.4 ≤ Mo ≤ 0.5%.

The results of calculations of permissible temperature differences, as well as heating and cooling rates depending on the pressure, are presented in Table 1. These values have been determined using the EN 12952-3 standard. Next, the temperature changes of the fluid during the start-up were determined for the drum and the outlet header. The heating of pressure components is carried out in such a manner that the total circumferential stresses on the hole edge, in the stress concentration point, caused by pressure and temperature do not exceed permissible values. Values of stress concentration coefficients were determined for elements taking into account the actual 3D geometry using the finite element method (FEM). These values made it possible to determine the permissible rates of temperature changes in the criteria elements. However, the FEM analysis indicates that the thermal stress concentration coefficients depend not only on the component geometry but also on the heat transfer coefficient on the inner surface. The optimum temperature and pressure changes during start-up of the pressure vessel should be determined concerning the total circumferential stress (caused by the difference in wall temperature and internal pressure) at the point of their concentration. The optimum temperature curves for the live steam superheater outlet header and boiler drum are presented in Fig. 2. Figure 3 shows the pressure changes in the OP-650-012 boiler header and boiler drum during optimum heating.

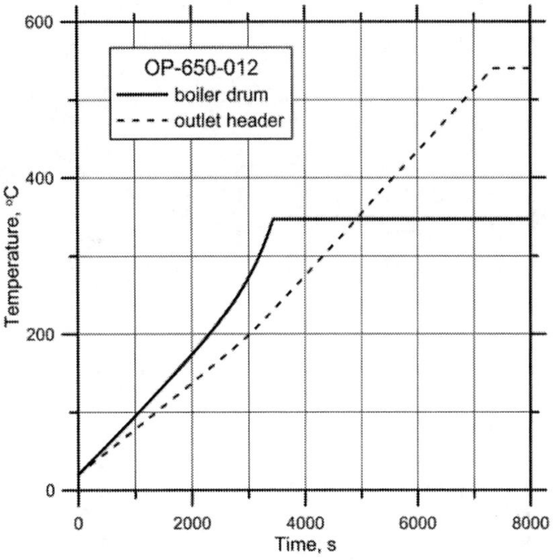

Fig. 2. Steam temperature changes for the OP-650-012 boiler drum and outlet header during start-up according to the proposed method.

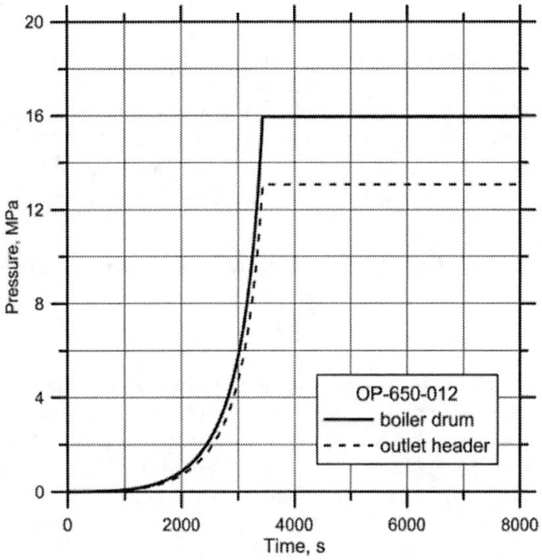

Fig. 3. Steam pressure changes for the OP-650-012 boiler drum and outlet header during start-up according to the proposed method.

The optimum outlet header heating time is equal to 2 hours and is two times higher than for boiler drum (Fig. 2). Pressure variations in the superheater outlet header as a function of time corresponding to a change of pressure in the drum. The pressure in the header is reduced by 15% compared with the value of pressure in the drum. It should be stressed that the effect of pressure on the optimum course of vapour temperature in the chamber is negligible since the thickness of the header wall is 100 mm, and the inner diameter is only 308 mm (Fig. 4). Stresses due to pressure at the points P_1 and P_2 are,

 https://doi.org/10.1051/matecconf/201824005007

therefore, very small and do not influence optimum temperature changes. Considering that the present start-up time of the boiler is 7.5 hours, it is possible to make it about three times shorter.

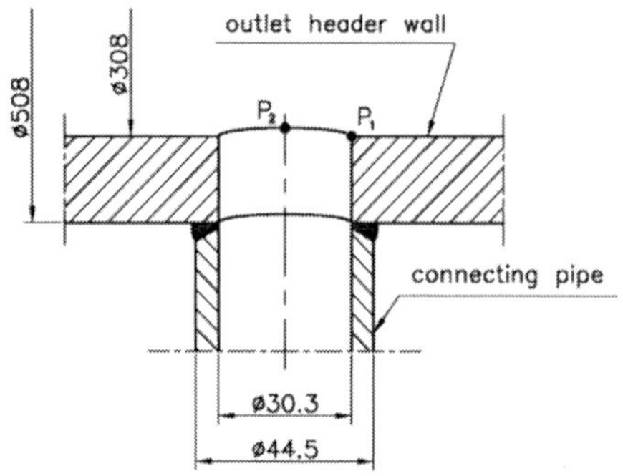

Fig. 4. Diagram of the analyzed connection between the junction of the outlet header with connecting pipe; P_1 – criterial point taken into account in the Standard EN 12952-3, P_2 – additional criterial point.

The boiler shutdown may be performed at any rate. During the shutdown, thermal and pressure-related stresses are summed up. Due to the reduction in the boiler pressure, the permissible cooling rates for the drum and outlet header may be very high. The boiler shutdown rate, i.e., a reduction in the steam pressure and the temperature is limited by factors other than thermal stresses. The problem of cooling optimization of boiler pressure components is less important, due to the large mass of water and metal in the boiler so that fast cooling of the boiler during normal operation is impossible.

4 Analysis of the temperature and stress field in the OP-650-012 boiler drum – constant heating rate

As shown in the previous chapter, the outlet header will be the limiting element of the boiler start-up. In this paragraph, the transient thermal-strength analysis of the OP-650-012 boiler drum was carried out for starting curves adjusted to the boiler outlet header heating time changes. The boiler start-up time from a cold state is 7300 seconds. The fluid temperature in the drum varies at a constant rate from the initial value 20°C to the nominal temperature of 340°C. The heat transfer coefficient is equal to 3000 W/(m^2·K). The FEM calculations were carried out using the Ansys 18.2 software. The mesh of elements used in the calculation is shown in Fig. 9. Figs. 5-8 shows the thermo-physical properties of steel grade 15NCuMNb, i.e.:

- thermal conductivity
- density
- heat capacity
- Young's modulus
- yield strength
- Poisons ratio
- linear thermal expansion coefficient

as a function of temperature.

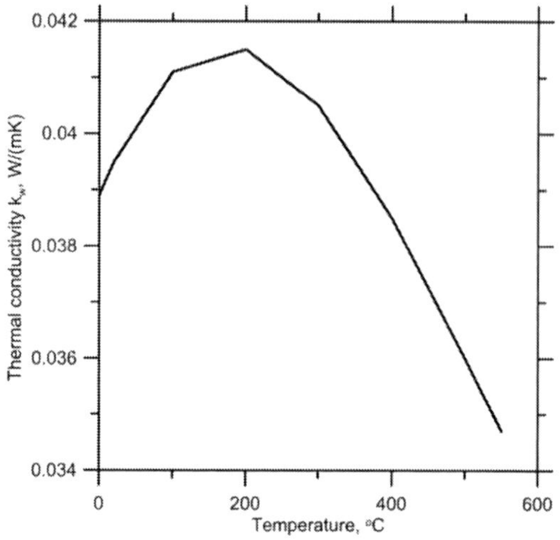

Fig. 5. Temperature dependence of thermal conductivity values for steel 15NCuMNb.

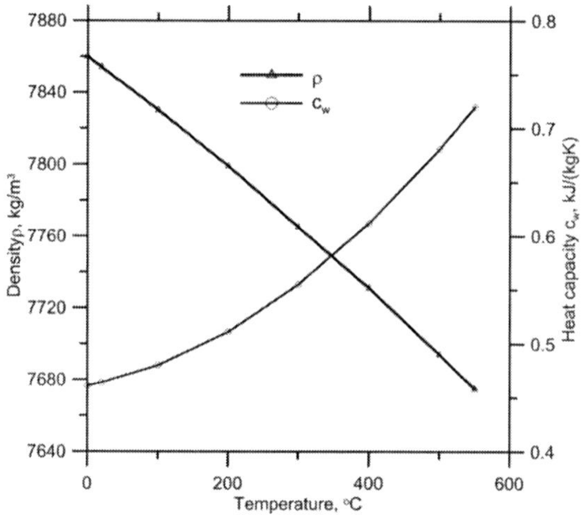

Fig. 6. Temperature dependence of density and heat capacity values for steel 15NCuMNb.

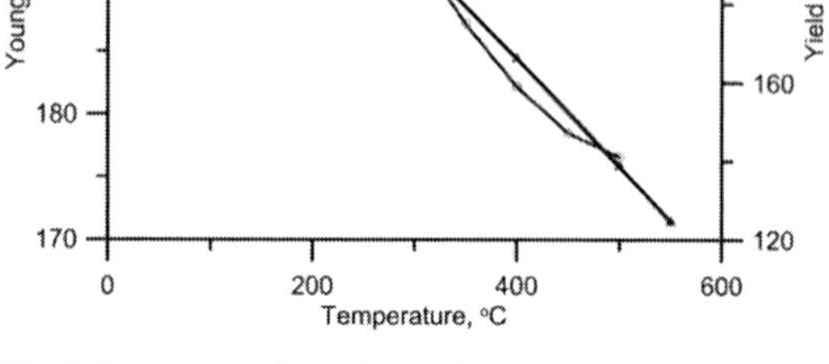

Fig. 7. Temperature dependence of Young's modulus and yield strength values for steel 15NCuMNb.

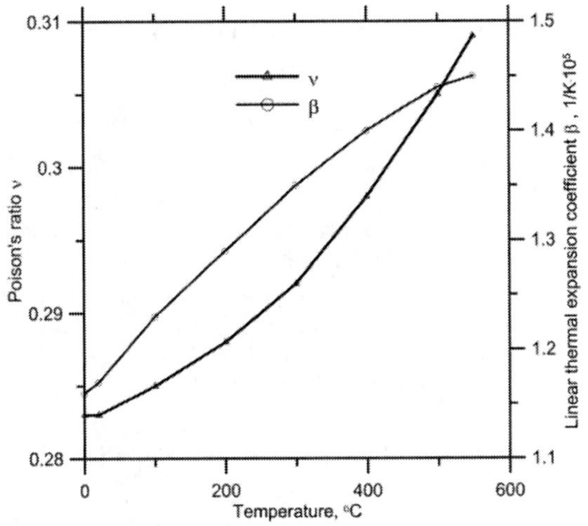

Fig. 8. Temperature dependence of Poisons ratio and linear thermal expansion coefficient values for steel 15NCuMNb.

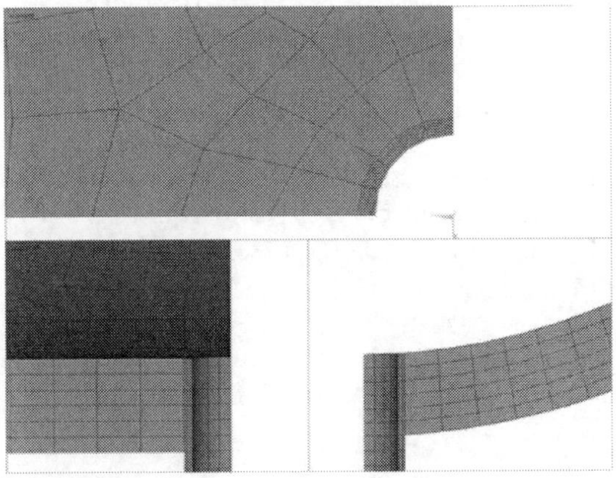

Fig. 9. Mesh of finite elements used to analyze the drum – down comer junction.

The EN 12952-3 standard was used to determine the permissible stresses limits for the startup (σ_{al} = -186 MPa) and shutdown (σ_{al} = 537 MPa) Fig. 11. The values of total circumferential stresses due to the thermal and pressure loads were calculated at the points P_1 and P_2 similarly as for the outlet header (Fig. 4) and compared with the permissible values in Fig. 11. Analysis of the results presented in Fig. 11 indicates that the stresses at points P_1 and P_2 are less than the limit values. The calculations take into account the pressure changes over time. Temperature changes on the inner and outer surface of the cylindrical part of the drum are shown in Fig 10. Figures 12 and 13 show the distribution of the temperature and von Misses stress field after 2000 seconds from the start of boiler start-up. Fig. 12 shows that the wall of the downcomer tube, which is much thinner than the wall thickness of the drum, heats up the fastest. The stress field analysis illustrated in Fig. 13 shows that the equivalent stresses reach their highest value at point P_2. The stress concentration at the edge of the hole is of a local character (it causes the stress field to be too high at a short distance from the hole).

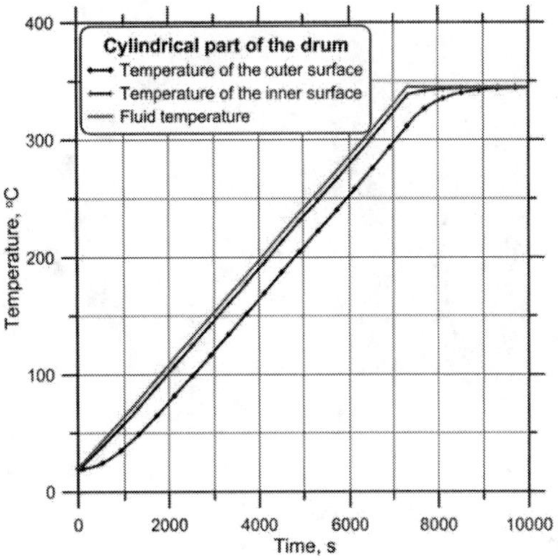

Fig. 10. Time changes in steam temperature during the OP-650-012 boiler drum start-up and in temperatures of the inside and outside surfaces of the cylindrical part of the drum.

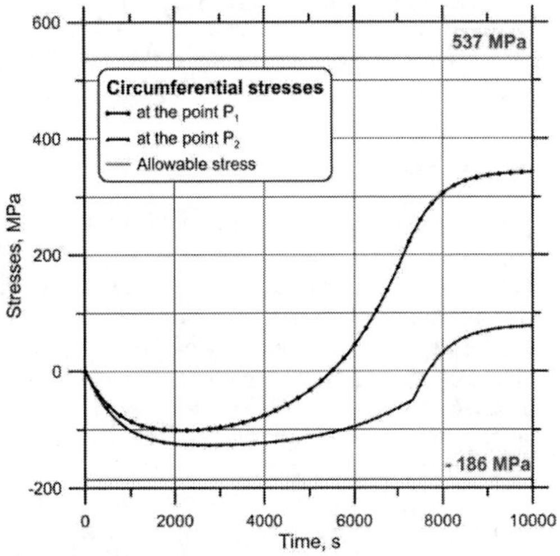

Fig. 11. Total circumferential stress due to thermal and pressure loads at points P_1 and P_2 during the OP-650-012 boiler start-up.

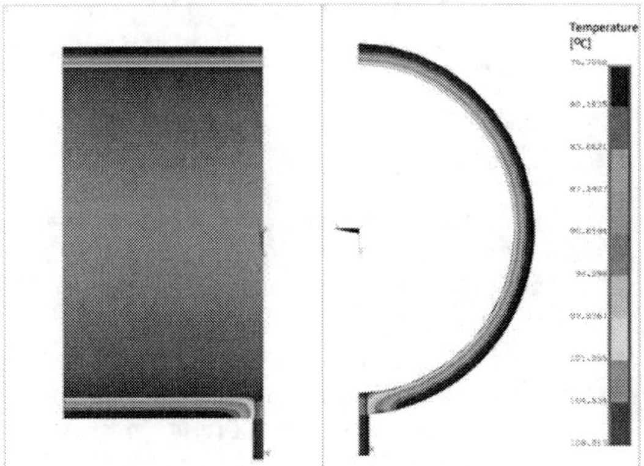

Fig. 12. Temperature distribution in the drum cross-section after 2000 s.

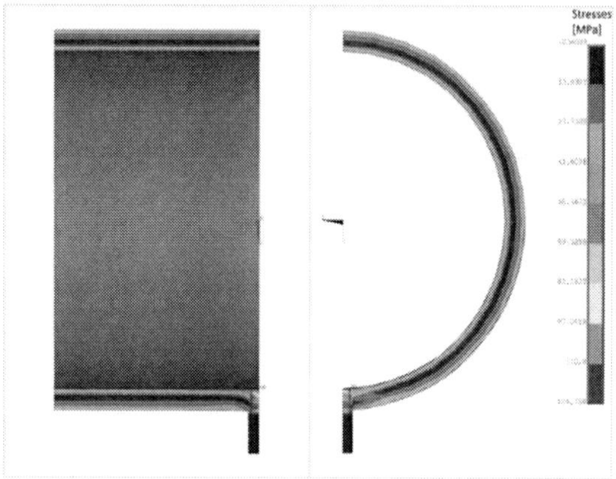

Fig. 13. Distribution of equivalent von Mises stress in time 4000 s.

The conducted calculations show that the existing temperature difference on the drum circumference influences slightly the stress condition of the joint between the drum and the outlet pipe, i.e., in the area of the highest stress concentration. Concluding, the scheduled temperature and pressure changes can be achieved practically.

Conclusions

The paper analyses the process of OP-650-012 boiler start-up. The method of determination of start-up curves, i.e. time changes in temperature and pressure of the working medium, was shown. Analyses and calculations were made for two criteria elements of the boiler, for the outlet header of the final super heater and boiler drum. In the analysis of start-up acceleration possibilities, the European Standard EN 12952-3, as well as our own method of start-up optimization, were used.

The results show that the start-up time can be reduced from 7.5 hours to just over 2 hours. For the boiler in question, the start-up time is limited by the stresses in the outlet header of the final superheater. When the start-up is carried out by the start-up curves determined for the outlet header, the maximum absolute value of circumferential stresses at the hole edge of the drum - downcomer junction is much lower than the allowable stress.

References

1. TRD 301, *Zylinderschalen unter innerem Überdruck. Technische Regeln für Dampfkessel* (Heymanns Beuth, Köln – Berlin, Germany, 143-185, 2001)
2. European Standard EN 12952-3, *Water-tube boilers, and auxiliary installations. Part 3: Design and calculation for pressure parts* (European Committee for Standardization, Brussels, 2012)
3. P. Dzierwa, J. Therm. Stresses **39**, 874-886 (2016)
4. J. Taler, P. Dzierwa, ASME J. Pressure Vessel Technology **137**, 011202-1-8 (2014)
5. J. Taler, P. Dzierwa, Int. J. Energ. Res. **35**, 897-908 (2011)
6. P. Dzierwa, M. Trojan, D. Taler, K. Kamińska, J. Taler, J. Therm. Sci. **25**, 380–388 (2016)
7. D. Taler, K. Kaczmarski, J. Therm. Sci. **25**, 549-557 (2016)
8. J. Taler, W. Zima, M. Jaremkiewicz, J. Therm. Stresses **39**, 386-397 (2016)
9. J. Taler, P. Dzierwa, D. Taler, M. Jaremkiewicz, M. Trojan, *Monitoring of thermal stresses and heating optimization including industrial applications* (Nova Science Publishers, New York, 2016)
10. J. Taler, D. Taler, K. Kaczmarski, P. Dzierwa, M.Trojan, T. Sobota, Energy **160** 500-519 (2018)
11. T. Sobota, Heat Transfer Eng. **39**, 1260-1271 (2018)
12. J. Taler, D. Taler, *Measurement of heat flux and heat transfer coefficient, Heat flux: processes, measurement techniques, and applications* (Nova Science Publishers, 1-103, New York, 2012)
13. M. Jaremkiewicz, D. Taler, T. Sobota, Int. J. Therm. Sci. **87**, 241-250 (2015)
14. M. Jaremkiewicz, Heat Mass Transfer **53**, 887-897 (2017)
15. M. Jaremkiewicz, J. Taler, Heat Transfer Eng. **39**, issues 13-14, 1227-1234 (2018)
16. M. Trojan, D. Taler, Fuel **150**, 75-87 (2015)

https://doi.org/10.1051/mateccconf/201824005008

Steady state CFD analysis of heat transfer coefficient in pressurised pipes of super-heater of OP210 steam boiler

Mariusz Granda*

Institute of Thermal Power Engineering, Cracow University of Technology, Poland

Abstract. The aim of the paper is Computational Fluid Dynamics (CFD) analysis of Wall Heat Transfer Coefficient (WHTC) of pressurized pipe as a part of super-heater of the OP210 boiler. The object of the investigation is convection from saturated steam to the wall of the pipe, which works under high pressure and high temperature. The analysis is an approach to obtain exact solutions of WHTC according to the third type boundary condition compared to direct results from CFD analysis. The paper consists of three-step approach typical for CFD analysis: (i) Pre-Processing, the most elaborated part of the analysis where knowledge about super-heaters, turbulence, velocity profile is important to 3D model, mesh and boundary condition definition. (ii) Simulation of steady state turbulent flow until convergence criteria are met. (iii) Post-Processing where different approaches to the WHTC are shown in comparison. Also, the investigation includes two different types of meshes (where a different number of inflation layers are used) and comparison between k-epsilon and Solid Shear Stress (SST) turbulence model.

1 Introduction

Super-heaters are significant accessories of steam boilers in terms of Rankine's cycle efficiency [1, 2]. To improve it, as high as possible parameters of steam should be used. Also, dehumidification of the vapour is very important due to the erosion of turbine blades [2]. In all of these cases, super-heaters are used. They are built from coils of seamless, bare tubes and headers. The first ones usually have 25 to 44 mm outside diameter, while the second ones 100 to 350 mm respectively [3]. There are two different types of super-heaters dependent on the mode of heat transfer. Convection and radiant super-heaters, where heat transfer is almost dependent on the parameters of the gas flow in the first case and highly dependent on the flue gas temperature in consequence of combustion in the second one. Dimensions are related to the operating pressure and total capacity.

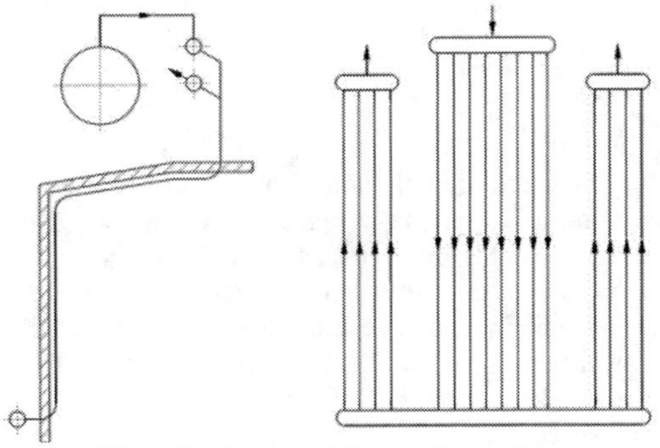

Fig. 1. Roof mounted design of super-heater.

Typical small power boilers, working under pressure from 10-12 MPa, where preheating and evaporation are most important in the heat absorption process, the surface of super-heater is respectively smaller [2]. Located on the roof, small, radiant furnace super-heater can be used (Fig.1)[3]. Higher operating pressure requires greater super-heaters, which are divided in stages (Fig.2), where steam is mixed in headers to avoid uneven temperature distribution (Fig. 3). Also, quench coolers can be used as a temperature regulator. As materials of super-heater work under high pressure and temperature, allowable stress and economic factors are most crucial during the engineering process. Moreover, wall temperature calculations should be prepared, so that the most appropriate material should be used. Under 420 °C carbon steel can be applied. Above it, till 540 °C, low-alloy ferritic steels should be used. From 500 °C till 600 °C ferritic-martensitic steel alloy is considered.

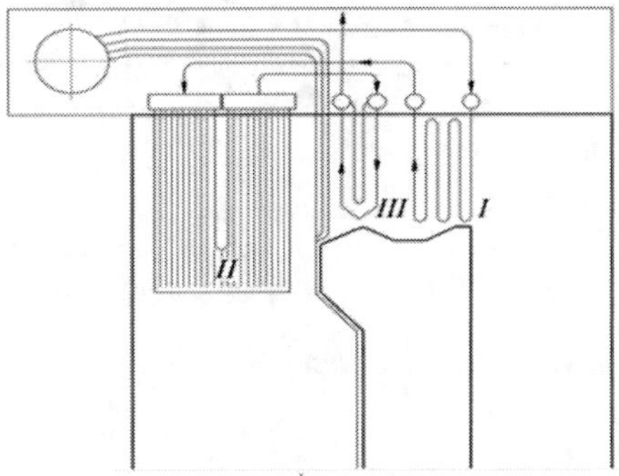

Fig. 2. Three-stage design of super-heater.

* Corresponding author: grandamariusz@gmail.com

Furthermore, austenitic stainless steel can work even up to 650 °C but this kind of material is very expensive and difficult to weld as well [4].

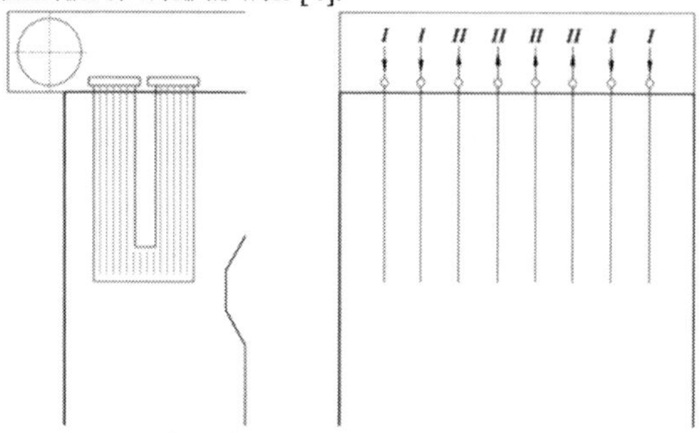

Fig. 3. Design of platen super-heater divided in stages.

2 Turbulence

2.1 General information

The flows occurring in nature can be divided into laminar, transitory and turbulent. The main and the most important criterion which describes every kind of flow is a dimensionless parameter called the Reynolds number (*Re*), that can be expressed by the relation of inertia forces to viscous forces:

$$Re=\rho UL/\mu, \quad (1)$$

where ρ is density, U is the mean speed of the flow, L is a characteristic dimension which is an internal diameter in case of cylindrical pipes and μ stands for dynamic viscosity respectively. Laminar flows, also called stratified are characterized by the fact that subsequent layers of the fluid slide on each other. The movement of the fluid elements is stable, and streamlines of the flow do not intersect. Unfortunately, most of the flows observed in nature are turbulent, where $Re \geq 2300$. A feature of such a flow is a chaotic movement of the fluid elements in a perpendicular direction to the general direction of the fluid motion, where the exchange of the fluid elements occurs between particular streamlines, and thus is characterized by the exchange of mass, momentum and energy. According to the Reynolds hypothesis, all parameters characterizing turbulent flow, such as speed, pressure and temperature, change over time around their time-averaged values [5, 6, 7]. Any flow value $f(x,y,z,t)$ can be expressed by the sum of the mean value $\bar{f}(x,y,z,t)$ and the fluctuation value $f'(x,y,z,t)$, as below:

$$f(x,y,z,t) = \bar{f}(x,y,z,t) + f'(x,y,z,t) \quad (2)$$

where the mean value is a time-averaged value:

$$\bar{f}(x,y,z,t) = \frac{1}{\Delta t}\int_t^{t+\Delta t} f(x,y,z,t)\,dt \quad (3)$$

Moreover, it should be noted, that the time interval must be chosen properly where the time-averaged value cannot change significantly. Further re-averaging of the time-averaged function value can be expressed as below:

$$\bar{f'}(x,y,z,t) = \frac{1}{\Delta t}\int_t^{t+\Delta t} f'(x,y,z,t)\,dt = 0 \quad (4)$$

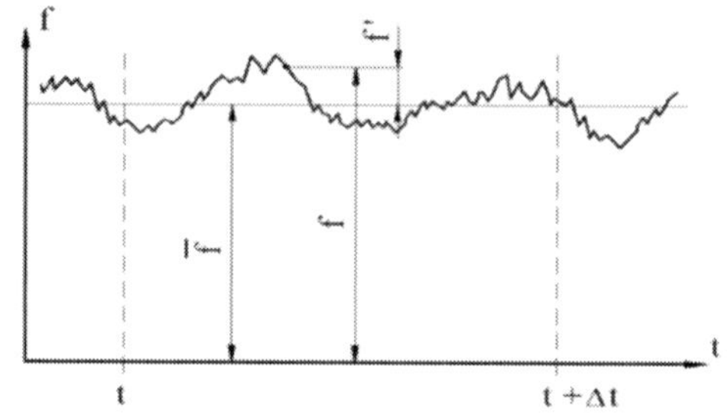

Fig. 4. Distribution of the flow value over time.

2.2 Turbulence simulation

2.2.1 Direct Numerical Simulation

Simulations of the turbulent flows can be carried out in terms of Direct Numerical Simulation method (DNS), which consists of direct solution of the Navier-Stokes equations, that enables accurate modelling of the phenomenon of turbulence. In this method, a mesh with elements smaller than individual swirls should be used. Also, the time step must be small enough to prevent crossing the mesh boundary by the analysed fluid element [8, 9]. These requirements are related to the substantial amount of computing resources and the long-time of problem solving. It translates into the enormous cost of this type of analysis.

2.2.2 Large Eddy Simulation

Large Eddy Simulation (LES) method is based on the simulation of large-scale turbulent movements with determined kinetic energy, and modelling of smaller swirls where viscosity forces are considerable [10]. This method does not require such a large amount of computing resources and is very accurate in reflecting the actual turbulent flow, so that it is more and more willingly used as a tool in the CFD simulations [10].

2.2.3 Reynolds Averaged Navier-Stokes Equations

Most engineering simulations of turbulent flows are based on the Reynolds Averaged Navier-Stokes (RANS) equations where any instantaneous value can be decomposed as described before. This method is not as time consuming as the previous ones and can be used for practical purposes. The RANS equations are non-linear, where each component of fluctuation generates the unknown and an additional equation must be solved. In this particular situation turbulence models are provided, and the problem can be solved. Turbulence models can be divided into two groups in terms of the used hypothesis. The first group is based on the Bossinesq theory, where the Reynolds stress tensor is related to the strain-rate tensor expressed by the dynamic eddy-viscosity [11, 12]. This group, which is referred to as the Eddy-Viscosity Models, contains single-equation, two-equation and multi-equation models. The most frequent used in industry are the two-equation models: k-ε, k-ω, SST k-ω.

The k-ε is the one of the most commercialized turbulence model, where turbulent kinetic energy k and turbulent dissipation rate ε are defined. The assumption of this model was its application to fully-developed turbulent flows with the high-Reynolds number [9, 11], so the k-ε model should be avoided in modelling of the low-Reynolds number flows with a sophisticated path, where high-pressure gradients occur.

Another popular turbulence model in CFD simulations is proposed by D.C. Wilcox. That is the k-ω model, where turbulent kinetic energy k and specific dissipation rate ω are employed. In this particular case, ω is the degree of turbulence dissipation related to the energy unit. On the basis of numerous investigations of flows along rough surfaces, obstacles with an adverse pressure gradient, this model is much more accurate in comparison to the others. However, with an increasing distance from the wall, a strong dependency on non-disturbed mean flow is observed, and the results of the simulation are not reliable [11].

The SST model is a hybrid of both k-ε and k-ω models, which is a smooth transition from the first one, used for limited flows, whose purpose is the reflection of turbulence near the boundary layer, to the second one, which has its advantages in modelling of fully-developed flows. A number of challenging researches, investigations on aerospace profiles, flows with an adverse pressure gradient, and obstacle flows have shown much better results compared to other models [11].

2.3 Velocity profile in turbulent flow

The theory about the velocity profile in laminar pipe flow is well-developed, but when it comes to turbulent flow, the velocity profile expressions are still based on experimental data. The velocity profile of turbulent flow is also axisymmetric, nevertheless it is almost flat with a sharp drop close to the wall. Taking into consideration the velocity profile from the wall to the axis, four regions can be described [13]. At first, the very thin, near to the wall, *viscous layer* with nearly linear velocity distribution can be specified. Flow in this sub-layer is almost identical to laminar flow. Next is *buffer layer* where turbulent effects are important but not as important as viscous effects. After the buffer layer, *overlap layer* can be defined. Viscous effects are still dominant. Last, *turbulent layer* where viscous effects no longer matter, fills up the rest part of the flow. The variety of flow characteristics shows difficulties in an analytical description of the velocity profile. However, taking into account the *viscous layer*, it can be said that the shear stress is directly proportional to the viscosity of the fluid [5, 6, 13]:

$$\tau = \mu \frac{du}{dy} \tag{5}$$

where τ is the shear stress, μ is the dynamic viscosity, u is the axial velocity and y is the distance from the wall. Considering that velocity distribution, where speed takes values from zero on the wall to a certain value and velocity gradient is constant, the previous equation can be simplified, as given:

$$\tau = \mu \frac{du}{dy} = \mu \frac{u}{y} = \rho\nu \frac{u}{y} \Longrightarrow \frac{\tau}{\rho} = \frac{\nu u}{y}. \tag{6}$$

Introducing friction velocity as $u_* = \sqrt{\tau/\rho}$, replacing it in the previous equation, velocity profile in viscous sub-layer can be described by dimensionless parameters u^+ and y^+ :

$$u_*^2 = \frac{\nu u}{y} \tag{7}$$

$$u^+ = \frac{u}{u_*} \text{ and } y^+ = \frac{y u_*}{\nu} \tag{8}$$

The name of this equation is the *law of the wall* [13], and linear velocity profile for the smooth wall, where the influence of the turbulent effect is insignificant is valid for $y^+ \leq 5$ (Fig 3). According to the experimental data velocity profile for the turbulent layer was also described by power or logarithmic equations as given:

$$u^+ = \frac{1}{\kappa} \ln(y^+) + B \tag{9}$$

$$\frac{u}{u_{max}} = \left(\frac{y}{R}\right)^{1/n} \text{ or } \frac{u}{u_{max}} = \left(1 - \frac{r}{R}\right)^{1/n} \tag{10}$$

$$\frac{\overline{u}}{u_{max}} = \frac{2n^2}{(n+1)(2n+1)} \tag{11}$$

where κ, B or n are experimental defined constants. Also n, is Reynolds dependent constant:

$$n = n(Re) \approx \frac{1}{\sqrt{\lambda}} \tag{12}$$

where λ is the Darcy friction factor. Both, power and logarithmic distribution are appropriate for the turbulent layer at range of $30 < y^+ < 350$, while the buffer layer cannot be described by mathematic equations.

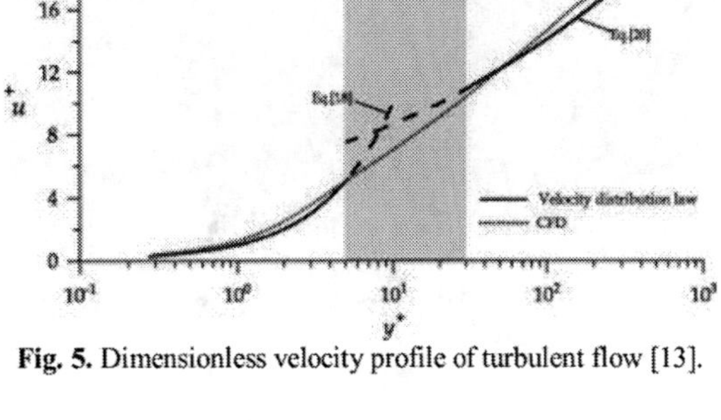

Fig. 5. Dimensionless velocity profile of turbulent flow [13].

3 CFD simulations

The subject of the analysis is the part of super-heater, that belongs to the furnace with a total capacity of 210 tons of steam per hour. The super-heater consists of 148, 38x4 mm 15HM steel pipes where a saturated steam flows under pressure of 10.5 MPa. The inlet temperature is 400 °C, and heat flux is set at 50000 W/m^2. The analysis is performed in Ansys CFX Academic environment to supply inner wall temperature and wall heat transfer coefficient through the pipe. The influence of the Earth's gravity is neglected.

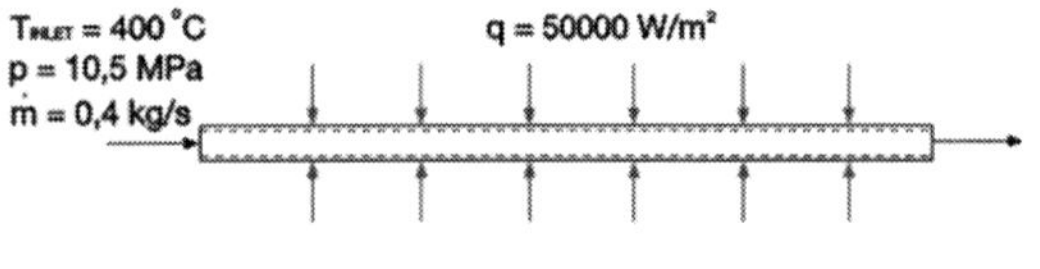

Fig. 6. Analysed problem.

In addition, due to consideration of fully developed flow, the length of 200 internal diameters is taken into account [7, 13]. To simplify the problem, the quad pipe model is created and consequently symmetry boundary condition is applied. Two different meshes are generated (Table 1, Fig. 7, Fig. 8). Properties of 15HM steel were provided and entered in the CFD-tool (Table 2) [14].

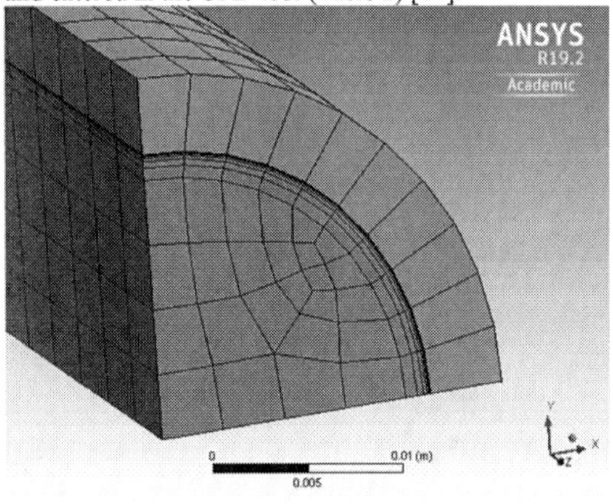

Fig. 7. Discrete model "Lite"

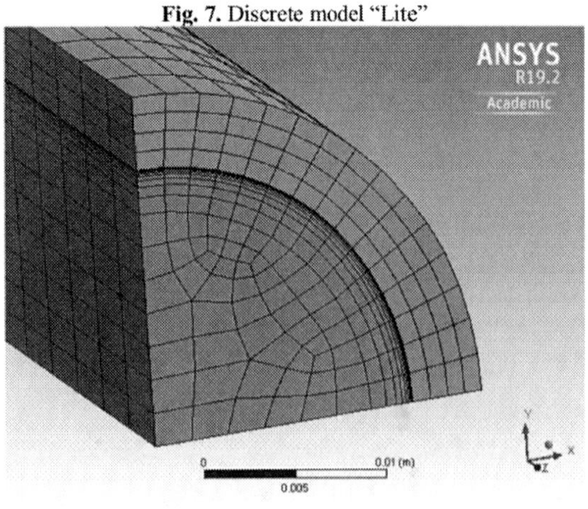

Fig. 8. Discrete model "Fine"

Table 1. Parameters of defined discrete models.

Discrete model	"Lite"	"Fine"
Number of Nodes	296k	576k
Number of Elements	256k	514k
Number of Inflation Layers	10	14

Table 2. Properties of 15HM steel.

	Steel 15HM 13CrMo44	$\rho\ [kg/m^3]$	$k\ [W/(mK)]$	$c_p [J/(kgK)]$
Temperature [°C]	20	7849	44,1	464
	100	7825	44,6	480
	200	7793	44,0	511
	300	7759	42,2	555
	400	7724	39,8	612
	500	7687	37,0	681
	600	7648	34,1	762

The CFD simulation is carried out on both discrete and turbulence (k-epsilon, SST) until the convergence criteria are met (10e-12).

4 Results

Determining the phenomenon of convection in the CFD analysis is much more complicated than it is assumed, as CFD-tools often accept inappropriate reference temperature during convection analysis. In order to examine the value of the wall heat transfer coefficient determined by the CFD-Post, the definition of the wall heat transfer coefficient (WHTC) according to the third type boundary condition should be referred to [5, 7, 15]:

$$WHTC = \frac{\dot{q}_{in}}{T_{in}-T_b} \tag{13}$$

where $\dot{q}_{in}$ is a heat flux provided to the inner wall, T_{in} is the temperature of the inner wall and most important value T_b is the bulk temperature. The bulk temperature reflects the average temperature of moving fluid in a particular cross section. To obtain real WHTC values, measurements at eleven equal-spaced cross section of inner wall temperature and bulk temperatures were gathered. The bulk temperature should have been measured as in the given equation:

$$T_b = \frac{\int_{A_w} uTdA}{\int_{A_w} udA} = \frac{2\pi \int_0^{r_w} uTrdr}{2\pi \int_0^{r_w} urdr} \tag{14}$$

and specific expression in CFD-Post was formulated:

***lengthInt**(Velocity u*Temperature*Y)@**radius** /*
lengthInt**(Velocity u*Y)@**radius

All data was gathered and presented in line charts (Fig. 9-16) to make comparison and conclusions.

https://doi.org/10.1051/matecconf/201824005008

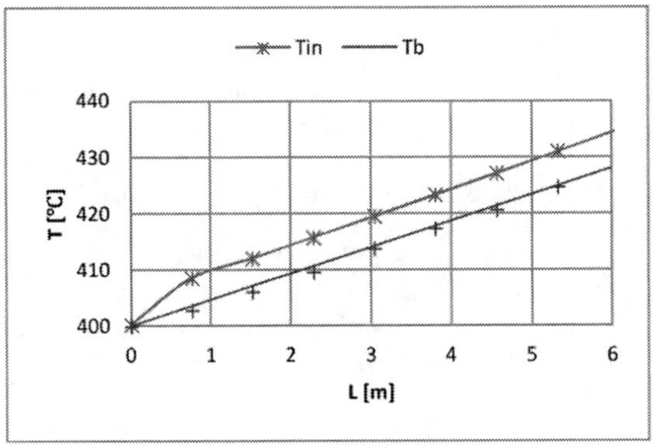

Fig. 9. Wall and bulk temperature in case of "Lite" and k-epsilon turbulence model.

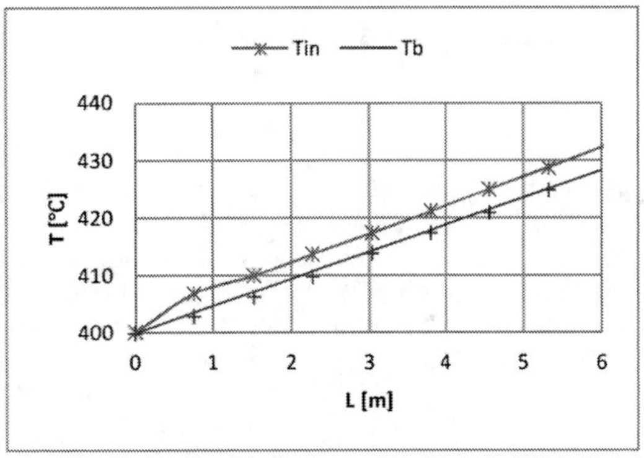

Fig. 10. Wall and bulk temperature in case of "Lite" and SST turbulence model.

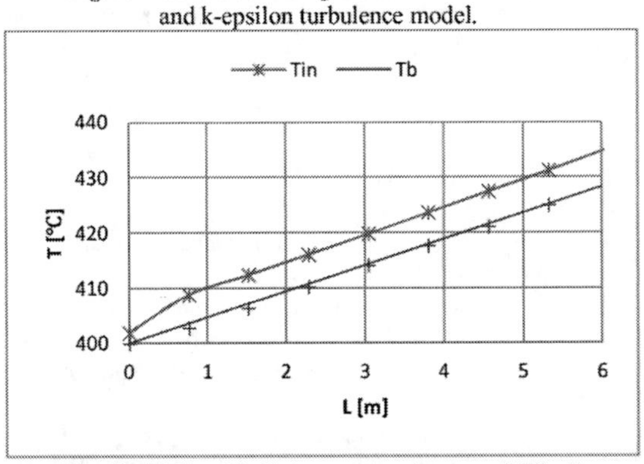

Fig. 11. Wall and bulk temperature in case of "Fine" and k-epsilon turbulence model.

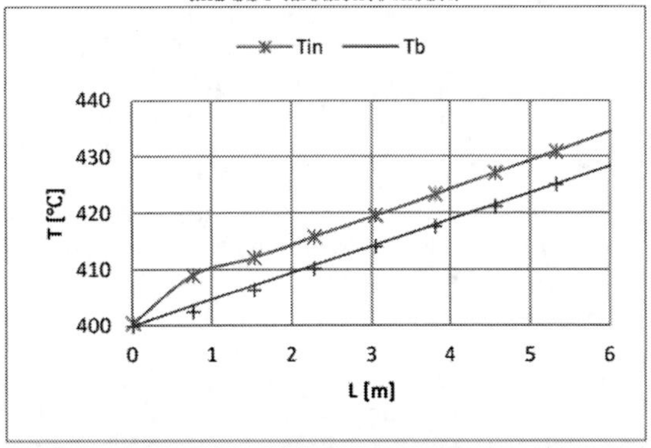

Fig. 12. Wall and bulk temperature in case of "Fine" and SST turbulence model.

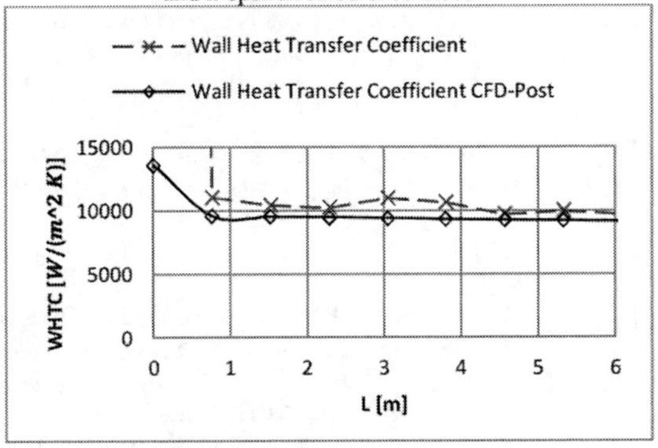

Fig. 13. WHTC comparison in case of "Lite" and k-epsilon turbulence model.

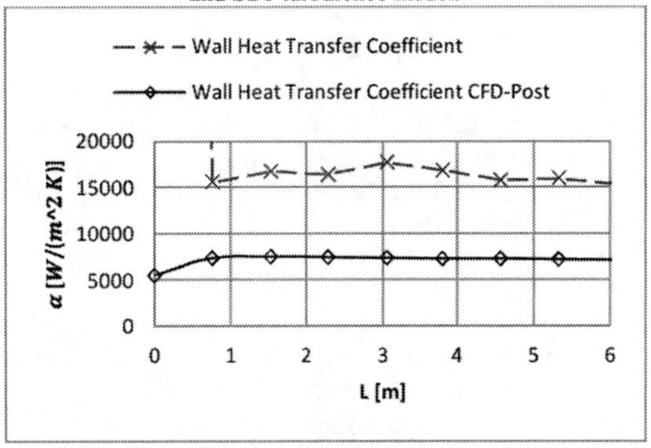

Fig. 14. WHTC comparison in case of "Lite" and SST turbulence model.

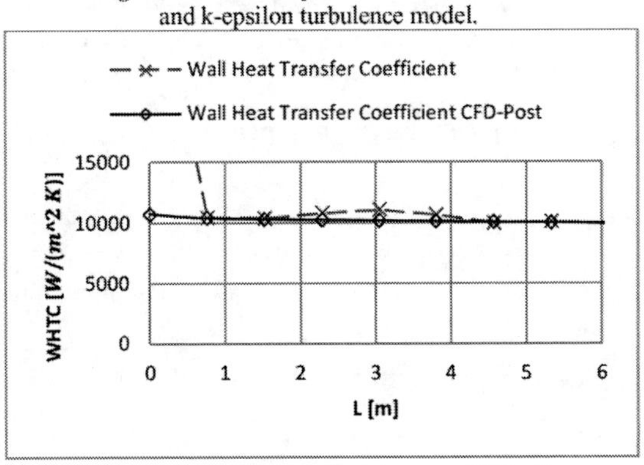

Fig. 15. WHTC comparison in case of "Fine" and k-epsilon turbulence model.

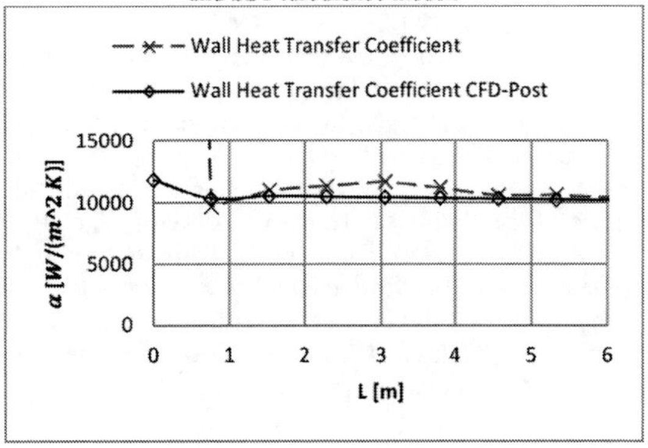

Fig. 16. WHTC comparison in case of "Fine" and SST turbulence model.

5 Discussion and conclusions

The distribution of the inner wall and the appropriate bulk temperature are presented in Fig. 9-12. Differences between these temperatures in every section, in terms of Eq. (13), have an influence on the Wall Heat Transfer Coefficient. Distribution of both, the exact and obtained directly from CFD-tool Wall Heat Transfer Coefficients are compared in Fig. 13-16. Moreover, the comparison between the k-epsilon and the SST is shown on the left (Fig. 9, 11, 13, 15) and the right (Fig. 10, 12, 14, 16) column respectively. Hardly any differences between the discrete models in case of the k-epsilon turbulence model can be observed. There are no significant discrepancies in temperature distribution that translates to nearly identical results of the WHTC. According to the SST turbulence model, the density of the mesh is a significant factor as far as the accuracy is concerned. Even the WHTC obtained by the shortest way is far away from the other values. Furthermore, the distribution of each value is much more chaotic due to the applied turbulence model. The k-epsilon turbulence model is the most appropriate in case of the fully developed flow of steam under high pressure and temperature. Solutions are obtained with right accuracy and without huge computational effort. What is more, a slight difference between approaches to the WHTC is noticeable because CFD-tools consider wall adjacent temperature as a reference temperature.

Widespread CFD-tools are a quite reliable source of information, although experience as well as adequate knowledge in the CFD analysis are needed. Unfortunately, the Wall Heat Transfer Coefficient in the CFX-Pre is simplified in relation to its definition that is directly proportional to the heat flux and inversely proportional to the difference of the wall and the bulk temperatures. As we know, the wall temperature is uncomplicated to define even during the investigation, but the bulk temperature is a much more complex problem. Values received by CFD-tools consider wall adjacent temperature, which is the temperature of the first node next to the wall, so Wall Heat Transfer Coefficient is fully mesh dependent, so if density tends to infinity, the WHTC tends too. Therefore, during discretization, appropriate y+ parameter should be taken into account. Moreover, there is no reason to use the SST turbulence model when the flow of steam under high pressure and temperature is under investigation. Based on the carried out analysis, it can be concluded that the value of the Wall Heat Transfer Coefficient is more or less 10,000 W/(m^{2}K). Furthermore, the roughness of the pipe is also worth attention as it has a huge influence on convection. In this particular case, the pipe was smooth, that means roughness was neglected and additional investigation in terms of this parameter should be carried out.

6 References

[1] M. Pawlik, F. Strzelczyk, *Elektrownie*, Wydawnictwo WNT, **4**, 43 (2014)
[2] J.B. Kitto, S.C, *Stultz, Steam its generation and use*, The Babcock & Wilox Company, **41**, 19.1-20, (2005)
[3] P. Orłowski, W. Dobrzański, E. Szwarc: *Kotły parowe*, Wydawnictwo WNT, **3**, 285-302 (1979)
[4] P. Ludowski, D. Taler, J. Taler, *Identification of thermal boundary conditions in heat exchangers of fluidized bed boilers*, Applied Thermal Engineering, **58**, 194-204, (2013)
[5] D. Taler, *Obliczenia i badania eksperymentalne wymienników ciepła I*, Wydawnictwo PK, (2016)
[6] K. Jeżowiecka-Kabsch, H. Szewczyk, *Mechanika płynów*, Oficyna wydawnicza Politechniki Wrocławskiej, 217-220 (2001)
[7] Y. A. Çengel, J.M. Cimbala, R.H. Turner, *Fundamentals of thermo-fluid science*, McGraw-Hill Higher Education, 869-891 (2012)
[8] Y. Kaneda, T. Ishihara, *High-resolution direct numerical simulation of turbulence*, Journal of Turbulence **7**, (2006)
[9] M. Paszko, K. Łygas, , *Współczesne metody modelowania przepływów turbulentnych w otoczeniu poruszającego się autobusu miejskiego*, Autobusy **12**, 1270 (2016)
[10] Y. Zhiyin, *Large-eddy simulation: Past, present and the future*, Chinese Journal of Aeronautics **28**, 11-24 (2014)
[11] M. Woelke, *Eddy Viscosity Turbulence Models employed by Computional Fluid Dynamic*, Wydawnictwa Naukowe Instytutu Lotnictwa, 92-109 (2007)
[12] P. Jamińska, *Wpływ sposobu zdefiniowania struktury wiatru w modelu turbulencji k-ε w wersji standard na rozkład współczynnika ciśnienia na powierzchni ścian prostopadłościanu*, Budownictwo i Architektura **10**, 93-97 (2012)
[13] N.M.C. Martins, N.J.G Carriço, H.M. Ramos, D.I.C. Covas, *Velocity-distribution in pressurized pipe flow using CFD: Accuracy and mesh analysis*, Computers & Fluids **105**, 218-230 (2014)
[14] J. Taler, P. Duda: *Rozwiązywanie prostych i odwrotnych zagadnień przewodzenia ciepła*, (2003)
[15] A. Neale, D. Derome, B. Blocken, J.E. Carmeliet, *CFD calculation of connective heat transfer coefficients and validation- Part I: Laminar flow*, (2006)
[16] CFX-Pre User's Guide, (2016)

https://doi.org/10.1051/matecconf/201824005009

Calculations of the pressure drop in the natural circulation boiler evaporator

Sławomir Grądziel[1,*], *Karol* Majewski[1]

[1] Institute of Thermal Power Engineering, Cracow University of Technology, al. Jana Pawła II 37, 31-864 Kraków, Poland

Abstract. The paper presents different models used to determine pressure losses in two-phase flows: the homogeneous model, the Lockhart-Martinelli, the Friedel and the Chisholm phase-slip models and the Martinelli-Nelson graphical method. The pressure losses are calculated for the evaporator of an OP-210 boiler with the output of 210×103 kg/h operating in one of the Polish power plants. The results obtained by means of the presented models are compared to each other.

1 Introduction

Two-phase flows occur in many industrial processes. In the power sector, the change in the state of aggregation takes place primarily in the power boiler evaporator and in the condenser. The appearance of the steam-water mixture involves a change in the thermal and flow properties. The thermal and flow conditions change considerably in the boiler evaporator, where the content of steam in the mixture varies with the amount of heat supplied along the medium flow path. The appearance of the steam-water mixture results in phenomena which cannot be observed in single-phase flows. The pressure drop should be determined taking account of interphase forces, the shape and size of bubbles, the velocity of individual phases (the split ratio) and many other factors [1, 3, 4].

The pressure drop can be calculated using many mathematical models. In literature, the steam-water mixture is treated as homogenous or it is considered under the phase-slip model assumptions. There are also models using graphical relations intended for the pressure drop determination. The paper presents results of the pressure drop in the OP-210 boiler evaporator determined by means of the homogeneous model, the split-phase models (the Lockhart-Martinelli, the Friedel, the Chisholm model) and the Martinelli-Nelson graphical method [1, 3, 4, 10].

2 Structure of two-phase flows in power boilers

Most power boiler evaporators in Poland are made of smooth tubes. The installed units are usually natural circulation boilers. For them, the steam dryness factor x at the risers outlet must not exceed 0.2. Due to that, the circulation multiplicity k cannot be smaller than 5:

$$x = \frac{1}{k} \tag{1}$$

The phenomena occurring in the boiler evaporator involve supercooled boiling and fully-developed nucleate boiling. The steam-water mixture flow structures in vertical channels are shown in Fig. 1.

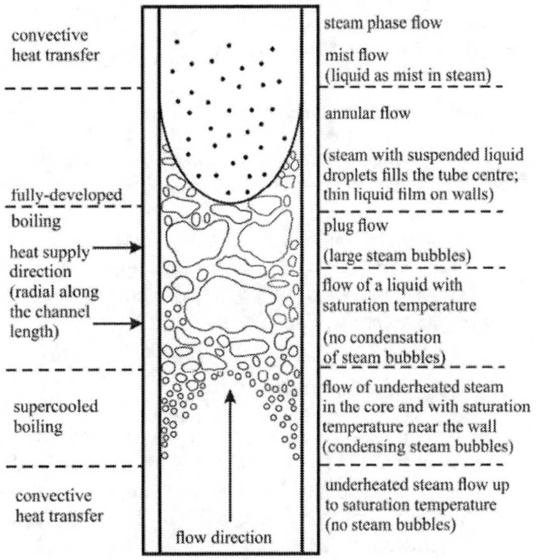

Fig. 1. Two-phase flow patterns for water-steam mixture in heated vertical channels [9].

Due to their complex flow structure, multi-phase flows are described by means of many different models

3 Models used for the pressure drop determination

* Corresponding author: gradziel@mech.pk.edu.pl

In the general case, the total pressure drop is found from the following relation [1-7]:

$$-\frac{dp}{dz}=-\frac{dp_f}{dz}-\frac{dp_a}{dz}-\frac{dp_g}{dz} \qquad (1)$$

where:

$\frac{dp_f}{dz}$ - frictional of pressure gradient, Pa/m,

$\frac{dp_a}{dz}$ - accelerational of pressure gradient, Pa/m,

$\frac{dp_g}{dz}$ - gravitational of pressure gradient, Pa/m.

In Equation (1), the pressure drops related to changes in momentum and potential energy are determined from the following relation:

$$\frac{dp_a}{dz}=\dot{G}^2\frac{d}{dz}\left[\frac{(1-x)^2}{\rho_L(1-\varepsilon)}+\frac{x^2}{\rho_L\rho_G}\right] \qquad (2)$$

$$\frac{dp_g}{dz}=g\left[\rho_G\varepsilon-\rho_L(1-\varepsilon)\right]sin\,\alpha \qquad (3)$$

where:

$\dot{G}$ – mass flux, kg/(m²s),

x – dryness factor,

ε – filling ratio,

ρ_L, ρ_G – density of the liquid and the gaseous phase, kg/m³,

α – tube inclination angle.

Determination of frictional pressure losses is a complex task. Therefore, many mathematical models have been developed to find the friction-related component. The mixture flow may be treated as homogeneous or as a slip flow. Using slip models involves calculating the two-phase flow multiplier, which makes it possible to find the friction-related pressure drop.

3.1 Homogenous model

The homogenous model application is related to the assumption that the fluid properties are averaged for the whole mixture and the gaseous and the liquid phase velocities are identical. The equality of the velocities of the two phases means that there is no slip between them. For the homogeneous model, the friction-related pressure drop is found from the following relation:

$$\frac{dp_f}{dz}=\frac{f_{TP}}{d_{in}}\frac{G^2}{2\rho_{TP}} \qquad (4)$$

where:

f_{TP} – two-phase friction factor,

$\dot{G}$ – mass flux, kg/(m²s),

ρ_{TP} – two-phase density, kg/m³.

The friction factor is found using classical equations, e.g. the Blasius formula, based on the Reynolds number found for averaged properties of the mixture [3, 4].

3.2 The Lockhart-Martinelli model

The Lockhart-Martinelli model is based on finding two-phase multipliers (the Lockhart-Martinelli multipliers), which make it possible to determine the pressure gradient related to frictional losses arising during the flow of the mixture [4]:

$$\frac{dp_{TP}}{dz}=\varphi_L^2\frac{dp_L}{dz} \qquad (5)$$

The two-phase multiplier used in equation (5) is found from the following relation:

$$\varphi_L^2=1+\frac{C}{X}+\frac{1}{X^2} \qquad (6)$$

The Martinelli parameters X and C used in equation (6) are described in [4].

3.3 The Friedel model

The Friedel model was developed based on 25 thousand measuring points. It is used for two-phase flows in vertical and horizontal tubes. The two-phase multiplier is found from the following relation [4]:

$$\varphi_{LO}^2=E+\frac{3.24FH}{Fr^{0.045}We^{0.035}} \qquad (7)$$

Two-phase multiplier φ_{LO}^2 concerns the flow of a mixture characterized by properties related to the liquid phase properties. Coefficients E, F, H and the Froude and the Weber numbers (Fr and We, respectively) used in equation (7) are described in [4].

3.4 The Chisholm model

The Chisholm model is an empirical method that can be used in a wide range of the steam pressure and dryness factor values. The two-phase multiplier is found from the following formula [4]:

$$\varphi_{LO}^2=1+\left(Y^2-1\right)\left[Bx^{0.5(2-n)}(1-x)^{0.5(2-n)}+x^{2-n}\right] \qquad (8)$$

where: Y^2 – the single-phase flow ratio between the gaseous and the liquid phase pressure gradients, B – coefficient dependent on the mass flow (mass flux), n – exponent describing the flow resistance coefficient. For the Blasius formula n=0.25.

The methods of determination of parameters Y^2 and B are described in [4].

3.5 • The Martinelli-Nelson graphical method

The Martinelli-Nelson method was developed on the example of flows of the steam-water mixture through horizontal ducts in the pressure range of 0.689÷20.7 MPa. The two-phase multiplier needed to establish friction-related losses is read from a chart developed based on different values of the steam pressure and dryness factor x (cf. Fig. 2).

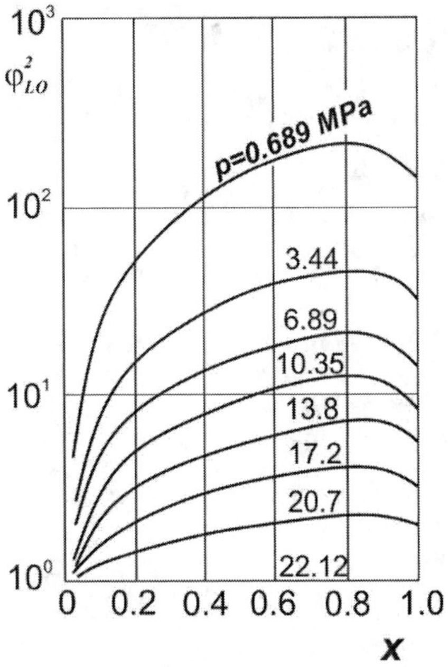

Fig. 2. Martinelli-Nelson correlation.

If the pressure value exceeds the critical point for water, the multiplier is $\varphi_{LO}^2 \approx 1$ [11]. The two-phase multiplier read from the chart enables determination of the friction-related pressure drop by means of the following formula:

$$\frac{dp_f}{dz} = \varphi_{LO}^2 \frac{dp_{f,LO}}{dz} \tag{9}$$

where:

$\frac{dp_{f,LO}}{dz}$ - frictional losses arising in the flow of a two-phase mixture with properties related to the properties of the liquid phase, Pa/m.

4 Comparison of the OP-210 boiler evaporator pressure drops determined by means of different methods

The OP-210 boiler is a single-drum natural circulation boiler. It generates superheated steam with the pressure of 9.8 MPa and temperature of 540°C. The boiler nominal output totals 210 x 103 kg/h. The pressure drop in the boiler evaporator was determined using the methodology described in [8]. The OP-210 boiler analysed circulation system, or contour, (with marked division into evaporation zones) is presented in Fig. 3.

The medium flow velocity and the degree of evaporation in the OP-210 boiler circulation contour were performed for the heat flux real values. The measurement methodology and the obtained results are presented in [2, 10].

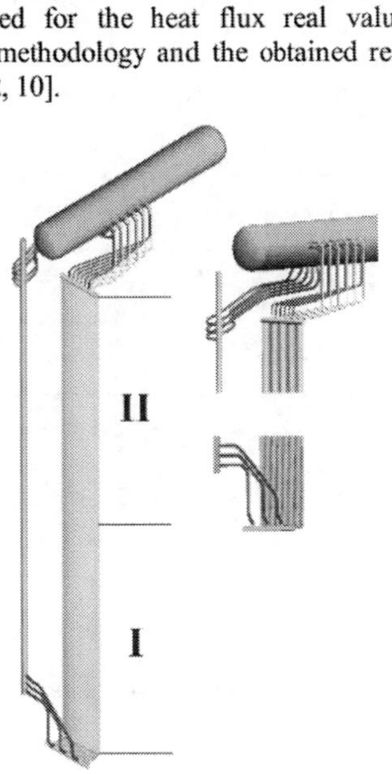

Fig. 3. Evaporator scheme of OP-210 power boiler.

Heating up to saturation parameters and partial evaporation of boiler water occur in the waterwall tubes. The calculation methodology is based on the evaporator division into a few zones [8]. Evaporation occurs in two zones so that the steam dryness factor at the outlet of the first and the second evaporation section is x_1=0.01 and x_2=0.07, respectively. A rise in the dryness factor value involves bigger friction-related pressure drops. The resistances obtained using the methodology described in [8], the homogenous model and the phase-slip models are compared in Table 1 and Fig. 4.

Table 1. Comparison of results.

		Section I	Section II
$\frac{dp_a}{dz}$, Pa/m		1546	2399
$\frac{dp_g}{dz}$, Pa/m		6019	3879
$\frac{dp_f}{dz}$, Pa/m	Model 1	142	208
	Model 2	174	260
	Model 3	318	968
	Model 4	211	375
	Model 5	194	348
	Model 6	110	396

Model 1 – literature-based [8]
Model 2 – homogenous model
Model 3 – the Lockhart-Martinelli model
Model 4 – the Friedel model
Model 5 – the Chisholm model
Model 6 – the Martinelli-Nelson graphical method

The calculation results indicate that the friction-related pressure drops take similar values for most of the models under consideration. Some differences occur for the Lockhart-Martinelli method, which was developed based on flow measurements of the air-water and air-oil mixture.

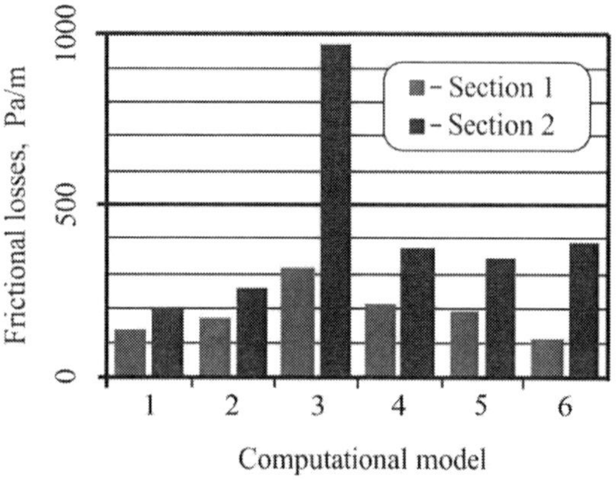

Fig. 4. Comparison of frictional losses established by means of different models: 1 – literature-based [8], 2 – homogenous model, 3 – the Lockhart-Martinelli model, 4 – the Friedel model, 5 – the Chisholm model, 6 – the Martinelli-Nelson graphical method

The obtained results indicate that frictional losses contribute to the total pressure drop only slightly. This is related to the small value of the steam dryness factor x and to the boiler evaporator vertical structure. Due to that, it is the changes in the mixture potential energy and momentum that have the dominant impact on the level of the total pressure drop.

5 Conclusions

It follows from the calculations that a rise in the gaseous phase content in the flowing mixture involves an increase in friction-related flow resistances. The applied models differ in their approach to the two-phase flow multiplier determination, which results in some differences. However, in the flow case under analysis, the content of the gaseous phase in the mixture is small and, therefore, the discrepancies between individual models are slight. The share of frictional pressure losses in the overall pressure drop is relatively small. For this reason, it does not really matter which model is selected to determine friction-related pressure losses – the selection has little impact on the total pressure drop in the circulation system of natural circulation boilers. The changes in the mixture potential energy and momentum have the dominant effect on the total pressure drop.

References

1. Dziubiński M., Prywer J., *Two-phase fluid mechanics* (WNT, Warszawa, 2009)
2. Grądziel S., *Modelling of thermal and flow phenomena occurring in evaporator of boiler with natural circulation* (Wydawnictwo Politechniki Krakowskiej, Kraków, 2012)
3. Hetsroni G., *Handbook of Multiphase System*, (Hemisphere Publishing Corporation, Washington, 1982)
4. Hewitt G., Shires G., Bott T., *Process heat transfer* (CRC Press, Begell House, 1994)
5. P. Ocłoń, M. Nowak, S. Łopata, J. of Therm. Sci., **23**, 2, pp. 177-186 (2014)
6. P. Ocłoń, M. Nowak, K. Majewski, AIP Conference Proceedings, **1558**, pp. 2419-2422 (2013)
7. P. Ocłoń, M. Nowak, B. Węglowski, T. Nabagło, P. Cisek, M. Jaremkiewicz, K. Majewski, J. of Appl. Comp. Sci., **22**, 1, pp. 111-135 (2014)
8. Kuznetsov N.W., Nitor W.W., Dubovski I.E., Karasina E.S., *Thermal Calculations of Steam Boilers. Standard Method* (Energy, Moscow, 1973)
9. Taler J., *Thermal and flow processes in large steam boilers. Modelling and monitoring* (PWN, Warszawa, 2011)
10. Zima W., Grądziel S., *Simulation of transient processesin heating surfaces of power boilers* (LAP LAMBERT Academic Publishing,Germany, 2013)
11. R.C.Martinelli, D.B.Nelson, Trans. Amer. Soc. Mech. Engrs., **70**, 6, pp. 695-702 (1948)

An analysis of high-temperature nuclear reactor coupled with gas turbine combined cycle

Marek Jaszczur[1]*, *Michał* Dudek[1], *Tomasz* Śliwa[2], *Zygmunt* Kolenda[1]

[1]AGH University of Science and Technology, Faculty of Energy and Fuels, Al. Mickiewicza 30, 30-059 Kraków, Poland
[2]AGH University of Science and Technology, Faculty of Drilling, Oil and Gas, Al. Mickiewicza 30, 30-059 Kraków, Poland

Abstract. At present many companies from the energy sector have to follow new regulations and concerns three crucial aspects of energy production: the impact on the environment, the efficiency of energy conversion and the cost of energy. From a technical point of view, the most efficient technology available today for electricity generation is based on a gas turbine combined cycle. In the present paper, an analysis of environmentally friendly, high-temperature gas nuclear reactor system coupled with gas turbine combined cycle technology has been investigated. The analysed system is one of the most advanced concepts and allow electricity generation with the higher thermal efficiency than could be offered by any currently existing nuclear power plant technology. The results show that it is possible to achieve thermal efficiency for nuclear power plant higher than 50% which is not only more than could be produced by any modern nuclear plant but it is also more than could be offered by most of the traditional power plants.

1 Introduction

Nuclear energy remains the most significant source of low carbon electricity production in the OECD and the second in the World [1]. Its importance as a current and future source of carbon-free energy must be recognised and should be treated on an equal footing with other low-carbon technologies. As a proven nuclear technology can play an important role in future energy systems for electricity production in many parts of the World [2].
In 2DS (2°C) scenario vision [1] for nuclear energy, it is assumed that the contribution of atomic technology makes significant progress in decarbonisation of the power systems. The 2DS vision focuses primarily on new generation nuclear technologies for electricity generation. However, it takes into account the potential of other energy applications such as combined heat and power, district heating, hydrogen and other fuels production or desalination [1].
In 2014, under construction were as much as 72 reactors and this was the highest amount for almost 3 decades. However, only 3 reactors were ready to operate. It is predicted according to the 2DS scenario, that China would account for the most significant increase in nuclear capacity from 17 GW in 2014 up to almost 250 GW until 2050 [1]. In the other countries mainly the Middle East, India, and the Russian Federation also significant increase in nuclear power capacity is predicted. It is also worth to notice that according to the 2DS scenario, nuclear capacity in most OECD countries, (except the Republic of Korea, Poland, Turkey and the UK) would decline or remain at a constant level at the same time.

In Table 1 nuclear power plant in 2018 under construction are presented accounting for different technologies: PWR (Pressurized Light-Water-Moderated and Cooled Reactor), PHWR (Pressurized Heavy-Water-Moderated and Cooled Reactor) and BWR (Boiling Light-Water-Cooled and Moderated Reactor) [4].
To achieve the goal of energy sector decarbonisation and in order to limit global temperature growth to just 2 degrees Celsius by the end of the century, a halving of global energy-related emissions up to the middle of XXI century is required. This requires an unprecedented transition in the energy production and a wide range of low-carbon energy technologies is needed to support this transition, including renewable energy technologies, energy efficiency gain, carbon capture and storage units.

Table 1. Nuclear Power Plant under construction in 2018.

Reactor Type	Number of Reactors under constructions	Total Net Electrical Capacity (MW_e)
PWR	48	50680
PHWR	4	2520
BWR	4	5253
HTGR	1	200

In this work, an analysis of high-temperature gas nuclear reactor (HTGR) system coupled with a gas turbine (GT) and steam turbine (ST) combined cycle technology has been presented. The proposed plant system includes one of the most advanced concepts and technology available today and allows electrical energy generation with the higher thermal efficiency than could be offered by any currently existing nuclear power plant. The considered

* Corresponding authors: jaszczur@agh.edu.pl

system consists of the HTGR as a primary source of heat, gas turbine, the heat recovery steam generation system (HRSG) and a steam turbine with a regeneration.

2 High-temperature nuclear reactor

A new generation high temperature gas-cooled nuclear reactors is the most innovative concept among current advanced nuclear reactor technologies available at present [3]. The U.S. Department of Energy (DoE) developed the very high-temperature gas-cooled reactor (VHTR) with a helium coolant for advanced applications and for further expand and safe use of nuclear energy [5,6]. This type of reactor posses several features that distinguish them from any available or currently under work type of reactors as well as provide significant operational advantages. In HTGR the fuel is in the form of small ceramic-coated particles tri-structural isotropic fuel (TRISO) capable of operating at very high temperatures and the helium is used as a coolant [7,8]. For this reason, the high-temperature nuclear reactor is designed to run continuously at an outlet temperature in the range 850-900°C. On the other hand, at the same time more advanced version of this reactor is also developed. The very high-temperature nuclear reactor able to operate with outlet temperature 900-1000°C.
Proposed solution provides the potential for high energy conversion efficiency and allows to obtain high-temperature process heat for various applications such as cement manufacture, hydrogen production [15], petroleum refineries, desulfurization of heavy oil, chemical industry and coal gasification (see Figure 1).

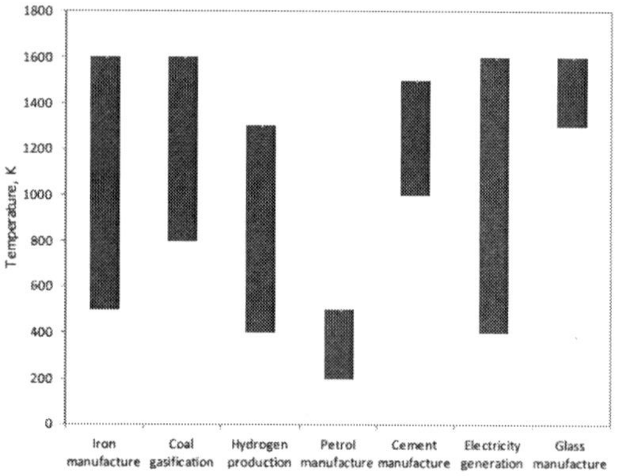

Fig. 1. Application for High-Temperature Nuclear Reactor's.

The example of selected High-Temperature Reactor's projects is shown in Table 2.

Table 2. Examples of HTR projects.

Program	Size MW_{th}	Temperature, °C	Energy Efficiency
MHTGR IGT [9,10]	200	850	48%
GTHTR300 [11,12]	600	850	45.8%
MPBR [10]	250	879	48%
GT-MHR [7,11,13]	600	850	>47%

2.1 Gas turbine combined cycle

The proposed gas turbine combined cycle (GTCC) employs single gas turbine (GT) and heat recovery steam generator that uses exhaust gas from a gas turbine to produce high-quality steam, which is supplied to a steam turbine (ST) [10]. In Figure 2 the advanced gas turbine combined cycle coupled with High-Temperature Nuclear Reactor and one pressure level Heat Recovery Steam Generator (1P HRSG) is presented while in Figure 3 the cycle with three pressure level Heat Recovery Steam Generator (3P HRSG) is also shown. The primary constraint in operation of the GTCC power plants is HRSG which is located directly after the gas turbine, where changes in temperature and pressure of the exhaust gases may cause significant thermal and mechanical stresses. The HRSG unit for combined cycle power plant contains three sections of heat exchanger modules for high, intermediate, and low-pressure steam respectively. In addition to that when the plant is operated in a common load-following operation mode, it can lead to sizeable thermal stress and in consequence may damage some components of the system. In order to solve this issue heat storage unit is implemented.

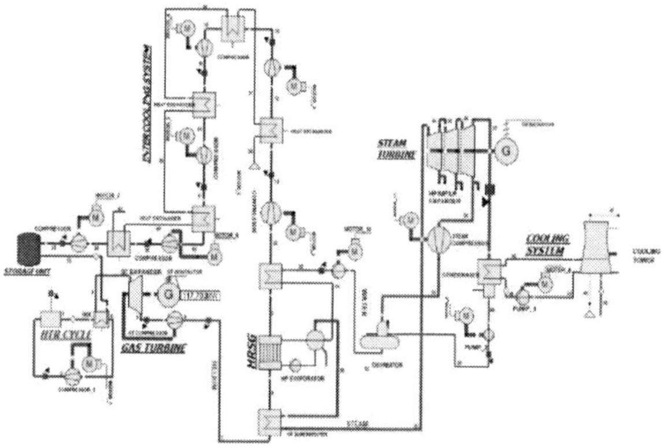

Fig. 2. Advanced Gas Turbine Combined Cycle coupled with HTR and one pressure level Heat Recovery Steam Generator.

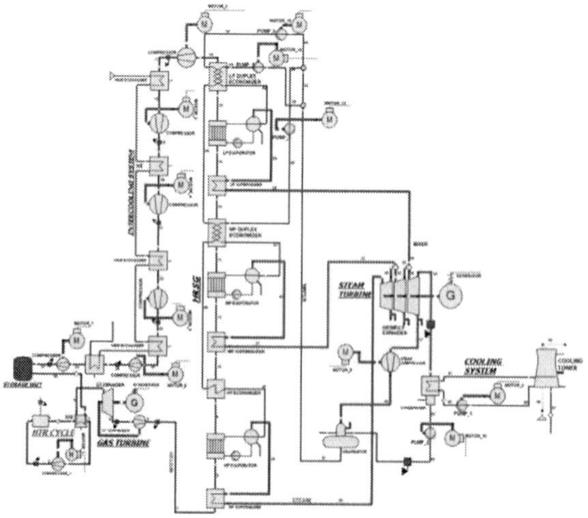

Fig. 3. Advanced Gas Turbine Combined Cycle coupled with HTR and three pressure level Heat Recovery Steam Generator.

2.2 Mathematical model

A mathematical model consist of the thermodynamic equations for High-Temperature Heat Exchanger (HTHE), compressor, gas turbine, steam turbine, evaporator, deareator, pumps, condenser and cooling tower. The calculations were made using software Ebsilon software, and the details about computer modelling as well as a model equations for all key components of the high-temperature nuclear reactor coupled with the gas turbine combined cycle are presented in other the authors work [14].

The heat recovery steam generator is considered as a one of the critical components of combined cycle power plants. Any change in its design directly affects the performance of the steam cycle and therefore the performance of the combined cycle power plant. The maximum heat transfer occurs in the evaporator section for the high-pressure level, and it occurs in the economiser section for the low-pressure level.

The heat from the gas turbine exhausted gases is recovered in a Heat Recovery Steam Generator to produce steam at the required pressure and temperature. This steam is then used to provide additional electricity using steam turbines. HRSGs are classified into single, dual, and triple pressure types depending on the number of drums in the boiler. As a consequence, the thermal efficiency of a power plant strongly depends on the HRSG design.

The thermal efficiency for all presented cases has been calculated according to the following formula:

$$\eta_{GTCC} = \frac{P_{GT} + P_{ST} - \sum_{i=1}^{n} N_{mi}}{Q_{HTR}} \tag{1}$$

where: P_{GT} is the gas turbine power, P_{ST} is steam turbine power, and $\sum N_{mi}$ are the pumps consumption, Q_{HTR} is the heat supplied from high-temperature nuclear reactor. Gas turbine power is given by the following equation [9]:

$$P_{GT} = \eta_{is}\eta_m\eta_e\eta_{mg}\dot{m}_{He1}cp_{He}\left[T_{IT} - T_{OUT}\right] - \frac{\eta_m \dot{m}_{He1} cp_{He}\left[T_{OUT} - T_{OUT\_1}\right]}{\eta_{isc}} \tag{2}$$

and power of steam turbine coupled with 1P (one pressure) level Heat Recovery Steam Generator is:

$$P_{ST} = \eta_{is}\eta_m\eta_e\eta_{mg}\dot{m}_s cp_s\left[T_{IT} - T_{OUT}\right] - u\dot{m}_s cp_s\left[T_u - T_{OUT}\right] \tag{3}$$

while for 2P (two pressure) level Heat Recovery Steam Generator) [9]:

$$P_{ST} = \eta_{is}\eta_m\eta_e\eta_{mg}\begin{pmatrix}\left[\dot{m}_{sHP}cp_{sHP}\left[T_{IT} - T_{OUTHP}\right] + \left(\dot{m}_{sHP} + \dot{m}_{sIP}\right)cp_{sIP}\left[T_{ITIP} - T_{OUTLP}\right]\right] \\ -u\left(\dot{m}_{sHP} + \dot{m}_{sIP}\right)cp_s\left[T_{ITLP} - T_{OUTLP}\right]\end{pmatrix} \tag{4}$$

and for 3P (three pressure) level Heat Recovery Steam Generator is as follows [8]:

$$P_{ST} = \eta_{is}\eta_m\eta_e\eta_{mg}\begin{pmatrix}\left[\dot{m}_{sHP}cp_{sHP}\left[T_{ITHP} - T_{OUTHP}\right] + \left(\dot{m}_{sHP} + \dot{m}_{sIP}\right)cp_{sIP}\left[T_{ITIP} - T_{OUTIP}\right]\right] \\ +\left(\dot{m}_{sHP} + \dot{m}_{sIP} + \dot{m}_{sLP}\right)cp_{sLP}\left[T_{ITLP} - T_{OUTLP}\right] \\ -u\left(\dot{m}_{sHP} + \dot{m}_{sIP} + \dot{m}_{sLP}\right)cp_s\left[T_{ITLP} - T_{OUTLP}\right]\end{pmatrix} \tag{5}$$

2.3 Assumption for calculations

The main assumptions for the thermodynamic cycle from Figure 2 and Figure 3 are shown in Table 3.

Table 3. Thermodynamic parameters used for calculation.

Parameters	Symbol	Value
Thermal power	Q_{HTR}	300 MW_{th}
Reactor outlet temperature	T_{ROUT}	850/900°C
Reactor coolant pressure	p_R	70 bar
Primary loop flow rate	$\dot{m}_{He}$	128 kg/s
Coolant pressure before Gas Turbine	P_{inGT}	70 bar
Coolant pressure after Gas Turbine	P_{ouGT}	35 bar
Gas Turbine Inlet Temperature	G_{TIT}	800/850°C
Coolant flow rate in gas cycle	$\dot{m}_{He1}$	104.5 kg/s
Gas expander mech. efficiency	η_m	99%
Gas expander isent. efficiency	η_{is}	90%
Steam temperature before ST	S_{TIT}	565 °C
Steam pressure before ST	S_{INP}	177 bar
Steam flow rate case1/case2	$\dot{m}_s$	48.9/57.7kg/s
Steam temperature after ST		28.96 °C
Steam turbine mech. efficiency	η_m	99.8%
Steam turbine isent. efficiency	η_{is}	88%
Gas Turbine power generation	P_{GT}	117.7 MW_e/ 123.1 MW_e
Compressor power consumption		6.1 MW
Steam Turbine power generation case1	P_{ST}	65.5/78.3/83.7MW_e
Steam Turbine power generation case2	P_{ST}	77.4/91.4/93.8MW_e
generator electrical efficiency	η_e	0.9856

3 Results and discussion

The results for an analysed Gas Turbine Combined Cycle coupled with High-Temperature Nuclear Reactor are shown in Figures 4-9. One may infer from the Figure 4 that the proposed system can generate electricity with a thermal efficiency in the range 44.8-51.4% depending on the case: 1P (one pressure) and 3P (three pressure) level Heat Recovery Steam Generator.

The use of the system based on the layout of Gas Turbine Combined Cycle coupled with High-Temperature Nuclear Reactor with one pressure heat recovery steam generation system from Figure 2 can generate of about 183.23 MW_e of electricity (see Figure 5 and Figure 6).

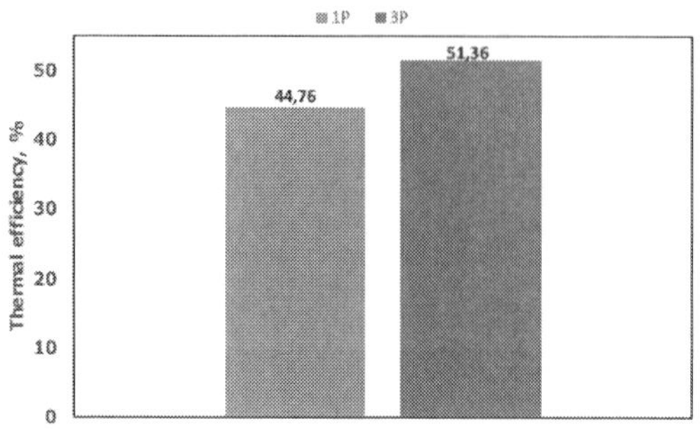

Fig. 4. The thermal efficiency for the 1P (one pressure) and 3P (three pressure) HRSG.

When the system is equipped with three pressure Heat Recovery Steam Generation, it is possible to obtain electrical power equal to 201.41 MW_e (see Figure 5 and Figure 6). One may obtain a significant growth in the electrical power when three pressure heat recovery steam generation unit is used.

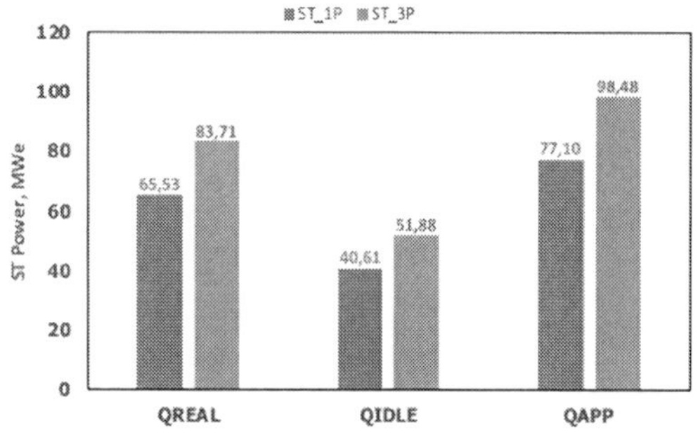

Fig. 5. The real, idle and apparent electrical power of steam turbine for the 1P (one pressure), and 3P (three pressure) HRSG.

The thermodynamic analysis has also been performed in order to investigate the effect of High-Temperature Nuclear Reactor temperature increase - from 800°C to 850°C – however, assuming the constant thermal power of the reactor equal to 300MW_{th} as well as constant temperature difference (about 50°C) in the primary heat exchanger. The results of this analysis are presented in Figures 6-8.

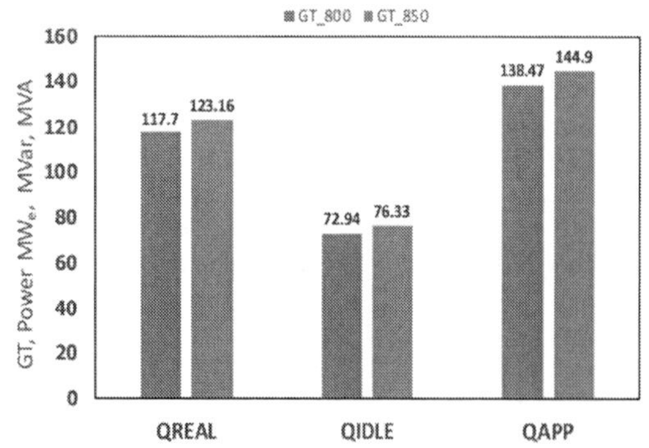

Fig. 6. The real, idle and apparent power for gas turbine and HTR output temperatures 800°C and 850°C.

It is clear that the thermodynamic cycle conditions depend on High-Temperature Nuclear Reactor parameters and steam parameters delivery to Steam Turbine (SST-3000 with a constant value of inlet pressure p_o=177 bar and inlet temperature T_o=565°C) as well as on the cycle configuration. The influence of the High-Temperature Nuclear Reactor temperature on the real, idle and apparent electrical power of the gas turbine is presented in Figure. 6.

In Figure 7 the real, idle and apparent electrical power of the Steam Turbine for the two different nuclear reactor outlet temperatures 800°C and 850°C and for the for the 1P (one pressure), and 3P (three pressure) Heat Recovery Steam Generator level are shown.

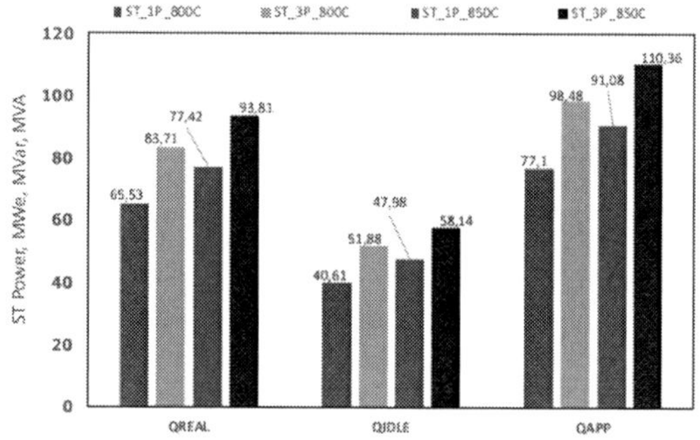

Fig. 7. The real, idle and apparent electrical power of steam turbine for cases 800°C, 850°C and for the 1P (one pressure), 3P (three pressure) HRSG levels.

The thermal efficiency for two different outlet temperatures of working fluid from High-Temperature Nuclear Reactor is shown in Figure. 8. As can be seen, the highest thermal efficiency can be observed for the system 3P (three pressure) HRSG levels system and reactor outlet temperature 800°C while for the system 1P (one pressure) HRSG system and temperature 850°C the efficiency is (due to reactor assumption) more than 10 percentage points lower.

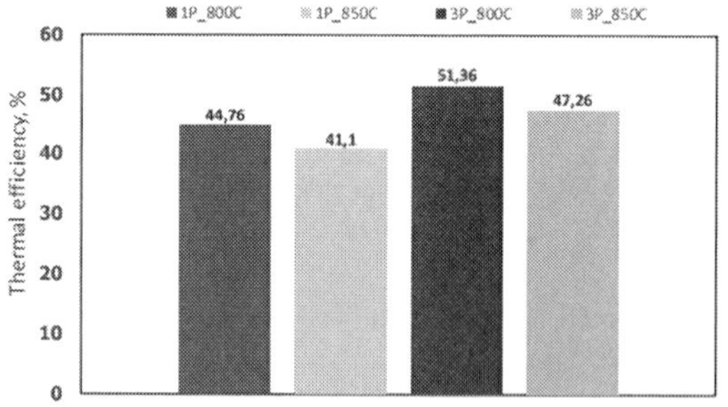

Fig. 8. The thermal efficiency comparison for the 1P (one pressure), 3P (three pressure) HRSG levels and for reactor outlet temperature 800°C, 850°C.

The value for the main thermodynamic parameters obtained from the calculation at the key points of analysed cycle (see numbering in Figure 2 and Figure. 3) and for the 1P (one pressure) and 3P (three pressure) HRSG levels are presented in Table 4, and 5.

Table 4. Thermodynamic parameters at selected points of the layout and for the reactor outlet temperature 850°C.

Lp.	1P				3P			
	p bar	t °C	m kg/s	h kJ/kg	p bar	T °C	m kg/s	h kJ/kg
1	70	850	128	4414	70	850	128	4414
4	69.5	800		4155	69.5	800		4155
5	35	566		2941	35	566		2941
6	36	577	104	3000	36	577	104	3000
10/15	35.9	311.2		1616	35.7	206.6		1072
21/26	42	104.8		544	42	99.7		517
23/28	177	565	48.9	3463	177	565	48.9	3463
31	-	-	-		100	444.9	61.78	3228
34	-	-	-		28	272.3	68.4	2929
27/35	0.04	28.9	42.1	2101	0.04	28.9	61.9	2061
32/41	1.7	115.1	48.9	483	0.5	81.3	68.4	340

Table 5. Thermodynamic parameters at selected points of the layout and for reactor outlet temperature 900°C.

Lp.	1P				3P			
	p bar	t °C	m kg/s	h kJ/kg	p bar	t °C	m kg/s	h kJ/kg
1	70	900	128	4673	70	900	128	4673
4	69.5	850		4414	69.5	850		4414
5	35	605		3144	35	605		3144
6	36	617	104.5	3205	36	617	104.5	3205
10/15	35.9	302.4		1570	35.7	206.7		1072
21/26	42	99.7		518	42	99.7		517
23/28	177	565	57.8	3463	177	565	57.8	3463
31	-	-	-		100	447.6	71.78	3235
34	-	-	-		28	276.9	74.67	2941
27/35	0.04	28.9	49.7	2101	0.04	28.9	67.6	2069
32/41	1.7	115.1	57.8	483	0.5	81.3	74.67	340

4 Conclusions

Proposed advanced gas turbine combined cycle coupled with the High-Temperature Nuclear Reactor is a desirable alternative to traditional and even modern energy production systems. It is based on one of the most advanced technologies which give the opportunity to produce electricity as well as technological heat for industrial and thermochemical processes without emissions of toxic or greenhouse gasses to the environment and with thermal efficiency higher than a traditional coal-fired power plant or nuclear power plants.

Presented system coupled with 1P (one pressure) or 3P (three pressure) level Heat Recovery Steam Generator is able to generate electrical energy at two stages: using Gas Turbine and Steam Turbine what have a significant impact on the system efficiency and system flexibility.

An increase in the reactor outlet temperature has a significant impact on the entire cycle thermal efficiency. However, under the current assumption (constant mass flow rate and constant reactor power) when increasing reactor temperature the energy efficiency conversion for the (1P, 3P) system is not growing due to the higher energy demand by the compressor.

Acknowledgement

The authors sincerely acknowledge financial support to the Polish Ministry of Science (Grant AGH No. 15.11.210.437 and Grant AGH No.11.11.210.312) and project GEMINI.

References

1. N. Energy, Technology Roadmap, Springer Reference (2015)
2. International Energy Agency, Energy Technology Perspectives, *Int. Energy Agency*, 412 (2015)
3. M. Dudek, Z. Kolenda, M. Jaszczur, W. Stanek, *J. Energy Resour. Technol.* **140**, 20910 (2018)
4. IAEA, Reference data series, nuclear power reactors in the world, **2**, 88 (2017)
5. A. C. Kadak, The Status of the US High-Temperature Gas Reactors, *Eng.* **2**, 119 (2016)
6. A. Bredimas, *Nucl. Eng. Des.* **271**, 41 (2014)
7. IAEA, High Temperature Gas Cooled Reactor Fuels and Materials, IAEA (2010).
8. K. Kok, Nuclear Engineering Handbook, (2009)
9. Z. Zhang, Y. Dong, F. Zhengming, Z. H. Wang, et al., *Eng.* **2**, 112 (2016)
10. O. Olumayegun, M. Wang, and G. Kelsall, Fuel, **180**, 694 (2016)
11. C. F. McDonald, *Appl. Therm. Eng.* **73**, 80 (2014)
12. G. Locatelli, M. Mancini, N. Todeschini, *Energy Policy*, **61**, 1503 (2013)
13. V. Zare, M. Hasanzadeh, *Energy Convers. Manag.* **128**, 227 (2016).
14. M. Dudek, M. Jaszczur, Q. Hassan, Z. Kolenda, *Int. Conf. on Computational Heat, Mass and Momentum Transfer*, Kraków, 2018.
15. I. Gutowska, Study on depressurized loss of coolant accident and its mitigation method framework at very high temperature gas cooled reactor, Oregon State University, Ph.D. thesis, (2015)

3-D simulation using for hydraulic calculation of the heat accumulator

Olha Kletska[1,*], *Anatoliy* Falendysh[1], *Arthur* Kagramanjan[1], and *Andrey* Onishchenko[1]

[1]Ukrainian State University of Railway Transport, Feuerbach Square 7, 61050 Kharkov, Ukraine

Abstract. The article is devoted to the peculiarities of hydraulic calculation of the heat accumulator in the software environment for 3D modeling - SolidWorks. Based on the results of the calculation, the values of the distribution of the velocities of the water flow and the distribution of the static pressure in the given plane of the heat accumulator were obtained, which eventually made it possible to identify the problem areas in the three-dimensional model and obtain the values of the pressure difference.

1. Introduction

One of the way to improve the ecological and economic indexes of the diesel locomotives that are in "a hot" reserve, is to use alternative sources of warming up such as a heat accumulator.. The functioning mode of this device is based on a latent heat of phase transitions of chemical substances, that is when the engine is running it is charged with heat, and during the storage of the locomotive in the "hot" reserve, the heat accumulated by the accumulator is spent for preheating the diesel locomotive.

For the purpose of optimisation of a choice of the additional equipment (a storage battery, the water pump) at the installation of the heat accumulator on a diesel locomotive hydraulic calculation on already constructed 3D sample pieces in program SolidWorks environment and in particular its module for simulation of a current of fluids and gases - Flow simulation is carried out.

Flow simulation software is fully integrated into SolidWorks for the calculation of liquid and gas flows inside and outside the SolidWorks model, and also calculates heat transfer from, to, and between these models by convection, radiation and thermal conductivity using computational fluid dynamics (CFD) technologies.

The correct choice of the water pump allows to pick up more precisely a storage battery that raises time of autonomous operation of a diesel locomotive and increases its ecological compatibility and fuel efficiency as a whole.

The problems in the field of energy accumulation based on the heat of phase transition are occupied scientists from the different countries such as W. Youssef, Y.T. Ge, S.A. Tassou «CFD modelling development and experimental validation of a phase change material (PCM) heat exchanger with spiral-wired tubes», Luca Constantin, Daniel Dragomir-Stanciu, Ionut Vasile Crismaru «Optimization of heat exchange in a heat accumulator with latent heat storage», Shilei Lu, Tianshuai Zhang, Yafei Chen «Study on the performance of heat storage and heat release of water storage tank with PCMs». However in one operation it is not mentioned hydraulic losses in examined devices and their influence on accessories choice at concrete field environment [1-4].

Nomenclature

ΔP Pressure difference (Pa)

2. Problem statement

The problem of environment protection is global for the whole world, therefore reduction of ejections of soiling substances with burnt gases is one of directions under its decision. Today the railway transport takes not the last place among sources of soiling substances in the presence of the big diesel park where as a heat-power plant uses an internal combustion engine.

Therefore, the first step in reducing pollutant emissions is to reduce reduction of fuel costs by diesel locomotives, both in the performance of shunting and trunk works, and when locomotives are in the "hot reserve". Due to some physical properties of construction and cost materials for the normal operation of a diesel locomotive, regardless of whether the driver's controller positions according to the specifications of the manufacturer, the temperature of water and oil should be not less than 40 °C. And the geographical position of Ukraine and the climatic conditions of operation of diesel locomotives require, to maintain them during heat-away the sludge at the required temperature level, by running the engines idling, or to warm them from an external source of energy - that is, warming up is an integral part of their operational cycle.

* Corresponding author: gurao@ukr.net

One of the way to improve the environmental and economic performance of diesel locomotives in a "hot" reserve is to use alternative heat sources such as a heat accumulator (HA). This device, the principle of which is based on the latent heat of phase transitions of chemical substances, that is, when the engine is running it is charged with heat, and during storage of the locomotive in the "hot" reserve, the heat accumulated by the accumulator is expended on heating the diesel locomotive.

Features of calculation of thermal accumulators with melting heat-accumulating material: To determine the main characteristics of a thermal battery with a phase transition, a detailed description of the processes occurring in the accumulator is required, which requires taking into account the mechanisms of sufficiently fine physical processes: radiation and convection in HA, volume changes in phase transformations, anisotropic properties of heat-storage material HSM) etc. On the basis of such calculations, the parameters at each point of the accumulator capacity are determined, which makes it possible to conduct its comprehensive studies. Simplified mathematical models [8] are often used to estimate the size and basic parameters of HA in which the constancy of the HSM and coolant properties, heat transfer in only one direction, neglect of the heat capacity of HSM and a others are often used. It make possible to receive the dependences for different types of accumulators, that allow to define their basic parametres.

It is known that solidification process HSM in capsular HA consists of the stages of the formation of phases section boundary of, its movement on a stroke of the heat-carrier and its collapse. The first two stages provide close to stable parameters at the output from the HA and, naturally, are of practical interest.

3. Analysis and modelling

To estimate the dimensions HA, the cylindrical element is considered under the assumptions that the properties of the coefficient of heat transfer and the velocity of heat carriers are constant, the axial heat transfer and the changes in the properties of the HSM during melting are absent, the speed of movement, and the shape of the phase boundary are constant.

Based on the results of calculations in the SolidWorks environment, a parametric model of a capsular-type heat accumulator with a melting heat-accumulating material based on barium hydroxide a $Ba(OH)\cdot 8H_2O$ (fig. 1) is constructed.

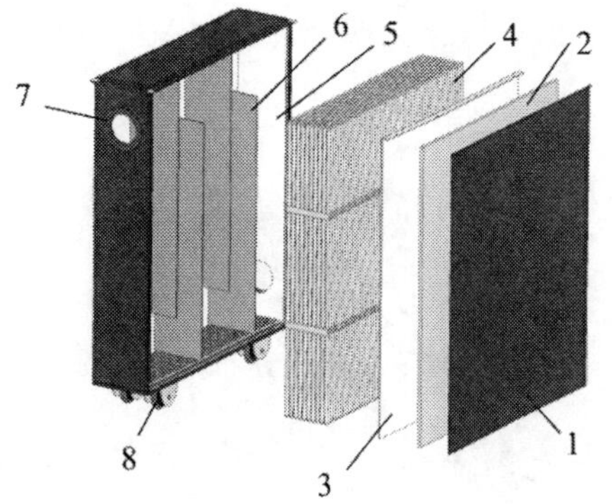

Fig. 1. Heat accumulator construction.
1. Lagging Sheet; 2. Thermal insulation; 3. A wall of the welded cage; 4. Pack of capsules; 5. The welded cage; 6. Transverse partition; 7. Outlet elbow; 8. Wheelbase.

Structurally, the model is a conventional heat exchanger made in the form of a rectangular parallelepiped. The device contains the leakproof welded cage from simple steel: with underwater and by-pass fitting pipes in which capsules with a working body are located. Capsules are hermetically sealed brass tubes. The package of capsules is assembled into a single matrix with a corridor arrangement of the latter. To improve the flow around the tubes and increase the heat transfer of the annular space, the body has transverse partitions. It should be noted that the design of the heat accumulator, that is shown in Fig. 1 is not final and can have different modifications [9].

To optimize the choice of additional equipment (storage battery, pumping water pump), when installing a heat accumulator on the diesel locomotive, a hydraulic calculation is carried out. Total hydraulic losses represent the sum of linear and local losses. The loss of friction is determined using formulas obtained as a result of theoretical or practical studies. Determining local losses is more difficult because for this, only experimental data are available. There is a wide variety of configurations of pipeline systems, therefore, there are various reasons for the occurrence of local resistances. The flow structure also can be quite complex. Therefore, the use of experimental data is limited. Flow simulation offers a different approach to solving similar problems and allows you to accurately determine the pressure loss in virtually any part of the pipeline system.

To solve this problem, this HA model was simplified and is a casing with an inlet and outlet branch pipe without thermal insulation and external cladding, wheelbase, with a set of tubes inside.

It is necessary to calculate the value of the total pressure loss, i.e. the difference between the pressure at the entrance to the heat accumulator and the pressure at the outlet from it, where the flow again becomes unperturbed.

To solve this type of problem, all the holes in the model must be closed with covers so that the internal space of the model is closed. At the inputs and outputs, the appropriate boundary conditions must be set. Covers are additional elements, with which you can close the holes (fig. 2).

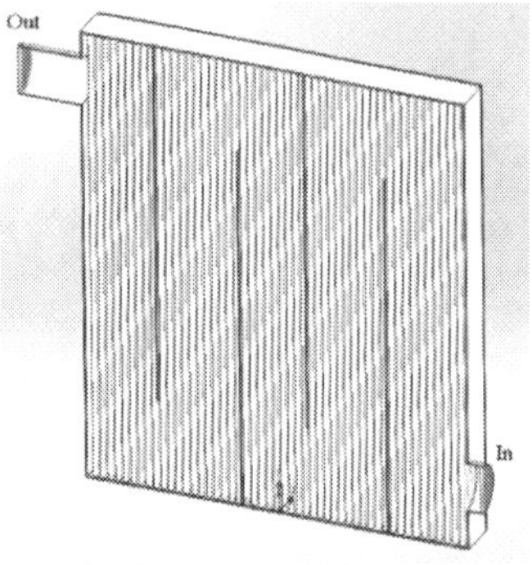

Fig. 2. Simplified HA model plugs at the input and output.

Boundary conditions or input data for calculation, the velocity at the entrance to the heat accumulator was chosen to be a numerical value of 0.15 m / s. In order that no errors occur in the specification of the boundary conditions, we set the "Pressure" condition on the outlet. Then the mass flow at this hole will be determined automatically so that the law of conservation of mass is satisfied. The static pressure was set equal to 101325 Pa, which means that the water flow pressure at the outlet from the HA is 1 atmosphere.

The only thing that remains unknown is the value of the total pressure at the inlet and outlet. The easiest and quickest way to determine these parameters is to set an appropriate calculation purpose.

As the global purpose the average mass flow was established. The inner surfaces of the covers at the entrance and exit from the HA were chosen and the boundary conditions were marked, in which the static pressure - local purposes - was selected. The Flow simulation project is ready for calculation. It will be completed when the average values of the mass flow, as well as the static pressure at the inlet and outlet of the HA, become steady. Flow simulation will automatically generate a calculated mesh. The entire calculation area will be divided into parts, which will later be subdivided into meshes. If there is a need, the meshes will be crushed further for a more detailed resolution of the geometry of the model. The results of the construction and the drawing of the mesh are presented below (tab. 1, fig. 3).

Table 1. Results of mesh construction (Number of cells).

Meshes	1326038
Meshes in the fluid environment	471790
Meshes in a solid	854248
Fractional meshes	417027
Irreguljarnye meshes	13
The truncated meshes	0

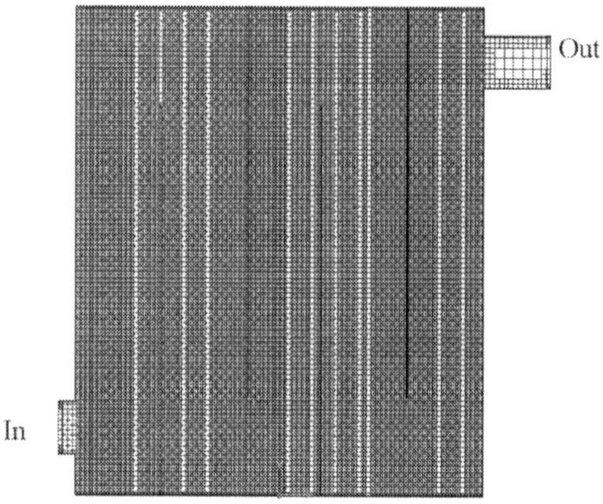

Fig. 3. Automatically generated model HA mesh.

4. Results and discussion

The results of calculation of the program are presented in fig. 4, 5, 6.

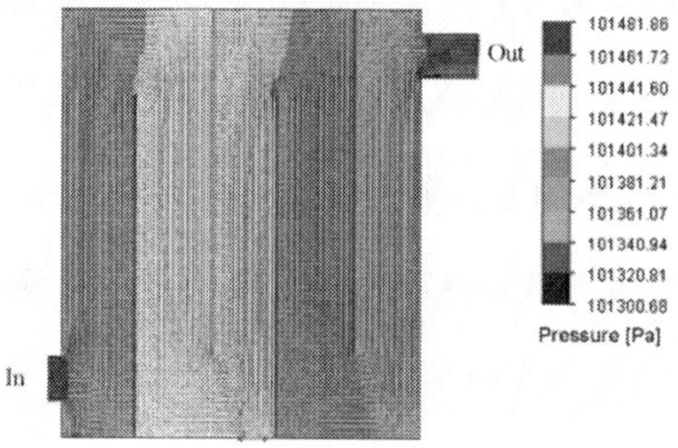

Fig. 4. Distribution of static pressure in the set plain HA.

It can be seen from the calculations that the pressure of the fluid flow decreases during the passage of the transverse partitions in the accumulator body (Fig. 4), since the presence of viscosity leads to a dissipation of energy, which eventually turns into heat. In streams of pure water, the final transition of mechanical energy to heat occurs when the individuality of the smallest vortex masses is lost, the isolated existence of which ceases due to the viscosity of the fluid.

The picture of the velocity distribution and the velocity vector is shown in fig. 5.

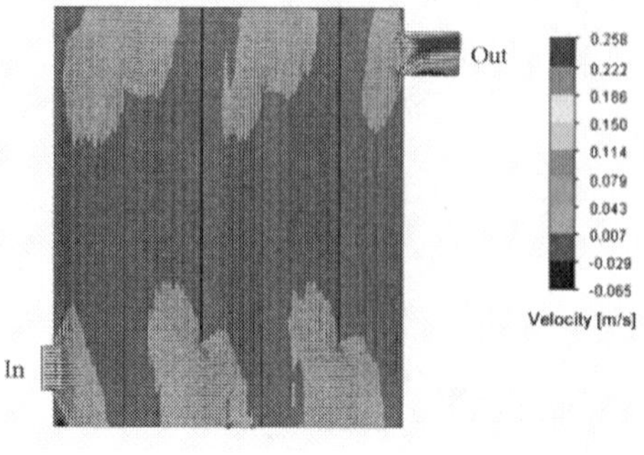

Fig. 5. Water stream movement in HA.

As in the first case, the velocity increases with the passage of the next transverse partition, which is accompanied by a drop in pressure. On the scale there are also velocities with a negative sign; these areas of blue colour show the zone of the reverse flow.

During the calculation of this task, a message appeared: "The appearance of the reverse flow at the boundary". It warns that at the hole where the boundary pressure condition was specified, in addition to the outflow, an inverse current is formed. In this problem, the reverse flow penetrates into the outgoing stream. When the reverse flow is formed, the accuracy of the results obtained is reduced. Moreover, in some cases, the convergence of goals can not be achieved at all (that is, the value of the goal will not become stable).

As the warning has not disappeared, it was necessary to discontinue calculation and increase the length of the outlet branch pipe at the exit of the heat accumulator so that there was no reverse flow at the border.

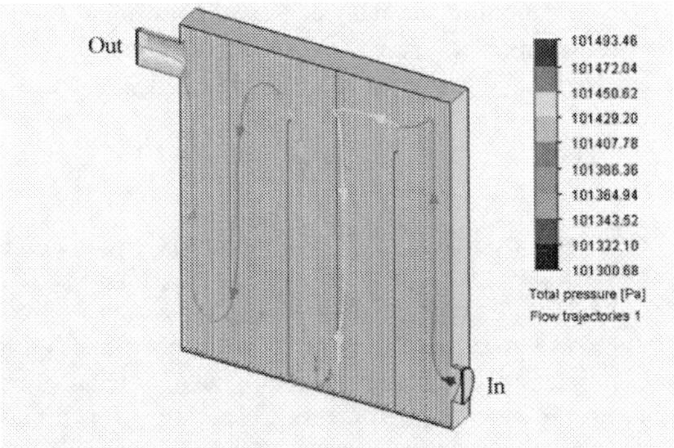

Fig. 6. Water stream trajectory in HA.

From the assaying of the received results follows that hydraulic losses of water pressure in the thermal accumulator amounted $\Delta P = 192{,}78$ Pa.

5. Conclusions

On the basis of hydraulic losses, we optimally select the water pump for pumping coolant through the cooling system of the locomotive, as well as the storage battery for its drive, that in turn increases the autonomous life of the diesel locomotive, which means that it increases its environmental friendliness and fuel efficiency.

Due to the fact that the design of the heat accumulator varies depending on the design features of the locomotive and the type of work performed by it, as well as the absence of typical HA projects, that is, the complete lack of experimental data, the use of Flow simulation allows the design engineer to obtain not only a numerical solution of the task, and visualization of the peculiarities of the "narrow" places of the design developed by him.

References

1. W. Youssef, Y.T. Ge, S.A. Energy Conversion and Management **157** (2018)
2. L. Constantin, D. Dragomir-Stanciu, I.V. Crismaru, Procedia Technology **19** (2015)
3. S. Lu, T. Zhang, Y. Chen, Energy and Buildings **158** (2018)
4. A.Z. Homich, O.I. Tupitcy'n, A.E. Simson, *E'konomiia topliva i teplotekhnicheskaia modernizatciia teplovozov* (M.: Transport, 1975)
5. V.V. Oshovskii`, D.I. Okhrimenko, A.Iu. Sy`soev, Seriia «Himiia i himicheskaia tekhnologiia». Vy`pusk No **15** (2010)
6. A.A. Aliamovskii`, *Komp`iuternoe modelirovanie v inzhenernoi`praktike* (SPb: BKHV - Peterburg 2005)
7. *SolidWorks Flow Simulation 2012*, Technical Reference (2012)
8. V.D. Levenberg, M.R. Tkach, V.A. Gol`strem, Akkumulirovanie tepla, pp. 49–74 (Kiev, Tekhnika, 1991)
9. A.O. Kagramanian, A.V. Onishchenko, Zaliznichnii` transport Ukraïni: naukovo-praktichnii` zhurnal, **1**, pp. 49-51 (2011)

Modeling of the heat transfer process in an air heat pump fanless evaporator

Piotr Kopeć[1,*], *Beata* Niezgoda-Żelasko[2]

[1,2] Institute for Thermal and Process Engineering, Cracow University of Technology, al. Jana Pawła II 37, 31-864 Cracow

Abstract. This paper analyses the mixed convection process in a fanless evaporator of an air heat pump. The text of the paper shows the authors' experimental studies results of the temperature distribution and the local values of heat transfer coefficients on the outer surface of vertical tubes with longitudinal fins for the case of mixed convection and fins of a specific shape of their cross-section (prismatic, wavy fins). The experimental studies include the air velocities w_a=2,3 m/s and the temperature differences between air and the refrigerant inside the heat exchanger tubes which is ΔT=24-40K. The results obtained were used for verification of CFD modeling of the heat transfer process for the discussed case of heat transfer and the geometry of the finned surface. The numerical analysis was performed for: the temperature distribution along the fin height, the tube perimeter and height, the distribution of local heat transfer coefficients on the finned tube perimeter and along its height. The simulated calculations were used to verify the method of determination of fin efficiency.

1 Introduction

Air is one of the types of bottom heat sources for heat pumps. The structure/design of an evaporator depends on the media involved in the heat transfer process and the manner air is obtained (natural or forced convection). In air heat pumps we can distinguish evaporators with forced air flow (fan exchangers) and fanless ones, working in conditions of free (natural) convection or mixed convection. A fanless evaporator is fabricated as a freely arranged system of vertical longitudinally-finned tubes [1]. The applicability of equations which may be used to calculate the air-side heat transfer coefficients depends on the flow type and geometry. For instance, the equations used for horizontal lamellated or transversely finned tubes include those proposed by: Schmidt [2], Norris-Spoford [3], Stasiulevicius [4] and Briggs-Young [5]. Papers [6] and [7] present the methods of measurements of local and mean heat transfer coefficients during transverse airflow around horizontal membrane tubes or tubes with two ribs (fins) under forced convection conditions. The authors of papers [8] and [9] looked at the heat transfer processes during forced transverse flow of exhaust gas around a horizontal bank of tubes with three ribs (fins).

The numerical analysis of natural convective heat transfer processes in natural convection processes on the surface of longitudinal fins is often performed for the flow inside longitudinally-finned tubes [10], [11]. Modeling of a heat transfer process can be used to solve reverse issues, in order to determine local mean heat transfer coefficients on membrane surfaces, after adopting the experimental study results as boundary conditions in solution of a system of equations of conservation (of momentum, energy, mass) [13]. The methods of this type are also used to search for solutions which will maximize the heat transfer coefficient [14], [15]. Experimental determination of mean or local values of heat transfer coefficients, especially for finned surfaces, is a complex measurement problem. This paper presents the heat transfer modeling process for fanless evaporators operated in the mixed convection conditions.

2 Experimental Studies

This paper presents the studies of heat transfer from the side of the bottom source of heat in a fanless evaporator of an air heat pump. The evaporator is made of vertical bimetallic tubes. The core of the evaporator tube is a copper tube. The outer part is an aluminium tube with drawn longitudinal fins having a specific shape, arranged centrally symmetrically along the circumference of the tube (see Figure 1). As the heat exchanger is operated in variable external conditions, the evaporators of that type can work in natural convection conditions and mixed convection conditions at low flow velocities (rates) and longitudinal or transverse flow around the tubes of the evaporator.

The experimental setup consisted of a system of two vertical finned tubes of the height of 2 m, whose axes were located at the distance of S_q=0.2306 m from each other. The experiments were carried out in a horizontal channel with a 2x0.464 m rectangular flow cross-section and the length of L_p=2.5 m (Figure 2). Airflow in the channel was forced using a radial fan. The heat exchanger tubes were fed with ice water at the

* Corresponding author: piotr.kopec@mech.pk.edu.pl

temperature of 7°C or with refrigerants (R407C and R507), having evaporation temperatures of: $-25 \leq T_R \leq -5$ (°C).

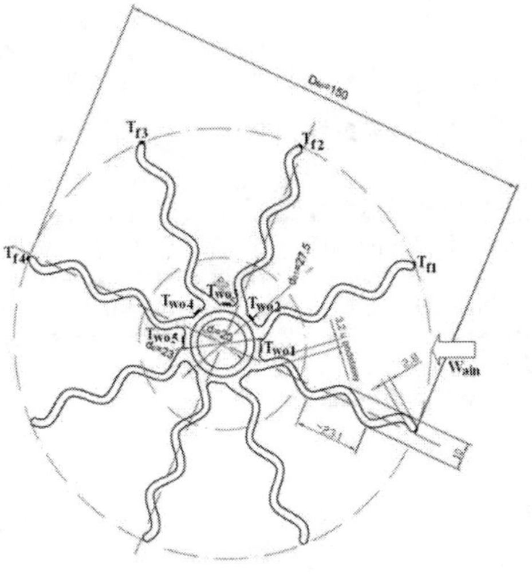

Fig. 1. Bimetallic tube cross-section.

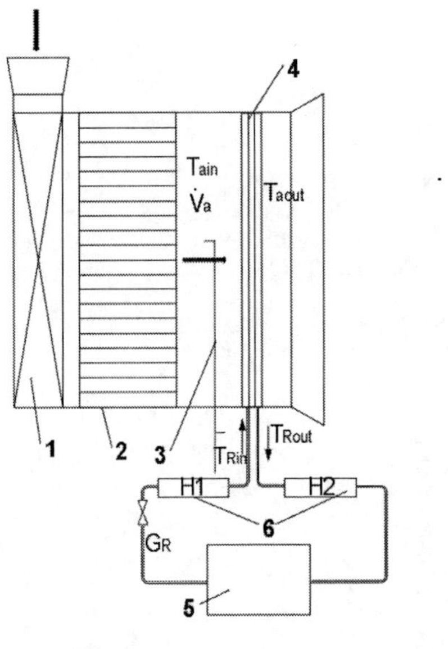

Fig. 2. Experimental setup: 1-radial fan, 2-flow stream equalizer, 3- Pitot tube, 4-studied tubes, 5-condensing unit, 6-electric heaters.

Calibrated NiCr-NiAl thermocouples were placed on the outer surface of one of the tubes to enable the measurement of the temperature of the outer tube surface and of the fin tips in the upper and lower part of the exchanger (Figure 1). The heat flux transferred from the air to the heat transfer surface was calculated using the heat balance equation for the cooling medium (water or refrigerant) [12].

2.1 Experimental studies results

During the experimental studies at the mentioned setup the temperature was measured of both bases and ends (tips) of the fins, in the top and bottom parts of the tube. The bottom means the measurement at the distance of 20 cm from the tube base, the top means the location at the distance of 170 cm from the tube base. On the basis of the data obtained, corresponding to the mixed convection conditions, the distribution of temperatures on the fin surface was determined.

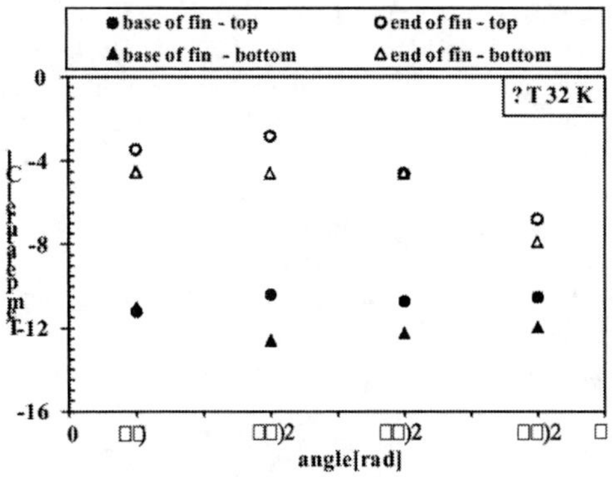

Fig. 3. Example of temperature distributions for mixed convection, ΔT=32K, w_a=2,3 m/s.

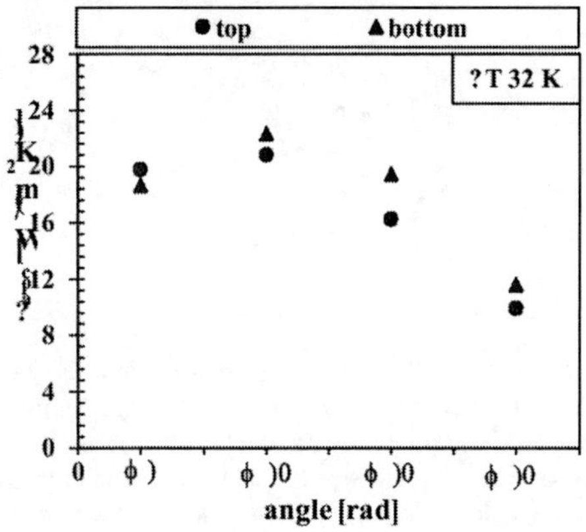

Fig. 4. Example of values of local heat transfer coefficients for mixed convection, ΔT=32K, w_a=2.3 m/s.

Figure 3 shows an example of temperature distributions on the surface of one tube of the air heat pump evaporator, operated in the mixed convention conditions.
Then these values were used to determine the values of the air-side heat transfer coefficients. In order to do that, the method of determining the local values of coefficients was used, as presented in the paper by Niezgoda-Żelasko, Żelasko [12]. Figure 4 shows an example of values of local heat transfer coefficients corresponding with the temperatures distribution shown in Figure 3.

 https://doi.org/10.1051/matecconf/201824005012

3 Numerical modelling of the heat transfer process

The heat transfer process between the finned tube and external air was modeled using the finite volume method and Ansys-Fluent software. The forced convection process modelling, the shape of the fins of the exchanger and the transverse direction of airflow required the modeling to be carried out in a 3D space. Due to the occurrence of the frontal surface (of inflow), the individual fins work in different ambient conditions. In this case the model could not be limited to one fin. An even number of fins in the exchanger allowed to assume a vertical symmetry axis and make a half of a real model. Air was modeled as a perfect incompressible gas. The dynamic viscosity coefficient and other physical parameters were assumed as constant values. Gravitational force was assumed along the model vertical axis. The mixed convection process modeling was carried out assuming the condition of the frontal velocity in the direction transverse to the exchanger under examination, the gravitational force being taken into account.

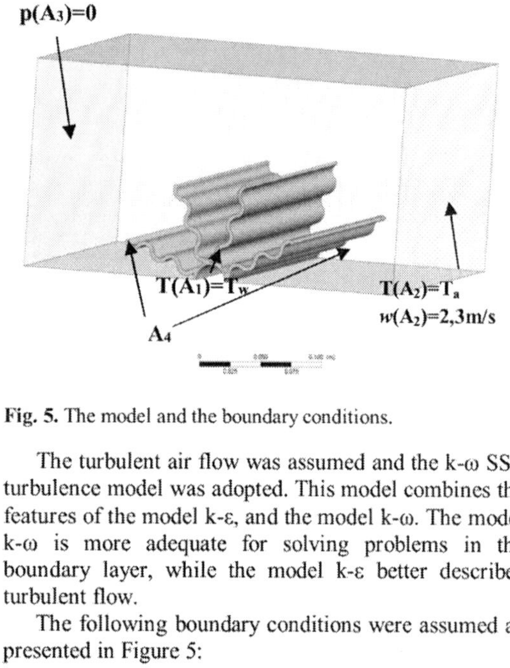

Fig. 5. The model and the boundary conditions.

The turbulent air flow was assumed and the k-ω SST turbulence model was adopted. This model combines the features of the model k-ε, and the model k-ω. The model k-ω is more adequate for solving problems in the boundary layer, while the model k-ε better describes turbulent flow.

The following boundary conditions were assumed as presented in Figure 5:

- constant temperature of the inner fin surface: $T(A_1)=T_w$,
- constant air temperature at the inlet: $T(A_2)= T_a$,
- condition of constant pressure at the air outlet $p(A_3)=0$,
- condition of velocity at the inlet $w(A_2)=2.3$ m/s,
- condition of symmetry for surface A_4,

The evaporator geometrics model was performed for real dimensions of the examined exchanger. Due to large dimensions of the exchanger, the number of elements of the calculation grid generated exceeded $9{\cdot}10^6$. The grid was hexagonal. In the areas of overlapping of the boundary conditions and in the area of air contact with the exchanger, the grid was refined. The thermodynamical parameters to be taken into consideration in the calculation model of the Fluent software corresponded to the operation parameters of the exchanger in real-life conditions.

3.1. Numerical analysis

The numerical analysis concerned the temperature distribution along the height of a fin for individual fins fixed on the perimeter of the exchanger tube as in Figure 1. CFD simulation was performed for the same heat transfer conditions and the temperature distribution along the studied fin was determined. The simulations were carried out for various values of the exchanger inner wall temperature.

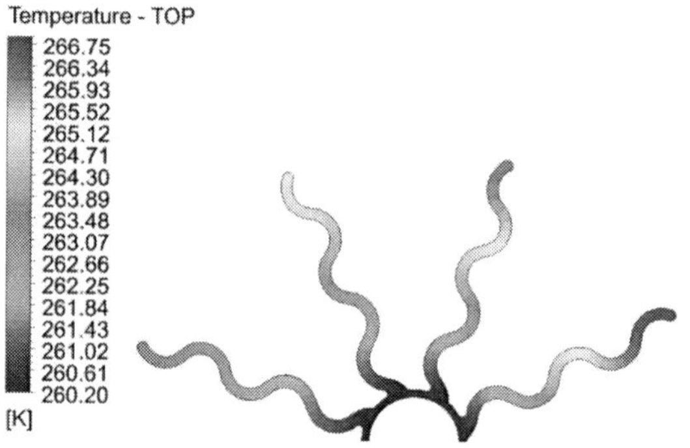

Fig. 6. Temperature distribution for CFD calculations, for mixed convection, ΔT=32K, w_a=2.3 m/s – top of the fin.

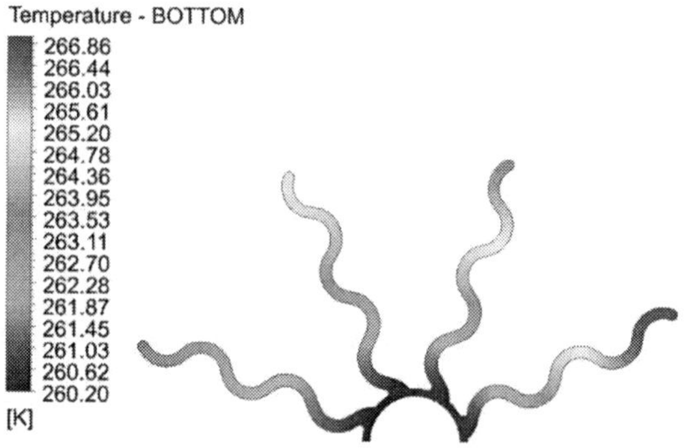

Fig. 7. Temperature distribution for CFD calculations, for mixed convection, ΔT=32K, w_a=2.3 m/s – bottom of the fin.

Figure 6 and 7 shows the temperature distribution in top and bottom parts of a fin, as obtained from the CFD modelling process. Figure 8 presents a comparison of the temperature values obtained at the base of fin and end of fin in its bottom and top parts, obtained from measurements and simulations. The direction of airflow and location of the fins in relation to it results in the difference of the temperature distribution for individual fins. This is confirmed by the temperature values obtained at the base and at the end of a fin, Figure 8.

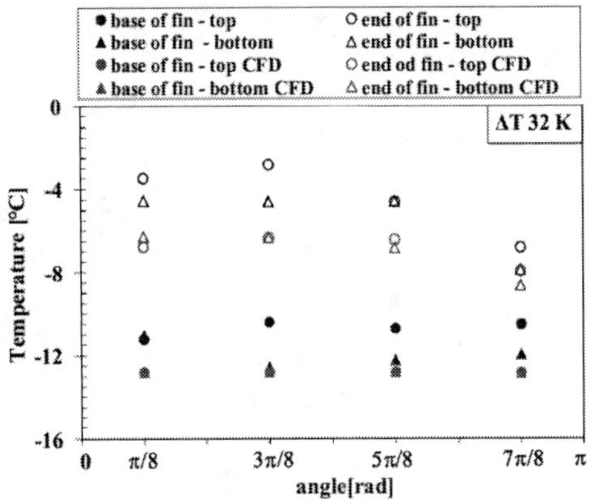

Fig. 8. Temperature distribution for CFD calculations, for mixed convection, ΔT=32K, w_a=2.3 m/s.

The second fin achieves the highest temperature – angle 3/8π. The maximum temperature value on the second fin results from the fact that it is almost perpendicular to inflowing air direction and consequently it is lipped by the largest stream of air volume. The results presented in Figure 8 evidence that in each of the cases discussed the qualitative distribution of temperature for measurements and simulation is similar. Various temperature values T_w will affect local values of heat transfer coefficients. Lower values T_w increase the temperature difference between ambient air and temperature T_w, which affects the higher values of heat transfer coefficients. Figure 9 presents an example of comparison of the local heat transfer coefficient values obtained.

The local heat transfer coefficients obtained by CFD modeling have similar values to the coefficients determined from experimental data.

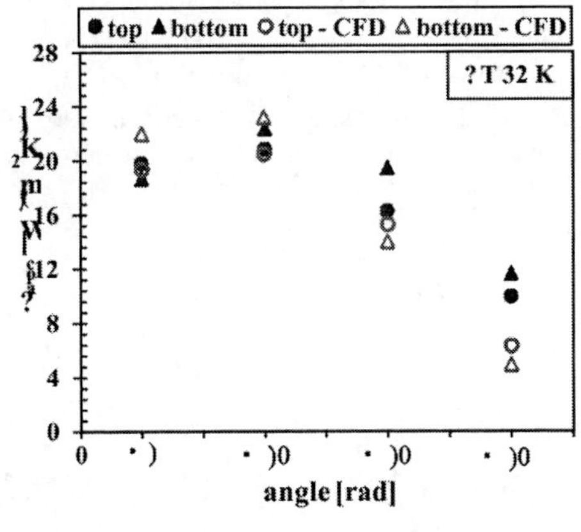

Fig. 9. Example of local heat transfer coefficient values obtained from CFD modeling for mixed convection, ΔT=32K, w_a=2.3 m/s.

The nature of changes in the coefficients is the same. The change of coefficients complies with the change in the fin temperature value. The highest local values are for the second fin, then for the first fin. These fins are situated on the side of the air inflow. In this area the air velocities are the highest, which determines the local heat transfer coefficients values.

3.1.1 Fin efficiency determination

The experimental studies and simulation calculations allowed to verify Schmidt's equations (1) for determination of fin efficiency. In the case of the discussed fin, the fin height "S" corresponds to the height of a straight fin having the surface area equivalent to the studied fin area. The temperature distribution obtained from experiment and from CFD modeling allowed to determine the verification efficiency of a fin from equation (2). Sample results of calculations for top and bottom of a fin are listed in Table 1.

$$\varepsilon_{fin-S} = \frac{\tanh(Sm)}{Sm} \qquad (1)$$

$$\varepsilon_{fin} = \frac{\overline{T}_{fin} - T_a}{\overline{T}_w - T_a} \qquad (2)$$

In mixed convection conditions individual fins feature various efficiencies, but not lower than 84%. The highest efficiencies for a given temperature difference between an internal surface of a heat exchanger and air are achieved by fins located on the leeward side, exceeding 91%. The discrepancies between the values obtained from simulation and the experimental parameters do not exceed 6%.

Table 1. Comparison of fins efficiencies.

ΔT=32K	Fin top			
angle	π/8	π3/8	π5/8	π7/8
ε_{fin-S} – (1)	0,874	0,874	0,874	0,874
$\varepsilon_{fin-exp}$ – (2)	0,847	0,847	0,875	0,924
ε_{fin} – CFD	0,896	0,889	0,889	0,916

ΔT=32K	Fin bottom			
angle	π/8	π3/8	π5/8	π7/8
ε_{fin-S} – (1)	0,865	0,865	0,865	0,865
$\varepsilon_{fin-exp}$ – (2)	0,847	0,847	0,863	0,919
ε_{fin} – CFD	0,880	0,860	0,909	0,964

4 Conclusions

The paper discussed an analysis of the heat transfer process in a fanless evaporator of an air heat pump. The temperature distribution on the outer surface of longitudinally finned tubes was analysed. The results obtained from experimental studies were compared to the temperature distribution obtained from numerical modeling of the heat transfer process, using finite volume method (Fluent). The discrepancies between the values obtained are minute. The relative difference of the measured and calculated temperatures along fin height was smaller than 2%.

The lowest temperature value of a fin end can be observed on the last fin – fixing angle 7/8π. This fin is located leeward. It is obstructed by preceding fins which consequently hinder its direct contact with the airflow. Additionally, the pressure is lower in this area. It results in air whirls and inflow of air of lower temperature, which is cooled down earlier by preceding fins. It should also be noted that the fin temperatures at the bottom of the exchanger are lower than the temperatures in its top part. This difference may result from the buoyant forces affecting the heat transfer process. The air giving up energy gets its temperature lower and increases its density. Apart from the air inflow, there is also an additional vertical airflow connected with the presence of buoyant force and whirls in the inter-fin area.

Using the temperature distribution, a comparison of fin efficiency calculation methods was carried out. The results obtained on the basis of these methods feature similar values and relative differences between the measured and calculated fin efficiencies do not exceed 4%.

Nomenclature

$\dot{G}$	refrigerant mass stream density, (kg/(m^{2}s))
S	substitute height of a straight fin having the area of a real fin, $S = A_{fin} / (n_{fin} \cdot L)$, (m)
S_q	tube spacing, (m)
s	fin thickness, (m)
m	parameter in formula (1) $m = \sqrt{(2 \cdot RCJ \cdot \alpha_{loc} / (s \cdot \lambda_{fin})}$, (1/m)
n	number of fins, (-)
L_p	channel length, (m)
RCJ	degree of process openness [16], (-)
T	temperature, (°C)
w	velocity, (m/s)

Symbols

Δ	increment
α	heat transfer coefficient, (W/(m^{2}K))
ε	fin efficiency, (-)
ϕ	measurement channel diameter, (m)
λ	thermal conduction coefficient, (W/(mK))

Subscripts

a	air
f	fin tip
fin	fin
loc	local value
R	refrigerant (cooling agent or water)
w	wall

References

1. B. Niezgoda-Żelasko, W. Zalewski, J. Żelasko, *Experimental research of the air-water heat pump with evaporator operating under free convection conditions* (Proceedings from the 44th Science and Technology Conference - Refrigerations Days, Poznań)
2. VDI-Wärmeatlas, 1991, Wärmeübertragung bei Querströmung um einzelne Rohrreihen und durch Rohrbündel, Gf 1- Gf 3, (Düsseldorf :VDI Verlag).
3. L. Kołodziejczyk, M. Rubik, *Refrigeration and air conditioning technology* (Warszawa: Arkady, 1976) (in Polish)
4. J. Stasiulevisius, V. Survila, Paper FC 6.5. Proc. 5th Int. Heat Trans. Conference, Tokyo (1974)
5. D. E. Briggs, E. H. Young, Chemical Engineering Programming Symposium, Ser. **59**. (1963)
6. J. Taler, T. Sobota, A. Cebula, Arch. of Therm. **26,** 1, pp 35-52 (2005)
7. J. Taler, A. Cebula, Mechanika, **202**, pp. 263-272 (2003)
8. M. Pronobis, S. Kalisz, R. Wejkowski, Heat and Mass Trans., **38,** pp. 343–350 (2002)
9. R. Wejkowski, M. Pronobis, *New kinds of convective tube bank - a comparison of physical and numerical modeling of heat transfer and pressure loss,* (Int. Conference: Combustion and Environment, Ostrava, Czech Republic, 2003)
10. I. M. Rustum, H. M. Soliman, Int. J. Heat Mass Trans., **33,** 7, pp. 343–350 (1990)
11. J. C. Chai, S. V. Patankar, Num. Heat Trans., **Part A 24,** pp. 67-87 (1993)
12. B. Niezgoda-Żelasko, J. Żelasko, Exp. Therm. And Fluid Sci., **57,** pp. 145-156 (2014)
13. J. Taler, T. Sobota, A. Cebula, Arch. of Therm., **26** 1, pp. 35-52 (2005)
14. S.C. Haldar, Int. J. of Therm. Sci., **49,** pp. 1977-1983 (2010)
15. A. Kumar, J. B. Joshi, A. K. Nayak, P.K. Vijayan, Int. J. of Heat and Mass Trans., **92,** pp. 507-522 (2016)
16. B. Niezgoda-Żelasko, W. Zalewski, *Refrigeration and air conditioning heat exchangers: thermal calculations* (Kraków, 2012) (in Polish)

The environmental impact of refuse derived fuel co-combustion with lignite

Piotr Krawczyk[1,*], *Krzysztof* Badyda[1], and *Aleksandra* Mikołajczak[1]

[1]Institute of Heat Engineering, Warsaw University of Technology, 21/25 Nowowiejska Street, 00—665 Warsaw, Poland

Abstract. Legal regulations on waste disposal require waste producers to limit landfilling and to find different ways of waste management, the preferred methods being recycling of material and energy potential. Currently, in Poland, the only consumers of refuse-derived fuels (RDF) are cement plants. However, their ability to utilize alternative fuel is far from the estimated potential. One solution would be to redirect the excess fuel to power and heat production facilities. Unfortunately, these sectors are facing a number of problems related to the thermal treatment of waste, mainly formal nature. Co-combustion of waste in power plants raises a lot of concern among their employees and local communities. Especially significant is the harmfulness of usage of fuel from waste for the people's lives or health, or the environment. The article compares the environmental footprints of the combustion process of: waste - based fuel (RDF) and lignite. The analysis was performed for a standard pulverized coal fired boiler. Comparative assessment was made by analyzing the total environmental impact of all combustion products of the two fuels. Final results have shown, that the environmental footprint of waste-based fuels can be similar or even smaller than traditional coal-based fuels.

1 Introduction

The issue of the waste to energy use is a current and extremely important issue from the implementation of sustainable development policy point of view. This challenge is particularly valid while referring to Refuse Delivered Fuel (RDF).

Like presented in [1], the potential of waste-based fuel production in Poland equals to 4,5-6 m. t/year. In Poland, in point of fact the only recipient of Refuse Derived Fuel (RDF) are cement plants. In the power plants and combined heat and power plants, the RDF is not used. The obstacle is formal and legal requirements established for the process of incineration / co-incineration of waste, emission of pollutants and the possibility of adverse changes in ash parameters.

It is estimated that Polish cement plants use approx. 1,000,000 Mg of waste-based fuels per year and can increase this amount by approx. 400,000 Mg [2, 3]. Possibilities for co-combustion of waste in this sector are limited. Cement plants will not be able to manage all the waste-based fuels produced in the country.

It is worth to mention, that the cement production peak occur in the summer, which causes the significant RDF demand reduction in the winter, when the cement plants work with the partial loads.

Basing on the facts presented above it can be claimed, that the part of the RDF producers will not be able to sell all produced waste-based fuel, unless the possibility of its use in the other sectors, like heat and/or power generation will be created

The energy use of waste would be possible to spread if the waste-based fuel could not have the waste status. However, the question arises whether it is possible to create such a fuel and carry out the procedure of losing the status of waste.

The conditions that have to be met in order to make the waste lost its status of waste and thus become a fuel are crucial. In legal acts [4], in addition to several technical, economic and technological conditions, the environmental condition can be found. Due to that the re-use of the object or waste substance should not lead to adverse effects on life, human health or the environment.

A literal confirmation of no harmfulness of fuel from waste is impossible. It is due to the fact that at the present moment, substances whose incineration "does not lead to negative effects on life, human health or the environment", even to a minimal extent, are not known. It should be emphasized that this applies not only to fuels from waste but also to any other currently used fuel. Therefore, it is necessary to determine the threshold of environmental nuisance at which we "consider" the fuel to be harmless to the environment. Such a point can be defined as, for example the ecological impact of the lignite burning.

In the other words, it can be assumed that in order to waste fuel to be deprived of the waste status it must meet environmental criteria, including indicators of environmental nuisance for human health and life determined by, for example lignite, which is after all allowed for the combustion as fuel.

* Corresponding author: piotr.krawczyk@itc.pw.edu.pl

The article presents own results of the elemental composition of wastes mixture investigation. The compounds were generated by being sorted from the over-sieve part of municipal waste unwanted substances from the fuel point of view, i.e. multi-material waste, ferrous and non-ferrous metals, non-flammable fractions (ash). Marked waste samples were prepared on a typical sorting line of municipal waste equipped with, among the others three optosegregators. The used sorting line is located in one of the Polish cities. The obtained results were used to assess the relative environmental nuisance of the mixture of waste used as fuel in comparison with lignite.

2. Characteristics of the compared fuels

The subjects to the analysis were: lignite and waste-based fuel. Technical parameters and elementary composition of the analyzed fuels are presented in Table 1.

Table 1. Results of elementary and technical analysis of analyzed fuels

Parameter	Unit	Lignite	Waste-based fuel
Total humidity	%	54.8	16.74
Volatile parts content	%	37.065	73.47
Higher Heating Value	J/g	17308	-
Lower heating value	J/g	7274.5	19009.6
Analytical humidity	%	14.9	2.99
Ash content	%	10.35	13.39
Carbon content	%	45.8	52.74
Hydrogen content	%	3.465	6.63
Total sulphur content	%	1.545	0.11
Nitrogen content	%	0.515	0.65
Chlorine content	%	0.0135	0.99
Fluorine content	%	0.06	0.0094
Sum of 9 heavy metals	mg/kg d.m.	224.3	261.2
Thallium	mg/kg d.m.	0.1	0.1
Cadmium	mg/kg d.m.	0.6	0.27
Mercury	mg/kg d.m.	0.416	0.35

The lignite with a chemical composition similar to the one presented in the table is mined in one of the Polish mines and used in the pulverized boilers.
The analyzed waste-based fuel was obtained on the standard MBP line of municipal waste segregation [5 - 7]. The substrate for the fuel production consists of municipal waste coming from the county town. During the installation work the so-called composite materials that were suspected of carrying heavy metals were tried to being get rid of. The analyzed fuel consisted mainly of four waste material fractions: plastics, waste paper, wood and textile materials in ratios resulting from their natural content in the tested waste stream.

3. Methodology of the fuel environmental harmfulness assessment

The environmental harmfulness of the given fuel combustion is the sum of the environmental harmfulness of all products of this process. It may be considered as the influence of a given substance on, among the others, environmental pollution, toxicity to living organisms and greenhouse effect [8 -13].
The properties of fuels significantly effect on the emission of harmful substances into the environment. These properties can be generally divided into two groups. The first group includes parameters affecting the technical suitability of the fuel for combustion: calorific value, moisture content, ash and combustible substances content and volatile parts. The second group consist of the chemical composition of the fuel and in particular the content of such harmful substances as sulfur, chlorine, fluorine and heavy metals. In addition to the harmful substances in the fuel, the emission in the exhaust gases depends on some combustion parameters such as: temperature distribution, boiler flue gas residence time, etc. Combustion products particularly sensitive to the process parameters are incomplete combustion products (i.e. CO) and nitrogen oxides.
The comparison of the particular fuels or the industrial installations operation due to their impact on the natural environment usually presents many difficulties. This is due to the need to take into account the specificity of both individual components and technological processes.
Descriptions of the methods for assessing the environmental impact of combustion installations are available in the literature. According to the approach proposed in [14], the entire combustion process should be analyzed comprehensively to determine appropriate indicators of nuisance for each possible combustion product. The total burden of all combustion products is then the sum of individual nuisances.
The indicator of the nuisance for the environment of the substance generated in the combustion process related to the unit of energy produced includes, among the others:

- calorific value of fuel;
- efficiency of the exhaust gases cleaning system (for each given components);
- coefficient of formation of a given component in the combustion chamber [kg/kg of fuel];
- the relative nuisance of a given component contained in the exhaust gas.

Table 2. Relative nuisance (toxicity) of the exhaust gases components [15].

Component	Value of toxicity coefficients
CO	1
Ash	5
TOC	5
HF	50
SO_2	1
NO_x	0,25
HCl	5
Cd, Tl, Hg	1000
Sb, As, Pb, Cr, Co, Cu, Mn, Ni, V, Sn	100

The relative nuisance of the given components contained in the exhaust depends on the type of substance and is usually related to the toxicity of the selected component,

e.g. SO_2. Their values based on [15] are presented in Table 2.
The total nuisance indicator, which can be calculated as the sum of nuisance indicators for individual substances arising in the combustion process, may be the basis for an objective comparison of the harmfulness of fuels. It takes into account not only the specificity of all components and their harmfulness, but also individual features of the installation.
According to the presented approach, when comparing ecological harmfulness of two fuels burned in the same installation, it is necessary to determine the same indicators characterizing the combustion installation, efficiency of exhaust gas cleaning etc. This way the result of the comparison does not depend on the characteristics of the combustion plant itself, and only indicators related to the physical and chemical properties of the fuel are concerned. In this situation, the individual environmental index of the substance "i" can be defined as:

$$JWU_i = Z_i \cdot W_K \cdot t_{sp\,i} \qquad (1)$$

where:
Z_i – component „i" content in the fuel (ash, sulphur, nitrogen, chlorine, fluorine, heavy metals) [mg/MJ];
W_{Ki} – conversion ratio of component "i" in the fuel to the harmful component in the exhaust gases - taking into account the existing purification system [%]
$t_{sp\,i}$ – the relative nuisance of component "i" contained in the exhaust gases [kg SO_2/kg i]
The total nuisance rate of a given fuel is the sum of individual harmfulness of its components and can be expressed as:

$$CWU = \sum Z_i \cdot W_K \cdot t_{sp\,i} \qquad (2)$$

Based on the presented approach the environmental nuisance of the analyzed fuels were compared.
For the purpose of the introduced method of fuels comparison the content of the analyzed elements in the fuel should be expressed as the amount of a given substance contained in the fuel per unit of chemical energy of this fuel, i.e. mg/MJ. For the purpose of these analysis, it was assumed that concentrations of all elements should be related to the unit of chemical energy in the working condition of the fuel (table 3).

Table 3. Comparison of the content of individual elements in lignite and waste-based fuel per unit of chemical energy.

Parameter	Sign	Lignite - working condition [mg/MJ]	Waste-based fuel – working condition [mg/MJ]
Ash content		13392	6044
Total sulphur content	St	1090,06	49.66
Nitrogen content	N	363.35	293.47
Chlorine content	Cl	9.53	446.97
Fluorine content	F	42.33	4.24
9 heavy metals total amount		10.225	11.441
Thallium	Tl	0.006	0.004
Cadmium	Cd	0.030	0.012
Mercury	Hg	0.023	0.015

Analyzing the data presented in the Table 3, it can be concluded that fuel from waste in relation to lignite is characterized by, among the others:
• lower ash content;
• lower sulfur content;
• lower nitrogen content;
• significantly higher chlorine content;
• lower content of fluorine;
• comparable content of the sum of "9" heavy metals (arsenic, lead, manganese, copper, chromium, nickel, vanadium, antimony, cobalt);

3.1 The conversion rates determination

As mentioned before, in order to determine the harmfulness (toxicity) of waste-based fuel in comparison with lignite, the total nuisance rate of the combustion process as defined above (equation 2) was used. The following fuel components were selected for the comparative analysis: ash, sulfur, nitrogen, chlorine, fluorine and heavy metals.
In the proposed approach, it was necessary to define the conversion factors of a given component in the fuel to the harmful component in the exhaust gases- taking into account the existing treatment system. In order to define them, the following data were used:
• for sulfur, nitrogen, chlorine, fluorine and ash – the measurements from one of the Polish lignite power plants;
• for heavy metals, including mercury - literature data [16].
The conversion rate determines what percentage of the weight of the component contained in the fuel (e.g. S) is converted to a harmful substance in the exhaust gases (e.g. SO_2) and passes through the exhaust clean-up system to the environment.
The conversion ratio for sulfur to sulfur dioxide emitted to the atmosphere was assumed at the level of 10%. This means that about 10% of the sulfur content introduced into the boiler with the fuel goes to the environment in the form of SO_2.
As a result of balancing the relevant components (elements) in the coal and flue-gases based on an actual data from the power plant, the following values of conversion rates were obtained:
•ash→ dust= 0,022 [%];
•N → NOx = 17,22 [%];
•Cl → HCl = 11 [%];
•F → HF = 3 [%].
The values of the N → NO_x conversion coefficient determined in the literature [17] for coal combustion takes values at the range of 12 - 20%.
To estimate the conversion rates for heavy metals, the mass balance of heavy metals were determined in the coal power plant system equipped with an electrostatic precipitator and a wet flue gas desulphurisation unit according to the BREF [18]
According to these data, the majority of heavy metals are removed from the exhaust gases with dust separation in the electrostatic precipitator. The only exception is mercury, which is in gaseous form and passes into the

environment in the amount of approx. 30% of its input mass introduced to the boiler.

Therefore, the following values of conversion ratios for heavy metals were assumed for the purposes of this assessment:

• The sum of "9" heavy metals - 0.1 [%];
• Thallium- 0.1 [%];
• Cadmium - 0.07 [%];
• Mercury - 30.6 [%].

4. Results

Using the determined conversion coefficients, the total nuisance index (CWU) was calculated as the sum of individual harmfulness of the fuel components for lignite and waste-based fuel. The results are presented in the tables 4 and 5.

Taking into account all the assumptions, the total environment nuisance indicator for the lignite equals to 216.2. For fuel from waste - 282.19. The comparison of both fuels shows that the harmfulness (nuisance) of waste-based fuel is about 30% higher than lignite (equation 3).

$$\frac{CWU_{waste-based\ fuel}}{CWU_{lignite}} = \frac{282.19}{216.2} = 1.3 \quad (3)$$

The analysis of the CWU structure for lignite indicates the sulphur and fluorine as the ones having the dominant role.

Table 4. The individual environmental index for lignite components and the total nuisance rate.

	Conversion ratio [%]		Relative nuisance (toxicity) of the exhaust gases components [kg SO_2/kg component]	Lignite - JWU
Dust	Ash→Dust	0.022	5	14.73
Sulphur (S)	$S \rightarrow SO_2$	10	1	109.006
Nitrogen (N)	$N \rightarrow NO_x$	17.22	0,25	15.64
Chlorine (Cl)	Cl→HCl	11	5	5.4
Fluorine (F)	F→HF	3	50	63.49
9 heavy metals total amount	9 heavy metals total amount	0.1	100	1.02
Thallium (Tl)	Tl→Tl	0.1	1000	0.006
Cadmium (Cd)	Cd→Cd	0.07	1000	0.021
Mercury (Hg)	Hg→Hg	30.6	1000	7.038
CWU				216.20

Table 5. The individual environmental index for waste-based fuel components and the total nuisance rate.

	Conversion ratio [%]		Relative nuisance of components [kg SO_2/kg component]	Waste-based fuel - JWU
Dust	Ash→Dust	0.022	5	6.64
Sulphur (S)	$S \rightarrow SO_2$	10	1	4.96
Nitrogen (N)	$N \rightarrow NO_x$	17.22	0,25	12.63
Chlorine (Cl)	Cl→HCl	11	5	245.83
Fluorine (F)	F→HF	3	50	6.36
9 heavy metals total amount	9 heavy metals total amount	0.1	100	1.14
Thallium (Tl)	Tl→Tl	0.1	1000	0.004
Cadmium (Cd)	Cd→Cd	0.07	1000	0.0084
Mercury (Hg)	Hg→Hg	30.6	1000	4.59
CWU				282.19

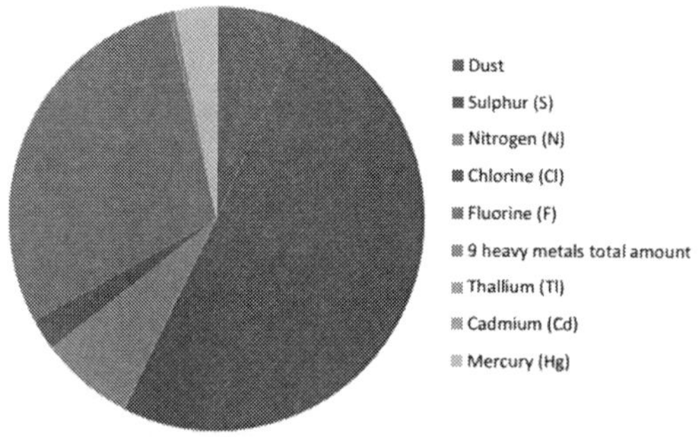

Fig. 1. The CWU structure for lignite.

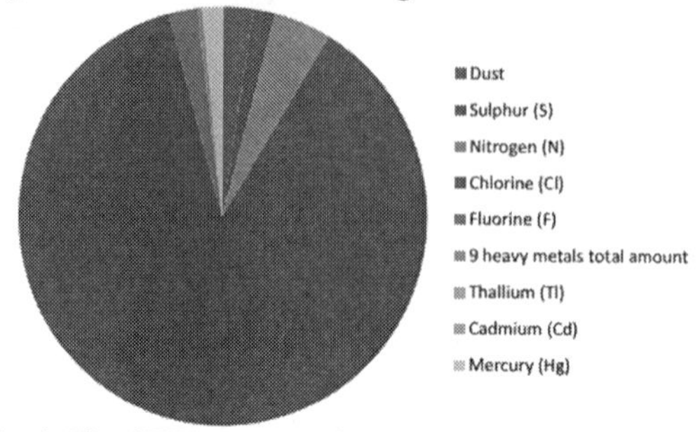

Fig. 2 The CWU structure for the waste-based fuel.

Analyzing the results obtained for the waste-based fuel, it can be concluded that the chlorine plays the major role in the CWU structure (approximately 87%).
During the preparation of the waste-based fuel at the sorting line, the optical separation was not used. This kind of segregation is implemented to remove PVC as the main carrier of chlorine in the waste. Hence the relatively high chlorine content in the analyzed waste-based fuel occur.
According to the results of the CWU comparison for lignite and waste-based fuel, reduction of chlorine content by approx. 30% in waste-based fuel (with unchanged other parameters), will result in equalization of both fuels indicators of nuisance. Basing on the authors experience, it can be concluded that applications of optical separation of PVC can lead to reduction of chlorine content in waste-based fuel by up to 70%.

5. Summary

The authors research results justified that the environmental impact of the waste-based fuel combustion may be comparable with the traditional fossil fuels combustion in the installation with the set parameters (including the configuration of the exhaust gas treatment process). Therefore, the presented results may be an argument in the discussion on the possible loss of the waste status by the fuel produced on the basis of municipal waste. It means that, from the legal regulations point of view, the waste-based fuel fulfill the condition that the use of the object or substance does not lead to negative consequences for life, human health or the environment.
What's more, the technical equipment of modern waste sorting plants allows for deep and very conscious sorting of waste. According to the obtained results, the waste fuels should maximize the ratios of four fractions: waste paper, textiles, wood, plastics, excluding PVC.
The analyzes were carried out with the assumption of the invariance of the combustion process parameters in the boiler when using both fuels. This proposition can be considered as legitimate when using co-combustion of fuel from waste with primary fuel, i.e. lignite in limited quantities (several percent of the mass fraction of waste-based fuel). Examples of such practices can be found, among the others in German energy sector. The Jänschwalde power plant co-combusts waste-based fuel on the large-scale. In 2012, 495,000 Mg of RDF was co-burned, which corresponds to 3.5% of the chemical energy of the fuel burned. Mostly used waste-based fuel is produced from mixed municipal waste.
Co-firing did not result in the reconstruction of the flue gas cleaning installation.

References

1. P. Krawczyk, J. Szczygieł, Rynek Energii, **6** (109), 6, (2013) (in Polish)
2. K. Badyda, P. Krawczyk, K. Pikoń, Energy, **100**, 6, (2016)
3. M. Kalisz, The power of eco-business, **5**, 6, (2013) (in Polish)
4. Directive 2008/98/EC of the European Parliament and of the Council of 19 November 2008 on waste and repealing certain directives (OJ L 312, 22.11.2008, pp. 3–30)
5. B. Bilitewski, G. Hardtle, K. Marek, *Podręcznik Gospodarki Odpadami,* (Seidel Przywecki, First edition, Warsaw 2003), (in Polish)
6. Polish Ministry of the Environment, Guidelines for the composting, fermentation as well as mechanical - biological waste treatment, 2008 (in Polish)
7. L. Sieja, Siła ekobiznesu, **7**, 5, (2013) (in Polish)
8. K. Pikoń, K. Gaska, Polish Journal of Environmental Studies, **18** (3A), 11, (2009)
9. J. Guinée, *Handbook on Life Cycle Assessment, Operational Guide to the ISO Standards, Eco-efficiency in industry and science*, (Volume 7, Kluwer Academic Publishers, New York, Boston, Dordrecht, London, Moscow, 2004)
10. Á. Restrepo, R. Miyake, F. Kleveston, E. Bazzo, Energy, **45**, 9, (2012)
11. K. Pikoń, Applied Energy, **75**, 8, (2003)
12. K. Pikon, M. Bogacka, J. Nadziakiewicz, W. Stanek, L. Czarnowska, SGEM 2014 Conference Proceedings, **4** (1), 8, (2014)
13. EPA. Emissions From Waste Incineration. Good Practice Guidance and Uncertainty Management in National Greenhouse Gas Inventories
14. J. Nadziakiewicz, K. Pikoń, Paliwa z Odpadów **3**, 5 (2001) (in Polish)
15. J. Niedziakiewicz, K. Wacławek, S. Stelmach, *Procesy termiczne utylizacji odpadów*, (Publisher of the Silesian University of Technology, 2012) (in Polish)
16. W. Kordylewski, *Spalanie i paliwa*, (Wroclaw University of Technology Publishing House, 2008) (in Polish)
17. R. Sekret, Emisja substancji szkodliwych w procesie fluidalnego spalania mieszanki biomasy i węgla brunatnego. Częstochowa listopad 2008 (in Polish)
18. Reference documents under the IPPC Directive and the IED (Industrial Emissions Directive, 2010/75/EU)

Modeling of a re-heat two-stage adsorption chiller by AI approach

Jaroslaw Krzywanski[1,*], *Karolina* Grabowska[1], *Marcin* Sosnowski[1], *Anna* Żyłka[1], *Karol* Sztekler[2], *Wojciech* Kalawa[2], *Tadeusz* Wójcik[2], and *Wojciech* Nowak[2]

[1]Jan Dlugosz University in Czestochowa, Faculty of Mathematics and Natural Sciences, Armii Krajowej 13/15, 42-200 Czestochowa, Poland
[2]AGH University of Science and Technology, Faculty of Energy and Fuels, Czarnowiejska 30, 30-059 Krakow, Poland

Abstract. A distinct advantage of adsorption chillers is their ability to be driven by heat of near ambient temperature. However the performance of the thermally driven adsorption systems is lower than that of other heat driven heating/cooling systems. It is the result of a poor heat transfer coefficient between the bed and the immersed heating surfaces of a built-in heat exchanger system. The aim of this work is to study the effect of thermal conductance values as well as other design parameters on the performance of a re-heat two-stage adsorption chiller. One of the main energy efficiency factors in cooling production, i.e. cooling capacity (CC) for wide-range of both design and operating parameters is analyzed in the paper. Moreover, the work introduces artificial intelligence (AI) approach for the optimization study of the adsorption cooler. The Adaptive Neuro – Fuzzy Inference System (ANFIS) was employed in the work. The developed ANFIS model can be applied for optimizations purposes and may constitute a submodel or a separate module in engineering calculations, capable to predict the CC of the re-heat two-stage adsorption chiller.

1 Introduction

The silica gel-water adsorption chillers constitute the economically viable and environmentally friendly technology, capable to convert waste thermal energy into useful cooling [1]. Their main disadvantages, are: the intermittence, bulkiness, and poor thermal conductivity of the used adsorbents leading to the low specific cooling power (SCP) and coefficient of performance (COP) [2].

Since the complexity of multi-bed adsorption chillers operation is still not sufficiently recognized the improvement of total efficiency of adsorption processes is still a challenging task [3 – 5].

Mathematical modelling is considered as an efficient research approach [6].
However sometimes additional assumptions should be made when modeling to get a trackable solution, the algorithms are often complicated and based on the solution of complex and time consuming sets of differential equations [7-9].

Artificial intelligence (AI) approach can be an alternative method for the above technics of data handling. The most convenient approach is to apply a non-iterative procedure, where one only needs to enter input parameters and call the performed the AI model.

The paper introduces the Adaptive Neuro – Fuzzy Inference System (ANFIS) for modelling of a re-heat two-stage adsorption chiller using low-temperature heat from cogeneration. The developed non-iterative model, using procedure similar to the one given in [10 - 14] allows to describe the behavior of the adsorption cooler and predict one of the main energy efficiency factors in cooling production, i.e. cooling capacity for wide-range of both design and operating parameters.

To our best knowledge we are the first to employ the Adaptive Neuro–Fuzzy Inference System to describe the behavior of the innovative adsorption chiller and predict one of the main energy efficiency factors in cooling production, i.e. cooling capacity (CC) for wide-range of both design and operating parameters.

2 Analysis and modeling

2.1 An object of investigations

The analyzed re-heat two-stage chiller consists of four adsorbent beds, one evaporator and one condenser as well as metallic tubes for heat transfer fluid and refrigerant flows [15, 16]. Silica gel is used as sorbent. The schematic diagram of a re-heat two-stage chiller is given in Fig. 1.

Long cycle (3400s) and short cycle (1300 s) are considered during the study. A long cycle time with heat source temperature at 60 °C is selected in this study. A detailed description of the chiller, including considered operating strategy, can be found elsewhere [15].

* Corresponding author: jkrzywanski@tlen.pl

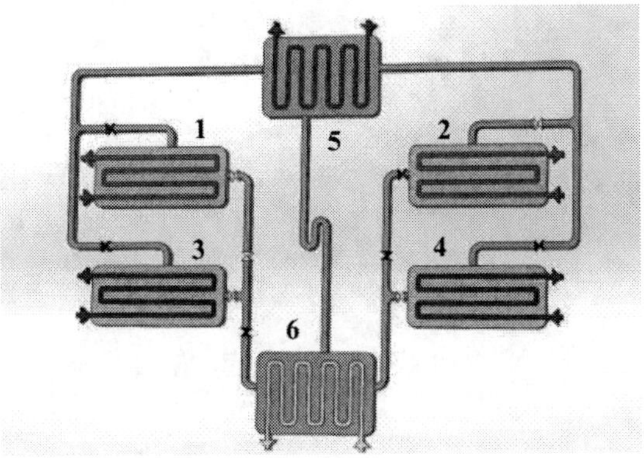

Fig. 1. The scheme of the adsorption chiller; 1-4 – adsorption beds, 5 – condenser, 6 – evaporator.

2.2 The Adaptive Neuro-Fuzzy Inference System

For the purpose of this paper the experimental results given in [15, 16] were used to derive and validate model parameters.

The Matlab® R2018a, with the Neuro-Fuzzy Toolbox is used to develop the model and predict the CC of the considered re-heat two stage adsorption chiller [17]. The following values: adsorbent mass (M_{sorb}), cycle time (t), hot water temperature (T), adsorbent-to-metal thermal capacitance ratio (C_{sorb}/C_{met}), overall evaporator and sorption element thermal conductance (hA_{evap}, hA_{sorb},) respectively are assumed as input parameters.

The total number of learning data used to develop model includes 80 samples of which 25 are independent checking data set, similar to the procedure given in [10, 11, 13].

The training the ANFIS model was performed for the following stopping criteria for training: number of training epochs equal to 3000 and the Error Tolerance = 0.1.

3 Results and discussion

The comparison between desired and calculated by the model results revealed that the calculated results are located within the range of ± 10 % of relative error, compared to the desired data.

In order to study the influence of a specific operating variable on the cooling capacity, other input parameters have to be fixed and keep unchanged during the analysis.

3.1 The effect of adsorbent mass on cooling capacity

The influence of adsorbent mass on cooling capacity can be found in Figure 2. As adsorbent mass increases from 4.7 kg to 32 kg. the cooling capacity increases due to the increase of bed sorption properties.

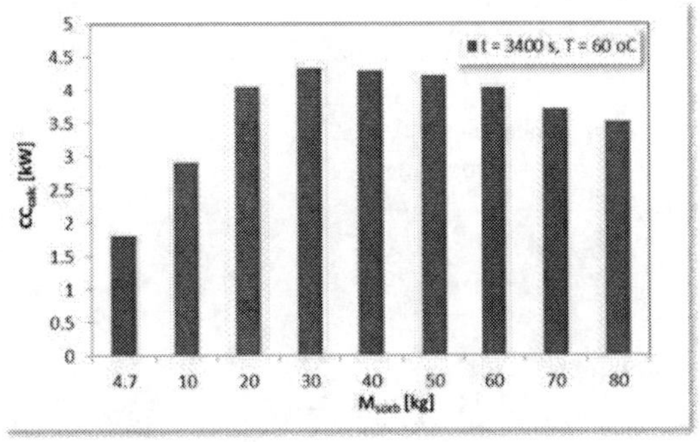

Fig. 2. The influence of adsorbent mass on cooling capacity.

3.2 The effect of adsorbent mass on cooling capacity

The influence of thermal conductance of sorption and element and evaporator on CC is depicted in Fig. 3. Three data series correspond to different values of thermal conductance of evaporator, namely: 1000, 2500 and 4000 W K^{-1}.

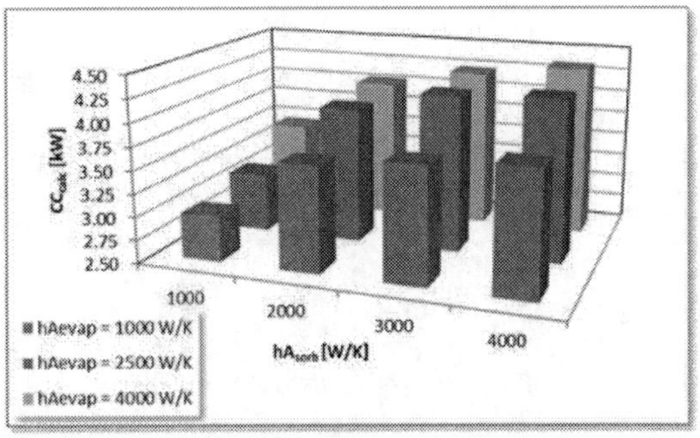

Fig. 3. The effect of sorption element and evaporator thermal conductance on cooling capacity.

The higher sorption element thermal conductance means better heat transfer conditions inside the sorption element and the increase in hA_{sorb} allows for better preparation of the bed for the adsorption process.

This also means higher amount of refrigerant inside the bed and better performance, leading to the increase in cooling capacity [15].

Similar explanation applies to the increase in the evaporator thermal conductance.

The performed calculations allows to determine optimal experimental and design parameters.

4 Conclusions

The paper deals with an innovative design in cooling production, a re-heat two-stage adsorption chiller.

The maximum relative error of the developed model is lower than 10 %.

The developed model constitutes an easy-to-use and powerful optimization tool which allows estimating the cooling capacity of the re-heat two-stage adsorption chiller.

Nomenclature

A	area	m^2
C	specific heat capacity	J/(kg K)
h	heat transfer coefficient	$W/(m^2 K)$
M	mass	kg
t	cycle time	s
T	hot water temperature	^oC

Subscripts

calc	calculated
evap	evaporator
met	metal
sorb	sorbent

Acronyms

AI	Artificial Intelligence
ANFIS	Adaptive Neuro-Fuzzy Inference System
CC	Cooling Capacity
COP	Coefficient of performance
SCP	Specific cooling power

Acknowledgment

Scientific work was performed within the confines of subsidies granted by the Faculty of Mathematics and Natural Sciences of Jan Dlugosz University in Czestochowa. The support is gratefully acknowledged.

References

1. X. Wang, Z. He, H.T. Chua, Int. J. Refrig. **52**, 32–41 (2015)
2. R.P. Sah, B. Choudhury, R.K. Das, Renew. Sustain. Energy Rev. **45**, 123–134 (2015)
3. Y.I. Aristov, B. Dawoud, I.S. Glaznev, A. Elyas, Int. J. Heat Mass Transf. **51**, 4966–4972 (2008)
4. B. Dawoud, Y. Aristov, Int. J. Heat Mass Transf. **46**, 273–281 (2003).
5. B.N. Okunev, A.P. Gromov, L.I. Heifets, Y.I. Aristov, Int. J. Heat Mass Transf. **51**, 246–252 (2008)
6. Y.I. Aristov, Int. J. Refrig. **32**, 675–686 (2009).
7. W. Muskała, J. Krzywański, T. Czakiert, W. Nowak, Rynek Energii **92**, 172-176 (2011)
8. W. Muskała, J. Krzywański, R. Rajczyk, M. Cecerko, B. Kierzkowski, W. Nowak, W. Gajewski, Rynek Energii **87**, 97-102 (2010)
9. T. Klajny, J. Krzywanski, W. Nowak, *Proc. of the 6th Int. Symp. On coal combustion, Conference: 6th International Symposium on Coal Combustion*, (Huazhong Univ Sci Technol, Wuhan, PEOPLES R. CHINA, Dec. 01-04, 148 – 53, 2007)
10. J. Krzywanski, K. Grabowska, F. Herman, P. Pyrka, M. Sosnowski, T. Prauzner, W. Nowak, Energy Convers. Manag. **153**, 313–22 (2017)
11. J. Krzywański, W. Nowak, Journal of Power Technologies **97**, 75-84 (2017)
12. J. Krzywanski, W. Nowak, Int. J. Heat Mass Transf. **94**, 327– 34 (2016)
13. J. Krzywanski, W. Nowak, J. Energy Eng. **142**, 1, Article Number 04015017 (2016)
14. J. Krzywanski, M. Wesolowska, A. Blaszczuk, A. Majchrzak, M. Komorowski, W. Nowak, Procedia Engineering **157**, 66 – 71 (2016)
15. M.Z.I. Khan, K.C.A. Alam, B.B. Saha, A. Akisawa, T. Kashiwagi, Appl. Therm. Eng. **27**, 1677-1685 (2007)
16. M.Z.I. Khan, K.C.A. Alam, B.B. Saha, Y. Hamamoto, A. Akisawa, T. Kashiwagi, Int. J. Therm. Sci. **45**, 511-519 (2006)
17. https://www.mathworks.com/help/fuzzy/

CFD modelling and validation of the rotary lobe compressed air expander

Michalina Kurkus-Gruszecka[1,*], *Piotr* Krawczyk[1], and *Piotr* Łapka[1]

[1]Warsaw University of Technology, Institute of Heat Engineering, Nowowiejska 21/25, 00-665 Warsaw, Poland

Abstract. The article covers the CFD analysis of the compressed air rotary lobe expander. Compressed air engines are commonly used in the explosion hazard zones, where the others power sources cannot be applied. The lobe construction seems to be cheaper solution in comparison with the classic turbine expanders at the corresponding power level.
In the paper the rotary lobe compressed air expander model was developed in the ANSYS CFX software. During implementation of the proposed numerical model the mesh motion problem was solved, which was challenging due to relatively small gaps between lobes and the housing as well as due to high rotary speed of lobes. What is more, the compressible fluid model was applied. Therefore, these conditions make the undertaken problem more complex and calculation more time consuming. Finally, the model was validated applying the catalogue data of the producer.
In the paper description of the developed model was presented. Then the results of simulations with their validation were shown which proved the correctness and accuracy of the model. The developed tool might be very useful for the further analysis of the expander including its optimization in terms of improvement of its efficiency or energy gain.

1 Introduction

The rotary lobe compressed air expander which is the object of the investigation in this paper in most cases is applied as the part of air motors. Air motors can successfully replace conventional, e.g. electric motors in explosive areas or in areas, where the suitable medium is available. The most common applications for air motors are elevators drives in the mines, coal mining locomotives, earth drilling rigs and units for winches [1,2]. Similar types of steam-based expanders can be applied in micro organic Rankine cycles (ORC).

The main advantage of gas expanders is the lower cost of application comparing to conventional turbine motors at the corresponding power level, combined with arduous and long-life systems. Other features determining the usage of air expanders are adjustable torque and output speed, the ability to stop under load without damage as well as a small number or lack of springs and pins [2]. Also, it is possible to install the air motor in wet, dirty and harsh environment due to its sealed housing and relatively compact structure.

In this paper the model of the rotary lobe compressed air expander was developed with the consideration of industrial solutions. As shown in Figure 1, the model used for the present simulation consists of two rotors moving inside the expanders housing. The torque is developed by the larger rotor situated in the lower part of the housing and then transferred to the rotating shaft by a synchronizing gear train. To preserve high efficiency of the expander, air gaps between the rotors and housing are relatively small which allows rotation process without physical contact. The frictionless work might result in the long maintenance free operation without downtime in industrial applications [1].

2 CFD model of the Rotary Lobe Expander

During the substantive preparations for modelling of the expander, no literature was found referring to rotary lobe expanders. However, literature on modelling related devices of a different type of operation or using a different medium is available. There are CFD models of rotary lobe pumps described in [3,4]. In particular the CFD modelling approach applied in [3] is similar to the one applied in the presented paper. Moreover, the relatively high cost of calculations in such simulations was also underlined. The turbulence and gas models applied for similar simulations are described in [5–9].There are also works related to models [10–12] and patents [13–17] of screw rotors, with lobes shape optimization taken into account. The lobes shape optimization is one of the next steps in the future development of the proposed rotary lobe expander model.

* Corresponding author: michalina.gruszecka@itc.pw.edu.pl

2.2 Geometry

In the first step a full 3D geometrical model of the expander was created. The model includes the fluid domain geometry, which corresponds to the housing of the expander and the geometries of two relevant rotors. The rotors were designed with different shapes (i.e. 2- and 5-lobes) having the proper proportions to allow synchronous operation, with consideration of face-clearance between lobes and the housing as well as center-clearance between the corresponding lobes.

Due to relatively high computing cost, the axial dimension of the expander was reduced to 1 mm, which corresponds to quasi-2D approach. This gives the opportunity to observe the main phenomena responsible for the expander operation. Also, it is associated with, e.g. neglecting pressure losses in the axial direction, which have an effect on the hereby presented results discussed further. The geometry of the rotary lobe air expander is presented in the Figure 1.

Figure 2 shows the geometry of the gas-filled expander chamber without rotating elements. In order to refine the grid, it was divided into five areas: the working area of the larger rotating circle, the working area of the smaller rotating circle, the central area the larger and smaller rotor engagement, the expander inlet as well as the outlet chamber and the outlet from the expander.

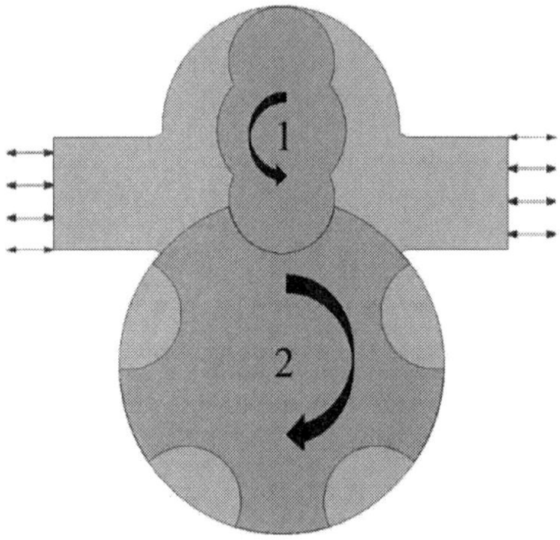

Fig. 1. Axial view of the rotary lobe expander geometry with rotor direction indicators. No. 1 – smaller rotor, no. 2 – larger rotor.

2.2 Mesh

The applied immersed solids method requires the generation of separate grids for the fluid domain and for solids. Thus, the three areas of discretization were designed: an area for the fluid domain and two areas for rotating wheels.

Fig. 2. Mesh of the fluid domain.

The density of the mesh was selected in relation to the values of gradients of local parameter such as pressure or velocity of the gas. The number of cells per unit volume was increased in specific areas, i.e. in the gaps between the walls of the expander and the lobes, in the area of meshing of the rotors. The grid was created while maintaining its quality criteria, e.g., aspect ratio and skewness on the reasonable levels.

The fluid domain was meshed with the sweep method (see Figure 2). The sweep method is adequate for complex geometries with little or no variation in a specific direction and requires definition of source and target boundary. The front plane of the 3D domain was selected as a source boundary.

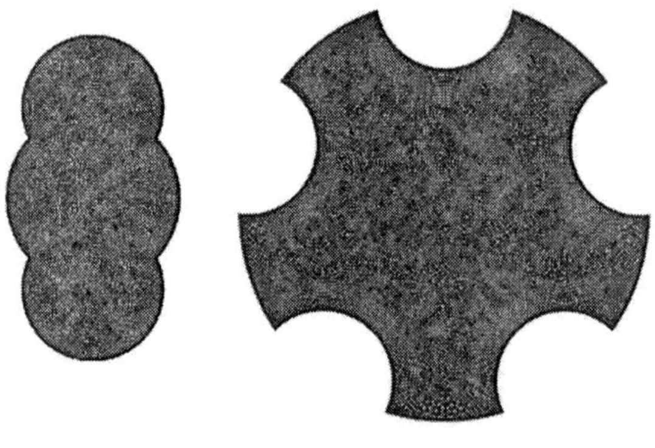

Fig. 3. Mesh of the smaller (on the left) and larger (on the right) rotors.

The density of the fluid domain grid was enhanced in the area of engagement of the rotors, which can be seen in Figure 2. The mesh properties for the fluid domain are presented in the Table 1. The grids for rotors were also created using the sweep method (see Figure 3), refined in areas of solid-fluid contact. The grid properties for the rotors are also presented in the Table 1.

Table 1. Meshes properties.

Parameter	Value		
	Fluid domain	Larger rotor	Smaller rotor
Number of cells	495,548	190,582	84,560
Number of nodes	744,897	287,340	127,602
Average aspect ratio	2.92	3.00	3.23
Average skewness	0.11	0.14	0.11
Average element quality	0.83	0.83	0.81

2.2 Computational domains, boundary conditions and simulations settings

In the ANSYS CFX the computational area was divided into three domains. The fluid flow region was defined as a fluid domain. The other two domains identified as solids were corresponding to two rotors. The boundary conditions for the inlet and outlet were set as fixed pressure values while for all remaining walls of the fluid area the no-slip wall was assumed.

Table 2. Parameters assumed in the simulations.

Parameter	Unit	No. of simulation		
		1	2	3
Speed of the larger rotating wheel	rev/min	1000	1000	1500
Speed of the smaller rotating wheel	rev/min	2500	2500	3750
Expanders inlet pressure	bar	4	8	3
Total simulation time	s	0.05	0.05	0.032
Time step	s	$5 \cdot 10^{-5}$	$5 \cdot 10^{-5}$	$3.2 \cdot 10^{-5}$
Number of time steps for a single simulation	-	1000	1000	1000
Number of revolutions of the smaller rotor during simulation	-	2.083	2.083	2
Number of revolutions of the larger rotor during simulation	-	0.833	0.833	0.8

In order to validate the rotary lobe air expander model, three simulations were conducted, and the obtained results were compared to the literature data. The value of pressure at inlet and rotation speed of the rotors took different values in the individual simulations. To maintain the appropriate stability and quality of calculations, the time step values were adapted to the rate of change of geometry which corresponds to the rotation speed. The number of time steps has been set to achieve the stabile solution and to observe the characteristics of the expander's operation. This refers to the execution of about two revolutions of the larger rotor.

In Table 2 the aforementioned parameters are presented. At the expander outlet fixed value of the pressure equal to 1 bar was assumed in the all simulations. Moreover, the inlet temperature was equal to 25□C and all walls of the fluid domain were adiabatic.

The catalogue data of the air expander producer does not provide detailed information about thermodynamic properties of the air (e.g. temperature or humidity). Therefore, the working medium applied in the presented simulations was the dry air with the thermophysical properties shown in the Table 3.

The flow inside the expander was assumed transient, compressible and turbulent. Therefore, ideal gas model together with the k-ω Shear Stress Transport (SST) turbulence model were applied during the simulations.

Table 3. Thermophysical properties of the working medium (dry air).

Parameter	Value	Unit
Molar mass	28.96	kg/kmol
Density	1.185	kg/m^3
Specific heat	1.0044	kJ/(kg·K)
Dynamic viscosity	$1.831 \cdot 10^{-5}$	kg/(m·s)
Thermal conductivity	$2.6 \cdot 10^{-2}$	W/(m·K)

2.2 Solver settings and residuals

As initially mentioned to investigate operation of the rotary lobe air expander transient calculations were carried out. The assumed main solver settings in the ANSYS CFX are presented below:

- Advection scheme: high resolution.
- Transient scheme: second order backward Euler.
- Residual type: root mean square (RMS).
- Convergence criteria: 10-4.

In order to conduct after-simulation analysis and to check the crucial parameters during the solution process the appropriate transient results were saved and transient monitors were created and registered. Selected variables which allowed analysis of the obtained results were saved for the whole computational domain for every time step. These variables were as follows: absolute pressure, total pressure in stationary frame of reference, velocity and temperature. Moreover, all together sixteen monitors were created for each simulation: two monitors for the average mass flow rate at the inlet and outlet to the fluid domain and fourteen monitors for the average pressure acting on the surface of half of each lobe, including both, smaller and larger rotors.

All simulations were carried out using the workstation with two processors (Intel® Xeon® CPU E5-2600 2.20 GHz) and with 64 GB RAM memory as well as utilizing of 8 to 16 computing nodes. The time required to carry out a single simulation equals approximately 26 h.

3 Results and discussion

In Figures 3-5 the average pressure acting on the surface of a half of each lobe, including both, smaller and larger rotors are presented. The obtained results are shown for simulation no. 2, which correspond to the inlet pressure of 8 bar and the rotation speed equals 1000 rev/min for the larger rotor. In each diagram a cyclical operation of the expander can be observed resulting in a regular change of the pressure on each monitored surface.

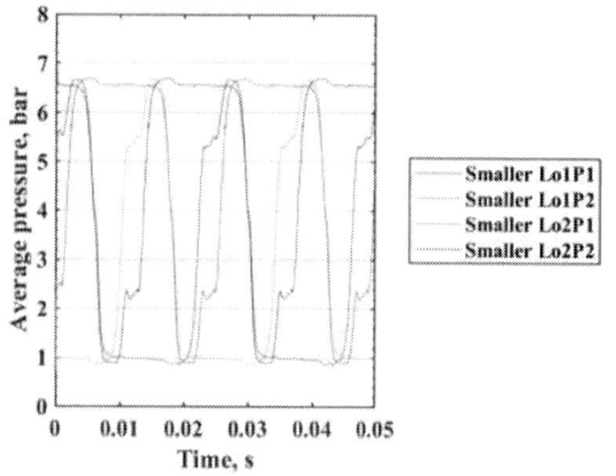

Fig. 4. The pressure variation for the smaller rotor for the pressure at the expander inlet of 8 bar and rotational speed of 1000 rpm.

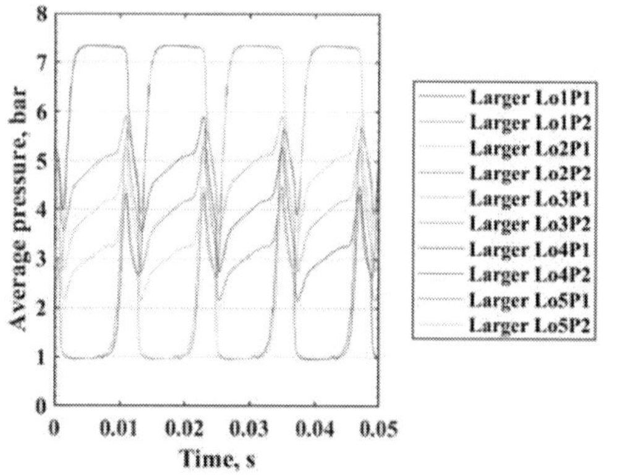

Fig. 5. The pressure variation for the larger rotor for the pressure at the expander inlet of 8 bar and rotational speed of 1000 rpm.

After obtaining the mean pressure values acting on the surface of the half of the lobe, the force acting in the direction perpendicular to the axis of rotation for each of the lobe was determined [17]. Then the torques acting on the individual rotors were determined as the sum of the component torques acting on the rotor from individual lobes. Such a procedure was necessary to obtain the actual value of the torque acting on the rotor, because at specific positions of the rotors, the force direction is unsuitable because it acts on the opposite direction to desired one.

The power of the modelled expander was calculated for the rotors as the product of the previously obtained torque values and the angular velocity – see Figure 6. The forces acting on the rotors and the power calculated on their basis show cyclicality resulting from the constant rotational speed of the rotors and the regular opening and closing of the space between the rotors and the walls of the expander.

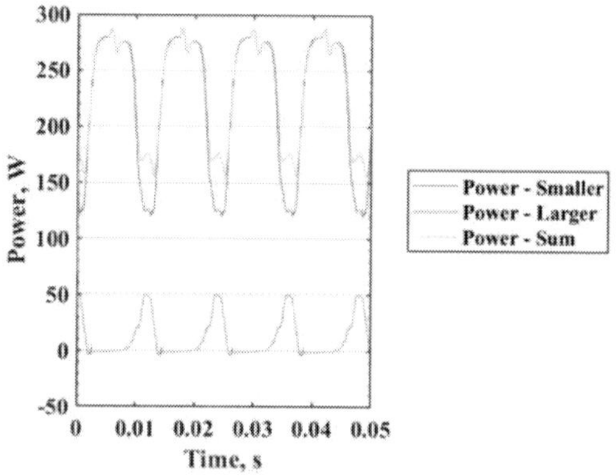

Fig. 6. Calculated expander power acting on rotors for the system thickness of 1 mm, inlet pressure of 8 bar and rotational speed of 1000 rpm.

The amplitude of the force and power values for the smaller rotor is much greater than for the larger rotor. This phenomenon occurs due to fewer lobes and uneven distribution of forces acting on the smaller wheel at the inlet and outlet of the expander.

Table 4. Results comparison with literature data.

Parameter	Unit	No. of simulation		
		1	2	3
Calculated average power on rotors	kW	9.97	23.45	9.86
Calculated average power of expander	kW	8.55	20.12	8.46
Expander's power according to literature data	kW	9.53	19.03	8.29
Error	%	10.28	5.73	2.05

To obtain the power value of the expander received at the output from the device $\dot{W}$, the calculated expander powers acting on the rotors $\dot{W}_R$ should be reduced by mechanical losses (mechanical efficiency of the gearing

https://doi.org/10.1051/matecconf/201824005015

η_g, mechanical efficiency of the smaller rotor η_{R1}, mechanical efficiency of the larger rotor η_{R2}) and losses resulting from medium leakages in the expander axial slots η_l. Therefore, the expander power was calculated based on the following formula:

$$\dot{W} = \dot{W}_R \eta_g \eta_{R1} \eta_{R2} \eta_l \tag{1}$$

The assumed final value of the expander's mechanical efficiency is 0.8579. As previously mentioned, the calculated power of expander was obtained for the thickness of 1 mm. To obtain the total power, the calculated value for the thickness of 1 mm was multiplied by the corresponding length of the expander presented in literature data. The comparison of calculated power with literature data is shown in Table 4.

The exemplary contours of velocity magnitude, contours of pressure with path lines and contours of pressure with velocity vectors are shown in Figures 10-15 for simulation no. 1 and 2. One can notice the leakage phenomena in the gaps between the rotors as well as smaller rotor and housing.

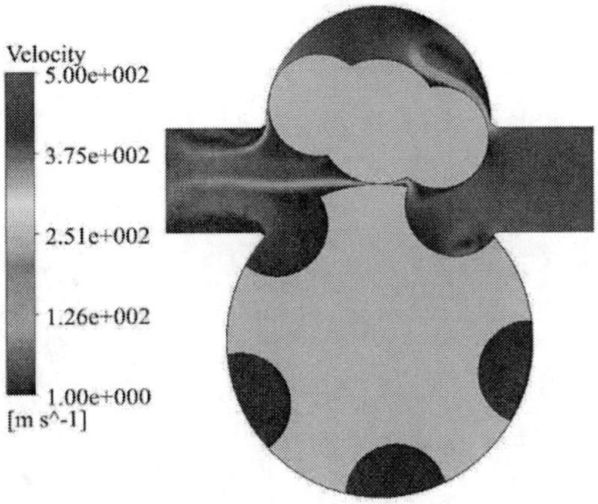

Fig. 7. Distribution of velocity magnitude in the expander for 0.005 s inlet pressure of 4 bar and rotational speed of 1000 rpm.

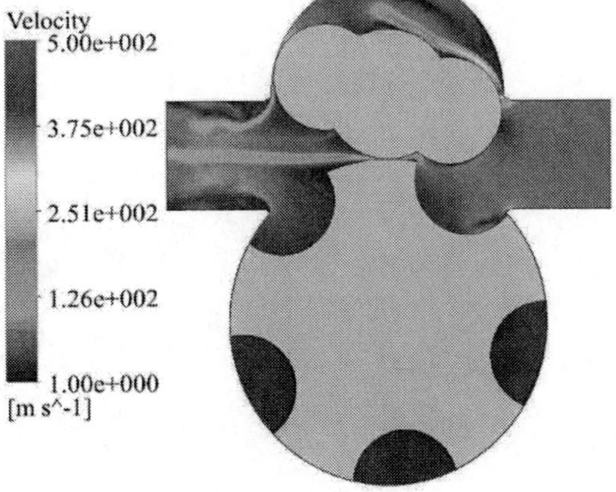

Fig. 8. Distribution of velocity magnitude in the expander for 0.005 s, inlet pressure of 8 bar and rotational speed of 1000 rpm.

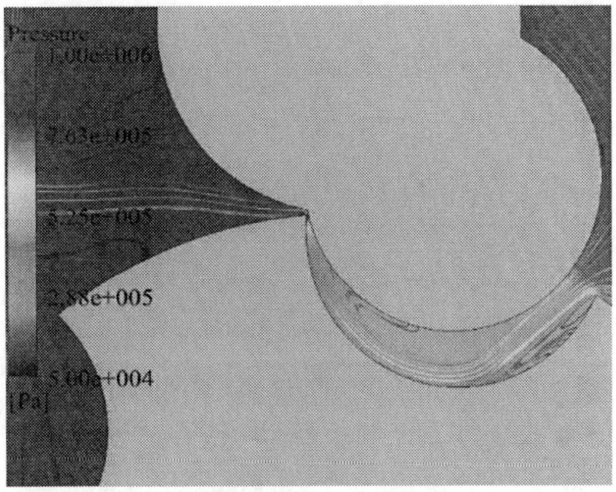

Fig. 9. Distribution of pressure and path lines in the gap between rotating wheels for 0.0025 s, inlet pressure of 8 bar and rotational speed of 1000 rpm.

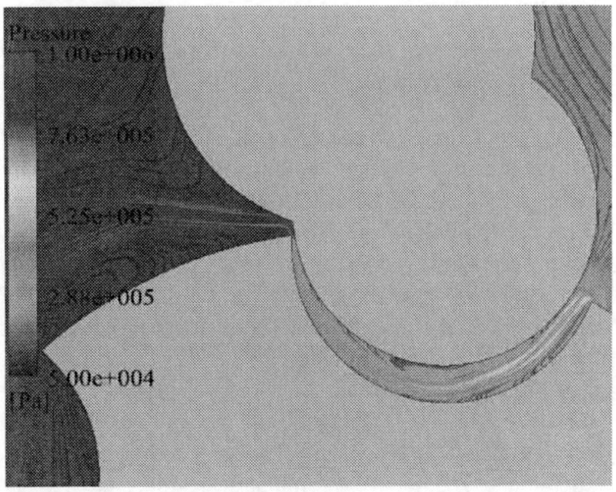

Fig. 10. Pressure distribution and path lines in the gap between rotating wheels for 0.0025 s, inlet pressure of 4 bar and rotational speed: 1000 rpm.

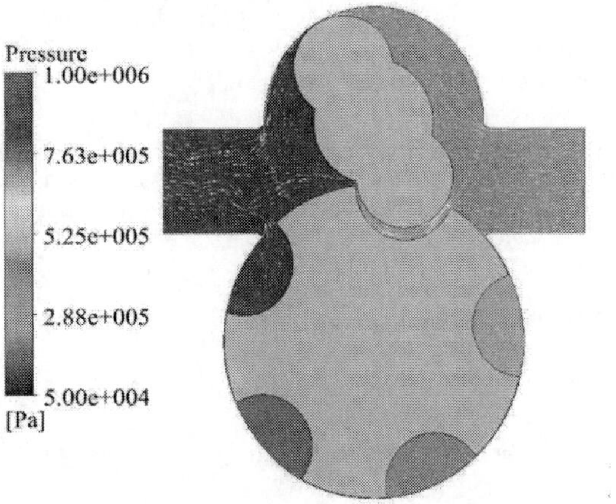

Fig. 11. Distribution of pressure and velocity vectors in the expander for 0.0025 s, inlet pressure of 4 bar and rotational speed of 1000 rpm.

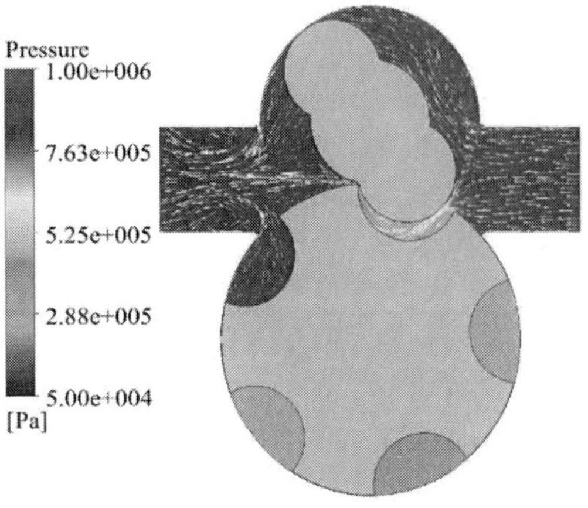

Fig. 12. Distribution of pressure and velocity vectors in the expander for 0.0025 s, inlet pressure of 8 bar and rotational speed of 1000 rpm.

2 Conclusions

In this paper the model of rotary lobe expander has been developed and then subjected to tests which validated the results obtained with data from literature. The model configuration and settings were selected in a way which ensured the highest possible accuracy and correctness of the obtained results as well as high computational efficiency. The validation activities were based on a device having a similar design for which geometrical and operational parameters are provided by the manufacturer. For the analysed cases (i.e. variable pressure of air inflow and variable rotational speed) satisfactory accuracy of simulated results with actual operational data was obtained. The maximal error was approximately 10 %.

It was shown that the developed model well simulates the operation of considered rotary lobe expander. Thus, it is ready to be used as a tool for conducting optimization in terms of geometry and parameters of the steam type expander.

This research is supported by the Programme of Research on the Electricity Sector which is co-financed by the European Union in the framework of the Smart Growth Operational Programme.

References

1. www.armakmotor.de
2. www.globe-benelux.nl
3. J. Schiffer and S. Klomberg, in *15th Int. Conf. Fluid Flow Technol.* (Budapest, 2012), pp. 1–8.
4. J. Vande Voorde, J. Vierendeels, and E. Dick, J. Comput. Appl. Math. **168**, 491 (2004).
5. ANSYS Inc., (2016).
6. P. A. C. Rocha, H. H. B. Rocha, F. O. M. Carneiro, M. E. Vieira da Silva, and A. V. Bueno, Energy **65**, 412 (2014).
7. F. R. Menter, Int. J. Comut. Fluid Dyn. **23**, 305 (2009).
8. D. Zhang, D. Pan, W. Shi, S. Wu, and P. Shao, Therm. Sci. **17**, 1551 (2013).
9. T.-H. Shih, W. W. Liou, A. Shabbir, Z. Yang, and J. Zhu, Compurers Fluids **24**, 227 (1995).
10. P. J. Singh and A. D. Onuschak, in *Int. Compress. Eng. Conf.* (1984), pp. 519–527.
11. W. He, Y. Wu, Y. Peng, Y. Zhang, C. Ma, and G. Ma, Appl. Therm. Eng. **51**, 662 (2013).
12. K. C. Ng, T. Y. Bong, and T. B. Lim, Heat Recover. Syst. CHP **10**, 119 (1990).
13. H. R. Nilsson, 2,975,963 (1961).
14. K. Kasuya, M. Fujiwara, M. Imai, T. Matsunaga, and Y. Takahashi, 4,401,420 (1980).
15. J. L. Bowman, 4,412,796 (1981).
16. H.-T. Lee, 4,890,992 (1990).
17. J. F. Gulich, *Centrifugal Pumps*, 3rd ed. (Springer, Berlin, 2014).

Review of audit calculation activities on the applicability of CFD software to nuclear safety problems

Gong-Hee Lee[1,2,*]

[1]Korea Institute of Nuclear Safety, Nuclear Safety Research Department, 34142 Daejeon, Republic of Korea
[2]University of Science and Technology, Nuclear and Radiation Safety Department, 34113 Daejeon, Republic of Korea

Abstract. The Korea Institute of Nuclear Safety (KINS) is tasked with the technical supporting for approving the safety of specific components or design modifications in nuclear power plant. When a licensee submits analysis to the Nuclear Safety & Security Commission (NSSC) for acceptance, KINS staff reviews this analysis and sometimes conducts independent audit calculations. Though recently licensing applications supported by using Computational Fluid Dynamics (CFD) software are increasing for nuclear safety problems, there is no CFD software which obtains a licensing from the domestic regulatory body until now. Additionally, there is no domestic regulatory guideline for the comprehensive evaluation of CFD software. Therefore, from the nuclear regulatory perspective, it is necessary to perform the systematic assessment and prepare the domestic regulatory guideline for checking whether valid CFD software is used for nuclear safety problems. In this paper, regulatory audit calculation activities on the applicability of CFD software to nuclear safety problems was briefly explained.

1 Introduction

The KINS is tasked with the technical supporting for approving the safety of specific components or design modifications in nuclear power plant. When a licensee submits analysis to the NSSC for acceptance, KINS staff reviews this analysis and at times conducts independent audit calculations. In some cases, CFD methods are used for these confirmatory studies which support safety evaluations, build confidence and reduce uncertainty in the final regulatory decisions. Typical examples are (1) thermal-flow analysis for high density spent fuel storage racks (2) estimation of moderator subcooling margin for Pressurized Heavy Water Reactor (PHWR).

Though recently licensing applications supported by using CFD software are increasing for nuclear safety problems, there is no CFD software which obtains a licensing from the domestic regulatory body until now. Additionally, there is no domestic regulatory guideline for the comprehensive evaluation of CFD software. Therefore, from the nuclear regulatory perspective, it is necessary to perform the systematic assessment and prepare the domestic regulatory guideline for checking whether valid CFD software is used for nuclear safety problems.

In this study, regulatory audit calculation activities on the applicability of CFD software to nuclear safety problems was briefly explained for the following items.

- Effect of asymmetric inlet flow rate on reactor core-inlet flow-distribution
- Single phase thermal stratification by emergency core cooling system injection
- Flow distribution inside a fuel assembly with twist-split type mixing vane

2 Case I: Effect of asymmetric inlet flow rate on reactor core-inlet flow-distribution

2.1 Overview

In general, an impeller of a Reactor Coolant Pump (RCP) installed in a Pressurized Water Reactor (PWR) is manually manufactured and therefore there may be a possibility of performance difference (about maximum 4% of design flow rate) per a RCP. Therefore, Kim et al. [1] measured reactor core-inlet flow-distributions under asymmetric inlet flow rate condition (case1 in Table 2) and compared it with those from symmetric inlet flow rate condition.

In this study, CFD analysis was conducted to verify the validity of the reactor internal flow distribution measured under asymmetric flow rate conditions by using ANSYS CFX and compared with the calculated results under the existing symmetric flow rate condition. In addition, we attempted to confirm whether asymmetrically entering flows due to the difference in RCP performance are properly mixed while passing through low support structures of a reactor, and provide core inlet flow rate distribution equivalent to symmetric flow rate conditions.

* Corresponding author: ghlee@kins.re.kr

2.2 Analysis model

APR+ Core Flow & Pressure Test Facility (ACOP), installed in the Korea Atomic Energy Research Institute (KAERI), is a 1/5 scaled-down model of APR+. It consists of a reactor vessel with two coolant loops (i.e., four cold legs and two hot legs). The scaling ratios applied to the test facility are summarized in Table 1.

Table 1. Summary of scaling parameters [2].

Parameters	APR+	ACOP
Temperature, ℃	310	60
Pressure, MPa	15	0.2
Density, kg/m^3	705.8	983.2
Viscosity, Ns/m^2	8.88×10^{-5}	4.66×10^{-4}
Length ratio	1	1/5
Area ratio	1	1/25
Volume ratio	1	1/125
Aspect ratio	1	1
Velocity ratio	1	1/2.17
Mass flow ratio	1	1/38.9
Core exit Re ratio	1	1/40.9
ΔP ratio	1	1/3.38

The internal structures of the reactor model (e.g., flow skirt and upper/lower core structures) had almost the same shapes as those in the original APR+, and satisfied geometrical similarity [1,2]. The core-inlet flow-rate distribution could be obtained by measuring the differential pressure and discharge coefficients at the venturi region of each core simulator. A total of 257 core simulators, which corresponded to the fuel assemblies, were installed in the reactor model. The upper head of the reactor, and some core-bypass flow-paths were neglected in the reactor model because these parts were expected to have little influence on the core-inlet flow-rate distribution. The criteria of the allowable data scattering for each core simulator inlet flow-rate distribution was ±1.5% [2].

The test matrix consists of three flow conditions, i.e., the symmetric or asymmetric flow conditions for 4-pumps operation, and the flow condition for 3-pumps operation. In this study, CFD simulation was conducted under the asymmetric flow conditions for 4-pumps operation (see Table 2). In Table 2, Asym. (case1~case3) correspond to asymmetric flow conditions for 4-pumps operation. In detail, case1 is real test condition and case2 & 3 are conditions for computation only.

Table 2. Inlet boundary conditions: flow rate (kg/s) at each cold leg (see Fig. 3(b) for the locations of each cold leg).

	Sym.	Asym. (case1)*	Asym. (case2)	Asym. (case3)
CL1A	135.0	136.1	138.1	140.9
CL1B	135.0	136.1	138.1	140.9
CL2A	**135.0**	**130.7**	**125.7**	**117.3**
CL2B	135.0	137.1	138.1	140.9
Total	540.0	540.0	540.0	540.0
Asymmetric	0%	5%	10%	20%

APR+ reactor internals are complex structures which support fuel assemblies, control rods and measuring instruments. The internal structures, especially those located in the upstream of the reactor core, may have a significant influence on the core-inlet flow-rate distribution; depending on both their shapes, and the relative distance between the internal structures and the core inlet [3]. Therefore, an exact representation of these internal structures is needed for CFD simulation of the core-inlet flow-rate distribution. However, such an approach requires a great deal of computing resources to analyze the real-flow phenomena inside a reactor.

In this study, as shown in Fig. 1, among the reactor internal structures located upstream of the reactor core, the real geometries of a flow skirt, Lower Support Structure Bottom Plate (LSSBP) and In-Core Instrumentation (ICI) nozzle support plate, were considered because these internal structures could significantly influence the flow-rate distribution at the core inlet. Meanwhile, to reduce total numbers of elements and thus minimize the required amount of computation, fuel assemblies and some internal structures (e.g., control-element guide tubes) were simply considered as each bulk volume (porous domain). Then, in order to reflect the velocity field and pressure drop occurring in the real-flow region; porosity and Isotropic Loss Models were applied to the porous domain.

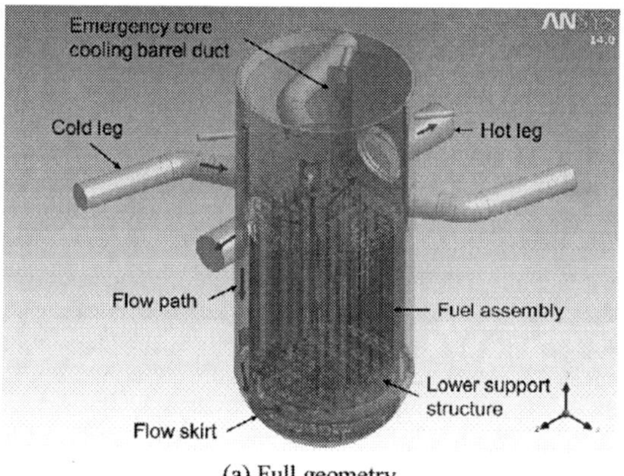

(a) Full geometry.

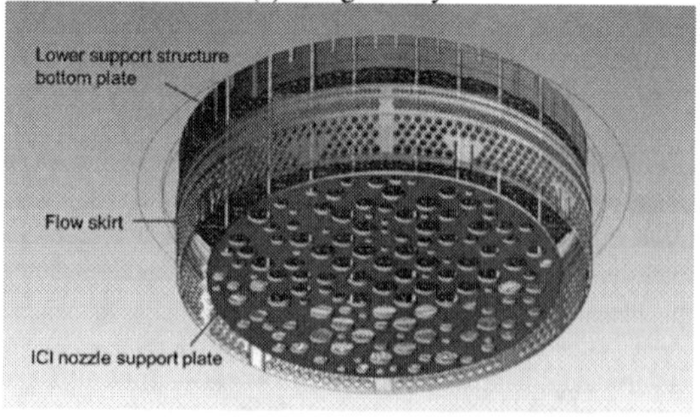

(b) Details of lower support structure.

Fig. 1. The computational domain.

Porosity is the ratio of the volume of fluid region to total volume; including both fluid and solid regions. It

 https://doi.org/10.1051/matecconf/201824005016

has an effect on flow acceleration in the porous domain. In this study, the porosity was determined by considering the real geometry of the reactor internal structures, and its magnitude was in the range of 0.5–0.75. A momentum source was used to model the momentum loss in the porous domain; which corresponds to a pressure drop in real reactor vessel. Loss coefficients were adjusted to match the magnitude of the pressure drop found in the porous domain, with those of the measurement.

2.3 Numerical modeling

The flow inside the scaled-down reactor model was assumed to be steady, incompressible, isothermal and turbulent. A high resolution scheme was used for both the convection-terms-of-momentum equations and - turbulence equations. The solution was considered 'converged' when the residuals of the variables were below 3×10^{-4}, and the variations of the target variables were small. Simulation was conducted with the commercial CFD software, ANSYS CFX R14.5.

The k-ε model, which is one of the most prominent Reynolds Averaged Navier-Stokes (RANS)-based turbulence models, was used to simulate the turbulent flow inside the scaled-down reactor. The reason is that this model has proven to be numerically stable and has offered a good compromise in terms of accuracy and robustness. In a previous study [4], turbulence models available in ANSYS CFX R13, for example k-ε model, Shear Stress Transport (SST) model, and SSG (Speziale, Sarkar and Gatski) Reynolds Stress model, were used to examine the turbulent flow inside the scaled-down reactor. Although the reactor internal-flow pattern differed locally; depending on the turbulence models used, the k-ε model showed the best agreement with the experimental data.

Fig. 2 shows the grid system for the computational domain that had the same size as the test facility. A hybrid mesh, made up of tetrahedrons, pyramids and prisms, was generated to prevent the oversimplification of the geometry, and to have more efficient mesh distribution. Prism layers were used to get higher resolution in the near-wall region. Total numbers of nodes were 7.3×10^7 and maximum y^+ in the downcomer was about 305.

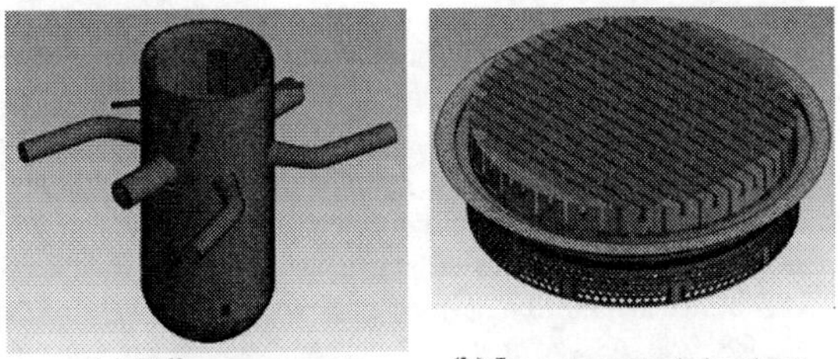

(a) Full geometry (b) Lower support structure and flow skirt

Fig. 2. Grid system.

By referring to the test conditions, as shown in Table 2, inlet flow-rates were imposed at each cold leg. Turbulence intensity at the inlet was assumed to be 5%. Light water at 60 ℃ was used as the working fluid. The 'average pressure over the whole outlet' option; with a relative pressure of 0 Pa, was used at each hot leg as an outlet-boundary condition. A no-slip condition was applied at the solid wall. To model the flow in the near-wall region, scalable wall functions were applied.

2.4 Results and Discussion

Fig. 3 shows the circumferential distribution of velocity (y-direction) in the downcomer (-0.6 m from the center of the cold leg). In the downcomer, velocity showed a non-uniform distribution; with peak negative magnitude between cold legs. In case of asymmetric flow condition, as the flow rate in a cold leg CL2A decreased, asymmetry of y-directional velocity profile between CL2A and other cold legs increased.

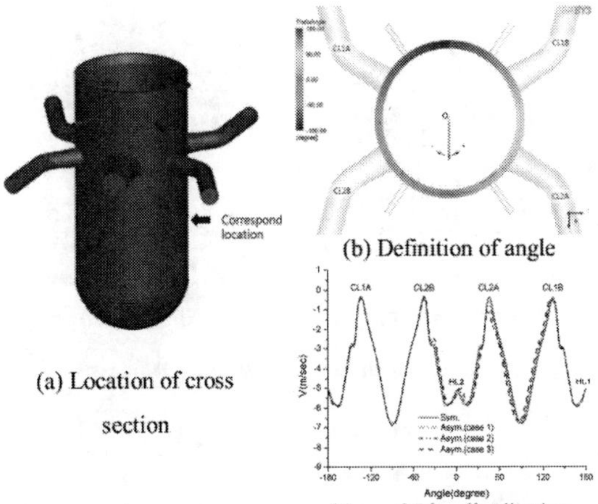

(a) Location of cross section (b) Definition of angle (c) y-velocity distribution

Fig. 3. Circumferential distribution of velocity (y-direction) in the downcomer (-0.6m from the center of cold leg).

Fig. 4 shows the normalized, mass-flow rate at the core-inlet plane. The magnitude is the ratio of the mass-flow rate per each fuel assembly, to the average mass-flow rate at the core-inlet plane.

The core inlet flow rate range calculated from the asymmetric flow rate conditions (case1~case3) was 80% ~ 141% of the average flow rate of the fuel assembly, which was the same as the calculation result at the symmetric flow rate condition. For the measured values as well, the core inlet flow rate range in the asymmetric flow rate condition was 86% ~ 126% of the average flow rate of the fuel assembly, which was the same as the measured flow rate at the symmetric flow rate condition.

Fig. 5 shows distribution of the normalized core-inlet mass-flow rate along the core centerline (A–A'). For calculation, though there was a minor difference (maximum 0.07) for certain fuel assemblies (no.74, 110 etc.), core inlet flow rate showed overall similar distribution regardless of asymmetric magnitude in an inlet flow rate. For measurement, although there was a minor difference (below 0.01) in a fuel assembly no.47, core inlet flow rate showed overall same distribution regardless of asymmetry in an inlet flow rate.

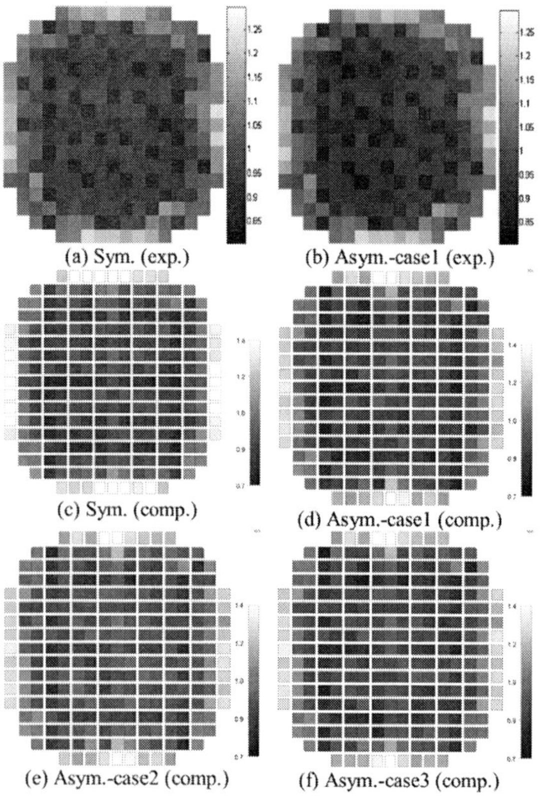

Fig. 4. Distribution of the normalized core inlet mass flow rate.

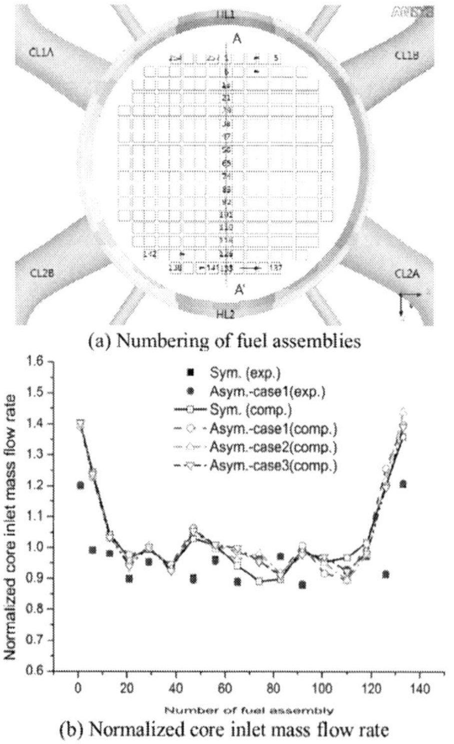

Fig. 5. Distribution of the normalized core inlet mass flow rate along core centerline (A-A').

In summary, asymmetrically entering flows due to RCP performance differences were properly mixed while passing through the internal structures installed at the lower part of the reactor, confirming that the core inlet flow distribution is equivalent to that under the symmetric flow condition.

3 Case II: Single phase thermal stratification by emergency core cooling system injection

3.1 Overview

Pressurized Thermal Shock (PTS) phenomenon is characterized by multi-dimensional and non-equilibrium flow conditions. For example, in case Emergency Core Cooling System (ECCS) is operated during Loss Of Coolant Accident (LOCA) in a PWR, PTS phenomenon occurs as cooling water is injected into a cold leg, mixed with the hot primary coolant, and then entrained into a reactor vessel. Insufficient flow mixing may cause temperature stratification and steam condensation. In addition, flow vibration may cause thermal stresses in the surrounding structures, which reduce the life of the reactor vessel. Because of this importance, the effect of PTS has been actively studied either experimentally or numerically [5]. Among these, the ROSA project [6,7] undertaken by the OECD/NEA includes six types of the separate or the integral effect tests. In this study, the calculation with ANSYS CFX R17.2 was performed for the Test 1「Temperature stratification and coolant mixing in unsteady state during ECCS injection」and then the predicted results were compared with the measured data.

3.2 Analysis model

The objective of Test 1 is to clarify the phenomenon of temperature stratification on the cold legs and the upper region of downcomer of the reactor vessel, which are important for the evaluation of PTS during the coolant injection by the ECCS, and to obtain the measured data for the main parameters with the aim of validating the simulation code and the numerical modeling. Fig. 6 shows the schematic diagram of the analysis model.

Test 1-1 (ST-NC-34 in JAEA) was performed as a separate effect test. The primary and secondary pressures under the single-phase natural circulation operating conditions were 15.5 MPa and 6.7 MPa, respectively, and the water level in the primary loop was maintained at 100% level. An inner diameter of a cold leg was 207 mm. The ECCS lines attached to the cold legs have different installation types and locations. While the ECCS line A was installed perpendicularly to the cold leg A, the ECCS line B formed a 45° angle with the cold leg B.

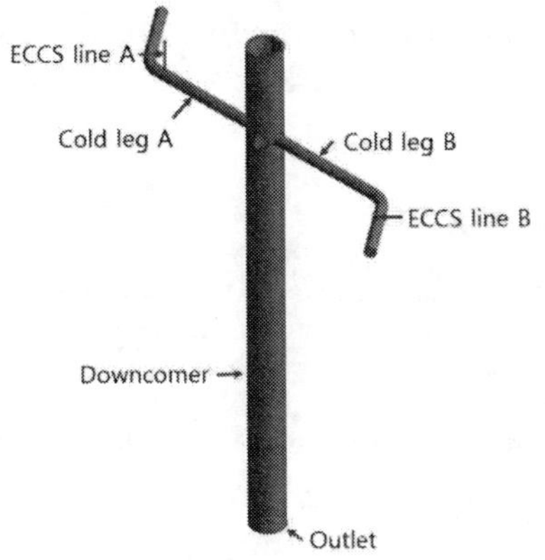

Fig. 6. Computational domain.

As shown in Fig. 8, 9 and 10(a), to investigate the thermal stratification phenomenon in the cold leg, 21 thermocouples were installed in 3 columns and 7 rows at two cross sections (TE2, TE3) in the cold leg between the ECCS line A and the downcomer. The nominal accuracy of the thermocouple measurements is ± 2.75 K. In this study, the measured and the calculated results were compared for the cases where the cold coolant was injected through the ECCS line A during time intervals of 2,091 s ~ 2,187 s under the natural circulation condition after the main coolant pump stopped.

3.3 Numerical modeling

In this study, the turbulent flow field inside the ROSA/LSTF (Large Scale Test Facility) was calculated under unsteady, single-phase, incompressible and non-isothermal conditions by using the commercial computational fluid dynamics software ANSYS CFX R17.2. High resolution scheme with quasi-second-order accuracy was used for the convection-terms-of-momentum and -turbulence equations. The second order backward Euler option, the default implicit time step method with the second order accuracy in ANSYS CFX R17.2, was used as the time discretization scheme. For the calculation of the unsteady flow, the calculation was performed during time intervals of 2,078 s ~ 2,218 s under the natural circulation condition with a time step of 0.1 s. It was determined that the solution was converged when the root mean square residuals of the individual equations were 10^{-5} or less at each time step. Baseline Reynolds stress model was used to simulate the turbulent flow inside the ROSA/LSTF. To model the flow in the near-wall region, the automatic wall treatment was applied.

Fig. 7 shows the grid system, generated by using ICEM-CFD software, for the computational domain that had the same size as the test facility. The grid shape was hexahedral, and the total number of elements used in the calculation was 1.99×10^6. In the computational domain, the maximum y^+ was about 411, and its location was near the region where the inner wall of the pressure vessel and the hot leg were bordered.

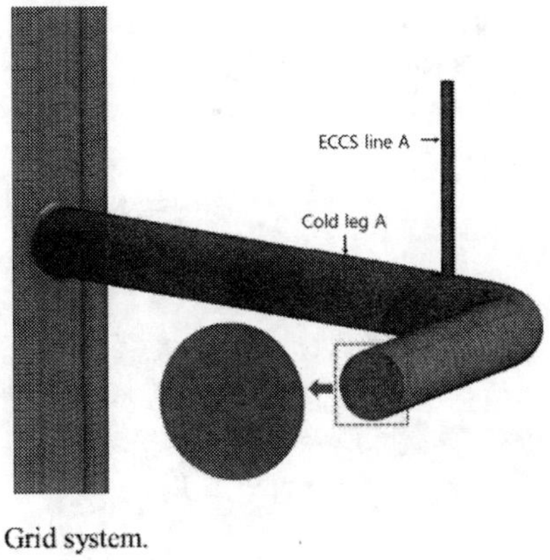

Fig. 7. Grid system.

The measured unsteady flow rate and temperature were applied at inlet of the cold leg A/B and the ECCS injection line A. For reference, the temperature difference between the hot coolant at the inlet of the cold leg A/B and the cold cooling water at the inlet of the ECCS injection line A was 200 K or more. On the other hand, the turbulence intensity at each inlet was assumed to be 5.0%. The average static pressure condition was applied at the outlet. No-slip and adiabatic conditions were applied at all wall interfaces including the pressure vessel.

3.4 Results and Discussion

Fig. 8 shows the streamline of the cold cooling water injected at inlet of the ECCS injection line A and the distribution of fluid temperature at the measurement sections (TE2, TE3) of the cold leg A.

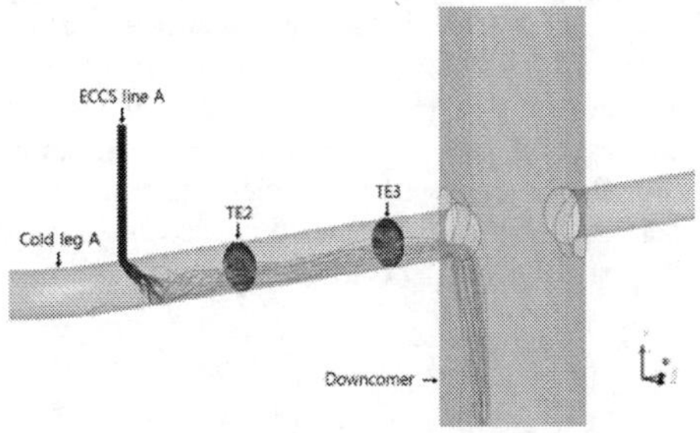

Fig. 8. Streamlines and contours of fluid temperature at cross-sectional plane TE2 and TE3 of the cold leg A (at t=2,128 s).

Cold cooling water injected at the inlet of the ECCS injection line A moved toward the lower part of the cold leg A and then flow mixing with the hot coolant occurred. The mixed flows proceeded along the cold leg A and then entered into the downcomer along the outer wall of the pressure vessel.

To identify the thermal stratification region, wall surface temperature distribution was shown in Fig. 9. The strongest thermal stratification was found at the location where the cold cooling water injected at the inlet of the ECCS injection line A contacted with the lower part of the cold leg A. In addition, as previously explained in Fig. 8, because the mixed flow moved along the lower part of the cold leg A and then entered into the downcomer along the outer wall of the pressure vessel, the thermal stratification phenomena occurred in these regions.

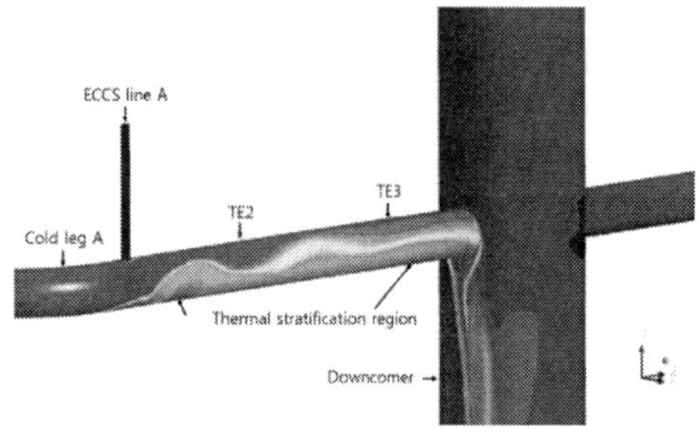

Fig. 9. Contour of wall surface temperature to demonstrate the thermal stratification region (at t=2,128 s).

Fig. 10 shows a comparison of the calculated and the measured temperature at three locations (TE2A1, TE2A4, TE2A6) in the cold leg A section (TE2) where are 1.6335 m from the pressure vessel center. The section (TE2) is located much closer to the ECCS injection line A in the downstream direction (farther away from the pressure vessel) in comparison with the section (TE3) shown in Fig. 8 and 9. Therefore, it is expected that the relatively weaker flow mixing between the cold cooling water injected at the inlet of the ECCS injection line A and the hot coolant occurs in comparison with the section (TE3), and, as a result, the measured temperature at three positions of the section (TE2) showed the obvious temperature gradient at three locations in the cross-section TE2 during the ECC water injection period. Though there were a little difference between the measured and computational results, the predicted fluid temperature was relatively in good agreement with the measured data during the ECC water injection period. Due to the non-disclosure limitations of measurement data, quantitative values of fluid temperature corresponding to the vertical axis of the graph (Fig. 10(b)~(d)) were not specified.

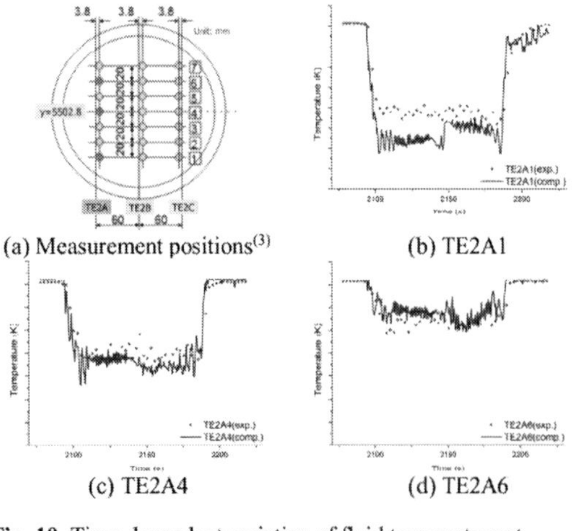

Fig. 10. Time-dependent variation of fluid temperature at cross-sectional plane TE2 in the cold leg A.

4 Case III: Flow distribution inside a fuel assembly with twist-split type mixing vane

4.1 Overview

In a PWR, the appropriate heat removal from the surface of fuel rod bundle is important for thermal margins and safety. A spacer grid that supports the fuel rods in a fuel assembly is equipped with mixing vanes that play a role in improving the heat transfer from the hot surfaces of the fuel rods to the coolant flow as the turbulence-enhancing devices.

As shown in Fig. 11, flow patterns generated by a spacer grid with mixing vanes generally consist of both the swirl flow inside subchannel and the cross flow at the fuel rod gaps. The swirl flow improves the heat removal at the fuel rod surface by mixing the hot water near the fuel rod surface with the relatively cold water at the subchannel center, while the cross flow contributes to mitigating the hot peaking of subchannels by exchanging the enthalpy between subchannels [9]. Therefore, the geometrical shape and arrangement of the mixing vane are important factors that determine the performance of the mixing vane.

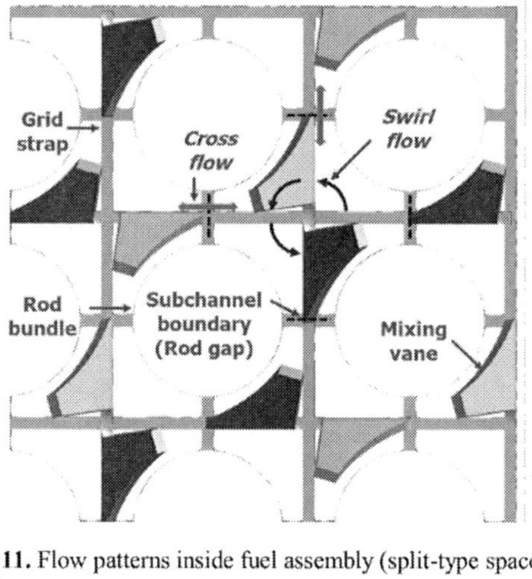

Fig. 11. Flow patterns inside fuel assembly (split-type spacer grid).

Because a spacer grid may cause rigorous mixing as well as greatly increased local turbulence levels inside the sub-channel, prediction of sub-channel flows, even in isothermal condition, is very difficult. In general, sub-channel analysis codes such as COBRA or VIPRE have been used to predict the flow and enthalpy distributions within fuel assemblies. However, these sub-channel codes rely on geometrically dependent mixing factors and empirical correlations to close the governing equations. The advantage of a CFD software for sub-channel flow predictions is that it does not rely to the same extent on these empiricisms. Therefore, CFD results have the potential for wider applicability to capture the essential features of the turbulent structures downstream of the spacer grid.

In this study, in order to examine the flow distribution inside a fuel assembly with twist-split type mixing vanes as a part of benchmark simulation for a Coordinated Research Project (CRP) on the application of CFD codes to Nuclear Power Plant (NPP) design, simulations were conducted with the commercial CFD software, ANSYS CFX R18.1. The predicted results were compared with the measured data from the Omni Flow Experimental Loop (OFEL) test facility.

4.2 Analysis model

As shown in Fig. 12, OFEL test facility [10] was used to measure the flow distribution inside a fuel assembly with twist-split type mixing vane. Test rig consisted mainly of a water storage tank, a centrifugal pump, a flow meter, and a test section. Test section was made up of a square housing, fixing plate, rod bundle, rod support, and spacer grid with twist-split type mixing vane. Length and outer diameter (D) of a fuel rod, and rod-to-rod pitch (P) were 2,000 mm, 25.4 mm, and 34.29 mm, respectively. Therefore, P/D was 1.35 which corresponded to the regular rod-bundle configuration.

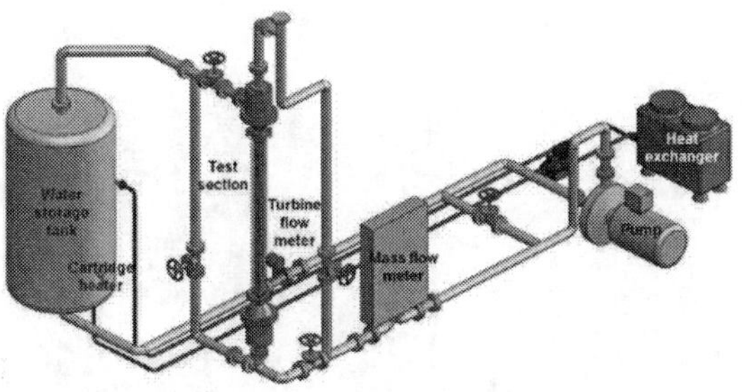

Fig. 12. Schematic diagram of OFEL facility [3].

As shown Fig. 13, velocity distribution inside the subchannel was measured at several cross-sectional planes by using either PIV (Particle Image Velocimetry) or LDV (Laser Doppler Velocimetry).
Inlet temperature and pressure condition of the working fluid were measured at 35℃ and 0.1 MPa, respectively. The bundle-average axial velocity was estimated to be 1.5 m/s.

4.3 Numerical modeling

The flow inside the fuel assembly was assumed to be unsteady, incompressible, isothermal and turbulent. A high resolution scheme for the convection-terms-of-momentum and -turbulence equations was used. 2nd Order Backward Euler scheme was used for the transient term. A time step of 0.001s was used with the maximum 10 iterations per time step. The solution was considered to be 'converged' when the residuals of variables were below 10^{-5} and the variations of the target variables were small. Simulation was conducted with the commercial CFD software, ANSYS CFX R18.1.

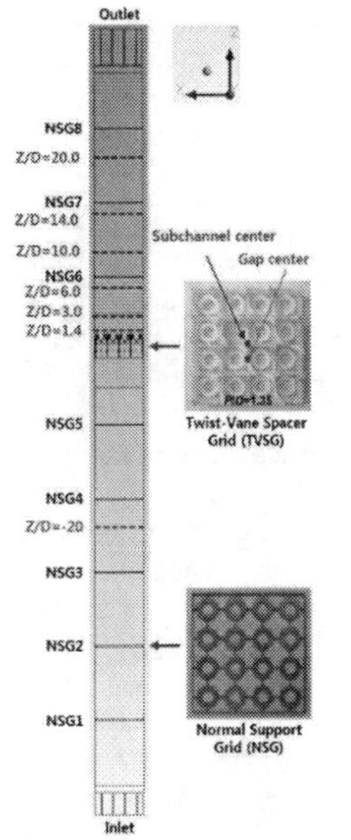

Fig. 13. Schematic diagram of test section and measurement positions for velocity components.

The ReNormalization Group (RNG) k-ε model was used to simulate the turbulent flow inside a fuel assembly. This model is based on renormalization group analysis of the Navier-Stokes equations. The transport equations for turbulence generation and dissipation are the same as those for the standard k-ε model, but the model constants differ. In general RNG k-ε model offers little improvement compared to the standard k-ε model.
Fig. 14 shows the grid system for the computational domain that had the same size as the test facility. A hybrid mesh, made up of tetrahedrons, wedges, pyramids and hexahedrons, was generated to prevent the oversimplification of the geometry, and to have more efficient mesh distribution. Prism layers were used to get higher resolution in the near-wall region. Total numbers of nodes 4.0×10^{7}.

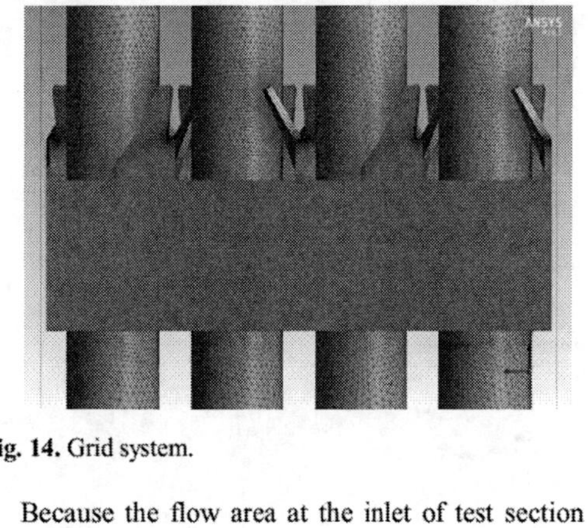

Fig. 14. Grid system.

Because the flow area at the inlet of test section is different from that in the flow passage, uniform velocity, calculated by considering the above-mentioned condition, was used as an inlet-boundary condition. The 'average pressure over the whole outlet' option; with a relative pressure of 0 Pa, was used as an outlet-boundary condition. A no-slip condition was applied at the solid wall. To model the flow in the near-wall region, the scalable wall function method was applied.

4.4 Results and Discussion

Fig. 15 shows the comparison of axial velocity profiles, measured and calculated along the z-axis (downstream of mixing vane) at both the subchannel center and the gap center of rod-to-rod. Both positions were indicated by the solid circles in Fig. 13. As shown in Fig. 15(a), defect (i.e. minimum) of the calculated axial velocity close to the mixing vane tip in the subchannel center agreed well with measurement. As the flow moved further downstream, the calculated axial velocity was slightly overestimated in the central sub-channel; compared with measurement. Additionally, at the gap center of rod-to-rod, though there was a little under-estimation in the velocity magnitude, the calculated axial velocity showed qualitatively similar profile in comparison with measurement (see Fig. 15(b)).

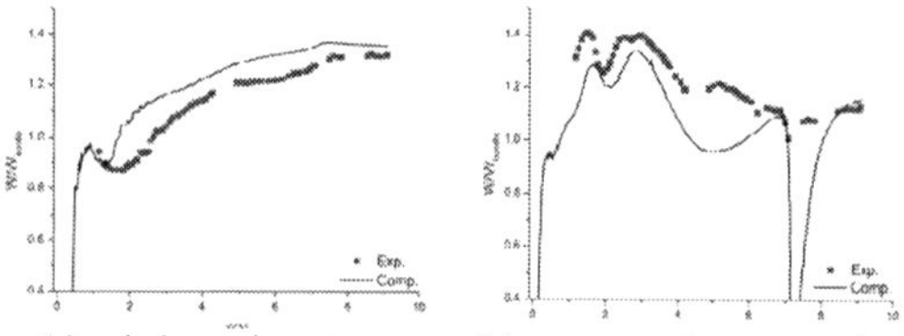

(a) subchannel center (b) gap center between rods

Fig. 15. Streamwise variation of axial velocity profile in the central subchannel.

Fig. 16 shows the lateral (or vertical) velocity (y-axis velocity component) distribution along a horizontal line in the central subchannel (see horizontal dashline in Fig. 17). As the flow moved further downstream, the positive (+) and negative (-) peak lateral velocity gradually decreased. This trend indicates that the strength of swirl flow gradually decays in the downstream region. The calculated velocity profiles showed similar patterns in comparison with measurement.

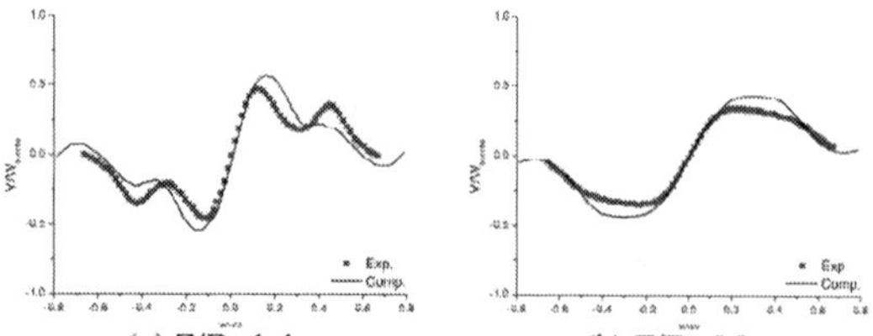

(a) Z/D=1.4 (b) Z/D=6.0

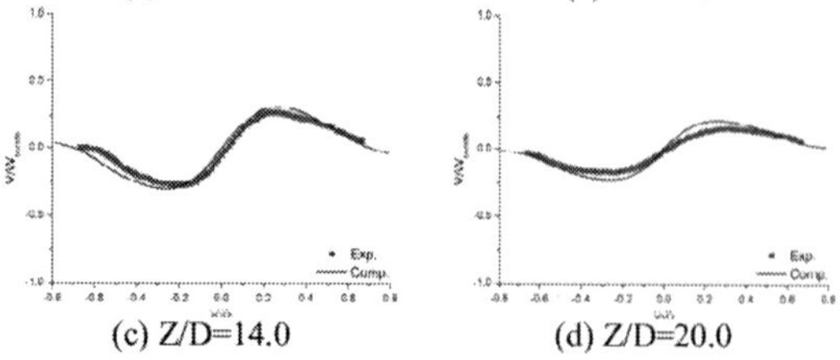

(c) Z/D=14.0 (d) Z/D=20.0

Fig. 16. Lateral velocity profile along the centerline in the central subchannel per the selected cross-sectional planes.

Fig. 17 shows the comparison of velocity vector in the central subchannel. At the location Z/D = 1.4, close to the mixing vane tip, a large elliptic vortex was generated in the subchannel center and two small secondary vortex appeared in the peripheral region near the upper and lower gaps between rods (see Fig. 17(a)). At the location Z/D = 3.0, major axis of a large elliptic vortex in the subchannel center rotated about 45° counter-clockwise and two small secondary vortex disappeared (see Fig. 17(b)). Finally, at the location Z/D = 6.0, the shape of central vortex became nearly circular (see Fig. 17(c)). The calculated velocity vector showed similar flow patterns in comparison with measurement.

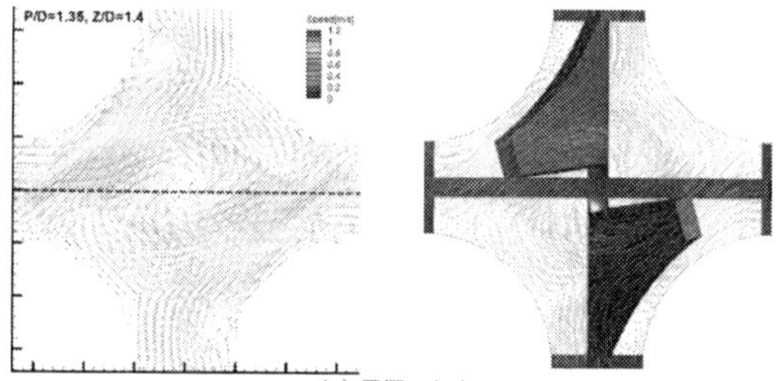

(a) Z/D=1.4

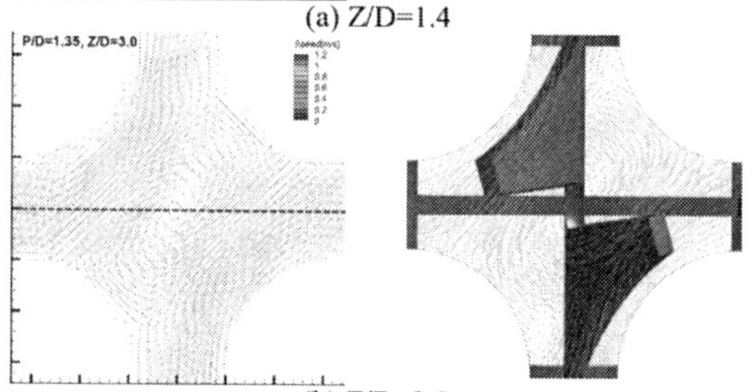

(b) Z/D=3.0

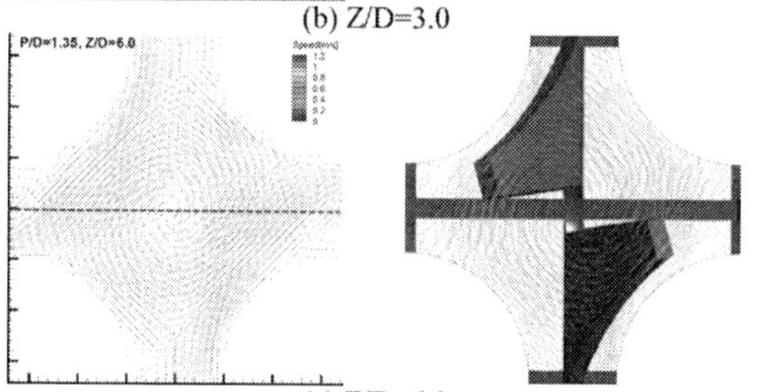

(c) Z/D=6.0

Fig. 17. Velocity vector in the central subchannel per the selected cross-sectional planes (left: experiment, right: computation)

5 Conclusions

In this study, regulatory audit calculation activities on the applicability of CFD software to nuclear safety problems was briefly explained. The major conclusions can be summarized as follows.

- Asymmetrically entering flows due to RCP performance differences were properly mixed while passing through the internal structures installed at the lower part of the reactor, confirming that the core inlet flow distribution is equivalent to that under the symmetric flow condition.

• The strongest thermal stratification was found at the location where the cold cooling water injected at the inlet of the ECCS injection line A contacted with the lower part of the cold leg A. Though there were a little difference between the measured and computational results, the predicted fluid temperature was relatively in good agreement with the measured data during the ECC water injection period.
• As the flow passing through a mixing vane moved downstream, vortex shape in the subchannel center changed from elliptic to circular type. The calculated velocity vector showed similar flow patterns in comparison with measurement.
• Although state of the art CFD software may solve reasonably these nuclear safety problems, it still has the limitation and a certain level of uncertainty in the calculation result. Main reasons may be grid quality, turbulence model and so on.
• Therefore, lessons learned from these audit calculations are being used to prepare the domestic regulatory guideline on the applicability of CFD software.

Disclaimer

The opinions expressed in this paper are those of the author and not necessarily those of the Korea Institute of Nuclear Safety (KINS). Any information presented here should not be interpreted as official KINS policy or guidance.

Acknowledgements

This paper was written by using the experimental data from OECD/NEA ROSA Project (CSNI2009/01). The authors also gratefully thank Dr. In, Wang-Kee in the KAERI for providing the OFEL experimental data and giving the valuable technical comments.

References

1. K.H. Kim, D.J. Euh, I.C. Chu, Y.J. Youn, H.S. Choi, T.S. Kwon, Nucl. Eng. Des. **265** (2013)
2. D.J. Euh, K.H. Kim, J.H. Youn, J.H. Bae, I.C. Chu, J.T. Kim, H.S. Kang, H.S. Choi, S.T. Lee, T.S. Kwon, Nucl. Eng. Tech. **44** (2012)
3. B.J. Lee, H.C. Jang, J.S. Cheong, S.J. Baek, Y.S. Park, J. Kor. Nucl. Soc. **33** (2001)
4. G.H. Lee, Y.S. Bang, S.W. Woo, A.J. Cheong, ICONE22-31255 (2014)
5. T. Farkas, I. Toth, Nucl. Eng. Des. **240** (2010)
6. H. Nakamura, T. Watanabe, T. Takeda, Y. Maruyama, M. Suzuki, Nucl. Eng. Tech. **41** (2009)
7. Japan Atomic Energy Agency, *Final Data Report of OECD/NEA ROSA Project* (Test 1-1, 2008)
8. ANSYS CFX, Release 17, ANSYS Inc.
9. S.-K. Chang, S. Kim, C.-H. Song, Nucl. Eng. Des. **279** (2014)
10. C.-H. Shin, C. Lee, C.-Y. Lee, W.-K., *LWR Fuel Development Division* (KAERI, 2015)

Verification of the heat transfer model for screw reactor

Anatoliy Levin*, *Alexander* Kozlov, *Denis* Svishchev, and *Maxim* Penzik

Melentiev Energy System Institute SB RAS, 130, Lermontov St., Irkutsk, 664033, Russia

Abstract. At present, the screw reactor pyrolysis is a promising and rapidly developing technology, because it helps obtain the end product, which is charcoal, tar, and synthesis gas. Additionally, pyrolysis can be considered as a preliminary stage for making the charcoal of the preset conversion for its further gasification when obtaining coal-derived gas. This research aims at establishing relationships between the operational parameters of pyrolysis and obtaining of the charcoal of preset conversion by means of mathematical modeling, full-scale experiment, and instrumental studies. Such comprehensive research ensures the accurate initial data, verification of the model, and optimal parameters for obtaining the product with preset properties as well as for further scaling of the reactor. In terms of design the pyrolyzer represents a recuperative heat exchanger where a mix of stack and recirculation gases is considered as a heat carrier. To prevent sintering of particles the screw transportation of fuel is used. The Comsol Multiphysics Software is used as a simulation environment. Heat exchange in the process of pyrolysis is simulated considering physical properties (porosity, permeability, etc.) of the medium. The constructed calculated mesh consisted of 604 thousand elements of three types (tetrahedrons, prisms and pyramids) and had the minimum size of 0.2 mm.

1 Introduction

According to the International Renewable Energy Agency (IRENA), in 2015 biomass generated 476933 GWh of electricity, which represents 8.6% of total power generated by renewable energy sources (5537516 GWh). At the same time, the growth of electricity produced by the biomass processing technologies from 2005 (195749 GWh) to 2015 was about 60% [1]. The main technology for obtaining electricity from biomass is thermochemical conversion. It accounts for about 75% of the electricity generated. The key reasons for the increase in the use of thermochemical conversion technologies were: increased energy consumption in developing countries, which led to the growth and significant fluctuations in world prices for hydrocarbons; considerations of energy security and general trends towards diversification of energy carriers in the fuel and energy balance; the new concept of distributed generation of energy, suggesting a reduction in the transport range by using local fuels; environmental safety that implies the use of renewable solid fuel resources, including waste [2, 3]. At present time, 160 projects for the design and construction of stations using the technology of biomass thermochemical conversion are being implemented by the international energy agency in the member countries of the Organization for Economic Co-operation and Development (OECD) [4]. However, the high-quality processing of biomass and waste has been only implemented as part of large installations with a unit capacity of more than 30 MW, and their application is unprofitable if extraction and transportation of fuel are costly [5]. To solve the problem of efficient use of such fuels, it is necessary to develop a set of technical solutions aimed at obtaining generator gas of acceptable quality and ensuring the production of electric and thermal energy at sufficiently low cost. The latter condition is the main impediment in resolving this issue in Russia and in other countries [6].

A substantional share of costs of the installed capacity as well as of operating costs of the realized conversion units is related to producer gas purification systems. In this regard, it is necessary to develop either existing gas cleaning systems or design alternative technologies for biomass thermochemical conversion, which by default imply the production of gas of acceptable quality for use in heat engines. One of such technologies is the technology of multi-stage gasification, which, due to the step-by-step organization of the process, allows for production of gas with a minimum content of tar products [7]. The key step of the multistage biomass gasification is pyrolysis, which, as a rule, is organized in a separate screw reactor, since it is pyrolysis that predetermines the further effectiveness of gasification. By varying the operating parameters of the pyrolysis process, one can vary both the composition of the producer gas and the yield of the target product (gas, tar, charcoal) [8]. As noted in [9], for all the universality of the screw reactor, its scaling seems to be a rather challenging task. This is confirmed by the fact that the multi-stage gas generator of the Viking type was scaled only up to 250 kW [10]. Apparently, the successful use and scaling of the screw pyrolyzer depend on a comprehensive study of the physicochemical properties of the biomass and the operating parameters of the process. It is also important to develop mathematical models that imitate the process under study and predict its progress. This work presents

* Corresponding author: lirt@mail.ru

the results of experimental research that was carried out at the laboratory stand of biomass pyrolysis in a screw reactor and provides results of mathematical modeling and verification of the model developed.

2 Experimental part

The experimental study was carried out at the laboratory stand of a screw pyrolyzer, the technological scheme of which is given at Fig. 1. The test material was wood pellets with the following charateristics: 6-8 mm in diameter; 2-5 cm long; ultimate and proximate analysis: C – 48 %, H – 4,6 %, O – 47,5%; devolatilization 81,7 %, ash content 2,2 %, moisture 8,2 %.

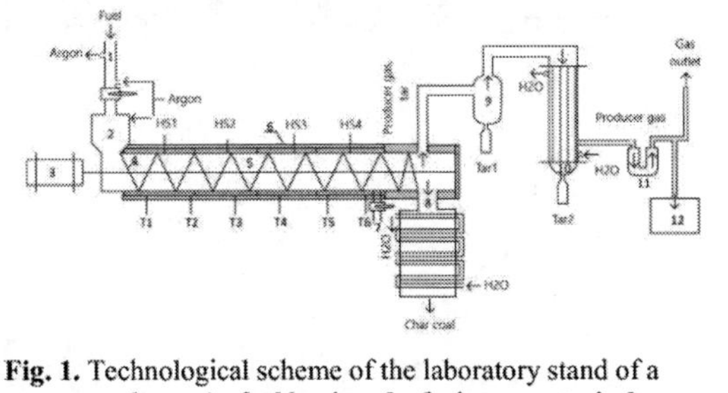

Fig. 1. Technological scheme of the laboratory stand of a screw pyrolyzer: 1– fuel bunker, 2– fuel storage tank, 3 – screw drive, 4 – screw, 5 – reactor, 6 – thermal isolation, 7 – sampler, 8– charcoal storage tank, 9– cyclon, 10 – heat exchanger, 11 – filter, 12 – gas chromatograph, T1-T6 – thermocouples, HS1-HS4 – heater sections.

The pellets in the reactor shaft (5) were moved by means of a screw (4) rotated by a variable speed drive (3). In the experiment, three rates of pellet transfer along the length of the reactor were used: 0.19, 0.37, and 0.6 mm/s. The wood pellets were heated by four sections of the electric heater HS1-HS4 located on the outer surface of the reaction chamber. Figure 2 shows the heating of the pyrolysis reactor at different fuel velocities.

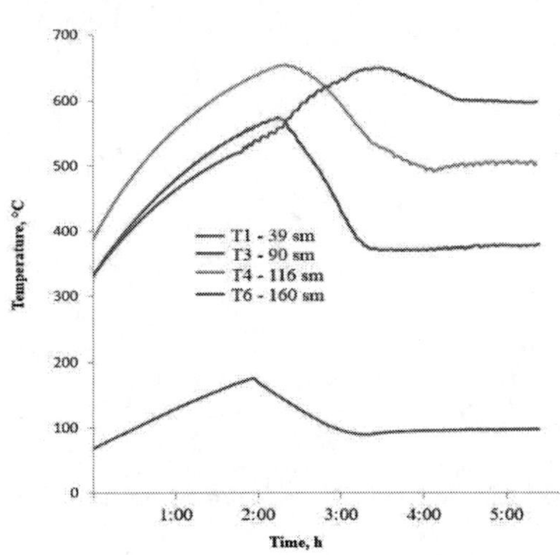

Fig. 2. Heating of the pyrolysis in different sections at the fuel velocity 0.37 mm/s

It can be seen from Fig. 2 that the initial warm-up stage lasts for two hours, and then we observe a decrease in the temperatures of the sections. It takes the screw pyrolyzer an hour to reach the stationary state.

To estimate the yield of producer gas, we used a balance for the inert gas-marker (argon), a certain amount of which was continuously supplied to the unit.

The hopper was purged with argon to remove air in the fuel mass. Then the fuel enters the storage tank (2) through an airproof gateway. To calculate the pellet consumption, the time of arrival of each portion of fuel into the storage tank was fixed.

The bulk of the reacted wood pellets enters the cooled charcoal storage tank (8). Sampling of charcoal to determine its technical characteristics (moisture, ash content, volatilization, elemental composition, heat of combustion) is carried out through the sampler (7). The producer gas enters the cyclone (9), in which solid particles and high boiling tar (Tar 1) are separated. The producer gas is cooled in the heat exchanger (10), where a low-boiling tar (Tar 2) is separated, and then passes through the filter (11). The yield of liquid products of the wood pyrolysis is estimated by the weight of the tar samples taken from the cyclone (9) and the heat exchanger (10). The thermocouples T1-T6 were used to measure the temperature of the pyroliser wall. The thermocouples that measured the layer temperature were immersed into the layer through special ports in the chamber wall. As a result of the experiment, we obtained the charcoal with a maximum conversion rate of 97% at the fuel velocity of 0.19 mm/s. We also drew the diaghram of the change in the bulk density of wood pellets depending on different degrees of conversion (Fig. 3).

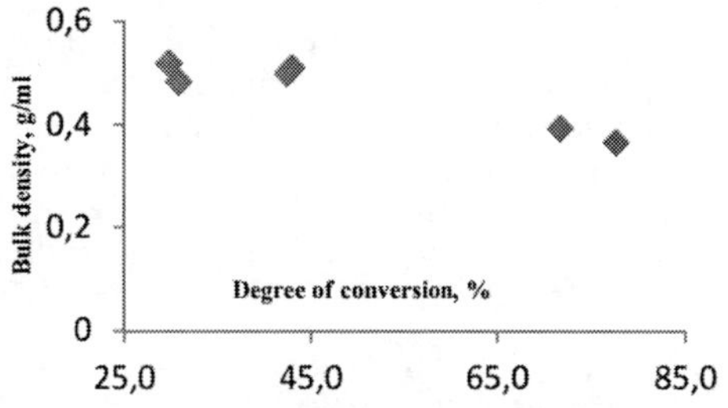

Fig. 3. Change in the bulk density of wood pellets depending on different degrees of pellets conversion.

Figure 3 shows that the bulk density varies insignificantly for samples with a degree of conversion up to 50%. The explanation for this is as follows: when pellets are pyrolyzed to a given degree of conversion, the geometric dimensions of the particles are preserved. It is worth noting that this dependence is further used in the mathematical model describing the pyrolysis process.

3 Mathematical model

We used the software based on the corresponding algorithms that ensure good convergence for the differential equations. The calculation mesh includes different number of cells:

- For gas part it includes hot gas from internal combustion engine and gas inside the screw above the porous medium.
- For a surface of solid elements, it includes all solid surfaces.
- For the others it includes solid volumetric parts such as screw, spiral, coating, etc.

The Comsol Multiphysics Software is used as a simulation environment. Heat exchange in the process of pyrolysis is simulated considering physical properties (porosity, permeability, etc.) of the medium [3]. The equation of a diffusion barrier for the plane separating the porous solid body from the gas space in the pyrolyzer is set as follows:

$$-nD_{s,i}\nabla c_{i,u} = \frac{D_{s,i}}{d_s}\left(c_{i,u} - c_{i,d}\right)$$
$$-nD_{s,i}\nabla c_{i,d} = \frac{D_{s,i}}{d_s}\left(c_{i,u} - c_{i,d}\right) \quad (1)$$

where d_s – barrier thickness, mm; D_{si} – diffusion coefficient, m^2/s. In the calculations we assumed a bed thickness of 5 mm and a diffusion coefficient of 10^{-6} m^2/s. The boundary conditions chosen to solve the problem are:

An inlet temperature of fuel is 20°C;
An inlet temperature of heating gases is 650°C;
A velocity of fuel in pyrolyzer 0,2÷3 mm/s;
An inlet velocity of heating gases is 0,5 m/s.

We calculated the heat required for the water evaporation by determining the concentration of the resulting vapors:

$$\nabla\cdot(-D_{H20}\nabla c_{H20}) + u\cdot\nabla c_{H20} = 70200e^{-\frac{34200}{8.314T}}. \quad (2)$$

Conversion of the solid fuel to coke was found using the equation

$$\nabla\cdot(-D_{cok}\nabla c_{cok}) + u\cdot\nabla c_{cok} = 54500e^{-\frac{96000}{8.314T}}. \quad (3)$$

The coefficients of the right-hand side of Eqs. (2), (3) were determined by the thermogravimetric analysis performed on solid fuel samples. It follows from Eq. (3) that the degree of fuel conversion is directly affected only by the temperature distribution, since the movement speed of solid particles is determined only by the rotation speed of the screw.

4 Analysis and modelling

As a result of parametric calculations within the above-mentioned range for the solid fuel velocity, we obtained distributions of the speed, temperature, and concentration fields of the substances formed. Figure 4 shows an example of spatial distribution of the coke concentration expressed in terms of the maximum achievable value (%). The spatial inhomogeneity of the conversion can be explained by the essential dependence of the temperature fields on the geometry of the screw, which has a high thermal conductivity. The influence of the fuel velocity on the solid fuel conversion is illustrated by Figure 5. It can be clearly seen that most of the wood bulk turns into coke halfway through the pyrolyzer.

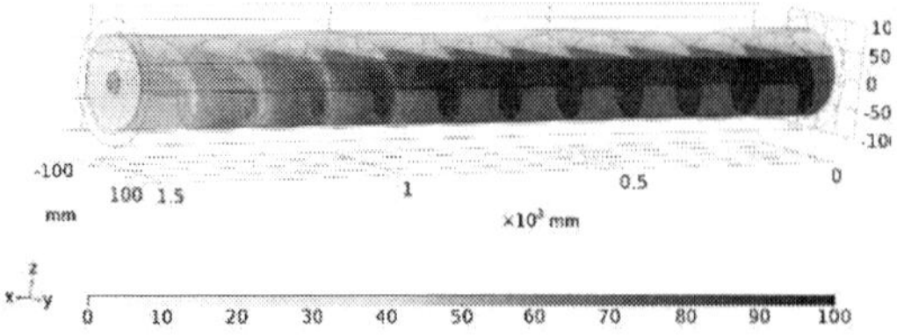

Fig. 4. Spatial distribution of the degree of wood bulk conversion into coke X_{coke} [%]

Fig. 5. Degrees of wood bulk conversion into coke for different movement speeds [mm/hour]:
1 – 106; 2 – 140; 3 – 176; 4 – 212; 5 – 248.

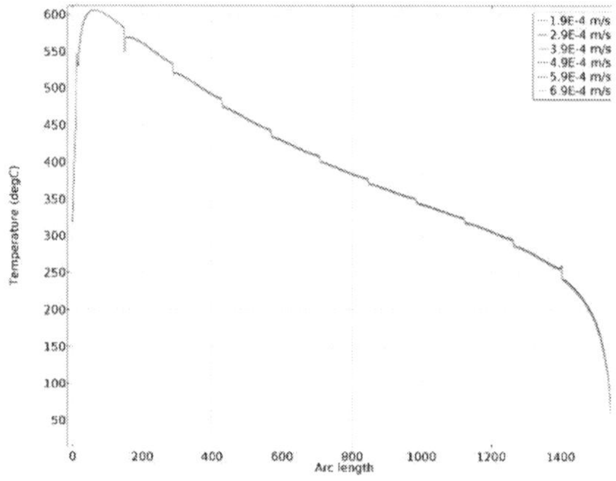

Fig. 6. Temperature distribution along the pyrolyzer for different movement speeds.

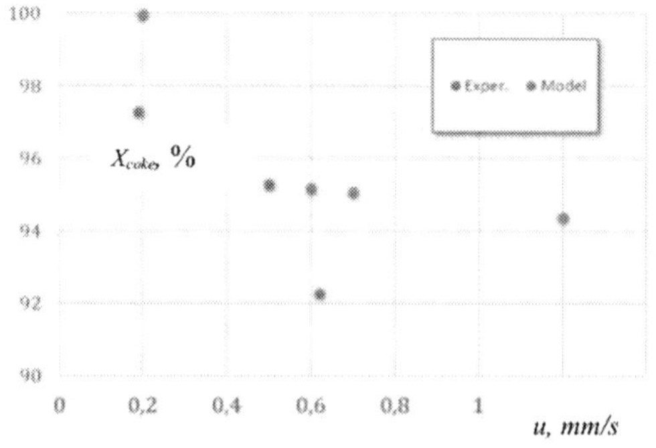

Fig. 7. Comparison between the charcoal conversion degrees obtained experiminally and via numerical modeling.

It can be seen from Figure 6 that the temperature profile is almost the same for all investigated solid fuel flow velocities through the screw reactor. As a result, the relationship between the final fuel conversion in the outlet section of the reactor and the rate of the medium should have the following form

$$X = X_0/u, \quad (3)$$

where X_0 is found from the experimental conditions. Equation (3) should hold for u=0÷0,01 m/s.

Figure 7 shows comparison of the experimental data obtained for different fuel velocities with the caluculated values. We can see a good agreement for the fuel velocities 0. 19 and 0,37 mm/s. If the fuel velocity is high (0.6 mm/s), the mathematical model derives the bigger fuel conversion degree than the one obtained experimentally. This can be explained as follows: in contrast to laboratory research, in the experimental facility the solid fuel particles form a close-packed backfill that prevents free convection of gases formed as a result of pyrolysis reactions. As a result, a diffusion barrier arises and limits the reaction rate.

5 Conclusions

The model of heat transfer and aerodynamics of the pyrolyzer was developed for express optimization calculations. Inclusion of the kinetic block in the model of heat transfer made it possible to determine the concentration fields of the formed substances (charcoal, steam, etc.).

Acknowledgements

This research was performed at the Melentiev Energy Systems Institute of SB RAS under support of the Russian Science Foundation (Grant No. 16-19-10227) using the High Temperature Circuit Multiple Access Research Center.

References

1. Brassard P., Godbout S., Raghavan V. Biosystems engineering **161**, 80- 92. (2017)
2. Funke A., Henrich E., Dahmen N., Sauer J. Energy Technology, **5(1),** 119-129. (2017)
3. http://www.irena.org/bioenergy
4. Molino A., Chianese S., Musmarra D. Journal of Energy Chemistry. **25,** 1, 10-25. (2016)
5. M. Castaldi, J. van Deventer, J.M. Lavoie, J. Legrand, A. Nzihou, Y. Pontikes, X. Py, C. Vandecasteele, P.T. Vasuedevan Waste and Biomass Valorization. **8,** 6, 1875-1884. (2017).
6. http://www.ieabioenergytask33.org
7. Hupa M., Karlstrom O., Vainio E. Proceedings of the Combustion Institute. **36**, 1, 113-134. (2017)
8. A. Kozlov, D. Svishchev, O. Marchenko, S. Solimin, V. Shamansky, A. Keiko // Setting the course for a biobased economy. Proceedings of the 25th European Biomass Conference / Ed. by L. Ek, H. Ehrnrooth, N .Scarlat, A. Grassi, P. Helm. - ETA-Florence Renewable Energies, 776-781. (2017).
9. Heidenreich S., Foscolo P.U. Progress in Energy and Combustion Science. **46**, 72-95. (2015).
10. Aramideh S., Xiong Q., Kong S.-Ch., Brown R.C. Fuel. **156**. 2. 234-242. (2015)
11. Skov I.R., Mathiesen B.V., Connolly D. *A review of biomass gasification technologies in Denmark and Sweden* (Department of Development and Planning, Aalborg University, 2013).
12. Levin A.A., Shamansky V.A., Kozlov A.N. Journal of Physics Conference Series, **754**, 02206. (2016)

First and second order dual phase lag equation. Numerical solution using the explicit and implicit schemes of the finite difference method

Ewa Majchrzak[1,*], and *Bohdan* Mochnacki[2]

[1]Silesian University of Technology, Konarskiego 18a, 44-100 Gliwice, Poland
[2]University of Occupational Safety Management, Bankowa 8, 40-007 Katowice, Poland

Abstract. In the paper the different variants of the dual phase lag equation (DPLE) are considered. As one knows, the mathematical form of DPLE results from the generalization of the Fourier law in which two delay times are introduced, namely the relaxation time τ_q and the thermalization one τ_T. Depending on the order of development of the left and right hand sides of the generalized Fourier law into the Taylor series one can obtain the different forms of the DPLE. It is also possible to consider the others forms of equation discussed resulting from the introduction of the new variable or variables (substitution). In the paper a thin metal film subjected to a laser pulse is considered (the 1D problem). Theoretical considerations are illustrated by the examples of numerical computations. The discussion of the results obtained is also presented.

1 Introduction

The macroscopic heat conduction model (the Fourier model) results from the assumption of instantaneous propagation of the thermal wave in the domain under considerations. It is known that this approach is not admittedly correct, but in the most of problems related to the heat conduction in the macro scale is fully satisfactory.

Seventy years ago Cattaneo [1] formulated an equation in which the delay time (relaxation time τ_q) of the heat flux in relation to the temperature gradient was taken into account. This hyperbolic PDE is known as the Cattaneo-Vernotte equation. In the case of the macroscopic problems the Cattaneo-Vernotte equation is used as a model of heat transfer processes for the non-typical materials with a complex internal structure (e.g. biological tissue) [2]).

The very high heating rates typical for the microscale heat transfer cause that the inclusion of the finite value of thermal wave velocity must be somehow taken into account.

The very popular model describing the microscale heat transfer is based on the dual phase lag equation (DPLE) – e.g. [3, 4]. The final form of DPLE results from the generalized form of the Fourier law in which both the relaxation time and the thermalization one are introduced (e.g. [5]). Depending on the number of terms in the Taylor series expansion of this law, the different forms of the dual phase lag equation can be obtained, in particular, the first and the second order DPLE and also the mixed variants.

Some simple problems described by the DPLE and supplemented by the appropriate boundary and initial conditions can be solved analytically. As an example the paper [6] can be mentioned. The solution concerns a thin metal film heating caused by a laser action, while this action is taken into account by the introduction of the artificial internal heat source.

The numerous approximate solutions of the different problems described by the first order DPLE can be found in literature, e.g. [7-9]. The numerical solutions concerning the second order DPLE (based on the finite difference method) are the subject of works [10- 14]. In the papers [12, 13] the mixed variants of DPLE are also considered.

Below, two variants of DPLE modified forms will be discussed. They are the result of a substitutions causing the appearance in the energy equation of the fragment which is identical as in the classical Fourier equation. The others components can be treated as the internal volumetric heat source. The computer program based on the explicit scheme of finite difference method is not complicated, additionally the stability condition has a simple form. At the stage of numerical modeling the 1D problem is considered. It results from the thermal aspects of the task discussed.

2 Governing equations

The starting point for the formulation of the dual phase lag equation (DPLE) is the generalized Fourier law in the form

* Corresponding author: ewa.majchrzak@polsl.pl

$$\mathbf{q}\left(X,t+\tau_q\right)=-\lambda\nabla T\left(X,t+\tau_T\right) \tag{1}$$

where $\mathbf{q}$ is a heat flux vector, λ is a thermal conductivity, T, X, t denote the temperature, spatial co-ordinates and time, τ_q [s] i τ_T [s] are the positive constants called the relaxation and thermalization times (lag times). To obtain the DPLE the left and right sides of equation (1) are developed into the Taylor series

$$\begin{gathered}\mathbf{q}(X,t)+\tau_q\frac{\partial\mathbf{q}(X,t)}{\partial t}+\frac{\tau_q^2}{2}\frac{\partial^2\mathbf{q}(X,t)}{\partial t^2}+\ldots=\\-\lambda\left[\nabla T(X,t)+\tau_T\frac{\partial\nabla T(X,t)}{\partial t}+\frac{\tau_T^2}{2}\frac{\partial^2\nabla T(X,t)}{\partial t^2}+\ldots\right]\end{gathered} \tag{2}$$

Depending on the assumed number of the components, the different forms of the DPLE can be obtained. Taking into account the first components of equation (2) and using the known diffusion equation one obtains [3, 4] (1D problem)

$$\begin{gathered}c\left[\frac{\partial T(x,t)}{\partial t}+\tau_q\frac{\partial^2 T(x,t)}{\partial t^2}\right]=\lambda\frac{\partial^2 T(x,t)}{\partial x^2}+\\\tau_T\lambda\frac{\partial}{\partial t}\left[\frac{\partial^2 T(x,t)}{\partial x^2}\right]+Q(x,t)+\tau_q\frac{\partial Q(x,t)}{\partial t}\end{gathered} \tag{3}$$

where c is a volumetric specific heat, Q is a capacity of volumetric internal heat sources. One can see, that the constant values of thermophysical parameters are taken into account and such an assumption is, as a rule, accepted. The more complex form of DPLE is associated with the introduction of the second-order derivatives, namely

$$\begin{gathered}c\left[\frac{\partial T(x,t)}{\partial t}+\tau_q\frac{\partial^2 T(x,t)}{\partial t^2}+\frac{\tau_q^2}{2}\frac{\partial^3 T(x,t)}{\partial t^3}\right]=\\\lambda\frac{\partial^2 T(x,t)}{\partial x^2}+\tau_T\lambda\frac{\partial}{\partial t}\left[\frac{\partial^2 T(x,t)}{\partial x^2}\right]+\\\lambda\frac{\tau_T^2}{2}\frac{\partial^2}{\partial t^2}\left[\frac{\partial^2 T(x,t)}{\partial x^2}\right]+Q(x,t)+\tau_q\frac{\partial Q(x,t)}{\partial t}+\\\frac{\tau_q^2}{2}\frac{\partial^2 Q(x,t)}{\partial t^2}\end{gathered} \tag{4}$$

The 'mixed' variants of equation discussed are also considered, for example

$$\begin{gathered}c\left[\frac{\partial T(x,t)}{\partial t}+\tau_q\frac{\partial^2 T(x,t)}{\partial t^2}+\frac{\tau_q^2}{2}\frac{\partial^3 T(x,t)}{\partial t^3}\right]=\\\lambda\frac{\partial^2 T(x,t)}{\partial x^2}+\lambda\tau_T\frac{\partial}{\partial t}\left[\frac{\partial^2 T(x,t)}{\partial x^2}\right]+\\Q(x,t)+\tau_q\frac{\partial Q(x,t)}{\partial t}+\frac{\tau_q^2}{2}\frac{\partial^2 Q(x,t)}{\partial t^2}\end{gathered} \tag{5}$$

Dual phase lag equation must be supplemented by the boundary conditions. Used in this paper the Neumann condition takes a the form

$$\begin{gathered}q_b(x,t)+\tau_q\frac{\partial q_b(x,t)}{\partial t}=\\\pm\lambda\left[\frac{\partial T(x,t)}{\partial x}+\tau_T\frac{\partial}{\partial t}\left(\frac{\partial T(x,t)}{\partial x}\right)\right]\end{gathered} \tag{6}$$

or (the second order model)

$$\begin{gathered}q_b(x,t)+\tau_q\frac{\partial q_b(x,t)}{\partial t}+\frac{\tau_q^2}{2}\frac{\partial^2 q_b(x,t)}{\partial t^2}=\\\pm\lambda\left[\frac{\partial T(x,t)}{\partial x}+\tau_T\frac{\partial}{\partial t}\left(\frac{\partial T(x,t)}{\partial x}\right)+\frac{\tau_T^2}{2}\frac{\partial^2}{\partial t^2}\left(\frac{\partial T(x,t)}{\partial x}\right)\right]\end{gathered} \tag{7}$$

where $q_b(x, t)$ is a given boundary heat flux, the sign "+" corresponds to $x = 0$ (left boundary), while sign "–" corresponds to $x = G$ (right boundary), G is the metal film thickness.

The initial conditions are in the form

$$t=0:\quad T(x,0)=T_p,\quad \left.\frac{\partial T(x,t)}{\partial t}\right|_{t=0}=u(x) \tag{8}$$

Additionally (for equations (4) and (5))

$$t=0:\quad \left.\frac{\partial^2 T(x,t)}{\partial t^2}\right|_{t=0}=v(x) \tag{9}$$

where T_p is an initial temperature, $u(x)$ and $v(x)$ are the known functions.

3 Modifications of DPLE

In the paper [8] a certain modification of the first order DPLE using the substitution technique is presented. In this paper we consider the second order DPLE (as in [12]) and the energy equation is extended by the introduction of the internal heat sources.

Let

$$T(x,t)=U(x,t)-\tau_q\frac{\partial T(x,t)}{\partial t}-\frac{\tau_q^2}{2}\frac{\partial^2 T(x,t)}{\partial t^2} \tag{10}$$

from which results that

$$\frac{\partial T(x,t)}{\partial t}=\frac{\partial U(x,t)}{\partial t}-\tau_q\frac{\partial^2 T(x,t)}{\partial t^2}-\frac{\tau_q^2}{2}\frac{\partial^3 T(x,t)}{\partial t^3} \tag{11}$$

and

$$\begin{gathered}\frac{\partial^2 T(x,t)}{\partial x^2}=\frac{\partial^2 U(x,t)}{\partial x^2}-\\\tau_q\frac{\partial}{\partial t}\left[\frac{\partial^2 T(x,t)}{\partial x^2}\right]-\frac{\tau_q^2}{2}\frac{\partial^2}{\partial t^2}\left[\frac{\partial^2 T(x,t)}{\partial x^2}\right]\end{gathered} \tag{12}$$

Introducing (11) and (12) into (4) gives

$$\frac{\partial U(x,t)}{\partial t}=a\frac{\partial^2 U(x,t)}{\partial x^2}+ a(\tau_T-\tau_q)\frac{\partial}{\partial t}\left[\frac{\partial^2 T(x,t)}{\partial x^2}\right]+ \frac{a}{2}(\tau_T^2-\tau_q^2)\frac{\partial^2}{\partial t^2}\left[\frac{\partial^2 T(x,t)}{\partial x^2}\right]+ \frac{1}{c}Q(x,t)+\frac{\tau_q}{c}\frac{\partial Q(x,t)}{\partial t}+\frac{\tau_q^2}{2c}\frac{\partial^2 Q(x,t)}{\partial t^2} \tag{13}$$

where $a = \lambda/c$ is a thermal diffusion coefficient.

This form of DPLE equation is interesting because the first two components (containing the function U) correspond to the classical diffusion equation, while the remaining ones can be treated as an internal heat source (thanks to which the formulation of the stability criterion for the explicit differential scheme used here is very simple). The boundary conditions should be also modified. In particular

$$\pm\lambda\frac{\partial U(x,t)}{\partial x}=\mp\lambda(\tau_T-\tau_q)\frac{\partial^2 T(x,t)}{\partial t\partial x}\mp \frac{\lambda}{2}(w_T\tau_T^2-w_q\tau_q^2)\frac{\partial^3 T(x,t)}{\partial t^2\partial x}+ q_b(x,t)+\tau_q\frac{\partial q_b(x,t)}{\partial t}+w_q\frac{\tau_q^2}{2}\frac{\partial^2 q_b(x,t)}{\partial t^2} \tag{14}$$

The last formula corresponds to the Neumann condition, of course.

The initial condition takes a form (c.f. equations (8) and (9))

$$t=0:\quad U(x,\,0)=T_p+\tau_q u(x)+\frac{\tau_q^2}{2}v(x) \tag{15}$$

Thus, at first the equation (13) supplemented by boundary conditions (13) and initial condition (15) should be solved. Next, using the formula (10), the temporary temperature field can be determined.

Another approach involves the introduction of two substitutions, namely [13]

$$V(x,t)=\tau_q T(x,t)+\frac{\tau_q^2}{2}\frac{\partial T(x,t)}{\partial t} \tag{16}$$

and

$$U(x,t)=T(x,t)+\frac{\partial V(x,t)}{\partial t} \tag{17}$$

Introducing formula (16) into (17) one obtains the dependence (10) and then equation (4) can be written in the fom

$$\frac{\partial U(x,t)}{\partial t}= a\frac{\partial^2}{\partial x^2}\left[T(x,t)+\tau_T\frac{\partial T(x,t)}{\partial t}+\frac{\tau_T^2}{2}\frac{\partial^2 T(x,t)}{\partial t^2}\right]+ \frac{1}{c}Q(x,t)+\frac{\tau_q}{c}\frac{\partial Q(x,t)}{\partial t}+\frac{\tau_q^2}{2c}\frac{\partial^2 Q(x,t)}{\partial t^2} \tag{18}$$

From (16) one has

$$\frac{\partial T(x,t)}{\partial t}=\frac{2}{\tau_q^2}\left[V(x,t)-\tau_q T(x,t)\right] \tag{19}$$

The last formula can be differentiated with respect to time and then

$$\frac{\partial^2 T(x,t)}{\partial t^2}=\frac{2}{\tau_q^2}\left[\frac{\partial V(x,t)}{\partial t}-\tau_q\frac{\partial T(x,t)}{\partial t}\right] \tag{20}$$

or (c.f. (17))

$$\frac{\partial^2 T(x,t)}{\partial t^2}=\frac{2}{\tau_q^2}\left[U(x,t)-T(x,t)-\tau_q\frac{\partial T(x,t)}{\partial t}\right] \tag{21}$$

By introducing (19) into (21) one obtains

$$\frac{\partial^2 T(x,t)}{\partial t^2}=\frac{2}{\tau_q^2}[U(x,t)+T(x,t)]-\frac{4}{\tau_q^3}V(x,t) \tag{22}$$

Taking into account the formulas (19) and (20), equation (18) can be written in the form

$$\frac{\partial U(x,t)}{\partial t}=\frac{a\tau_T^2}{\tau_q^2}\frac{\partial^2 U(x,t)}{\partial x^2}+\frac{a(\tau_q-\tau_T)^2}{\tau_q^2}\frac{\partial^2 T(x,t)}{\partial x^2}+ \frac{2a\tau_T(\tau_q-\tau_T)}{\tau_q^3}\frac{\partial^2 V(x,t)}{\partial x^2}+\frac{1}{c}Q(x,t)+ \frac{\tau_q}{c}\frac{\partial Q(x,t)}{\partial t}+\frac{\tau_q^2}{2c}\frac{\partial^2 Q(x,t)}{\partial t^2} \tag{23}$$

As in the case of equation (13), the first two components containing the function U correspond to the classical diffusion equation, while the remaining ones can be treated as an artificial internal heat source.

The equation (23) is supplemented by the modified boundary conditions

$$\pm\lambda\frac{\partial U(x,t)}{\partial x}=\mp\lambda\frac{(\tau_q-\tau_T)^2}{\tau_T^2}\frac{\partial T(x,t)}{\partial x}\mp \frac{2\lambda(\tau_q-\tau_T)}{\tau_T\tau_q}\frac{\partial V(x,t)}{\partial x}+ \frac{\tau_q^2}{\tau_T^2}\left[q_b(x,t)+\tau_q\frac{\partial q_b(x,t)}{\partial t}+\frac{\tau_q^2}{2}\frac{\partial^2 q_b(x,t)}{\partial t^2}\right] \tag{24}$$

and the initial condition (15).

After solving the problem (23), (24), (15), the ordinary differential equation (17) with the initial condition (c.f. formula (16))

$$t=0:\quad V(x,0)=\tau_q T_p+\frac{\tau_q^2}{2}u(x) \qquad (25)$$

is solved.

Finally, the solution of the ordinary differential equation (16) with the initial condition $T(x,0) = T_p$, allows one to determine the temporary temperature field.

4 Finite Difference Method

The problems presented previously have been solved using the different variants of the finite difference method (explicit and implicit schemes [12, 16]).

All above presented variants of DPLE have been modeled using the in-house computer programs based on the FDM, while two selected solutions will be shown and compared here. In particular, the thermal processes proceeding in the domain of thin metal film subjected to the laser pulse are considered. The first solution applies equation (23) with the internal volumetric heat source resulting from the laser action. The energy equation is supplemented by the Neumann conditions (in the adiabatic version) and the appropriate initial conditions. At the stage of numerical modeling the explicit scheme of the FDM has been used. One can see, that equation (23) has a form which is convenient for numerical modeling. It contains the simple differential operators without the mixed derivatives. The same physical problem has been solved using the equation (4) and the adequate boundary-initial conditions. In this version the implicit scheme of the FDM has been applied.

The domain considered is covered with a geometrical mesh Ω_h: $0 = x_0 < x_1 < \ldots < x_{j-1} < x_j < x_{j+1} < \ldots < x_n = G$ with a constant step h, while along the time line the points Ω_t: $0 = t^0 < t^1 < \ldots\ t^{f-2} < t^{f-1} < t^f < \ldots < t^F < \infty$ with a constant step Δt are distinguished. The Cartesian product $\Omega_h \times \Omega_t$ creates a time-spatial differential mesh.

The following approximation of equation (23) is proposed:

$$\begin{aligned}\frac{U_j^f-U_j^{f-1}}{\Delta t}&=a\frac{\tau_T^2}{\tau_q^2}\frac{U_{j-1}^{f-1}-2U_j^{f-1}+U_{j+1}^{f-1}}{h^2}+\\ &a\frac{\left(\tau_q-\tau_T\right)^2}{\tau_q^2}\frac{T_{j-1}^{f-1}-2T_j^{f-1}+T_{j+1}^{f-1}}{h^2}+\\ &a\frac{2\tau_T\left(\tau_q-\tau_T\right)}{\tau_q^3}\frac{V_{j-1}^{f-1}-2V_j^{f-1}+V_{j+1}^{f-1}}{h^2}+\\ &\frac{1}{c}\left[Q_j^{f-1}+\tau_q\left(\frac{\partial Q}{\partial t}\right)_j^{f-1}+\frac{\tau_q^2}{2}\left(\frac{\partial^2 Q}{\partial t^2}\right)_j^{f-1}\right]\end{aligned} \qquad (26)$$

from where

$$\begin{aligned}U_j^f&=\left(1-\frac{2a\Delta t\tau_T^2}{\tau_q^2h^2}\right)U_j^{f-1}+\frac{a\Delta t\tau_T^2}{\tau_q^2h^2}\left(U_{j-1}^{f-1}+U_{j+1}^{f-1}\right)+\\ &\frac{a\Delta t\left(\tau_q-\tau_T\right)^2}{\tau_q^2h^2}\left(T_{j-1}^{f-1}-2T_j^{f-1}+T_{j+1}^{f-1}\right)+\\ &\frac{2a\Delta t\tau_T\left(\tau_q-\tau_T\right)}{\tau_q^3h^2}\left(V_{j-1}^{f-1}-2V_j^{f-1}+V_{j+1}^{f-1}\right)+\\ &\frac{\Delta t}{c}\left[Q_j^{f-1}+\tau_q\left(\frac{\partial Q}{\partial t}\right)_j^{f-1}+\frac{\tau_q^2}{2}\left(\frac{\partial^2 Q}{\partial t^2}\right)_j^{f-1}\right]\end{aligned} \qquad (27)$$

The well known stability criterion for parabolic equations determines the critical time step.

The FDM form of the boundary conditions is the following ($x = 0$):

$$\begin{aligned}\lambda\frac{U_1^{f-1}-U_0^{f-1}}{h}&=-\lambda\frac{\left(\tau_q-\tau_T\right)^2}{\tau_T^2}\frac{T_1^{f-1}-T_0^{f-1}}{h}-\\ &\lambda\frac{2\left(\tau_q-\tau_T\right)}{\tau_T\tau_q}\frac{V_1^{f-1}-V_0^{f-1}}{h}+\\ &\frac{\tau_q^2}{\tau_T^2}\left[q_b(x,t)+\tau_q\frac{\partial q_b(x,t)}{\partial t}+\frac{\tau_q^2}{2}\frac{\partial^2 q_b(x,t)}{\partial t^2}\right]_0^{f-1}\end{aligned} \qquad (28)$$

and ($x = G$):

$$\begin{aligned}-\lambda\frac{U_n^{f-1}-U_{n-1}^{f-1}}{h}&=\lambda\frac{\left(\tau_q-\tau_T\right)^2}{\tau_T^2}\frac{T_n^{f-1}-T_{n-1}^{f-1}}{h}+\\ &\lambda\frac{2\left(\tau_q-\tau_T\right)}{\tau_T\tau_q}\frac{V_n^{f-1}-V_{n-1}^{f-1}}{h}+\\ &\frac{\tau_q^2}{\tau_T^2}\left[q_b(x,t)+\tau_q\frac{\partial q_b(x,t)}{\partial t}+\frac{\tau_q^2}{2}\frac{\partial^2 q_b(x,t)}{\partial t^2}\right]_n^{f-1}\end{aligned} \qquad (29)$$

Finally

$$\begin{aligned}U_0^{f-1}&=U_1^{f-1}+\frac{\left(\tau_q-\tau_T\right)^2}{\tau_T^2}\left(T_1^{f-1}-T_0^{f-1}\right)+\\ &\frac{2\left(\tau_q-\tau_T\right)}{\tau_T\tau_q}\left(V_1^{f-1}-V_0^{f-1}\right)-\\ &\frac{h\tau_q^2}{\lambda\tau_T^2}\left[q_b(x,t)+\tau_q\frac{\partial q_b(x,t)}{\partial t}+\frac{\tau_q^2}{2}\frac{\partial^2 q_b(x,t)}{\partial t^2}\right]_0^{f-1}\end{aligned} \qquad (30)$$

and

$$U_n^{f-1} = U_{n-1}^{f-1} - \frac{\left(\tau_q - \tau_T\right)^2}{\tau_T^2}\left(T_n^{f-1} - T_{n-1}^{f-1}\right) - \frac{2\left(\tau_q - \tau_T\right)}{\tau_T \tau_q}\left(V_n^{f-1} - V_{n-1}^{f-1}\right) - \frac{h\tau_q^2}{\lambda \tau_T^2}\left[q_b(x,t) + \tau_q \frac{\partial q_b(x,t)}{\partial t} + \frac{\tau_q^2}{2}\frac{\partial^2 q_b(x,t)}{\partial t^2}\right]_n^{f-1} \tag{31}$$

Next one solves equation (17) (the Euler method)

$$\frac{V_j^f - V_j^{f-1}}{\Delta t} = U_j^{f-1} - T_j^{f-1} \tag{32}$$

or

$$V_j^f = V_j^{f-1} + \Delta t\left(U_j^{f-1} - T_j^{f-1}\right) \tag{33}$$

The last stage of the algorithm results from equation (19)

$$\frac{T_j^f - T_j^{f-1}}{\Delta t} = \frac{2}{\tau_q^2}\left(V_j^{f-1} - \tau_q T_j^{f-1}\right) \tag{34}$$

or

$$T_j^f = \left(1 - \frac{2\Delta t}{\tau_q}\right)T_j^{f-1} + \frac{2\Delta t}{\tau_q^2}V_j^{f-1} \tag{35}$$

Now, the FDM algorithm (implicit scheme) for the model based on equation (4) will be discussed. The following approximation of the energy equation is proposed

$$\frac{T_j^f - T_j^{f-1}}{\Delta t} + \tau_q \frac{T_j^f - 2T_j^{f-1} + T_j^{f-2}}{(\Delta t)^2} + \frac{\tau_q^2}{2}\frac{T_j^f - 3T_j^{f-1} + 3T_j^{f-2} - T_j^{f-3}}{(\Delta t)^3} = a\left(\frac{\partial^2 T}{\partial x^2}\right)_j^f + \frac{a\tau_T}{\Delta t}\left[\left(\frac{\partial^2 T}{\partial x^2}\right)_j^f - \left(\frac{\partial^2 T}{\partial x^2}\right)_j^{f-1}\right] + \frac{a\tau_T^2}{2(\Delta t)^2}\left[\left(\frac{\partial^2 T}{\partial x^2}\right)_j^f - 2\left(\frac{\partial^2 T}{\partial x^2}\right)_j^{f-1} + \left(\frac{\partial^2 T}{\partial x^2}\right)_j^{f-2}\right] + \frac{1}{c}Q_j^f + \frac{\tau_q}{c}\left(\frac{\partial Q}{\partial t}\right)_j^f + \frac{\tau_q^2}{2c}\left(\frac{\partial^2 Q}{\partial t^2}\right)_j^f \tag{36}$$

where

$$\left(\frac{\partial^2 T}{\partial x^2}\right)_j^s = \frac{T_{j-1}^s - 2T_j^s + T_{j+1}^s}{h^2} \tag{37}$$

and $s = f$, $s = f-1$ or $s = f-2$.

After mathematical manipulations one has

$$AT_{j-1}^f + BT_j^f + AT_{j+1}^f = C_j^f \tag{38}$$

where

$$A = -\frac{a\left[2(\Delta t)^2 + 2\tau_T \Delta t + \tau_T^2\right]}{2h^2(\Delta t)^2}$$

$$B = \frac{2(\Delta t)^2 + 2\tau_q \Delta t + \tau_q^2}{2(\Delta t)^3} - 2A$$

$$C_j^f = \frac{2(\Delta t)^2 + 4\tau_q \Delta t + 3\tau_q^2}{2(\Delta t)^3}T_j^{f-1} - \frac{2\tau_q \Delta t + 3\tau_q^2}{2(\Delta t)^3}T_j^{f-2} + \frac{\tau_q^2}{2(\Delta t)^3}T_j^{f-3} - \frac{a\tau_T(\Delta t + \tau_T)}{h^2(\Delta t)^2}\left(T_{j-1}^{f-1} - 2T_j^{f-1} + T_{j+1}^{f-1}\right) + \frac{a\tau_T^2}{2h^2(\Delta t)^2}\left(T_{j-1}^{f-2} - 2T_j^{f-2} + T_{j+1}^{f-2}\right) + \frac{1}{c}Q_j^f + \frac{\tau_q}{c}\left(\frac{\partial Q}{\partial t}\right)_j^f + \frac{\tau_q^2}{2c}\left(\frac{\partial^2 Q}{\partial t^2}\right)_j^f \tag{39}$$

Here one can see, that the FDM equations corresponding to (4) are much more complicated compared to equation (23), among other things, three levels of time must be introduced.

Approximation of boundary conditions (7) is the following
(x=0):

$$\lambda \frac{T_1^f - T_0^f}{h} = -\frac{\lambda \tau_T}{\Delta t}\left(\frac{T_1^f - T_0^f}{h} - \frac{T_1^{f-1} - T_0^{f-1}}{h}\right) - \frac{\lambda \tau_T^2}{2(\Delta t)^2}\left(\frac{T_1^f - T_0^f}{h} - \frac{T_1^{f-1} - T_0^{f-1}}{h} + \frac{T_1^{f-2} - T_0^{f-2}}{h}\right) + \left[q_b(x,t) + \tau_q \frac{\partial q_b(x,t)}{\partial t} + \frac{\tau_q^2}{2}\frac{\partial^2 q_b(x,t)}{\partial t^2}\right]_0^f \tag{40}$$

and (x=G)

$$-\lambda \frac{T_n^f - T_{n-1}^f}{h} = \frac{\lambda \tau_T}{\Delta t}\left(\frac{T_n^f - T_{n-1}^f}{h} - \frac{T_n^{f-1} - T_{n-1}^{f-1}}{h}\right) + \frac{\lambda \tau_T^2}{2(\Delta t)^2}\left(\frac{T_n^f - T_{n-1}^f}{h} - 2\frac{T_n^{f-1} - T_{n-1}^{f-1}}{h} + \frac{T_n^{f-2} - T_{n-1}^{f-2}}{h}\right) + \left[q_b(x,t) + \tau_q \frac{\partial q_b(x,t)}{\partial t} + \frac{\tau_q^2}{2}\frac{\partial^2 q_b(x,t)}{\partial t^2}\right]_n^f \tag{41}$$

Finally

$$-DT_0^f + DT_1^f = 2\tau_T(\Delta t + \tau_T)\left(T_1^{f-1} - T_0^{f-1}\right) - \tau_T^2\left(T_1^{f-2} - T_0^{f-2}\right) + E\left[q_b(x,t) + \tau_q \frac{\partial q_b(x,t)}{\partial t} + \frac{\tau_q^2}{2}\frac{\partial^2 q_b(x,t)}{\partial t^2}\right]_0^f \tag{42}$$

and

$$DT_{n-1}^f - DT_n^f = \\ -2\tau_T\left(\Delta t+\tau_T\right)\left(T_n^{f-1}-T_{n-1}^{f-1}\right)+\tau_T^2\left(T_n^{f-2}-T_{n-1}^{f-2}\right)+ \\ E\left[q_b\left(x,t\right)+\tau_q\frac{\partial q_b\left(x,t\right)}{\partial t}+\frac{\tau_q^2}{2}\frac{\partial^2 q_b(x,t)}{\partial t^2}\right]_n^f \tag{43}$$

where

$$D=2\left(\Delta t\right)^2+2\tau_T\Delta t+\tau_T^2 \\ E=\frac{2h\left(\Delta t\right)^2}{\lambda} \tag{44}$$

The transition from t^{f-1} to t^f ($f \geq 3$) requires the solution of the system of equations (42), (38), (43) with a three-band main matrix that is fastest solved using the Thomas algorithm.

5 Results of computations

The thin metal film (G = 100 nm, 1 nm = 10^{-9} m) made of gold is considered. Thermophysical parameters of gold equal: λ = 315 W/(m K), c = 2.4897 MJ/(m^3 K), τ_q = 8.5 ps, τ_T = 90 ps (1 ps = 10^{-12} m) [4].

The surface x = 0 of the domain is subjected to the laser pulse. The source function associated with the laser irradiation has the following form [16, 17]

$$Q(x,t)=\sqrt{\frac{\beta}{\pi}}\frac{1-R}{t_p\,\delta}\,I\,\exp\left[-\frac{x}{\delta}-\beta\frac{(t-2t_p)^2}{t_p^2}\right] \tag{45}$$

where I is the laser intensity, t_p is the characteristic time of laser pulse, δ is the optical penetration depth, R is the reflectivity of the irradiated surface and β= 4 ln2.

For x = 0 and x = G the no-flux boundary conditions are assumed: $q_b(0, t) = q_b(G, t)$=0.

The parameters determining the capacity of internal heat source are equal to I_0 = 13.7 J/m^2, t_p = 0.1 ps, δ = 15.3 nm, R = 0.93. The initial temperature of domain equals T_p = 300 K, while the initial values of functions $u(x)$ = 0, $v(x)$ = 0 (c.f. equations (8), (9)). Differential mesh parameters: n = 100, Δt = 0.00002 ps.

In Figure 1 the comparison of results obtained using the algorithms discussed is presented. Solid lines correspond to the explicit scheme of FDM for the equation (23) with the boundary conditions (24) and initial condition (15), while the symbols correspond to the implicit scheme of FDM for the equation (4) with the boundary conditions (7) and initial conditions (8), (9). In this Figure the temperature histories at the points x = 0 (irradiated surface), x = 20 nm and x = 40 nm are shown. As can be seen, the results are almost the same. For example, for the first and the second variants of FDM, the maximum temperatures at the irradiated surface are equal to 306.536 K and 306.535 K, respectively.

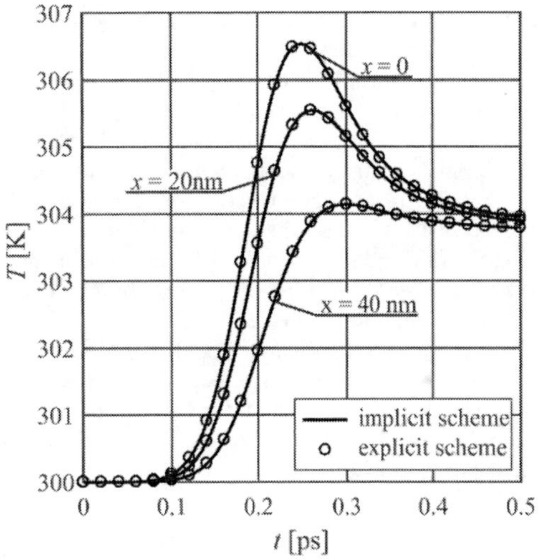

Fig. 1. Temperature history at the points x = 0, x = 20 nm and x = 40 nm

It is worth checking what are the differences in the solutions obtained by means of the first and second order DPLE. In Figure 2 the temperature histories at the irradiated surface for laser intensities 13.7 J/m^2 and 50 J/m^2 (t_p = 0.1 ps) are shown. The differences are essential. Moreover, for higher laser intensity these differences are greater. For I = 13.7 J/m^2 the maximum temperatures are equal to 313.14 K and 306.54 K, respectively, while for I = 50 J/m^2 the maximum temperatures are equal to 347.94 K and 323.85 K. The calculations were also carried out for higher laser intensities (100, 200 and 300 J/m^2) and this trend was confirmed, namely I = 100 J/m^2: maximum temperatures: 395.90 K and 347.70 K; I = 200 J/m^2: maximum temperatures: 491.79 K and 395.41 K; I = 300 J/m^2: maximum temperatures: 587.68 K and 443.11 K.

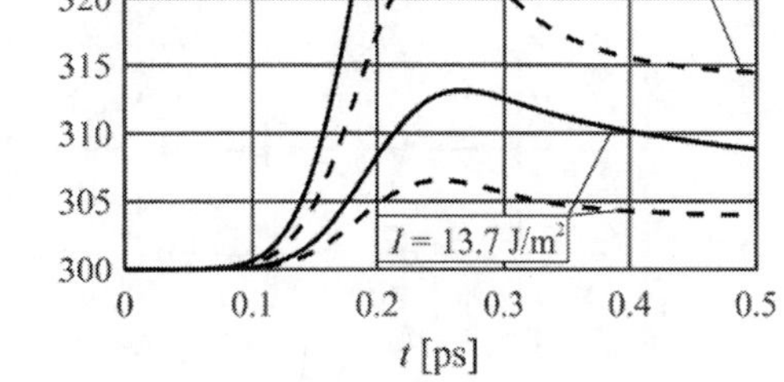

Fig. 2. Temperature histories for the first and second order DPL at the point x = 0 (t_p = 0.1 ps)

Figure 3 illustrates the heating/cooling curves at the irradiated surface for a shorter laser pulse t_p = 0.05 ps and laser intensities 13.7 J/m^2 and 50 J/m^2, respectively. For I = 13.7 J/m^2 the maximum temperatures are equal to 315.12 K and 308.28 K, respectively, while for I = 50 J/m^2 the maximum temperatures are equal to 355.17 K and 330.22 K. As previously, for the higher laser intensity the differences between the solutions obtained using the first and second order DPL are greater. For the other intensities: I = 100 J/m^2: maximum temperatures: 410.34 K and 360.44 K; I = 200 J/m^2: maximum temperatures: 520.68 K and 420.88 K; I = 300 J/m^2: maximum temperatures: 631.02 K and 481.33 K.

It should be noted that the results presented in Figures 2 and 3 have been obtained using the implicit scheme of the FDM.

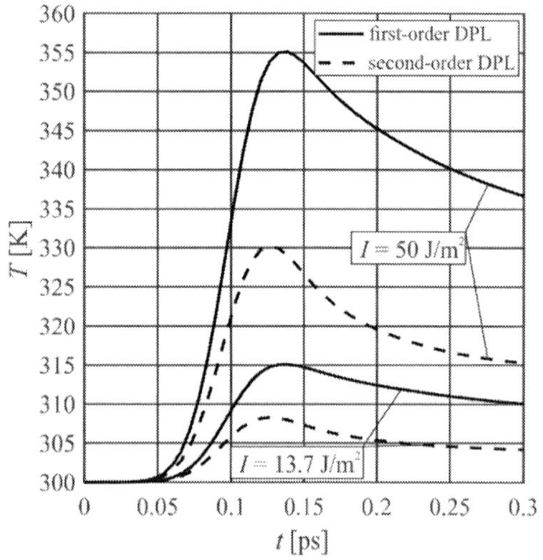

Fig. 3. Temperature histories for the first and second order DPL at the point x = 0 (t_p = 0.05 ps)

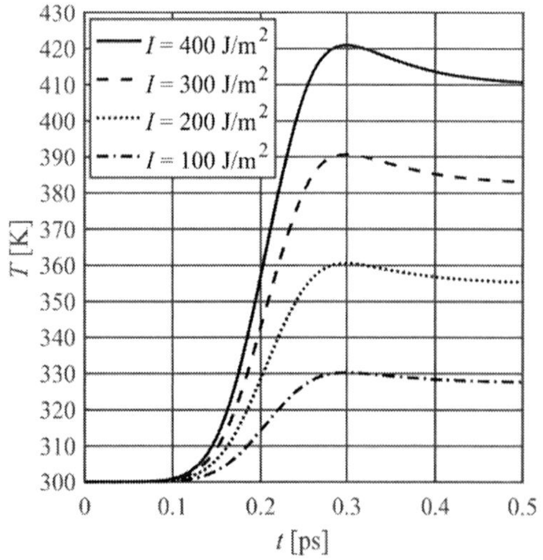

Fig. 4. Temperature histories for the second order DPL at the point x = 0 (t_p = 0.1 ps)

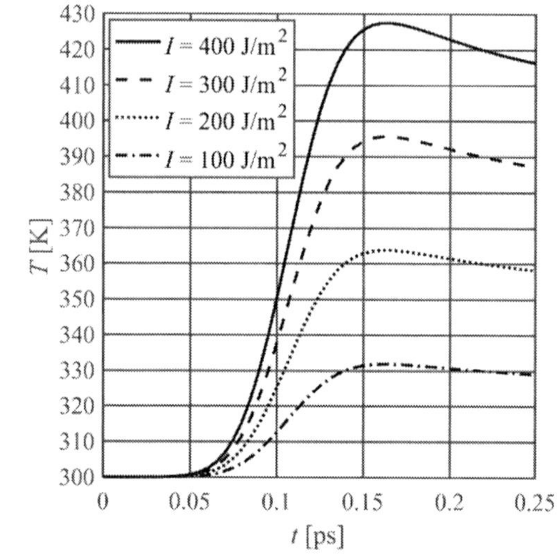

Fig. 5. Temperature histories for the second order DPL at the point x = 0 (t_p = 0.05 ps)

Figures 4 and 5 illustrate the temperature histories at the irradiated surface for the second order DPL obtained using the explicit scheme of FDM and for two values of laser pulse t_p = 0.1 ps and t_p = 0.05 ps. As mentioned previously, for shorter laser pulse the temperatures are higher: I = 100 J/m^2 the maximum temperatures are equal to 347.94 K and 360.45 K; I = 200 J/m^2 the maximum temperatures: 395.41 K and 420.91 K; I = 300 J/m^2 the maximum temperatures: 443.12 K and 481.36 K; I = 400 J/m^2 the maximum temperatures: 490.08 K and 541.82 K.

6 Conclusions

In the paper the second order dual phase lag models are considered. The results of numerical computations show that obtained solutions are practically the same, despite that the final forms of the DPLE are different. The models resulting from the introduction of certain substitutions are more complicated than the classical one, but on the other hand, the algorithms using the FDM are essentially simpler. The results obtained have been compared with the solutions of the first order DPLE and the differences are clearly visible (see: previous chapter).

Acknowledgement

The paper and research are financed within the project 2015/19/B/ST8/01101 sponsored by The National Science Centre (Poland).

References

1. M.C. Cattaneo, Atti De Seminario Matematico e Fisico Della Universi ta di Modena **3**, 3, 3-21 (1948)
2. W. Roetzel, N. Putra, S.K. Das, Int. J. Therm. Sci. **42**, 541-552 (2003)

3. Z.M. Zhang, *Nano/Microscale Heat Transfer*, (McGraw-Hill, 2007)
4. D.Y. Tzou, *Macro to Microscale Heat Transfer: The Lagging Behavior* (John Wiley & Sons, Ltd, 2015)
5. A. Faghri, Y. Zhang, J. Howell, *Advanced Heat and Mass Transfer*, (Global Digital Press, 2010)
6. M. Ciesielski, J. Appl. Math. Comput. Mech. **16**, 1, 33-40 (2017)
7. E. Majchrzak, B. Mochnacki, Int. J. Multiscale Comput. Eng. **12**, 1, 65-77 (2014)
8. W. Dai, R. Nassar, Numer. Meth. Part. D. E. **16**, 441-458 (2000)
9. H. Wang, W. Dai, R. Melnik, Int. J. Therm. Sci. **45**, 1179-1196 (2006)
10. M.A. Castro, F. Rodriques, J. Cabrera, J.A. Martin, J. Comput. Appl. Math. **201**, 432-440 (2016)
11. D. Deng, Y. Jiang, D.L. Liang, Appl. Math. Comput. **309**, 31-48 (2017)
12. E. Majchrzak, B. Mochnacki, J. Theor. Appl. Mech. **56,** 2, 393-402 (2018)
13. J. Cabrera, M.A. Castro, F. Rodriquez, J.A. Martin, Math. Comput. Model. **57**, 1625-1632 (2013)
14. D. Deng, Y. Jiang, D. Liang, Appl. Math. Comput. **309**, 31-48 (2017)
15. E. Majchrzak, B. Mochnacki, J. Appl. Math. Comput. Mech. **15**, 3, 89-96 (2016)
16. J.K. Chen, D.Y. Tzou, J.E. Beraun, Int. J. Heat Mass Trans. **49**, 307-316 (2016)
17. E. Majchrzak, B. Mochnacki, J.S. Suchy, J. Theor. Appl. Mech. **47,** 2, 383-396 (2009)

Review of the literature discussing the problem of uneven thermal stresses in the furnace chamber

Marek Majdak[1,*]

[1] Institute of Thermal Power Engineering, Cracow University of Technology, al. Jana Pawła II 37, 31-864 Kraków, Poland

Abstract. Uneven thermal stresses occurring in the walls of the furnace chamber in high power boilers are the reason for the occurrence of significant thermal stresses. The occurrence of excessive stress leads to serious damage to the waterwall tubes, which result in the necessity of more frequent repairs of the boiler. Therefore, finding a way to reduce stress values is one of the key challenges in improving productivity and increase the service life of boilers.
In this paper will be presented the review of the literature concerning the problem of uneven temperature distribution on the surface of the combustion chamber walls. There will also be presented studies that discuss the methods of modifying the furnace chamber to reduce the values of the stresses occurring in it.

1. Literature review

Flow-through boilers for supercritical parameters allow obtaining parameters exceeding the values characteristic for the critical point of water, which results in a factor for which there is no boundary between the liquid and gas phases. In boilers of this type, it is particularly important to care for the safety of hydraulic components, which is determined by controlling the temperature of the flow elements, the absence of pulsation of the medium flow, installation of protection to prevent the occurrence of stagnation or backflow in the boiler [1].

To precisely determine the working conditions of the walls of the furnace chamber, numerous analyzes were carried out based on the division of waterwall tubes and the factor flowing through them into finite elements. Determining the working conditions of furnace chamber walls using direct measurements is particularly difficult due to the high temperatures at which the measuring devices must be operated. To stabilize the measurement conditions and protect the measuring devices against damage numerous elements attached to the tested walls of the firing chambers are used. The design and principle of operation of one of these elements is described in [2].

The main problem related to the changing conditions of waterwall tube work is the identification of waterwall tube parameters on the thickness of their cross-section in the event of a transient thermal state of their operation. One of the methods of identifying these parameters in online mode is the method proposed in [3], allowing to determine the temperature distribution of the material from which the furnace chamber walls are made, and the temperature distribution of the medium flowing through the waterwall tubes based on heat transfer equations solved using methods of elementary balances.

Another method of solving the inverse problem in cylindrical elements, i.e. determining the temperature distribution on the cross-section of a thick-walled element based on the temperature measured on its external surface, taking into account the variable material parameters in which the analyzed elements are made and taking into account variable factor parameters, has been proposed in [4]. The temperature distribution in the wall of the element was determined using the simple method, taking into account the dependence of the thermophysical properties of the material and the temperature of the medium inside the tubes.

Another issue that should be taken into account in order to properly model the temperature distribution and stresses occurring in the waterwall tubes mounted on the furnace chamber walls is the problem of determining the dynamics of changes in material parameters due to a rapid change in their working conditions - rapid temperature change at the inlet to the waterwall tube or step change heat flux falling on the wall of the furnace from the outside. One of the solutions developed for this problem, based on the use of a differential scheme in explicit and implicit form, has been presented in [5].

The boiler evaporator dynamics model, based on the solution of balance equations describing the principles of energy, mass and momentum conservation, has been presented in [6]. This model is a model with distributed parameters, described by differential equations with separated variables. By using this model, it is possible to determine the exact temperature and stress distribution in the boiler of the power boiler, taking into account the thermophysical properties of the material from which the evaporator is made and time parameters of the medium that passes through it.

The mathematical model allowing for the analysis of heat exchange processes occurring in boiler furnace

* Corresponding author: marek.majdak@mech.pk.edu.pl

chamber waterwalls for supercritical parameters was also presented in [7]. In the article is presented a methodology for solving equations of energy, mass and momentum balance, based on the model proposed in [5] and computational verification of obtained results together with comparing them with other strict solutions for transients available in the literature.

In the literature, there are also available analyzes of the thermal load distribution at the height of the furnace chamber. In [8-10], the results of calculations of the thermal load distribution determined by measurements of the face temperature of the waterwall tube are presented using a measurement insert that allows obtaining more accurate measurement results, assuming a constant temperature of the medium flowing inside the analyzed waterwall tube.

The paper [11] presents the results of numerical computations performed for the supercritical boiler's furnace chamber waterwalls with a steam output of 2400×10^3 kg/h. The model is based on the solution of equations describing the conservation of mass, momentum and energy. The aim of the calculations is to determine the distribution of mass flow, enthalpy and the fluid pressure in tubes, using the Runge-Kutta method to solve the system of ordinary differential equations.

Another method of modifying the heat exchange process in high-power power boilers is the use of internally rifled tubes. The articles [12-14] present the results of calculations and numerical simulations allowing for the comparison of the heat exchange process in smooth tubes and internally rifled tubes, which have an increased internal surface area.

A comprehensive analysis of the heat exchange phenomenon occurring in the furnace of a fluidized bed boiler with a circulating bed, taking into account the phenomena occurring between adjacent particles moving inside the bed, and between the bed and the boiler furnace wall is presented in [15]. The study also discusses the heat transfer between the waterwall tubes and the fluid flowing through them, as well as the influence of tube grooving on the heat exchange intensity.

The problem of the analysis of phenomena related to heat exchange in furnace chambers of fluidized bed boilers has also been described in [16]. The authors of the study discussed the problem with consideration of the division into two processes – the transfer of heat between the gas and the particles of the bed, and the transfer of heat between the fluidized bed and the surface of the tubes immersed in the bed.

An extensive analysis of the calculation of thermo-hydraulic operating conditions of a fluidized bed boiler with a circulating bed operating on supercritical parameters is presented in [17-19]. Authors presented the assumptions of the mathematical model allowing analysis of the boiler's operation taking into account the variable mass flow of the circulating factor in it, and presented changes in the temperature distribution of the circulating medium and metal from which the furnace chamber is made in selected circulation contours depending on the boiler load and stream size mass of the medium flowing through the boiler.

The design and working conditions of the fluidized bed boiler, in which the Benson type evaporator was used, i.e., the evaporator built of vertical waterwall tubes, allowing to determine the mass flow of the medium flowing through the individual waterwall tubes, depending on the thermal load affecting them, are described in [20]. The use of the presented method allows to reduce hydraulic losses and reduce the stresses occurring in the boiler furnace chambers, mainly during boiler operation under a reduced mass stream of the heat-receiving medium.

The author of the publication [21] presents in it some of the most common causes of failure of pressure elements of boilers, with particular emphasis on the surfaces of firing chamber waterwalls. One of the most important causes of failure, which is based already in the wrong design of the furnace chamber, is the problem of overheating of waterwall tubes to a temperature exceeding the limit value. As a result of exceeding the permissible temperature, the material's creep strength decreases dramatically - this phenomenon can be significantly reduced by using the proposed layout of waterwall tubes with different pitch distribution on the width of the furnace chamber.

The studies [22-24] present new models of the division of the boiler waterwall tube with the adjacent fin on the control volume. The temperature distributions for the medium and selected cross-sectional points of waterwall tubes at various heights were presented, based on data regarding working conditions for one of the Polish power plants operating on supercritical parameters.

The comparison of the influence of changes in the parameters of the fluidized bed boiler with a circulating bed on the phenomenon of heat exchange occurring in the combustion chamber is presented in [25]. The authors of the study focused on the comparison of known computational models and the determination of the impact of fluidized bed density on heat exchange in the boiler's furnace chamber.

The paper [26] discusses the process of remodeling the EKM-1,3 grate boiler for an OF-1,3 fluidized boiler. The authors of the article focused on the discussion of installed and modernized boiler components, as well as on the analysis of the results of tests carried out, in a wide range of operating parameters, in a fixed and unchanged boiler operating condition. The test results can be used during the design process of the fluidized bed boiler as well as during its operation.

The description of the structure and principles of operation of the world's first fluidized bed boiler for supercritical parameters with a vertical arrangement of the evaporator tubes was presented in [27]. The authors of the article focused on discussing the process of setting up the installation, determining the parameters of the fuel for which combustion was designed, as well as discussing individual elements of the system that makes up the water-steam cycle. The parameters of the boiler's evaporator's operation with the vertical arrangement of the waterwall tubes with other implementations of combustion chamber waterwalls occurring in the flow boilers have been compared, thanks to which it was shown that the proposed implementation is more advantageous in terms of boiler operation parameters, which has a positive effect on fuel consumption as well as reduction of emission of pollution and greenhouse gases into the atmosphere.

One of the items presenting cross-sectional methods of analyzing the phenomenon of heat exchange taking place in the furnace of the fluidized bed boiler is [28]. In this publication, the author analyzes in detail the phenomenon of heat exchange in a simplified - two-dimensional and three-dimensional fluidized bed model, discusses the methods of experimental monitoring of this phenomenon and discusses the results obtained by using a numerical model of fluidized bed simulation compared to the results of measurements of fluidized bed parameters using thermographic and infrared radiation.

References

1. V. A. Lokshin, D. F. Peterson, A. L. Schwarz, *Standard methods of hydraulic design for power boilers* (CRC Press, Berlin, Germany, 1988)
2. B. Węglowski, *Identification of the thermal operating conditions of boilers combustion chamber walls* (in polish) (Wydawnictwo Politechniki Krakowskiej, Cracow, Poland, 1995)
3. W. Zima, *The inverse problem of heat conduction in pressure elements of boilers* (in polish) (Wydawnictwo Politechniki Krakowskiej, Cracow, Poland, 1996)
4. S. Grądziel, *Identification of transient thermal state of thick-walled boiler elements* (in polish) (Wydawnictwo Politechniki Krakowskiej, Cracow, Poland, 1997)
5. W. Zima, *Simulation of the transient processes in power boilers steam superheaters* (in polish) (Wydawnictwo Politechniki Krakowskiej, Cracow, Poland, 2004)
6. S. Grądziel, *Modeling of flow-thermal phenomena occurring in the boiler's evaporator with natural circulation* (in polish) (Wydawnictwo Politechniki Krakowskiej, Cracow, Poland, 2012)
7. S. Łopata, S. Grądziel, *Energy machines and devices* (in polish) 365-382 (Wydawnictwo Politechniki Krakowskiej, Cracow, Poland, 2013)
8. S. Łopata, S. Grądziel, *Energy machines and devices* (in polish) 71-90 (Wydawnictwo Politechniki Krakowskiej, Cracow, Poland, 2013)
9. J.Taler, P.Duda, B.Węglowski, W.Zima, S.Grądziel, T.Sobota, D.Taler,, Fuel **88**, 305-311 (2009)
10. S. Grądziel, Archivum Combustionis **35,** 75-88 (2015)
11. S. Grądziel, K. Majewski, Chemical and Process Engineering **37**, 2, 199-213 (2016)
12. S. Grądziel, K. Majewski, Procedia Engineering **157**, 44-49 (2016)
13. K. Majewski, S. Grądziel, Chemical and Process Engineering **37**, 2, 251-260 (2016)
14. S. Grądziel, K. Majewski, Journal Heat Transfer Engineering **39**, 13-14, 1243-1250 (2018)
15. P. Basu, *Circulating Fluidized Bed Boilers* (Springer, London, UK, 2015)
16. S. Wiśniewski, T. S. Wiśniewski, *Heat exchange* (in polish) (WNT, Warsaw, Poland, 2013)
17. J. Pan, G. Wu, D. Yang, Applied Thermal Engineering **32**, 41-48 (2012)
18. J. Pan, G. Wu, D. Yang, Applied Thermal Engineering **82**, 225-236 (2015)
19. L. Wang, D. Yang, Z. Shen, K. Mao, J. Long, Applied Thermal Engineering **95**, 42-52 (2016)
20. S. J. Goidich, R. J. Docherty, K. P. Meizer,. Modern Power Systems **32**, 4, 29-32 (2012)
21. A. Jasiński, *Diagnostics of water boilers - main causes of failure of pressure elements* (in polish) (Energopomiar, Gliwice, Poland, 2013)
22. W. Zima, M. Nowak-Ocłoń, Procedia Engineering **157**, 200-206 (2016)
23. W. Zima, M. Nowak-Ocłoń, P. Ocłoń, Energy **92**, 117–127 (2015)
24. W. Zima, M. Nowak-Ocłoń, P. Ocłoń, Energy **148**, 809-823 (2018)
25. P. Basu, P. K. Nag, Chemical Engineering Science **51**, 1, 1-26 (1996)
26. W. Dobrzański, K. Jurek, W. Szwarc, *Experimental research of the experimental OF-1,3 fluidized bed boiler* (in polish) (Wydawnictwo Politechniki Warszawskiej, Warsaw, Poland, 1987)
27. S. J. Goidich, R. J. Docherty, K. P. Melzer, *The World's First Supercritical FW-BENSON Vertical PC Boiler – The 750 MWe Longview Power Project* (Foster Wheeler, 2011)
28. A. Patil, *Heat transfer in gas-solid fluidized beds* (Technische Universiteit Eindhoven, Eindhoven, Holland, 2015)

Thermal flow analysis of vertical combustion chamber waterwall tubes operation

Marek Majdak[1,*], *Sławomir* Grądziel[1]
[1] Institute of Thermal Power Engineering, Cracow University of Technology, al. Jana Pawła II 37, 31-864 Kraków, Poland

Abstract. In this paper will be presented an algorithm that allows to determine the temperature distribution in a vertical pipe through which is flowing the medium with high temperature and pressure. The method of pipe division into control volumes presented in paper allows to determine the temperature distribution on the pipe cross-section and to determine changes in its value at the pipe height. The issue will be solved by taking into account the variability of the parameters of the material from which the pipe is made and the parameters of the fluid, depending on the temperature. The applied algorithm allows to determine the change in the value of individual parameters in time, due to which it is a useful tool for analysis of the operation of the system in the initial stage of its work, as well as with variable parameters of flow.

Nomenclature

A – cross-sectional area, m^2
c – specific heat, J/kgK
d – diameter, m
g - standard gravity, m/s^2
h – enthalpy, kJ/kg
k – thermal conductivity, W/mK
m – mass flow of the medium, kg/s
p – pressure, Pa
r– radius, m
t – temperature of the medium, °C
υ – velocity of the medium, m/s
z – height, m

Greek letters

α – heat transfer coefficient, W/m^{2}K
β – the angle of inclination of the waterwall tube, °
$\Delta\varphi$– characteristic angle for the finite volume under consideration, rad
Δz – height of finite volume, m
θ – wall temperature, °C
ρ – density, kg/m^3
τ – time, s

Subscripts

in – inner
m – medium
o – outer

Introduction

One of the most important operational problems occurring in high-power boilers is damage to the heat exchange surface in the combustion chamber, which in the case of boilers fired with solid fuels result in a higher frequency of maintenance shutdowns and the need to perform costly and difficult to correct repairs.

During boiler operation, waterwall pipes are exposed to large and uneven stresses, which in the boiler start-up phase result from the variability of boiler operation parameters over time. In the later period of boiler operation, stresses affecting waterwall pipes result from uneven thermal load of the surface through which heat exchange is implemented, as well as from variable flow conditions inside the pipes - depending on the type of power boiler, the pressure, mass flow and state of aggregation of the heat receiving medium - dependences allowing to determine the distribution of these parameters are presented in [1]. In many studies, it has been emphasized that the heat load changes at the width of the furnace chamber, so changing the pitch of the vertical waterwall pipes may allow to reduce the thermal stresses occurring in them. Another solution aimed at reducing thermal stresses is the use of internally rifled tubes instead of smooth tubes - a comparison of the heat exchange process for both types of tubes is presented in [2-4]. Detailed analysis of the operating parameters of individual boiler waterwall fragments may allow to reduce the frequency of damage occurring in them resulting from thermal stresses exceeding the permitted values.

Problem analysis

The problem of optimizing the arrangement of vertical waterwall tubes in the boiler furnace chamber will be analyzed (Fig. 1). The analysis will examine the thermal stresses occurring in the walls of the furnace chamber of the boiler, made in the form of vertical waterwall tubes

* Corresponding author: marek.majdak@mech.pk.edu.pl

connected with fins. The proposed mathematical model will take into account the variability of the thermophysical parameters of the medium and the metal from which the waterwall tubes are made, depending on the temperature. To determine the parameters of the factor circulating inside the waterwall pipes IAPWS IF97 water and steam tables have been implemented to the created mathematical model. Dependencies of thermophysical properties of steel, from which screen tubes were made, have been approximated based on available parameter tables determined by material manufacturers. For the analysis, it will be assumed that the boiler operates on supercritical parameters, and the heat load changes on the width of the furnace chamber, according to the results of tests carried out for fluidized bed boilers operating on supercritical parameters. The simplified distribution of the heat flux density in the cross-section of the furnace chamber is shown below (Fig. 2).

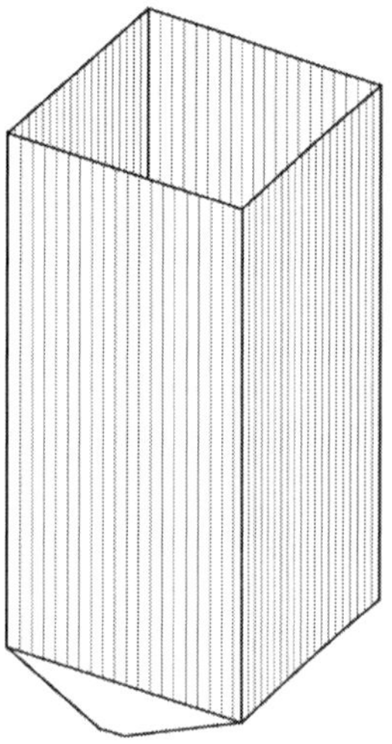

Fig. 1 The contour of the combustion chamber with vertical waterwall tubes

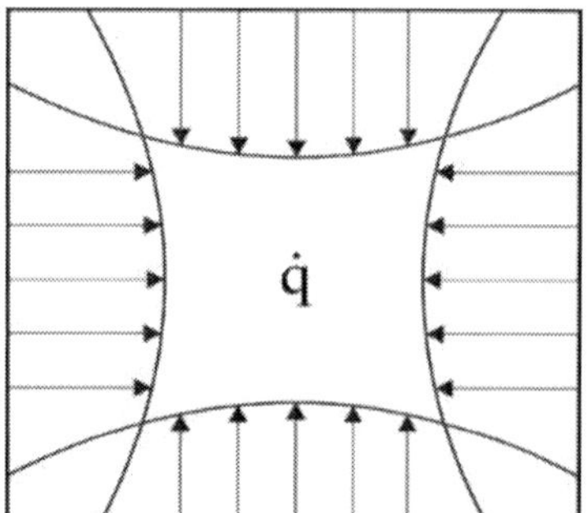

Fig. 2 The heat flux density distribution in the cross section of the combustion chamber

Optimization of the screen pipe arrangement is aimed at reducing the thermal stresses occurring in the furnace chamber walls, associated with significant temperature differences of the medium flowing through waterwall tubes occurring in boilers operating on supercritical parameters - in boilers with natural water circulation this phenomenon does not occur, because the temperature distribution of the medium in the width of the chamber is practically uniform.

The basic principles on which the calculation algorithm is based on the assumptions presented in [5] and [6-7]. The division of the analyzed waterwall tubes into control volumes by the method proposed in [6-7] has been presented in (Fig. 3). By using this, it is possible to determine the temperature distribution over the cross section of the pipe and the analysis of changes its value on the height of the tube. On the basis of the determined temperature distribution, an analysis of thermal stresses occurring in the waterwall tube will be carried out.

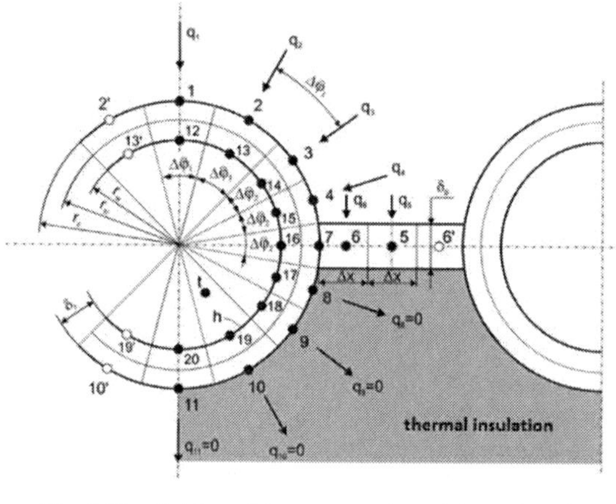

Fig. 3 The division of the cross-section of the analyzed waterwall tubes into control volumes

Equations of the energy balance for selected control volumes, allowing the creation of a matrix system describing the temperature distribution in a single screen tube are presented below:

$$c_{1,j}\rho_{1,j}\frac{\Delta\hat{\varphi}_1}{2}\left(r_o^2-r_m^2\right)\Delta z\frac{d\theta_{1,j}}{d\tau}=k_{1,j}\frac{\theta_{2,j}-\theta_{1,j}}{\Delta\hat{\varphi}_1 r_o}\Delta r\Delta z+ k_{1,j}\frac{\theta_{12,j}-\theta_{1,j}}{\Delta r}\Delta\hat{\varphi}_1 r_m\Delta z+k_{1,j}\frac{\Delta\hat{\varphi}_1}{2}\left(r_o^2-r_m^2\right)\frac{\theta_{1,j-1}-\theta_{1,j}}{\Delta z}+ k_{1,j}\frac{\Delta\hat{\varphi}_1}{2}\left(r_o^2-r_m^2\right)\frac{\theta_{1,j+1}-\theta_{1,j}}{\Delta z}+q_{1,j}\Delta\hat{\varphi}_1 r_o\Delta z \quad (1)$$

$$c_{11,j}\rho_{11,j}\frac{\Delta\hat{\varphi}_1}{2}\left(r_o^2-r_m^2\right)\Delta z\frac{d\theta_{11,j}}{d\tau}=k_{11,j}\frac{\theta_{10,j}-\theta_{11,j}}{\Delta\hat{\varphi}_1 r_o}\Delta r\Delta z+ k_{11,j}\frac{\theta_{20,j}-\theta_{11,j}}{\Delta r}\Delta\hat{\varphi}_1 r_m\Delta z+k_{11,j}\frac{\Delta\hat{\varphi}_1}{2}\left(r_o^2-r_m^2\right)\frac{\theta_{11,j-1}-\theta_{11,j}}{\Delta z}+ k_{11,j}\frac{\Delta\hat{\varphi}_1}{2}\left(r_o^2-r_m^2\right)\frac{\theta_{11,j+1}-\theta_{11,j}}{\Delta z} \quad (2)$$

$$c_{12,j}\rho_{12,j}\frac{\Delta\hat{\varphi}_1}{2}\left(r_m^2-r_{in}^2\right)\Delta z\frac{d\theta_{12,j}}{d\tau}=k_{12,j}\frac{\theta_{13,j}-\theta_{12,j}}{\Delta\hat{\varphi}_1 r_{in}}\Delta r\Delta z+ k_{12,j}\frac{\theta_{1,j}-\theta_{12,j}}{\Delta r}\Delta\hat{\varphi}_1 r_m\Delta z+k_{12,j}\frac{\Delta\hat{\varphi}_1}{2}\left(r_o^2-r_m^2\right)\frac{\theta_{12,j-1}-\theta_{12,j}}{\Delta z}+ k_{12,j}\frac{\Delta\hat{\varphi}_1}{2}\left(r_o^2-r_m^2\right)\frac{\theta_{12,j+1}-\theta_{12,j}}{\Delta z}+\alpha_j\left(t_j-\theta_{12,j}\right)\Delta\hat{\varphi}_1 r_{in}\Delta z \quad (3)$$

$$c_{20,j}\rho_{20,j}\frac{\Delta\hat{\varphi}_1}{2}\left(r_m^{\ 2}-r_{in}^{\ 2}\right)\Delta z\frac{d\theta_{20,j}}{d\tau}=k_{20,j}\frac{\theta_{19,j}-\theta_{20,j}}{\Delta\hat{\varphi}_1 r_{in}}\Delta r\Delta z+$$
$$k_{20,j}\frac{\theta_{11,j}-\theta_{20,j}}{\Delta r}\Delta\hat{\varphi}_1 r_m\Delta z+k_{20,j}\frac{\Delta\hat{\varphi}_1}{2}\left(r_o^{\ 2}-r_m^{\ 2}\right)\frac{\theta_{20,j-1}-\theta_{20,j}}{\Delta z}+ \tag{4}$$
$$k_{20,j}\frac{\Delta\hat{\varphi}_1}{2}\left(r_o^{\ 2}-r_m^{\ 2}\right)\frac{\theta_{20,j+1}-\theta_{20,j}}{\Delta z}+\alpha_j\left(t_j-\theta_{20,j}\right)\ \Delta\hat{\varphi}_1 r_{in}\Delta z$$

The next step in creating an algorithm allowing to determine the temperature distribution and distribution of stresses occurring on the furnace chamber wall was the implication of the equations describing the change of parameters flowing in the waterwall tubes given in [5] and their integration with the adopted division of analyzed elements into the control volumes described above. The applied equations of mass conservation, momentum balance, and energy balance for the factor flowing inside the analyzed waterwall tubes, presented in [5] in the form of differential equations with separated variables, are presented below:

$$\frac{\partial \dot{m}}{\partial z}=-A\frac{\partial \rho}{\partial \tau} \tag{5}$$

$$\frac{\partial}{\partial z}\left(\frac{\dot{m}^2}{A^2\rho}+p\right)=-\frac{1}{A}\frac{\partial \dot{m}}{\partial \tau}-\frac{\partial p_t}{\partial z}-\rho g sin\beta \tag{6}$$

$$\frac{\partial h}{\partial z}=\frac{\rho A}{\dot{m}}\left(-\frac{\partial h}{\partial \tau}+\frac{4\alpha\left(\theta-t\right)}{d_{in}\rho}\right) \tag{7}$$

$$\frac{dh_j^\tau}{dz}=\frac{\rho_j^{\tau-\Delta\tau}A}{\dot{m}_j^{\tau-\Delta\tau}}\left(-\frac{\partial h_j^\tau-\partial h_j^{\tau-\Delta\tau}}{\Delta\tau}+\frac{4\alpha_j^{\tau-\Delta\tau}\left(\theta_j^{\tau-\Delta\tau}-t_j^{\tau-\Delta\tau}\right)}{d_{in}\rho_j^{\tau-\Delta\tau}}\right) \tag{8}$$

$$\rho_j^{\ \tau}=f\left(h_j^{\ \tau},p_j^{\ \tau-\Delta\tau}\right) \tag{9}$$

$$\frac{d\dot{m}_j^{\ \tau}}{dz}=-A\frac{\rho_j^{\ \tau}-\rho_j^{\ \tau-\Delta\tau}}{\Delta\tau} \tag{10}$$

$$\frac{d}{dz}\left(\frac{\left(\dot{m}^2\right)_j^\tau}{A^2\rho_j^{\ \tau}}+p_j^{\ \tau}\right)=-\frac{1}{A}\frac{\dot{m}_j^{\ \tau}-\dot{m}_j^{\ \tau-\Delta\tau}}{\Delta\tau})- \tag{11}$$
$$\frac{dp_t}{dz}-\rho_j^{\ \tau}gsin\varphi$$

$$t_j^{\ \tau}=f\left(h_j^{\ \tau},p_j^{\ \tau}\right) \tag{12}$$

The proposed calculation algorithm, based on the above-described dependences, allows to determine the temperature distribution of the waterwall tube and the factor flowing through it, taking into account all the thermophysical properties of the medium and the furnace chamber wall in individual finite volumes. Below, on (Fig. 4 - 7), the temperature distributions for selected cross-sectional points of the waterwall tube will be presented, determined at different heights above the inlet of the medium to the waterwall tube.

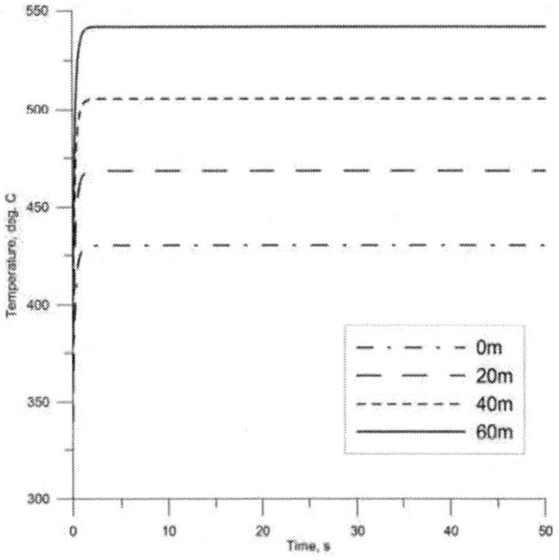

Fig. 4 Wall temperature distribution for point "1" depending on the height above the inlet of the tube

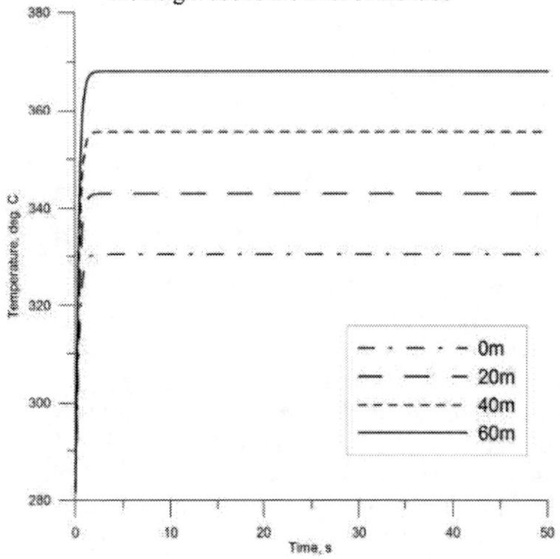

Fig. 5 Wall temperature distribution for point "11" depending on the height above the inlet of the tube

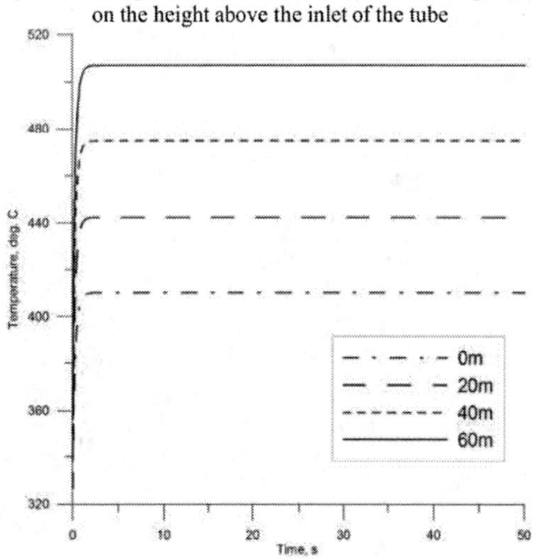

Fig. 6 Wall temperature distribution for point "12" depending on the height above the inlet of the tube

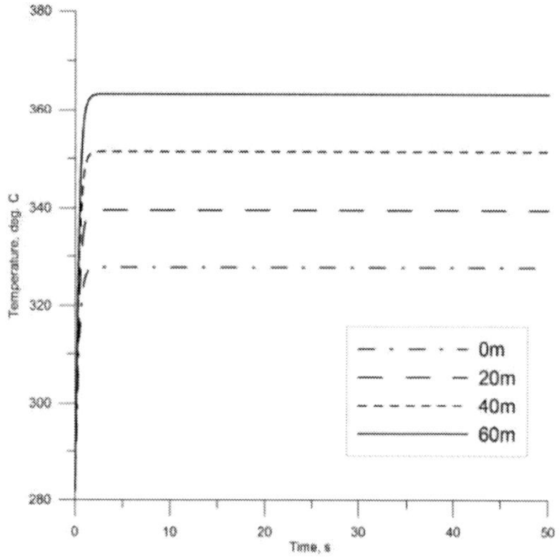

Fig. 7 Wall temperature distribution for point "20" depending on the height above the inlet of the tube

Summary

The paper presents the basic assumptions of the created algorithm allowing for the analysis of temperature distributions occurring in vertical waterwall tubes installed in the boiler's furnace chamber. Numerical calculations will have to take into account the variability of thermophysical parameters of waterwall tubes and the factor flowing in them, depending on the temperature in individual finite volumes. The numerical model will also take into account the difference in the mass flow of the medium flowing through the individual waterwall pipes and the change of the heat load value falling on individual waterwall tubes to accurately reflect the operating conditions of the boiler's evaporator. The results obtained using the numerical model will be the input data for the analysis of stresses occurring in vertical waterwall tubes.

References

1. S. Grądziel, K. Majewski, Chemical and Process Engineering **37**, 2, 199-213 (2016)
2. S. Grądziel, K. Majewski, Procedia Engineering **157**, 44-49 (2016)
3. K. Majewski, S. Grądziel, Chemical and Process Engineering **37**, 2, 251-260 (2016)
4. S. Grądziel, K. Majewski, Journal Heat Transfer Engineering **39**, 13-14, 1243-1250 (2018)
5. S. Grądziel, *Modeling of flow-thermal phenomena occurring in the boiler's evaporator with natural circulation* (in polish) (Wydawnictwo Politechniki Krakowskiej, Cracow, Poland, 2012)
6. W. Zima, M. Nowak-Ocłoń, P. Ocłoń, Energy **92**, 117–127 (2015)
7. W. Zima, M. Nowak-Ocłoń, P. Ocłoń, Energy **148**, 809-823 (2018)

Numerical modelling of retrofitting an industrial boiler type OP-230 for predicting NOx reduction

Przemysław Motyl[1,*], *Jan* Łach[2]

[1]Kazimierz Pulaski University of Technology and Humanities in Radom, Faculty of Mechanical Engineering, Malczewskiego 29, 26-600 Radom, Poland
[2]Bialystok University of Technology, Faculty of Engineering Management, Ojca Tarasiuka 2, 16-001 Kleosin, Poland

Abstract. This study focuses on the indirect co-firing CFD simulations of reductions in NOx emissions relative to the coal combustion. For the given biomass-derived syngas composition and 15% syngas heat input, two arrangements of nozzles dedicated to the syngas injection into the furnace of an older PC-fired boiler of type OP-230 with low-emission burners on the front wall were chosen to investigate the impact of their location on the reduction in NOx emissions. It was shown that the highest reduction achieved by means of the improvement of the air staging via the dual-fuel technique can reach up to 40% (Case 1: nozzles at the elevation of the bottom row of burners) and up to 32% (Case 2: nozzles above the upper level of the burners). The modelling results also showed that one can meet the requirements for co-firing of coal with waste derived syngas. The boiler can be retrofitted for the fulfilment of the prospective environmental protection regulations relating to levels of NOx emissions by combining the described technology with SNCR methods. It is also very important that the indirect co-firing enables reduction of carbon dioxide emissions proportionally to the syngas heat input.

1 Introduction

The direct and indirect biomass co-firing with coal are widely regarded as environmentally friendly technologies that can provide the commonly known benefits although they also have their disadvantages [1-3]. This study is conducted to determine whether a PC-fired boiler of type OP-230 with low-emission burners on the front wall can be effectively retrofitted via indirect biomass co-firing for the fulfilment of the prospective environmental protection regulations relating to levels of NOx emissions. The problem is of modifying the standard air staging technique, originally adapted for lowering NOx emissions slightly below 400 mg/Nm3, by biomass-derived syngas injection above or below the existing coal burners into the furnace through dedicated nozzles offering the largest flexibility in volumetric flow rate. This solution leads to the improvement of the air staging via the dual-fuel technique and its integration with low-NOx coal burners relative to the combustion of coal. The expected reduction in NOx emissions depends on arrangement of syngas nozzles, syngas composition and heat input replacement. The last factor determines the amount of avoided CO2 emissions. The various attractive ways to modify low-NOx emission burners and the air distribution with the use of OFA, ROPA and SOFA nozzles may be found in [4-5]. The overview of NOx reduction technology is also done e.g. in [6-7]. Experiences with the commercial implementation of indirect co-firing technology were acquired mainly in Finland (the Kymijärvi PC-fired CHP plant, Kymijärvi II power plant and the Vaskiluodon Voima plant in Vaasa) but also in Austria (the Zeltweg PC-fired power plant), Belgium (the installation at Ruien) and the Netherlands (the installation at Geertruidenberg) [3,8-9]. A large part of currently conducted research is focused on co-firing simulations with help of ANSYS Fluent CFD software [10] and other CFD software packages [11]. For example, the detailed numerical results for several options of co-firing the paper rejects syngas in a coal-fired boiler were reported in [11]. They are focused on the NOx and CO emissions and furnace LOI depending on syngas heat input and furnace stoichiometry. The highest NOx reduction of 46% relative to the coal combustion corresponded to 23% heat replacement. The simulations of the indirect biomass co-firing in a 600MW tangential PC-fired boiler with 14% heat input injection through the lowest layer burner conducted in [12] showed first of all that NOx emission can be reduced about 50-70% relative to the baseline case (no syngas). The results of similar simulations of co-firing the sewage sludge syngas in a PC-fired OP-650 boiler are given in [13]. The technical feasibility of co-firing biodegradable waste-derived LCV syngas as much as 10, 20 and 40% by heat input with coal in a 230t/h dual PC-gas fired boiler was investigated in [14]. Examining two extreme calorific values, it was discovered that the higher the syngas heat content, the greater coal substitution without the serious impact on boiler operation. The results of above studies give only some

* Corresponding author: p.motyl@gmail.com

idea of the magnitude of NOx reduction relative to the baseline achieved under indirect co-firing different syngas with coal in selected PC-fired boilers. Therefore, for a given boiler and a type of coal and syngas, the reduction in NOx emissions cannot be predicted a priori from known hitherto data. It can be determined through proper research. This justifies the use of our own numerical simulations to conduct the evaluation of reductions in NOx concentrations in the furnace of a PC-fired boiler of type OP-230 with low-emission burners on the front wall. The comparison of the results obtained with the prospective standards of NOx emissions will answer the question formulated above. Simultaneously, it will turn out whether the syngas derived from agro biomass and fuels from waste containing such troublesome components as chlorine and alkali can be co-fired with coal.

2 Formulation of the problem

The basic data of the boiler are as follows: size - 8m×26.7m×8.66m, coal consumption Bc=8.242kg/s, primary and secondary air volume flow rate (and temperature) - 14.48 and 25.96Nm3/s (378 and 603 K), respectively. The secondary air swirl is in line with the burner manufacturer recommendations. The boiler is equipped with (i) eight low-emission burners located at three levels on the front wall and characterized by the concentric dust nozzle supplying a fuel-dust-air mixture into the dust separator in which the mixture is divided into independent streams with alternately high and low concentration of dust coal favouring the formation of a reduction zone; (ii) twelve OFA nozzles installed on both opposite walls; and (iii) two bottom air nozzles located on the side walls in the region near the bottom of the main burners. The air staging system with OFA nozzles ensures complete combustion. The amount of air supplied to the boiler is controlled by monitoring flue gas oxygen level at the outlet from the furnace.

Table 1. Basic properties of coal

Item	Value
Proximate analysis [wt%]	
Moisture	12.4
Volatile matter	22.5
Fixed carbon	52.5
Ash	12.6
Ultimate analysis [wt%] DAF	
Carbon	80.77
Hydrogen	5.87
Oxygen	10.93
Nitrogen	1.73
Sulfur	0.29
Chlorine	0.40
LHV [MJ/kg]	23.44

Table 2. Composition of co-fuel [15] (standard temperature 20°C and pressure 1013hPa)

Components [%mol]	
Item	**Value**
CO2	22.93
CH4	9.27
CO	30.91
H2	36.89
LHV [MJ/kg]	11.97

For the 100% boiler load, six of eight available burners were active in our simulations. The necessary settings were fixed according to the Fluent packet documentation. The mesh system consists of 990 000 polyhedral cells. The hybrid Finite-Rate/Eddy-Dissipation model [10] was used to describe the combustion of coal and the co-firing process. The kinetic data are taken from [15-16]. The standard k-ε turbulence model enables the turbulent Reynolds equations to be closed. The Discrete Ordinates radiation model was used to simulate the radiation heat transfer. The combustion of coal is the baseline case (Case 0). The coal chemical composition is given in Table 1. We limit ourselves to study of co-firing one syngas whose chemical composition is presented in Table 2. In order to evaluate its impact on the reduction in NOx emissions, two arrangements of nozzles were simulated: (i) four nozzles at the elevation of the bottom row of burners (Case 1); and (ii) two nozzles on the front wall above the upper level of the burners (Case 2). These two simulations were conducted under the assumption that the heat replacement is equal to 15%. The co-fuel was fed to the furnace through nozzles of a diameter providing a gas velocity within 25 to 30m/s to ensure gas supply to the center of the combustion chamber. The indicated two arrangements of nozzles resulted in different oxygen concentration and temperature profile in the furnace. The results achieved are compared with those for the baseline (Case 0). The gas supply was accompanied by a change of air distribution in the furnace. The reduced amount of combusted coal dust allowed to move the air from the burners to the OFA nozzles. To simplify a comparative analysis it was assumed that the syngas temperature at the inlet to the boiler was the same in Case 1 and Case 2 and equal to 1000 K.

3 Results and discussion

The first stage of numerical simulations consisted of reproducing the coal combustion conditions (Case 0) using the inputs taken from the recommendations of the boiler and coal burner manufacturers. The average values: T=1340K, λ=1.23, O2=3,6% dry, and NOx=380mg/Nm3 (6%O2, dry) at the furnace exit (Figs. 1 and 2) are within the range declared in the boiler manual and are taken as the reference points for the indirect co-firing analysis (Case 1 and Case 2).

https://doi.org/10.1051/matecconf/201824005021

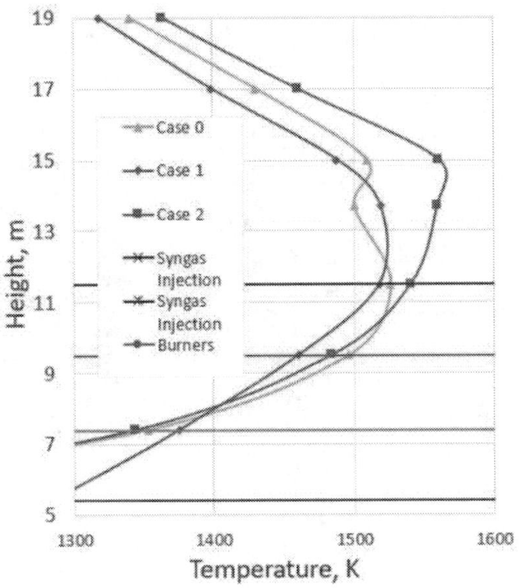

Fig. 1. Mean temperature distribution along the furnace elevations.

Fig.1 shows two peaks of temperature which are slightly displaced towards the rear wall because the burners are located on the front wall. The first peak above the burners is due to the course of the combustion process at region of burners. Next, the following drop in temperature reaches the OFA nozzles. The coal burnout results in slight increase in temperature above OFA nozzles corresponding to the second peak. The results of the second stage of simulations are related to Case 1 and Case 2. On one hand, the syngas low heat content relative to coal and increased moisture content resulting from methane and hydrogen combustion (Table 2) cause a lowering of flame temperature (i.e. flame cooling effect) in the burner zone as shown in Figs. 1 to 3. On the other hand, the syngas co-firing increases the temperature near the co-fuel nozzles.

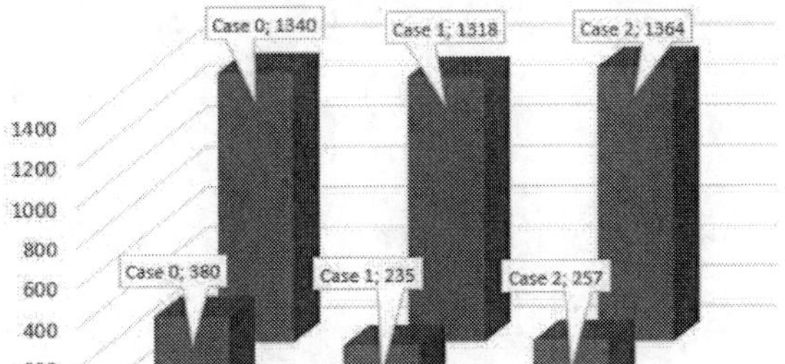

Fig. 2. Mean temperature and NOx concentration at the furnace exit.

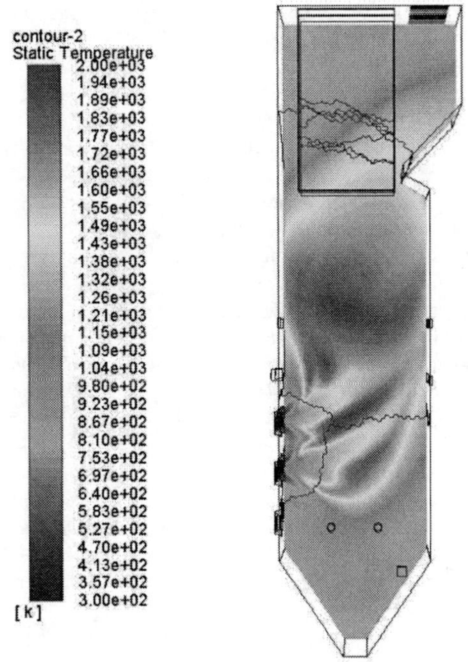

Fig. 3. Temperature profile in the furnace (Case 2)

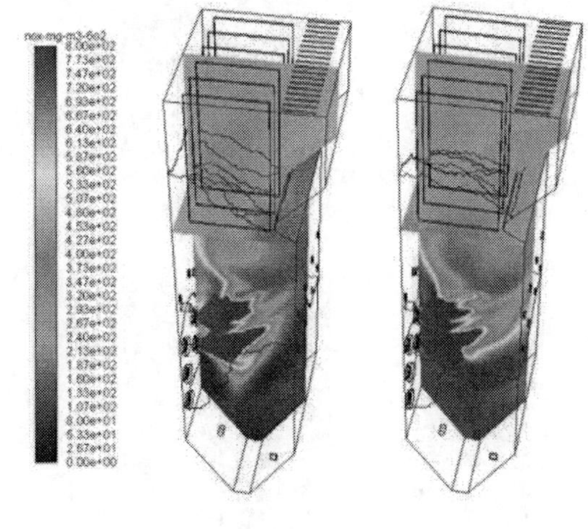

Fig. 4. NOx values [mg/Nm3] for the Case 1 and Case 2.

Comparing the mean temperatures at the furnace exit (Fig. 2), we can conclude that the amount of co-fuel injection in Case 2 is limited because such an injection at the elevation of the upper row of burners while increasing the amount of air in OFA nozzles raises the local temperature in the burnout zone as illustrated in Fig.3. Further increase of gas flow rate can cause an adverse impact on the boiler operation. The NOx concentration along the entire height of the furnace is higher for Case 2 than for Case 1 (Fig. 4) which results from higher combustion temperature (Fig.1). This does not apply to the local fuel riche zone in which the NO is

https://doi.org/10.1051/matecconf/201824005021

removed by hydrocarbons introduced into the furnace with the syngas injection in Case 2. The NOx concentration distribution obtained with Case 1 (Fig. 4) results from: (i) stable oxygen deficient conditions (Fig.5) in the zone of burners what is also illustrated in Fig.6; and (ii) limited successive uniform increase in oxygen concentration (Fig.5) up to levels of OFA nozzles and small amounts of NO production in the burnout zone; both due to the flame cooling effect. The syngas velocity ensuring its injection to the center of combustion chamber does not significantly affect the oxygen depletion near the screens. Decreasing the excess air in the main combustion zone has only a small impact on increasing the exposure of screens to low-oxygen and high-temperature corrosion in the region from burners to the level of OFA nozzles. Figs. 7 and 8 show that Case 1 and Case 2 ensure a long residence time of the co-fuels (more than two seconds) in the combustion chamber and not less than 2 seconds residence at the temperature above 1123K, respectively. The syngas can be therefore derived from gasification of agricultural residues biomass or containing such troublesome components as chlorine and alkali.

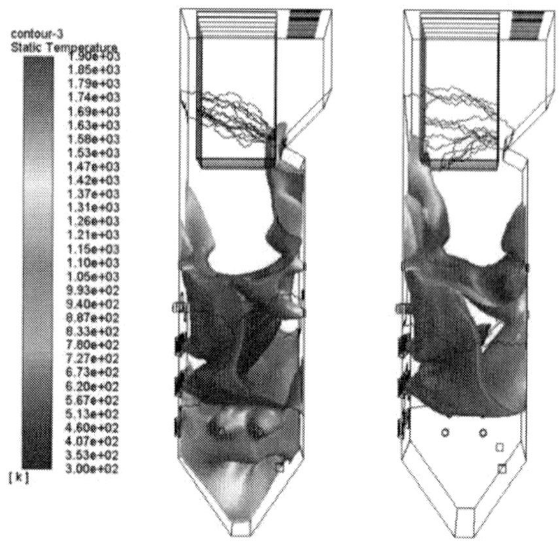

Fig. 6. Substoichiometric zone in the furnace (Case 1 - on the left, Case 2 – on the right) colored according to the temperature-color scale.

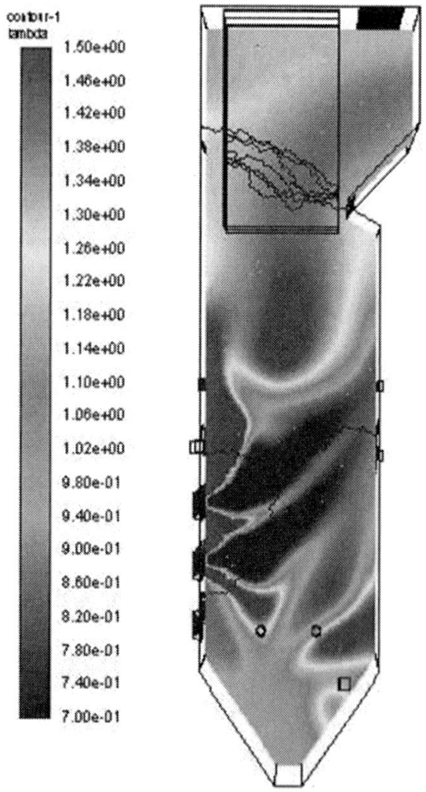

Fig. 5. Excess air ratio distribution in the furnace (Case 1).

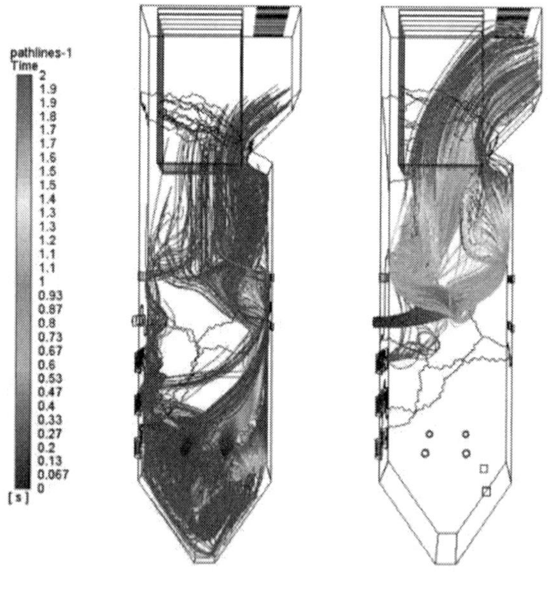

Fig. 7. Pathlines released from the dedicated syngas nozzles (Case 1 - on the left, Case 2 – on the right).

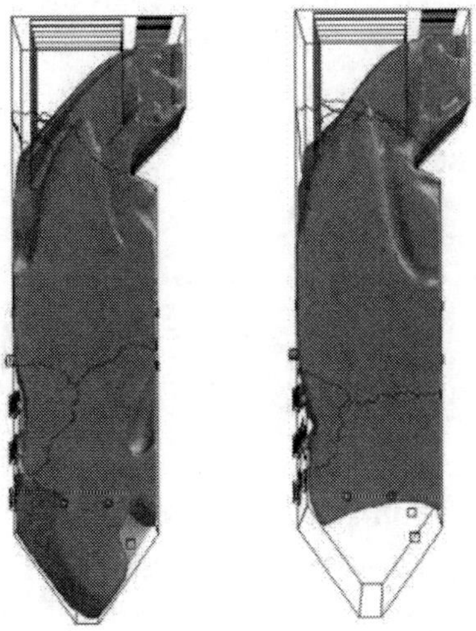

Fig. 8. Temperature region above 1123K (Case 1 - on the left, Case 2 – on the right).

4 Conclusions

Our CFD simulations were directed at the evaluation of possibilities, limitations and benefits of the co-firing environmentally friendly biomass-derived syngas with coal in an older PC-fired boiler of type OP-230 with low-emission burners on the front wall still operating in Poland as heating units in combined heat and power (CHP) district energy plants. Particular attention was focused on the evaluation of reductions in NOx emissions relative to the coal combustion. For the biomass-derived syngas composition (Table 2) and 15% heat input replacement, two arrangements of nozzles dedicated to the syngas injection into the furnace were chosen to investigate the impact of their location on the reduction in NOx emissions. Ensuring the optimal operating conditions of the boiler, it was shown that the highest reduction achieved by means of such an improvement of the air staging via the dual-fuel technique can reach up to 38% (Case 1) and up to 32% (Case 2). The modelling results also showed that one can meet the requirements for co-firing of coal with syngas produced by gasification of agricultural and agro industrial residues [17,18] as well as solid recovered fuels (SRFs) containing such troublesome components as, for example, chlorine and alkali Note that there is particular interest in gasification of agro biomass renewing every year. Fortunately, the mentioned troublesome component remain in the gasifier as slag and ash that are removed from there, therefore does not enter the boiler. Thus, it is easier to use ashes obtained from co-firing as additive for concrete production. The gasification of agro biomass and fuels formed from waste (SRFs, RDFs, REFs) is probably the forward looking technology that can even play a key role in the use of these combustible substances in the future energy sector. In the latter case, we would have to deal with a way to reduce the amount of waste deposited in landfills. The prospective environmental protection regulations relating to NOx emissions can be fulfilled by combining the described technology with SNCR methods. It is also very important that the indirect co-firing enables reduction of carbon dioxide emissions proportionally to the syngas heat input.

Acknowledgements

Research have been carried out in the framework of 3190/23/P and S/WZ/1/2015 works financed by Ministry of Science and Higher Education from the funds for science.

References

1. J.U. Watts, A.N. Mann, D.L. Russel, *An overview of NOx control technologies demonstrated under the Department of Energy's Clean Coal Technology Program* (U.S. DOE, Federal Energy Technology Center, 2000)
2. C.D. Holland, *A Summary of NOx Reduction Technologies* (Texas Institute for Advancement of Chemical Technology, 2002)
3 R. Fernando, *Experience of indirect cofiring of biomass and coal* (IEA Clean Coal Centre, London, UK, 2002)
4. C. Zhou, Y. Wang, Q. Jin, Q. Chen, Y. Zhou, J. ENERGY INST. (to be published)
5. S. Chen, B. He, D. He, Y. Cao, G. Ding, X. Liu, Z. Duan, X. Zhang, J. Song, X. Li, Energy **122**, 287-300 (2017)
6. R.D. Bell, F.P. Buckingham, *An overview of technologies for reduction of oxides of nitrogen from combustion furnace* (https://www.mpr.com/uploads/news/nox-reduction-coal-fired.pdf)
7. R.K. Srivastava, R.E. Hall, S. Khan, K. Culligan, B.W. Lani, J. Air & Waste Manage. Assoc. **55**, 9, 1367-1388 (2005)
8. R. Fernando, *Co-gasification and indirect cofiring of coal and biomass* (IEA Clean Coal Centre, London, UK, 2009)
9. M. Blomberg, *Fuel conversion at the Vaasa power plant: Large-scale biomass gasification plant integrated to a coal-fired boiler* (http://biofuelregion.se/wpcontent/uploads/Mauri-Blomberg-Vaskilouden-Voima-Oy.pdf)
10. ANSYS® Academic Research Fluent Manual, Release 18
11. K.T. Wu, H.T. Lee, C.I. Juch, H.P. Wan, H.S. Shim, B.R. Adams, S.L. Chen, Fuel **83**, 1991-2000 (2004)
12. C. Dong, Y. Yang, R. Yang, J. Zhang, Appl. Energy **87**, 2834-2838 (2010)
13. W.P. Adamczyk, S. Werle, A. Ryfa, Appl. Therm. Eng. **73**, 343-350 (2014)
14. S. Kalisz, M. Pronobis, D. Baxter, Energy **33**, 1770-1778 (2008)
15. R. Chacartegui, M. Torres, D. Sánchez, F. Jiménez, A. Muñoz, T. Sánchez, Fuel Process. Technol. **92**, 213-320 (2011)
16. L. Chen, S.Z. Yong, A. Ghoniem, Prog. Energy Combust. Science **38**, 156-214 (2012)
17. A. Wasiak, O. Orynycz, MPER, **8**, 1, 60-68 (2017)
18. A. Wasiak, O. Orynycz, CMES'17, **15**, (2007)

A model of water-spray cooling effect on a continuous casting process

Hocine Mzad[1,*], *Abdessalam* Otmani[1], *Kamel* Bey[1], and *Stanisław* Łopata[2]

[1]Department of Mechanical Engineering, Badji Mokhtar University of Annaba, P.O. Box 12, DZ-23000, Annaba, Algeria
[2]Institute of Thermal Power Engineering, Cracow University of Technology, Poland

Abstract. The intention of this study is to give an idea about the influence of water-spray cooling on the solidification process of the liquid metal which enables to locate the shear region. The effect of spray heat transfer coefficient (h_{spray}) during the liquid-to-solid transition through the cooled zone temperature and the metal latent heat of solidification are highlighted. A gray iron continuous casting process subjected to water-sprays cooling was simulated using the commercial code COMSOL MULTIPHYSICS 5.2. The obtained results show the great influence of h_{spray} on the location of transition region as well as the relationship between h_{spray}, wall outer temperature, latent heat dissipation, and the solidification time.

Keywords: Metal, Melting, Water-spray, Simulation, Heat transfer.

1 Introduction

The final metal structure is strongly affected by the set up cooling mode throughout liquid metal solidification. Continuous casting transforms molten metal into solid on a continuous basis and includes a variety of important commercial processes. These processes are the most efficient way to solidify large volumes of metal into simple shapes for subsequent processing.

Pascon et al. [1] developed a thermomechanical 2D model of the strand in the mould using the finite element code LAGAMINE. The approach uses the generalized plane strain state that allows a variable thickness of the slice in time, and provides one equation for the vertical force equilibrium.

Water is used to cool the mould in the initial stages of solidification, and then below the mould, where it is in direct contact with the newly solidified surface of the metal. A study by Tebbal and Mzad [2], that contributes to obtain a better quality of metal and steel with a possible minimum of energy and water expenses, lead to an empirical correlation defining the dispersion profile of a water jet beneath liquid sprayers.

Water cooling affects the product quality by controlling the heat removal rate that creates and cools the solid shell and generating thermal stresses and strains inside the solidified metal. The heat extraction coefficient during secondary cooling depends greatly on the surface temperature of the ingot, as represented by boiling water-cooling curves [3].

Several mathematical models are proposed for cooling a metal solidifying in a continuous casting mould, the purpose is to develop such a control to obtain the best possible steel quality, that is, minimizes the defects in the final product. More details about the history of the models are available in [4, 5].

The rate of heat transfer from a plate due to impinging of an array of water jets was investigated numerically through the solution of heat conduction equation, Mzad and Tebbal [6]. To highlight the parameters affecting spray cooling, the simulation was carried out for eight different sprayers in the range of ejected fluid pressure between 1 bar and 3 bars.

Aiming to reduce the occurrence of the surface and internal defects in the products, Ji and Yang [7] developed a dynamic secondary cooling control system. To accurately predict and control temperature in real time during the continuous casting process, a fast, accurate transient solidification and heat transfer model was developed, which provide feedback to a control system.

The disposition of nozzles that create overlapping jet zones results in dispersion profiles that depend on the experimental conditions. In order to understand the fluid behaviour in the overlap region, a research work [8] suggests a hydrodynamic investigation using a program of simulation based on the experimental correlations of pulverized water jets. The influence of the hydrodynamic parameters and the nozzle conception, by varying the

* Corresponding author: h_mzad@yahoo.fr

water pressure, is highlighted which gives a very wide operating range.

In view of an increase of the casting speed under the constraint that the metallurgical length should be maintained constant, Bellet et al. [9] focuses on two points; the impact of local stress generated along the surface and subsurface of the cast product when passing by water spray sprinklers and the automatic determination of the water flow rate to be imposed to each sprinkler.

Recent developments in spray cooling applied to the second cooling are reviewed and evaluated by Raudensky et al. [10]. The quantification of heat transfer performance of spray cooling in continuous casting is then studied and the leading correlations developed for the heat transfer coefficient related to major spray cooling parameters are selected and compared.

A new Nusselt number correlation for spray cooling at large Reynolds numbers and high surface temperatures for water sprays impinging perpendicularly onto a flat plate is presented in Hnizdil's et al. research work [11]. A large set of experimental data on spray cooling of hot surfaces with water has been analysed, including the water temperature effects.

Mzad and Khelif [12] conducted a numerical investigation on heat transfer characteristics of water spray impingement cooling of stationary beryllium-copper alloy plate. The material is exposed to non-stop intensive cooling beneath a full cone spray nozzle. The only controlling parameter taken during the numerical study was water spray pressure in the working range of 1 to 4 bar.

In this paper, a numerical simulation of a continuous casting cooling process was performed based on a new model of temperature variation in the metal transition zone. Obviously, we will analyse the influence of the heat transfer coefficient (h_{spray}) on some of thermal parameters affecting the solidification process like mould wall temperature and metal latent heat of solidification/fusion.

2 Problem modelling

A steady state continuous casting process is modelled and simulated using the commercial code COMSOL MULTIPHYSICS 5.2. The metal is first melted in a furnace and poured into a ladle. From the ladle the hot metal is transferred into the tundish. The hot metal is poured into the continuous casting machine from the tundish. The mould in the continuous casting process is water cooled; this helps speed up the solidification of the metal casting. The continuous casting does not completely harden in the mould; it does spend enough time in the water cooled mould to develop a protective solidified skin of an adequate thickness on the outside.

The thin shell like solidified metal withdraws from the mould and passes through a straightening roller. In the chamber the strand is water sprayed, which prevents porosity. After solidification, predetermined lengths of strands are cut into pieces using mechanical shears or travelling oxyacetylene torches (Fig. 1).

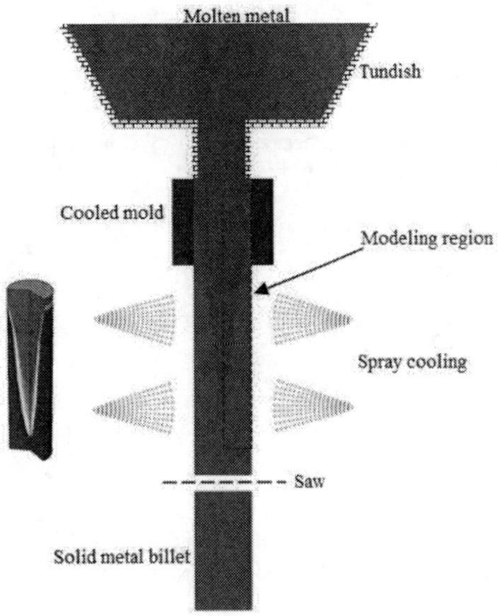

Fig. 1. Schematization of the studied case.

By assumption, in order to simplify the problem, the flow field of the liquid metal was not considered and it was assumed that there was no volume change during solidification. It has also been assumed that the casting speed, of 2.5 mm/s, is constant and uniform throughout the studied domain. In this work we have supposed an axisymmetric 2D geometry of the continuous casting domain composed of two rectangular sections. One represents the metal inside the mould and the second represents the spray cooled region outside the mould. In this region radiation cooling takes place. The considered material is a gray cast iron with the thermal properties shown in table 1.

Table 1. Gray cast iron thermophysical properties.

Parameter and designation	Value
Latent heat of fusion (ΔH_f), J/kg	240×10^3
Solidus temperature T_s, K	1356
Liquidus temperature T_l, K	1463
Specific heat (C_{p_s}) at T_s, J/(kg.K)	660
Specific heat (C_{p_l}) at T_l, J/(kg.K)	950
Thermal conductivity (k_s) at T_s, W/(m.K)	29
Thermal conductivity (k_l) at T_l, W/(m.K)	26
Density (ρ_s) at T_s, kg/m^3	6964
Density (ρ_l) at T_l, kg/m^3	6368
Estimated viscosity (μ_l) at T_l, Pa.s	0.0143

The heat conduction equation which describes the heat transfer of the studied case is written as follows:

$$\rho C_p \frac{\partial T}{\partial t} + \rho C_p u \nabla T + \nabla q = Q \quad (1)$$

Hypothetically, C_{p_s} is supposed constant in solid phase, but in liquid phase C_{p_l} is evaluated using the following formula:

$$C_{p\_l} = C_{p\_s} + \frac{\Delta H}{T_m} \quad (2)$$

The steady state heat equation of continuous casting process is merely written:

$$\rho C_p u \Delta T + \Delta(-k\Delta T) = 0 \quad (3)$$

The density and the specific enthalpy are respectively:

$$\rho = \theta\rho_s + (1-\theta)\rho_l \quad (4)$$

$$H = \frac{1}{\rho}\left[\theta\rho_s H_s + (1-\theta)\rho_l H_l\right] \quad (5)$$

During the liquid to solid transformation, a significant amount of latent heat is released and the total amount of heat released per unit mass during the transition is given by the enthalpy variation ΔH.

Obviously, the specific heat varies before, during and after the transition:

$$\Delta C_p = \frac{\Delta H}{T} \quad (6)$$

After some formal transformations:

$$C_p = \frac{1}{\rho}\left(\theta_1\rho_S C_{p\_s} + \theta_2\rho_l C_{p\_l}\right) + \Delta H_f \frac{\partial \alpha_m}{\partial T} \quad (7)$$

where:

$$\alpha_m = \frac{(1-\theta)\rho_l - \theta\rho_s}{2\left[\theta\rho_S + (1-\theta)\rho_l\right]} \quad (8)$$

The thermal conductivity varies allowing the expression:

$$k = \theta k_s + (1-\theta)k_l \quad (9)$$

The convection heat flux transferred to the mould by the liquid metal can be estimated using Newton equation:

$$q_0 = h(T_{ext} - T) \quad (10)$$

where, h = 800 W/(m^2.K).

Similarly, convection at the outer cooled zone is computed using the equation 10. T_{ext} and T are respectively external and mould wall temperatures. In this situation, h became spray heat transfer coefficient.

As mentioned before, radiation occurs during the continuous casting process, and it is described by Stefan-Boltzmann equation in which the temperature T is expressed in Kelvin:

$$q_0 = \varepsilon\sigma\left(T_{ext}^{\ 4} - T^4\right) \quad (11)$$

where σ symbolise the Stefan-Boltzmann constant and ε the material emissivity (ε = 0.8).

3 Results and discussion

The study consists of water-spray cooling simulation. Figure 2 illustrates the result of this mode of cooling. One can distinguish a complete solidification of the liquid metal before reaching the cutting zone, which is sought in continuous casting processes.

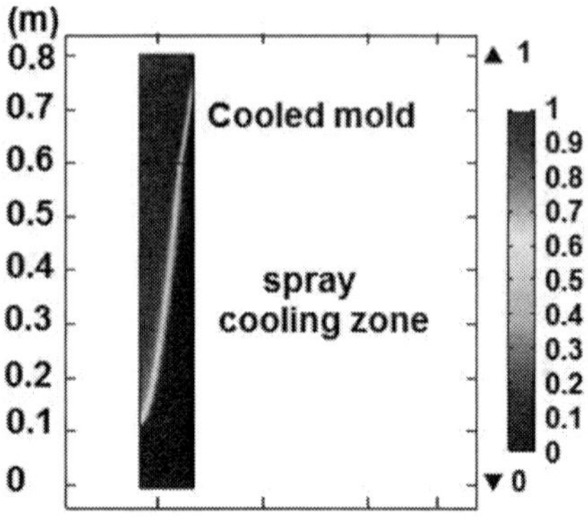

Fig. 2. Spray cooling effect on metal solidification.

The spray cooling process is very efficient since it accelerates the process through a large amount of heat evacuation without water wasting, which imply time and energy saving.

In this section we draw attention to the influence of heat transfer coefficient (h_{spray}) on the cooling rate of the metal. Even if the values adopted during the simulation are below the actual values [6], the obtained results show very significantly the effectiveness of water spray cooling (Fig. 3). Indeed, the increase in h_{spray} accelerates heat transfer.

If we consider the level L = 0.3 m, the wall temperature drops to 35% corresponding to h_{spray} value of 1500 W/(m^2.K). If h_{spray} value is higher, of the order of 6000 W/(m^2.K), the decrease in wall temperature is almost 70%.

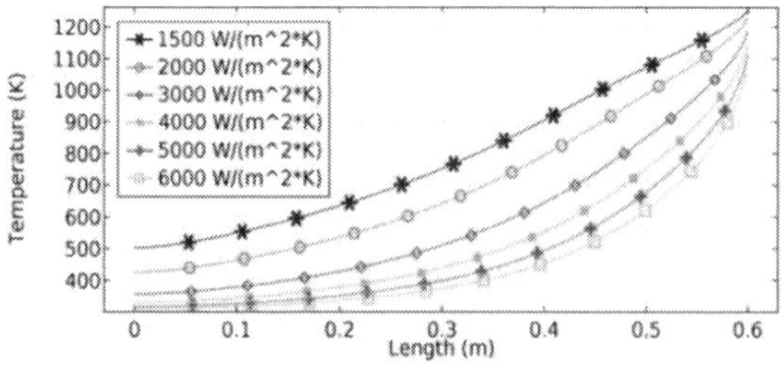

Fig. 3. Influence of spray heat transfer coefficient on wall temperature.

The computed values of latent heat of solidifying are obtained showing different sites of heat evacuation depending on the initial value of h_{spray} (Fig. 4). Visibly, the dissipation of the latent heat is slower when h_{spray} decreases, since the wall temperature remains relatively high.

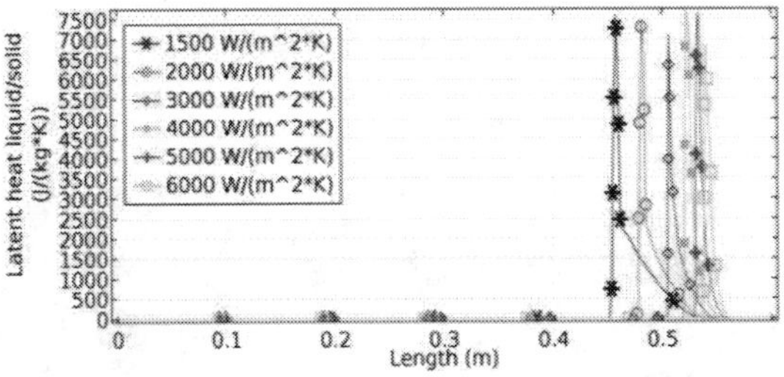

Fig. 4. Influence of spray heat transfer coefficient on latent heat.

The solidification is faster for higher heat transfer coefficients, e.g for h_{spray} = 6000 W/(m^2.K) metal solidification begins at L = 0.5615 m and finish at L = 0.537 m, whereas, for h_{spray} = 1500 W/(m^2.K) the process begins at the level L = 0.532 m and finish at L = 0.451 m (Fig. 5). This investigation is an attempt to take account of the evolution of latent heat and to determine its effect on the rate of solidification which implicitly has a remarkable influence on the location of the transition region (Fig. 6). The metal solidifies gradually from the outer walls, and the rate of advance of crystallization is thought to play an important part in the segregation of impurities from the liquid metal.

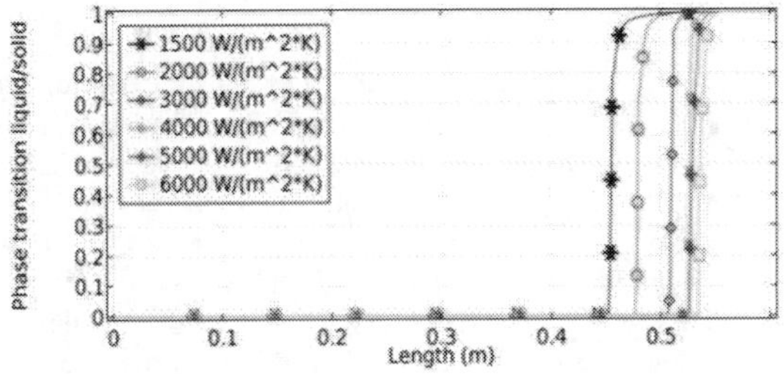

Fig. 5. Influence of spray heat transfer coefficient on phase transition.

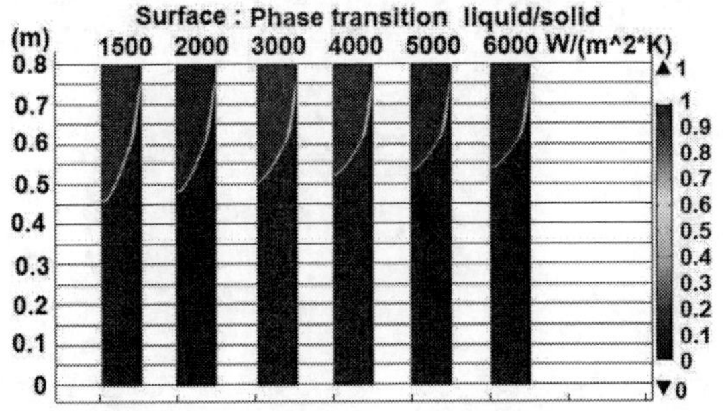

Fig. 6. Influence of spray heat transfer coefficient on transition localization.

4 Final conclusions

The use of water spray cooling through horizontal nozzles in continuous casting process allows a uniform and homogeneous solidification of the metal which ensures a cutting zone entirely solid.

The localization of the liquid-to-solid transition zone is strongly affected by spray heat transfer coefficient. Difference between latent heats of melting and solidifying for the considered metal is probably due to different heat capacities for undercooled melt and solid state.

It is assumed that the performed simulations will be possible to use in optimization of real casting process, but it would be very useful also, due to an insufficient experimental and theoretical data in the accessible literature in case of phase transitions, to perform more significant experiments.

References

1. F. Pascon, A.M. Habraken, M. Bourdouxhe, F. Labory, Mechanics and Industry **1**, 61-70, in French (2000)
2. M. Tebbal, H. Mzad, Forsch. im Ingenieurwesen, **68**, 126-132 (2004)
3. J. Sengupta, B.G. Thomas, M.A. Wells, Metallurgical and Materials Transactions **A, 36**, 187-204 (2005)
4. A.V. Lotov, G.K. Kamenev, V.E. Berezkin, K. Miettinen, Applied Mathematical Modelling **29**, 653-672 (2005)
5. S.V. Lukin, N.I. Shestakov, T.I. Strashko, A.V. Zverev, Russian Metallurgy **3**, 184-188 (2007)
6. H. Mzad, M. Tebbal, Heat Mass Transfer **45**, 287-295 (2009)
7. Z. Ji, J. Yang, *3rd Int. Conf. on Intelligent Networks and Intelligent Systems* (ICINIS), (Shenyang, China, 2010)
8. H. Mzad, M. Elguerri, Atomization and Sprays **22**, 447-460 (2012)
9. M. Bellet, L. Salazar-Betancourt, O. Jaouen, F. Costes, *8th European Continuous Casting Conference* (Graz, Austria, 2014)
10. M. Raudensky, A.A. Tseng, J. Horsky, J. Kominek, Metallurgical Research and Technology **113**, 509 (2016)
11. M. Hnizdil, M. Chabicovsky, M. Raudensky, T.W. Lee, J. of Flow Control Measurement and Visualization **4**, 104-113 (2016)
12. H. Mzad, R. Khelif, Procedia Engineering **157**, 106-113 (2016)

https://doi.org/10.1051/matecconf/201824005023

Study of *Chlorella vulgaris* sedimentation process

Agnieszka Patyna[1], *Małgorzata* Płaczek[1,*], and *Stanisław* Witczak[1]

[1] Department of Process Engineering, Faculty of Mechanical Engineering, Opole University of Technology, Opole, Poland

Abstract. The paper reports the results of *Chlorella vulgaris* sedimentation process including description of cultivation condition of microalgal biomass. The process of algae cultivation was carried out in photobioreactor comprising systems of carbon dioxide supply, mixing and artificial LED illumination. The growth of microalgae was determined alternatively in three ways by measuring the amount of dry mass over time, counting the cells and measurement of optical density by use of a spectrophotometer. Algae biomass with different concentration was subjected to the separation process by gravity. This led to the determination of the characteristic of sedimentation process for different concentrations and cell sizes. The experimental results indicate that sedimentation process offers a tool with a potential application for microalgae harvesting.

1 Introduction

Algal biomass is more and more commonly considered as a raw material with potential application in production of biofuels as well as electricity and heat generation. Due to the depletion of oil resources as well as an increase in environmental pollution, more and more extensive research is being carried out with regard to the potential application of algae biomass as a source of renewable energy [1]. Algae also contain a wide range of nutrients, so they can play the role of a source of food for humans and farm animals. The cultivation of algae does not need to occupy large areas, and thus the production rate of their biomass is much higher than traditional crops. In addition, growing algae in closed photobioreactors can provide high biomass concentrations, as a result of which the cost of algae production can be reduced. Nevertheless, the reactors themselves require the input of relatively expensive materials and consume considerable amounts of energy. The assessment of the economic feasibility of the cultivation also needs to take into account the costs of lighting and CO_2 supply, energy for mixing as well as costs of nutrient. The method applied for separation of this type of biomass, which also determines the efficiency of its cultivation also forms a significant consideration. In spite of this, research works dealing with the production conditions of this type of algae are more numerous, as algal biomass both offers an alternative to renewable energy sources as well as finds a potential application in removing CO_2 from flue gases and in sewage treatment. All of this prove demonstrates the potential applicability of this type of mass in environmental protection.

At present, the separation of algae from cultivation medium and their dewatering form the most energy-consuming stages of the cultivation. As a result, research focusing on the reduction of the overall costs of their production forms a current need. To support this statement, we can say the energy needed for centrifuging algae from 1 liter of culture medium is equal to 13.8 MJ/kg of dry mass on average, and in the case of algal strains with a high lipid content, it can reach as much as 26.2 MJ/kg of dry matter. The cultivation of algae applied in the production of biodiesel is carried out on a large scale and the costs associated with their isolation can incur considerable costs. It is estimated that the cost of algae dewatering forms a 20-30% of the total expenses needed for the production of biofuels [2]. The whole process is also complex due to the microscopic size of the algae cells (2-200 μm) [3] and it is frequently carried out over several stages, as present by the example in Figure 1.

.

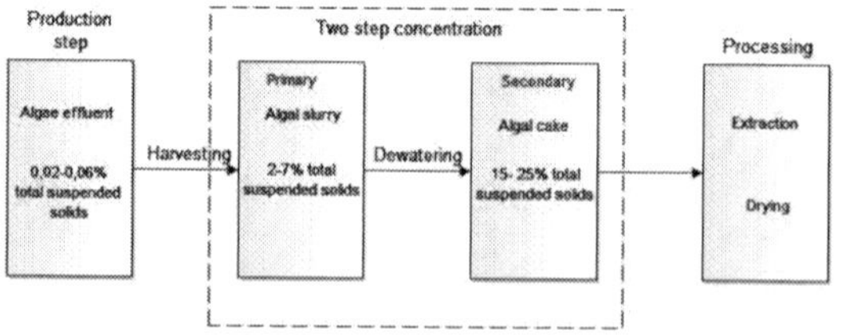

Fig. 1. Diagram of algae cultivation for energy purposes [4].

There are a variety of studies into the extraction of algae accompanying water treatment processes and focusing on the effect of various parameters on this process. Coagulation and flocculation techniques are commonly applied for this purpose. It is also possible to perform the separation of algae from the cultivation medium in a similar manner. The technique applied for

* Corresponding author: m.placzek@po.opole.pl

the separation of biomass from liquid depends primarily on the strain of algae applied in the process and the concentration of the suspension. Initial flocculation and coagulation can also be carried out with the purpose of improving the efficiency of the existing types of mechanical processes (centrifugation, filtration, sedimentation) [5,6,7,8]. In the selection of the technique applied to select algae concentration, the aspect of great importance is associated with the quality of the final product. The use of additional chemicals in the production of algae is not recommended e.g. in the food industry, where high product quality is required.

Among the above methods applied for harvesting of algae, sedimentation plays an important role since this method guarantees low operating costs. However, this approach is very slow (the velocity of gravitational settling of algal cells is equal on average to 0.1-2.6 cm/h) and performs better for larger cells, such as *Spirulina* [3]. Nevertheless, the technique is cheap and offers an easy control the process.

A preliminary flocculation process is carried out in order to improve the sedimentation efficiency, resulting in an increase in the size of the settling particles. Flocculation in this case increases cell dimensions, as a result of which better conditions are provided for the settling of algae. Unfortunately, this requires the use of chemical compounds ($Al_2(SO4)_3$, $FeCl_3$, $Fe_2(SO_4)_3$) [3,4,8]. Many types of algae may also be prone to autoflocculation, the intensity of which depends on the culture conditions. Mechanisms of this are most commonly associated with:

- electrostatic interaction between the negatively charged cell surface and positively charged ions in the culture liquid (neutralization of charges),
- development of salt deposits and extracellular polymers substances (EPS), as well as sweep floc mechanism involving the interaction of microalgae cells with bacteria,
- formation of positive bridge complexes between cells connected by extracellular polymer compounds (bridging).

The potential for the occurrence of various mechanisms of autoflocculation is schematically shown in Fig. 2.

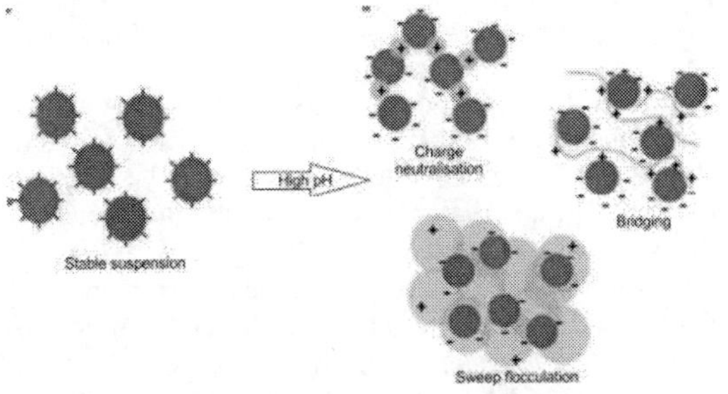

Fig. 2. Examples of autoflocculation types: charge-neutralization, bridging and sweep floc mechanism [9].

Other factors that can influence the autoflocculation process are also worth more extensive discussion. Many studies indicate the dependence of the production of extracellular polymer substances on the level of illumination and temperature [10,11]. Also the concentration of oxygen dissolved in the culture liquid may result in the formation of large and compacted floccoli [12]. The increase in EPS production is also related to the aging of cells in batch cultures, in which the best conditions for flocculation are known to exist at the end of the exponential growth phase and in the stationary phase [13].

2 Results and discussion

The study of sedimentation was carried out in laboratory conditions in order to determine the effects of the variable concentrations of microalgae cells on the velocity and efficiency of the settling process. Two strains of single-cell *Chlorella vulgaris* BA 002 and BA 167 algae were used as the study material in the tests (strains were acquired from the Culture Collection of Baltic Algae of the Institute of Oceanography at the University of Gdańsk). These strains differ slightly in their size and morphology. Algae were cultivated in a stirred tank photobioreactor (Fig. 3) using the standard F/2 production medium based on distilled water at the constant temperature of 27°C.

Fig.3. Experimental setup applied for research of productivity of *Chlorella vulgaris* microalgae.

A summary with details of the production medium based on the procedure described in [14] is found in Table 1. The composition of the production medium plays a crucial role since it considerably determines the productivity of the algal culture and the size and composition of the substances found in the algae (such as lipids).

Table 1. Composition of cultivation medium used in the study.

Medium components	Concentration [g/dm^3]
$FeSO_4 \cdot 7H_2O$	0.042
K_2HPO_4	0.53
$MgSO_4 \cdot 7H_2O$	0.34
Citric acid $\cdot 2H_2O$	1.094
Urea	1.1
$CaCl_2 \cdot 2H_2O$	0.08
Na_2SO_4	0.2
Glucose	40

The algal slurry samples harvested during the culture provided data needed to determine its productivity and individual cell size distribution. As a result of microscopic tests carried out using the Motic AE2000-T inverted microscope with digital camera Moticam 5+ with image analysis software, the study found that both algae strains have a similar range of individual cell sizes, as shown in Fig. 4 and Fig. 5. The mean size of cells for both suspensions was 4 μm.

The microscope was also used to determine the number of cells per unit of the medium volume. Each sample with a given number of cells was tested simultaneously to determine the absorbance value and the dry mass content. For this purpose, the Thermo Scientific Orion AquaMate 8000 UV-Vis Spectrophotometer and a RADWAG moisture analyzer MA.X2.A with accuracy of mass measurement of 0.1 mg were used.

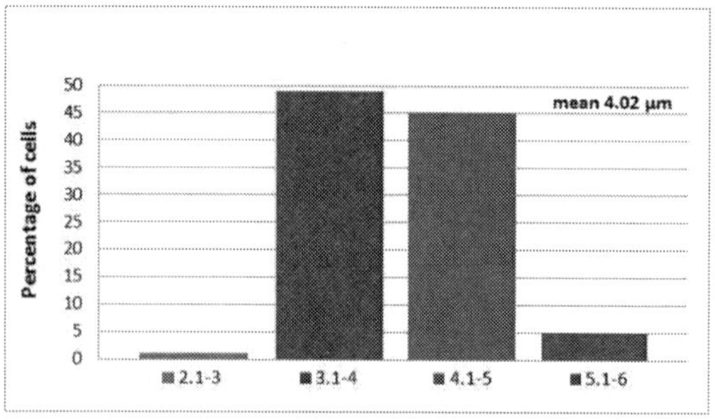

Fig. 4. Distribution of particle size (cells) for *Chlorella vulgaris* BA 002 algae.

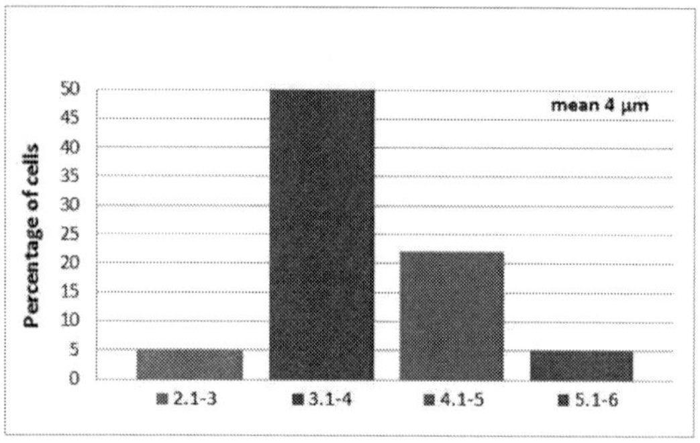

Fig. 5. Distribution of particle size (cells) for *Chlorella vulgaris* BA 167 algae.

The sedimentation tests were carried out in measuring cylinders with a capacity of 100 cm^3. Each cylinder contained a reference scale applied to record the results in millimeters. Tests were carried out for various initial concentrations of the suspension, as shown in Figure 6 for a specific example.

As we can see, the process is accompanied by the occurrence of an interface between the clear liquid and the slurry at the top of the cylinder and between the slurry and sediment in its bottom. In this way, three zones are created with different concentrations of microalgae. During the tests, measurements were made at the time intervals of 5, 15 and 30 minutes depending on the settling velocity of the suspension. At the same time, samples were taken from each of the analyzed zones (i.e. clear liquid zone, interface and the sediment) with the purpose of determining the concentrations of microalgae by applying microscopic observations.

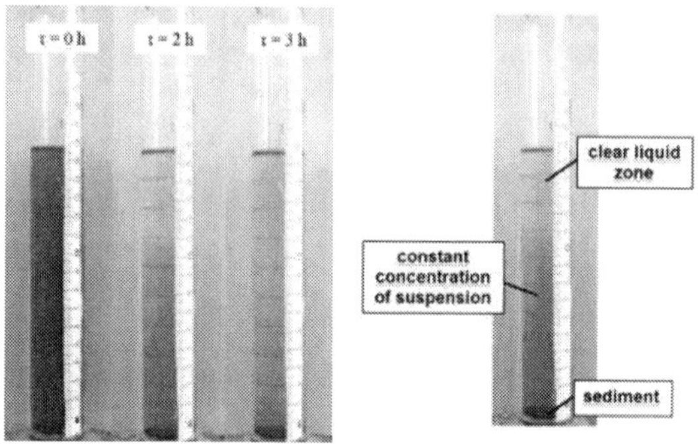

Fig. 6. Suspension with the concentration of $3.25 \cdot 10^5$ cells/cm^3 throughout the settling process.

These observations as well as the measurements of dry mass provided the varying concentration of algal cells in each zone throughout the sedimentation process. Examples with microscope slides obtained for suspensions with various concentrations are found in Fig. 7.

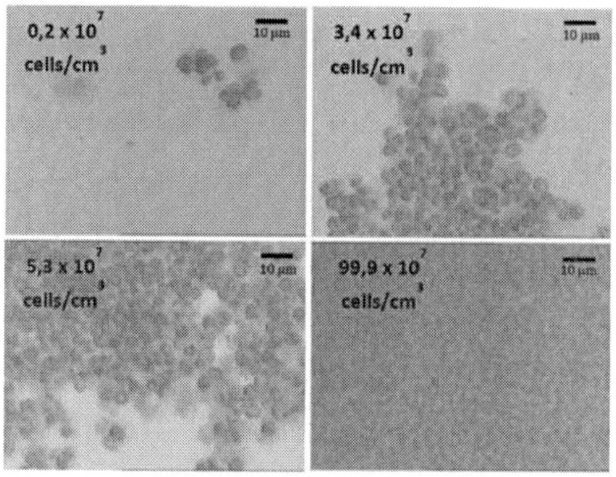

Fig. 7. Examples of microscopic preparations for *Chlorella vulgaris* BA 002 algal suspensions with various concentrations.

As we can see from the microscope photographs, in each zone there is a different number of cells, and in the sludge interface zone, at the level close to the bottom of the cylinder, a variation in the concentration occurs. On the basis of dewatering and drying of samples with different concentrations of algae, a relation was developed to account for the volumetric concentration of algae with their mass concentration, as well as with the absorbance value. These values were subsequently used in assessing the efficiency of sedimentation. The relation between the various sizes for *Chlorella vulgaris* BA 002 strain is summarized in Table 2.

On the basis of the experimental data, sedimentation curves were developed to determine the variations in the location of specific interfaces. Due to the fact that the thickness of the sediment was very small, i.e. equal to 1-3 mm on average, a decision was made that the study will only focus on the zone between the clear liquid and suspension with constant concentration (it was done with the purpose of clarity of the drawings).

Table 2. Measured values of *Chlorella vulgaris* BA 002 algal suspension.

Number of cells in unit volume [cells/cm^3]	Concentration [g/dm^3]	Absorbance [-]
0.5 x 10^7	0.437	0.185
1.0 x 10^7	0.733	0.456
10 x 10^7	6.063	5.342
50.0 x 10^7	29.750	27.06

Examples with the results of sedimentation process of the *Chlorella vulgaris* BA 002 and BA 167 algae strains are presented in Figs. 8 and 9.

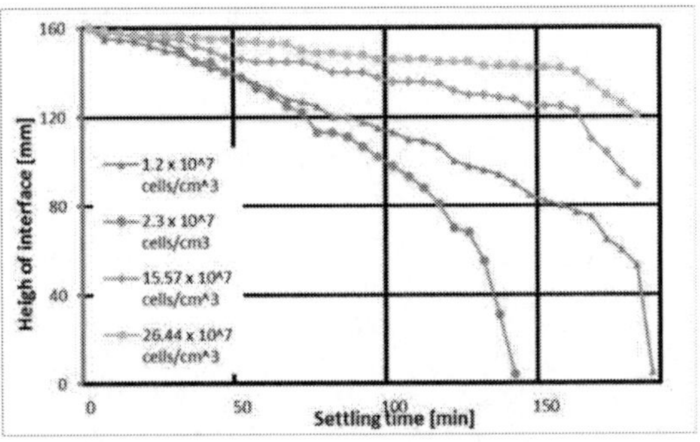

Fig.8. Algae settling curves – *Chlorella vulgaris* BA 002.

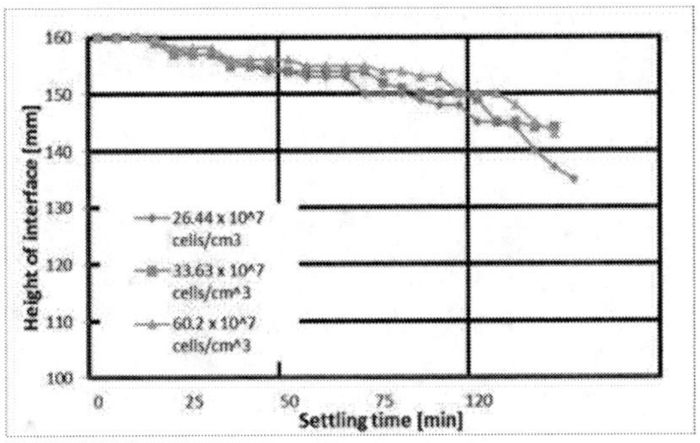

Fig.9. Algae settling curves – *Chlorella vulgaris* BA 167.

The analysis of the experimental data demonstrates that the velocity of formation of the clear liquid zone is variable over time and is relative to the initial value of the concentration of a specific suspension. Throughout the initial phase of the settling period, the course is linear, but after about 120 minutes the settling rate increases. Microscopic observations of the sizes of particle agglomerates formed as a result of autoflocculation were carried out to determine the variable conditions of the suspension during the settling process. Dino-Lite Edge AM7515MT8A Digital Microscope was used for the study. Examples of the results of the particle size formed are shown in Fig.10. We can also note here that the average size of agglomerates in this case was 20-50 μm and it was many times greater from individual algal cells, whose size is 3-5 μm. The studies conducted in this area demonstrated that the effect of algae agglomeration does not occur in the conditions when very low concentrations are present in the suspension and is limited in the extent at its high concentrations.

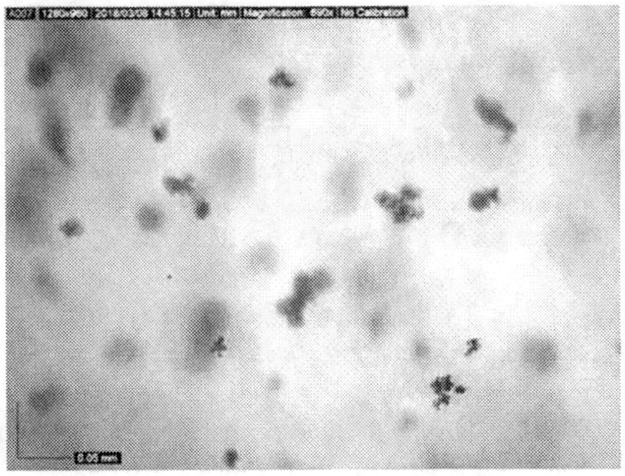

Fig. 10. Microscope image of algal suspension – *Chlorella vulgaris* BA 002 with the initial concentration of 1.3·10^7 cells/cm^3, (911 mg/dm^3).

The consequence of the occurrence of different algal cell sizes during settling process is associated with various settling velocities, as shown in Fig.11.

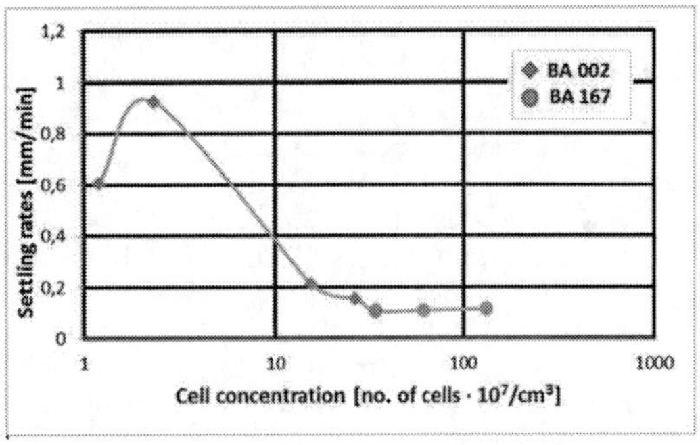

Fig. 11. Settling rates of *Chlorella vulgaris* algae for BA 002 and BA 167 species.

From the course of the curve in this figure, as can conclude that algal cells have lower settling velocities at both low and high concentrations compared to the conditions marked by their mean concentrations. The singularities in terms of settling velocities result from the interactions of the algal cells, which could be likely to either repel each other (as the particles can carry a negative electrostatic charge) or agglomerate, depending on the chemical state of the culture medium. The explanation of this phenomenon requires further systematic research.

3 Conclusions

The present study confirmed the potential applicability of settling processes for harvesting algae from the cultivation medium. The analysis showed that the settling velocity is significantly affected by the initial concentration of the suspension and the conditions in which the cultivation is performed. The acceleration of the process can involve the use of specific concentrations of the suspension and the application of the phenomenon of autoflocculation. A more comprehensive description of the process needs further studies in the area.

References

1. E. Suali, R. Sarbatly, Renewable and Sustainable Energy Reviews **16** (2012)
2. E. Molina Grima, E. H. Belarbi, F. G. Acien Fernandea, A. Robles Medina, Y. Chisti, Biotechnology Advances **20** (2003)
3. I. Rawat, R. Ranjith Kumar, T. Mutanda, F. Bux, Applied Energy **103** (2013)
4. N. Uduman, Y. Qi, M.K. Danquah, G.M. Forde, A. Hoadley, Journal of Renewable and Sustainable Energy **2**, 012701 (2010)
5. K. Dong-Heui, K. Mi-Sug, Water Science & Technology **72**, 5 (2015)
6. T.M. Mata, A.A. Martins, N. S. Caetano, Renewable and Sustainable Energy Reviews **14** (2010)
7. L. Duu-Jong, L. Guan-Yu, C. Yin-Ru, C. Jo-Shu, International Journal of Hydrogen Energy **37** (2012)
8. L. Brennan, P. Owende, Renewable and Sustainable Energy Reviews **14** (2010)
9. N.T. Tran, J.R. Seymour, N. Siboni, C.R. Evenhuis, B. Tamburic, Algal Research **26** (2017)
10. F.M. Lupi, H.M.L. Fernandes, I. Sa-Correia, J.M. Novais, J. Appl. Phycol. **3** (1991)
11. J. Moreno, M.A. Vargas, H. Olivares, J. Rivas, M.G. Guerrero, Journal Biotechnology **60** (1998
12. B.M. Wilén, P. Balmér, Water Research **33** (1991)
13. X. Zhang, P. Amendola, J.C. Hewson, M. Sommerfeld, Q. HU, Bioresource Technology **116** (2012)
14. R.A. Andersen, *Algal Culturing Techniques*, Elsevier Academic Press, New York 2005

Numerical prediction of steady state temperature based on transient measurements

Ewa Pelińska-Olko[1,*], *Marek* Lewkowicz[1]

[1] Wrocław University of Science and Technology, Faculty of Mechanical and Power Engineering, 27 Wyb. Wyspiańskiego st, 50-370 Wrocław, Poland

Abstract. We show how to use numerical analysis of short-time range experimental data for predicting the limit steady-state value of the investigated parameter. In this article the approach has been applied to a specific, although typical, thermal problem: determining the average steady-state temperature of a heater in the convective and radiative heat exchange with the environment. First, we describe a heat exchange experiment aimed at obtaining temperature experimental data in both short and long time range. Then we present a methodology for applying two methods, i.e., neural networks and least squares approximations, for obtaining predictions about the steady-state temperature values based on short time experimental data. The aim of the study is to compare the predictions to each other and to the long time experimental values, with the aim of determining the applicability range of the two methods.

Nomenclature

A, B, D - coefficients of the least squares approximation
$b(i), B(i), G(i)$ - error functions for bases indexed by i
T_e - the environment temperature
$\bar{T}_g$ - the average heater temperature
T_∞ - the limit temperature
n - the number of measurement points
$y(x)$ - test function
$\bar{y}, \bar{y}_i$ - average values of functions

1 Introduction

All physical systems do evolve with time. However, if the system is in steady state, i.e., the dependence of its parameters on time is negligible, then the description of physical phenomena in the system simplifies considerably. That's why many material properties (in particular the thermal ones), such as heat transfer coefficient, Pecklet coefficient, or emissivity, are determined experimentally through measurements performed in steady state. In the experimental practice, steady state is achieved as a result of prolonged maintenance of the system under constant external conditions. Time needed to reach steady state can be very long. For example, for thermal systems, in order to achieve a steady state temperature field, heating times of several hours or days may be required. Such long-lasting experiments become expensive and otherwise troublesome. The main motivation for this paper is an attempt to shorten such long-lasting experiments without substantial loss in accuracy of final results. We compare two numerical procedures predicting the limit steady-state temperature values based on measurements performed in much shorter time. The first method is based on artificial neural networks while the second one uses least squares approximations.

The paper is organized as follows. We start with a description of a test rig providing experimental temperature values to be numerically tested. Next, a short description of neural networks used in the paper is given, followed by a rapid review of the least squares approximation method. Numerical analysis of the experimental data is preceded by a test application of the methods to artificial, computer-generated data. Having passed successfully the test, each method is applied to real data. Finally, the steady-state predictions are obtained and compared.

2 Obtaining experimental data

A test bench described below has been designed in order to obtain several series of experimental temperature data in a time range long enough to be considered as reaching the steady state. Initial portions of the data will serve as input for our numerical procedures. Their output - the limit predictions - will be compared with the limit experimental data.

2.1 Test rig

The main part of our test bench is a heater, which is an aluminum cuboid sized 3 by 3 by 4 cm, containing a

* Corresponding author: ewa.olko@pwr.edu.pl

https://doi.org/10.1051/matecconf/201824005024

resistance wire. The heater is placed in a closed cubical chamber made of a 5 mm thick aluminum plate, with outside dimension 15 cm. The heater is suspended approximately in the center of the chamber. Three thermocouples are placed on its surface. The heater (5) with thermocouples inside the chamber (3) are shown in the right part of Fig.1. The chamber is placed inside the desiccator (1) connected to vacuum pump (2). The heater and the interior of the chamber are painted with black paint of known emissivity $\varepsilon = 0.97$. The temperature is measured by 6 thermocouples (6), previously scaled to each other. Three of them were placed on the surface of the heater, the remaining three - on the aluminum chamber surface. The readings of each triple was averaged, which gave two streams of space-average values: one for the heater and one for the chamber. The voltage is maintained at the desired level by a voltage stabilizer (4).

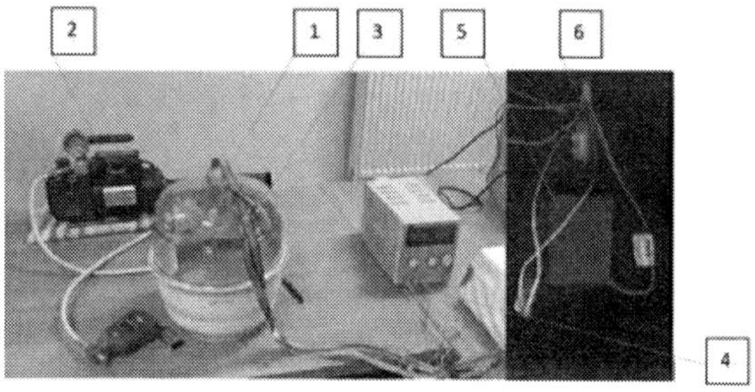

Fig. 1. The test rig.

2.2 Experimental data

The data from the test-bench were saved on the disk, using in-house software coded in Delphi. The measurements were performed in a continuous mode using a Keithley 2000 multimeter. The temperature in the specified places of the station was recorded in the form of strings of voltage values and placed in files that could be read using Excel. Each experiment was performed in two stages. First, the heater was heated for some time at the atmospheric pressure. Then which the air was removed from the desiccator with help of the vacuum pump, until the absolute pressure inside the desiccator reached about 300 Pa. The heat stream dissipated by the heater was still the same, but the lack of air inside the desiccator caused an increase in temperature on the tested surfaces, especially on the surface of the heater. The measurements were performed until temperature stabilized again. The duration of the external conditions at the set level was long in this case. Fig. 2 presents a sample temperature distribution in K, on the heater and on the surface of the closed chamber.

Time, on the abscissa axis, is presented in the form of number of the consecutive measuring points. The measuring point is the moment in which the meter measured the temperature and then saved this information to a file. The average time between two following measuring points is 10.88s. The moment of switching on the vacuum pump about 600th measuring point divides the temperature graph into two parts. The first part represents heating in conditions in which free convection dominated at atmospheric pressure. The second part of the graph shows the temperature change during further heating of the heater under the conditions of absolute pressure in the desiccator at the level of 300Pa. Under such conditions, the heat exchange through radiation was dominant.

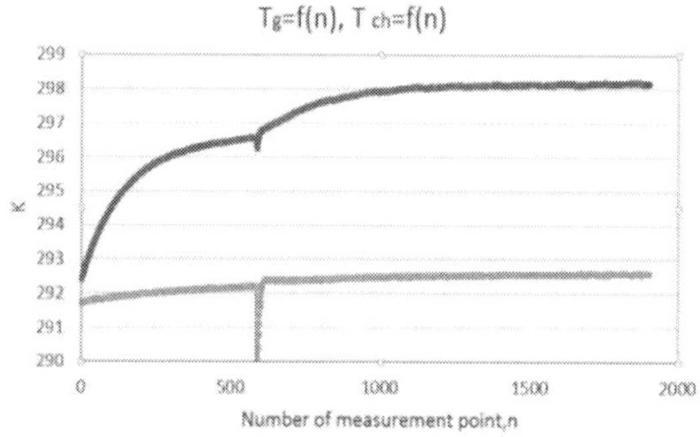

Fig. 2. The temperature of the heater and on the inner surface of the closed chamber in the heating process.

3 Theoretical background

3.1 Artificial neuron network

Artificial neural networks have found many applications. The first of them are from the 1950s. Very simple objects were modeled using the first networks and the network itself was optimized. A breakthrough in the work on artificial networks was the development of the backward propagation algorithm in the 80's, which was rediscovered three times . Networks are currently applied to a wide range of areas. For example, they are used in the regulation of technological processes to optimize regulators, including PID regulators, for complex objects such as boilers, heating nodes [1], [2], gas turbines. Large objects are modeled using neural networks too. In [3], for example, the authors modeled a furnace, in [4] - a power block, and in [5] - a steam cooler. Neural networks are also used in predicting various parameters characterized by high randomness of measured values. In [6], [7] the authors successfully used neural networks to make long-term temperature and air humidity predictions of relevance in housing. Works using neural networks often describe, to a very limited extent, the architecture of the network used, the way of signal processing, but very rarely one can find a description of the procedures for obtaining an appropriate teaching base. Artificial neural networks have been programmed on the model of neuronal cells in the brain, in which the signal is delivered by means of several protrusions, the so-called dendrites, processed in a cell and sent out with exactly one output, the so-called axon. A similar approach is used here. The input data is provided by the so-called input layer to the hidden layer [8]. The calculation results are recovered in the so-called output layer. Our program uses the back propagation algorithm. The accuracy of the returned results is influenced by the structure of the network (weights, architecture) and the number of the hidden neurons [9], [10], [11]. The number of neurons is usually determined by trial and error. In this work we used a program written in R version 3.4.2 X86 = 64-w 64-mgw32 / x64, which works with the nnet database containing the Fit Neural Network module [12]. The neural network used for calculations has one hidden layer, created by 6 neurons.

3.2 The least squares approximation

The method of least squares is a standard approach in data fitting. The best fit in the least-squares sense minimizes the sum of squared residuals, i.e., the differences between the observed (measured) values, and the values provided by a model. The simplest choice for the model is a linear combination of fixed, properly chosen functions. The fit is accomplished by finding the coefficients of the combination, satisfying the least squares paradigm, and this leads to a linear system of equations. Our data are obtained from a physical process of heat exchange between a body and environment, presumably satisfying Newton's Law of Cooling :

$$\frac{dQ}{dt} = hS(T(t) - T_e). \qquad (1)$$

If the body is treated as an internal heat source with constant power Q and heat capacity C, then the heat balance :

$$Q = C\frac{dT}{dt} + hS(T(t) - T_e) \qquad (2)$$

leads to a solution :

$$T(t) = \frac{Q}{hS} + T_e - De^{-\frac{hS}{C}t}. \qquad (3)$$

As h, C are unknown, we represent the solution in the form:

$$T(t) = A - De^{-Bt} \qquad (4)$$

and seek the coefficients A, B, D according to the least squares principle. Thus the model is linear in A, D but non-linear in B. For experimental data $(t_i, T_i), i = 1..n$ it is a standard procedure to seek the best-fit A, B, D coefficients by equating to zero the three partial derivatives of the error function (5) whose minimum is sought :

$$\sum_{i=0}^{n}(A - De^{-Bt_i} - T_i)^2. \qquad (5)$$

This leads to a system of equations (6) :

$$\begin{cases} \sum\left(A - De^{-Bt_i} - T_i\right) = 0 \\ \sum\left(A - De^{-Bt_i} - T_i\right)e^{-Bt_i} = 0 \\ \sum\left(A - De^{-Bt_i} - T_i\right)e^{-Bt_i}t_i = 0 \end{cases} \qquad (6)$$

The first two equations enable one to express A and D explicitly in terms of B. Substitution to the third equation gives a single non-linear equation in B:

$$R_0S_1Y_1 + R_2S_1Y_0 + nR_1S_2 = R_0S_2Y_0 + nR_2Y_1 + R_1S_1^{\,2} \qquad (7)$$

where the notation has been used : $Y_0 = \sum T_i$,
$Y_1 = \sum T_i e^{-Bt_i}$, $S_1 = \sum e^{-Bt_i}$, $S_2 = \sum e^{-2BT_i}$,
$R_0 = \sum e^{-Bt_i}t_i$, $R_1 = \sum e^{-Bt_i}t_iT_i$, $R_2 = \sum_{\tau=0}^{N} e^{-2Bt_i}t_i$.

This highly non-linear equation can be solved numerically and provides the value of B, which lets us find the values of D and A. Thus the long-term prediction of the temperature is made on the basis of:

$$\lim_{t\to\infty} T(t) = \lim_{t\to\infty}\left(A - De^{-Bt}\right) = A. \qquad (8)$$

We also use a simplified approach, in which the approximating function has the form :

$$T(t) = D\left(1 - e^{-Bt}\right). \qquad (9)$$

This corresponds to $A = D$ and is applicable if the data has been shifted to satisfy $t_1 = 0; T_1 = 0$.

4 Modelling based on neural network

4.1 Artificial data

In the first step, we checked the operation of the network for an analytic function with an explicitly defined form: $y(x) = \frac{x}{50 \cdot 2.73^{-5}}\frac{1}{x+1} + 296$; x=1,2,.... The character of the curve was similar to the experimental one, in the temperature range on the right as in Fig.2. The calculated value of the horizontal asymptote is the limit value considered as a steady state value and is: $\lim_{x\to\infty} y = 299.0328$.

The values of the y function were generated in Excel program in a special way. The domain of the y function contained 1,300 cells. The parameter $\overline{y}$ is the arithmetic average value of the y function calculated in Excel from the last 50 of the 1,300 cells. The difference between the asymptotic value and the last one is: $\lim_{x\to\infty} y - \overline{y} = 0.0047$.

It can therefore be considered as a fixed value. The method of using the network to predict y in a steady state consisted of two stages. In the first one, the network was taught on data generated from a specific subset of the domain of the function y. This dataset is hereinafter referred to as the *teaching base*. The teaching base may have a different amount of data, which is referred to as *different length of the teaching base*. In this paper, all teaching bases have a common origin, here it is the point (1, 297). The first database marked as $i = 1$ contained data from the first 100 cells. After the teaching process, the program generated the so-called Testing Vector, which generated the last 50 predicted y-values from which the average value $\bar{y}_i$ was calculated, different for each teaching base. Having $\bar{y}$ and $\bar{y}_i$, three error functions were constructed as:

$$b(i) = \bar{y} - \bar{y}_i, \quad (10)$$

$$B(i) = \bar{y} - \frac{1}{i}\sum_{k=1}^{i} \bar{y}_k, \quad (11)$$

$$G(i) = \bar{y} - \left(\prod_{k=1}^{i} \bar{y}_k\right)^{1/i} \quad (12)$$

where i determines the length of the teaching base. Table 1 contains the values of the error functions for the analytically constructed function y, for different teaching bases.

Table 1. Comparison of the error functions values for the different teaching bases using analytical data.

i	$b(i)$	$B(i)$	$G(i)$
1	0.09604388	0.09604388	0.09604388
2	0.03404388	0.06504388	0.06504549
3	-0.09477612	0.01177055	0.01178111
4	-0.11781612	-0.02062612	-0.02061293
5	0.01704388	-0.01309212	-0.01308119
6	-0.13231612	-0.03296279	-0.03295038
7	0.01604388	-0.02596183	-0.02595071
8	0.00704388	-0.02183612	-0.02182618
9	0.00404388	-0.01896056	-0.01895162
10	0.00804388	-0.01626012	-0.01625196
11	0.00404388	-0.0144143	-0.01440683
12	0.00504388	-0.01279279	-0.01278589
13	0.00024388	-0.01178997	-0.01178358

We can observe that the values of the function $B(i)$ and $G(i)$ are practically the same (Figure 3). The function $b(i)$ monotonically decreases from $i = 7$, to the value of 0.00024 for $i = 13$. The rule is the longer teaching base, the greater compliance of the network values with the real data. $B(i)$ and $G(i)$ reach values below $0.02(i = 3)$ much faster, but their value stabilizes at level about 0.012 only. It should be highlighted, that the temperature measured with the accuracy of 0.02K is considered to be sufficient in the vast majority of measurement practice.

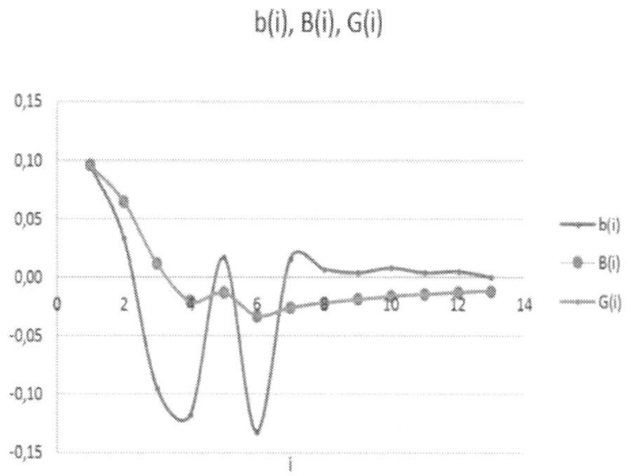

Fig. 3. The dependences of the error functions on i for an analytical data.

4.2 Radiative data

Neural networks had been used in a real experiment to determine the average temperature of the heater in the steady state, in conditions of a heat exchange dominated by radiation. The evolution of the heater temperature is shown in the right part of Figure 2. The total time of heating was about 8.5 h, while the heating under conditions close to the vacuum lasted about 3.93 hours. Speed of temperature changes in the range of the last 50 measurement points stabilizes at the level $\pm 0.036K/h$.

We recognized this temperature as the one in steady-state. The average temperature in the steady state $\bar{T}_g$ is equal to 298.194K. The network predicted the temperature values for the last 50 measurement points of the experiment. The average of these values $\bar{T}_{gi}$ was compared with the corresponding average temperature $\bar{T}_g$, calculated for the actual experimental data. Teaching bases were constructed similarly as previously. We can observe on the base of b(i) value (Table 2) the average steady-state temperatures for the various teaching bases used by the network. We can observe that almost all temperature values (except four items, for $i = 1,2,3,4$) are very close to $\bar{T}_g$.

Additionally the values $b(i)$, $B(i)$, $G(i)$ are increasing in compare to the analytical case . However, the shape of these functions is similar. For $i \geq 5$ the values of error functions begin to monotonically go to zero.

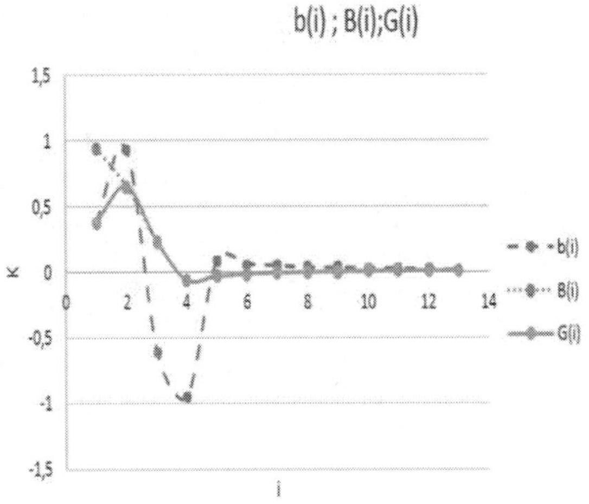

Fig. 3. The dependences of the error functions on i for an analytical data.

Table 2. Comparison of the error functions using experimental-radiative data for different bases.

i	$b(i)$	$B(i)$	$G(i)$
1	0.37498296	0.37944983	0.374982961
2	0.93198296	0.65348300	0.653613300
3	-0.61581700	0.23038300	0.231070200
4	-0.94835700	-0.06430000	-0.063349943
5	0.07898296	-0.03565000	-0.034877927
6	0.05982296	-0.01973000	-0.019092357
7	0.04698296	-0.01020000	-0.009652129
8	0.03398296	-0.00468000	-0.004197394
9	0.03198296	-0.00061000	-0.000177138
10	0.02898296	0.00235300	0.002739001
11	0.02598296	0.00450100	0.004852163
12	0.02110548	0.00587500	0.006196429
13	0.01435032	0.00619100	0.006487703

It is noteworthy that all values $b(i)$ for $i \geq 1$ are greater than $B(i)$ and $G(i)$ for each teaching base. In the analytical case, the value of $b(i)$ was the greatest for small i, but for $i \geq 6$ definitely less than the values of the other error functions. The data used in both cases, i.e. analytical and experimental one, differed by the amplitude of the oscillatory distortions and the place of their occurrence. We also have tested the methodology using bases of equal length placed in different places, but the results were very divergent and did not allow us to draw any conclusions regarding the steady-state temperature. Research in the area will be continued.

4.3 Mixed convective-radiative data

The artificial neural network has been used to determine the temperature value in the distant future, which we consider as the time in steady-state. For this purpose we used the data from the first part of the experiment Fig.2. in which the convective heat exchange was dominating. As before, the teaching bases were constructed as multiples of data from 100 measuring points with a common beginning. To determine the average temperature on the heater, the equations of error functions have to be transformed. If $b(i), B(i), G(i)$ are negligible the average temperature in distant time can be determined as $\overline{T}_g$ in form:

$$\overline{T}_g = \overline{T}_{g_i} \tag{13}$$

$$\overline{T}_g = \frac{1}{i}\sum_{k=1}^{i}\overline{T}_{gk} \tag{14}$$

$$\overline{T}_g = \left(\prod_{k=1}^{i}\overline{T}_{gk}\right)^{1/i} \tag{15}$$

where: $\overline{T}_{gi}$ is the average temperature on the heater provided by the network, for different teaching bases as in Table 3.

Table 3. Temperature in steady state for experimental mixed data expressed by means of error functions elements.

i	$\overline{T}_{gi}$	$\frac{1}{i}\sum_{k=1}^{i}\overline{T}_{gk}$	$\left(\prod_{k=1}^{i}\overline{T}_{gk}\right)^{1/i}$
1	296.125	296.125	296.125
2	296.153	296.139	296.132
3	307.946	300.074	300.023
4	296.470	299.173	299.131
5	296.592	298.657	298.622
6	296.733	298.337	298.306

We can see the values: $\overline{T}_{gi}$. $\frac{1}{i}\sum_{k=1}^{i}\overline{T}_{gk}$. $\left(\prod_{k=1}^{i}\overline{T}_{gk}\right)^{1/i}$ for $i = 3$ returned by the network are much higher than the others. This is probably due to the larger change in the slope of the experimental curve and the specific placement of the teaching base ending. Therefore, this value was excluded in further considerations. After the exclusion $\overline{T}_{gi=3}$ calculated in Table 3, the values of all error functions had been reduced. After recalculating the corresponding values and taking into account that $\frac{1}{i}\sum_{k=1}^{i}\overline{T}_{gk}$ and $\left(\prod_{k=1}^{i}\overline{T}_{gk}\right)^{1/i}$ differ minimally, the range of the returned value $\overline{T}_g$ is:

$$\overline{T}_g \in \left(\overline{T}_{gi=5}; \left(\prod_{k=1}^{5}\overline{T}_{gk=5}\right)^{1/5}\right)\text{K} \tag{16}$$

These are parts of the error functions introduced above. Therefore $\overline{T}_g \in (296.41;296.73)K$, which means the temperature increase is between $\Delta T_g \in (3.98;4.3)$.

5 Modelling based on least squares

5.1 Artificial data

Similarly as for the neural network method, the capability of the least squares approach can be tested with artificial computer-generated data obtained for random t_i and :

$$T_i = A - De^{-Bt_i} \tag{17}$$

where we have fixed, realistic $A;B;D$. The least squares algorithm described above recovers the coefficients with excellent precision.

5.2 Mixed convective-radiative data

The numeri-cal procedure solving the non-linear equation in B and calculating A and D is applied to the 580 data $(t_i;T_i)$ obtained experimentally for the mixed convective-radiative heat exchange described above. Replacing t_i with $t_i - t_1$ and T_i with $T_i - T_1$, one can obtain $t_1 = T_1 = 0$ and use approximation of the form $D(1 - e^{-Bt_i})$. The approximation for the whole range of data gave prediction $\lim T = 4.18138$. The actual (experimental) value at the end of the experiment was $T_{580} = 4.15188$. Note that the linear trend coefficient during the experiment varied from 0.00015 to 0.00010 K/s. The last value corresponds to 0.36 K per hour and one can expect that a comparable raise of the temperature would take place if the experiment was prolonged. Therefore one can suspect that the prediction was underestimated. In order to verify this suspicion our prediction method was applied to several data ranges forming initial segments of the full range. The predictions of the limit values of T obtained for the initial segments containing multiples of 50 experimental data are shown in the first two columns of Table 4. For example, the prediction based on the first 200 data (from the whole set containing 580 data) was 4.02 K. This is certainly an under-estimation, since that value of the temperature difference has been actually obtained as soon as at the 470th measurement. Taking (as the approximation basis) the final segments instead of the initial ones we get predictions shown the last two columns of Table 4. The final segment beginning at 1 is the full range and gives prediction 4.15 as before. Shorter final segments give higher predictions, which confirms our previous observation that temperature grows faster there than it would follow from the theoretical expectation based on Newton's Law.

Table 4. Long-term temperature predictions based on several initial and final data segments.

End of segment	Prediction	Beginning of segment	Prediction
50	3.30	1	4.15
100	3.84	50	4.15
150	3.96	100	4.16
200	4.02	150	4.17
250	4.06	200	4.18
300	4.08	250	4.19
350	4.09	300	4.21
400	4.11	350	4.22
450	4.13	400	4.23
500	4.15	450	4.26

5.3 Pure radiative data

The second set of the experimental data was obtained for pure radiative heat exchange. The experiment lasted longer and the trend line coefficient reached 0.00001 K/s, which is 10 times smaller than in the first experiment. This corresponds to 0.036 K per hour growth and lets us consider the process to be stabilized. We agree that the steady-state (limit) value is $T_\infty = 1.39188$, the average of the last 100 experimental values. Averaging was applied in order to eliminate oscillations. First we note that the approximation by a function $D(1 - e^{-Bt_i})$in the full range seems to be slightly better than that for convection, as the standard deviation is now 0.016 and average error is 0.013, compared to 0.023 and 0.019 for the previous set of data. It is crucial to verify the quality of the approximation based on initial segments of different length. The relevant information is collected in Table 5. The length of the segment is given in the first column; it is followed by the coefficients B and C for the approximating function of the form $D(1 - e^{-Bt_i})$. Thus D is the forecast of the approximated data at infinity. This should be compared with $T_\infty = 1.39188$. It can be observed that the initial lengths greater than 500 give forecasts which deviate from T_∞ by at most 0.01 K. For the initial segment of length 400, the error is 0.017, and for 300 it is as big as 0.06. Shorter segments give unacceptably large forecast errors. The last column gives the forecast based on the approximation of the more general form $A - De^{-Bt_i}$. It turns out that the forecasts for long intervals differ from the previous ones by mK. For shorter intervals the differences are bigger, reaching 0.03 K, always in favor of the second approximation form.

Table 5. Coefficients B and limit temperature predictions.

length	Coefficient B	$T_{\infty(B,D)}$	$T_{\infty(A,B,D)}$
1300	0.00039747	1.38359	1.38113
1200	0.00040014	1.38016	1.37694
1100	0.00040198	1.37773	1.37358
1000	0.00040221	1.37739	1.37217
900	0.00040096	1.37917	1.37262
800	0.00039796	1.38381	1.37553
700	0.00039586	1.38728	1.37607
600	0.00039355	1.39130	1.37529
500	0.00038530	1.40726	1.38432
400	0.00036519	1.45529	1.42144
300	0.00034789	1.50449	1.44393
250	0.00029926	1.68746	1.62360
200	0.00020915	2.27300	2.48332

6 Conclusions

Sample temperature distribution for the mixed convective-radiative and almost purely radiative heat exchange processes have been obtained experimentally in order to provide data needed for the analysis performed in the paper. The steady-state temperature, i.e., its value at distant time, for the chosen distribution has been gained in two ways. One uses artificial neutral network and the other is based on the least squares algorithm.

Each method used initial data segments of different length and provided an independent forecast of the limit steady-state temperature value. For long initial segments the methods gave temperature forecasts highly compatible with each other as well as with the experimental steady-state temperature value. As expected, for shorter intervals the forecasts were less compatible. In any case, the application of the approximation makes it possible to obtain a reliable forecast of the steady-state limit temperature with accuracy of approximately 0.03 K based on experiment lasting 3-4 hours, while reaching the same limit temperature experimentally required about 8-10 hours. Shortening of the experiment time allows one to reduce costs and other inconvenience connected with prolonged experiments.

The predictions of steady-state temperature obtainable by least squares are less prone to input distortion than those from neural networks. These issues are the subject of further work.

References

1. J. Lichota, Rynek Energii **4,** 61-67 (2009)
2. J. Lichota, M. Skałecki, *35-th International Technical Conference of Clean Coal & Fuel Systems* (Florida, USA 2010)
3. H. Rusinowski, W. Stanek, Energ. Convers. Manage. **48,** 2802-2809 (2007).
4. S. Lu, B. Hoog, Int. J. Elec. Power **22,** 67-78 (2000)
5. M. Mohammadzaheri, L. Chen, A. Ghaffar, J. Willison, Simul. Model. Pract. Th. **17,** 398-407 (2009)
6. A. Ruano, E. Crispim, E. Conceição, M. Lúcio Energ. Buildings **38**, 682-694 (2006)
7. L. Mba, P. Meukam, A. Kemajou, Energ. Buildings **121,** 32-34
8. R. Tadeusiewicz, *Sieci neuronowe* (Warszawa 1992)
9. M. Eldracher, *The 1992 ACM/SIGAPP Symposium on Applied Computing 1,* 1098-1104 (ACM Press, 1992)
10. E. Cantu-Paz, *Genetic and Evolutionary Computation Conference,* 790-800 (Springer, Berlin, Heidelberg 2003)
11. N. Pavlidis, D. Tausolis,V. Plagianakos, G. Nikiforidis, M. Vrahatis, *IJCNN 2005. Proceedings 2005 IEEE International Joint Conference*, **4**, 2190-2194 (2005)
12. https://cran.rproject.org/web/pacages/nnet/nnet.pdf

Comparison of dust emissions from an individual wood-fired boiler and dust emissions in the production of the same amount of heat in a combined heat and power plant

Amadeusz Rak[1], and *Leszek* Kulesza[1*]

[1]Cracow University of Technology Faculty of Environmental Engineering ul. Warszawska 24, 31-155 Kraków, Poland

Abstract. Comparison of dust emissions from an individual wood fired boiler (of various types) and of dust emissions during generation of the same amount of heat in a heat and power station were the aim of this work. For the purpose of the calculations the dust emissions from the heat and power station were assumed as the maximum permissible dust emissions, in line with the regulation of the Minister of Environment of 4 November 2014 on emission standards for certain types of installations, fuel combustion sources and waste incineration and co-incineration devices. The emissions of dusts for boilers were assumed based on available publications. For the purposes of the analysis, three boilers of 15 [kW] power were considered. Based on the obtained results it can be stated that the amounts of dust emitted by selected wood-fired boilers are not significantly different from the amount of dust emitted by a heat and power station for the same effect on the recipient's side. Heating boilers of powers lower than 15 [kW] exist and for some of them the dust emissions are even smaller than for the boilers discussed herewith. The discussed boilers have heating efficiency of 86 %. If their efficiency were to be higher, the concentration of the dust would have been lower. The solid fuel (wood) fired boilers considered in this work are not the best ones on the market with respect the dust emissions. The ongoing development of the fuel combustion technology allows expecting even further reduction of emissions of dusts.

1 Introduction

A high number of boilers of low power and low energy efficiency is operated in Poland. They are emitting high volumes of products of incomplete combustion. In order to limit the emission of solid combustion products and to increase the energy efficiency of boilers, the heat sources have to be properly designed and operated.

The main method for the generation of the primary energy is combustion of fuels. This process is however also one of the key sources of emissions to air. Even though the existing technologies in the power industry allow to meet the legal requirements on the emissions to air, there are still existing vast possibilities to further reduce the emissions of air pollutants by means of improved air protection technologies [1].

Comparison of dust emissions from an individual wood fired boiler (of various types) and of dust emissions during generation of the same amount of heat in a heat and power station were the aim of this work. As in reality the distance between a recipient and a heat and power station varies, the considerations assumed utilization of heat from a heating network and electric heating. Dust emissions of individual wood-fired boilers of 15 [kW] power in a single-family house were compared with dust emissions from a heat and power station during generation of the same amount of heat. An analysis of the results and conclusions conclude the paper.

2 Legal requirements for dust emissions from heat and power stations

Legal requirements on dust emissions for heat and power stations have been established by the regulation of the Minister of Environment of 4 November 2014 [2]. It contains emission standards for sulphur dioxide, nitrogen oxides, carbon monoxide and dust. The requirements for dusts have been referenced in table 1.

Table 1. Emission standards for combustion of solid and liquid fuels, excluding gas engines.

Nominal heating power of the source in [MW]	Dust emission standards in [mg/m^3]		
	for 6% oxygen content in the off-gases		for 15% oxygen content in the off-gases for gas turbines and 3% for the remaining sources
	biomass and peat	other solid fuels	liquid fuels
1	2	4	5

* Corresponding author: lkulesza@wp.pl

≥ 50 and ≤100	30	30	30
≥ 100 and ≤300	20	25	25
> 300	20	20	20

Reference: Regulation of the Minister of Environment of 4 November 2014 on emission standards for certain types of installations, fuel combustion sources and waste incineration and co-incineration devices. (Official Journal, position 1546 of 7 November 2014), p. 22. [2]

3 Emissions of air pollutants from small-scale boilers manufactured in Poland

Table 2 presents, based on literature references, results of measurements of concentrations of pollutants produced during combustion of fuels in various heating boilers manufactured in Poland. All of the boilers (except for one, which was then omitted in further considerations) have heating efficiency exceeding 85%. The tests were conducted in accredited laboratories and the results were provided in the technical documentation and sales folders.

Table 2. Comparison of pollutant concentrations produced by various boilers [3].

Boiler type	Power [kW]	Fuel	Dust Concentration at 10% O_2 [mg/m^3]
FUWI chips fired	15	woodchips	46
FUWI/Bio-P	29	woodchips	21
FUWI chips fired	30	woodchips	32
Cichewicz /Futura Picus	14 ÷60	woodchips	90
HEF/Eko-Plus "U" Pellets	18 ÷50	woodchips	55
SAS/ Agro-Eco	15 ÷100	woodchips	50÷75
Termo-Tech/Biomaster	10 ÷100	woodchips	80
MCI Sp. z.o.o./ R-Eco Pellet	24,6	woodchips	25
Koterm/ KWMP 4	31 ÷60	woodchips	20÷60
Defro Duo/ Defro Duo	15 ÷42	woodchips	45÷70
Zębiec/KP	15 ÷60	woodchips	90
Elekromet/ Eko-PE	20 ÷35	woodchips	45÷80
Elekromet/ Eko-PE	20	woodchips	45
Hiton/Hiton-25	15 ÷25	woodchips	145
Elekromet/ Eko-PE	35	woodchips	80
Sigma Dual - chips	20 ÷25	woodchips	-
Sigma Dual - wood	20	wood	-
Cichewicz/ FB25	20	sawdust briquettes	485
Buderus KP	18 ÷38	wood logs	150
Hiton/ Moderator	20	oats	560
Moderator/Smok	60	biomass briquettes	495

Reference: Marek Juszczak, Low power, biomass-fired source of heat: energy-environmental efficiency for selected fuels, published by Poznan University of Technology, Poznan 2016, p. 44. [3]

4 Comparative calculations of dust emissions for a boiler of 15 [kW] power fired with beech chips and for a heat and power station

For the purposes of the analysis, three boilers of 15 [kW] power were considered: woodchips-fired FUWI, Termo-Tech/Biomaster and Hiton/Hiton-25, of efficiency 86%. Stoichiometric calculations assuming the net calorific value of the woodchips W_d = 14836 [kJ/kg], indicated that the required feed of woodchips is 4,23 [kg/h] and the stream of off-gases is 31,46 [m^3/h].

The power of the comparative heat and power station was assumed as 100 [MW]. The efficiency of heat generation was assumed as 85 %, of heat distribution as 86,7 %, of electricity generation as 38 %, and of distribution of electricity as 92,7 %. Based on stoichiometric calculations, with the assumed net calorific value of bituminous coal W_d = 21000 [kJ/kg], the required feed of the coal was:
For heating with district heating - 3,49 [kg/h]
and stream of off-gases 33,61 [m^3/h].
For electric heating – 7,30 [kg/h]
and stream of off-gases 70.30 [m^3/h].

The resulting stream of off-gases carries the amount of dusts equalling the standards listed in table 1 for heat and power stations, i.e. 30 [mg/m^3] and for small-sized boilers equalling the one from table 2 - i.e. 46, 80 and 145 [mg/m^3] for each boiler type respectively. The results have been elaborated in table 3.

5 Analysis of results and conclusions

For the purpose of the calculations the dust emissions from the heat and power station were assumed as the maximum permissible dust emissions, in line with the regulation of the Minister of Environment of 4 November 2014 on emission standards for certain types of installations, fuel combustion sources and waste incineration and co-incineration devices. In reality, this emission might be lower, meaning the calculated amount of dust emissions from the heat and power station might be lower. The emissions of dusts for boilers were assumed based on the publication of Marek Juszczak [3].

Based on the obtained results it can be stated that **the amounts of dust emitted by selected wood-fired boilers are not significantly different from the**

amount of dust emitted by a heat and power station for the same effect on the recipient's side. In addition, the FUWI boiler generates a significantly smaller amount of dust in comparison to the heat and power station when the building is heated using electric heating.

Heating boilers of powers lower than 15 [kW] exist and for some of them the dust emissions are even smaller than for the boilers discussed herewith. One of them is boiler Futura Pellets of heating power equal 5,4 [kW] which generates 6 [mg/m^3] of dust.

Table 1. Compilation of results.

Selected boilers				
Boiler type	Power [kW]	Fuel	Dust concentration at 10% O_2 [mg/m^3]	Amount of dust generated over the course of one hour [g/h]
FUWI woodchips fired	15	Woodchips	46	1,447
Temo-Tech/ Biomaster	15	Woodchips	80	2,517
Hilton/ Hilton-25	15	Woodchips	145	4,562
Energy	Power [MW]	Fuel	Dust concentration at 10% O_2 [mg/m^3]	Amount of dust generated over the course of one hour [g/h]
Heat and power station (heat)	100	Bituminous coal	30	1,008
Heat and power station (electricity)	100	Bituminous coal	30	2,109

Reference: Own calculations.

The discussed boilers have heating efficiency of 86 %. If their efficiency were to be higher, the concentration of the dust would have been lower. An example here is boiler Pellets Unit ETA of 15 [kW] heating power and heating efficiency 93,5 % which emits 19 [mg/m^3] of dust.

The solid fuel (wood) fired boilers considered in this work, as one can conclude from the previous paragraph, are not the best ones on the market with respect the dust emissions. The ongoing development of the fuel combustion technology allows expecting even further reduction of emissions of dusts.

Coupling of a single-family house to a heat distribution network generates costs - for the construction of the district heating substation and the fee for the hook-up. Part (or even the entirety) of these costs can be covered by the utilities company. It is dependent however on its policy.

When a given building is to be heated using electric heating, it must be powered with three-phase electric power and the wiring inside the building has to be adapted as well. These operations also generate costs.

For a small boiler, all data is taken from materials from the bibliography; there is no description of the sensors there. For the boiler of the combined heat and power plant, the maximum permissible dust emission values were adopted for the calculations (measurement data were not used). Dust emitted from small boilers propagates at low altitude - a maximum of a dozen or so meters. Dust emitted from the chimney of the heat and power plant propagates at the height of several dozen meters.

References

1. J.Telenga-Kopyczyńska, J.Konieczyński, *Perspektywy ograniczenia emisji pyłu, ditlenku siarki i tlenków azotu w wyniku zastosowania BAT w elektroenergetyce i ciepłownictwie w Polsce,* (1th edition, Instytut Chemicznej Przeróbki Węgla, Zabrze, Instytut Podstaw Inżynierii Środowiska PAN, Zabrze, 2008)
2. Rozporządzenia Ministra Środowiska z dnia 4 listopada 2014 r. w sprawie standardów emisji dla niektórych rodzajów instalacji, źródeł spalania paliw oraz urządzeń spalania lub współspalania odpadów, Dziennik. Ustaw, poz. 1546 z dnia 7 listopada 2014r.
3. M. Juszczak, *Źródło ciepła małej mocy zasilane biomasą: efektywność energetyczno-ekologiczna dla wybranych paliw* (1th edition, Wydawnictwo Politechniki Poznańskiej, Poznań, 2016)

Prediction of a methane circular pool fire with fireFoam

Camilo Sedano[1], *Omar* López[1,*], *Alexander* Ladino[2], and *Felipe* Muñoz[3]

[1]Department of Mechanical Engineering, Universidad de los Andes, 111711 Bogotá, Colombia
[2]Department of Mechanical Engineering, Universidad del Valle, Cali, Colombia
[3]Department of Chemical Engineering, Universidad de los Andes, 111711 Bogotá, Colombia

Abstract. In the present work, the *fireFoam* solver was used with Large Eddy Simulation (LES) and the Eddy Dissipation Concept (EDC) for modelling a medium-scale methane pool fire. A convergence analysis performed, showed that a 2 Million elements three-dimensional mesh, is good enough to attain good numerical results. By comparing the numerical results obtained, with the experimental ones, as well as numerical results from previous studies, it was proven that the *fireFoam* solver is able to obtain satisfactory results.

1 Introduction

Developments in fire safety engineering have focused on fire detection, suppression, structures' heating and smoke filling rates [1]. Classically, fire phenomena have been studied through experimental techniques, which can be highly expensive. Also, due to the random and destructive nature of fires, experimental results can be nearly impossible to replicate. Due to this, computational and numerical modelling of fire phenomena has received an increasing interest in recent years.

All the same, several difficulties arise when modelling fire dynamics. In particular, different phenomena are taking place simultaneously, such as: turbulent flow and turbulent mixture processes, thermodynamics, heat transfer (radiation) and chemical kinetics [2]. Furthermore, the fact that each of these phenomena considers different time and length scales, thus implying that there are several assumptions that must be applied in order to couple these. This is why, different models which simplify the combustion, chemical kinetics and fluid dynamics processes must be used [3].

In conflagrations (defined as an uncontrolled fire spread [2]), these simplifications lead to two main types of models: zone and field models. The first one separates the space in which the fire takes place into two zones: one where the products of the combustion process remain and another where the reactants remain, until consumed by the reaction. On the other hand, field models use Computational Fluid Dynamics (CFD) for reactive flows, in order to solve the coupled equations of Navier-Stokes with chemical kinetics (mixture fracture solution) [4].

Although field models have a higher mathematical complexity than zone models, the first are able to adjust to almost any geometric domain and constraints. Therefore, this field has recently drawn an increasing interest, in order to develop reliable models which are accurate and exact enough to predict fires. Some known field models are the Flame Surface Density model developed by Trouvé [5], the Partially Stirred Reactor model of Chen [6] or the Laminar Flamelet Model initially proposed by Peters [7].

The Eddy Dissipation Concept (EDC) is one of the main combustion models used within field models. In it, Magnussen recommends a way to "relate the rate of combustion to the rate of dissipation of eddies" [8], by assuming that the rate of reaction depends on the mean concentration of a reacting specie, turbulent kinetic energy and its' dissipation rate [9]. The EDC began as a mode able to consider turbulent and momentum mixing, while taking into account the chemical kinetics. Most recently the interest in this model has shifted to develop an EDC which can use Large Eddy Simulation (LES) turbulence models instead of the Reynolds Averaged Navier-Stokes (RANS) models traditionally used [9-12].

In recent years there have been several studies regarding the simulation of pool fires. The experiment proposed by Tieszen in 2002 [13] has been one of the main cases studied. The simulations by Xin in 2008 [14] and Cheung in 2009 [15] using the Fire Dynamics Simulator (FDS) used Tieszen experiment as base case. Their results mainly show the flow's velocity fields with great details of the fluctuating velocity results. Another example is the study of Novozhilov and Koseki [16] who use the FIRE software to simulate the results obtained in different experimental analysis including the experiment by Weckman and Strong [17]. They report results for the evolution of temperature with time, as well as normalized flame heights and burning rates for pool fires of different fuels and sizes. Finally, in 2014 Chen [18] recreates computationally the results obtained by Weckman and Strong [17], in an attempt to test the capabilities of the LES models developed in the solver *fireFoam*. Chen developed a new model for radiative

* Corresponding author: od.lopez20@uniandes.edu.co

https://doi.org/10.1051/matecconf/201824005026

emission, obtaining different results for the flame heights, heat generation, temperature and emissive contours. However, some of these results do not agree completely with the experimental ones.

Hence, the main objective of this work is to predict the dynamics of a medium-scale pool fire and study the influence of the computational domain's discretization especially in the vertical (flame propagation) direction, when using LES turbulence modeling with the EDC as the combustion model. As test case, the experiment proposed by Tieszen et al in 2002 [13] was selected and simulated.

2 Modelling

2.1. Experimental setup

The experiment developed by Tieszen et al. consisted of a 1 m methane pool fire in the FLAME facility [13], was used as base case for the present numerical study. The research focused on taking velocity measurements using Particle Image Velocimetry (PIV), thus showing contours for average velocities in the axial and cross-stream directions, velocity fluctuations and root-mean squares contours. The PIV technique was used in order to obtain detailed results which could be easily compared to numerical solutions using LES turbulent models [13]. This case has been well documented in literature, since different authors have used it as a case study with which they could validate the numerical results obtained using different solvers (FDS [15], SFLM [19], CSE Model [20]).

The experimental setup consists of a 1 m diameter burner located at the center of a 6.1m cubic facility, 2.45 m above the floor. The methane was injected at a speed of 0.097 m/s to assure a total heat release rate of 2.07 MW. Further details can be obtained from the original article of the experiment [13].

2.2 Computational setup

The simulations were carried out using OpenFOAM – 2.4.0 (OF). This version was compiled with the *fireFoam* solver developed between FMGlobal and CFD Direct. This way, the combustion and radiation models to be used with the solver, can be implemented with the turbulent models available in OF. Also, the model can then be solved using the finite volume methods available in these libraries. The *fireFoam* solver uses the PIMPLE algorithm in order to solve the Navier-Stokes equations for compressible flows. This algorithm allows combining both the SIMPLE and PISO algorithm, such that it can use different correction equations for the pressure and velocity [9].

2.3 Computational models

To run the simulations, two LES turbulence models were selected and compared. First, the Spectral Eddy Viscosity (SEV) [21] model was used in order to perform a mesh convergence analysis. This way, a mesh could be selected such that it could get the most accurate results, without incurring in higher computational cost. This mesh would then be used with the Smagorinsky turbulence model such that a comparison between the two models could be fairly done by using the exact same mesh.

First, the SEV model proposed by Menon was implemented. This turbulence model "accounts for the total energy transfer" by assuming that energy and energy transfer are in phase [21]. Since this model consists of only one equation for the turbulent kinetic energy, less computational cost is required for the solution of the model. Although this assumption is big enough to deviate the results from the exact equations [21], previous computational studies [18] have proven good agreement between the numerical and experimental results.

On the other hand, the Smagorinsky model suggests that the viscous stresses are proportional to the flows' strain rate [22]. Its main advantage is the fact that it uses an algebraic solution while other models (such as SEV) use a differential solution. This implies that the Smagorinsky model should also incur in a low computational cost. This model has been one of the main turbulence models used in the simulation of conflagrations [15, 19].

The Eddy Dissipation Concept proposed by Magnussen was chosen to model the combustion. The EDC models the turbulent mix as a function of the turbulent kinetic energy and its dissipation, the latter which is equal to the reaction rate. This leads to an expression of the mass fraction in terms of the two turbulent properties [23], such that the model only solves one additional transport equation for the fuel's mass fraction [9].

Regarding the chemical kinetics solution, the *irreversibleinfiniteReaction* model was used which is available in the OF library. This model takes into account only one reaction (equation 1) to solve the chemistry of the case at hand, basically a one equation equilibrium model. By using a one reaction model, the chemical kinetics solution is greatly simplified since it limits the number of species (five in this case) thus reducing the amount of iterations and reaction rate equations the solver must solve, therefore reducing the global computational cost of the simulation.

$$CH_4 + 2O_2 + 7.5N_2 \rightarrow CO_2 + 2H_2O + 7.5N_2 \quad (1)$$

The finite volume Discrete Ordinates Model (fvDOM) was also included in the simulation in order to model the radiation heat transfer. In the present work, the model was implemented considering the gaseous phase as a gray gas participating media, while the Radiative Transfer Equation (RTE) is solved through a "loose coupling" method. This means that the RTE is solved using the solution of the previous time step (lagged in time). Then, the divergence of the radiative heat field is used as a source term for the energy equation which is solved in the PIMPLE (PISO) algorithm [24].

The simulations were set to have an adjustable time step dependent on the Courant-Friedrich-Lewy number (CFL). By using a CFL of 0.35, the simulation would adjust the time step such that the CFL fluctuates around 0.35. This way the entire model depends on the specified CFL rather than on a fixed time step, which will allow to solve the problem depending on the velocity of the fluid flow. Finally, the PISO algorithm will be solved using a Gauss Linear scheme, with an implicit Euler time discretization.

2.3 Computational domain, mesh and boundary conditions

The computational domain is a 3 m cube, according to the numerical simulation performed by Cheung et al [15], to make sure that the boundary conditions imposed by the facility are far enough from the flame, as well as disregarding the possible wall effects. The mesh was generated using OF's *blockMesh* utility, to create hexahedral elements. A total of 90 cells across the diameter of the pool fire were used [18], thus giving a total of 270 cells across the domain due to the restrictions from the geometry's vertices. As a first approach, four different meshes were generated in order to study the influence of the number of elements in the flame propagation (axial) direction. All of these were refined towards the bottom face, such that there are more elements near the burner's exit. Table 1 shows the main details of these meshes.

Table 1. Mesh descriptions.

Mesh	No. Cells Base	No. of Cells Vertical	Total Cells [Millions]	Min. cell volume [m^3 x10^{-7}]	Max. Aspect Ratio
Coarse (C)	270	40	1.62	32.14	28.91
Medium (M)	270	50	2.03	25.72	23.13
Fine (F)	270	60	2.43	21.43	19.28
Super-Fine (SF)	270	70	2.84	18.37	16.53

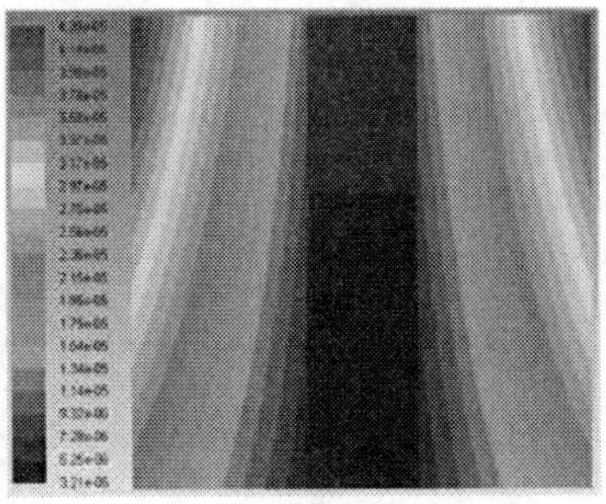
a)

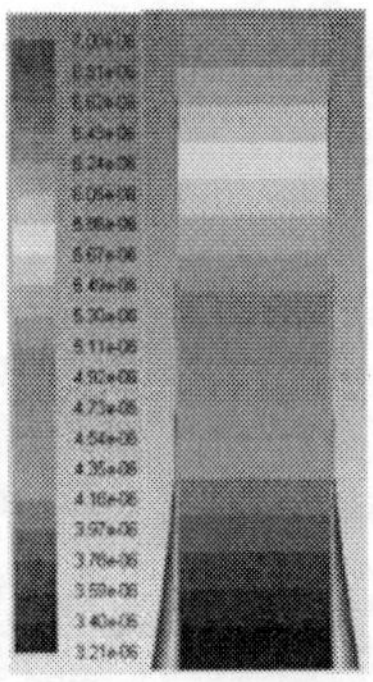
b)

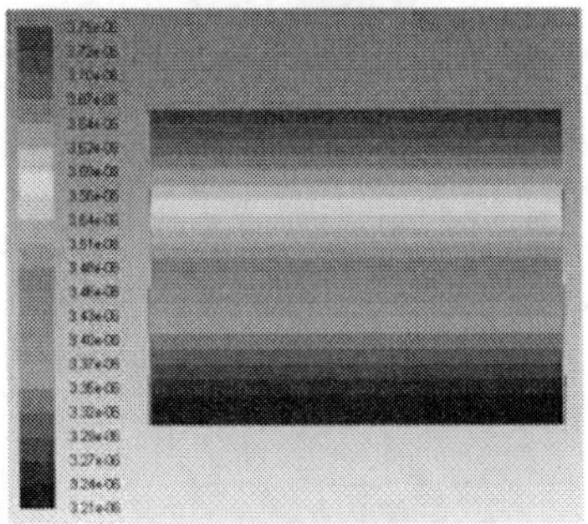
c)

Fig. 1. Cell volume for mesh C. a) Midplane; b) Center region of midplane; c) Lower burner zone (30cm width x 30 cm height).

As shown by Figure 1 c), the smallest cell volumes are close to the burner's exit. This was intended in order to better capture the details of the combustion process of the reactive flow. Since the minimum cube-root-volume (CRV) for the coarse mesh is of 0.015 m, this could work as a first approach towards the size of the filter of the LES turbulence model to be used. Figure 2 shows the visualization and some details of Mesh C (1.62 million cells).

a)

b)

c) 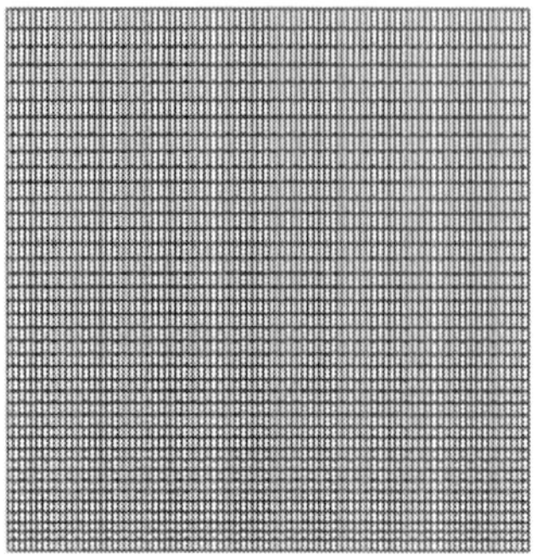

Fig. 2. Mesh a) Top view; b) Top view detail; c) Side view.

Four different boundaries were defined: the base, sides, outlet and an <inlet patch on the "base" face. The boundary conditions were set according to Table 2 and Table 3.

Table 2. Boundary conditions for pressure and velocity.

Boundary	Pressure	Velocity
Sides	calculated	*pressureInletOutletVelocity*
Outlet	calculated	*inletOutlet*
Inlet	calculated	0.097 m/s
Base	calculated	*zeroGradient*

Table 3. Boundary conditions for temperature, turbulent kinetic energy and mass fraction of methane and oxygen.

Boundary	T	k	CH_4	O_2
Sides	*inletOutlet*			
Outlet				
Inlet	275 K	1.41E-06 m^2/s^2	1	0
Base	*zeroGradient*			

The *zeroGradient* condition does not allow the variable to have any gradients in the normal direction to that face. The *inletOutlet* condition depends on the direction of the flow. When the flow leaves the domain, it behaves as a *zeroGradient* condition; however when the flow enters the domain (backflow), it performs as a constant value condition. The inlet values used for the *inletOutlet* boundary condition were: U = 0 m/s, T = 300 K, k = 1.41E-6 m^2/s^2, CH_4 mass fraction = 0 and O_2 mass fraction = 0. The *pressureInletOutletVelocity* backflow, instead of having a constant value for backflow, it is calculated from the pressure obtained from the internal cell value [25]. The kinetic turbulent energy was calculated according to equation 2:

$$k = 1.5(UI)^2 \qquad (2)$$

Where I is the turbulent intensity (I = 0.01). At the "Outlet" boundary, k also has a value of 1.41E-6 m^2/s^2 when performing as an inlet condition. For the Smagorinsky turbulence model, instead of specifying a value for the turbulent kinetic energy k, the boundary condition must be specified for the kinematic viscosity μ. In this case, a *zeroGradient* condition was used for this property at all boundaries.

3 Results and Discussion

3.1 Spectral Eddy Viscosity model

The results shown are based on time averages taken in a 15 s time frame (the last 15 s of the simulation).

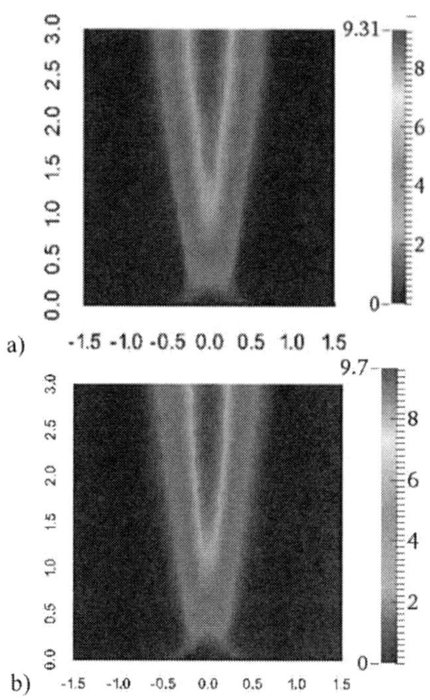

Fig. 3. Velocity contours [m/s] for a) C mesh and b) SF mesh. (Domain scale in meters).

Figure 3 shows that SF mesh obtains average velocities up to 9.7m/s while the coarser mesh can only achieve values up to 9.3 m/s. Also, looking at the geometrical characteristics of the contours, the higher velocities (i.e. the red contours) begin at heights close to the 1.15 m mark for the C mesh, while for the SF mesh, these contours begin close to 1.3 m heights. The biggest difference however, appears at the bottom region of the domain. The 3 m/s contours are clearly more detailed in the finer mesh. This also appears to affect the cone-shaped geometry between the 0.5 m and 0.5 m widths up to the 0.3 m height. These differences between the contours of the different meshes are rather interesting, therefore data for axial and cross-stream velocity profiles at different heights were used to observe the convergent behavior of the simulations, as well as to compare them against the experimental results obtained by Tieszen et al

https://doi.org/10.1051/matecconf/201824005026

[13] and some numerical results from previous studies [15, 19].

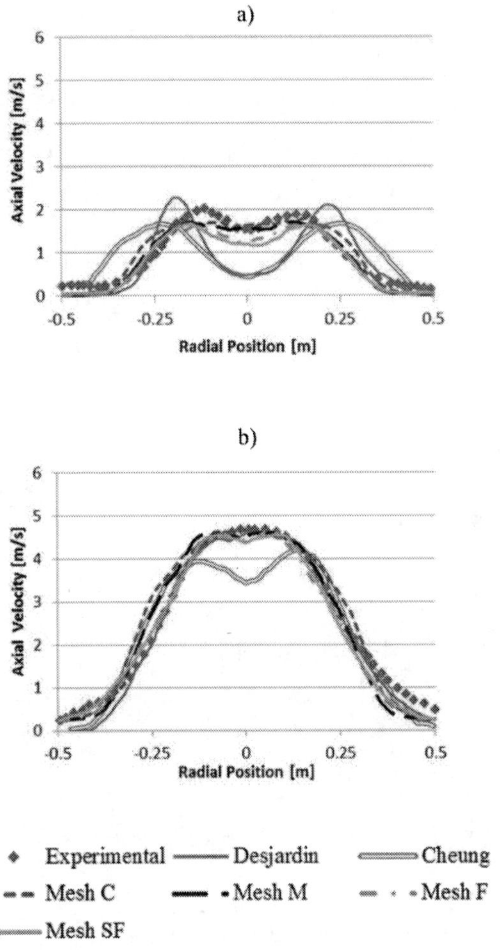

Fig. 4. Axial velocity [m/s] vs radial positon [m] at different heights; a) 0.2 m; b) 0.6 m.

From Figure 4 it can be observed that the general behavior of the simulations is very similar to the experimental results, especially for the 0.6 m height. Also, it is worth noting that the present implementation with fireFoam is able to obtain better results than the ones obtained by Cheung with FDS [15] or Desjardin with SFLM [19]. Finally, it is clear that there is a convergent behavior and particularly that Mesh M is able to obtain very accurate results without incurring in higher computational costs. To this regard, the maximum average axial velocity was selected as the main convergence criterion.

Table 4 shows the maximum average axial velocities for the four meshes at four different heights, as well as the average error for each mesh, when comparing to the experimental results. That is, the average of the error between the numerical result and the experimental result, for each height. This shows that Mesh M is the one with the least average error, thus proving to be the most accurate one. Also, since it is the second coarsest mesh, it does show an advantage regarding computational cost. However, it also shows that all four meshes overestimate the top velocity at the 0.8 m height.

Table 4. Mesh convergence. Maximum axial velocity [m/s] at different heights.

Data	0.2 m	0.4 m	0.6 m	0.8 m	Error
Experimental	2.02	3.54	4.64	5.23	-
Mesh C	1.79	3.38	4.67	5.76	6.73%
Mesh M	1.71	3.47	4.61	5.67	6.63%
Mesh F	1.66	3.38	4.66	5.63	7.63%
Mesh SF	1.67	3.36	4.57	5.58	7.71%

Taking into account that the principal transport phenomenon is buoyancy, it is expected that the flow is dragged towards the centerline of the fire. This means, that there should be a symmetry towards the center plane (radial position 0 m), since the vertical transport drags the surrounding fluid creating the "necking" region at the bottom regions of the domain due to the entrainment phenomenon. Figure 4 does show this symmetric behavior expected, while Figure 3 shows the necking region at the bottom of the domain. This behavior can also be observed in Figure 5 which shows line contours for axial and cross-stream velocities for the bottom parts of the domain.

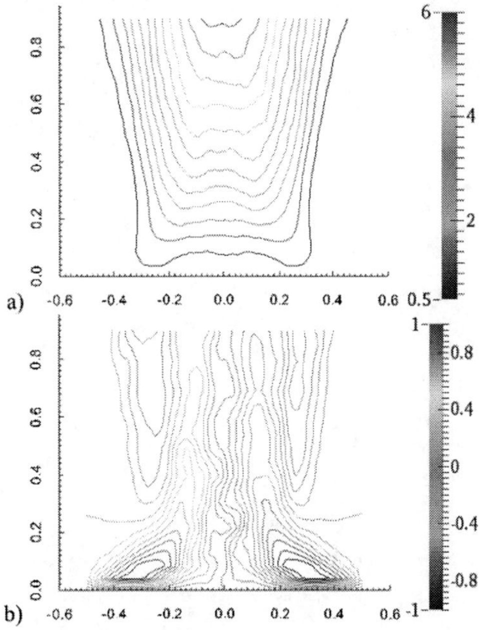

Fig. 5. Velocity line contours [m/s] for Mesh M; a) Axial velocity; b) Radial velocity (scales in meters).

Figure 5 a) shows the necking behavior expected from the poolfire, when looking at the curvature of the 0.5 m/s contour, which expands outwards from 0.3 m to

0.5 m widths. This matches the high velocity contours shown in Figure 5 b), which in turn show certain symmetry towards the centerline. Furthermore, both images help validate the use of the *fireFoam* software, since they show a very good resemblance with the experimental results obtained by Tieszen et al [13]. First, in Figure 5 a) the contours between 2.5 m/s and 3.5 m/s are close to the 0.4 m height. Second, the high velocity contours in Figure 5 b) (velocities above 0.8 m/s) go up to 0.2 m heights and between 0.2 m and 0.4 m wide. Similarly, the lower velocities (between -0.1 m/s and 0.1 m/s) are condensed in the central part of the domain, and towards the upper boundary.

3.2 Smagorinsky model

Previous research showed different studies [15, 19] which used the Smagorinsky LES model to account for the turbulent flow. Therefore, in the present study will compare the Spectral Eddy Viscosity proposed in the previous section and the Smagorinsky model. To do so, Mesh M (2.03M elements) was selected, taking into account that the results showed that this mesh was able to get accurate results without incurring in higher computational costs.

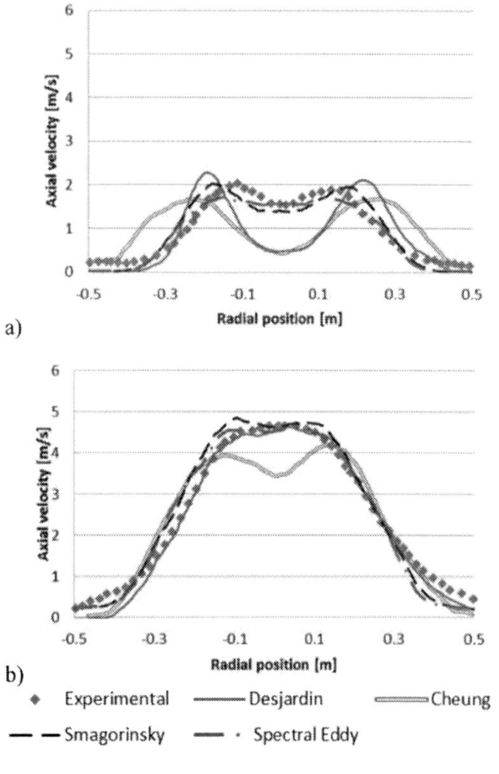

Fig. 6. Turbulent models comparison. a) Axial velocity at 0.2m height; b) Axial velocity at 0.6 m height.

Figure 6 shows the results obtained for the velocity in both the axial and cross-stream direction, against the radial position at different heights. From these images it is evident that the Smagorinsky model is able to obtain better results than the Spectral Eddy Viscosity, when being compared to the experimental results. However, it is especially interesting to observe that the present implementation in *fireFoam* of the Smagorinsky model is able to obtain better results than other computational tools such as FDS [15] and SFLM [19].

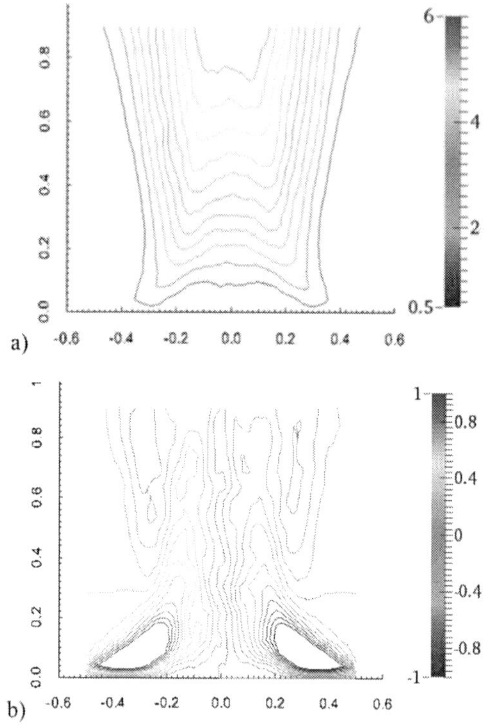

Fig. 7. Smagorinsky – Velocity contours [m/s]; a) Axial velocity; b) Radial velocity (scales in meters).

Looking at Figure 7, and by comparing it to Figure 5, the Smagorinsky model does prove to obtain more detailed results. For example, the curvature of the 0.5 m/s axial velocity contour is more distinct towards the boundaries of the domain, as well as the curvature towards the centerline. This in turn, allows the other contours to have a peak close to the 0 m mark, which shows a better agreement with the experimental results. Also, looking at Figure 7 b) the higher velocity contours (red and blue contours) seem to be closer to each other, while maintaining the 0 m/s contour in the central part of the domain. This shows that the Smagorinsky model is able to better detail the phenomena at hand.

Finally, Table 5 shows that the simulation on the coarse (C) mesh took approximately 107 hours of CPU time running 16 parallel processes in an Intel Xeon processor E5-2695 v2, while the finer (SF) mesh took 285 hours of CPU with the same number of parallel processes. This again proves the necessity of the different models used, considering the complexity of the combustion process. Especially the turbulent and chemical kinetics models used are necessary to significantly reduce the computational cost. It is also interesting to notice that the implementation with the

Smagorinsky turbulent model (which uses the Medium mesh) also has a shorter computing time as the Spectral Eddy Viscosity implementation for the same mesh. This proves that not only the Smagorinsky model achieves better results, but also incurs in fewer computational costs.

Table 5. Computational cost.

Mesh	Processors	Time [h]
Coarse (C)	16	107
Medium (M)	16	298
Fine (F)	16	334
Super-Fine (SF)	16	385
Smagorinsky (M)	16	202

4 Conclusions

In the present work, simulations were carried out in order to replicate the experimental results obtained by Tieszen et al in a medium-scale methane pool fire, through the use of the open source code *fireFoam*. The numerical results show a good agreement with the experimental results particularly with axial (flame propagation direction) velocities and velocity line contours. Although, the numerical results for line contours are not geometrically exact to the experimental ones, the general behavior of these show that the models that were implemented can simulate effectively a pool fire. The regions with high velocities are similar in both cases (experimental and numerical), for both axial and cross-stream velocities. In particular, it is interesting to see the curvature of the axial line contours, which is a clear effect of buoyancy and entrainment, due to the high velocity areas shown in the cross-stream line contours. Also, mesh convergence was analyzed by running simulations with four different mesh sizes. Based on the convergence analysis, it is appropriate to affirm that a mesh with 90 divisions across the pool fire's diameter and 50 divisions in the axial direction (medium size mesh) is fine enough to obtain reliable results. By comparing the results of the 2.03M elements mesh using two different turbulent models, it was possible to confirm that the Smagorinsky LES model is able to achieve better results, while doing so in less computing time. Taking all of this into account, the *fireFoam* tool has proven to be an effective solver for the numerical simulations of medium scale pool fires. This can in turn lead to suggest the use of the computational tool for diffusion turbulent flames phenomenon of larger scales.

The authors would like to thank the Engineering schools of both, Universidad de los Andes and Universidad del Valle (Grant CI2868), for providing the resources and support for the development of the present research. Also, to the HPC support team at the Universidad de los Andes, for providing the cluster computing service in which the simulations took place.

References

1. G. Hekestad, Fire Safety J., pp. 25-32 (1984)
2. V. Novozhilov, *Prog. in Energy and Combustion Sci.*, **27**, pp. 611-666 (2001)
3. S. Turns, *An Introduction to Combustion*, McGraw-Hill, Singapore (2000)
4. B. Merci and T. Beji, *Fluid Mechanics Aspects of Fire and Smoke Dynamics in Enclosures*, CRC Press, Chennai, India (2016)
5. A. Trouvé and T. Poinsot, *Proc. of the Summer Program 1992* (1992)
6. J.-Y. Chen, Combustion Sci. and Tech., **122**, pp. 63-94 (1997)
7. N. Peters, *Prog. in Energy Combustion Sci.*, **10**, pp. 319-339 (1986)
8. B. Magnussen, and B.H. Hjertager, *Intl. Symposium on Combustion*, **16**, pp. 719-729 (1976)
9. H. K. Versteeg, and W. Malalasekera, *An introduction to Computational Fluid Dynamics. The Finite Volume Method*, Pearson Prentice Hall, Essex (2007)
10. S. Pope, New J. of Physics, **35**, pp. 1-24 (2004)
11. K. McGrattan, et al., Fire Safety J., **30**, pp. 161-178 (1998)
12. Z. Chen et al., Fire Safety J., **64**, pp. 12-26 (2014)
13. S. Tieszen et al., Fire, Combustion and Flame, **129**, pp. 378-391 (2002)
14. Y. Xin et al., Combustion and Flame, **153**, pp. 499-509 (2008)
15. S.C. Cheung and G. Yeoh, Intnl. J. of Thermal Sci., **48**, pp. 2187-2202 (2009)
16. V. Novozhilov and H. Koseki, Combustion and Sci. Tech., **176**, pp. 1283-1307 (2010)
17. E.J. Weckman and A.B. Strong, Combustion and Flame, **105**, pp. 245-266 (1996)
18. Z. Chen et al., Intnl. J. of Heat and Mass Transfer, **70**, pp. 389-408 (2014)
19. P.E. Desjardin, Combustion Sci. and Tech., **177**, p.p. 1883-1916 (2005)
20. S. Ferraris et al., Fire Safety Sci., **8**, pp. 963-974 (2005)
21. S. Menon et al., Computers & Fluids, **25**, pp. 165-180 (1996)
22. J. Smagorinsky, Monthly Weather Review, **91**, pp. 99-164 (1963)
23. B. Magnussen, *ECCOMAS Thematic Conference on Computational Combustion*, Lisbon (2005)
24. J. Tencer and J.R. Howell, J. of the Brazilian Society of Mechanical Sciences and Engineering, **119** (2015)
25. NEXTfoam, *Boundary Conditions – OpenFOAM-2.3.0* (2014)

Numerical investigation of passive stabilization of the surface temperature using PCM and metal foam

Mirosław Seredyński[1,*]

[1] Institute of Heat Engineering, Faculty of Power and Aeronautical Engineering, Warsaw University of Technology, 21/25 Nowowiejska Str., 00-665 Warsaw, Poland

Abstract. Stabilization of the temperature of PVs is of great importance, due to strong relation between operating temperature and its efficiency. Due to low thermal conductivity of PCM, intensification of heat transfer to the operating material is needed. The four fully passive PV's temperature stabilization systems, based on the phase change material (PCM) are numerically investigated in this paper. Apart from the natural convection promotion, intensification of heat transfer is done with aluminium fins and aluminium foam. The simplified computational model based on the equilibrium formulation of energy transport equation, taking into account viscous and inertial fluid flow resistances in the porous material, is formulated and solved with the general purpose software - ANSYS Fluent. Proposed model is succesfully verified by comparing the results with available in literature numerical solutions.Simulations outcomes are presented, the temperature and liquid fraction distributions in proposed geometry configurations, temperature variations determined in selected points and averaged on the heated surface temperature plots. Presented results help to assign the best configuration.

1 Introduction

The temperature stabilization of electronic devices and photovoltaics is very important due to restricted operational requirements and considerable impact of temperature on their efficiency. As was reported by many researchers, e.g. Radziemska [1], the overall efficiency of a PV panel decreases by 0.5 pp for every 1°C rise above 25°C.

The temperature stabilization methods can be roughly divided into active and passive ones. The latter is not dependent on the external power supply system, so it is highly recommended for systems which operating must be infallible and recovery of heat is not needed.

Among methods of passive cooling of PVs those based on phase change materials (PCM) increase their popularity. Basic properties of such materials is a melting temperature and latent heat. The phase change temperature must be tuned to temperature of upper source of heat and its difference should be as small as possible to minimize exergy losses and to assure high quality of temperature stabilization. The common materials considered for PV temperature stabilization are organic pcm's, which are chemically stable, non expensive, safe and does not cause corrosion in metals. Their main drawback is a low thermal conductivity, close to 0.2 W/(mK), so some effort is needed to intensify the heat transfer between the hot surface and the heat storage material [2].

In the extensive survey of cooling techniques for PVs presented by Sathe and Dhoble [3], those based on the pcm's were considered among the novel ones. Various solutions vere mentioned: based on natural convection promotion, utilization of fins and developed heat transfer area.

The passive temperature stabilization system for the building-integrated photovoltaics based on the pcm was analyzed experimentally and numerically by Huang et al. [4]. Three system configurations were studied: plain aluminium plate system, PV/PCM system without and with internal fins. Height of fins was lower than the depth of the container, so heat was not transferred immediately, through fins, to the rear wall. Authors presented validation of the numerical model for all considered cases with performed experiment, where the constant insolation 1000W/m^2 was imposed on the vertical wall, and finally good conformity was reported.

Analysis of the role of internal finning of the PCM container exposed to high thermal radiation flux on the temperature limiting was investigated by the same authors (Huang et al.) [5] and [6]. It was concluded, wider fins prevent from overheating for longer time [5] and lower distances between fins considerably reduce the front temperature [6].

Enhancement of heat transfer using metal foams (porosity 95% and 10 PPI) saturated with PCMs, compared to pure PCMs, was investigated experimentally and numerically by Zhao et al. [7]. The sample was heated electrically from the bottom with a constant heat flux. Detailed analysis of charging and discharging of the composite showed the considerable increase in rate of the processes, 3-10 times for charging

* Corresponding author: msered@itc.pw.edu.pl

and 2 times for discharging. The numerical model of heat transfer in the composite based on the non-equilibrium energy balance formulation neglected effect of natural convection in the molten PCM.

Li et al. [8] studied numerically and experimentally melting process in the high-porosity copper metal foam filled with PCM. The effect of porosity and pore density on the wall temperature and the nonuniformity of temperature in the foam were investigated. Authors discussed the enhanced effective conductivity of foam matrix but also its fluid flow suppressing effect. The proposed computational model included non-equilibrium heat transfer, both viscous and inertial resistance for fluid flow and natural convection in molten PCM. Experiments outcomes revealed the temperature of the heated wall is more efficiently lowered, the porosity of the composite is lower and the pore density is lower.

The composite made of graphite matrix impregnated with PCM was used by Atkin and Farid [9] for temperature stabilization of PV cells. This solution was compared to other stabilization methods, based on passive cooling with fins. The composite material used as the heat sink, the graphite foam infused with PCM, had much higher thermal conductivity, equal to 16.6 W/(mK), than the conductivity of pure PCM, equal to 0.25 W/(mK). However, due to structure of the heat storage material, heat conduction was the only heat transfer mechanism. Experiment was conducted under artificial conditions, the insolation of maximum intensity 920 W/m^2, was supplied with halogen lamps. Authors observed the best performance for the system with graphite-PCM composite as the heat storage material and additional finned heat sink at the rear side. For 12 hours simulated daylight load the PV peak efficiency was higher by 2pp then for the reference case.

In the proposed paper four configurations of the passive temperature stabilization system are numerically investigated. Considered cases involve geometry with or without aluminium fins, with porous aluminium fins or porous aluminium layer attached to the heating surface. The numerical simulations are performed with the general purpose ANSYS Fluent software. The special treatment of mass, momentum and energy transport through porous zone, considered in two cases, is necessary so effective transport parameters in the two phase domain, like thermal conductivity and permeability are estimated. Results of numerical simulations in the simplified, two dimensional geometry, enable assessment of the heat transfer rate from the hot surface to heat container and the related temperature stabilization. Additionally, the mechanisms of heat transfer to PCM are discussed, based on the average temperature history of the hot surface.

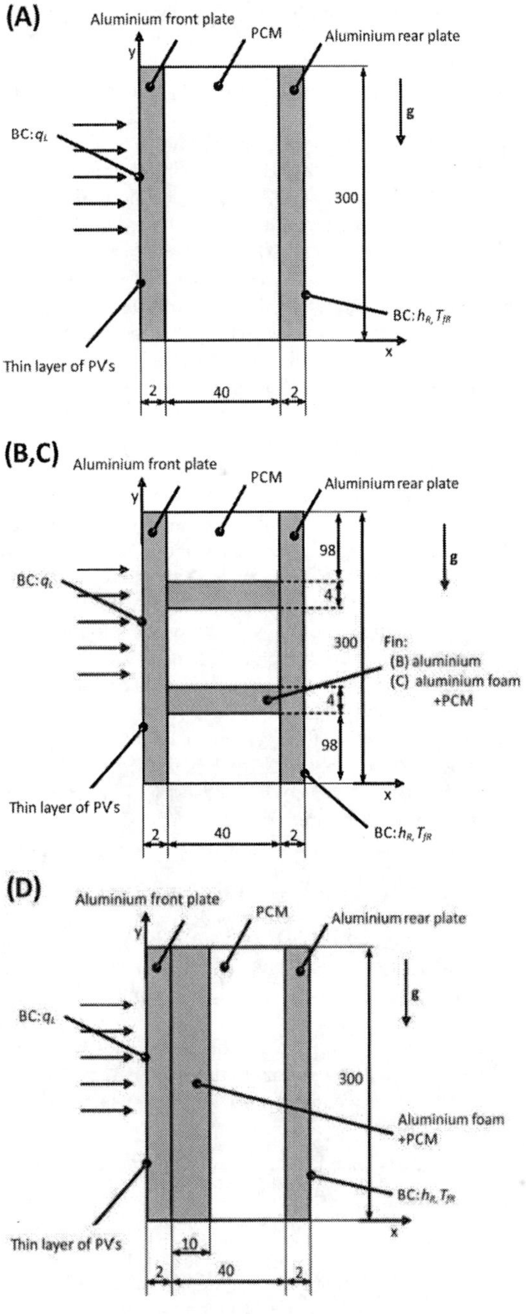

Fig. 1. Schematic of considered cases: (A) the reference geometry, (B) with two fins made of Al, (C) with two fins made of Al foam and (D) with Al foam layer attached to the hot plate. All dimensions are given in mm, not to scale.

2 Problem statement

In this paper the role of natural convection in molten PCM as well as enhancement of heat transfer with fins and well heat conducting porous foam are investigated.

The four configurations of the passive temperature stabilization system, based on the PCM are considered (Fig. 1). Between the front and rear plates made of aluminium, the PCM is stored. The total width and height of the system are 44 mm and 300 mm, respectively. In the two configurations additional fins made of aluminium (case B) or aluminium foam (case C) are present. In the case D, the layer of aluminium foam of 10 mm thickness is attached to the front aluminium plate. For cases C and D, the open pores and high porosity foam is used, completely filled with PCM, so both diffusive and convective heat transfer can develop in it. The perfect contact between vertical aluminium plates and fins or foam layer is assumed. At the external surface of the front plate the thin layer of PVs is fastend. The thickness of the PVs layer as well as thermal contact resistance at its back surface are neglected. At the surface of PVs, called subsequently the hot surface, the external heat flux is imposed, which reflects the insolation close to its highest value. The heat flux density, q_L, is uniform and constant in time, equal to $1000\,W/m^2$. The same value was assumed by several authors in their numerical or experimental research (e.g. [4]). In this simplified analysis the other components of heat transfer from the hod wall, namely the convective and radiative terms are neglected. Horizontal boundaries of the system are assumed to be perfectly insulated. On the external surface of the rear plate, called subsequently the cold surface, the third kind boundary condition is imposed, heat transfer coefficient, h_R, is 5 W/(m^{2}K) and ambient air temperature, T_R, is 293.15 K.

3 Computational model

The general purpose CFD software, ANSYS Fluent, has been adopted to perform numerical simulations of heat and mass transfer in the PVs temperature stabilization system. Heat transfer enhancement methods, like finning, using of well heat conducting porous foams saturated with pcm, are taken into account, so heat and mass transport phenomena through a porous zone should be taken into account.

The considered mass, momentum and energy transport equations are following:

$$\nabla \cdot \mathbf{V} = 0 \tag{1}$$

$$\frac{\partial(\rho \mathbf{V})}{\partial t} + \nabla \cdot (\rho \mathbf{V}\mathbf{V}) = \nabla \cdot (\mu_l \nabla \mathbf{V}) + \mathbf{B} \tag{2}$$

$$-\varepsilon_l \varepsilon_F \nabla p - \frac{\varepsilon_l \varepsilon_F \mu_l}{K}\mathbf{V} - \frac{\varepsilon_l \varepsilon_F C_I}{K^{1/2}}\rho|\mathbf{V}|\mathbf{V}$$

$$\frac{\partial(\rho c T)}{\partial t} + \nabla \cdot (\rho c \mathbf{V} T) = \tag{3}$$

$$= \nabla \cdot (k \nabla T) + \rho L \frac{\partial g_s}{\partial t} - \rho L \nabla \cdot (g_l \mathbf{V}_l)$$

where the effective material properties: density, ρ, specific heat, c, viscosity, μ, thermal conductivity, k, latent heat, L are presented without indices. Momentum balance equation (2) is written in semiheuristic volume-averaged general form [10], valid for both bulk liquid and porous zone. The one domain approach is adopted, where the highly non-uniform two-phase medium is replaced with the artificial one, which properties are determined as averaged across the phases occupying the representative elementary volume (REV). The nominal properties of PCM and aluminium are listed in the table 1. It is also assumed, the intrinsic properties of aluminium foam are the same as of the solid aluminium, however in reality those properties can slightly vary. The assumption of the equilibrium heat transfer (eq. 3) is justified with the relatively low heat fluxes and high density of pores of the foam and is sufficient for preliminary analysis.

The source term, B, in the eq. 2, express the buoyancy forces in the molten PCM. The simple Boussinesq model is taken

$$\mathbf{B} = -\varepsilon_l \varepsilon_F \rho_{ref} \beta \mathbf{g}(T - T_{ref}), \tag{4}$$

where the reference density, ρ_{ref}, and reference temperature, T_{ref}, are equal to the nominal density of the PCM and the initial temperature of the system, respectively. Product of foam porosity, ε_F, and volumetric liquid fraction of PCM, ε_l, is equal to the volumetric fraction of liquid phase with respect to the total mixture volume. The same term appears in the pressure gradient, the Darcian and Forcheimer damping terms in eq. 2.

To estimate the permeability, K, and the intertial resistance coefficient, C_I, more detailed informations on the metal foam microstructure must be supplied. Presented model and simulations are based on the assumption, the pore number per inch (PPI) is 10. Other microstructure parameters, like pore diameter, dp, and fiber diameted, df, are taken from Li et al. [8]. Tortuosity coefficient, χ, and related characteristic length d_k, are derived with formulae [8]

$$\chi = 2 + 2\cos\left[\frac{4\pi}{3} + \frac{\cos^{-1}(2\varepsilon_F - 1)}{3}\right], \tag{5}$$

$$d_k = \frac{2d_p}{3-\chi}. \tag{6}$$

Permeability and inertial coefficient are derived with formulae [8]

$$K = \frac{(\varepsilon_l \varepsilon_F)^2 d_k^2}{36(\chi - 1)\chi}, \tag{7}$$

$$C_I = \frac{0.00212(1 - \varepsilon_l \varepsilon_F)^{-0.132}}{\sqrt{K}}\left(\frac{d_f}{d_p}\right)^{-1.63}, \tag{8}$$

During the phase change the temperature of PCM changes in some temperature range, namely between solidus and liquidus temperatures, so the process is not isothermal as for pure substances or mixtures at the eutectic compositions. It means the zone of coexistence of solid and liquid phases appear during phase change process. Generally the properties of this mixture depend on intrinsic properties of phases and the fraction of the solid phase. It is assumed here that all material properties of solid and liquid PCM are the same and temperature independent (table 1), so all effective properties of PCM are constant. In two cases, C and D, the permeable metal foam fully saturated with PCM occupies part of the computational domain, so effective transport properties should be determined there. Effective density, specific heat and latent heat are calculated as simple averages based on the mixture theory:

$$\rho = (1-\varepsilon_F)\rho_F + \varepsilon_F \rho_{PCM}, \quad (9)$$

$$c = (1-g_F)c_F + g_F c_{PCM}, \quad (10)$$

$$L = \varepsilon_F L_{PCM}. \quad (11)$$

The effective thermal conductivity of the foam saturated with PCM was determined with formula adopted from [11]:

$$k = 0.35\left[\varepsilon_F k_{PCM} + (1-\varepsilon_F)k_F\right] + \frac{0.65}{\left(\frac{\varepsilon_F}{k_{PCM}} + \frac{1-\varepsilon_F}{k_F}\right)}. \quad (12)$$

Table 1. Material properties.

	Unit	PCM	Aluminium
Density, ρ	kg/m³	770	2 675
Specific heat, c	J/(kgK)	2 000	903
Thermal conductivity, k	W/(mK)	0.2	211
Viscosity, μ	kg/(ms)	0.02	n.a.
Volumetric expansion coefficient, β_T	1/K	0.001	n.a.
Melting temperature, T_M	°C	27-29	n.a.
Latent heat, L	J/kg	220 000	n.a.

The considered transport equations (eq. 1-3) supplied with the closure relations (eq. 4-12) were solved with the ANSYS Fluent software. The non-stationary equations were integrated with the fully implicit scheme with time step 0.01s. The second order upwind scheme was used for all equations. The pressure and velocity fields were coupled with the PISO scheme.

4 Verification

The implementation of proposed model in the commercial software Fluent was werified with numerical results presented by Beckermann and Viskanta [12]. The considered case is the melting of gallium in the porous matrix of glass beads in a square cavity heated from the left wall. All material properties and boundary conditions were adopted from the referred paper [12]. Both viscous resistance and inertial resistance terms were taken into account according to formulations presented in [12]. Domain was discretized with the uniform mapped grid of 10 000 control volumes. The fully implicit non-iterative time advancing PISO scheme was used with uniform time sted 0.01s. The second order upwind scheme was used for each balance equation. Gradients were determined with the Least Squares Cell Based scheme. The comparison of phase interface predited for several time instants with ANSYS Fluent (solid lines) and numerical (dashed lines) taken from the paper are presented in Fig. 2. Results reveal very good conformity.

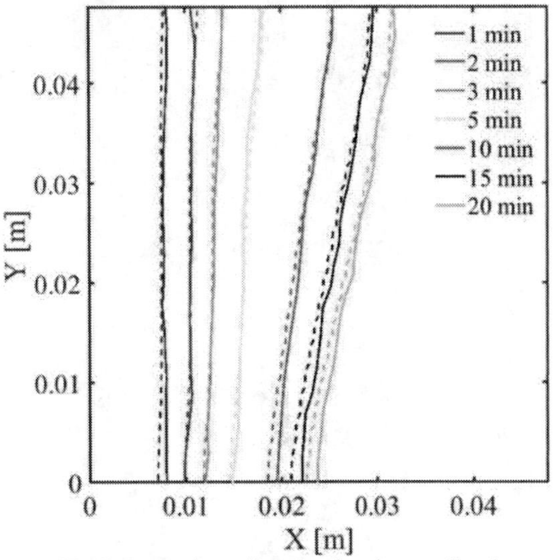

Fig. 2. Comparison of interface locations predicted numerically, present work – solid lines and the reference data [12] – dashed line.

5 Results and discussion

The presented model (eq. 1-3) supplied with closure relations (eq. 4-8) and definitions of effective metal properties (eq. 9-12) was used to simulate the response of systems A-D to thermal loading as were detailed in the previous chapters.

The comparison of temperature distributions (Fig. 3) after 1h of heating reveals that fins made of aluminium (case B) or aluminium foam (case C) considerably promote heat transfer to the back side of the container. So the maximum temperature of the front plate is visibly reduced to 309.1 K for the case B and 312.78 K for the case Ccompared to 318.66 K for the case A. Presence of the aluminium foam slab attached to the front plate (case D) also reduces the maximum temperature but it is slightly higher than in cases B and C. Presence of fins (cases B and C) considerably equalize temperature distribution in the PCM, heat is more efficiently dissipated in the bottom part of the container. Similar effect of temperature equilization is observed in the case D. Temperature across the metal foam layer is nearly constant, but it increases towards its upper part. Beyond the foam layer the single vortex is formed, similar to those present in the case A, but rate of melting in the upper part (case D) is lower than observed in the case A (Fig. 4). Contrary, the predicted melting rate in the bottom part is the highest for case D (Fig. 4).

This tendency for temperature equalization is confirmed in plots of temperature variation in time (Fig. 5), determined at three points, P_1, P_2, P_3, located in the halt height of the domain (Table 2).

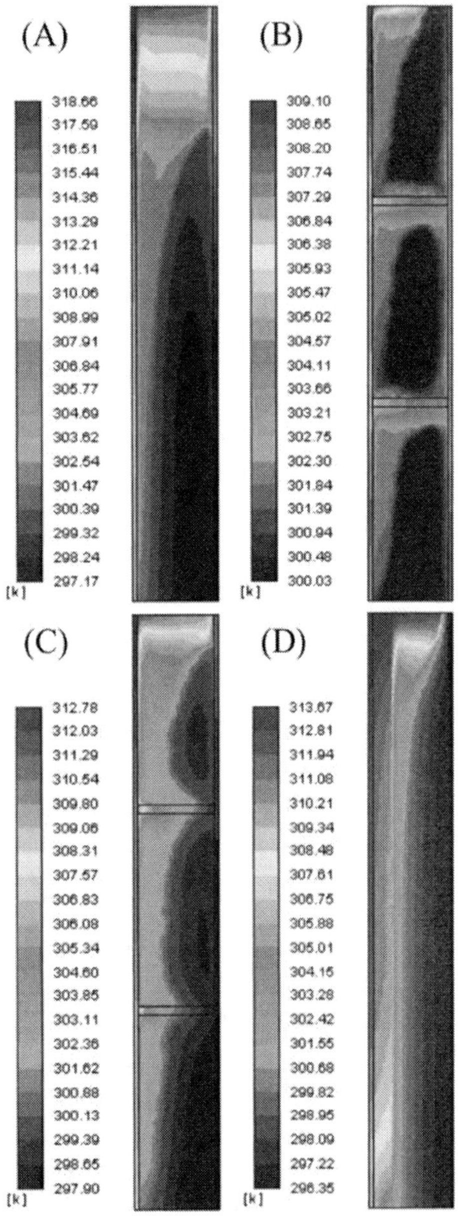

Fig. 3. Comparison of temperature distributions in the PCM container after 1 h of heating.

The highest surface temperature is obtained for the case A (blue solid line). The existence of the evident maximum in temperature, predicted at point P_1, is observed for cases A and D, where single vortex forms in the molten PCM. It is related with the conductive only heat transfer mode present in the starting period of heating, where the width of the molted PCM layer is too low for development of natural convection. It is particularly visible in the central and bottom part of the container. The time period, when the heat transfer is convective only, is reduced considerably in cases B and C due to more efficient heat transfer to the rear part of the container and additional PCM melting close ti fins and rear plate. After the convective flow develops, the temperature in P_1 is visibly reduced. Additionally, for the case D, the temperature in P_1 is initially reduced due to effective heat transport through the composite layer.

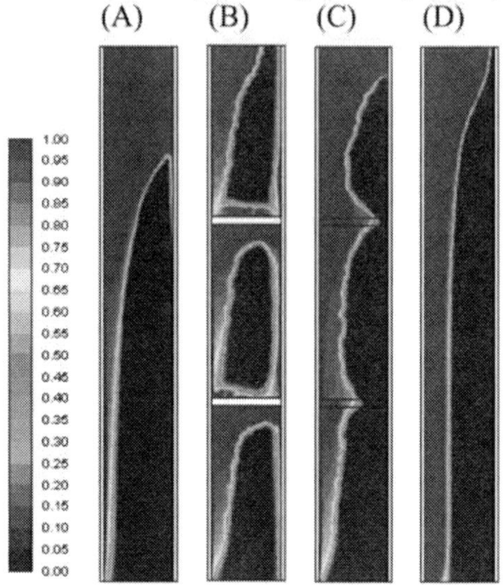

Fig. 4. Comparison of liquid fraction distributions in the PCM container after 1 h of heating.

Table 2. Coordinates of probe points. The origin of the global Cartesian coordinate system is in the left, bottom corner of the domain.

	X [mm]	Y [mm]
P_1	4	150
P_2	10	150
P_3	22	150

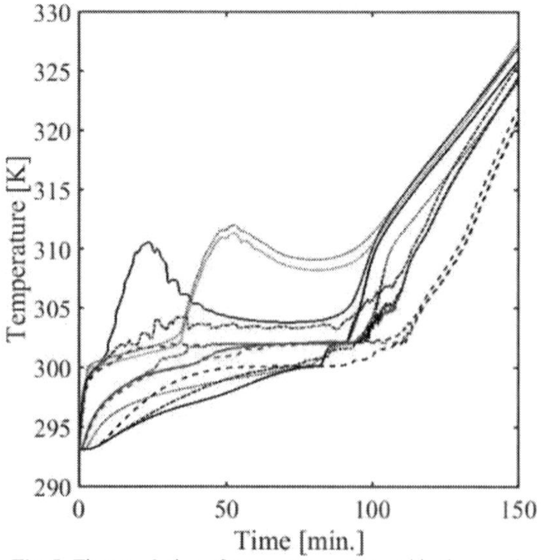

Fig. 5. Time evolution of temperature measured in three probe points: P_1 – blue line, P_2 – red line and P_3 – black line. For case A – solid line, case B – dashed line, case C – dash-dot line and case D – dotted line.

Variations in time of the average temperatures calculated on the heating wall are presented in Fig. 6. The shapes of curves are similar to those predicted in P_1 (Fig. 5). The local maximum in temperature is evident for cases A and D, but the latter is much more lower so the temperature stabilization is better, especially in the first period of heating. The best temperature stabilization configuration is the B case.

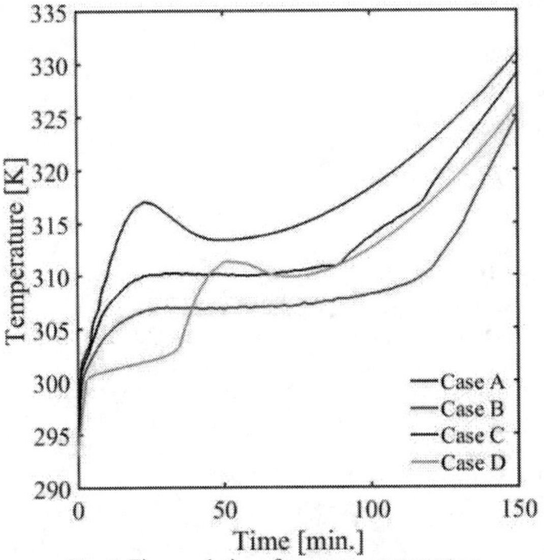

Fig. 6. Time evolution of an average temperature of the heating surface.

6 Conclusions

In the proposed paper the performance of four configurations of the heat sink used for temperature stabilization are discussed. The dominant heat transfer mode can be easily deduced based on the average temperature predicted on the heating wall. The retarded heat transport to PCM due to long time period necessary to develop natural convection in the molten PCM manifests with maximum in the temperature plot. It can be avoided as fins are present (cases B and C), the rate of melting is higher due to simultaneous melting close to the fins and the rear wall. It is worth noting, the fins connecting the front and rear walls are highly effective, melting occurs also close to the real wall (Fig. 5). Taking into account the average temperature on the heated wall the best considered arrangement, assuring the lowest average front temperature, is the B case. The issue of validity of the equilibrium model for heat transport will be addressed in the future research. Based on the werification of the model with respect to positively validated numerical data presented in [12], the utilization of the equilibrium model is justified.

References

1. E. Radziemska, Renew. Energ. **28**, 1 (2003)
2. A. Hasan, S.J. McCormack, M.J. Huang, B. Norton, Sol. Energy **84**, 1601 (2010)
3. T.M. Sathe, A.S. Dhoble, Renew. Sust. Energ. Rev. **76**, 645 (2017)
4. M.J. Huang, P.C. Eames, B. Norton, Int. J. Heat Mass Tran. **47**, 2715 (2004)
5. M.J. Huang, P.C. Eames, B. Norton, Sol. Energy **80**, 1121 (2006)
6. M.J. Huang, P.C. Eames, B. Norton, N.J. Hewitt, Sol. Energ. Mat. Sol. C. **95**, 1598 (2011)
7. C.Y. Zhao, W. Lu, Y. Tian, Sol. Energy **84**, 1402 (2010)
8. W.Q. Li, Z.G. Qu, Y.L. He, W.W. Tao, Appl. Therm. Eng. **37**, 1 (2012)
9. P. Atkin M.M. Farid, Sol. Energy **114**, 217 (2015)
10. M. Kaviany, Principles of heat transfer in porous media, 2nd edition, Springer, New York (1995).
11. A. Bhattacharya, V.V. Calmidi, R.L. Mahajan, Int. J. Heat Mass Tran. **45**, 1017 (2002)
12. C. Beckermann, R. Viskanta, Int. J. Heat Mass transfer, **31**, 35 (1988)

Numerical research of solidification dynamics with anisotropy and thermal fluctuations

Przemysław Smakulski*[1], *Jun* Ishimoto[2], and *Sławomir* Pietrowicz[1]

[1]Department of Thermodynamics, Theory of Machines and Thermal Systems, Wrocław University of Science and Technology, Wybrzeże Wyspiańskiego 27 St. 50 - 370 Wrocław, Poland
[2]Institute of Fluid Science, Tohoku University, 2-1-1, Katahira, Aoba-ku, Sendai 980-8577, Japan

Abstract. The influence of thermal fluctuations and anisotropy on the solidification process of a pure component is analyzed. It helps to understand the unstable freezing process where complicated structures such as dendrites could be formed due to a supercooling event reached during the cooling of the biological material. The study consists of mathematical modeling, validation with an analytical solution, and shows the influence of thermal noises on phase front dynamics. The analysis has been modeled in the framework of the Phase Field Method with Cahn-Hilliard formulation of a free energy functional [1]. The phase front is described by the Eulerian approach of fluid fields and formulated as a Phase Field scalar variable (order parameter) with a predefined, diffused boundary thickness. The results describe an influence scale onto directional phase front propagation dynamics, and how significant are stochastic thermal noises in micro-scale freezing.

Nomenclature

η - order-parameter, -;
M_1 - mobility parameter;
$F(\eta, T)$ - free energy functional, J;
$f_0(\eta, T)$ - homogenous free energy density of a single component, J/m^3;
κ - positive gradient energy coefficient, J/m;
ω - constant barrier height, J/m^3;
σ - liquid/solid surface tension, J/m^2;
T_m - melting point temperature, K;
μ_k - interface kinetic coefficient, m/s K;
w_{min} - interface thickness, m;
L - latent heat of fusion, J/m^3;
T - actual temperature, K;
α - thermal diffusion, m^2/s;
c_p - specific heat per unit volume, J/m^3 K;
δ - magnitude of anisotropy, -;
B - anisotropy symmetry constant, -;
χ - an amplitude of fluctuations, - ;
r - is a random number between the range $-1<r<1$, -.

1 Introduction

Cell/tissue protection is becoming an integral part in many current medical fields like a transplantation, preservation of female fertility, or tissue engineering [1]. The principle of cryopreservation is based on lowering the metabolic activity. The biological materials chemical reactions, which sum is metabolism, could be reduced by decreasing it's temperature [1]. In the paper [2] the author estimated that for normothermic animals, metabolic activity decreases 1,5 to 2 times for every 10 K decrease in temperature. Thus, the lower the temperature is, the longer material could be stored.

It was shown in [1], when choosing the temperature below 143 K (−130°C) material viability is remained over the years [3]. Unfortunately, cryopreservation methodology has to deal with several undesirable events, which could occur during the freezing and thawing process. This includes osmotic stresses, recrystallization of the ice, cryoprotectant agents toxicity or Intracellular Ice Formation (IIF). Among all of those listed events, IIF was found as a most dangerous for preserving cells and tissues [4]. It stands out as a formation of the ice inside a cell, and causes damage of organelles. The IIF occurs when a cell contains unbound (i.e. freezable) supercooled water, and the actual temperature of the cell interior is below the nucleation point [5]. To protect the cell interior before undesirable ice nucleation, the freezing process has to be performed in a way to avoid this event. One well known method of freezing biological material is the (so-called) slow freezing method, where the extracellular icefront is propagated relatively slowly to dehydrate the cell interior before reaching the nucleate temperature of water [6].

Many protocols exist concerning the slow freezing method. Some of them were completed after years of experiments. Alternatively, the research has been going in the direction of numerical simulations since 1984, where the first mathematical description of cell dehydration was proposed by Mazur [7]. In this paper, Mazur developed relative volume water content of the cell, taking as a parameter the cooling rate of freezing, permeability of the membrane and the actual temperature of the cell interior. By this simple method, it was possible to determine whether cooling avoids IIF or not. Recient mathematical descriptions of the cell freezing process deal with many variables aspects concerning freezing itself. The model presented by Jaeger et al. [8,9] is a two-dimensional finite element approach. The main emphasis was placed to the osmotic response of the cell membrane to an uneven flux of water during ice freezing propagation. This work presents an effect of the external ice propagation front, which changes the local concentration of the salt due to solidification phase movement. For modeling a complicated system of the phases an Arbitrary Lagrangian-Eulerian method (ALE) has been chosen, where an numerical mesh is adapted to the movment of the biological cell boundary. Kinetics of the water was modeled as a response of unequal salt concentration difference between extracellular and intracellular space. Later mathematical models were mainly focused on unstable freezing ice patterns due to the high subcooling of the solution. An example could also be cited, the two consecutive works of Udaykumar and Mao [10] and Yang and Udaykumar [11], in which numerical calculations of solidification itself were presented in a binary mixture solution (water-salt). The research was carried out on FEM fixed cartesian mesh by the immersed phase boundary method with the supply of level-set method. As the results show, the dendritic pattern of the freezing is shown due to instabilities caused by large subcooling.

Thermal instabilities together with the anisotropic solidification process is the main objective analized in the article. To investigate the influence of the dendritic pattern on the directional freezing front, the mathematical model based on Phase Field Method (PFM) has been proposed. This approach allows for the detection of the solidification pattern on the basis of primary free energy minimization in the numerical domain. This part is a background of the wider project concerning numerical research of solidification front influence on biological cell structure.

2 Analysis and modelling

Considering only the moving boundary problem two basic approaches have to be pointed out: Eulerian and Lagrangian.

Lagrangian methods maintain the interface as discontinuity and explicitly follow the evolution of tracking mesh points. In the opposite extreme, Eulerian methods deals with fixed grid and implicit volume fraction tracking. Unquestionable advantage of using the Eulerian approach is related to the formulation of the problems involving free boundaries, where deformation of the free surface is so large that Lagrangian methods cannot be used [12]. From the Eulerian tracking methods, one, the PFM was choosen for further analysis. PFM formulation was introduced by Cahn and Hilliard in 1958 [13] and its principles were based on Ginzburg and Landau free energy functional [14]. The evolution of the interface is represented by implicitly given phase-field variable called order parameter (η) [14] and has diffusive characteristics across the phase boundary. The main idea of diffuse interface is associated with the free energy minimization [14]. Thanks to the diffusive-interface approach, PFM could track unstable and complex solid-liquid topologies without the necessity of explicit implementation of boundary conditions on the phase border [15]. Thus, all variables are smooth through the two-phase medium and remain continuous over the numerical domain. The basic form of PFM governing equation is shown below:

$$\frac{\partial \eta}{\partial t} = -M_1 \frac{\delta F(\eta, a_1, \dots, a_i)}{\delta \eta} \tag{1}$$

The order parameter usually represents microscopic variables, like magnetic spins, but in transition between solid and liquid, the order parameter has a macroscopic character and is associated with density difference. Where solid/liquid phase transition is considered, each phase in the system is represented by the finite value of the phase-field parameter, in that case; η = 1 corresponds to solid, and η = 0 to liquid phase. In PFM the free energy functional of the single component liquid/solid transition is described as:

$$F(\eta, T) = \int_V \left[f_0(\eta, T) + \frac{\kappa}{2} (\nabla \eta)^2 \right] dV \tag{2}$$

In literature, this representation is called a Cahn-Hilliard free energy functional. The RHS parts of the above presented formulation describe the actual local density of free energy a $f_0(\eta, T)$ in homogenous system. The related gradient term is responsible for the diffusive character of the interface. The local free energy density of a pure component is assumed to be:

$$f_0(\eta, T) = \omega\, g(\eta) + h(\eta)\, \Delta f_0(T) \tag{3}$$

where two additional functions, the double-well function $g(\eta) = \eta^2(1-\eta)^2$, and interpolating function $h(\eta) = \eta^3(6\eta^2 - 15\eta + 10)$ are implemented directly. Double-well function of that form ensuring the stable minimum $\partial f_0/\partial \eta = 0$ for $\eta = 1$ and 0, where the second, $h(\eta)$ function provides smooth interpolation of the local free energy difference between two phases. The final form of the PFM equation has been presented in the following form:

$$\frac{\partial \eta}{\partial t} = -M_1 [\omega\, g'(\eta) + h'(\eta)\, \Delta f_0(T) + \kappa^2 \nabla^2 \eta] \tag{4}$$

$$\omega = \frac{\sigma}{w_{min}} \quad (5)$$

$$M_1 = \frac{T_m \cdot \mu_k}{1.364 \cdot L \cdot w_{min}} \quad (6)$$

$$\kappa = 1.364 \cdot \sigma \cdot w_{min}, \quad (7)$$

The local free energy difference $\Delta f_0(T)$ is represented by Gibbs free energy per unit volume [16].

$$\Delta f_0(T) = \Delta g = L - T\frac{L}{T_m} \quad (8)$$

As it can be seen, local free energy density depends on the latent heat of fusion and actual temperature level. This consideration allows for the tracking of the phase front implicitly without additional implementation of the melting temperature on the border. It also consistents with the Gibbs-Thomson equation at vanishing interface thickness – the sharp-interface limit condition. The temperature itself provides energy field potential which establishes the direction of the phase boundary force. For the temperature equal to T_m the relation above is equal to zero, where for $T < T_m$ local free energy difference is positive, and for $T > T_m$ obtains a negative value. Temperature field is derived from the first law of thermodynamics, and is represented by:

$$\frac{\partial T}{\partial t} = \alpha \nabla^2 T + \frac{L}{c_p} h'(\eta) \frac{\partial \eta}{\partial t} \quad (9)$$

An additional term related to latent heat of fusion was implemented as a source term interpolated by function $h(\eta)$ which comes from PFM. It was assumed, that the Peclet number in cell freezing is Pe << 1. For such a low value of Pe number, motion of the fluid inside the described systems is transported mainly due to diffusion without any convection effects.

Morphologically complex structures begin to grow until constrained by surface tension and interface kinetic effects. The anisotropy of the surface free energy plays a significant role in the dynamics of dendritic growth [17]. The first approach to describe the instability phenomena during solidification was presented by Mullins and Sekerka [10,18]. Interfacial energy anisotropy in the model was introduced based on the assumption that scalar gradient coefficient κ depends on the interface orientation [19]. Function of that coefficient in 2-D Cartesian space has the following form [19,20]:

$$\kappa(\theta) = \kappa_0(1 - \delta\cos(B\,\theta)) \quad (10)$$

where $\theta = \tan^{-1}(\eta_y/\eta_x)$ is the angle between the normal to the interface and reference axis. Subscripts in η_x, η_y represent the order parameter derivatives at x and y axis, and the κ_0 is a isotropic constant coefficient. It was shown by Kobayashi [14], that increasing the anisotropy magnitude parameter (δ) significantly influences the solid structure morphology. Roughly speaking, the patters for higher values of anisotropy magnitude result on more individual dendrites formation, with weak side-branching. It should be noticed, that the mechanisms of side-branching in the crystal form are mainly governed by the thermal fluctuation occurring near the dendrite tip. These stochastic noise forces are added into the model, to consider microstructure evolution changes [15]. The stochastic noise is included into the main governing equation. Final form of the order-parameter is equal to:

$$\frac{\partial \eta}{\partial t} = -M_1[16\,\chi\,r\,g(\eta)\,f'(T) + \kappa(\theta)^2\nabla^2\eta] \quad (11)$$

The random "r" number is simulated on C++ language by a pseudo random number generator. The described mathematical formulation has been implemented by numerical packages OpenFOAM 2.3.0 [21].

3 Validation of the model

Validation was performed in two steps. First of all, the stable directional solidification process was compared with the one-dimensional "classic" Stefan problem formulation. The analytical solution of that problem was solved for pure water using the von Neumann approach without any thermal fluctuations or initial subcooling. The results are shown in Fig. 1. Secondly, the mathemathical model for instability of the freezing pattern was compared with a pure component by the directional freezing of pure metal (nickel) cast [19]. The obtained PFM results are presented in Fig. 2. The comparison data pattern could be found in the above listed article.

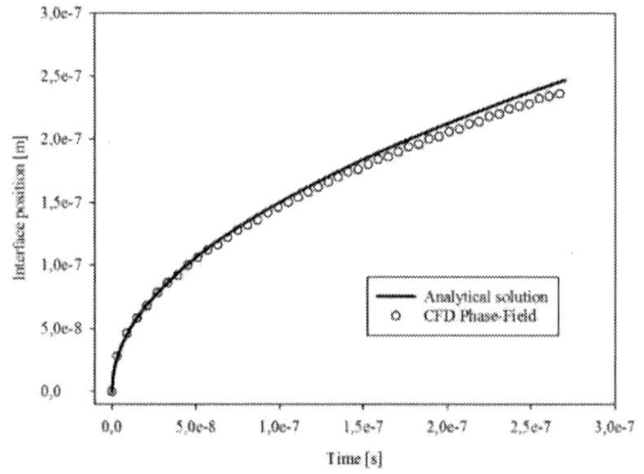

Fig. 1. Analytical and numerical solution for the one dimensional Stefan problem.

In both cases, the developed numerical description of a unnstable solidification front showed very good agreement with the analytical solution of the Stefan problem, or dendritic growth pattern for the nickel cast. Some deviation in the matter of one-dimensional Stefan solution colud be explained due to the diffusive character of the solidification front itself in PFM. In the Stefan problem, solidification front is assumed to be a sharp-edge.

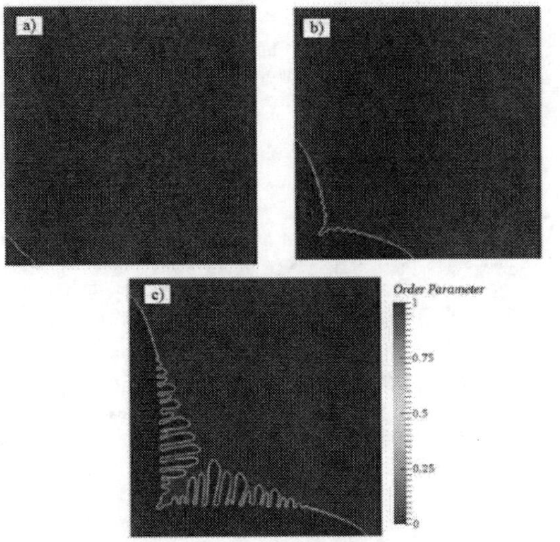

Fig. 2. Results of the calculations for the pure component with anisotropy and thermal fluctuations, the simulation time is equal to: a) 1e-8 s; b) 5e-8 s; c) 1e-7 s

4 Results and discussion

The morphology changes during the unstable solidification of pure water were checked by this method. In case of pure water, the 6th-fold symmetry of anisotropy was assumed [14]. The nucleation seed of the radius $2.5 \cdot 10^{-8}\, m$ was inserted in the middle of bulk liquid water which represets a solid sphere of a pure substance immersed into an undercooled melt. The numerical domain was equal to $2\ x\ 2\ \mu m$. The single cell grid size was assumed to be about $\Delta x = 2.5 \cdot 10^{-9} m$ which results in 640 000 mesh cells in total. Initial water temperature inside the numerical domain was set to the uniform value of 245.84 K. It corresponds to undercooling of $\Delta T = 27.3\, K$. The pure water properties used in the numerical calculations are presented in Tab. 1. The result of the simulation is studied in Fig. 3.

In Fig. 3 the snowflake like dendritic structure can be observed. Very characteristic is the sharp, visible dendrites pattern which follows anisotropy directions, with additional secondary branching from the main direction of anisotropy. This is a characteristic breakdown of the solid–liquid interface due to the Mullins–Sekerka instability [10,22]. It is worth mentioning, that dendrite growth is only due to thermal fluctuations, which is an invaluable advantage of taking into consideration free energy based on PFM. No initial sinusoidal perturbation on the phase boundary was applied. The temperature field propagates together with the solid-liquid interface. The very visiblie relation between order-parameter and temperature iso-lines dependency is shown in Fig. 4. The melting point temperature for pure water is set exactly on the interface border, where order-parameter achieve a value close to 1 (see Fig. 3 for comparison), which corresponds to 100% solid representation.

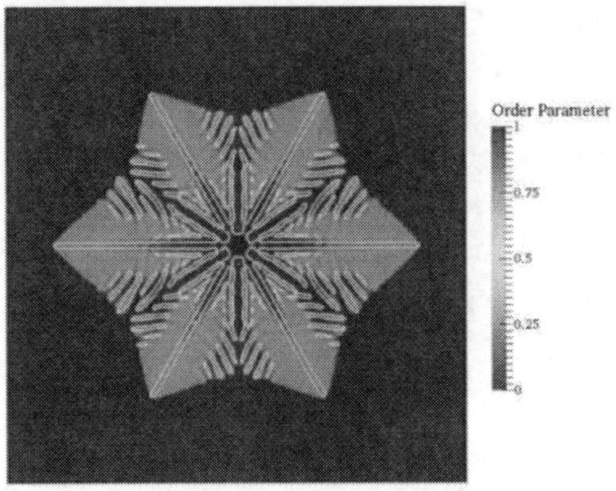

Fig.3. Pure water unstable solidification with anisotropy and thermal noise – order parameter, the simulation time is equal to 1e-7 s.

Table 1. Physical properties of water used during numerical calculations

Symbol	Value	Units
T_m	273.15	K
L	$3.33 \cdot 10^{8}$	J/m^3
σ	0.01598	J/m^2
cp	$4.2 \cdot 10^{6}$	J/m^3K
α	$1.4 \cdot 10^{-7}$	m^2/s
w_{min}	$1 \cdot 10^{-8}$	m
μ_k	5	$m/s\,K$
δ	0.1	-
B	6	-
χ	0.6	-

The value of order-parameter between 0 and 1 represents the region which is called “the mushy zone” – a mixture of solid and liquid. Therefore, for that region the temperature of the melt is lower than the temperature for pure ice, taking into consideration present boundary condition. Depending on the direction of dendrites growth, through time, the temperature field forms a strong thermal gradient. This feature could have an impact to cell freezing protocols like a red blood cell, which size is approx. 6-8μm in diameter [21]. To show thermal gradients of temperature profile, Fig. 5 represents temperature distribution profile over probe lines. Those proble lines were placed horizontaly - in the direction of main dendrite structure gowth; and diagonally – in the directional approx. 45deg. inclined to the previous line (in between the main dendrites directions). The zero point for both probes was set in the center of the snow-like frozen water. As a result, it can be seen, that there are differences in temperature propagation over time The fastest temperature diffusion is in the direction of main dendrite growth orientation (horizontal probe). The temperature stip gradient profile reaches the position of 0.6 μm after 1e-7 s for diagonal position, where the same temperature jump in Fig. 5 is reached after 7e-8 s for the main dendrite tip (horizontaly positioned line).

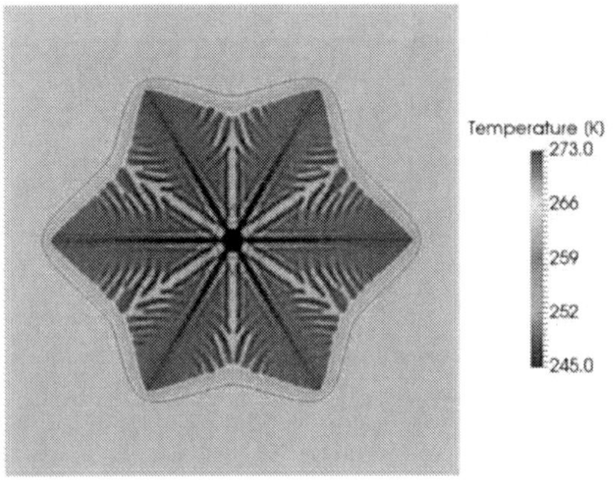

Fig. 4. Temperature isolines for the simulation time equals to 1e-7 s.

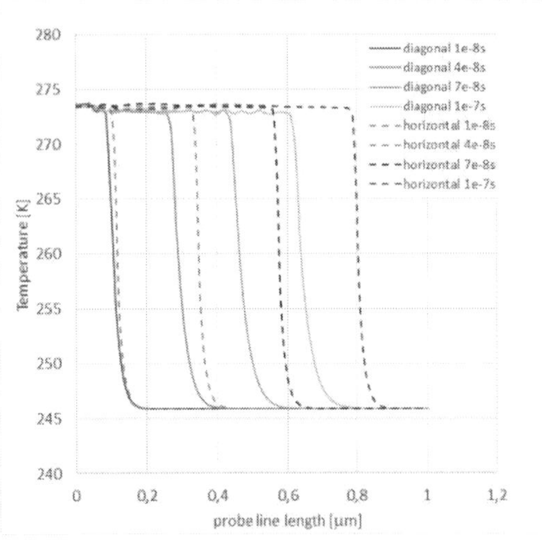

Fig. 5. Temperature distribution vs. time, depending on direction of data acquisition.

Conclusions

The new numerical approach for the directional freezing of biological cells, by using the PFM, has been presented and validated. Our interest is in the computation of dendritic solidification used in cryopreservation of cells and tissue. There is strong evidence, that the fluctuations and thermal noises have an impact on the solidifiction process under subcooling conditions. The temperature profile, as well as the direction of the solid front depends on anisotropy freezing and thermal fluctuations. Moreover, the resultant temperature field shows a strong relationship with the direction of solidifying instabilities during the time of solidification. The primary goal of this paper is to present a numerical technique for capturing a diffusive solid interface during the growth process. The microsegregation of solid interface is a usually an ignored issue due to the simplification of the cryopreservation models. The present model is a primiary step to develop a more adequate mathematical description of slow freezing methodology.

The authors wish to thank the Japan Society for the Promotion of Science (JSPS) and Ministry of Education, Culture, Sports, Science and Technology - Japan (MEXT) for their financial support of this project.

References

1. A.I. Zhmakin, *Fundamentals of Cryobiology*, (Springer Berlin Heidelberg, Berlin, Heidelberg, 2009)
2. B. Rubinsky, Heart Fail. Rev. **8,** 277–284 (2003)
3. E. Erdemli, S. Özkavukcu, J. Ankara Med. Sch. **24,** 187–196 (2003).
4. D. Li, B. Chen, W.J. Wu, G.X. Wang, Y.L. He, Appl. Therm. Eng. **73**, 1489–1500 (2014)
5. P. Mazur, Cryobiology. **14,** 251–272 (1977)
6. L. Bahari, A. Bein, V. Yashunsky, I. Braslavsky, PLoS One. **13,** 1–17 (2018)
7. P. Mazur, Am. J. Physiol. **247,** C125–C142 (1984)
8. M. Jaeger, M. Carin, M. Medale, G. Tryggvason, Biophys. J. **77,** 1257–67 (1999)
9. M. Jaeger, M. Carin, J. Comput. Phys. **179,** 704–735 (2002)
10. H.S. Udaykumar, L. Mao, Int. J. Heat Mass Transf. **45,** 4793–4808 (2002)
11. Y. Yang, H.S. Udaykumar, J. Comput. Phys. **210,** 55–74 (2005)
12. C.. Hirt, B.. Nichols, J. Comput. Phys. **39,** 201–225 (1981)
13. J.W. Cahn, J.E. Hilliard, J. Chem. Phys. **31,** 688–699 (1959)
14. R. Kobayashi, Phys. D. **63,** 410–423 (1993)
15. N. Moelans, B. Blanpain, P. Wollants, Comput. Coupling Phase Diagrams Thermochem. **32,** 268–294 (2008)
16. G.F. Naterer, *Heat transfer in Single and Multiphase Systems* (CRC Press LLC, Boca Raton, 2003)
17. G.B. McFadden, A.A. Wheeler, R.J. Braun, S.R. Coriell, R.F. Sekerka, Phys. Rev. E. **48,** 2016–2024 (1993)
18. W.W. Mullins, R.F. Sekerka, J. Appl. Phys. **35,** 444–451 (1964)
19. J. Choi, S.K. Park, H.Y. Hwang, J.Y. Huh, Acta Mater. **84,** 55–64 (2015)

20. I. Loginova, G. Amberg, J. Ågren, Acta Mater. **49,** 573–581 (2001)
21. C.J. Greenshields, *OpenFOAM User Guide, OpenFOAM Found. Ltd.* (2015) 230.
22. J.S. Langer, Rev. Mod. Phys. **52** (1980).
23. M. L. Turgeon, *Clinical Hematology: Theory and Procedures* (Lippincott Williams & Wilkins, 2004)

Heating of thick-walled steam header – an experimental study

Tomasz Sobota[1,*]

[1]Cracow University of Technology, Institute of Thermal Power Engineering, Faculty of Mechanical Engineering, Al. Jana Pawła II 37, 31-864 Cracow, Poland

Abstract. The paper presented the description of the experimental installation for testing of the computer systems for on-line monitoring of the power boilers operation. The results of the experiment can be used for calculation of temperature and thermal stresses distribution in thick walled pressure elements based on the solution of the inverse heat conduction problem.

1 Introduction

The Polish power plants, in the vast majority, operate power units installed 25 years ago or earlier. Therefore, in the near future thermal power plants facing the need to solve two major problems:

1. Assessment of the degree of usage of the major power plant equipment including boilers, turbines and steam pipelines. Proper estimation of pressure elements damage ratio of power units allows to decide on exchange used elements or decommissioning whole power unit.
2. The increase in the efficiency of electricity generation, including the efficiency of steam boilers already in use and newly built. On-line monitoring of the power boilers operation will reduce the unit cost of electricity production (the cost of 1 kWh of electric energy generation).

Institute of Thermal Power Engineering Cracow University of Technology developed and implemented in several domestic power plants two computer systems for on-line monitoring power boiler operation - Thermal Limitation Block (BOT) and Efficiency Limitation Block (BOS) [1-9].

Thermal Limitations Block allows for safe and quick start-up and shutdown of power boilers, resulting in reduced start-up losses while the high durability of boiler pressure parts is maintained. Block Efficiency Limitations (BOS) allow for power boiler operation with high efficiency and automatic cleaning of the heating surfaces of the combustion chamber and the steam super heater with sootblowers.

Further implementation of the developed systems in the domestic power plants, will enable to increase its competitiveness to foreign electricity producers. In order to further development of the monitoring system experimental installation named Installation for testing computer systems for on-line monitoring of power boilers operation. This installation enables to conduct the research and development in conditions similar to those that exist in steam boilers installed in the power plants [10-14].

2 Description of the experimental installation

In order to further development and improvement of the computer systems for on-line monitoring of power boiler operation in the laboratory of the Division of Thermal Power Engineering Cracow University of Technology modern experimental installation was built (Fig. 1). During its construction, particular emphasis was placed on ensuring that the conditions are similar to those that may occur in pressure components installed in power plants. In addition to metal temperature, temperature, mass flow rate, and pressure of the steam were measured as a function of time.

Fig. 1. View of the experimental installation for testing the computer system for on-line monitoring thick-wall pressure components

To measure temperature distribution on the outer surface of the steam header pre-calibrated thermocouples were

* Corresponding author: tsobota@mech.pk.edu.pl

used. The experimental installation consists of the following elements:

- steam generation unit with steam output capacity 700 kg/h, equipped with a three-stage oil burner,
- installation for boiler feed water treatment,
- tray-type (cascade-type) deaerator,
- blow down flash-vessel for heat recovery,
- steam pressure reduction station 10 bar /4 bar /2 bar,
- boiler control system,

steam header made of martensitic high alloy P91 steel.

2.1. Steam header

The steam header was made out of the martensitic high alloy P91 steel with the following physical properties: specific heat, $c = 486$ J/(kg·K), thermal conductivity, $k = 29$ W/(m·K), density, $\rho = 7750$ kg/m^3, modulus of elasticity, $E = 2{,}28 \cdot 10^{11}$ Pa, Poisson's ratio, $\nu = 0{,}29$, coefficient of thermal expansion $\beta = 0{,}098 \cdot 10^{-4}$ 1/K. The outer diameter d_{out}, wall thickness δ and the length L are: 355 mm, 50 mm and 3765 mm, respectively (Fig. 2).

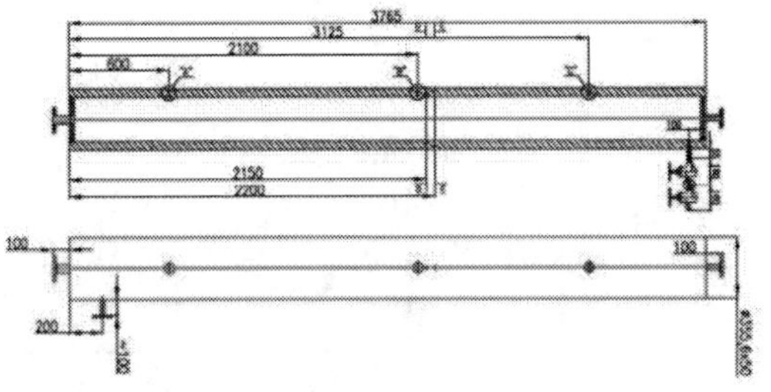

Fig. 2. Diagram of the steam header; A, B, C – location of industrial thermometers, przekrój D-D – the location of the 13 K-type thermocouples on the half of the outer circumference, przekrój E-E – temperature measurement on the wall thickness at four different depths

a) b)

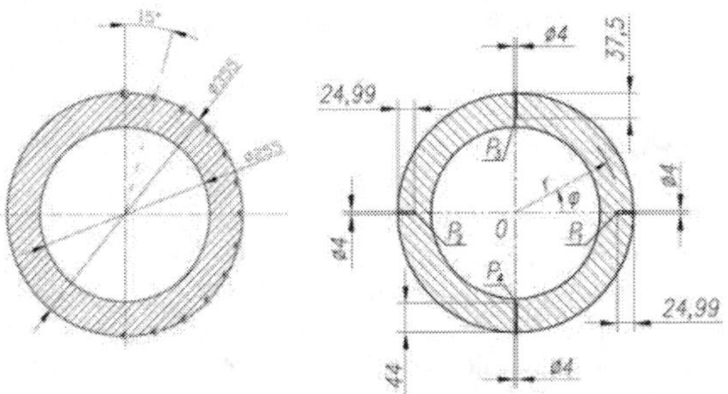

Fig. 3. Location of temperature sensors on the steam header a) thermocouples at the outer surface of the steam header, b) location of interior thermocouples used for experimental validation of the inverse procedure

Thirteen thermocouples NiCr-NiAl (K-type) were mounted every 15° on the half of the outer circumference at the distance of 2150 mm from the inlet of steam (Fig. 3a). The temperature measurement on the outer surface of the steam header will be used to determine the temperature distribution on wall cross-section and stress distribution on the inner and outer surfaces of the header using the inverse heat conduction methods.

To validate results of calculation based on the inverse procedure, temperature measurements carried out with four K-type thermocouples, with outer diameter of 3 mm, installed at the interior locations: $\delta_1 = \delta_2 = d_{out} - 0{,}50\delta$, $\delta_3 = d_{out} - 0{,}75\delta$ and $\delta_4 = d_{in} + 6$ mm (Fig. 3b) are used.

Fig. 4. View of the instrumented steam header before the foundation of the thermal insulation

At the top forming the steam header, at a distance of 600 mm (point A), 2100 mm (point "B"), and 3125 mm (point "C") from steam inlet (Fig. 2), 3 through-holes were drilled and threaded, in which industrial thermometers in massive housings made of steel 1H18N9T was placed. Implemented with their assistance saturated steam temperature measurement will be used to develop and to test methods for measuring the temperature of the medium flowing at high speed. The developed in the future method allows determining the actual temperature of the medium, on which the inertia caused by the massive thermometer housing will not affect.

3 Results of the experiment

The measurements were carried out in a transient state. The steam mass flow rate through the steam header was varying during the experiment. During the experiment, the following data were collected using data acquisition system: the temperature at the outer surface of the steam header, wall temperature at the three different depth and steam mass flow rate and its pressure at the header inlet. As the result of conducted experiment time distributions of the temperature at the steam header's outer surface half of the circumference (Fig. 5), internal wall temperature (Fig. 6) and steam temperature, pressure and mass flow rate (Fig. 7) during start-up, operation and shut down were collected.

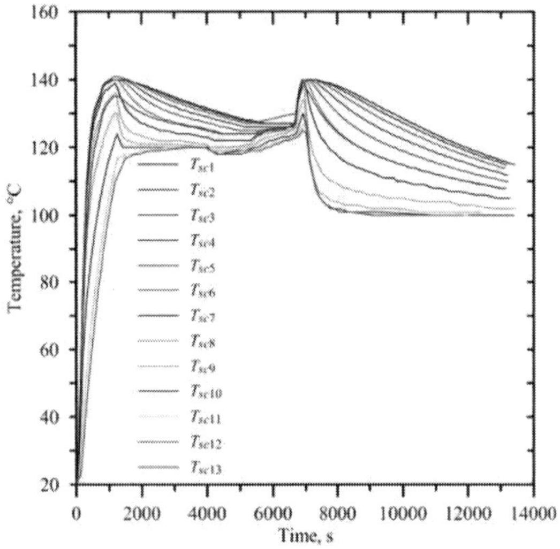

Fig. 5. Time changes of the measured temperature at the steam header's outer surface

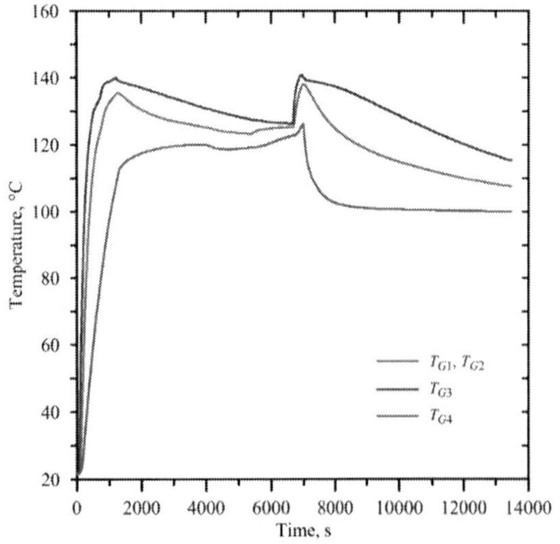

Fig. 6. Measured time changes of the wall temperature at the interior points

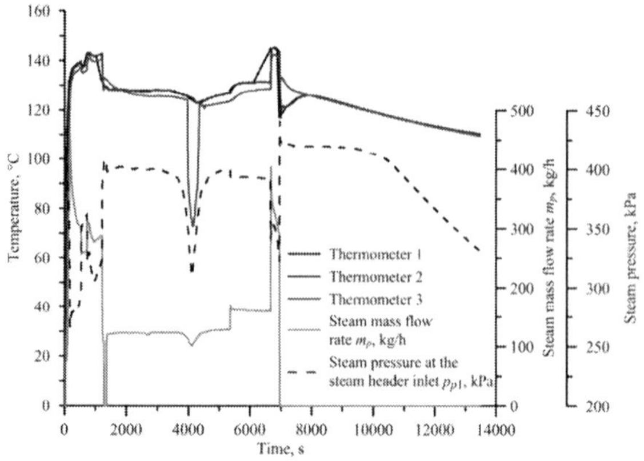

Fig. 7. Time changes of steam temperature, pressure and mass flow rate during the experiment

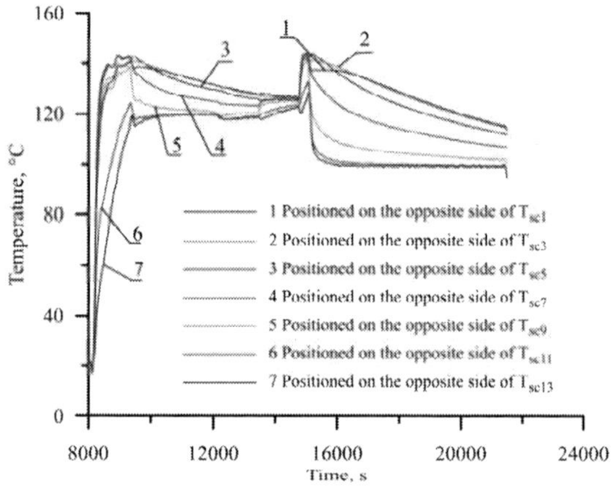

Fig. 8. Time changes of the calculated temperature at the steam header's inner surface

The results of experiments can be used for calculation of temperature and thermal stresses distribution in thick walled pressure elements based on the solution of the inverse heat conduction problem [1-3].

Conclusion

The paper presented the description of the experimental installation for testing of the computer systems for on-line monitoring of the power boilers operation. The results of the experiment are also presented. Developed at the Division of the Thermal Power Engineering Cracow University of Technology and tested in the described in the paper installation computer systems allow for quick to start-up and shutdown of the power boilers, their safe operation with high efficiency and automatic cleaning of heating surfaces of the combustion chamber and the steam super heaters with sootblowers.

References

1. J. Taler, D. Taler, K. Kaczmarski, P. Dzierwa, M. Trojan, T. Sobota, Energy, **160**, 500-519 (2018)
2. T. Sobota Tomasz, Heat Transfer Eng., **39**(13-14), 1260-1271 (2018)
3. T. Sobota, E3S Web of Conferences, **13**, (2017)
4. T. Sobota, E3S Web of Conferences, **14**, (2017)
5. J. Taler, B Węglowski, W. Zima, P. Duda, S. Grądziel, T. Sobota, A. Cebula, and D. Taler, Proc. IMechE **222** Part A: J. Power and Energy, 11-24 (2008)
6. J. Taler, S. Gądziel, T. Sobota, D. Taler, Archiwum Energetyki, **XXXVIII**(1), 97-122 (2008)
7. S. Lubecki, D. Taler, T. Sobota, Archiv. of Thermodynamics, **29** (4), 87-96 (2008)
8. J. Taler, Sz. Lubecki, T. Sobota, Energetyka, **2-3**, 680-681 (2011)
9. J. Taler, B. Węglowski, T. Sobota, D. Taler, M. Trojan, P. Dzierwa, M. Jaremkiewicz, M. Pilarczyk, Proc. of the ASME 2016 Power Conference POWER2016, (2016)
10. T. Sobota, Testing station for computerized systems for continuous monitoring of power boilers' operation, in *Modern Power Systems and Units*, J. Taler (Ed.) (Wydawnictwo Politechniki Krakowskiej, Kraków, 577-586, 2007)
11. M. Jaremkiewicz, D. Taler, T. Sobota, Int. J. of Thermal Sci., **87**, 241-250 (2015)
12. M. Jaremkiewicz, D. Taler, T. Sobota, App. Thermal Eng., **29**, pp. 3374–3379 (2009)
13. J. Taler, B. Węglowski, D. Taler, T. Sobota, P. Dzierwa, M. Trojan, P. Madejski, M. Pilarczyk, Energy, **92**(1), 153-159 (2015)
14. J. Taler, D. Taler, T. Sobota, P. Dzierwa, Archiv. of Thermodynamics, **32** (3), 103-116 (2011)

Determination of heat losses through building partitions

Tomasz Sobota[1,*], *Jan* Taler[1]

[1]Cracow University of Technology, Institute of Thermal Power Engineering, Faculty of Mechanical Engineering, Al. Jana Pawła II 37, 31-864 Cracow, Poland

Abstract. An numerical algorithm which can be use to calculate heat stream density transferred through wall with air void is presented . Some results of influence of thermal isolation on heat stream density transferred through the wall are presented too . The algorithm shown in this paper has general character and can be use to calculation loss of heat through the walls of block flats, furnaces , cold rooms and others objects and devices.

1 Introduction

The need to reduce energy consumption and maintain an unchanged temperature inside the building imposes specific requirements with regard to the thermal insulation of external walls and roofs, and thus - the materials used to insulate them.
The choice of the method used to insulate a specific building depends on many factors such as:

1. insulated wall structure,
2. technical condition of the wall ,
3. the required thermal and humidity conditions to be met in the insulated building,
4. the financial capacity of the investor. [4]

Increasing the thermal insulation of external walls results in a reduction in energy demand of 15 to 30% depending on the existing characteristics of the building [3] and the investment outlays, depending on their scale, can be recouped within 1 to 5 years. [1] Also the use of new insulating materials and increased insulation thickness leads to a significant reduction of heat losses through the partitions, and consequently to a reduction in the demand for thermal power in rooms. [2]

Many buildings have external walls made as so-called slotted walls (double), where the air gap, acting as thermal insulation, runs vertically through the height of at least one storey and divides the wall into two: internal (load-bearing) and external (elevation).

Although the air has a low coefficient of heat conduction, the air gap is not a good insulator, which is due to the complex way of transferring heat through the gap. There is both heat exchange through free convection (caused by the temperature difference between the supporting and elevation wall), radiation and conduction. The authors of many publications, e.g. [5-7] in determining the heat flux transmitted through the vertical air gap where natural convection occurs, use a substitute heat conduction coefficient, which is based on the criterion numbers Gr , Pr :

$$\lambda_k = \lambda \cdot \frac{0.0236 \cdot (\mathrm{Gr} \cdot \mathrm{Pr})^{1.393}}{(\mathrm{Gr} \cdot \mathrm{Pr}) + 1.01 \cdot 10^4}$$

However, the equivalent heat conduction coefficient does not take into account heat exchange through radiation.

In this paper we will present an algorithm for calculating the density of heat flowing through a wall with an air gap. It was assumed that the total density of the heat flux passing through the air gap is the sum of three components from convection, conduction and radiation $q = q_\alpha + q_\lambda + q_r$.

2 Algorithm of calculation of heat loss through walls of buildings

The calculation of the density of the heat stream penetrating from the inside of the room with the temperature T_w to the environment with the temperature T_z will be carried out in an iterative way. This is due to non-linear equations describing heat exchange through natural convection and radiation. At a known internal temperature, the density of the heat flux penetrating through the wall is selected so that the calculated external temperature is equal to the actual external temperature. From the mathematical point of view, a non-linear algebraic equation with one unknown q is solved.

To solve such an equation, the method of setting the interval has been used in this paper. The calculation of the surface temperature of the air gap from the external environment is also iterative. The surface temperature of the air gap is calculated using the straight iteration method (fixed point method).

At the assumed value q, the surface temperature of the inner wall is calculated first (Fig. 1).

$$T_1 = T_w - \frac{q}{h_w} \tag{1}$$

where T_1 - internal wall surface temperature, T_w - internal room temperature, q [W/m2] - heat flux density (unit

* Corresponding author: tsobota@mech.pk.edu.pl

heat loss) and h_w - heat transfer coefficient from internal air to wall.

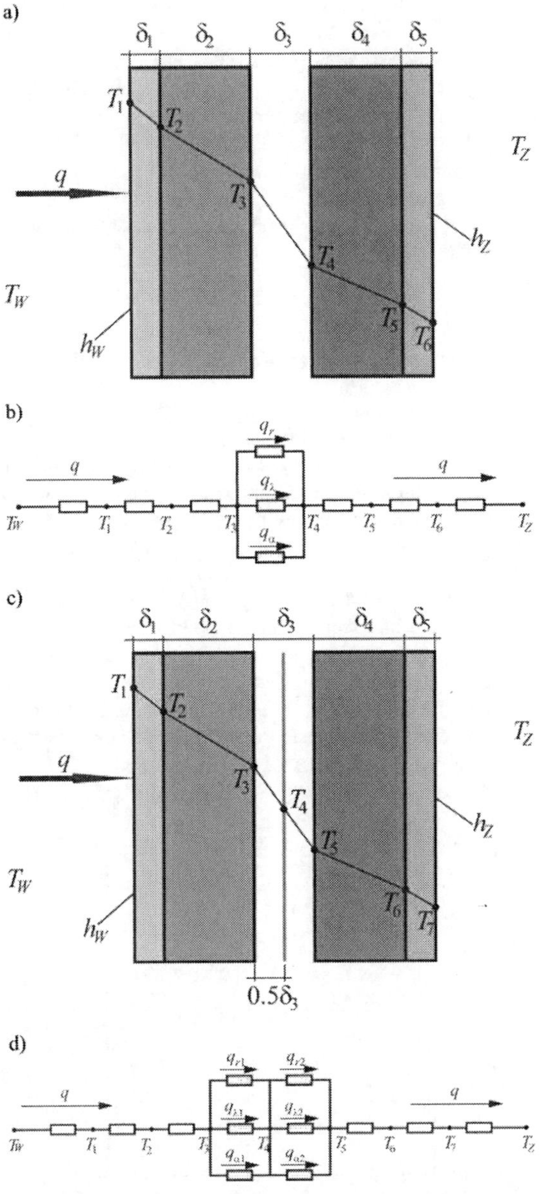

Fig. 1. Scheme of the wall with air void used to computations: a) wall with air void, b) heat flow through the wall diagram, c) wall with air void and radiation shield, d) heat flow through the wall with air void and radiation shield diagram

Then the temperature T_2 at the plaster-wall boundary is calculated.

$$T_2 = T_1 - \frac{q\cdot\delta_1}{k_1} = T_w - \frac{q}{h_w} - \frac{q\cdot\delta_1}{k_1} = T_w - q\cdot\left(\frac{1}{h_w}+\frac{\delta_1}{k_1}\right) \quad (2)$$

where δ_1 - plaster layer thickness, k_1 - thermal conductivity of the plaster. The temperature T_3 on the supporting wall surface is calculated consecutively.

$$\begin{aligned} T_3 &= T_2 - \frac{q\cdot\delta_2}{k_2} = \\ &= T_w - \frac{q}{h_w} - \frac{q\cdot\delta_1}{k_1} - \frac{q\cdot\delta_2}{k_2} = \\ &= T_w - q\cdot\left(\frac{1}{h_w} + \frac{\delta_1}{k_1} + -\frac{\delta_2}{k_2}\right) \end{aligned} \quad (3)$$

where δ_2 - load-bearing wall thickness, k_3 - heat conduction coefficient of the internal wall. Then the temperature T_4 on the internal surface of the façade wall is iteratively calculated. The total heat flux density through the air gap is determined by the formula:

$$q = q_\lambda + q_\alpha + q_r \quad (4)$$

where

- the component of the heat flux transferred by conduction,

$$q_\lambda = \frac{k_3}{\delta_3}\cdot(T_3 - T_4) \quad (5)$$

- convective component of heat flux density ,

$$q_\alpha = h_k(T_3 - T_4) \quad (6)$$

- radiation componen of heat flux.

$$q_r = \frac{\sigma\cdot(T_3^4 - T_4^4)}{\frac{1}{\varepsilon_1}+\frac{1}{\varepsilon_2}-1} \quad (7)$$

After substitution (5) - (7) to (4) the following is obtained

$$q = h_k\cdot(T_3 - T_4) + \frac{k_3}{\delta_3}\cdot(T_3 - T_4) + \frac{\sigma\cdot(T_3^4 - T_4^4)}{\frac{1}{\varepsilon_1}+\frac{1}{\varepsilon_2}-1} \quad (8)$$

from where it follows

$$(T_3 - T_4) = \frac{q}{h_k + \frac{k_3}{\delta_3} + \frac{\sigma\cdot(T_3 - T_4)\cdot(T_3^2 + T_4^2)}{\frac{1}{\varepsilon_1}+\frac{1}{\varepsilon_2}-1}} \quad (9)$$

The temperature T_4 is calculated from the following equation using the simple iteration method

$$T_4^{(n+1)} = T_3 - \frac{q}{h_k + \frac{k_3}{\delta_3} + \frac{\sigma\cdot(T_3 - T_4^{(n)})\cdot\left[T_3^2 + \left(T_4^{(n)}\right)^2\right]}{\frac{1}{\varepsilon_1}+\frac{1}{\varepsilon_2}-1}} \quad (10)$$

where n = 0, 1, 2, … is the number of subsequent iteration, δ_3 - thickness of the air gap, k_3 thermal conduction of the air, h_k - heat transfer coefficient of the air in the air gap, ε_1, ε_2 - emissivity of the surface of load-bearing and elevation wall, $\sigma = 5{,}67\cdot10^{-8}$- Stefan - Boltzmann constant, W/(m^{2}K^4). As the first approximation (n = 0) can be taken as $T_4^{(0)} = T_3$ - 4K.

The iterative process ends when

$$\left|T_4^{(n+1)} - T_4^{(n)}\right| \le \varepsilon \quad (11)$$

where ε is the specified tolerance.

The following dependencies were used to determine the convective heat transfer coefficient α_k: [7]

- for Gr ≤ 2000

$$\mathrm{Nu} = 1 \tag{12}$$

- for 2000 ≤ Gr ≤ 200000

$$\mathrm{Nu} = \frac{0.18\mathrm{Gr}^{0.25}}{\left(\frac{L}{\delta_3}\right)^{\frac{1}{9}}} \tag{13}$$

- for 200000 < Gr ≤ 10^7

$$\mathrm{Nu} = \frac{0.065\mathrm{Gr}^{\frac{1}{3}}}{\left(\frac{L}{\delta_3}\right)^{\frac{1}{9}}} \tag{14}$$

where Gr is the Grashoff number $\mathrm{Gr} = \frac{g\rho^2\beta x^3}{\eta^2}(T_w - T_\infty)$ and Nu is the Nusselt number, L is air gap height.

The temperature T_5 at the boundary between the external wall and external plaster is then calculated

$$T_5 = T_4 - \frac{q\cdot\delta_4}{k_4} \tag{15}$$

where δ_4 - external wall thickness, k_4 - thermal conductivity of the external wall and T_6 - temperature on the external side of the wall.

$$T_6 = T_5 - \frac{q\cdot\delta_5}{k_5} = T_4 - q\cdot\left(\frac{\delta_4}{k_4} + \frac{\delta_5}{k_5}\right) \tag{16}$$

where δ_5 - thickness of external plaster, k_5 - thermal conductivity coefficient of external plaster. The last step is to calculate the ambient temperature T_z , which is used for comparison with the actual ambient temperature.

$$T_z = T_6 - \frac{q}{h_z} \tag{17}$$

where h_z is the heat transfer coefficient of the outside air.

The iterative calculation of q shall be carried out using the interval method until the criterion

$$\left|T_z - T_{z,rz}\right| \leq \varepsilon \tag{18}$$

where $T_{z,rz}$ is the actual ambient temperature. Usually, only a few iterations are followed by a heat flux density at which the unevenness is met (18). The presented calculation algorithm is general in nature, e.g. there is no small difference between temperatures T_3 and T_4 and can be used to calculate heat losses through walls of residential buildings, furnaces, cold rooms and other objects and devices.

3 Calculation examples

In order to present the efficiency of the calculation algorithm described above, unit heat losses through the wall of residential buildings of different constructions will be determined. It will be shown that the exchange of heat through radiation in the air gap between the load-bearing wall and the façade wall has a very unfavourable impact on the insulation of the wall, i.e. it contributes significantly to increasing the value of the heat-transfer coefficient through the wall. Examples of calculations were carried out for the following data:

- internal plaster of thickness $\delta_1 = 0.015$ m ,
- internal wall of thickness $\delta_2 = 0.250$ m, made of perforated (cored) brick with thermal conductivity $k_2 = 0.47$ W/(mK),
- air gap of thickness $\delta_3 = 0.060$ m,
- external wall of thickness $\delta_4 = 0.120$ m made of solid brick with thermal conductivity $k_4 = 0.75$ W/(mK),
- external plaster of thickness $\delta_4 = 0.015$ m.

The heat transfer coefficient h_z from the surrounding depends on the wind speed w and is described by the formula [8]

$$h_z = 7.15\cdot w^{0.78} \tag{19}$$

4 Discussion of the obtained results

In the building industry, air void is used to reduce the loss of heat flux density permeating through external walls. The use of this technique allows to limit the density of the heat flux required to maintain the desired temperature in the indoor room by about 17%-19% in relation to the normal wall (for example, used in the calculations). In [4] the recommended minimum thickness of the air gap is 0.04 m. As you can see from Fig. 2. the best effect can be obtained when the width of the air gap is within the range 0.04 ≤ δ_3 ≤ 0.07 m, because the curve representing the dependence of the density of the heat flux as a function of the thickness of the air gap is the steepest in this range. However, with an increase in the thickness of the air gap above δ_3 = 0/07 m, the course of the curve is significantly softened (Fig. 2).

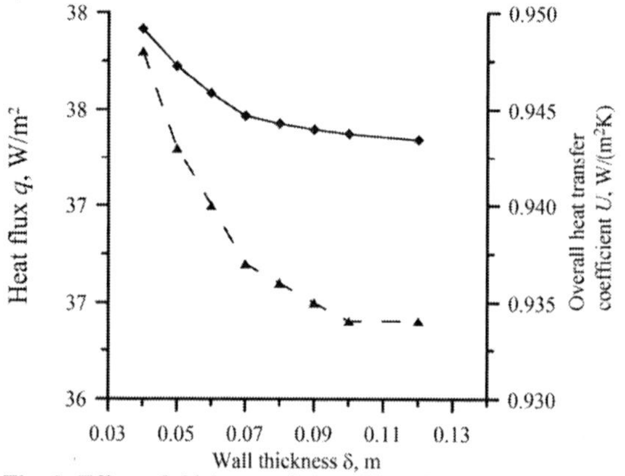

Fig. 2. Effect of thickness of air void on heat stream density and overall heat transfer coefficient

The determined values of the overall heat transfer coefficient U are of the range 0.934 W/(m^{2}K) to 0.948 W/(m^{2}K) and are on average 18% lower than for a normal wall (U = 1.150 W/(m^{2}K)) made of the same materials.

In the air gap, the density of the heat flux is transmitted through three paths: convection, radiation and conduction (Fig. 1). The following table shows the shares of the individual heat transfer methods in the total heat transfer stream.

Table 1. Font styles for a reference to a journal article.

Method of heat transfer	Percentage of total heat flux
Conduction	7.8%
Convection	19.51%
Radiation	72.69%

These shares were determined for the case when the wall with an air void was not insulated. As you can see, almost three quarters of the heat flux is transmitted through radiation in the gap. Reducing the share of radiation in the total heat flux density passing through the partition by using screens inside the air gap can be one of the ways to reduce heat loss in the room. In the case of using a single screen with the same emissivity as the masonry, the density of the heat flux transmitted by the radiation is reduced twice. In order to determine the effect of the screen on the total heat flux density that penetrates through the partition, numerical calculations have been made for the following case. In the wall with an air void as presented in Section 3. The air gap was divided into two parts by a screen made of polished aluminium foil. The diagram of a building envelope with an aluminium foil screen and the corresponding thermal resistance system is shown in Fig. 1c and Fig. 1d. As a result of this procedure, the share of radiation in the total heat flux density was reduced to 11% in the first part of the air gap and to 11.6% in the second part, and the of heat flux transferred through the building wall decreased from 37.583 W/m^2 in the case of a wall with an unshielded air gap, to 21.194 W/m^2 after the use of aluminium foil.

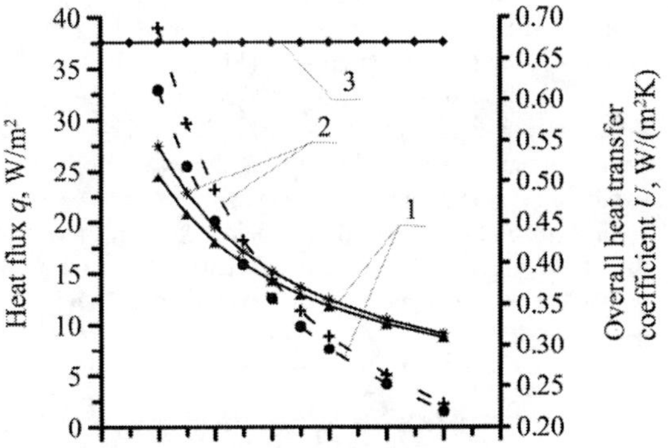

Fig. 3. Effect of additional thermal isolation (foamed polystyrene λ = 0,034 W/(m^{2}K)) of the wall with air void on heat stream density and overall heat transfer coefficient.

The next stage of calculations was an attempt to determine the effect of additional external insulation of the wall with air void on the heat flux density and a comparison with the case of thermal insulation of the same thickness but placed between the supporting and elevation wall - Fig. 3. As can be seen in Fig. 3 the very application of insulation with minimum thickness (in both cases), gives visible effects in the form of reduced density of the heat flux. For example, for a 0.06 m thick thermal insulation, in both cases a reduction of heat flux density by more than 60% was achieved. For the thickness of the foamed polystyrene layer equal to 0.06 m one obtains q = 14.27 W/m^2 when the wall is insulated from the outside and q = 15.19 W/m^2 when the insulation is inside the partition. In the case of external insulation, the losses are reduced by

$$\frac{(15,19-14,27)}{15,19}\cdot 100\% = 6,1\ \% .$$

It can be noticed that insulating the wall from the outside is more advantageous in terms of thermal insulation. For technological reasons, it is more advantageous to build walls with internal insulation (smaller thickness of the wall, there is no need to use metal anchors), however, the argument for a wall with an air void with additional insulation from the outside is in addition to reducing heat losses also a more favourable distribution of temperatures T_3 and T_4 on the surfaces of the air gap and the temperature of the façade wall T_5.

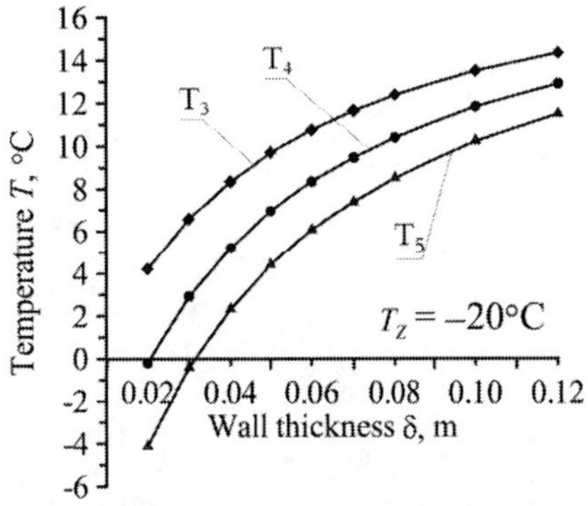

Fig. 4. Effect of thickness insulating material on T_3 i T_4 temperatures on the air void surfaces and T_5 temperature outside the wall

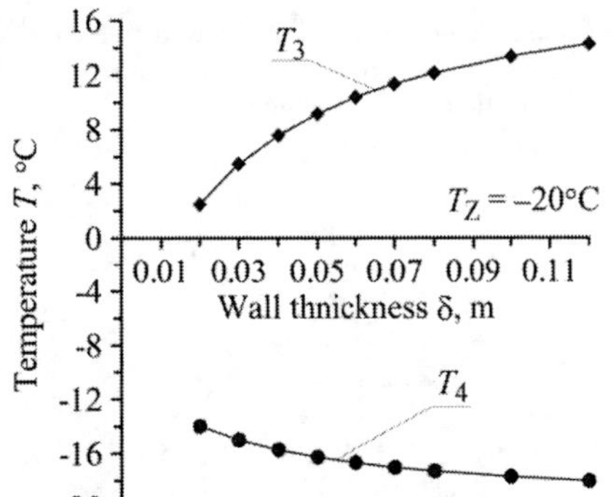

Fig. 5. Changes of T_3 i T_4 temperatures (Fig. 1a) when the air void is filled by different thickness insulating material

As shown in Fig. 4. for this type of partition there is almost no freezing of the external wall. On the other hand, if the insulation is placed inside the wall, the façade wall in the winter period is still working at negative temperatures, regardless of the ambient temperature, if only the latter is lower than zero (Fig. 6). In order to show the benefits of thermal insulation of external partitions, the wall thickness distribution for the following models of building partitions is shown in Fig. 7 wall with a non-heat-insulated air gap, wall with a screened air void and wall with a screened air gap with thermal insulation from the outside.

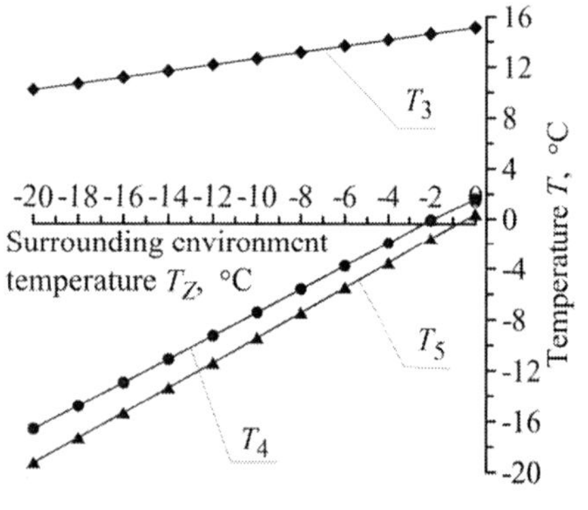

Fig. 6. Changes of T_3 i T_4 temperatures (Fig. 1a) when the air

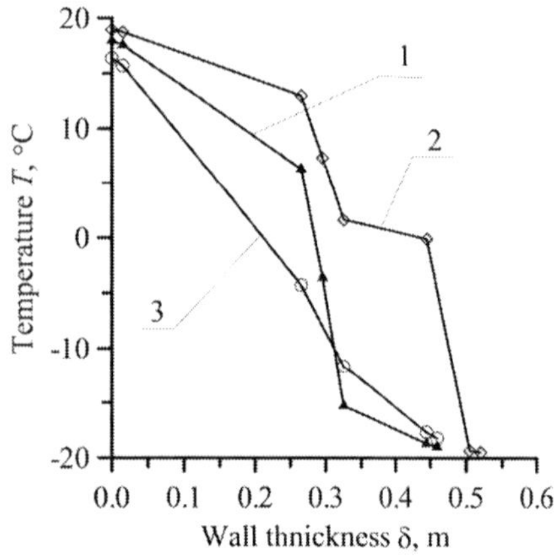

Fig. 7. Wall temperature distribution: 1 – wall with air void equipped with a radiation shield, 2 – wall with air void equipped with a radiation shield and outer thermal insulation, 3 – wall with air void

a)

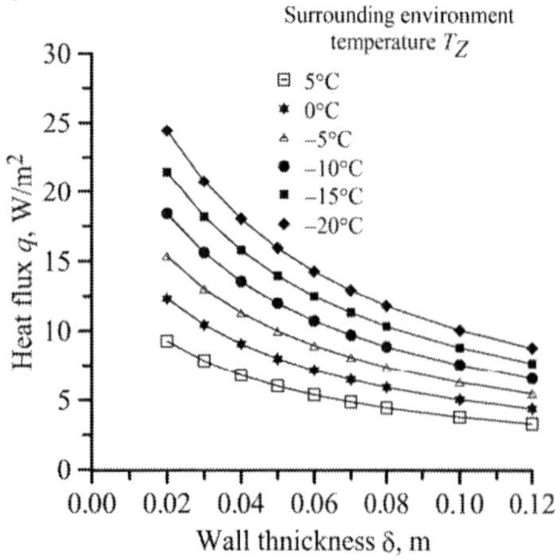

b)

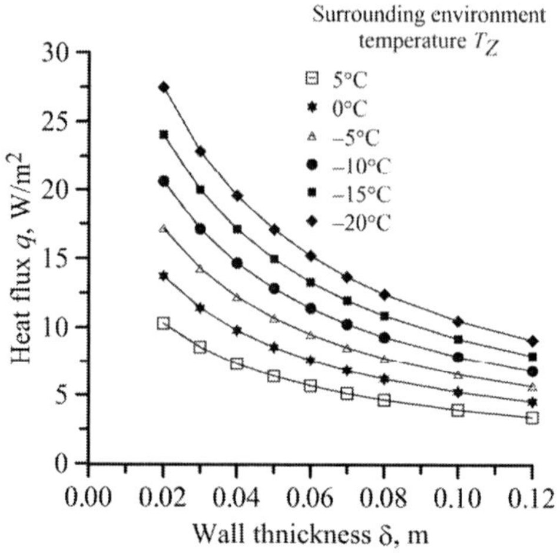

Fig. 8. Wall temperature distribution: 1 – wall with air void equipped with a radiation shield, 2 – wall with air void equipped with a radiation shield and outer thermal insulation, 3 – wall with air void

Fig. 8 shows the change of unit heat loss at the assumed temperature inside the room, at different outside temperatures, as a function of the thickness of the applied insulation. In the range of ambient temperatures from –5°C to –20°C these curves are very steep, which means that a slight increase in the thickness of the applied insulation contributes to a significant reduction in the loss of heat flux density, and at the same time lowers the costs of space heating. Assuming an extreme case, that the outside temperature in winter is –20°C and insulating the wall with a layer of 0.10 m foamed polystyrene, whose thermal conductivity k = 0.034 W/(mK), it is possible to reduce the heat flux transferred through the wall of the order of 3.7 times in relation to the non-insulated wall (Fig. 8a). The heat flux for such insulated wall is q = 10.06 W/m^2 and the overall heat transfer coefficient U = 0.252 W/(m^{2}K) – Fig. 8a. Similar results were achieved in the case of thermal insulation inside the partition (Fig. 8b).

The insulation material also has a great influence on the reduction of heat flux density. It should be characterized by resistance to moisture and the lowest possible coefficient of heat conduction.

5 Final remarks

The general algorithm of calculation of heat losses through building partitions presented in the paper allows to determine heat losses taking into account convection and radiation in the air gap. It enables determination of thermal losses through building partitions and the distribution of temperatures on its thickness much more precisely than in the case of algorithms taking into account only the heat conduction resistance of individual layers.

References

1. Regulation of the Minister of Infrastructure and Construction of the Republic of Poland on technical conditions to be met by buildings and their location. Journal of Acts, **1422** (2015)
2. J. Kwiatkowski, P. Matwiejuk, District Heating, Heating, Ventilation, **9**, pp. 325-329 (1994)
3. R. Rabjasz, District Heating, Heating Ventilation, **6**, pp. 250 – 256 (1995)
4. J. Panas (Ed.), *New construction foreman's guide* (ARKADY, Warsaw 2007)
5. F. Hell, *Grundlagen der Wärmeübertragung*, (VDI - Verlag , Düsseldorf ,1979)
6. W.P. Isachenko, W.A. Osipowa, A.S. Sukomiel, *Tieplopieredacha* (Energiya, Moscow, 1965)
7. J.R. Welty, *Engineering heat transfer* (John Wiley & Sons Inc., New York, 1974)
8. H. Recknagel, E. Sprenger, E.-R.Schramek, *Heating and air conditioning manual including refrigeration and hot water supply* (EWFE, Gdańsk, 1994)

Pipeline heating and cooling

Tomasz Sobota[1,*], *Dawid* Taler[2], and *Szczepan* Lubecki[3]

[1]Cracow University of Technology, Institute of Thermal Power Engineering, Faculty of Mechanical Engineering, Al. Jana Pawła II 37, 31-864 Cracow, Poland
[2]Cracow University of Technology, Institute of Thermal Engineering and Air Protection, Faculty of Environmental Engineering, ul. Warszawska 24, 31-155 Cracow, Poland
[3]BIPROTECH LLC, ul. Kamieńskiego 47, 30-644 Cracow, Poland

Abstract. The work aims to determine the undetermined temperature distribution of the medium and the pipeline wall using the finite difference method. Time courses of the temperature of the flowing medium and pipeline walls caused by a step change in temperature of the medium at the pipeline inlet, obtained by the numerical method, were compared with the courses calculated using strict analytical formulas. The numerical method of determining transient courses of the temperature of medium and pipeline wall can be used in the analysis of heating and cooling of heating or power pipelines with any changes in time of mass flow of the flowing medium or temperature of the medium at the inlet to the pipeline.

1 Introduction

During start-up and shutdown of power units, heating and cooling of pipelines, in particular, the pipeline connecting the boiler with the turbine, is an important issue. The temperature course of the medium is important not only because of the thermal stresses in the pipeline wall but also because of the stresses in the fittings installed on the pipeline [1-16]. In the case of steam or gases, the temperature changes along the pipeline length are higher due to the lower density and specific heat of the medium. Large wall thicknesses and considerable wall lengths also contribute to a significant reduction in the temperature of superheated steam in power pipelines.

2 Mathematical formulation of the problem

The heating of the pipeline by a flowing factor is described in the energy balance equation for the fluid [17]

$$\tau_{cz}\frac{\partial T_{cz}}{\partial t}+\frac{1}{N_{cz}}\frac{\partial T_{cz}}{\partial z^{+}}=T_{sc}\big|_{r=r_w}-T_{cz} \tag{1}$$

and wall

$$\frac{1}{r}\frac{\partial}{\partial r}\left(r\frac{\partial T_{sc}}{\partial r}\right)+\frac{1}{L_r^2}\frac{\partial^2 T_{sc}}{\partial\left(z^{+}\right)^2}=0\,. \tag{2}$$

Boundary and initial conditions are in the form

$$-\lambda\frac{\partial T_{sc}}{\partial r}\bigg|_{r=r_w}=\alpha_{cz}\left(T_{cz}-T_{sc}\big|_{r=r_w}\right)$$
$$\frac{\partial T_{sc}}{\partial r}\bigg|_{r=r_z}=0 \tag{3}$$

$$T_{cz}\big|_{z=0}=T_0+f\left(t\right)$$
$$\frac{\partial T_{sc}}{\partial z^{+}}\bigg|_{z=0}=\frac{\partial T_{sc}}{\partial z^{+}}\bigg|_{z=L_r}=0 \tag{4}$$

$$T_{cz}\big|_{t=0}=T_0\,,\qquad T_{sc}\big|_{t=0}=T_0 \tag{5}$$

where Eqs. (1-5) have the following designations: T_{cz} i T_{sc} – fluid temperature and wall temperature, respectively in °C, t – time in s, $z^{+}=z/L_r$ – dimensionless Cartesian coordinate, L_r – steam pipeline length in m, T_0 – constant initial fluid temperature in °C, $f(t)$ – fluid temperature changes in time at pipeline inlet in °C. The following formulae define the number of heat transfer units N_{cz}, and the fluid time constant τ_{cz}

$$N_{cz}=\frac{h_{cz}A_{in}}{\dot{m}_{cz}c_{pcz}}=\frac{h_{cz}U_{in}L_r}{A_{cz}w_{cz}\rho_{cz}c_{pcz}}$$
$$\tau_{cz}=\frac{m_{cz}c_{pcz}}{h_{cz}A_{in}}=\frac{A_{cz}L_r\rho_{cz}c_{pcz}}{h_{cz}A_{in}} \tag{6}$$

where h_{cz} denotes heat transfer coefficient on the inner surface of the pipeline in W/(m^{2}K), $A_{in}=U_w\cdot L_r$ – inner surface area of the pipeline in m^2, $\dot{m}_{cz} = A_{cz}w_{cz}\rho_{cz}$ – fluid mass flow rate in kg/s, ρ_{cz} – fluid density in kg/m^3, w_{cz} – velocity of the fluid flowing the pipeline in m/s, c_{pcz} – fluid specific heat in J/(kgK), $U_{in} = \pi\ d_{in}$ – inner circumference of the pipeline in m, $A_{cz} = \pi\ (d_{in})^2/4$ – cross-sectional area of the channel occupied by the fluid in m^2.

The initial boundary problem described by Eqs. (1-5) was solved with the control volume method.

The heating process is analysed on the example of a steam pipeline connecting a steam power boiler with a turbine. The pipeline wall is divided into eight layers. The control volumes in the area of heating medium - superheated steam comes into contact with the cells located on the internal radius of the pipe r_w. Due to axial symmetry, only half of the pipeline was analysed.

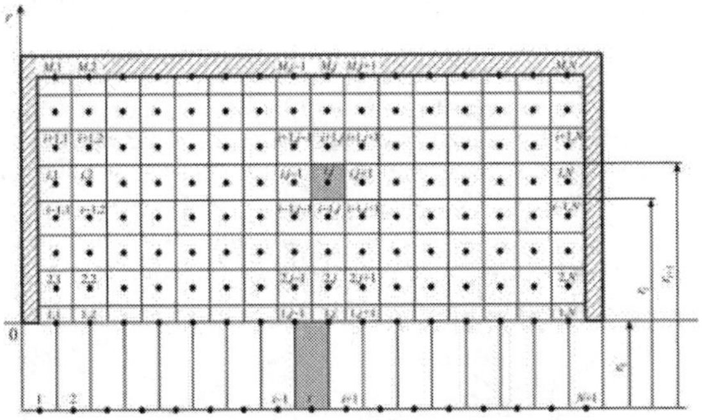

Fig. 1. Diagram of the pipeline division into control volumes

Temperature changes of the fluid and the pipeline wall describe the elementary balance equations recorded for the division shown in Figure 1. Below are presented exemplary heat balance equations for the cell (i,j) lying inside the wall and the cell (i) for the fluid:

$$
\begin{aligned}
& c_{sc}\rho_{sc}\pi\left(r_{i+1}^2 - r_i^2\right)\Delta z \frac{T_{sc,(i,j)}^{n+1} - T_{sc,(i,j)}^{n}}{\Delta t} = \\
& = \pi\left(r_{i+1}^2 - r_i^2\right)\frac{T_{sc,(i,j-1)}^{n} - T_{sc,(i,j)}^{n}}{\Delta z} + \\
& + \pi\left(r_{i+1}^2 - r_i^2\right)\frac{T_{sc,(i,j+1)}^{n} - T_{sc,(i,j)}^{n}}{\Delta z} + \\
& + 2\pi r_i \Delta z \frac{T_{sc,(i-1,j)}^{n} - T_{sc,(i,j)}^{n}}{\Delta r} + \\
& + 2\pi r_{i+1} \Delta z \frac{T_{sc,(i+1,j)}^{n} - T_{sc,(i,j)}^{n}}{\Delta r}
\end{aligned}
\tag{7}
$$

$n = 0, 1, \ldots; i = 1, \ldots, M; j = 1, \ldots, N,$

$$
\begin{aligned}
& \tau_{cz}\frac{T_{cz,i}^{n+1} - T_{cz,i}^{n}}{\Delta t} + \frac{1}{N_{cz}}\frac{T_{cz,i}^{n} - T_{cz,i-1}^{n}}{\Delta z} = \\
& = \frac{T_{sc,(1,j-1)}^{n} + T_{sc,(1,j)}^{n}}{2} - T_{cz,i}^{n}
\end{aligned}
\tag{8}
$$

$n = 0, 1, \ldots; i = 2, \ldots, N+1.$

In the Eqs. (7-8) the following notations were assumed: Δt - time step in s, Δz - spatial step in m, r - radial step in m, Δr - radial step in m, k_{sc} - thermal conductivity of the pipeline wall material in W/(mK), ρ_{sc} - density of the pipeline wall material in kg/m3, c_{sc} – specific heat of the pipeline material in J/(kgK), r_{in} - internal radius of the pipeline in m. The superscripts indicate the time step number. The assumption that the analyzed pipeline is ideally insulated on the external surface was made.

In order to ensure the stability of calculations of T_{cz} factor temperature and T_{sc} wall temperature using formulae (7-8), the Courant condition [17] must be met.

$$\frac{w_i \cdot \Delta t}{\Delta z} \le 1, \qquad i = 1, \ldots, N. \tag{9}$$

and the condition of stability of the solution of the heat conduction equation using the explicit method [6].

In this case, the pipeline heating can be analyzed, in which the time changes in temperature of the flowing medium at the inlet to the channel is described by any function $f(t)$. In a particular case, when $f(t) = \Delta T_{cz}$. This problem has a strict analytical solution [18]:

$$\frac{T_{cz} - T_0}{\Delta T_{cz}} = e^{-(\xi+\eta)}U(\xi,\eta), \tag{10}$$

$$\frac{T_{sc} - T_0}{\Delta T_{cz}} = e^{-(\xi+\eta)}U(\xi,\eta) - e^{-(\xi+\eta)}I_0\left(2\sqrt{\xi\eta}\right), \tag{11}$$

where I_0 is the modified Bessel function of the zero order and the U is the function defined by the formula

$$U(\xi,\eta) = \sum_{n=0}^{\infty}\sum_{k=0}^{n}\frac{\eta^n \xi^k}{n!k!}. \tag{12}$$

In Eqs. (10-12) the following designations were assumed:

$$
\begin{aligned}
& \xi = \frac{z \cdot N_{cz}}{L_r}, \\
& \eta = \frac{t - z/w_{cz}}{\tau_M} = \frac{t - \left(z\,\tau_{cz} N_{cz}\right)/L_r}{\tau_M}.
\end{aligned}
\tag{13}
$$

Eqs. (11-13) will be used to assess the accuracy of the finite difference method. In order to obtain a high accuracy of the analytical solution, the double series (12) was taken as $n > 400$ in the computer calculations, as the series is slowly converging.

3 Numerical simulation of pipeline heating

The steam pipeline to be analysed is made of 13HMF steel. Its dimensions are as follows: length $L_r = 48$ m, internal diameter $d_{in} = 0.217$ m, and wall thickness $g_{sc} = 0.028$ m. The physical properties of the pipe material are: density $\rho_m = 7650$ kg/m^3, specific heat $c_m = 519$ J/(kgK) and thermal conductivity $\lambda_m = 35$ W/(mK). It has been assumed that superheated steam with temperature $T_{cz} = 540$°C and pressure $p_{cz} = 11$ MPa flows suddenly into the pipeline. The steam thermal properties are – density $\rho_{cz} = 28{,}492$ kg/m^3, specific heat $c_{cz} = 2484$ J/(kgK) [19].

The initial temperature of the pipeline and fluid is T_0 = 300°C. The value of heat transfer coefficient on the inner surface of the pipeline, for the preset flow velocity of the fluid w_{cz}, was calculated from the Dittus-Boelter correlation formula. The distribution of fluid temperature T_{cz} and pipeline wall temperature T_{sc} as a function of time was determined from the solution of the a system of elementary balance equations and analytical solution described by Eqs. (10-11).

The analysis of Figs. 2 and 3 shows that the elementary balance method gives results with very good accuracy. It can be noticed that at higher steam speeds both the medium and the wall reach the steady state faster. The method of elementary balances can be applied for any changes in time of the medium temperature at the inlet and at temperature and pressure dependent thermal properties of the medium and wall material. In such cases, it is difficult or impossible to find an analytical solution.

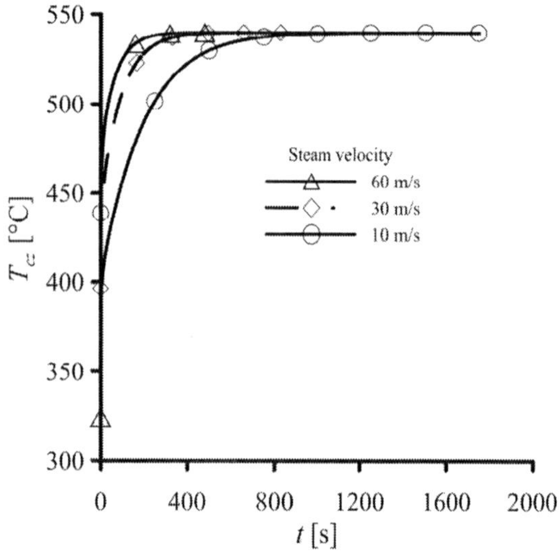

Fig. 2. Time changes of steam temperature for different steam flow velocities at the outlet from the pipeline; the points are the values obtained by a strict analytical solution and the solid line temperature runs determined by the finite difference method

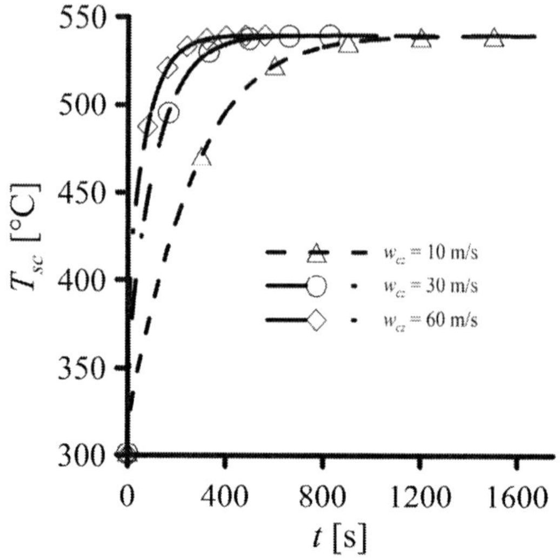

Fig. 3. Change in wall temperature over time at the outlet from the pipeline at different steam velocities; the points are the values obtained by a strict analytical solution, and the solid line temperature runs determined by the finite difference method

Figure 4 shows changes in temperature difference between the inner and outer surface of the pipeline at the end of the pipeline.

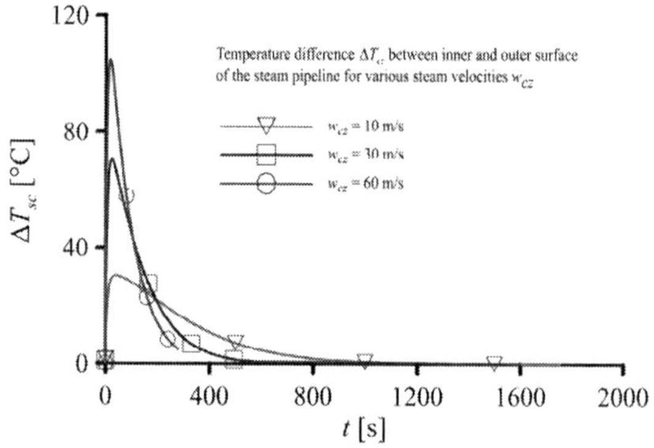

Fig. 4. The temperature difference between the inner and outer surface of the pipeline wall

The analytical solution, allowing to determine the change of heating medium temperature and pipeline wall temperature was derived assuming that the pipeline wall is a body with concentrated thermal capacity and without radial heat dissipation inside the wall. Therefore, in the next stage, the wall temperature as a function of time obtained from the tight solution (8) was compared with the average wall temperature of the pipeline (fig. 5) determined according to the following formula:

$$\overline{T}_{sc}(z) = \frac{2}{r_z^2 - r_w^2} \int_{r_w}^{r_z} T(r,z) r dr . \qquad (14)$$

where the temperature distribution $T(r, z)$ is obtained from a numerical solution.

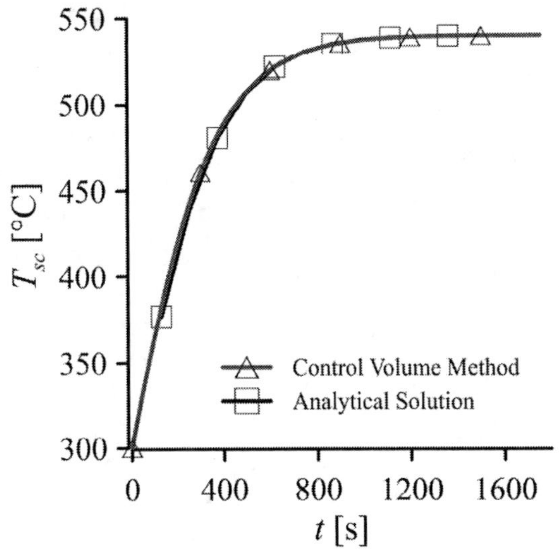

Fig. 5. Comparison of the wall temperature distribution at the end of the pipeline obtained from the analytical solution (8) with the value of the mean temperature calculated from Eq. (11) for the case when the steam flow velocity w_{cz} = 10 m/s.

4 Simulation of pipeline heating with FLUENT program

FLUENT [20] was used to determine the distribution of the temperature field in the cross-section of the steam pipeline under analysis. A two-dimensional model of the pipeline was built, which included axial symmetry.

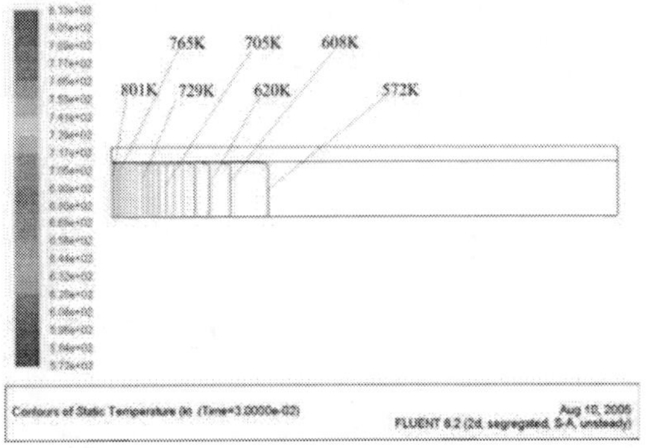

Fig. 6. The temperature distribution of steam obtained with the fluent program at time t = 0.03 s

Then as boundary conditions were set: steam velocity and temperature at the pipeline inlet, wall temperature, heat transfer coefficient on the internal surface of the pipe and ideal thermal insulation on the outer surface of the pipeline, i.e. $dT/dr = 0$. The distribution of the fluid temperature, at different times points during pipeline heating, is shown in Figures 6 and 7. The simulation was carried out for the following data: T_{cz} = 540°C, T_{sc} = 300°C, w_{cz} = 60 m/s, h_{cz} = 4269 W/(m^{2}K).

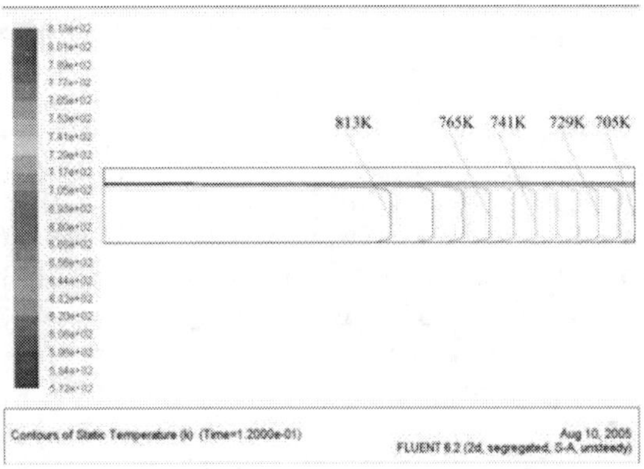

Fig. 7. Temperature distribution of steam obtained with the Fluent program at time t = 0.03 s

Conclusion

The explicit differential method presented in this paper can be used for the analysis of pipeline heating and cooling. A comparison of the calculated time histories of factor and pipeline wall temperature with a stepwise increase in factor temperature at the pipeline inlet with the histories obtained using strict analytical formulae shows that the accuracy of the differential method is very good. The finite difference method can be applied to any temporary changes in temperature of the medium at the inlet to the pipeline and depending on the temperature of thermo-physical properties of the medium and wall.

References

1. D.J. Littler (Ed.), *Modern Power Station Practice. Volume G – Station Operation and Maintenance. Third Edition* (Pergamon Press, Oxford, 1991)
2. L. Cwynar, *Rozruch kotłów parowych* (WNT, Warszawa 1981)
3. J. Taler, D. Taler, K. Kaczmarski, P. Dzierwa, M. Trojan, T. Sobota, Energy, **160**, 500-519 (2018)
4. T. Sobota. Heat Transfer Eng., **39**(13-14), 1260-1271 (2018)
5. T. Sobota, E3S Web of Conferences, **13**, (2017)
6. T. Sobota, E3S Web of Conferences, **14**, (2017)
7. J. Taler, B Węglowski, W. Zima, P. Duda, S. Grądziel, T. Sobota, A. Cebula, and D. Taler, Proc. IMechE **222** Part A: J. Power and Energy, 11-24 (2008)
8. J. Taler, S. Gądziel, T. Sobota, D. Taler, Archiwum Energetyki, **XXXVIII**(1), 97-122 (2008)
9. S. Lubecki, D. Taler, T. Sobota, Archiv. of Thermodynamics, **29** (4), 87-96 (2008)
10. J. Taler, Sz. Lubecki, T. Sobota, Energetyka, **2-3**, 680-681 (2011)
11. J. Taler, B. Węglowski, T. Sobota, D. Taler, M. Trojan, P. Dzierwa, M. Jaremkiewicz, M. Pilarczyk,

Proc. of the ASME 2016 Power Conference POWER2016, (2016)
12. T. Sobota, Testing station for computerized systems for continuous monitoring of power boilers' operation, in *Modern Power Systems and Units*, J. Taler (Ed.) (Wydawnictwo Politechniki Krakowskiej, Kraków, pp. 577-586 2007)
13. M. Jaremkiewicz, D. Taler, T. Sobota, Int. J. of Thermal Sci., **87**, 241-250 (2015)
14. M. Jaremkiewicz, D. Taler, T. Sobota, Appl. Thermal Eng., **29**, 3374–3379 (2009)
15. J. Taler, B. Węglowski, D. Taler, T. Sobota, P. Dzierwa, M. Trojan, P. Madejski, M. Pilarczyk, Energy, **92**(1), 153-159 (2015)
16. J. Taler, D. Taler, T. Sobota, P. Dzierwa, Archiv. of Thermodynamics, **32** (3), 103-116 (2011)
17. P.H. Oosthuizen, D. Naylor, *Introduction to Convective Heat Transfer Analysis* (McGraw-Hill, New York, 1999)
18. E.P Serov, B.P. Korolkov, *Dinamika parogienieratorov*. (Energoizdat, Moscow, 1981)
19. *ASME Steam Tables. Sixth Edition* (The ASME, New York, 1993)
20. *FLUENT 6*.0 (Fluent Inc., Lebanon USA, 2004)

Transient heat transfer at fluid flow in a thick-walled pipeline

Tomasz Sobota[1,*], *Dawid* Taler[2], and *Szczepan* Lubecki[3]

[1]Cracow University of Technology, Institute of Thermal Power Engineering, Faculty of Mechanical Engineering, Al. Jana Pawła II 37, 31-864 Cracow, Poland
[2]Cracow University of Technology, Institute of Thermal Engineering and Air Protection, Faculty of Environmental Engineering, ul. Warszawska 24, 31-155 Cracow, Poland
[3]BIPROTECH LLC, ul. Kamieńskiego 47, 30-644 Cracow, Poland

Abstract. The work aims to determine the transient temperature distribution of the medium and the pipeline wall using the finite difference method. Time courses of the temperature of the flowing medium and pipeline walls caused by a step change in temperature of the medium at the pipeline inlet, obtained by the numerical method, were compared with the courses calculated using strict analytical formulas. The numerical method of determining the transient distribution of temperature of medium and pipeline wall can be used in the analysis of heating and cooling of heating or steam pipelines with any changes in time of mass flow rate of the flowing medium or temperature of the medium at the inlet to the pipeline..

1 Introduction

During start-up and shutdown of power units, heating and cooling of pipelines, in particular, the pipeline connecting the boiler with the turbine, is an important issue. The temperature course of the medium is important not only because of the thermal stresses in the pipeline wall but also because of the stresses in the fittings installed on the pipeline [1-16]. In the case of steam or gases, the temperature changes along the pipeline length are higher due to the lower density and specific heat of the medium. Large wall thicknesses and considerable lengths also contribute to a significant reduction in the temperature of superheated steam in pipelines.

2 Mathematical formulation of the problem

The temperature changes of the medium and the pipeline wall are described by the following equations

$$\tau_{cz}\frac{\partial T_{cz}}{\partial t}+\frac{1}{N_{cz}}\frac{\partial T_{cz}}{\partial z^{+}}=T_{sc}-T_{cz}, \tag{1}$$

$$\tau_{M}\frac{\partial T_{sc}}{\partial t}=T_{cz}-T_{sc} \tag{2}$$

The boundary condition and initial conditions are in the form

$$T_{cz}\big|_{z=0}=T_0+f(t), \tag{3}$$

$$T_{cz}\big|_{t=0}=T_0, \tag{4}$$

$$T_{sc}\big|_{t=0}=T_0. \tag{5}$$

In Eqs. (1-5) the following designations have been adopted: T_{cz} i T_{sc} – fluid and wall temperature in °C, t - time in s, $z^{+} = z/L_r$ – - dimensionless Cartesian coordinate, L_r – steam pipeline length in m, T_0 – constant initial temperature of the fluid in °C, $f(t)$ – fluid temperature changes in time at pipeline inlet in °C.
The number of heat transfer units N_{cz}, the fluid time constant τ_{cz} and steam pipeline wall time constant τ_M are defined by the following formulae:

$$N_{cz}=\frac{h_{cz}A_{in}}{\dot{m}_{cz}c_{pcz}}=\frac{h_{cz}U_{in}L_r}{A_{cz}w_{cz}\rho_{cz}c_{pcz}}, \tag{6}$$

$$\tau_M=\frac{m_m c_m}{h_{cz}A_w}=\frac{A_m L_r \rho_m c_m}{h_{cz}A_{in}}, \tag{7}$$

$$\tau_{cz}=\frac{m_{cz}c_{pcz}}{h_{cz}A_{in}}=\frac{A_{cz}L_r\rho_{cz}c_{pcz}}{h_{cz}A_{in}}, \tag{8}$$

where h_{cz} denotes heat transfer coefficient on the inner surface of the pipeline in W/(m^{2}K), $A_{in} = U_w{\cdot}L_r$ – inner surface area of the pipeline in m^2, $\dot{m}_{cz} = A_{cz}w_{cz}\rho_{cz}$ – fluid mass flow rate in kg/s, ρ_{cz} – fluid density in kg/m^3, w_{cz} – velocity of the fluid flowing the pipeline in m/s, c_{pcz} – fluid specific heat in J/(kgK), $m_m = A_m L_r \rho_m$ – mass of the pipeline in kg, ρ_m – density of the pipeline material in kg/m^3, c_m – specific heat of the pipeline material in J/(kgK), $U_{in} = \pi d_{in}$ – inner circumference of the pipeline in m, $A_{cz} = \pi (d_{in})^2/4$ – cross-sectional area of the channel occupied by the fluid in m^2, $A_m=\pi\left(d_{out}^2-d_{in}^2\right)/4$ – cross-sectional area of the pipeline wall w m^2.
The Eqs. (1-2) have been derived with the assumption that the pipeline wall is thermally insulated on the external surface. Also, it has been assumed that the

* Corresponding author: tsobota@mech.pk.edu.pl

distribution of temperature and fluid velocity in the cross-section of the pipeline does not depend on the radius. Due to the complexity of the problem described by the Eqs. (1-5), numerical methods, e.g. presented in the paper [17], must be used to determine the transient temperature of the medium and the pipeline. The initial-boundary value problem (1-5) was solved with the explicit finite differences method. Fig. 1 shows a diagram of the differential grid for the calculation of the temperature distribution of the medium and the pipeline.

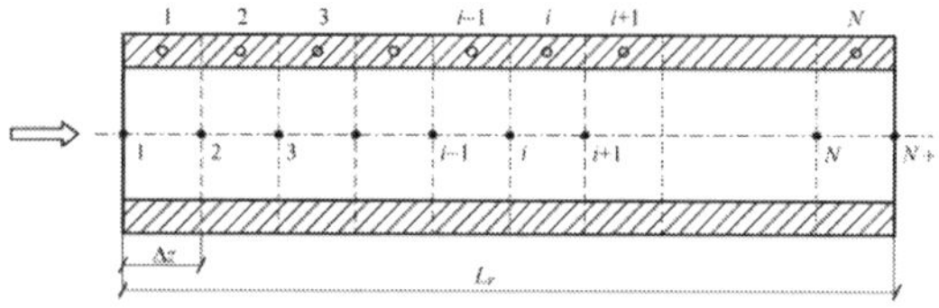

Fig. 1. Diagram of the finite difference grid

Applying the explicit finite differences method to the solution of the initial boundary problem (1-5), it is obtained

$$\tau_{cz}\frac{T^{n+1}_{cz,i+1}-T^{n}_{cz,i+1}}{\Delta t}+\frac{1}{N_{cz}}\frac{T^{n}_{cz,i+1}-T^{n}_{cz,i}}{\Delta z^{+}}= =T^{n}_{sc,i}-\frac{T^{n}_{cz,i}+T^{n}_{cz,i+1}}{2}, \tag{9}$$

$$\tau_M\frac{T^{n+1}_{sc,i}-T^{n}_{sc,i}}{\Delta t}=\frac{T^{n}_{cz,i}+T^{n}_{cz,i+1}}{2}-T^{n}_{sc,i} \tag{10}$$

$$T^{0}_{cz,i}=T^{0}_{sc,i}=T_0\,, \tag{11}$$

$$T^{n}_{cz,1}=f(n),\quad i=1,\ldots,N \quad n=0,1,2,\ldots, \tag{12}$$

where $t_n = n\cdot\Delta t$, Δt – denotes the time step in s, and z $\Delta z^+ = \Delta z/L_r$ is dimensionless co-ordinate.

After solving the equations (9-10) in respect to $T^{n+1}_{cz,i+1}$ and $T^{n+1}_{sc,i}$ it is obtained

$$T^{n+1}_{cz,i+1}= =T^{n}_{cz,i+1}-\frac{\Delta t}{\tau_{cz}N_{cz}}\left(\frac{T^{n}_{cz,i+1}-T^{n}_{cz,i}}{\Delta z^{+}}-T^{n}_{sc,i}+\frac{T^{n}_{cz,i}+T^{n}_{cz,i+1}}{2}\right) \tag{13}$$

$$T^{n+1}_{sc,i}=T^{n}_{sc,i}+\frac{\tau_M}{\Delta t}\left(\frac{T^{n}_{cz,i+1}+T^{n}_{cz,i}}{2}-T^{n}_{sc,i}\right) \tag{14}$$

$i = 1, \ldots, N,\quad n = 0, 1, 2, \ldots$

In order to ensure stability of calculations using Eqs. (13-14), the Courant condition [4, 5] must be met

$$\frac{w_i\cdot\Delta t}{\Delta z}\le 1,\quad i=1,\ldots,N, \tag{15}$$

where w_i is the fluid velocity in the i-th node.

In a particular case, when $f(t) = \Delta T_{cz}$ the problem described by the Eqs. (1-5) has a strict analytical solution [18]:

$$\frac{T_{cz}-T_0}{\Delta T_{cz}}=e^{-(\xi+\eta)}U(\xi,\eta)\,, \tag{16}$$

$$\frac{T_{sc}-T_0}{\Delta T_{cz}}=e^{-(\xi+\eta)}U(\xi,\eta)-e^{-(\xi+\eta)}I_0\left(2\sqrt{\xi\eta}\right), \tag{17}$$

where I_0 is the modified Bessel function of the zero order and the U is the function defined by the formula

$$U(\xi,\eta)=\sum_{n=0}^{\infty}\sum_{k=0}^{n}\frac{\eta^n\xi^k}{n!k!}\,. \tag{18}$$

In Eqs. (16-18) the following designations were adopted:

$$\xi=\frac{z\cdot N_{cz}}{L_r}$$
$$\eta=\frac{t-z/w_{cz}}{\tau_M}=\frac{t-(z\,\tau_{cz}N_{cz})/L_r}{\tau_M} \tag{19}$$

Eqs. (16-19) will be used to assess the accuracy of the finite difference method. In order to obtain a high accuracy of the analytical solution, the double series (18) was taken as $n > 400$ in the computer calculations, as the series is slowly converging.

3 Numerical simulation of steam pipeline heating

The steam pipeline to be analysed is made of 13HMF steel. Its dimensions are as follows: length L_r = 48 m, internal diameter d_{in} = 0.217 m, and wall thickness g_{sc} = 0.028 m. The physical properties of the pipe material are: density ρ_m = 7650 kg/m^3, specific heat c_m = 519 J/(kgK) and thermal conductivity λ_m = 35 W/(mK). It has been assumed that superheated steam with temperature T_{cz} = 540°C and pressure p_{cz} = 11 MPa flows suddenly into the pipeline. The steam thermal properties are – density ρ_{cz} = 28,492 kg/m^3, specific heat c_{cz} = 2484 J/(kgK) [19].

The initial temperature of the pipeline and fluid is T_0 = 300°C. The value of heat transfer coefficient on the inner surface of the pipeline, for the preset flow velocity of the fluid w_{cz}, was calculated from the Dittus-Boelter correlation formula. The distribution of fluid temperature T_{cz} and pipeline wall temperature T_{sc} as a function of time was determined from the solution of the system of differential Eqs. (13-14) and analytical Eqs. (16-17).

The analysis of Figs. 2 and 3 shows that the finite difference method gives results with very good accuracy. It can be noticed that at higher steam speeds both the medium and the wall reach the steady state faster. Finite difference method can be applied for any changes in time of the medium temperature at the inlet and at temperature and pressure dependent thermal properties of medium and wall material. In such cases, it is difficult or impossible to find an analytical solution.

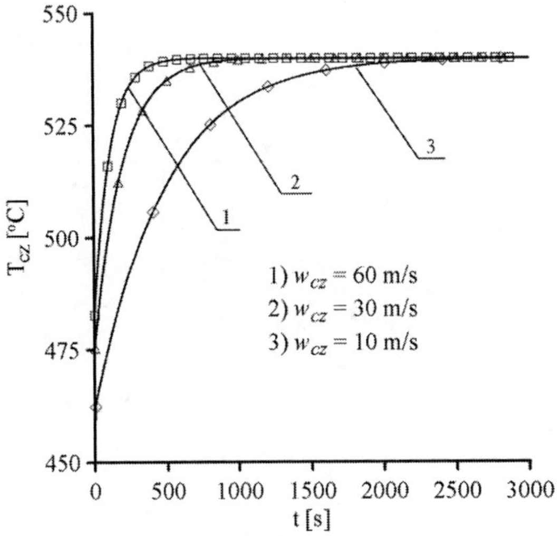

Fig. 2. Water steam temperature variation over time at the outlet from the pipeline for different steam flow velocities; the points are the values obtained by a strict analytical solution and the solid line temperature runs determined by the finite difference method

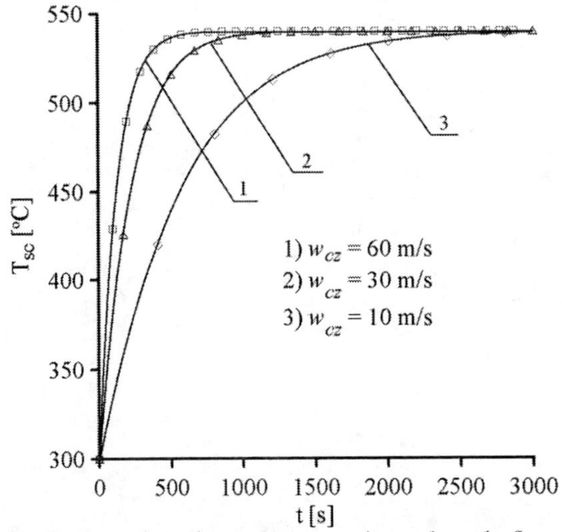

Fig. 3. Change in wall temperature over time at the outlet from the pipeline at different steam velocities; the points are the values obtained by a strict analytical solution, and the solid line temperature runs determined by the finite difference method

An example of transient courses of fluid and wall temperature caused by the linear increase of fluid temperature with constant speed v_T = 20 K/min from initial temperature T_0 = 300°C to final temperature T_k = 540°C is shown in Figs. 4 and 5. Calculations were carried out with the described explicit finite difference method. The analysis of Figs 4 and 5 shows that the pipeline heating speed is strongly influenced by the steam mass flow rate.

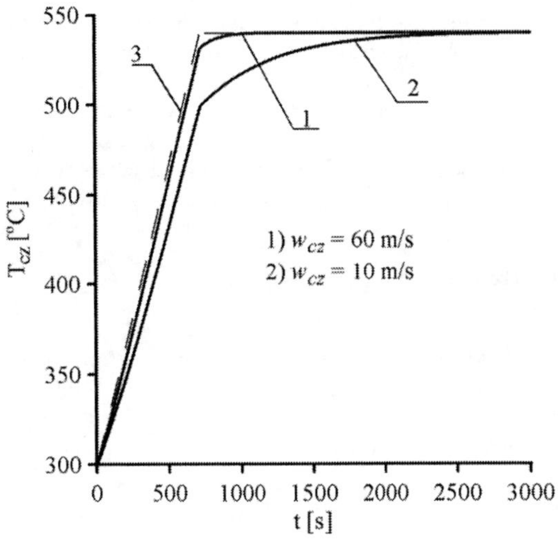

Fig. 4. Fluid temperature change over time at the outlet from the pipeline at different steam flow velocities and temperature change at the inlet to the pipeline described by the stepwise line – Eq. (3)

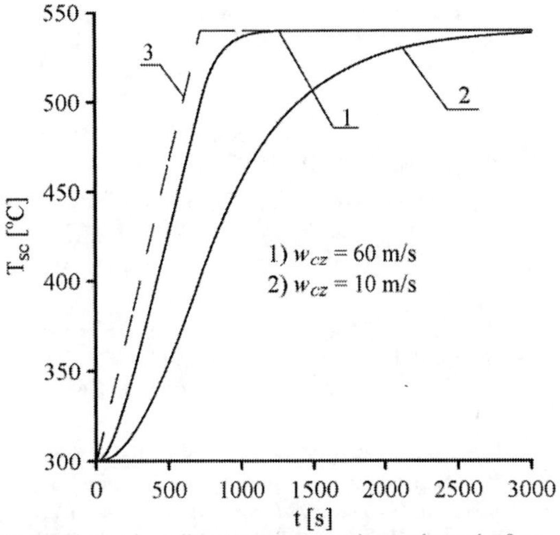

Fig.5. Change in wall temperature over time at the outlet from the pipeline at different steam flow velocities and temperature change at the inlet to the pipeline described by the stepwise line – Eq. (3)

If the steam flow velocity increases, the time of steam flow through the pipeline decreases and the heat transfer coefficient h_{cz} on the internal surface of the pipeline increases. Heat transfer coefficient increases from h_{cz} = 312,7 W/(m^{2}K) at w_{cz} = 10 m/s to h_{cz} = 1311 W/(m^{2}K) at w_{cz} = 60 m/s. The wall time constant τ_M decreases more than fourfold and the heating of the pipeline wall is much faster at w_{cz} = 60 m/s.

Conclusion

The explicit finite difference method presented in this paper can be used for the analysis of pipeline heating and cooling. A comparison of the calculated time histories of fluid and pipeline wall temperature with a stepwise increase in fluid temperature at the pipeline inlet with the histories obtained using strict analytical formulae shows that the accuracy of the differential method is very good. The finite difference method can be applied to any temporary changes in temperature of the medium at the inlet to the pipeline and depending on the temperature of thermo-physical properties of the medium and wall.

References

1. D.J. Littler (Ed.), *Modern Power Station Practice. Volume G – Station Operation and Maintenance. Third Edition* (Pergamon Press, Oxford, 1991)
2. L. Cwynar, *Rozruch kotłów parowych* (WNT, Warszawa 1981)
3. J. Taler, D. Taler, K. Kaczmarski, P. Dzierwa, M. Trojan, T. Sobota, Energy, **160**, 500-519 (2018)
4. T. Sobota. Heat Transfer Eng., **39**(13-14), 1260-1271 (2018)
5. T. Sobota, E3S Web of Conferences, **13**, (2017)
6. T. Sobota, E3S Web of Conferences, **14**, (2017)
7. J. Taler, B Węglowski, W. Zima, P. Duda, S. Grądziel, T. Sobota, A. Cebula, and D. Taler, Proc. IMechE **222** Part A: J. Power and Energy, 11-24 (2008)
8. J. Taler, S. Gądziel, T. Sobota, D. Taler, Archiwum Energetyki, **XXXVIII**(1), 97-122 (2008)
9. S. Lubecki, D. Taler, T. Sobota, Archiv. of Thermodynamics, **29** (4), 87-96 (2008)
10. J. Taler, Sz. Lubecki, T. Sobota, Energetyka, **2-3**, 680-681 (2011)
11. J. Taler, B. Węglowski, T. Sobota, D. Taler, M. Trojan, P. Dzierwa, M. Jaremkiewicz, M. Pilarczyk, Proc. of the ASME 2016 Power Conference POWER2016, (2016)
12. T. Sobota, Testing station for computerized systems for continuous monitoring of power boilers' operation, in *Modern Power Systems and Units*, J. Taler (Ed.) (Wydawnictwo Politechniki Krakowskiej, Kraków, pp. 577-586 2007)
13. M. Jaremkiewicz, D. Taler, T. Sobota, Int. J. of Thermal Sci., **87**, 241-250 (2015)
14. M. Jaremkiewicz, D. Taler, T. Sobota, Appl. Thermal Eng., **29**, 3374–3379 (2009)
15. J. Taler, B. Węglowski, D. Taler, T. Sobota, P. Dzierwa, M. Trojan, P. Madejski, M. Pilarczyk, Energy, **92**(1), 153-159 (2015)
16. J. Taler, D. Taler, T. Sobota, P. Dzierwa, Archiv. of Thermodynamics, **32** (3), 103-116 (2011)
17. P.H. Oosthuizen, D. Naylor, *Introduction to Convective Heat Transfer Analysis* (McGraw-Hill, New York, 1999)
18. E.P Serov, B.P. Korolkov, *Dinamika parogienieratorov*. (Energoizdat, Moscow, 1981)
19. *ASME Steam Tables. Sixth Edition* (The ASME, New York, 1993)

The influence of adsorption chillers on CHP power plants

Karol Sztekler[1*], *Wojciech* Kalawa[1], *Sebastian* Stefański[1], *Jarosław* Krzywanski[2], *Karolina* Grabowska[2], *Marcin* Sosnowski[2], and *Wojciech* Nowak[1]

[1]AGH University of Science and Technology, Faculty of Energy and Fuels, 30-059 Cracow, Poland
[2]Jan Dlugosz University in Czestochowa, Faculty of Mathematics and Natural Sciences, 42-200 Czestochowa, Poland

Abstract. The simultaneous production of electricity, heat and cooling, the so-called trigeneration, allows for substantial savings in the chemical energy of fuels. More efficient use of the primary energy contained in fuels translates into tangible earnings for power plants while reductions in the amounts of fuel burned, and of non-renewable resources in particular, certainly have a favourable impact on the natural environment. The main aim of the above-described project was to analyse the influence of use adsorption contacted to conventional CHP power plant. An adsorption chiller is an item of industrial equipment that is driven by low grade heat and intended to produce chilled water and desalinated water. Adsorption chillers ACH can by used for utilization heat from many industrial process where temperature medium is too low for use absorption chillers. In this article modelled the cycle of a conventional heat power plant integrated with an adsorption chiller-based plant. Multi-variant simulation calculations were performed using IPSEpro simulation software.

1 Introduction

Polish electricity production is mainly based on conventional fuels, particularly hard coal and lignite (jointly 79%); sources with smaller shares in electricity production include: RES 13%, gaseous fuels 5%, biomass and others 3%. In 2016, electricity power production in Poland 166.6 TWh while its consumption was 159.1 TWh. In 2016, the total thermal capacity installed at the licenced heat producers was 54,259.8 MW, and the generation capacity was 53,434.7 MW of which heat supplied to consumers connected to the network was 235,545.9 TJ. With regard to heat production, the predominant fuel utilised for this purpose is coal 75%; followed by gaseous fuels 7.2% and RES 7.6%. Production of heat in cogeneration units used 2,230,263,334 GJ energy from coal fuels [1-3].

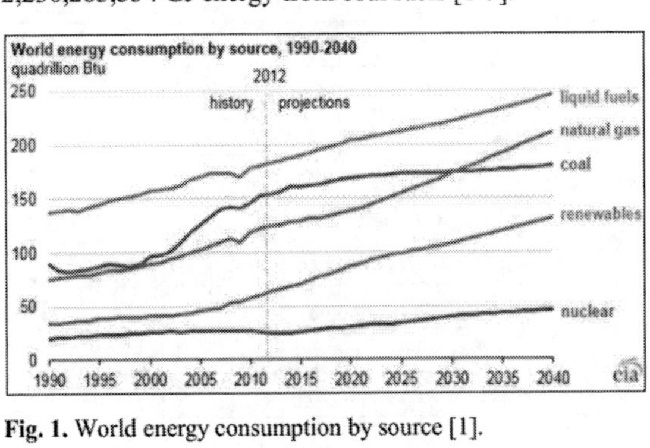

Fig. 1. World energy consumption by source [1].

According to figure number one fossil fuels still will be dominate fuels use for production energy.

The use of fossil fuels implies a lot of adverse effects on the environment, such as global warming, acid rain, smog as well as soil and water pollution. Counteracting this phenomenon involves increasing the efficiency of electricity production and searching for more efficient methods of conversion of the chemical energy of a fuel to electricity or replacing existing fuels with ones that are more environmentally friendly. One of the ways to increase the efficiency of the use of chemical energy from fuels with simultaneous reduction of the negative effect on the environment is generation through CHP (Combined Heat and Power) or through polygeneration, i.e. generation of an extra utility beside heat and electricity, e.g. cooling or compressed air. This paper presents the possibility of increasing the effectiveness of cogeneration units through the application of adsorption chillers fed by heat from the terminal unit of the district heat network. The greater quantity of heat generated from combusting fuels will be effectively utilised to, inter alia, feed chillers and the cooling produced will be used for example for air conditioning. This will contribute to a reduction in energy consumption for cooling purposes while at the same time reducing the amount of heat that remains unused in the process of generating electricity at conventional power plants, and consequently to savings in fossil fuels and improving the condition of the environment we live in [4-6].

* Corresponding author: sztekler@agh.edu.pl

https://doi.org/10.1051/matecconf/201824005033

2 Modelling process

2.1 Adsorption chillers

The operating principle of an adsorption cooling system (Figure 2) consists of evaporating the refrigerant in the evaporator under reduced pressure.

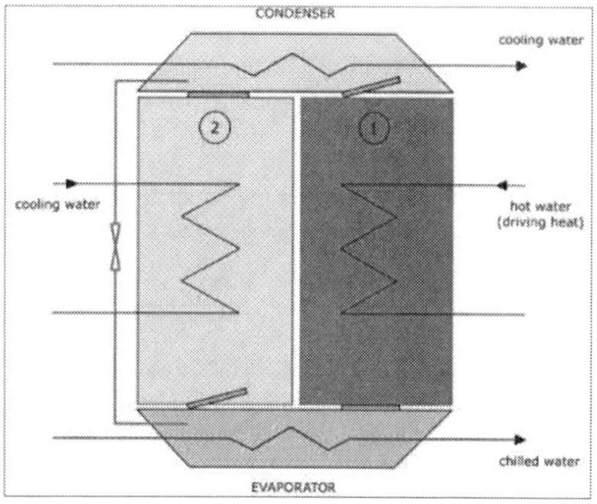

Fig. 2. Adsorption cooling-desalination system [4].

The process takes place without the use of a mechanical compressor but takes advantage of the thermal effect of adsorption and desorption. The system operates on a cyclical basis and the process of adsorption and desorption (bed regeneration) takes place alternately. Refrigerant vapours discharged from the evaporator are adsorbed in an adsorption bed, which is accompanied by the generation of heat which is absorbed by the cooling water [13]. In order to enable further operation of the bed, it is necessary to remove the adsorbate from the adsorbent (desorption). This process requires the application of heat from an external source. Refrigerant vapours released in the desorption process are discharged into the evaporator. In a closed system, the condensate from the condenser is fed back to the evaporator. It is possible to apply an adsorption chiller with an open system and using water as the refrigerant. The constant feeding of water from an external source results in producing condensate that is devoid of the substances dissolved in the feed water. Therefore, they can be used in the utilisation of waste heat in power plants and in combined heat and power plant [7-10]. For analysis will choose Adsorption chiller BryAir ADC Series D-FRAME F-200. The main parameters of ADC chiller present in table 1.

Table 1. Technical specification of BryAir chiller ADC Series F-FRAME F-200 [11].

Cooling capacity	900 kW
Electricity consumption	1.3-2.4 kW
Refrigerant type	Water
Heat for regeneration Temperature range	52°C-93°C
Temperature drop for regeneration process	$\Delta t = 6{,}6°C$
Chilled water	3°C-20°C
Condenser water	10°C-39°C

The value of the regeneration capacity of a chiller working based on the silica gel/water pairing is variable and strongly dependent on the hot water inlet temperature. Based on the graph in Fig. 3.

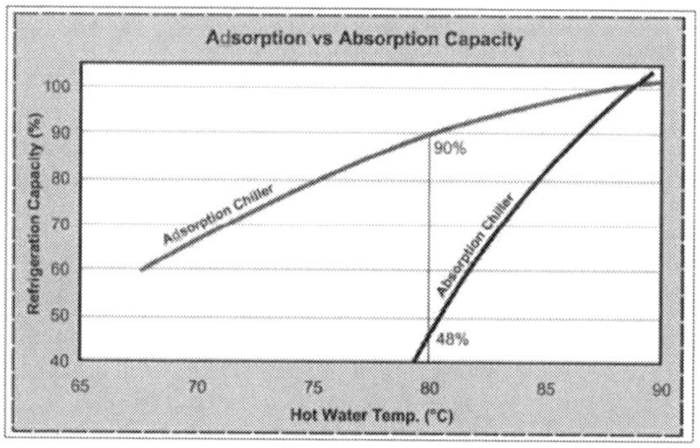

Fig. 3. Influence hot water temperature on refrigeration capacity adsorption-absorption chillers [11].

Adsorption cooling systems are being dynamically improved. Their range of application is expanding, and in combination with renewable energy sources they can make a cheap source of cooling. Commercialisation of the solutions found in adsorption cooling industry is possible if the most essential disadvantages are eliminated, i.e. the small cooling capacity in relation to the mass of the installation and the fluctuations in the cooling performance resulting from the cyclic character of operation.

2.2 CHP power plant

The thermodynamic cycle of a cogeneration system considered based on the example of the CHP plant in Czestochowa with a power of 64MW was defined numerically in the IPSEpro software based on the design parameters. The main parameters of the system have been presented in Table 2.

Table 2. Main parameters of the CHP plant [12].

CHP operating parameters	
Nominal output of superheated steam	280 t/h
Fresh steam temperature	515°C
Fresh steam pressure in the boiler	111Bar
Feed water temperature	230°C
Electrical power of the generator	64 MWe
Heating capacity	121,25 MW
Supply water temperature	86°C
Return water temperature	57°C

Apart from generating power, a CHP plant provides end users with large amounts of heat that can also be effectively utilised for coolant production. In order to increase the energy efficiency of the power plant, a proportion of the heat generated in the cogeneration

https://doi.org/10.1051/matecconf/201824005033

process will be used to generate coolant through the use of adsorption chillers for this purpose.

To maintain a high level of electricity production, the heat and power plant must also produce heat. In the summer heat it is usually not used, so you need to get an additional demand for its reception or "release" them into the atmosphere in wet or dry towers, as it happens in condensing power plants. Figure 4 shows integration adsorption chiller with CHP power plant.

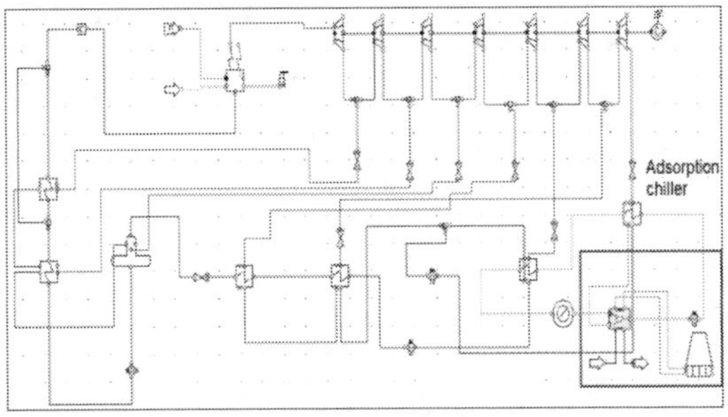

Fig. 4. CHP power plant integration with adsorption chiller.

Traditionally, during the summer, heat is used to heat utility water, but it is only 10 to 15% of the total heat output. Thus, the adsorption system increases the efficiency of the heat and power plant in the summer season, i.e. outside the heating season. Therefore, in the summer period it is possible to maintain the level of efficiency from the heating period (85-90%) if we obtain a sufficient number of cold consumers. Table 3 shows results this analysis [11].

Table 3. Summary of the most important data

Name of parameter	Symbol	Value	Unit
Return water from the district heating network	T_z	57	°C
CHP electrical capacity	N_{el}	64,49	MW
CHP heat capacity	$\dot{Q}_s$	121,25	MW
Heat for preparation utility water	$\dot{Q}_{UW}$	18,18	MW
Maximum heat influx into the adsorption chiller	$\dot{Q}_z$	103,07	MW
Maximum cooling capacity	$\dot{Q}_{Ch}$	37,1	MW
COP (Coefficient Of Performance) Bry Air chiller ADC	COP	0,36	-
EUF_T (indicator of chemical energy use of fuel in the trigeneration system)	EUF_T	99,8	[%]
District water temperature drop in the chiller	ΔT_z	6.6	°C
District water mass flow for heat utility water	$\dot{m}$	149,85	kg/s
District water mass flow for adsorption chillers	$\dot{m}$	849,15	kg/s

3 Summary

The main objective of this paper was to examine the effect of the application of adsorption chillers on the improved energy efficiency of a conventional power plant. Actions aimed at increasing and improving it, as well as the savings in non-renewable energy resources, are becoming an increasingly important aspect of the operation of power plants. An indicator that clearly demonstrates the relative energy effect of the application of trigeneration technology, i.e. the utilisation of the heat capacity of the cogeneration unit for additional production of coolant. During summer CHP power plants is operating only for production heat for preparation utility water but it is only 10 to 15% of the total heat output. The consequence of this situation is the limitation of electricity production because in CHP systems, the production of electricity is dependent on heat production. Using adsorption chillers, the CHP system operates at the same load as in winter and additionally produce chilled water for air conditioning. In this case maximum cooling capacity is about 37,1 MW. This type of solution can be implemented if we find, cold consumers for example, large refrigeration centers in office buildings, data storage centers (server rooms) and shopping centers, whose demand for cold is amounts to several MW.

Acknowledgements

The first author acknowledges the support provided by Faculty of Energy and Fuels, AGH University of Science and Technology, Cracow, Poland. The article was funded from government money Faculty of Energy and Fuels number 11.11.210.216

References

1. Directive 2012/27/EU of the European Parliament and of the Council of 25 October 2012 on energy efficiency, amending Directives 2009/125/EC and 2010/30/EU and repealing Directives 2004/8/EC and 2006/32/EC *OJ L* **315**
2. OECD, *Environmental Outlook to 2050* (2012)
3. P. Byrne *et al., Renewable and Sustainable Energy Rev.* **47** 703 (2015)
4. L. Zisheng, W. Ruzhu, *Applied Energy* **174** 224 (2016)
5. K. Sztekler, M. Komorowski, M. Tarnowska, Ł. Posak, *E3S Web of Conferences* **10**, 00095 (2016)
6. K. Sztekler, K. Wojciechowski, M. Komorowski, M. Tarnowska, *E3S Web of Conferences* **14**, 01031 (2017)
7. R. Z. Wang, Y. X. Xu, J. Y. Wu, M. Li, H. B. Shou, *Appl. Therm. Eng.* **22** 603 (2002)
8. R. Z. Wang, M. Li, Y. X. Xu, J. Y. Wu, *Solar Energy* **68** 189 (2000)
9. X. J. Zhang, R. Z. Wang, *Renewable Energy* **27** 401 (2002)
10. G. Najeh, G. Slimane, M. Souad, B. Riad, E.G. Mohammed, *Case Studies in Thermal Engineering* **8** 337 (2016)

11. Bry-Air technical data, https://www.bryair.com /products-solutions/adsorption-chillers
12. Internal technological documentation from power plants Kozienice, Poland
13. Thu K, Yanagi H, Saha B B and Ng K C 2013 *Int J. of Heat and Mass Transfer* **65** 662

https://doi.org/10.1051/matecconf/201824005034

Modeling of the boiler economizer

Marcin Trojan[1,*], and *Mariusz* Granda[1]

[1]Institute of Thermal Power Engineering, Cracow University of Technology, Cracow, Poland

Abstract. The boiler economizer is a tube heat exchanger located in the final part of the convective duct. In the economizer, water flowing into the boiler is preheated by flue gases. The paper presents the boiler economizer mathematical model with distributed parameters, which can be used to simulate its operation. The developed mathematical model makes it possible to determine temperatures of the tube and working medium of the boiler economizer. In addition, the non-linear mathematical model of the entire boiler allows to analyze the influence of ash fouling of individual boiler heating surfaces on the economizer operation. The proposed model can also be used for monitoring heat and flow parameters of the economizer in on-line mode.

1 Introduction

The subject of this paper is the boiler economizer mathematical model which can be applied in the economizer design and operation calculations.

Literature in the field of hydraulic and thermal boiler calculations is huge. Both analytical methods and modeling CFD are used for mathematical modeling of processes in boilers [1,2]. Computational modeling is an excellent way to optimize boiler design and performance. However, analytical methods are more suitable to monitor the operation of the boiler on-line. The reason for this is the computation time of CFD simulations. A lot of attention is paid in the literature to modeling of steam superheaters. The complexity of heat transfer processes in the steam superheaters causes some difficulties in mathematical modeling of superheaters. The steam superheater cannot be calculated using the method based on the logarithmic mean temperature difference between the fluids (the LMTD method) or the ε-NTU method (effectiveness – the number of transfer units). It is caused by a large dependence of the water steam specific heat on pressure and temperature [3-5]. In works [6, 7] a standard method for calculating steam boilers was presented. In this method, superheaters are calculated as common heat exchangers assuming constant physical properties of the liquid.

Jan Taler et al. [8] proposed a transient mathematical model for the combustion chamber for the optimization of the plant start-up time. In the work [9] also discuss the optimal boiler start-up by using mathematical models of the critical pressure components of a steam boiler.

Corrosion of economizer tubes is the subject of many works [10, 11]. The available literature contains a little information about the economizer modeling. In the work [12] models of economizer with smooth ducts and economizer ducts embossed with turbulence inducing ribs were presented. Stevanovic et al. [13] proposed the numerical model which allows to demonstrate how the high-pressure economizer can be used to raise the primary control reserve in coal-fired thermal power plants.

2 Mathematical model of the boiler economizer

Having passed through steam superheaters, flue gases are directed onto the boiler economizer, located in the final part of the convective duct. The economizer is a tube heat exchanger where water flowing into the boiler is preheated by flue gases. Thereby, the flue gas temperature is reduced (flue gas waste heat recovery).

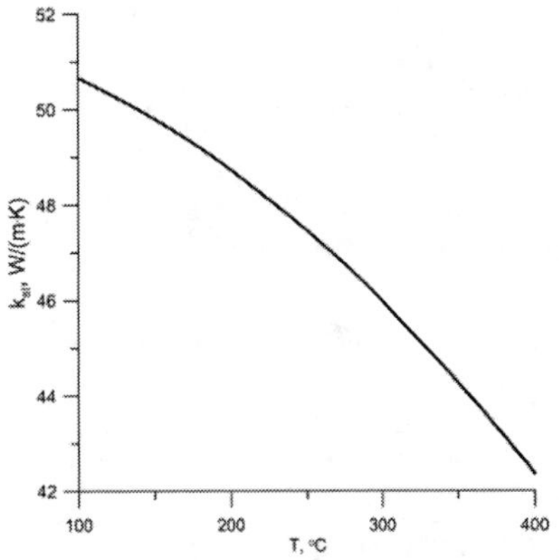

Fig. 1. Temperature-dependent changes in the heat conductivity coefficient for steel 20.

The OP-210M boiler has a two-stage economizer made of tubes arranged in a staggered configuration. The economizer tubes are made of Russian steel 20, whose heat conductivity coefficient is approximated using the relation

* Corresponding author: marcin.trojan@pk.edu.pl

$$k_w = 35.762267 - 6.01214 \cdot 10^{-7}\, T^{2.5} \tag{1}$$

where k_w is in W/(m·K) and T in °C.

Temperature-dependent changes in the heat conductivity coefficient for steel 20 are shown in Fig. 1. The economizer first and second stage are made of tubes connected with fins along the entire length (membrane heating surface). Due to the temperature field symmetry, the economizer tubes can be treated as longitudinally finned ones (Fig. 2).

Fig. 2. The economizer longitudinally finned tube.

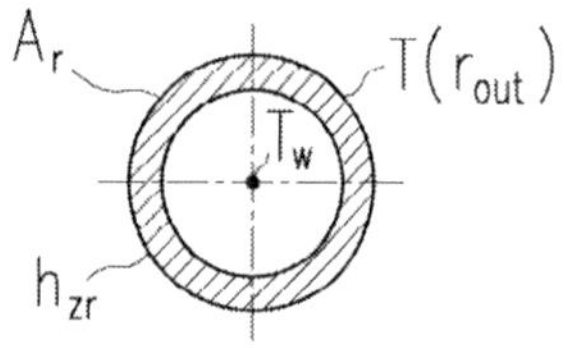

Fig. 3. Equivalent smooth tube taking account of fins located on the flue gas side.

Basic data concerning the economizer first and second stage are listed in Table 1.

Table 1. Basic data of the economizer first and second stage.

Basic data	stage I	stage II
Surface area of tubes with no fins, m^2	733.5	1107.5
Number of smooth tubes	76	54
Tube length, m	96	204
Number of tube rows	12	25
Tube pitch (transverse pitch s_1 x longitudinal pitch s_2), mm	90 x 50	130 x 50
Outer diameter x tube wall thickness, mm	32 x 5	32 x 5
Fin width x fin thickness, mm	68 x 4	68 x 4
Tube and fin material	B18	B18
Flue gas flow free cross-section, m^2	17.7	21.8
Flow nature	Counterflow	Counterflow

The economizer mathematical modeling is similar to modeling the steam superheater. However, it has to be taken into account that the working medium flowing in tubes is water. It should also be emphasized that due to high specific heat of this medium, the heat flux that has to be supplied to water from flue gases is very high. The presence of fins on the superheater tubes will be taken into consideration by introducing an equivalent (weighted) heat transfer coefficient on the flue gas side.

The weighted heat transfer coefficient on the flue gas side related to the outer surface of a smooth tube (with no longitudinal fins) is determined from the condition of equality between the heat flux absorbed by the outer surface of the smooth tube (assuming that the heat transfer coefficient is equal to the weighted one) and the total heat flux absorbed by the fins and the smooth tube surface in between the fins (Fig. 2 and Fig. 3)

$$\begin{aligned} &h_{zr} A_r \left[T\left(r_{out}\right) - T_g\right] = \\ &= h_g A_{rm}\left[T\left(r_{out}\right) - T_g\right] + h_g A_{fin}\eta_{fin}\left[T\left(r_{out}\right) - T_g\right] \end{aligned} \tag{2}$$

The symbols in Eq. (2) are as follows: h_{zr} – weighted heat transfer coefficient, flue gas side, (W/m^{2}K), h_g – heat transfer coefficient, flue gas side, (W/m^{2}K),
η_{fin} – fin efficiency, A_r – smooth tube outer surface area (m^2), A_{rm} – smooth tube surface area in between the fins (m^2), A_{fin} – fin surface area (m^2), $T\left(r_{out}\right)$ – tube outer surface temperature (°C), T_g – flue gas temperature (°C).

After transformations, Eq. (2) gives the following relation for the weighted heat transfer coefficient on the flue gas side

$$h_{zr} = h_g\left[\frac{A_{rm}}{A_r} + \eta_{fin}\left(h_g\right)\frac{A_{fin}}{A_r}\right] \tag{3}$$

The efficiency of a straight fin with constant thickness is expressed as [14-17]

$$\eta_{fin} = \frac{\mathrm{tgh}\left(m\, H_z\right)}{m\, H_z} \tag{4}$$

where parameter m is described using the following equation

$$m = \sqrt{\frac{2h_g}{k_{fin}\,\delta_{fin}}} \tag{5}$$

and the fin height is found from the relation (Fig. 2)

$$H_z = \frac{s_{fin}}{2} - r_{out}\cos\varphi \tag{6}$$

Surface areas A_{fin}, A_r, A_{rm}, are defined by the following formulae

$$A_{fin} = 4\left(\frac{s_{fin}}{2} - b\right) = 4\cdot H_z \tag{7}$$

$$A_r = 2\,\pi\,r_{out} \tag{8}$$

$$A_{rm} = (2\,\pi - 4\,\varphi)\,r_{out} \tag{9}$$

Angle φ is calculated from (Fig. 2)

$$\varphi = \arcsin\left(\frac{\delta_{fin}}{2\,r_{out}}\right) \tag{10}$$

Using the weighted heat transfer coefficient h_{zr}, the economizer is calculated as if it was made of smooth tubes.

For the staggered tube arrangement, the Nusselt number is found from the following relation [18]

$$Nu = 0.196\,\sigma_1^{-0.247}\sigma_2^{-0.051}\,\mathrm{Re}^{0.674}\,\mathrm{Pr}^{0.44} \tag{11}$$

where

$$\sigma_1 = \frac{s_1}{d} \text{ and } \sigma_2 = \frac{s_2}{d} \tag{12}$$

The water side Nusselt number was calculated using Gnielinski's correlation [19]

$$\mathrm{Nu} = \frac{\frac{\xi}{8}(\mathrm{Re}-1000)\mathrm{Pr}}{1+12.7\sqrt{\frac{\xi}{8}}\left(\mathrm{Pr}^{2/3}-1\right)}\left[1+\left(\frac{d_{in}}{L}\right)^{2/3}\right]\left(\frac{\mathrm{Pr}}{\mathrm{Pr}_{in}}\right)^{0.11} \tag{13}$$

$4\cdot10^3 \le \mathrm{Re} \le 10^6 \quad 0.5 \le \mathrm{Pr} \le 200$

where the friction factor for smooth tubes is given by the Filonienko equation [20]

$$\xi = \left(1.82\log\mathrm{Re} - 1.64\right)^{-2} \tag{14}$$

The heat transfer coefficient at the tube inner surface for transition regime from laminar to turbulent and for turbulent flow can also be determined using the correlation proposed by Taler [21]

$$\mathrm{Nu} = \mathrm{Nu}_{m,q}(\mathrm{Re}=2300) + \frac{\frac{\xi}{8}(\mathrm{Re}-2300)\mathrm{Pr}^{1.008}}{1.08+12.39\sqrt{\frac{\xi}{8}}\left(\mathrm{Pr}^{2/3}-1\right)} \times\left[1+\left(\frac{d_{in}}{L}\right)^{2/3}\right]\left(\frac{\mathrm{Pr}}{\mathrm{Pr}_{in}}\right)^{0.11}, \tag{15}$$

$2300 \le \mathrm{Re} \le 10^6, \quad 0.1 \le \mathrm{Pr} \le 1000, \quad \frac{d_{in}}{L} \le 1$

where: d_{in} - inner diameter of the tube, L - tube length.

The mean Nusselt number for laminar flow Nu_m with respect to tube length L can be expressed for uniform wall heat as [21]

$$\mathrm{Nu}_{m,q} = \left(\mathrm{Nu}_{m,q,1}^{4.8290} + \mathrm{Nu}_{m,q,2}^{4.8290}\right)^{(1/4.8290)} \quad 1\cdot10^{-6} \le \frac{1}{\mathrm{RePr}}\frac{x}{d_{in}} \le 1.0 \tag{16}$$

The symbol $Nu_{m,q,1}$ in Eq. (16) denotes the mean Nusselt number for hydrodynamically and thermally fully developed flow

$$\mathrm{Nu}_{m,q,1} = \frac{48}{11} = 4.364 \tag{17}$$

The symbol $Nu_{m,q,2}$ denotes the mean Nusselt number for hydrodynamically and thermally fully developed flow over the plate with linear temperature profile in the fluid and constant heat flux at the wall surface

$$\mathrm{Nu}_{m,q,2} = 3^{1/3}\Gamma(2/3)\left(\mathrm{Re\,Pr}\frac{d_{in}}{L}\right)^{1/3} = 1.9530\left(\mathrm{Re\,Pr}\frac{d_{in}}{L}\right)^{1/3}, \quad \frac{1}{\mathrm{Re\,Pr}}\frac{L}{d_{in}} \le 0.0005 \tag{18}$$

where the symbol Γ designates the gamma function [22].

Eq. (18) is valid only for the initial section of the entrance region when the parameter $\frac{1}{\mathrm{Re\,Pr}}\frac{L}{d_{in}}$ is small. The coefficient of determination r^2 and the absolute maximum error $|\varepsilon_{\max}|$ for formula (16) are $r^2 = 0.99979$ and $|\varepsilon_{\max}| \le 3.9\%$.

The economizer was modeled in the same way as the superheaters, using the developed distributed parameter mathematical model proposed in [23, 24]. The general assumptions made in the development of a numerical model of the boiler economizer are as follows: the water and gas flow is one dimensional, the physical properties of fluids are functions of temperature, axial heat conduction in the tube wall and fluid is negligible, the temperature and flue gas velocity are constant over the channel cross-section before the economizer, and heat transfer coefficients on the inner and outer tube surfaces are uniform. By using the partial differential equations describing the space and time changes of steam T_s, tube wall T_w, ash layer T_a and flue gas temperatures T_g can be obtained

- the steam temperature at the outlet of the control volume (Fig. 4)

$$T_{s,i+1} = T_{w1,i} - \left(T_{w1,i} - T_{s,i}\right)\exp\left(-\Delta N_{s,i+\frac{1}{2}}\right), \quad i = 1,\dots,N \tag{19}$$

where the symbol $\Delta N_{s,i+\frac{1}{2}}$ denotes the number of transfer units on the steam side for the i-th control volume, defined as

$$\Delta N_{s,i+\frac{1}{2}} = \frac{2\pi\, r_{in}\, h_{s,i}\, \Delta x}{\dot{m}_s\, \overline{c}_{p_{s,i}}} \tag{20}$$

- the system of three nonlinear algebraic equations for tube wall temperatures $T_{w1,i}$, $T_{w2,i}$ and $T_{w3,i}$

$$T_{w1,i} = \frac{1}{h_{s,i}\, d_{in} + \dfrac{k_w\left(T_{w1,i}\right) + k_w\left(T_{w2,i}\right)}{2}\dfrac{d_c}{\delta_w}} \times \left[h_{s,i}\, \overline{T}_{s,i}\, d_{in} + \frac{k_w\left(T_{w1,i}\right) + k_w\left(T_{w2,i}\right)}{2}\frac{d_c}{\delta_w} T_{w2,i}\right], \quad i = 1,\dots,N \tag{21}$$

$$T_{w2,i} = \frac{1}{\dfrac{k_w\left(T_{w1,i}\right) + k_w\left(T_{w2,i}\right)}{2}\dfrac{d_c}{\delta_w} + \dfrac{k_a}{\delta_a} d_s} \times \left[\frac{k_w\left(T_{w1,i}\right) + k_w\left(T_{w2,i}\right)}{2}\frac{d_c}{\delta_w} T_{w1,i} + k_a \frac{d_s}{\delta_a} T_{w3,i}\right], \quad i = 1,\dots,N \tag{22}$$

$$T_{w3,i} = \frac{1}{\left[h_{g,i}\left(d_o + 2\delta_a\right) + k_a \dfrac{d_s}{\delta_a}\right]} \times \left[h_{g,i}\left(d_o + 2\delta_a\right)\overline{T}_{g,i} + k_a \frac{d_s}{\delta_a} T_{w2,i}\right], \quad i = 1,\dots,N \tag{23}$$

where:
$d_c = \left(d_{in} + d_o\right)/2 = r_{in} + r_o$, $\quad d_s = d_o + \delta_a = 2r_o + \delta_a$.

- the flue gas temperature at the outlet of the control volume

$$T''_{g,i} = T_{w3,i} - \left(T_{w3,i} - T'_{g,i}\right)\exp\left(-\Delta N_{g,i}\right), \quad i = 1,\dots,N \tag{24}$$

where the symbol $\Delta N_{g,i}$ denotes the number of transfer units on the gas side for the i-th control volume, defined as

$$\Delta N_{g,i} = \frac{2\pi\left(r_o + \delta_a\right)\Delta x\, h_{g,i}}{\dot{m}_g\, \overline{c}_{p_{g,i}}} \tag{25}$$

where Δx is the control volume length.

The resulting system of nonlinear algebraic equations (19), (21) – (23), and (24) for the temperatures at nodes in steam, wall, and flue gas areas was solved by the Gauss-Seidel method.

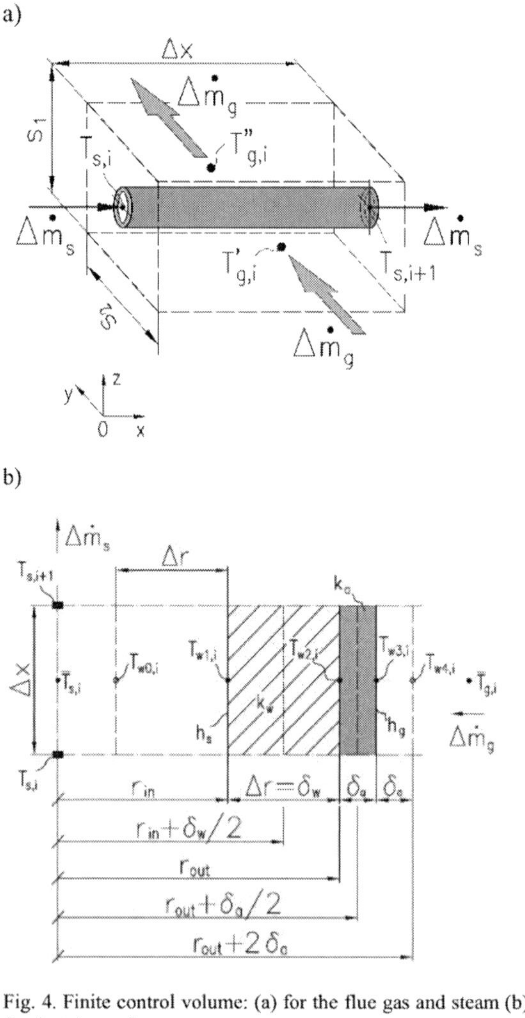

Fig. 4. Finite control volume: (a) for the flue gas and steam (b) for the tube wall.

3 Results of modeling the operation of the economizer second stage

From the economizer first stage, the boiler feed water is directed to the economizer second stage. In terms of the flue gas flow direction, the heater is located in the flue gas duct downstream the economizer first stage.

The economizer second stage is made of 54 rows, with twenty-five tubes with outer diameter $d_o = 32$ mm and wall thickness $\delta_{ECO} = 5$ mm each (Fig. 5).

A mathematical model of the device was developed. The flow arrangement and division of the economizer second stage into finite volumes is shown in Fig. 6. Considering the flow arrangement, the economizer second stage can be classified as a parallel-cross-flow heat exchanger. The economizer tubes are in a staggered configuration. The tube arrangement is shown in Fig. 5.

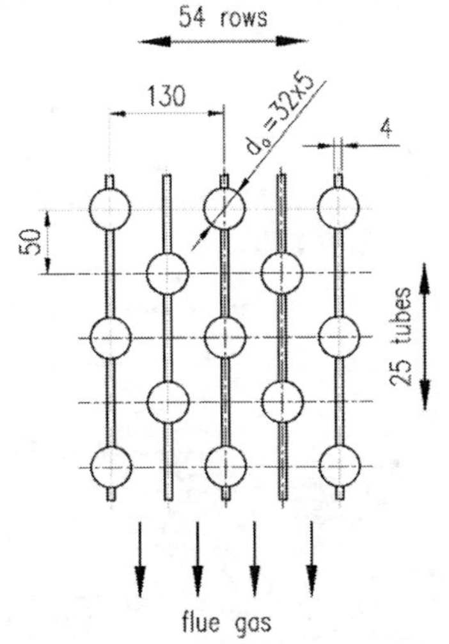

Fig. 5. Staggered tube arrangement in the economizer second stage.

The symbols A1-AN, B1-BN and C1-CN in Fig. 6 denote the temperature of the inner and outer tube surfaces, and the outer temperature of the ash deposit, respectively:

A1=RPWD11(1), RPWD12(1), RPWD13(1); A2=RPWD11(2), RPWD12(2), RPWD13(2);

AI=RPWD11(I), RPWD12(I), RPWD13(I); AN-1=RPWD11(N-1), RPWD12(N-1), RPWD13(N-1);

AN=RPWD11(N), RPWD12(N), RPWD13(N);

B1=RPWD21(1), RPWD22(1), RPWD23(1); B2=RPWD21(2), RPWD22(2), RPWD23(2);

BI=RPWD21(I), RPWD22(I), RPWD23(I); AN-1=RPWD21(N-1), RPWD22(N-1), RPWD23(N-1);

BN=RPWD21(N), RPWD22(N), RPWD23(N);

C1=RPWD251(1), RPWD252(1), RPWD253(1); C2=RPWD251(2), RPWD252(2), RPWD253(2);

CI=RPWD251(I), RPWD252(I), RPWD253(I);

CN-1=RPWD251(N-1), RPWD252(N-1), RPWD253(N-1);

CN=RPWD251(N), RPWD252(N), RPWD253(N).

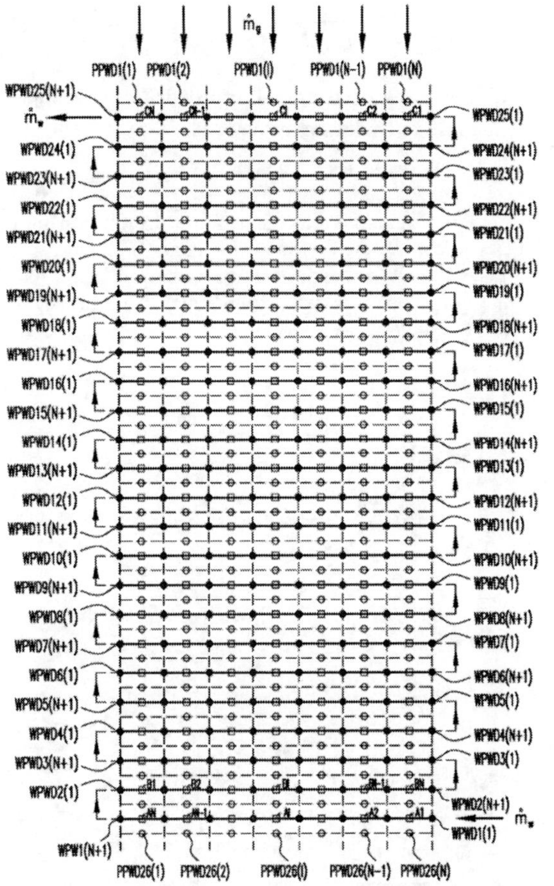

Fig. 6. Division of cross-parallel-flow the second stage of economizer with twenty-five passes into finite volumes: PPWD1(I), PPWD2(I), … , PPWD26(I) – flue gas temperature, RPWD11(I), RPWD12(I), RPWD13(I), … , RPWD251(I), RPWD252(I), RPWD253(I) – temperature of the inner and outer tube surfaces, and the outer temperature of the ash deposit, respectively, WPWD1(I), WPWD2(I), … , WPWD25(I) – water temperature.

The following data are assumed in the economizer second stage model: $\delta_{fin} = 4$ mm, $s = 130$ mm, flue gas duct width $a_d = 7700$ mm, flue gas duct height $b_d = 75 \cdot 45$ mm $= 3375$ mm. The flue gas flow velocity is determined in the tube bank free cross-section. Considering the flue gas duct dimensions, the free cross-section surface area is $A_d = 18{,}8$ m^2. Considering the gaps between the economizer and the duct walls, the free cross section area of $A_d = 21{,}8$ m^2 is assumed, according to the data provided by the manufacturer. The equivalent thickness of the radiative layer is found from formula

$$s_z = \frac{3.6V}{A} = 3.6 \cdot \frac{\left[2s_1s_2 - \frac{\pi d_o^2}{2} - 2\left(2s_2 - d_o\right)\delta_{fin}\right]}{2\pi d_o + 2\left(2s_2 - d_o\right)} \quad (19)$$

It totals $s_z = 115.9$ mm. The economizer second stage flue gas and water mass flows are $\dot{m}_g = 62.7$ kg/s and $\dot{m}_w = 49.5$ kg/s, respectively.

Fig. 7 presents the temperature distribution of flue gases flowing through the economizer first stage in two cases: (1) all the boiler heating surfaces are clean and (2) the boiler operates with steam superheaters affected by ash fouling.

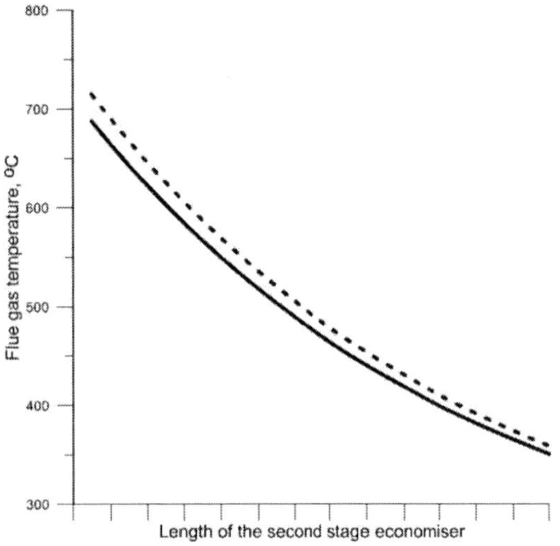

Fig. 7. Temperature distribution of flue gases flowing through the economizer second stage (continuous line – clean boiler, dashed line – individual steam superheater stages affected by ash fouling).

If the boiler heating surfaces are clean, the flue gas temperature upstream the economizer second stage is $687.7\,^{o}C$. The water temperature at the economizer second stage inlet is $239.6\,^{o}C$ and equal to the water temperature at the economizer first stage outlet. The drop in the flue gas temperature for the boiler clean heating surfaces totals $336.7\,^{o}C$, which means that the temperature of flue gases downstream the economizer second stage is $351.0\,^{o}C$, and this is equal to the flue gas temperature at the economizer first stage inlet. The water temperature at the economizer outlet (Fig. 8) is in this case $314.3\,^{o}C$; consequently the increment in the temperature of water flowing through the economizer second stage totals $74.7\,^{o}C$. The biggest increment in the water temperature, $4.4\,^{o}C$, is obtained in tube 25, which is in the zone of the flue gas highest temperatures.

If the steam superheater individual stages are affected by ash fouling, the flue gas temperature upstream the economizer second stage is $715.0\,^{o}C$, which is a value higher by $27.3\,^{o}C$ compared to the case with the boiler operation with clean steam superheaters. The water temperature at the economizer inlet is $240.5\,^{o}C$. The flue gas temperature downstream the economizer second stage is $358.9\,^{o}C$, which means a drop by $356.1\,^{o}C$. Passing through the economizer second stage, water is heated to the temperature of $318.9\,^{o}C$. The biggest increment in the water temperature, $4.5\,^{o}C$, is obtained in tube 25, which is in the zone of the flue gas highest temperatures.

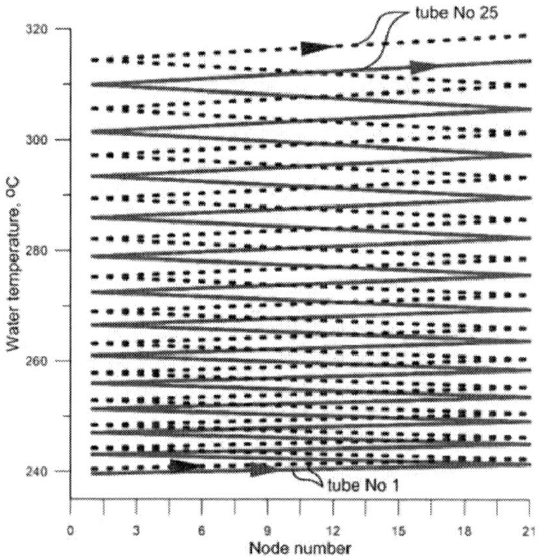

Fig. 8. Temperature distribution of water flowing through the economizer second stage (continuous line – clean boiler, dashed line – individual steam superheater stages affected by fouling).

4 Conclusions

The proposed mathematical model is suitable for modeling multi-pass boiler economizers and various flow systems. Due to the short calculation time, the method can be used for monitoring heat and flow parameters of the economizer in on-line mode. The developed mathematical model makes it also possible to determine temperatures of the tube outer and inner surfaces and external fouling for all heating surfaces of the boiler.

The developed mathematical model allows to determine parameters of working mediums and the temperature of the wall, which is essential in design calculations. The knowledge of the metal temperature under the different boiler loads enables a correct selection of the steel grade. Determining the tube walls temperature also makes it possible to avoid overheating of the tubes, e.g. at the boiler low loads.

The proposed method can be used to model plate fin and tube heat exchangers and other cross-flow tube exchangers, which are often used for example in air-conditioning and refrigeration.

Nomenclature

A_d	free cross-section surface area in flue gas duct, m
A_{fin}	fin surface area, m^2
A_{rm}	smooth tube surface area in between the fins, m^2
A_r	smooth tube outer surface area, m^2
d_{in}	inner diameter, m
d_{out}	outer diameter, m
h_g	heat transfer coefficient, flue gas side, W/m^{2}K
h_{zr}	weighted heat transfer coefficient, flue gas side, W/m^{2}K
H_z	fin height, m
k_a	ash deposit thermal conductivity, W/(m·K)
k_{fin}	fin thermal conductivity, W/(m·K)
k_w	tube material thermal conductivity, W/(m·K)
L	tube length, m
Nu	Nusselt Number
Pr	Prandtl Number
r_{in}	inner radius of the tube, m
r_{out}	outer radius of the tube, m
Re	Reynolds Number
s_z	equivalent thickness of the radiative layer, m
T	temperature, °C
T_g	flue gas temperature, oC
T_s	steam temperature, °C
T_w	wall temperature, °C
$T'_{g,i}$	flue gas temperature at the inlet to the control volume, °C
$T''_{g,i}$	flue gas temperature at the outlet of the control volume, °C
Γ	gamma function
δ_a	ash deposit thickness, m
δ_{fin}	fin thickness, m
η_{fin}	fin efficiency
ξ	friction factor for smooth tubes

References

1. Raghavan V., *Combustion Technology. Essentials of Flames Burners* (John Wiley & Sons Ltd., UK, 2016)
2. Ranade V.V., Gupta D.F., *Computional Modeling of Pulverized Coal Fired Boilers* (CRC Press – Taylor & Francis Group, Boca – Raton, 2015)
3. Hewitt G.F., Shires G.L., Bott T.R., *Process Heat Transfer*, CRC Press - Begell House: Boca Raton, 1994)
4. Kakaç S., Liu H., *Heat Exchangers: Selection, Rating, and Thermal Design, second ed.* (CRC Press -Taylor & Francis Group: Boca Raton, 2002)
5. Shah R.K., Sekulić D.P., *Fundamentals of heat exchanger design* (Wiley: Hoboken, 2003).
6. Kuznetsov N. V., Mitor V. V., Dubovskij I. E., Karasina E. S., *Thermal Calculations of Steam Boilers. Standard Method* (Energy, Moscow, 1973) (in Russian)
7. Lokshin V. A., Peterson D. F., Schwarz A. L., *Standard Methods of Hydraulic Design for Power Boilers*, (Hemisphere, Washington, 1988)
8. Taler J., Dzierwa P., Taler D., Harchut P., Energy, **92,** pp. 160-170 (2015)
9. Dzierwa P., Journal of Thermal Stresses, **39,** pp. 874-886 (2016)
10. Qua H.C., Khaw C.C., Tan C.S., Wang X., Ooi J.B., Engineering Failure Analysis, **18,** 8, pp. 2201-2210 (2011)
11. Sagayaraj D., Suresh S., Chandrasekar M., International Journal of Minerals, Metallurgy and Materials, **16,** 5, pp. 534-539 (2009)
12. Lowrey S., Sun Z., International Journal of Refrigeration, **60,** pp. 38–53 (2015)
13. Stevanovic V.D., Ilic M., Djurovic Z., Wala T., Muszynski S., Gaji I., Energy, **147,** pp. 782-798 (2018)
14. Çengel Y.A., *Introduction to Thermodynamics and Heat Transfer 2nd Edition* (McGraw-Hill, Hoboken, 2008)
15. Kraus A.D., Aziz A., Welty J., *Extended surface heat transfer* (Wiley, New York, 2001)
16. Kreith F., Manglik R.M., Bohn M.S., *Principles of heat transfer* (Seventh Edition, Cengage Learning, Stamford, USA, 2011).
17. Taler J., Duda P., *Solving Direct and Inverse Heat Conduction Problems* (Springer, Berlin, 2006)
18. Pronobis M., Kalisz S., Wejkowski R., Heat and Mass Transfer, **38,** pp. 343-350 (2002)
19. Gnielinski V., International Chemical Engineering, **16,** pp. 359-368 (1976)
20. Filonienko G. K., Teploenergetika 1, **4,** pp. 40-44 (1954), (in Russian)
21. Taler D., International Journal of Thermal Sciences, **108,** pp. 108–122 (2016)
22. Thompson W.J., *Atlas for Computing Mathematical Functions*, New York, Wiley (1997).
23. Trojan M., Taler D., Fuel **150,** pp. 75–87 (2015)
24. Taler D., Trojan M., Dzierwa P., Kaczmarski K., Taler J., International Journal of Thermal Sciences, **129,** pp. 320–333 (2018)

Influence of selected cycle components parameters on the supercritical CO_2 power unit performance

Marcin Wołowicz*, *Jarosław* Milewski, and *Gabriel* Ziembicki

Institute of Heat Engineering, Warsaw University of Technology, Nowowiejska 21 Street 00-665 Warsaw, Poland

Abstract. The paper presents the influence of selected components parameters on the performance of supercritical carbon dioxide power unit. For this analysis mathematical model of supercritical recompression Brayton cycle was created. The analysis took into consideration changes in the net cycle power and efficiency for different compressor inlet temperatures. The results were obtained for a fixed minimum pressure of 7.4 MPa and fixed recompression split ratio. The studies conducted in this paper included also consideration of sensitivity of the cycle efficiency to a change in recuperators heat transfer area. In order to determine how each recuperator influences the cycle performance, an analysis of efficiency dependence on the recuperators area was made. Another parameters that were investigated are to a change in turbine and compressors isentropic efficiency and their influence on the cycle efficiency. In the reference cycle, isentropic efficiencies were set up as 88% for both the main and recompression compressor, and 90% for the turbine. Since isentropic efficiency is a sort of measure of broadly defined quality of a turbine or compressor, including airfoil shape, sealing, etc., it may be a significant cost factor that should be considered during cycle design. Therefore, a sensitivity analysis of cycle efficiency to both compressors and turbine isentropic efficiencies was conducted.

1 Introduction

In the 1960s, Feher [1] studied various gases' properties for a purpose of finding the most suitable one for a supercritical thermodynamic cycle. Carbon dioxide was proposed as a working fluid due to several reasons. First, its physical properties e.g. critical pressure, which is significantly lower compared to water, allows lower operating pressure. Then, CO_2 thermodynamic and transport properties are well known, hence cycle analysis is based on reasonably firm data. Finally, carbon dioxide is abundant, non-toxic and has relatively low cost. The analysis proved that CO_2 supercritical cycle offers several desirable features as high thermal efficiency (the investigated cycle reached a thermal efficiency of 55% under ideal conditions), low volume to power ratio and no blade corrosion and cavitation.

2 Supercritical CO_2 cycles

The most basic and compact supercritical CO_2 cycle is a simple Brayton cycle. It is simple and offers relatively good efficiency. However, there is still potential to improve its performance. The biggest reduction in efficiency of the supercritical Brayton cycle comes from the large irreversibility in the recuperator [2]. So called compound cycles can overcome this problem and can perform significantly better than the regular supercritical Brayton cycles.

Supercritical CO_2 cycles can be divided in three groups: pre-compression cycles, partial cooling cycles and recompression cycles.

The pre-compression Brayton cycle is one of the ways to increase the generation within the cycle and reduce the pinch-point problem. As shown in Fig. 1 the cycle is similar to normal Brayton cycle with a small modification.

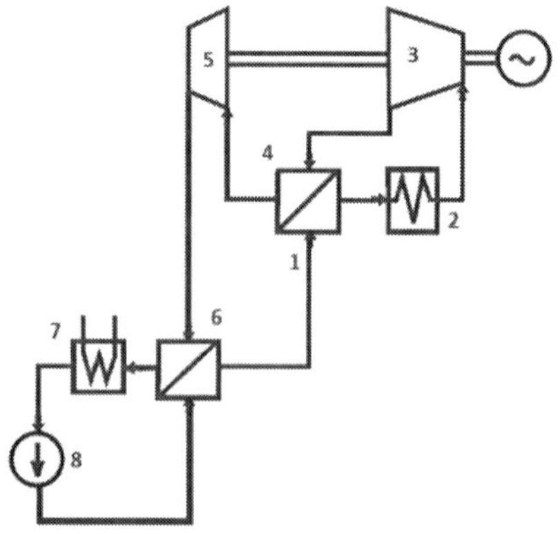

Fig. 1. Schematic of the pre-compression Brayton cycle [2].

First, the working fluid is compressed and then heated in the high temperature recuperator (1) using exhaust heat from the turbine. The fluid passes to a heat source (2) where the heat is added and then expands in the turbine

* Corresponding author: marcin.wolowicz@itc.pw.edu.pl

(3). The remaining exhaust heat is extracted from the fluid in the high temperature recuperator (1). The difference from the normal Brayton cycle is that in the middle of recuperation process, when the hot fluid temperature approaches to the heated fluid temperature, a compressor (5) is introduced that compresses the fluid to higher pressure. As the fluid pressure rises, so does its temperature and specific heat. Thus, regeneration process can continue and more available heat is returned to the heated fluid. This extra heat reduces average temperature at which heat is rejected from the cycle and increases the average temperature at which heat is added to the cycle. This results in an efficiency improvement of 6% over a Brayton cycle that would otherwise suffer from the pinch point problem [3].

Another cycle layout that aims at reducing Brayton cycle drawbacks is the partial cooling cycle presented in Fig. 2. In general, its operation differs from the previously described cycle by two adjustments. The first is that only a fraction of the working fluid is compressed in the low temperature compressor (pump). The rest is compressed in the recompression compressor that is introduced before the pre-cooler and after the pre-compression compressor. The second difference is the introduction of another pre-cooler before the pre-compression compressor. This way, similar to the pre-compression cycle, more heat is available for the regeneration process.

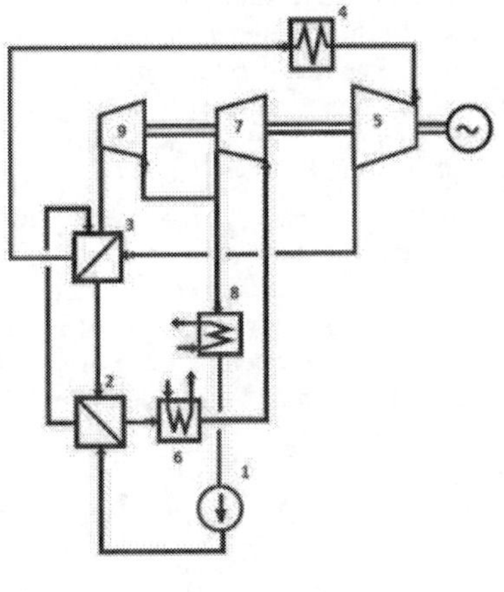

Fig. 2. Schematic of the partial cooling Brayton cycle [2].

After compression in the main compressor (1), a fraction of the working fluid is heated in the low temperature recuperator (2) and merged with the flow from the re-compressing compressors that is at the same conditions. The fluid is then heated in the high temperature recuperator (3) and in the heat source (4) in turn and then enters the turbine (5). After the expansion process the fluid returns its heat in the high and low temperature recuperator (2,3). Then it passes to the pre-cooler (6) where it is cooled to the pre-compressor inlet temperature, and subsequently compressed in the pre-compressor (7). A part of the pre-compressed fluid is sent to the pre-cooler (8) and the main compressor. The rest is recompressed in the second recompressing compressor (9) to the high temperature recuperator inlet conditions, and then is merged with the stream from the main compressor. This move eliminates the pinch point problem, since due to the lower mass flow rate on the high pressure side of the low temperature recuperator, the mass flow weighted heat capacity of the streams is about equal and a pinch point does not occur.

The cycle improves its efficiency by reducing the average temperature of heat rejection so that the efficiency improvement is bigger than that for the pre-compression cycle.

Although the partial cooling cycle seems to be attractive due to its efficiency benefits, the complication of a cycle layout may prove detrimental to the economic. Therefore, another cycle is introduced, a recompression cycle, which is simpler than both partial cooling and pre-compression cycle. The general layout of the cycle is shown in Fig. 3.

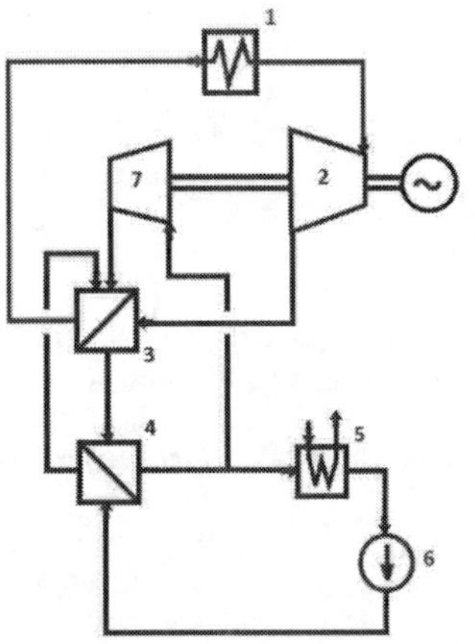

Fig. 3. Schematic of the recompression Brayton cycle [2].

The advantage of this cycle is that it completely eliminates one precooler and pre-compressing compressor from the cycle. After the regeneration process in the high temperature recuperator (3) the fluid is heated in the heat source (1) and passes to the turbine (2). Then in enters successively the high and low temperature recuperators (3,4) and returns its heat to the fluid at the high pressure side. The fluid flow is then split into two streams. The first is sent directly to the recompression compressor, where it is compressed to the same pressure conditions as the CO_2 leaving the main compressor and merged with it in the high pressure recuperator. The second flow is cooled in the precooler (5), compressed in the main compressor (6) and heated in the recuperators.

The effect of recompression is sufficient to overcome a pinch point problem. Owing to the decreased mass flow rate at the high pressure side of the low temperature recuperator, the mass flow weighted heat capacity of the streams is about equal at both sides and a pinch point does not occur. The recompression cycle is, along with the pre-compression cycle, the simplest among the surveyed cycles. In addition, at the desired operating conditions of turbine inlet pressure and temperature (20

MPa and 550°C), it proves to achieve the highest efficiency of all examined cycles [2, 4].

3 Model description

The supercritical CO_2 Brayton power cycle with recompression was selected for further examination due to its relatively high power conversion efficiency compared to other cycle layouts proposed in literature. In this cycle configuration a portion of CO_2 flow is recompressed without rejection of heat. Splitting the flow allows to compensate for the difference between specific heats of the high and low pressure CO_2 flows in the low temperature recuperator.

The cycle evaluation was performed by using the GateCycle application. The design of the cycle assumes a supercritical CO_2 inlet flow with a given parameters, which is generated by a generic heat source. The CO_2 is expanded through the turbine to produce electric power. The flow at a lower temperature and pressure then passes through high and low temperature recuperators successively, where it is further cooled. Then, the flow is split into two streams. One stream is cooled down in the cooler that provides additional cooling to the fluid before it enters the main compressor. The second stream after being split is compressed in the recompression compressor and then mixed with the first flow which has similar state in terms of pressure and temperature. The merged stream passes to the high temperature recuperator, where the heat is recovered and then returns to the heat source.

As a first step of analysis, a reference cycle was developed to be used as a basis for further examination. Realistic component operating conditions and design parameters were selected to be consistent with real case application. The reference cycle assumes a heat source which is able to generate constant CO_2 flow at temperature of 550°C and mass flow rate of 10,000 t/h. At the bottom of the cycle the flow temperature and pressure are 31.25°C and 7.4 MPa and are very close, but still slightly above the CO_2 critical point (30.98°C and 7.373 MPa). The split ratio is initially assumed as 0.28, which means that this part of the flow is sent directly to the recompression compressor without being cooled before to the lowest cycle temperature. Besides the main parameters values, the following assumptions have been made:

1. the system operates at steady state conditions,
2. expansion and compression processes are adiabatic but non-isentropic,
3. appropriate isentropic efficiencies were employed for both turbine and compressor,
4. effectiveness for heat exchangers are used for low and high temperature recuperators,
5. pressure drops due to friction in all the heat exchangers and pipelines are negligibly small,
6. changes in kinetic and potential energies in each component are negligible.

4 Obtained model analysis

4.1 Compressor inlet temperature

Fig. 4 shows changes in the net cycle power and efficiency for different compressor inlet temperatures.

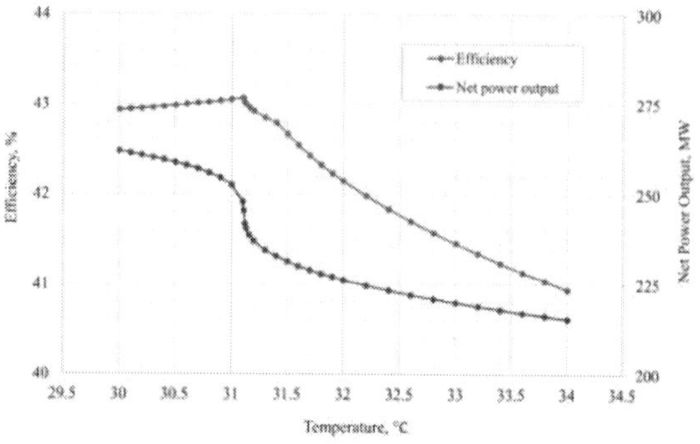

Fig. 4. Efficiency and net power output for different compressor inlet temperatures (fixed split flow ratio).

The results were obtained for a fixed minimum pressure of 7.4 MPa and fixed recompression split ratio. A rapid variation in both cycle net power and efficiency can be seen in the near critical region, when carbon dioxide moves from supercritical gas state to supercritical liquid state passing through the peak in specific heat capacity rate observable in the critical or pseudocritical point. Compressor inlet temperature exhibits the optimum point at the value of critical temperature where the efficiency is maximum. For temperature of 31.1°C efficiency reaches the maximum value of 43.08%, which is about 0.2 percentage points more than efficiency of the reference cycle, where compressor inlet temperature is 31.25°C. However, the magnitude of efficiency gain may not be sufficient in the face of increased cooler duty. Due to the peak in specific heat capacity rate in the critical temperature, required cooler area necessary to cool the working fluid to temperature of 31.1°C is about 11% higher than that in the reference cycle.

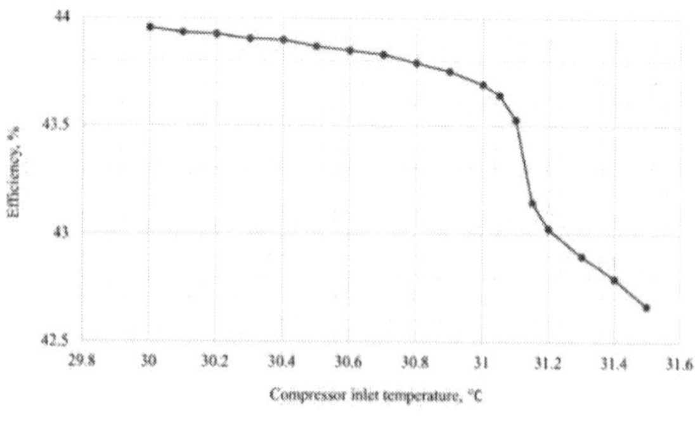

Fig. 5. Cycle efficiency as a function of compressor inlet temperature (split flow ratio dynamically optimized for each temperature).

One may consider the existence of efficiency peak and decrease along with decreasing minimal cycle temperature somewhat odd in the light of widely known thermodynamic dependencies that predict higher efficiencies for lower temperature of heat removal. The correlation, however, does not violate thermodynamic laws - lower efficiency for lower temperature results from absence of the optimization of parameters, whose optimum values depend on compressor inlet temperature.

In the examined cycle, recompression split flow ratio is the parameter which plays a crucial role in the observed phenomenon. Fig. 5 shows efficiency dependence on minimal cycle temperature when the condition of optimum split ratio is satisfied. The graph was obtained by plotting efficiency lines for various recompression split flow ratios at different temperatures levels shown in Fig. 6. Then the maximum efficiency points for each temperature level where collected and illustrated in the graph as a variable dependent on temperature. The shape of efficiency curve presented in Fig. 5 looks more realistic - the abrupt surge is still present, but then, as the temperature decreases, so does the efficiency, which results mainly from decrease in compressor work.

An optimum point for net cycle power does not exist. As the temperature of CO_2 before the compressor declines, the compression work is also reduced, what leads to the higher net cycle power.

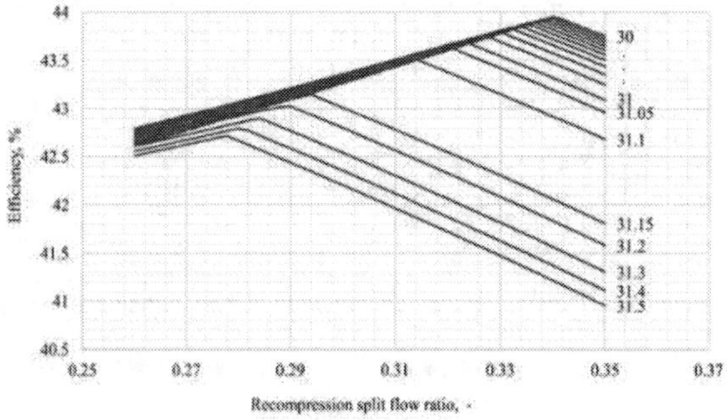

Fig. 6. Optimum recompression split flow ratio for compressor inlet temperatures ranging from 30°C to 31.5°C.

4.2 Heat exchangers area

In order to determine how each recuperator influences the cycle performance, an analysis of efficiency dependence on the recuperators area was made. Cycle efficiency was calculated for both recuperators area varying independently of each other from -50% to 50% of its initial value. In the reference cycle recuperators effectiveness was set to 90%, which translates into heat transfer area of 612,363 m^2 and 482,667 m^2 for the low temperature and high temperature recuperators respectively.

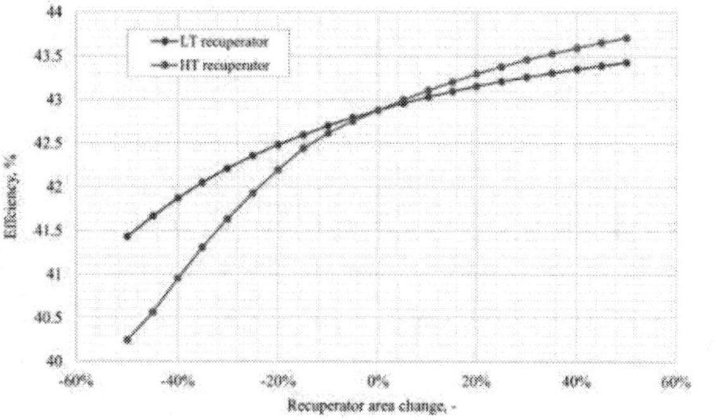

Fig. 7. Sensitivity of the cycle efficiency to a change in recuperators heat transfer area.

As shown in Fig. 7, the rule that the cycle efficiency grows when recuperator area is increased and falls when recuperator area is reduced applies to both recuperators. However, the efficiency line for high temperature recuperator is steeper, which means that it is easier to obtain efficiency change manipulating high temperature recuperator area.

The general tendency revealed by Fig. 7 can be explained by broader dependence between efficiency and recuperation level in the supercritical CO_2 cycle. As the recuperator area increases, more heat can be transferred and more heat can be return to the cold fluid. Thus, less heat is lost in the cooler and less heat has to be supplied to the cycle by a heat source. As a result the ratio of net cycle power to added heat, which is in fact the cycle efficiency, increases.

4.3 Turbine and compressors isentropic efficiency

In the reference cycle, isentropic efficiencies were set up as 88% for both the main and recompression compressor, and 90% for the turbine. Since isentropic efficiency is a sort of measure of broadly defined quality of a turbine or compressor, including airfoil shape, sealings, etc., it may be a significant cost factor that should be considered during cycle design. Therefore, a sensitivity analysis of cycle efficiency to both compressors and turbine isentropic efficiencies was conducted. In order to present and compare how turbine and compressor quality influences the cycle performance, isentropic efficiency of these devices was changed independently in the range from 78% to 95%.

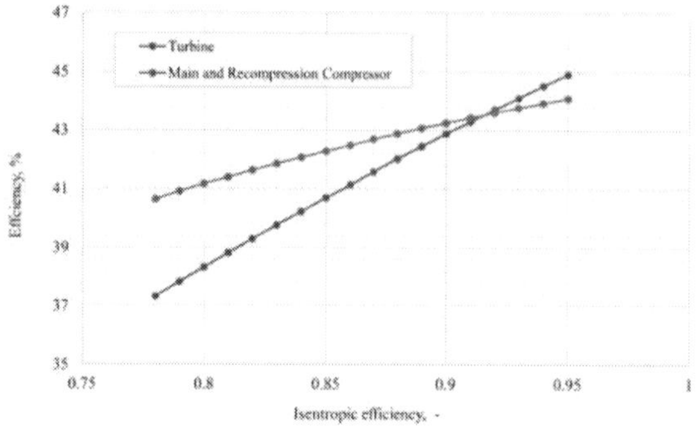

Fig. 8. Sensitivity of the cycle efficiency to a change in turbine and compressors isentropic efficiency.

Calculated efficiency lines are presented in Fig 8. Since both main compressor and recompression compressor are used to realize the same process, their efficiencies were changed simultaneously and are presented by the continuous green line. Thus, the slopes of both lines can be used to determine which process (compression or expanding) has more impact on the recompression cycle. A significant difference occurs between these two processes in case of compression, 1 percentage point change in isentropic efficiency change translates into slightly more than 0.2 percentage points of net cycle efficiency, whereas for expanding process the change of net efficiency is almost 0.45 percentage points.

5 Conclusions

Due to a strong influence of the compressor inlet temperature on optimal recompression split flow ratio, these parameters were altered and optimized simultaneously to mitigate a distortion of efficiency curve. Optimization of the minimal cycle temperature is a good example of such distortion. For a fixed recompression ratio, decrease in the compressor inlet temperature in the area under the critical point causes fall in efficiency, which may seem unnatural. The results obtained when recompression ratio was dynamically optimized for each temperature level indicate that the cycle efficiency rises along with decreasing compressor inlet temperature above and under the critical point as well. There is an abrupt efficiency surge of about 0.7 percentage points in the near critical region caused by sudden fall of compression work. However, similar rise also occurs in the cooler heat transfer area, because of increased amount of heat that has to be removed from the cycle. This fact may be a deciding factor in choosing the lowest cycle temperature, which, considering additional cost of the cooler, probably should be set as close as possible to the critical temperature.

For the reference cycle the optimum recompression split flow ratio is 0.28, which means that this part of the flow is recompressed without being cooled before to the minimum cycle temperature. However, increase in recompression rate always demands more heat transfer area to satisfy assumed heat exchanger effectiveness, therefore an optimum split flow ratio should be a trade-off between efficiency gain and additional capital cost. During design and optimization of a specific system, it is crucial to take into consideration that recompression flow ratio is very sensitive to other cycle parameters, in particular compressor inlet temperature. For instance, for temperature range from 30°C to 31.5°C, the recompression ratio that guarantees the highest efficiency varies from 0.28 to 0.34.

Since achieving a given isentropic efficiency in two different machines as compressor and turbine is not of the same level of difficulty, and their isentropic efficiencies influence overall net cycle efficiency differently, it is possible that shifting the effort of efficient design from one device to another may be more effective in terms of cost and performance.

Acknowledgements

This scientific work was financially supported as a research project number 2015/19/D/ST8/02780 by the National Science Centre, Poland

References

1. E, Feher, Energy Conversion **8**, pp. 85-90 (1968)
2. V. Dostal, M.J. Driscoll., P. Hejzar,, A *Supercritical Carbon Dioxide Cycle for Next Generation Nuclear Reactors, Advanced Nuclear Power Technology Program*, (Thesis (Sc. D.) - Massachusetts Institute of Technology, 2004)
3. G. Angelino, 1968, ASME Paper No. **68-GT-23** (1968)
4. M. Kulhánek, V. Dostál, Thermodynamic Analysis and Comparison of Supercritical Carbon Dioxide Cycles, Supercritical CO_2 Power Cycle Symposium, pp. 1 – 7 (2011)

Pulverized coal gasification with steam and flue gas

Robert Zarzycki[1,*]

[1]Czestochowa University of Technology, Department of Energy Engineering, Brzeźnicka 60A, 42-201 Częstochowa, Poland

Abstract. The study presents the concept and numerical calculations of the coal dust gasification in the entrained flow reactor with power of 16 MW_t. The gasification process in the reactor can be performed in the atmosphere of O_2, CO_2 and H_2O. The combustible gases obtained during gasification are composed mainly of CO and H_2 and can be used to feed pulverized coal-fired boilers. Integration of the reactor (reactors) for coal dust gasification with the pulverized coal-fired boiler allows for improved flexibility, especially in the range of low loads if stabilization of coal dust combustion in pulverized-fuel burners or support for their work with ignition burners fed with gas or light fuel oil is necessary. The concept of the gasification reactor assumes strong eddy motion of the coal dust, which substantially allows for elongation of the time of fuel remaining in the reactor and obtaining a high reaction level. The concept of the entrained flow reactor presented in this study and the results of numerical calculations can be helpful for development of the devices with greater powers which in the nearest future should be integrated in the systems of pulverized coal-fired boilers in order to reduce their minimum load without using the ignition burners.

1 Introduction

Ensuring uninterrupted supplies of electricity and heat is mainly obtained in energy processes by combustion of fossil fuels (hard and brown coal, petroleum and natural gas). With the development of nuclear energy and renewable energy technologies, the demand for non-renewable fuels is gradually decreasing. Despite numerous benefits, such as low costs of electricity production, low level of waste and insignificant demand for fuel and huge resources of the fuel, nuclear energy has also some drawbacks, such as high costs of construction of nuclear power plants, generation of waste that is very dangerous to humans and the environment, and risk of failures. The risk of failures, terrorist attack and nuclear disasters that have occurred (Fukushima, Chernobyl) substantially limited the trust to the nuclear energy. Many countries with nuclear power plants plan to shut them down and return to the use of fossil fuels.
A dynamic development of renewable energy based mainly on the use of wind power allows for limitation of the use of fossil fuels. The major problem with renewable energy is the lack of stability of energy supplies, whose production depends substantially on atmospheric conditions (wind). In such cases, conventional power plants (coal-fired) have to be prepared for the replenishment of the electricity deficit in the electric power system. In many cases, operation of high-performance power units with minimal power (minimum load) or temporary shutting down the units is necessary due to the surplus of electricity in the electric power system.

Therefore, it is critical that modern power units are characterized by a low value of minimum load and high dynamics of changes in the load [1]. Technology of fluidized-bed furnaces allows for operation of the unit in the range of 40-100%, with minimal power of the unit limited by the necessity of maintaining the fluidization process inside the furnace [2]. In the case of pulverized-fuel boilers, minimal load depends on the design and specifications of the boiler and ranges from 40% to 60%. This is connected with stabilization of the coal dust process at low boiler loads (lower temperatures inside the combustion chamber) and ensuring the minimal water circulation rate in the radiant tubes of the boiler. Reduction of the pulverized-fuel boiler power below the minimal level requires starting the ignition burners in order to ensure a stable coal dust combustion process. The use of the ignition burners fuelled by light fuel oil, mazut or natural gas increases the costs of steam power unit operation and emissions of harmful substances.

The increase in flexibility of steam power units equipped in pulverized-fuel boilers can be achieved by replacing ignition fuels (fuel oil, mazut, natural gas) with the gas generated during coal gasification. This solution allows for utilization of the basic fuel in the power plant (hard coal and brown coal) for production of the syngas. Installation of burners in pulverized-fuel boilers fuelled with synthesis gas allows for stabilization of the combustion of coal dust at low loads of the power unit and maintaining the boiler in the state of hot reserve using gas burners fuelled by synthesis gas. The use of burners fuelled with synthesis gas during power unit start up is also possible (replacing the ignition burners). This proposal of the use of synthesis gas allows for limitation of consumption of expensive ignition fuels, especially if the steam power unit has to be operated over the broad range of changes in the load.

* Corresponding author: zarzycki@is.pcz.czest.pl

The achievement of such a goal requires a reactor design that ensures control over the coal gasification process. In the case of actual power facilities, it is important that unit power of the reactor should allow for replacing of one or several ignition burners.

The study presents the concept and numerical computations of the reactor operation to perform coal gasification process in the atmosphere of recirculating flue gas and water steam.

2 The concept of reactor for coal gasification

Coal gasification process can be performed using one of the three types of reactors (Fig. 1):

- moving bed reactors;
- fluidized-bed reactors;
- entrained-flow reactors.

Moving bed reactors (Fig. 1a) (Lurgi) are designed to gasify solid fuels with grain size ranging from 5 to 80 mm. Time of fuel gasification process depends on the gasification medium (oxygen, air, water steam) and pressure at which the process is performed (environmental pressure, overpressure) and ranges from several minutes to several hours. The most popular reactors include counter-current flow of fuel and gasifying agent with their power reaching 350 MW_t. Heat needed for the gasification process originates from partial combustion of the charge. Gasification in a stationary bed requires fuel and low agglomerating capacity of coal. The main problem connected with operation of moving bed reactors is the disposal of fragmented fuel and liquid hydrocarbons [4-9].

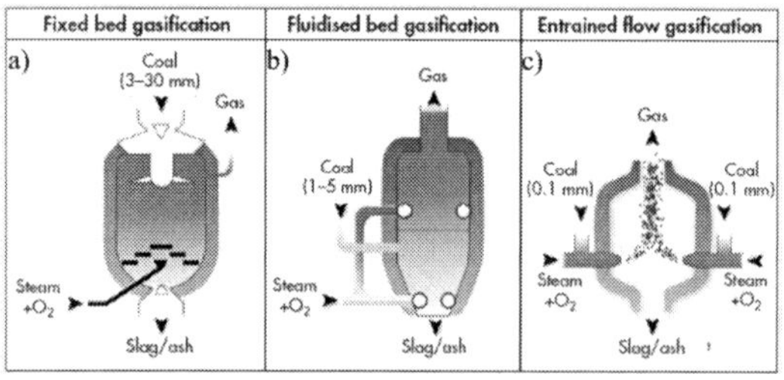

Fig. 1. Gasification process for coal [3].

In fluidized-bed reactors (Fig. 1b) (Winkler), it is necessary to prepare fuel through its fragmentation in the range of 0.5-6 mm in order to ensure efficient fluidizing. Time of fuel remaining in the reactor ranges from 10 to 100 s while its temperature has to be lower than the softening point of the ash contained in the coal. Unit powers of fluidized-bed reactors reach 700 MW_t. Gasifying agents include oxygen, air and water steam introduced from the bottom. The major problem with operation of fluidized-bed reactors is poor conversion of the C element resulting from low process temperatures limited by the ash softening point [4-9].

In entrained flow reactors (Fig. 1c) (Koppers-Totzek), fragmented coal is transported in the streams of the gasifying agent (oxygen and water steam mixture). With fragmentation of fuel to grains of below 0.1 mm and the temperature during the gasification process ranging from 1200-1600°C, time of fuel grain remaining amounts to several seconds, which allows for reaching high unit gas efficiency and reactor power of 700 MW_t. One benefit of this type of generators is the opportunity to use caking coals as raw materials and no temperature limits caused by ash composition. The main problems connected with operation of entrained flow reactors is to ensure continuous supply of the gasifying agent and fuel with the specific ratio and cooling of the gas generated during gasification [4-9].

The biggest potential connected with the achievement of high efficiencies of synthesis gas, rate of changes in efficiency and opportunity to control coal gasification process is observed for entrained flow reactors for coal gasification [9]. They allow for coal dust gasification in the atmosphere of recirculated flue gas and in the presence of water steam. Ash vitrification can also be performed [10-16].

Fig. 2 presents a design of entrained-flow reactor for coal gasification with power output of 16 MW_t. The reactor is composed of seven main components presented in Fig. 2. The **C1** dome contains one burner nozzle with diameter of 50 mm and four nozzles for supplying gases with specific content to the reactor. Below the **C1** dome there is **C2** with diameter of 2 m, composed of ten cylindrical segments with the height of 0.5 m each. Each segment has eight nozzles with diameter of 50 mm at the angle of 45°. The nozzles were installed so that their distance from the symmetry axis of the chamber **C2** in each lower segment is incremented by 50 mm. This design allows for a strong eddy motion of the gas and fuel supplied to **C2**. Below the **C2** chamber there is the **C3** chamber with the height of 1 m and reducing diameter to 1.8 m, where three levels of nozzles with diameter of 50 mm are located. Each level contains eight nozzles (distributed every 90°). This system of nozzles ensures a strong eddy motion of the gas and fuel. Below the chamber **C3** there is the **C4** chamber with special design, composed of ten segments with the height of 0.5 m each. Each segment has four levels of nozzles distributed every 90° with diameter of 50 mm. Side walls of each segment in the **C4** chamber were shaped so that the nozzles are installed tangentially to the chamber walls. The nozzles are also inclined at the angle of 5°.

The chamber **C5**, composed of two truncated cones is located in the lower part, with a step between each other aimed to extend the time of fuel remaining in the boiler. Nozzles are tangentially installed in the side walls of the **C5** chamber, at five steps, distributed every 90°. Below them there is the **C6** chamber, cylindrically-shaped, with diameter of 3 m and tangentially installed channel used to discharge gases from the reactor. A plunger is installed inside the **C6** chamber. Below this point there is the **C7** chamber composed of two truncated cones and a cylindrical chamber. The chambers **C6** and **C7** act as a cyclone that allows for separation of solid particles from the gas generated in the reactor.

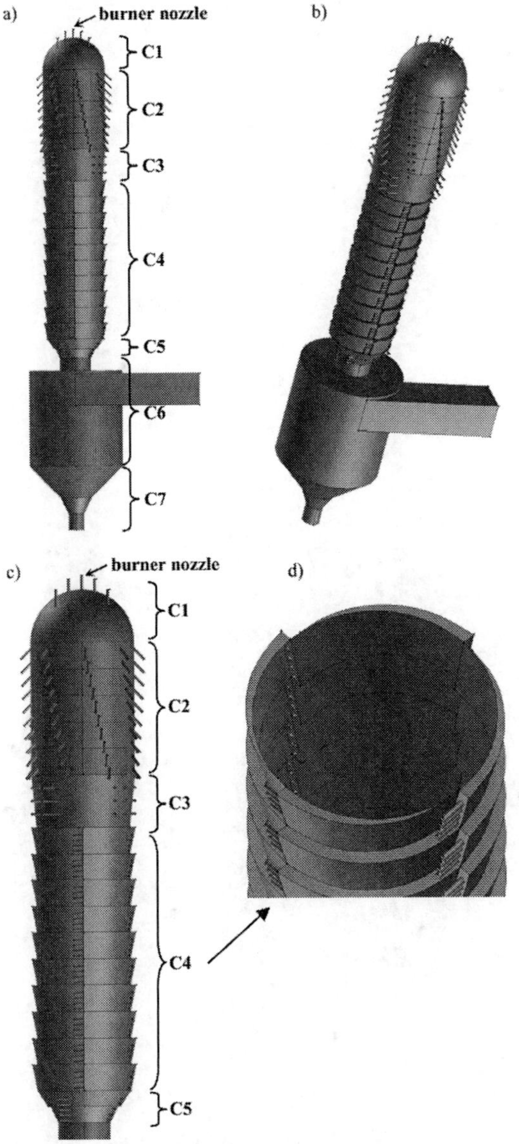

Fig. 2. Entrained flow reactor for coal gasification.

3 Numerical analysis of coal dust gasification process

Geometry and the grid used for computation of the coal dust gasification reactor (Fig. 2) was developed using the GAMBIT software. The ANSYS FLUENT 14 software was used to compute coal dust gasification process.

Calculations of coal dust and gas flow were performed using the Reynolds Stress turbulence model. Modelling of flow of coal dust was based on the Discrete Phase Model, whereas coal dust combustion was performed using Species Transport model, which allows for modelling chemical reactions both in the solid phase and gaseous phase [17-22]. The Single-Rate fuel devolatilization model was employed. Calculations were based on the radiation model termed Discrete Ordinate (DO). A numerical model of the process allows for computation of combustion, fuel devolatilization and fuel gasification in the atmosphere of air, O_2, CO_2 and water vapour.

Calculations of coal dust gasification process for the dust with mean size of 0.25 mm were carried out for coal dust with physicochemical parameters as presented in Table 1.

Table 1. Results of technical and elemental analysis of the fuel used for simulations (dry state).

M [%]	A [%]	VM [%]	FC [%]	C [%]	H [%]	N [%]	O [%]	HHV [MJ/kg]
5	5	45	45	85	10	1	4	24.7

A - ash
M - moisture
VM - volatile matter
FC - fixed carbon
HHV - higher heating value

For simplification purposes, it was adopted that the fuel does not contain sulphur. Reaction rate constants were derived from the studies [17,18].

Calculations for combustion and gasification of coal dust were described with seven reactions:

- Reaction of volatile matter (VM) combustion

$$VM = C_{1.74}H_{5.95}O_{0.15}N_{0.0428}$$

$$VM + 2.28\ O_2 = 1.74\ CO + 2.79\ H_2O + 0.0214\ N_2 \quad (1)$$

- Reaction of oxidation of carbon oxide

$$CO + 0.5\ O_2 = CO_2 \quad (2)$$

- Reaction of oxidation of fixed carbon (FC)

$$C_{(s)} + 0.5\ O_2 = CO \quad (3)$$

- Boudouard's reaction

$$C_{(s)} + CO_2 = 2\ CO \quad (4)$$

- Synthesis of water gas

$$C_{(s)} + H_2O = CO + H_2 \quad (5)$$

- Hydrogen oxidation reaction

$$H_2 + 0.5\ O_2 = H_2O \quad (6)$$

- Methane oxidation reaction

$$CH_4 + 2\ O_2 = CO_2 + 2\ H_2O \quad (7)$$

Numerical computations were performed for two variants:

- **V1** - gasification in the atmosphere of O_2/CO_2,
- **V2** - gasification in the atmosphere of O_2/H_2O.

In both cases (**V1** and **V2**), the conditions of fuelling the **C1** and **C2** chambers were identical. The burner located in the **C1** chamber was fuelled by the O_2/CH_4 mixture at the rate of 15 m/s with volumetric ratios of 80/20%. The nozzles were fuelled by the mixture of O_2/CO_2 at the speed of 5 m/s with volumetric ratios of 80/20%. The **C2** chamber was fuelled by the O_2/CO_2 mixture with the speed of 5 m/s, with volumetric ratio of 70/30%. With this gas, coal dust was also supplied to the **C2** chamber. Total flux of the coal dust was 0.65 kg/s, whereas 0.11 kg/s was supplied at the highest step, and reduced for each step down by 0.01 kg/s. The lowest step of the **C2** chamber was fuelled at a rate of 0.02 kg/s.

For the **V1** variant, the **C3** chamber was fuelled with the O_2/CO_2 mixture with the rate of 7 m/s and volumetric ratio of 50/50%. The **C4** chamber was fuelled with the gas with every second step of nozzles with the parameters similar to the **C3**. The chamber **C5** was fuelled by the O_2/CO_2 mixture at a rate of 10 m/s, with volumetric ratio of 50/50% only in the upper cone.

The difference between **V1** and **V2** variants is connected with replacement of the carbon dioxide supplied to **C3**, **C4** and **C4** (variant **V1**) with water steam (variant **V2**).

Temperature of gas mixtures supplied with nozzles to the reactor was identical for both cases (**V1** and **V2**), i.e. 300°C.

The results of the gasification process in the O_2/CO_2 atmosphere (variant **V1**) are presented in Fig. 3. Fig. 3a presents distribution of the peripheral gas rate in the vertical cross-section resulting from the supply of the gas with tangentially designed nozzles. The highest levels of the peripheral component are observed near the reactor walls in the chamber **C4** and **C5**. Concentration of the fuel supplied to the reactor with nozzles installed in the **C2** chamber is presented in Figs. 3b, 3c and 3d. With the nozzles installed in the **C2** chambers at the angle of 45° and their shift with respect to axes at individual steps, the fuel supplied to the chamber is decomposed in the form of the rings observed inside the **C2** chamber along its height (Fig. 3b, Fig. 3c).

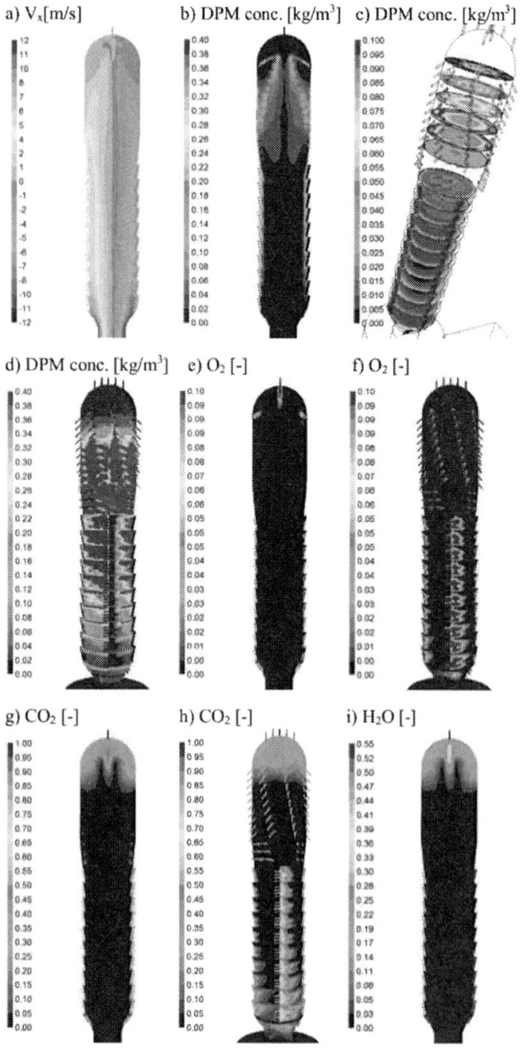

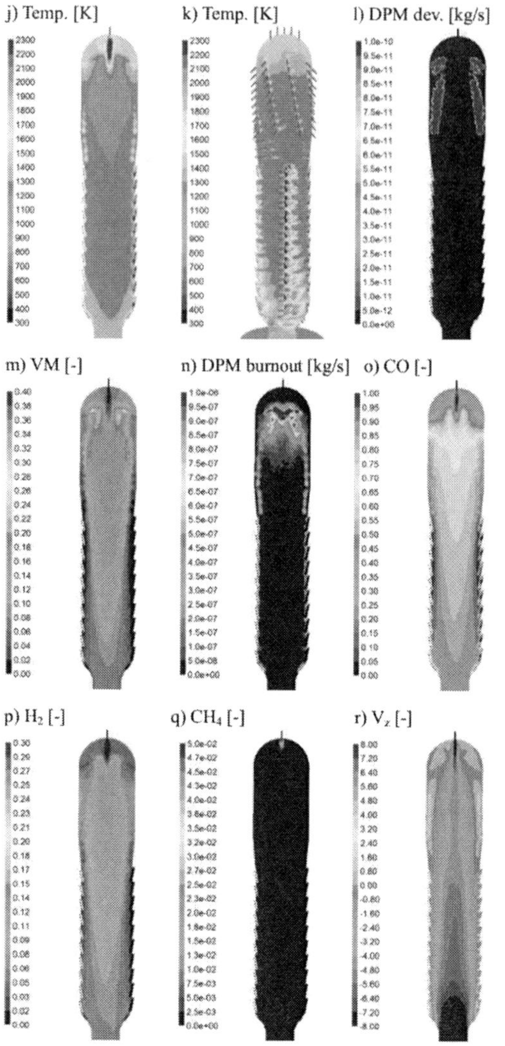

Fig. 3. Distribution of selected process values (**V1**): a) peripheral velocity component, b) fuel concentration in the vertical cross-section, c) fuel concentration, d) fuel concentration on the furnace walls, e) O_2 concentration, f) O_2 concentration on the furnace walls, g) CO_2 concentration, h) CO_2 concentration on the furnace walls, g) H_2O concentration, j) temperature distribution in the vertical cross-section, k) temperature distribution near furnace walls, l) devolatilization rate distribution in the vertical cross-section, m) volatile matter concentration, n) fuel burning rate, o) CO concentration, p) H_2 concentration, q) CH_4 concentration, r) vertical velocity component.

With the effect of the centrifugal force caused by swirling gas (Fig. 3a), the fuel supplied to the **C2** chamber is transported near the reactor walls (Fig. 3c). The gas supplied tangentially to the **C3** and **C4** chambers at a rate of 7 m/s increases the value of the centrifugal force and, consequently, the fuel remains only near the reactor walls.

This phenomenon is very beneficial due to the extension of time of fuel remaining near the specifically shaped walls of the **C4** chamber, whereas the gasifying agent is supplied only near the walls, which allows for

https://doi.org/10.1051/matecconf/201824005036

gasification of the fuel in this area. The special shape of the **C4** chamber helps intensify mixing of gas and the fuel swirling inside the reactor (Fig. 4), allowing for an efficient gasification process. Process conditions were set so that fuel concentration at the outlet from the **C5** chamber was equal zero. This means that total stream of the fuel supplied to the reactor is gasified in its interior. In the analysed variant, the gasification agents were O_2 and CO_2, with distributions presented in Figs. 3e, 3f, 3g and 3h. Distributions of O_2 concentration in the vertical cross-section of the reactor (Fig. 3e) show substoichiometric conditions inside the reactor whereas elevated O_2 concentration is observed only near the nozzles that fuel the C_4 chamber. Analysis of CO_2 distribution reveals elevated concentration of this gas in the **C1** chamber and in the upper part of the **C2** chamber. This is connected with the necessity of combustion (reaction 1) of a specific amount of gases generated through fuel devolatilization (Fig. 3l, 3m) due to the necessity of ensuring heat for the performance of endothermic processes of coal gasification. Furthermore, CO_2 is generated through combustion of the O_2/CH_4 mixture supplied with the burner nozzle. Due to the gasification process (reaction 4), the CO_2 concentration is noticeably reduced, and in the central part of the **C2** chamber, it equals nearly zero. CO_2 concentration near reactor walls is presented in Fig. 3h. Similar to O_2, its elevated concentration is observed only near the nozzles that supply the O_2/CO_2 mixture. This suggests intensive processes of mixing and coal gasification (reaction 4). Figure 3i illustrates H_2O concentration. Presence of H_2O is due to the combustion of the O_2/CH_4 mixture (reaction 7) in the burner installed in the **C1** chamber and humidity contained in the fuel. Similar to CO_2 distribution, H_2O concentration was significantly reduced in the initial part of the the **C2** chamber, which suggests the gasification process (reaction 5) using water steam, which results in the increase in H_2 concentration (Fig. 3p).

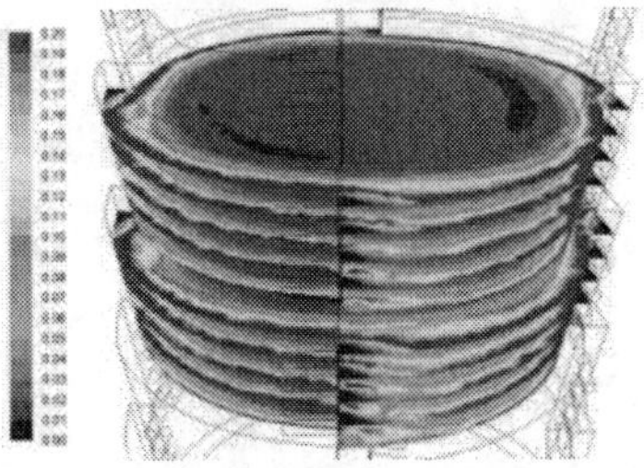

Fig. 4. Distribution of fuel concentration in **C4** (**V1**).

Figures 3j and 3k illustrate distributions of temperature inside the reactor and near its walls, respectively. The highest temperature values are observed near the burner (chamber **C1**) and near the walls of the **C4** chamber. Gasification processes require ensuring adequate amount of heat to perform them and maintain optimal thermal conditions due to the pattern of coal gasification reaction. Mean temperatures in the reactor are around 1000-1200°C near the reactor walls (**C4** chamber). In the locations where the O_2/CO_2 mixture is supplied, local temperature reaches the level near 2000°C. This allows for a fast heating of the fuel in this zone and intensification of the gasification processes (reactions 3, 4 and 5). Local temperature of 2000°C also allows for vitrification of ash remaining after the gasification process. It is possible in this case to discharge the molten ash with the combustible gases obtained through gasification to the **C6** and **C7** chambers. The fuel supplied to the **C2** chamber is devolatilized, which can be observed in Fig. 3l, which illustrates fuel devolatilization rate. This results in the distribution of concentration of volatile matter (VM), as presented in Fig. 3m.

The highest VM concentration occurs in the upper part of the **C2** chamber and is insignificantly reduced with the flow towards the **C4** and **C5** chambers. Near the reactor walls, VM concentration is noticeably decreased, which results from partial combustion of volatile matter (reaction 1) and dilution with gasification products (reaction 3, 4 and 5) generated near the reactor walls. The parameter which reflects the rate of fuel conversion in the reactor is presented in Fig. 3n. The highest fuel conversion rates are observed near the burner and near the locations of supply of the gasifying agent (O_2/CO_2). The main product of the gasification process for the **V1** is CO, whose distribution is presented in Fig. 3o. Near the burner nozzle (**C1** chamber), CO concentration is ca. 20-40%, whereas down the stream, the CO content in the reactor axis rises noticeably to the level of 60-80%. The highest values of the CO concentrations are observed near the wall of the **C3** and **C4** chambers, with elevated fuel concentration (see Figs. 3b, 3c, 3d) and places where the gasification factor of O_2/CO_2 is supplied. The other gasification product is H_2, with its distribution presented in Fig. 3p. Presence of this gas component results on the one hand from gasification of coal dust in the presence of water steam generated due to combustion of CH_4 and humidity contained in fuel. Furthermore, H_2 is partially obtained from combustion of a part of volatile matter (reaction 1) and then gasification of fuels in the generated stream of water vapour (reaction 5). Due to the superstoichiometric conditions of O_2/CH_4 burner operation, methane supplied to the do reactor is quickly burnt (Fig. 3q). Figure 3r presents the vertical component of gas velocity in the reactor. With adequate installation of the nozzles in side walls of the reactor, it is possible to obtain flow of gas towards the **C1** chamber near the walls, whereas in the reactor axis, the flow is noticeably oriented towards the **C5** chamber. This organization of gas flow is beneficial due to the improved mixing of gases in the reactor and, importantly, allows for extension of the time of remaining the tiniest fuel grains through their flow towards the **C1** chamber. This allows for a substantial limitation of the steam of fuel, which, with the molten slag, could flow to the **C6** and **C7** chambers.

The selected results of computations of the gasification process in the presence of water steam (variant **V2**) are presented in Fig. 5. Distribution of temperature for **V2** variant is similar to the distribution for **V1** in chambers **C1**, **C2**, **C3**. Slightly higher temperatures are observed in the **C4** and **C5** chambers,

 https://doi.org/10.1051/matecconf/201824005036

resulting from the difference in the demand for heat at gasification in the presence of CO_2 (8) and H_2O (9).

- the Boudouard reaction

$$C_{(s)} + CO_2 = 2\,CO \qquad +172\text{ MJ/kmol} \quad (8)$$

- the water gas reaction

$$C_{(s)} + H_2O = CO + H_2 \quad +131\text{ MJ/kmol} \quad (9)$$

Analysis of the distribution of CO concentration for **V2** (Fig. 5b) reveals that the highest values of CO concentrations occur near the **C2** chamber walls whereas the CO concentration decreases down the stream. Analysis of the H_2 distribution (Fig. 5c) reveals a noticeable increase in H_2 concentration near the walls of the **C3** and **C4** (places of O_2/H_2O mixture supply).

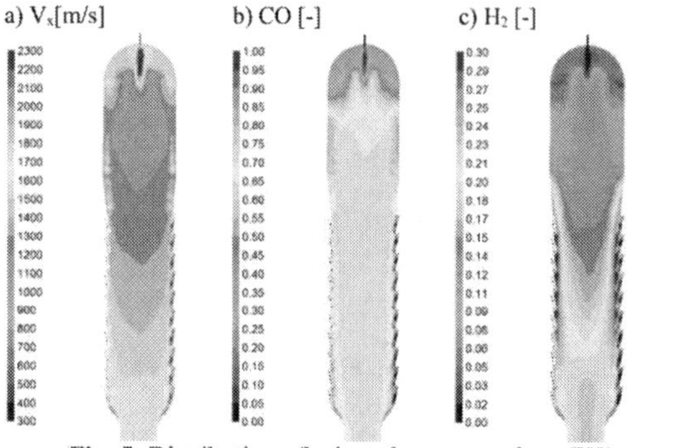

Fig. 5. Distribution of selected process values (**V2**): a) temperature distribution in the vertical cross-section, b) CO concentration, c) H_2 concentration.

Table 2 compares composition of gases that leave the **C5** chamber. It can be noticed that for the **V1** variant, process gas is composed of 86% CO and 5.65% H_2. In the case of **V2** (gasification in the presence of water steam) the CO concentration reaches 68.5% whereas H_2 concentration exceeds 21%. Values of other gas components for both cases of **V1** and **V2** were similar.

Table 2. Selected process parameters at the outlet from **C5** for coal dust gasification (**V1**, **V2**).

Variant	CO [%]	H_2 [%]	CO_2 [%]	H_2O [%]	VM [%]	O_2 [%]	CH_4 [%]
V1	86.30	5.65	1.25	0.20	6.52	1.61E-04	1.19E-05
V2	68.67	21.37	1.52	2.58	5.84	1.39E-03	1.35E-05

4 Conclusions

The concept and the design of coal dust gasification reactor presented in the study allows for performing the process of coal dust gasification in the atmosphere of O_2/CO_2 and O_2/H_2O. Division of reactor into zones allows for controlling the gasification process. Process of fuel gasification is performed both in the central part of the reactor (the **C2** chamber) and near its walls (**C3**, **C4** chambers). Strong eddy motion of the fuel with the special design of the **C4** chamber allows for intensification of the process of mixing between the fuel and gasification agent and extension of the time of fuel remaining in the reactor. The proposed solution also allows for ash vitrification in the lower part of the **C4** chamber and in the **C5** chamber. The computations of the fuel gasification process revealed opportunities to obtain the product containing 86.3% CO and 5.6% H_2 for the **V1** variant and 68.7 % CO and 21.4 H_2 for the **V2** variant. These gases can be directly used for fuelling pulverized-fuel boilers in order to improve their flexibility, stabilization of the coal dust combustion process at low boiler loads and replacing the ignition burners during boiler start-up.

The gasification process in the atmosphere of O_2/CO_2 and O_2/H_2O allows for the control of temperature inside the reactor and separation of the processes of coal gasification and fly ash vitrification. With changes in the composition of the gasifying agent, it is possible to obtain a product (synthesis gas) with expected composition.

During designing and construction of new pulverized coal-fired boilers, they can be equipped in coal gasification reactors using the solution presented in this study. A compact design of the reactor allows for its installation near a pulverized coal-fired boiler. Using the coal dust as a fuel supplied to a coal gasification reactor allows for a substantial limitation of consumption of fuel oil or gas during operation of the pulverized coal-fired boiler. The design of the reactor should be integrated especially with pulverized coal-fired boilers where oxy-fuel combustion processes are performed.

The installation of the coal gasification reactors in a pulverized coal-fired boiler system allows for:

- increasing the flexibility and dynamics of blocks with dust boilers,
- increase in thermal efficiency of boilers,
- reduction of the minimum technological operation of the dust boiler without the need to start the firing torches,
- implementation of the combustion and/or gasification process of solid fuels,
- carrying out the process of ash refining (e.g. during the period of cheaper night pricing).

The proposed solution constitutes an interesting technological option aimed at retrofitting and modernization of 200 MW_e class of dust mills.

Acknowledgements

Ministry of Science and Higher Education Republic of Poland BS/PB-404-301/11

References

1. R. Zarzycki, R. Kobyłecki, Z. Bis, Mechanic no **2**, (2017)
2. Bis Z., *Kotły fluidalne: teoria i praktyka*, (Wydawnictwo Politechniki Częstochowskiej, Częstochowa, 2010)
3. http://www.bine.info, Power plants with coal gasification, projekt info 09/06.
4. Borowiecki T., Kijeński J., Machnikowski J., Ściążko M., *Czysta energia, produkty chemiczne i paliwa z węgla - ocena potencjału rozwojowego,*

(Instytut Chemicznej Przeróbki Węgla, Zabrze, 2008)
5. Tomeczek J., *Zgazowanie węgla* (Skrypty centralne nr 1551/4, Politechnika Śląska, Gliwice, 1991)
6. A.J. Minchener, Fuel, vol. 84, (2005)
7. A.G. Collot, International Journal of coal Geology, vol. **65**, (2016)
8. R. Zarzycki, R. Kobyłecki, M. Kratofil, Z. Bis, Energy Policy Journal, Tom 17, zeszyt 4, (2014)
9. Higman C., Van Der Burgt M., *Gasification*, (Elsevier Science, USA, 2003)
10. R. Zarzycki, Z. Bis, Journal of Thermal Science, Volume **26**, Issue 2, (2017)
11. R. Zarzycki, Z. Bis, R. Kobyłecki, Procedia Engineering, Vol.**157**, (2016)
12. R. Zarzycki, Z. Bis, Procedia Engineering, Vol.**157**, (2016)
13. R. Zarzycki, Z. Bis, 12 th International Conference on Fluidized Bed Technology, (2017)
14. R. Zarzycki, Z. Bis, E3S Web of Conferences 13, 05003, (2017)
15. R. Zarzycki, E3S Web of Conferences 13, 05004, (2017)
16. R. Zarzycki, 12 th International Conference on Fluidized Bed Technology, (2017)
17. L. Chen, M. Gazzino, A.F. Ghoniem, 35th International Technical Conference on Coal Utilization & Fuel Systems. Clearwater, Florida, (2010)
18. D. Toporov, P. Bocian, P. Heil, A. Kellermann, H. Stadler, S. Tschunko, M. Förster, R. Kneer, , Combustion and Flame, vol. **155**, (2008)
19. M. Vascellari, G. Cau, Proceeding of CCT2009 Fourth International Conference on Clean Coal Technologies. Dresden, Germany, (2009)
20. P. Warzecha, A. Bogusławski, Archivum Combustionis, Vol. **12**, nr 3, (2012)
21. R. Zarzycki, M. Kratofil, D. Pawłowski, M. Ścisłowska, R. Kobyłecki, Z. Bis, Energy Policy Journal, Tom 16, zeszyt 3, (2013)
22. ANSYS Fluent Theory Guide, (2011)

Cyclone furnace as a way for mercury removal from lignite

Robert Zarzycki[1,*], and *Michał* Wichliński[1]

[1]Czestochowa University of Technology, Department of Energy Engineering, Brzeźnicka 60A, 42-201 Częstochowa, Poland

Abstract. The power plants will have to meet the requirements regarding permissible Hg emission levels by 2021. A particularly unfavourable situation is observed for front of lignite, which contains more mercury than hard coal, and in addition its combustion also causes more CO2 emissions. The paper presents the proposal to use the cyclone furnace in the process of heat treatment for the process of the release of mercury from lignites. The structure and methods of operation of the cyclone furnace were presented. The paper also discusses the methodology of the process of numerical modelling of combustion and gasification of coal dust. The results showed that changing a fuel stream fed to the cyclone furnace allows for controlling the temperature and degree of fuel devolatilization. This leads to heating of the fuel to the desired temperature due to the expected level of mercury removal from the fuel. This is very important especially in the case of fuels containing significant amounts of moisture, such as lignite. The experiments performed in the study confirmed the results of numerical calculations for heating fuel and showed the possibility of over 90% removal of mercury from the fuel. The results confirmed the laboratory workplace rotary kiln.

1 Introduction

Processes of fossil fuel combustion for the purposes of electricity and heat generation contribute to the degeneration of the natural environment. The contaminants removed from flue gas include dust, sulphur oxides and nitrogen oxides. Technologies are currently being developed to separate carbon dioxide from flue gas in order to store it underground [1, 2]. Limitations of heavy metal emissions (such as mercury) are expected to be introduced in the nearest future. Mercury is an element which is more difficult to be removed from flue gas compared to the previously mentioned compounds as its content in carbon and, consequently, in flue gas, is insignificant. Medium content of mercury in Polish energy coals is around 80 ng/g [3, 4]. This insignificant mercury content in carbon, combined with huge amounts of burnt coal, results in emissions of mercury at the level of 50 Mg mercury per year per one power plant [5]. Total mercury emissions from anthropogenic sources in Poland is 11 Mg/year [6]. Of this number, over 9 Mg is caused by combustion of carbon fuels. Around 1.1 Mg is emitted as a result of non-ferrous metals production and 0.7 Mg - from cement production [7].

Limitation of mercury emissions can be reached by flue gas cleaning or through removal of mercury from coal before combustion. Removal of mercury vapour from flue gas (post-combustion methods) is burdensome since the amount of sorbents used is very high and content of mercury in the flue gas stream is very small. Furthermore, fly ash is also contaminated with activated carbon, mostly used to capture mercury vapours together with the adsorbed mercury and non-reacted sorbent. A promising option is to remove mercury from fuel before combustion (pre-combustion methods) which allows for cleaning the coal and than use it in various areas of the industry. One of these methods is low-temperature thermal processing, with its efficiency reaching over 90% at temperature of ca. 400°C, which was demonstrated in numerous publications [8-10]. Another concept of mercury removal from fuel was proposed in the paper [11]. Overheated steam was used as a heat carried in this solution. The concept related to fluidized-bed furnaces and allowed for limitation of waste caused by low-temperature processing.

This study presents the proposal of thermal processing of fuel using the cyclone furnace.

2 Cyclone furnace

Cyclone furnaces are understood to mean energy equipment where a strong eddy flow of gases is used to support the process of solid, liquid or gaseous fuel combustion [12, 13]. Cyclone furnaces also allow for solid fuel gasification in the atmosphere of flue gases or water vapour [14-18] and vitrification of fly ashes [19, 20]. Apart from these function, cyclone furnaces can be used for thermal processing of fuel that consists in its

* Corresponding author: zarzycki@is.pcz.czest.pl

drying and controlled devolatilization to the expected level. In this case one can perform the controlled process of fuel heating that allows for heating coal grains to the required temperature while releasing a certain amount of volatile parts and removing mercury in the form of vapour. Details of the assumptions of low-temperature fuel processing for the needs of mercury removal have been discussed in studies [8-10].

The cyclone furnace used in the experimental and numerical examinations (Fig. 1) is composed of two chambers. Fragmented fuel is supplied by tangentially designed nozzles to the upper chamber PC2. The process of heating, drying and initial fuel devolatilization is performed in this chamber. The PC1 chamber is located below the upper chamber and used to continue the process of heating, fuel devolatilization, combustion and gasification. In order to ensure undisturbed flow of fuel from the PC2 to PC1 chamber, a part in the form of the plunger that separates the flow of fuel grains from PC2 to PC1 from the flow of gas from PC1 to PC2 was installed. The PC1 chamber is made of several cylinders with decreasing diameters that form the steps at the places of changes in PC1 chamber diameter. A set of nozzles was designed tangentially to the cylinder generatrix of the PC1 chamber in order to introduce the driving gas at high speed. The aim of the tangentially supplied driving gas is firstly to generate a strong eddy motion of fuel grains and to ensure adequate composition (air, oxygen, carbon dioxide, water vapour) due to performance of the combustion process, gasification and thermal fuel processing aimed to control heating, fuel devolatilization and mercury removal. With PC1 chamber diameter that decreases with the vertical flow of coal grains, the centrifugal force that allows for maintaining a substantial fuel mass in strong eddy current increases while time of remaining the grains necessary to heat the fuel to the required temperature increases (thermal processing of the fuel).

This concept of the cyclone furnace allows for the controlled heating and devolatilization of the fuel swirling in the furnace. The heat needed for the process of heating and devolatilization is obtained from combustion of a certain part of gases generated during fuel devolatilization. The control of thermal processing of fuel can be performed by:

- changing the stream of fuel mass supplied to the cyclone furnace at invariable parameters of driving gas: possibility to achieve the conditions of fuel combustion in the furnace, gasification and controlled heating and fuel devolatilization;
- change in the speed of the driving gas (PC1 chamber): changes in the centrifugal force (control of time of fuel remaining in the furnace) and changes in the amount of heat generated from oxidation of volatile matter;
- change in the composition of the driving gas (PC1 chamber): possibility to control temperature inside the PC1 chamber by changing the stream of oxygen needed for oxidation of volatile matter (change in the amount of heat generated).

The above presented methods allow for determination of optimal process parameters that ensure the required conditions of thermal processing of fuel due to the expected level of mercury removal.

During the experimental examinations, the process of thermal processing of fuel was controlled by changing the stream of fuel mass over a wide range from combustion conditions through gasification to the state when stable thermal fuel processing conditions are obtained: partial fuel devolatilization that ensures the heat for maintaining the process of fuel heating to the assumed temperature.

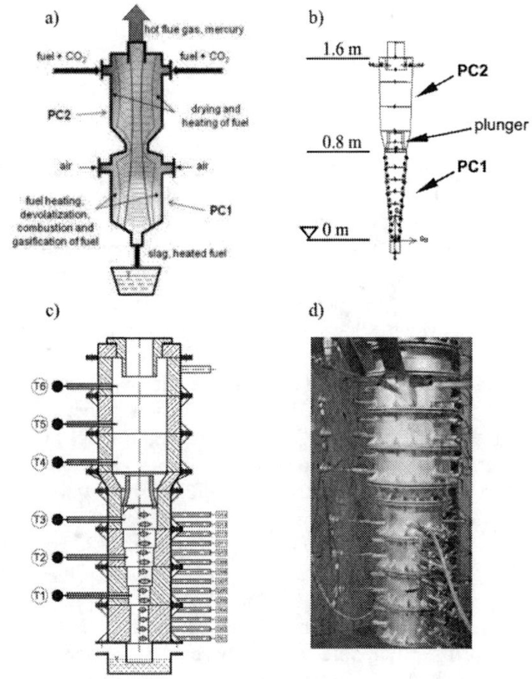

Fig. 1. Design a), geometry b), cross-section c), view of cyclone furnace d).

3 Experimental examinations and numerical analysis of thermal processing of fuel in a cyclone furnace

The below presented results of numerical computations and experimental examinations were performed for brown coal with physicochemical parameters as presented in Table. 1. The examinations were conducted for fuel with medium grain size (1.5 mm). The fuel was supplied to the PC2 chamber (upper) with two tangent channels in the CO_2 stream at the speed of 3 m/s and temperature of 25°C. The stream of fuel supplied was changed from 2 to 16 g/s. Air with temperature of 25°C was supplied to the PC1 chamber at the speed of 3 m/s, being the driving gas used to induce strong eddy current of fuel in the PC1 chamber. For the fuel stream of 2 g/s, air-fuel ratio is λ=1.2 and ensures complete fuel combustion.

https://doi.org/10.1051/matecconf/201824005037

For the above presented process conditions, numerical analysis of the combustion process and fuel devolatilization were performed for variable stream of fuel mass.

Table 1. Physicochemical fuel parameters (all the data in the analytical state).

M [%]	A [%]	VM [%]	HHV [kJ/kg]	C [%]	H [%]	N [%]	S [%]	Hg [ng/g]
25.3	40.2	38.8	13112	33.2	4.4	0.6	2.5	180.7

A - ash
M - moisture
VM - volatile matter
HHV - higher heating value

Geometry and the grid used for computation of the cyclone furnace was developed using the Gambit software (Fig. 1b). Computation of the fuel grain flow, devolatilization process and combustion of volatile matter was performed by means of the ANSYS FLUENT 14 software. Calculations of fuel grain and gas flow were performed using the Reynolds Stress turbulence model. Modelling of flow of coal grains was based on the Discrete Phase Model, whereas coal dust combustion was performed using Species Transport model, which allows for modelling chemical reactions both in the solid phase and gaseous phase [21-25]. The Single-Rate fuel devolatilization model was employed. Calculations were based on the radiation model termed Discrete Ordinate (DO). A numerical model of the process allows for computation of combustion, fuel devolatilization and fuel gasification in the atmosphere of air, O_2, CO_2 and water vapour.

Calculations of the process of heating, devolatilization, gasification and combustion of fuel with replacement diameter of 1.5 mm were carried out for the fuel with physicochemical parameters as presented in Table 1. For simplification purposes, it was adopted that the fuel does not contain sulphur. Reaction rate constants were derived from the studies [22, 23].

Calculations for combustion and gasification of coal dust were described with the following six reactions:

- Reaction of volatile matter (VM) combustion
  $$VM = C_{0.13}H_{2.93}O_{1.55}N_{0.0422}$$
  $$VM + 0.02\ O_2 = 0.13\ CO + 1.46\ H_2O + 0.0211\ N_2 \quad (1)$$
- Reaction of oxidation of carbon oxide
  $$CO + 0.5\ O_2 = CO_2 \quad (2)$$
- Reaction of oxidation of fixed carbon (FC)
  $$C_{(s)} + 0.5\ O_2 = CO \quad (3)$$
- Boudouard's reaction
  $$C_{(s)} + CO_2 = 2\ CO \quad (4)$$
- Synthesis of water gas
  $$C_{(s)} + H_2O = CO + H_2 \quad (5)$$
- Hydrogen oxidation reaction
  $$H_2 + 0.5\ O_2 = H_2O \quad (6)$$

The numerical model takes into consideration the heat transfer through furnace walls to its surroundings.

Actual thickness of individual components of the laboratory furnace (Fig. 1d) [26, 27] with coefficients of heat penetration and conduction was also taken into consideration for the materials used. It was adopted that air temperature around the furnace was 25°C.

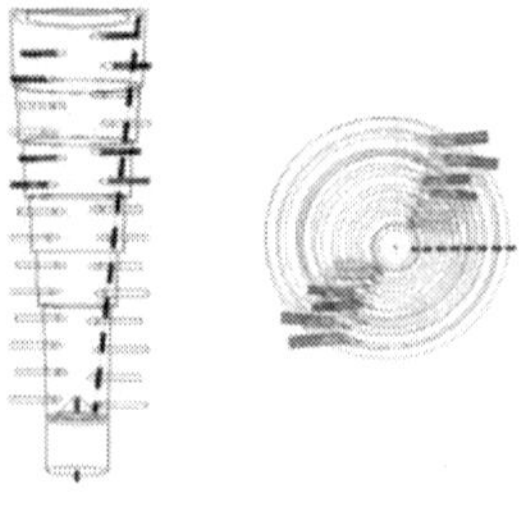

Fig. 2. Location of the line used to determine profiles of temperature, fuel concentration and fuel devolatilization rate.

Figures 3, 4 and 5 present the time-averaged profiles of temperature, fuel concentration and fuel devolatilization rate for selected streams of fuel mass supplied to the cyclone furnace (from 2 to 16 g/s). Profiles (Fig. 3, 4, 5) were determined along the dashed line visible in Fig. 2. It is located at the distance from 10 to 25 mm from the PC1 chamber. Figures 6, 7 and 8 presented time-averaged distributions of temperature, fuel concentration and fuel devolatilization rate for the three selected fuel streams: 2, 6 and 10 g/s. Analysis of the temperature profiles presented in Fig. 3 reveals that for the fuel stream of 2 g/s, temperature at the level of z = 0.3 m reaches the level of 1300°C, which suggests the intensive fuel combustion in this zone of the PC1 chamber. This fact is confirmed by profiles of fuel concentration and fuel devolatilization rate (Figs. 4, 5) for the analysed case (2 g/s) and temperature distribution (Fig. 6a). While increasing the stream of the fuel supplied to the cyclone furnace, a temperature decline (Fig. 3 and 6) is observed, connected with substoichiometric conditions of the process and necessity to supply greater amount of heat to the heating and devolatilization process. Increasing the stream of the fuel supplied is noticeable in concentration profiles (Fig. 4) and concentration distributions (Fig. 7) An increase in fuel concentration can be observed near the PC1 chamber walls (Fig. 7b and 7c) compared to the case of reference (Fig. 7a, 2 g/s).

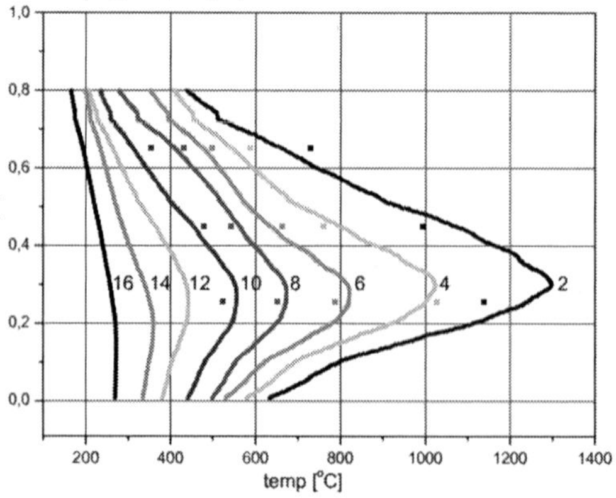

Fig. 3. Time-averaged temperature profiles near the PC1 chamber, points - results of experimental studies (T1, T2, T3).

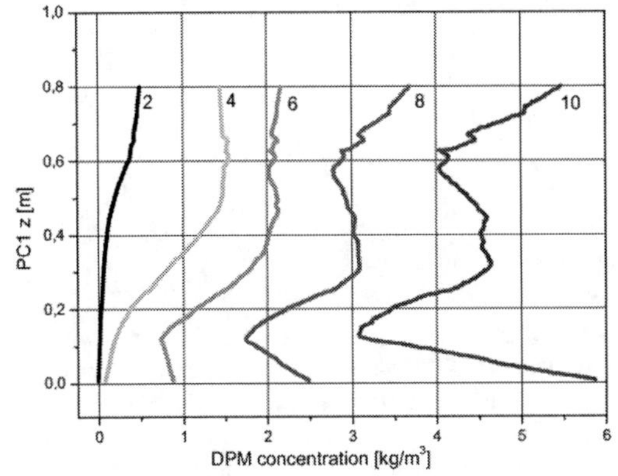

Fig. 4. Time-averaged profiles of fuel concentration near PC1 chamber walls.

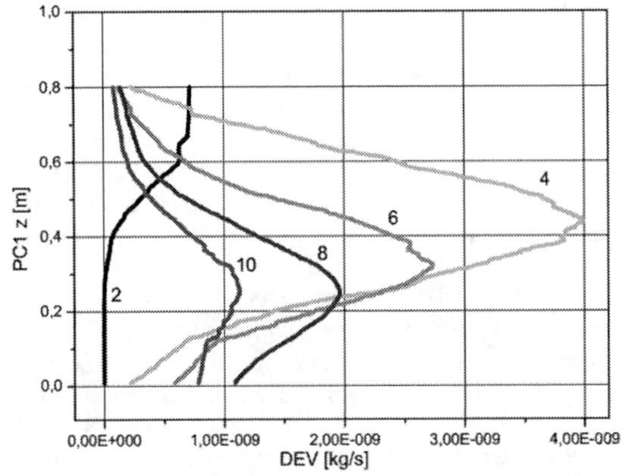

Fig. 5. Time-averaged profiles of fuel devolatilization rate near PC1 chamber walls.

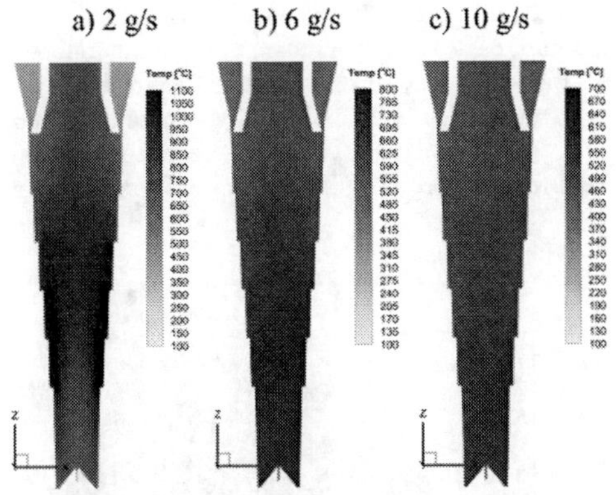

Fig. 6. Time-averaged temperature distributions in the vertical cross-section of the PC1 chamber.

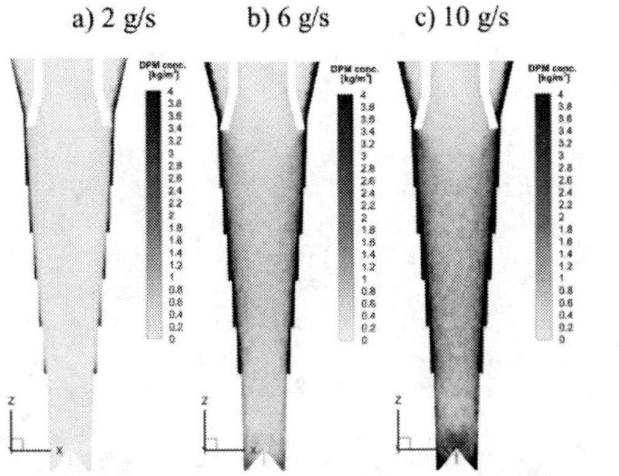

Fig. 7. Time-averaged fuel concentration distributions in the vertical cross-section of the PC1 chamber.

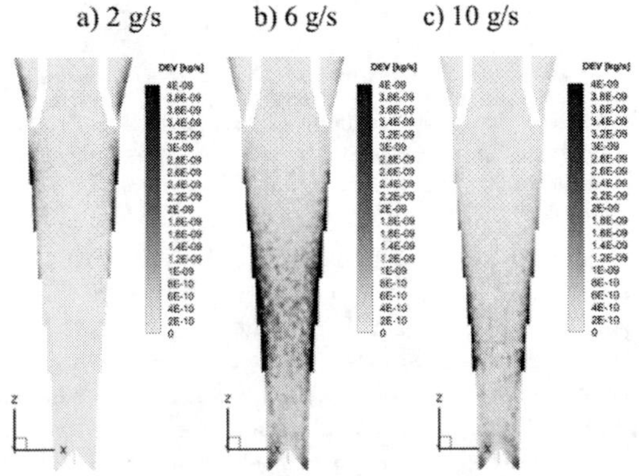

Fig. 8. Time-averaged fuel devolatilization rate distributions in the vertical cross-section of the PC1 chamber.

The increase in the fuel mass stream also requires greater energy (momentum) generated by the driving gas in order to maintain a strong eddy motion of the fuel. If the swirling fuel mass exceeds the boundary value of the driving gas stream, further stable swirling is impossible and fuel flows towards the lower part of the furnaces, which is reflected by the increased fuel concentration near the coordinate $z = 0$ m where the container that allows for collecting the heated and thermally processed fuel is located. The increase in the stream of the supplied fuel impacts significantly on its devolatilization rate (Fig. 5, 8). Temperature (Fig. 3, 6) and devolatilization rate (Fig. 5, 8) are closely related since the increase in temperature leads to the increase in devolatilization rate and therefore allows for generation of an adequate amount of volatile parts whose combustion allows for maintaining adequate temperature inside the PC1 chamber. The increase in volatile matter in the analysed cases that results from the increased devolatilization rate (Figs. 5, 8) and increased stream of volatile matter does not translate into increased temperature (Figs. 3, 6) due to maintaining constant stream of driving gas (oxidant). The increase in fuel stream requires increased heat stream for the purposes of heating the increased fuel mass.

In conclusion of the presented results of the numerical computations, maintaining constant stream of driving gas (oxidant) and variable fuel stream can ensure the conditions that allow for the controlled heating and fuel devolatilization for the fuel flowing to the container located in the lower part of the PC1 chamber.

Fig. 9a presents the instantaneous image of the fuel combustion process (2 g/s) in the PC1 chamber, Fig. 9b presents the interior of the PC1 chamber covered with the molten slag after the fuel combustion process. Fig. 9b illustrates the location of the thermocouples and the internal plunger that divides PC1 and PC2 chambers.

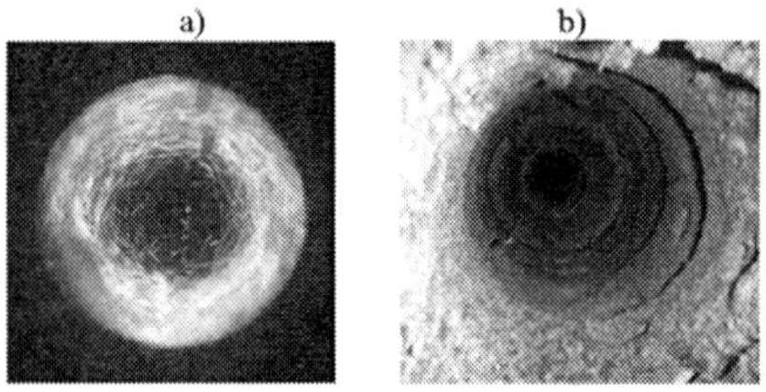

Fig. 9. The view of the fuel burnt in the PC1 chamber (2 g/s) a), interior of the PC1 chamber covered with molten slag after the combustion process b).

Analysis of the combustion and fuel devolatilization process was also performed using the cyclone furnace model (Figs. 1c, 1d) for five fuel rates (2, 4, 6, 8, 10 g/s). Experimental conditions were consistent with the conditions of numerical computations. The obtained results of the experimental examinations allowed for determination of mean values of temperature T1, T2 and T3 (Fig. 1.c) along the height of the PC1 and were added to the Figure 3 in the form of points. Comparison of the temperature distributions obtained from numerical computations and experimental examinations revealed that maximal differences between the values did not exceed 100°C. The experimental examinations for variable stream of fuel mass allowed for determination of the value of temperature to which fuel can be heated for the fuel mass stream. Table 2 presents mean values of fuel temperature which, after thermal processing in the cyclone furnace was supplied to the container located in the lower part of the furnace.

Table 2. Fuel temperature after thermal processing in the cyclone furnace.

Fuel flow rate [kg/s]	2	4	6	8	10
Fuel temperature [°C]	-	612	363	292	255

4 Laboratory examinations of the process of mercury removal in rotary furnace station

The laboratory examinations of the process of mercury removal from brown coal were conducted on a rotary furnace station (Fig. 10).

Fig. 10. Rotary furnace station.

The examinations were performed for the temperatures of 200-400°C. These temperatures were chosen because the temperature of 200°C allows for removal of a substantial amount of mercury from brown coal [10]. Furthermore, maximal temperature was determined at the level which guarantees removal of over 90% of mercury from coal but it is low enough not to cause a substantial degradation of brown coal as a fuel. The examinations of the rotary furnace station were performed in the air atmosphere. Time of remaining the samples in the furnace for individual temperature was chosen so that the sample temperature at the outlet from the furnace was 200, 250, 300, 350 and 400°C for individual tests. After obtaining the required temperature by the samples leaving the rotary furnace, they were cooled in the atmosphere of nitrogen and next the examinations of mercury content were performed in the coal samples. The difference between the content of mercury in the samples before and after thermal processing allowed for computation of the amount of mercury removed for individual temperatures. Mercury content was evaluated using the Lumex RA-915+ spectrometer with the RP-91 device. The methodology of measurements of mercury content in solids using the Lumen spectrometer has been extensively discussed in previous publications [3].

The results indicated that it is possible to remove around 18% of mercury from brown coal already at the temperature of 200°C whereas at the temperature of 250°C, this efficiency reached over 50%. In the case of higher temperatures (300 and 350°C), degree of mercury removal was 78% and over 90%, respectively. With maximal temperature of 400°C, mercury removal degree was 98% (Fig. 11).

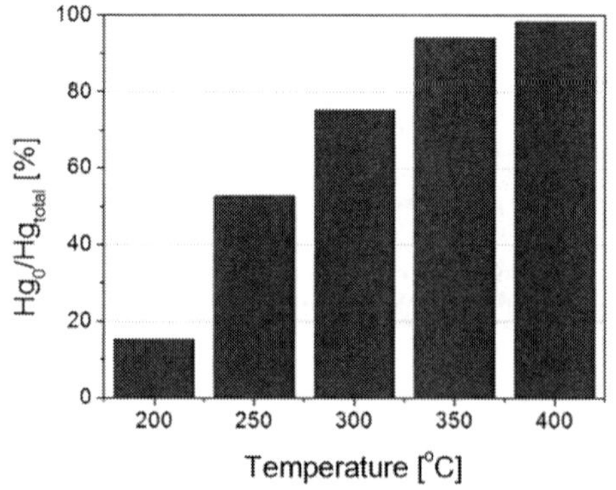

Fig. 11. Amount of mercury removed at individual temperatures for brown coals in rotary furnace station.

5 Examinations of mercury content in coal samples

After completion of the examinations in the rotary furnace station, the temperature ranges were determined for the cyclone furnace. Also in this case, the temperature reached by individual coal samples leaving the cyclone furnace station was significant (Table 2). Mercury content was also evaluated for coal brown samples using the Lumex spectrometer. The results are presented in Fig. 12.

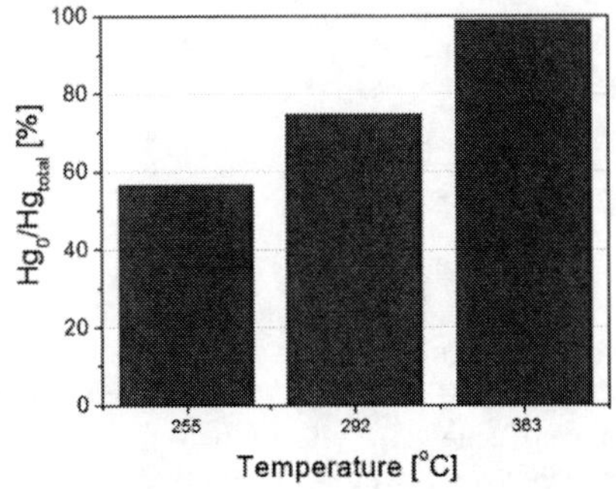

Fig. 12. Amount of mercury removed at individual temperatures for brown coals in cyclone furnace station.

In the cyclone furnace, the amount of mercury removed was nearly 60% for the initial temperature of 255°C (10 g/s). For higher temperatures of 292 (8 g/s) and 363°C (6 g/s), degree of mercury removal was very high i.e. over 75%, and nearly 99%, respectively.

In addition to the degree of mercury removal, fuel degradation degree which occurs at a specific process temperature is also important. Degree of fuel degradation was computed as a difference in the content of volatile matter before and after thermal processing. For the examinations performed on the rotary furnace station, the loss of volatile matter is presented in Fig. 13. For the lowest temperature (200°C), the amount of lost volatile matter was 15%, whereas for the temperature of 250°C, this was slightly over 15%. For the temperature of 300 and 350°C, the amount of lost volatile parts was 20 and 27%, respectively. For the highest temperature (400°C) the amount of lost volatile matter was over 32%.

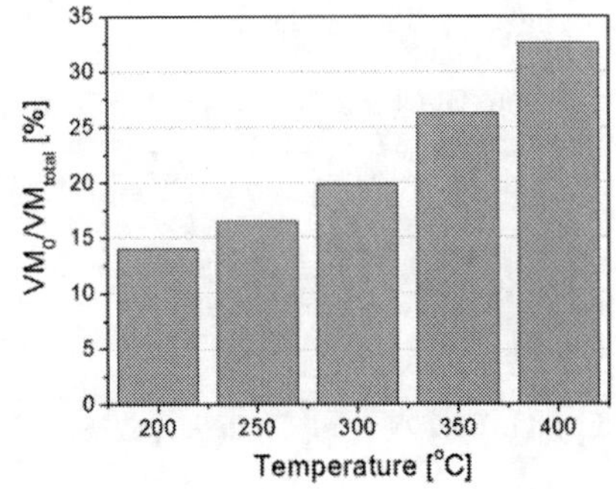

Fig. 13. Amount of volatile matter lost at individual temperatures for brown coals in rotary furnace station.

Similar examinations of the amount of lost volatile matter was performed for the cyclone furnace station. The results are presented in Fig. 14.

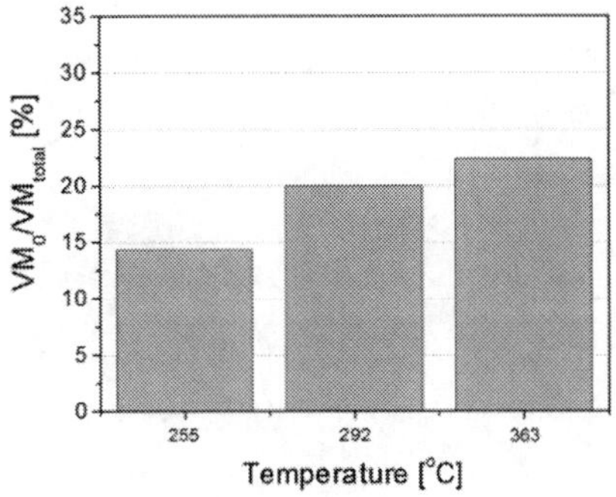

Fig. 14. Amount of volatile matter lost at individual temperatures for brown coals in the cyclone furnace station.

For the cyclone furnace, the amount of lost volatile matter was insignificantly lower than for the rotary furnace station. This is likely to have been caused by more efficient processes of heat transfer inside the cyclone furnace, which translated into faster reaching the expected temperature by the sample, which reduced the loss of volatile matter. For the initial temperature (255°C) (10 g/s) the amount of lost volatile matter was over 14%. For higher temperatures of 292 (8 g/s) and 363°C (6 g/s), this was 20% and 22%, respectively.

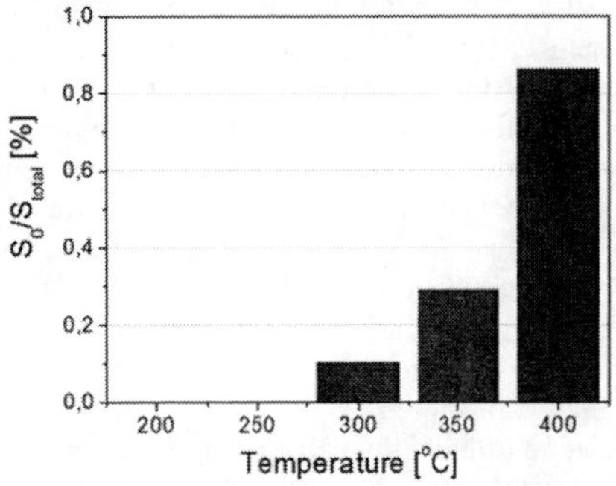

Fig. 15. Changes in sulphur content at individual temperatures for brown coals in rotary furnace station.

Apart from the evaluation of lost volatile matter, the change in the sulphur content in the brown coal was also examined for individual temperatures. Sulphur content was lower for the temperature range of 300-350°C both for rotary furnace and cyclone furnace. The results for the rotary furnace station are presented in Fig. 15. For initial temperatures of 200 and 250°C, no changes in sulphur content was found for the samples after thermal processing. A noticeable reduction in sulphur content was observed in brown coal samples for higher temperatures of 300, 350 and 400°C. Sulphur content for the highest temperature (400°C) was reduced by over 0.8%. Similar results were observed for coal subjected to thermal processing in the cyclone furnace For these samples, changes in sulphur content (0.1%) were observed only for the temperatures of over 292°C (8 g/s). For higher temperature of 363°C (6 g/s), this was around 0.2% (Fig. 16).

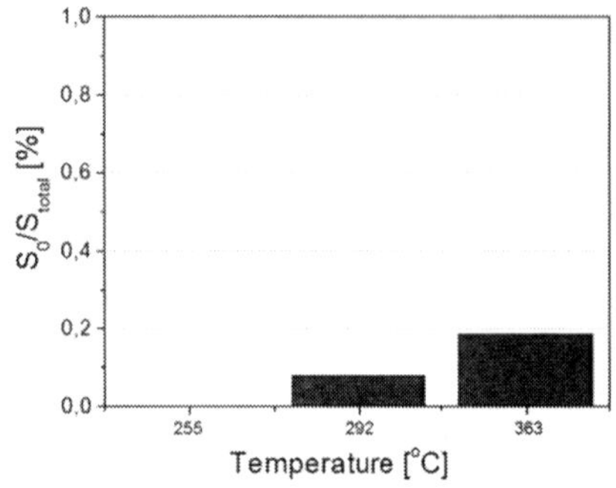

Fig. 16. Changes in sulphur content at individual temperatures for brown coals in cyclone furnace station.

6 Conclusions

The solution for the process of fuel thermal processing using cyclone furnace allows for controlled thermal processing of fuel and mercury removal from the fuel.
With strong eddy motion of fuel, cyclone furnaces allow for intensive mixing and heating of the fuel, thus allowing for high efficiency of thermal processing. Consequently, the cyclone furnace can be scaled to the dimensions that allow for thermal processing of fuel for the purposes of actual energy furnaces. The level of removed mercury can be controlled by changing the stream of fuel supplied to the furnace, changing the aerodynamics (velocity of the driving gas) or oxygen concentration in the driving gas. These parameters translate into the value of temperature inside the cyclone furnace and time of remaining (thermal processing) of fuel, which results in heating the fuel to the desired temperature that ensures the required level of the removed mercury.

The gases obtained during thermal processing, composed of the flue gas, volatile matter and water vapour and containing mercury vapour after cooling and condensation of humidity can be supplied to the absorber that allows for removing mercury vapour from gases. Cleaned gases can be next supplied to the furnace in order to be burnt. This technology of thermal processing of fuel and mercury removal does not lead to chemical energy loss and reduced performance of power units.
The results of the experiments conducted on the cyclone furnace station for the temperatures of 255-363°C showed that the amount of Hg was from nearly 60% to 99%. These results are similar to those obtained for the rotary furnace station, which demonstrated the opportunities for reducing mercury emissions from 50 to over 90% in the temperature range of 250-350°C. The amount of lost volatile matter for brown coal after processing on the cyclone furnace ranged from 15% (for temperature of 255°C) to 22% (for temperature of 363°C). Similar results were obtained for the rotary furnace station, with values ranging from 17% (temperature of 250°C) to 27% (temperature of 350°C). Thermal processing of coal also resulted in a reduction of sulphur content, which in the case of the cyclone furnace was from 0.1% to 0.2% for temperatures of 292 and 363°C. These results were confirmed in the rotary furnace, with sulphur content for these temperatures reducing from 0.1% to 0.3% for the temperatures of 300 and 350°C.

The results obtained for mercury content after thermal processing in the cyclone furnace lead to the conclusion that it is possible to remove substantial amounts of mercury already at the temperature of ca. 300°C (75%), with the loss of volatile matter at the level not exceeding 20%.

Acknowledgements

Ministry of Science and Higher Education Republic of Poland BS/PB-404-301/11

References

1. B. Metz, O. Davidson, H. de Coninck, M. Loos, L. Meyer, Cambridge University Press, New York (2005)
2. R. Zarzycki, M. Panowski, Journal of Energy Resources Technology **140**, 3 (2018)
3. M. Wichliński, R. Kobyłecki, Z. Bis, Archives of Environmental Protection **39**, 2 (2013)
4. T. Olkuski, Energy Policy Journal **10**, 2 (2017)
5. A. Głodek, J. M. Pacyna, Atmospheric Environment **43**, 35 (2009)
6. A. Głodek, D. Panasiuk, J. M. Pacyna, Water Air Soil Pollut **213** (2010)
7. AMAP/UNEP, *Technical Background Report for the Global Mercury Assessment 2013*, (UNEP Chemicals Branch, Switzerland, 2013)
8. A. Iwashita, S. Tanamachi, T. Nakajima, H. Takanashi, A. Ohki, Fuel **83**, 6 (2004)
9. T. Dziok, A. Strugała, A. Rozwadowski, A. Okońska, Chemical Industry **93**, 12 (2014)
10. M. Wichliński, R. Kobyłecki, Z. Bis, Fuel Processing Technology **119** (2014)
11. R. Zarzycki, M. Wichliński, Energy Policy Journal **17**, 4 (2014)
12. R. Zarzycki, Z. Bis, Journal of Thermal Science **26**, 2 (2017)
13. R. Zarzycki, Z. Bis, R. Kobyłecki, Procedia Engineering **157** (2016)
14. Robinson W. D., *The Solid Waste Handbook : A Practical Guide*, (John Wiley and Sons Ltd, New York, 1986)
15. P. He, S. Luo, G. Cheng, B. Xiao, L. Cai, J. Wang, Renewable Energy **37**, 1(2012)
16. R. Zarzycki, Z. Bis, Procedia Engineering **157** (2016)
17. R. Zarzycki, Z. Bis, 12 th International Conference on Fluidized Bed Technology (2017)
18. R. Zarzycki, Z. Bis, E3S Web of Conferences **13**, 05003 (2017).
19. R. Zarzycki, E3S Web of Conferences **13**, 05004 (2017)
20. R. Zarzycki, 12 th International Conference on Fluidized Bed Technology (2017)
21. L. Chen, M. Gazzino, A.F. Ghoniem, 35th International Technical Conference on Coal

Utilization & Fuel Systems. Clearwater, Florida (2010)
22. D. Toporov, P. Bocian, P. Heil, A. Kellermann, Stadler H., Tschunko S., Förster M., Kneer R., Combustion and Flame **155** (2008)
23. M. Vascellari, G. Cau, Proceeding of CCT2009 Fourth International Conference on Clean Coal Technologies, Dresden, Germany (2009)
24. P. Warzecha, A. Bogusławski, Archivum Combustionis **12**, 3 (2012)
25. ANSYS Fluent Theory Guide (2011)
26. R. Zarzycki, R. Kobyłecki, M. Kratofil, M. Ścisłowska, D. Pawłowski, Z. Bis, Energy Policy Journal **17**, 3 (2014)
27. R. Kobyłecki, R. Zarzycki, Z. Bis, Gas Power Generation **2** (2016)

Proposed liquid CO_2 cycle refrigeration system for heat hazard control

Xiaowei Zhai[1,2], *Yu* Xu[1,2*], *Zhijin* Yu[1,2], and *Kai* Wang[1,2]

[1]College of Safety Science and Engineering, Xi'an University of Science and Technology, 710054 Xi'an, China
[2]Shaanxi Key Laboratory of Prevention and Control of Coal Fire, 710054 Xi'an, China

Abstract. Liquid CO_2 can absorb heat via phase change and generates cryogenic CO_2 which can effectively solve the problem of thermal damage in the deep coal mining process. The CO_2 cycle refrigeration device system is designed to effectively cool down the working surface of the mine in which CO_2 is cyclically utilized. COMSOL Multiphysics simulation software is used to characterize the CO_2 cycle refrigeration system in mine of the heat transfers process between CO_2 and the air flow in tunnel. The results show that the reduction of steady air flow temperature reach at 8 °C in the tunnel by CO_2 cycle refrigeration system before the air flow into work face. we analyzed the influence of main parameters on refrigeration system and gets the results:1) The refrigeration system get higher cooling efficiency of cryogenic CO_2 when the ventilation velocity of local fan is increased, and the temperature of outlet CO_2 and steady air flow in tunnel has increased; 2) Increasing the CO_2 flow, the refrigeration effect of the system is enhanced obviously, but CO_2 refrigeration capacity utilization ratio is reduced; 3) Increasing the length of helical tube would led to use CO_2 refrigerating capacity more efficiently; 4) The cooling effect of the cooling system can be improved obviously by lowering the the CO_2 cooling temperature.

1 Introduction

With the increases in the depths of mines, high-temperature mines have been more and more now [1]. A high-temperature environment directly affects mineworkers in work condition and work efficiency, and is easy to cause insecure behavior and lack of enthusiasm [2]. In addition, it will seriously endanger the health of workers that working in a high temperature environment for a long time [3]. Therefore the heat harmful problem of high temperature mine needs to be solved urgently.

Various cooling technologies have been used to prevent heat hazards and ensure safe and efficient mining. Based on the principle of vortex tube energy separation, Ma Li et al. studied the portable cooling technology and method in high temperature mine for individual cooling [4]. Chen W et al. proposed a split-type vapor compression refrigerator [5], using air as the refrigerant, pneumatic refrigerator was cooled down in a local freezer and it combined with cooling clothing to cool the workers in high temperature working environments. Deng bao developed an air-to-air conditioning system with compressors, condensers, evaporators, etc., to cool down the working face [6]. However, due to the small heat capacity of the air, the thermal damage of the working face was not effectively controlled. SUN Xikui et al. studied the influencing factors of radiation cooling technology of ice water cooling, and established a refrigeration system that can produce 37.5t of ice and 62.7t cold water at 2 °C per hour [7]. Guo Pingye et al. has designed a refrigeration systems which used the cold water as the cold source, and established evaluation method for mine refrigeration system to optimizes the refrigeration system and improve the cooling effect of the working face [8,9].

However, this system needs a lot of cold water from the ground as refrigeration. Furthermore, there are impurities in the water, and the cooling system pipes form scales in the process of transporting water after a long run time, which increases the resistance of the water transportation and increase the system operation burden. The heat exchange efficiency is reduced because of the decrease in the heat conductivity of the pipes caused by the scale of water, which leads to a distinct reduction in the cooling effect of the refrigeration system.

Due to its low price, accessibility, large heat capacity and being environmental friendly, the use of CO_2 as a refrigerant has attracted widely attention in recent years [10]. Previous studies have focused on the transcritical cycle and the relationship between the storage environment of carbon dioxide and refrigeration capacity is conducted [11-14]. However, the processes of heat exchange between the CO_2 and air flow or thermal diffusion in mine are less reported. These are the critical processes for the refrigeration system which ultimately determine the cooling effect for the work face, refrigeration efficiency and working cost of the refrigeration system. In this paper, a model of mine CO_2 cycle cooling system is developed, and the parameters of

* Corresponding author: xy1235813@sina.com

the local fan air volume, CO_2 velocity, CO_2 helical tube length and CO_2 cooling temperature are investigated.

2 Cooling method of CO_2 cycle refrigeration in mine

The CO_2 cycle refrigeration system is designed to cool down the working face of a high temperature mine. The system includes compressor, gas cooler, throttle device, monitoring device, evaporator, pipeline, heat-exchanger and local fan. The low temperature CO_2 gas is performed as cold source and cyclically utilized. If the liquid CO_2 is cooled in the heat exchange system directly, it will easy to the tube jam in pipelines for the liquid CO_2 phase change produce dry ice. So the liquid CO_2 phase transition is carried out in a special evaporator. In the evaporator device, as the pressure reduce, the liquid CO_2 absorb heat by change of state and forms low temperature CO_2, then the low temperature CO_2 enters the helical tube. The helical tube is located in the local air duct and contact with the wind flow provided by the local fan in the tunnel, which makes the low temperature CO_2 exchange heat with the wind flow to form the cold air flow. The cold air flow is sent to the working face in the last by the duct to create comfortable working conditions for workers. Fig.1 shows a schematic of CO_2 cycle refrigeration system.

In the Fig. 2, after the heat exchange, the CO_2 gas returns to the ground through the pipeline and forms liquid CO_2 by compression system and gas cooler. Then the liquid CO_2 through the pipeline into vaporization system which are installed on the underground and vaporizes into low-temperature CO_2 gas. In order to ensure the safety of the system, a safety valve is installed on the ground of the conveying pipe. When the CO_2 leaks occur in the pipe, the safety valve will be opened and the CO_2 in the pipeline wiil be released into the air on the ground. The CO_2 is circularly used in the CO_2 cycle refrigeration system for sustainable cooling the work face in mine, which greatly reduces the cost of materials required for refrigeration. At the same time, the CO_2 has more larger heat capacity than water and air resulting it has a better cooling effect at the same mass of water or air. The heat exchange between CO_2 and air is the key for the refrigeration system. To analyze the process of heat exchange between CO_2 and the air flow as well as the influence of main parameters on the cooling effect, a cooling model of CO_2 refrigeration in mine is established in this paper.

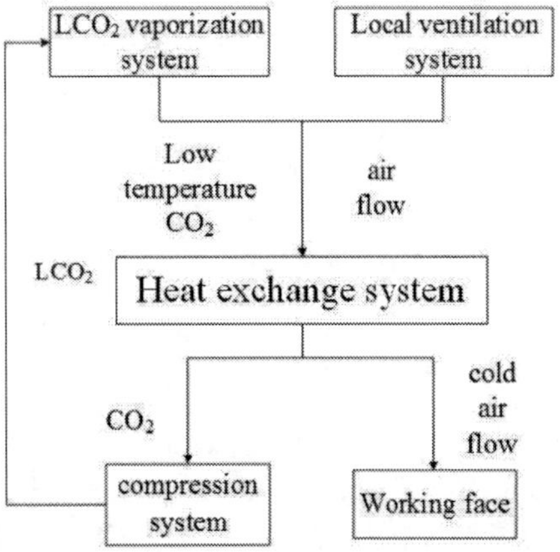

Fig. 1. Schematic of CO_2 cycle refrigeration system.

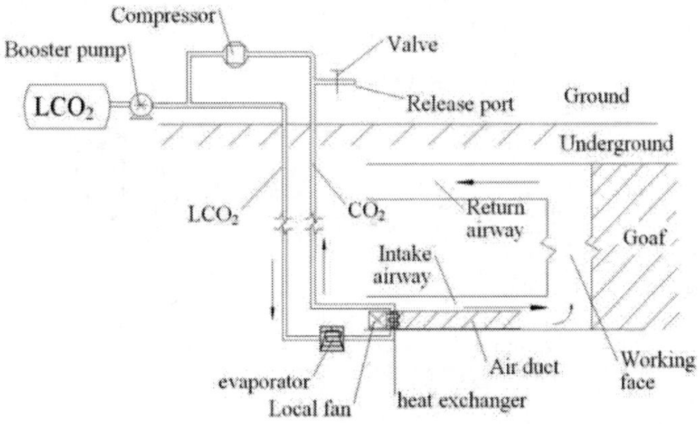

Fig. 2. The device of CO_2 cycle refrigeration system.

3 Modeling and simulation

3.1 Flow equation

Considering the balance of hydrostatic energy, kinetic energy and potential energy, one can be obtain the fluid flow equation which is used to describe the motion of a fluid in a pipeline under the action of gravity, viscous resistance and pressure.

$$\begin{cases} -\nabla p - f_D \dfrac{\rho}{2d_h} u|u| + F = 0 \\ \nabla \cdot (A\rho u) = 0 \end{cases} \tag{1}$$

where p is the pressure, Pa; ρ is the density, kg/m^3; A is the cross-sectional area, m^2; u is the tangential fluid

https://doi.org/10.1051/matecconf/201824005038

velocity, m/s; d_h is the pipe diameter, m; and F is the volume force, N/m^3.

In Eq.(1), f_D is the *Darcy* friction factor, which is related to the Reynolds number, pipe surface roughness and pipe diameter. It can be expressed as:

$$f_D = 8\left[\left(\frac{8}{Re}\right)^{12} + \left(B+C\right)^{-1.5}\right]^{1/12} \quad (2)$$

where the B is defined as:

$$B = \left[-2.457In\left(\left(\frac{7}{Re}\right)^{0.9} + 0.27\left(\frac{e}{d}\right)\right)\right]^{16} \quad (3)$$

C is defined as:

$$C = \left(\frac{37530}{Re}\right)^{16} \quad (4)$$

where Re is the Reynolds number; e is the surface roughness; d is the pipe diameter, m.

3.2 Energy conservation equation

The energy equation of pipeline flow is used to describe the heat exchange and thermal diffusion between the fresh air flow and the CO_2 pipeline.

$$\rho AC_p u \cdot \nabla T = \nabla \cdot Ak\nabla T + f_D \frac{\rho}{2d_h}\left|u\right|^3 + Q_{wall} \quad (5)$$

where C_p is the heat capacity at constant pressure, J/(kg·K); T is the temperature, K; and k is the thermal conductivities, W/(m·K). Q_{wall} is the a source of heat exchange between the wall and the surrounding wall, W/m; which can be can be expressed as:

$$Q_{wall} = hZ\left(T_{ext} - T\right) \quad (6)$$

where Z is the circumference of the CO_2 transportation pipeline, *m*; h is the overall coefficient of heat exchange, W/(m^2·K); T_{ext} is the external temperature of the CO_2 conveying steel tube, K; and T is the temperature of CO_2, K.

3.3 Physical model and numerical simulated

Based on the Section 2.1and 2.2, as a background of CO_2 cycle refrigeration system in a mine, a simulated model of heat-exchange system in tunnel was constructed at a ratio of 1:1, as depicted in Fig. 3. The physical model was a cuboid tunnel with width of 4 m, height of 4 m and length of 30 m. The tunnel of inlet wind speed is set to 1m/s, and the temperature set as 313.15 K. There is a duct with a radius of 0.6 m in the tunnel, which is supplied by a local fan with wind air of 5 m/s. A helical steel tubes for conveying low temperature CO_2 gas are installed inside the wind tubing for heat transfer between air flow and CO_2, which is 3.6 m long, radius of the outer ring with 0.4 m, radius of the inner ring with 0. 05 m. The CO_2 flow of transmission pipeline inlet is set to 4 m/s and the inlet temperature is set to 233.15 k. The physical model of heat-exchange system in tunnel mesh illustration consisting of 10,000 elements. To achieve the coupled process, all the governing equations can be implemented into and numerical solved by COMSOL multiphysics based on the finite element method, as determination of wind speed and temperature distribution under steady state, and the transient solver was properly selected.

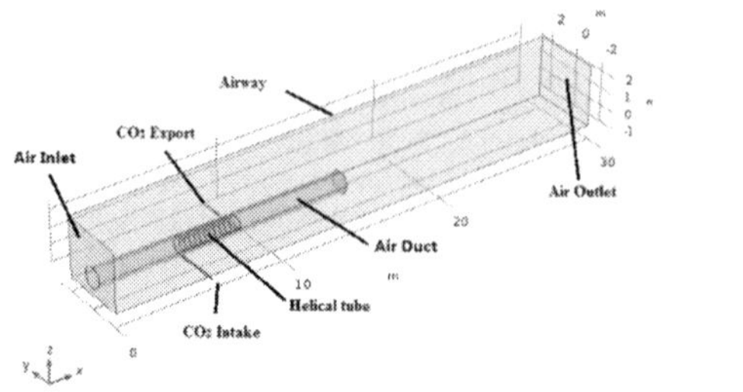

Fig. 3 The model of heat-exchange system in tunnel.

4 Simulated result

In the Fig. 4, we can see that a high speed wind flow is gradually mixed with the air flow after it comes out of the local wind duct, and the air speed is gradually distributed evenly in the tunntl. In the process of wind flow movement, the wind flow in the duct exchanges heat with the helical pipeline which used to carried CO_2 , forming a large amount of cold air flow. Then, the high speed cold wind flow enters the tunnel gradually mixes with the fresh air flow and the wind flow tend to stable gradually, at the same time the temperature of wind flow decrease. Finally, the stable air flow temperatures in the tunnel are reduced obviously. We can see the temperature distribution in the tunnel in Fig. 5.

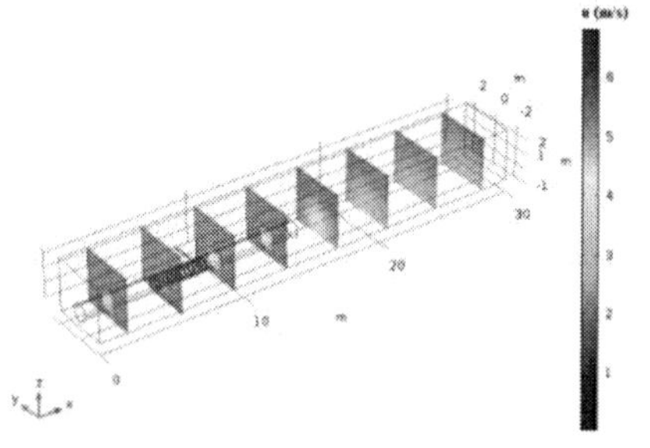

Fig. 4 Velocity distribution in the model.

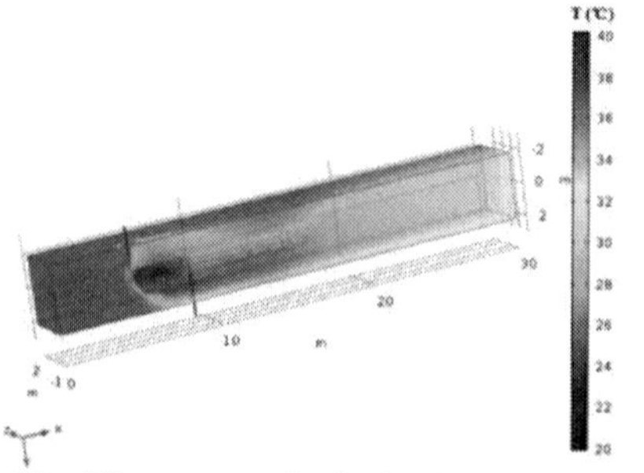

Fig. 5 Temperature distribution in the model.

The velocity and temperature distribution in the central axis of the duct, as shown in Fig. 6, has nonlinear change characteristics. When the wind flow in the duct is in contact with the CO_2 conveying pipeline, the wind flow velocity increases due to the resistance of the pipe, and the heat exchange occurs between the air flow and the CO_2 pipeline cause of the wind flow temperature fast reduce. At that time the cold wind flow temperature is rapidly lower to 24 °C. When the wind flow in the duct enters the tunnel, the velocity of the cold air flow decreases and the temperature increases gradually and stably, because of cold air flow is mixed with the hot air flow gradually. The temperature of the air flow is stabilized at 32.54 °C in the tunnel.

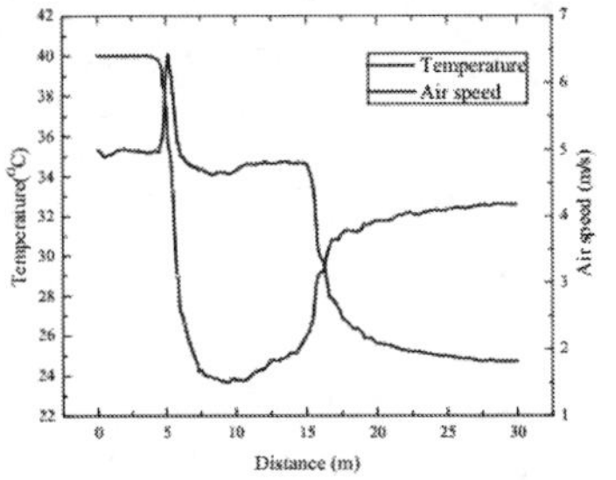

Fig. 6 Temperature and wind speed curve of the central axis of the duct.

5 Analysis of cooling effect

5.1 The effect of the ventilation speed

To analyse the effect of ventilation speed on CO_2 cycle refrigeration system, we set the flow velocity in the CO_2 pipeline is 4 m/s, and changed the wind speed in the duct of the model to obtain the refrigeration effect under different local fan air speed. When the wind speed is 2 m/s, the wind flow temperature is 18.73 °C in duct, and the final steady air flow temperature is 33 °C in the tunnel.

With the increase of ventilation speed, the average heat exchange capacity of per cubic flow decreases, so the outlet wind temperature in the duct increases obviously. However, due to the increase of the total cold air flow, the temperature of cold air flow dilution by the hot air in tunnel is weakened, and the steady air flow temperature in the tunnel change little compared to low ventilation speed.

In Fig. 7(b) shows that with the increase of ventilation speed, the heat exchange between CO_2 and air increase, and the temperature of CO_2 at the outlet pipeline increases. However, the temperature of steady air flow increases because of the increase of cold air volume, but the air flow temperature in the tunnel is very small after dilution.

To obtain the functional relationship between the cooling effect in tunnel and the ventilation speed. The formula of steady air flow temperature and ventilation speed are fitted by using ORIGIN software. The expression of function relation between steady air flow temperature $T(u)$ and ventilation speed are obtained: $T(u)=32.326-0.375u+0.0835$. The temperature of steady air flow in the tunnel is proportional to the square of local fan air speed.

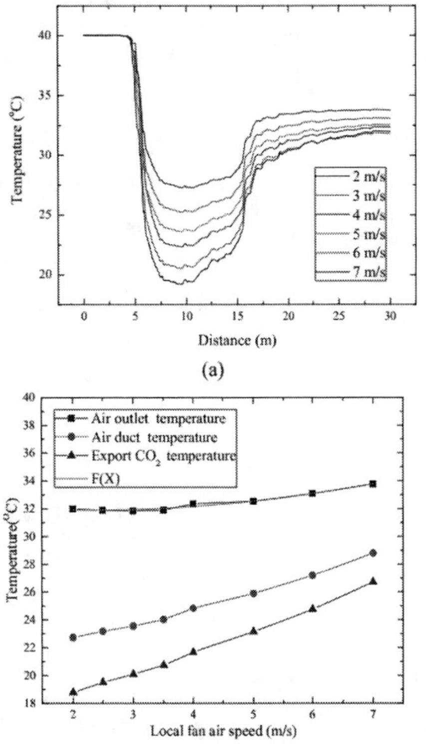

(b)

Fig. 7. The effect of ventilation speed u on the CO_2 cycle refrigeration system, where the subfigure (a) correspond to the ventilation speed of local fan at u = 2, 3, 4, 5, 6 and 7 m/s with the temperature curve of the central axis of the duct, respectively.

5.2 The effect of the CO_2 flow

Fig. 8 illustrates the effect of the CO_2 flow l on the CO_2 cycle refrigeration system. The local ventilation speed is set at 5 m/s, and the CO_2 velocity in the pipeline is set at 2, 3, 4, 5 and 6 m/s, respectively. It can be seen from Fig. 8(a) that with the increase of CO_2 velocity, the cooling capacity of the system increases, and the temperature of air flow decrease more obviously. When the CO_2 velocity is 6 m/s, the temperature of cold wind in the duct reaches to 20.39 °C, and the steady air flow temperature in the tunnel decreases to 30.39 °C.

Compared with the increase of ventilation speed, the temperature descends more obvious by increases CO_2 velocity. According to figure. 8(b), with the increase of CO_2 flow velocity, the CO_2 temperature in outlet pipeline decreases after heat transfer, so the cooling utilization ratio of CO_2 decreases, but the air flow temperature in tunnel drop more obvious. The formula of steady air flow temperature and CO_2 flow velocity are fitted by ORIGIN software. Then, the expressions of steady air flow temperature $T(l)$ and CO_2 current is: $T(l)=42.11-3.469l+0.267l^2$.

Therefore, when refrigeration effect of the system improved by increase of the CO_2 flow velocity, the length of heat transfer pipe or the air speed of ventilation should be increased in order to increase the cooling utilization ratio of cryogenic CO_2 and the cost be saved.

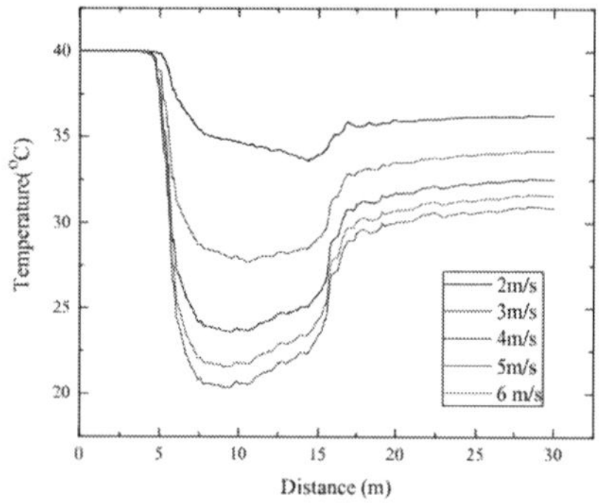

(a)

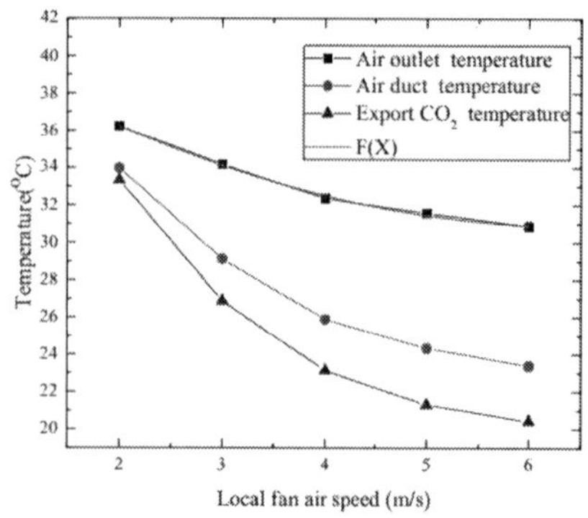

(b)

Fig. 8. The effect of CO_2 flow l on the CO_2 cycle refrigeration system, where the subfigure (a) correspond to the CO_2 velocity in pipeline at l = 2, 3, 4, 5 and 6 m/s with the temperature curve of the central axis of the duct, respectively.

5.3 The effect of the helical tube length

The helical pipes are used to heat exchange with wind flow directly and transport CO_2. It can be observed in the Fig. 9 that with the increase of the helical tube length, the heat exchange between CO_2 and air flow increase results in the increase of CO_2 temperature after heat exchange and more efficient use of CO_2 refrigerating capacity. When the length of helical tube was increased from 1.8 m to 4.2 m, the temperature of CO_2 in outlet pipeline increased from 22.01 to 23.25 °C due to the sufficiently heat exchange between CO_2 and air flow. Meanwhile the air flow temperature at outlet of tuyere from 27.76 °C reduced to 25.50 °C and air temperature in the laneway is also reduced by 1.34 °C degrees centigrade than that of the original.

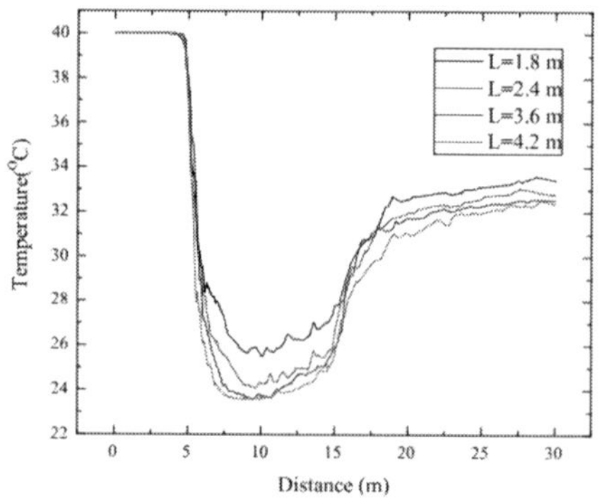

(a)

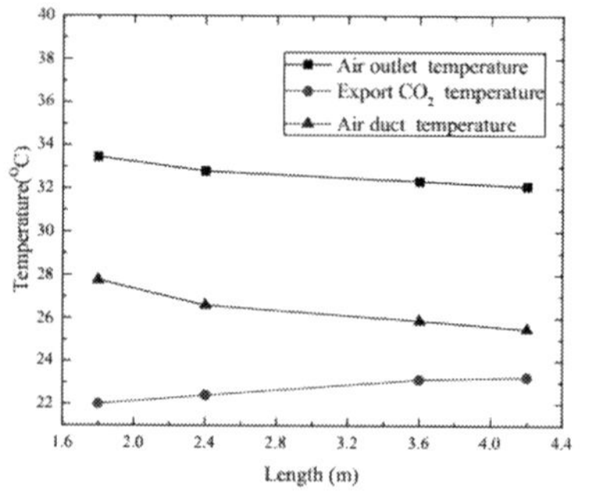

(b)

Fig. 9. The effect of CO_2 helical tube length on the CO_2 cycle refrigeration system, where the subfigure (a) correspond to the CO_2 helical tube length at L= 1.8, 2.4, 3.6 and 4.2 m with the temperature curve of the central axis of the duct, respectively.

5.4 The effect of CO_2 the cooling temperature

In the evaporator, the heat is absorbed by liquid CO_2 phase change which forms low temperature CO_2, and then the low temperature CO_2 enters the heat exchanger and exchange heat with the air flow to form the cold air flow. Before CO_2 entering the heat exchanger, The CO_2 temperature have an important impact on the CO_2 cycle refrigeration system. We can be seen from the Fig.10, as the CO_2 cooling temperature raise, the temperature of wind flow in duct, CO_2 in outlet pipeline and steady air in tunnel are all increased because of CO_2 refrigerating capacity decreases. When the CO_2 cooling temperature is -45 °C, the temperature of steady air flow in tunnel drop 8 °C after passing through the CO_2 refrigeration system. However, when CO_2 cooling temperature is at the -30 °C, the steady air flow temperature is only reduced 6.72 °C. Thus, the temperature of CO_2 formed by the phase transition of the liquid CO_2 in the evaporator is an important parameter affecting the refrigeration system. Before CO_2 enters the heat exchanger, the heat preservation measures should be taken on the CO_2 pipeline to prevent cooling loss to improve the cooling effect of the cooling system on the working surface in the mine.

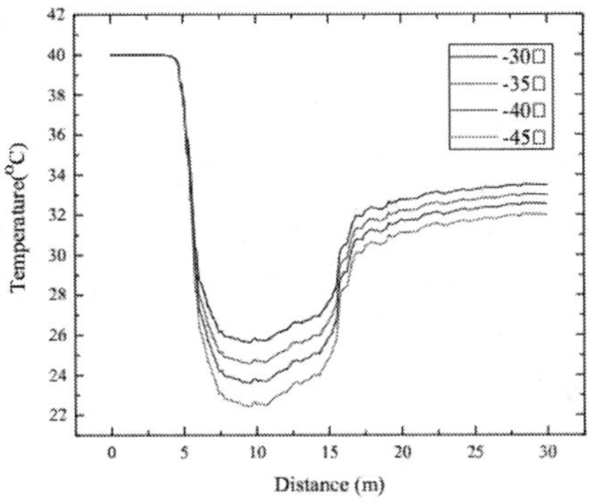

(a)

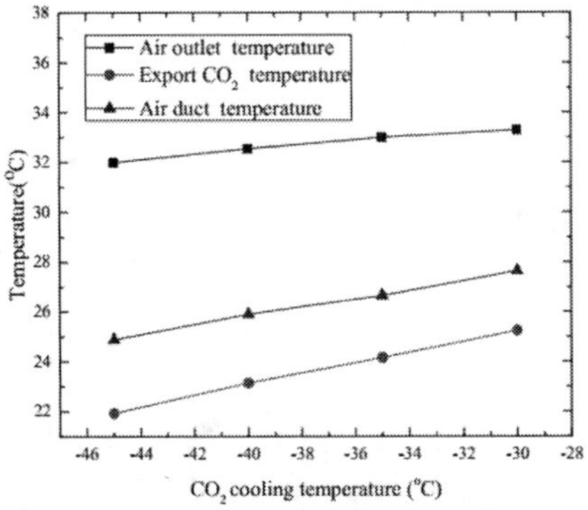

(b)

Fig. 10. The effect of CO_2 cooling temperature on the CO_2 cycle refrigeration system, where the subfigure (a) correspond to the CO_2 cooling temperature T= -30, -35, -40 and -45 °C with the temperature curve of the central axis of the duct, respectively.

6 Conclusion

In this work, a fully coupled steady model of compositional gas flow in tunnel and the non-equilibrium heat transfer between the CO_2 and the gas is developed to characterise the CO_2 cycle refrigeration system in mine of the heat transfer process between CO_2 and the air flow in the tunnel. Based on the results of this study, the following conclusions can be drawn:

1) The wind flow provided by the local fan is cooled by CO_2 heat exchange system and then the high speed wind flow is gradually mixed with the air flow form low temperature steady cold air flow in tunnel.

2) When the the CO_2 flow in pipeline is constant, the refrigeration efficiency would be improved with the increase of ventilation speed of local fan, and the temperature of steady air flow in tunnel and CO_2 in outlet pipeline also has increased.

3) Under the condition of wind speed of local fan is constant, the increase of CO_2 velocity is beneficial to enhance the refrigeration capacity of the system, but the efficiency of CO_2 refrigeration will be reduce. Therefore, the length of the heat exchanger tube and air speed of the local fan should be increased when the CO_2 flow velocity is increased.

4) With the increase of the helical tube length, the heat exchange recovery between CO_2 and the air flow increase results in the temperature of CO_2 in outlet pipeline increases and the use of CO_2 refrigerating capacity more efficiently.

5) The temperature of cooling CO_2 have an important impact on the cycle refrigeration system. The air outlet temperature, export CO_2 temperature and air duct temperature would increased with the increase of the CO_2 cooling temperature. To improve the cooling effect of the cooling system on the working surface in the mine, the heat preservation measures should be taken on the CO_2 pipeline to prevent cooling capacity loss.

References

1. M.C. He, H.P. Xie, S.P. Peng, YD. Jiang. J. Rock Mechanics & Engineering **24**, 16, 2803-2813 (2005)
2. Z.G. Su, Z.A. Jiang, Z.Q. Sun. J. Procedia Earth & Planetary Science **1**, 1, 414-419 (2009)
3. A.G. Chen, A.J. Prof. China Safety Science Journal **645**, 889-892 (2004)
4. L. Ma, L.R. Zhang, B. Li, H. Wen, Z.P. Wang. J. Coal Technology **33**, 11, 278-281 (2014)
5. W. Chen, S. Liang, J. Liu. Appl. Thermal Eng. **105**, 425-435 (2016)
6. B. Deng, H. Zhang, Y.D. Song, Z. Jiang, Y. Shi. J. Refrigeration and Air Conditioning, **3**, 301-304 (2014)
7. X.K. Sun, X.H. Li, W.M. Cheng. J. Mining & Safety Engineering, **26**, 1, 105-109 (2009)
8. P.Y. Guo. Int. J. Mining Science and Technology **23**, 3, 453-456. (2013)
9. P.Y. Guo, Y. Wang, M. Duan, D. Pang, N. Li. Int. J. Mining Science and Technology **25**, 4, 649-654 (2015)
10. A. Pearson. Int. J. Refrigeration **28**, 8, 1140-1148. (2005)
11. L. Cecchinato, M. Chiarello, M. Corradi. Applied Energy **87**, 6, 2095-2101 (2010)
12. A. Subiantoro, K.T. Ooi. Int. J. Refrigeration **33**, 4, 675-685 (2010)
13. J. Sarkar. Adv Res Mech Eng **1**, 22–29 (2010)
14. Y. Cai, C. Zhu, Y. Jiang, H. Shi. Energy Conversion & Management. **89**, 92–98 (2015)

AUTHOR INDEX

AUTHOR INDEX

AUTHOR INDEX